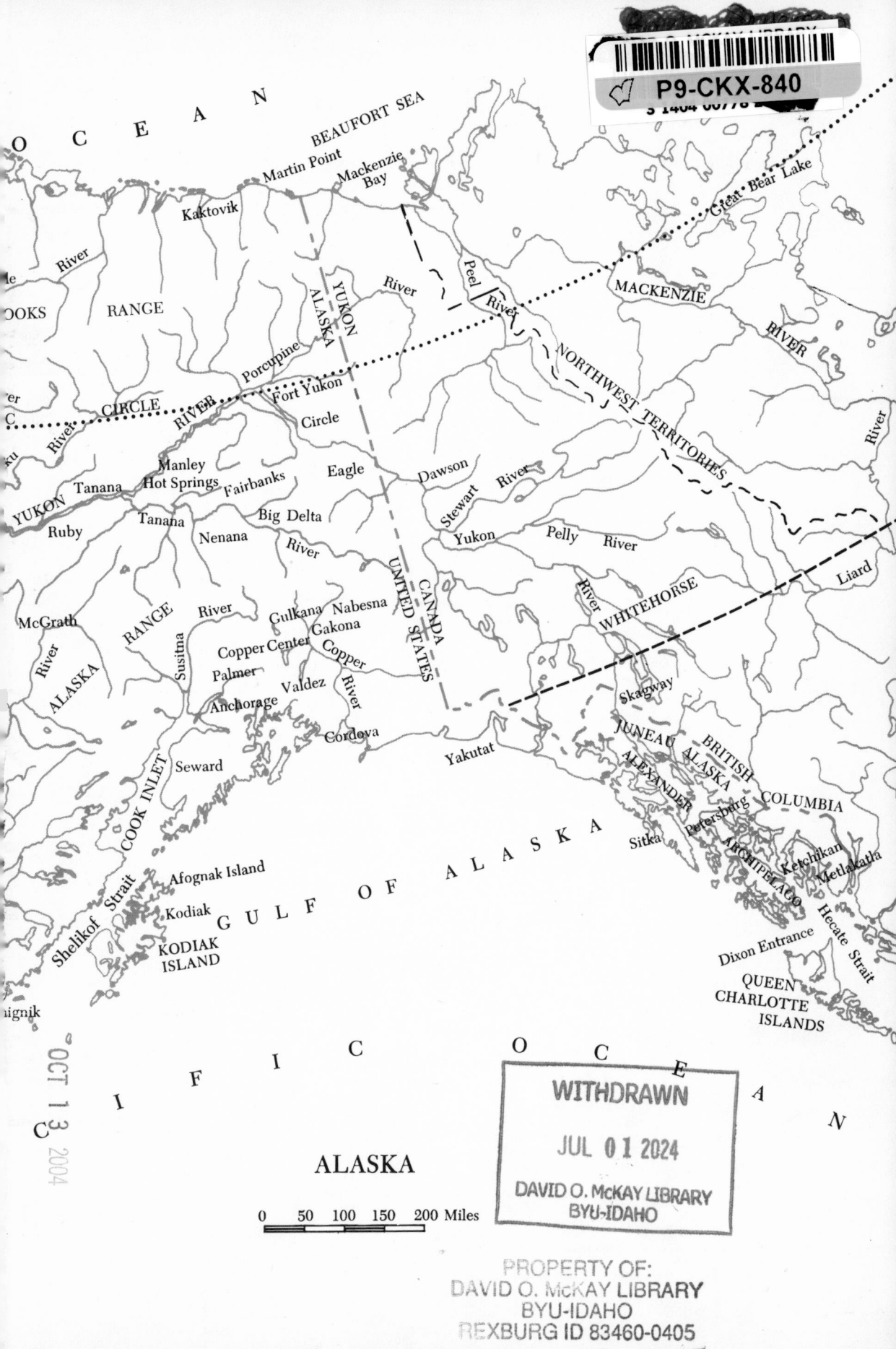

ALASKA

0   50   100 150 200 Miles

P9-CKX-840

WITHDRAWN
JUL 0 1 2024
DAVID O. McKAY LIBRARY
BYU-IDAHO

PROPERTY OF:
DAVID O. McKAY LIBRARY
BYU-IDAHO
REXBURG ID 83460-0405

OCT 13 2004

WITHDRAWN

Anderson's
FLORA
of ALASKA
and Adjacent Parts of Canada

Anderson's
FLORA
of ALASKA
and Adjacent Parts of Canada
Stanley L. Welsh
Brigham Young University Press   Provo, Utah

Library of Congress Catalog Card Number: 74-16260
International Standard Book Number: 0-8425-0705-1
Brigham Young University Press, Provo, Utah 84602
© 1974 by Brigham Young University Press. All rights reserved
Printed in the United States of America
74   2.5M   0579

Library of Congress Cataloging in Publication Data

Anderson, Jacob Peter.
  Anderson's Flora of Alaska and adjacent parts of
Canada.

  "Volume is completely new as to text, keys, and
illustrations."
  Bibliography: p. 673.
  1. Botany—Alaska. 2. Botany—Yukon Territory.
3. Botany—British Columbia. I. Welsh, Stanley L.,
ed. II. Title.
QK146.A52   581.9'798   74-16260
ISBN 0-8425-0705-1

Jacob Peter Anderson, 1874–1953

Maxine Morgan Williams

For Jacob Peter Anderson, Richard W. Pohl, and Duane Isely

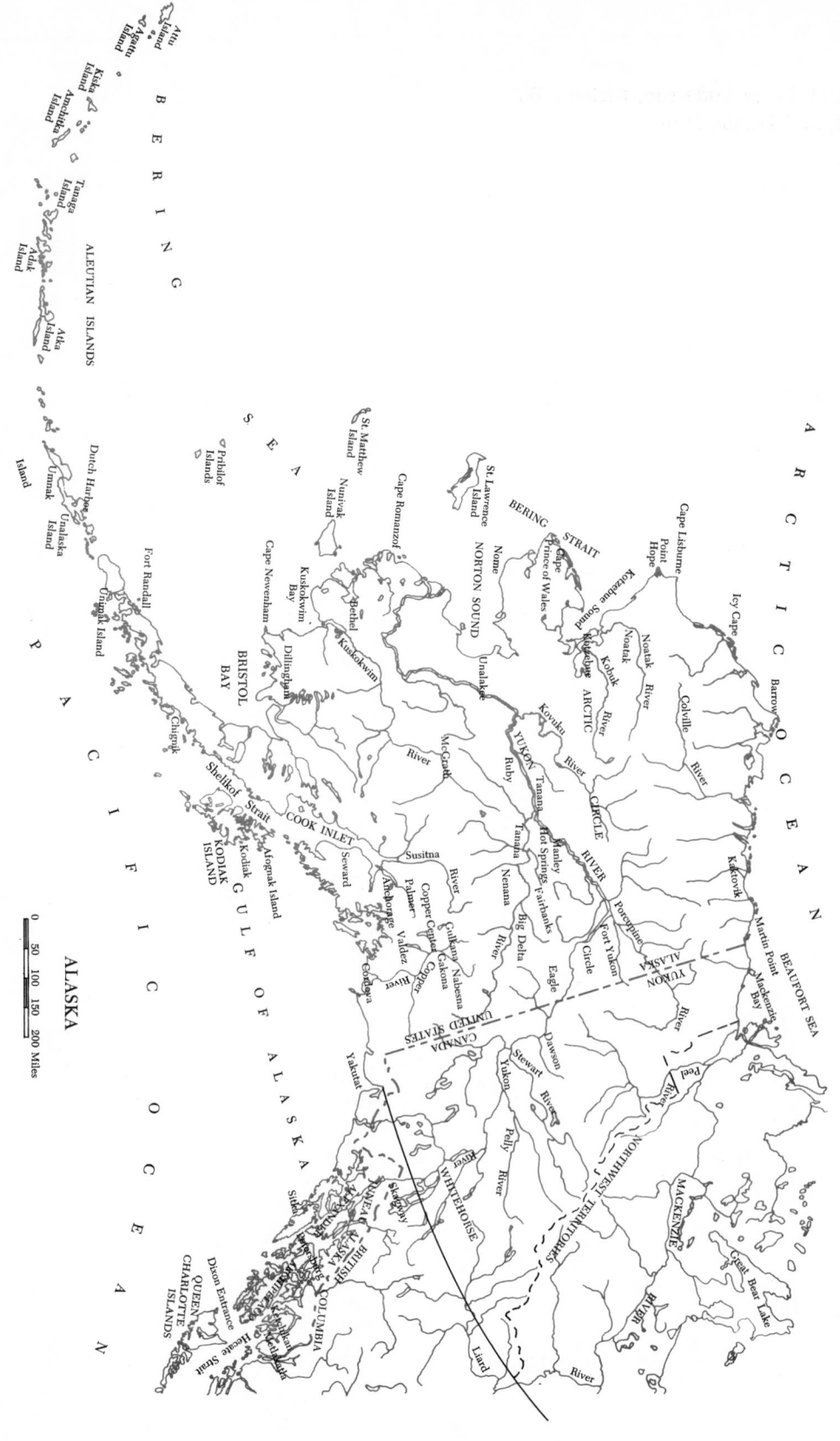

Contents

Foreword

When Dr. J. P. Anderson died in 1953, his labor of love, a field manual of the Flora of Alaska based upon his personal collections and vast field experience, existed only in the form of nine preliminary articles on the flora which appeared in the Iowa State Journal of Science. These articles were not in the form which Dr. Anderson would have wished, but he felt that he must put on paper what he knew, before his time ran out. He had started to revise these articles when he died. Supplies of reprints were soon exhausted. Efforts to secure outside funds to renew work on the Alaskan Flora project proved fruitless. Therefore, in 1959 we republished Dr. Anderson's articles in book form, to make them available to the many persons who sent us requests for his works. We hoped that it might be possible to produce an up-to-date revision of the work, based upon Dr. Anderson's herbarium and more recent field studies.

We were indeed fortunate to secure the interest and cooperation of one of the most active American field botanists, Dr. Stanley Welsh, in preparing the new edition. In 1965, with the aid of funds provided by the Anderson Botany Bequest and salary assistance from Brigham Young University, Dr. Welsh began an extensive program of field studies of the flora of Alaska, Yukon, and the Northwest Territories. This campaign, which has continued to the present, has greatly extended our knowledge of the flora of the Western Arctic. The specimens collected have been added to the Anderson Herbarium at Iowa State, making it the most comprehensive collection of Alaskan plants in existence. In addition to his field studies, Dr. Welsh has spent extended periods of time in Ames, studying the Anderson collection.

The present volume is completely new as to text, keys, and illustrations. It is entirely Dr. Welsh's work, but it has been based upon the Anderson Herbarium and designed to retain the "flavor" of Dr. Anderson's original. It will serve as a work of reference not only for the professional, but also for many others who are interested in the largely unspoiled flora of our last frontier.

R. W. Pohl
Iowa State University
Ames, Iowa
5 Feb. 1974

Preface

Jacob Peter Anderson was born on 7 April 1874, in Glenwood, Sevier County, Utah. He was reared in the Midwest where he received his formal education at the University of Nebraska and at Iowa State University, graduating from Iowa State in 1913. On 1 February 1914 he began a journey which was to lead to his life's work on the Alaska flora.

On his arrival in Alaska he immediately began to collect plants. By 1924 he had collected more than 2,200 numbers of vascular plants and 1,100 of other plant species, mainly parasitic fungi. However, the entire collection was destroyed by fire in November of 1924. Duplicates of some of this early collection are located in the U. S. National Herbarium, New York Botanical Garden, and Iowa State University Herbarium. Many of the specimens were represented only in the Anderson herbarium and these were lost entirely. In 1925 he began a second collection, which ultimately amounted to more than 11,000 of his own specimens, and more than that number received in exchange from other workers.

In October of 1941, Dr. Anderson moved to Ames, Iowa. During his tenure at Iowa State University, he compiled a treatment of the *Flora of Alaska and Adjacent Parts of Canada*. The work was published in nine fascicles. Fortunately, all of them were completed before Anderson's death at the age of seventy-nine in 1953. His collections of plants, known as the J. P. Anderson Herbarium of Arctic and Boreal Plants, were bequeathed according to the terms of his will to Iowa State University where they are preserved as a part of the University Herbarium. Dr. Anderson left a modest bequest to Iowa State University to be used in helping to finance a revision of his *Flora*.

My own interest in Alaskan plants began in 1957 with my graduate studies at Iowa State University. The office which I shared with another graduate student was the one used previously by Dr. Anderson. Lining the hallway and in a room near the office was the Anderson collection. My research dealt with genera of legumes with circumboreal representation, and I consulted the Anderson collection with increasing frequency to determine the distribution of our American representatives in the northwestern part of the continent. Some of the species were found under names currently being used in North American plant taxonomy. Others were filed under names that were not consistent with either American or Eurasian materials with which they were conspecific.

Because of the apparent need for more nearly uniform taxonomic treatment, I undertook revisions of some genera during the 1960s. Both *Astragalus* (1963) and *Oxytropis* (1967) were revised during that period. In February of 1965, I was contacted by Dr. Richard W. Pohl, director of the herbaria at Iowa State University. Dr. Pohl suggested that I undertake the revision of Anderson's *Flora*.

I spent most of the field season of 1965, 1966, and 1968 in Alaska and adjacent Canada, supported by financial aid from the Anderson bequest. I was able to travel most of the highways in Alaska and Yukon, and take a trip to Point Barrow, Point Hope, Kotzebue, and Nome in 1966. I collected at Juneau, Skagway, and Haines during the 1966-68 period. My collections during those years yielded almost 4,000 separate numbers and several thousand duplicates. The first set of plants was selected for deposit with the Anderson collection at Iowa State University. Brigham Young University was chosen the depository for the second set. Major sets of duplicates were distributed to New York Botanical Garden, University of North Carolina, University of Alaska, and Komarov Botanical Institute.

In 1969, I spent a few days collecting by helicopter from an oil exploration camp in northern British Columbia. This allowed for a reevaluation of plant distributions and for the inclusion of some few species not previously reported for that region. Some 500 specimens were collected in British Columbia during 1969. Oil exploration camps at Sam Lake, Yukon, in 1970 and at Canoe and Loon lakes, NWT in 1971 and 1973 were centers for operation into the British, Barn, and Richardson mountains. Excursions were also taken into the Coastal Plains of Northern Yukon, Mackenzie River Delta, and Old Crow Flats. Some 1,800 specimens were collected during the three years. Travel in the region by helicopter presented an opportunity to visit many sites which had not been collected in previously. The trips by air made possible two reports on the flora in small parts of the included area. The reports (1971a, 1971b) include descriptions of the areas studied and contain annotated lists of taxa collected.

The revision of Anderson's *Flora* has been based on much new material provided by contemporary collectors. At the outset of my study, I was contacted by Maxcine Morgan Williams, a longtime resident of Alaska and a botany enthusiast with a continuing interest in Alaskan plants encouraged by association during the 1930s with

J. P. Anderson. During the 1960s, she was engaged in a self-appointed task of collecting representatives of the flora to add to the herbarium of the Juneau Botanical Club. That herbarium was built in response to a need felt by a group of interested amateur botanists mainly in the Juneau area. The club and its members had collected widely in Alaska and Yukon. The herbarium, consisting of several thousand specimens, was housed during the early period of my study in the home of Mrs. Amy Rude in Juneau. Recently, the collection has been permanently deposited in the Alaska State Museum in Juneau.

Mrs. Williams has contributed almost 2,000 critically collected specimens for this study. These have been taken along the highways in southern Alaska and Yukon, and from such remote regions as the Arctic Coast, Seward Peninsula, Anaktuvuk Pass, islands of the Bering Sea, the Aleutians, and Alaska Peninsula.

Through Mrs. Williams I was introduced to several careful botanical collectors. Mrs. Aline Strutz of Anchorage had been collecting native Alaskan plants for several years. She has provided more than 1,000 specimens, mainly from southern Alaska, but also from the Arctic Coast, Seward Peninsula, and the Aleutians. Mrs. Helen Schmuck of Nenana has provided many interesting specimens from that region. Perhaps the finest collection of plants from the southeastern portion of Alaska was provided by Mr. Dennis Jacques during the summer of 1972. These important specimens contained several new records of range extensions. Mrs. Katherine Reed has traveled widely in interior and southwestern Alaska. Several hundred of her collections support the basic work in this treatment.

Dr. William Mitchell of Palmer, Alaska, aided me during the first year of fieldwork. He also determined critical specimens of grass species and verified numerous others. The support of the Alaska Agricultural Experimental Station in providing drying equipment and storage space aided materially in this study. I was also given free access to the herbarium of the experiment station.

Drs. Richard W. Pohl and Duane Isely, both of Iowa State University, assisted me at every point during my investigation. Financial aid, space, and facilities of the herbarium at Iowa State were provided generously. During 1966-67, I devoted the entire year to study of the specimens in the Anderson herbarium at Ames, Iowa, and during the winter of 1972-73 I returned to that institution for some weeks to conclude the treatment of the monocots.

I was able to examine certain materials from the herbaria of the University of Alaska, Gray Herbarium, New York Botanical Garden, U. S. National Museum, and from the Riksmuseum in Stockholm.

To all persons noted above and to the curators of the various herbaria from which specimens have been obtained, I am grateful. Several others have contributed to this work in various ways. It is with special gratitude that I single out Dr. Glen Moore, my companion during the trips northward in 1966 and 1968. Dr. Moore is a plant collector of unlimited energy. He collected plants long after I was so tired I could not even place plants into the plant press.

Mrs. Kay Thorne must be recognized for her zeal in providing accurate illustrations of representative species from the region. Her talent is clearly seen in her fine work.

Dr. Keith Rigby made it possible for me to visit the oil exploration camps as a guest of Union Oil Company of Canada. I am grateful to Dr. Rigby and to the officers of the Union Oil Company. Their support in the field made possible the exploration of regions not previously accessible. Several of the geological staff members at the exploration camps became enthusiastically involved in providing specimens from sites where they were conducting geological studies. To each of them I am grateful.

The facilities of the herbarium of Brigham Young University have been used extensively in this work. Also, that herbarium is the main repository of specimens from the latter years of my fieldwork and for all specimens received as gifts or for determination during the course of this study. Therefore, I am especially grateful to the officers of Brigham Young University for their support in this revision. The late Dr. Dana Stocks provided support and encouragement throughout the project.

Dr. George Argus provided the treatment of the willows. He deserves special thanks for this task.

Finally, I wish to express thanks to my wife, Stella, for her long-suffering, help, encouragement, and support without which this work could not have been completed.

Stanley L. Welsh
Provo, Utah
September 1, 1973

Introduction

Alaska, Yukon Territory, and adjacent British Columbia is a vast and magnificent region. Together, Alaska and Yukon Territory comprise roughly 775,000 square miles, and the included portion of northwestern British Columbia comprises approximately 25,000 square miles, bringing the total for the area treated in this manual to approximately 800,000 square miles.

Because of its great size and the attendant difficulties of travel within it, the region has been opened to botanical exploration slowly and with much hardship. Early botanists to visit the region were associated mainly with expeditions engaged in more mundane tasks than those involved with the advancement of botanical knowledge. The expeditions moved by ship along the coasts and later into the navigable rivers which drain from the land mass. Most of the specimens were collected along the seacoasts.

The advent of the twentieth century did not find conditions for travel in interior Alaska much improved. Winter was the most auspicious time for moving long distances overland but was the least opportune time for plant collecting. Travelers during the growing season went by boat, foot, or horse, and were plagued by poor trails and hordes of mosquitoes.

A romantic aura soon developed about Alaska and about Arctic North America generally. This romanticism has stirred the hearts of men for generations, and they found their way north on a series of pretexts. Most sought for gold, but others were in search of experience and self-satisfaction. Among the adventurers were persons with some botanical training. From these men a trickle of specimens began to arrive at the major herbaria. The volume of specimens increased in size between the years 1900 and 1940. The increase in botanical collecting was due primarily to the activities of a few persons. Among them there are two names which stand out: J. P. Anderson and Eric Hultén. Dr. Anderson was a resident of Alaska from 1914 to 1941. He traveled widely and amassed a large and quite complete collection of flowering plants, vascular cryptogams, and rust fungi. Dr Hultén traveled in the Aleutian Islands in the 1930s where he collected materials that formed the basis of his *Flora of the Aleutian Islands*. A. E. Porsild of Canada collected in Alaska in the 1930s and in the Yukon in the 1940s. His collections have added substantially to the knowledge of portions of the region previously unknown.

The impact of world events during the past three decades has focused attention upon Alaska as an important military base and as a strategic site for long-distance flights. Alaska-Yukon has been the scene of increasing activities of all types since the 1940s. Highways have been constructed which tie several of the principal cities and towns together, and which connect them with the rest of Canada and the United States. Travel by air has opened the entire country to exploration. Thousands of people have gone north to seek their fortunes, find "peace of mind," enjoy the scenery, settle, or hunt and fish. The gold which flowed from this tremendous region during the historic past flows northward now in an ever increasing amount.

In parallel development with this increase in activity was a growing awareness of the flora of Alaska, Yukon, and British Columbia. Through funds provided by various agencies, several botanical investigations have been carried out, usually as part of some project of broader scope. Most of the field work has been done by young botanists who did not mind the rigors of field work in this boreal wilderness. Information taken from herbarium labels indicates that most of the investigators were in the field for only one or two collecting seasons. The contributions made by these collectors are very important, more so than the length of their tenure in the field would seem to indicate. Large numbers of specimens were collected from previously unexplored regions. Ranges were extended, and new taxa were discovered.

Specimens from the early years of this botanical interest found their way to Anderson, Hultén, and Porsild. Their interpretations were added to those of all earlier collectors and helped to form the basic body of knowledge on which both Anderson's and Hultén's floras were based. The mid- to late 1940s formed the turning point from the strictly exploratory phase of Arctic American botany. The publications of Anderson and Hultén added impetus to the botanical renaissance. Now it was possible for a person with training in basic botany to determine most of the plants with which he came in contact.

Members of the Juneau Botanical Club, an organization of amateur botanists inspired by J. P. Anderson, began an active drive to build a collection. This herbarium now consists of several thousand specimens and is housed in Juneau. The University of Alaska has amassed a very large collection of arctic and boreal plants from Alaska and Yukon. The Anderson herbarium is housed at Iowa State University in Ames, Iowa. It has continued to grow, even though its founder has passed away. The tremendous collection of arctic plants in the National Museum of Canada continues to receive specimens from throughout Canada and Alaska. The very large collection of specimens from the American Arctic in the Naturhistoriska Riksmuseet at Stockholm rivals those in North American herbaria in size and coverage.

The present work is an attempt to bring together current information on indigenous and the common cultivated plants of Alaska and Yukon. The presentation differs from traditional floras in that the treatment is in alphabetical order by families, genera, and species, with only the major categories being in phylogenetic order. Thus, the work begins with vascular cryptogams and conifers. The dicotyledoneae follows and comprises the bulk of the treatment. The monocot families are at the end of the work. The reason for this modification is principally to allow for ease of use to those students of the flora who spend so much time looking for bits of information on individual species.

DIVISION TRACHEOPHYTA

Plants with well-developed vascular system: xylem and phloem, leaves, either macrophylls or microphylls, and roots; reproduction by means of spores or seeds, the latter borne in cones or flowers.

1a. Plants with microphylls (small, scale-like leaves, usually with a single vein); reproduction by means of spores; flowers or woody cones lacking. (2)

1b. Plants with macrophylls (large leaves, usually with more than a single vein); reproduction by spores or seeds, the latter borne in flowers or cones. *Subdivision PTEROPSIDA*. (3)

2a. Stems not jointed; leaves green and imbricated, not whorled or forming a sheath at the node; plants either aquatic and grasslike or terrestrial and resembling mosses or miniature conifers; sporangia borne in axils of leaves or modified leaves in cone. *Subdivision LYCOPSIDA* .... p. 1

2b. Stems jointed and fluted; leaves not green, reduced to a whorl of connate scales at the nodes; plants neither grass- nor moss- nor coniferlike; sporangia borne under peltate, scalelike structures in terminal cones. *Subdivision SPHENOPSIDA* .... p. 7

3a. Plants fernlike, with broad leaves, reproducing by spores; flowers and cones lacking. *Class FILICINAE* .... p. 11

3b. Plants generally not fernlike, reproducing by seeds; flowers or cones present. (4)

4a. Seeds borne naked on the surface of a scale, these crowded together on an axis and forming a cone; flowers lacking. *Class GYMNOSPERMAE* .... p. 29

4b. Seeds borne in ripening carpels; plants with flowers. *Class ANGIOSPERMAE*. (5)

5a. Cotyledons 2; stems mostly increasing in diameter by means of a cambium between the xylem and phloem; leaves mostly net-veined; flower parts in 4s or 5s. *Subclass DICOTYLEDONEAE* .... p. 39

5b. Cotyledons 1; stems usually lacking a cambium, or if a cambium is present, it produces entire vascular bundles; leaves mostly parallel-veined; flower parts usually in 3s. *Subclass MONOCOTYLEDONEAE* .... p. 484

Subdivision LYCOPSIDA

Plants with solid stems; leaves microphyllous, scalelike, subulate, or grasslike, arranged spirally, basally, or opposite; sporophylls leaflike, with axillary sporangia or with the sporangia adnate to the sporophyll base, commonly aggregated into a terminal strobilus (except in *Isoetes* and *Lycopodium selago*); spores all alike (homosporous) or of two types (heterosporous).

1a. Plants aquatic, submersed in ponds or lakes, or occasionally growing on

exposed mud; leaves grasslike, long and slender, from broadly clasping bases; sporangia borne imbedded in the leaf base. ISOETACEAE (*Isoetes*)....
.... p. 2

1b. Plants terrestrial, growing in various habitats, but usually not in ponds or lakes; leaves scalelike or subulate; sporangia borne in terminal strobili or at leaf bases along the stem. (2)

2a. Plants homosporous; sterile leaves lacking a ligule; strobili terete. LYCOPODIACEAE (*Lycopodium*) .... p. 3

2b. Plants heterosporous; sterile leaves with a ligule at the base; strobili quadrangular (terete in *S. selaginoides*). SELAGINELLACEAE (*Selaginella*) .... p. 6

ISOETACEAE
Quillwort Family

Plants perennial, aquatic, amphibious, or sometimes terrestrial herbs; stems cormlike, with leaves clustered in a close spiral at the summit of the stem; leaves simple, elongate, dilated basally, the blade hollow and transversely septate, the outermost sterile, the next innermost bearing megasporangia, and the next innermost bearing microsporangia; sporangia solitary, enclosed in a cavity on the ventral side of the leaf base; ligule (a small flap of tissue) borne above the sporangial cavity; spores of two types, microspores and megaspores.

ISOETES L.

Stems very short; leaves more or less cylindrical, elongate; sporangia borne at the base of the leaves, usually covered by a vellum, a thin flap of tissue.

Pfeiffer, N. E. 1922. Monograph of the Isoetaceae. Ann. Mo. Bot. Garden 9:79–232.

Isoetes echinospora Dur.
Quillwort

Stems 2-lobed; leaves 2.5–10(13) cm long, tapering from the base to a long, slender

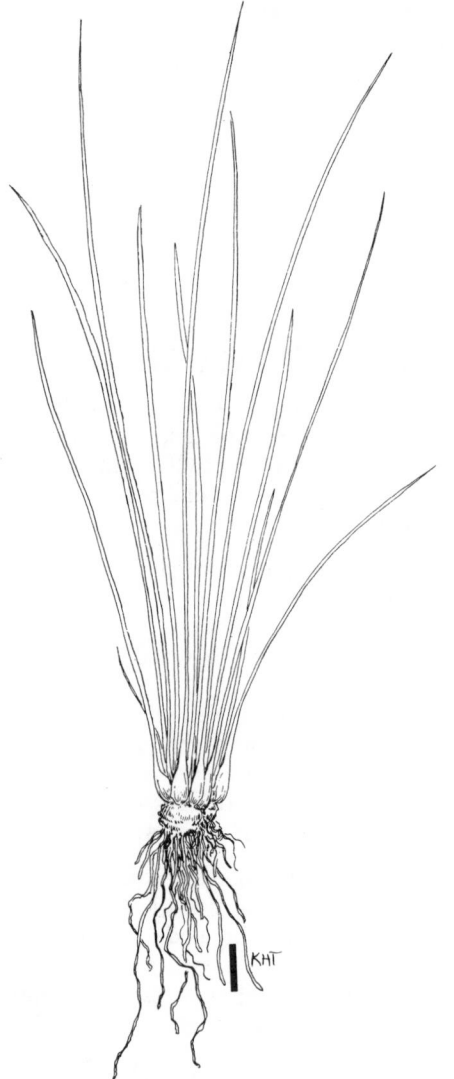

Isoetes echinospora Dur. (× 0.7).

tip, straight or curved; ligule deltoid; sporangia commonly 3–5 mm long; megaspores 0.25–0.6 mm broad, white, covered with short spines, the spines sharp, blunt, or some of them bifid.

Ponds and lakes; in widely scattered localities in Alaska south of Arctic Circle, but mostly coastal, and in southern Yukon; eastward to the Atlantic and south to California, Utah, Colorado, Minnesota, and

New Jersey; circumboreal (*I. echinospora* Dur.; *I. muricata* Dur.; *I. braunii* Dur.; *I. maritima* Underw.; *I. braunii* var. *maritima* [Underw.] Pfeiffer; *I. macounii* A. A. Eat.; *I. asiatica* [Makino] Makino; *I. flettii* [A. A. Eat.] Pfeiffer; *I. muricata* var. *braunii* [Dur.] Reed; *I. muricata* ssp. *muricata* [Underw.] Hultén; *I. echinospora* var. *truncata* Eaton in Gilbert; *I. truncata* [Eaton] Clute). The large number of synonyms indicates the degree of confusion as regards this assemblage of specimens. A monographic study is indicated.

LYCOPODIACEAE
Clubmoss Family

Plants perennial, terrestrial; stems elongate, upright or trailing, simple or branching from the apex, lacking axillary buds, leafy throughout or nearly so; leaves simple, 1-veined, spirally arranged or less commonly opposite, mostly small and scale- or needlelike; sporophylls like the vegetative leaves or unlike them, usually aggregated into a terminal strobilus, (except in *L. selago*); sporangia solitary, axillary or adnate to the ventral surface of the sporophyll, all alike; ligule lacking; spores all alike.

LYCOPODIUM L.

A single genus with characteristics of the family.

Wilce, J. H. 1965. Section Complanata of the genus *Lycopodium*. Nova Hedw. Beih. 19:1–233.

1a. Sporophylls appearing like the vegetative leaves, elongate, commonly several times longer than wide. (2)
1b. Sporophylls distinctly different from the vegetative leaves, commonly 1–2 times longer than wide. (3)

2a. Sporangia borne in the axils of ordinary foliage leaves, not forming strobili; stems perennial and evergreen, arising in small, erect clusters; plants widespread, frequently gemmiferous (bearing budlike gemmae among the leaves). *L. selago*
2b. Sporangia borne in the axils of sporophylls aggregated into a terminal sessile cone; stems annual, prostrate, with erect branches bearing strobili; plants of extreme southeastern Alaska. *L. inundatum*

3a. Strobili borne on slender peduncles. (4)
3b. Strobili borne sessile on vegetative branches. (5)

4a. Stems appearing flattened and wing margined; leaves 4-ranked, those of the upper and lower stem surfaces differing from the decurrent marginal ones. *L. complanatum*
4b. Stems appearing terete, not margined; leaves in about 10 ranks, all alike, not readily separable in upper or lower stem surfaces. *L. clavatum*

5a. Leaves commonly more than 4 mm long, borne in 6–10 ranks. (6)
5b. Leaves less than 4 mm long, borne in 4–5 ranks. (7)

6a. Prostrate stems subterranean; erect stems much branched, resembling a small conifer; leaves mostly less than 5 mm long; plants restricted to southern Alaska and Yukon. *L. obscurum*
6b. Prostrate stems above ground; erect stems simple, or with few long, erect-ascending branches, not especially coniferlike; leaves mostly more than 5 mm long; plants of broad distribution. *L. annotinum*

7a. Vegetative branches flattened; leaves in 4 ranks, the decurrent marginal ones unlike those of upper and lower surfaces; plants of coastal and interior localities. *L. alpinum*
7b. Vegetative branches terete; leaves in (4)5 ranks, all alike; plants of coastal southern Alaska. *L. sitchense*

Lycopodium alpinum L.
Alpine Clubmoss; Ground Fir

Stems perennial, the horizontal above

ground or barely covered, producing erect, dichotomously branched, aerial stems, the fertile ones 5–16 cm long, the vegetative ones generally shorter, at least some more or less flattened; leaves 4-ranked, more or less dimorphic, those on the lower surface trowel-shaped, the free tips commonly 2–3 mm long on fertile branches, commonly 6-ranked and all alike; strobili solitary or few, sessile or essentially so, 1–2.5 cm long; sporophylls greenish or yellowish, becoming brown in age, ovate, acuminate, slightly erose.

Meadows in tundra and heathlands, and in woods; in most of Alaska south of the Brooks Range, and in southern Yukon; eastward to northeastern North Amer-

Lycopodium annotinum L. (× 0.5).

ica; circumboreal (*Diphasium alpinum* [L.] Rothm.).

Lycopodium annotinum L.
Stiff Clubmoss

Stems perennial, the horizontal above ground, producing erect or ascending, simple or dichotomously branched aerial stems, the fertile ones commonly 5–20(25) cm long, the vegetative ones subequal to the fertile or slightly shorter, terete; leaves 8-ranked, all about alike, mostly 4–10 mm long, oblong-elliptic to narrowly lanceolate, on aerial stems serrulate or entire, minutely spinulose-tipped; strobili solitary, sessile, 1.2–3.5 cm long; sporophylls straw-colored or greenish, ovate, acuminate, with scarious, denticulate margins.

Ground layer in woods, thickets, heathlands, and tundra; in much of Alaska from the Brooks Range southward, and most of the Yukon; eastward to the Atlantic and south to Oregon, Colorado, Minnesota, Pennsylvania, and Virginia; circumboreal. There are three very common species of *Lycopodium* in our region, *annotinum, clavatum*, and *selago*. Perhaps the most abundant is *L. annotinum*. Two poorly defined varieties, joined by intermediates, are present.

1a. Plants yellowish, with slender, entire-margined leaves, mostly known from interior sites (*L. pungens* La-Pyl.; *L. annotinum* ssp. *pungens* [La-Pyl.] Hultén). *L. annotinum* var. *pungens* (LaPyl.) Desv.

1b. Plants green, seldom yellowish, with broader, serrulate-margined leaves, widely distributed (includes *L. annotinum* var. *alpestre* Hartm.; *L. annotinum* ssp. *alpestre* [Hartm.] Löve & Löve). *L. annotinum* var. *annotinum*

Lycopodium clavatum L.
Running Clubmoss

Stems perennial, the horizontal above ground, producing erect or ascending, simple- or few-branched, aerial stems, the fertile ones mostly 6–20(25) cm tall, the vegetative commonly much shorter than the fertile, terete; leaves mostly 10-ranked, all about alike, mostly 4–11 mm long, lance-linear to oblong, entire, long aristate-attenuate apically, the tip commonly twisted or bent, 1–3 mm long; strobili 1–5, borne on peduncles 0.5–10(15) cm long, 1.8–4.5 cm long; sporophylls straw-colored or greenish, ovate, long aristate-acuminate, with scarious, denticulate margins.

Woods, heathlands, and tundra; in most of Alaska south of the Brooks Range, and most of the Yukon; eastward to the Atlantic, and south to California, Idaho, Michigan, and North Carolina; circumboreal (*L. clavatum* var. *monostachyon* Grev. & Hook.; *L. clavatum* ssp. *monostachyon* [Grev. & Hook.] Sel.). Proposed segregates do not appear to be maintainable either by means of single morphological features or by combinations of them, although there is a tendency for specimens of the interior to have short peduncles.

Lycopodium complanatum L.
Ground Cedar

Stems perennial, the horizontal below ground or rarely above ground, producing erect, freely branched, aerial stems 6–25 (30) cm tall, the vegetative branches commonly overtopped by the fertile ones, flattened; leaves 4-ranked (6-ranked on some stems), the lateral ones opposite, continuous with the stem margin, the decurrent base longer than the free apex, the tip 1–2 mm long, ovate to triangular-acuminate, those of the upper side large and with an elevated base, those of the lower side with only the free tip evident; strobili 1–3, borne on peduncles 1–6.5 cm long, 0.6–2.7 cm long; sporophylls straw-colored or greenish, turning brown in age, ovate, acuminate, erose.

Woods, thickets, and heathlands; in much of Alaska south of the Brooks Range, except for the southwestern and extreme southeastern portions, and in southern Yukon; east to the Atlantic, south to Wash-

ington, Idaho, Montana, Minnesota, New England; circumboreal (*Diphasium complanatum* [L.] Rothm.).

Lycopodium inundatum L.
Bog Clubmoss

Stems annual, the horizontal above ground, producing erect or ascending, simple, aerial stems 2–10(15) cm tall, all branches fertile, more or less flattened; leaves 8–10 ranked, all about alike, mostly 4–8 mm long, lance-linear, entire, gradually attenuate to a soft tip; strobili solitary, sessile, 1.5–3.5(4) cm long; sporophylls green, lance-attenuate, resembling foliage leaves, entire.

Pond and lake shores, and bogs; in extreme southeastern Alaska and adjacent British Columbia; and disjunctly in northeastern North America; Europe, Asia (*Lepidotis inundata* [L.] C. Boerner; *Plananthus inundatus* [L.] Beauv.).

Lycopodium obscurum L.
Tree Clubmoss

Stems perennial, the horizontal deeply placed below ground producing erect, much-branched, aerial stems 7–20(25) cm tall which resemble small conifers, terete or flattened; leaves 6–8-ranked, all about alike, mostly 2–5 mm long, elliptic to oblong, entire, spinulose-tipped; strobili solitary, sessile at the ends of branches, 1–2.5 (3) cm long; sporophylls greenish, becoming yellowish brown, cordate, abruptly acuminate, the hyaline margins more or less erose.

Woods, bogs, heathlands, and tundra; in widely spaced localities in interior central Alaska, on the Alaska Peninsula, the western Aleutians, and extreme southeastern portion of the Panhandle, and in central western Yukon; eastward to the Atlantic and southward to Washington, Idaho, Montana, Michigan, West Virginia, and North Carolina; Asia (*L. dendroideum* Michx.; *L. obscurum* var. *dendroideum* [Michx.] D. C. Eaton).

Lycopodium selago L.
Fir Clubmoss

Stems perennial, the horizontal, below ground portion very short, producing erect or ascending, simple or dichotomously branched aerial stems 2–25(30) cm long, commonly tufted, terete; leaves 8-ranked, all about alike, subappressed and 3–5 mm long or widely spreading and 4–10 mm long, narrowly lanceolate to lance-attenuate, entire or nearly so, attenuate to acuminate; gemmae (bulbils) present in the axils of some upper leaves; strobili not apparent, the sporophylls green to yellowish, essentially like the vegetative leaves, entire.

Alpine and arctic tundra, heathlands, and woods; almost throughout Alaska and Yukon; eastward to the Atlantic and southward to Oregon, Montana, Michigan, and North Carolina; circumboreal. Several infraspecific taxa have been proposed for specimens that represent morphological extremes within *L. selago*. All are connected by a continuous series of intermediates, and the control appears to be ecological rather than genetic. The named segregates include: var. *appressum* Desv. (ssp. *appressum* [Desv.] Hultén), leaves ovate-lanceolate, appressed, and incurved; var. *miyoshianum* Makino (ssp. *miyoshianum* [Makino] Calder & Taylor; ssp. *chinense* [Christens.] Hultén), with strongly spreading to reflexed lance-attenuate leaves; and var. *selago*, with lance-attenuate, ascending leaves.

Lycopodium sitchense Rupr.
Alaskan Clubmoss

Stems perennial, the horizontal above ground or slightly covered, producing erect, dichotomously branched aerial stems, most of which are vegetative and only 3–10 cm long, the fertile branches much longer than the vegetative ones, terete; leaves commonly 5-ranked, all about alike, the free tips commonly 2–4 mm long, on fertile branches commonly 6-ranked; strobili solitary, or less commonly paired, sessile or essentially so, 1–2.5 cm long; sporophylls yellowish green, becoming brown in age, ovate, acuminate, slightly erose.

Alpine meadows, heathlands, and woods; in coastal and insular southern Alaska and southern Yukon; disjunctly eastward to the Atlantic and southward to Oregon, Idaho, Montana, Michigan, and New Hampshire; Asia (*L. sabinaefolium* var. *sitchense* [Rupr.] Fern.; *L. sabinaefolium* ssp. *sitchense* [Rupr.] Calder & Taylor). This entity is closely allied to *L. alpinum* from which it can be identified only with difficulty, mainly on the basis of terete stems, and leaves alike in 5–6 ranks.

SELAGINELLACEAE
Spikemoss Family

Plants perennial, terrestrial; stems trailing, with erect, ascending, or prostrate dichotomous branches, lacking axillary buds, leafy throughout; leaves simple, 1-veined, spirally arranged, mostly scale- or needle-like; sporophylls similar to the vegetative leaves, aggregated into terminal strobili; sporangia solitary, axillary, of two types, usually the uppermost producing numerous tiny microspores, and the lower 1–4 large megaspores; ligules present.

SELAGINELLA Beauv.

A single genus, with characters of the family.

Tryon, R. M. 1955. *Selaginella rupestris* and its allies. Ann. Mo. Bot. Gard. 42:1–99.

1a. Leaves without a dorsal groove, the margin spinulose-toothed; sporophylls loosely aggregated, mostly over 2.5 mm long; strobili terete. *S. selaginoides*

1b. Leaves with a dorsal groove, the margin merely ciliate, not toothed; sporophylls densely aggregated, mostly less than 2.5 mm long; strobili quadrangular. (2)

2a. Plants loosely spreading; setae at leaf apices white; distributed widely in Alaska and western Yukon and in northern British Columbia. *S. sibirica*

2b. Plants densely clump-forming; setae at leaf apices becoming yellowish; reported by Tryon (1955) to be present in southeastern Alaska, the report requires verification. *S. densa* Rydb. var. *standleyi* (Maxon) Tryon

Selaginella selaginoides (L.) Link
Low Selaginella

Stems of two types, the vegetative ones more or less prostrate and mat-forming, the fertile ones ascending, mostly (1)3–10(15) cm long; leaves spirally arranged, commonly 1.5–3 mm long, lanceolate to elliptic, spinulose-toothed marginally, sharply attenuate; strobili terete, solitary, 2–4 cm long, sessile, not strongly differentiated; sporophylls similar to vegetative leaves, 2.5–5 mm long, prominently spinulose-toothed marginally.

Woods, thickets, and heathlands, in moist sites; in much of Alaska south of the Brooks Range and in southern Yukon and in the Mackenzie; disjunctly eastward to the Atlantic and south to Nevada, Idaho, Wyoming, Minnesota, and New Hampshire; circumboreal.

Selaginella sibirica (Milde) Hieron.
Northern Selaginella

Stems prostrate, with erect or ascending branches, mostly 0.5–1.5 (2) cm long, forming open, loose mats; leaves densely imbricate, commonly 1–2(2.5) mm long, lance-attenuate to narrowly oblong, ciliate marginally, decurrent basally, abruptly white-setose apically, the setae forming conspicuous tufts at the tips of vegetative branches; strobili solitary, terminating short, ascending to erect branches, mostly 0.5–1.5 cm long, quadrangular; sporophylls dissimilar to vegetative leaves, definitely 4-ranked, commonly 3–4 times as wide as the leaves, ovate, shortly setose, ciliate or lacking cilia. Ridge tops, rock outcrops, and less commonly in thickets in open woods; in much of Alaska (except for the northern coast and southeastern portion), and in northern to southwestern Yukon and northern British Columbia; eastward to the Mackenzie; Asia (*S. rupestris* f. *sibirica* Milde; *S. schmidtii* var. *krauseorum* Huron). This entity is apparently closely related to *S. densa* Rydb., whose range approaches that of *S. sibirica* from the south.

Subdivision SPHENOPSIDA

Plants with solid or hollow, jointed, longitudinally ribbed stems; leaves microphyllous, whorled, small, and scalelike; strobili spikelike, bearing numerous stalked, peltate scales with sporangia on the lower surface; spores numerous, all alike.

EQUISETACEAE
Horsetail Family

Stems jointed, ribbed, with a persistent meristem at the base of each internode; leaves whorled, scalelike, connate-sheathing; branches (if any) breaking through base of the sheath, alternating with ridges of internodes above; peltate sporangiophores, each with 5–10 sporangia beneath a polygonal cap; spores spherical, each with 4 spirally wound elaters.

EQUISETUM L.

Plants rhizomatous, perennials; stems annual or perennial and evergreen, with silicified cell walls; strobili borne on photosynthetic stems, or on specialized non-photosynthetic stems.

Hanke, R. L. 1963. A taxonomic monograph of the genus *Equisetum*—subgenus Hippochaete. Nova Hedw. Beih. 8:1–123.

1a. Stems evergreen, usually unbranched, or, if so, then lacking regular whorls of branches; cones with peduncles seldom exceeding the subtending sheath, apiculate. (2)

1b. Stems annual, not evergreen, often of 2 types, some usually with regular whorls of branches; cones with at least some peduncles much surpassing the subtending sheath, rounded apically. (4)

2a. Leaf sheaths 3-toothed; stems lacking a central cavity. *E. scirpoides*
2b. Leaf sheath (3)4–20-toothed or more; stems with a central cavity. (3)

3a. Stems 5–10(12)-ridged, commonly 1–3(4) dm tall; teeth of leaf sheaths persistent. *E. variegatum*
3b. Stems (14)16–20-ridged or more, commonly 2–10 dm tall or more; teeth of sheaths jointed, deciduous. *E. hyemale*

4a. Plants bearing strobili in summer; fertile and sterile stems alike, green; ridges of stem smooth or minutely cross-wrinkled, lacking tubercles or spicules. (5)
4b. Plants bearing strobili in spring (less commonly in summer); fertile and sterile stems not alike; ridges of stem with tubercles or spicules, or almost smooth, but not cross-wrinkled. (6)

5a. Stems deeply 5–10-grooved, the central cavity about 1/6 the diameter of the stem; teeth of leaf sheaths (2.5) 3–7 mm long, conspicuously white-margined. *E. palustre*
5b. Stems shallowly 9–25-grooved, the central cavity more than 1/2 the diameter of the stem; teeth of leaf sheaths 1.5–3.5 mm long, not or scarcely white-margined. *E. fluviatile*

6a. Fertile stems whitish, pinkish, yellowish, or brownish, soon withering; ridges of stem with minute bumps or cross-ridges; branches of sterile stems not again branched. *E. arvense*
6b. Fertile stems becoming green and branched, persistent; ridges of stem with spicules, tubercles, or transverse ridges; branches of sterile stems often again branched. (7)

7a. Leaf sheaths flaring upwards, the chestnut brown teeth cohering in several broad lobes; branches usually again branched. *E. sylvaticum*
7b. Leaf sheaths subcylindrical, the teeth free or nearly so; branches usually unbranched. *E. pratense*

Equisetum arvense L.
Meadow Horsetail

Stems annual, of 2 types, the sterile ones (0.5)1–5(6) dm tall, 1–5 mm thick, 10–12-ridged, the ridges with minute bumps and cross-ridges, the central cavity about 1/4 the stem diameter, the stomates in 2 broad bands, not sunken, the sheaths 5–10 mm long, greenish, with teeth 1–3 mm long, persistent, separate or some united, brown or blackish, the margins sometimes pale and hyaline; branches in regular whorls, 3–4-ridged, solid, usually not branched again; fertile stems whitish, pinkish, brownish, or yellowish, borne in springtime, soon withered, 0.6–3 dm tall, 3–8 mm thick, the sheaths 10–20 mm long, with teeth 5–9(11) mm long, some connate; strobilus 5–35 mm long or more, with peduncles much longer than the subtending sheath, blunt apically.

Arctic and alpine tundra, heathlands, thickets, and woods; throughout Alaska and Yukon; widely distributed in North America; circumboreal (*E. boreale* Bong.; *E. arvense* ssp. *boreale* [Bong.] A. Löve; *E. arvense* var. *boreale* [Bong.] Ledeb.). This is one of the more persistent weedy plants of cultivated land in Alaska and Yukon.

Equisetum fluviatile L.
Swamp Horsetail

Stems annual, all about alike, mostly 3–10 dm tall, 3–8 mm thick, shallowly 9–25-grooved, the ridges smooth or with minute, transverse wrinkles, the central cavity about 3/4 the stem diameter, the stomates in 1 band in each groove, not sunken; sheaths 4–10 mm long, greenish, the teeth 1.5–3.5 mm long, persistent, black or blackish, not or only narrowly hyaline-margined;

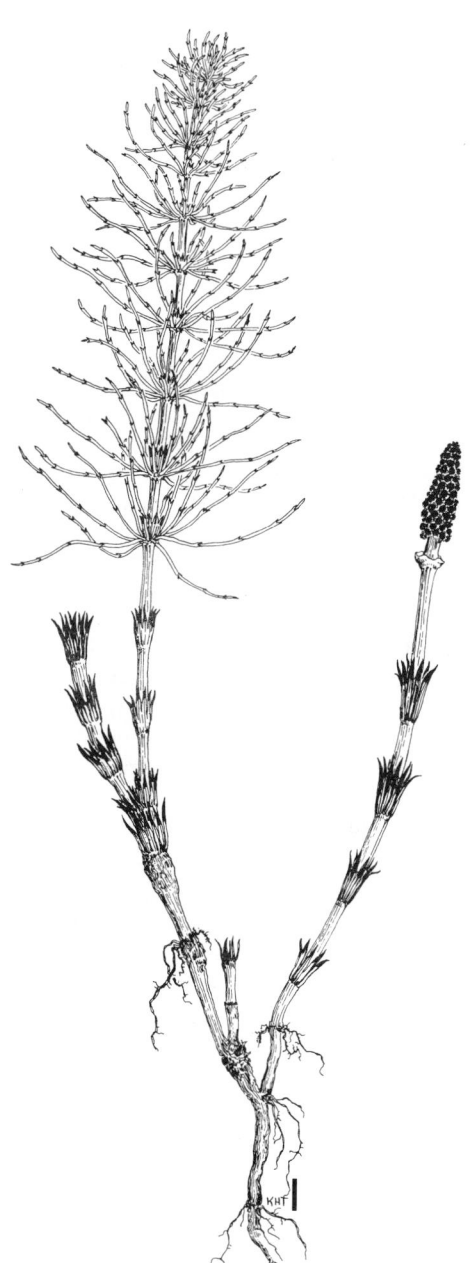

Equisetum arvense L. ($\times$ 0.4).

branches lacking, or few to many and whorled, 4–6-ridged, not again branched; strobilus 5–20 mm long, subsessile or with peduncles slightly surpassing the subtending sheath, blunt apically.

Marshes and tidal flats; in much of Alaska (except for northern coastal regions and the Aleutians) and most of the Yukon; eastward to the Atlantic and southward to Washington, Idaho, Wyoming, Minnesota, and Pennsylvania; circumboreal. *E. fluviatile* has been reported to produce poisoning in horses in south-central Alaska.

Equisetum hyemale L.
Common Scouring-rush

Stems perennial, evergreen, all alike, commonly 2–10 dm tall or more, 4–10 mm thick or more, with (14)16–20 ridges or more, the ridges with 2 rows of tubercles or 1 row of transverse ridges, the central cavity about 3/4 the stem diameter, the stomates in 2 rows in each groove, not sunken; sheaths 3–10(15) mm long, usually with 2 black bands separated by a grayish band at maturity, the teeth 2–4 mm long, deciduous, black, hyaline-margined, jointed to the sheath; strobilus 10–25 mm long, subsessile or with peduncles subequal to the subtending sheath, stoutly apiculate.

Streambanks, seeps, and marshes; in the southeastern third of Alaska (and westernmost Aleutians) and southern Yukon; widespread in North America; Eurasia.

Equisetum palustre L.
Marsh Horsetail

Stems annual, all about alike, 1.5–6(8) dm tall, 1–3 mm thick, deeply 5–10-grooved, the ridges minutely transversely wrinkled, the central cavity less than 1/3 the stem diameter, the stomates in a broad band in each groove, not sunken; sheaths 4–10 mm long, greenish, the teeth (2.5)3–7 mm long, persistent, blackish or brownish, with conspicuous hyaline margins; branches lacking or few to many and whorled, 5–6-ridged, not again branched; strobilus 8–20 mm long, subsessile or with peduncles much surpassing the subtending sheath, blunt apically.

Stream banks, marshes, muskegs, and pond margins; in most of Alaska (except for northern coastal regions, the Aleutians and the Panhandle), and much of the Yukon; eastward to the Atlantic and south to Washington, Idaho, Montana, Nebraska, and Pennsylvania; circumboreal.

Equisetum pratense Ehrh.
Meadow Horsetail

Stems annual, of 2 types, the sterile ones 1–5 dm tall, 1–4 mm thick, 10–18-ridged, the ridges with blunt tubercles or transverse ridges, the central cavity 1/3 to 1/2 the stem diameter, the stomates in 2 bands in the grooves, not sunken; sheaths (1.5) 2–6 mm long, subcylindrical, greenish white, the teeth 1–4 mm long, persistent, usually separate, with narrow brownish median, and with broad hyaline margins; branches in regular whorls, mostly 3-ridged, solid, not again branched; fertile stems pale, borne in springtime, later becoming green and producing whorls of simple branches, 1–5 dm tall, 2–3 mm thick, with teeth 3–5 mm long, mostly distinct; strobilus 10–20 mm long, with peduncles much longer than the subtending sheath, blunt apically.

Woods and thickets, and less commonly in meadows; in most of Alaska, except for coastal northern and southern portions, and much of the Yukon; eastward to the Atlantic and south to Montana, Iowa, and New Jersey; circumboreal.

Equisetum scirpoides Michx.
Dwarf Scouring-rush

Stems perennial, evergreen, all alike, commonly 0.3–1.5 dm tall, 0.5–1 mm thick, with 6 ridges, each ridge with a single row of tubercles, the stomates in 2 rows in each major groove; central cavity lacking; sheaths (1)1.5–5 mm long, black or the base green or pale, the teeth 3, with hyaline margins, about 1–2 mm long; strobilus 3–5 mm long, subsessile or shortly pedunculate, apiculate.

Moist woods, heathlands, and tundra, often in moss; in most of Alaska, except for

Equisetum sylvaticum L.
Woodland Horsetail

Stems annual, of 2 types, the sterile ones 1.5–5(7) dm tall, 1–5 mm thick, 10–18-ridged, the ridges each with 2 rows of spreading or recurved spicules, the central cavity more than 1/2 the diameter of the stem, the stomates in 2 bands in the grooves, not sunken; sheaths 3–12(20) mm long, flaring, green, whitish, or brown (especially apically), the teeth 2–10 mm long, persistent, connate into several broad, brown lobes with hyaline margins; branches regularly whorled, 4–5-ridged, commonly again branched; fertile stems brownish or pale, borne in springtime, later becoming green and producing whorls of green compound branches, about the same size as the sterile branches but with larger sheaths and teeth; strobilus 1–3 cm long, with peduncle much longer than the subtending sheath, blunt apically.

Woods, heathlands, and meadows; in most of Alaska south of the Brooks Range, (except for the Aleutians), and in southern Yukon; eastward to the Atlantic and southward to Idaho, South Dakota, Iowa, and Kentucky; circumboreal.

Equisetum variegatum Schleich.
Variegated Scouring-rush

Stems perennial, evergreen, all alike, commonly (0.5)1–4 dm tall, 1–2(4) mm thick, with 5–12 ridges, each ridge with 2 rows of tubercles, the central cavity 1/4 to 1/3 the diameter of the stem, the stomates in 2 rows in each groove, sunken below the epidermis; sheaths (1)2–4 mm long, the base not easily distinguished, flared, black or blackish apically, the teeth 1–2(3) mm long, with conspicuous white-hyaline margins; strobilus (0.3)0.7–1 cm long, subsessile or shortly pedunculate, prominently apiculate.

Muskegs, streambanks, woods, and tundra; almost throughout Alaska and Yukon; eastward to the Atlantic and south to Washington, Utah, Illinois, and Pennsylvania; circumboreal.

Subdivision PTEROPSIDA
Class FILICINAE

Plants with solid, usually below ground stems; leaves (fronds) macrophyllous, alternate, mostly large and compound, or variously lobed or divided, commonly with petioles (stipes); sporangia often borne in sori on vegetative or modified leaves, usually on the lower surface or along the margins; spores numerous, all alike.

1a. Sporangia borne naked, on erect spike- or paniclelike segment of normal frond arising from base of blade or from petiole. OPHIOGLOSSACEAE.... .... p. 11
1b. Sporangia borne on margin or underside of the frond, usually with fertile and sterile portions of blade similar, dimorphic in some. (2)

2a. Sori marginal, enclosed by a 2-valved margin of frond; leaf blade membranous, one cell thick; sporangia with oblique annulus. HYMENOPHYLLACEAE .... p. 11
2b. Sori on the underside of the frond or submarginal; leaf blade several cells thick; sporangia with a vertical annulus POLYPODIACEAE .... p. 14

HYMENOPHYLLACEAE
Filmy Fern Family

Plants small, delicate, with slender, threadlike, long-creeping rhizomes; leaves alternate, and widely spaced along the rhizome, the blades only one cell thick; sori marginal, in cup- or purselike openings in the leaf margin; sporangia with an oblique annulus; spores tetrahedral.

HYMENOPHYLLUM L.

A single genus, with characteristics outlined for the family.

Hymenophyllum wrightii van den Bosch

Plants delicate; rhizomes threadlike, branched, forming loose mats; stipes to 2 cm long, glabrous, narrowly winged upwards; blades triangular-ovate, 1–3 cm long, and to half as wide, once to twice pinnately parted, the rachis winged; leaflets (pinnae) few, divided or the upper simple; indusia broadly ovate, the sori at tips of veins.

Moist rocks and tree trunks, in wet, maritime regions; in coastal and insular southeastern Alaska and British Columbia; Asia (*Mecodium wrightii* [van den Bosch] Copeland). This mosslike fern has long been overlooked in our flora (see Taylor, T. M. C. 1967. *Mecodium wrightii* in British Columbia and Alaska. Amer. Fern. Journ. 57[1]:1–6). The plant is known from Alaska (gametophytes only) from Biorka Island, 56° 52′ N., 135° 33′ W. (UBC). The species is known from sporophyte material from Queen Charlotte Island.

OPHIOGLOSSACEAE
Adder's-tongue Family

Plants more or less succulent, with short, tuberous, erect rhizomes; leaves 1 per stem, rarely more, simple or compound, nodding in bud, not circinate; fertile portion of frond distinct, borne erect, arising from the stipe; sporangia borne naked, on the panicle- or spikelike fertile segment, lacking indusia; sporangia without an annulus, opening by a regular split; spores tetrahedral, very numerous.

Clausen, R. T. 1938. A monograph of the Ophioglossaceae. Mem. Torrey Club 19(2):1–177.

1a. Blade simple, the veins reticulate; sporangia borne on a spikelike fertile stalk; known in our region only from Kodiak Island. *Ophioglossum*

1b. Blade compound or divided, the veins not reticulate; sporangia borne in paniclelike clusters; widely distributed. *Botrychium*

BOTRYCHIUM Sw.

Leaves with sterile blades pinnately or ternately compound, pinnatifid, or bipinnatifid, the veins not netted; fertile segment pinnate or bipinnate, appearing paniculate.

1a. Sterile leaf blade bipinnate to tripinnate, broadly deltoid; buds hairy; plants often more than 20 cm tall. (2)

1b. Sterile blade pinnatifid to pinnate-pinnatifid, oblong-lanceolate, often longer than broad; buds glabrous; plants usually less than 20 cm tall. (3)

2a. Fertile leaf segment arising far down the petiole, nearly at ground level; sterile blade evergreen. *B. multifidum*

2b. Fertile leaf segment arising at or near the base of the sterile blade; sterile blades deciduous. *B. virginianum*

3a. Sterile blade once pinnate or ternate, the lobes rounded or again slightly lobed; midvein of lobes lacking. (4)

3b. Sterile blade pinnate or bipinnatifid, the segments pinnatifid, the lobes various, the midvein of lobes distinct. (5)

4a. Sterile blade petiolate, the few lobes poorly developed; reported from Cannery Bay, Kodiak Island. *B. simplex* E. Hitchc.

4b. Sterile blade sessile or nearly so, the several lobes well developed; our most common and most widely distributed grapefern. *B. lunaria*

5a. Sterile blade deltoid, about as wide as long; fertile segment reflexed in bud. *B. lanceolatum*

5b. Sterile blade ovate to ovate-oblong, longer than wide; fertile segment erect in bud. *B. boreale*

Botrychium boreale (Fries) Milde
Northern Grapefern

Plants stout, fleshy, 4–20(26) cm tall, glabrous, yellow-green; sterile blade ovate to oblong in outline, sessile or nearly so, attached at or above the middle of the plant, 1.5–6 cm long, 1.2–4.5 cm broad, pinnate, the pinnae again pinnatifid, the ultimate lobes rounded to obtuse to acutish, not generally broadly rounded; fertile segments exceeding the sterile, the stalk 1–4(7) cm long, the fertile portion 0.8–6 cm long, simple or branched.

Alpine meadows and heathlands; in much of the southeastern one-third of continental Alaska, and in coastal and insular southern Alaska, widely distributed in the Yukon; south to Oregon and Nevada; Greenland, Eurasia (*B. lunaria* var. *boreale* Fries; *B. boreale* ssp. *obtusilobum* [Rupr.] R. T. Clausen; *B. crassinervium* var. *obtusilobum* Rupr.; *B. boreale* ssp. *obtusilobum* [Rupr.] R. T. Clausen; *B. boreale* var. *obtusilobum* [Rupr.] Brown).

Botrychium lanceolatum (Gmel.) Angstr.
Lance-leaved Grapefern

Plants fleshy, 6–20(35) cm tall, glabrous, green; sterile blade deltoid in outline, sessile or nearly so, attached near the top of the plant, 1–6 cm long, 1–8 cm broad, ternate, the pinnae again pinnatifid and usually longer than broad, the ultimate lobes obtuse, not broadly rounded; fertile segments subequal to or longer than the sterile, the stalk 0.5–3 cm long, the fertile portion 0.6–5 cm long, branched.

Alpine meadows and heathlands; in coastal and insular southern Alaska, and southern Yukon; eastward to the Atlantic and south to Washington, Utah, Wisconsin, and Pennsylvania (*Osmunda lanceolata* Gmel.).

Botrychium lunaria (L.) Swartz
Moonwort

Plants fleshy, (3)5–24(28) cm tall, glabrous, green; sterile blade oblong in out-

shaped, and often overlapping, frequently broadly rounded; fertile segment longer than the sterile, the stalk 1–9 cm long, the fertile portion (0.5)1–7 cm long, branched.

Woods, muskegs, open slopes, and meadows; in most of Alaska south of the Brooks Range, and much of the Yukon; east to the Atlantic and south to California, Arizona, Colorado, Michigan, and Maine; circumboreal. Two morphologically separable entities are represented among our materials. The following key will allow segregation of most specimens.

1a. Pinnae crowded, tending to overlap, usually more than 6 mm broad; our common phase. *B. lunaria* var. *lunaria*

1b. Pinnae widely spaced, not overlapping, usually less than 5 mm broad; uncommon (*B. minganense* Victorin; *B. lunaria* var. *minganense* [Victorin] Dole; *B. lunaria* f. *minganense* [Victorin] Clute; *B. onondagense* Underw.). *B. lunaria* var. *onondagense* (Underw.) House.

Botrychium multifidum (Gmel.) Trevis
Leathery Grapefern

Plants fleshy, commonly 10–30(50) cm tall, hairy, becoming glabrescent in age, green; sterile blade deltoid in outline, stalked, attached near the stem base, 3–16(30) cm broad, ternately 2–4 times compound, ultimate segments crenate, obtuse, persistent into the second season; fertile segment longer than the sterile, the stalk 4–19(25) cm long, the fertile portion (3)5–15(20) cm long, branched.

Open woods, meadows, and muskegs; in coastal and insular southern Alaska; east to the Atlantic and south to California, Idaho, Wyoming, Iowa, and North Carolina.

Botrychium virginianum (L.) Swartz
Rattlesnake Fern

Plants slender, 25–60(75) cm tall, hairy when young, becoming glabrescent in age,

Botrychium lunaria (L.) Swartz (× 0.8).

line, sessile or shortly stalked, attached at various heights along the plant, 1.2–7 cm long, 0.7–3 cm wide, pinnate, the pinnae entire, crenate, or incised, broadly fan-

green; sterile blade deltoid in outline, sessile or nearly so, attached above the middle of the plant, (5)8–17(20) cm long, (5)12–22(35) cm wide, ternately 2–4 times compound, the ultimate segments crenate, obtuse, deciduous; fertile segment longer than the sterile, the stalk 3.5–18 cm long, the fertile portion 2–10 cm long, branched.

Open grassy slopes and woods; in coastal and insular southern Alaska; east to the Atlantic and south to California, Arizona, and Florida. Our materials belong to ssp. *europaeum* (Angstr.) R. T. Clausen.

OPHIOGLOSSUM L.

Leaves with sterile blade simple, entire, the veins conspicuously netted; fertile segment simple, unbranched, spikelike.

Ophioglossum vulgatum L.
Adder's-tongue

Plants mostly 10–30 cm tall, glabrous; sterile blade elliptic to ovate, sessile or nearly so, attached well above the ground level, entire, rounded to obtuse apically, mostly 3–10 cm long and 1–4 cm broad; fertile segment longer than the sterile, mostly 3–10 cm long, the fertile spike 1–4 cm long, unbranched, with two rows of sessile or embedded sporangia along one side.

Open slopes; in our area known only from Unalaska Island; disjunctly southward to Washington and in eastern North America; circumboreal. *O. vulgatum* has not been collected in Alaska in recent times, and might well be extinct at Unalaska. It is to be expected elsewhere, however. Our material has been segregated as var. *alaskanum* (Britt.) C. Chr.

POLYPODIACEAE
Common Fern Family

Plants small to large; stems consisting of short to elongate rhizomes; leaves alternate, petiolate, erect to spreading or pendulous, circinate in bud (forming characteristic fiddleheads), simple or variously compound, the blade several cells thick; fertile and sterile leaves alike, or dissimilar in some genera; sporangia borne in sori, these marginal or on the lower surface, naked or more or less covered by an indusium, the sporangia opening by means of a vertical annulus.

1a. Leaves conspicuously dimorphic, the fertile with much contracted pinnae. (2)

1b. Leaves not dimorphic, the fertile and vegetative much alike. (4)

2a. Plants with both fertile and vegetative leaves once pinnately compound, the pinnae of fertile leaves involute, entire, with a continuous narrow sorus; plants of coastal and insular southern Alaska. *Blechnum*

2b. Plants with both fertile and vegetative leaves more than once pinnately compound, or with the pinnae pinnatifid, with sori various, but not as above; plants of various distribution. (3)

3a. Vegetative leaves commonly less than 20 cm tall, often exceeded by the fertile leaves. *Cryptogramma*

3b. Vegetative leaves commonly more than 50 cm tall, much longer than the fertile leaves. *Matteuccia*

4a. Sori borne along the margin of the lower surface of the leaves, covered by the inrolled indusiumlike margin. (5)

4b. Sori borne on the lower surface of the frond, not covered by the margin, naked or with true indusia. (6)

5a. Leaves twice to thrice pinnately compound, with a main central axis; sori interrupted along the margin; blades and rhizomes hairy, not scaly; stipe thick, tough, yellow to brown. *Pteridium*

5b. Leaves twice compound with a dichotomously branching rachis; sori interrupted along the margin; blades glabrous, the stipe (petiole) and rhizome scaly, not hairy; stipe slender, brittle, black. *Adiantum*

6a. Indusia lacking, even on young leaves, the sori naked. (7)

6b. Indusia present, at least on young leaves, the sori more or less protected. (10)

7a. Leaves once pinnately compound or pinnatifid, the pinnae entire or merely toothed, not again pinnatifid, evergreen; petioles articulated to the rhizome. *Polypodium*

7b. Leaves twice or thrice pinnately or ternately compound or pinnatifid, the pinnae lobed or again compound, herbaceous; petioles continuous with the rhizome. (8)

8a. Leaves at least thrice pinnately compound, densely crowded on the short, stout rhizome, lanceolate in outline. *Athyrium*

8b. Leaves twice pinnately compound, or the pinnae merely pinnatifid, or if thrice pinnately or ternate-pinnately compound, then broadly deltoid in outline. (9)

9a. Leaves more or less ternately compound, the pinnae once to twice pinnately compound or pinnatifid, the blade broadly deltoid, often broader than long, glabrous (glandular in *G. robertianum*); pinnae articulated to the rachis. *Gymnocarpium*

9b. Leaves pinnately compound, the pinnae pinnatifid, the blade ovate to deltoid in outline, longer than broad, more or less hairy at least when young, the lower pinnae not articulated to the rachis. *Thelypteris*

10a. Indusium elongate, flaplike, attached along the sorus on the side next to the edge of the leaf segment, sometimes curved apically but not kidney-shaped. (11)

10b. Indusium peltate, kidney-shaped, or attached below the sorus, not as above. (12)

11a. Plants small, the leaves commonly less than 20 cm long; leaves once pinnately compound, with toothed pinnae about as broad as long. *Asplenium*

11b. Plants large, the leaves commonly more than 30 cm long; leaves twice to thrice pinnately compound, the compound or pinnatifid pinnae several times longer than broad. *Athyrium*

12a. Leaves evergreen, more or less leathery, often sharply toothed; indusium peltate, attached at the center and opening all around. *Polystichum*

12b. Leaves deciduous, or less commonly evergreen, not leathery or sharply toothed; indusium not peltate. (13)

13a. Indusium orbicular to kidney-shaped, attached at the sinus; plants mostly coarse, with conspicuously scaly petioles; leaves often more than 2.5 dm long (except in *Dryopteris fragrans*). (14)

13b. Indusium hoodlike, or platelike and borne beneath sorus; plants mostly slender and delicate, with scaly or glabrous petioles; leaves commonly less than 2.5 dm long. (15)

14a. Leaves once pinnately compound, the pinnae deeply pinnatifid, pubescent along the upper surface and usually also along the sides of the rachis with unicellular hairs; vascular bundles in petiole 2. *Thelypteris*

14b. Leaves twice to thrice pinnately compound, merely scaly or minutely glandular along the rachis; vascular bundles in petiole 3 or more. *Dryopteris*

15a. Indusium hoodlike, attached at one side, partly under the sorus and commonly pushed back as the sorus expands; veins reaching the leaf margin; leaves not accompanied by conspicuous, persistent, petiole bases. *Cystopteris*

15b. Indusium platelike, borne symmetrically beneath the sorus, splitting into radiating, hairlike segments; veins not reaching leaf margin; leaves accompanied by conspicuous, persistent, petiole bases. *Woodsia*

ADIANTUM L.

Plants medium sized, delicate, mesophytic ferns, arising from short, creeping rhizomes beset with brownish or blackish scales; leaves deciduous, the petioles shining, wiry, each with one vascular bundle, the ultimate segments petiolulate, much broader than long, dichotomously veined; sori on reflexed margin, interrupted.

Adiantum pedatum L.
Northern Maidenhair Fern

Rhizome thick, beset with brown scales; leaves solitary or few, mostly 20–70 cm long, the petioles purplish black, 10–40

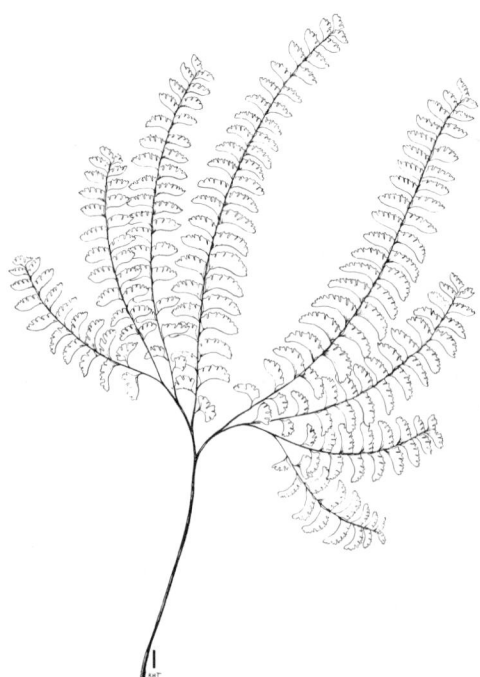

Adiantum pedatum L. (× 0.25).

cm long, the blade glabrous, with main pinnae dichotomously branched, shorter than the petioles and more or less parallel with the ground, 3–60 cm long and about as wide; pinnae commonly 8 or more, with the central ones the longer; pinnules broader than long, with a straight lower margin and a curved, lobed upper margin bearing the sori.

Moist sites in woods, and thickets; in coastal and insular southern Alaska from Unalaska eastward; east to the Atlantic and south to California, Utah, Oklahoma, and Georgia; Asia (*A. pedatum* var. *aleuticum* Rupr.; *A. pedatum* ssp. *aleuticum* [Rupr.] Calder & Taylor).

ASPLENIUM L.

Plants small to medium sized, delicate, mesophytic, evergreen ferns; rhizomes short, scaly; leaves once pinnately compound, the pinnae toothed or subentire, the veins not reaching the margin; petiole wiry, slender, green to black, with an X-shaped vascular bundle at the base; sori elongate, each borne on a veinlet; indusium flaplike, attached along the vein on the side of the sorus towards the pinnae margin.

1a. Leaf rachis green or greenish. *A. viride*
1b. Leaf rachis dark brown to purplish brown. *A. trichomanes*

Asplenium trichomanes L.
Maidenhair Spleenwort

Rhizomes very short; leaves clustered, 5–20 (22) cm long, glabrous, evergreen, associated with persistent leaf bases of previous seasons; petiole slender, curved, shining, dark reddish brown; blade oblong to linear in outline, the rachis colored like the stipe; pinnae mostly 13-35 opposite or offset pairs, 2–8 mm long and 1–6 mm broad, toothed; sori elongate, with conspicuous indusium.

Rock outcrops, crevices, and talus slopes; in insular southeastern Alaska, where evidently rare; disjunctly eastward to the At-

Asplenium trichomanes L. (× 0.4).

lantic and south to Oregon, Arizona, Texas, Alabama, and Georgia; Eurasia.

Asplenium viride Huds.
Green Spleenwort

Rhizomes short; leaves clustered, 2–15 cm long, glabrous or sparsely glandular-hairy, usually not evergreen, usually with persistent leaf bases of previous seasons; petioles slender, brown to purplish brown at the base, becoming greenish upwards; blade oblong in outline, the rachis green or greenish; pinnae mostly 3–20 opposite or subopposite pairs, 2–9 mm long and 1–6 mm wide, crenate; sori elongate, with conspicuous indusium.

Wet rock outcrops and crevices, commonly on limestone; mostly in coastal and insular southeastern Alaska, but also known from coastal western and south-central Alaska and western Yukon. This small fern is easily overlooked and should be expected at other sites as well.

ATHYRIUM Roth

Plants medium to large sized, mesophytic, deciduous ferns; rhizomes short, scaly, ascending, and covered with persistent, flattened petiole bases of preceding seasons; leaves 2–4 times pinnately compound, the pinnae once to thrice compound or pinnatifid, the veins reaching the margin; petiole coarse, flattened and black basally becoming herbaceous upwards, scaly, with two vascular bundles at the base, these anastomosing upwards; sori round to elongate; indusium hyaline, flaplike, attached along the vein on the side of the sorus towards the margin of the segment, or lacking.

1a. Leaf blades finely dissected, the pinnules narrowly lanceolate in outline, acute apically; indusium lacking; plants uncommon in coastal and insular south-central and southeastern Alaska. *A. distentifolium*

1b. Leaf blades rather coarsely dissected, the pinnules lance-oblong to lanceolate in outline, more or less rounded apically to acute, indusium present; plants common in much of coastal and insular Alaska and less commonly northward to the 66th parallel in Alaska and Yukon. *A. filix-femina*

Athyrium distentifolium Tausch ex Opiz
Alpine Lady Fern

Rhizomes short, with leaves arranged in a vaselike tuft; leaves mostly 20–80 cm long, glabrous, deciduous, usually borne with persistent leaf bases of previous seasons; petioles coarse, blackish and scaly basally, becoming greenish or straw-colored and sparsely scaly upwards; blade 15–60 cm long, lance-elliptic in outline, 2–4 times pinnately compound or pinnatifid; pinnae mostly 15–25 pairs; pinnules numerous, lanceolate in outline, acute; sori round, less than 1 mm wide; indusium lacking.

Moist alpine thickets; in coastal and insular south-central and southeastern Alaska; southward to California, Nevada, and Colorado, and disjunctly in northeastern North America; circumboreal (*A. distentifolium* ssp. *americanum* [Butters] Hultén; *Aspidium alpestre* Hoppe). Our material belongs to var. *americanum* [Butters] Cronq.

Athyrium filix-femina (L.) Roth
Lady Fern

Rhizomes short, with leaves arranged in a vaselike tuft; leaves mostly (20)30–130 cm long, glabrous, deciduous, usually with persistent leaf bases of the previous seasons; petioles coarse, blackish and scaly basally, becoming greenish or straw-colored and sparsely scaly upwards; blade (15)25–100 cm long, lanceolate to lance-elliptic in outline, 2–3 times pinnately compound or pinnatifid; pinnae mostly 20–35 pairs; pinnules numerous, oblong to lanceolate in outline, rounded to acute; sori oblong to horseshoe-shaped; indusium straight or curved, often toothed.

Woods, thickets, meadows, and stream banks; in coastal and insular southern Alaska, and in the interior as far north as the 66th parallel in Alaska, and in southern Yukon; eastward to the Atlantic and south to California, Texas, and Florida; circumboreal (*A. cyclosorum* Rupr.; *A. filix-femina* var. *cyclosorum* Rupr.; *A. filix-femina* ssp. *cyclosorum* [Rupr.], C. Chr. in Hultén; *A. felix-femina* var. *sitchense* Rupr.). It does not seem reasonable to segregate our materials from typical *A. filix-femina* of Europe.

BLECHNUM L.

Plants medium sized, mesophytic, evergreen ferns; rhizomes short, scaly, ascending, and covered with persistent leaf bases of preceding seasons; leaves of two types, fertile and vegetative, once pinnately compound or pinnatifid, the veins reaching the margin on sterile leaves; petiole slender, brown, scaly, with 3 or more vascular bundles at the base, fertile leaves with slender pinnae, each with a pair of veins parallel to the midvein; sori continuous, parallel to the midvein; indusium membranous, attached along the margin and opening towards the midvein.

Blechnum spicant (L.) Roth
Deer Fern

Rhizomes short, with strongly dimorphic leaves clustered at the apex; vegetative leaves mostly 10–75 cm long; petioles 0.5–18(25) cm long, scaly basally, dark purplish brown, blades elliptic to oblong in outline, once pinnately compound, pinnae numerous, entire or serrulate, oblong, rounded to obtuse apically; fertile leaves longer than the vegetative ones, commonly 25–90 cm long, with petioles mostly 12–37 cm long, the fertile pinnae numerous, commonly longer and narrower than the vegetative ones, linear, usually acute; sori continuous; indusium conspicuous, brown-hyaline.

Moist woods and thickets; in coastal and insular southern Alaska; southward to California; Europe, Asia.

CRYPTOGRAMMA R. Br.

Plants small, mesophytic, deciduous or evergreen ferns; rhizomes short or somewhat elongate, scaly; leaves of two types, fertile and vegetative, mostly (1)2–3 times pinnately compound or pinnatifid, the veins reaching the margin, at least on vegetative leaves; petiole slender, green, straw-colored, or purplish black, scaly, with a single vascular bundle; fertile leaves longer than the vegetative ones, with relatively long, narrow pinnules; sori more or less continuous along the margin of pinnules, the margin revolute and forming a false indusium.

1a. Rhizomes short, densely leafy and often clothed with old leaf bases; leaves tufted, evergreen; plants mostly of coastal and insular southern Alaska, less commonly in the interior. *C. crispa*

1b. Rhizomes elongate, sparsely leafy, lacking old leaf bases; leaves deciduous; plants mostly of interior Alaska and Yukon, less commonly in coastal western and southeastern Alaska. *C. stelleri*

Cryptogramma crispa (L.) R. Br. ex Hook.
Parsley Fern; Rock-brake

Rhizomes short, compactly branched, clothed with scales, old leaf bases, and tufted leaves; vegetative leaves (3)7–25 cm long, with scaly, straw-colored to greenish petioles 1.5–17 cm long and ovate to ovate-lanceolate blades 2–3 times pinnately compound, the pinnae commonly 5–11, twice pinnately compound, the ultimate segments toothed; fertile leaves longer than the vegetative ones, the fertile pinnae about as many as the vegetative, the ultimate segments linear to narrowly oblong; sori more or less continuous, covered by the revolute margin of the ultimate segments.

Outcrops, open slopes, thickets, and woods; in much of Alaska south of the 65th parallel, and in southern Yukon; southward to Mexico and New Mexico; disjunctly eastward to the Great Lakes; Europe, Asia (*C. acrostichoides* R. Br. ex Richards.; *C. crispa* var. *sitchensis* [Rupr.] C. Chr.). Our material belongs to var. *acrostichoides* (R. Br.) C. B. Clarke. Attempts to segregate our material into two varieties have not been successful. The var. *sitchensis* is supposed to differ in having broad, sterile blades which are finely dissected, and in having obovate, evidently toothed, tertiary lobes. Only arbitrary separation seems possible, and it appears best to treat our specimens as belonging to a single highly variable entity.

Cryptogramma stelleri (Gmel.) Prantl
Slender Cliff-brake

Rhizomes slender, creeping, scaly, with leaves scattered along the length of the rhizome; vegetative leaves 5–15 cm long, with scaly purplish brown petioles 1–9 cm long and ovate to oblong blades 1–2 times pinnately compound, the pinnae commonly 3–5, these usually merely pinnatifid, the ultimate segments crenately toothed; fertile leaves commonly surpassing the vegetative ones, mostly 6–19 cm long, with petioles 3–9 cm long, these colored much like

Cryptogramma crispa (L.) R. Br. (× 0.5).

the vegetative petioles, the ultimate segments lanceolate to lance-linear; sori continuous, the revolute margin membranous, translucent.

Wooded slopes, rock outcrops, and crevices; from the Seward Peninsula eastward through central Alaska to west-central Yukon and the Mackenzie; disjunctly eastward to the Atlantic and south to Washington, Nevada, Colorado, Iowa, and West Virginia; Asia.

CYSTOPTERIS Bernh.
Plants small to medium sized, mesophytic,

delicate, deciduous ferns; rhizomes short or elongate, scaly; leaves clustered or scattered, all about alike, mostly 2–4 times pinnately or ternate-pinnately compound or pinnatifid, the veins reaching the margin; petioles slender, scaly, brown to green or straw-colored, not jointed, with 2 vascular bundles; sori round, borne along veins on the lower side of the blade; indusium attached under the sorus, the free tip hoodlike, arching over the sorus, often pushed back as the sorus enlarges, soon withering.

Blasdell, R. F. 1963. A monographic study of the fern genus *Cystopteris*. Mem. Torrey Bot. Club 21(4):1–102.

1a. Leaf blade lanceolate, commonly more than twice longer than broad, the lowermost pinnae not larger than the others. *C. fragilis*

1b. Leaf blade deltoid-ovate, commonly about as broad as long, the lowermost pinnae much larger than the others. *C. montana*

Cystopteris fragilis (L.) Bernh.
Fragile-fern

Rhizomes short, thickish, densely scaly, with leaves more or less tufted; leaves 4–35 (40) cm long, the petioles scaly and brownish at the base, becoming greenish or straw-colored above or less commonly brownish throughout, 1.5–17 cm long; blade lanceolate to elliptic in outline, 4–20 cm long and 1.5–7 cm broad, 2–3 times pinnately compound, glabrous or nearly so, the pinnae once to twice pinnately compound or pinnatifid, the ultimate segments toothed or lobed.

Woods, thickets, heathlands, meadows, rock outcrops, and talus slopes; in most of Alaska and Yukon except for the northern coastal plain; widely distributed in North America; circumboreal and circumaustral (*Polypodium fragile* L.; *C. dickieana* R. Sim; *C. fragilis* var. *dickieana* [R. Sim] Moore; *C. fragilis* ssp. *dickieana* [R. Sim] Hyl.).

Cystopteris montana (Lam.) Bernh.
Mountain Fragile-fern

Rhizomes slender, long and creeping, scaly, with leaves arising singly; leaves 12–40 cm long, the petioles scaly for some distance above the base, brown basally becoming greenish or straw-colored upwards, 8–30 cm long; blade ovate to deltoid in outline, 4–15 cm long, 4–17 cm wide, 3–4 times ternate-pinnately compound, glabrous or nearly so, the lower pair of pinnae much larger than the upper ones.

Moist sites in woods, heathlands, and meadows, often on limestone; in west-central and interior Alaska and southern Yukon, and less commonly in coastal southern Alaska; disjunctly eastward to the Atlantic and south to British Columbia, Montana, Colorado, and Quebec; Eurasia (*Polypodium montanum* Lam.).

DRYOPTERIS Adans.

Plants medium sized to large, mesophytic, deciduous or evergreen ferns; rhizomes short, scaly, covered with persistent leaf bases of previous seasons; leaves clustered, all about alike, mostly 2–3 times pinnately compound or pinnatifid, the veins reaching the margin; petioles coarse, scaly nearly or quite to the blade, brown to green, with 3–more vascular bundles; sori roundish, borne along veins on the lower side of the blade; indusium roundish or reniform, attached at the sinus.

1a. Leaves densely clustered, evergreen, mostly less than 30 cm long, the blades oblong in general outline, the lowermost pinnae commonly smaller than those at middle of blade; plants of talus slopes and rock outcrops. *D. fragrans*

1b. Leaves loosely clustered, deciduous, mostly more than 30 cm long, the blades ovate-lanceolate in outline, the lowermost pinnae larger than the others; plants of woods. *D. austriaca*

Dryopteris austriaca (Jacq.) Woynor ex Schinz & Thell.

Spinulose Shield-fern

Rhizome thick, ascending, scaly, with loosely tufted leaves; leaves deciduous (15)30–90 cm tall or more, the petioles dark brown basally, green or straw-colored above, scaly almost or quite to the blade, 6–35 cm long; blade lanceolate to ovate in outline, 10–50 cm long, 8–36 cm wide, thrice pinnately compound or pinnatifid, the ultimate segments softly spinulose-toothed, the lowermost pinnae commonly larger and unequally pinnate-pinnatifid, with the first pair of second-order pinnae unequal in size, the basally directed one much the larger; indusium glabrous, or sometimes glandular.

Moist woods, thickets, and stream banks; in most of Alaska south of the Brooks Range (except for southeastern continental portions) and in southern Yukon; eastward to the Atlantic and south to California, Idaho, Wyoming, Iowa, Tennessee, and North Carolina; circumboreal (*D. dilatata* [Hoffm.] Gray; *Aspidium spinulosum* var. *americanum* Fisch. ex Kunze; *D. dilatata* ssp. *americana* [Fisch.] Hultén). This is one of the truly common species of fern in Alaska.

Dryopteris fragrans (L.) Schott

Fragrant Shield-fern

Rhizomes very coarse, densely scaly, with densely tufted leaves of the current and previous seasons; leaves evergreen, (5)10–28 cm long, the petioles brown basally, green or straw-colored above, densely scaly, 1–12 cm long; blade oblong to oblong-oblanceolate in outline, (3)6–25 cm long, 1–4.5 cm wide, once to twice pinnately compound or pinnatifid, the ultimate segments with rounded teeth, the lowermost pinnae smaller than those near the middle of the blade, the second-order pinnae not unequal in size; indusium glandular margined.

Rock outcrops, cliffs, and talus slopes; from low to high elevations in most of Alaska except for coastal southern and northern portions, and most of the Yukon; eastward to the Atlantic and south to northern British Columbia, Minnesota and New York; Eurasia (*D. fragrans* var. *remotiuscula* Kom.).

GYMNOCARPIUM Newman

Plants medium sized, delicate, mesophytic, deciduous ferns; rhizome elongate, slender, sparsely scaly, the leaf bases of previous seasons usually not persistent; leaves scattered, all about alike, ternate, the pinnae pinnately once to twice compound, the lowermost pair jointed to the rachis, the veins reaching the margin; petioles slender, scaly below the middle, brown at the base, green to straw-colored upwards, with 2 vascular bundles; sori roundish, borne along the veins on lower side of the blade; indusium lacking.

Wagner, W. H.   1966.   New data on the North American oak ferns, *Gymnocarpium*. Rhodora 68(774):121–38.

1a.   Leaves glandular along the rachis and blade, the terminal division of the blade larger than the lateral segments; plants mostly of central and east-central Alaska and west-central Yukon. *G. robertianum*

1b.   Leaves glandless or nearly so, the terminal division of the blade only slightly larger than the lateral segments; plants widespread in Alaska and Yukon south of the 65th parallel. *G. dryopteris*

Gymnocarpium dryopteris (L.) Newman

Oak-fern

Rhizomes slender, the leaves scattered; leaves (12)20–43 cm long, the petioles brown at the base (less commonly throughout), green to straw-colored upwards, sparsely scaly near the base or up to about the middle, 8–35 cm long, 4–22 cm broad, ternate, the three main segments again once to twice pinnately compound or pinnatifid, the terminal segment larger than or subequal to the lateral segments, glabrous or nearly so, not glandular; sori submarginal.

Woods, thickets, heathlands, and meadows; in much of Alaska and Yukon south of the 65th parallel; eastward to the Atlantic and south to Oregon, Arizona, New Mexico, Kansas, and West Virginia; circumboreal (*Phegopteris dryopteris* [L.] Fee; *Dryopteris linnaeana* C. Chr.; *Thelypteris dryopteris* [L.] Slosson in Rydb.; *Dryopteris disjuncta* [Rupr.] Morton).

Gymnocarpium robertianum (Hoffm.) Newman
Limestone Oak-fern; Scented Oak-fern

Rhizomes slender, with leaves scattered; leaves 8–35 cm long, the petioles brown at the base, straw-colored to green upwards, sparsely scaly near the base or upwards to about the middle, 6–25 cm long; blade deltoid to ovate in outline, 4–12 cm long, 3–11 cm broad, ternate, the three main segments again once to twice pinnatifid or pinnately compound, the terminal segment commonly much larger than the lateral ones, minutely glandular-puberulent; sori submarginal.

Woods; in central and east-central and less commonly in south-central Alaska, and west-central Yukon; eastward to Newfoundland and south to British Columbia, Iowa, and North Carolina. The apparent hybrid between *G. dryopteris* and *G. robertianum* has been designated as *G.* × *heterosporum* W. H. Wagoner (Scamman 913, Wiseman G H). They may be identified by the blackish, strongly irregularly shaped spores.

MATTEUCCIA Todaro

Plants large, elegant, mesophytic, deciduous ferns; rhizomes short, erect or ascending, coarse, scaly, covered with persistent leaf bases of previous seasons; leaves clustered, of two types, fertile and vegetative, once pinnately compound, the pinnae pinnatifid, the veins reaching the margin; petiole coarse, brown and flattened basally, becoming straw-colored to green above, scaly, with two vascular bundles; fertile leaves much shorter than the vegetative, with slender pinnules; sori elongate, covered by the revolute margin of the pinnule; indusium hoodlike, soon withering.

Matteuccia struthiopteris (L.) Todaro
Ostrich-fern

Rhizomes coarse, erect, clothed with scales and old leaf bases and tufted leaves; vegetative leaves 60–120 cm long, with short, sparsely scaly petioles 8–25 cm long, and elliptic to oblong blades once pinnately compound with pinnatifid pinnae, the ultimate segments more or less revolute and entire to remotely toothed and more or less long-hairy; fertile leaves shorter than the vegetative ones, commonly 30–50 cm long with petioles mostly 15–30 cm long, the fertile pinnae numerous, much smaller than the vegetative ones, dark brown to blackish with age.

Woods; in central, south-central, and southwestern continental Alaska; eastward to Newfoundland and south to British Columbia, South Dakota, Missouri, and Virginia; Eurasia (*Struthiopteris filicastrum* All.; *Onoclea struthiopteris* [L.] Hoffm.).

POLYPODIUM L.

Plants small to medium sized, evergreen, mesophytic ferns; rhizomes elongate, thickish, clothed with scales, not beset with old leaf bases; leaves scattered, all about alike, once pinnately compound or merely pinnatifid, the pinnae serrulate, the veins apparently not reaching the margin; petioles straw-colored to greenish, lacking scales, with 3 vascular bundles at the base, jointed to the rhizome; sori round or oval, each borne at the end of a veinlet on the lower side of the blade; indusium lacking.

Lang, F. A. 1971. The *Polypodium vulgare* complex in the Pacific Northwest. Madroño 21:235–54.

Lloyd, R. M., and F. A. Lang. 1964. The *Polypodium vulgare* complex in North America. Brit. Fern. Gaz. 9(5):168–77.

1a. Pinnae commonly 3.5–5 times or more longer than broad, at least some often more than 30 mm long, acute to attenuate or less commonly rounded apically; plants of coastal and insular southern Alaska. *P. glycyrrhiza*

1b. Pinnae commonly 2–3(3.5) times longer than broad, mostly less than 30 mm long, rounded to acute apically; plants of interior Alaska and Yukon, rarely in coastal south-central Alaska. (2)

2a. Sori oval, placed about halfway between the midvein and margin; rhizome scales with midvein colored about like the lateral portions; plants of rather broad distribution in Alaska and western Yukon. *P. hesperium*

2b. Sori round, placed along the margin; rhizome scales with midrib darker than lateral portions; plants known only from west-central Yukon. *P. virginianum*

Polypodium glycyrrhiza D. C. Eaton

Licorice-root; Licorice Fern

Rhizomes thickish, licorice flavored, densely scaly, the scales uniformly colored; leaves (4)10–45 cm long, the petioles straw-colored to greenish throughout, lacking scales, mostly (1)3–15 cm long; blade lanceolate to elliptic or oblong in outline, (2.5)5–30(35) cm long, (1.5)3–11 cm broad, the longest pinnae mostly (7)15–60 mm long and (2)3–9 mm broad, commonly 3.5–5 (or more) times longer than broad; sori round, ordinarily borne nearer the midvein than the margin.

Tree trunks, mossy logs, cliffs, rocks, and banks, in woods and thickets; in coastal and insular southern Alaska from the westernmost Aleutians eastward through the Panhandle; southward to California (*P. vulgare* L. var. *occidentale* Hook.; *P. vulgare* ssp. *occidentale* [Hook.] Hultén).

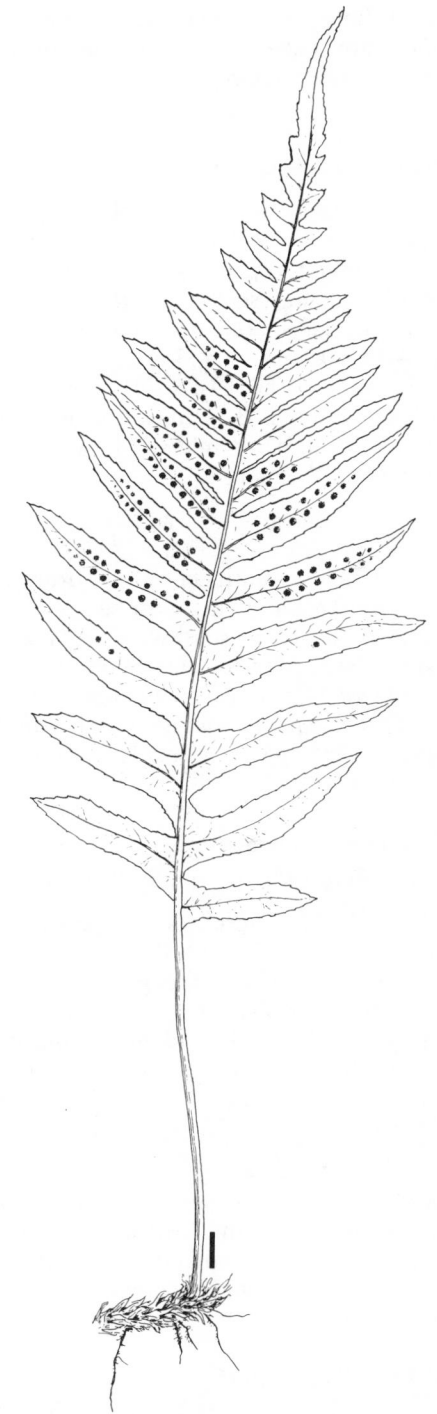

Polypodium glycyrrhiza D. C. Eaton (× 0.5).

Hybrids between *P. glycyrrhiza* and *P. hesperium* are known from regions to the south, but have not yet been reported from Alaska.

Polypodium hesperium Maxon

Rhizome thickish, sweet, densely scaly, the scales uniformly colored; leaves mostly 5–30 cm long, the petioles straw-colored to greenish, lacking scales, mostly 1–10 cm long; blade lance-oblong to oblong in outline, 4–15(20) cm long, 1–5 cm wide, the longest pinnae (6)10–25(30) mm long, 4–8 mm wide, 2–3.5 times longer than wide; sori oval, borne midway between the midvein and the margin.

Rocks and crevices; in widely scattered locations in interior central Alaska, and less commonly south-central (Turnagain Arm) Alaska, and in west-central and southwestern Yukon; southward to Mexico, Arizona, New Mexico, and South Dakota (*P. vulgare* L. var. *columbianum* Gilbert; *P. vulgare* ssp. *columbianum* [Gilbert] Hultén). This species is evidently rare in our region.

Polypodium virginianum L.
Virginia Polypody; Common Polypody

Rhizome thickish, acrid, densely scaly, the scales with a dark central stripe; leaves mostly 10–35 cm long, the petioles straw-colored to greenish, mostly 5–15 cm long; blade lance-oblong to oblong in outline, 5–20 cm long, 2–6 cm wide, the longest pinnae 10–30(37) mm long, 4–7(8) mm wide, commonly 2–3.5(4) times longer than wide; sori round, nearly marginal.

Shaded, mossy rocks; in west-central Yukon and northeastern British Columbia; eastward to Newfoundland and south to Alberta, Arkansas, and Georgia. This species is evidently rare in our region.

POLYSTICHUM Roth

Plants small to large sized, evergreen, mesophytic ferns; rhizomes short, thick, erect to ascending, clothed with scales and leaf bases; leaves tufted, all about alike, once or twice pinnately compound, the pinnae or pinnules spinulose-toothed (except in *P. aleuticum*), the veins apparently not reaching the margin; petioles brown, or straw-colored upwards, sparsely to copiously scaly or finally lacking scales, with mostly 4–5 vascular bundles; sori round, borne on the veins on the lower side of the blade; indusium arising from the center of the sorus, centrally or excentrically attached, peltate.

1a. Leaves twice pinnately compound, the pinnae, or some of them, conspicuously cleft or pinnately compound. (2)

1b. Leaves once pinnately compound, the pinnae merely toothed, or obliquely lobed. (3)

2a. Basal pinnule on the uppermost side of the pinnae conspicuously longer than the next adjacent pinnule; rachis of leaf with 1–2 buds on the lower side near the tip. *P. andersonii*

2b. Basal pinnule on the uppermost side of the pinnae subequal to the next adjacent pinnule; rachis of leaf lacking buds. *P. braunii*

3a. Leaves to about 15 cm long, the blade papery; pinnae with rounded teeth, lacking aristate to spinulose processes; plants known only from Atka Island. *P. aleuticum*

3b. Leaves mostly 20–90 cm long or more, the blade leathery; pinnae with spinulose teeth; plants of coastal and insular southern Alaska, but not known from Atka Island. (4)

4a. Pinnae once to twice longer than broad, the spinulose teeth more or less spreading; leaves mostly less than 30 cm long; plants of broad distribution in coastal and insular southern Alaska. *P. lonchitis*

4b. Pinnae commonly more than 4 times longer than broad, the spinulose teeth appressed-ascending; leaves often more than 30 cm long; plants of the southern portion of the Alexander Archipelago. *P. munitum*

Polystichum aleuticum C. Chr. ex Hultén
Aleutian Holly-fern; Aleutian Shield-fern

Rhizome obliquely ascending, clothed with scales and persistent leaf bases of the previous seasons; leaves to 15 cm long, the petioles brown, slender, mostly 5–8 cm long, at first sparsely scaly, finally naked; blade oblong to narrowly oblanceolate in outline, once pinnately compound, to 10 cm long and 1.5–2 cm wide, pinnae ovate to ovate-oblong, equally bilobed basally, to 10 mm long and 4 mm broad, the median lobe with few, deltoid, acute nonaristate teeth; rachis and midvein scaly; indusium excentrically peltate, erose, glabrous.

Habitat not recorded, known only from the type locality on Atka Island. This species, like others which appear only in the westernmost Aleutians, seems to be more closely related to Asiatic than to American species. It seems likely that further work will demonstrate this entity as being conspecific with some Asiatic species. The plants are about the size and form of *Woodsia alpina*.

Polystichum andersonii Hopkins
Anderson's Holly-fern

Rhizome erect or ascending, densely clothed with scales and persistent leaf bases; leaves mostly 30–120 cm long, the petioles brown to straw-colored, coarse, mostly 6–35 cm long, scaly throughout; blade elliptic to lance-oblong in outline, twice pinnately compound or pinnatifid, 25–90 cm long, 6–20 cm wide, characteristically bearing one or more buds in the axils of the upper pinnae; pinnae lance-oblong, the largest 30–100 mm long and 10–35 mm broad at the base, the teeth spinulose-aristate, the basal pinnule on the upper side of at least the lower pinnae longer than the adjacent pinnule; rachis

and midveins scaly; indusium peltate, erose, the teeth more or less glandular.

Moist woods and thickets; in coastal and insular southern Alaska, where distribution is bipartite, in the westernmost Aleutians and in the Panhandle. It is to be sought in the intervening area (*P. braunii* ssp. *andersonii* [Hopkins] Calder & Taylor). Putative hybrids between *P. andersonii* and *P. munitum* are known in southeastern Alaska.

Polystichum braunii (Spencer) Fée
Prickly Shield-fern

Rhizome erect, densely clothed with scales and persistent leaf bases; leaves mostly 30–100 cm long, the petioles brown to straw-colored, coarse, mostly 6–15 cm long, scaly throughout; blade elliptic to lance-elliptic or oblanceolate, twice pinnately compound or pinnatifid, 25–80 cm long, 5–20 cm wide, not bearing buds; pinnae lance-oblong, the largest 40–120 mm long and 15–25 mm broad at the base, the teeth aristate-spinulose, the basal pinnule on the upper side of the lower pinnae subequal to or slightly longer than the next adjacent pinnule; rachis and midvein scaly; indusium peltate, often erose.

Woods and thickets; in coastal and insular southeastern Alaska from the westernmost Aleutians eastward through the Panhandle, and southward to British Columbia; east to Newfoundland and south to Wisconsin and Pennsylvania (*P. alaskense* Maxon; *P. braunii* var. *alaskense* [Maxon] Hultén; *P. braunii* ssp. *alaskense* [Maxon] Calder & Taylor). *P. braunii* is reported to hybridize with *P. lonchitis*, and a specimen from Juneau (Stonehouse 53, ISC) seems to represent such a plant.

Polystichum lonchitis (L.) Roth
Holly-fern

Rhizome erect or ascending, densely covered with scales and persistent leaf bases; leaves tufted, mostly 15–50(60) cm long, the petioles dark brown basally, straw-colored upwards, coarse, scaly throughout; blade oblong to elliptic or narrowly oblanceolate in outline, once pinnately compound, 10–50 cm long, 2–7.5 cm wide, not bearing buds; pinnae obliquely ovate to obliquely lanceolate, often curved and unequally lobed at the base, the largest 10–40 mm long, 5–14 mm broad, at least the lower less than twice longer than broad, the teeth spinulose, spreading-ascending; rachis and midvein scaly; indusium erose-dentate.

Woods and thickets and on rocky slopes and outcrops; in coastal and insular southern Alaska from the westernmost Aleutians eastward through the Panhandle; eastward to the Atlantic and south to California, Arizona, New Mexico, and Michigan.

Polystichum munitum (Kaulf.) Presl
Western Sword-fern; Dagger-fern

Rhizome ascending, densely clothed with scales and persistent leaf bases; leaves tufted, mostly (20)50–100(150) cm long, the petioles brownish to straw-colored or greenish, coarse, scaly throughout; blade lance-elliptic to oblong in outline, once pinnately compound, mostly 30–100 cm long, 4–30 cm wide, not bearing buds; pinnae obliquely oblong-lanceolate, straight or curved, unequally lobed at the base, the largest 20–150 mm long and 8–15 mm broad, even the lowermost ones more than 3 times longer than broad, the teeth spinulose, appressed or incurved; rachis and midvein scaly; indusium peltate, fringe-margined.

Woods; in the southern portion of the Alexander Archipelago; southward to California, Idaho, and Montana.

PTERIDIUM Gled.

Plants medium to large sized, mesophytic, deciduous ferns; rhizomes elongate, deeply subterranean, covered with hairs but not with scales; leaves scattered, pinnately twice to thrice compound, the pinnae once to twice compound or pinnatifid, the veins confluent along the margin; petiole coarse, black basally, becoming straw-colored upwards, scaleless, the base covered by a feltlike mass, with several vascular bundles, these anastomosing upwards; sori continuous along the margin of pinnae or pinnules, protected by the inrolled indusial margin of the leaf.

Pteridium aquilinum (L.) Kuhn in von der Decken
Western Bracken

Rhizomes creeping, wide-spreading; leaves scattered, mostly 30–150(200) cm long, hairy, deciduous; petioles mostly 20–100 cm long, blackish and covered with a feltlike mass basally, glabrous and straw-colored upwards; blade 20–100 cm long, deltoid to triangular in outline, usually 3 times pinnately compound, the basal pair of pinnae often larger and more strongly dissected than the others; pinnules widely spreading, oblong to linear or triangular, with revolute margins; sori marginal, protected by the revolute margin.

Woods, muskegs, and meadows; in coastal and insular southeastern Alaska; widely distributed in North America; Europe, Asia, and Southern Hemisphere (*P. aquilinum* ssp. *lanuginosum* [Bong.] Hultén). Our material belongs to var. *pubescens* Underw.

THELYPTERIS Schmidel

Plants medium sized, mesophytic, deciduous ferns; rhizomes slender and creeping or short and stout, sparsely scaly; leaves scattered or tufted, once pinnately compound, the pinnae merely pinnatifid, or twice pinnately compound or pinnatifid, the veins extending to the margin; petiole slender, brown at base, becoming straw-colored upwards, scaly throughout or lacking scales upwards, with 2 vascular bundles at the base, these anastomosing upwards, sori round, each borne along a vein on the lower leaf surface; indusium small or lacking, kidney-shaped, attached at the sinus, usually ciliate.

1a. Leaf blades triangular-ovate, the lowermost pinnae the largest; ultimate segments ciliate and more or less hairy on the upper and lower surfaces; plants rather broadly distributed in southern Alaska and in southeastern Yukon. *T. phegopteris*

1b. Leaf blades elliptic to oblanceolate, the lower pinnae smaller than those at the middle of the blade; ultimate segments not ciliate, not or only slightly hairy above, glandular-hairy to glabrous beneath; plants of coastal and insular southern Alaska. *T. limbosperma*

Thelypteris limbosperma (All.) H. P. Fuchs
Mountain Wood-fern

Rhizomes short, stout, horizontal to ascending, sparsely scaly, clothed with persistent leaf bases; leaves deciduous, tufted, mostly 25–90 cm long, the petioles brown basally (or less commonly brown throughout), straw-colored upwards, scaly almost or quite to the blade, 5–25 cm long; blade elliptic to oblanceolate in outline, 20–75 cm long, once pinnately compound, the pinnae merely pinnatifid (sometimes deeply so); lowermost pinnae much smaller than those at the middle of the blade; sori small, borne near the revolute margin; indusium kidney-shaped, small, soon deciduous.

Woods, meadows, heathlands, and alpine tundra; in coastal and insular southern Alaska from the Aleutians eastward through the Panhandle; south to Washington; Europe, Asia, (*Polypodium limbospermum* All.; *Polypodium oreopteris* Ehrh.; *Dryopteris oreopteris* [Ehrh.] Maxon; *Thelypteris oreopteris* [Ehrh.] Slosson in Rydb.; *Lastrea limbosperma* [All.] Holub.).

Thelypteris phegopteris (L.) Slosson in Rydb.
Northern Beech-fern; Long Beech-fern

Rhizomes slender, long-creeping, sparsely scaly, not clothed with leaf bases; leaves deciduous, scattered, mostly 10–50 cm long, the petioles brown basally, straw-colored upwards, sparsely scaly and more or less hairy almost throughout, 7–33 cm long; blade triangular-ovate in outline, mostly 5–20 cm long and 5–18 cm wide, once to twice pinnately compound, or the pinnae merely pinnatifid; lowermost pinnae the largest; sori small, borne very near the margin; indusium none.

Woods, thickets, and meadows; in much of Alaska south of the 65th parallel, except for central and eastern Alaska, and disjunctly in southeastern Yukon; eastward to the Atlantic and south to Oregon, Iowa, Michigan, and North Carolina; circumboreal (*Polypodium phegopteris* L.; *Dryopteris phegopteris* [L.] C. Chr.; *Phegopteris polypodioides* Fee; *Phegopteris dryopteris* [L.] Fee; *Lastrea phegopteris* [L.] Bory.).

WOODSIA R. Br.

Plants small to medium sized, mesophytic to subxerophytic deciduous ferns; rhizomes short, horizontal or ascending, scaly, beset with persistent leaf bases; leaves clustered, all about alike, once to twice pinnately compound, the veins not reaching the margin; petioles slender, scaly or hairy or both, dark brown to brown or straw-colored, jointed near the base (except in W. *scopulina*), with 2 vascular bundles at the base, these anastomosing upwards; sori round, borne along veins on the lower side of the blade, sometimes confluent in age; indusium borne beneath and enclosing the sorus with hairlike lobes.

Brown, D. F. M. 1964. A monographic study of the fern genus *Woodsia*. Nova Hedwegia 16:1–154.

1a. Petioles white-hairy, not jointed at the base; plants known from coastal south-central and southeastern Alaska, and southwestern Yukon. *W. scopulina*

1b. Petioles more or less scaly, seldom white-hairy, jointed near the base, the joint appearing as a thickened, dark ring; plants broadly distributed. (2)

2a. Leaves glabrous, the blades lacking hairs or scales, except for the indusial hairs; pinnae as broad as long. *W. glabella*

2b. Leaves with hairs and scales in addition to the indusial hairs; pinnae mostly longer than broad. (3)

3a. Pinnae more or less densely scaly and hairy on the lower surface, the largest with 3–more lobes per side, conspicuously longer than broad. *W. ilvensis*

3b. Pinnae more or less hairy and sparsely scaly or lacking scales entirely on the lower surface, the largest commonly with 2–3(4) lobes per side, only slightly longer than broad. *W. alpina*

Woodsia alpina (Bolton) S. F. Gray
Alpine Woodsia

Rhizome short, thickish, scaly, clothed with persistent leaf bases; leaves 5–16 cm long, the petioles scaly, brownish, jointed near the base, 1–9 cm long, the persistent base below the joint 0.5–2 cm long; blade oblong to linear in outline, 3–9 cm long, 0.5–1.6 cm wide, once pinnately compound; pinnae pinnatifid, with commonly 2–3(4) lobes on each side, the largest 4–10 mm long and 3.5–6 mm broad, longer than broad or about as broad as long, the lower surface more or less hairy and sparsely scaly or the scales none; sori usually confluent; indusium of delicate, hairlike lobes.

Rock crevices, outcrops, and talus; in most of continental Alaska and Yukon; eastward to the Atlantic and south to British Columbia and New York; circumboreal (*Acrostichum alpinum* Bolton).

Woodsia glabella R. Br.
Smooth Woodsia

Rhizome short, thickish, scaly, clothed with persistent leaf bases; leaves 1.5–10 (15) cm long, the petioles scaly at the brownish base, jointed near the base, 0.5–

6 cm long, the persistent base below the joint 0.5–2.5 cm long; blade oblong-oblanceolate to linear in outline, 2–10 cm long, 0.3–1.4 cm wide, once pinnately compound; pinnae pinnatifid or simple, commonly with 1–2 lobes per side, the largest 3–6 mm long and 3–6 mm wide, as broad as long, glabrous and scaleless; sori often confluent; indusium of few, delicate, hairlike lobes.

Moist rock outcrops, crevices, and talus; in most of continental Alaska and Yukon; east to the Atlantic and south to British Columbia and New York; circumboreal.

Woodsia ilvensis (L.) R. Br.
Rusty Woodsia

Rhizome short, thickish, scaly, clothed with persistent leaf bases; leaves (4)6–15 (22) cm long, the petioles scaly throughout, jointed near the base, mostly 2–7(9) cm long, the persistent base below the joint 0.5–4.5 cm long; blade lance-oblong to oblong or elliptic, (3)4–11(12.5) cm long, 0.8–2.5(3) cm wide, once pinnately compound; pinnae pinnatifid, commonly with 3–4 (sometimes more) lobes per side, the largest (3)5–15(17) mm long and 3–8 mm wide, usually longer than broad, hairy and scaly on the lower surface; sori often confluent; indusium of many, slender, hairlike lobes.

Moist rocks, cliffs, and crevices; in much of continental and western coastal Alaska south of the Brooks range, and disjunctly in scattered localities in the Yukon; east to the Atlantic and south to British Columbia, Alberta, Illinois, and North Carolina; circumboreal (*Acrostichum ilvense* L.). *Woodsia ilvensis* is known to hybridize with *W. alpina* and *W. glabella.*

Woodsia scopulina D. C. Eaton
Rocky Mountain Woodsia

Rhizome short, thickish, scaly, clothed with persistent leaf bases; leaves 10–30 cm long, the petioles scaly at the base, pilose and more or less glandular, not jointed, mostly 6–20 cm long; blade lance-oblong, 5–12(14) cm long, 1.7–4.5 cm wide, once to twice pinnately compound; pinnae once compound or merely pinnatifid, usually with several lobes per side, the largest 10–22 mm long and 5–13 mm wide, longer than broad, more or less glandular and hairy with broad, flattened scalelike hairs on the lower surface; sori more or less distinct; indusium of flat, toothed or scalelike segments, more or less concealed by the sorus.

Open, rocky slopes; in coastal, south-central and southeastern Alaska and southern Yukon; east to Quebec and south to California, New Mexico, and South Dakota.

Subdivision PTEROPSIDA

Class GYMNOSPERMAE

Trees and shrubs, with scalelike or needle-like evergreen or deciduous leaves; cones unisexual, both male and female cones borne on the same (monoecious), or less commonly on separate plants (dioecious); male (staminate) cones or strobili soft, smaller than the female (ovulate) ones, deciduous; female cones with usually numerous, spirally arranged scales, with seeds borne on the upper surface.

1a. Leaves alternate or subopposite, 2-ranked; ovulate cones reduced to a single seed with a fleshy, finally red-colored, cup-shaped aril; plants dioecious. TAXACEAE .... p. 38
1b. Leaves opposite, whorled, or spirally arranged, not 2-ranked; ovulate cones well developed, woody or fleshy, but not as above; plants monoecious or dioecious. (2)

2a. Plants shrubs or trees with opposite or whorled, scalelike or linear-subulate leaves; cones mostly less than 15 mm long, the scales opposite or whorled, woody or fleshy. CUPRESSACEAE .... p. 29
2b. Plants mostly trees, with leaves spirally arranged or borne in clusters;

cones mostly more than 15 mm long, the scales spirally arranged, woody to papery. PINACEAE .... p. 31

CUPRESSACEAE
Cypress Family

Plants monoecious trees or shrubs; leaves evergreen, opposite or whorled, scalelike or awl-shaped and needlelike; staminate cones small, terminal or axillary, the microsporophylls decussate; ovulate cones terminal with commonly 2–12 opposite or whorled scales, bearing 1-several ovules, dry or fleshy at maturity.

1a. Plants low-growing, prostrate to ascending shrubs; branchlets not arranged in flat sprays; cones fleshy, berrylike, indehiscent. *Juniperus*
1b. Plants trees, or if shrubs then erect; branchlets arranged in flat sprays; cones woody or leathery, eventually dehiscent. (2)

2a. Cones reflexed, longer than broad, the scales laterally attached; plants of the southern portion of the Alexander Archipelago. *Thuja*
2b. Cones erect, as broad as long or broader, the scales peltately attached; plants of south-central and southeastern coastal regions. *Chamaecyparis*

CHAMAECYPARIS Spach

Trees, with opposite, decussate, scalelike leaves, the lateral leaves folded, the upper and lower ones flattened or rounded; branchlets arranged in more or less flattened sprays; staminate cones ellipsoid to ovoid; ovulate cones subglobose, with 3–6 pairs of peltate scales, each bearing 2–5 ovules; cones dehiscent, woody to leathery, dark brown to gray brown, glaucous.

Chamaecyparis nootkatensis (D. Don) Spach
Alaska Cedar

Trees, mostly 10–30 m tall, with spreading branches and drooping branchlets; ulti-

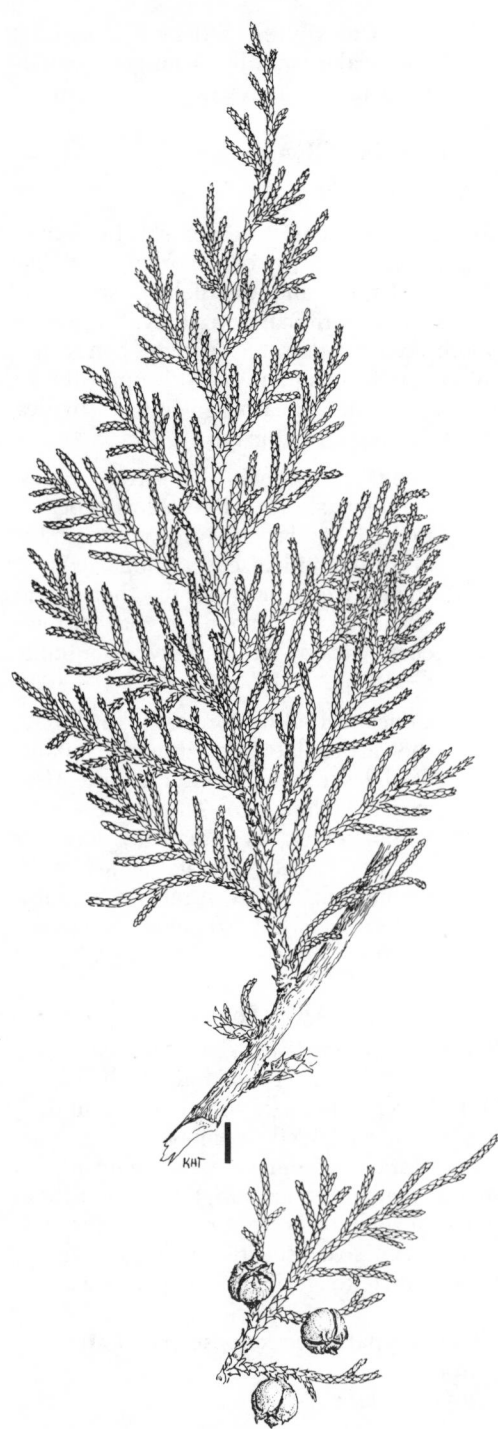

mate branchlets subterete or slightly flattened, 1.5–2.5 mm broad; bark gray brown, scaly, peeling in long narrow strips; wood yellow in color; leaves 1–2.5(3) mm long, closely imbricate, persisting on the branchlet for several years; cones globose, 0.6–1.2 cm long.

Coastal woodlands, from near sea level to subalpine sites; in coastal and insular southern Alaska from the Prince William Sound eastward through the Panhandle; southward to California (*Cupressus nootkatensis* D. Don in Lamb.).

JUNIPERUS L.

Shrubs, with opposite or whorled, scalelike or awl-shaped leaves; branchlets subterete or angular, not in flattened sprays; staminate cones subcylindrical to subglobose; ovulate cones subglobose with mostly 3–8 opposite or whorled scales, becoming fleshy and berrylike at maturity; cones indehiscent, fleshy, green or glaucous and turning brownish, blackish, or purplish at maturity.

1a. Leaves all awl-shaped, jointed to the branchlet, commonly in whorls of 3; stems usually ascending; plants widely distributed in Alaska and Yukon. *J. communis*
1b. Leaves all scalelike, or some of them awl-shaped, decurrent basally, not jointed to the branchlet, usually opposite; stems decumbent to procumbent or prostrate; plants of southeastern, continental Alaska and southern Yukon. *J. horizontalis*

Juniperus communis L.
Common Juniper

Stems commonly ascending, mostly 0.5–2 m long; leaves jointed to the stem, mostly in whorls of 3, awl-shaped, 0.3–1(1.5) cm long, spinulose-tipped, usually marked with a white band on the upper surface, dark green on the lower surface; cones maturing the second season, green, ripening bluish black, 0.5–1 cm broad, usually 1-seeded.

Chamaecyparis nootkatensis (D. Don) **Spach** (× 0.6).

Juniperus communis L. (× 0.4).

Bluffs, alluvial fans, open to dense woods, rock outcrops, and alpine tundra; in much of continental and coastal south-central to southeastern Alaska and much of the Yukon; east to the Atlantic and south to California, New Mexico, and Georgia; circumboreal (*J. communis* var. *saxatilis* Pallas; *J. nana* Willd.; *J. sibirica* Burgsd.; *J. communis* ssp. *nana* [Willd.] Syme). Our material belongs to var. *montana* Ait.

Juniperus horizontalis Moench
Creeping Juniper

Stems commonly decumbent to procumbent or prostrate, mostly 0.5–1.5 m long; leaves decurrent, opposite, scalelike or awl-shaped, 0.1–0.4(0.6) cm long, acute to spinulose-tipped, lacking a white band on the upper surface; cones maturing the first season, green, ripening blue purple or blue black, glaucous, 0.5–1 cm broad, mostly 3–5 seeded.

Bluffs, alluvial fans, woods, and terraces; in continental south-central Alaska and southern Yukon; eastward to the Atlantic and south to British Columbia, Wyoming, Colorado, Nebraska, Illinois, and New York.

THUJA L.

Trees with opposite, decussate, scalelike leaves, the lateral leaves folded, the upper and lower rounded or flattened; branchlets arranged in distinctive, flattened sprays; staminate cones subglobose; ovulate cones reflexed, oblong-ellipsoid, with 4–6 pairs of laterally attached scales, the middle scales each bearing 2–3 ovules; cones dehiscent, woody, green, turning tan or brownish at maturity.

Thuja plicata Donn ex D. Don in Lamb.
Western Red Cedar

Trees mostly 20–40 m tall, with widely spreading branches and drooping or pendulous branchlets; ultimate branchlets 1–2.5 mm wide, distinctly compressed; bark reddish brown peeling in long slender strips; leaves mostly 1–2(3) mm long, closely imbricate, persisting for several years and enlarging; cones oblong-ellipsoid, when closed 0.4–0.6 cm broad, ultimately dehiscent, the scales spreading-ascending.

Coastal forests, from near sea level to subalpine; in the southern half of the Alexander Archipelago; southward to California, Idaho, and Montana.

PINACEAE
Pine Family

Plants monoecious trees; leaves evergreen or deciduous, needlelike, linear to oblong, borne singly or in clusters of 2–5, or densely aggregated on short lateral shoots; cones solitary, axillary or terminal; staminate cones small, soft, with spirally arranged

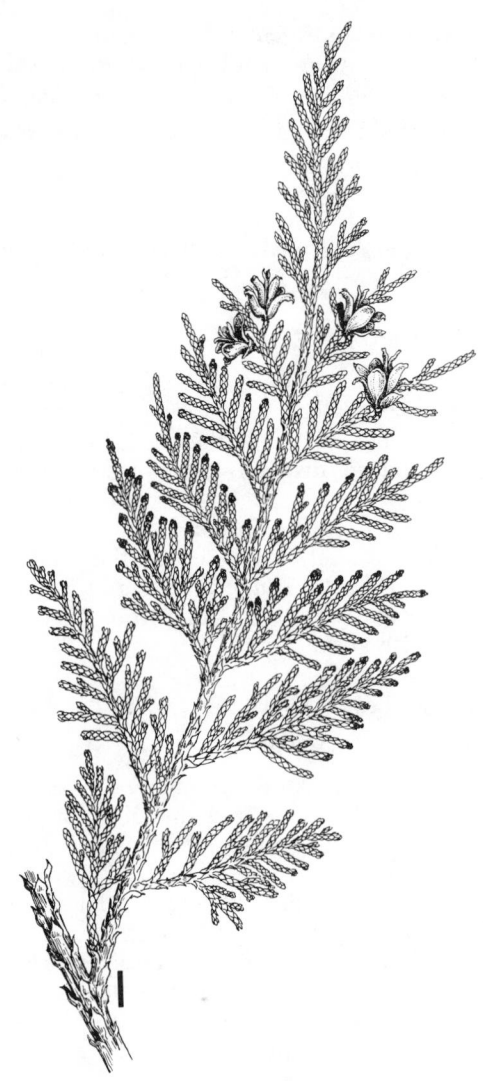

Thuja plicata Donn (× 0.5).

microsporophylls; ovulate cones small to large, the several to many scales spirally arranged, each subtended by a bract and bearing 2 ovules; cones woody to leathery or papery.

1a. Leaves borne in dense clusters on short spur branches, deciduous; cones less than 2 cm long, persistent. *Larix*

1b. Leaves borne singly or in clusters of 2–5, persistent; cones mostly more than 2 cm long, persistent or not. (2)

2a. Leaves born in clusters of 2–5, these fitting together to form a cylinder. *Pinus*

2b. Leaves borne singly, not fitting together to form a cylinder. (3)

3a. Branches smooth where needles have fallen, the leaves deciduous to the base; cones erect on the branches, the scales deciduous from the apex, with the axis alone persisting on the tree. *Abies*

3b. Branches rough where needles have fallen, the leaves deciduous above the persistent base; cones pendulous or reflexed, falling whole, the scales persistent. (4)

4a. Leaves quadrangular (somewhat flattened in *Picea sitchensis*), commonly sharply pointed. *Picea*

4b. Leaves flattened, blunt apically. *Tsuga*

ABIES Mill.

Evergreen, spirelike or conical trees, with thin, grayish bark often bulged by resin vesicles when young, dark gray and thick in age; leaves borne singly, spirally arranged, flat, blunt, narrowed to a short, stout petiole, wholly deciduous, the leaf scar nearly circular; winter buds blunt, resin covered; staminate cones catkinlike, cylindrical; ovulate cones stiffly erect, cylindrical, maturing in one season, the scales shed singly at maturity, the slender axis persistent on the branches for several years.

1a. Branches appearing spraylike, the leaves horizontally spreading; leaves with 2 broad lines of white stomata on the lower surface only; plants of the southern Alexander Archipelago. *A. amabilis*

1b. Branches not spraylike, the leaves directed upwards on both surfaces; plants of south-central and continental southeastern Alaska, southern Yukon, and adjacent British Columbia. *A. lasiocarpa*

Abies amabilis (Dougl.) Forbes
Silver Fir

Trees to 25 m tall or more; twigs pubescent; needles green and lacking stomates above, with two broad silvery bands of stomates on the lower surface, flattened, tending to form flattened sprays, the needles horizontally spreading or appressed along the top of the twig, mostly 10–30 mm long; ovulate cones purple, mostly 8–10(12) cm long and 3–4(6) cm thick.

Coastal forests; in the southern third of the Alexander Archipelago; southward to California (*Picea amabilis* Dougl.).

Abies lasiocarpa (Hook.) Nutt. (× 0.4).

Abies lasiocarpa (Hook.) Nutt.
Subalpine Fir

Trees to about 10 m tall, rarely more, or dwarfed and shrublike in northern localities or in alpine sites; twigs pubescent; needles bluish green, stomatiferous on both surfaces, flattened, all tending to turn upwards, not spraylike, mostly (5)10–25(30) mm long; ovulate cones deep purple, mostly 6–10 cm long and 2.5–3.5 cm thick.

Forests; in south-central and continental southeastern Alaska, southern Yukon, and adjacent British Columbia; southward to Oregon, Arizona, and New Mexico (*Pinus lasiocarpa* Hook.).

LARIX Adans.

Deciduous trees with scaly bark; leaves of two types, thin scalelike, hair-tipped, bractlike leaves at the tips of short branches, and green needles, these borne in clusters on short, spur shoots or spirally arranged on long shoots, all deciduous; winter buds blunt; staminate cones cylindrical, catkinlike; ovulate cones curved upwards, ovoid, maturing in one season, the entire cone more or less persistent, not falling apart at maturity.

Larix laricina (DuRoi) K. Koch
Tamarack

Trees to 10(18) m tall; twigs glabrous, covered with decurrent leaf bases; needles linear, 5–22 mm long, less than 0.5 mm wide, borne in clusters of 7–45 on short, spur shoots, blunt apically, or spirally arranged and apiculate on elongate shoots; ovulate cones mostly 10–20 mm long and 7–20 mm thick, purplish, drying brownish.

Muskegs; in west-central to east-central Alaska, and central and southeastern Yukon; east to the Atlantic and south to British Columbia, Minnesota, and West Virginia (*Pinus laricina* DuRoi; *Larix alaskensis* Wight). Our material has been assigned to var. *alaskensis* (Wight) Raup.

Larix laricina (DuRoi) K. Koch (× 0.4).

PICEA A. Dietr.

Evergreen trees with scaly bark; needles borne singly, spirally arranged, quadrangular and square in cross section or more or less flattened, often sharply acute, deciduous above the base (quickly deciduous from cut branches), the twigs roughened with persistent peglike leaf bases; winter buds blunt; staminate cones cylindrical, catkinlike; ovulate cones curved downward, ovoid to cylindrical, maturing in one season, the cones more or less persistent, not falling apart at maturity.

1a. Decurrent leaf bases of young twigs ciliate and often pubescent; lower bud scales sharply acuminate, often pubescent dorsally; cones mostly 15–30 mm long, persistent on the tree, the scales erose-denticulate. *P. mariana*

1b. Decurrent leaf bases of young twigs glabrous, seldom sparsely ciliate; lower bud scales obtusish to sharply acuminate, glabrous dorsally; cones mostly 25–90 mm long, seldom persisting, the scales entire or erose-crenulate. (2)

2a. Needles almost flat on the upper surface, much wider than thick; cone scales erose-denticulate; plants of coastal and insular southern Alaska from Kodiak eastward. *P. sitchensis*

2b. Needles 4-angled and 4-sided, only some of them sometimes much wider than thick; cone scales entire or subentire, occasionally splitting; plants widespread, less common or absent in coastal regions. *P. glauca*

Picea glauca (Moench) Voss
White Spruce

Trees to 25(30) m tall; bark thin, scaly, silvery brown or blackish; young branchlets covered by decurrent, glabrous or rarely sparsely ciliate leaf bases; needles 4-angled, blue green, standing out on all sides of the branchlet or commonly pointing upwards, mostly 6–15 mm long, more or less sharply pointed, stomatiferous on all sides; lowermost bud scales obtusish to shortly acuminate, merely ciliate or only rarely pubescent dorsally; cones 25–65 mm long, only tardily persistent, the scales subentire to entire, often splitting at irregular intervals, rounded to truncate or emarginate apically.

Interior forests; in most of continental Alaska south of the crest of the Brooks Range and in the Yukon south of 69°15′N;

eastward to the Atlantic and south to British Columbia, Wyoming, South Dakota, Wisconsin, and New York (*Pinus glauca* Moench; *Picea glauca* var. *porsildii* Raup; *P. glauca* var. *albertiana* [Brown] Sarg.). This is the common coniferous tree in interior forests, the important lumber tree of commerce in the interior. The trees average only about 8–10 m tall and 2–3 dm in diameter, but specimens on terraces in the interior grow to 25 m and 5–7 dm in thickness. It tends to intergrade at points of

Picea glauca (Moench) Voss (× 0.6).

contact with both *P. mariana* and *P. sitchensis*. The infraspecific status of our materials requires clarification, and until further work is done it seems best to treat our specimens in a single polymorphic entity.

Picea mariana (Mill.) Britt., Sterns & Pogg
Black Spruce

Trees to 10(15) m tall; trunk mostly less than 2 dm thick; bark thin, scaly, usually gray or blackish; young branchlets covered by decurrent, ciliate, and often hairy leaf bases; needles 4-angled, bluish green, standing out on all sides or mostly pointed upwards, usually 5–10(12) mm long, sharp or blunt apically, stomatiferous on all sides; lowermost bud scales acuminate, ciliate and commonly pubescent dorsally; cones 15–30 mm long, persistent, the scales erosedenticulate, broadly rounded apically.

Poorly drained muskegs and wet meadows; in much of continental Alaska south of the Brooks Range and in coastal south-central Alaska, and much of the Yukon; eastward to the Atlantic and south to British Columbia, Pennsylvania, and New Jersey (*Abies mariana* Mill.). Black spruce is the characteristic tree of poorly drained sites. Dead or senile trees often have dense clusters of branchlets near the top. The presence of hairs on the branchlets has been used as a principal diagnostic feature. However, the character is not absolute, varying from dense to sparse. Some specimens show evidence of hybridization with *P. glauca*.

Picea sitchensis (Bong.) Carr.
Sitka Spruce

Trees to 50 m tall or more; trunk to 20 dm in diameter or more; bark thin, scaly, grayish brown to purplish; young branchlets covered by decurrent, nonciliate, glabrous leaf bases; needles flattened, much wider than thick, green to bluish green, standing out in all directions, usually 10–25 mm long, sharp apically, with 2 broad stomatiferous bands on the lower surface; lowermost bud scales obtuse to abruptly or gradually acuminate, ciliate and rarely hairy dorsally; cones 45–90(100) mm long, only tardily deciduous, the scales erose-denticulate, rounded to emarginate.

Coastal forests; in southern Alaska from Kodiak Island eastward through the Panhandle; southward to California (*Pinus sitchensis* Bong.). Sitka spruce was planted at Unalaska in 1805 where it still persists. Plantings should be expected elsewhere in Alaska and the Yukon. This is an important lumber tree of coastal forests. There are indications of hybridization between Sitka spruce and white spruce in south-central Alaska.

PINUS L.

Evergreen trees with scaly bark; leaves of two types, thin scalelike leaves subtending the base of short spur branches, and green needles borne in clusters of 2–5 on spur branches, the spur branches borne spirally on the twigs; winter buds acutish, resinous; staminate cones ovoid to cylindrical; ovulate cones variously arranged but not erect in age, ovoid to lance-ovoid, maturing in one or two seasons, more or less persistent, woody, not falling apart at maturity.

1a. Needles borne in clusters of 2; trees native or cultivated in southern Alaska and southern Yukon. *P. contorta*

1b. Needles borne in clusters of 5; trees cultivated in south-central Alaska (Anchorage) and perhaps elsewhere, native to western United States. Bristlecone Pine. *P. aristata* Engelm.

Pinus contorta Dougl. ex. Loud.
Lodgepole Pine

Trees 2–12(20) m tall or more; trunk mostly 2–4(7) dm in diameter; bark thin scaly, brownish to blackish; needles in 2s, green to yellow green, semicylindrical, though concave on the upper surface, mostly 15–75 mm long, persistent for several years; cones woody, ovoid to lance-ovoid, lopsided, mostly 30–60 mm long, persisting on the tree for many years and

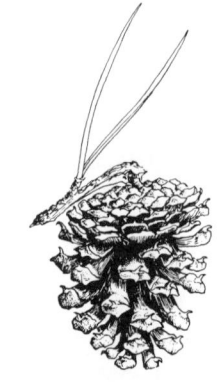

Pinus contorta Dougl. (× 0.4).

opening irregularly, the scales terminated by a flattened umbo, this bearing a sharp, curved spine.

Muskegs; in coastal southeastern Alaska, and on terraces, alluvial fans, glacial moraines, and mountain slopes in southern Yukon and adjacent British Columbia; southward to California, Nevada, Utah, and Colorado. This species is cultivated in Anchorage and perhaps in other sites. The indigenous plants have been treated as belonging to two varieties separable by means of the following key:

1a. Trees mostly less than 5 m tall, shrubby; bark dark brown to blackish, furrowed in old age; plants in coastal muskegs. *P. contorta* var. *contorta*

1b. Trees mostly more than 6 m tall, slender; bark thin, brownish, not furrowed in age; plants of interior sites (*P. contorta* ssp. *latifolia* [Engelm.] Critchfield). *P. contorta* var. *latifolia* Engelm.

TSUGA Carr.

Evergreen trees, with scaly to deeply furrowed bark; needles borne singly, spirally arranged, flat, rounded apically, deciduous above the base (quickly deciduous from cut branches), the twigs roughened with persistent, peglike leaf bases; winter buds blunt; staminate cones ovoid, pedunculate; ovulate cones usually pendulous, ovoid to cylindrical, maturing in one season, more or less persistent, falling as a unit.

1a. Needles and branchlets tending to spread and form flat sprays, the needles with 2 broad stomatiferous bands beneath, lacking stomates above; cones 15–25 mm long. *T. heterophylla*

1b. Needles and branchlets tending to spread in all directions, not forming flat sprays, the needles stomatiferous on both sides; cones mostly 30–65 mm long. *T. mertensiana*

Tsuga heterophylla (Raf.) Sarg.
Western Hemlock

Trees to 50 m tall or more; trunk to 10 dm or more in diameter; bark becoming thick, strongly furrowed, brown; young branchlets covered with short hairs and long, multicellular hairs; needles flat, without a central thickening, yellow green above, with 2 broad white stomatiferous bands below, mostly 3–15(22) mm long,

Tsuga heterophylla (Raf.) Sarg. (× 0.4).

arranged in flat sprays; cones 15–25 mm long, oblong-ovoid, more or less persistent, the scales obtuse to rounded, puberulent.

Coastal forests; in south-central and southeastern Alaska; southward to California, Idaho, and Montana (*Abies heterophylla* Raf.). There are reports of hybrids between *T. heterophylla* and *T. mertensiana*. This is an important forest tree in southeastern Alaska.

Tsuga mertensiana (Bong.) Carr.
Mountain Hemlock

Trees to 20 m or sometimes more; trunk to about 5 or 6(10) dm in diameter; bark thick, strongly furrowed, dark brown to gray brown or black; young branchlets pubescent with both short and long, multicellular hairs; needles flat above, more or less keeled below, or less commonly somewhat keeled above, yellowish green to blue green, almost equally stomatiferous on both surfaces, mostly 8–20(25) mm long, tending to stand out on all sides of the branchlets, not forming flat sprays; cones mostly (25)30–65 cm long, cylindrical, more or less persistent, the scales broadly rounded, puberulent.

Coastal to montane forests; in south-central and southeastern Alaska; southward to California, Nevada, Idaho, and Montana (*Pinus mertensiana* Bong.).

TAXACEAE
Yew Family

Plants dioecious, shrublike trees; leaves evergreen, spirally arranged, needlelike; staminate cones small, globular, axillary; ovulate cones much reduced, the ovules solitary, each surrounded by a fleshy disc (aril), the aril ripening and brightly colored at maturity.

TAXUS L.

A single genus with characteristics of the family.

Taxus brevifolia Nutt. (× 0.4).

Taxus brevifolia Nutt.
Western Yew

Small, shrublike trees to 5(9) m tall or more; bark thin, scaly; branchlets glabrous, terete; leaves flat, yellowish green, mostly 10–18 mm long and 1–2 mm wide, sharply acuminate apically, 2-ranked on the branchlets, persistent for several years; seeds 5–6 mm long, surrounded by a fleshy, red aril.

Coastal forests; in the southernmost islands of the Alexander Archipelago; south to Oregon, Idaho, and Montana.

Subdivision PTEROPSIDA
Class ANGIOSPERMAE
Subclass DICOTYLEDONEAE

1a. Plants woody; trees, shrubs, or subshrubs. KEY I. .... p. 39
1b. Plants herbaceous. (2)

2a. Perianth consisting of a single whorl (arbitrarily called sepals), or none. Key II. .... p. 40
2b. Perianth consisting of 2 whorls (sepals and petals). (3)

3a. Corolla of petals united, at least at the base. KEY V. .... p. 44
3b. Corolla of separate petals. (4)

4a. Stamens numerous, more than twice as many as the petals. KEY III ....
.... p. 42
4b. Stamens few, not more than twice as many as the petals. KEY IV. .... p. 42

KEY I. PLANTS WOODY; TREES, SHRUBS, OR SUBSHRUBS.

1a. Plants parasitic on branches of trees; rooting in the host (*Tsuga*), usually yellow green. VISCACEAE (*Arceuthobium*) .... p. 484
1b. Plants not parasitic on branches of trees, rooting in soil. (2)

2a. Leaves opposite. (3)
2b. Leaves alternate. (8)

3a. Leaves and branchlets stellate-pubescent; perianth in a single whorl; fruit a red drupe. ELAEAGNACEAE (*Shepherdia*) .... p. 221
3b. Leaves and branchlets glabrous or pubescent, but not stellate; perianth of sepals and petals, or if sepals only, then the fruit not a drupe. (4)

4a. Leaves palmately veined and lobed or parted; fruit a double samara. ACERACEAE (*Acer*) .... p. 46
4b. Leaves pinnately veined, simple or compound; fruit drupaceous or capsular. (5)

5a. Ovary superior; stamens 2–5; fruit a capsule. (6)
5b. Ovary inferior; stamens 4–5; fruit drupaceous. (7)

6a. Plants erect, deciduous shrubs; stamens 2; fruit a 2-valved capsule. OLEACEAE (*Syringa*) .... p. 295
6b. Plants decumbent subshrubs; stamens 5; fruit a 2–3-valved capsule. ERICACEAE .... p. 223

7a. Petals united; stamens 5 (rarely 4). CAPRIFOLIACEAE .... p. 69
7b. Petals separate; stamens mostly 4. CORNACEAE (*Cornus*) .... p. 173

8a. Perianth consisting of a single whorl (arbitrarily called sepals), or none (See also Chenopodiaceae.) (9)
8b. Perianth consisting of both sepals and petals. (14)

9a. Plants evergreen, trailing; leaves with narrow involute blades. EMPETRACEAE (*Empetrum*) .... p. 222
9b. Plants deciduous; leaves mostly with expanded blades. (10)

10a. Leaves and branchlets stellate-pubescent; flowers solitary or in axillary clusters. ELAEAGNACEAE (*Elaeagnus*) .... p. 221
10b. Leaves and branchlets variously pubescent or glabrous, but not stellate; flowers in catkins. (11)

11a. Buds with a single visible scale; dioecious subshrubs, shrubs, or trees. SALICACEAE (*Salix*) .... p. 402
11b. Buds with few to several visible scales; dioecious or monoecious shrubs or trees. (12)

12a. Leaves ovate, once-serrate; dioecious trees; fruit a 2–4-valved capsule. SALICACEAE (*Populus*) .... p. 401
12b. Leaves various in shape, but if ovate to lanceolate then doubly serrate, or else the plants monoecious shrubs; fruit a nutlet or waxy drupe. (13)

13a. Leaves narrowly oblanceolate to spatulate, toothed only near the

apex; fruit a waxy drupe. Myrica-
ceae (*Myrica*) .... p. 292

13b. Leaves ovate to lanceolate or orbicular, toothed from near the base; fruit
a nutlet, this often winged. Betu-
laceae .... p. 51

14a. Corolla of united petals. (15)

14b. Corolla of separate petals. (17)

15a. Flowers in dense, involucrate heads;
leaves mostly lobed or divided. Compositae (*Artemisia*) .... p. 121

15b. Flowers variously displayed, but not
in involucrate heads; leaves mostly
not lobed or divided. (16)

16a. Stigma 3-lobed; ovary 3-loculed; fruit
a 3-valved capsule. Diapensiaceae
(*Diapensia*) .... p. 219

16b. Stigma capitate; ovary 2–10-loculed;
fruit a capsule, berry, or drupe. Eri-
caceae .... p. 223

17a. Stamens numerous, more than twice
as many as the petals. Rosaceae....
.... p. 369

17b. Stamens few, not more than twice as
many as the petals. (18)

18a. Plants depressed shrubs or subshrubs;
leaves simple, pinnately veined, very
small. (19)

18b. Plants erect or ascending shrubs;
leaves palmately veined, often lobed,
moderate to large. (20)

19a. Sepals 3, petaloid; petals 3; stamens
3; fruit a fleshy, black berry; leaves
deeply grooved beneath, congested.
Empetraceae (*Empetrum*) .... p. 222

19b. Sepals 5; petals 5, reflexed; stamens
5; fruit a red berry; leaves flat, not
congested on the stem. Ericaceae
(*Oxycoccus*) .... p. 231

20a. Leaves with spines along the petioles
and main veins; stems coarse and
spiny; inflorescence terminal. Arali-
aceae (*Oplopanax*) .... p. 50

20b. Leaves lacking spines along the petioles and main veins; stems slender,
armed or unarmed; inflorescence
lateral. Saxifragaceae (*Ribes*)..........

.... p. 427

KEY II. PERIANTH CONSISTING OF A SINGLE WHORL, OR NONE.

1a. Plants parasitic on branches of trees,
rooting in the host (*Tsuga*), usually
yellowish green. Viscaceae (*Arceuthobium*) .... p. 484

1b. Plants not parasitic on branches of
trees, rooting in the soil. (2)

2a. Leaves alternate or basal and alternate. (3)

2b. Leaves opposite or whorled, at least
on the lower part of the stem (See
also Chenopodiaceae). (18)

3a. Leaves odd-pinnately compound, or
palmately lobed, the lobes serrate;
stipules present, not sheathing. Rosaceae (*Sanguisorba, Alchemilla*)......
.... p. 369

3b. Leaves various, but not as above;
stipules absent, or if present, then
sheathing. (4)

4a. Flowers lacking a perianth, the naked
staminate and pistillate flowers borne
in cup-shaped involucres (cyathia);
plants with milky juice. Euphorbi-
aceae (*Euphorbia*) .... p. 237

4b. Flowers with a perianth, variously
arranged, but not as above; plants
with watery or milky juice. (5)

5a. Flowers sessile, in involucrate heads;
anthers united into a tube around the
style. Compositae .... p. 106

5b. Flowers variously arranged, but usually not in heads; anthers not joined.
(6)

6a. Flowers perigynous, 4-merous; stamens 4 or 8, inserted in the notches of
a disk which almost fills the center
of the flower; leaves orbicular, with
6–10 broadly rounded teeth. Saxifragaceae (*Chrysosplenium*) .. p. 422

6b. Flowers epigynous, perigynous, or
hypogynous, usually not 4-merous;
stamens not inserted on a disk (except in Umbelliferae and Santalaceae); leaves not as above. (7)

7a. Ovary inferior; plants with perfect flowers. (8)

7b. Ovary superior or plants dioecious. (10)

8a. Leaves simple, well distributed along the stem; flowers in axillary or terminal cymes; fruit drupaceous. SANTALACEAE .... p. 420

8b. Leaves mostly once to thrice compound; flowers in compound umbels; fruit a schizocarp or berry. (9)

9a. Styles 2; fruit a schizocarp; leaves several to many. UMBELLIFERAE........ .... p. 467

9b. Styles 5; fruit a purplish berry; leaves usually solitary. ARALIACEAE (*Aralia*) .... p. 50

10a. Pistils several to many per flower; stamens usually 10–many. RANUNCULACEAE .... p. 347

10b. Pistils 1 per flower; stamens usually 10 or less (more than 10 in Papaveraceae). (11)

11a. Plants aquatic, usually more or less submerged; perianth lacking. CALLITRICHACEAE (*Callitriche*) .... p. 65

11b. Plants terrestrial, sometimes growing on moist soil; perianth present. (12)

12a. Perianth showy; sepals and petals both present in bud, the sepals caducous and represented only by scars during anthesis; plants with milky juice. PAPAVERACEAE (*Papaver*) .... p. 304

12b. Perianth not or not especially showy; plants various, but not as above. (13)

13a. Ovary 2-loculed; stamens 2, 4, or 6; fruit a silicle. CRUCIFERAE (*Lepidium*) .... p. 210

13b. Ovary 1-loculed; stamens 2–9; fruit an achene or utricle. (14)

14a. Plants with a stipular sheath (ocrea) above each node; perianth segments mostly 6. POLYGONACEAE .... p. 314

14b. Plants lacking stipules; perianth segments 1, 4, 5, or 6. (15)

15a. Leaves all basal; plants scapose perennials; flowers borne in cup-shaped involucres. POLYGONACEAE (*Eriogonum*) .... p. 315

15b. Leaves chiefly cauline; plants annual or perennial; flowers not borne in involucres. (16)

16a. Plants neither scurfy-pubescent nor with spinose bracts or sepals; flowers in small, axillary clusters; styles 1. URTICACEAE (*Parietaria*) .... p. 477

16b. Plants commonly scurfy-pubescent, or with spinose bracts and sepals; styles 2–5. (17)

17a. Bracts subtending flowers spinose; plants not scurfy. AMARANTHACEAE (*Amaranthus*) .... p. 48

17b. Bracts subtending flowers not spinose; plants commonly scurfy. CHENOPODIACEAE .... p. 99

18a. Plants aquatic; leaves simple, or the immerse ones finely dissected. (19)

18b. Plants terrestrial; leaves various, but not as above. (20)

19a. Leaves simple, or if dissected then ovary 4-loculed; stigmas 4. HALORAGACEAE .... p. 248

19b. Leaves 2–3 times palmately divided; ovary 1-loculed; stigmas 1. CERATOPHYLLACEAE .... p. 99

20a. Flowers sessile, in involucrate heads; anthers united into a tube around the style. COMPOSITAE .... p. 106

20b. Flowers not in involucrate heads; anthers not united. (21)

21a. Leaves pinnately parted or compound; inflorescence of terminal or axillary, many-flowered cymes. VALERIANACEAE (*Valeriana*) .... p. 478

21b. Leaves palmately lobed or parted, or simple and neither lobed nor parted; inflorescence various. (22)

22a. Leaves palmately lobed or parted; cauline leaves in 1–2 whorls of 3. RANUNCULACEAE (*Anemone*) .. p. 349

22b. Leaves simple, not lobed or parted; cauline leaves several or lacking, not in whorls of 3. (23)

23a. Plants with stinging hairs; flowers inconspicuous, in axillary spikes. URTICACEAE (*Urtica*) .... p. 477
23b. Plants lacking stinging hairs; flowers not in axillary spikes. (24)

24a. Flowers in terminal umbels, subtended by 4 large petaloid bracts (the whole resembling a single, terminal flower). CORNACEAE (*Cornus*) .. p. 173
24b. Flowers variously arranged, but not subtended by petaloid bracts. (25)

25a. Ovary inferior, 2-loculed; flowers in cymes. RUBIACEAE (*Galium*) .. p. 398
25b. Ovary superior, 1-loculed; flowers axillary, or in racemes. (26)

26a. Styles 2–5; calyx segments distinct. CARYOPHYLLACEAE .... p. 74
26b. Styles 1; calyx segments united. PRIMULACEAE (*Glaux*) .... p. 338

KEY III. COROLLA OF SEPARATE PETALS; STAMENS MORE THAN TWICE AS MANY AS THE PETALS.

1a. Plants aquatic; leaves immersed or floating, cordate or peltate; carpels united or separate (in *Brasenia*). NYMPHAEACEAE .... p. 293
1b. Plants terrestrial, or if aquatic and with separate carpels, the leaves not peltate. (2)

2a. Ovary inferior, or partly so. ROSACEAE .... p. 369
2b. Ovary superior. (3)

3a. Sepals 2–3, caducous; flowers mostly 2.5–6 cm broad; plants with milky juice. PAPAVERACEAE (*Papaver*) .... p. 304
3b. Sepals mostly 5 or more (caducous in some Ranunculaceae); flowers mostly less than 2.5 cm broad; plants lacking milky juice. (4)

4a. Staminal filaments united into a tube surrounding the styles. MALVACEAE (*Malva*) .... p. 290
4b. Staminal filaments not united into a tube. (5)

5a. Stamens, petals, and sepals all attached to the margin of a hypanthium; flowers perigynous. ROSACEAE.... .... p. 369
5b. Stamens, petals, and sepals all attached to the base of the ovary; flowers hypogynous. RANUNCULACEAE .... p. 347

KEY IV. COROLLA OF SEPARATE PETALS; STAMENS FEW, NOT MORE THAN TWICE AS MANY AS THE PETALS.

1a. Pistils 3 or more. (2)
1b. Pistils 1 (rarely 2). (5)

2a. Plants succulent, fleshy; sepals, petals, and pistils 4–5 each. CRASSULACEAE (*Sedum*) .... p. 175
2b. Plants not succulent; pistils often more numerous than the sepals and petals. (3)

3a. Plants aquatic; leaves peltate. NYMPHAEACEAE (*Brasenia*) .... p. 293
3b. Plants terrestrial, or if aquatic, then the leaves not peltate. (4)

4a. Stamens inserted on a hypanthium; flowers perigynous. ROSACEAE............ .... p. 369
4b. Stamens inserted at the base of the ovary; flowers hypogynous. RANUNCULACEAE .... p. 347

5a. Ovary inferior, or partly so. (6)
5b. Ovary superior. (11)

6a. Plants aquatic; leaves mostly whorled, simple or finely dissected; stamens 1 or 4–8. HALORAGACEAE .... p. 248
6b. Plants terrestrial, sometimes growing in wet places. (7)

7a. Inflorescence a compound umbel; fruit a schizocarp or berry; leaves mostly ternate or finely divided. (8)
7b. Inflorescence a raceme or panicle, or a simple umbel (or the flowers solitary); fruit not a schizocarp or berry; leaves not as above. (9)

8a. Styles 2; fruit a schizocarp; **leaves several.** UMBELLIFERAE .... p. 467

8b. Styles 5; fruit a **purplish berry;** leaves solitary (rarely 2–3). ARALI-ACEAE (*Aralia*) .... p. 50

9a. Inflorescence a simple umbel subtended by 4 large petaloid bracts. CORNACEAE (*Cornus*) .... p. 173

9b. Inflorescence racemose or paniculate (or the flowers solitary), not subtended by petaloid bracts. (10)

10a. Corolla of 2 or 4 petals; stamens 2 or 8. ONAGRACEAE .... p. 295

10b. Corolla of 5 petals; stamens 5 or 10. SAXIFRAGACEAE .... p. 421

11a. Styles 2–5 (appearing as 6–10 in Droseraceae), distinct to near the base. (12)

11b. Styles 1, sometimes lobed or divided at the apex, or the style obsolete. (17)

12a. Leaves with long, glandular, sensitive hairs which trap insects. DROSER-ACEAE (*Drosera*) .... p. 219

12b. Leaves variously pubescent or glabrous, but not as above. (13)

13a. Leaves opposite. CARYOPHYLLACEAE .... p. 74

13b. Leaves alternate, or basal and alternate, or if opposite (as in some Portulacaceae), then the sepals 2. (14)

14a. Leaves strictly basal; flowers in heads, subtended by broad, scarious bracts. PLUMBAGINACEAE (*Armeria*) .... p. 310

14b. Leaves cauline, at least some (all basal in some Saxifragaceae); flowers not in heads nor with broad scarious bracts subtending the inflorescence. (15)

15a. Petals blue, caducous; fruit a 10-loculed capsule. LINACEAE (*Linum*) .... p. 289

15b. Petals variously colored, but not blue; fruit less than 10-loculed. (16)

16a. Sepals 2; stamens opposite the petals or fewer (5–3). PORTULACACEAE .... p. 328

16b. Sepals 5; stamens alternate with the petals or more numerous (except 3 in *Tolmiea*). SAXIFRAGACEAE .... p. 421

17a. Sepals 2–3. (18)

17b. Sepals 4, 5 or more. (20)

18a. Corolla regular. PORTULACACEAE .... p. 328

18b. Corolla irregular. (19)

19a. Sepals 3, with 2 small and green, the third petaloid and produced backward into a recurved spur. BALSAMI-NACEAE .... p. 50

19b. Sepals 2, equal or nearly so, lacking a spur (the corolla sometimes spurred). FUMARIACEAE (*Corydalis*) .... p. 238

20a. Flowers irregular. (21)

20b. Flowers regular. (23)

21a. Flowers papilionaceous, the upper petal (banner) enclosing the others in bud; fruit a legume; leaves mostly compound. LEGUMINOSAE .... p. 259

21b. Flowers not papilionaceous; fruit a capsule; leaves simple. (22)

22a. Flowers solitary, the petals violet to yellow or white. VIOLACEAE (*Viola*) .... p. 480

22b. Flowers few to numerous, the petals pink, greenish yellow, or creamy white. PYROLACEAE (*Pyrola*) .. p. 345

23a. Leaves compound, mostly basal, deeply pinnatifid. GERANIACEAE (*Erodium*) .... p. 245

23b. Leaves simple, may be pinnately or palmately divided. (24)

24a. Sepals (bracts), petals (in reality sepals), and stamens 4; stamens inserted at edge of a disk. ROSACEAE (*Alchemilla*) .... p. 371

24b. Sepals and petals 4 or 5; stamens 2, 5, or 6, or if 4, then not inserted at edge of a disc. (25)

25a. Sepals and petals 4; stamens 6(4 plus 2), 4, or rarely 2. CRUCIFERAE .. p. 177

25b. Sepals and petals (4)5(7); stamens (4)5–10(12). (26)

26a. Leaves palmately lobed, cleft, or divided, stipulate; carpels tailed at maturity, separating from each other as 1-seeded, indehiscent segments. Geraniaceae (*Geranium*) .... p. 245

26b. Leaves simple pinnate, lacking stipules; carpels not as above. (27)

27a. Flowers solitary (rarely paired), axillary; hypanthium cone shaped; petals pink or white, less than 4 mm long. Lythraceae (*Lythrum*) .. p. 290

27b. Flowers various, solitary to several, subscapose or racemose, not axillary; hypanthium lacking or broad and flat, not cone shaped; petals variously colored, mostly more than 4 mm long. (28)

28a. Stamens 5, alternating with broad, dissected, glandular-tipped staminodia; ovary 1-loculed. Saxifragaceae (*Parnassia*) .... p. 426

28b. Stamens 5–more; staminodia lacking; ovary more than 1-loculed. Pyrolaceae .... p. 343

KEY V. COROLLA OF PETALS UNITED AT LEAST NEAR THE BASE.

1a. Ovary inferior, or partly so. (2)
1b. Ovary superior (see Menyanthaceae). (9)

2a. Stamens more than 5; anthers opening by terminal pores. Ericaceae.... p. 223

2b. Stamens 5 or less (if more as in *Adoxa*, the sepals 2–3); anthers not opening by terminal pores. (3)

3a. Stamens united by their anthers. (4)
3b. Stamens separate. (5)

4a. Flowers sessile, in involucrate heads; stamens adnate to the corolla. Compositae .... p. 106

4b. Flowers not in involucrate heads; stamens not adnate to the corolla. Campanulaceae (*Lobelia*) .... p. 69

5a. Basal leaves ternately compound, long-petioled; stem leaves in a single, opposite pair, ternately lobed. Adoxaceae (*Adoxa*) .... p. 47

5b. Basal leaves not long-petioled, ternate; stem leaves simple, or pinnately lobed, divided, or compound. (6)

6a. Leaves alternate; flowers mostly large and more or less bell shaped. Campanulaceae (*Campanula*).......... p. 66

6b. Leaves opposite or whorled; flowers mostly small and not bell shaped. (7)

7a. Leaves pinnately lobed or compound; stamens 2–3(4); flowers irregular; plants herbaceous, frequently ill scented. Valerianaceae (*Valeriana*) .... p. 478

7b. Leaves simple, or pinnately compound; stamens 4–5; flowers regular or irregular; plants herbaceous or woody, mostly not ill scented. (8)

8a. Plants woody shrubs, vines, or trailing subshrubs; leaves opposite (rarely whorled), or perfoliate; fruit 1–several-seeded. Caprifoliaceae.......... p. 69

8b. Plants herbaceous; leaves opposite or whorled; fruit 2-seeded. Rubiaceae (*Galium*) .... p. 398

9a. Stamens more than 5 (mostly twice as many as the corolla lobes). (10)

9b. Stamens 5 or less (if more, then the same number as the corolla lobes). (13)

10a. Corolla segments distinctly united, saucer, cup, or tube shaped. (11)

10b. Corolla segments not markedly united, usually connate only near the base, or only part of the segments united. (12)

11a. Anthers opening by terminal pores. Ericaceae .... p. 223

11b. Anthers opening by a longitudinal slit. Pyrolaceae .... p. 343

12a. Petals 5; fruit a legume or a loment. Leguminosae .... p. 259

12b. Petals 4, in 2 unlike pairs; fruit a capsule with 2 parietal placentae. FUMARIACEAE (*Corydalis*) ...... p. 238

13a. Plants parasitic, devoid of chlorophyll; leaves small, scalelike. ORO-BANCHACEAE .... p. 302

13b. Plants usually not parasitic, having chlorophyll; leaves usually well developed. (14)

14a. Corolla irregular. (15)
14b. Corolla regular. (18)

15a. Ovary with 1 ovule per locule, appearing 4-loculed, 4-lobed; fruit of 4 indehiscent 1-seeded nutlets; stems mostly square; plant often with a mintlike odor. LABIATAE .... p. 252

15b. Ovary with more than 1 ovule per locule, usually not 4-loculed nor 4-lobed; stems mostly round; plants lacking a distinctive odor. (16)

16a. Plants aquatic; leaves often dissected and bearing small bladders; corolla spurred; stamens 2. LENTIBULARIA-CEAE (*Utricularia*) .... p. 288

16b. Plants usually terrestrial, or if aquatic then other than above. (17)

17a. Plants scapose; scapes bearing a single, blue to purple flower. LEN-TIBULARIACEAE (*Pinguicula*) .... p. 287

17b. Plants not scapose; inflorescence few- to many-flowered. SCROPHULARIACEAE .... p. 444

18a. Plants with milky juice; pistils 2, separate at the base, united by the styles. APOCYNACEAE (*Apocynum*)...... .... p. 49

18b. Plants usually lacking milky juice; pistils 1. (19)

19a. Sepals 2; styles 3, or 3-lobed; stamens 2–5. PORTULACACEAE ...... p. 328

19b. Sepals (4)5 or more, or connate with (4)5 or more lobes; styles various; stamens various. (20)

20a. Stamens as many as the corolla lobes and opposite them. (21)

20b. Stamens as many as the corolla lobes and alternate with them, or fewer. (22)

21a. Flowers in headlike cymules, subtended by broad, straw-colored or pinkish, translucent bracts. PLUM-BAGINACEAE (*Armeria*) .... p. 310

21b. Flowers variously arranged, solitary, in umbels, or in racemes, not subtended by broad translucent bracts. PRIMULACEAE .... p. 333

22a. Corolla small (2 mm broad or less), scarious, veinless; capsule opening by a lid; leaves mostly basal; inflorescence a dense spike. PLANTAGINACEAE (*Plantago*) .... p. 307

22b. Corolla various, but not as above; leaves mostly cauline; inflorescence mostly other than a spike. (23)

23a. Ovary 4-lobed, appearing 4-loculed; fruit consisting of 4 nutlets at maturity. BORAGINACEAE. .... p. 56

23b. Ovary not 4-lobed, the locules 1, 2–3; fruit not consisting of nutlets. (24)

24a. Style 3-cleft; ovary 3-loculed, or the stigma 3-lobed; fruit a 3-valved capsule. (25)

24b. Style not 3-cleft; ovary 1–2-loculed; fruit not as above. (26)

25a. Plants dwarf perennials with pedunculate, subscapose, single-flowered inflorescences; peduncle naked or single-bracted and usually with 3 bracts at the base of the calyx; stigma merely 3-lobed. DIAPENSIACEAE (*Diapensia*) .... p. 219

25b. Plants various, but if dwarf perennials, the inflorescence not as above; styles 3-cleft. POLEMONIACEAE .... p. 310

26a. Ovary 1-loculed. (27)
26b. Ovary with 2 or more locules. (29)

27a. Leaves mostly basal, from a stout rhizome, trifoliolate, or broadly orbicular to reniform; scapes bearing compound cymes or racemes. MENY-ANTHACEAE .... p. 291

27b. Leaves cauline, at least some, **neither** trifoliolate nor broadly orbicular, or if orbicular, then with few, large teeth. (28)

28a. Leaves opposite, entire; styles 1 or none. GENTIANACEAE .... p. 239
28b. Leaves alternate or basal and alternate, or if opposite, then pinnatifid; styles 2 or 2-cleft apically. HYDROPHYLLACEAE .... p. 250

29a. Plants aquatic; leaves strictly basal, long-petioled, the blades elliptic to spatulate, entire. SCROPHULARIACEAE (*Limosella*) .... p. 451
29b. Plants terrestrial, sometimes growing in wet soil; leaves not strictly basal, or if long-petioled, then not entire. (30)

30a. Styles 2 or 2-cleft apically; fruit a capsule. HYDROPHYLLACEAE (*Phacelia*) .... p. 250
30b. Styles 1; stigma entire or merely lobed apically; fruit a capsule or berry. (31)

31a. Principal leaves basal, long-petiolate, the blades orbicular to reniform, palmately veined, with few large teeth or lobes. HYDROPHYLLACEAE (*Romanzoffia*) .... p. 251
31b. Principal leaves cauline, mostly short-petiolate or sessile, the blades mostly ovate to lanceolate, pinnately veined, entire to serrate or pinnately lobed. (32)

32a. Stamens 2; flowers pale blue to blue. SCROPHULARIACEAE (*Veronica*) ....
.... p. 461
32b. Stamens usually 4–5, or others represented by staminodia; flowers variously colored. SOLANACEAE ....
.... p. 465

ACERACEAE
Maple Family

Trees or shrubs; leaves opposite, palmately lobed; flowers regular, perfect or imperfect, or polygamous, borne in terminal or lateral racemes or panicles; sepals 4–5; petals separate, 4–5, or lacking; stamens 4–10, usually 8, inserted at the inner or outer edge of a glandular disk; pistils 1, the ovary superior, 2-loculed; styles 2; fruit a double samara.

ACER L.
Small trees or shrubs; leaves palmately lobed and veined; flowers in axillary racemes or panicles; petals shorter than the sepals, or lacking.

1a. Inflorescence elongate, the flowers more than 10; fruit bristly-hairy; leaves often over 12 cm broad (at least some); reported from extreme southeastern Alaska (requires verification). *A. macrophyllum* Pursh
1b. Inflorescence short, the flowers commonly 10–fewer; fruit glabrous or nearly so; leaves less than 11 cm broad. (2)

2a. Leaves 7–9-lobed; stamens inserted at the inner edge of the glandular disk; plants uncommon in southeastern Alaska. *A. circinatum*
2b. Leaves 3–5-lobed; stamens inserted at the outer edge of the glandular disk; plants indigenous to southeastern Alaska, or cultivated introductions. (3)

3a. Leaves 3-lobed, the central lobe much longer than the lateral ones; cultivated ornamental. *A. ginnala*
3b. Leaves 3–5-lobed, the central lobe not much larger than the lateral ones; indigenous to southeastern Alaska, sometimes cultivated. *A. glabrum*

Acer circinatum Pursh
Vine Maple

Shrub or small tree, to 5 m tall; stems reddish purple to brownish; leaf blades 3–7.5 cm long, 5–11 cm broad, 7–9-lobed, acuminate apically, cordate basally; flowers 1–few, 6–8 mm broad, borne in 2-leaved shoots; sepals 4–5, purplish; petals white, shorter than the sepals; stamens 8, inserted at the inner edge of the glandular disk; samara wings widely (almost horizontally) spreading.

Cultivated; reportedly indigenous in southeastern Alaska; south to California.

Acer ginnala Maxim
Amur Maple

Shrub or small tree, 1–4 m tall; stems greenish to reddish; leaf blades 3–9 cm long, 2–7 cm broad, 3-lobed, the central lobe much longer than the lateral ones, acuminate apically, subcordate to truncate basally; flowers few to several, borne in long-peduncled panicles; sepals 4–5, greenish; petals yellowish white; stamens 8–10,

Acer glabrum Torr. (× 0.3).

inserted at the outer edge of the glandular disk; samara wings nearly parallel.

Cultivated ornamental; in southern Alaska; introduced from Asia.

Acer glabrum Torr.
Douglas Maple

Shrub or small tree, 1–5 m tall; stems reddish to purplish; leaf blades 2–12 cm long, 1.5–12 cm broad, 3–5-lobed, acuminate apically, subcordate basally; flowers few to several, 6–8 mm broad, borne on 2-leaved shoots; sepals 4–5; petals subequal to the sepals, or lacking; stamens 8–10, inserted at the outer edge of the glandular disk; samara wings not widely spreading.

Woods and thickets; in southeastern Alaska (cultivated in southern and southeastern portions); southward to California and eastward to New Mexico, Nebraska, and Alberta (*Acer douglasii* Hook.; *A. glabrum* ssp. *douglasii* [Hook.] Wesmael). Our material belongs to var. *douglasii* (Hook.) Dippel.

ADOXACEAE
Moschatel Family

Small, perennial herbs, from scaly rhizomes; stems erect, with a single pair of opposite, palmately veined and lobed leaves, and 1–3 basal, long-petiolate, ternately compound leaves; flowers regular, perfect, borne in heads; sepals 2–3(4); petals united, the corolla rotate, 4–5-lobed; stamens 8–10; anthers 1-loculed; filaments inserted in pairs in the sinuses of the corolla lobes; pistils 1, the ovary partly inferior, 4–5-loculed, each with a single ovule; styles 4–5, short; fruit a dry berry with 4–5 nutlets.

ADOXA L.

Inconspicuous, rhizomatous herbs; leaves ternately compound or palmately lobed, petiolate; flowers inconspicuous, yellowish green, borne in heads; fruit a dry berry.

Adoxa moschatellina L.

Moschatel

Plants 0.6–1.8 dm tall, with a musky odor, from rhizomes with fleshy scales; basal leaves 4–16 cm long, the blades ternately compound, the 3 leaflets deeply 3-lobed, the lobes again lobed, green and glabrous; cauline leaves opposite, 0.8–4 cm long, the blade palmately lobed, the lobes again lobed or toothed; peduncles 2–5 cm long, glabrous; flowers sessile in heads; sepals 2–3, 1–1.5 mm long, green; petals 4–5, 1–2 mm long, yellowish green; fruit 2–3 mm long.

Woods, thickets, talus slopes, and meadows; in much of Alaska south of the Brooks Range, except for coastal and insular portions, and in southern Yukon; eastward to the Atlantic and southward to Colorado, Iowa, and New York; circumboreal.

AMARANTHACEAE

Amaranth Family

Herbaceous plants; leaves simple, alternate or some opposite, not mealy; flowers inconspicuous, perfect, monoecious or dioecious, borne in dense racemes or spikes, or in axillary clusters, subtended by spinose bracts; sepals 5; corolla lacking; stamens usually 5; pistils 1, the ovary superior, 1-loculed; styles 2–3; fruit a 1-seeded capsule or utricle.

AMARANTHUS L.

Annual, weedy herbs; stems prostrate to erect; leaves alternate, simple; flowers in axillary or terminal clusters, small and inconspicuous, subtended by spinose bracts; fruit 1-seeded, the seeds black, shiny.

1a. Flowers in small axillary clusters; stems prostrate or ascending. *A. graecizans*

1b. Flowers in terminal and axillary spikes; stem erect. *A. retroflexus*

Amaranthus graecizans L.

Redroot

Annual, monoecious herbs, the stems prostrate to ascending, to 5 dm long or more, glabrous or pubescent; leaf blades commonly 1–2.5 cm long, 0.3–1.5 cm broad, obovate to elliptic, entire to undulate; petioles 2–20 mm long; flowers in small, axillary clusters; bracts spinose, 1.5–3 mm long; sepals spinose, 1.5–3 mm long on pistillate flowers, smaller and scalelike on staminate flowers; seed lens shaped, 1.3–1.7 mm long, black, shiny.

Introduced weed in southeastern Alaska, where it evidently does not persist; widespread in the northern hemisphere.

Amaranthus retroflexus L.

Redroot Pigweed

Annual, dioecious herbs, the stems erect, 3–5 dm tall or more, puberulent; leaf blades 2–8 cm long, 0.5–5 cm broad, lanceolate to ovate or oblanceolate, entire or

Adoxa moschatellina L. (× 0.6).

undulate; petioles 5–60 mm long; flowers in terminal and axillary spikes; bracts spinose, 3–5 mm long; sepals of pistillate flowers 2.5–4 mm long; seeds lens shaped, 0.8–1.2 mm long, black, shiny.

Introduced weed in southeastern Alaska, where it evidently does not persist; widespread in the northern hemisphere.

APOCYNACEAE
Dogbane Family

Perennial, rhizomatous herbs with milky juice; leaves opposite or rarely whorled, simple, entire; flowers few to numerous, borne in axillary or terminal cymes, perfect, regular; sepals 5, distinct to near the base; petals united, the corolla 5-lobed, campanulate; stamens 5, alternate to the corolla lobes; anthers appressed to the style; pistils 1, the ovaries 2, distinct below the solitary style, superior; fruit a follicle; seeds with a terminal tuft of hair.

APOCYNUM L.

Perennials, with erect stems and opposite, simple leaves; flowers small, showy; sepals 5, lanceolate, scarious-margined; corolla campanulate, white to pink; ovaries distinct, connate above into a single style, separating at maturity into 2 separate, many-seeded follicles.

Apocynum androsaemifolium L.
Dogbane

Plants with milky juice, the stems 2–7 dm tall, glabrous or hairy; leaf blades 2.5–8 cm long, 1.2–4.5 cm broad, lance-ovate to elliptic or oblong, abruptly acuminate apically, obtuse to rounded basally, entire, commonly glabrous and green above, paler and usually pubescent beneath; petioles 2–4 mm long, hairy; flowers small, showy; sepals 2–3 mm long; corolla pinkish, 6–8 mm long, the lobes shorter than the tube; follicles 5–12 cm long, to 0.5 cm thick; seeds with long hair.

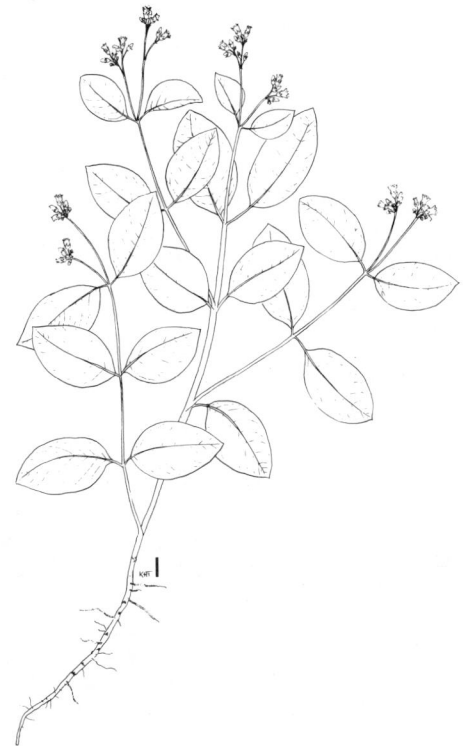

Apocynum androsaemifolium L. (× 0.3).

Open hillsides and woods; in southeastern quarter of Alaska (except for the Panhandle) and in southern Yukon; widely distributed in North America.

ARALIACEAE
Ginseng Family

Perennial herbs or shrubs, with branches armed or unarmed; leaves ternately compound, or palmately veined and lobed, alternate; flowers perfect or polygamous; sepals represented by 5 teeth or inconspicuous; petals 5, separate; stamens 5, alternate with the petals; pistils 1, the ovary inferior, 2–5-loculed; styles 5; fruit a berry.

1a. Plants herbaceous, unarmed; leaves ternately compound; flowers in umbels. *Aralia*
1b. Plants woody, armed; leaves simple,

palmately lobed; flowers in panicles. *Oplopanax*

ARALIA L.

Perennial, unarmed herbs; leaves alternate or basal, ternately compound; flowers perfect, borne in umbels; sepals 5, minute; petals 5, separate, early deciduous; styles 5, distinct nearly to the base; ovary 5-loculed; fruit a berry.

Aralia nudicaulis L.
Wild Sarsaparilla

Rhizomatous herbs, the flowering stems arising from short, vertical, woody branches; leaves 1(2–3) per stem, basal, 2–5 dm long, ternately compound, the leaflets commonly 9, sometimes more, 5–15 cm long, 2–8 cm broad, oblong to elliptic, acuminate apically, oblique basally, serrate; inflorescence of (2)3–7 umbels; flowers 5–6 mm long, numerous, the pedicels 5–15 mm long; sepals minute; petals greenish white, 2–3 mm long; berries purplish, 6–8 mm long.

Woods; at Liard Hot Springs, British Columbia, and extreme southeastern Yukon; southward to Washington and east to the Atlantic.

OPLOPANAX Miq.

Shrubs, with armed branches, petioles, and leaf blades; leaves alternate, palmately veined and lobed, deciduous; flowers perfect or imperfect, borne in panicles; sepals 5, small; petals 5, separate; styles 2, distinct to near the base; ovary 5-loculed; fruit a berry.

Oplopanax horridum (J. E. Smith) Miq.
Devil's Club

Shrubs, 0.5–3 m tall or more; stems coarse, armed with stiff spines; leaves more than 1 per stem, blades palmately veined and lobed, 5–35 cm long and 10–45 cm broad or more, lobes doubly serrate, acuminate,

cordate basally, more or less pubescent, spiny along the veins; petioles about as long as the blades, armed with spines; inflorescence a panicle; flowers 5–6 mm long, numerous, the pedicels 1–4 mm long; sepals minute; petals 2–3 mm long, greenish white; berries red, 5–8 mm long.

Woods, stream banks, rock outcrops, and meadows; in coastal southern Alaska; south to California, Idaho, and Montana, and disjunctly from Michigan to Ontario (*Panax horridum* J. E. Smith in Rees; *Echinopanax horridum* [J. E. Smith] Decne. & Planch).

BALSAMINACEAE
Jewelweed Family

Annual, succulent herbs; stems erect or ascending from tap or fibrous roots; leaves alternate, simple; flowers irregular, perfect, axillary, solitary or few in racemes; sepals 3, the upper one petaloid and strongly spurred; petals 5, or united and appearing as 3; stamens 5, the filaments and anthers closely covering the ovary; pistils 1,

Oplopanax horridum (J. E. Smith) Miq. (× 0.3).

ovary superior, 5-loculed; fruit an elastically dehiscent capsule.

IMPATIENS L.

Annual herbs, with simple, alternate leaves; flowers showy, yellow; sepals 3, dissimilar, the upper one petaloid and spurred; petals 5, appearing as 3; stamens 5; ovary 5-carpelled and 5-loculed (1-loculed by breakdown of the partitions); fruit a capsule.

Impatiens noli-tangere L.
Western Touch-me-not

Annual herbs, the stems 2–7 dm tall, erect or ascending, succulent; leaf blades 1.5–9 (12) cm long and 0.7–4.5 cm broad, ovate to elliptic, acute to obtuse apically, obtuse to rounded basally, serrate to crenate-serrate, glabrous; petioles 3–50 mm long; lower sepals 4–5 mm long, acuminate, the spurred sepal yellow, spotted purplish, 15–25 mm long, the spur strongly recurved; petals yellow, spotted purplish; capsules to 25 mm long.

Moist woods, stream banks, and seeps; in southern Alaska (except for the southwestern portion); southward to Oregon and Idaho; Eurasia (*I. occidentalis* Rydb.).

BETULACEAE
Birch Family

Monoecious trees or shrubs; leaves alternate, simple, deciduous; flowers borne in catkins, each scale subtending 3–6 flowers; perianth none or minute; staminate catkins drooping at anthesis, usually much longer than the pistillate; stamens commonly 1–4, the flowers adnate to the bract; pistillate catkins ovoid to cylindrical; pistils 1, the ovary naked or inferior, (1)2-loculed and (2)4-ovuled; styles 2; fruit a 1-seeded nutlet or samara.

1a. Pistillate catkins woody, persistent and conelike; leaves distinctly doubly serrate. *Alnus*
1b. Pistillate catkins papery, the scales deciduous and falling apart at maturity, the rachis often persistent for some time; leaves distinctly doubly serrate or only once serrate. *Betula*

ALNUS Mill.

Trees or shrubs; leaves alternate, simple, doubly serrate to lobed, the teeth often acuminate; flowers in catkins, borne with or before the leaves; staminate catkins 1–3, the bracts peltate, subtending 3–6 flowers, each flower with 2–4 stamens; pistillate catkins conelike, 1–several, the persistent woody bracts subtending 2 flowers; fruit a winged or wingless nutlet.

1a. Catkins borne on leafy shoots, arising from buds of the current season; nutlets with wings about half as wide as the body or more; our most common alder. *A. crispa*
1b. Catkins borne on naked twigs of the previous season; nutlets wingless or with wings less than half as broad as the body. (2)

2a. Leaves conspicuously bicolored, much paler beneath than above; fruits narrowly winged; plants of southeastern Alaska. *A. rubra*
2b. Leaves not at all, or only slightly bicolored, about as green above as below; fruits wingless; plants broadly distributed in southern Alaska and Yukon. *A. incana*

Alnus crispa (Ait.) Pursh
Green Alder

Shrubs or rarely small trees; stems commonly 1–5 m tall and 0.2–2 dm thick; bark grayish to brownish; winter buds acute; twigs densely glandular, lacking hairs except sometimes on the pistillate inflorescence; leaf blades 2–11.5 cm long, 1.5–10 cm broad, ovate to elliptic, acute to acuminate apically, obtuse to rounded, truncate, or subcordate (rarely acute) basally, finely doubly serrate with numerous, acuminate teeth, and often shallowly lobed, about the same color below as above, usually glabrous above, hairy in the vein axils

Alnus crispa (Ait.) Pursh (× 0.3).

and sometimes generally beneath; petioles 5–20 mm long, glabrous or sometimes hairy; pistillate catkins arising from buds of the current season, 3–8(12) per cluster, 8–20 mm long, 7–12 mm thick; wing of nutlet half as broad as the body or more.

Tundra, heathlands, thickets, woods, mountain slopes, bogs, stream banks, and terraces; in most of Alaska; eastward to Labrador and southward to California, Colorado, Minnesota, and North Carolina; Eurasia. This is Alaska's most common and most widespread alder. Two morphological variations are evident in our specimens. There are intermediates, and definitive segregation of all specimens seems impos-

sible. The following key will allow identification of most specimens.

1a. Leaf margins shallowly lobed as well as doubly serrate; plants more common in coastal southern Alaska, but not restricted there (*A. sinuata* [Reg.] Rydb.; *A. fruticosa* Rupr., in part; *A. fruticosa* var. *sinuata* [Reg.] Hultén; *A. sitchensis* [Reg.] Sarg.; *A. crispa* ssp. *sinuata* var. *laciniata* Hultén; *A. crispa* var. *sinuata* [Reg.] Breitung). *A. crispa* var. *laciniata* Hultén.

1b. Leaf margins not at all, or only slightly lobed, merely doubly serrate; plants more common in the interior, and in coastal western Alaska, but not restricted there. *A. crispa* var. *crispa*

Alnus incana (L.) Moench
Mountain Alder

Shrubs or small trees; stems commonly 1–5(10) m tall and 0.2–2 dm thick; bark grayish to brownish; winter buds blunt; twigs puberulent and commonly glandular as well; leaf blades 2.5–9(11) cm long, 1.5–6.5(10) cm broad, ovate to elliptic, oblong or oval, obtuse to rounded or less commonly acute or abruptly acuminate, obtuse to rounded or subcordate basally, doubly serrate with numerous teeth, and shallowly lobed, the lower leaf surface usually paler than the upper but not markedly so, hairy to glabrous above, usually hairy along the veins beneath; petioles 5–20(30) mm long, puberulent; pistillate catkins arising from branches of the previous season, 3–9 per cluster, 9–15 mm long, 8–10 mm thick; wing of nutlet lacking.

Woods, stream banks, bogs, and terraces; in much of Alaska and Yukon south of the 68th parallel (except for the Aleutians; uncommon in southwestern Alaska); eastward to Nova Scotia, and south to California, New Mexico, Arizona, and Pennsylvania; Eurasia (*Betula alnus* var. *incana*

L.; *A. tenuifolia* Nutt.; *A. incana* ssp.
tenuifolia [Nutt.] Breitung; *A. incana* var.
virescens authors, not Wahl.; *Betula rugosa* DuRoi). Plants from Alaska and Yukon have been assigned to ssp. *rugosa*
(DuRoi) R. T. Clausen var. *occidentalis*
(Dippel.) C. L. Hitchc.

Alnus rubra Bong.
Red Alder

Trees, to 10 m tall or more, and 1–5 dm
thick or more; bark gray; winter buds
blunt; twigs glandular, lacking hairs; leaf
blades 4.5–12(15) cm long, 2.5–10 cm
broad, ovate to elliptic, acute to acuminate
or obtuse apically, acute to obtuse basally,
doubly serrate with numerous teeth, and
shallowly lobed, the margin distinctly
revolute, paler and often reddish beneath,
glabrous or nearly so above, hairy along
the veins and sometimes generally beneath;
petioles 8–17 mm long, glabrous or sparingly hairy; pistillate catkins arising from
branches of the previous season, 3–9 per
cluster, 10–20 mm long, 6–12 mm thick;
wing of nutlet less than half as broad as
the body.

Stream banks and woods; in southeastern
Alaska; southward to California (*A. ore-
gona* Nutt.).

BETULA L.

Trees or shrubs; leaves alternate, simple,
deltoid to ovate, oval, orbicular, elliptic, or
obovate, serrate to crenate, crenate-serrate,
or doubly serrate; flowers in catkins, borne
with or before the leaves; staminate catkins 1–4 per bud, pendulous or spreading
in flower; pistillate catkins usually solitary,
erect in flower; staminate flowers in clusters of 3, the stamens 2; bracts of pistillate
catkins 3-lobed, deciduous; pistillate flowers 2–3 per bract; fruit a samara.

Dugle, J. R. 1966. A taxonomic study of
western Canadian species in the genus
Betula. Can. Jour. Bot. 44:929–1007.

1a. Plants low to moderate-sized shrubs;
leaf blades 0.5–2.5 cm long (to 4 cm
long on juvenile shoots), oval to orbicular or less commonly ovate to
obovate, crenate to crenate-serrate,
with usually 10 teeth per side or less.
B. glandulosa

1b. Plants moderate to large shrubs or
trees (rarely low shrubs); leaf blades
(1)1.5–8 cm long, ovate to deltoid,
elliptical, or obovate, crenate-serrate
to serrate or doubly serrate, with usually 10–40 teeth per side. (2)

2a. Bark exfoliating, white, reddish, yellowish, or purplish; trees; leaves acuminate to attenuate to acute; samara
wings broader than the body. *B. papyrifera*

2b. Bark not exfoliating, dull to shiny,
gray to yellowish brown or reddish
brown; plants shrubs or small trees;
leaves obtuse to acute or abruptly
acuminate; samara wings subequal to
the body width or narrower. (3)

3a. Bark shiny; leaves sharply and usually doubly serrate; samara wing about
as broad as the body. *B. occidentalis*

3b. Bark dull; leaves once (rarely twice)
crenate-serrate; samara wing narrower than the body. *B. glandulifera*

Betula glandulifera (Reg.) Butler

Shrubs or small, treelike shrubs, commonly
(0.5)1–3 m tall (rarely more), with 1–
several trunks to 1(3) dm in diameter;
bark not exfoliating, gray to brown or purplish, dull, the lenticels not or seldom conspicuous; twigs puberulent to glabrous,
bearing yellowish, crystalline, resin glands;
leaf blades 1–3.5(4) cm long, 0.8–3 cm
broad, obovate to broadly elliptic or ovate,
rounded to obtuse or acute apically, cuneate to obtuse and entire basally, once
crenate-serrate (or some serrations double),
with commonly 10–20(25) teeth per side,
not hairy in the lower vein axils, minutely
hairy to glabrous on the margins near the
base; petioles 3–12 mm long, glabrous or

puberulent; pistillate catkins 12–23(30) mm long, 6–10 mm thick, the bracts glabrous or puberulent dorsally, ciliate; samara wing narrower than the body.

Heath, muskegs, stream banks, woods, meadows, and tundra, common just below tree line in mountainous areas; from central western Alaska, eastward through central and southern Alaska, and southern to northern Yukon; eastward to Quebec, and southward to Oregon, Wyoming, North Dakota, Minnesota, and Michigan (*B. pumila* var. *glandulifera* Reg.; *B. obovata* Butler; *B. hallii* Howell; *B. glandulosa* var. *glandulifera* [Reg.] Gleason; *B. beeniana* A. Nels.; *B. kenaica* authors, not Evans).

Keying out here, and included in the description, are several different phases, some of which have been demonstrated to be hybrids. The specimens from eastern Alaska and southern Yukon may be distinguished by the following key. Specimens from western and southwestern Alaska may represent additional hybrids. More work on Alaskan specimens is indicated.

1a. Leaves broadly ovate, crenate-serrate, with 15–25 teeth per side; plants of eastern Alaska and southern Yukon (*B. glandulosa* × *resinifera* authors, in part), a hybrid between *B. glandulosa* and *B. occidentalis*. *B.* × *eastwoodae* Sarg.

1b. Leaves elliptic to obovate, crenate to crenate-serrate, with 10–20 teeth per side. (2)

2a. Leaves elliptic, 2–2.5 cm long, crenate, usually with 10–15 teeth per side; wing of samara about one-half as wide as the nutlet; a hybrid between *B. glandulosa* and *B. glandulifera*. *B.* × *sargentii* Dugle

2b. Leaves obovate, 2–4 cm long, crenate to crenate-serrate, with 14–20 teeth per side; wing of samara more than one-half as wide as the nutlet; plants of eastern Alaska and southern Yukon. *B. glandulifera* (Reg.) Butler

Betula glandulosa Michx.
Glandular Birch

Shrubs, commonly 0.3–1.5(2) m tall, with 1–several main stems; bark not exfoliating, gray to brown or purplish, the lenticels not or seldom conspicuous; twigs puberulent to glabrous, bearing yellowish, crystalline, resin glands; leaf blades 0.5–2.5 cm long (to 4 cm long on some juvenile shoots), 0.4–2 cm broad (to 3.5 cm broad on juvenile shoots), oval to orbicular, broadly elliptic, or obovate, or less commonly ovate, rounded to obtuse or rarely acute apically, cuneate to rounded or subcordate basally, once crenate to crenate-serrate (or some serrations double) with commonly 10 teeth per side or less (sometimes to 15 per side), not hairy in the lower vein axils, minutely hairy to glabrous on the margins near the base; petioles 1–9 mm long, glabrous or puberulent; pistillate catkins 7–20(25) mm long, 3–8 mm thick, the bracts commonly glabrous dorsally, ciliate; samara wings narrower than half the body width.

Tundra, heathlands, open woods, terraces, bogs, stream banks and bars; in most of Alaska and Yukon; east to Newfoundland, and south to California, Utah, Colorado, and New York; Greenland, Asia. *B. glandulosa* is our most common birch. It apparently hybridized with *B. glandulifera* and possibly with other entities as well. There are two more or less distinctive, but apparently intergrading taxa within the complex. These have received different interpretations by various workers. The following key will serve to distinguish most specimens.

1a. Leaves with entire, more or less cuneate to obtuse bases; bracts of catkins often with a distinctive, dorsal, glandular hump at maturity; samara wings often broader upwards. *B. glandulosa* var. *glandulosa*

1b. Leaves with teeth almost or quite to the base; bracts of catkins seldom if ever with a distinctive, dorsal, glandular hump at maturity; samara wings broadest near the middle (*B. sibirica* Ledeb.; *B. exilis* Sukacz.; *B. nana* ssp. *exilis* [Sukacz.] Hultén). *B. glandulosa* var. *sibirica* (Ledeb.) Blake

Betula occidentalis Hook.
Water Birch

Shrubs or small trees, commonly 3–6 m tall and with several trunks to 2 dm thick or more; bark not or scarcely exfoliating, reddish brown to brown, shining, marked with pale, longitudinal lenticels; twigs pubescent to glabrous, bearing yellowish to reddish, crystalline, resin glands; leaf blades (1)1.5–5 cm long, (0.7)1–4 cm broad, ovate, acute or abruptly acuminate apically, obtuse to rounded or less commonly cuneate to truncate basally, sharply and often doubly serrate, with commonly 15–25 teeth per side, not hairy in the lower vein axils, minutely hairy to glabrous on the margins near the base; petioles 5–15 mm long, glabrous or puberulent; pistillate catkins 15–40 mm long, 4–10 mm thick, the bracts puberulent and ciliate; samara wing subequal to the width of the nutlet.

Stream banks, terraces, and slopes; in central and eastern Alaska and southern Yukon; east to Mackenzie and south to California, Utah, Colorado, and South Dakota (*B. fontinalis* Sarg.; *B. papyrifera* ssp. *occidentalis* [Hook.] Hultén). Water birch is known to hybridize with both *B. glandulosa* and *B. papyrifera*.

Betula papyrifera Marsh.
Paper Birch

Trees, commonly 5–20 m tall or more, with usually 1 main trunk to 5 dm thick or more; bark exfoliating, creamy white to reddish, yellowish, or purplish brown, marked with pale, elongate, horizontal lenticels; twigs puberulent to glabrous, bearing yellowish to whitish, crystalline, resin glands, or these poorly developed or lacking; leaf blades 2.5–8 cm long, 1.3–7 cm broad, ovate to lance-ovate, acuminate to acute or sometimes obtuse apically, cuneate to obtuse, rounded, truncate, or subcordate basally, sharply and often doubly serrate and sometimes slightly lobed as well, with usually 15–20 teeth per side, hairy to glabrous in the lower vein axils, hairy to glabrous on the margins near the base and often over the upper surface; petioles 8–40 mm long, glabrous or puberulent; pistillate catkins 15–35(40) mm long, 5–15 mm thick, the bracts glabrous or puberulent dorsally; samara wing broader than the body.

Woods, stream banks, slopes, and terraces; in most of southern Alaska and southern Yukon; eastward to Labrador and southward to Washington, Colorado, Nebraska, Iowa, Illinois, and North Carolina. Our material has been treated as belonging to at least three species: *B. resinifera; B. kenaica;* and *B. papyrifera*. These entities have been segregated by means of presence or absence of crystalline glands; occurrence of hair in the lower vein axils, or its absence; and by such tenuous characteristics as leaf shape and color, and bark color. Leaf shape has not proved useful in any attempts at segregation. Bark color varies from specimen to specimen in the same population and seems not to be definitive. More recently, Dugle, 1966, has demonstrated that *B. resinifera* has a chromosome number of $2n=28$, while the chromosome number of *B. papyrifera* varies from $2n=56$–84, or rarely 28. Glands and pubescence appear to be the most distinctive diagnostic features. Thus, it is possible to recognize two sympatric varieties in our flora.

1a. Twigs lacking crystalline, yellowish glands (or rarely with a few); lowermost vein axils on the lower leaf surface usually hairy (*B. kenaica* Evans; *B. papyrifera* var. *kenaica* [Evans] Henry; *B. papyrifera* var. *commutata* [Reg.] Fern, the usually dark-barked coastal phase). *B. papyrifera* var. *papyrifera*

1b. Twigs moderately to densely clothed

with crystalline, yellowish glands; vein axils hairy or glabrous (*B. resinifera* Britt. ex Britt. & Rydb.; *B. alaskana* Sarg., not Lesquereau, a fossil plant; *B. neoalaskana* Sarg.; *B. papyrifera* var. *humilis* authors, not [Reg.] Hultén). *B. papyrifera* var. *neoalaskana* (Sarg.) Raup.

BORAGINACEAE

Borage Family

Annual, biennial, or perennial herbs; leaves alternate, simple, often bristly-hairy; flowers perfect, regular, frequently in scorpioid cymes; calyx usually 5-lobed; corolla united, 5-lobed, commonly with appendages at the throat; stamens 5, alternate with the corolla lobes, adnate to the corolla; ovary superior, 2-carpellate, usually deeply 4-lobed and appearing 4-loculed; style entire or 2-cleft, arising between the lobes of the ovary; fruit at maturity separating into 4 nutlets.

1a. Plants tall, slender perennials; flowers blue to red purple, 6–8 mm long, the tube 2.5–3.5 mm long; nutlets spreading in fruit, bearing numerous, barbed prickles. *Cynoglossum*
1b. Plants annual or perennial, but not large and coarse; flowers blue, white, yellow, or orange, various in size; nutlets not spreading in fruit, armed or unarmed. (2)

2a. Corolla yellow to orange. *Amsinckia*
2b. Corolla blue or white. (3)

3a. Corolla tubular or tubular-funnelform, the tube mostly more than 5 mm long; flowers nodding. *Mertensia*
3b. Corolla campanulate to rotate, the tube less than 5 mm long; flowers variously arranged, but usually not nodding. (4)

4a. Plants procumbent annuals; fruiting calyx much enlarged, folded; flowers solitary, axillary. *Asperugo*
4b. Plants erect or ascending, annual or perennial; fruiting calyx moderately

or not at all enlarged; flowers several to many, not axillary. (5)

5a. Fruiting pedicels reflexed. *Hackelia*
5b. Fruiting pedicels erect or spreading, or the flowers subsessile. (6)

6a. Flowers white; plants annual, or if perennial, then restricted to central Yukon Valley (see *Cryptantha*). (7)
6b. Flowers blue (white in albino forms); plants annual, biennial, or perennial. (8)

7a. Nutlets smooth; plants annual, or wrinkled and plants perennial. *Cryptantha*
7b. Nutlets with warts or ridges; plants annual. *Plagiobothrys*

8a. Plants slender annuals or biennials; nutlets armed with hooked prickles. *Lappula*
8b. Plants perennial; nutlets unarmed (except in *Eritrichium splendens*). (9)

9a. Plants depressed pulvinate, seldom over 7 cm tall, or if taller then the limb commonly over 10 mm broad; nutlets with an oblique upper face encircled by an upturned flange or rim; distribution mostly arctic-alpine. *Eritrichium*
9b. Plants sometimes caespitose, but not pulvinate, mostly over 7 cm tall; nutlets without an upturned flange or rim encircling the upper face; distribution often other than above. *Myosotis*

AMSINCKIA Lehm. Nom. cons.

Annual, stiffly hairy herbs with decumbent to erect, elongate stems; leaves alternate, lanceolate to oblong or linear, sessile or the lower ones petiolate, stiffly hairy; flowers numerous, borne in spicate scorpioid cymes; calyx 5-lobed, moderately enlarging in fruit, the lobes distinct almost to the base; corolla tubular, 5-lobed, yellow to yellow orange or orange, the throat with or without appendages; stamens 5, inserted on the corolla tube; nutlets keeled ventrally and often dorsally as well, ridged and warty dorsally, unarmed, the scar small, located near the lower end of the ventral keel.

1a. Corolla tube lacking appendages at the throat; stamens attached above the middle of the corolla tube; our common species. A. menziesii

1b. Corolla tube with hairy appendages at the throat; stamens attached below the middle of the corolla tube; rare in our region. A. lycopsoides

Amsinckia lycopsoides Lehm. ex Fisch. & Mey.
Fiddleneck

Annual, the stems 1.5–6 dm tall or more, simple or branched, spreading-hairy, the stiff hairs with pustular bases; leaves 1.5–8 cm long and 0.8–1.5 cm broad, lanceolate to oblong or linear, stiffly hairy, sessile and clasping basally or the lower ones petiolate; sepals 5–10 mm long in fruit, stiffly hairy; corolla 6–9 mm long, yellow to orange, the throat with hairy appendages, the limb 3–6 mm broad; stamens inserted below the middle of the tube; nutlets 2.5–3 mm long, ovoid, wrinkled and warty dorsally, the dorsal keel poorly or not at all developed.

Weedy species; in southwestern Alaska (Attu Island), and to be expected elsewhere; adventive from western North America where it occurs from southern British Columbia southward to California.

Amsinckia menziesii (Lehm.) Nels. & Macbr.

Annual, the stems 1.5–8 dm tall, simple or branched, spreading hairy with stiff, pustular-based hairs; leaves 1.5–10 cm long, 0.5–2.5 cm broad, lanceolate to oblong, elliptical, or linear, stiffly hairy, sessile and clasping basally, or the lower ones petiolate; sepals 5–10 mm long in fruit, stiffly hairy; corolla 4–8 mm long, yellow, the throat

open, lacking appendages, glabrous; stamens inserted above the middle of the tube; nutlets 2.5–3 mm long, ovoid, wrinkled and warty dorsally, the dorsal keel often prominent in at least the upper portion.

Weedy species of disturbed soils; in southern Alaska and southern Yukon, where it may represent a recent introduction from the range of the species in western Canada and United States (*Echium menziesii* Lehm.; A. lycopsoides authors, not Lehm. ex Fisch. & Mey., at least in part).

ASPERUGO L.

Annual, retrorsely hairy herbs, with weak, clambering, elongate stems; leaves alternate, or some opposite or rarely whorled, oblanceolate to elliptic or lanceolate, petiolate, stiffly short-hairy; flowers solitary in bract or leaf axils, or in the forks of branches; calyx 5-lobed, enlarging in fruit, the lobes about as long as the tube, toothed near the base; corolla campanulate, 5-lobed, blue, the throat with appendages; stamens 5, inserted on the corolla tube; nutlets not keeled, smooth but reticulately patterned, unarmed, the scar small, located near the upper end of the ventral surface.

Asperugo procumbens L.
Catchweed

Annual, the stems 1–5 dm long or more, retrorsely hairy with short, pustular-based hairs; leaves 1.5–10 cm long, 0.4–2.5 cm broad, oblanceolate to elliptic or lanceolate, stiffly short-hairy, petiolate, gradually reduced and subsessile upwards; calyx 2.5–8 mm long, much expanded in fruit, stiffly short-hairy; corolla 2–3 mm long, blue, the throat with appendages, the limb 2–3 mm broad; stamens inserted near the middle of the tube; nutlets 2.5–3 mm long, ovoid, smooth but reticulately patterned.

Weedy species in disturbed soils in southern Alaska (to be expected elsewhere); native to Eurasia.

CRYPTANTHA Lehm.

Annual or perennial, stiffly hairy herbs, with erect or ascending stems; leaves alternate or mainly basal, oblanceolate to spatulate, oblong, or lanceolate, sessile or petiolate, stiffly hairy; flowers numerous, borne in simple to compound, scorpioid cymes, or some of them solitary in leaf or bract axils; calyx 5-lobed, slightly to moderately enlarging in fruit, the lobes distinct or nearly so; corolla tubular, 5-lobed, white, the throat with appendages; stamens 5, inserted on the corolla tube; nutlets smooth and shining or slightly wrinkled, the scar elongate, running along much of the ventral surface.

1a. Plants annual; corolla about 1 mm broad; nutlets smooth, shining. *C. torreyana*

1b. Plants perennial; corolla 4–6 mm broad; nutlets slightly wrinkled, dull. *C. shackletteana*

Cryptantha shackletteana Higgins

Perennial, caespitose, arising from a branching subwoody caudex and stout taproot, the stems 0.7–1.8 dm tall, strigose and spreading-hairy with slender, nonpustular hairs; leaves basal and cauline, 2–13 cm long, 0.1–0.5 cm broad, linear, strigose on both surfaces and with few, obscurely pustulate hairs dorsally; sepals 3–5 mm long (7–10 mm long in fruit), stiffly hairy; corolla, including limb, 5–6 mm long, 5–6 mm broad, white, the appendages yellow; nutlets 3.3–3.6 mm long, lance-ovoid, roughened and ridged on both surfaces, the margins smooth, the scar about as long as the ventral surface, open.

Steep, south-facing greenstone slope at Mission Bluff near Eagle, Alaska; endemic. *C. shackletteana* is apparently most closely related to *C. spiculifera* (Piper) Payson, from which it is disjunct by several hundred miles.

Cryptantha torreyana (Gray) Greene

Annual, the stems 0.8–4 dm tall, strigose and spreading-hairy with slender, nonpustular hairs; leaves all cauline, 1–3.5 cm long, 0.2–0.6 cm broad, oblong to linear, pubescent with coarse, spreading, pustular-based hairs or the hairs lacking swollen bases, short-petiolate to subsessile; sepals 4–8 mm long in fruit, stiffly hairy; corolla, including limb, 3–4 mm long, about 1 mm broad, white, the appendages yellowish; nutlets 1.5–2.2 mm long, ovoid, smooth throughout, shining, the lateral edges rounded, the scar running the length of the ventral surface, mottled brownish.

Weedy species; in southeastern Alaska (Skagway), where it is probably adventive from the range of the species in the western United States and Canada.

CYNOGLOSSUM L.

Perennial, stiffly hairy herbs, with erect or ascending stems; leaves alternate, elliptic to oblong, the lowermost long-petiolate, the upper ones auriculate-clasping, stiffly hairy; flowers numerous, borne in long-pedunculate, racemose or paniculate, more or less scorpioid cymes; calyx 5-lobed, slightly enlarging in fruit, the lobes distinct to below the middle; corolla short-funnelform, 5-lobed, blue to red purple, the throat with appendages; stamens 5, inserted on the corolla tube; nutlets armed throughout with stout, barbed prickles, the scar broad, above the middle of the ventral surface.

Cynoglossum boreale Fern.
Wild Comfrey

Perennial, the stems 3–8 dm tall, pubescent with coarse, spreading or descending hairs (at least below); leaves dimorphic, the lower ones 10–30 cm long, the blades 2–7 cm broad, elliptic to oblong, petiolate, the upper ones gradually reduced, becoming sessile and auriculate-clasping; inflorescence borne on a long, naked peduncle;

sepals 1.5–2.5 mm long, stiffly hairy; corolla, including limb, 3–5(8) mm long, the limb 4–7 mm broad, blue, the appendages yellowish; nutlets 3.5–5 mm long, prickly throughout, radially spreading, the scar large and broad near the upper end of the ventral surfaces.

Woods; at Liard Hot Springs, British Columbia; east to Newfoundland and south to Connecticut, Michigan, and Manitoba.

ERITRICHIUM Schrad.

Perennial, softly hairy, pulvinate-caespitose or merely caespitose herbs, with villous to glabrate, erect stems, or the plants acaulescent; flowers few to many, borne in contracted, cymose clusters, or in slightly expanded, false panicles; calyx 5-lobed, not enlarging in fruit, the lobes distinct or nearly so; corolla tubular, 5-lobed, blue, the throat with appendages; stamens 5, inserted on the corolla tube; nutlets with an oblique upper face encircled by an upturned flange or rim.

Wright, William F. 1902. The genus *Eritrichium* in North America. Bull. Torrey Club 29:407–14.

1a. Flowers 4–7 mm broad; leaves pilose. *E. nanum*
1b. Flowers 9–13 mm broad; leaves strigose. *E. splendens*

Eritrichium nanum (Vill.) Schrad.
Arctic Forget-me-not

Perennial, pulvinate-caespitose, acaulescent or short-caulescent herbs, arising from a branching caudex and taproot, the caudex branches clothed with persistent leaves; flowering stems lacking, or more commonly 0.2–0.7(1) dm tall, softly villous to glabrate, erect; leaves 0.4–1 cm long, all or mostly in compact basal rosettes, oblong to oblanceolate, obovate, or ovate, long-pilose with slender, white hairs; cauline leaves alternate, long-pilose; flowers few to many, sessile, or on pedicels to 2 mm long; calyx lobes 2–3 mm long, densely villous; corolla, including limb, 3–6 mm long, the limb 4–7 mm broad, blue, the appendages commonly yellow; nutlets 0.5–1 mm long, obconic, the oblique upper face encircled by an upturned flange or rim, this toothed.

Arctic and alpine tundra and heathlands; in much of Alaska (except for the southeastern portion) and in northern Yukon; in the Rocky Mountains southward to Oregon and New Mexico; Eurasia. Our material has previously been treated as belonging to two species. The differences between the entities are slight and possibly represent only ecological races. It seems best therefore, to treat them as portions of a single species.

1a. Flower cluster sessile; plants acaulescent, mostly of northern and western Alaska and islands of the Bering Sea, and in northern Yukon (*E. chamissonis* DC.). *E. nanum* var. *chamissonis* (DC.) Herder
1b. Flower cluster borne on a leafy stem, elevated above the basal rosette; plants widely distributed (*E. aretioides* [Cham.] DC.). *E. nanum* var. *aretioides* (Cham.) Herder

Eritrichium splendens Kearney in Wight
Showy Forget-me-not

Perennial, caespitose, caulescent herbs, arising from a branching caudex and taproot, the caudex branches clothed with persistent leaf bases; flowering stems (0.3) 0.4–2(3) dm tall, appressed-strigose, erect or ascending; basal leaves 1.5–6 cm long, 0.2–0.4 cm broad, narrowly oblanceolate, tapering to a narrow petiole; cauline leaves 1–4 cm long, linear to narrowly oblong or narrowly oblanceolate, appressed-strigose; flowers few to many; pedicels (2)4–15 mm long; calyx lobes 2.5–3.5 mm long, strigose; corolla, including limb, 5–8 mm long, the limb 9–13 mm broad, blue, the appendages yellowish (?); nutlets 1.5–2.5 mm long, obliquely obconic, the upper face encircled with an upturned flange of irregular, spinulose teeth, the upper face minutely tuberculate or spinulose, glabrous.

Arctic alpine; in central Alaska, from near the Bering Strait eastward to central Yukon; endemic. This entity requires additional study. It should be compared, when more material has been collected, with the Asian *E. rupestre* (Pallas) Bunge, with which it might be identical.

HACKELIA Opiz

Annual or biennial, erect, spreading-hairy herbs, with erect or ascending stems; leaves alternate, oblanceolate to spatulate or the upper ones narrowly lanceolate to elliptic or oblong, the lowermost petiolate; flowers numerous, borne in compound, scorpioid cymes; pedicels deflexed in fruit; calyx 5-lobed, not enlarging in fruit, the lobes distinct to the base or nearly so; corolla tubular, 5-lobed, blue or white, the throat with appendages; stamens 5, inserted on the corolla tube; nutlets with barbed prickles around the margin of the dorsal surface, the scar rather broad near the base of the ventral keel.

Johnston, I. M. 1923. Studies in the Boraginaceae. 1. Restoration of the genus *Hackelia*. Contr. Gray Herb. n.s. 68:43–48.

Hackelia deflexa (Wahl.) Opiz

Annual or biennial, caulescent herbs, the stems 1.5–5 dm tall or more, spreading hairy below, strigose above; leaves 1–8 cm long (or more), 0.2–2 cm broad, the lower ones oblanceolate to spatulate and petiolate, the upper ones lanceolate to elliptic or oblong and sessile, pubescent with spreading, pustular-based hairs; flowers numerous, short-pedicellate; pedicels deflexed in fruit; sepals 1–1.5 mm long, strigose; corolla 1.5–2.5 mm long; nutlets 2–3 mm long, the marginal prickles in 1 row, distinct almost to the base, the dorsal surface minutely tuberculate and short-hairy, the sides minutely hairy.

Rocky slopes and roadsides; in southern Alaska (Glenn Highway, mile 97) and in southern Yukon (Rancheria, Alaska Highway, mile 710); eastward to Quebec and southward to Washington, Idaho, Colorado, Iowa, and Vermont (*H. leptophylla* [Rydb.] Johnst., *H. jessicae* authors, not [McGregor] Brand.).

LAPPULA Moench

Annual or biennial (perennial?), erect, strigose herbs with erect or ascending stems; leaves alternate, oblanceolate to spatulate or the upper ones oblong to linear, the lowermost petiolate; flowers numerous, borne in compound, scorpioid cymes; pedicels short and erect or ascending in fruit; calyx 5-lobed, not enlarging in fruit, the lobes distinct to the base or nearly so; corolla tubular, 5-lobed, blue or white, the throat with appendages; stamens 5, inserted on the corolla tube; nutlets with 1 or more rows of barbed prickles around the dorsal surface, the rather narrow scar along most of the ventral keel.

Lappula echinata Gilib.
Stickseed

Annual or biennial, caulescent herbs, the stems (0.6)1.5–4.5 dm tall, strigose throughout or with some spreading hairs; leaves 0.6–8 cm long, 0.2–1 cm broad, the lower ones narrowly oblanceolate to spatulate, the upper ones oblong, elliptic or lance-elliptic, pubescent with ascending, more or less pustular-based hairs; flowers numerous, subsessile to short-pedicellate; pedicels erect or ascending in fruit; sepals 2–3.5 mm long, hairy; corolla, including limb, 3–4 mm long, the limb 2–3.5 mm broad, blue or white; nutlets 2.5–4 mm long, the marginal prickles in 2–3 rows (the outer rows often much shorter than the inner), distinct to the base or nearly so, the dorsal surface tuberculate, not hairy.

Weedy species of roadsides, railroads, and dry hillsides; in southern Alaska and Yukon, where probably indigenous; widely distributed in North America; Eurasia (includes *L. myosotis* Moench). The writer has not been able to distinguish *L. occidentalis* (Wats.) Greene (*L. redowskii* authors, not [Hornem.] Greene) from among our material. The distinguishing feature of this entity is that its nutlets have only a single row of marginal prickles.

MERTENSIA Roth. Nom. Cons.

Perennial, glabrous or hairy herbs, with procumbent, decumbent, ascending, or erect stems; leaves alternate, lanceolate, elliptic, oblong, oblanceolate, or spatulate, sessile, or the lower ones petiolate, glabrous or hairy; flowers several to many, borne in cymes; calyx 5-lobed, cleft to below the middle, moderately enlarging in fruit; corolla funnelform or tubular, 5-lobed, blue or sometimes pink or white, enlarged at the throat and distinguished into tube and limb, the throat with appendages; stamens 5, attached to the corolla tube; nutlets wrinkled and often tuberculate (rarely weakly spinose) dorsally, or smooth and shiny, the scar large, triangular, near the base of the ventral keel.

Macbride, J. F. 1916. The true Mertensias of western North America. Contr. Gray Herb. n.s. 48:1–20.

Williams, L. O. 1937. A monograph of the genus *Mertensia* in North America. Ann. Mo. Bot. Gard. 24:17–159.

1a. Plants fleshy, glabrous or nearly so, procumbent to decumbent, maritime, throughout coastal Alaska and coastal northern Yukon. *M. maritima*
1b. Plants not or only slightly fleshy, hairy over much of the herbage, erect or ascending, or decumbent only at the base, not maritime. (2)

2a. Pedicels glabrous; plants 1–2.2 dm tall; corolla limb subequal to the tube; distributed north of the Brooks Range. *M. drummondii*
2b. Pedicels hairy; plants 2–10 dm tall;

Lappula echinata Gilib. (× 0.4).

corolla limb longer than the tube; distributed mostly south of the Brooks Range. *M. paniculata*

Mertensia drummondii (Lehm.) G. Don
Drummond Bluebell

Perennial, the stems 1–2.2 dm long, few to several from a branched, subterranean or superficial caudex and taproot, decumbent to ascending or erect, glabrous; leaves not or only slightly fleshy; basal leaves 2–13 cm long, 0.2–0.9 cm broad, the blade narrowly elliptic to oblanceolate, long-petiolate; cauline leaves reduced and subsessile upwards, pustular above and below, and sometimes setose as well; flowers several to many, borne in subcorymbose, contracted cymes, or with some solitary in the leaf axils; pedicels glabrous; sepals 3.5–6 mm long, sparingly strigose to glabrous dorsally and ventrally, appressed ciliate; corolla blue (or white?), 9–15 mm long, the tube 4.5–6.5 mm long, the limb campanulate, subequal to the tube; nutlets not seen.

Arctic tundra; north of the Brooks Range and eastward to Mackenzie; endemic.

Mertensia maritima (L.) S. F. Gray
Oysterleaf

Perennial, the stems 1–10 dm long, several to many, arising from a branched, often subterranean caudex and taproot, procumbent to decumbent, glabrous throughout; leaves fleshy, 0.5–12 cm long, 0.4–4.5 cm broad, oblanceolate to obovate, elliptic, or ovate, abruptly acuminate to acute, obtuse, or rounded, glabrous, or the upper surface with pustular outgrowths, the lower ones petiolate, becoming smaller and subsessile to sessile upwards; flowers several to many, borne in subcorymbose, contracted to open cymes, or with some solitary in leaf axils; pedicels glabrous; sepals 2–4 mm long in flower, to 9 mm long in fruit, glabrous; corolla blue or rarely white, or pink in bud, 3.5–11(12) mm long, the tube 1–3 mm long, the limb campanulate, longer than the tube; nutlets 2.5–4 mm long, smooth or slightly wrinkled.

Beaches, spits, bars, and less commonly river banks; in coastal and insular Alaska, and in coastal northern Yukon; eastward to Newfoundland and south to Massachusetts; circumpolar. This very attractive seashore plant is separable into two more or less distinctive phases in Alaska. The separation is not absolute because transitional specimens do occur, and individual specimens vary.

1a. Corollas 3.5–8 mm long; sepals commonly 3 mm long or less in flower and less than 5 mm long in fruit; our common phase (including the Aleutian Islands). *M. maritima* var. *maritima*
1b. Corollas 7–11(12) mm long; sepals commonly 3–4 mm long in flower and often over 5 mm long (to 9 mm) in fruit; plants of the Aleutians (*M. asiatica* [Takeda] Macbr.). *M. maritima* var. *asiatica* (Takeda) Welsh

Mertensia paniculata (Ait.) G. Don
Tall Bluebell

Perennial, the stems 2–10 dm long, few to several arising from a caudex and taproot, erect or ascending, or less commonly sprawling, hairy to glabrate; leaves not fleshy, the cauline ones 3–15 cm long, 1-7 cm broad, lanceolate to ovate, elliptic, or linear-lanceolate, hairy above and beneath or glabrous on the lower surface, acuminate to acute, the lower ones petiolate, becoming smaller and sessile or subsessile above; flowers few to many, borne in contracted to open, paniculate cymes, or with some solitary in the leaf axils; pedicels strigose; sepals (2)3.5–5(8) mm long, pubescent ventrally and dorsally, or the dorsal surface glabrous, ciliate; corolla blue (pink in bud), pink, or white, (10)11–19 mm long, the tube (4)5–8 mm long, the limb campanulate, longer than the tube; nutlets 2.5–5 mm long, wrinkled or spinulose-tuberculate dorsally and along the margins.

Woods, thickets, meadows, terraces, gravel bars, and stream banks; over much of Alaska south of the Brooks Range and most of the Yukon; eastward to Quebec, and southward to Iowa, Montana, Idaho, and Oregon. The forms of *M. paniculata* assembled in Alaska and Yukon have been interpreted previously as belonging to several taxa, most of them at specific level. Mor-

and pointing backward; plants of broad distribution (*M. alaskana* Britt.; *M. paniculata* var. *alaskana* [Britt.] L. O. Williams). *M. paniculata* ssp. *paniculata*

MYOSOTIS L.

Perennial or annual, spreading-hairy to strigose herbs with erect, ascending, or decumbent stems; flowers several to many, borne in scorpioid cymes which elongate in fruit; calyx 5-lobed, slightly enlarging in fruit, the lobes distinct to near or below the middle; corolla tubular, 5-lobed, blue or less commonly white, the throat with appendages; stamens 5, inserted on the corolla tube; nutlets smooth and shining, the scar nearly basal, the margin narrow and upturned.

1a. Calyx closely appressed-strigose; plants often decumbent or stoloniferous at the base; cultivated and escaping. *M. scorpioides*

1b. Calyx with some spreading or spreading-ascending, often hooked hairs; plants erect or ascending from a caudex or taproot, not stoloniferous; indigenous perennials. *M. sylvatica*

Myosotis scorpioides L.
Forget-me-not

Perennial (or annual?), often stoloniferous herbs, the stems decumbent to ascending or erect, 2–6 dm tall, strigose with stiff, appressed hairs; leaves 1.5–8 cm long, 0.3–2 (2.8) cm broad, oblanceolate to elliptic or oblong; flowers erect, showy; fruiting pedicels spreading, equaling or surpassing the calyx; calyx 2–4 mm long, strigose throughout with appressed, stiff hairs, the lobes subequal to the tube or shorter; corolla blue, 5–10 mm broad; style equaling or longer than the nutlets; nutlets 2–2.5 mm long, smooth.

Cultivated ornamental; escaping in southern and western Alaska and southern Yukon; adventive from Europe. We follow Cronquist, 1959, Vasc. Plants, Pacific N. W.

Mertensia paniculata (Ait.) G. Don (× 0.3).

phological similarities seem to override the differences and thus a more conservative interpretation is indicated.

1a. Nutlets spinulose-tuberculate, at least along the margins; sepals glabrous dorsally; lower leaf surface appressed-pubescent with forward pointing hairs; plants of western and southwestern Alaska (*M. paniculata* var. *eastwoodiae* [Macbr.] Hultén). *M. paniculata* ssp. *eastwoodiae* (Macbr.) Welsh

1b. Nutlets merely wrinkled, not at all spinulose-tuberculate; sepals hairy dorsally or less commonly glabrous; lower leaf surface spreading hairy or glabrous, or the hairs appressed

4:231, in using the epithet *scorpiodes* in place of *M. palustris* (L.) Lam.

Myosotis sylvatica Hoffm.

Forget-me-not

Perennial, fibrous-rooted herbs from a branching caudex, the stems ascending to erect, 0.3–4.5 dm tall, spreading-hairy throughout; leaves 1–11 cm long, 0.2–1.2 cm broad, oblanceolate to elliptic or oblong; flowers erect, showy; fruiting pedicels spreading or ascending, equaling or surpassing the calyx; calyx 2–5 mm long, bearing some spreading to ascending, often hooked hairs, the lobes longer than the tube; corolla blue or sometimes white, the

Myosotis sylvatica Hoffm. (× 0.3).

limb 4–10 mm broad; style shorter than the nutlets; nutlets 1–2 mm long, smooth.

Meadows, thickets, spits, talus slopes, and rock outcrops; through most of Alaska and Yukon; east to Mackenzie and south to Idaho, Wyoming, and South Dakota; Eurasia (*M. alpestris* F. W. Schmidt; *M. alpestris* ssp. *asiatica* Vesterg. ex Hultén). This very attractive plant is the official state flower of Alaska.

PLAGIOBOTHRYS Fisch. & Mey.

Annual, strigose herbs with prostrate to ascending or erect stems; leaves alternate, or at least the lower ones opposite, oblong to linear or narrowly oblanceolate, petiolate to sessile, strigose to spreading-hairy; flowers numerous, borne in simple to compound, scorpioid cymes, or some solitary in leaf axils; calyx 5-lobed, slightly to moderately enlarging in fruit, the lobes distinct to below the middle; corolla tubular, 5-lobed, white, the throat with appendages; stamens 5, inserted on the corolla tube; nutlets roughened or wrinkled, the small scar near the base of the ventral keel.

Johnston, I. M. 1923. Studies in the Boraginaceae. 4. A. synopsis and redefinition of *Plagiobothrys*. Contr. Gray Herb. n.s. 68:57–80.

————. 1932. Studies in the Boraginaceae IX. 1. The Allocarya section of *Plagiobothrys* in the western United States. Contr. Arn. Arb. 3:5–82.

1a. Corolla limb 5–10 mm broad; plants erect; inflorescence lacking foliose bracts; plants rare, in southeastern Alaska. *P. figuratus*

1b. Corolla limb 1–4 mm broad; plants prostrate to ascending or erect; inflorescence with few to numerous, foliose bracts. (2)

2a. Hairs of sepals spreading-ascending; sepals less than twice longer than the nutlets; plants of southeastern quarter of Alaska and southern Yukon. *P. scouleri*

2b. Hairs of sepals mostly appressed; sepals commonly 2–3 times longer than the nutlets; plants of southwestern Alaska and Aleutians. *P. orientalis*

Plagiobothrys figuratus (Piper) Johnst. ex Peck

Annuals, the stems 1–4 dm tall, erect, strigose to spreading-hairy with slender, nonpustular hairs; leaves all cauline, alternate or the lower ones opposite, 0.5–8 cm long, 0.1–0.5 cm broad, linear to oblong, pubescent like the stem or some hairs pustular based; branches of inflorescence often in pairs, lacking foliose bracts; sepals 3–4 mm long, stiffly ascending-hairy; corolla white, 5–10 mm broad, the appendages yellow; nutlets 1.2–1.5 mm long, ovoid, wrinkled or roughened, the scar near the base of the ventral keel.

Weedy species of disturbed soils; in southeastern Alaska; adventive in our region; indigenous from southern British Columbia south to Oregon (*P. hirtus* [Greene] Johnst. var. *figuratus* [Piper] Johnst.).

Plagiobothrys orientalis (L.) Johnst.

Annuals, the stems mostly 1–2 dm long, prostrate to ascending or rarely erect, appressed-hairy with slender, nonpustular hairs; leaves all cauline, alternate or the lower ones opposite, 1–7 cm long, 0.3–0.8 cm broad, oblong to linear, pubescent like the stem, or some hairs pustular based; branches of inflorescence not in pairs, bearing numerous, foliose bracts; sepals 3–6 mm long, appressed-hairy; corollas white, 2.5–4 mm broad; nutlets 2–2.5 mm long, ovoid, wrinkled, the scar near the base of the ventral keel.

Weedy species of disturbed soils; in southwestern Alaska and in the Aleutian Islands; Asia (*Heliotropium orientale* L.; *Allocarya orientalis* [L.] Brand).

Plagiobothrys scouleri (H. & A.) Johnst.

Annuals, the stems 0.5–2 dm long, prostrate

to ascending or erect, appressed-hairy with slender, nonpustular hairs; leaves all cauline, alternate or the lower ones opposite, 1–5 cm long, 0.1–0.4 cm broad, linear to oblong or narrowly oblanceolate, pubescent like the stem or with some pustular-based hairs; branches of inflorescence not or only rarely in pairs, bearing few to several foliose bracts; sepals 2–3 mm long, ascending-hairy; corollas white, 1–3(4) mm broad; nutlets 1.5–2 mm long, ovoid, wrinkled, the scar near the base of the ventral keel.

Weedy species of disturbed soils; in the southeastern quarter of Alaska and southern Yukon; south to California and New Mexico (*P. cusickii* [Greene] Johnst.; *P. cognatus* [Greene] Johnst.). Possibly, this entity is adventive in our region.

CALLITRICHACEAE
Water-starwort Family

Slender, annual, monoecious herbs; leaves simple, opposite, entire, floating or submerged; flowers imperfect, solitary in leaf axils; calyx and corolla lacking; staminate flower with a single stamen; pistillate flower with a single pistil, the ovary superior, 2-loculed (or 4-loculed by intrusion of false septae); styles 2, slender; fruit a schizocarp.

CALLITRICHE L.

Slender, inconspicuous, aquatic or amphibious annuals, with slender, submerged or floating stems, or the stems prostrate on mud or on wet banks; flowers inconspicuous, axillary, subtended by hornlike bracts or lacking them; perianth lacking; fruit splitting at maturity into 2–4 segments.

Fassett, N. C. 1951. *Callitriche* in the New World. Rhodora 53:137–55, 161–82, 185–94, 209–22.

1a. Stems commonly less than 7 cm tall; fruits scarcely or not at all winged, the tiny pitlike depressions on the fruit surface not arranged in rows; plants evidently rare. *C. anceps*

1b. Stems usually more than 7 cm tall; fruits winged near the apex or wingless, the tiny pitlike depressions arranged in rows or not. (2)

2a. Leaves all linear, 1-veined, their bases not joined by a ridge or wing; fruit grayish, usually winged apically, the pitlike depressions irregularly arranged. *C. hermaphroditica*

2b. Leaves linear and 1-veined, or the upper ones broadened and several-veined; fruit usually tan to pale reddish brown, more or less winged apically, the pitlike depressions arranged in rows. *C. verna*

Callitriche anceps Fern.

Plants slender, usually less than 7 cm long; leaves 4–9 mm long, linear, 1-veined and bidentate apically, or the upper ones oblanceolate and 1–3 veined; flowers not subtended by bracts; fruit brownish, to 1 mm long, scarcely or not at all winged, the pitlike depressions not in definite rows.

Shallow ponds; in western and southern Alaska; south to Washington and Utah; eastern North America.

Callitriche hermaphroditica L.

Plants slender, 0.5–3 dm long; leaves 0.5–3 cm long, all linear, 1-veined, and bidentate apically; flowers not subtended by bracts; fruits grayish, 1–2 mm long, narrowly winged apically, the pitlike depressions not in definite rows.

Shallow ponds; in widely scattered localities in Alaska and Yukon; east to Newfoundland and south to California, New Mexico, and New York; Eurasia (*C. autumnalis* L.).

Callitriche verna L.
Vernal Water-starwort

Plants slender, 0.5–3 dm long; leaves 0.5–2.5 cm long, linear, 1-veined, and biden-

tate apically, or the upper ones oblanceolate to obovate and 1–3-veined; flowers subtended by bracts; fruits tan to pale reddish brown, 1–2 mm long, more or less winged apically, the pitlike depressions arranged in rows.

Shallow ponds and lakes; over much of southern Alaska and Yukon; widely distributed in North America; Eurasia (*C. palustris* authors, not Hegelm.; *C. bolanderi* Hegelm.; *C. heterophylla* Pursh ssp. *bolanderi* [Hegelm.] Calder & Taylor; *C. heterophylla* var. *bolanderi* [Hegelm.] Fassett). This is our most common species of water-starwort.

CAMPANULACEAE
Bellflower Family

Plants herbaceous, perennial (or some biennial); leaves alternate or basal, simple; flowers regular or irregular, perfect; sepals 5; petals 5, united; stamens 5, alternate with the corolla lobes, the anthers connate or distinct; pistils 1; ovary inferior, 2–5-loculed; styles 1; fruit a capsule.

1a. Corolla regular; filaments and anthers separate. *Campanula*

1b. Corolla irregular; filaments and anthers united. *Lobelia*

CAMPANULA L.

Perennial or biennial herbs, the stems decumbent to ascending or erect; leaves alternate or chiefly basal, linear to elliptic, oblanceolate, lanceolate, or obovate; flowers solitary, or 2–several in racemose or paniculate clusters; sepals 5, the lobes linear to oblong or subulate; corolla 5-lobed, regular, blue or rarely white; stamens 5, distinct; ovary 3(5)-loculed; styles 1, the stigmas 3(5)-lobed; capsules opening by pores.

1a. Styles exserted from the corolla; corolla lobes spreading or recurved, equaling or longer than the tube; plants of southeastern Alaska, or of east-central Alaska and southern Yukon. (2)

1b. Styles shorter than or equaling the corolla, not or only slightly exserted; corolla lobes erect to ascending, shorter than or rarely equaling the tube; plants of various distribution. (3)

2a. Leaves sharply serrate; plants of southeastern Alaska. *C. scouleri*

2b. Leaves subentire to irregularly serrate; plants of east-central Alaska and southern Yukon. *C. aurita*

3a. Ovary and sepals glabrous; flowers 1–10 or more; cauline leaves commonly 10–30. *C. rotundifolia*

3b. Ovary and sepals more or less hairy; flowers solitary or less commonly 2–6; cauline leaves 10 or less. (4)

4a. Leaves entire or merely serrulate; corolla 4–10 mm long; flowers solitary. *C. uniflora*

4b. Leaves crenate to sharply serrate; corolla (14)15–35 mm long; flowers solitary or less commonly 2–6. (5)

5a. Leaves merely crenate; calyx sinuses subtended by reflexed, attenuate appendages; calyx lobes lance-oblong to lanceolate, entire; corolla lobes lance-ovate; plants of the Aleutian Islands. *C. chamissonis*

5b. Leaves sharply serrate with few to several teeth, or some subentire; calyx sinuses not subtended by appendages; calyx lobes lance-attenuate, often toothed or lobed; corolla lobes ovate to broadly triangular; plants broadly distributed. *C. lasiocarpa*

Campanula aurita Greene
Yukon Bellflower

Perennial, from elongate rhizomes, with 1–several decumbent to ascending or erect stems, the stems mostly 1–3.5 dm tall, pubescent in decurrent lines and scabrous, or glabrous and merely scabrous; leaves mainly cauline, usually 10–20, 0.8–3.5 cm long, 0.1–0.6 cm broad, linear to oblong, elliptic, or oblanceolate, entire to irregularly serrulate, glabrous throughout or ciliate, especially near the base; flowers solitary or more commonly 2–7; pedicels scabrous; calyx and ovary glabrous; calyx lobes lance-attenuate, entire or toothed; calyx sinuses not subtended by appendages; corolla blue, 12–18 mm long, the lobes lanceolate, spreading, longer than the tube; styles exserted from the corolla but shorter than it; anthers 3–5 mm long; capsules obconic to subcylindrical.

Open slopes, woods, and rock outcrops; in central eastern Alaska, southern Yukon, and northern British Columbia; eastward to Mackenzie; endemic.

Campanula chamissonis Federov
Hairy-flowered Harebell

Perennials, from slender to thickish rhizomes, with usually 1 erect or ascending stem, the stems commonly 0.5–1.2 dm tall, more or less hairy, not scabrous, often purplish; leaves mainly basal, these 1–8 cm long, 0.3–1.2 cm broad, the blades elliptic to obovate, crenate to crenate-serrate, glabrous throughout or more or less hairy, distinctly petiolate; cauline leaves 3–10, reduced upwards; flowers solitary; pedicels hairy to glabrous; calyx and ovary hairy; calyx lobes lance-oblong to lanceolate, entire; calyx sinuses subtended by reflexed, attenuate appendages; corolla blue, 20–35 mm long, the lobes lanceolate, erect or ascending, shorter than the tube; styles shorter than the corolla tube; anthers 3–5 mm long; capsules obconic to subcylindrical.

Meadows and hillsides; in the Aleutian Islands; Asia (includes *C. pilosa* Gray; *C. dasyantha* authors, not Bieb.).

Campanula lasiocarpa Cham.
Mountain Harebell

Perennials, from a slender rhizome, with 1–several, ascending to erect stems, the stems 0.2–2 dm tall, glabrous or sparingly

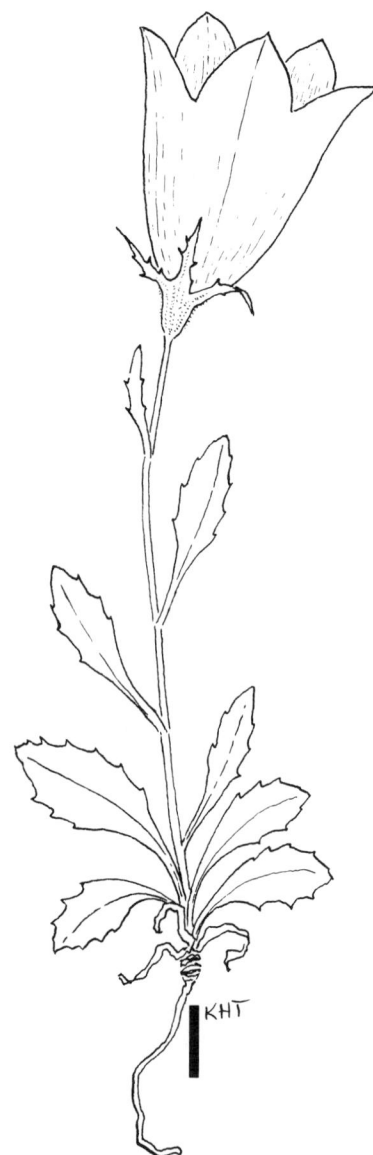

Campanula lasiocarpa Cham. (× 1.0).

hairy, especially upwards, not scabrous; leaves mainly basal, these 1–6 cm long, 0.2–1.3 cm broad, the blades elliptic to oblanceolate, obovate, or oblong, sharply serrate to subentire, glabrous or ciliate along the conspicuous petioles; cauline leaves 3–10, reduced upwards; flowers solitary, or less commonly 2–6; calyx and ovary hairy;

calyx lobes lance-attenuate, usually lobed or toothed; calyx sinuses lacking appendages; corolla blue (rarely white), 15–35 mm long, the lobes ovate or broader, erect or ascending, much shorter than the corolla tube; anthers 3–5 mm long; capsules obconic to subcylindrical.

Alpine meadows, rock outcrops, open woods, terraces, and muskegs; in much of Alaska and Yukon; eastward to Mackenzie and southward to Washington; Asia. This is the most common harebell in interior Alaska.

Campanula rotundifolia L.
Bluebells of Scotland

Perennials, from a slender rhizome or from a caudex, with 1–several, decumbent to ascending or erect stems, the stems 1–5 dm tall or more, glabrous and smooth or minutely scabrous in some; leaves mainly cauline, commonly 10–22, 0.7–8 cm long, 0.1–1.5 cm broad, more or less dimorphic, the lower ones often with broad, cordate-ovate, lanceolate, or elliptic blades and distinctive petioles, the upper ones reduced and linear to lance-linear, oblong, narrowly serrulate, subentire, or entire, the margin scabrous; flowers solitary, or 2–15; pedicels glabrous; calyx and ovary glabrous; calyx sinuses lacking appendages; corolla blue (rarely white), 15–30 mm long, the lobes ovate or broader, erect or ascending, much shorter than the tube; styles shorter than the corolla, but usually exceeding the tube; anthers 5–7 mm long; capsules obconic, nodding.

Rock outcrops, terraces, stream banks, and open slopes; in coastal and insular southern and southwestern Alaska, except for the Aleutians, and rarely in southern Yukon; east to Newfoundland and south to California and Mexico; Eurasia (*C. latisepala* Hultén; *C. latisepala* var. *dubia* Hultén; *C. latisepala* × *rotundifolia*). In Alaska *C. rotundifolia* demonstrates considerable variation, especially in leaf size and shape; the broader leaved plants were designated as *C. latisepala,* but since these intergrade insensibly with the narrow-leaved phases, only arbitrary separation appears possible. Thus, it seems best to treat all phases as portions of one highly polymorphic species.

Campanula scouleri Hook. ex DC.
Scouler Harebell

Perennials, from slender rhizomes, the stems 1–several, decumbent to ascending or erect, 1–4 dm tall, glabrous to short-hairy; leaves mainly cauline, usually 4–10, 1–6 cm long, 0.4–2 cm broad, ovate to lanceolate, sharply serrate, glabrous, the lower ones distinctly petiolate, reduced upwards; flowers solitary, or commonly 2–5; pedicels glabrous; calyx lobes lance-attenuate, entire; calyx and ovary glabrous; calyx sinuses lacking appendages; corolla blue, 8–12 mm long, the lobes lanceolate, spreading or recurved, longer than or merely subequal to the tube; styles longer than the corolla, exserted; capsule subglobose.

Woods; in southeastern Alaska (where apparently rare); southward to California.

Campanula uniflora L.
Arctic Harebell

Perennials, from elongate rhizomes, the stems 1–several, decumbent to erect, 0.3–2(2.8) dm tall, glabrous, or more or less hairy in decurrent lines; leaves mainly cauline, usually 4–10, 0.5–4(4.5) cm long, 0.1–0.5(0.7) cm broad, elliptic to oblanceolate or linear, entire to crenulate, glabrous or nearly so, or the upper ones hairy, scabrous marginally; flowers solitary; pedicels glabrous or nearly so; calyx lobes lance-attenuate, entire; calyx and ovary hairy; calyx sinuses lacking appendages; corolla blue to purplish, 4–10 mm long, the lobes ovate, erect, about as long as the tube; styles shorter than the corolla, not inserted; anthers 1–2 mm long; capsules subcylindrical.

Tundra and heath; in widely disjunct localities through much of Alaska and western Yukon; southward to Colorado; Eurasia.

LOBELIA L.

Perennial herbs, the stems erect or nearly so; leaves alternate, linear, oblanceolate, or spatulate; flowers few to many in a raceme; sepals 5, the lobes linear to subulate; corolla 5-lobed, irregular, blue to white; stamens 5, the filaments and anthers connate; ovary 2-loculed; styles 1, the stigma 2-lobed; capsules opening near the apex.

Lobelia kalmii L.

Perennials, from fibrous roots, the stems 1–several, erect, glabrous or sparingly hairy below, 1–4 dm tall; leaves more or less dimorphic, the lower ones 1–3 cm long, 0.3–0.6 cm broad, spatulate to oblanceolate, the cauline leaves linear to oblanceolate, glabrous or sparingly hairy along the petioles; flowers few to many; pedicels minutely short-hairy, usually 2-bracted; sepals lance-attenuate, glabrous; ovary glabrous; corolla blue to white, usually with a white or yellow eye, 7–12 mm long, the lower lip longer than the tube; capsule obconic, opening at the apex.

Marl fen, at Liard Hot Springs, British Columbia; eastward to Newfoundland and southward to Washington, Montana, Minnesota, and Pennsylvania (*L. kalmii* var. *strictiflora* Rydb.; *L. strictiflora* [Rydb.] Lunell).

CAPRIFOLIACEAE
Honeysuckle Family

Subshrubs, shrubs, woody vines, or small trees; leaves opposite (rarely some whorled), simple or compound, lacking stipules or rarely with them; flowers irregular or regular, perfect, arranged in cymes, or in axillary, pedunculate pairs, or in head- or spikelike clusters; calyx 5 (3–4)-lobed; corolla of united petals, typically 5-lobed; stamens 4–5; pistils 1; ovary inferior, 1–5-loculed; styles 1; fruit a berry or drupe.

1a. Corolla rotate, regular; style short or absent. (2)

1b. Corolla tubular, often irregular; style elongate. (3)

2a. Leaves pinnately compound; flowers numerous, borne in terminal, compound, paniculate cymes on elongate leafy branches. *Sambucus*

2b. Leaves simple, palmately lobed; flowers several to many, borne in terminal, contracted, paniculate cymes on short, axillary branches. *Viburnum*

3a. Stamens 4; stems trailing; flowers in pairs on long, erect peduncles. *Linnaea*

3b. Stamens 5; stems erect, ascending, or less commonly trailing; flowers variously arranged, but if in axillary pairs the stems not trailing. (4)

4a. Flowers often less than 10 mm long, the corollas white or tinged pinkish, paired in the leaf axils and often in terminal, spikelike clusters as well; plants commonly 1 m tall or less, of southeastern Alaska. *Symphoricarpos*

4b. Flowers commonly more than 10 mm long, the corollas white, pink, orange, or yellowish, borne in pedunculate, axillary pairs, or in terminal, spike- or headlike clusters; plants often more than 1 m tall, of southern Alaska, southern Yukon, and northern British Columbia. *Lonicera*

LINNAEA L.

Subshrubs, the stems prostrate, rooting at the nodes; leaves opposite, simple; flowers borne in pairs, on elongate, axillary peduncles; calyx 5-lobed; corolla 5-lobed, regular or nearly so, funnelform, pink or pinkish; stamens 4, in 2 pairs, attached near the corolla base; ovary 3-loculed, with only 1 functional; style elongate; fruit dry, 1-seeded.

Linnaea borealis L.
Twin-flower

Stems mostly 1–12 dm long, trailing, mod- erately to sparingly hairy, becoming glabrous in age; flowering stems erect or ascending, commonly 0.1–1.5 dm long; leaves 0.5–2 cm long, 0.3–1.5 cm broad, elliptic to obovate or orbicular, or sometimes ovate, entire or few-toothed, rounded to obtuse apically, shortly petiolate, long-hairy to glabrous; peduncles axillary, 1.5–6.5 cm long, stipitate-glandular and often villous, bearing a pair of small bracts at the apex; flowers paired (rarely more); pedicels mostly 0.5–2 cm long, hairy like the peduncles; ovary glandular-hairy; calyx 2–5 mm long; corolla 6–12 mm long, pink or pinkish, hairy within; fruit 1.5–3 mm long, hairy.

Woods, thickets, meadows, open dry slopes, terraces, and stream banks; in much of Alaska and Yukon; eastward to the Atlantic and southward to California, Arizona, New Mexico, South Dakota, Indiana, and West Virginia; circumpolar. Our materials include two varieties which may be distinguished by the following key.

1a. Leaves orbicular to obovate, or rarely elliptic; corolla tube flaring from within the calyx, mostly 10 mm long or less; plants of southwestern and interior Alaska. *L. borealis* var. *borealis*

1b. Leaves commonly elliptic to obovate, or less commonly orbicular; corolla tube flaring at or above the tip of the calyx teeth, often over 9 mm long; plants of southeastern Alaska and southern Yukon, and less commonly of interior Alaska (*L. americana* Forbes; *L. borealis* var. *americana* [Forbes] Rehd.; *L. borealis* ssp. *americana* [Forbes] Hultén; *L. longiflora* [Torr.] Howell; *L. borealis* ssp. *longiflora* [Torr.] Hultén). *L. borealis* var. *longiflora* Torr.

LONICERA L.

Shrubs, the stems erect, or vines with twining stems; leaves opposite, simple; flowers borne in axillary, pedunculate pairs, or in terminal, head- or spikelike clusters; calyx 5-lobed; corolla 5-lobed, 2-lined, ir-

Linnaea borealis L. (× 0.5).

regular, funnelform, swollen near the base, white, pink, orange, or yellowish; stamens 5, attached near the corolla base; ovary 2–3-loculed, each locule functional; style elongate; fruit a berry, several-seeded.

1a. Flowers in terminal head- or spikelike clusters; uppermost leaves ordinarily connate-perfoliate; corollas commonly 18–25 mm long, orange to red orange; plants of northern British Columbia. *L. dioica*

1b. Flowers borne in axillary, pedunculate pairs; uppermost leaves not connate-perfoliate; corollas commonly less than 18 mm long, yellow, white, or pink; plants of various distribution. (2)

2a. Corollas yellow, often tinged purplish, glandular-hairy; flowers subtended by large, foliose bracts; plants indigenous in southeastern Alaska. *L. involucrata*

2b. Corollas white to pink, glabrous; flowers not subtended by foliose bracts; plants cultivated in southern Alaska. *L. tatarica*

Lonicera dioica L.
Honeysuckle

Stems twining, 0.5–1.5 m long, sparingly hairy to glabrous; leaves 2.5–8 cm long, 2–6 cm broad, elliptic to obovate, entire, obtuse apically, obtuse to acute basally, glabrous and green above, glaucous and hairy beneath, often ciliate, sessile or shortly petiolate; uppermost leaves usually connate perfoliate; peduncles 0.5–2.5 cm long, glabrous; flowers several, borne sessile in terminal, spike- or headlike clusters; ovary glabrous; calyx less than 1 mm long; corolla 18–25 (30) mm long or more, orange or orange red, pubescent externally; fruit 8–12 mm long, red.

Woods; at Liard Hot Springs in northern

British Columbia; eastward to Quebec and southward to North Carolina, Oklahoma, and southern British Columbia (*L. glaucescens* Rydb.). Our plants belong to var. *glaucescens* (Rydb.) Butters.

Lonicera involucrata (Richards.) Banks ex Spreng.
Black Twinberry

Shrubs 0.5–2 m tall, the young stems glabrous, quadrangular; leaves 2.5–16 cm long, 1.5–7.5 cm broad, elliptic to oblanceolate or broadly lanceolate, entire, acuminate apically, acute to rounded basally, green and glabrous above or hairy along the midvein, pale and more or less hairy beneath, short-petiolate; peduncles 0.5–4 cm long; flowers borne in sessile pairs, subtended by broad, greenish or purplish black, foliose, glandular-hairy bracts commonly 1 cm long or more; ovary glandular-hairy; calyx less than 1 mm long; corollas 12–18 mm long, yellow (sometimes tinged purplish), pubescent externally; fruit 8–12 mm long, black.

Woods and stream banks; in southeastern Alaska; east to Quebec and south to California, New Mexico, and Mexico (*L. ledebourii* Esch.).

Lonicera tatarica L.
Tatarian Honeysuckle

Shrubs 1–3 m tall or more, the young stems glabrous, more or less terete; leaves 1.5–6 cm long, 0.5–3 cm broad, lanceolate to oblong, entire, attenuate to acuminate or acute apically, obtuse to rounded or subcordate basally, green and glabrous above, pale and glabrous beneath, short-petiolate; peduncles 3–15 mm long; flowers borne in sessile pairs, subtended by narrow, scarious bracts; ovary glabrous; calyx less than 1 mm long; corollas 7–15 mm long, white to pink, glabrous; fruit 4–6 mm long, orange or red.

Cultivated ornamental; in southern Alaska; native to Eurasia. Several horticultural forms are cultivated, including var. *alba*

Loesel., with white flowers, and var. *rosea* Reg., with flowers rose outside and pink inside. The plants are hardy and flower in late June and early July.

SAMBUCUS L.

Shrubs, with coarse, erect or ascending stems; leaves opposite, pinnately compound, sometimes stipulate; flowers numerous, borne in terminal, compound, paniculate cymes; calyx 5-toothed; corolla 5-lobed, rotate, regular, white to cream; stamens 5, attached to the corolla tube; ovary 3–5-loculed, each locule functional; style very short or lacking; fruit a 3–5-seeded berry.

Sambucus racemosa L.
Red Elderberry

Shrubs 1–2(3) m tall, with subherbaceous, pithy twigs; leaves 5–35 cm long, the leaflets (3)5–7, 2.5–15 cm long, 1–6 cm broad, lanceolate to elliptical, oblanceolate, or oblong, acuminate apically, unequally obtuse to acute basally, serrate, petiolulate, green and glabrous or hairy only along the midvein above, pale and hairy along the veins beneath; flowers numerous. borne in terminal, paniculate cymes with a distinct central rachis extending beyond the lowermost branches, on leafy lateral shoots; calyx lobes less than 1 mm long; corollas white to cream, 4–6 mm broad, the lobes much longer than the tube; fruit 4–6 mm long, red.

Woods, stream banks, hillsides, and thickets; in coastal and insular southern Alaska and less commonly in the interior; eastward to the Atlantic and southward to northeastern United States, California, Arizona, and New Mexico; Eurasia. Our material has been designated as ssp. *pubescens* (Michx.) House var. *arborescens* (T. & G.) Gray.

SYMPHORICARPOS Duhamel

Shrubs, with slender, erect, or ascending stems; leaves opposite, simple; flowers solitary to more commonly 2–several, borne in short, terminal or axillary spikes; calyx 5-lobed; corolla 5-lobed, more or less irregular, funnelform, white to pink; ovary 4-loculed, with only 2 functional; style elongate; fruit a 2-seeded berry.

Jones, G. N. 1940. A monograph of the genus *Symphoricarpos*. Jour. Arn. Arb. 21:201–52.

Symphoricarpos albus (L.) Blake
Snowberry

Shrubs, mostly 0.5–1.5 m tall, the young stems glabrous, terete or nearly so; leaves 1.5–7 cm long, 0.8–4 cm broad, ovate to elliptic, entire or coarsely toothed, rounded to obtuse apically, rounded to obtuse basally, green and glabrous above, pale and sparingly hairy to glabrous beneath, short-petiolate; peduncles 3–30 mm long; flowers solitary, or more commonly 2–several in axillary or terminal spikes, each closely subtended by a pair of minute bracts; ovary glabrous; corollas 5–7 mm long, white to pink, densely hairy within; fruit 6–12 mm long, white.

Woods and thickets; in coastal and insular southeastern Alaska; eastward to Quebec and southward to California, Utah, Colorado, Nebraska, and Virginia (*S. rivularis* Suksd.; *S. albus* ssp. *laevigatus* [Fern.] Hultén). Plants from Alaska belong to var. *laevigatus* (Fern.) Blake.

VIBURNUM L.
Shrubs, with erect or ascending stems; leaves opposite, simple, palmately lobed and veined; flowers several to many, borne in axillary, compound, corymbose cymes; calyx 5-lobed; corolla 5-lobed, rotate, regular, white to cream; stamens 5, attached to the corolla tube; ovary 3-loculed, only 1 locule functional; style short; fruit a 1-seeded drupe.

Viburnum edule (Michx.) Raf.
Highbush Cranberry

Shrubs to 2 m tall, with slender, sub-

herbaceous twigs; leaves 2.5–15 cm long, 1.5–13 cm broad, elliptic to orbicular or lanceolate, palmately lobed and veined, rounded to obtuse basally, serrate, glabrous and green above, pale and usually spreading-hairy along the veins beneath; flowers few to many, borne in corymbose cymes, on short, axillary branches with 1 pair of leaves; calyx less than 1 mm long; corollas white to cream, 4–7 mm broad, the lobes much longer than the tube; fruit 8–15 mm long, red or orange.

Woods, thickets, stream banks, open slopes, cold bogs, and lake shores; in most of Alaska (except for the Aleutians and extreme northern portions) and Yukon; eastward to Newfoundland and south to Oregon, Idaho, Colorado, and Pennsylvania

Viburnum edule (Michx.) Raf. (× 0.3).

(*V. pauciflorum* Pylaie). The plant, including its fruit, has a distinctive, musky odor. The fruit is used for making jams and jellies.

CARYOPHYLLACEAE
Pink Family

Annual, biennial, or perennial herbs; leaves opposite, simple, entire or minutely serrulate; flowers usually perfect, regular, solitary or borne in cymes; sepals 4–5, separate or united; petals 4–5, separate or lacking; stamens usually as many as, or twice as many as, the sepals; pistils 1; ovary superior, 1-loculed, or incompletely 3–5 loculed; styles 2–5; fruit a capsule.

1a. Leaves with distinctive scarious stipules. (2)
1b. Leaves without scarious stipules. (3)

2a. Styles and valves of capsules 3; leaves opposite (sometimes with clustered, axillary leaves as well). *Spergularia*
2b. Styles and valves of capsules 5; leaves whorled. *Spergula*

3a. Sepals distinct or nearly so. (4)
3b. Sepals united, forming a tube. (7)

4a. Petals 2-cleft, -parted, or -lobed, (lacking in some *Stellaria* species). (5)
4b. Petals rounded to emarginate (lacking in *Arenaria chamissonis*). (6)

5a. Styles usually 5; capsule cylindrical, membranous, often 1–2 times longer than the calyx, dehiscent by 10 teeth. *Cerastium*
5b. Styles usually 3; capsule ovoid to short-cylindrical, seldom as long as the calyx, dehiscent by 6–8 teeth. *Stellaria*

6a. Styles usually 5, alternate with the sepals; valves of capsule opposite the sepals. *Sagina*
6b. Styles usually 3, if more, then opposite the sepals; valves of capsule alternate with the sepals. *Arenaria*

7a. Flowers immediately subtended by 1–2 pairs of more or less connate bracts. *Dianthus*
7b. Flowers not closely subtended by bracts, or if so, then bracts not connate. (8)

8a. Styles commonly 2; plants introduced. (9)
8b. Styles 3–5; plants native, or less commonly introduced. (10)

9a. Calyx 3–5 mm long, 5-veined, conspicuously membranous below the sinuses; plants cultivated and escaping. *Gypsophila*
9b. Calyx 10 mm long or more, 10–20-veined, not or only slightly membranous below the sinuses; plants weedy species of disturbed sites. *Vaccaria*

10a. Calyx teeth longer than the tube; petals red, mostly 20–30 mm long; plants adventive, uncommon. *Agrostemma*
10b. Calyx teeth shorter than the tube; petals white to pink, mostly less than 20 mm long; plants native (except *Silene noctiflora*). (11)

11a. Styles usually 5; capsules dehiscent by 5 (or 10, by splitting of the primary) valves. *Lychnis*
11b. Styles usually 3; capsules dehiscent by 6 (or rarely by 8–10) valves. *Silene*

AGROSTEMMA L.

Annual herbs, with erect stems; leaves opposite, lacking stipules; flowers solitary or several, borne in an open cyme; calyx of 5 united sepals, the tube 10-ribbed, shorter than the subfoliaceous teeth; petals 5, showy, the distinctive blade borne on a slender claw, lacking appendages; stamens 10, inserted at the base of the ovary; styles 5, alternating with the sepals; ovary 1-loculed; capsule 5-valved.

Agrostemma githago L.
Corn Cockle

Plants mostly 1.5–4 dm tall, the stems

finely white-hairy; leaves lance-linear, mostly 2–8 cm long and 0.2–0.6 cm broad; peduncles (or pedicels) 5–15 cm long; calyx tube 10–15 mm long, becoming thickened and conspicuously 10-ribbed in age; calyx lobes 15–35 mm long, lance-linear; petals red, commonly 20–30 mm long; capsule about equaling the calyx tube.

Introduced weedy species of disturbed sites; in southern Alaska; widely distributed in North America.

ARENARIA L.

Annual, biennial, or perennial herbs, with prostrate to erect stems; leaves opposite, linear to ovate, lacking stipules; flowers solitary, or few to many, borne in open or contracted cymes; sepals 5, distinct or nearly so, 1–3-veined; petals 5, white, the blade entire to slightly emarginate, but not bilobed, showy to inconspicuous, or sometimes lacking; stamens usually 10, inserted at the base of a glandular disk; styles usually 3; ovary 1-loculed; capsule opening by 3 or 6 teeth.

Maguire, B. 1951. Studies in the Caryophyllaceae—V. *Arenaria* in America north of Mexico. A conspectus. Am. Midl. Nat. 46:493–511.

———. 1958. *Arenaria rossii* and some of its relatives in America. Rhodora 60: 710.

1a. Plants with fleshy stems and leaves; flowers more or less unisexual; distribution maritime. *A. peploides*

1b. Plants not at all fleshy, or if slightly so, then the plants with perfect flowers and seldom if ever maritime. (2)

2a. Leaves ovate to lanceolate, oblanceolate, or broadly elliptical, acute to rounded apically. (3)

2b. Leaves linear to oblong or subulate, acute, abruptly acute, or blunt apically. (6)

3a. Plants densely caespitose, compact,

the internodes all obscured by the thatch of persistent leaves; flowers lacking petals, subsessile or borne on pedicels to 0.5 cm long. *A. chamissonis*

3b. Plants moderately caespitose to open, the internodes readily apparent (at least some); flowers with petals, on pedicels often over 0.5 cm long. (4)

4a. Plants caespitose, usually less than 10 cm tall, from a branching caudex and taproot, not rhizomatous or stoloniferous; leaves less than 1 cm long. *A. humifusa*

4b. Plants not caespitose, often over 10 cm tall, from elongate, horizontal rhizomes or stolons; leaves often more than 1 cm long (at least some). (5)

5a. Sepals broadly spreading or ascending in flower, usually tinged purplish; capsule inflated at maturity, 7–10 mm long. *A. physodes*

5b. Sepals erect (seldom spreading) in flower, usually greenish; capsule not inflated at maturity, 5 mm long or less. *A. lateriflora*

6a. Leaves linear-filiform, commonly 2–5 cm long; capsules dehiscent by 6 teeth. (7)

6b. Leaves subulate to oblong or linear, commonly less than 1(2) cm long; capsules dehiscent by 3 teeth. (8)

7a. Flowers borne in open cymes, pedicellate; plants of southeastern third of Alaska and southern Yukon. *A. capillaris*

7b. Flowers borne in contracted, subcapitate clusters, the pedicels short or lacking; plants reported for southern Yukon. *A. congesta*

8a. Sepals acute to acuminate; plants slender, annual, biennial, or short-lived perennials, from weak taproots. (9)

8b. Sepals broadly obtuse; plants slender to coarse perennials with strong taproots. (11)

9a. Plants glandular-puberulent, especially in the inflorescence, or rarely glabrous; leaves 3-veined. *A. rubella*

9b. Plants glabrous throughout; leaves 1-veined. (10)

10a. Inflorescence 1-flowered; sepals often tinged purplish. *A. rossii*

10b. Inflorescence commonly 2–7-flowered; sepals seldom purplish. *A. stricta*

11a. Leaves more or less flattened, strongly ciliate marginally, 3-veined; capsules mostly 10–15 mm long. *A. macrocarpa*

11b. Leaves seldom flattened (except in some *A. laricifolia*); capsules 10 mm long or less. (12)

12a. Cauline leaves in (2)3–4(6) pairs; basal leaves sharply acute to blunt apically; sepals erect or ascending in flower, seldom purplish throughout; flowers often 2 per stem. *A. laricifolia*

12b. Cauline leaves in (0)1–2(3) pairs; basal leaves blunt apically (acutish in *A. sajanensis*); sepals ascending to spreading or less commonly erect in flower, often purplish throughout; flowers solitary (rarely in pairs). (13)

13a. Sepals (2)3–8 mm long, spreading to ascending in flower; petals commonly 8 mm long or more; plants common, widely distributed. *A. arctica*

13b. Sepals (2)2.5–4(5) mm long, ascending to erect in flower; petals commonly less than 8 mm long; plants uncommon, known from widely disjunct localities. (14)

14a. Petals oblong, about equaling the sepals, or slightly longer; sepals 2–4.2 mm long; plants rare. *A. sajanensis*

14b. Petals obovate, up to twice longer than the sepals; sepals 2.5–5 mm long; plants uncommon. *A. obtusiloba*

Arenaria arctica Steven
Arctic Sandwort

Perennial plants, forming loose to dense mats 0.5–4 dm broad or more, arising from a taproot and branching caudex, the caudex branches prostrate, clothed with persistent leaves; flowering stems 0.1–1 dm tall, erect or ascending; basal leaves 0.3–1.5 cm long, commonly less than 0.1 cm broad, oblong to linear, 1-veined, abruptly obtuse apically, ciliate in the lower portion, often somewhat glandular, or glabrous; cauline leaves 1–2 pairs, 0.3–1 cm long, broader than the basal ones; flowers erect, solitary; pedicels glandular-hairy; sepals (2)3–8 mm long, spreading to ascending in flower, often purplish throughout, obtuse apically, glandular-hairy to glabrate; petals white, rarely purplish in age, (5)8–12 mm long, obovate; capsules 6–10(12) mm long, 3-valved; seeds 0.8–1 mm long, reniform, warty.

Arctic and alpine tundra and heathlands; over much of Alaska and western and northern Yukon and adjacent Mackenzie; Eurasia (*Minuartia arctica* [Steven] Asch. & Graebn.). The large-petaled phases of *A. arctica* are among the most beautiful of our sandworts. This species is remarkably variable, however, and small-flowered plants are common. *A. arctica* apparently intergrades with *A. obtusiloba* on the one hand, and with *A. laricifolia* on the other. It has not been possible to place all specimens positively. In addition, putative hybrids between *A. arctica* and the more easily distinguished *A. macrocarpa* are known. Population analysis and breeding studies are indicated.

Arenaria capillaris Poir.
Beautiful Sandwort

Perennial plants, forming dense clumps, arising from a taproot and branching caudex, the caudex branches erect, ascending or decumbent, clothed with persistent leaves; flowering stems (0.3)0.8–2(2.5) dm tall, erect or ascending; basal leaves grasslike, 1.5–7(10) cm long, less than 0.1 cm broad, linear-filiform, 1-veined, sharply acute apically, minutely ciliate to glabrous; cauline leaves 2–5 pairs, 0.8–4 cm

Arenaria arctica Steven (× 1.0).

long, similar to the basal ones, transitional with the scarious-margined bracts upwards; flowers erect, 1–3(5); pedicels glabrous; sepals 3–6.5 mm long, erect or nearly so in flower, the scarious margins sometimes purplish, obtuse to abruptly acute apically, glabrous; petals white, 5–10 mm long, obovate; capsules 5–8 mm long, 6-valved; seeds 1.5–1.8 mm long, reniform, minutely warty.

Open hillsides, talus slopes, and gravelly terraces; in the eastern half of Alaska and southern Yukon; southward to Oregon, Nevada, and Montana; Eurasia (*A. nardifolia* authors, not Ledeb.; *A. capillaris* ssp. *formosa* Maguire, not *A. formosa* Fisch.). Our material belongs to var. *capillaris*. It differs from var. *americana* (Maguire) Davis by having the inflorescence glabrous, and by having longer sepals. The var. *americana* approaches our region from the south.

Arenaria chamissonis Maguire
Matted Sandwort

Perennial plants, forming dense clumps to 1 dm broad or more, arising from a taproot and branching caudex, the caudex branches erect or ascending, or less commonly prostrate, clothed with persistent leaves; flowering stems 0.4(1) cm tall or less, erect; basal leaves 0.3–0.8 cm long, 0.1–0.2 cm broad, oblanceolate to obovate or elliptic, 1(3)-veined, acute or abruptly acuminate to obtuse apically; cauline leaves (0)1 pair, to 2 mm long; flowers erect, solitary; pedicels glabrous; sepals 2–4 mm long, erect in flower, the scarious margins not purplish, acute apically, glabrous; petals lacking; capsules 2.5–4 mm long, 6-valved; seeds 1.2–1.4 mm broad, reniform, almost smooth.

Arctic and alpine tundra and heathlands; from the Seward Peninsula northward to Cape Lisburne, and disjunctly eastward in the Alaska Range; Asia (*Cherleria dicranoides* [Cham. & Schlecht.] Hultén, not *A. dicranoides* H. B. K.).

Arenaria humifusa Wahl.
Low Sandwort

Perennial plants, forming compact clumps to 1 dm broad, arising from a taproot and branching caudex, the caudex branches prostrate to ascending, not obscured by persistent leaves; flowering stems 0.1–0.4(0.6) dm tall, erect or nearly so; basal leaves 0.3–0.8 cm long, 0.1–0.2 cm broad, oblanceolate to elliptic, 1(3)-veined, acute apically; cauline leaves commonly 2–3(4) pairs, 0.3–0.5 cm long; flowers erect, solitary; the pedicels pubescent; sepals 3–4 mm long, erect in flower, the margins scarious, obtuse to acutish apically, more or less glandular-hairy; petals white, commonly 3–4 mm long, narrowly oblong; capsules 4–6 mm long, 6-valved (or tardily splitting and apparently 3-valved); seeds subreniform, 0.7–0.9 mm long, almost smooth.

Peaty soils, in alpine sites and in open woods; in widely disjunct localities from the Seward Peninsula eastward to the Alaska Range and in southern Yukon; eastward to Newfoundland and southward to British Columbia and Alberta; Europe (*A. longipedunculata* Hultén). This plant is evidently rare in our region.

Arenaria laricifolia L.
Larch-leaved Sandwort

Perennial plants, forming loose to (less commonly) dense clumps to 4 dm broad, arising from a taproot and branching caudex, the caudex branches prostrate to ascending, clothed with persistent leaves; flowering stems 0.2–1.2(1.7) dm tall, erect or ascending; basal leaves 0.1–1.7(2) cm long, to 0.1 cm broad, linear-subulate to linear or narrowly oblong, 1 (or obscurely 3)-veined, sharply acute to blunt apically, minutely ciliate and often glandular-hairy, or less commonly glabrous; cauline leaves (2)3–4(6) pairs, 0.2–1.4 cm long, becoming smaller upwards; flowers erect, 1–2 (rarely more); pedicels glandular-hairy; sepals 3.5–6 mm long, erect or steeply ascending, the scarious margins seldom purplish, obtuse apically, glandular-hairy; petals white, 6–10 mm long, obovate; capsules 6–8 mm long, 3 (rarely 5)-valved; seeds subreniform, 0.8–1 mm long, warty on the back.

Gravelly or sandy soils in arctic or alpine tundra, or in heathlands and less commonly in open woods; over much of continental Alaska and Yukon; Europe (*Minuartia laricifolia* [L.] Schinz & Thell; *M. yukonensis* Hultén). This entity has suffered from broadly differing interpretations. Portions of it (those specimens with obtuse leaves) have been included within *A. arctica*. There is some indication that *A. laricifolia* intergrades with both *A. arctica* and *A. obtusiloba*. Further work is necessary before an understanding of this complex will be possible. However, at least two more or less distinctive and evidently sympatric entities appear worthy of taxonomic recognition.

1a. Leaves acute apically, often linear-subulate, mostly obscurely 3-veined and glandular-pubescent. *A. laricifolia* var. *laricifolia*

1b. Leaves obtuse, linear to narrowly oblong, commonly 1-veined and glabrous to glabrate or merely ciliate, rarely glandular-hairy (*A. obtusiloba* authors, not [Rydb.] Fern., in part). *A. laricifolia* var. *hulténii* Welsh

Arenaria lateriflora L.
Blunt-leaved Sandwort

Perennial, rhizomatous or stoloniferous plants, the stems arising singly or in small clumps, the stolons or rhizomes lacking conspicuous, persistent leaves; flowering stems 0.4–2(2.5) dm tall, erect or ascending; basal leaves smaller than the upper ones; cauline leaves 3–7 pairs, 0.3–2.6 cm long, 0.1–1.1 cm broad, the largest ones near the middle of the stem, lanceolate to elliptic or lance-oblong, pinnately 3–5-veined, acute to obtuse apically, minutely hairy below, especially along the midrib, ciliate, often with sterile, axillary shoots; flowers erect, solitary or 2–5 in terminal cymes; pedicels puberulent to glabrate or glabrous; sepals 2–3 mm long, erect or spreading in flower, with whitish, scarious margins, obtuse apically, glabrous or ciliate near the base; petals white, 3.5–8 mm long, obovate; capsules 2–4 mm long, 6-valved; seeds subreniform, 0.8–1.1 mm long, black, shining.

Dry slopes, meadows, terraces, open woods, and thickets; in much of Alaska and Yukon, mostly south of the 68th parallel; eastward to Newfoundland and southward to New Mexico, Missouri, and Rhode Island; Eurasia (*Moehringia lateriflora* [L.] Fenzl.).

Arenaria macrocarpa Pursh
Long-podded Sandwort

Perennial plants, forming loose to dense mats 0.5–4(10) dm broad or more, arising from a taproot and branching caudex, the caudex branches prostrate to ascending or erect, clothed with persistent leaves; flowering stems 0.1–0.6 dm tall, erect or ascending; basal leaves 0.4–1.5 cm long, to 0.2 cm broad, lance-subulate to lance-oblong, 3-veined, acute to obtuse apically, ciliate; cauline leaves (0)1–6 pairs, 0.3–1 cm long, often broader than the basal ones;

flowers erect, solitary; pedicels glandular-pubescent with multicellular hairs; sepals 4–6.5 mm long, ascending to erect in flower, often purplish, obtuse apically, glandular-hairy to glabrate; petals white, 8-13 mm long, obovate; capsules (5)10–16 mm long, 3–4-valved; seeds 0.8–1.2 mm long, subreniform, strongly warty.

Arctic and alpine tundra and heathlands; over most of Alaska, and less commonly in the Yukon; Eurasia (*Minuartia macrocarpa* [Pursh] Ostenf.; *A. macrocarpa* var. *rosea* Hultén). This entity is distinctive, but apparently hybridizes with *A. arctica* and perhaps also with *A. laricifolia*.

Arenaria obtusiloba (Rydb.) Fern.
Alpine Sandwort

Plants perennial, forming loose to dense mats 0.5–4 dm broad, arising from a taproot and branching caudex, the caudex branches prostrate to ascending or erect, clothed with persistent leaves; flowering stems 0.1–0.8 dm tall, erect or ascending; basal leaves 0.3–0.8 cm long, commonly less than 0.1 cm broad, oblong to linear, 1-veined, abruptly obtuse apically, ciliate almost or quite to the apex, or glabrous; cauline leaves (0)1–2 pairs, 0.2–0.8 cm long, commonly broader than the basal ones; flowers erect, solitary; pedicels glandular-hairy; sepals 2.5–4(5) mm long, spreading-ascending to nearly erect in flower, often purplish, the apex obtuse, glandular-hairy to glabrate; petals white, 4–6(10) mm long, obovate; capsules 4–8 mm long, 3–4-valved, the seeds 0.8–1 mm long, smooth or nearly so.

Arctic and alpine tundra; uncommon in widely disjunct localities in Alaska and in the Yukon; southward to Oregon and New Mexico and disjunctly eastward to Labrador and Greenland (*Minuartia obtusiloba* [Rydb.] House). Our material appears to represent a series of morphological intermediates between *A. arctica* and *A. laricifolia*. None of the Alaskan specimens I have examined is strictly identical with plants from the southern Rocky Mountains. These latter specimens may be distinguished by the nearly erect sepals and more or less acute leaves which tend to be glabrous. The southern Rocky Mountain specimens seem to be more closely related to *A. laricifolia.* Further work is indicated.

Arenaria peploides L.
Sea-beach Sandwort

Plants perennial, forming loose to dense mats, commonly 1–8 dm broad or more, arising from a deep-seated taproot and horizontal, substoloniferous or subrhizomatous caudex branches, these not obscured by persistent leaves; flowering stems 0.5–5 dm long or more, prostrate to ascending or erect; lowermost leaves smaller than those of the middle stem; cauline leaves 3–10 pairs or more, 0.5–5 cm long, 0.3–2 cm broad, elliptic to oblong, oblanceolate or obovate, pinnately several- to many-veined, fleshy, acute to acuminate apically, glabrous, often with sterile, axillary shoots; flowers erect, solitary in the upper leaf axils, some imperfect on at least some plants; pedicels glabrous; sepals 4–8 mm long, spreading in flower, greenish, glabrous, acute apically; petals white or greenish white, 2–3 mm long, obovate; capsules 8–12 mm long, subglobose, 3–5-valved; seeds 3–4 mm long, shining.

Maritime beaches; in coastal Alaska and northern Yukon; eastward to Nova Scotia and southward to Oregon and Virginia; circumboreal (*Honkenya peploides* [L.] Ehrh.). Our specimens have been segregated into two main entities.

1a. Main leaves commonly 2–4 times longer than broad; flowering stems usually more than 1.5 dm long; plants of coastal southern Alaska (*Honkenya peploides* ssp. *major* [Hook.] Hultén). *A. peploides* var. *major* Hook.
1b. Main leaves commonly twice longer than broad or less; flowering stems usually less than 1.5 dm long; plants of coastal western and northern Alas-

ka and Yukon (*H. peploides* var. *diffusa* [Hornem.] Ostenf.; *H. peploides* ssp. *diffusa* [Hornem.] Hultén). *A. peploides* var. *peploides*

Arenaria physodes Fisch. ex DC.
Merckia

Perennial, rhizomatous, or stoloniferous plants, the stems arising singly or in small to large clumps and more or less mat-forming, the stolons or rhizomes without conspicuous, persistent leaves; flowering stems 0.2–2 dm long or more, prostrate to ascending or erect; lowermost leaves smaller than those of the middle stem; main cauline leaves 3–7 pairs or more, 0.4–2.5 cm long, 0.3–1.2 cm broad, lanceolate to elliptic or ovate, obscurely pinnately veined, more or less fleshy, acute to acuminate apically, pubescent below, especially on the midvein or glabrous, ciliate, sometimes with sterile, axillary shoots; flowers erect, solitary, terminal or axillary; pedicels pubescent with multicellular hairs; sepals 3–6 mm long, spreading to ascending in flower, with whitish or purplish scarious margins, obtuse to more or less acute apically, glabrous or sparsely villous, commonly purplish; petals white, 4–6 mm long, oblanceolate; capsules 7–10 mm long, subglobose, 6-lobed by invagination of the sutures, 6-valved; seeds 1.2–1.5 mm long, minutely roughened, yellowish brown.

Arctic and alpine tundra, heath, open woods, and stream gravels; over most of Alaska and Yukon; east to Mackenzie; Asia (*Merckia physodes* [Fisch.] Fisch. ex Cham. & Schlecht.; *Wilhelmsia physodes* [Fisch.] McNeill).

Arenaria rossii R. Br. ex Richards.
Ross Sandwort

Plants short-lived perennials, biennials, or rarely annuals, forming cushions or mats 0.5–2 dm broad, arising from a taproot and branching caudex, the caudex branches prostrate to ascending, clothed with persistent leaves; flowering stems 0.2–1.5 dm tall, erect or nearly so; basal leaves 0.3–1 cm long, less than 0.1 cm broad, linear to lance-linear, 1-veined, obtuse apically, glabrous; cauline leaves (0)2–3 pairs, 0.3–1.2 cm long, similar to the basal ones, often with short, sterile, axillary branches; flowers erect, solitary; pedicels glabrous; sepals (1.5)2–3.5 mm long, spreading or spreading-ascending in flower, more or less purplish throughout, acute apically; petals white, 2–3.5 mm long, oblanceolate or lacking; capsules 1.5–2.5 mm long, 3-valved; seeds 0.6–1 mm long, reddish brown, minutely roughened.

Arctic and alpine tundra and heathlands; in much of Alaska north of the 64th parallel and in most of the Yukon; southward to Washington, Idaho, and Colorado; Asia, Greenland (*Minuartia rolfii* Nannfeldt). Three more or less intergrading varieties are present in Alaska and Yukon.

1a. Petals lacking; plants rare and local in southeastern Yukon. *A. rossii* var. *apetala* Maguire
1b. Petals present; plants of various distribution. (2)

2a. Sepals commonly 1.5–2.5 mm long, obtusish, usually 1-veined; petals surpassing the sepals; plants densely tufted, rare, known from disjunct sites in eastern Alaska and Yukon. *A. rossii* var. *rossii*
2b. Sepals commonly 2–3.5 mm long, acute, commonly 3-veined; petals equaling or shorter than the sepals; plants loosely tufted, common in Alaska and Yukon (*Minuartia elegans* var. *orthotrichoides* [Schischk.] Hultén; *M. rossii* var. *elegans* [Cham. & Schlecht.] Hultén). *A. rossii* var. *elegans* (Cham. & Schlecht.) Welsh

Arenaria rubella (Wahl.) J. E. Smith
Reddish Sandwort

Plants short-lived perennials, biennials, or rarely annuals, forming cushions or mats 0.4–3 dm broad, arising from a taproot and branching caudex, the caudex branches prostrate to ascending or erect, clothed with persistent leaves; flowering stems 0.2–1.5 dm tall, erect or ascending; basal leaves 0.3–1(1.2) cm long, less than 0.1 cm broad, lance-linear, 3-veined, acute apically; glandular-puberulent to glabrate; cauline leaves 2–6 pairs, 0.1–1.4 cm long, similar to the basal ones but reduced upwards, often with short, sterile, axillary branches; flowers erect, solitary or more commonly 2–4; pedicels glandular-hairy; sepals (2)3–4.5 mm long, spreading in flower, commonly more or less purplish throughout, acute to acuminate apically; petals white, 2.5–5 mm long, oblanceolate; capsules 3–5 mm long, 3–4-valved; seeds 0.4–0.6 mm long, reddish brown, minutely roughened.

Gravelly soils, rock outcrops, and talus slopes in arctic and alpine tundra, heath, and open woods; in most of Alaska (except for the Aleutian Islands) and Yukon; eastward to Greenland and southward to California, Nevada, New Mexico, and Gaspé; Eurasia (*A. verna* authors, not L.; *A. propinqua* Richards.; *A. quadrivalvis* R. Br.; *Minuartia rubella* [Wahl.] Graebn. ex Asch. & Graebn.). The glabrous form is known as f. *epilis* (Fern.) Polunin.

Arenaria sajanensis Willd.
Sajan Sandwort

Perennial plants, forming dense to moderately dense tufts 0.3–1.5 dm broad, arising from a taproot and branching caudex, the caudex branches prostrate to ascending or erect, clothed with persistent leaves; flowering stems 0.1–0.5 dm tall, erect or nearly so; basal leaves 0.2–1 cm long, less than 0.1 cm broad, linear, 1-veined, acutish to obtuse apically, glabrous, or ciliate in the lower portion, cauline leaves (0)2–3 pairs, 0.2–0.8 cm long, often broader than the basal ones; flowers erect, solitary, or in pairs, the pedicels glandular-puberulent; sepals 2–4.2 mm long, ascending to erect in flower, sometimes purplish at the tips, obtuse apically, glandular-hairy to glabrate; petals white, 2–4.5 mm long, oblong to narrowly oblanceolate; capsules 2.5–5 mm long, 3-valved; seeds 0.6–0.8 mm long, reddish brown, smooth.

Alpine tundra and heathlands; from widely disjunct sites in Alaska and Yukon; eastward to Greenland; Eurasia (*Stellaria biflora* L., not *Arenaria biflora* L.; *Minuartia biflora* [L.] Schinz. & Thell.).

Arenaria stricta Michx.
Rock Sandwort

Plants short-lived perennials (or biennials?), forming loose to moderately dense clumps 1–5 dm broad, arising from a taproot and branching caudex, the caudex branches commonly ascending, clothed with persistent leaves; flowering stems 0.6–2.5 dm tall, prostrate to erect, glabrous; basal leaves 0.5–1.8 cm long, less than 0.1 cm broad, linear to oblong, 1 (or obscurely 3)-veined, obtuse to acutish apically, glabrous; cauline leaves (0)2–6 pairs, 0.2–1.4 cm long, reduced upwards, often with short, sterile, axillary branches; flowers erect, solitary or more commonly 2–7; pedicels glabrous; sepals 3–4 mm long, erect or ascending in flower, seldom or not at all purplish throughout, acute to acuminate apically; petals white, 2–3.5 mm long, oblanceolate to oblong; capsules 3–4.5 mm long, 3(4 or 5)-valved; seeds 0.5–0.8 mm long, dark reddish brown, roughened.

Dry rocky tundra, moist clayey banks, dry sandy slopes, and open woods; in central, eastern Alaska and southern Yukon; eastward to Labrador and southward to Virginia, Arkansas, Texas, Wyoming, and Oregon; Eurasia (*Minuartia stricta* [Sw.] Hiern; *M. dawsonensis* [Britt.] Mattf.). Our materials are assignable to ssp. *dawsonensis* (Britt.) Maguire. They have been accorded widely differing interpretations. Both *Arenaria dawsonensis* Britt. and *A. uliginosa* Schleich. have been treated at specific levels within our region. A careful examination of herbarium specimens has failed to demonstrate consistent differences which would allow meaningful segregation of the specimens into two entities. It seems best to treat specimens from Alaska and Yukon conservatively.

CERASTIUM L.

Annual, biennial, or perennial herbs, with prostrate to erect stems; leaves opposite, ovate to lanceolate, oblong, or linear, lacking stipules; flowers solitary and terminal, or more commonly in terminal, open or congested cymes; sepals 5, distinct to the base, 1(3–5)-veined; petals 5, white, showy to inconspicuous, the blade notched or bilobed, or sometimes lacking; stamens commonly 10, inserted at the base of the ovary; styles usually 5, opposite the sepals; ovary 1-loculed; capsule opening by 10 teeth.

Hultén, E. 1956. The *Cerastium alpinum* complex. A case of world-wide introgressive hybridization. Svensk. Bot. Tidskr. 50:411–95.

1a. Plants 2–6.5 dm tall or more; leaves lance-linear, 2–10 cm long; petals (15)18–25 mm long; known from northern and east-central Alaska and western Yukon. *C. maximum*

1b. Plants 0.3–2.5(3, or seldom more) dm tall; leaves various in shape, but if lance-linear, then commonly less than 2 cm long; petals usually less than 12 mm long; distribution various. (2)

2a. Petals subequal to the sepals; plants annual, biennial, or short-lived, adventive perennials. (3)

2b. Petals longer than the sepals; plants indigenous perennials. (4)

3a. Pedicels (0)1–4 mm long; sepals 3–5 mm long; plants annual. *C. viscosum*

3b. Pedicels (at least some) commonly 5–10(20) mm long; sepals 4.5–6.5 mm long; plants biennial or short-lived perennials. *C. vulgatum*

4a. Leaves linear to lance-linear, oblong, or narrowly elliptical (the uppermost cauline leaves sometimes broader);

sterile branches arising in most leaf axils. *C. arvense*

4b. Leaves lanceolate to elliptical, oblong, or oblanceolate; sterile branchlets lacking, or present only in the lowermost axils. (5)

5a. Stems and pedicels moderately to densely pubescent with spreading, yellowish hairs, commonly 1–1.5 mm in diameter or more; plants of coastal and insular, southern and southwestern Alaska. *C. fischerianum*

5b. Stems and pedicels variously pubescent but seldom as above, commonly 1 mm broad or less, or if more, then usually not of coastal southern Alaska. *C. beeringianum*

Cerastium arvense L.
Field Chickweed

Perennial plants, forming loose to dense mats or clumps 1–4 dm broad, arising from stoloniferous or subrhizomatous, prostrate stems; flowering stems 0.6–3(4) dm long, decumbent to ascending or erect, from glabrous to descending or spreading-hairy on the middle internodes, becoming densely glandular above and in the inflorescence; leaves (0.3)1–2.5 cm long, 0.1–0.4(0.7) cm broad, linear to lance-linear, oblong, or narrowly elliptical (or the uppermost leaves sometimes lanceolate), 1-veined, more or less pubescent above and beneath, ciliate, acute apically; most cauline leaves with sterile, axillary shoots; flowers erect, commonly 3–6 in an open, more or less dichotomous cyme; pedicels 0.4–3 cm long, glandular-hairy; bracts scarious-margined; sepals 4–6.5 mm long, glandular-hairy, scarious-margined, the innermost broadly so; petals white, 8–12 mm long, deeply bilobed; capsules 6–10 mm long, cylindrical; seeds 0.8–1.2 mm long, yellowish or reddish brown.

Alpine tundra, heath, woodlands, and meadows; in the southern third of Alaska and southern Yukon; widely distributed in North America; Eurasia (*C. campestre* Greene). Our specimens include some apparent intermediates between *C. arvense* and *C. beeringianum*. Otherwise, they are fairly uniform.

Cerastium beeringianum Cham. & Schlecht.
Bering Chickweed

Perennial plants, forming loose to dense mats or clumps, 0.4–4 dm broad or more, arising from taproots and more or less stoloniferous, prostrate stems; flowering stems 0.3–2.5(3.5) dm tall, decumbent to erect, the lower internodes glabrous to moderately, descending-hairy, becoming moderately to densely hairy and often glandular above; leaves 0.3–3.8 cm long, (0.2)0.3–0.8 cm broad, lanceolate to lance-oblong, or less commonly oblong to linear, 1-veined, more or less pubescent on both surfaces (or rarely glabrous except along the vein), ciliate, acute to obtuse or rounded apically; cauline leaves lacking sterile, axillary shoots, or bearing them only in the lowermost axils; flowers erect, commonly 3–6 in an open, more or less dichotomous cyme; pedicels 0.2–3.5 cm long, glandular-hairy or merely spreading-hairy; uppermost bracts narrowly scarious-margined; sepals 3–7 mm long, pubescent like the pedicels, scarious-margined, the innermost broadly so; petals white, 6–12 mm long, deeply bilobed; capsules 8–12 mm long, cylindrical; seeds 0.7–1.1 mm long, yellowish to brownish.

Arctic and alpine tundra, heath, woods, meadows, and open slopes; almost throughout Alaska and Yukon; eastward to Mackenzie and southward to California, Arizona, and Colorado; Asia (*C. arcticum* authors, not Lange; *C. bailynickii* Tolm.; *C. jenisejense* Hultén; *C. regelii* authors not Ostenf.?; *C. scammaniae* Polunin). This is our most common and most variable species. It apparently intergrades with both *C. arvense* and *C. fischerianum*. Two poorly defined varieties are present in Alaska.

1a. Leaves sparingly pubescent or glabrous, except for along the midrib; pubescence of pedicels not glandular; bract margins not at all or only slightly scarious; plants of the Aleutian Islands (*C. aleuticum* Hultén). *C. beeringianum* var. *aleuticum* (Hultén) Welsh

1b. Leaves sparingly to moderately pubescent; pubescence of pedicels often glandular; bract margins somewhat scarious; plants of broad distribution (*C. beeringianum* var. *grandiflorum* Hultén). *C. beeringianum* var. *beeringianum*

Cerastium fischerianum Ser. ex DC.
Fischer Chickweed

Plants perennial, forming loose mats or clumps, commonly 1–5 dm broad or more, arising from taproots and stoloniferous or subrhizomatous, horizontal stems; flowering stems (0.3)0.8–5 dm tall, decumbent to ascending or erect, or sprawling, the lower internodes sparingly to moderately spreading- or descending-hairy, becoming more densely pubescent upwards, with spreading, yellowish, often glandular hairs; leaves 0.7–5 cm long, 0.3–1.5 cm broad, lanceolate to oblong, elliptic, or oblanceolate, 1-veined, more or less pubescent on both surfaces, ciliate, acute to obtuse or rounded apically; stems lacking sterile, axillary shoots, or bearing them only in the lowermost axils; flowers erect (deflexed in fruit), commonly 2–10 in an open, more or less dichotomous cyme; pedicels 0.3–3.5 cm long, pubescent with yellowish, spreading, frequently glandular hairs; uppermost bracts not or only slightly scarious-margined; sepals (5)6–10 mm long, pubescent like the pedicels, scarious-margined, the innermost broadly so; petals white, 7–12 mm long, deeply bilobed; capsules 10–14 mm long, subcylindrical; seeds 1–1.5 mm long, reddish brown.

Hillsides, lake shores, and gravelly spits; in coastal and insular southern and south-western Alaska; Asia (*C. unalaschkense* Takeda). This entity is ordinarily easily distinguished by its large size and conspicuous, yellowish pubescence. However, it apparently intergrades to a limited extent with *C. beeringianum*. Additional study is indicated.

Cerastium maximum L.
Great Chickweed

Plants perennial, the stems solitary, or few in clumps, arising from subrhizomatous branches; flowering stems commonly 2–7 dm tall, erect or ascending, the lower internodes moderately pilose, becoming glandular upwards; leaves 2–10 cm long, 0.3–1.2 cm broad, lance-attenuate, 1 (or obscurely several)-veined, more or less pubescent on both surfaces, short-ciliate, attenuate apically; stems with sterile, axillary shoots or less commonly lacking them; flowers erect, showy, commonly 3–10, in an open or congested cyme; pedicels 0.2–2.5 (6) cm long, glandular-pubescent; uppermost bracts not scarious-margined; sepals 8–11(12) mm long, moderately to sparingly glandular-hairy, scarious-margined; petals white, (15)18–25 mm long, deeply bilobed; capsules 15–22 mm long, tapering to the apex; seeds 2–2.5 mm long, yellowish brown.

Open woods, gravel bars, and terraces; in central eastern and northern Alaska and northern and western Yukon; Asia. This is our most beautiful and most distinctive *Cerastium* species.

Cerastium viscosum L.
Mouse-ear Chickweed

Plants annual, arising from slender taproots, the stems 0.5–5 dm long, decumbent to erect, moderately to densely villous and commonly glandular as well; leaves 0.3–3 cm long, 0.2–1.3 cm broad, obovate or oblanceolate to elliptic, lanceolate, or ovate, 1-veined, pubescent above and beneath, long-ciliate, acute to obtuse or rounded apically; stems lacking sterile, axillary shoots (often with axillary flowering shoots); flowers erect or spreading, inconspicuous, commonly several to many in a congested cyme; pedicels (0)1–2 mm long, spreading-hairy; uppermost bracts scarcely if at all scarious-margined; sepals 3–5 mm long, pubescent like the upper stem, scarious-margined; petals white, 3–4(5) mm long, bilobed; capsules 6–9 mm long, cylindrical; seeds 0.3–0.5 mm long, brown.

Adventive, weedy species of old clearings, homesteads, gardens, and roadsides; known from many disjunct localities in Alaska and Yukon; widespread in North America; native to Eurasia (*C. glomeratum* Thuill.).

Cerastium vulgatum L.

Plants biennial or short-lived perennial, forming clumps mostly 1–4 dm broad, arising from taproots and often from horizontal, stoloniferous branches; flowering stems 0.5–4 dm long, decumbent to erect, moderately (sparsely) villous and often glandular as well; leaves 0.7–2.5 cm long, 0.2–1 cm broad, lanceolate to ovate, lance-oblong, or oblanceolate, 1-veined, coarsely pubescent on both surfaces, long-ciliate, obtuse to rounded or acute; stems lacking sterile, axillary shoots (but often with axillary flowering shoots); flowers erect or spreading, inconspicuous, commonly several to many in a congested to open cyme; pedicels commonly 3–10 mm long or more, spreading-hairy; uppermost bracts distinctly scarious-margined; sepals 5–7 mm long, spreading-hairy, scarious-margined; petals white, 5–7 mm long, bilobed; capsules 7–10 mm long, cylindrical; seeds 0.6–0.8 mm long, reddish brown.

Adventive weedy species of disturbed soils; in widely scattered sites in Alaska and Yukon; common in North America; native of Eurasia (*C. caespitosum* Gilib.; *C. fontanum* Baum ssp. *triviale* [Link] Jalas).

DIANTHUS L.

Perennial herbs, with erect or ascending stems; leaves opposite, lance-linear to oblong or linear, lacking stipules; flowers solitary, or 2–4, borne in branching cymes; sepals 5, united to near the apex, each several-veined; petals 5, pink to pink purple, showy, the blade spreading at right angles to the claw, toothed; stamens 10, connate, adnate to the base of the petals; styles usually 2; ovary 1-loculed, stipitate; capsule opening by 4–5 teeth.

Dianthus repens Willd.
Northern Pink

Plants 0.5–1.7(2.5) dm tall, from a taproot and branching caudex, the stems erect or ascending, glabrous; leaves 1–4.5 cm long, 0.1–0.3 cm broad, lance-linear to oblong or linear, 1-veined, acute to attenuate apically, glabrous; flowers solitary, or 2–4, erect, closely subtended by 2–4 bracts, the peduncles glabrous; calyx urn-shaped, 10–14 mm long, purplish, glabrous, the teeth 2–4 mm long; petals pink to pink purple, 18–25 mm long, the blades 7–13 mm long, toothed apically; capsules 12–17 mm long, 4–5-valved; seeds 1.2–2 mm long, brownish, flattened.

Rock outcrops and talus slopes; in widely scattered sites over much of Alaska (except for the southeastern and southwestern portions); Eurasia.

GYPSOPHILA L.

Annual plants, with erect, much-branched stems; leaves opposite, lanceolate to linear or oblong, lacking stipules; flowers numerous, borne in open cymes; sepals 5, united to near the middle, 1-veined, scarious below the sinuses; petals 5, white to pinkish, the blades rounded or emarginate apically, but not bilobed; stamens 10, inserted at the base of the ovary; styles 2; ovary 1-loculed; capsule opening by 4–6 valves.

Gypsophila elegans Bieb.
Baby's-breath

Plants 1.5–3.5 dm tall, from a taproot, the stems erect, glabrous; leaves 1–5 cm long

Dianthus repens Willd. (× 0.8).

or more, 0.2–1 cm broad, narrowly lanceolate to oblong, 1-veined, acute to attenuate apically, glabrous, 0.5–4 cm long; calyx campanulate, 2–3 mm long, often purplish, glabrous, the teeth 1–1.5 mm long; petals white or pinkish, 5–8 mm long; capsules

3–4 mm long, 4–6-valved; seeds 1–1.4 mm long, brownish black.

Cultivated and escaping in central Alaska; native to Eurasia.

LYCHNIS L.

Perennial herbs with erect or ascending stems; leaves opposite or chiefly basal, lance-linear to oblong, lacking stipules; flowers solitary, or 2–5 in congested or more or less open cymes; sepals 5, united to near the apex, each 2–4-veined; petals 5, white to pink or reddish, the blade spreading at right angles to the claw (when exserted from the calyx), retuse or 2–4-lobed; stamens 10, connate, adnate to the base of the petals; styles 5(4); ovary 1-loculed (or incompletely 5-loculed), stipitate; capsules opening by 5(4) or 10(8) teeth. Our *Lychnis* species have been variously treated as included within *Melandrium* or as part of an expanded *Silene*. Neither course is acceptable; the basic segregation of the species is still the same. Wholesale transfer of species to other genera does allow botanists a chance for presenting an array of confusing new names and combinations (see Hultén, 1973, p. 482).

Hultén, E. 1973. Supplement to flora of Alaska and neighboring territories. Bot. Not. 126:459–512.

Maguire, B. 1950. Studies in the Caryophyllaceae—IV. A synopsis of the North American species of the subfamily Silenoideae. Rhodora 52:233–45.

1a. Flowers (1–2)3–6, borne erect in more or less congested cymes; calyx 3–6 mm broad, cylindrical or slightly constricted apically; stems 0.5–3(4) dm tall, strongly glandular. *L. triflora*

1b. Flowers commonly solitary (sometimes 2–6), borne nodding or erect, when more than 1, borne in more or less open cymes; calyx commonly more than 6 mm broad, subcylindrical to distinctly urn-shaped; stems variable in height, moderately to obscurely or not at all glandular. (2)

2a. Plants 2.3–6 dm tall; flowers 1–6, usually in open cymes, commonly erect; petals exserted, white; plants of central, northern, and eastern Alaska and western Yukon. *L. taylorae*

2b. Plants 0.3–2.5(3.5) dm tall; flowers solitary (less commonly 2–3, in more or less contracted cymes), nodding and petals included or shortly exserted, or erect and petals exserted; broadly distributed. (3)

3a. Flowers nodding in bud, the petals included within, subequal to, or slightly longer than the calyx, calyx bladdery-inflated, urn-shaped. *L. apetala*

3b. Flowers erect or spreading in bud, the petals exserted; calyx inflated, cylindrical or more or less urn-shaped. *L. furcata*

Lychnis apetala L.
Nodding Lychnis

Plants 2.3–6 dm tall, from a taproot and and branching caudex, the stems erect or ascending, sparingly white-hairy below, becoming glandular-villous above with multicellular hairs, the cross walls purplish; basal leaves 1.5–6 cm long, 0.2–0.9 cm broad, oblanceolate to narrowly elliptical, 1-veined, acute to obtuse apically, glabrous to glabrate, ciliate; cauline leaves 2–3 pairs, reduced upwards, the upper ones often pubescent with multicellular hairs; flowers solitary, or rarely 2–3, nodding in bud, spreading to erect in fruit; pedicels villous with multicellular, glandular hairs, the cross walls purplish; calyx urn-shaped to campanulate, purplish veined and commonly more or less hairy like the pedicels, 10–19 mm long, (5)6–15 mm broad, the teeth 1.5–3 mm long, ciliate; petals pinkish or purplish (or white?), included within the calyx or exserted 1–4 mm, bilobed; capsules erect, 12–18(22) mm long, opening by 5–6 (or 10–12, by

splitting of the primary) teeth; seeds 1.5–2.5 mm long, with a broad, more or less inflated winged margin.

Meadows, rock outcrops, open slopes, lake shores, in tundra and heathlands, and in open woods; in much of Alaska (except for the southeastern portion), and most of the Yukon; eastward to the Atlantic Ocean and southward to Utah and Colorado; circumboreal (*Melandrium apetalum* [L.] Fenzl.; *M. macrospermum* Porsild; *Silene macrosperma* [Porsild] Hultén; *M. soczavaeanum* Schischk.; *L. soczavaeanum* [Schischk.] J. P. Anders.; *Silene wahlbergella* Chawdhuri). The species is represented in our region by ssp. *arcticum* (Fries) Hultén. An entirely glabrous phase is known as var. *glabra* Reg. *L. apetala* is closely related to *L. furcata* with which it apparently intergrades.

Lychnis furcata (Raf.) Fern.
Arctic Lychnis

Plants 0.3–2.5(3.5) dm tall, from a taproot and branching caudex, the stems erect or nearly so, sparingly white-villous below, becoming glandular-villous above with multicellular hairs, the cross walls often purplish; basal leaves 1.5–5.5 cm long, 0.2–0.8 cm broad, oblanceolate to narrowly elliptical, 1-veined, acute to obtuse apically, glabrous to glabrate, ciliate; cauline leaves 2–3 pairs, reduced upwards, the upper ones sometimes pubescent with multicellular hairs; flowers solitary, or sometimes 2–3, erect or spreading in bud, erect in fruit; pedicels villous with multicellular, glandular hairs, the cross walls often purplish; calyx campanulate to urn-shaped, purplish veined, and commonly more or less hairy like the pedicels, 10–17 mm long, (5)6–15 mm broad, the teeth 1.5–3.5 mm long, ciliate; petals white to pinkish, exserted (2)3–7 mm, bilobed; capsules 12–18 mm long, opening by 5–6 (10–12) teeth; seeds 0.9–1.5 mm long, with a narrowly winged margin.

Meadows, rock outcrops, terraces, and

flood plains, in tundra and heathlands; in most of Alaska (except for the southwestern portion), and in the Yukon; eastward to Labrador; circumpolar (*L. affinis* authors, not J. Vahl ex Fries; *Melandrium affine* authors; *L. furcata* ssp. *elatior* [Reg.] Maguire; *M. affine* var. *brachycalyx* [Raup] Hultén).

Lychnis taylorae Robins.
Taylor Lychnis

Plants 2.3–6 dm tall, from a taproot and branching caudex, the stems erect, sparingly to moderately white-hairy or glabrous below, becoming glandular-villous above with multicellular hairs, the cross walls purplish, or rarely glabrous throughout; basal leaves 1.5–8 cm long, 0.2–0.8 cm broad, linear-oblanceolate to oblanceolate, 1-veined, acute to obtuse apically, glabrous above and beneath or pubescent along the vein, ciliate or glabrous marginally; cauline leaves 2–5 pairs, reduced upwards, the upper ones glandular-ciliate to villous, or the surfaces glabrous; flowers solitary or more commonly 2–6, erect in bud and in fruit; pedicels glandular-villous with multicellular hairs, the cross walls purplish or glabrous; calyx campanulate or somewhat constricted apically, purplish veined and more or less hairy like the pedicels, or glabrous, 7–11 mm long, (3)5–8 mm broad, the teeth 1–2.5 mm long, ciliate; petals white, exserted 2–5 mm, bilobed; capsules erect, 10–14 mm long, opening by 5–6(10–12) teeth; seeds 1–1.5 mm long, with a moderate winged margin.

Open woods and rock outcrops; in northern, central, and eastern Alaska and western Yukon; east to Mackenzie; endemic (*L. funstoni* Wight ex Mertie, name only; *Melandrium taylorae* [Robins.] Tolm.; *M. taylorae* var. *glabrum* Hultén; *L. furcata* ssp. *elatior* var. *glabra* [Hultén] Maguire; *L. furcata* ssp. *elatior* [Reg.] Maguire, in part; *Silene taylorae* [Robins.] Hultén). This entity appears to be closely related to *L. furcata,* but its tall stature,

more numerous and smaller flowers, and smaller capsules appear to be definitive.

Lychnis triflora R. Br.

Plants 0.5–3(4) dm tall, from a taproot and branching caudex, the stems erect or ascending, white-hairy and more or less glandular throughout, conspicuously glandular above, the hairs multicellular, with purplish cross walls; basal leaves 1.5–6 cm long, 0.2–0.8 cm broad, narrowly oblanceolate, 1-veined, acute to obtuse apically, commonly hairy on both surfaces, ciliate; cauline leaves 2–4 pairs, reduced upwards, the upper ones glandular-hairy; flowers solitary, or more commonly 3–6, erect in bud and in fruit; pedicels and peduncles glandular-villous with multicellular hairs, the cross walls purplish; calyx cylindrical or slightly constricted apically, purplish veined, more or less hairy like the pedicels, 6–11 mm long, 3–6 mm broad, the teeth 1–3 mm long, ciliate; petals white or pinkish, exserted 2–5 mm, bilobed; capsules 8–11 mm long; opening by 5–6(10–12) teeth; seeds 0.6–1 mm long, slightly or not at all winged.

Open slopes, terraces, woods, and muskegs; in the eastern and northern portions of Alaska and southern Yukon; eastward to Mackenzie; Greenland, Asia. Our materials belong to var. *dawsonii* Robins (*L. triflora* ssp. *dawsonii* [Robins.] Maguire; *L. dawsonii* [Robins.] J. P. Anders.; *Melandrium ostenfeldii* Porsild; *M. taimyrense* Tolm., at least for our material; *Silene sorensenis* [B. Boi.] Bocq.). This entity belongs to a section of *Lychnis* with very confused nomenclature. More work is indicated.

SAGINA L.

Annual or perennial herbs, with prostrate, ascending, or erect stems; leaves opposite, linear, basally connate, apiculate apically, lacking stipules; flowers solitary, or 2–several in cymes; sepals 4–5, distinct, obscurely 1(3)-veined; petals 4–5, usually shorter than the sepals, white, the blades entire to slightly emarginate; stamens 4–5(8–10), inserted at the base of a glandular disk; styles 4–5, alternate with the sepals; ovary 1-loculed; capsule opening by 4–5 valves, these opposite the sepals.

1a. Plants annual, lacking basal rosettes, scarcely if ever with clustered secondary leaves in their axils; known only from coastal, southern Alaska. *S. occidentalis*

1b. Plants biennial or perennial, with more or less well-developed basal rosettes, and commonly with clustered secondary leaves in the leaf axils; distribution various. (2)

2a. Sepals ordinarily purplish margined, 1.3–2.1 mm long; flowers erect in fruit; plants 1–5(10) cm tall; plants of wide distribution in coastal Alaska and Yukon, and less commonly some distance inland. *S. intermedia*

2b. Sepals not purplish margined, if less than 2 mm long, then the flowers commonly reflexed in fruit; plants of southern Alaska and Yukon. (3)

3a. Sepals 1.5–2 mm long; flowers commonly reflexed in fruit. *S. saginoides*

3b. Sepals 2.5–3.2 mm long; flowers seldom reflexed in fruit. *S. crassicaulis*

Sagina crassicaulis Wats.
Beach Pearlwort

Plants biennial or perennial, forming diffuse clumps to 2 dm broad, arising from taproot and branching caudex, the caudex branches prostrate to ascending; flowering stems 0.2–1.1 dm tall, decumbent to erect; basal leaves 0.8–3.5 cm long, less than 0.1 cm broad, linear, 1-veined, apiculate apically, glabrous; cauline leaves 2–6 pairs, usually with clustered, secondary, axillary leaves; flowers erect, solitary and terminal or lateral, or 2–3 and the lower ones lateral; sepals 2.5–3.2 mm long, erect, greenish, with scarious margins, obtuse apically, glabrous or with some stipi-tate-glandular hairs; petals white, 1.5–2.5 mm long, or lacking; capsules 3–5 mm long, 4–5-valved, erect at maturity; seeds 0.3–0.4 mm long, reddish brown.

Moist, sandy or gravelly beaches, sea cliffs, and salt marshes; in coastal, southern Alaska; southward to California; Asia (*S. maxima* Gray var. *crassicaulis* [Wats.] Hara; *S. litoralis* Hultén; *S. crassicaulis* var. *litoralis* [Hultén] Hultén).

Sagina intermedia Fenzl ex Ledeb.
Snow Pearlwort

Plants biennial or perennial (or annual?), forming dense to diffuse clumps to 1(1.5) dm broad, arising from taproots and branching caudex, the caudex branches commonly ascending; flowering stems 0.1–0.5(1) dm tall, decumbent to erect; basal leaves 0.4–1(1.5) cm long, less than 1 mm broad; linear, 1-veined, apiculate apically, glabrous; cauline leaves 1–8 pairs, usually with clustered, secondary, axillary leaves; flowers erect, solitary and terminal or axillary, or 2–3 and the lower ones lateral; sepals 1.3–2.1 mm long, erect, greenish, the margins purplish or rarely merely scarious, obtuse apically, glabrous; petals white, 1–1.5 mm long, or lacking; capsules 1.5–3.5 mm long, 4–5-valved, erect at maturity; seeds 0.3–0.4 mm long, reddish brown.

Moist rock outcrops, beaches, spits, and hillsides; throughout coastal Alaska and northern Yukon, and less commonly inland; eastward to Labrador; Greenland, Eurasia.

Sagina occidentalis Wats.
Western Pearlwort

Plants annual, forming solitary stems or freely branched clumps, arising from a taproot, the stems 0.5–1.5 dm long, erect or ascending; basal leaves not in rosettes, mostly 1–3 cm long, less than 0.1 cm broad, linear, 1-veined, apiculate apically, glabrous; cauline leaves 2–6 pairs, scarcely ever with clustered, secondary leaves in the axils; flowers erect, usually several per stem; sepals 1.5–2.5 mm long, erect, greenish, the margins scarious, glabrous or glandular-hairy; petals white, 1.2–2.6 mm long; capsules 2–5 mm long, 5-valved, erect at maturity; seeds 0.2–0.3 mm long, reddish brown.

Moist soils; in coastal southern Alaska, where it is uncommon and possibly adventive from coastal western Canada and United States.

Sagina saginoides (L.) Britt.
Arctic Pearlwort

Plants biennial to perennial (or rarely annual?), forming compact to more or less diffuse clumps 0.3–1.5(2) dm broad, arising from a taproot and branching caudex, the caudex branches prostrate to ascending; flowering stems 0.2–0.7(1.4) dm tall, decumbent to erect; basal leaves 0.4–1.5 cm long, usually less than 0.1 cm broad, linear, 1-veined, apiculate apically, glabrous, cauline leaves 2–6 pairs, usually with clustered, secondary, axillary leaves; flowers erect, solitary and terminal or lateral, or 2–3 and the lower ones lateral; sepals 1.5–2 mm long, erect, greenish, the margins scarious, glabrous, obtuse to rounded apically; petals white, 1–1.5 mm long; capsules 2.5–3 mm long, 4–5-valved, commonly recurved at maturity; seeds 0.2–0.4 mm long, reddish brown.

Moist soil, mud flats, alpine meadows, open woods, and rock outcrops; in coastal, southern Alaska and for some distance in the interior of southern Alaska and Yukon; eastward to the Atlantic Ocean and south to California, Mexico, Michigan, and Quebec; circumboreal (*S. linnaei* Presl.). This entity seems to approach *S. intermedia*, and when the purplish margins of the sepals are not evident, or when specimens lack mature, recurved fruit, they are difficult to assign to one species or the other.

SILENE L.

Annual or perennial herbs, with prostrate, decumbent, ascending, or erect stems; leaves opposite, lanceolate to lance-oblong or linear, lacking stipules; flowers solitary or few to many in open to contracted cymes; sepals 5, united to near the apex, each 2 (or more)-veined; petals 5, white to pink, red, or purplish, the blade ascending or spreading at right angles to the claw, entire or several-lobed apically, often appendaged at juncture of blade and claw; stamens 10, adnate to the base of the petals; styles commonly 3 (less commonly 4–5); ovary 1-loculed (or incompletely 3–5-loculed), stipitate; capsule opening by 6 (rarely 8–10) teeth.

Hitchcock, C. L., and B. Maguire. 1947. A revision of the North American species of *Silene*. Univ. Wash. Publ. Bio. 13:1–73.

1a. Plants annual, coarse, erect; calyx 13–23 mm long; leaves 0.7–3 cm broad; introduced weedy species. S. *noctiflora*

1b. Plants perennial, slender or mat-forming, decumbent to ascending or erect; calyx commonly 13 mm long or less (except in *S. douglasii*); leaves mostly less than 0.7 cm broad; indigenous species. (2)

2a. Plants forming dense mats; leaves linear to oblong, 3–12 mm long, and less than 2 mm broad; petals pink. S. *acaulis*

2b. Plants forming loose clumps or the stems solitary; leaves narrowly lanceolate to lance-oblong, commonly over 2 mm broad (at least some); petals commonly white. (3)

3a. Calyx bright pink purple; stems erect, or decumbent at the base only. S. *repens*

3b. Calyx green, or scarious with green along the veins, or with purplish teeth; stems sprawling-ascending or decumbent. (4)

4a. Calyx (10)12–15 mm long; petal appendages 1–2(3) mm long; plants of northern British Columbia. S. *douglasii*

4b. Calyx 5–11 mm long; petal appendages 0.3–1 mm long; plants of southern Alaska and Yukon. (5)

5a. Calyx 5–8 mm long; petal appendages
less than 0.5 mm long; seeds smooth,
shiny, black. *S. menziesii*

5b. Calyx 8–11 mm long; petal appendages 0.5–1 mm long; seeds roughened, dull. *S. williamsii*

Silene acaulis L.
Moss Campion

Perennial, dioecious, or less commonly per-
fect-flowered plants, forming dense mats
or cushions, 0.3–4 dm broad or more,
arising from a taproot and branching caudex, the caudex branches prostrate to ascending or erect, clothed with persistent
leaves; flowering stems lacking or to 0.5
dm tall, erect; basal leaves 0.5–3.5 cm
long, to 1.5(2) mm broad, linear to lance-
linear, 1(3)-veined, connate-sheathing basally, acute apically, ciliate with descending
hairs, glabrous on both surfaces; cauline
leaves lacking, or sometimes with 1 (rarely
2) reduced pair near the stem base; flowers
erect, solitary, borne sessile, or on glabrous
pedicels 0.5–5 cm long; calyx cylindrical to
campanulate, 4.5–8.5 mm long, often purplish tinged, glabrous, the teeth 1–2 mm
long, ciliate; petals pink to pink purple, or
rarely white, 8–12 mm long, the blade subentire to emarginate, the appendages 0.5–1
mm long or lacking; capsules 3-locular, 4–
10 mm long; seeds 0.8–1.2 mm long, light
brown.

Arctic and alpine tundra, heathlands, and
less commonly in open woods; in most of
Alaska and Yukon; east to the Atlantic and
south to Arizona, New Mexico. and New
Hampshire; circumpolar. Two weak, and in
our region, apparently completely intergrading varieties are present.

1a. Calyx commonly 4–6(6.5) mm long;
petal blades mostly emarginate; flowering stems usually 1.5 cm long or
less; plants of broad distribution, our
most common phase (ssp. *arctica* Löve

& Löve). *S. acaulis* var. *exscapa* (All.)
DC.

1b. Calyx commonly 7–8.5 mm long; petal
blades entire or nearly so; flowering
stems often 1.5 cm long or more;
plants mostly of mountains of western central and eastern Alaska, and
southern Yukon (ssp. *subacaulescens*
[Williams] Hultén). *S. acaulis* var.
subacaulescens (Williams) Fern. & St.
John

Silene douglasii Hook.
Douglas Campion

Perennial, perfect plants, forming loose
clumps, arising from a branching caudex
and stout taproot; flowering stems 1–4 dm
long or more, decumbent to ascending;
basal leaves smaller than the main cauline
ones, often withered by flowering time;
cauline leaves mostly 2–8 cm long, 0.2–0.7
(1.2) cm broad, oblanceolate to narrowly
lanceolate, 1 (several)-veined, slightly connate basally, acute apically, hairy above
and beneath and rarely somewhat glandular; flowers erect, solitary, or more commonly 2–several in open or contracted
cymes; pedicels 0.3–3.5 cm long, descending-hairy; calyx campanulate, (10)12–15
mm long, greenish or purplish, pubescent
to nearly glabrous, the teeth 1.5–3 mm long,
ciliate; petals white to greenish or pinkish,
12–16 mm long, the blades bilobed, the appendages 1–2(3) mm long; capsules 1-loc-
uled, 8–12 mm long (including stipe),
the seeds 1.1–1.4 mm long, reddish brown,
roughened.

Dry slopes; in northern British Columbia;
disjunctly south to California, Nevada, and
Utah. This entity is evidently rare in northern British Columbia.

Silene menziesii Hook.
Menzies Campion

Perennial, dioecious (or rarely perfect?)
plants, forming loose clumps, arising from
elongate rhizomes; flowering stems 0.5–3
dm long, or more, sprawling to decumbent,

ascending or less commonly erect; basal leaves smaller than the main cauline ones, often withered by flowering time; cauline leaves 1.5–7 cm long, 0.3–2.2 cm broad, lanceolate to elliptic or oblong, 1 (several)-veined, not or only slightly connate basally, acuminate to attenuate or acute apically, glandular-pubescent above and beneath, especially along the veins, glandular-ciliate; flowers erect, solitary, or more commonly 2-several in open cymes; pedicels 0.4–3 cm long, glandular-hairy; calyx campanulate, 5–8 mm long, purplish or greenish, pubescent and glandular, the teeth 1.5–2.5 mm long, ciliate; petals white, 6–10 mm long, the blades bilobed, the appendages 0.4 mm long or less; capsules 1-loculed, 5–8 mm long (including stipe); seeds 0.7–0.9 mm long, black, shiny.

Open woods and meadows; in coastal, south-central and southeastern Alaska, and in southern Yukon; south to California and New Mexico. Two varieties are present in our region.

1a. Lower internodes pubescent with 3–7-celled, nonglandular hairs. *S. menzeisii* var. *menzeisii*
1b. Lower internodes pubescent with short glandular hairs. *S. menziesii* var. *viscosa* (Greene) A. S. Hitchc. & Maguire

Silene noctiflora L.
Night-flowering Catchfly

Annual, perfect or occasionally dioecious plants, arising from taproots, the stems 2–8 dm tall, erect; leaves 2–10 cm long, 0.6–2.5(4) cm broad, lanceolate to elliptic or the lower ones oblanceolate, several-veined, slightly connate basally, acute to attenuate or obtuse apically, hairy above and beneath, ciliate; flowers erect, solitary or few- to several-veined, slightly connate basally, acute to attenuate or obtuse apically, hairy above and beneath, ciliate; flowers erect, solitary or few to several in open or contracted cymes; pedicels 0.3–1 cm long, glandular-hairy; calyx tubular, 12–25

mm long, greenish or pinkish, or scarious between the veins, glandular-hairy, the teeth 4–9 mm long, ciliate; petals white to pinkish, 20–35 mm long, the blades bilobed, the appendages 0.5–1.5 mm long; capsule 3-locular, 1.5–2.5 cm long (including stipe); seeds 0.8–1.2 mm long, brown, roughened.

Weedy species of disturbed soils; in southern Alaska (to be expected elsewhere); widely distributed in North America; introduced from Eurasia.

Silene repens Pers.
Pink Campion

Plants perfect, perennial, forming clumps, arising from elongate rhizomes; flowering stems 0.7–3.5 dm tall, sparingly or not at all branched or branching only basally; basal leaves withered or lacking at flowering time; cauline leaves 1.5–6 cm long, 0.2–0.8 cm broad, lance-linear, linear, or narrowly oblong, 1-veined, not basally connate, attenuate apically, pubescent above and beneath, especially along the veins, ciliate; flowers erect or ascending, commonly 2–15, in a contracted, paniculate cyme; pedicels 0.1–1 cm long, white-villous; calyx tubular, 10–14 mm long, bright pink purple or seldom greenish, villous with whitish, multicellular hairs, the teeth 1–2.5 mm long, ciliate; petals white to pinkish, 15–20 mm long, the blades bilobed, the appendages 0.7–1.2 mm long; capsules 3(4)-locular, 10–14 mm long (including stipe); seeds 0.8–1 mm long, brown, roughened.

Dry, grassy slopes, open woods, rock outcrops, and meadows; in interior, central, northern, and eastern Alaska and most of the Yukon; eastward to Mackenzie, southward to British Columbia, and disjunctly south in Idaho, Wyoming, and Montana. Our materials belong to ssp. *repens* (ssp. *purpurata* [Greene] C. L. Hitchc. & Maguire). Occasional specimens lack the distinctive coloration of the calyx and might be confused with *S. williamsii*. However, the longer petals, little-branched stem, and more or less congested inflorescence appear definitive for S. *repens*.

Silene williamsii Britt.
Williams Campion

Plants perfect, perennial, forming loose clumps, arising from elongate rhizomes; flowering stems 0.8–4(4.5) dm tall, much branched, sprawling to ascending or erect; basal leaves withered or lacking at flowering time; cauline leaves 1.5–6 cm long, 0.2–1.3 cm broad, lance-attenuate to lance-linear, 1-veined, not basally connate, attenuate apically, pubescent and often glandular above and beneath, ciliate; flowers erect to spreading, commonly 3–many, in open leafy cymes; pedicels 0.3–1.5 (2.5) cm long, glandular-villous; calyx tubular, 8–11 mm long, greenish to scarious or seldom purplish apically, the teeth 1.5–3 mm long, ciliate; petals white, 9–13 mm long, the blades bilobed, the appendages 0.5–1 mm long; capsules 8–10 mm long (including a stipe about 1 mm long); seeds 0.7–1.2 mm long, reddish brown, roughened.

Open woods, grassy slopes, roadsides, airstrips, and rock outcrops; in central and eastern Alaska and western Yukon; endemic. This species is apparently most closely related to S. *menziesii*, which it grossly resembles and with which it is sympatric in southern Yukon. The larger flowers, commonly smaller leaves, and reddish brown, roughened seeds appear to segregate specimens of S. *williamsii* (S. *menziesii* ssp. *williamsii* [Britt.] Hultén).

SPERGULA L.

Annual herbs, with prostrate to ascending or erect stems; leaves whorled, linear, not connate basally, acute apically, stipulate; flowers few to many, borne in open terminal cymes; sepals 5, distinct, obscurely 1-veined; petals 5, white, shorter to longer than the sepals, the blades entire; stamens 10 (or sometimes only 5); styles usually 5, alternate with the sepals; ovary 1-loculed; capsule opening by 5 valves, these opposite the sepals.

Spergula arvensis L.
Spurry

Plants annual, forming diffuse clumps to 10 dm broad or more, arising from taproots; stems 0.8–8 dm long, prostrate to ascending or erect; leaves whorled, 1–5 cm long, less than 1 mm broad, linear, 1-veined, acute apically, glandular-hairy to glabrate; stipules membranous; flowers erect, spreading or descending, few to many in terminal cymes (sometimes some axillary); sepals 2–6 mm long, erect, greenish with scarious margins, obtuse to acute apically, glandular-hairy; petals white, 2–5 mm long; capsules 2.5–6 mm long; seeds 1.2–1.5 mm long, black, minutely roughened.

Weeds of disturbed soils; in southern Alaska; adventive from Europe.

SPERGULARIA (Pers.) J. & C. Presl Nom. Cons.

Annual (or biennial?) herbs, with prostrate, decumbent, ascending, or erect stems; leaves opposite (or clustered), linear, not connate basally, acute to apiculate apically; stipules membranous, connate-sheathing; flowers few to many, borne in open, leafy, terminal cymes; sepals 5, distinct, obscurely 1-veined; petals 5, white to pinkish, shorter than the sepals, the blades entire; stamens 2–10; styles 3; ovary 1-loculed; capsules opening by 3 valves.

1a. Leaves glabrous or nearly so, lacking clustered, axillary leaves or seldom with them; stamens 2–5; seeds usually conspicuously wing-margined; our most common sand-spurry. S. *canadensis*

1b. Leaves more or less hairy, usually with clustered, axillary leaves; stamens usually 10 (rarely 6–9); seeds not wing-margined; rare. S. *rubra*

Spergularia canadensis (Pers.) G. Don.
Canada Sand-spurry

Plants annual, forming small clumps, arising from taproots, the stems 0.4–3 dm long, erect or ascending; leaves 1–4.5 cm long, mostly 0.1–0.2 cm broad or less, linear, 1-veined, acute apically, glabrous or nearly so, usually lacking clustered, secondary axillary leaves; flowers erect, spreading or reflexed, few to many, in leafy cymes; pedicels sparingly to moderately glandular-hairy, or glabrous; sepals 2–4 mm long, greenish, with scarious margins, obtuse apically, glabrous or glandular-hairy; petals white to pinkish, 1–3 mm long; stamens 2–5; capsules 3–5 mm long, 3-valved; seeds 0.8–1.2 mm long, light brown, usually with a distinctive membranous wing.

Sea beaches and saline marshes; in coastal southern Alaska, from Kodiak Island eastward; southward to California and from Newfoundland to New York (*S. marina* authors, not [L.] Griseb.). Our material belongs to var. *canadensis*.

Spergularia rubra (L.) J. & C. Presl
Purple Sand-spurry

Plants annual (rarely biennial?), forming small to large clumps, arising from taproots, the stems 0.5–3 dm long, prostrate to ascending; leaves 0.5–2 cm long, usually less than 0.1 cm broad, linear, 1-veined, mucronate apically, more or less glandular-hairy, usually with clustered secondary, axillary leaves; flowers erect, spreading or deflexed, few to many in leafy cymes; pedicels glandular-hairy; sepals 3–4.5 mm long, greenish, with scarious margins, obtuse apically, glandular-hairy; petals pinkish, 2–3.5 mm long; stamens (6–9)10; capsules 3–5 mm long, 3-valved; seeds 0.4–0.6 mm long, dark brown, not winged.

Weedy species of disturbed sites; apparently introduced in coastal southern Alaska, and in the interior (Fairbanks), to be expected elsewhere; adventive from Europe.

STELLARIA L.

Annual or perennial herbs, with prostrate to decumbent, ascending, or erect stems; leaves opposite (rarely whorled), linear-lanceolate to ovate, lacking stipules; flowers solitary, in the leaf axils, or few to many and borne in axillary or terminal cymes; sepals (4)5, distinct, obscurely 1–3-veined; petals (4)5, white, deeply to shallowly bilobed, sometimes reduced or lacking; stamens usually 10, inserted at the base of the ovary; styles usually 3 (rarely 4–5); ovary 1-loculed; capsule opening by 6 (rarely by 8 or 10) teeth.

Fernald, M. L. 1940. *Stellaria calycantha*. Rhodora 42:254–59.

Hultén, E. 1943. *Stellaria longipes* Goldie and its Allies. Bot. Not. 1943:251–70.

Porsild, A. E. 1963. *Stellaria longipes* Goldie and its Allies in North America. Nat. Mus. Can. Bul. 186:1–35.

1a. Lower leaves distinctly petiolate; pubescence of stems, pedicels, and petioles in longitudinal lines; introduced annual weed of disturbed soil. *S. media*

1b. Lower leaves sessile or shortly petiolate; pubescence of stems, pedicels and petioles uniform or lacking, not in longitudinal lines; plants indigenous, broadly distributed. (2)

2a. Stems distinctly villous; sepals often ciliate marginally with numerous, short hairs; plants from interior, western and northern Alaska and Yukon. *S. longipes*

2b. Stems glabrous or only sparingly villous; sepals not ciliate, or if so, then the hairs long, few, and near the base only. (3)

3a. Leaves 3–8 mm long, 1–2 mm broad; stems often 0.4 mm broad or less; sepals 2.5–4 mm long. *S. crassifolia*

3b. Leaves often over 8 mm long or over 2 mm broad or both, or if less (as in some *S. humifusa* and *S. umbellata*), then the sepals mostly 4–5 mm long and the plants drying brownish or the flowers in axils of scarious bracts. (4)

4a. Leaves linear or narrowly oblong, minutely toothed marginally (use at least 30×); flowers few to many, borne in open, scarious-bracted cymes. *S. longifolia*

4b. Leaves lance-linear to lance-attenuate, lanceolate, lance-ovate, or oblong, minutely toothed or entire marginally; flowers solitary or few to several in leafy or scarious-bracted cymes. (5)

5a. Leaves lance-ovate to lanceolate, entire, shortly petiolate; flowers solitary in leaf axils; plants of coastal southern Alaska. *S. crispa*

5b. Leaves lance-attenuate to oblong, or less commonly lanceolate, sessile, or if shortly petiolate, then the margin minutely toothed. (6)

6a. Leaf margin minutely toothed (use a strong lens), and often ciliate with multicellular hairs; leaves commonly 3–9 mm broad; flowers axillary, or in leafy bracted cymes. *S. calycantha*

6b. Leaf margin entire; leaves 1–3 mm broad, or if 3–5 mm broad, then the flowers commonly solitary. (7)

7a. Leaves lance-attenuate, commonly 5–10 times longer than broad or more; flowers solitary, or 2–5 in open cymes. *S. longipes*

7b. Leaves lanceolate, commonly less than 4 times longer than broad; flowers solitary or rarely 2–3. (8)

8a. Sepals 2.5–3 mm long; flowers (1) 2–3, borne in subumbellate clusters, or the uppermost pedicel subtended by minute bracts; plants of broad distribution. *S. umbellata*

8b. Sepals 4–9 mm long; flowers solitary in leaf axils, or rarely 2; plants of various distribution. (9)

9a. Flowers borne in the axils of scarious bracts; sepals (6)7–9 mm long; plants of the southeastern quarter of Alaska. *S. alaskana*

9b. Flowers borne in the axils of foliage leaves; sepals mostly (3.5)4–6 mm long. (10)

10a. Stems elongate, with 1–several flowers along its length; leaves fleshy, not leathery or shining; plants broadly distributed. *S. humifusa*

10b. Stems more or less contracted, commonly 1-flowered; leaves shining, leathery, not or only somewhat fleshy; plants of the Aleutians, or rarely elsewhere. *S. ruscifolia*

Stellaria alaskana Hultén
Alaska Starwort

Perennials from elongate rhizomes, forming small to moderate clumps; flowering stems 0.3–1 dm tall, glabrous; leaves (0.5) 0.8–1.9 cm long, 0.2–0.7 mm broad, lanceolate to elliptic, acute to acuminate or attenuate apically, sessile, glabrous, entire; flowers solitary, or less commonly 2, borne in the axils of scarious bracts; pedicels 1–20 mm long, glabrous; sepals (6.5)7–9 mm long, 3-veined, scarious-margined, not ciliate; petals white, 3–6 mm long; capsules straw-colored, 6–8 mm long, opening by 6 teeth; seeds 0.8–1.2 mm long, light brown, roughened.

Rock outcrops, talus slopes, and moraines in alpine tundra; in interior (vicinity of McKinley National Park and eastward) and southeastern Alaska; endemic.

Stellaria calycantha (Ledeb.) Bong.

Perennials from elongate rhizomes, forming moderate to large clumps; flowering stems 0.6–5 dm long or more, glabrous, scabrous, or sparsely short-villous; leaves 0.8–5 cm long, 0.2–0.9 cm broad, lanceolate to lance-attenuate, narrowly oblong or lance-linear, attenuate, acute, or acuminate apically, sessile or rarely shortly petiolate, glabrous, minutely serrulate and often long-ciliate marginally; flowers solitary and axillary, or more commonly with open, terminal cymes, the bracts scarious-margined to green throughout; pedicels mostly 4–4.5

mm long, glabrous or scabrous; sepals (1.5)2–4.5 mm long, obscurely 1–3-veined, scarious margined, not ciliate; petals white, 1–4 mm long, or lacking; capsules straw-colored to purplish, 3.5–6 mm long, opening by 6 teeth; seeds 0.6–1 mm long, reddish brown, smooth to slightly roughened.

Wet meadows, river banks, thickets, open woods, and roadsides; in most of Alaska from the Brooks Range southward and in southern Yukon; eastward to the Atlantic and southward to California, Utah, Wyoming, Michigan, and New York; Eurasia. The complex of forms segregated herein as varieties have been treated as two or more species or subspecies by previous authors. The entities have been segregated on the basis of flower size, size of leaves, degree of branching of the inflorescence, and on whether the stem is scabrous or not. Several trends are recognizable, and these trends form the basis of the varieties herein proposed, because each entity passes almost completely into all of the others. The following arbitrary key will serve to characterize most specimens.

1a. Sepals 1.5–2.7 mm long; inflorescence open, flowers several to many; plants of interior Alaska and Yukon, or less commonly in coastal, western and southern Alaska (S. calycantha ssp. interior Hultén; S. calycantha var. isolepis Fern., in part). S. calycantha var. calycantha

1b. Sepals (2)2.5–4.5 mm long; inflorescence various; plants of coastal and insular southern Alaska. (2)

2a. Inflorescence open, several- to many-flowered; stems scabrous or glabrous; plants intergrading with both var. calycantha and var. bongardiana. S. calycantha var. sitchana (Steud.) Fern.

2b. Inflorescence compact, 1–few-flowered; stems scabrous or rarely glabrous (S. borealis Bigel. var. bongardiana Fern.). S. calycantha var. bongardiana (Fern.) Fern.

Stellaria crassifolia Ehrh.
Fleshy Starwort

Perennials from slender rhizomes, forming small to large mats or clumps; flowering stems 0.3–3 dm long, glabrous; leaves 0.2–0.8(1) cm long, to 2 mm broad, lance-attenuate to lance-linear, acute to attenuate apically, sessile or nearly so, glabrous, entire; flowers solitary, or few to several in open cymes, the bracts green throughout or with very narrow, scarious margins; pedicels commonly 3–40 mm long, glabrous; sepals 2.5–4 mm long, 3-veined, scarious-margined, not ciliate; petals white, 2.5–5 mm long (or more?); capsule straw-colored, 4–5 mm long, opening by 6 teeth; seeds 0.7–1 mm long, reddish brown, roughened.

Moist soils, in muskegs, open woods, and on lake shores; over much of Alaska (except for the Aleutian Islands) and Yukon; eastward to Newfoundland and south to Idaho and Colorado; Eurasia.

Stellaria crispa Cham. & Schlecht.
Crisp Starwort

Perennials from slender rhizomes, forming small to large mats; flowering stems 1–4.5 dm tall, glabrous; leaves 0.4–2.6 cm long, 0.2–1.5 cm broad, lanceolate to ovate, acuminate to attenuate apically, shortly petiolate to subsessile, glabrous, entire; flowers solitary in leaf axils; pedicels mostly 5–30 mm long, glabrous; sepals 3–4.5 mm long, obscurely 3-veined, scarious-margined, not ciliate; petals usually lacking; capsules straw-colored or brownish, 3.5–6 mm long, opening by 6 teeth; seeds 0.7–1 mm long, brown, roughened.

Wet soil in woods, on stream banks, and on beaches; in coastal and insular southern Alaska, and rarely in southern Yukon, southward to California and Montana.

Stellaria humifusa Rottb.
Low Chickweed

Perennials from slender rhizomes, forming

small to large mats or clumps, the stems 0.2–3.8 dm long, glabrous; leaves 0.4–1.5 cm long, 0.1–0.5 cm broad, lanceolate to elliptic, acute to obtuse apically, sessile or nearly so, glabrous or with a few cilia along the margins, entire, fleshy (wrinkled and often brownish when dried); flowers solitary in the axils of foliage leaves; pedicels commonly 0.5–3 cm long, glabrous; sepals 4–5 mm long, 3-veined, scarious-margined; petals white, 4–6 mm long, capsules 4–5 mm long, straw-colored, opening by 6 teeth; seeds 0.8–1 mm long, brown, slightly roughened.

Spits, bars, lake shores, beaches, and marshes; throughout coastal Alaska and Yukon; eastward to Labrador and south to Oregon. This entity has been confused with S. *crassifolia*, but has fleshy leaves that wrinkle and tend to turn brownish when dried, and the sepals are ordinarily longer.

Stellaria longifolia Muhl. ex Willd.
Long-leaved Starwort

Perennials from elongate rhizomes, forming loose clumps; flowering stems 1–3.5 dm long, glabrous or scabrous; leaves 0.8–4 cm long, 0.1–0.3 cm broad, linear to narrowly elliptic, attenuate to acute apically, sessile, glabrous or sparingly ciliate, minutely serrulate marginally; flowers few to many in open cymes, the bracts scarious throughout; pedicels commonly 3–3.5 mm long, glabrous or scabrous; sepals 2–3 mm long, obscurely 3-veined, scarious-margined, not ciliate; petals white, 2–3.5 mm long; capsules 3–6 mm long, greenish or yellowish, opening by 6 teeth; seeds 0.7–0.8 mm long, brown, slightly roughened.

Wet meadows, marshes, muskegs, and roadsides; in the southeastern quarter of Alaska and southern Yukon; eastward to Newfoundland and southward to California, New Mexico, and South Carolina; Eurasia. S. *longifolia* approaches S. *calycantha* in many of its technical characteristics, but differs in the very narrow leaves and long-branched, more open cymes.

Stellaria longipes Goldie
Long-stalked Starwort

Perennials from slender rhizomes, forming small to large clumps or mats; flowering stems 0.3–3.2 dm long, villous, glabrate, or glabrous; leaves 0.4–2.6(4) cm long, 0.1–0.4 cm broad, lanceolate to lance-attenuate, sessile, sparingly villous, especially along the margin near the base, or glabrous or glabrate, entire, smooth and shiny; flowers solitary, or few to several in open cymes, the bracts scarious, scarious-margined, or green throughout, ciliate or not; sepals 3–5 mm long, 3-veined, scarious-margined, glabrous or hairy dorsally, marginally ciliate or not; petals white, 3–8 mm long; capsules 4–6 mm long, often purplish, opening by 6 teeth; seeds 0.6–0.9 mm long, brown, slightly roughened.

Stream banks, lake shores, open woods, and sandy slopes; in most of Alaska and Yukon; eastward to Newfoundland and south to California, Arizona, New Mexico, Minnesota, and New York; Eurasia. The *longipes* complex has been treated as consisting of a series of morphologically variable entities, often recognized at specific level. The primary diagnostic criteria include pubescence of sepals, leaves, and stems, presence or absence of scarious bracts in the inflorescence, and number of flowers. An examination of numerous specimens of the complex from our region reveals that every possible combination of pubescence position with bract type and with flower number can be demonstrated. Most specimens which lack scarious bracts are 1-flowered. In the case of other species wherein cymose inflorescences are reduced to a single flower, the subtending, primary bracts are foliose. Only when secondary branches are developed do typical, scarious (or only scarious-margined) bracts develop. Thus, it has been possible to designate taxa within the complex only arbitrarily and a conservative treatment is indicated. The following key will allow one to recognize those races which occur most frequently.

1a. Sepals pubescent dorsally, or ciliate, or both. (2)
1b. Sepals glabrous, not ciliate. (3)
2a. Sepals ciliate, not pubescent dorsally (*S. ciliatosepala* Trautv.; *S. ciliatosepala* var. *arctica* [Schischk.] Hultén; *S. edwardsii* R. Br.). *S. longipes* var. *edwardsii* (R. Br.) Gray
2b. Sepals ciliate and pubescent dorsally (*S. laxmannii* authors, not Fisch.). *S. longipes* var. *laeta* (Richards.) Wats.
3a. Flowers subtended by scarious or scarious-margined bracts (at least some), commonly with more than 1 flower per stem (*S. stricta* Richards.; *S. longipes* var. *stricta* [Richards.] Rydb.; *S. vestita* Greene; *S. longipes* var. *vestita* [Greene] Polunin). *S. longipes* var. *longipes*
3b. Flowers not subtended by scarious bracts, arising in the axils of ordinary foliage leaves or green bracts, commonly with only 1 flower per stem (*S. monantha* Hultén; *S. monantha* var. *altocaulis* Hultén). *S. longipes* var. *altocaulis* (Hultén) C. L. Hitchc.

Stellaria media (L.) Cyrill.

Common Chickweed

Annuals or biennials (?), from taproots, forming loose mats or clumps, the stems mostly 1–5 dm long, decumbent and rooting at the nodes, pubescent with multicellular hairs in longitudinal lines; leaves 1–5(6) cm long, 0.5–2.5(3) cm broad, the blades ovate to elliptic or broadly lanceolate, abruptly acuminate, glabrous above and beneath, ciliate, at least the lower ones with distinctive petioles to 2 cm long, the petioles pubescent; flowers axillary, or commonly few to several in short, leafy bracted cymes; pedicels 2–60 mm long, hairy like the stems; sepals 3–6 mm long, 3-veined, scarious-margined, ciliate basally, hairy dorsally; petals white, 2.5–6 mm long; capsules 4–8 mm long, straw-colored or greenish, opening by 6 valves; seeds 0.8–1.2 mm long, brown, uniformly warty.

Weed of disturbed soil, usually associated with activities of man; mostly in southern Alaska and southern Yukon; adventive from Eurasia. This is one of the most aggressive weeds of cultivated land in Alaska.

Stellaria ruscifolia Pallas ex Schlect.

Perennials from elongate rhizomes, forming small to moderate clumps; flowering stems 0.3–1.2 dm tall, glabrous; leaves 0.4–1.5 cm long, 0.2–0.6 cm broad, ovate-lanceolate to lanceolate, acuminate to acute apically, sessile, glabrous, entire; flowers solitary in axils of foliage leaves, the pedicels 5–40 mm long, glabrous; sepals 4–6 mm long, obscurely 3-veined, scarious-margined, not ciliate; petals white, 5–7 mm long; capsules 4–6 mm long, straw-colored, opening by 6 teeth; seeds 0.8–1.2 mm long, brown, roughened.

Gravelly sites; in the Aleutian Islands and in insular southeastern Alaska (Hidden Glacier); Asia. This material is designated as ssp. *aleutica* Hultén. It approaches, if not passes into, *S. alaskana*. The more coastal range, slightly smaller flowers, and apparent lack of scarious bracts seem to be the most important distinguishing features. Some plants from the north of the Brooks Range, herein treated as *S. longipes* var. *altocaulis* (e. g. Lake Peters, Spetzman 801, ISC), approach *S. ruscifolia* in some characteristics (notably in short, thick, coriaceous leaves). The complex is in need of additional study.

Stellaria umbellata Turcz.

Perennials from slender rhizomes, forming small clumps or mats; flowering stems 0.5–1 dm long, glabrous; leaves 0.3–0.9 cm long, 0.1–0.3 cm broad, elliptic to lanceolate, acute apically, sessile or nearly so, glabrous, entire; flowers (1)2–3, subumbellate in the axils of scarious bracts, or the uppermost pedicel subtended by minute, paired, scarious bracts; pedicels 7–20 mm long, glabrous; sepals 2.5–3 mm long, obscurely 3-veined, scarious-margined, not ciliate; petals lacking; capsules 3–4.5 mm long, straw-colored, opening by 6–8 teeth; seeds 0.5–0.7 mm long, brownish, roughened.

Alpine tundra; disjunct in central eastern to northwestern Alaska; Asia.

VACCARIA Medic.

Annual herbs, with erect or ascending stems; leaves opposite, lanceolate to oblong, lacking stipules; flowers few to numerous in open cymes; sepals 5, united nearly to the apex; calyx tube 10-veined and sharply angled; petals 5, pinkish, the blade retuse; stamens 10; styles 2 (rarely 3); ovary 1-loculed; capsules opening by 4 teeth.

Vaccaria pyramidatus Medic.
Cow-herb

Plants annual, from taproots, the stems 1.5–8 dm tall, glabrous; leaves 3–8 cm long, 0.5–2.5 cm broad, lanceolate to oblong, 1–3-veined, attenuate to acute apically, connate-clasping and sessile to short-petiolate basally, glabrous; flowers few to many, erect or spreading; pedicels commonly 5–60 mm long, glabrous; calyx tubular, 11–15 mm long in flower, purplish, glabrous, inflated and strongly 5-angled in fruit, the teeth 1–2 mm long; petals pink, the blades 5–8 mm long, retuse; seeds 1.7–2.1 mm long, reddish black.

Weedy species of disturbed soils; in southern Alaska and southern Yukon; adventive from Europe (*Saponaria vaccaria* L.; *S. segetalis* Neck.; *V. segetalis* [Neck.] Garcke ex Asch.).

CERATOPHYLLACEAE

Hornwort Family

Plants herbaceous, submersed aquatics; leaves whorled, palmately dissected; flowers imperfect, solitary and sessile in leaf axils, each subtended by a calyxlike involucre of 8–14 bracts, the perianth lacking; stamens 12–16; pistil 1, the superior ovary with a single carpel; fruit an achene.

CERATOPHYLLUM L.

A single genus with characteristics of the family.

Ceratophyllum demersum L.
Hornwort

Stems elongate, to 10 dm long or more, freely branched and forming large masses; leaves whorled, 5–12 per node, each 2–3 times palmately divided, 1–3 cm long, the segments linear and flat, antrorsely toothed; achenes 4–5 mm long, ellipsoidal, provided with 2 basal spines.

Ponds; in interior eastern and western Alaska and northern Yukon and adjacent Mackenzie; widespread in North America; Eurasia.

CHENOPODIACEAE

Goosefoot Family

Herbaceous plants; leaves simple, alternate or opposite, often scurfy (mealy) with collapsed, globular hairs; flowers inconspicuous, monoecious, dioecious, polygamous, or perfect, in axillary or terminal cymes or panicles; sepals 1–5; corolla lacking; stamens 1–5, opposite the calyx lobes, or fewer; pistils 1, the ovary superior, 1-loculed, 1-ovuled; styles 2–5; fruit a utricle.

1a. Leaves scalelike, opposite; stems fleshy, jointed; branches opposite; plants of saline soils. *Salicornia*

1b. Leaves not scalelike, at least the upper ones alternate; stems not fleshy and jointed; branches usually alternate; plants of various habitats. (2)

2a. Plants densely long-hairy, subshrubs, perennial; known from southwestern Yukon. *Eurotia*

2b. Plants glabrous or variously hairy, herbaceous, annual; distribution various. (3)

3a. Leaves linear or subulate, often somewhat fleshy and almost terete. (4)

3b. Leaves with well-developed blades. (5)

4a. Leaves fleshy, often nearly terete;

perianth lobes 5; plants of saline soils. *Suaeda*

4b. Leaves subulate, neither fleshy nor terete; perianth lobes 1; plants of sandy habitats (sometimes saline). *Corispermum*

5a. Perianth of a single bractlike segment, not enclosing the fruit; stamens 1; plants ordinarily less than 20 cm tall; leaves elliptic to oblong, usually with 1 pair of lateral lobes near the middle. *Monolepis*

5b. Perianth 3–5-lobed, at least partially enclosing the fruit, or the fruit enclosed by connate subtending bracts; stamens 3–5; plants various, but seldom as above. (6)

6a. Flowers perfect; perianth regularly 5-lobed; fruit not enclosed by connate bracts. *Chenopodium*

6b. Flowers imperfect, the pistillate mostly without a perianth and subtended by connate bracts, the staminate with a 3–5-lobed perianth. (7)

7a. Stigmas commonly 2–3; leaves often scurfy, variously shaped, toothed, lobed, or entire; plants indigenous or introduced. *Atriplex*

7b. Stigmas commonly 4–5; leaves glabrous, triangular-ovate, sometimes hastately lobed; cultivated and escaping. Spinach. *Spinacia oleracea* L.

ATRIPLEX L.

Annual, monoecious (dioecious) herbs; leaves simple, entire or toothed, often hastately lobed, alternate or the lower ones opposite, more or less mealy or glabrous; flowers imperfect, inconspicuous, borne in spikes or panicles, the staminate with 3–5-lobed perianth and 3–5 stamens, the pistillate lacking perianth (or rarely with one), but subtended and enclosed by 2 basally connate bracts with entire or toothed margins; stigmas usually 2; pericarp membranous, the seed usually erect.

1a. Leaf blades of at least the main leaves mostly over 4 cm broad; fruiting bracts about as broad as long, ovate to orbicular; plants often more than 5 dm tall; adventive, weedy species of interior sites. *A. hortensis*

1b. Leaf blades usually less than 4 cm broad (rarely more); fruiting bracts longer than broad, lanceolate, lance-attenuate, or triangular; plants usually less than 5 dm tall; indigenous to coastal, western and southern Alaska, or adventive. *A. patula*

Atriplex hortensis L.
Garden Orache

Plants monoecious, annual, mostly 5–10 dm tall or more, the stems erect or decumbent, simple or branched, mealy to glabrous; leaves petiolate, the lower ones often opposite, the upper alternate, the blades 5–18 cm long, 2–15 cm broad, triangular-ovate to lanceolate, rounded to obtuse apically, cuneate to cordate or somewhat hastate basally, entire or irregularly toothed, more or less mealy when young, glabrous in age; flowers borne on spikes arranged in terminal and axillary panicles; staminate flowers with a 3–5-lobed perianth and horizontal seed, the majority with 2 ovate to orbicular, basally connate bracts mostly 6–10 mm broad at maturity, the margins entire or toothed, the enclosed fruit and seed erect.

Adventive, weedy species; interior Alaska (to be expected elsewhere); native to Asia. Garden orache has been grown for use as a potherb. Our specimens might represent plants which have escaped from cultivation.

Atriplex patula L.
Spearscale

Plants monoecious (dioecious), annual, mostly 0.5–5(8) dm tall, the stems decumbent to erect, simple or more commonly freely branched, more or less mealy, or glabrous; leaves petiolate to subsessile or sessile (?), the lower opposite, the upper alternate, the blades 1–13 cm long, 0.2–5 cm broad, lanceolate to lance-linear, oblong, or linear, the apex acute to obtuse or rounded, the base cuneate to obtuse and sometimes hastately lobed, entire or toothed; flowers in terminal or axillary spikes or panicles; staminate flowers with 5-lobed perianth; pistillate flowers without a perianth, enclosed by 2 lanceolate, lance-attenuate, or triangular, basally connate bracts mostly 3–20 mm long and 2–7 mm broad at maturity, the margins entire or toothed near the base.

Usually in saline soils; coastal western and southern Alaska, or less commonly some distance inland; broadly distributed in North America; Eurasia. Four more or less distinctive but intergrading varieties are present in our region. It appears that at least some portion of the variation is due to differential response of a series of genotypes to different habitats. Plants collected from below high tide level appear to be the basis of *A. drymarioides* Standl. They grade within a few feet into typical var. *obtusa*.

1a. Leaves linear, usually less than 4 mm broad; pistillate bracts linear to lance-linear. *A. patula* var. *zosteraefolia* (Hook.) C. L. Hitchc.
1b. Leaves lanceolate to oblong, seldom less than 4 mm broad; pistillate bracts lanceolate to lance-attenuate or triangular. (2)

2a. Leaf blades triangular-hastate (at least some); pistillate bracts usually with toothed margins and tuberculate projections on the back; adventive (?) somewhat weedy plants of disturbed sites. *A. patula* var. *patula*
2b. Leaf blades lanceolate to oblong, not at all or only some of them hastately lobed; pistillate bracts entire or toothed only near the base, not or only slightly tuberculate on the back. (3)

3a. Leaf blades 0.2–2(3) cm broad, lanceolate to linear-lanceolate, or oblong; pistillate bracts not or only slightly thickened basally; plants of

western and southern Alaska (*A. gmelinii* C. A. Mey.; *A. drymarioides* Standl.). *A. patula* var. *obtusa* (Cham.) C. L. Hitchc.
3b. Leaf blades mostly (2)3–5 cm broad, broadly lanceolate to lanceolate; pistillate bracts somewhat basally thickened at maturity; plants of southern Alaska (*A. alaskensis* Wats.). *A. patula* var. *alaskensis* (Wats.) Welsh

CHENOPODIUM L.

Annual, perfect herbs; leaves simple, entire or toothed, often hastately lobed, alternate, mealy or glabrous; flowers perfect, inconspicuous, borne in terminal and axillary spikes, panicles, or cymes; perianth usually 5-lobed, the lobes enclosing the fruit, sometimes becoming fleshy at maturity; stamens 5 or fewer; stigmas 2–3; pericarp membranous, more or less adherent to the horizontal or less commonly erect seed.

Aellen, P., and T. Just. 1943. Key and synopsis of the American species of the genus *Chenopodium* L. Am. Midl. Nat. 30:47–76.

1a. Seeds commonly erect in the ovary (see also *C. glaucum*); leaves hastate to sagittate basally (less commonly merely cuneate), not whitish or grayish mealy beneath. (2)
1b. Seeds commonly horizontal in the ovary; leaves various but sometimes hastate basally, often grayish or whitish mealy beneath. (3)

2a. Calyx not fleshy or red at maturity; flower clusters mostly less than 4 mm broad; plants of southern Yukon. *C. rubrum*
2b. Calyx fleshy and reddish at maturity; flower clusters commonly more than 4 mm broad; plants of broad distribution. *C. capitatum*

3a. Stems decumbent to prostrate; leaves green above, whitish mealy beneath, dentate to subentire; seeds erect in at least some fruits; plants known from central eastern Alaska. *C. glaucum*

3b.   Stems usually erect; leaves various; seeds horizontal; plants of various distribution. (4)

4a.   Leaves commonly bright green above and beneath, slightly or not at all mealy, coarsely dentate, the apices acuminate to attenuate; plants known from the vicinity of Dawson City, Yukon. *C. hybridum*

4b.   Leaves commonly grayish mealy, at least beneath, variously toothed to subentire, the apices merely acute to obtuse or rounded. (5)

5a.   Leaves lance-oblong to linear, entire or some of them slightly hastately lobed; plants of southern Yukon. *C. leptophyllum*

5b.   Leaves lanceolate to triangular-lanceolate, at least some of them hastately lobed or toothed; plants broadly distributed in southern Alaska and Yukon. *C. album*

Chenopodium album L.

Lamb's Quarters

Plants (1)2–10 dm tall or more, the stems erect or ascending, simple or branched, sparingly to densely mealy; leaves mostly 2–12 cm long, the blades 0.3–5 cm broad, ovate to triangular or lanceolate, shallowly to deeply toothed, or sometimes entire, cuneate basally, acute to obtuse or rounded apically, sparsely to moderately grayish mealy, at least beneath, the petioles mostly 0.3–5 cm long; flowers clustered in spikes borne in terminal and usually also lateral panicles; perianth lobes free to below the middle, mealy, becoming keeled and enclosing the fruit; pericarp adherent to the horizontal seed; fruit 1–1.5 mm broad.

Adventive weedy species of disturbed soils; in southern Alaska and Yukon; introduced from Eurasia (*C. berlandieri* Moq. ssp. *zschackei* Murr. & Zobel).

Chenopodium album L. (× 0.5).

Chenopodium capitatum (L.) Asch.
Strawberry Spinach

Plants 1–8 dm tall, the stems erect or ascending, simple or more commonly branched, glabrous; leaves mostly 2–12 (20) cm long, the blades 1–5(10) cm broad, triangular-hastate, shallowly to deeply toothed or subentire, hastately lobed basally, acute to obtuse apically, often turning reddish, glabrous; the petioles mostly 0.5–6(14) cm long; flowers clustered in axillary, capitate spikes, the lower clusters subtended by foliose bracts, the upper ones bractless or with reduced bracts; perianth lobes free to below the middle, not mealy, becoming fleshy and reddish, shorter than the fruit; pericarp adherent to the erect seed; fruit about 1 mm long.

In sandy and gravelly soils along stream courses, lake shores, and highways; in the southeastern third of Alaska and in southern Yukon; south to California and New Mexico and east to Quebec and Massachusetts.

Chenopodium glaucum L.

Plants 0.5–3 dm tall, the stems prostrate to erect, commonly much branched, mealy to glabrate or glabrous; leaves 0.2–2(3) cm long, the blades 0.9–1(1.2) cm broad, triangular to oblong or lanceolate, coarsely toothed to subentire, obtuse to cuneate basally, rounded to obtuse apically, whitish mealy beneath, greenish above, the petioles 0.1–1 cm long; flowers clustered in short, axillary spikes and in terminal spikes or panicles; perianth lobes free almost to the base, more or less mealy, not fleshy at maturity, shorter than the fruit; pericarp free of the usually horizontal (sometimes vertical) seed; fruit 0.6–0.8 mm broad.

Moist sites; vicinity of Manley Hot Springs, Alaska; disjunctly eastward to Quebec and southward to Oregon, Arizona, and New Mexico; Eurasia, Africa, Australia (C. glaucum var. pulchrum Aellen). Our ma-

terials have been treated as belonging to ssp. salinum (Standl.) Fedde.

Chenopodium hybridum L.

Plants 1.5–10 dm tall or more, the stems erect, simple or branched, glabrous except above; leaves 3–15 cm long or more, the blades 1.5–12 cm broad, ovate to deltoid-ovate or broadly lanceolate, coarsely toothed to lobed, rounded to subcordate basally, acuminate apically, glabrous or sparingly mealy, the petioles mostly 1–7 cm long; flowers clustered in loose, large, terminal panicles and sometimes in reduced axillary spikes, the inflorescence more or less mealy and sometimes glandular; perianth lobes free almost to the base, mealy, not fleshy at maturity, shorter than the fruit; pericarp adherent to the horizontal seed; fruit 1.5–2 mm broad.

Weedy species of disturbed soils; known in our region only from the vicinity of Dawson City, Yukon (to be expected elsewhere); probably adventive, native to Eurasia (C. gigantospermum Aellen ex Fedde; C. hybridum ssp. gigantospermum [Aellen] Hultén).

Chenopodium leptophyllum (Moq.) Wats.
Narrow-leaved Goosefoot

Plants 1.5–7 dm tall, the stems decumbent to erect, simple or branched, commonly grayish mealy; leaves 1–4 cm long, the blades 0.2–6 mm broad, lance-oblong to linear, entire or shallowly lobed, cuneate basally, acute to obtuse apically, moderately to densely grayish mealy beneath, greenish above, the petioles 0.1–0.8 cm long; flowers clustered in terminal panicles or spikes, or in axillary spikes, the inflorescence grayish mealy; perianth lobes free almost to the base, mealy, not fleshy at maturity, equaling or longer than the fruit; pericarp loosely adherent to the horizontal seed; fruit 0.8–1.5 mm broad.

Disturbed soils; in southern Yukon (to be expected elsewhere); adventive from western North America where it occurs from British Columbia eastward, and southward to Mexico. Our limited material is assignable to var. *leptophyllum*.

Chenopodium rubrum L.

Plants 1–5 dm tall or more, the stems decumbent to erect, usually branched, glabrous or somewhat hairy above; leaves 2–10 cm long, the blades 0.5–4 cm broad, triangular to ovate or lanceolate, coarsely toothed or hastately lobed to subentire, obtuse to cuneate or hastate basally, obtuse to acute apically, not mealy (at least at maturity), greenish or turning reddish, the petioles 0.5–5 cm long; flowers clustered in axillary spikes, and often in a terminal panicle, the inflorescence not mealy; perianth lobes free to below the middle, not mealy nor becoming fleshy and reddish at maturity, shorter than the fruit; pericarp not adherent to the erect seed; fruit about 1 mm long.

Moist, saline meadows; in southern Yukon; eastward to Newfoundland and southward to California; Eurasia. Our material belongs to var. *humile* (Hook.) Wats.

CORISPERMUM L.

Annual, perfect herbs; leaves simple, alternate, entire, linear, glabrous or sparingly stellate-hairy; flowers perfect, inconspicuous, borne in axillary and terminal spikes, the subtending bracts scarious-margined; perianth consisting of a single papery scale (rarely 2–3), not enclosing the fruit; stigmas 2; pericarp adherent to the strongly flattened, erect seed.

Corispermum hyssopifolium L.
Bugseed

Plants 1–3 dm long or more, the stems decumbent to erect, commonly much branched, stellate-hairy to glabrate; leaves 1–6 cm long, 0.1–0.4 cm broad, linear to lance-linear, entire, narrowed basally to an indistinct petiole, cuspidate apically, glabrous or stellate-pubescent near the base,

transitional upward with the bracts of the inflorescence; bracts of inflorescence lanceolate to ovate-acuminate, with broad, scarious margins, 5–9 mm long, enclosing the flowers and obscuring the rachis of the spike; spikes mostly 1–3 cm long: fruit strongly flattened and winged, obovate to orbicular, 3–4 mm long.

Sandy soils along streams; in central eastern Alaska; evidently rare in our region, where it might be indigenous; widely distributed in western North America where it is reported to be adventive from Eurasia.

EUROTIA Adans.

Perennial, polygamous or dioecious subshrubs; leaves simple, entire, alternate, stellate-hairy, and spreading-hairy; flowers unisexual, inconspicuous, borne in axillary clusters or in terminal spicate inflorescences; staminate flowers with a 4-lobed perianth and 4 stamens; pistillate flowers lacking a perianth; stigmas 2; pericarp free from the erect seed.

Eurotia lanata (Pursh) Moq.
Winterfat

Plants 0.8–4 dm tall or more, the decumbent to erect stems arising from a woody base, stellate-hairy with longer, straight hairs intermixed, the hairs white or fading yellowish in age; leaves 1–4 cm long, 0.1–0.4 cm broad, linear to narrowly lanceolate, entire and somewhat revolute, stellate-hairy, sessile above, short-petiolate below; flowers borne in dense, axillary clusters or more or less spicate along branch tips; pistillate flowers 2–4 in the axils, subtended by 2 connate bractlets which enclose the fruit at maturity; staminate flowers in spicate axillary clusters, the perianth segments 4, 1.5–2 mm long; fruiting bracts 3–6 mm long.

Steep mountain slope; vicinity of Kluane Lake (Slims River), Yukon; disjunctly southward from Saskatchewan to Washington and southward to California, Arizona, New Mexico, and Texas.

MONOLEPIS Schrad.

Annual, polygamo-monoecious herbs; leaves simple, hastately lobed to entire, alternate, mealy to subglabrous, fleshy; flowers unisexual, inconspicuous, borne in axillary clusters; perianth consisting of 1 bractlike scale (rarely 2–3, or lacking), not enclosing the fruit; stigmas 2; pericarp reticulately patterned or warty, adherent to the erect seed.

1a. Leaves entire; perianth segment 1 mm long or less, rounded apically; plants rare in northern Alaska. *M. spathulata*
1b. Leaves with a pair of lateral lobes near the middle; perianth segments 1–2 mm long, acute apically; plants of southern Alaska and Yukon. *M. nuttalliana*

Monolepis nuttalliana (Schult.) Greene
Poverty-weed

Plants 0.5–2(3) dm tall, the stems ascending to erect, simple or much branched, mealy to subglabrous; leaves 1–5 cm long, the blades 0.1–1.5 cm broad, lanceolate to elliptic or oblong, with 1 pair of lateral lobes near the middle, reduced and sometimes entire upwards, sparsely mealy to glabrate, the petioles 0.1–2 cm long; flowers borne in dense, sessile, axillary clusters; perianth segments 1–2 mm long, more or less acute apically; pericarp pitted, usually pale; fruit 1–1.5 mm broad.

Weedy species of worked soils; in southern Alaska and southern Yukon; disjunctly southward from British Columbia to California and New Mexico and eastward to Manitoba and Missouri.

Monolepis spathulata Gray

Plants 0.5–1.5 dm tall, the stems prostrate or ascending, simple or branched from the base, sparsely mealy to glabrous; leaves 0.5–1.8 cm long, the blades 0.1–0.3 cm broad, oblong to oblanceolate, entire, reduced upwards, glabrous or nearly so, the petioles 0.1–0.8 cm long; flowers borne in dense, sessile, axillary clusters; perianth segments 0.5–1 mm long, more or less rounded apically; pericarp minutely warty, often reddish; fruits 0.5–1 mm broad.

Dry sandy hilltop; in northern Alaska (Sadlerochit River); disjunctly southward from Oregon to Mexico, Idaho, and Nevada.

SALICORNIA L.

Annual or perennial herbs from taproots or rhizomes; leaves simple, scalelike, opposite, connate, glabrous; flowers perfect, borne sessile, in opposite groups of 3, sunken in depressions of thickened, terminal spikes, subtended by scalelike bracts; perianth consisting of 4 connate segments, free at the tip around a slitlike opening, enclosing the fruit; stigmas commonly 2; pericarp thin, free from the erect, retrorsely-pubescent seed.

1a. Plants annual, from slender taproots, the stems commonly erect; central flower much above the lateral ones. *S. europaea*
1b. Plants perennial, from elongate rhizomes, the stems often trailing and sometimes rooting at the nodes; central flower not much above the lateral ones. *S. pacifica*

Salicornia europaea L.
Slender Glasswort

Plants 0.1–2.5 dm tall, from slender taproots, the stems fleshy, erect or ascending, commonly branched, jointed, glabrous; leaves scalelike, often with a scarious margin; spikes 0.5–5 cm long, the joints mostly 1.5–4 mm long and 1–2 mm thick; central flower much above the lateral ones; fruit dehiscent, the seeds falling separately.

Saline soils, on tidal flats and lake shores; in coastal, southern Alaska and in southern Yukon; widely distributed in temperate regions of the northern hemisphere (*S. herbacea* L.).

Salicornia pacifica Standl.

Pacific Samphire

Plants 0.5–3 dm long or more, from elongate rhizomes, the stems prostrate to ascending or erect, sometimes rooting at the nodes, jointed, glabrous; leaves scalelike, often with a scarious margin; spikes 1–4 cm long, the joints mostly 1.5–2.5 mm long and 2–3 mm thick; central flower only slightly above the lateral ones; fruit dehiscent, the seeds falling separately or sometimes adherent to the calyx.

Saline tidal flats and beaches; in maritime, southeastern Alaska; southward to Baja California and eastward to Utah. Our plants have been treated previously as belonging to the same taxon as the plants of coastal eastern North America and of Europe. Their ultimate status must await additional study.

SUAEDA Forsk. Nom. Cons.

Annual herbs from slender taproots; leaves simple, alternate, fleshy, subterete, glabrous; flowers perfect or unisexual, clustered in the axils of the leaves, or sometimes terminal and spikelike; perianth lobes 4–5, fleshy, basally connate, enclosing the fruit at maturity; stigmas usually 2; pericarp free from the commonly horizontal shining seed.

Suaeda maritima (L.) Dumort

Low Sea-blite

Plants 0.3–2.5(3) dm tall, the stems decumbent to erect, commonly branched; leaves 0.5–3 cm long, 0.1–0.2 mm broad, linear to oblong, subterete, fleshy, often glaucous; bracts of inflorescence like the leaves but much smaller upwards; perianth lobes 0.8–1.4 mm long, equal or nearly so; pericarp free from the horizontal, shining seed; fruit 1.3–1.6 mm broad.

Maritime beaches and tidal flats or saline soils; in the interior of Alaska and Yukon; southward to Washington and disjunctly to northeastern North America; Europe.

Salicornia pacifica Standl. (× 0.4).

COMPOSITAE
Composite Family

Herbs or shrubs; leaves alternate or opposite, simple, pinnatifid, or compound; flowers in involucrate heads, these solitary or several to many in corymbose or cymose clusters; heads few- to many-flowered, surrounded by bracts which form a cup-shaped, cylindrical, or urn-shaped involucre; flowers all with ligulate corollas, all with tubular corollas, or with both, and when both are present, with the tubular corollas (disk corollas) forming the central disk, and the ligulate corollas (rays) forming an outer, radiating row; calyx crowning the summit of the ovary and modified as a pappus of capillary bristles, scales, awns, or a short crown, or lacking; petals 5, united; stamens 5, alternate with the corolla lobes, filaments free, the anthers united, forming a tube around the style, or nearly distinct; pistils 1, the ovary inferior, 1-loculed; styles 1, 2-cleft; fruit an achene. *Note*: The individual flowers in some genera have no bracts at the base of the ovary and after the achenes have fallen the receptacle is naked; in other genera each flower has a scalelike or bristly bract at the base of the ovary and the receptacle is said to be chaffy.

1a. Corollas all raylike, the flowers all perfect; plants with milky juice. KEY I. .... p. 107
1b. Corollas not of ray flowers only, some or all of them tubular; plants mostly with watery juice. (2)

2a. Ray flowers present. KEY II. .... p. 107
2b. Ray flowers lacking or vestigial; heads discoid, all flowers tubular. KEY III. .... p. 108

KEY I. COROLLAS ALL RAYLIKE; PLANTS WITH MILKY JUICE.

1a. Pappus of plumose bristles, of bristles and scales, of scales, of minute awns, or lacking. (2)
1b. Pappus of simple bristles only (these sometimes barbellate). (5)

2a. Pappus lacking, or of a few minute scales. *Lapsana*
2b. Pappus of well-developed bristles. (3)

3a. Plants with leafy stems; leaves merely toothed. *Picris*
3b. Plants subscapose; leaves pinnatifid to pinnately lobed. (4)

4a. Receptacle chaffy; achenes long-beaked. *Hypochaeris*
4b. Receptacle not chaffy; achenes beakless. *Leontodon*

5a. Plants scapose; heads solitary. (6)
5b. Plants with at least some cauline leaves; heads few to several (rarely solitary). (8)

6a. Achenes spinulose or with ridges near the top of the body, or less commonly smooth, tipped by a slender beak. *Taraxacum*
6b. Achenes smooth or nearly so, beaked or beakless. (7)

7a. Pappus bristles brownish, more or less united and falling as a unit; achenes beakless. *Apargidium*

7b. Pappus bristles whitish, distinct, falling separately; achenes beaked. *Agoseris*

8a. Main cauline leaves hastate-deltoid, sharply toothed, the petioles broadly winged; corollas white. *Prenanthes*
8b. Main cauline leaves various in shape, not hastate-deltoid, or much reduced upwards; corollas yellow or blue, rarely white. (9)

9a. Achenes more or less flattened; stems leafy, mostly over 3 dm tall. (10)
9b. Achenes subterete; stems with leaves much reduced upwards (except in some *Hieracium*), often less than 3 dm tall. (11)

10a. Achenes beaked; flowers yellow or blue; involucres cylindrical to ovoid-cylindrical. *Lactuca*
10b. Achenes beakless; flowers yellow; involucres broadly campanulate to hemispheric. *Sonchus*

11a. Plants from taproots; pappus white or whitish. *Crepis*
11b. Plants from short, rhizomatous caudex; pappus tan to brown. *Hieracium*

KEY II. RAY FLOWERS PRESENT.

1a. Rays white to pink, purple, blue, or red, but not yellow. (2)
1b. Rays yellow or orange. (10)

2a. Plants acaulescent or nearly so; rays white to pinkish; involucral bracts with hyaline, fringed margins; known only from southern Yukon. *Townsendia*
2b. Plants caulescent (acaulescent in some *Erigeron*); rays various; involucral bract margins usually not both hyaline and fringed; plants of various distribution. (3)

3a. Pappus of capillary bristles; receptacle naked. (4)
3b. Pappus none, or a mere crown, or of scales or awns; receptacle naked or chaffy. (6)

4a. Basal leaves long-petiolate, cordate, sagittate, or palmately lobed; stem leaves much reduced, bladeless. *Petasites*

4b. Basal leaves various, but not as above; stem leaves reduced or not. (5)

5a. Involucral bracts in 3–more rows, the outer ones progressively shorter, or subequal, green throughout or scarious near the base; rays comparatively broad; style branches more than 0.5 mm long; plants often with rhizomes. *Aster*

5b. Involucral bracts in 1–2 rows, subequal, green or scarious throughout; rays mostly very narrow; style branches 0.5 mm long or less; plants seldom with rhizomes (except in *E. peregrinus*). *Erigeron*

6a. Plants scapose; pappus none. *Bellis*

6b. Plants with at least some cauline leaves; pappus present or absent. (7)

7a. Rays short, 2–4 mm long. *Achillea*

7b. Rays mostly over 10 mm long. (8)

8a. Receptacle chaffy or bristly, at least towards the middle. *Anthemis*

8b. Receptacle naked. (9)

9a. Leaves entire to dentate or pinnatifid with broad lobes. *Chrysanthemum*

9b. Leaves pinnately dissected, with linear to filiform lobes. *Matricaria*

10a. Leaves opposite. (11)

10b. Leaves alternate (sometimes opposite below in *Helianthus* and *Madia*). (12)

11a. Pappus of capillary bristles; plants perennial. *Arnica*

11b. Pappus of retrorsely barbed awns; plants annual. *Bidens*

12a. Pappus of capillary bristles. (13)

12b. Pappus of awns, scales, a short crown, or none. (15)

13a. Involucral bracts in 1 series, sometimes with 1–2 rows of very short bractlets at the base. *Senecio*

13b. Involucral bracts in 2–more series, unequal. (14)

14a. Plants from a rhizome or caudex; roots numerous; heads many, small, mostly 3–8 mm high. *Solidago*

14b. Plants with taproots; heads few to several, mostly over 7 mm high. *Haplopappus*

15a. Receptacle chaffy throughout; pappus of caducous, scalelike awns, or a short crown. (16)

15b. Receptacle naked or with a single row of chaffy scales between ray and disk flowers. (18)

16a. Leaves pinnatifid, the narrow segments again toothed. *Anthemis*

16b. Leaves not pinnatifid into narrow segments, entire to toothed or lobed, simple, ovate to lanceolate or oblong. (17)

17a. Leaves ovate to lanceolate, petiolate. *Helianthus*

17b. Leaves oblong to lance-oblong, sessile and more or less clasping. *Chrysanthemum*

18a. Leaves once to thrice pinnatifid. *Tanacetum*

18b. Leaves not pinnatifid, entire, linear. *Madia*

KEY III. RAY FLOWERS LACKING OR VESTIGAL; HEADS DISCOID, ALL FLOWERS TUBULAR.

1a. Pappus lacking, or of scales, awns, or a short crown. (2)

1b. Pappus of capillary or plumose bristles. (6)

2a. Leaves opposite; plants annual. *Bidens*

2b. Leaves alternate; plants perennial or annual. (3)

3a. Inflorescence spicate, racemose, or paniculate, the heads rarely capitate, several to numerous. *Artemisia*

3b. Inflorescence corymbiform or capitate, or the heads solitary. (4)

4a.   Leaves entire, toothed, or pinnately lobed, the bases sheathing. *Cotula*
4b.   Leaves pinnately dissected. (5)

5a.   Receptacle flat, or merely convex. *Tanacetum*
5b.   Receptacle strongly conic, pointed. *Matricaria*

6a.   Leaves spiny, thistlelike; receptacle densely bristly. *Cirsium*
6b.   Leaves not spiny; receptacle not bristly (except in *Saussurea*). (7)

7a.   Basal leaves long-petiolate, cordate, sagittate, or palmately lobed; stem leaves much reduced, bladeless. *Petasites*
7b.   Basal leaves various, but not as above; stem leaves reduced or not. (8)

8a.   Flowers pinkish or blue to purple; pappus bristles plumose; receptacle usually bristly. *Saussurea*
8b.   Flowers mostly yellow or white; pappus bristles rarely plumose; receptacle naked. (9)

9a.   Basal leaves large, the blades to 20 cm broad or more, reniform, the base cordate. *Cacalia*
9b.   Basal leaves large or small, but not both reniform and cordate at the base. (10)

10a. Involucral bracts in 1 series, sometimes with 1–2 rows of very short bractlets at the base. *Senecio*
10b. Involucral bracts in 2–more series, unequal. (11)

11a. Plants with taproots; heads all with outer flowers pistillate and the central ones perfect. *Gnaphalium*
11b. Plants fibrous rooted, often rhizomatous or stoloniferous; heads dioecious or nearly so. (12)

12a. Basal leaves conspicuous; cauline leaves much reduced; mostly pistillate plants known. *Antennaria*
12b. Basal leaves early deciduous; cauline leaves little reduced, numerous; pistil-

late plants commonly with a few staminate flowers. *Anaphalis*

ACHILLEA L.

Perennial, rhizomatous, aromatic herbs with watery juice; stems erect or ascending, simple and subentire to incised or dissected; heads several to many, borne in compact to open corymbose cymes; involucral bracts imbricate in several rows, chaffy, the margin scarious and hyaline; receptacle chaffy; ray flowers present, usually 3–12, pistillate, fertile, white, pink, or pink purple; disk flowers mostly 10–more, perfect, fertile; pappus none; style branches flattened; achenes compressed, callus margined, glabrous, beakless.

1a.   Leaves subentire to shallowly toothed; rays over 3 mm long; plants cultivated and escaping. *A. ptarmica*
1b.   Leaves incised or dissected; rays usually less than 3 mm long; plants indigenous or uncommon in cultivation. (2)

2a.   Leaves pinnately dissected, the divisions again dissected; rachis about as broad as thick. *A. millefolium*
2b.   Leaves incised, the divisions merely toothed; rachis several times broader than thick. *A. sibirica*

Achillea millefolium L.
Yarrow

Rhizomes horizontal, the stems ascending to erect, 0.2–10 dm tall, villous-tomentose, simple or branched from the upper axils; leaves 2–20 cm long, reduced upwards, pinnately dissected, the divisions again dissected; heads borne in hemispheric or flat-topped, corymbose cymes; involucre 4–6 mm high, the bracts dark to light margined, villous to glabrate; rays usually about 5, 2–3.5 mm long, white to pink or pink purple; disk flowers 10–20; achenes 1–2 mm long.

Gravelly or sandy soils in numerous habitats; in much of Alaska (except for the ex-

treme northern portion) and Yukon; widely distributed in North America; circumboreal. The *millefolium* complex is difficult both taxonomically and nomenclatural-
ly. Several phases have been recognized
at specific rank, or at varietal or subspecific level in one or more species. In Alaska
and Yukon, three more or less completely
intergrading entities have been recognized:
an introduced ornamental (ssp. *millefolium*, n=27), and the other two indigenous
(ssp. *lanulosa*, n=18, and ssp. *borealis*,
n=27).

1a. Plants cultivated; ultimate leaf segments lanceolate to ovate. *A. millefolium* ssp. *millefolium*
1b. Plants indigenous; ultimate leaf segments linear to narrowly lanceolate.
(2)

2a. Involucral bracts with blackish or
brownish margins; plants widespread
in Alaska and Yukon; our common
Achillea. A. millefolium ssp. *borealis*
(Bong.) Breitung
2b. Involucral bracts with pale margins;
plants mostly of interior Alaska (*A.
occidentalis* Raf. ex Rydb.). *A. millefolium* ssp. *lanulosa* (Nutt.) **Piper**

Achillea ptarmica L.
Sneezeweed

Rhizomes horizontal to ascending, the
stems ascending to erect, 1.5–6 dm tall, villous-tomentose, simple to much branched;
leaves 1.5–5 cm long, subentire to sharply
serrate, little reduced upwards; heads few
to many, borne in corymbose cymes; involucre 4–5 mm high, the bracts with dark
to pale margins, villous to glabrate; rays
6–15, 3–5 mm long, usually white; disk
flowers numerous; achenes 1–2 mm long.

Cultivated ornamental; escaping in southern Alaska; Eurasia.

Achillea sibirica Ledeb.
Siberian Yarrow

Rhizomes horizontal to ascending, the
stems erect, 3–12 dm tall, villous-tomentose, simple or branched from the upper
axils; leaves 2–15 cm long, incised, the divisions merely toothed, gradually reduced
upwards; heads borne in corymbose cymes;
involucres 3.5–5 mm long, the bracts with
dark to pale margins, villous to glabrate;
rays usually 7–12, 1–3 mm long; disk flowers 15–30; achenes 1–2 mm long.

Woods, river banks, roadsides, often in
gravelly soil; in interior Alaska and Yukon,
east to Manitoba, and disjunctly to the
Gaspé Peninsula, and south to British Columbia and Saskatchewan (*A. multiflora*
Hook.).

AGOSERIS Raf.

Perennial scapose herbs with milky juice
from taproots; leaves all basal, entire to
pinnately lobed or merely toothed; heads
solitary on a scape; involucral bracts in 2–
several series, herbaceous or the inner
ones hyaline or nearly so; receptacle usually naked; corollas all raylike, perfect, yellow to orange, often drying pinkish or purplish; pappus of capillary bristles; style
branches semicylindrical; achenes angular
or terete, prominently nerved, glabrous,
usually beaked.

1a. Achenes with a slender beak commonly over half as long as the body
of the achene; flowers brownish
orange when fresh, often drying purplish. *A. aurantiaca*
1b. Achenes with a short, stout beak less
than half as long as the body, or the
beak lacking; flowers yellow, often
drying pinkish. *A. glauca*

Agoseris aurantiaca (Hook.) Greene
Mountain Dandelion

Plants 1–4(5) dm tall, from a simple or
branched caudex; leaves 5–30 cm long,
0.5–3 cm broad, narrowly oblanceolate, entire to toothed or lobed, villous to glabrate; scapes villous-tomentose to nearly
glabrous, often persistent; involucres 15–
25 mm long, the outer bracts villous and

ciliate, often purple-spotted; corollas brownish orange, often drying purplish; achene body 4–8 mm long, the beak slender, from about half as long to longer than the body.

Meadows and hillsides; in southern Yukon and southeastern Alaska (Skagway); east to Alberta and south to California and New Mexico (*A. gracilens* [Gray] Kuntze).

Agoseris glauca (Pursh) Raf.

Plants 1–3.5 dm tall, from a simple or branched caudex; leaves 5–25 cm long, 0.3–5 cm broad, narrowly oblanceolate, entire, or less commonly toothed or lobed, sparsely villous; scapes villous-tomentose, often persistent; involucres 20–25 mm tall, the outer bracts villous and ciliate, sometimes purple-spotted; corollas yellow, often drying pinkish; achene body 4–10 mm long, the beak stout, to half as long as the body.

Open slopes and meadows; in extreme northern British Columbia and southeastern Alaska (Skagway), to be expected in the Yukon; east to Manitoba and Minnesota, and south to Arizona (*A. scorzonerae-folia* [Schrad.] Greene). Our material belongs to var. *dasycephala* (T. & G.) Jeps.

ANAPHALIS DC.

Perennial, dioecious or polygamo-dioecious, rhizomatous herbs with watery juice; stems ascending to erect, simple or branched from the uppermost axils; leaves simple, alternate, only gradually reduced upwards, entire; heads several to many, borne in hemispheric or flattopped corymbose cymes; involucral bracts imbricate in several rows, chaffy, scarious, white or with a dark, triangular, basal spot; receptacle naked; corollas of disk flowers only, imperfect, whitish, the pistillate heads sometimes bearing some central staminate flowers, the pistillate corollas tubular-filiform, the staminate corollas tubular-funnelform; pappus of capillary bristles; styles branches somewhat flattened; achenes very small, roughened, glabrous to sparingly hairy.

Anaphalis margaritacea (L.) B. & H.
Pearly Everlasting

Plants 1.5–8 dm tall, the stems white villous-tomentose; leaves only gradually reduced upwards, 2.5–12 cm long, 0.5–2 cm broad, narrowly lanceolate to oblong, elliptic, or oblanceolate, sessile, entire, flat to slightly revolute, white tomentose below, commonly less pubescent and greenish above; heads showy, the involucres 4–7 mm high, 5–10 mm broad, the bracts pearly white, with a dark, triangular base, glabrous; achenes to about 1 mm long.

Talus slopes and open woods; in coastal and insular southern Alaska, from the Aleutian Islands eastward through the Panhandle; widely distributed in North America; Asia (*Gnaphalium margaritaceum* L.).

ANTENNARIA Gaertn.

Perennial, dioecious herbs with stolons, caudices, or rhizomes, the juice watery; stems ascending to erect, usually simple; leaves simple, alternate and basal, the cauline generally reduced upwards; heads solitary to many, borne in corymbose cymes; involucral bracts imbricate in several rows, scarious, at least marginally, often colored; receptacle naked; corollas of disk flowers only, imperfect, whitish or tawny; pistillate corollas tubular-filiform, the pappus of numerous capillary bristles; staminate corollas tubular-funnelform, the pappus of few clavate or barbellate, usually flattened bristles; style branches slightly flattened; achenes terete to slightly compressed, glabrous or papillose.

Porsild, A. E. 1950. The genus *Antennaria* in northwestern Canada. Can. Field-Nat. 64:1–25.

———. 1965. The genus *Antennaria* in eastern arctic and subarctic America. Saertyrk Botanisk Tidssk. 61:22–55.

1a. Upper leaf surface glabrous, bright green (or turning brownish in age); flowering stems usually less than 1.5

dm tall; plants of the Aleutian Islands. *A. dioica*

1b. Upper leaf surface densely to sparsely tomentose, usually grayish or silvery, sometimes green, but if so then the flowering stems usually over 1.5 dm tall, or plants not of the Aleutian Islands. (2)

2a. Heads solitary (rarely 2–3); plants usually less than 1(1.9) dm tall. (3)

2b. Heads commonly 3–several per peduncle (rarely solitary); plants often more than 1.5 dm tall (except in *A. alpina*). (4)

3a. Scarious portion of pistillate involucral bracts brownish or blackish green, the apices usually acute to acuminate as well as erose. *A. monocephala*

3b. Scarious portion of pistillate involucral bracts white or tan, the apices usually rounded to obtuse in outline as well as erose. *A. shumaginensis*

4a. Upper leaf surfaces green, sparsely tomentose, soon glabrate; plants of southeastern Yukon. *A. neglecta*

4b. Upper leaf surfaces usually grayish tomentose, or if greenish then plants of different distribution. (5)

5a. Largest leaves 4–20 cm long, distinctly 3(5)-veined; plants (1)2–6 dm tall. *A. pulcherrima*

5b. Largest leaves 0.5–3(4) cm long, with only 1 more or less distinct vein; plants 0.5–2.5(4) dm tall. (6)

6a. Scarious portion of involucral bracts bright pink to pale pink or white. *A. rosea*

6b. Scarious portion of involucral bracts brownish green or blackish green or dirty tan to whitish. (7)

7a. Involucral bracts usually abruptly acuminate to acute apically (at least the inner ones), the scarious terminal portion brownish green to blackish green throughout. *A. alpina*

7b. Involucral bracts mostly obtuse to rounded apically, the scarious terminal portion merely dirty tan, often whitish apically. *A. umbrinella*

Antennaria alpina (L.) Gaertn.
Alpine Pussytoes

Plants caespitose from a caudex or mat-forming and stoloniferous, 0.2–2.3 dm tall; basal leaves 0.5–3(4) cm long, 0.1–0.5 cm broad, cuneate-oblanceolate to spatulate, acute to obtuse or rounded apically, grayish and tomentose on both surfaces or greenish and subglabrous above; heads (1) 2–5(9), borne in subcapitate to somewhat open cymes; pistillate involucres 4.5–8(10) mm high, villous-tomentose below, the scarious tips of the bracts uniformly blackish green or brownish green, at least the inner ones acuminate to acute, often erose; staminate involucres 4–5 mm high, the scarious tips of the bracts often pale apically; achenes glabrous or papillose.

Arctic and alpine tundra and heathlands; over most of Alaska and Yukon; east to Labrador and Greenland; circumboreal. The entities herein interpreted as belonging to *A. alpina* have been treated previously at specific or subspecific level. Both sexual and apomictic entities are involved, and segregation at specific level has been based on differences in size of the pistillate involucre, size, shape, and pubescence of basal leaves, on whether staminate plants are known or not (a fact not always demonstrable in specimens in the herbarium or in the field), and on the presence or absence of stolons. Despite the use of these and other features, the delimitation of the purposed entities has been to a large extent almost completely arbitrary. Thus, it seems best to recognize our materials as belonging to a single polymorphic species with several more or less distinctive varieties.

1a. Plants stoloniferous, the basal rosettes borne normally on prostrate or ascending stolons, mat-forming; known from southeastern Yukon. (2)

1b. Plants not stoloniferous, the basal rosettes borne on short caudex branches, clump-forming; plants of broad distribution. (3)

2a. Involucres mostly 5–7 mm high; heads 4–9; achenes with papillae (*A. atriceps* Fern.; *A. media* authors, not Greene; *A. pedunculata* Porsild). *A. alpina* var. *stolonifera* (Porsild) Welsh

2b. Involucres mostly 7–9 mm high; heads 1–3; achenes glabrous (*A. frieseana* var. *megacephala* [Fern.] Hultén; *A. megacephala* Fern.). *A. alpina* var. *megacephala* (Fern.) Welsh

3a. Basal leaves 1–3(4) cm long, narrowly cuneate-oblanceolate, with well-developed petioles, the blades often greenish above; plants mostly of western and northern Alaska, but also in central Alaska and northern Yukon (*A. alaskana* Malte; *A. frieseana* [Trautv.] Ekm.; *A. frieseana* ssp. *alaskana* [Malte] Hultén; *A. neoalaskana* Porsild). *A. alpina* var. *frieseana* Trautv.

3b. Basal leaves 0.5–2 cm long, oblanceolate to spatulate, the petioles not well developed, the blades usually densely white tomentose above; plants of broad distribution in Alaska and Yukon (*A. densifolia* Porsild; *A. ekmaniana* Porsild; *A. subcanescens* Ostenf. ex Malte; *A. frieseana* ssp. *compacta* [Malte] Hultén). *A. alpina* var. *compacta* (Malte) Welsh

Antennaria dioica (L.) Gaertn.

Plants mat-forming, stoloniferous, 0.3–1.5 (2) dm tall; basal leaves 0.8–3 cm long, 0.3–0.9 cm broad, cuneate-spatulate, rounded to obtuse or cuspidate apically, whitish or grayish tomentose beneath, glabrous or soon glabrate and green above (or turning brownish in age); heads 2–5 (8), borne in a subcapitate or slightly open cyme; pistillate involucres 7–11 mm high, villous-tomentose below, the scarious tips of the bracts white to pink, rounded to obtuse apically, entire or erose; staminate involucres 6–7 mm high, the scarious tip like the pistillate ones; achenes glabrous.

Gravelly soils; in the western Aleutian Islands; Eurasia (*Gnaphalium dioicum* L.; *A. insularis* Greene).

Antennaria monocephala DC.

Plants mat-forming, with short or elongate stolons, 0.2–1.6 dm tall; basal leaves 0.5–2 cm long, 0.2–0.5 mm broad, cuneate-oblanceolate to -spatulate, rounded to obtuse or cuspidate apically, grayish tomentose to glabrate below, glabrous or glabrate and green above; heads solitary; pistillate involucres 4–8 mm high, tomentose to glabrate below, the scarious tips of the bracts blackish or brownish green, at least the inner ones acute to acuminate apically, usually erose; staminate involucres 3.5–6 mm high, the inner bracts often pale apically; achenes glabrous or papillose.

Arctic and alpine tundra and heathlands; through most of Alaska and Yukon; east to Labrador and Greenland (*A. angustata* Greene; *A. monocephala* ssp. *angustata* [Greene] Hultén; *A. monocephala* var. *exilis* [Greene] Hultén; *A. monocephala* var. *latisquamea* Hultén; *A. philonipha* Porsild; *A. monocephala* ssp. *philonipha* [Porsild] Hultén; *A. pygmaea* authors, not Fern.). *A. monocephala* is closely related to *A. alpina,* but is distinguished from that entity by its monocephalic habit and usually by the green upper leaf surfaces.

Antennaria neglecta Greene

Plants mat-forming, with well-developed stolons, 1.5–4 dm tall; basal leaves 2–5 cm long, 0.7–2 cm broad, cuneate-oblanceolate, acute to obtuse and cuspidate apically, whitish tomentose below, green and sparsely tomentose to glabrous above, obscurely 3-veined; heads 5–9, borne in a compact to open cyme; pistillate involucres 6–10 mm high, tomentose below, the scarious tips of the bracts white to pale tan, rounded to obtuse or the inner ones narrow and acute, erose; staminate plants unknown in our range; achenes papillose.

Open woods; in southeastern Yukon; eastward to Newfoundland and south to California, Arizona, and Virginia (*A. howellii* Greene; *A. neglecta* ssp. *howellii* [Greene] Hultén). Our materials belong to var. *howellii* (Greene) Cronq.

Antennaria pulcherrima (Hook.) Greene

Plants rhizomatous and stoloniferous, 1–6 dm tall; basal leaves 4–20 cm long, 0.4–2 cm broad, narrowly oblanceolate, tapering basally to a slender petiole, acute to attenuate apically, thinly grayish tomentose beneath and above, 3(5)-veined; heads 4–20; pistillate involucres 8–11 mm high, tomentose below, the scarious tips of the bracts white to pale tan, or brownish green, at least the inner ones narrowly lance-attenuate to acute, entire; staminate involucres 5–8 mm high, tomentose below, the scarious tips of the bracts pale tan to whitish, broad, rounded apically, entire or erose; achenes glabrous.

Open woods, flood plains, terraces, and mountain slopes; in the eastern half of Alaska and most of Yukon; eastward to Newfoundland, and south to Utah and Colorado (*A. pulcherrima* var. *angustisquama* Porsild; *A. carpatica* var. *pulcherrima* Hook.).

Antennaria rosea Greene

Plants mat-forming, stoloniferous, 0.5–4 dm tall; basal leaves 0.5–2.5 cm long, 0.2–0.9 mm broad, cuneate-oblanceolate to spatulate, acute to obtuse apically, whitish or grayish tomentose beneath, tomentose to glabrate and less commonly greenish above, indistinctly 1-veined; heads 3–20, borne in a capitate to open cyme; pistillate involucres 5–9 mm high, tomentose below, the scarious tips of the bracts bright pink, pale pink, white, or merely whitish, obtuse

to rounded apically, usually erose; staminate plants unknown in our region; achenes usually glabrous.

Open woods, meadows, flood plains, and terraces; in much of the eastern half of Alaska and southern Yukon; east to Ontario and south to California, Arizona, and New Mexico (*A. alborosea* Porsild & Porsild; *A. breitungii* Porsild; *A. elegans* Porsild; *A. incarnata* Porsild; *A. laingii* Porsild; *A. leuchippi* M. P. Porsild; *A. nitida* Greene; *A. oxyphylla* Greene; *A. rosea* var. *nitida* [Greene] Breitung; and *A. subviscosa*

Antennaria rosea Greene (× 0.4).

Fern., at least in part). As interpreted herein, *A. rosea* is represented by a series of highly variable populations interspersed with local, more or less uniform populations. The more stable populations have been treated previously as belonging to several species. The characteristics by which they have been segregated are often trivial and it seems best to consider them as part of a polymorphic species. Further work is indicated.

Antennaria shumaginensis Porsild

Plants caespitose to mat-forming, the stolons short, 0.8–1.5(1.9) cm tall; basal leaves 1–3 cm long, 0.3–0.6 cm broad, spatulate to obovate, obtuse to acute and cuspidate apically, whitish tomentose below, glabrous and greenish above, or more or less tomentose above; heads 1–2; pistillate involucres 6–7 mm high, tomentose below, the scarious tips of the bracts pale tan, attenuate apically, erose; staminate involucres somewhat smaller than the pistillate, the bracts pale; achenes glabrous or slightly papillose.

Western Pacific Coast region (Popof Island) and Bristol Bay; endemic. It seems probable that *A. shumaginensis* is only a phase of *A. monocephala* and might be best recognized at some rank under that entity.

Antennaria umbrinella Rydb.

Plants mat-forming, stoloniferous, 0.5–2 dm tall; basal leaves 0.5–2.5 cm long, 0.2–0.6 cm broad, cuneate or rounded and sometimes cuspidate apically, grayish tomentose beneath or rarely glabrate, grayish tomentose to glabrate above, indistinctly 1-veined; heads 3–6, borne in subcapitate to open cymes; pistillate involucres 5–8 mm high, tomentose below, the scarious tips of the bracts from dirty brownish green to pale tan or the innermost almost white, acute to rounded, usually erose; staminate plants unknown in our region; achenes glabrous.

Open woods and dry alpine slopes; in much

of Alaska (except for the extreme western and northern portions) and in the Yukon; east to Hudson Bay and south to California, Arizona, and Colorado (*A. isolepis* Greene; *A. pallida* E. Nels.). *A. umbrinella* is transitional between *A. rosea* and *A. alpina*. The extreme forms approach those species in bract shape, color, and stature.

ANTHEMIS L.

Annual or short-lived perennial, taprooted, aromatic herbs with watery juice; stems erect, commonly branched; leaves alternate, once to thrice pinnately dissected; heads solitary on the uppermost branches; involucral bracts imbricate in several rows, chaffy, the margins scarious or hyaline; receptacle hemispheric, chaffy at least near the middle; ray flowers present, white or yellow, usually 10–more, sterile; disk flowers numerous, perfect, fertile; pappus none or a short crown; style branches flattened; achenes subterete to compressed, not callous margined, glabrous, beakless.

1a. Rays yellow; pappus a short crown; disk commonly more than 12 mm broad. *A. tinctoria*
1b. Rays white; pappus lacking; disk commonly less than 10 mm broad. *A. cotula*

Anthemis cotula L.
Mayweed

Plants annual, 1–5 dm tall; stems simple or branched, ill-scented; leaves 1–6 cm long, twice to thrice pinnatifid, the ultimate segments lance-oblong, sparsely villous and glandular-dotted; heads borne solitary at the ends of the uppermost branches; ray flowers commonly 10–20, white, sterile, 5–10 mm long; disk flowers numerous; disk 7–10(12) mm broad; receptacle chaffy only in the middle, the bracts narrowly subulate; achenes slightly flattened, glandular, the pappus lacking.

A weed of disturbed soils; in southern Alaska and Yukon; introduced from Europe.

Anthemis tinctoria L.

Yellow Camomile

Plants short-lived perennials, 2.5–6 dm tall; stems simple or branched; leaves 1.5–7 cm long, once to twice pinnatifid, the segments oblong in outline, merely toothed or lobed, villous-tomentose below, glabrous or glabrate above, sparsely glandular-dotted; heads borne solitary at the ends of the uppermost branches; ray flowers 20–35, yellow, fertile, 7–14 mm long; disk flowers numerous; disk 12–15 mm broad or more; receptacle chaffy throughout, the bracts narrow and with yellow awn tips; achenes compressed; pappus a short crown.

Cultivated ornamental; escaping in southeastern Alaska; native of Europe.

APARGIDIUM T. & G.

Perennial, scapose herbs with milky juice, from taproots; leaves all basal, entire or pinnately lobed or merely toothed; heads solitary on a scape; involucral bracts in several rows, herbaceous, or the inner ones with hyaline margins; receptacle naked; corollas of ray flowers only, perfect, yellow (drying whitish or pinkish); pappus of brownish capillary bristles, connate at the base; style branches semicylindrical; achenes cylindrical, 10–12-ribbed, glabrous, truncate.

Apargidium boreale (Bong.) T. & G.

Plants 1–5 dm tall, from a simple or branched caudex; leaves 5–25 cm long, 0.3–1.2 cm broad, narrowly lanceolate to lance-oblong, entire to toothed or lobed, glabrous, often persistent; involucres 10–15 mm high, the bracts black, strigulose, not ciliate; corollas yellow, often drying whitish or pinkish; achenes 5–6 mm long, beakless.

Wet meadows; in southern and southeastern Alaska; southward to California (*Apargia borealis* Bong.; *Microseris borealis* [Bong.] Schulz-bip.).

ARNICA L.

Perennial herbs from rhizomes or caudices, with watery juice; stems erect, simple or branched above; leaves opposite, or the uppermost alternate, simple, entire or toothed; heads solitary, or 3–9(11) in corymbose clusters; involucral bracts subequal or evidently biseriate, herbaceous; receptacle naked, convex; ray flowers present, yellow or orange, several to many, fertile, or rarely lacking (in *A. parryi*); disk flowers numerous, perfect, fertile; pappus of barbellate or subplumose capillary bristles; style branches flattened; achenes cylindrical, 5–10-nerved, pubescent to glabrate or glabrous, often glandular.

Maguire, B. 1943. A monograph of the genus *Arnica*. Brittonia 4:386–510.

1a. Anthers purple; involucral bracts callous tipped. (2)

1b. Anthers yellow; involucral bracts not callous tipped. (3)

2a. Heads nodding; rhizomes naked; leaves chiefly basal. *A. lessingii*

2b. Heads erect; rhizomes densely clothed with persistent leaf bases; leaves chiefly cauline. *A. unalaschcensis*

3a. Cauline leaves (4)5–9 pairs; pappus brownish; heads often 5–more per main stem (see also *A. diversifolia*). (4)

3b. Cauline leaves (0)1–4(5) pairs; pappus brownish or white; heads 1–4(5) per main stem. (5)

4a. Involucral bracts acuminate to attenuate, lacking an apical tuft of hair. *A. amplexicaulis*

4b. Involucral bracts merely acute to abruptly rounded (rarely acuminate), bearing an apical or subapical tuft of hair. *A. chamissonis*

5a. Leaves (at least the lower) cordate, ovate, or broadly lanceolate, often cordate, truncate, or obtuse basally, seldom cuneate. (6)

5b. Leaves narrowly lanceolate to lance-oblong or lanceolate, usually cuneate basally. (8)

6a. Pappus brownish, subplumose; leaves usually obtuse to subcuneate basally. *A. diversifolia*

6b. Pappus white, merely barbellate; leaves usually subcordate, truncate or obtuse basally (at least the lower). (7)

7a. Achenes glabrous, or glabrous near the base; blades of basal leaves much longer than the petioles, or the petioles lacking; plants of southern and southeastern Alaska and southern Yukon. *A. latifolia*

7b. Achenes uniformly, though sometimes sparingly hairy; blades of basal leaves subequal to, or shorter than, the petioles; plants of southeastern Alaska and southern Yukon. *A. cordifolia*

8a. Pappus brownish, subplumose; plants of southeastern Yukon. (9)

8b. Pappus white, merely barbellate; plants of various distribution. (10)

9a. Heads with ray flowers absent or reduced; immature heads nodding. *A. parryi*

9b. Heads with ray flowers present; immature heads erect. *A. mollis*

10a. Pubescence at peduncle apex white; heads often more than 1, usually erect; achenes uniformly hairy. *A. alpina*

10b. Pubescence at peduncle apex yellowish; heads rarely more than 1, usually nodding; achenes glabrous throughout, or only in the lower portion, or sometimes hairy throughout. *A. louiseana*

Arnica alpina (L.) Olin
Alpine Arnica

Plants 1–5 dm tall, the stems erect or ascending, simple or branched only in the inflorescence, sparsely to densely villous and sometimes also glandular, especially above; basal leaves 3.5–22 cm long, 0.5–2 cm broad, narrowly lanceolate to lance-elliptic, 3–5-veined, pilose and often glandular, tapering to a slender petiole, usually entire; cauline leaves (0)2–4(5) pairs, becoming smaller upwards, sessile or the lower ones petiolate, usually entire, acute to attenuate; heads solitary, or 3–5, usually erect, the peduncle apex usually densely white-villous; involucres 11–18 mm high, the bracts lanceolate, acute to acuminate, pilose and usually glandular, the tip ciliate, not with a conspicuous apical tuft of hair; rays usually 10–15, yellow, the teeth conspicuous; achenes 4–7.5 mm long, uniformly hairy, often glandular; pappus white, barbellate.

Arctic and alpine tundra or more commonly in open woods; in eastern half of Alaska and most of the Yukon; circumpolar. The specimens treated under *A. alpina* are polymorphic, and have been assigned to a series of infraspecific taxa.

1a. Leaves (at least the lowermost) remotely but conspicuously serrate-dentate; plants of central eastern Alaska and southern Yukon (includes *A. lonchophylla* Greene). *A. alpina* var. *lonchophylla* (Greene) Welsh

1b. Leaves entire (rarely inconspicuously serrate-dentate in var. *vestita*); plants of various distribution. (2)

2a. Heads solitary; disk achenes 4–6 mm long; plants less than 2.5 dm tall, occurring in northern Alaska and Yukon, and in alpine sites in southern Yukon (*A. alpina* ssp. *angustifolia* [Vahl] Maguire). *A. alpina* var. *angustifolia* (Vahl) Fern.

2b. Heads 1–5(7); disk achenes various; plants commonly more than 2.5 dm tall, or if shorter and the heads solitary, then arctic-alpine in distribution in southern Yukon, or disk achenes over 6 mm long. (3)

3a. Disk achenes 6–7.5 mm long; heads solitary; leaves and stems conspicuously tomentose; plants 0.5–2 dm tall, known from southeastern Yukon (*A. alpina* ssp. *tomentosa* [Macoun] Maguire). *A. alpina* var. *tomentosa* (Macoun) Cronq.

3b. Disk achenes 4–5.5 mm long; heads 1–7; leaves and stems pilose; plants 1.5–5 dm tall; widely distributed in eastern Alaska and southern Yukon. (4)

4a. Involucral bracts and usually the peduncles densely whitish villous, the bract surface obscured by hair; plants of the Copper River drainage in south-central Alaska. *A. alpina* ssp. *attenuata* (Greene) Maguire var. *vestita* Hultén

4b. Involucral bracts and peduncles thinly villous or hirsute, not obscured by the hair; plants widely distributed in eastern Alaska and southern Yukon. *A. alpina* ssp. *attenuata* var. *linearis* Hultén

Arnica amplexicaulis Nutt.

Plants 2–8 dm tall, the stems erect or ascending, simple and/or branching only in the inflorescence, sparsely villous with multicellular hairs; basal leaves much smaller than the cauline ones, often withered by flowering time; cauline leaves (4) 5–7(8) pairs, lanceolate to lance-elliptic, the largest ones near or slightly below the middle of the stem, with 3–5 main veins, tapering to a slender base, or merely sessile, distinctly to obscurely serrate-dentate or subentire; heads 1–5(11), the peduncle apex sparingly villous with usually yellowish hairs, often mixed with stipitate glands; involucres 9–15 mm high, the bracts narrowly lanceolate, acute to acuminate, pilose and sometimes glandular, the tips ciliate, lacking conspicuous apical tuft of hair; rays usually 8–14, yellow, the teeth conspicuous; achenes 4–6 mm long, sparsely hairy with simple and compound hairs, sometimes glandular; pappus brownish, subplumose.

Mountain slopes, woods, and subalpine sites; in coastal southern Alaska and southern Yukon; south to California and east to Montana.

1a. Leaves distinctly serrate-dentate, all sessile or subsessile; heads usually 3–7; plants of southeastern Alaska and southwestern Yukon. *A. amplexicaulis* var. *amplexicaulis*

1b. Leaves indistinctly serrate to entire, the lower cauline leaves with slender, winged petioles; heads usually 1; plants from Glacier Bay westward to Kodiak and Katmai, and less commonly in southern Yukon (*A. amplexicaulis* ssp. *prima* Maguire). *A. amplexicaulis* var. *prima* (Maguire) B. Boi.

Arnica chamissonis Less.

Plants 2–8 dm tall, the stems erect or ascending, simple or more commonly branched above, sparsely to densely tomentose or villous with multicellular hairs, and often glandular as well; basal leaves 3–11(15) cm long, 0.3–1.2 cm broad, lanceolate to oblong or oblanceolate, with 3–5 main veins, pilose or tomentose, tapering to a slender petiole, subentire to distinctly toothed, smaller than the cauline ones and often withered by flowering time; cauline leaves (4)5–8(9) pairs, lanceolate to lance-elliptic, the largest ones near or slightly below the middle of the stem, the lower ones usually petiolate, the upper sessile, subentire to distinctly toothed; heads (1)3–9, the peduncle apex sparingly to densely villous with whitish hairs, often intermixed with glands; involucres 9–15 mm high, the bracts lanceolate, obtuse, acute or less commonly acuminate, sparsely to densely pilose, ciliate, the tips with a conspicuous tuft of whitish hair; rays usually 10–16, yellow, the teeth conspicuous; achenes 4–6 mm long, hairy and glandular to glabrate; pappus brownish, barbellate to subplumose.

Open slopes, meadows, and woods; in southern Alaska from the Aleutians to the

Panhandle and in southern Yukon; south to California and New Mexico and east to Ontario.

1a. Plants whitish pilose and tomentose, known from southern Yukon (*A. foliosa* var. *incana* Gray; *A. chamissonis* ssp. *foliosa* var. *incana* [Gray] Hultén; *A. chamissonis* ssp. *incana* [Gray] Maguire; *A. chamissonis* f. *incana* [Gray] B. Boi.). *A. chamissonis* var. *incana* (Gray) Hultén

1b. Plants not both pilose and tomentose, greenish, of various distribution. (2)

2a. Pappus often subplumose; hairs at apex of peduncle with conspicuous cross-walls; plants widely distributed in southern Alaska and southern Yukon. *A. chamissonis* ssp. *chamissonis*

2b. Pappus usually barbellate; hairs at apex of peduncle lacking conspicuous cross-walls; plants of southeastern Yukon. *A. chamissonis* ssp. *foliosa* (Nutt.) Maguire

Arnica cordifolia Hook.

Plants (1.5)2–6 dm tall, the stems erect or ascending, simple or branched above, sparsely villous with multicellular hairs, and often glandular as well; basal leaves smaller than the cauline, often withered by flowering time, the petioles longer than the blades; cauline leaves 2–4(5) pairs, the blades 2–10 cm long, 2–6 cm broad, cordate-ovate to orbicular or the uppermost lanceolate, the largest ones below the middle of the stem, the lower leaves with petioles longer than the blades, the upper ones sessile or subsessile, serrate-dentate to subentire; heads 1–3 (or rarely more, when lateral branches arise from the lower axils), the peduncle apex villous with whitish hairs, often intermixed with glands; involucres 14–20 mm high, the bracts lanceolate to oblong, acuminate to acute, sparsely to densely pilose and often glandular-ciliate, the tip bearing a moderate tuft of hair; rays usually 10–15, yellow, the teeth conspicuous; achenes 4–5.5 mm long, uniformly hairy and often glandular; pappus white, barbellate.

Open to dense woods; in southeastern Alaska and southern Yukon; south to California and east to New Mexico, Colorado, Nebraska, and Michigan. Our materials belong to var. *cordifolia*.

Arnica diversifolia Greene

Plants 1.5–4 dm tall, the stems erect or ascending, simple or branched above, sparsely villous with multicellular hairs and often glandular as well, or subglabrous; basal leaves smaller than the cauline, often withered by flowering time, borne on slender petioles shorter than the blades; cauline leaves 2–4(5) pairs, the blades 2–8 cm long, 0.8–6 cm broad, or the uppermost narrowly lanceolate, the largest ones at the middle of the stem or below, becoming sessile or subsessile above, usually irregularly serrate; heads 1–3, or more, the peduncle apex sparsely to moderately villous with whitish hairs and often with glands; involucres 10–15 mm high, the bracts lanceolate, acuminate to acute, sparsely to densely pilose and often glandular, ciliate, the tip lacking a tuft of hairs; rays usually 12–15, yellow, shallowly toothed; achenes 5–7 mm long, glabrous or sparsely and uniformly hairy; pappus brownish, subplumose.

Open woods; in southern and southeastern Alaska and southern Yukon; southward to California. *A. diversifolia* is apparently rare in our region. Possibly it represents an occasional hybrid between *A. amplexicaulis* and *A. latifolia*.

Arnica latifolia Bong.

Plants 1–6 dm tall, the stems erect or ascending, simple or branched above, sparsely villous with multicellular hairs and often glandular; basal leaves smaller than the cauline, usually withered by flowering time, the petioles usually shorter than the blades; cauline leaves 2–4(5) pairs, the blades 2–10 cm long, 0.5–7 cm broad, cordate-ovate to lanceolate, the largest ones at the middle or below, the lower leaves with petioles shorter than the blades, the upper ones sessile or subsessile, usually conspicuously serrate-dentate; heads 1–5, the peduncle apex sparsely to moderately villous with whitish or yellowish hairs and often glandular as well; involucres 9–17 mm high, the bracts lanceolate acuminate to acute, sparsely pilose and often glandular, ciliate, lacking an apical tuft of hair; rays usually 8–12, yellow, shallowly toothed; achenes 5–8 mm long, glabrous or sparsely hairy, or glabrous in the lower portion; pappus white, barbellate.

Open woods, meadows, and heathlands; in southern Alaska and southern Yukon; south to California and east to Colorado, Wyoming, and Alberta.

Arnica lessingii Greene

Plants 0.5–3.5 dm tall, the stems erect or ascending, simple or rarely branched above, moderately to densely villous with multicellular hairs, not glandular; basal leaves smaller than the cauline ones, often withered at flowering time, or all leaves appearing basal, the petioles broadly winged; cauline leaves 3–5(6) pairs (sometimes all near the stem base), 1.5–14.5 cm long, 0.5–2.5 cm broad, lanceolate to elliptic or oblanceolate, obtuse to acute or rounded apically, the lower ones tapering to a winged petiole, or all sessile, entire to serrate; heads solitary, usually nodding, the peduncle apex moderately villous with brownish hairs; involucres 12–17 mm high, the bracts lanceolate to elliptic, abruptly obtuse, with a broad, callous tip, pilose, ciliate, lacking an apical tuft of hair; rays mostly 8–14, yellow, the teeth conspicuous; anthers purplish; achenes 5–6.5 mm long, sparsely hairy to glabrous; pappus brownish, barbellate to subplumose.

Arctic and alpine tundra, heath, and less commonly in open woods; over much of Alaska and Yukon; east to Mackenzie and south to British Columbia; Asia (A. lessingii ssp. norbergii Hultén & Maguire).

Arnica louiseana Farr.

Plants 0.5–4 dm tall, the stems erect or ascending, simple or rarely branched above, sparsely to moderately villous with multicellular hairs and sometimes also glandular; basal leaves 1.2–8 cm long, 0.4–1.8 cm broad, lanceolate to oblong or oblanceolate, obscurely to conspicuously 3–5-veined, pilose to glabrous, tapering to a short, broad petiole or subsessile, entire to serrate; cauline leaves (0)1–4 pairs, 2–9 cm long, 0.4–2.5 cm broad, larger than the basal leaves, becoming smaller upwards, all sessile or the lower ones with short, winged petioles, obtuse to rounded or less commonly acute; heads solitary or rarely 3, usually nodding, the peduncle apex densely villous with yellowish hairs; involucres 10–18 mm high, the bracts lanceolate, attenuate to acuminate or less commonly obtuse, sparsely to densely pilose and often glandular, sometimes tinged red, the tips ciliate, lacking a conspicuous tuft of hair; rays 9–14(16), yellow, the teeth conspicuous; achenes 4–6 mm long, glabrous or glabrous only in the lower portion, or pubescent throughout and sometimes glandular; pappus white, barbellate.

Arctic and alpine tundra, heathlands, and open woods; over much of Alaska and Yukon; east to Gaspé Peninsula and south to British Columbia and Alberta; Asia. Two more or less distinctive but intergrading phases of A. louiseana are present in our region.

1a. Involucral bracts sparsely pilose to glabrous apically, distinctly reddish; plants of broad distribution (A. brevifolia Rydb.; A. illiamnae Rydb.; A. mendenhallii Rydb.; A. nutans Rydb.; A. snyderi Raup; A. louiseana ssp. frigida [C. A. Mey.] Maguire; A. frigida C. A. Mey. ex Iljin). A. louiseana var. frigida (C. A. Mey.) Welsh
1b. Involucral bracts densely pilose al-

most or quite to the apex, obscurely or not at all reddish tinged; plants of central and eastern Alaska (*A. louiseana* ssp. *frigida* var. *pilosa* Maguire). *A. louiseana* var. *pilosa* Maguire

Arnica mollis Hook.

Plants 1.5–6 dm tall, the stems erect, simple or branched above, sparsely to moderately villous with multicellular hairs and sometimes also glandular; basal leaves smaller than the cauline and often withered at flowering time; cauline leaves 2–4 pairs, the largest ones usually near the stem base, lanceolate to elliptic or oblanceolate, all sessile or the lower ones tapering to broad petioles, serrate to entire; heads 1–5, the peduncle apex sparsely to moderately villous with whitish hairs and also glandular; involucres 10–18 mm high, the bracts lanceolate, acuminate, pilose and often glandular, ciliate, lacking an apical tuft of hairs; rays usually 14–20, yellow, the teeth short; achenes 5–7 mm long, uniformly hairy; pappus brownish, subplumose.

Alpine meadows and mountain slopes; in south-central Alaska and southeastern Yukon (?); south to California and east to Colorado.

Arnica parryi Gray

Plants 2–6 dm tall, the stems simple or branched above, sparsely to moderately villous with multicellular hairs and often with stipitate glands; basal leaves smaller than the cauline ones, often withered at flowering time; cauline leaves 2–4 pairs, the largest near the base, lanceolate, 3–20 cm long, 0.7–4 cm broad, acute, tapering basally into a slender petiole, becoming sessile or subsessile above, entire or toothed; heads 1–3 or more, nodding in bud, the peduncle apex moderately villous and also glandular; involucres 10–14 mm high, the bracts lanceolate, acute, sparsely pilose and glandular, ciliate, lacking apical tuft of hair; rays usually lacking; achenes 4–6 mm long, uni-

formly hairy, glabrous below, or glabrous; pappus brownish, barbellate to subplumose.

Alpine meadows and open woods; in southeastern Yukon; south to California and east to Colorado.

Arnica unalaschcensis Less.

Plants 0.6–4 dm tall, the stems erect, simple, moderately to densely villous with multicellular hair, not glandular; basal leaves smaller than the cauline, often withered at flowering time; cauline leaves 2–5 pairs (seldom all near the base), 2.5–15 cm long, 0.6–3.5 cm broad, lanceolate to elliptic or oblanceolate, obtuse to rounded or rarely acute apically, all sessile or the lower ones tapering to broad petioles, serrate to entire; heads solitary, erect, the peduncle apex moderately villous with whitish hair; involucres 10–18 mm high, the bracts lanceolate, abruptly obtuse, with a callous tip, sparsely pilose and glandular, ciliate, lacking an apical tuft of hair; rays mostly 12–18, yellow, the teeth conspicuous; anthers purplish; achenes 3.5–5 mm long, sparsely hairy and often glandular; pappus brownish, subplumose.

Aleutian Islands and islands of the Bering Sea; eastern Asia (*A. obtusifolia* Less.).

ARTEMISIA L.

Perennial herbs, subshrubs, or shrubs from taproots or rhizomes, the juice watery; stems spreading to ascending or erect, simple or much branched; leaves alternate or basal, entire to toothed, lobed, or divided; heads several to numerous, borne in spicate, racemose, or paniculate clusters, rarely subcapitate; involucral bracts imbricate in several rows, dry, at least the inner with scarious margins; receptacle naked or beset with long hairs, often glandular; corollas of disk flowers only, perfect, or sometimes the central ones sterile, the marginal merely pistillate; marginal corollas tubular, the central ones tubular-funnelform; pappus lacking, or a short crown;

style branches flattened; achenes subterete or angular, glabrous.

Hall, H. M., and F. E. Clements. 1923. Genus *Artemisia*. pp. 31-156. *In*: The phylogenetic method in taxonomy. Carn. Inst. Wash. Pub. No. 326.

Hultén, E. 1954. *Artemisia norvegica* Fr. and its allies. Nytt Magasin Botanikk 3:63–82.

Keck, D. D. 1946. A revision of the *Artemisia vulgaris* complex in North America. Proc. Cal. Acad. Sci. IV. 25:421–68.

1a. Plants 0.3–1.5(1.7) dm tall, caespitose; heads in dense apical clusters; distributed in western and northern Alaska and northern Yukon. (2)
1b. Plants (1)1.5–10 dm tall or more, or if less, then densely white-pilose, or heads not as above, caespitose or not; heads in elongate spikes, racemes, or panicles; distribution various. (6)

2a. Flowering stems arising from elongate, spreading, woody branches of the previous season; lower leaves typically once to twice ternately divided. *A. alaskana*
2b. Flowering stems arising directly from the caudex; lower leaves various. (3)

3a. Central flowers of head lacking normal ovaries, abortive; corollas glabrous. *A. campestris*
3b. Central flowers of head with normal ovaries, fertile; corollas glabrous or hairy. (4)

4a. Corolla distinctly hairy near the apex; basal leaves pinnatifid or ternately divided, silvery-pilose with appressed hairs, lacking conspicuous apical tuft of long silvery hair. *A. glomerata*
4b. Corolla glabrous; basal leaves entire, or once to twice pinnatifid or ternately divided, variously hairy. (5)

5a. Leaves greenish, only sparsely strigulose, once to twice ternately divided; flowers purplish or only rarely yellowish. *A. globularia*

5b. Leaves silvery-pilose with long hairs, bearing conspicuous tufts at the lobed apices, merely 3-lobed; flowers yellow. *A. senjavinensis*

6a. Flowering stems arising from more or less elongate, spreading, woody branches of the previous season, seldom directly from the caudex; lower leaves 0.3–2.5(5) cm long, the ultimate segments linear to oblong. (7)
6b. Flowering stems arising directly from the caudex or rhizome apex, herbaceous or only slightly woody below; lower leaves, when present, usually more than 3 cm long. (8)

7a. Leaves of sterile shoots and lower stems 0.3–1.5(2.5) cm long, the segments less than 0.8 mm broad; receptacle often beset with long, white hairs. *A. frigida*
7b. Leaves of sterile shoots and lower stem 1.5–2.5(5) cm long, the segments 0.6–2(3) mm broad; receptacle glabrous. *A. alaskana*

8a. Leaves all entire, or the lower ones sometimes toothed or lobed, glabrous and green above and beneath, or white-hairy on both surfaces. (9)
8b. Leaves deeply incised, pinnatifid, or ternately divided, variously pubescent. (10)

9a. Leaves green above and beneath; central flowers of head with normal ovaries; plants of southern Alaska and southern Yukon. *A. dracunculus*
9b. Leaves white-hairy above and beneath, or somewhat greenish above; central flowers of head with abortive ovaries; plants of southern Yukon. *A. ludoviciana*

10a. Leaves primarily or all cauline, tomentose below (glabrous in *A. biennis*), at least when young, green or greenish above; plants with rhizomes, or annual and taprooted. (11)
10b. Leaves primarily basal, the cauline reduced upwards, silvery-villous to

strigulose or glabrate, scarcely tomentose, more or less uniformly colored above and beneath; plants from caudices or only occasionally from rhizomes. (13)

11a. Plants annual or biennial, from taproots; leaves green, essentially glabrous; adventive weeds. *A. biennis*

11b. Plants perennial, rhizomatous; leaves white-tomentose beneath, green to tomentose above. (12)

12a. Leaves bipinnately divided (at least some), the ultimate segments again toothed, usually less than 5 cm long; plants of northern British Columbia. *A. michauxiana*

12b. Leaves once to twice pinnately parted, the ultimate segments entire, mostly more than 5 cm long; plants of broad distribution. *A. tilesii*

13a. Main cauline leaves twice to thrice pinnately divided, green on both surfaces, glabrous or sparsely villous (densely so in some *A. norvegica*); central flowers of head with normal ovaries. (14)

13b. Main cauline leaves simple or merely once pinnatifid or once to twice palmately divided, uniformly silvery-silky to strigulose, seldom greenish; central flowers of head with normal or abortive ovaries. (15)

14a. Basal leaves (5)10 cm long or more, the main primary divisions mostly more than 2 cm long; heads with peduncles 1 cm long or less; plants of central eastern Alaska. *A. laciniata*

14b. Basal leaves less than 10 cm long, the main primary divisions mostly less than 2 cm long, or if the basal leaves are larger, then the peduncles 1–5 cm long or more; plants of broad distribution. *A. norvegica*

15a. Plants less than 1 dm tall, densely white-hairy; plants known only from the western Aleutians. *A. aleutica*

15b. Plants mostly 1–5 dm tall, densely white-hairy to strigulose and somewhat greenish; plants widely distributed, but not of the Aleutians. (16)

16a. Involucral bracts sparsely to densely pilose; corollas glabrous or nearly so, yellow, the central flowers with normal ovaries; leaves or leaf segments lacking apical tufts of hair; plants disjunct in western, northern, central, and southern Alaska and southern Yukon. *A. furcata*

16b. Involucral bracts glabrous to sparsely or less commonly densely hairy; corollas glabrous to long-hairy, yellow, or more commonly tinged reddish or brownish, the central flowers with abortive ovaries; leaves or leaf segments commonly with long, apical tufts of hair; plants widely distributed in Alaska and Yukon. *A. campestris*

Artemisia alaskana Rydb.
Alaska Wormwood

Plants with a stout caudex or woody crown, shrubby at the base; flowering stems arising from short, prostrate to ascending, woody offsets, rarely if ever directly from the caudex, (1.5)2–5.5(6) dm tall, white-tomentose, turning brownish in age; leaves of basal offsets 1.5–5 cm long, once or more commonly twice ternately divided, the segments flat, 0.6–2(3) mm broad, oblong to broadly linear, lacking stipulelike divisions near the base; leaves of flowering stems reduced upwards, merely ternate or finally entire, tomentose above and below or somewhat greenish above; inflorescence paniculate, or less commonly merely racemose; heads numerous, nodding, sessile or some on peduncles to 3 cm long; involucres 3.5–5 mm high, 6–9 mm broad, the bracts tomentose, with brownish or hyaline margins; marginal flowers pistillate, fertile; central flowers perfect, fertile, the corollas glabrous (often glandular), yellowish; receptacle glabrous; achenes glabrous.

Artemisia alaskana Rydb. (× 0.4).

Open, dry hillsides, flood plains, and lake shores; in the eastern two-thirds of Alaska (except for extreme northern and southern portions) and in southern to northern Yukon and adjacent Mackenzie; south to northern British Columbia; Asia (includes *A. bigelovii* authors, not Gray; *A. cana* authors not Pursh; *A. krushiana* authors, not Besser; *A. tyrellii* Rydb.). *A. alaskana* has been mistaken for several other species in our region, but the woody offsets and truly tomentose stems and leaves distinguish it from other species, except for *A. frigida,* which it closely resembles.

Artemisia aleutica Hultén
Aleutian Wormwood

Plants with simple or branched caudex and stout taproot, the caudex branches short, clothed with persistent leaf bases, the stems arising directly from the caudex, 0.5–1 dm tall, villous-tomentose; leaves of basal rosettes 1.5–5 cm long, twice palmately divided, the ultimate segments narrowly oblong to oblanceolate, acute, densely white-villous (the hair turning brownish in age) on both surfaces, the petioles often greatly expanded; cauline leaves similar to the basal, greatly reduced and merely ternate above; inflorescence spicate or racemose; heads numerous, erect or spreading, borne sessile or on peduncles to 1.5 cm long; involucres 5–7 mm high, 6–8 mm broad, the bracts white-villous, glabrate in age, greenish or brownish, the margin hyaline; marginal flowers pistillate, fertile; central flowers sterile, the ovary abortive, the corolla hairy, purplish; receptacle glabrous; achenes glabrous.

Open slopes; in western Aleutian Islands; endemic. This entity is apparently most closely related to *A. campestris.*

Artemisia biennis Willd.
Biennial Wormwood

Plants annual or biennial, with taproots, the stems 3–10 dm tall or more, glabrous; basal leaves commonly withered by flowering time; cauline leaves well developed, mostly 5–15 cm long, once pinnately divided, the segments oblong to oblanceolate, again toothed, essentially glabrous, green; inflorescence spicate or in spicate panicles; heads numerous, crowded, sessile or subsessile, erect or nearly so; involucres 2–3 mm high, 2–4 mm broad, the bracts glabrous, greenish to yellowish, the margin hyaline; marginal flowers pistillate, fertile; central flowers perfect, fertile, the corollas glabrous; receptacle and achenes glabrous.

Weed of disturbed soils; in central eastern Alaska; disjunctly southward and eastward through much of North America; Europe. *A. biennis* is probably adventive in Alaska.

Artemisia campestris L.

Plants with simple or branched caudex and stout taproot, the caudex branches short, clothed with persistent leaf bases, the stems arising directly from the caudex, (0.7)1–5(7) dm tall, villous-tomentose to glabrate or glabrous; leaves of basal rosettes 2–17(22) cm long, twice to thrice pinnately or palmately divided, the ultimate segments linear to narrowly oblong, acute, sparsely strigulose and greenish to densely silky-pilose on both surfaces; cauline leaves similar to the basal, reduced upwards, becoming merely ternate or entire; inflorescence subcapitate when young, spicate, racemose or paniculate; heads numerous, erect, borne sessile or on peduncles to 1(2) cm long; involucres 2.5–5 mm high, 2–7 mm broad, the bracts glabrous to sparsely pilose, or less commonly densely silky-hairy, greenish to yellowish or brownish, the margins hyaline; marginal flowers pistillate, fertile; central flowers sterile, the ovary abortive, the style often undivided, the corolla sparsely hairy to glabrous; receptacle and achenes glabrous.

Arctic and alpine tundra, heathlands, and open woods; in much of Alaska and Yukon; east to the Atlantic and south to Oregon, Arizona, Michigan, and Vermont; circumboreal (*A. borealis* Pallas). Our highly variable material belongs to ssp. *borealis* (Pallas) H. & C.

1a. Inflorescence spicate, racemose, or compactly paniculate; involucral bracts glabrous to sparsely pilose; plants sparsely strigulose to loosely villous, distribution broad (*A. borealis* ssp. *purshii* [Besser] Hultén; *A. borealis* var. *purshii* Besser; *A. campestris* var. *purshii* [Besser] Cronq.; *A. manca* Rydb.; *A. richardsonia* Besser; *A. spithamaea* Pursh). *A. campestris* var. *borealis* (Pallas) M. E. Peck

1b. Inflorescence compactly to openly paniculate; involucral bracts glabrous or densely villous; plants subglabrous to strigulose or densely, appressed silky-hairy, known from south-central Alaska and southern Yukon. (2)

2a. Plants sparsely strigulose to subglabrous, mostly 2–5(7) dm tall; involucres glabrous or only sparsely pilose; known from southern Yukon and central eastern Alaska. *A. campestris* var. *canadensis* (Michx.) Welsh

2b. Plants densely, appressed silky-hairy, mostly 2.5–4 dm tall; involucres densely to moderately pilose, glabrate in age; known from south-central and southwestern Alaska. *A. campestris* var. *strutziae* Welsh

Artemisia dracunulus L.

Plants with stout rhizomes, the roots fibrous, the stems 4.5–9 dm tall or more, glabrous; leaves primarily cauline, 1–7 cm long, 0.1–0.9 cm broad, entire, or some of them cleft, glabrous or nearly so, green on both surfaces; inflorescence paniculate; heads numerous, nodding, borne sessile or on short peduncles; involucres 1.8–4 mm high, 2.5–6 mm broad, the bracts glabrous, greenish or yellowish, with hyaline margins; marginal flowers pistillate and fertile, the style bilobed; central flowers sterile, the ovary abortive, the style branches often undivided, the corolla glabrous (often glandular); receptacle and achenes glabrous.

Open, dry slopes; in south-central Alaska and southern Yukon; east to Manitoba and south to Mexico, Texas, and Illinois; Eurasia (*A. dracunculoides* Pursh; *A. glauca* Pallas).

Artemisia frigida L.
Prairie Sagewort

Plants with stout caudex and woody crown, usually shrubby at the base; flowering stems arising from short, prostrate or ascending, woody offsets, seldom directly from the caudex, 0.5–4.5 dm tall, white-tomentose to strigulose (turning brownish in age); leaves of basal offsets much like the stem leaves, 0.5–1.5(2.5) cm long, twice to thrice ternately (or subpinnately) divided into linear segments less than 0.5 (0.8) mm broad, often with stipulelike divisions near the base, whitish pilose-tomentose throughout, or fading brownish; inflorescence paniculate or less commonly borne sessile or on very short peduncles; involucres 2–3.5 mm high, 4–6 mm broad, the bracts pilose-tomentose, with brownish-scarious margins; marginal flowers pistillate, fertile; central flowers perfect, fertile, the corolla glabrous (often glandular), yellow or tinged reddish; receptacle long-hairy to glabrous or nearly so; achenes glabrous.

Steep, open slopes, lake shores, flood plains, and sandy terraces; in the eastern two-thirds of Alaska and much of the Yukon; east to Quebec and south to Arizona and Kansas; Asia. *A. frigida* is more or less uniform in our region, except that materials from eastern and south-central Alaska, and from western Yukon lack hair on the receptacle, or possess only a few hairs, and appear to have slightly larger heads than normal. In both of these features the plants approach *A. alaskana*, and might indicate hybridization of the two entities.

1a. Receptacle densely long-hairy; heads often less than 5 mm broad; plants of broad distribution. *A. frigida* var. *frigida*

1b. Receptacle glabrous, or with 1–few hairs; heads often more than 5 mm broad; plants from southwestern Yukon and from south-central and eastern Alaska (*A. rupestris* L. ssp. *woodii* Neilson). *A. frigida* var. *williamsiae* Welsh

Artemisia furcata Bieb.

Plants with a simple or more commonly branched caudex and stout taproot, the caudex branches short, clothed with persistent

leaf bases, the stems arising directly from the caudex, 0.7–3.6 dm tall, strigulose to glabrate; leaves of basal rosettes 2–10(12) cm long, (1)2–3 times palmately divided, the ultimate segments narrowly oblong to linear, acute to obtuse, sparsely to densely strigulose above and beneath or less densely so and greenish above; cauline leaves pinnately divided, ternate, or the upper ones entire; inflorescence racemose or spicate; heads few to numerous, erect or spreading, or less commonly some nodding, borne sessile or on peduncles to 2(3) cm long; involucres 3–6 mm high, 4.5–8 mm broad, the bracts sparsely to moderately pilose, greenish, the margins brownish-scarious; marginal flowers pistillate, perfect; central flowers perfect, fertile, yellowish or seldom purplish, glabrous or nearly so; receptacle glabrous; achenes glabrous.

Arctic and alpine tundra, often on talus slopes; on islands of the Bering Sea, western, southern and interior Alaska, and southern Yukon; disjunctly south to Washington; Asia (*A. trifurcata* Steph. ex Spreng.; *A. hyperborea* Rydb.).

1a. Basal leaves mostly 1(less commonly 2–3) times palmately or pinnately divided; involucres moderately to densely villous; plants of northern and western Alaska. *A. furcata* var. *furcata*

1b. Basal leaves commonly 2–3 times palmately compound; involucres sparingly to moderately villous (rarely densely so); plants of southern Alaska and southern Yukon (*A. heterophylla* Besser). *A. furcata* var. *heterophylla* (Besser) Hultén

Artemisia globularia Cham. ex Besser

Plants with a simple or branched caudex and stout taproot, the caudex branches short, clothed with persistent leaf bases, the stems arising directly from the caudex, 0.3–1.5(1.7) dm tall, villous-tomentose; leaves of basal rosettes 1–4.5 cm long, ternately or palmately once to twice divided, the ultimate segments oblong to linear, uniformly though sparsely strigulose and greenish to densely whitish-strigulose; cauline leaves 1–few, usually once ternate, much reduced; inflorescence subcapitate; heads several, borne sessile, or the lowermost on short peduncles; involucres 3.5–6 mm high, 6–11 mm broad, the bracts pilose, with brownish-scarious margins; marginal flowers pistillate, fertile; central flowers perfect, fertile, the corollas glabrous (often glandular), purplish or rarely yellow; receptacle glabrous; achenes glabrous.

Arctic and alpine tundra; in northwestern, west-central and southwestern Alaska, and less commonly from the mountains of the interior; Asia (*A. globularia* var. *lutea* Hultén; *A. globularia* f. *lutea* [Hultén] B. Boi.; *A. norvegica* ssp. *globularia* [Besser] H. & C.; *Ajania globularia* [Besser] Poljak.).

Artemisia glomerata Ledeb.

Plants with simple or more commonly branched, often subterranean caudex, the caudex branches short or elongate (when underground), the upper parts clothed with persistent leaf bases, the stems arising directly from the caudex branches, 0.5–1.5 dm tall, pilose to villous-tomentose; leaves of basal rosettes 0.5–3(4.5) cm long, once to twice ternately divided, the segments oblong-lanceolate to linear, uniformly whitish-strigulose throughout; cauline leaves few, simple or once ternately or pinnately divided; inflorescence subcapitate to densely spicate, or the lowermost heads racemose; heads several to many, sessile or the lower ones on peduncles to 1.5 cm long; involucres 3–4 mm high, 3.5–5 mm broad, the bracts pilose dorsally, with brownish-scarious margins; marginal flowers pistillate, fertile; central flowers perfect, fertile, the corollas hairy, yellowish; receptacle and achenes glabrous.

Arctic and alpine tundra; in northern and northwestern Alaska, and islands of the Bering Sea, and disjunctly in southwestern Alaska and in northern Yukon (*A. norvegica* ssp. *glomerata* [Ledeb.] H. & C.; *Ajania glomerata* [Ledeb.] Poljak.).

Artemisia laciniata Willd.

Plants with simple or branched caudex and stout taproot, the caudex branches short, clothed with persistent leaf bases, the stems arising directly from the caudex, 2–5 dm tall, strigulose to spreading-hairy or glabrate (especially below); leaves of basal rosette 5–20 cm long, twice to thrice pinnately divided, the segments lance-acute, sparsely hairy to strigulose or pilose and greenish on both surfaces; cauline leaves once to twice pinnately divided, or the uppermost entire; inflorescence paniculate, racemose, or merely spicate; heads several to numerous, spreading to nodding, borne sessile or on short peduncles to 1 cm long; involucres 3–5 mm high, 4–8 mm broad, the bracts glabrous to sparsely pilose, greenish or yellowish, the margins hyaline; marginal flowers pistillate, fertile; central flowers perfect, fertile, the corollas hairy, yellowish; receptacle and achenes glabrous.

Open, dry hillsides; in central Alaska; Eurasia (*A. laciniatiformis* Kom.; *A. macrobotrys* Ledeb., at least in part; *A. tanacetifolia* authors, not L.). This entity is apparently closely related to *A. norvegica*, but can be distinguished by the well-developed basal leaves and usually smaller heads.

Artemisia ludoviciana Nutt.

Plants with stout rhizomes, the roots fibrous, the stems 3–10 dm tall, white-to-mentose; leaves primarily cauline, 3–9 cm long, 0.3–1 cm broad, entire or merely lobed, white-tomentose above and beneath; inflorescence paniculate; heads numerous, erect or spreading, borne sessile or on short peduncles; involucres 2.5–4.5 mm high, 3–4 mm broad, the bracts tomentose, with scarious margins; marginal flowers pistillate, perfect; disk flowers perfect, fertile, the corollas glabrous, yellow; receptacle and achenes glabrous.

Open slopes; in northern British Columbia (at Lake Bennett), where it might be adventive; otherwise from southern British Columbia south to Mexico and east to Arkansas, Illinois, and Ontario (*A. gnaphalodes* Nutt.; *A. ludoviciana* var. *gnaphalodes* [Nutt.] T. & G.).

Artemisia michauxiana Bess. in Hook.

Plants with rhizomes or with a caudex and stout taproot, the stems 2–5(6) dm tall, white-tomentose to glabrous and green; leaves primarily cauline, 1.5–5 cm long, once to twice pinnately divided, bright green above, white-tomentose beneath; inflorescence paniculate; heads numerous nodding or finally erect, borne sessile, or on slender peduncles to 1.5 cm long; involucres 3–4 mm high, (3)3.5–5 mm broad, the bracts glabrous or nearly so, greenish or yellowish, with hyaline margins; marginal flowers pistillate, fertile; inner flowers perfect, fertile, the corollas glabrous, yellowish or purplish; receptacle glabrous; achenes glabrous.

Open slopes; in northern British Columbia (Lake Atlin vicinity) and south to Utah, Nevada, and Wyoming. This species should be sought in southern Yukon.

Artemisia norvegica Fries

Plants with simple or branched caudex and stout taproot, the caudex branches short, clothed with persistent leaf bases, the stems arising directly from the caudex, 1–7(10) dm tall, glabrous to sparsely villous and often reddish, or less commonly densely pilose; leaves of basal rosettes mostly 3.5–10 cm long (rarely to 20 cm long on sterile shoots), twice to thrice pinnately divided, the segments lance-attenuate to acute, glabrous or nearly so and green on both surfaces or sparsely to moderately whitish-villous; cauline leaves similar to the basal, the lower ones often longer than the basal, becoming smaller upwards, often with stipulelike divisions near the base; inflorescence paniculate, race-

mose, or less commonly spicate; heads few to numerous, nodding (or the upper ones erect), borne sessile or on peduncles to 6(10) cm long; involucres 3.5–7 mm high, 7–13(16) mm broad, the bracts glabrous or nearly so and greenish, with brownish-scarious margins, or less commonly whitish-pilose; marginal flowers pistillate, fertile; central flowers perfect, fertile, the corollas long-hairy from near the base, or rarely glabrous, yellowish, often tinged reddish; receptacle and achenes glabrous.

Arctic and alpine tundra, heathlands, or less commonly in open woods; over most of Alaska and Yukon; east to Mackenzie, and south to California and Colorado; Eurasia (*A. arctica* Less.; *A. chamissoniana* Besser ex Hook.; *A. longepedunculata* Britt. & Rydb.). Two more or less completely intergrading varieties are present in our materials. The extremes are remarkably distinct and seem to require taxonomic recognition.

1a. Involucres commonly sparsely to densely whitish-pilose (sometimes glabrous); primary cauline leaves 5.5 cm long or less, often pilose with long hairs (sometimes glabrous); corollas glabrous to densely hairy; distributed in northern and northwestern Alaska and northern Yukon (*A. comata* Rydb.; *A. arctica* ssp. *comata* [Rydb.] Hultén; *A. trifurcata* authors, not Steph. ex Spreng.). *A. norvegica* var. *comata* (Rydb.) Welsh

1b. Involucres glabrous; primary cauline leaves often over 5.5 cm long, glabrous or nearly so; corollas usually hairy from near the base; widely distributed in Alaska and Yukon (*A. arctica* var. *beringensis* Hultén; *A. norvegia* ssp. *beringensis* [Hultén] Hultén). *A. norvegica* var. *saxatilis* (Besser) Jeps.

Artemisia senjavinensis Besser

Plants densely caespitose, with branching caudex and stout taproot, the caudex branches short to elongate, ascending to erect, clothed with persistent leaves, the stems arising directly from the caudex, 0.4–1 dm tall, white villous-tomentose (becoming brownish in age); leaves of basal rosettes 0.4–1(1.4) cm long, apically 3-lobed, white villous-tomentose; cauline leaves several, ternately divided or pinnately divided, or the uppermost entire; inflorescence subcapitate or the lowermost heads on short peduncles; heads several to many, erect or spreading; involucres 2.5–4.5 mm high, 3.5–5 mm broad, the bracts densely white-pilose, with brownish-scarious margins; marginal flowers pistillate, fertile; central flowers perfect, fertile, the corollas glabrous, yellow; receptacle and achenes glabrous.

Open gravelly slopes, in alpine tundra or heath; in extreme western Alaska, from the Seward Peninsula and vicinity of Kotzebue Sound; Asia (*A. androsacea* Seem.; *Ajania senjavinensis* [Besser] Poljak.).

Artemisia tilesii Ledeb.

Plants with stout rhizomes, the roots fibrous, the stems (0.5)2.5–12 dm tall or more, loosely tomentose to glabrate, often purplish tinged; leaves primarily cauline, (2)6–15 cm long, once or twice pinnatifid, or the upper ones sometimes entire, the lobes ascending, typically green and sparsely tomentose to glabrate above and densely whitish-tomentose beneath; inflorescence paniculate; heads several to many, nodding or spreading, or less commonly erect, borne sessile or on short to elongate peduncles; involucres 3–5 mm high, 8–10 mm broad, the bracts sparingly to densely cobwebby-tomentose, brownish or greenish, the margins scarious; marginal flowers pistillate, fertile; central flowers perfect, fertile, the corollas glabrous (often glandular), yellow, often tinged reddish; receptacle and achenes glabrous.

Arctic and alpine tundra, heathlands, and open woods; through most of Alaska and Yukon; east to Hudson Bay and south to Oregon and Montana; Asia (*A. hookeriana* Besser ex Hook.). Three more or less poorly defined varieties are recognizable among materials from Alaska and Yukon.

1a. Leaves subpalmately divided, the main lateral lobes often more than 5 cm long; plants of the Aleutian Islands (*A. unalaskensis* Rydb.; *A. unalaskensis* var. *aleutica* Hultén). *A. tilesii* var. *aleutica* (Hultén) Welsh

1b. Leaves pinnately divided, the main lateral lobes usually less than 3 cm long; plants of broad distribution. (2)

2a. Inflorescence compact, with few to many flowers, often overtopped by the upper leaves; involucral bracts 12–28; plants mostly of northern and insular western Alaska, and northern Yukon, or of alpine sites. *A. tilesii* var. *tilesii*

2b. Inflorescence compact to much branched, many-flowered, not overtopped by leaves; involucral bracts mostly 11–13; plants of broad distribution (*A. gormanii* Rydb.; *A. tilesii* ssp. *gormanii* [Rydb.] Hultén; *A. elatior* [T. & G.] Rydb.; *A. tilesii* var. *elatior* T. & G.; *A. tilesii* ssp. *elatior* [T. & G.] Hultén; *A. tilesii* ssp. *unalaschcensis* [Besser] Hultén). *A. tilesii* var. *unalaschcensis* Besser

ASTER L.

Perennial rhizomatous herbs, or less commonly from caudices, with watery juice; stems decumbent to ascending or erect, simple or branched; leaves alternate, simple, entire or toothed; heads solitary or few to several in corymbose clusters; involucral bracts strongly imbricate to subequal, herbaceous throughout, or with scarious margins near the base; receptacle flat or merely convex, naked; rays blue, purple, pink, or white, few to numerous, pistillate; disk flowers numerous, perfect, fertile, yellow or tinged reddish or purplish; pappus of capillary bristles; style branches flattened, oblong to lanceolate, mostly more than 0.5 mm long; achenes 2–several-nerved, pubescent or glabrous.

1a. Lower cauline leaves long-petiolate, the blades ovate-lanceolate to lanceolate, with subcordate to truncate or abruptly obtuse bases; plants of northern British Columbia. A. ciliolatus

1b. Lower cauline leaves long-petiolate to sessile, the blades attenuate to acute or sessile and auriculate-clasping basally, not both long-petiolate and with blades subcordate to truncate or obtuse basally; plants of various distribution. (2)

2a. Involucral bracts glabrous dorsally, the margin usually ciliate; peduncles not glandular (see also A. alpinus). (3)

2b. Involucral bracts hairy and often glandular dorsally; peduncles hairy or glandular, or both. (6)

3a. Leaves oblong to lanceolate, usually more than 10 mm broad; ray flowers blue, purple, or pink; plants of various distribution. A. subspicatus

3b. Leaves linear, less than 10 mm broad; ray flowers white or pale purplish; plants of southern east-central Alaska and southern Yukon. (4)

4a. Plants annual; rays lacking on marginal flowers; plants known from southern Yukon. A. brachyactis

4b. Plants perennial; rays present and conspicuous on marginal flowers; plants of southern Alaska and southern Yukon. (5)

5a. Stems softly pubescent throughout; middle and upper stem leaves merely sessile, not auriculate-clasping; plants of open woods in southern Yukon and east-central Alaska. A. commutatus

5b. Stems pubescent only above, often the hair in decurrent lines below the leaf bases; middle and upper cauline leaves more or less auriculate-clasping; plants usually of cold bogs, in the southeastern quarter of continental Alaska and southern Yukon. A. junciformis

6a. Plants with heads solitary; involucral bracts subequal; achenes distinctly flattened, 2-nerved; ray flowers white or pink; known from southern and northern Yukon (rarely southern Alaska). A. alpinus .

6b. Plants with usually more than 1 head; involucral bracts various; achenes not much flattened, several-nerved; ray flowers mostly blue, purple, or pink. (7)

7a. Involucral bracts and peduncles merely hairy, not glandular, usually tinged purplish; narrower lower portion of disk-corolla tube equaling or surpassing the uppermost expanded portion; plants widely distributed in Alaska and Yukon. A. sibiricus

7b. Involucral bracts and peduncles glandular, greenish or purplish; narrower lower portion of disk-corolla tube shorter than the uppermost expanded portion; plants of various distribution. (8)

8a. Leaves mostly sharply serrate, ovate to elliptic or obovate; involucral bracts strongly imbricate; plants of northern British Columbia. A. conspicuus

8b. Leaves entire or serrate, lanceolate to oblong or linear; involucral bracts moderately imbricate to subequal; plants of southern Yukon, southeastern Alaska, and northern British Columbia. (9)

9a. Leaves linear-lanceolate to narrowly oblong, the lower ones merely sessile, less than 10 mm broad; plants less than 3 dm tall, known only from southwestern Yukon. A. yukonensis

9b. Leaves lanceolate to lance-oblong, the lower ones attenuate basally, often petiolate, more than 10 mm broad; plants 4–9 dm tall, known from southern Yukon, northern British

Columbia, and southeastern Alaska. *A. modestus*

Aster alpinus L.
Alpine Aster

Plants from a caudex or short rhizome, the stems arising singly, or several together, erect or ascending, (0.3)0.8–3 dm tall, villous with multicellular hairs; basal leaves 2–8 cm long, 0.3–1.6 cm broad, oblanceolate, obtuse to rounded apically, tapering to the petiole, entire, 3-veined, sparsely spreading-hairy above and below, ciliate; cauline leaves reduced upwards, becoming oblong to linear and sessile, not auriculate-clasping; heads solitary, the peduncles villous with multicellular hairs; involucres 6–10 mm high, 14–20(24) mm broad, the bracts oblong, hairy and often glandular dorsally or rarely almost glabrous, in several subequal series, greenish throughout; rays white to pink or blue, mostly 8–15 mm long, 1–2 mm broad; pappus white to tan; achenes flattened, 2-nerved, pubescent.

Open slopes, woods, sandy flats, lake shores, tundra, and clay terraces; in southern and northern Yukon and in southwestern Alaska; east to Mackenzie and south to Colorado; Eurasia. American material belongs to var. *vierhapperi* (Onno) Cronq. (*A. alpinus* ssp. *vierhapperi* Onno).

Aster brachyactis Blake

Plants annual, from a taproot, commonly branching from most leaf axils, the stems erect or ascending, 1–5 dm tall or more, glabrous; basal leaves usually absent by flowering time; cauline leaves 3–12 cm long, 0.1–0.9 cm broad, linear to narrowly oblong, glabrous above and beneath, ciliate; heads several to numerous, in a paniculate inflorescence; involucre 5–11 mm high, 6–12 mm broad, the bracts lance-linear, glabrous dorsally, short-ciliate or glabrous marginally, subequal to somewhat imbricate, green at the tip, the base whitish; outer pistillate flowers with tubular co-

rollas lacking rays; pappus whitish, surpassing the corollas.

Moist, usually saline soils, often near lake shores or ponds; in southern Yukon; southward to Washington and Wyoming and eastward to Missouri and Minnesota.

Aster ciliolatus Lindl. ex Hook.

Plants with elongate rhizomes, the stems arising singly or less commonly several together, erect, 2–10 dm tall or more, pubescent with spreading, multicellular hairs or glabrous; basal leaves usually absent by flowering time, long-petiolate; lower cauline leaves 8–20 cm long, the blades mostly 5–10 cm long and 2–4.5 cm broad, ovate-lanceolate, subcordate to truncate or obtuse basally, acute to acuminate apically, pubescent like the stem or glabrous, serrate, the margin scabrous, the petioles broadly winged; cauline leaves gradually reduced upwards, finally lanceolate, sessile and often subentire, not auriculate-clasping; heads numerous in corymbose panicles, the peduncles spreading-hairy to glabrous; involucres 5–8 mm high, 5–12 mm broad, the bracts lance-oblong, glabrous dorsally, short-ciliate, imbricate in several series, green at the tip, the base whitish; rays blue, 8–15 mm long, 1–1.5 mm broad; pappus whitish; achenes several-nerved, glabrous or pubescent.

Open woods; in northern British Columbia; east to Quebec and south to North Dakota, South Dakota, and Montana (*A. lindleyanus* T. & G.; *A. laevis* authors, not L.).

Aster commutatus (T. & G.) Gray

Plants from short rhizomes, the stems arising singly or several together, erect, mostly 3–5 dm tall, soft-hairy throughout; basal leaves linear, sessile, usually withered by flowering time; cauline leaves 2–3 cm long, 0.1–0.2(0.3) cm broad, linear, entire, sessile, not auriculate-clasping, short-pubescent above, glabrous beneath, revolute; heads several to many, in paniculate clusters, the peduncles hairy, not glandular; involucres 4–7 mm high, 6–10 mm broad, the bracts lanceolate, glabrous, in 2 subequal series, green throughout or the inner ones white at the base; rays white or pale pinkish, mostly 6–10 mm long, 1–1.5 mm broad; pappus white; achenes several-nerved, hairy.

Open woods; in east-central Alaska and southern Yukon; southward to Arizona, New Mexico, and Kansas and eastward to Minnesota (A. *elegantulus* Porsild).

Aster conspicuus Lindl. in Hook.

Plants with elongate rhizomes, the stems arising singly, erect, 3–10 dm tall, strongly glandular above and sometimes also hairy; basal leaves smaller than the middle cauline ones, usually withered by flowering time, short-petiolate to subsessile; cauline leaves mostly 5–15(18) cm long and 1–6 (8) cm broad, ovate to broadly elliptic, serrate, acute to obtuse basally, not auriculate-clasping, sessile or nearly so, acute to acuminate apically; heads few to many, in corymbose panicles, the peduncles glandular; involucres 9–12 mm high, 9–15 mm broad, the bracts lanceolate, densely glandular, imbricate in several series, green at the tip, whitish at the base; rays blue to violet, mostly 10–15 mm long; pappus whitish; achenes several-nerved.

Open woods; in northern British Columbia; southward to Oregon, Idaho, and Wyoming and eastward to Saskatchewan.

Aster junciformis Rydb.

Plants from short to elongate rhizomes, the stems arising singly, erect, 2–6 dm tall, glabrous or glabrate below, pubescent in lines only, or throughout above; basal leaves linear to oblong, sessile, or nearly so, usually withered by flowering time; cauline leaves 2–7 cm long, 0.2–0.7 cm broad, linear to oblong, entire, sessile, often more or less auriculate-clasping, glabrous above and below, scabrous-ciliate; heads few to many, in corymbose panicles, the peduncles

villous, not glandular; involucres 5–7 mm high, 9–14 mm broad, the bracts oblong, glabrous, imbricate in several series, greenish throughout or more commonly whitish, at least near the base and often almost throughout; rays white to pale violet, 7–15 mm long; pappus white; achenes several-nerved, hairy.

Cold bogs and wet meadows; in the southeastern third of continental Alaska and southern Yukon; east to Quebec and south to New Jersey, Minnesota, Colorado, and Idaho (A. *junceus* authors, not Ait.).

Aster modestus Lindl. ex Hook.

Plants with elongate rhizomes, the stems arising singly, erect, mostly 4–9(10) dm tall, sparsely to densely villous with multicellular hairs and glandular above, glabrate to glabrous below; basal leaves smaller than the middle cauline ones, usually withered by flowering time; cauline leaves 3.5–12 cm long, 0.7–2.5(4) cm broad, lanceolate to elliptic, serrate to subentire, sessile, sometimes more or less auriculate-clasping, acuminate to acute apically, glabrous above, sparsely villous beneath; heads few to many, in corymbose panicles, the peduncles glandular; involucres 7–11 mm high, 12–20 mm broad, the bracts narrowly lanceolate, glandular dorsally, not ciliate, subequal, herbaceous throughout, or the inner ones whitish near the base; rays violet or purple, mostly 8–15 mm long, 1.2–1.8 mm broad; pappus whitish or yellowish; achenes several-nerved, hairy.

Open woods and mountain slopes, often along streams; in southeastern Alaska, southern Yukon, and northern British Columbia; south to Oregon, Idaho, and Montana and east to Ontario and Minnesota (A. *unalaschkensis* var. *major* Hook.; A. *major* [Hook.] Porter).

Aster sibiricus L.
Siberian Aster

Plants with elongate rhizomes, the stems arising singly or few to several together, ascending to erect, 0.5–4.5(5) dm tall, sparsely to moderately villous with multicellular hairs, rarely glabrate; basal leaves smaller than the middle cauline ones, usually withered by flowering time; cauline leaves 2–8(10) cm long, 0.5–3.5 cm broad, lanceolate to elliptic, oblong, or oblanceolate, sharply serrate to entire, sessile or the lower ones short-petiolate, not auriculate-clasping, acute to acuminate or less commonly obtuse apically, hairy above and beneath, or glabrate to glabrous above; heads solitary or more commonly 2–several in corymbose clusters, the peduncles villous, not glandular; involucres 7–15 mm high, 13–24 mm broad, the bracts oblong to lance-oblong, hairy dorsally, not glandular, ciliate, in several subequal or distinctly imbricate series, greenish throughout or commonly suffused with purple; rays purple, mostly 8–15 mm long, 1.5–2.5 mm broad; pappus brown, yellowish, or rarely whitish; achenes several-nerved, hairy.

Open woods, hillsides, stream banks, terraces, and meadows; over most of Alaska (except southeastern portion) and Yukon; south to Oregon, Idaho, and Wyoming; Eurasia (*A. arcticus* Eastw.; *A. montanus* Richards., not Nutt.; *A. richardsonii* Spreng.; *A. salsuginosus* Richards.).

Aster subspicatus Nees

Plants with elongate rhizomes or caudex, the stems arising singly or several together, erect, mostly 2–8 dm tall, sparsely villous with multicellular hairs to glabrate; basal leaves smaller than the cauline ones, usually withered by flowering time; cauline leaves 2–12 cm long, 0.5–2.5 cm broad, lanceolate to oblong, elliptic or oblanceolate, entire to serrate, the lower ones often petiolate, the middle and upper ones sessile and commonly auriculate-clasping, acute to attenuate apically, glabrous above and beneath, scabrous-ciliate; heads solitary or more commonly few to several in corymbose clusters, the peduncles villous; involucres 7–9 mm high, 13–18 mm broad,

the bracts oblong-lanceolate, glabrous dorsally, short-ciliate, imbricate in several series, green at the tip, the margins near the base brownish or yellowish; rays purple, mostly 10–20 mm long, 1.5–2.5 mm broad; pappus brownish or purplish, rarely whitish; achenes several-nerved, hairy.

Moist woods, meadows, and roadsides; in southern and southeastern Alaska; south to California and east to Alberta, Montana, and Idaho (*A. douglasii* Lindl. ex Hook.; *A. foliaceous* Lindl. ex DC.). It has not been possible to segregate specimens into two entities on a consistent basis. The characteristics used as diagnostic features appear to be variable and their interpretation arbitrary. Thus, it seems best to recognize all phases as belonging to one polymorphic species.

Aster yukonensis Cronq.
Yukon Aster

Plants with elongate rhizomes and usually also with a caudex, the stems commonly arising several together, ascending, mostly 0.5–2.5 dm tall, sparsely to moderately villous with multicellular hairs, becoming glandular upwards; basal leaves similar to the cauline, sessile or subpetiolate; cauline leaves 1.4–7 cm long, 0.2–0.4 cm broad, narrowly oblong to linear, entire, glabrous above and beneath, or the upper ones sparsely hairy and also glandular, sessile and auriculate-clasping (at least above), acute to acuminate apically or less commonly some of them obtuse; heads solitary or few in corymbose clusters, the peduncles glandular and villous; involucres 7–10 mm high, 12–16 mm broad, the bracts oblong, glandular dorsally and along the margins, in several subequal series, greenish throughout or with yellowish-scarious margins near the base; rays blue to purple, 6–10 mm long, 1.5–2.5 mm broad; pappus purplish to whitish; achenes several-nerved, hairy.

Gravelly slopes, roadcuts, and lake shores; along the south end of Kluane Lake, Yukon, and disjunctly, the Brooks Range of Alaska; endemic.

BELLIS L.

Plants scapose herbs, with fibrous roots, the juice watery; stems leafless, simple; leaves all basal, simple, petiolate, toothed to entire; heads solitary; involucral bracts in 2 subequal series, herbaceous throughout; receptacle conic to hemispheric, naked; rays white, pink, or purple, numerous, pistillate; disk flowers numerous, perfect, yellow; pappus lacking; style branches flattened; achenes flattened, usually 2-nerved, pubescent.

Bellis perennis L.
European Daisy

Plants 0.2–2 dm tall; leaves all basal, with short to long petioles, the blades 0.7–3(4) cm long, 0.5–2.5 cm broad, obovate to oval or orbicular, dentate to entire, obtuse to rounded or emarginate apically, pubescent above and beneath with coarse, spreading hairs; scapes pubescent with ascending hairs; heads solitary; involucres 5–6 mm high, 9–15 mm wide, the bracts ovate to broadly lanceolate, rounded to obtuse apically, sparsely hairy dorsally, often suffused with purple; rays white to pink or purple, mostly 8–10 mm long, 1.5–2.5 mm broad; pappus lacking; achenes flattened.

Cultivated ornamental; escaping and persisting in southern Alaska (Unalaska); introduced from Europe.

BIDENS L.

Annual herbs with fibrous roots, or rooting along the lower stem, the juice watery; stems usually erect, commonly branched; leaves opposite, simple or pinnately compound; heads few to several in cymose inflorescences; involucral bracts in 2 series, the outer herbaceous, the inner somewhat petaloid and striate; receptacle flat or slightly convex, chaffy throughout, the chaff similar to the inner involucral bracts;

ray flowers present, yellow, neutral or pistillate, or lacking; disk flowers numerous, perfect, fertile, yellow; pappus of (1)2–4 awns or teeth, these retrorsely barbed, persistent; style branches flattened; achenes flattened, pubescent, usually 2–4-awned.

Sherff, E. E. 1937. The genus Bidens. Field Mus. Pub. Bot. 16:1–709.

1a. Leaves simple, the middle and upper ones usually sessile or subsessile; plants from interior Alaska. B. cernua
1b. Leaves pinnately compound, with 3–5 leaflets, petiolate; plants from southeastern Alaska. B. frondosa

Bidens cernua L.
Bur-marigold

Plants mostly 1–5(8) dm tall, the stems sparsely spreading-hairy to glabrous; leaves simple, 1.5–10(15) cm long and 0.5–3(4) cm broad, narrowly lanceolate to lance-ovate, coarsely serrate to subentire, glabrous; heads nodding in age; outer involucral bracts 5–8, green, foliaceous, unequal, spreading or reflexed, the inner bracts erect, mostly 6–8 mm long; rays 6–8, yellow, or lacking; achenes mostly 5–7 mm long, tan, the 2–4 awns retrorsely barbed.

Moist soil; in interior Alaska, where it is possibly adventive; widely distributed in the Northern Hemisphere.

Bidens frondosa L.

Plants mostly 2–8(10) dm tall, the stems glabrous or nearly so; leaves petiolate, pinnately compound with 3–5 leaflets, 2–6(8) cm long and 0.5–2 cm broad, serrate; heads erect in age; outer involucral bracts 5–8, green, subfoliaceous, subequal, erect or spreading, the inner bracts erect, mostly 5–7 mm long; rays usually lacking; achenes mostly 5–9 mm long, dark brown or black, the 2 awns barbed.

Moist soil; in southwestern Alaska (Annette Island), where it is possibly adventive; widespread in North America.

CACALIA L.

Perennial, rhizomatous herbs, with watery juice; stems erect, simple, flexuous; leaves alternate, simple, palmately lobed and toothed, petiolate; heads several, borne in narrow, racemelike panicles; involucral bracts subequal, in 1 series, chaffy, the margin hyaline; receptacle flat, naked; ray flowers lacking; disk flowers 3–6, perfect, fertile; pappus of capillary bristles; style branches flattened; achenes subterete, beakless, glabrous.

Cacalia auriculata DC.

Plants 4–12 dm tall, the stems erect, flexuous, glabrous or sparsely pubescent; leaves petiolate, the blades 3–10(13) cm long, 4.5–16(21) cm broad, reniform, palmately lobed and veined, the lobes dentate, glabrous or nearly so; inflorescence narrowly paniculate; heads several, spreading or ultimately nodding; involucres 7–10 mm long, 5–8 mm broad, more or less constricted near the middle, the bracts usually 5, straw-colored, lance-oblong, each constricted near the middle, the margin hyaline; disk flowers yellow, usually 3–6; pappus white; achenes glabrous.

Slopes and ravines, often where wet; in the Aleutian Islands; Asia (*C. auriculata* ssp. *kamtschatica* [Maxim.] Hultén). Our material belongs to var. *auriculata*.

CHRYSANTHEMUM L.

Perennial herbs, from short to elongate rhizomes, or from a caudex, the juice watery; stems decumbent to ascending or erect, simple or less commonly branched; leaves alternate or principally basal, entire to pinnatifid or palmately lobed; heads solitary or few in an open corymbose inflorescence; involucral bracts imbricate in 2–4 series, greenish, with margins brownish-scarious; receptacle flat or convex, naked; ray flowers white or yellow, numerous, pistillate, fertile; disk flowers numerous, perfect, fertile, yellow; pappus lacking or a

short crown; style branches flattened; achenes several nerved, beakless, glabrous.

1a. Leaves entire, chiefly basal, the subscapose stems with 1–few, often scarious-margined, linear leaves; plants of northern Alaska and Yukon. *C. integrifolium*

1b. Leaves pinnatifid, or palmately lobed, the distinctly leafy stems with only the upper leaves entire or linear, or all of them lobed; plants of various distribution. (2)

2a. Rays yellow; plants introduced weeds, known from the Aleutians. *C. segetum* L.

2b. Rays white; plants indigenous or introduced. (3)

3a. Basal leaves palmately or subpalmately lobed, usually present at flowering time; peduncle apex tomentose; plants often less than 3 dm tall, occurring throughout coastal Alaska. *C. arcticum*

3b. Basal leaves pinnately lobed or cleft, or merely toothed, usually withered by flowering time; peduncle apex glabrous or nearly so; plants often more than 3 dm tall, weeds of southern Alaska. *C. leucanthemum*

Chrysanthemum arcticum L.
Arctic Daisy

Plants rhizomatous or with a caudex, the stems erect or ascending, or decumbent basally, 0.6–4(7.5) dm tall, glabrous or sparsely tomentose, especially above; basal leaves 1.5–16 cm long, the blades 0.4–4.5 cm broad, palmately or subpalmately lobed, cuneate to obovate, oval or ovate, usually persistent at flowering time, glaucous or green, glabrous or tomentose when young; cauline leaves reduced upwards, finally entire; heads usually solitary (seldom 2), the peduncle apex tomentose; involucres 7–13 mm high, 15–27 mm broad, the bracts oblong to lance-oblong, the middle greenish, the broad, brownish margins

Chrysanthemum arcticum L. (× 0.4).

scarious, glabrous or tomentose; rays white, fading brownish or bluish, 10–25 mm long, 2.5–6 mm broad; achenes subterete, about 10-ribbed.

Marshes, tidal flats, stream banks, meadows, and beaches; throughout coastal Alaska and Yukon; east to Labrador; Eurasia (*Leucanthemum arcticum* [L.] DC.; *L. gmelinii* Ledeb.; *C. arcticum* ssp. *gmelinii* [Ledeb.] Kita.; *Dendranthema arcticum* [L.] Tsvel.; *L. hultenii* Löve & Löve; *D. hultenii* [Löve & Löve] Tsvel.). It has not been possible to distinguish morphologically those plants designed as ssp. *polare* Hultén, except on a wholly arbitrary basis. They apparently intergrade completely with the southern materials. Indeed, individual specimens of the *polare* type are found throughout the southern range of the species.

Chrysanthemum integrifolium Richards.

Plants rhizomatous, or with a well-developed caudex, the caudex branches clothed with a thatch of persistent leaves, the stems simple, erect, 0.5–1.6(2) dm tall, pilose to glabrate; leaves primarily basal, 0.5–3.5 cm long, 0.1–0.2 cm broad, linear to narrowly oblong, entire, sparsely to moderately pilose, or glabrate in age, long-ciliate; cauline leaves 1–few, linear, often with brownish-scarious margins near the apex; heads solitary, the peduncle apex villous; involucres 4–7 mm high, 10–14 mm broad, the bracts oblong, the middle greenish, the broad, brownish margin scarious, glabrous or sparsely pilose; rays white, fading yellowish or brownish, 8–15 mm long, 3–5 mm broad; achenes subterete, about 5-ribbed.

Arctic and alpine tundra and heathlands; in much of Alaska and Yukon north of the 64th parallel; south to British Columbia and east to the Canadian Arctic Archipelago; Asia (*Leucanthemum integrifolium* [Richards.] DC.; *Dendranthema integrifolium* [Richards.] Tsvel.).

Chrysanthemum leucanthemum L.
Ox-eye Daisy

Plants rhizomatous, the stem simple or branched above, erect, mostly 2–8 dm tall, glabrous or sparsely villous with multicellular hairs; basal leaves 4–15 cm long, 0.7–2.5 cm broad, pinnately lobed or cleft, or merely toothed, oblanceolate to spatulate, usually withered by flowering time, glabrous or sparsely villous; cauline leaves well developed, reduced upwards, finally entire; heads solitary or less commonly 2–4, the peduncle apex glabrous or nearly so; involucres 7–11 mm high, 15–22 mm broad, the bracts narrowly lanceolate, greenish or brownish, the lateral margins narrow, brownish, flaring and broad at the tip; rays white, fading brownish, 12–18 (20) mm long, 3–5 mm broad; achenes subterete, about 10-ribbed.

Disturbed soils; in southern and southeastern Alaska, and to be expected elsewhere; widely distributed in North America; adventive from Eurasia (*Leucanthemum vulgare* Lam.; *L. leucanthemum* [L.] Rydb.).

CIRSIUM Mill.

Perennial or biennial herbs with rhizomes or taproots, the juice watery; stems erect, simple or branched; leaves alternate, simple, pinnatifid or merely pinnately lobed or coarsely toothed, often decurrent, spiny; heads (1) few to several, borne in compact or corymbose cymes; involucral bracts imbricate in several rows, at least some spine tipped; receptacle flat to hemispheric, densely bristly; disk flowers alone present, perfect or imperfect, pink purple to white, with slender tube and long, narrow lobes; pappus of plumose, rarely barbellate bristles; style branches connate, except at the tip, with a tuft of hair at the base of the branches; achenes moderately 4-angled, glabrous.

1a. Plants rhizomatous perennials, dioecious; heads mostly 1–2(2.5) cm high; weedy species of southern and southeastern Alaska. *C. arvense*
1b. Plants taprooted biennials, perfect; heads usually more than 2.5 cm high; distribution various. (2)
2a. Upper surface of leaves spinulose-hispid; peduncles and stems commonly spiny-winged with conspicuous decurrent leaf bases; weedy species of southeastern Alaska. *C. vulgare*
2b. Upper surface of leaves cobwebby-tomentose, villous, or glabrous, not spinulose-hispid; peduncles and stems variously or not at all winged; distribution various. (3)
3a. Involucral bracts lanceolate to lance-ovate (at least the outer and middle), the inner ones often with dilated, scarious, fringed tips; plants of southeastern Yukon. *C. foliosum*
3b. Involucral bracts lance-linear, usually none of them with dilated, scarious, fringed tips; plants of southeastern Alaska or of the Aleutian Islands. (4)
4a. Leaves mostly 1.5–2 times longer than broad, ovate to obovate in outline; plants of the Aleutian Islands. *C. kamtschaticum*
4b. Leaves mostly 4–10 times longer than broad, oblong to lance-oblong in outline; plants of southeastern Alaska. *C. edule*

Cirsium arvense (L.) Scop.
Canada Thistle

Perennial, rhizomatous herbs, the stems mostly 5–10 dm tall, glabrous or sparingly tomentose; leaves 4–15 cm long, 2–6 cm broad, shallowly to deeply pinnatifid or merely lobed, the margin spiny, glabrous to tomentose above and beneath; heads several to many, unisexual; involucres 1–2 (2.5) cm high, the bracts lance-ovate, at least the outer ones and often all of them spine tipped, tomentose to glabrous; corollas pink purple to white; pappus of pistillate heads longer than the corollas, that of staminate heads shorter than the corollas; achenes 3–5 mm long.

Roadsides, abandoned farms, cultivated land, and other disturbed sites; in southern and southeastern Alaska, and to be expected elsewhere; widely introduced in North America; adventive from Eurasia.

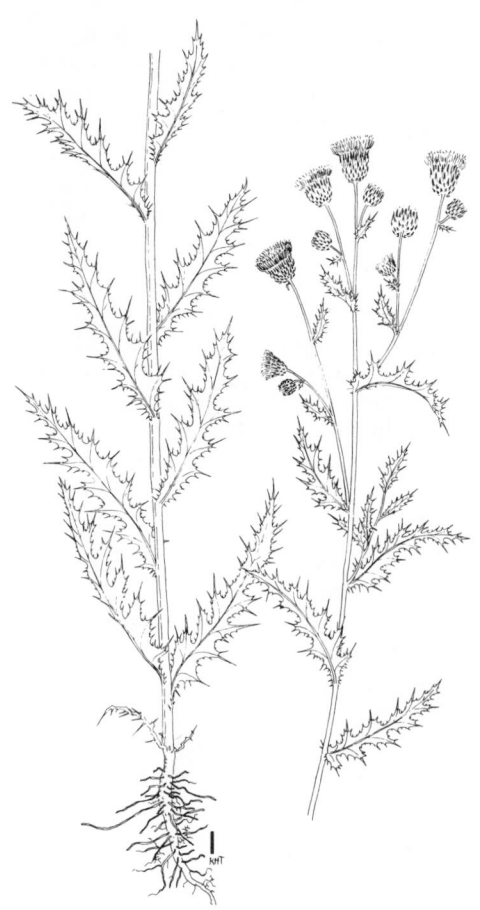

Cirsium arvense (L.) Scop. (× 0.3).

Cirsium edule Nutt.
Edible Thistle

Biennial or short-lived perennial, with tap-roots, the stems mostly 4–12 dm tall, spar-ingly to moderately cobwebby-tomentose; leaves mostly 5–20 cm long and 1.5–5 cm broad, shallowly to deeply pinnatifid, gla-brous, except on the midrib below, or spar-ingly cobwebby on both surfaces, spiny; heads solitary or few, perfect; involucres 2–4 cm high, the bracts lance-linear, at least the outer ones spine tipped, moderate-ly cobwebby-tomentose; corollas pink purple.

Streamsides and wet meadows; in south-eastern Alaska (Hyder); south to Oregon.

Cirsium foliosum (Hook.) DC.
Leafy Thistle

Perennials with taproots, the stems mostly 3–10 dm tall, sparsely to moderately cob-webby-tomentose; leaves mostly 5–30 cm long and 1.5–6 cm broad, shallowly to deeply pinnatifid or merely toothed, spar-ingly tomentose on both surfaces, or less so beneath, spiny; heads (1) few to sever-al, clustered at the stem apex; involucres 2–3.5 cm high, the bracts lanceolate or lance-ovate (at least the outer and middle ones), the outer ones spine tipped, the in-ner ones often with dilated, scarious, fringed tips; corollas pink to white.

Meadows and thickets; in southeastern Yu-kon; east to Saskatchewan and south to California, Arizona, and Colorado.

Cirsium kamtschaticum Ledeb. ex DC.
Kamchatka Thistle

Perennials with taproots, the stems mostly 4–10(20) dm tall, sparingly villous; leaves mostly 8–20(25) cm long, 4–17 cm broad, broadly lobed, the sinuses extending two-thirds to three-fourths the distance to the midrib, moderately decurrent, sparingly villous on the main veins above and be-neath, otherwise glabrous, spiny; heads solitary or few; involucres 2–3 cm high, the bracts lance-linear, usually all spine tipped, more or less tomentose; corollas pink purple.

Meadows and beaches; in westernmost Aleutian Islands; Asia.

Cirsium vulgare L.
Bull Thistle

Biennials with taproots, the stems mostly 4–15 dm tall, spiny to tomentose, conspicu-ously winged, with decurrent leaf bases; leaves mostly 5–25 cm long and 2–8 cm broad, pinnatifid, spinulose-hispid above, tomentose and whitish or sometimes green-ish beneath, spiny; heads solitary or few to several; involucres 2.5–4 cm high, the bracts all spine tipped, more or less tomentose; corollas pink purple.

Disturbed soils; in southeastern Alaska; a weedy species of wide distribution in North America; adventive from Eurasia (*C. lanceolatum* authors, not Hill.).

COTULA L.

Perennial (or annual?) herbs with fibrous roots, the juice watery; stems decumbent to ascending or erect; leaves alternate, pinnatifid to entire; heads solitary at the ends of branches, pedunculate; involucral bracts in 2 subequal series, dry, yellowish, with narrow to broad scarious margins; receptacle flat to conic, naked except for the persistent stipe bases; corollas of disk flowers only, the marginal ones pistillate, fertile, the central ones perfect and fertile or rarely sterile; pappus lacking; style branches flattened; achenes stipitate, flattened or merely 2-nerved.

Cotula coronopifolia L.
Mud-disk

Plants 0.5–2(3) dm tall, the stems often trailing and rooting at the nodes, glabrous or nearly so; leaves 2–6 cm long, 0.2–0.6 (1) cm broad, oblong, pinnatifid to entire, sessile, the base sheathing around the stem; heads solitary on naked peduncles; involucres 3–5 mm high, 7–11 mm broad, the bracts lanceolate to lance-oblong or oblong, truncate to rounded apically, yellowish, the margins hyaline, glabrous; marginal flowers flattened, long-stipitate, those of the central flowers merely 2-nerved, short-stipitate, the stipes persistent on the receptacle.

Beaches and tidal flats; in southeastern Alaska; widely distributed along the beaches of the world; adventive from South Africa.

CREPIS L.

Annual, biennial, or perennial herbs from taproots, with milky juice; leaves basal, or cauline and alternate, entire to pinnatifid or merely toothed; heads few to numerous, in compact or open corymbose clusters; involucral bracts in 1–2 series, the outer ones much shorter than the inner, herbaceous or scarious, the margin hyaline; receptacle flat or nearly so, naked; corollas of ray flowers only, perfect, yellow; pappus of capillary bristles; style branches semicylindrical; achenes terete or nearly so, beaked or beakless, prominently 10–20-ribbed, glabrous.

1a. Plants annual or biennial; cauline leaves auriculate-clasping; introduced weedy species of southern Alaska and southern Yukon. (2)
1b. Plants perennial; cauline leaves (if any) not auriculate-clasping; plants not weedy, indigenous, of broad distribution. (3)
2a. Achenes 1.5–2.5 mm long, commonly pale brown to straw-colored; inner involucral bracts glabrous within. *C. capillaris*
2b. Achenes 2.5–4 mm long, dark purplish brown at maturity; inner involucral bracts hairy within. *C. tectorum*
3a. Plants in open tufts, the stems easily visible, usually more than 1 dm tall; achenes with the beak about one-fourth as long as the body, the ribs minutely roughened. *C. elegans*
3b. Plants in dense tufts, usually less than 1 dm tall; achenes beakless or the beak less than one-fifth as long as the body, the ribs smooth. *C. nana*

Crepis capillaris (L.) Wallr.
Slender Hawk's-beard

Annual or biennial herbs, the stems erect, simple or branched, mostly 2–6 dm tall, sparsely spreading-hairy; basal leaves 3–20 cm long, 0.5–3 cm broad, lanceolate to oblanceolate, denticulate to pinnatifid, or bipinnatifid, glabrous or pubescent with stiff, spreading hairs, especially along the lower midvein, petiolate; cauline leaves reduced upwards, sessile and auriculate-clasping;

heads (1) several to numerous, mostly 20–60-flowered, borne in an open inflorescence; involucres 5–8 mm high, 5–14 mm broad, the inner bracts lance-attenuate, 8–16, tomentose, often glandular-hairy, glabrous within, the outer bracts lance-linear; achenes to 2.5 mm long, pale brown to straw-colored at maturity, not beaked.

Weedy species of disturbed soils; in central and southeastern Alaska; introduced at widely scattered sites in North America; adventive from Europe.

Crepis elegans Hook.
Elegant Hawk's-beard

Perennial herbs, the stems much branched, 0.8–3 dm tall, elongate, glabrous; basal leaves mostly 3–10 cm long, 0.6–2 cm broad, ovate to lanceolate, elliptic, or oblong, pinnatifid or subhastately lobed, glabrous, petiolate; cauline leaves reduced upwards, sessile, not auriculate-clasping; heads numerous, mostly 6–10-flowered, borne in an open inflorescence; involucre 7–10 mm high, 2–4 mm broad, the inner bracts narrowly oblong, 8–10, greenish or blackish, glabrous, the outer bracts oblong to oblanceolate; achenes 4.5–5.5 mm long, the beak 0.8–1.5 mm long, the ribs minutely roughened.

Stream banks, terraces, gravel bars, and lake shores; in the southeastern third of continental Alaska and most of the Yukon; east to Saskatchewan and south to British Columbia and Montana.

Crepis nana Richards.
Dwarf Hawk's-beard

Perennial herbs, the stems much branched, mostly 0.2–1.1(1.5) dm tall, contracted, usually obscured by the leaves, glabrous; basal leaves 1–7.5(9) cm long, 0.2–1.2 cm broad, the blades spatulate to orbicular, ovate, or rarely cordate, glabrous, petiolate; cauline leaves similar to the basal, not auriculate-clasping; heads several to numerous, mostly 9–12-flowered, borne in a compact cushionlike inflorescence; involucre

7–12 mm high, 2–6 mm broad, the inner bracts narrowly oblong, 8–12, greenish or blackish, glabrous, the outer bracts lance-attenuate; achenes 4–6 mm long, the beak less than 0.5 mm long, or lacking, the ribs smooth.

Gravelly flood plains, bars, terraces, lake shores, and talus slopes; over much of Alaska, except for the southwestern portion, and over most of the Yukon; east to Labrador and Newfoundland and south to California, Utah, and Quebec; Asia. Two varieties are recognizable.

1a. Leaves entire or nearly so; plants broadly distributed. *C. nana* var. *nana*
1b. Leaves pinnatifid; plants of northern Alaska (*Youngia americana* Babcock). *C. nana* var. *lyratifolia* (Turcz.) Hultén

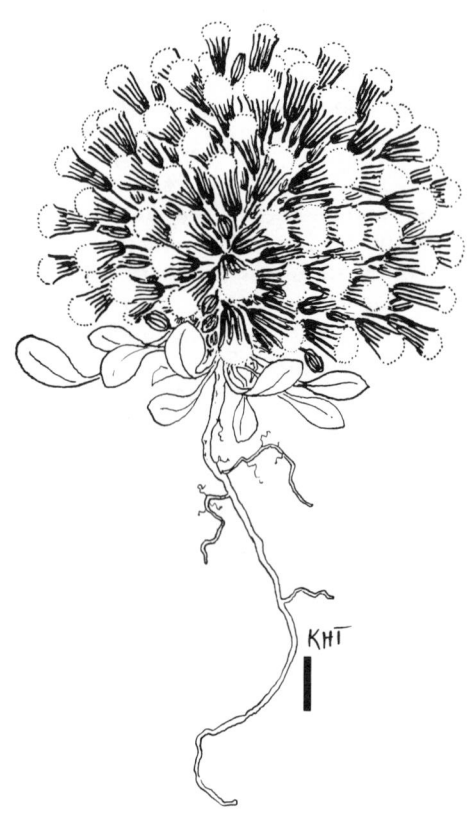

Crepis nana Richards. (× 0.8).

Crepis tectorum L.

Annual herbs, the stems erect, simple or branched, mostly 1.5–5 dm tall, appressed-hairy to glabrate; basal leaves 2–15 cm long, 0.3–3.5 cm broad, oblanceolate to oblong, denticulate to pinnatifid, glabrous to villous, petiolate; cauline leaves reduced upwards, becoming sessile and often auriculate-clasping; heads (1) few to numerous, mostly 20–60-flowered, borne in sub-corymbose clusters; involucres 6–11 mm high, 4–12 mm broad, the inner bracts 12–15, lance-attenuate, tomentose and often glandular-hairy as well, strigose within; outer bracts subulate; achenes 2.5–4 mm long, dark purplish brown at maturity, not or scarcely beaked.

Disturbed sites; in southern Alaska, southern Yukon, and northern British Columbia; adventive from Europe.

ERIGERON L.

Biennial or perennial herbs, from a caudex and taproot or less commonly from a rhizome, with watery juice; stems decumbent to ascending or erect, simple or branched; leaves alternate or all basal, simple, entire, toothed, or ternately dissected; heads solitary or few to numerous, in corymbose or racemose clusters; involucral bracts subequal to imbricate, herbaceous throughout, or scarious throughout or only near the apex; receptacle flat or convex, naked; ray flowers pink, purple, blue, or white, numerous, pistillate, rarely rayless; disk flowers numerous, perfect, fertile, yellow, rarely otherwise; pappus of capillary bristles; style branches flattened, lanceolate to ovate, less than 0.5 mm long; achenes 2–several-nerved, usually pubescent.

Cronquist, A. 1947. Revision of the North American species of *Erigeron*, north of Mexico. Brittonia 6:121–302.

1a. Involucres moderately to densely villous or tomentose with whitish, yellowish, or purplish hairs; heads solitary, rarely 2–4; plants mostly less than 2.5 dm tall. (2)

1b. Involucres sparsely to moderately hairy or subglabrous, or spreading-hairy, not villous; heads solitary or few to numerous; plants of various size, often over 2.5 dm tall (except in *E. hyssopifolius*). (8)

2a. Pappus pinkish or purplish; plants forming small mats, the caudex branches slender, prostrate. *E. purpuratus*

2b. Pappus whitish to tan, not pinkish or purplish; plants not forming mats, the caudex branches stout, erect or ascending. (3)

3a. Leaves and stems (and involucres) densely white villous-tomentose; plants mostly 0.7–1 dm tall, of northern and northwestern Alaska. *E. muirii*

3b. Leaves and stems variously pubescent, but not densely white-villous; involucres villous with whitish or purplish hairs; plants of various size and distribution. (4)

4a. Ray corollas 3–6 mm long, 0.3–0.8 (1) mm broad, inconspicuous. (5)

4b. Ray corollas 7–15 mm long, 1–2 mm broad, showy. (6)

5a. Hairs of the involucre and upper stem with purplish black cross-walls; involucral bracts purplish black or greenish. *E. humilis*

5b. Hairs of involucre with colorless cross-walls, or some of them with reddish purple cross-walls; involucral bracts reddish purple. *E. uniflorus*

6a. Hairs of the involucre with black or purplish black cross-walls; plants mostly 1 dm tall or less. *E. hyperboreus*

6b. Hairs of the involucre with colorless cross-walls, or some of them reddish purple; plants 0.4–2.5(4) dm tall. (7)

7a. Cauline leaves linear-lanceolate to oblong, acuminate; ray flowers mostly 45–75. *E. yukonensis*

7b. Cauline leaves lanceolate to ovate, merely acute or less commonly acuminate; ray flowers mostly 100–125. *E. grandiflorus*

8a. Leaves once to twice ternately divided, all basal or nearly so; stems scapose or subscapose. *E. compositus*
8b. Leaves entire or merely toothed: cauline leaves well developed; stems leafy. (9)

9a. Rays erect, inconspicuous, very narrow, about equaling or only slightly exceeding the pappus; pappus longer than the disk corollas. (10)
9b. Rays spreading, conspicuous, narrow to broad, much surpassing the pappus; pappus shorter than the disk corollas. (11)

10a. Cauline leaves narrowly lanceolate to oblong, oblanceolate, or less commonly linear; rayless pistillate flowers present between the ray and disk flowers; inflorescence corymbose, the peduncles curved-ascending, or the heads solitary. *E. acris*
10b. Cauline leaves linear to oblong; rayless pistillate flowers lacking; inflorescence racemose, the peduncles erect or nearly so, or the heads solitary. *E. lonchophyllus*

11a. Leaves numerous, all nearly alike, commonly less than 3 cm long and 0.4 cm broad; plants of Yukon. *E. hyssopifolius*
11b. Leaves neither numerous nor alike, the basal ones larger and broader, or differently shaped from the others, often over 3 cm long, or over 0.4 cm broad, or both; plants of various distribution. (12)

12a. Stems few to several from a branching caudex (seldom single), usually decumbent-ascending, mostly less than 2.5 dm tall; cauline leaves linear or linear-lanceolate to narrowly oblong or oblanceolate, merely sessile. (13)

12b. Stems 1–few from a caudex or rhizome, erect or less commonly decumbent-ascending, often more than 2.5 dm tall; cauline leaves usually lanceolate to lance-oblong and more or less auriculate-clasping. (14)

13a. Basal leaves evidently 3-veined, often over 5 mm broad (at least some); plants locally common in eastern Alaska and southern Yukon. *E. caespitosus*
13b. Basal leaves 1-veined, commonly less than 5 mm broad; plants of southwestern Yukon. *E. pumilus*

14a. Ray corollas mostly 1.5–3 mm broad; leaves glabrous above and beneath, or sparsely hairy, especially along the veins. *E. peregrinus*
14b. Ray corollas 0.2–1.3 mm broad; leaves moderately hairy above and beneath, the hairs spreading. (15)

15a. Involucres (5)6–9 mm high; rays mostly 125–175, 8–15 mm long, 0.6–1.2 mm broad; plants widely distributed in eastern Alaska and Yukon. *E. glabellus*
15b. Involucres 4–6 mm high; rays 150 or more, 5–10 mm long, 0.2–0.6 mm broad; plants of northern British Columbia and disjunctly in northern Yukon. *E. philadelphicus*

Erigeron acris L.
Fleabane Daisy

Plants biennial or perennial, with simple or few-branched caudex and a taproot, the stems 1–7.5 dm tall, sparsely to moderately spreading-hairy or subglabrous; basal leaves 2–14 cm long, 0.3–1.6 cm broad, oblanceolate to spatulate, entire or toothed, acute to obtuse or rounded apically tapering to a petiole basally, glabrous on both surfaces or spreading-hairy, ciliate or not; cauline leaves narrowly lanceolate to oblong or oblanceolate, or less commonly linear, acute to attenuate apically, sessile; heads solitary or more commonly few to many in corymbose clusters, the peduncles curved-ascending; involucres 5–11 mm high, 8–22 mm broad, the bracts linear-lanceolate, long-attenuate, glandular or hairy, or both; pistillate flowers numerous, the outer ones (rays) with pink to purplish or white corollas 2–4 mm long and 0.2–0.5 mm broad, the inner ones rayless; disk corollas shorter than the white to pinkish pappus; achenes 2-nerved, hairy.

Gravel bars, terraces, stream banks, and lake shores, often in open woods; in most of Alaska and Yukon; east to the Atlantic and south to California, Colorado, and Maine; Eurasia. Our material includes three more or less distinctive varieties.

1a. Peduncles and involucres nearly or quite glandless, more or less spreading-hairy (*E. alpinus* var. *elatus* Hook.; *E. elatus* [Hook.] Greene). *E. acris* var. *elatus* (Hook.) Cronq.
1b. Peduncles and involucres more or less glandular, also often spreading-hairy. (2)

2a. Plants usually 3 dm tall or more, with several to many heads (*E. elongatus* Ledeb., not Moench.; *E. kamtschaticus* DC.; *E. angulosus* Gand. var. *kamtschaticus* [DC.] Hara; *E. acris* var. *asteroides* authors, not [Andrz.] DC.; *E. acris* ssp. *politus* [Fries] Schinz & Keller). *E. acris* var. *kamtschaticus* (DC.) Herder
2b. Plants mostly 1–2.5 dm tall, with solitary or few heads (*E. nivalis* Nutt.; *E. jucundus* Greene; *E. debilis* [Gray] Rydb.). *E. acris* var. *debilis* Gray

Erigeron caespitosus Nutt.

Plants perennial, with well-developed, usually branched caudex and stout taproot, the stems 0.6–3 dm tall, decumbent-ascending, or less commonly erect, moderately pubescent with stiff, white, multicellular, spreading or descending hairs; basal leaves 2.5–14 cm long, 0.2–0.8(1) cm broad, narrowly oblanceolate to spatulate, entire, rounded to obtuse or acute apically, tapering to a slender petiole basally, 3-veined, hairy like the stems; cauline leaves oblong to linear-lanceolate, numerous, gradually reduced upwards, sessile; heads solitary, or few to several, the peduncles suberect; involucres 4–7 mm high, 10–15 mm broad, the bracts lance-oblong, attenuate, pubescent with white, spreading hairs and often glandular as well; rays mostly 30–100, white to blue or pink, 5–15 mm long, 1–2 mm broad; disk corollas longer than the whitish to tan pappus; achenes 2-nerved, hairy.

Dry gravelly flats, open slopes, terraces, and lake shores; in central eastern Alaska and southern Yukon; disjunctly southward from British Columbia to Arizona and Nebraska.

Erigeron compositus Pursh

Plants perennial, with a branched caudex and taproot, the stems scapose or subscapose, 0.5–2 dm tall, decumbent to ascending or erect, pubescent with spreading hairs and often glandular as well; basal leaves 1–8 cm long, once to twice ternately divided, glabrous or spreading-hairy, especially along the petioles; cauline leaves greatly reduced, simple, entire, linear; heads solitary; involucre 5–8 mm high, 10–15 mm broad, the bracts lance-oblong, attenuate, purplish apically, pubescent with white, spreading hairs, more or less glandular; rays mostly 20–60, white, pink, or blue, 4–10 mm long, 1–2 mm broad, or absent; disk corollas exceeding the whitish to tan pappus; achenes 2-nerved, hairy.

Open slopes, rock outcrops, and lake shores; in the eastern half of mainland Alaska and disjunctly in the northwestern part, and in most of the Yukon; east to Greenland and south to California, Arizona, South Dakota, and Quebec. Two very poorly defined varieties are present in our region.

1a. Leaves mostly once ternate (*E. trifidus* Hook.; *E. pedatus* Nutt.; *E. gormanii* Greene). *E. compositus* var. *discoideus* Gray

1b. Leaves mostly twice to thrice ternate; our common variety (*E. multifidus* Rydb.). *E. compositus* var. *glabratus* Macoun

Erigeron glabellus Nutt.

Plants biennial or perennial, with simple or branched caudex and taproot, the stems 1.5–5 dm tall, erect or nearly so, moderately to densely pubescent with white, spreading, multicellular hairs; basal leaves 2–12 cm long, 0.3–3 cm broad, oblanceolate, entire or toothed, acute to obtuse or rounded apically, tapering to a broad petiole basally, only the midvein prominent, spreading-hairy above and beneath; cauline leaves reduced upwards, lance-oblong to lance-linear, several to numerous, finally sessile; heads solitary, or few to numerous, the peduncles erect or nearly so; involucres 5–9 mm high, 12–22 mm broad, the bracts lance-oblong, attenuate, pubescent with white, spreading hairs, not glandular; rays mostly 125–175, blue to pink or white, 8–15 mm long, 0.6–1.2 mm broad; disk corollas exceeding the white to tan pappus; achenes 2-nerved, hairy.

Gravel bars, terraces, stream banks, open woods, and meadows; in central and eastern Alaska and Yukon; south to Utah and east to Wisconsin (*E. glabellus* ssp. *pubescens* [Hook.] Cronq.; *E. asper* var. *pubescens* [Hook.] Breitung; *E. turneri* Greene). Our material belongs to var. *pubescens* Hook.

Erigeron grandiflorus Hook.

Plants perennial, with simple or branched caudex and taproot, the caudex sometimes rooting, the stems 0.5–2.5 dm tall, sparsely to moderately pubescent with spreading, multicellular hairs; basal leaves (1)2–9 cm long, 0.4–1 cm broad, oblanceolate, acute to obtuse apically, tapering to a petiole basally, hairy above and beneath, obscurely 3–5-nerved; cauline leaves lanceolate to ovate, few to several, reduced upwards, sessile; heads solitary, the peduncles erect; involucres 8–10 mm high,

18–24 mm broad, the bracts lance-linear, long-attenuate, the tips purplish, densely villous with whitish, multicellular hairs, the cross-walls not or only seldom purplish; rays mostly 100–125, blue to pink, 10–15 mm long, 1–2 mm wide; disk corollas exceeding the whitish to tan pappus; achenes 2-nerved, densely hairy.

Cliffs and talus slopes and tundra; in central and eastern Alaska and northern and southern Yukon; east to Mackenzie and south to British Columbia and Alberta. A new species, *E. hultenii* Spongberg (Rhodora 75:801. 1973) has been described from near Anchorage. It is said to be intermediate between *E. peregrinus* and *E. grandiflorus*.

Erigeron humilis Grah.

Plants perennial, with simple or few-branched caudex and taproot, the stems 0.3–2 dm tall, moderately to densely villous, the hairs with purplish black cross-walls, at least above; basal leaves 1–8 cm long, 0.3–1.1 cm broad, oblanceolate to spatulate, rounded to obtuse apically, tapering to a slender or broad petiole basally, entire, sparsely hairy to glabrous above and beneath, ciliate; cauline leaves narrowly oblong, acute to attenuate apically, sessile; heads solitary (rarely 2), the peduncles erect; involucres 6–9 mm high, 12–18 mm broad, the bracts lance-oblong, attenuate, purplish black or greenish, villous with long, multicellular hairs, the cross-walls purplish black; rays mostly 50–150, white to pink or purplish, mostly 3–6 mm long, 0.3–1 mm broad; disk corollas subequal to or exceeded by the white to tan pappus; achenes 2-nerved, hairy.

Arctic and alpine tundra, often in gravelly soil; in much of Alaska and Yukon; south to Montana and east to the Atlantic; circumpolar (*E. unalaschkensis* [DC.] Vierh.; *E. pulchellus* var. *unalaschkensis* DC.).

Erigeron hyperboreus Greene

Plants perennial, with simple or branched caudex and taproot, the stems 0.4–1 dm tall, moderately to densely villous, the hairs with purplish black cross-walls, at least above; basal leaves 1–4 cm long, 0.1–0.4 cm broad, narrowly oblanceolate, acute or obtuse apically, tapering to the base or subsessile, entire, hairy above and beneath; cauline leaves oblong to linear, attenuate; heads solitary, the peduncles erect; involucres 6–8 mm high, 14–18 mm broad, the bracts lance-oblong, attenuate, purplish throughout or only apically, villous with long, multicellular hairs, the cross-walls purplish black; rays mostly 40–60, blue, 8–12 mm long; disk flowers exceeding the whitish pappus; achenes 2-nerved, hairy.

Arctic and alpine tundra; in western, northern, and south-central Alaska, and northwestern Yukon (?); east to Mackenzie; endemic (*E. alaskanus* Cronq.; *E. alaskanus* f. *albiflorus* Jordal; *E. hyperboreus* f. *albiflorus* [Jordal] B. Boi.)

Erigeron hyssopifolius Michx.

Plants perennial, with simple or branched, slender caudex and fibrous roots, the stems 1.2–3 dm tall, strigulose to glabrous with white hairs; leaves all essentially alike, 0.7–2.5 cm long, 0.1–0.4 cm broad, linear to linear-oblanceolate, acute apically, tapering to the base or subsessile, often with short leafy shoots in the axils; heads solitary or rarely more, the peduncles erect; involucres 4–6 mm high, 6–12 mm broad, the bracts lance-linear, attenuate, greenish, strigose with whitish hairs; rays 20–50, white to pink or pink purple, 4–8 mm long; disk flowers exceeding the yellowish pappus; achenes 2-nerved, hairy.

Moist sites; in north-central Yukon; east to Mackenzie and Newfoundland, and southward to Vermont.

Erigeron lonchophyllus Hook.

Plants biennial or short-lived perennial, with simple or branched caudex and a taproot, stems mostly 1–5 dm tall, sparsely to moderately spreading-hairy; basal leaves 0.7–10 cm long, 0.1–1.2 cm broad, narrowly oblanceolate to spatulate or oblong, entire, acute to obtuse or rounded, tapering to a petiole, glabrous or sparsely hairy above and below, usually not ciliate; cauline leaves linear to oblong-lanceolate or oblong, acute apically, sessile, often long-ciliate; heads solitary or more commonly few to several in racemose clusters, the peduncles erect or nearly so, not curved near the base; involucres 5–9 mm high, 8–18 mm broad, the bracts lance-linear, attenuate or merely acute, hairy, not glandular; pistillate flowers all with rays, numerous, mostly 2–3 mm long and 0.2–0.5 mm broad, pink or white; disk corollas shorter than the white or whitish pappus; achenes 2-nerved, hairy.

Dry, open slopes, gravel bars, terraces, and wet bogs; in southeastern two-thirds of mainland Alaska and southern Yukon; south to California, Utah, and New Mexico, and east to Quebec and South Dakota.

Erigeron muirii Gray

Plants perennial, with simple or branched caudex and stout taproot, the stems 0.5–1 dm tall, densely villous-tomentose with long, tangled, multicellular, whitish hairs; basal leaves 1–4 cm long, 0.2–0.7 cm broad, oblanceolate to spatulate, rounded to obtuse apically, tapering to the base, entire, pubescent like the stem; cauline leaves few, linear to lance-linear, sessile; heads solitary, the peduncle erect; involucres 6–9 mm high, 11–20 mm broad, the bracts lance-oblong to lanceolate, acute to attenuate, purplish, pubescent like the stems; rays mostly 50–100, white or pink (blue?), 8–13 mm long, 1–1.5 mm broad; disk corollas surpassing the whitish pappus; achenes 2-nerved, hairy.

Rock outcrops, meadows, and alluvial fans; in northern Alaska and northern Yukon; endemic (*E. grandiflorus* ssp. *muirii* [Gray] Hultén).

Erigeron peregrinus (Pursh) Greene

Plants perennial, with rhizomes or caudex, the stems 0.6–6 dm tall or more, sparingly to moderately villous with multicellular hairs, or glabrous; basal leaves 1.5–15(20) cm long, 0.4–2.5(3.5) cm broad, oblanceolate to spatulate, entire or toothed, rounded to obtuse or acute apically, tapering to a petiole basally, glabrous above and beneath or sparingly pubescent, ciliate, usually with 3–5 main veins; cauline leaves lanceolate to oblong, elliptic, or oblanceolate, sessile (at least above); heads solitary (rarely more), the peduncles erect; involucres 6–12 mm high, 12–30 mm broad, the bracts lance-oblong, attenuate, sparsely to moderately pubescent with white (rarely purplish) hairs, villous-ciliate, sometimes glandular; rays mostly 30–80, pink purple, pink, or white, 8–16 mm long or more, 1.5–3 mm broad; disk corollas exceeding the whitish to tan pappus; achenes 4–7-nerved, hairy.

Wet meadows and open woods; throughout coastal, southern Alaska and less commonly some distance in the interior; south to California, Utah, and New Mexico (*Aster peregrinus* Pursh; *A. unalaschkensis* Less; *A. tilesii* Wilkstr.). Our material is somewhat variable. Two subspecies have been recognized.

1a. Involucral bracts hairy dorsally or merely ciliate, not glandular; plants common, widely distributed in southern Alaska (*E. peregrinus* var. *dawsoni* Greene). *E. peregrinus* ssp. *peregrinus*

1b. Involucral bracts densely glandular dorsally, rarely also hairy; plants uncommon, in southeastern Alaska (*E. callianthemus* Greene). *E. peregrinus* ssp. *callianthemus* (Greene) Cronq.

Erigeron philadelphicus L.

Plants biennial or short-lived perennial (or annual?), with usually simple caudex and tap or fibrous roots, the stems 2–7 dm tall, pubescent with long, spreading hairs or glabrous; basal leaves mostly 3–15 cm long, 0.5–3.5 cm broad (rarely larger), oblanceolate or obovate, coarsely toothed to sub-

entire, rounded to obtuse apically, sparsely to moderately hairy above and beneath; heads solitary, or more commonly few to many, in a corymbose inflorescence; involucre 4–6 mm high, 6–15 mm broad, the bracts lance-oblong, acute to acuminate, with broad, hyaline margins, pubescent with spreading hairs or glabrous; rays mostly 150 or more, pink to pink purple or white, 5–10 mm long, 0.2–0.6 mm broad; disk corollas surpassing the whitish pappus; achenes 2-nerved, hairy.

Open woods; in northern British Columbia (Liard Hot Springs) and disjunctly in northwestern Yukon; widely distributed in North America.

Erigeron pumilus Nutt.

Plants perennial, with a well-developed, usually branched caudex and stout taproot, the stems 0.5–2.5 dm long, ascending to erect, copiously spreading-hairy; basal leaves (2)3–6 cm long, 0.1–0.4(0.6) cm broad, linear-oblanceolate, entire, acute apically, tapering to a slender petiole basally, 1-veined, hairy like the stems; cauline leaves linear to linear-oblanceolate, numerous, gradually reduced upwards, becoming sessile or subsessile; heads solitary or few to several, the peduncles suberect; involucres 4–7 mm high, 8–14 mm broad, the bracts lance-oblong, attenuate, pubescent with white, spreading hairs and often glandular as well; rays mostly 50–100, white to blue or pink, 5–12 mm long, 0.7–1.5 mm broad; disk corollas longer than the whitish to tan pappus; achenes 2-nerved, hairy.

Dry, open slopes; in southwestern Yukon; disjunctly southward from Washington and British Columbia south to California, Arizona, and New Mexico, and eastward to Kansas and Saskatchewan.

Erigeron purpuratus Greene

Plants perennial, with usually well-developed, horizontal caudex branches and taproot, the stems 0.2–1 dm tall, spreading-hairy and finely glandular; basal leaves 0.8–3.5 cm long, 0.1–0.5 cm broad, narrowly oblanceolate to oblong or spatulate, entire or 3-toothed, villous to glabrate or glabrous; cauline leaves greatly reduced, the stems subscapose; heads solitary, the peduncles variously curved or straight, stout; involucres 7–10 mm high, 9–20 mm broad, the bracts lance-linear, attenuate, mostly purplish throughout, moderately to densely villous with long, multicellular hairs, at least some with purplish black cross-walls; rays mostly 60–90, pink or white, 4–6 mm long, 0.5–1 mm broad; disk corollas subequal to or shorter than the pinkish or purplish pappus; achenes 2-nerved, hairy to nearly glabrous.

Gravel bars, terraces, flood plains, rock outcrops, and roadsides; in much of the eastern and northern portions of Alaska and southwestern and northern Yukon; south to British Columbia; endemic (*E. denalii* A. Nels.).

Erigeron uniflorus L.

Plants perennial, with simple or less commonly branching caudex and taproot, the stems 0.3–2 dm tall (rarely taller), spreading-hairy; basal leaves 1.5–8 cm long, 0.2–0.7 mm broad, oblanceolate, villous to glabrate or glabrous, entire; cauline leaves few, narrowly lanceolate to oblong or linear, sessile; heads solitary, borne on erect, stout peduncles; involucres 7–12 mm high, 12–26 mm broad, the bracts lance-linear, attenuate, purplish, more or less densely woolly-villous with long, multicellular hairs, the cross-walls colorless or sometimes purplish; rays mostly 100–200, pink to purplish or white, 3–6 mm long, 0.3–0.6 mm broad; disk corollas shorter than the whitish to tan pappus; achenes 2-nerved, more or less hairy.

Meadows, hills, and alluvial deposits; in arctic and alpine tundra, known from widely disjunct localities over much of Alaska and southern Yukon; east to Quebec and Greenland (*E. eriocephalus* Vahl.

ex Hornem.). Our material belongs to ssp. *eriocephalus* (Vahl.) Cronq.

Erigeron yukonensis Rydb.

Plants perennial, with a branching or simple caudex and taproot, the stems 0.5–2.5 (4) dm tall, sparsely to moderately spreading-hairy; basal leaves 1–16 cm long, 0.2–0.8 cm broad, narrowly oblanceolate to oblong or linear, acute or acuminate apically, tapering to a petiole basally, hairy above and beneath, 1–3-nerved; cauline leaves lance-linear to oblong, few to several, reduced upwards, sessile; heads solitary or less commonly 2–4, the peduncles erect; involucres 6–10 mm high, 10–19 mm broad, the bracts lance-linear, attenuate, the tips purplish, sparsely to moderately villous with whitish, multicellular hairs, the cross-walls colorless; rays mostly 45–75, pink or pink purple, 10–15 mm long, (1.1)1.5–2.2 mm broad; disk flowers exceeding the whitish to tan pappus; achenes 2-nerved, hairy.

Arctic and alpine tundra; in northern Alaska and northern to southern Yukon; east to Mackenzie (*E. glabellus* ssp. *pubescens* var. *yukonensis* [Rydb.] Hultén).

GNAPHALIUM L.

Annual, perfect herbs, with taproots, the juice watery; stems erect or decumbent to ascending, simple or much branched; leaves simple, entire, alternate, not reduced upwards; heads several to many, borne in terminal and axillary, corymbose clusters; involucral bracts slightly to conspicuously imbricate, chaffy, scarious; receptacle naked; corollas of disk flowers only, the outer ones tubular-filiform and pistillate, the inner ones tubular-funnelform and perfect, yellowish or whitish; pappus of capillary bristles; style branches somewhat flattened; achenes small, nerveless, glabrous.

Gnaphalium uliginosum L.
Cudweed

Plants 0.3–2.5 dm tall, usually branched,

the stems densely white-tomentose; leaves 0.8–5 cm long, 0.1–0.4 cm broad, narrowly oblanceolate to oblong, white- or grayish-tomentose, not reduced upwards; heads several to many, in bracteate, terminal and axillary clusters; involucres 2–3.5 mm high, woolly at the base, the bracts brownish; achenes about 0.5 mm long.

Weed of waste places, cultivated land, and greenhouses; in southern Alaska and southern Yukon; adventive from Eurasia.

HAPLOPAPPUS Cass. Nom. Cons.

Perennial herbs, from caudex and taproot, the juice watery; stems erect from ascending to prostrate caudex branches, simple, subscapose; leaves alternate, chiefly basal, simple, entire; cauline leaves few or lacking; heads solitary; involucral bracts more or less imbricate in several rows, scarious along the sides; receptacle flat to convex, naked; rays yellow, few to several, pistillate; disk flowers numerous, perfect, fertile, yellow; pappus of capillary bristles; style branches flattened; achenes hairy.

Haplopappus macleanii Brandegee

Plants mat-forming, the caudex branches slender, spreading to ascending, clothed with persistent leaves, the subscapose flowering stems 0.3–0.6 dm tall, minutely glandular; basal leaves 0.5–1 cm long, less than 0.1 cm broad, linear to acute apically, glabrous above and beneath, ciliate; cauline leaves 1–2 or lacking; heads solitary; involucres mostly 6–8 mm high and 11–13 mm broad, the bracts oblong to lance-oblong, attenuate, greenish along the center of the apex, otherwise scarious, minutely glandular; rays 5–12, yellow, about 6 mm long; disk flowers longer than the tan to whitish pappus; achenes densely hairy, about 3 mm long.

Dry, open slopes; in southern Yukon; endemic.

HELIANTHUS L.

Annual herbs with taproots or perennials with rhizomes, the juice watery; stems erect, simple or less commonly branched; leaves alternate or opposite, simple, toothed, petiolate; heads solitary or few; involucral bracts imbricate in several series; receptacle flat to convex, chaffy throughout; ray flowers present, yellow, neuter; disk flowers numerous, perfect, fertile, usually purplish; pappus of 2 deciduous awns; style branches flattened; achenes moderately compressed, glabrous or nearly so.

1a. Leaves mostly opposite; plants rhizomatous perennials; reported from the vicinity of Dawson City, Yukon. *H. laetiflorus* Pers.

1b. Leaves mostly alternate (or only the lowermost opposite); plants annuals with taproots, reported from central and southern Alaska. *H. annuus*

Helianthus annuus L.
Annual Sunflower

Plants annual, the stems mostly 3–10 dm tall or more, sparsely to moderately spreading-hairy with course, multicellular hairs; leaves alternate or the lowermost opposite, 5–40 cm long, the blades 3–15 cm broad, ovate to lance-ovate, lanceolate, or deltoid, acute to acuminate or abruptly obtuse apically, acute to truncate or cordate basally; heads solitary or few to several; involucres 1.3–2.5 cm high (or higher), the bracts ovate-acuminate, coarsely hairy, ciliate; rays numerous, yellow, conspicuous; achenes large, edible.

Rare to occasional weed, or in cultivation; in central and southern Alaska; native to western United States.

HIERACIUM L.

Perennial herbs with milky juice, from subrhizomatous caudex and fibrous roots; leaves alternate or primarily basal, simple, entire or toothed; heads few to many, in corymbose clusters; involucral bracts imbricate to subequal, green or greenish black, or the inner ones with scarious margins;

receptacle naked; corollas of ray flowers only, perfect, yellow, white, or orange red; pappus of white to brownish capillary bristles; style branches semicylindrical; achenes terete or angular, usually tapering to the base, beakless.

1a. Cauline leaves numerous, well developed, the basal ones usually deciduous by flowering time; involucral bracts glabrous or nearly so; plants of northern British Columbia and southeastern Yukon. *H. umbellatum*

1b. Cauline leaves few, much reduced upwards, the basal ones well developed, present at flowering time; involucral bracts sparsely to densely black-hairy; plants of broad distribution. (2)

2a. Flowers red orange; plants rhizomatous and often stoloniferous; introduced in southeastern Alaska. *H. aurantiacum* L.

2b. Flowers white or yellow; plants not both rhizomatous and stoloniferous; variously distributed. (3)

3a. Flowers white; involucral bracts and peduncle apex only sparsely long-hairy; plants of southeastern Alaska and southwestern Yukon. *H. albiflorum*

3b. Flowers yellow; involucral bracts and/or peduncle apex moderately to densely villous with long hairs; plants of various distribution. (4)

4a. Upper part of stem and involucre with grayish or yellowish brown hairs, these mostly 1.5–4 mm long, not at all or only sparsely stipitate-glandular. *H. triste*

4b. Upper part of stem and involucre with grayish hairs mostly less than 1.5 mm long and often conspicuously stipitate-glandular as well. *H. gracile*

Hieracium albiflorum Hook.
White Hawkweed

Stems mostly 3–6 dm tall or more, moder-ately to densely pubescent with long, spreading hairs near the stem base, glabrous or nearly so above; basal leaves mostly 4–12(15) cm long and 0.7–2(3) cm broad, oblanceolate, entire or denticulate, obtuse (though mucronate) apically, tapering basally to a broad petiole, sparsely to moderately long-hairy above and beneath; cauline leaves mostly 3–4, becoming sessile and much smaller above; heads several to many, the peduncles sparsely long-hairy and also glandular; involucres 7–11 mm tall, 6–12 mm broad, the bracts imbricate in several series, sparsely long-hairy to glabrous, or often glandular; corollas white; achenes 2–3 mm long.

Open woods; in southeastern Alaska and southwestern Yukon; east to Saskatchewan and south to California, Utah, and Colorado.

Hieracium gracile Hook.
Slender Hawkweed

Stems mostly 1.5–4 dm tall, moderately short-villous and glandular to glabrate basally, moderately villous and stipitate-glandular above, sometimes also with non-glandular hairs mostly less than 1.5 mm long; basal leaves 3.5–15 cm long, 0.4–2.5 cm broad, oblanceolate, entire or denticulate, obtuse to rounded apically, tapering basally to a slender petiole, sparsely hairy, glandular, or glabrous on both surfaces, ciliate or not; cauline leaves 2–3, much reduced above, usually all petiolate; heads solitary or more commonly few to many, the peduncles white-hairy and with blackish stipitate glands, sometimes also long black-hairy; involucres 6–10 mm high, 5–15 mm broad, the bracts subequal or of several lengths, moderately pubescent with blackish hairs; corollas yellow; achenes 2.5–3.5 mm long.

Alpine meadows and open slopes; in southern Alaska, from the eastern Aleutian Islands eastward, and in southern Yukon; east to Mackenzie and south to California and New Mexico; South America (*H. gra-*

cile var. *yukonense* Porsild; *H. triste* ssp. *gracile* [Hook.] Calder & Taylor). Our material is variable, especially in number of heads, stature, and in the nature of the basal leaves. Despite this, it seems best to treat all of our specimens as belonging to var. *alaskanum* Zahn.

Hieracium triste Willd.
Woolly Hawkweed

Stems mostly 0.3–3(4.5) dm tall, sparsely to moderately short- to long-hairy below, moderately to densely villous and scarcely or not at all glandular above, the hairs mostly 1.5–5 mm long; basal leaves 1–16 cm long, 0.4–3 cm broad, oblanceolate to obovate, entire or denticulate, rounded to obtuse apically, tapering basally to a slender petiole, sparsely hairy to glabrous above, usually all petiolate; heads solitary or more commonly 2–5(7), the peduncle villous with long, grayish or blackish hairs, these underlain by shorter, whitish hairs, and less commonly with some stipitate glands; involucres 7–10 mm high, 5–14 mm broad, the bracts subequal or with some shorter ones, usually obscured by long hairs; corollas yellow; achenes 2–3 mm long.

Alpine meadows, stream banks, and talus slopes; in coastal and insular southern Alaska, and less commonly in south-central Alaska and southern Yukon; south to British Columbia; Asia (*H. triste* var. *tristiforme* Zahn). Putative hybrids between *H. triste* and *H. gracile* have been reported. Their true nature requires careful investigation. The genus is noted for its apomictic members.

Hieracium umbellatum L.

Stems mostly 3–9 dm tall, sparsely villous and long-hairy basally or glabrous, short-hairy above; basal leaves much reduced, usually withered by flowering time; cauline leaves numerous, mostly 2.5–12 cm long and 0.5–2 cm broad, lance-oblong, acute apically, merely sessile or somewhat clasping basally, denticulate to subentire, more or less pubescent with short subconic hairs, especially near the margin; heads commonly 5–many, the peduncles pubescent with short white hair, and sometimes also with short subconic hair; involucres 7–12 mm high, 6–22 mm broad, the bracts conspicuously imbricate, glabrous or nearly so; corollas yellow; achenes 2–3 mm long.

Moist woods; in southeastern Yukon and northern British Columbia; east to Labrador and south to Oregon, Idaho, Colorado, and Michigan; Eurasia (*H. scabriusculum* Schwein.; *H. canadense* authors, not Michx.). *H. umbellatum* is closely related to *H. canadense*, and apparently passes into that entity in more southern regions.

HYPOCHAERIS L.

Perennial, subscapose herbs from taproots, the juice milky; leaves primarily basal, simple, pinnately lobed to pinnatifid, the cauline leaves small and bractlike; heads solitary, or few in a branching inflorescence; involucral bracts in several series, greenish black, the inner ones with hyaline margins; receptacle chaffy; corollas of ray flowers only, perfect, yellow, or purplish on the lower surface; pappus of plumose, capillary bristles; style branches semicylindrical; achenes several-nerved, subterete, minutely roughened, long-beaked.

Hypochaeris radicata L.
Cat's-ears

Plants 1.5–5 dm tall, the stems simple or branched above, glabrous or spreading-hairy below; basal leaves 3–16(25) cm long, 0.5–3.5(5) cm broad, oblanceolate, pinnately toothed or pinnatifid, sparsely to moderately spreading-hairy above and below, rounded to obtuse apically, tapering to a broad petiole basally; cauline leaves alternate, minute or lacking; heads solitary, or more commonly 2–5, terminating the branches, the peduncles glabrous; involucres 5–15 mm high, 7–20 mm broad, the bracts glabrous or stiffly hairy along the midribs; corollas numerous, longer than the bracts; achenes 4–7 mm long, the beak mostly 2–3 mm long.

Weedy species of disturbed soils; in the Aleutian Islands (Atka) and in southeastern Alaska; introduced from Europe.

LACTUCA L.

Annual, biennial, or perennial herbs from taproots, the juice milky; leaves chiefly cauline, alternate, simple, entire to lobed or pinnatifid; heads several to numerous; involucral bracts in 2 or more series, green or greenish, often tinged with red or purple, the inner ones often with hyaline margins; receptacle naked; corollas all raylike, yellow, blue, or white, perfect; pappus of capillary bristles; style branches semicylindrical; achenes compressed, several-nerved, beaked, glabrous or hispid.

1a. Plants perennial; involucres in fruit mostly 12–18 mm high; pappus white; leaves not spinulose along the lower midvein; known from central Alaska. *L. tatarica*

1b. Plants annual or biennial; involucres in fruit mostly 6–12(13) mm high; leaves spinulose along the lower midvein, at least some of them, or the pappus brownish; distribution various. (2)

2a. Pappus brownish; achene beak shorter than half the length of the body or lacking; corollas usually bluish or white; plants indigenous to southeastern Alaska. *L. biennis*

2b. Pappus white; achene beak subequal to the body or longer; corollas yellow (drying blue); plants adventive in central Alaska. *L. serriola*

Lactuca biennis (Moench) Fern.
Tall Blue Lettuce

Plants annual or biennial, the stems 6–20 dm tall, glabrous, not spinulose; leaves mostly 7–20(40) cm long and 2.5–15(20)

cm broad, pinnatifid or merely toothed, acute apically, wing-petioled and more or less clasping basally, glabrous and often spinulose along the lower midvein; heads several to many, with 13–30 flowers or more; involucres mostly 9–12(13) mm high and 8–15 mm broad, the bracts often purplish tinged; corollas bluish to white, or sometimes yellow; pappus brownish; achenes 4–5 mm long, beakless or with a short, stout beak.

Open woods, meadows, and roadsides; in southeastern Alaska and northern British Columbia; east to Newfoundland and south to California, Colorado, and North Carolina (*L. spicata* authors, not *Sonchus spicatus* Lam.).

Lactuca serriola L.
Prickly Lettuce

Plants annual or biennial, the stems 3–15 dm tall, glabrous, often spinulose; leaves mostly 5–20(30) cm long and 1–5(8) cm broad, pinnately lobed or merely toothed, acute or abruptly acuminate apically, sagittate-clasping basally, glabrous, spinulose along the lower midrib; heads numerous, with 12–25 flowers; involucres mostly 8–12(13) mm high and 3–6 mm broad, the bracts sometimes reddish tinged; corollas yellow (fading blue); pappus white; achenes mostly 4–6 mm long, the beak subequal to the body or longer.

Waste places, evidently rare; in central Alaska (Manley Hot Springs), and to be expected elsewhere; adventive from Europe (*L. virosa* authors, not L.). The cultivated lettuce, *L. sativa* L., is also grown in Alaska; it seldom flowers.

Lactuca tatarica (L.) C. A. Mey.

Plants perennial, the stems mostly 2–10 dm tall, glabrous, not spinulose; leaves mostly 5–12(15) cm long and 0.5–4 cm broad, entire to lobed or more or less pinnatifid, acute to attenuate apically, sessile and more or less clasping basally, glabrous, not spinulose; heads few to many, with 20–50

Lactuca serriola L. (× 0.3).

flowers; involucres mostly 12–18 mm high and 6–12 mm broad, the bracts often reddish tinged; corollas blue; pappus white; achenes 4–6 mm long, the beak stout.

Meadows and open woods; in central eastern Alaska and disjunctly south to California and east to Missouri and Minnesota (*Sonchus pulchellus* Pursh; *L. pulchella* [Pursh] DC.; *L. tatarica* ssp. *pulchella* [Pursh] Stebb.).

LAPSANA L.

Annual herbs from taproots, the juice milky; leaves alternate, simple, subentire to toothed or lyrate-pinnatifid; heads numerous; involucral bracts in 2 series, the inner ones large and keeled, the outer minute, greenish; receptacle naked; corollas of ray flowers only, perfect, yellow; pappus none; style branches semicylindrical; achenes subterete, several-nerved, tapering to both ends, beakless.

Lapsana communis L.
Nipplewort

Plants mostly 3–10 dm tall, the stems erect, simple or branched, pubescent with stipitate glands or glabrous; leaves mostly 5–12 cm long and 2–5(7) cm broad, the blades subentire to toothed, or the lower ones lyrate-pinnatifid, sparsely hairy to glabrous above and below; heads numerous, the peduncles glabrous or nearly so; involucres 5–8 mm high, 3–9 mm broad, the bracts glabrous, the inner ones keeled, the outer ones very small; flowers mostly 10–14; achenes 3–5 mm long.

Disturbed soils; in southeastern Alaska; adventive from Eurasia.

LEONTODON L.

Perennial herbs with a caudex and fibrous roots, the juice milky; leaves all basal, simple, pinnately lobed to entire; heads solitary, or few to several in a branching inflorescence; involucral bracts in several series, greenish black, the inner ones with hyaline margins; receptacle not chaffy; corollas of ray flowers only, perfect, yellow; pappus of plumose, capillary bristles; style branches semicylindrical; achenes several-nerved, subterete, not beaked.

Leontodon autumnale L.
Fall Dandelion

Plants 1–5 dm tall or more, the stems simple or branched above, glabrous throughout or more commonly hairy at the apex; leaves 4–30 cm long and 0.5–4 cm broad or more, oblanceolate, pinnately lobed to entire, glabrous or moderately hairy, long-attenuate apically, tapering to a broad petiole basally; heads solitary or 2–5, terminating the branches, the peduncles bracteate, glabrous; involucres 7–13 mm high, 9–20 mm broad, the bracts stiffly hairy; corollas numerous, longer than the bracts; achenes 4–7.5 mm long, beakless.

Waste places; in central Alaska (to be expected elsewhere); adventive from Europe.

MADIA Mol.

Annual herbs from taproots, the juice watery; stems erect, simple to much branched; leaves alternate or the lower ones opposite, simple, entire; heads 1–many, borne in compact clusters; involucral bracts in 1 series, equal, greenish, enfolding the ray achenes; receptacle flat, with a single row of bracts between the ray and disk flowers, otherwise naked; ray flowers present or sometimes lacking in some heads, yellow, pistillate and fertile; disk flowers perfect, fertile or sterile, yellow; pappus none; style branches flattened; achenes compressed, glabrous, beakless.

Madia glomerata Hook.
Tarweed

Stems 1–7 dm tall, simple or branched, pubescent with coarse, ascending to spreading, multicellular hairs mixed with slender, unicellular hairs; leaves 2–7 cm broad, 0.1–0.4 cm broad, linear to lance-linear or narrowly oblong, hairy above and beneath, long-ciliate; heads 1–many in small clusters, short-pedunculate to sessile; involucres 6–9 mm high, 3–5 mm broad, the bracts enclosing the ray achenes, greenish, pubescent and glandular; ray flowers yellow, mostly 1.5–2.5 mm long, usually 1–3, or lacking in some heads; disk corollas hairy; pappus none; achenes glabrous, mostly 3–4 mm long.

Weedy species of disturbed roadsides and waste places; in southern Alaska and southern Yukon; possibly introduced from its range in British Columbia, east to Saskatchewan, and south to California, Arizona, and Colorado.

MATRICARIA L.

Annual, biennial, or short-lived perennial herbs, from tap or fibrous roots, the juice watery; stems erect or ascending, simple to much branched; leaves alternate or basal and alternate, pinnately once or twice dissected; heads solitary, or few to several in an open cluster; involucral bracts imbricate in 2–3 series, greenish, with hyaline or brownish, scarious margins; receptacle hemispheric to conic, naked; ray flowers white, numerous, pistillate and fertile, or lacking; disk flowers numerous, perfect, fertile, yellow or greenish yellow; pappus a short crown; style branches flattened; achenes several-nerved, beakless, glabrous.

1a. Ray flowers lacking; heads usually several; stems branching above. *M. matricarioides*

1b. Ray flowers present, conspicuous, rarely absent; heads solitary, or sometimes 2–several: stems branching from the caudex or above. (2)

2a. Involucral bracts with dark brownish or brownish black margins; primary leaf divisions usually less than 1.5 cm long, the ultimate segments lance-oblong; plants indigenous to coastal western, northwestern, and northern regions. *M. ambigua*

2b. Involucral bracts with pale brownish margins; primary leaf divisions often over 1.5 cm long, the ultimate segments linear. *M. maritima*

Matricaria ambigua (Ledeb.) Krylof
Arctic Chamomile

Plants perennial, the stems 0.8–2.5(3.5) dm tall, branching from the caudex, or less commonly above, glabrous; basal leaves well developed, forming a more or less persistent tuft, 1.5–7 cm long, the primary divisions 0.3–1.3 cm long, the ultimate segments lance-oblong, glabrous or ciliate;

cauline leaves similar to the basal ones; heads solitary, or less commonly 2–several, the peduncles glabrous; receptacle convex; involucres 5–9 mm high, 12–24 mm broad, the bracts lanceolate to ovate, with dark brown to brownish black, scarious margins, glabrous; ray flowers numerous, white, 9–22 mm long, 1.5–4.5 mm broad, or rarely lacking; achenes 4-angled.

Matricaria ambigua (Ledeb.) Krylof (× 0.5).

Tundra and heath; along the coast in western, northwestern, and northern Alaska, and northern Yukon; east to Baffin Land; Eurasia (*M. inodora* f. *phaeocephala* Rupr.; *M. maritima* var. *phaeocephala* [Rupr.] Hyl.; *Chrysanthemum grandiflorum* Hook.; *M. inodora* var. *grandiflora* [Hook.] Ostenf.; *Tripleurospermum phaeocephalum* [Rupr.] Pobed). This entity is closely related to *M. maritima,* and might be best treated at some infraspecific level within that species. If the name-bearing synonym, *Pyrethrum ambiguum* Ledeb., proves to be misapplied, the correct specific epithet might well be *M. phaeocephala* (Rupr.) Stefans.

Matricaria maritima L.

Plants annual or biennial, rarely perennial, the stems 2–6 dm tall, usually branching above, glabrous; basal leaves usually withered by flowering time; cauline leaves 2–12 cm long, the primary divisions mostly 10–25 mm long, the ultimate segments linear, glabrous; heads commonly few to several, the peduncles glabrous; receptacle convex; involucres 5–7 mm high, 13–18 (21) mm broad, the bracts lanceolate, with usually pale brown scarious margins, glabrous; ray flowers numerous, white, 10–17 mm long, 2.5–4 mm broad; achenes 4-angled.

Cultivated ornamental; escaping in east-central Alaska and possibly elsewhere; adventive from Europe (*M. inodora* L.; *Tripleurospermum inodorum* [L.] Schulz-Bip.; *M. maritima* ssp. *inodora* [L.] Clapham; *M. maritima* var. *agrestis* [Knaf.] Willm.).

Matricaria matricarioides (Less.) Porter
Pineapple Weed

Plants annual, the stems 0.7–4.5 dm tall, branching above, or less commonly throughout, glabrous, ill scented; basal leaves usually withered by flowering time; cauline leaves 1–8 cm long, the primary divisions 4–17 mm long, the ultimate segments lance-oblong, glabrous; heads few to numerous, the peduncles glabrous; receptacle conic; involucres 2.5–4.5 mm long, 7–12 mm broad, the bracts ovate to obovate, with broad, hyaline margins, glabrous; ray flowers lacking; achenes obscurely 4-nerved.

Weed of disturbed soils; widely scattered in Alaska and in the Yukon; south to California and Arizona, and widely introduced elsewhere (*M. suaveolens* [Pursh] Buch., not L.).

PETASITES Mill.

Perennial herbs from rhizomes, the juice watery; stems erect or ascending, decumbent only at the base, merely bracteate, the bracts alternate, lacking true foliage leaves, branching only in the inflorescence; foliage leaves arising directly from the rhizome, simple, toothed or lobed, long-petiolate; heads several to many, borne in corymbose clusters; involucral bracts in 1 series, subequal, or with a few shorter, outer bracts, greenish or commonly suffused with red or purple; receptacle flat, naked; outer flowers pistillate, the corollas white, with the rays developed or not; disk flowers tubular, 5-lobed, perfect, fertile, white or tinged red; pappus of capillary bristles, elongating in fruit; style branches short, often undivided; achenes 5–10-nerved, glabrous, beakless.

1a. Leaves shallowly to conspicuously dentate, usually with 20 or more teeth per side, more or less hastately lobed basally. *P. sagittatus*
1b. Leaves distinctly lobed, or if merely dentate, then the teeth usually less than 15 per side, sagittate to cordate basally, seldom or not hastate. *P. frigidus*

Petasites frigidus (L.) Fries
Arctic Sweet Coltsfoot

Plants with elongate rhizomes; flowering stems 0.6–4.5(5) dm tall, bearing few to many alternate bracts, more or less white-tomentose; foliage leaves arising directly from the rhizome, the blades mostly 2–18 cm long from sinus to apex and 3–25 cm broad, cordate to reniform, cordate-ovate, or more or less orbicular, conspicuously lobed, or if merely toothed, then the teeth usually less than 15 per side, green and tomentose to glabrate or glabrous above, densely to sparsely tomentose beneath, the petioles mostly 2–30 cm long; heads several to many, the peduncles moderately stipitate-glandular and often tomentose; involucres 6–12 mm high, 8–20 mm broad, the bracts oblong to lanceolate, pubescent basally with glandular, multicellular hairs, the cross-walls often colored; rays small to conspicuous, white.

Petasites frigidus (L.) Fries (× 0.3).

Moist sites in tundra, heath, or woods; throughout Alaska and Yukon; east to the Atlantic and south to California, Michigan, and Massachusetts; Eurasia. *P. frigidus* is highly polymorphic, and because of extensive hybridization among the various entities within *P. frigidus,* and between *P. frigidus* and *P. sagittatus,* classification is difficult. Further work is indicated. Three poorly defined varieties are present within our region.

1a. Leaves shallowly lobed to merely toothed, the blades often less than 6 cm long; hybridizing with, and passing into the other varieties; plants common over much of Alaska and Yukon. *P. frigidus* var. *frigidus*

1b. Leaves evidently lobed, the sinuses extending one-fourth the distance to the midrib or leaf base, the blades often more than 6 cm long; distribution various. (2)

2a. Leaves palmately lobed and veined, the sinuses extending one-half or more the distance to the leaf base; plants of southern to northern Yukon (*Tussilago palmata* Ait.; *Nardosmia palmata* [Ait.] Hook.; *N. speciosa* Nutt.). *P. frigidus* var. *palmatus* (Ait.) Cronq.

2b. Leaves palmately to pinnately lobed and veined, the sinuses seldom extending half the distance to the midrib or leaf base; plants of broad distribution in Alaska and Yukon (*P. alaskanus* Rydb.; *P. arcticus* Porsild; *P. hyperboreoides* Hultén; *P. hyperboreus* Rydb.; *P. vitifolia* Greene). *P. frigidus* var. *nivalis* (Greene) Cronq.

Petasites sagittatus (Banks) Gray
Arrowleaf Sweet Coltsfoot

Plants from elongate rhizomes; flowering stems mostly 2–6.5 dm tall, bearing few to many, alternate, scarious to subherbaceous, often brownish, tomentose to glabrate bracts, more or less white-tomentose; foliage leaves arising directly from the rhizome, the blades mostly 4–23 cm long from sinus to apex, 2.5–21 cm broad, hastate to cordate, merely toothed with usually 20 or more teeth per side, green and glabrous or more or less tomentose above, moderately to densely tomentose beneath, the petioles 0.5–40 cm long; heads several to many, the peduncle tomentose and stipitate-glandular or glandless; involucres 7–10 mm high, 12–20 mm broad, the bracts oblong to lanceolate, pubescent basally with multicellular, glandular hairs, the cross-walls mostly colorless; rays small, white.

Moist sites usually in bogs or swamps; in the southeastern third of Alaska (rarely elsewhere), and southern Yukon; east to Labrador and south to Washington, Idaho, Montana, and Colorado (*Tussilago sagittata* Banks ex Pursh).

PICRIS L.

Biennial herbs from taproots, the juice milky; leaves basal and cauline, simple, alternate, toothed; heads several to many, borne in corymbose clusters; involucral bracts in several series, greenish black, the inner ones with scarious margins; receptacle naked or merely short-hairy; corollas all raylike, perfect, yellow; pappus of plumose, capillary bristles; style branches semicylindrical; achenes 5–10-ribbed, subterete, minutely roughened, narrowed at the apex, not beaked.

Picris hieracioides L.

Plants mostly 3–8 dm tall, the stems erect, simple or branching from the middle and upper nodes, moderately to densely pubescent with stiff, spreading, reddish to purplish black hairs, especially below; basal leaves 6–15 cm long, 1–4 cm broad, oblanceolate, toothed, obtuse to acute apically, narrowed to a broad petiole basally, sparingly to moderately hairy like the stems; cauline leaves well developed, lanceolate, becoming sessile upwards and more or less auriculate-clasping; heads several to many, mostly 10–16 mm high and 8–15 mm broad, the bracts lance-oblong, attenuate, coarsely hairy, especially along the midveins; corollas numerous, longer than the bracts; achenes about 4 mm long, beakless.

Beaches and slopes; in the western Aleutian Islands (Attu); Eurasia. Our material belongs to var. *alpina* Koidz. (*P. kamtschatica* Ledeb.; *P. hieracioides* ssp. *kamtschatica* [Ledeb.] Hultén; *P. hieracioides* var. *kamtschatica* [Ledeb.] B. Boi.).

PRENANTHES L.

Perennial herbs from tuberous roots, the juice milky; leaves chiefly cauline, alternate, simple, hastate-deltoid, sharply toothed, with broadly winged petioles (becoming sessile upwards); heads few to many, in corymbose clusters; involucral bracts subequal, or with a few shorter, outer ones, green or tinged reddish, the inner ones with hyaline margins; receptacle naked; corollas of ray flowers only, perfect, white or pinkish; pappus of capillary bristles; style branches semicylindrical; achenes terete, 5–15-nerved, beakless, glabrous.

Prenanthes alata (Hook.) D. Dietr.
Rattlesnake Root

Plants 2–8 dm tall, the stems simple to much branched, especially above, sparingly pubescent with short, contorted, multicellular hairs; leaves chiefly cauline, 3.5–17 cm long, the blades 0.6–7 cm broad or more, the main ones hastate-deltoid, sharply toothed, the petioles broadly winged, minutely puberulent to glabrous above and below; heads few to many, the peduncles moderately villous; involucres 6–14 mm high, 3–13 mm broad, the bracts lance-oblong, sparingly villous; pappus brownish; rays mostly 10–15, white or pinkish, mostly 5–10 mm long; achenes 4–6 mm long.

Stream banks, open woods, sea cliffs, beaches, and open slopes; in insular and coastal southern Alaska, from the Aleutians eastward through the Panhandle; south to Oregon (*Sonchus hastatus* Less.; *Nabalus*

hastatus [Less.] Heller; *P. lessingii* Hultén; *Nabalus alatus* Hook.).

SAUSSUREA DC. Nom. Cons.

Perennial, rhizomatous or subrhizomatous herbs with watery juice; stems erect, unarmed; leaves alternate, simple, entire to toothed or lobed, unarmed; heads few to several, borne in corymbose clusters; involucral bracts imbricate in several series, green throughout, or the inner ones scarious at the base, often suffused with purple; receptacle flat or convex, naked or with slender bristles; corollas of disk flowers only, perfect, pink purple or white, falling as a unit; style branches semicylindrical to flattened, the style with a ring of hair immediately below the branches; achenes more or less compressed, about 5-nerved, beakless, glabrous.

1a. Cauline leaves ovate to lanceolate, cordate to truncate or obtuse basally; plants usually 5 dm tall or more, known from southeastern Alaska. S. *americana*

1b. Cauline leaves linear to elliptic, oblong or narrowly lanceolate, acute to attenuate basally; plants usually less than 5 dm tall, of various distribution. (2)

2a. Involucral bracts all lance-acuminate, about equal in length; receptacle naked; plants of southwestern, western, and northwestern Alaska. S. *nuda*

2b. Involucral bracts varying in shape, the outer ovate to broadly lanceolate, usually much shorter than the inner; receptacle usually bristly; plants of broad distribution. S. *angustifolia*

Saussurea americana D. C. Eaton

Plants mostly 5–9 dm tall, the stems glabrous or nearly so; leaves mostly 5–12 cm long and 1–5 cm broad, the blades ovate to lanceolate or triangular, basally cordate to truncate or obtuse (or abruptly acuminate), acuminate apically, dentate, sparsely

to moderately villous-tomentose, especially
on the veins beneath, green above; heads in
a compact corymbose cluster, the peduncles more or less viscid; involucres 10–13
mm high, 8–12 mm broad, the bracts imbricate in 3–4 series, the outer ones ovate,
about 1.5–2 times longer than broad,
tomentose to glabrate; flowers pink purple
or white; receptacle naked or more or less
bristly.

Open woods; in southeastern Alaska and
northern British Columbia; south to Oregon
and Idaho.

Saussurea angustifolia (Willd.) DC.

Plants 0.3–4.5 dm tall, the stems sparingly
to moderately (densely) villous-tomentose;
leaves mostly 2.5–13 cm long and 0.2–0.8
(1.1) cm broad, narrowly oblong to elliptic, lanceolate, or oblanceolate, attenuate
to acute basally, acute or attenuate apically, entire or denticulate, often revolute,
glabrous or sparingly to moderately villous or tomentose, especially below and
sometimes also glandular, green above;
heads few to several in a compact corymbose cluster, the peduncles sparingly to
densely villous; involucres 10–15 mm high,
10–16 mm broad, the bracts imbricate in
2–4 series, the outer ones ovate to lanceolate or lance-acuminate, mostly 1.5–2(4)
times longer than broad, pilose to villous
or tomentose, or less commonly glabrous;
flowers pink purple or white; receptacle
usually bristly.

Meadows, bogs, and open woods; over
much of Alaska and the Yukon; east to
Mackenzie and south to Saskatchewan;
Asia. Our abundant material is variable in
stature, viscidity, pubescence, and shape
of involucral bracts. Three more or less
completely intergrading varieties occur in
our region.

1a. Outer involucral bracts usually (1)
1.5–2(2.5) times longer than broad;
stems well developed, often over 1.5
dm tall; leaves mostly less than 0.8

Saussurea angustifolia (Willd.) DC. (× 0.5).

cm broad; widely distributed, our
most common *Saussurea*. *S. angustifolia* var. *angustifolia*

1b. Outer involucral bracts usually 2–
4 times longer than broad; stems
usually less than 1.5 dm tall; leaves
variable, but with some often over

0.8 cm broad; distribution various. (2)

2a. Leaves more or less tomentose, slightly if at all glandular-viscid, the leaf bases commonly with a white tomentum; plants commonly 0.3–0.9 dm tall, mostly from interior Alaska and Yukon (*S. viscida* var. *yukonensis* [Porsild] Hultén). **S. angustifolia var. yukonensis** Porsild

2b. Leaves tomentose throughout, or more commonly merely villous and glandular-viscid along the margins or on the lower midvein; leaf bases white-tomentose or not; plants commonly 0.6–1.6 dm tall, mostly of insular and coastal western Alaska. **S. angustifolia var. viscida** (Hultén) Welsh

Saussurea nuda Ledeb.

Plants commonly 1–3 dm tall, the stems sparingly to moderately tomentose, or merely villous; leaves 4–20 cm long, 0.6–2.5 cm broad, oblanceolate to elliptic or oblong, acute to attenuate basally, acute to acuminate apically, entire or denticulate, glabrous or sparingly tomentose on both surfaces, green above and below; heads few to several, in a compact, corymbose cluster, the peduncles more or less tomentose and often villous as well; involucres 9–13 mm high, 14–18 mm broad, the bracts imbricate in 2–3 series, the outer ones lance-acuminate, subequal to the inner ones, villous or villous-tomentose; flowers pink purple or white; receptacle commonly naked.

Tundra and heathlands; in coastal southwestern, western, and northwestern Alaska; Asia.

SENECIO L.

Annual, biennial, or perennial herbs, with rhizomes, caudices, or taproots, the juice watery; stems erect, ascending, or decumbent at the base; leaves alternate, simple, entire, toothed, or lobed to pinnatifid; heads solitary, or few to many, in corymbose cymes; involucral bracts in 1 series, often with smaller bractlets at the base, green throughout, or the margins scarious or hyaline, or variously colored; receptacle flat or convex, naked; ray flowers yellow or orange, or reddish; pappus of capillary bristles; style branches flattened; achenes subterete, 5–10-nerved, glabrous or pubescent.

Barkley, T. M. 1962. A revision of *Senecio aureus* Linn. and allied species. Trans. Kansas Acad. Sci. 66:318-408.

1a. Leaves pinnately to subpalmately lobed, the lobes 5 cm long or more; plants of the western Aleutian Islands. *S. cannabifolius*

1b. Leaves variously lobed to toothed or entire, but not as above; plants of various distribution. (2)

2a. Involucral bracts and peduncle apex pubescent with purplish or brownish, multicellular hairs. *S. atropurpureus*

2b. Involucral bracts and peduncle apex glabrous, or pubescent with yellowish multicellular hairs, or white-tomentose. (3)

3a. Stems, leaves, and involucres sparingly to moderately or densely tomentose; involucres (20)25–55 mm broad; cauline leaves well developed; plants of beaches in coastal southern, western, and northwestern Alaska. *S. pseudo-arnica*

3b. Stems, leaves and involucres variously pubescent or glabrous; involucres commonly less than 25 mm broad (except in some *S. fuscatus*, which has poorly developed cauline leaves); habitat seldom as above. (4)

4a. Main cauline leaves petiolate, the blades hastate-triangular to ovate or lanceolate, serrate-dentate, the base hastately lobed, truncate, or abruptly obtuse to acute. *S. triangularis*

4b. Main cauline leaves sessile, or if petiolate, then the blades lanceolate to oblong or oblanceolate, entire, denticulate, serrate, or pinnatifid, attenuate to acute basally. (5)

5a. Cauline leaves well developed, not much reduced upwards, irregularly pinnatifid; plants introduced, more or less weedy. (6)

5b. Cauline leaves much reduced upwards, variously toothed, lobed, or entire; plants indigenous, perennial (annual or biennial in S. *congestus*). (8)

6a. Upper leaves short-petiolate to sessile, not clasping; leaf lobes usually not toothed; plants short-lived perennials, reported from the vicinity of Tok (introduced from western United States). S. *eremophilus* Richards.

6b. Upper leaves auriculate-clasping; leaf lobes again toothed. (7)

7a. Heads rayless; plants glabrous or glabrate; involucral bracts black tipped. S. *vulgaris*

7b. Heads with minute rays; plants viscid-hairy; involucral bracts not black tipped. S. *viscosus*

8a. Involucral bracts and peduncle apex moderately to densely pubescent with yellowish or whitish, multicellular hairs; rays yellow; cauline leaves linear-lanceolate, entire, or the lower ones broader and toothed. S. *yukonensis*

8b. Involucral bracts and peduncle apex glabrous, or tomentose, lacking multicellular hairs, or if pubescent with multicellular hairs, then the rays yellow orange to orange (see S. *fuscatus*) or the stem also villous with multicellular hairs (see S. *congestus*). (9)

9a. Plants annual or biennial; stems and peduncles villous with multicellular hairs. S. *congestus*

9b. Plants perennial; stems and peduncles glabrous or merely tomentose. (10)

10a. Leaf pubescence of 2 types, of coarse, flattened, multicellular hairs overlain by a white tomentum (seldom merely tomentose); ray flowers orange or yellow orange when present; involucral bracts usually purplish throughout and more or less pubescent with whitish or yellowish hairs. S. *fuscatus*

10b. Leaf pubescence of only 1 type, merely white-tomentose, or lacking; involucral bracts green throughout or only the tips black or purplish, glabrous or sparingly tomentose. (11)

11a. Heads solitary (rarely 2); blades of basal leaves about as broad as long or broader, merely crenate (rarely lobed); plants glabrous or nearly so. S. *resedifolius*

11b. Heads 2–many (rarely solitary); blades of basal leaves longer than broad, dentate, crenate, or lobed; plants glabrous, or more or less tomentose. (12)

12a. Leaves sharply denticulate, the main cauline ones ovate to lanceolate, lance-oblong, elliptic, oblanceolate, or triangular, often more than 6 cm long. (13)

12b. Leaves crenate, serrulate, or pinnatifid, the main cauline ones linear to lance-oblong, oblanceolate, or elliptic, often less than 6 cm long. (14)

13a. Involucral bracts black tipped; basal leaves with broadly winged petioles; plants widespread in Alaska and Yukon. S. *lugens*

13b. Involucral bracts with dark colored tips, but not black tipped; basal leaves with slender petioles; plants of southern Yukon. S. *sheldonensis*

14a. Heads typically lacking rays. (15)

14b. Heads typically bearing rays. (16)

15a. Disk corollas yellow; heads commonly 8–20; leaves thin; plants of lowland regions in southeastern quarter of Alaska and southern Yukon. S. *indecorus*

15b. Disk corollas orange or reddish orange; heads commonly 1–6; leaves subsucculent; plants alpine or subalpine, known from southern Yukon. *S. pauciflorus*

16a. Plants usually conspicuously tomentose at flowering time; heads commonly (1)2–6. *S. hyperborealis*

16b. Plants glabrous, or tomentose only at the base, or in the leaf axils, or in the inflorescence at flowering; heads often 5 or more. (17)

17a. Blades of basal leaves obovate or orbicular to oblanceolate or elliptic, shallowly lobed to crenate or subentire; stems immediately below the inflorescence mostly 1–2.5 mm in diameter; plants of southern Yukon. *S. streptanthifolius*

17b. Blades of basal leaves oblanceolate, or less commonly obovate or orbicular, crenate or crenate-serrate, or subentire; stems immediately below the inflorescence 0.5–1.5(2) mm in diameter; plants from the southeastern quarter of Alaska and southern Yukon. *S. pauperculus*

Senecio atropurpureus (Ledeb.) Fedtsch.

Plants perennial, rhizomatous, 0.3–2.5 dm tall, the stems erect or ascending, glabrous or sparingly white-tomentose below, becoming villous or villous-tomentose above with purplish or brownish, multicellular hairs, the cross-walls purplish; basal leaves 1.5–11 cm long, 0.2–2 cm broad, oblanceolate to obovate, entire or dentate-serrate, rounded to obtuse or less commonly attenuate or acute apically, acute to obtuse basally, petiolate or sessile, glabrous or white-tomentose; cauline leaves becoming smaller upwards, lanceolate, ovate, or lance-oblong, entire or toothed; heads solitary, the peduncle apex villous or villous-tomentose with purplish or brownish multicellular hairs; involucres 7–12 mm high, 10–24 mm broad, the bracts lance-oblong, acute, green throughout or the tips purplish tinged, the margins hyaline, sparingly or more commonly moderately to densely villous or villous-tomentose with purplish or brownish, multicellular hairs; rays yellow, mostly 5–17 mm long and 2.5–6 mm broad, or vestigial or lacking; achenes glabrous.

Arctic and alpine tundra and heathlands, often in moist sites; over most of Alaska, except for the Aleutians and the southeastern Panhandle, and over much of the Yukon; east to Labrador; Eurasia. The forms of *A. atropurpureus* with ray flowers well developed are among the most showy of ragworts in Alaska. The species is a polymorphic one, consisting of a series of more or less distinctive populations. Four intergrading varieties are recognizable among our materials.

1a. Basal leaves conspicuously petiolate, mostly (1.5)3.5–11 cm long and 0.2–1 cm broad, irregularly dentate or denticulate; plants from the Seward Peninsula, and disjunctly eastward along the Alaska Range and St. Elias Mts. *S. atropurpureus* var. *dentatus* (Gray) Hultén

1b. Basal leaves sessile or short-petiolate, mostly 1.5–4 cm long and 0.6–2 cm broad, entire or irregularly serrate-dentate; plants of various distribution. (2)

2a. Upper portion of stem and involucre densely tomentose with brownish hairs; leaves irregularly serrate-dentate to entire; plants from disjunct localities, mostly in interior Alaska and Yukon (*S. kjellmanii* Porsild; *S. atropurpureus* ssp. *tomentosus* [Kjellm.] Hultén). *S. atropurpureus* var. *tomentosus* (Kjellm.) Hultén

2b. Upper portion of stem and involucre sparingly to moderately villous with usually purplish (apparently blackish) hairs; leaves entire or sometimes denticulate; distribution various. (3)

3a. Ray flowers well developed, mostly 12–17 mm long and 4–6 mm broad;

peduncles often less than 3 cm long; plants of northwestern coastal Alaska and islands of the Bering Sea. *S. atropurpureus* var. *atropurpureus*

3b. Ray flowers mostly less than 12 mm long or less than 4 mm broad, or both, or vestigial, or lacking; peduncles usually more than 3 cm long; plants of broad distribution in Alaska and Yukon (*Cineraria frigida* Richards.; *S. atropurpureus* ssp. *frigidus* [Richards.] Hultén). *S. atropurpureus* var. *ulmeri* (Steffen) Porsild

Senecio cannabifolius Less.

Plants perennial, with elongate rhizomes, 0.5–1.5 m tall, or more, the stems glabrous to sparingly tomentose, especially above; basal leaves petiolate, usually withered by flowering time; cauline leaves 10–20 cm long, 8–15 cm broad, pinnately or subpalmately cleft, the lobes usually over 5 cm long, serrate, glabrous or more commonly hairy, especially along the midveins beneath, petiolate or the upper ones subsessile, usually auriculate at the base; heads numerous in corymbose cymes, the peduncles tomentose; involucres 4–6 mm long, 5–7 mm broad, the bracts in 2 series, the outer ones much shorter than the inner, green throughout, or the margins scarious, pubescent to glabrate; rays yellow, 7–12 mm long, 1.5–2 mm broad; achenes glabrous.

Mountain slopes and meadows; in the westernmost Aleutian Islands (Attu); Asia (*S. palmatus* [Pallas] Ledeb.). Our specimens belong to var. *cannabifolius*.

Senecio congestus (R. Br.) DC.
Marsh Fleabane

Plants annual or biennial, with fibrous roots, 0.5–10 dm tall or more, the stems sparingly to moderately or densely villous, or villous-tomentose with whitish or yellowish, multicellular hairs, especially above; basal leaves 3–20 cm long, 0.5–3 cm broad, oblong to elliptic or narrowly oblanceolate, pinnately lobed or toothed to subentire, glabrous or villous, petiolate or sessile, often lacking by flowering time; cauline leaves 2–20 cm long, 0.3–5 cm broad, lanceolate to oblong, pinnately lobed or toothed, villous to glabrate, gradually reduced upwards, sessile; heads several to many, the peduncles moderately to densely villous or villous-tomentose with yellowish or pale brownish hairs; involucres 5–13 mm high, 5–14 mm broad, the bracts in 1 series, lance-linear, attenuate, greenish or yellowish throughout, or sometimes suffused with purple, the margins hyaline, glabrous or sparingly to densely villous or villous-tomentose with yellowish hairs; rays yellow, mostly 3–9 mm long and 1–2 mm broad, or lacking; achenes glabrous.

Moist sites, along streams, in seeps, and in bogs; in most of Alaska (except for southwestern and southeastern portions) and Yukon; eastward to Labrador and south to Alberta and Iowa; Eurasia. Three poorly defined varieties may be distinguished by the following arbitrary key.

1a. Leaves all pinnatifid, the lobes often 1 cm long or more; inflorescence usually densely long villous-tomentose; plants mostly from western Alaska, rarely elsewhere. *S. congestus* var. *laceratus* (Ledeb.) Fern.

1b. Leaves merely toothed to subentire, or only the lower ones pinnatifid, or if all pinnatifid, then the inflorescence not densely long villous-tomentose; plants of various distribution. (2)

2a. Inflorescence congested; peduncles and often upper leaves and involucres obscured by a villous tomentum; plants mostly from northern Alaska and northern Yukon. *S. congestus* var. *congestus*

2b. Inflorescence open or expanding; pedicels variably hairy, but seldom as above; upper leaves and involucres mostly sparingly hairy; plants widely distributed in interior Alaska and Yukon. (*S. palustris* [L.] Hook., not

Vell; *S. congestus* var. *tonsus* Fern.).
S. congestus var. *palustris* (L.) Fern.

Senecio fuscatus (Jord. & Fourr.) Hayek

Plants perennial, from a short, subrhizomatous caudex or from a rhizome, 0.5–3 dm tall, the stems sparingly to densely white-tomentose, becoming villous or villous-tomentose with long, whitish or yellowish, multicellular hairs above; basal leaves 3–10 cm long, 0.6–2.5(3.5) cm broad, oblanceolate to obovate, entire, or less commonly toothed, rounded to obtuse or acute apically, subsessile to broadly winged-petiolate or less commonly with slender petioles, tomentose or the tomentum deciduous, underlain with flattened, multicellular hairs (these rarely lacking); cauline leaves lance-oblong or lance-attenuate to oblong or oblanceolate, entire or toothed, gradually reduced upwards, sessile; heads solitary, or 2–4, the peduncles white-tomentose and often villous-tomentose with yellowish multicellular hairs as well; involucres 6–10 mm high, 10–24 mm broad, the bracts in 1 series, lance-linear, attenuate, suffused purplish throughout (rarely greenish), tomentose to glabrate or sparingly to densely villous with yellowish multicellular hairs; rays orange, yellow orange, or sometimes yellowish or purplish, 9–20(30) mm long, 1–3.5 mm broad; achenes hairy.

Arctic and alpine tundra and heathlands; widely distributed in western, northern and central Alaska, and in northern and western Yukon; east to Mackenzie; Eurasia (*S. denalii* A. Nels.; *S. lindstroemii* [Ostenf.] Porsild.) This entity is polymorphic, varying in shape of basal leaves, flower color, and amount and position of multicellular hairs. Further work may demonstrate the necessity of recognizing one or more infraspecific taxa.

Senecio hyperborealis Greenm.
Northern Groundsel

Plants perennial, from a caudex or short rhizome, 0.5–3(3.6) dm tall, the stems sparingly to moderately white-tomentose, or sometimes glabrate; basal leaves 2–8 cm long, 0.3–2 cm broad, the blades oblanceolate to elliptic or less commonly orbicular to obovate, lobed, crenate, or subentire, rounded to obtuse apically, tapering basally to slender or winged petiole, tomentose on both surfaces, or more so beneath, sometimes glabrate or glabrous above and beneath; cauline leaves oblanceolate to oblong, entire or more commonly pinnatifid, gradually reduced upwards, sessile, or the lower ones petioled; heads solitary, or 2–6 (rarely more), the peduncles white-tomentose or glabrate; involucres 5–7 mm high, 5–15 mm broad, the bracts in 1 series, or with 1 or more short outer ones, lance-oblong, attenuate, green throughout or suffused with purple, especially near the tip, sparingly tomentose or glabrous; rays yellow, 5–11 mm long, 1.5–3 mm broad, or sometimes lacking; achenes glabrous or pubescent.

Mountain slopes, terraces, and floodplains; over much of Alaska (except for extreme northern, southwestern, and southeastern portions) and in the Yukon; east to Mackenzie and south to British Columbia and Alberta (*S. conterminus* Greenm.; *S. ogotorukensis* Packer).

Senecio indecorus Greene

Plants perennial, from a caudex, 1–8 dm tall, the stems glabrous; basal leaves 1.5–12 cm long, 0.6–3 cm broad, the blade elliptic to obovate or oblanceolate to ovate or lanceolate, crenate or serrate, rounded to obtuse or acute apically, abruptly or gradually acute to acute basally, glabrous above and below; petioles slender, usually tomentose near the base; cauline leaves oblanceolate to oblong or elliptic, often coarsely toothed or finally pinnatifid, gradually reduced upwards, finally sessile and more or less auriculate-clasping; heads 4–many, borne in open, subumbellate corymbs, the peduncles glabrous or sparsely tomentose, bearing   (0)1–several   small   bractlets;

involucres 7–11 mm high, 9–20 mm broad, the bracts in 1 series, or with a few short outer scales, lance-linear, green throughout or the tips suffused with purple, or purplish throughout, glabrous or sparsely tomentose; rays lacking, or when rarely present, then yellow and less than 5 mm long; disk flowers yellow; achenes glabrous.

Open woods, slopes, and roadsides; in the southeastern quarter of continental Alaska and southern Yukon; east to Quebec and south to California, Idaho, Wyoming, and Michigan (*S. pauciflorus* var. *fallax* Greenm.).

Senecio lugens Richards.

Plants rhizomatous perennials, 1–8 dm tall, the stems glabrous or sparsely tomentose or villous; basal leaves 3–21 cm long, 0.5–4.5 cm broad, oblanceolate, elliptic, or ovate-lanceolate, denticulate, or subentire, rounded or more commonly acute apically, tapering basally to a winged or slender petiole, glabrous or sparingly tomentose, especially along the margin; cauline leaves lance-oblong to lanceolate, denticulate to subentire, sessile or the lower ones petiolate, gradually reduced upwards; heads few to many, borne in compact to open cymes, the peduncles tomentose, bracteate; involucres 5–9 mm high, 8–12 mm broad, the bracts lance-oblong, in 2 series, the outer ones much shorter, green, the margins scarious below the tips, black or purplish black, glabrous or sparingly tomentose; rays yellow, 7–15 mm long, 1.5–4 mm broad; achenes glabrous.

Meadows, slopes, bogs, and open woods; in most of Alaska (except for extreme southwestern and southeastern portions) and throughout the Yukon; eastward to Northwest Territory and south to Washington, Wyoming, Iowa, and Manitoba.

Senecio pauciflorus Pursh

Plants perennial, from a caudex, 1–4 dm tall, the stems glabrous or glabrate; basal leaves 1.5–10 cm long, 1–4 cm broad, the blade elliptic-ovate or ovate, or subreniform, subcordate, truncate, or abruptly cuneate basally, sparsely tomentose to glabrate or glabrous; petioles slender; cauline leaves oblong or lance-oblong, variously toothed or lobed, or finally almost entire, gradually reduced upwards, becoming sessile; heads usually 1–6 (sometimes more), borne in subumbellate corymbs, the peduncles glabrous or tomentose; involucres 6–10 mm high, 8–17 mm broad, the bracts in 1 series, or with a few short outer scales, lance-linear, green, or more commonly tinged purplish, glabrous or sparsely tomentose; rays lacking, or when rarely present, then yellow or yellow orange and 5–7 mm long; disk flowers orange or reddish; achenes glabrous.

Alpine meadows, open woods, lake shores, and river terraces; in southern Yukon; eastward to Labrador and south to California, Idaho, Wyoming, and Quebec.

Senecio pauperculus Michx.

Plants perennial, from a subrhizomatous caudex, 1.5–5.5(7) dm tall, the stems glabrous or glabrate; basal leaves 1.5–8(10) cm long, 0.4–1.4(2.5) cm broad, the blades oblanceolate to elliptic or oblong, crenate, serrate, or subentire, rounded to obtuse apically, tapering basally to a slender petiole, glabrous or glabrate; cauline leaves oblanceolate to oblong or lance-linear, toothed or pinnatifid, gradually reduced upwards, sessile or the lower ones petiolate; heads 2–10 or more, the peduncles glabrous or glabrate, bracteate; involucres (3)4–8 mm high, 6–15 mm broad, the bracts in 1 series, or with some short outer ones, lance-oblong, green, or suffused with purple near the tip, the margins hyaline, glabrous, or sparingly tomentose; rays yellow, 5–10 mm long, 2–3 mm broad, or rarely absent; disk flowers yellow; achenes glabrous or hairy.

Meadows, open woods, and roadside gravels; in the southeastern quarter of mainland Alaska and southern Yukon; east to Labrador and south to Oregon, New

Mexico, and Georgia (S. *pauperculus* var. *balsamitae* [Muhl.] Fern.). This entity is apparently closely related to S. *hyperborealis,* S. *indecorus,* and S. *streptanthifolius.*

Senecio pseudo-arnica Less.

Plants perennial, 0.5–6 dm tall or more, rhizomatous, the rhizomes ascending to erect, the stems coarse, glabrous or glabrate below, more or less tomentose above; basal leaves usually smaller than the cauline ones, often withered by flowering time; cauline leaves (2.5)5–22 cm long, 1–10 cm broad, oblanceolate, obovate or oblong, serrate to subentire, rounded to obtuse apically, cuneate or attenuate basally, broadly petioled to sessile, more or less tomentose above and beneath, or glabrate above, not reduced above, or reduced in size only in the inflorescence; heads solitary or less commonly 2–5, the peduncles tomentose or glabrous; involucres 10–20 mm high, 20–45 mm broad, the bracts in 1 row, or some of the outer ones of variable shape and size, lance-oblong to lance-linear, attenuate, more or less tomentose, scarious or green, or suffused with purple; rays yellow, 10–25 mm long, 3–7 mm broad; achenes glabrous.

Beaches and tidal flats; in coastal and insular western, southwestern, southern, and southeastern Alaska; south to British Columbia; disjunctly in coastal eastern North America; Asia.

Senecio resedifolius Less.

Plants perennial, from a caudex or short rhizome, 0.4–2.3 dm tall, the stems erect or ascending, glabrous or rarely tomentose near the base; basal leaves 1.5–7(9) cm long, 0.4–1.8 cm broad, the blades ovate, lanceolate, or orbicular, or sometimes subreniform or subcordate, rounded or obtuse apically, abruptly acute or cuneate basally, crenate, subentire or more or less pinnatifid, glabrous above and below, petiolate, rarely pubescent near the petiole base; cauline leaves greatly reduced upwards, often becoming entire or subentire and ses-

sile upwards; heads solitary (sometimes 2), the peduncles glabrous or sparingly tomentose, often suffused with purple; rays yellow or yellow orange, 7–15 mm long, 1.5–4 mm broad, or lacking; achenes glabrous.

Arctic and alpine tundra and heathlands; in most of Alaska and in western and northern Yukon; disjunctly eastward to Newfoundland and southward to Washington, Montana, and Wyoming.

Senecio sheldonensis Porsild
Sheldon Groundsel

Plants perennial, with short rhizome or caudex, mostly 3–4 dm tall, the stems erect, glabrous; basal leaves smaller than the main cauline ones, usually withered by flowering time; cauline leaves 2–10 cm long, 0.7–3 cm broad, lance-ovate to lanceolate, denticulate, obtuse to acute apically, attenuate basally, glabrous, the lower ones broadly to narrowly petiolate, the upper ones sessile; heads 3–4, the peduncles glabrous; involucres 7–10 mm high, 8–16 mm broad, the bracts in 1 row, lance-oblong, attenuate, green, the tips dark colored; rays yellow, about 10 mm long and 1 mm broad; achenes glabrous.

Open woods and meadows; in southern Yukon and northern British Columbia; endemic.

Senecio streptanthifolius Greene

Plants perennial, from a caudex, mostly 1.5–4 dm tall, the stems glabrous or sometimes tomentose, especially in the basal leaf axils; basal leaves 2–8 cm long, 0.4–3 cm broad, the blades oblanceolate to obovate or orbicular, crenate, serrate, dentate, or subentire, rounded to obtuse apically, acute to attenuate basally, glabrous, petiolate; cauline leaves much reduced upwards, becoming sessile and often auriculate-clasping; heads 4–10 or more, the peduncles glabrous or sparingly tomentose (the bractlet axils sometimes densely tomentose); involucres 4–8 mm high, 4–10 mm broad, the bracts in 1 series or sometimes 1 or more short outer ones present, lance-oblong, green, or suffused with purple, glabrous; rays yellow, mostly 5–10 mm long and 1–2 mm broad; achenes glabrous.

Open woods and meadows; in southern Yukon and northern British Columbia; east to Mackenzie and south to California, Utah, and Colorado (*S. cymbalarioides* Nutt. not Buek.; *S. cymbalarioides* var. *borealis* [T. & G.] Greenm.).

Senecio triangularis Hook.

Plants perennial from a caudex or rhizome, 3–12 dm tall or more, the stems glabrous to sparingly short-villous, the pubescence more evident above; basal leaves smaller than the main cauline ones, usually withered by flowering time; cauline leaves 4–20 cm long and 0.6–8 cm broad (or more), the lower ones petiolate, with blades broadly to narrowly triangular, or triangular-hastate to triangular-cordate, the upper ones becoming sessile, smaller and narrower, dentate or serrate-dentate to denticulate, or less commonly some subentire, glabrous or glabrate above, usually short-villous along the veins beneath; heads 3–many, the peduncles sparingly villous; involucres 8–12 mm high, 9–16 mm broad, the bracts in 1 series, or with 1–more short to elongate outer ones, lance-oblong, green, the basal margins scarious, the tips black and tufted-hairy, otherwise sparingly hairy to glabrate; rays yellow, mostly 8–14 mm long and 2–6 mm broad; achenes glabrous.

Alpine and subalpine meadows, woods, and thickets; in southern Alaska, mostly near the coast; from the Alaska Peninsula eastward, and in southern Yukon; east to Mackenzie and south to California, New Mexico, and Saskatchewan.

Senecio viscosus L.
Viscid Groundsel

Plants annual or biennial, with taproots, 1–6 dm tall, the stems glandular-viscid throughout; basal leaves smaller than the main cauline ones, often withered by flowering time; cauline leaves not much reduced upwards, mostly 2.5–10 cm long and 1–6 cm broad, deeply pinnately lobed, the segments again toothed or lobed, pubescent like the stems; heads few to numerous, the peduncles glandular-viscid; involucres 7–10 mm high, 6–16 mm broad, the bracts in 1 series, or with few outer shorter ones, lance-subulate, green with scarious margins, the tips not black, glandular-viscid; ray flowers yellow, mostly 1–4 mm long and 0.5–1 mm broad; achenes hairy.

Roadsides and open woods; in the vicinity of the Dyea-Liarsville road in southeastern Alaska; widely established in eastern North America; adventive from Europe.

Senecio vulgaris L.
Common Groundsel

Plants annual or biennial, with fibrous or taproots, 1–5.5 dm tall, the stems glabrous or sparingly villous; basal leaves smaller than the main cauline ones, often withered by flowering time; cauline leaves not much reduced upwards, 2–10 cm long, 0.5–4.5 cm broad, irregularly pinnatifid, the lobes again toothed, glabrous, or more or less villous, especially along the veins beneath, the lower ones petiolate, the upper ones becoming sessile and auriculate-clasping; heads few to numerous, the peduncles sparingly villous to glabrous; involucres 5–8 mm high, 4–10 mm broad, the bracts in 2 series, the outer ones short and black-tipped, lance-linear, green with scarious margins, the tips black, glabrous; ray flowers lacking; achenes hairy.

Weedy species of disturbed sites; in the southeastern quarter of Alaska (including the Panhandle) and in southern Yukon; adventive from Europe.

Senecio yukonensis Porsild
Yukon Groundsel

Plants perennial, rhizomatous, 1.5–3.5 dm tall, the stems sparingly villous to glabrate below, becoming more densely villous above; basal leaves 2–10 cm long, 0.4–1.5 cm broad, oblanceolate, elliptic, or oblong, denticulate to subentire, obtuse to acute apically, tapering to a broad or narrow petiole basally, moderately to sparingly tomentose or glabrate, or glabrous on both surfaces; cauline leaves much reduced upwards, becoming sessile, linear, and entire or subentire; heads usually (1)2–6, the peduncles moderately to densely villous-tomentose with yellowish multicellular hairs; involucres 6–12 mm high, 9–25 mm broad, the bracts in 1 row, lance-oblong to lance-linear, greenish, the margins hyaline, moderately to densely villous; rays yellow, 7–12 mm long, 2–3 mm broad, or lacking; achenes glabrous.

Meadows and heathlands; from widely disjunct locations in most of Alaska north of the 62nd parallel, and in most of the Yukon; endemic (*S. alaskanus* Hultén).

SOLIDAGO L.

Perennial herbs with rhizome or caudex, the roots fibrous, the juice watery; stems decumbent-ascending to erect, simple or branched above; leaves alternate, simple, entire or toothed; heads several to many, borne in compact to open panicles; involucral bracts imbricate in several series, scarious, or the tips green; receptacle flat or convex, naked; rays yellow, few to several, pistillate, fertile; disk flowers several to numerous, perfect, fertile, yellow; pappus of capillary bristles; style branches flattened, oblong to lanceolate; achenes several-nerved, pubescent or rarely glabrous.

1a. Plants with well-developed rhizomes, usually over 3.5 dm tall; involucres commonly 4 mm high or less. *S. canadensis*

1b. Plants with short rhizomes or a caudex, usually less than 3.5 dm tall; involucres commonly 4 mm high or more. (2)

2a. Petioles of lower leaves long-ciliate with multicellular hairs; flowers usually in a single corymbose cluster or with one or more additional clusters borne on elongate lateral branches; plants broadly distributed in Alaska and Yukon. *S. multiradiata*

2b. Petioles of lower leaves not ciliate with multicellular hairs, glabrous or merely scabrous; flowers commonly in several clusters, the lateral ones on short branches. *S. spathulata*

Solidago canadensis L.
Canada Goldenrod

Plants rhizomatous, the caudex poorly developed or lacking, the stems 3–10 dm tall or more, short-hairy throughout or glabrous below; basal leaves usually withered by flowering time; cauline leaves 2–14 cm long, 0.5–2.5 cm broad, lance-elliptic to lance-oblong, acute apically, tapering to the base, serrate to entire, rough on both surfaces, short-ciliate; heads several to numerous, in terminal, paniculate clusters, secund or not; involucres 3–4(5) mm high, 3–6 mm broad, the bracts lance-oblong, more or less glandular; rays yellow, 1–3 mm long; achenes hairy.

Open woods, in meadows, and along roadsides; in coastal and insular southern Alaska, and disjunctly from interior Alaska and Yukon; widely distributed in North America.

1a. Involucres mostly 3–3.5(4) mm long, the inner bracts usually more than twice longer than the outer ones; inflorescence elongate; plants of interior sites (*S. elongata* Nutt.). *S. canadensis* var. *salebrosa* (Piper) Jones

1b. Involucres mostly 3.5–4(5) mm long, the inner bracts usually less than twice longer than the outer ones; inflorescence generally compact; plants of coastal regions (*S. lepida* DC.). *S. canadensis* var. *subserrata* (DC.) Cronq.

Solidago multiradiata Ait.
Northern Goldenrod

Plants with short rhizomes or a caudex,

the stems (0.5)1.5–4(7) dm tall, sparingly to moderately short-villous, especially above; basal leaves 1.5–20(27) cm long, 0.4–2(2.5) cm broad, oblanceolate, obtuse to acute apically, tapering to a winged petiole basally, entire or less commonly serrate, glabrous above and below, short-ciliate on the blade, long-ciliate with multi-cellular hairs on the petiole margins near the base, often persistent at flowering time; cauline leaves gradually reduced upwards, becoming sessile and sometimes clasping; heads several to numerous, in terminal or lateral, corymbose panicles, not secund, the lateral branches (when present) usually more than 2 cm long; involucres 5–8 mm high, 7–14 mm broad, the bracts lance-oblong to lance-linear, attenuate, not glandular; rays yellow, 4–6.5 mm long; achenes hairy.

Open woods, meadows, and slopes; almost throughout Alaska and Yukon; east to Newfoundland and south to California and New Mexico; Asia (*S. multiradiata* var. *scopulorum* Gray; *S. multiradiata* var. *arctica* [DC.] Fern.).

Solidago spathulata DC.

Plants from a caudex or with short rhizomes and conspicuous fibrous roots, the stems 1.5–6 dm tall, glabrous or sparingly short-hairy above; basal leaves 3–15 cm long, 0.8–1.4(2) cm broad, narrowly oblanceolate, obtuse to acute apically, tapering to a winged petiole basally, toothed to subentire, glabrous above and below or sparingly hairy, short-ciliate on the blade, the petiole margins glabrous, usually persistent at flowering time; heads numerous, borne in contracted spicate panicles, not secund, the lateral branches usually less than 2 cm long; involucres 3.5–5.5 mm high, 5–7 mm broad, the bracts oblong to lance-oblong, obtuse or acute, more or less glutinous; rays yellow, 1.5–3 mm long; achenes hairy.

Open woods, river banks, terraces, and meadows; in the southeastern third of continental Alaska and southern Yukon; east

to Quebec and south to California, Arizona, and New Mexico (*S. decumbens* Greene; *S. oreophila* Rydb.; *S. decumbens* var. *oreophila* [Rydb.] Fern.; *S. yukonensis* Gand.).

SONCHUS L.

Annual or perennial herbs from taproots or deep-seated, rhizomelike roots, the juice milky; leaves chiefly cauline, alternate, simple, entire to lobed or pinnatifid; heads few to several; involucral bracts imbricate in several series, green or greenish (drying brownish), the inner ones with hyaline margins; receptacle naked; corollas of ray flowers only, yellow, perfect; pappus of capillary bristles; style branches semi-cylindrical; achenes compressed, several- to many-nerved, beakless, glabrous.

1a. Plants perennial, spreading by rhizomelike roots; involucres more than 14 mm long in fruit, pubescent with stipitate glands. *S. arvensis*

1b. Plants annual, from taproots; involucres less than 14 mm long in fruit, glabrous, or less commonly with stipitate glands. (2)

2a. Leaves sharply and narrowly toothed, and sometimes lobed; achenes not transversely wrinkled, merely longitudinally nerved. *S. asper*

2b. Leaves sharply and broadly toothed or merely toothed, and lyrate-pinnatifid; achenes transversely wrinkled and longitudinally nerved. *S. oleraceus*

Sonchus arvensis L.
Perennial Sow-thistle

Plants perennial, with deep-seated, rhizomelike roots, the stems 4–10 dm tall or more, pubescent with coarse, stipitate glands, at least above, often glabrous below; leaves 5–40 cm long, 0.8–10 cm broad, more or less pinnatifid, auriculate-clasping basally, acute to obtuse apically, prickly-margined, heads few to several, the peduncles stipitate-glandular; involucres 14–20 mm high and 10–30 mm broad in fruit, the bracts lance-oblong to lance-linear, pubescent like the peduncles; rays yellow, mostly 10–20 mm long; achenes 2.5–3.5 mm long, several-nerved and transversely wrinkled.

Weedy species of disturbed soils; in southeastern Alaska; adventive from Europe, widely distributed in North America.

Sonchus asper (L.) Hill
Spiny Sow-thistle

Plants annual, from taproots, the stems 3–10 dm tall, pubescent with coarse, stipitate glands, at least above, often glabrous below; leaves 3–15 cm long, 1–5 cm broad, merely lobed or lobeless, auriculate-clasping basally, acute to acuminate or less commonly obtuse apically, the margins armed with slender, sharp prickles; heads few to several, the peduncles stipitate-glandular or glabrous; involucres 9–14 mm long and 10–16 mm broad in fruit, the bracts lance-oblong or lance-linear, glabrous or with few stipitate glands; rays yellow, mostly 5–10 mm long; achenes 2–3 mm long, several-nerved, not transversely wrinkled.

Weedy species of disturbed sites; in widely scattered localities in Alaska (to be expected in the Yukon); adventive from Europe, widely distributed in North America.

Sonchus oleraceus L.
Common Sow-thistle

Plants annual, from taproots, the stems 2–10 dm tall or more, glabrous throughout or less commonly with stipitate glands above; leaves 4–20 cm long, 0.6–10 cm broad, more or less lyrate-pinnatifid, auriculate-clasping basally, acute to obtuse apically, irregularly and broadly toothed, the teeth weakly prickly; heads few to several, the peduncles glabrous or sometimes stipitate-glandular; involucres 10–13 mm high and 8–20 mm broad in fruit, the

bracts lance-linear to lance-oblong, glabrous or with few stipitate glands; rays yellow, mostly 8–12 mm long; achenes 2–3 mm long, several-nerved and transversely wrinkled.

Weedy species of disturbed habitats; in widely disjunct sites in Alaska (to be expected in the Yukon); adventive from Europe, widely distributed in North America.

TANACETUM L.

Perennial, aromatic herbs from rhizomes, or from a caudex, the juice watery; stems erect, simple or branched above; leaves alternate, chiefly cauline, pinnately once to thrice dissected; heads solitary or 2–numerous in corymbose clusters; involucral bracts imbricate in several series, greenish, with brownish, scarious margins; receptacle flat or convex, naked; ray flowers yellow, several to numerous, pistillate, fertile, or lacking; disk flowers numerous, perfect, fertile, yellow; pappus a short crown; style branches flattened; achenes several-nerved, commonly glandular, beakless.

1a. Heads numerous, lacking rays; leaves once to twice pinnatifid, the ultimate segments often more than 2 mm broad; plants cultivated and escaping. *T. vulgare*

1b. Heads 1–4 (rarely more), with yellow rays; leaves twice to thrice pinnatifid, the ultimate segments mostly less than 1.5 mm broad; plants indigenous. *T. bipinnatum*

Tanacetum bipinnatum (L.) Schulz-bip

Plants rhizomatous, the stems 0.8–5(6) dm tall or more, sparingly to moderately villous, or densely so above; basal leaves 3–30 cm long, usually persisting only on sterile branches, twice to thrice pinnately divided, the ultimate divisions usually less than 1.5 mm broad, sparingly to moderately villous; cauline leaves similar to the basal, gradually reduced upwards; heads solitary or

Tanacetum bipinnatum (L.) Schulz-bip (× 0.5).

2–4 (rarely more), the peduncles moderately to densely villous; involucres 5–10 mm high, 8–22(24) mm broad, the bracts ovate to lanceolate or oblanceolate, with hyaline to brownish, scarious margins;

rays yellow, 2–7 mm long, 2–5 mm broad; achenes glandular.

River banks, floodplains, and open woods; in much of Alaska (except for southeastern portion) and western Yukon; east to Newfoundland and south to Quebec; Eurasia.

1a. Heads ordinarily solitary, mostly 18–20 mm broad, the involucral bracts with dark brown, scarious margins; rays 4–7 mm long; plants mostly of western and northern Alaska and Yukon—rarely in the interior (*Artemisia kotzebuensis* Besser; *T. kotzebuense* [Besser] Besser). *T. bipinnatum* ssp. *bipinnatum*

1b. Heads ordinarily 2–4 or more, mostly 8–17 mm broad, the involucral bracts with hyaline to brownish margins; rays mostly 1–4 mm long; plants of interior eastern Alaska and western Yukon (*T. huronense* Nutt.; *Chrysanthemum bipinnatum* ssp. *huronense* [Nutt.] Hultén). *T. bipinnatum* ssp. *huronense* (Nutt.) Welsh

Tanacetum vulgare L.

Common Tansy

Plants rhizomatous, the stems 4–10 dm tall or more, glabrous or nearly so; basal leaves commonly withered by flowering time; cauline leaves 7–20 cm long, once to twice pinnately divided, the ultimate divisions usually more than 1.5 mm broad, glabrous or nearly so, only slightly reduced upwards; heads numerous, the peduncles sparingly hairy to glabrous; involucres 2–4 mm high, 4–8 mm broad, the bracts lanceolate to lance-oblong, with hyaline to brownish margins; rays lacking; achenes glandular.

Cultivated and escaping; in southeastern Alaska; adventive from the Old World (*Chrysanthemum vulgare* [L.] Bernh.; *T. boreale* Fisch.).

TARAXACUM Hall

Perennial scapose herbs with milky juice, from taproots; leaves all basal, pinnatifid to subentire; heads solitary on a scape; involucral bracts in 2 series, herbaceous, the outer ones shorter and sometimes spreading or reflexed, the inner ones often dilated or appendaged apically, usually with broad, hyaline or scarious margins, at least basally; receptacle naked; corollas of ray flowers only, perfect, yellow; pappus of capillary bristles; style branches semicylindrical; achenes angular or terete, prominently nerved or ribbed, usually spinulose or with ridges near the body apex, glabrous, beaked.

Sherff, E. E. 1920. North American species of *Taraxacum*. Bot. Gaz. 70:329–59.

1a. Inner involucral bracts commonly dilated or bearing appendages apically; outer bracts usually appressed or ascending; achenes straw-colored to olive-drab or brownish; plants indigenous, our most common species. *T. ceratophorum*

1b. Inner involucral bracts usually not dilated or appendaged apically, or if so, then the achenes reddish, or the outer bracts reflexed, or both; plants indigenous or adventive. (2)

2a. Outer bracts reflexed or spreading; achenes straw-colored to olive-drab or brownish; involucres often over 15 mm high; weedy, adventive plants of disturbed sites. *T. officinale*

2b. Outer bracts appressed or ascending (reflexed in some *T. laevigatum*); achenes red to brownish red or blackish; involucres often less than 15 mm high; nonweedy, indigenous plants of undisturbed sites (except *T. laevigatum*). (3)

3a. Achene bodies black or grayish black; plants mostly of northern and western Alaska, and disjunctly in the interior. *T. lyratum*

3b. Achene bodies red or brownish red; plants broadly distributed, or else weeds of disturbed sites. (4)

4a. Inner bracts commonly dilated or appendaged apically; achenes obscurely or not at all quadrangular; introduced plants of disturbed soils. *T. laevigatum*

4b. Inner involucral bracts scarcely or not at all dilated or appendaged apically; achenes tending to be sharply quadrangular; indigenous species of undisturbed sites. *T. eriophorum*

Taraxacum ceratophorum (Ledeb.) DC.
Horned Dandelion

Plants mostly 0.5–5 dm tall, from simple or branched caudex and taproot; leaves 4–35 cm long, 0.7–6 cm broad, subentire to toothed, or more commonly pinnately lobed to pinnatifid, the terminal lobe usually broader than the others, tapering basally to a more or less winged petiole; scapes sparingly villous, sometimes moderately so below the head; involucres 13–22 mm tall in flower, the outer bracts ovate to lanceolate, appressed or ascending, the inner ones lance-oblong, attenuate, the apex dilated or appendaged, rarely only slightly so; rays yellow (sometimes fading bluish); achene bodies 3–7 mm long, straw-colored to olive-drab or brownish, the beak usually 2–4 times longer than the body; pappus white.

Tundra, heath, and woods; indigenous in much of Alaska and Yukon; east to the Atlantic, and south to Massachusetts, New Mexico, and California; circumboreal (*T. andersonii* Hagl.; *T. angulatum* Hagl.; *T. arietinum* Hagl.; *T. aureum* Hagl.; *T. caligans* Hagl.; *T. chamissonis* Greene; *T. chlorostephum* Hagl.; *T. collinum* DC.; *T. demissum* Hagl.; *T. dumetorum* Greene; *T. eurylepium* Dahlst. ex Ostenf.; *T. eyerdamii* Hagl.; *T. fabbeanum* Hagl.; *T. festivum* Hagl.; *T. flavovirens* Hagl.; *T. hypochoeropse* Hagl.; *T. integratum* Hagl.; *T. kodiakense* Hagl.; *T. lacerum* Greene; *T. latilimbatum* Hagl.; *T. leptoglossum* Hagl.; *T. leptopholis* Hagl.; *T. maurolepium* Hagl.; *T. microceras* Hagl.; *T. mitratum* Hagl.; *T. multesimum* Hagl.; *T. ochraceum*

Hagl.; *T. oncophorum* Hagl.; *T. ovinum* Greene; *T. paralium* Hagl.; *T. patagiatum* Hagl.; *T. pellianum* Porsild; *T. phalolepis* Hagl.; *T. pribilofense* Hagl.; *T. scotostigma* Hagl.; *T. signatum* Hagl.; *T. speirodon* Hagl.; *T. sublacerum* Hagl.; *T. trigonolobum* Dahlst.). The large number of synonyms is an indication of the great diversity of this polymorphic species. At least some portion of the variation is attributable to the apomictic nature of the entity. Recognition of the various apomictic races as species has been discussed by previous authors (notably by M. L. Fernald. 1933. Rhodora 35:369–86). The continued recognition of such races as species seems both impractical and unnecessary. Possibly some of the races should be recognized at infraspecific level, perhaps as formae. Certainly, further work is indicated.

Taraxacum eriophorum Rydb.

Plants mostly 0.4–1.5 dm tall, from simple or branched caudex and taproot; leaves 2.5–13 cm long, 0.4–2 cm broad, merely lobed or more commonly pinnatifid, the terminal lobe usually broader than the lateral ones, tapering basally into a slender petiole; scapes glabrous or rarely sparingly villous, especially below the head; involucres 8–14 mm tall in flower, the outer bracts ovate to lanceolate, appressed or ascending-spreading, the inner ones lance-oblong to oblong, the apex slightly or scarcely dilated, unappendaged; rays yellow (sometimes bluish on the outer surface); achene bodies mostly 3–4 mm long, red to brownish red, quadrangular, the beak shorter than the body or to 1–2 times longer; pappus white.

Arctic and alpine tundra, heathlands, and less commonly in open woods; in much of Alaska (and in Yukon?); southward to Wyoming; Asia (*T. callorhinorum* Hagl.; *T. carneocoloratum* A. Nels.?; *T. chromocarpum* Hagl.; *T. kamtschaticum* Dahlst.; and *T. lateritium* Dahlst.). Specimens of *T. eriophorum* approach *T. ceratophorum* on the one hand and *T. lyratum* on the

other. It seems possible that the assemblage herein interpreted as *T. eriophorum* might well have arisen from integration of some basic *lyratum-ceratophorum* stock.

Taraxacum laevigatum (Willd.) DC.

Plants mostly 1–4 dm tall, from simple or branched caudex; leaves mostly 5–25 cm long and 1–4 cm broad, commonly deeply pinnatifid throughout, the terminal lobe seldom much broader than the lateral ones, tapering basally to a more or less winged petiole; scapes sparingly villous to glabrous; involucres 10–20 mm tall in flower, the outer bracts ovate to lanceolate, appressed to ascending, or sometimes reflexed, the inner ones lance-oblong, attenuate, the apex not much dilated, rarely appendaged; rays yellow (sometimes fading bluish); achene bodies mostly 3–4 mm long, red to brownish red, the beak from shorter than to 1–3 times longer than the body; pappus white.

Weedy species of waste places; evidently uncommon in both Alaska and Yukon; adventive from Eurasia (*T. erythrospermum* Andrz. ex Bess.; and *T. scanicum* Dahlst.).

Taraxacum lyratum (Ledeb.) DC.

Plants mostly 0.3–1 dm tall, from simple or branched caudex and taproot; leaves 1–8 cm long, 0.3–1.5 cm broad, pinnately lobed to pinnatifid, the terminal lobe broader than the lateral ones, or less commonly subentire, tapering basally to a slender or more or less winged petiole; scapes glabrous or nearly so; involucres 7–14 mm tall, the outer bracts lanceolate to ovate, appressed or ascending-spreading, the inner ones lance-oblong to oblong, scarcely or slightly dilated, rarely appendaged; rays yellow (fading bluish); achene bodies 3–6 mm long, black or grayish black (brownish to straw-colored when young), the beak shorter than or slightly longer than the body; pappus white.

Arctic and alpine tundra; in northern, western, and mountainous interior Alaska and

Yukon; southward to Nevada, Arizona, and Colorado; Asia (*T. alaskanum* Rydb.; *T. hyperarcticum* Dahlst.; *T. phymatocarpum* Vahl ex Hornem.; and *T. sibiricum* Dahlst.).

Taraxacum officinale Weber ex Wiggers
Common Dandelion

Plants mostly 0.5–6 dm tall (or more), from simple or branched caudex and taproot; leaves 5–40 cm long, 1–10 cm broad, pinnately lobed to pinnatifid, the terminal lobe broader than the lateral ones, tapering basally into a more or less winged petiole; scapes sparingly villous to subglabrous, or often moderately to densely villous below the head; involucres 15–25 mm high in flower, the outer bracts lance-acuminate, reflexed, the inner ones lance-attenuate, not or scarcely dilated apically, rarely appendaged; rays yellow, or bluish externally; achene bodies 3–4 mm long, straw-colored to olive-drab, the beak usually 2–4 times longer than the body; pappus white.

Weedy species of disturbed soils; in southern Alaska and southern Yukon; adventive from Eurasia (*T. cinericolor* Hagl.; *T. dahlstedii* Lindb. f.; *T. decorifolium* Hagl.; *T. retroflexum* Lindb. f.; *T. undulatum* Lindb. f.; and *T. vagans* Hagl.).

TOWNSENDIA Hook.

Perennial herbs, from taproot and simple or branched caudex, the juice watery, acaulescent or nearly so; leaves alternate or principally basal, entire; heads solitary, borne sessile or nearly so; involucral bracts imbricate in several series, usually with a green median stripe and fringed margins; receptacle flat, naked; rays white to pink, pistillate, fertile, several to many; disk flowers numerous, perfect, fertile, yellow; pappus of disk flowers of numerous, rigid, bristlelike scales, that of the ray flowers often shorter and scalelike; style branches flattened; achenes 2-nerved, beakless, pubescent.

Townsendia hookeri Beaman

Plants mostly 2–5 cm tall, acaulescent or nearly so, arising from a taproot and stout caudex; leaves 0.5–3 cm long and 0.1–0.2 (0.3) cm broad, narrowly oblanceolate, strigose; heads solitary, sessile or nearly so; involucres 12–16 mm high and 10–15(20) mm broad, the bracts narrowly lanceolate, attenuate, sparingly hairy to glabrate, the hyaline margin fringed; rays white to pink, 10–15 mm long; achenes hairy.

Dry slopes; in southern Yukon; disjunctly in southern Canada and western contiguous United States.

CORNACEAE
Dogwood Family

Shrubs, subshrubs, or herbs; leaves simple, opposite or apparently whorled, entire; flowers perfect, regular, arranged in cymes or umbels, the inflorescence sometimes subtended by a whorl of petaloid leaves; sepals 4, minute; petals 4, distinct; stamens 4, alternate with the petals; pistils 1, the ovary inferior, 2–4-loculed; styles 1, arising from a glandular disk; fruit a drupe.

CORNUS L.

Plants woody or herbaceous; leaves subpalmately to pinnately veined, entire; flowers arranged in corymbose cymes or in umbels, the inflorescence subtended by usually 4 petaloid bracts, or the bracts lacking; sepals 4, minute; petals 4, whitish, greenish, or purplish; stamens 4; ovary inferior, 2-carpellate; fruit a drupe.

1a. Plants shrubs to 20 dm tall or more, inflorescence not subtended by an involucre. *C. stolonifera*
1b. Plants apparently herbaceous, to 2.5 dm tall; inflorescence subtended by petaloid involucre. (2)

2a. Petals, sepals, and ovaries dark purplish; lateral veins arising from the base of the leaf or nearly so. *C. suecica.*

2b. Petals, sepals, and ovaries usually greenish or yellowish (the petals sometimes suffused purplish); lateral veins arising from the midvein in the lower third of the leaf. *C. canadensis*

Cornus canadensis L.
Bunchberry

Plants rhizomatous, herbaceous, the stems 0.5–2.5 dm tall, minutely strigose with malpighian hairs, leafless or with 1 or less commonly 2 pairs of foliage leaves below the terminal pair or whorl of 4–6 leaves, these 2–6.5 cm long, and 0.8–4 cm broad, elliptic to obovate, entire, sparsely strigose above, glabrous or nearly so and paler beneath; flowers borne in a solitary, umbellate cyme, subtended by 4 white, yellowish, or pinkish to purplish bracts, these 0.9–2 cm long, 0.6–1.5 cm broad, deciduous in fruit; sepals minute, yellowish to reddish; petals 1–1.5 mm long, usually bicolored (yellowish and purplish), at least one bearing a slender awn to 1 mm long; ovary usually greenish when young, densely white-hairy, the style shorter than the stamens; drupes 6–9 mm long, red.

Moist woods and heathlands; in much of Alaska and Yukon south of the 68th parallel; east to Labrador and Greenland, and south to California, New Mexico, Minnesota, and Pennsylvania. *C. canadensis* forms hybrids with *C. suecica* L. The hybrids have been treated variously as species or as infraspecific taxa within *C. canadensis.* If a binomial is to be applied to those morphological intermediates, it should be *C.* × *unalaschkensis* Ledeb. (*C. canadensis* var. *intermedia* Farr; *C. intermedia* [Farr] Calder & Taylor).

Cornus stolonifera Michx.
Red-osier Dogwood

Shrubs to 20 dm tall or more, the stems often reddish or purplish when young, strigose to strigulose, soon glabrate or glabrous; leaves opposite, petiolate, the blades 3–12 cm long, 1.5–8 cm broad, oblong to

Cornus canadensis L. (× 0.4).

elliptic, hairy above and beneath, paler beneath; flowers borne in corymbose cymes, not subtended by showy bracts; sepals minute, yellowish; petals 2–4 mm long, white, all lacking awns; ovary greenish when young, white-hairy, the style shorter than the stamens; drupes 6–8 mm long, white to bluish.

Moist open woods; in the southeastern third of mainland Alaska, in the Panhandle, and in southern Yukon; widely distributed in North America. Two more or less distinctive varieties are present within our region.

1a. Branchlets and leaves merely strigose; styles 2–3 mm long; petals 3–4 mm long; plants mostly of southeastern Alaska. *C. stolonifera* var. *occidentalis* (T. & G.) C. L. Hitchc.

1b. Branchlets and leaves mostly strigulose; styles mostly 1–2 mm long; petals 2.5–3(3.5) mm long; plants of interior Alaska and Yukon (*C. stolonifera* var. *interior* [Rydb.] St. John; *C. stolonifera* var. *baileyi* [Coult. & Evans] Drescher). *C. stolonifera* var. *stolonifera*

Cornus suecica L.
Swedish Cornel

Plants rhizomatous, herbaceous, the stems 0.4–2.6 dm tall, minutely strigose with malpighian hairs, commonly with 1–2(3) pairs of foliage leaves below the terminal pair or whorl of 4–6 leaves, these 1.1–3.5 cm long, 0.7–2.2 cm broad, ovate to elliptic or oblanceolate, entire, sparsely strigose above, glabrous or glabrate and paler beneath; flowers borne in a solitary, umbellate cyme, subtended by 4 white, yellowish, pinkish, or purplish bracts, these 0.6–1.4 cm long, 0.4–1 cm broad, deciduous in fruit; sepals minute, purplish; petals 1–1.5 mm long, dark purplish, at least 1 petal with a slender awn to 1 mm long; ovary dark purplish when young, sparsely to densely white-hairy, the style about as long as the stamens; drupes 7–12 mm long, red.

Woods, heath, and alpine tundra; widely distributed in south-central, southwestern, and western Alaska as far north as Kotzebue Sound, on the islands of the Bering Sea, and less commonly in southeastern Alaska; Eurasia, Greenland.

CRASSULACEAE
Stonecrop Family

Plants perennial or annual, succulent, herbaceous; leaves alternate or opposite, simple, fleshy, entire or toothed; flowers perfect or imperfect, regular, borne in cymose clusters; sepals 4–5, distinct or nearly so; petals 4–5, distinct or connate basally; stamens 4, 8, or 10; pistils 4–5, distinct or nearly so, superior, each 1-carpelled, the styles 1 per pistil; fruit a follicle.

1a. Plants perennial; leaves alternate; stamens 8 or 10. *Sedum*

1b. Plants annual; leaves opposite; stamens 4. *Tillaea*

SEDUM L.

Plants perennial, with taproot and short to elongate, subrhizomatous caudex branches, the stems decumbent to ascending or erect; leaves alternate, subterete to flattened, entire or toothed; flowers showy, borne in subcapitate to paniculate cymes; sepals usually 5, persistent; petals usually 5, yellow, greenish yellow, or dark purple; pistils usually 5; follicles erect.

1a. Flowers dark purple, rarely greenish yellow; leaves flattened, often irregularly toothed; petals 2–3 mm long; stamens exserted. *S. rosea*

1b. Flowers yellow; leaves subterete or flattened, entire; petals 6–12 mm long; stamens not exserted. (2)

2a. Leaves lanceolate to elliptic in outline, subterete to slightly flattened; petals distinct; plants of southern Yukon and southeastern Alaska. *S. lanceolatum*

2b. Leaves spatulate, obovate, or oblanceolate in outline, flattened; petals connate at the base; plants of southeastern Alaska. *S. oreganum*

Sedum lanceolatum Torr.
Stonecrop

Caudex branches slender; stems decumbent basally, producing sterile basal branches with imbricate, fleshy leaves; flowering stems 0.3–2.5 dm tall; leaves lanceolate to elliptic, 0.4–2 cm long, 0.1–0.4 cm broad, subterete or somewhat flattened, entire; sepals 2.5–6 mm long; petals 6–9 mm long, yellow, distinct, persistent; stamens shorter than the petals; follicles erect, united at the base, the styles erect or spreading.

Open woods and grasslands; in southern Yukon and southeastern Alaska; south to California and east to Alberta, South Dakota, Nebraska, and New Mexico (S. stenopetalum authors, not Pursh). Our material belongs to var. lanceolatum.

Sedum oreganum Nutt. ex T. & G.

Caudex branches slender; stems decumbent basally, producing sterile, basal branches with imbricate, fleshy leaves; flowering stems 0.6–1.5 dm tall; leaves spatulate to obovate or oblanceolate, 0.5–2 cm long, 0.2–0.7 cm broad, flattened, entire; sepals 2.5–4.5 mm long, persistent; petals 8–12 mm long, yellow, united near the base; stamens shorter than the petals; follicles erect, distinct or nearly so, the styles erect.

Open rocky slopes; in southeastern Alaska; southward to California.

Sedum rosea (L.) Scop.
Roseroot

Caudex branches stout, clothed with persistent, scalelike leaves; stems erect or ascending, not or only rarely with sterile, basal branches, 0.3–3(4.5) dm tall; leaves ovate to elliptic or oblanceolate, reduced and scalelike below, becoming larger upwards, 0.4–4 cm long, 0.2–1.1(1.5) cm broad, flattened, irregularly toothed to entire; sepals 1.2–2 mm long, persistent, distinct; stamens longer than the petals, the filaments purple or yellow, lacking in pistillate flowers; follicles erect, distinct, the styles spreading.

Arctic and alpine tundra, heath, and less commonly in open woods; in most of Alaska and Yukon; south to California,

Nevada, and Colorado and east to Newfoundland and Maine; Greenland, Eurasia.

1a. Petals, staminal filaments, and anthers yellow or greenish yellow; plants uncommon in our region (*Rhodiola rosea* L.). *S. rosea* var. *rosea*
1b. Petals and usually the staminal filaments dark purple; plants common

Sedum rosea (L.) Scop. (× 0.7).

in our region (*S. frigidum* Rydb.; *S. rosea* ssp. *integrifolium* [Raf.] Hultén var. *frigidum* [Rydb.] Hultén; *S. alaskanum* [Rose] Rose ex Hutchins.). *S. rosea* var. *integrifolium* (Raf.) Berger

TILLAEA L.

Plants diminutive annuals, from taproots, the stems prostrate to ascending or erect; leaves opposite, flattened, entire; flowers minute, borne singly or clustered in the leaf axils; sepals usually 4, persistent; petals usually 4, whitish; pistils commonly 4; follicles erect.

Tillaea aquatica L.
Pygmyweed

Annuals, the stems prostrate and more or less stoloniferous to ascending or erect, 2–8 cm tall; leaves linear to narrowly oblanceolate, 3–6 mm long, 0.2–1 mm broad, flattened, entire, connate-sheathing basally; sepals 0.5–1 mm long, connate about half the length; petals 1–2 mm long, whitish, membranous; stamens much shorter than the petals; follicles erect, distinct, the styles erect.

Mud flats and ephemeral pools; in coastal southern Alaska; south to Mexico, and eastward disjunctly to the Atlantic; circumboreal (*Crassula aquatica* [L.] Schonl.).

CRUCIFERAE
Mustard Family

Herbaceous annual, biennial, or perennial plants; leaves alternate (or basal and still alternate), simple to compound; flowers perfect, regular or nearly so (irregular in *Iberis*), arranged in racemes, spikes, or corymbs; sepals 4, soon deciduous; petals 4 or lacking; stamens 6(4 and 2), 4, or 2; pistils 1; ovary superior, 2-loculed or only 1-loculed by abortion of the partition; fruit a silique or silicle. *Note*: Mature fruit is often necessary for accurate determination.

1a. Plants with cauline leaves auriculate-, cordate-, or sagittate-clasping. KEY I. .... p. 177

1b. Plants with cauline leaves sessile or petiolate, but not clasping, or the leaves all basal. (2)

2a. Plants aquatic or maritime, fleshy, or terrestrial and the petals lacking. KEY II. .... p. 178

2b. Plants terrestrial, seldom aquatic or maritime, rarely fleshy; petals present, not rudimentary or lacking. (3)

3a. Flowers yellow. KEY III. .... p. 178

3b. Flowers white, pink, or purplish. KEY IV. .... p. 179

KEY I. PLANTS WITH CAULINE LEAVES CLASPING.

1a. Flowers yellow. (2)

1b. Flowers white, pink, or purplish. (6)

2a. Cauline leaves deeply lyrate-pinnatifid. (3)

2b. Cauline leaves entire to sinuate-dentate, at least the upper ones not pinnatifid. (4)

3a. Pods 1–3 times longer than broad; stems spreading-hairy, at least below. *Rorippa*

3b. Pods more than 5 times longer than broad; stems glabrous or glabrate. *Barbarea*

4a. Stem leaves ovate to lanceolate, glabrous or nearly so; pods several times longer than broad. *Brassica*

4b. Stem leaves lanceolate to linear, rough-hairy; pods 1–2 times longer than broad. (5)

5a. Seeds and ovules several per locule; valves of pod with a distinct central vein extending to the apex, not reticulately pitted. *Camelina*

5b. Seeds 1 (ovules usually 2) per locule; valves of pod conspicuously reticulately pitted, not with a central vein. *Neslia*

6a. Pods several times longer than broad; petals white, pinkish, or purplish. (7)

6b. Pods mostly less than twice longer than broad; petals white to cream, or rarely pink to purplish. (8)

7a. Plants glabrous and glaucous, 1.5 dm tall or less, annual. *Arabidopsis*

7b. Plants hairy, at least below, or more than 2 dm tall, biennial or perennial. *Arabis*

8a. Petals cream; pods obovoid, the valves with a distinct central vein extending to the apex. *Camelina*

8b. Petals white; pods strongly compressed. (9)

9a. Plants pubescent, at least below; fruit cuneate-obcordate. *Capsella*

9b. Plants glabrous; fruit obovate to orbicular, often winged and samaralike. *Thlaspi*

KEY II. PLANTS AQUATIC OR MARITIME, FLESHY, OR TERRESTRIAL AND THE PETALS LACKING.

1a. Plants strictly aquatic. (2)

1b. Plants terrestrial or maritime, but not aquatic. (3)

2a. Leaves simple, linear, more or less terete. *Subularia*

2b. Leaves compound or pinnatifid, the blade well developed, flat. *Rorippa*

3a. Plants fleshy maritime herbs; basal leaves with long, slender petioles and reniform to oval blades, or cuneate-spatulate and coarsely dentate apically. (4)

3b. Plants seldom fleshy or maritime, but if so (compare *Cakile*), the leaves not as above. (5)

4a. Basal leaves with slender petioles, the blades reniform to oval; flowers white; silicles 4–7(10) mm long. *Cochlearia*

4b. Basal leaves cuneate-spatulate; flowers yellow; silicles (6)8–25 mm long. *Draba*

5a. Petals lacking; silicles flattened at

right angles to the partition, about as broad as long. *Lepidium*

5b. Petals present; silicles with 2 parts, breaking into an upper and a lower segment at maturity, longer than broad. *Cakile*

KEY III. PLANTS TERRESTRIAL; PETALS PRESENT; FLOWERS YELLOW.

1a. Fruit not over 3 times longer than broad. (2)

1b. Fruit over 3 times longer than broad. (5)

2a. Fruit distinctly compressed parallel to the partition. (3)

2b. Fruit terete or quadrangular, slightly if at all flattened. (4)

3a. Fruit oval to orbicular in outline; seeds 1–2 in each locule. *Alyssum*

3b. Fruit elliptic to oblong in outline; seeds several in each locule. *Draba*

4a. Plants glabrous, or with simple hairs only. *Rorippa*

4b. Plants pubescent with stellate hairs. *Lesquerella*

5a. Plants at least somewhat hairy, with some or all hairs stellate or branched. (6)

5b. Plants glabrous, or with simple hairs only. (8)

6a. Fruit at maturity strongly compressed parallel to the partition, often twisted. *Draba*

6b. Fruit terete or quadrangular, not strongly compressed. (7)

7a. Cauline leaves pinnate, bipinnate, or deeply pinnatifid. *Descurainia*

7b. Cauline leaves entire or sinuate-dentate. *Erysimum*

8a. Fruit indehiscent, the sutures lacking, the beak 1–3 cm long. *Raphanus*

8b. Fruit dehiscent, the sutures evident, the beak (if present) less than 1(1.5) cm long. (9)

9a. Flowers mostly less than 4 mm broad;

beak of fruit less than 3 mm long. *Sisymbrium*

9b. Flowers mostly 4–8 mm in diameter; beak of fruit over 2 mm long. *Brassica*

KEY IV. PLANTS TERRESTRIAL; PETALS PRESENT; FLOWERS WHITE, PINK, OR PURPLE.

1a. Fruit body not over 3 times longer than broad (somewhat longer in *Cakile*). (2)

1b. Fruit body 3–many times longer than broad. (8)

2a. Plants succulent maritime annuals; fruit transversely 2-segmented. *Cakile*

2b. Plants of various habitats, usually not maritime, not succulent; fruit not 2-segmented. (3)

3a. Fruit flattened at right angles to the partition. (4)

3b. Fruit not flattened, or flattened parallel to the partition. (5)

4a. Flowers large and showy, irregular (2 petals much longer than the others); fruit 4–6 mm broad; plants cultivated and escaping. *Iberis*

4b. Flowers inconspicuous, tiny, often lacking petals; fruit less than 4 mm broad; weedy plants of waste places. *Lepidium*

5a. Styles 2–3 mm long; plants erect annuals 3–10 dm tall or more. *Berteroa*

5b. Styles less than 2 mm long; plants perennial or rarely annual, commonly less than 3 dm tall. (6)

6a. Fruit oval, with thin winglike margin. *Alyssum*

6b. Fruit elliptical to oblong, longer than broad, lacking a winglike margin. (7)

7a. Leaves palmately lobed or pinnatifid, at least some. *Smelowskia*

7b. Leaves variously toothed or entire, but not as above. *Draba*

8a. Plants at least somewhat hairy, with some or all hairs stellate or branched. (9)

8b. Plants glabrous, or with simple hairs only (glandular-pubescent in *Parrya*). (15)

9a. Fruit at maturity strongly compressed parallel to the partition. (10)

9b. Fruit terete or quadrangular, not strongly compressed. (12)

10a. Petals bright pink to pink purple; siliques 25 mm long or more, 2–3 mm broad. *Erysimum*

10b. Petals white or yellowish, 2–7 mm long; siliques 5–45 mm long, less than 1 mm broad, or if broader, then less than 20 mm long. (11)

11a. Fruit less than 20 mm long, often twisted, elliptic to oblong or lanceolate. *Draba*

11b. Fruit over 20 mm long, not twisted, oblong. *Arabis*

12a. Plants pulvinate-caespitose perennials less than 15 cm tall, or if decumbent to erect and more than 15 cm tall, then the siliques hairy; flowers white, rarely purplish; usually alpine. (13)

12b. Plants decumbent to erect, not caespitose, often over 15 cm tall; flowers usually pink to purple or white. (14)

13a. Plants grayish hairy; basal and cauline leaves both 3–5-lobed (at least some); silicles glabrous. *Smelowskia*

13b. Plants seldom grayish hairy; leaves entire or merely dentate; siliques often hairy. *Braya*

14a. Petals white, rarely pinkish, 7 mm long or less; plants indigenous. *Arabis*

14b. Petals pink to pink purple, rarely white, more than 10 mm long; plants cultivated ornamentals. *Hesperis*

15a. Fruit indehiscent, the sutures lacking, the beak 1–3 cm long. *Raphanus*

15b. Fruit dehiscent, the suture distinct,

the beak (if present) less than 5 mm long. (16)

16a. Fruit terete or nearly so; plants of wet places, rooting at the nodes. *Rorippa*

16b. Fruit distinctly flattened; plants of various habitats, if growing in wet places, then not rooting at the nodes (except in some *Cardamine*). (17)

17a. Plants glandular-pubescent throughout, rarely glabrous; siliques 4–7 mm broad. *Parrya*

17b. Plants glabrous or pubescent, but not glandular; siliques mostly less than 4 mm wide. (18)

18a. Leaves compound, or if simple, the plants usually less than 10 cm tall (see *C. bellidifolia*); flowers white, pink, or purplish, or varicolored in populations. *Cardamine*

18b. Leaves simple, often entire (the basal ones pinnatifid in *Arabis*); plants often over 10 cm tall, or if less than the fruit only 8–12 mm long or the partition imperfect or lacking; flowers usually white. (19)

19a. Flowers in the axils of cauline leaves (bracts); stems 2–5 cm long, pubescent; partition lacking. *Aphragmus*

19b. Flowers in bractless racemes; stems often more than 5 cm long, pubescent to glabrous; partition present or lacking. (20)

20a. Plants glabrous, glaucous; leaves usually entire, ovate to lanceolate; partition lacking or perforate. *Eutrema*

20b. Plants pubescent to glabrous; leaves entire or toothed, usually oblanceolate; partition complete, not perforated. (21)

21a. Siliques glabrous; pedicels and rachis glabrous; leaves mostly cauline. *Arabis*

21b. Siliques often pubescent; pedicels and rachis at least somewhat hairy; leaves chiefly basal. *Braya*

ALYSSUM L.

Plants stellate-pubescent perennials or annuals from taproots; leaves alternate, simple, entire, tapering to the base, not clasping; flowers in racemes, the pedicels not subtended by bracts; sepals 4, deciduous; petals 4, yellow or white; stamens 6, each filament (or some of them) with a whitish process near the base; style slender, the stigma capitate; fruit a silicle, less than twice longer than broad, broadly elliptic to oval in outline, compressed parallel to the septum, the valves nerveless; seeds 1–2 per locule.

1a. Plants annual; petals white or pale yellow; styles less than 1 mm long. *A. alyssoides*

1b. Plants perennial; petals yellow; styles 1–2 mm long. *A. americanum*

Alyssum alyssoides L.
Alyssum

Plants annual, the herbage stellate-pubescent, the stems 0.5–3 dm long; leaves spatulate to oblanceolate, 0.8–3.7 cm long, 0.2–0.6 cm broad, stellate and green on both surfaces; pedicels spreading-ascending, 2–5 mm long; flowers 3–5 mm long, the sepals 2.8–4 mm long, green, stellate-pubescent, persistent; petals white or pale yellow, surpassing the sepals; silicles 3–5 mm long and about as broad, the valves stellate-pubescent; styles persistent, mostly 0.4–0.8 mm long.

Roadside in south-central Alaska (Rapids Lodge, Hodgson 98, July 1954, US), and to be expected elsewhere; widespread in North America; adventive from Europe.

Alyssum americanum Greene
American Alyssum

Plants perennial, decumbent, the herbage stellate-pubescent, the stems (0.5)1–3 dm long; leaves spatulate, 0.5–1.6 cm long, 0.1–0.3(0.5) cm broad, stellate on both surfaces, greenish above, silvery beneath; pedicels spreading, (3)4–8 mm long;

Alyssum americanum Greene (× 0.5).

flowers 3–4 mm long, the sepals 2–2.5 mm long, greenish, stellate-pubescent, deciduous; petals yellow, surpassing the sepals; silicles 3–5 mm long, 2.5–3.5(4.5) mm broad, the valves stellate-pubescent; styles persistent, 1–2 mm long.

Talus slopes; along the central Yukon Valley, from Rampart, Alaska, eastward to west-central and northern Yukon, and in the Porcupine River Valley; Asia.

APHRAGMUS Andrz.

Plants pubescent with short, simple hairs, or glabrous, rhizomatous; leaves mainly basal, simple, entire, tapering to the base, not clasping; flowers in short, subumbellate racemes, the pedicels short, subtended by foliose bracts (cauline leaves); sepals 4, deciduous; petals 4, white or purplish; stamens 6, the filaments lacking processes; style short, the stigma capitate; fruit a silique, mostly 3–5 times longer than broad, oblong-ellipsoid in outline,

slightly compressed, the septum lacking, the valves obscurely 1-nerved; seeds (2) 4–10.

Schulz, O. E. 1923. *Aphragmus.* pp. 197-98. *In*: A. Engler. Pflanzenreich IV. 105. Heft 84:1–388.

Aphragmus eschscholtziana Andrz. ex DC.

Plants ascending to erect, the stems 0.2–0.6 dm long; basal leaves 1.5–4 cm long, the blades narrowly ovate to spatulate, 0.4–0.8 cm wide, tapering to a slender petiole, glabrous; cauline leaves (bracts) 0.5–1.2 cm long, (0.1)0.2–0.5 cm broad, entire, glabrous, borne atop a naked stem; pedicels 1–2.5(6) mm long, erect, glabrous or minutely pubescent; sepals 1.5–1.8 mm long, obtuse, glabrous; petals 1.8–2.2 mm

Aphragmus eschscholtziana Andrz. (× 0.9).

long, white or purplish, longer than the sepals; siliques 6–12(15) mm long, 2–3 mm broad, the valves glabrous; styles 0.2–0.5 (0.7) mm long.

Tundra and heath, often in wet areas; in widely scattered sites ranging from the Seward Peninsula and Aleutian Islands to southern Alaska and southwestern Yukon.

ARABIDOPSIS (DC.) Schur

Plants glabrous, glaucous annuals from taproots; leaves alternate, simple, entire, auriculate-clasping basally; flowers in racemes, the pedicels not subtended by bracts; sepals 4, deciduous; petals 4, white; stamens 6, the filaments lacking processes; style short, the stigma slightly bilobed; fruit a silique, many times longer than broad, linear in outline, subterete or slightly compressed, the valves obscurely 1-nerved; seeds numerous.

Arabidopsis salsuginea (Pallas) N. Busch

Plants erect or the branches ascending, glabrous and glaucous, 0.5–1.5 dm tall; leaves oblong-lanceolate, 0.3–1.8 cm long, 0.1–0.4 cm broad, auriculate-clasping basally, acute to obtuse apically, glaucous and glabrous; pedicels spreading, 1–9 mm long; flowers 1.8–2.5 mm long, the sepals pinkish, 0.8–1.2 mm long, glabrous; siliques 7–16 mm long, to 1 mm broad; styles less than 0.5 mm long, persistent.

Saline soils; known from southwestern Yukon; east to Mackenzie and from Alberta east to Manitoba and south to Colorado (*Thellungiella salsuginea* [Pallas] Schultz).

ARABIS L.

Plants glabrous or variously pubescent, biennial or perennial herbs from taproots; leaves alternate or basal, simple, entire or toothed, or the basal lyrate-pinnatifid, the cauline ones auriculate-clasping or merely sessile or subpetiolate; flowers in racemes, the pedicels not subtended by bracts; sepals 4, deciduous; petals 4, white, pink,

or purplish; stamens 6, the filaments lacking processes; styles short, the stigma obscurely bilobed; fruit a silique, many times longer than broad, linear, compressed, the valves mostly 1-nerved; seeds numerous.

Rollins, R. C. 1941. A monographic study of *Arabis* in western North America. Rhodora 43:289–325, 348–411, 425–81.

1a. Cauline leaves attenuate basally or merely sessile, not at all auriculate-clasping. (2)

1b. Cauline leaves auriculate-clasping basally (except in some *A. holboellii*). (3)

2a. Basal leaves lyrate-pinnatifid to entire; cauline leaves attenuate basally; plants widespread in our region. *A. lyrata*

2b. Basal leaves entire; cauline leaves acute basally, sessile; plants known from southern Yukon. *A. nuttallii*

3a. Pedicels and siliques ascending to erect, and often tightly appressed (see also *A. hookeri*). (4)

3b. Pedicels and siliques spreading-ascending to deflexed. (7)

4a. Stems densely hirsute with long, spreading hairs, except above; cauline leaves often denticulate; seeds in 1 row in each locule; plants rather broadly distributed in southern Alaska and southern to northern Yukon. *A. hirsuta*

4b. Stems glabrous or glabrate, or densely hirsute only near the base; cauline leaves seldom denticulate; seeds in 2 rows in each locule; plants local in southern Alaska and Yukon. (5)

5a. Plants 0.4–2(2.5) dm tall; petals pink purple; plants from southwestern Yukon. *A. lyallii*

5b. Plants 3–6 dm tall or more; petals white or pinkish; plants variously distributed. (6)

6a. Basal leaves pubescent with malpighian hairs or glabrous; petals 7–10 mm long. *A. drummondii*

6b. Basal leaves pubescent with simple, bi- or trifurcate hairs; petals 4–6 mm long. *A. glabra*

7a. Stems densely hirsute with long, spreading hairs, except ·above; pedicels spreading-ascending, the **pods** erect or ascending. *A. hookeri*
7b. Stems variously pubescent to glabrate but not hirsute with long, spreading hairs; pedicels ascending to deflexed, the pods spreading-ascending, spreading, or deflexed. (8)

8a. Plants usually less than 2 dm tall, the stems arising from a caudex; racemes usually with 10(20) flowers or less; plants from southwestern Yukon. *A. lemmonii*
8b. Plants usually over 2.5 dm tall, the stems arising from a taproot; racemes usually more than 10-flowered; plants rather broadly distributed in the southeastern quarter of Alaska and southern Yukon. (9)

9a. Pubescence of lower stems and leaves mostly bi- or trifurcate; siliques spreading, the pedicels ascending to spreading, usually not abruptly curved near the base. *A. divaricarpa*
9b. Pubescence of lower stems and leaves mostly more than 3-pronged; siliques deflexed, the pedicels deflexed, abruptly curved near the base. *A. holboellii*

Arabis divaricarpa A. Nels.

Plants biennial, the stems solitary or few from a taproot, (2)2.5–7 dm tall, glabrous from near the base or pubescent with appressed bi- or trifurcate hairs, glabrous above; basal leaves entire or serrate, 1.2–4 (6) cm long, 0.3–0.8 cm broad, oblanceolate, tapering basally into a slender petiole; cauline leaves commonly longer than the internodes, 1.3–4 cm long, 0.2–0.6 cm broad, oblong to lanceolate, usually entire, sessile and auriculate, the upper ones usually glabrous, the lower ones mostly pubescent; racemes many-flowered, much elongating

in fruit; pedicels 5–12 long in fruit, ascending to spreading, usually not abruptly bent near the base; sepals 3–5 mm long, pinkish, glabrous or sparsely hairy with several-branched hairs; petals 6–9 mm long, pink to purplish; siliques 30–80 mm long, 1.5–2.5 mm broad, glabrous, spreading, straight or slightly curved, compressed; seeds often in 2 rows, winged.

Disturbed soils; in the southeastern quarter of mainland Alaska and southern Yukon; disjunctly eastward to the Atlantic and south to California, Utah, and Colorado.

Arabis drummondii Gray
Drummond Rockcress

Plants biennial or short-lived perennial, the stems solitary or 2–several from a taproot or caudex, 3–7 dm tall, usually glabrous throughout; basal leaves entire or rarely denticulate, 1.8–7 cm long, 0.3–1.2 cm broad, oblanceolate, tapering basally into a slender petiole, glabrous or pubescent with malpighian hairs; cauline leaves usually longer than the internodes, 1.5–6 cm long, 0.4–1.5 cm broad, oblong to lanceolate, usually entire, sessile and auriculate, usually glabrous; racemes mostly many-flowered, elongated in fruit; pedicels 7–15 mm long in fruit, erect; sepals 3–4.5 mm long, greenish or pinkish, glabrous; petals 7–10 mm long, white or pinkish; siliques 30–80 mm long, 2–3 mm broad, glabrous, erect, straight, strongly compressed; seeds in 2 rows, broadly winged.

Disturbed soils along streams and roadsides; known from widely disjunct localities in southern Alaska and Yukon, where it is evidently only locally common; eastward to the Atlantic and southward to California, Utah, and Colorado.

Arabis glabra (L.) Bernh.
Tower Mustard

Plants biennial or short-lived perennial, the stems solitary to several from a taproot, 3–12 dm tall or more, spreading-hairy (rarely glabrous) near the base, glabrous above; basal leaves toothed or entire, 3–12 cm long, 0.8–2.5 cm broad, oblanceolate, tapering to a broad petiole, pubescent with stalked bi- or trifurcate hairs, or some of the hairs simple; cauline leaves usually longer than the internodes, 2.5–12 cm long, 0.4–3.5 cm broad, lanceolate, usually entire, sessile and auriculate, glabrous or pubescent with simple and stalked, bi- or trifurcate hairs; racemes many-flowered, elongated in fruit; pedicels 4–10 mm long in fruit, erect; sepals 3–4 mm long, green or pinkish, glabrous; petals white to pink, 4.5–6 mm long; siliques 15–90 mm long, 1–1.5 mm broad, glabrous, erect, straight, compressed; seeds in 2 rows, narrowly winged.

Open woods; in southeastern Alaska, southern Yukon, and adjacent British Columbia; southward to California, and eastward to the Atlantic (*Turritis glabra* L.).

Arabis hirsuta (L.) Scop.
Hairy Rockcress

Plants biennial (rarely short-lived perennial), the stems solitary or few to several from a taproot, 2–6 dm tall or more, spreading-hairy with simple or forked hairs at least near the base, glabrous above; basal leaves toothed to entire, 1.5–9 cm long, 0.5–2.5 cm broad, oblanceolate, tapering to a broad petiole, pubescent with simple and forked hairs, often purplish beneath; cauline leaves longer or shorter than the internodes, 1–5(7) cm long, 0.3–2 cm broad, lanceolate, often serrate-dentate, sessile and auriculate, usually pubescent (at least beneath); racemes several- to many-flowered, elongating in fruit; pedicels 0.5–10 mm long in fruit, erect; sepals 2–4 mm long, greenish, glabrous; petals white to pinkish, 4–8 mm long; siliques 25–90 mm long, 1–2 mm broad, glabrous, erect, straight or slightly curved, compressed; seeds in 1 row, winged or wingless.

Flood plains, roadsides, or rocky outcrops; in southern half of Alaska and southern to

northern Yukon; east to the Atlantic and south to California, Arizona and New Mexico; circumboreal.

1a. Petals 3–5 mm long; plants from interior Alaska and Yukon (*A. hirsuta* ssp. *pycnocarpa* [Hopkins] Hultén). *A. hirsuta* var. *pycnocarpa* (Hopkins) Rollins

1b. Petals mostly more than 5 mm long; plants of coastal southern Alaska (*A. hirsuta* ssp. *eschscholtziana* [Andrz.] Hultén). *A. hirsuta* var. *eschscholtziana* (Andrz.) Rollins

Arabis holboellii Hornem.
Holboell Rockcress

Plants biennial, the stems solitary or few to several from a taproot, 2–10 dm tall, pubescent with appressed or stalked, branched hairs near the base, glabrous above; basal leaves entire or toothed, 1–6 cm long, 0.2–1 cm broad, oblanceolate, densely to moderately pubescent with branched hairs, tapering basally to a slender petiole; cauline leaves usually longer than the internodes, sessile and auriculate or merely sessile, usually pubescent (at least the lower ones); racemes many-flowered, elongating in fruit; pedicels 5–14 mm long in fruit, reflexed, abruptly bent near the base; sepals 2.5–5.5 mm long, greenish or pinkish, pubescent or glabrous; petals 5–10 mm long, pink to white; siliques 30–80 mm long, 1–2.5 mm broad, glabrous, deflexed, straight or slightly curved, compressed; seeds in 1 row or irregular, narrowly winged.

Dry, gravelly or sandy hillsides, flats, or roadsides, often in open woods; in the southeastern quarter of Alaska and southern Yukon; east to Greenland and south to California, Utah, Colorado, Nebraska, and Michigan.

1a. Cauline leaves merely sessile, not auriculate; plants from southwestern Yukon. *A. holboellii* var. *pendulocarpa* (A. Nels.) Rollins

1b. Cauline leaves auriculate-clasping; distribution various. (2)

2a. Upper cauline leaves usually glabrous, not or only slightly revolute, the basal leaves usually greenish; siliques 2–2.5 mm broad; plants uncommon. *A. holboellii* var. *holboellii*

2b. Upper cauline leaves usually pubescent, revolute, the basal leaves commonly grayish; siliques 1–2 mm broad; plants common. *A. holboellii* var. *retrofracta* (Grah.) Rydb.

Arabis hookeri Lange
Hooker Rockcress

Plants biennial, the stems solitary or few to several from a taproot, 2–6 dm tall, hirsute at the base with long, spreading hairs, becoming glabrate or glabrous above; basal leaves denticulate, 3–9 cm long, 0.4–1.2 cm broad, narrowly oblanceolate, tapering basally to a slender petiole, pubescent with stalked, stellate hairs; cauline leaves longer or shorter than the internodes, 2–5 cm long, 0.2–0.6 cm broad, oblong to lance-oblong, often denticulate, sessile and auriculate, usually pubescent; racemes many-flowered, elongating in fruit; pedicels 5–16 mm long in fruit, ascending; sepals 1.5–2 mm long, pinkish, pubescent; petals 2.5–4 mm long, white; siliques 18–40 mm long, 1–1.5 mm broad, glabrous, erect, straight, slightly compressed contrary to the partition; seeds in 2 rows or irregular, not winged.

Dry hillsides and roadsides, usually in open woods; in the southeastern quarter of mainland Alaska and southern Yukon; east to Greenland. This entity has several technical characters which indicate only slight relationship to other species of *Arabis*. On the basis of these features, *A. hookeri* is often placed in other genera. However, the traditional view of placing it with *Arabis* is followed herein because of confusion as to generic limits of related genera (*Turritis mollis* Hook.; *Arabidopsis mollis* [Hook.] Schulz; *Halimolobos mollis* [Hook.] Rollins).

Arabis lemmonii Wats.
Lemmon Rockcress

Plants perennial, the stems few to several from a branching caudex, mostly 1–2 dm tall, pubescent at the base with branching hairs or rarely glabrous, becoming glabrate or glabrous above; basal leaves entire or shallowly toothed, 0.8–2.5 cm long, 0.2–0.5 cm broad, oblanceolate, tapering basally to a slender petiole, pubescent with branching hairs; cauline leaves shorter or longer than the internodes, 0.5–1.5 cm long, 0.2–0.3 mm broad, oblong to lanceolate, usually entire, sessile and auriculate, glabrous or the lower ones pubescent; racemes mostly less than 10-flowered (to 20-flowered), elongating in fruit; pedicels 2–6 mm long, ascending to spreading; sepals 2–3.5 mm long, pinkish, glabrous or pubescent; petals pink to pink purple, 5–7 mm long; siliques 20–50 mm long, 1.5–2.5 mm broad, glabrous, ascending to spreading or slightly deflexed, straight or curved, compressed; seeds in 1 row, winged.

Alpine talus; in southeastern continental Alaska and southwestern Yukon; disjunctly from northern to southern British Columbia, southward to California and east to Montana and Colorado.

Arabis lyallii Wats.
Lyall Rockcress

Plants perennial, with 1–several stems from a branching caudex, 0.4–2.5 dm tall, glabrous or sparsely hairy near the base; basal leaves entire, 0.8–2(3) cm long, 0.1–0.3 cm broad, oblanceolate, tapering basally to a slender petiole, glabrous or pubescent near the base with simple or bifurcate hairs; cauline leaves shorter or longer than the internodes, 0.5–1.5 cm long, 0.1–0.2 cm broad, oblong to lance-oblong, entire, sessile and auriculate, glabrous or the lower ones ciliate; racemes mostly 10-flowered or less (to 15-flowered), elongating in fruit; pedicels 3–10 mm long, ascending to erect; sepals 3–4 mm long, purplish or greenish; petals pink purple, 6–10 mm long; siliques 20–60 mm long, 1–2 mm broad, glabrous, erect or ascending, straight or curved; seeds in 1 row, winged.

Slopes west of Kluane Lake, in southwestern Yukon; disjunctly southward from British Columbia to California and eastward to Alberta, Wyoming, and Utah.

Arabis lyrata L.

Plants short-lived perennials (biennials), the stems usually several from a taproot, 0.5–3.5 dm tall, glabrous throughout or sparsely hirsute near the base, or rarely hairy throughout; basal leaves entire, toothed, or lyrate-pinnatifid, 1.2–6(12) cm long, 0.3–1.5(2.4) cm broad, oblanceolate, tapering basally to a slender petiole, glabrous or pubescent with trifurcate hairs; cauline leaves shorter or longer than the internodes, 1–5(8.5) cm long, 0.3–1.2(2.5) cm broad, spatulate or oblanceolate, entire, toothed, or lobed, sessile or tapering to a broad petiole, glabrous; racemes few- to many-flowered, elongated in fruit; pedicels 5–12 mm long, ascending to spreading; sepals 2–3 mm long, greenish or pinkish, glabrous or pubescent; petals white to pinkish, 4.5–7 mm long; siliques 12–45(50) mm long, 1–1.5 mm broad, glabrous, ascending, straight or curved, compressed; seeds in 1 row, not winged.

Tundra, heath, or woods; over most of Alaska and Yukon; east to the Atlantic, and south to Washington and Alberta; Asia. The bulk of our material belongs to var. *kamchatica* Fisch. ex DC. (*A. lyrata* ssp. *occidentalis* [Wats.] Piper; *A. arenicola* [Richards.] Gelert). The var. *lyrata* is present in widely scattered localities and may be distinguished by the presence of trifurcate hairs on the lower leaf surface.

Arabis nuttallii Robins. ex Gray
Nuttall Rockcress

Plants perennial, the stems solitary or few

from a taproot, 1–3 dm tall, hirsute below, glabrous above (or throughout); basal leaves usually entire, 1–3.5 cm long, 0.3–1 cm broad, oblanceolate, tapering basally to a slender petiole, glabrous or long-hairy above, long-hairy below and ciliate; cauline leaves shorter or longer than the internodes, 0.5–1.5 cm long, 0.2–0.5 cm broad, lanceolate to oblong, entire, sessile or nearly so, not auriculate, strongly ciliate and hirsute below; racemes few- to many-flowered, elongated in fruit; pedicels 4–11 mm long; sepals 3–4 mm long, greenish, glabrous or pubescent; petals white or pinkish, 5–8 mm long; siliques 12–20 mm long, 1–1.5 mm broad, glabrous, ascending, compressed; seeds in 1 row, wingless.

Known in our area only from southern Yukon (Whitehorse); otherwise from southern Alberta and Washington south to Wyoming, Utah, and Nevada.

BARBAREA R. Br.

Plants glabrous to sparsely hirsute biennials (annuals) from taproots; leaves alternate, lyrate-pinnatifid to pinnately compound, the cauline leaves auriculate-clasping; flowers in racemes, the pedicels not subtended by bracts; sepals 4, deciduous; petals 4, yellow; stamens 6, the filaments lacking processes; style short, the stigma obscurely bilobed; fruit a silique, several to many times longer than broad, linear, only slightly compressed, the valves 1-nerved; seeds many.

Barbarea orthoceras Ledeb.
Wintercress

Plants erect, the stems 1.5–10 dm tall, glabrous; leaves lyrate-pinnatifid to pinnately compound or rarely reduced to the terminal lobe, mostly 4–10(15) cm long and 1–2.5(4) cm broad, glabrous or the petiole and lower lobes sparsely hirsute; cauline leaves reduced upwards, auriculate-clasping; racemes many-flowered, the pedicels 2–4 mm long, ascending; sepals 2.5–3.5 mm long, yellowish, glabrous; petals 4–5.5

mm long, yellow; siliques 15–50 mm long, 1.5–2.5 mm broad, glabrous, erect or ascending, slightly compressed, straight or somewhat curved; seeds in 1 row, pitted.

Moist soil along streams, seeps, beaches, and roadsides; in much of Alaska and Yukon south of the 68th parallel; east to the Atlantic and south to California and Colorado; Eurasia (*B. americana* Rydb.).

BERTEROA DC.

Plants stellate-pubescent annuals from taproots; leaves alternate and basal, simple, entire, reduced upwards and sessile, not auriculate; flowers in racemes, the pedicels not subtended by bracts; sepals 4, deciduous; petals 4, white; stamens 6, the filaments lacking processes; style long, slender, the stigma bilobed; fruit a silicle, 1–3 times longer than broad, compressed parallel to the septum, the valves 1-nerved or nerveless; seeds several.

Berteroa incana (L.) DC.

Plants erect, the stems 3–10 dm tall or more, stellate-hairy; basal leaves oblanceolate, 3–5 cm long, entire, petiolate, usually withered by flowering time; cauline leaves becoming reduced upwards, sessile or short-petiolate below; racemes many-flowered, the pedicels erect or ascending, 4–10 mm long; sepals 2–3 mm long, greenish to whitish, hairy; petals white, 4–6 mm long, deeply bilobed; silicles 5–7 mm long, 2–3 mm broad, moderately inflated, stellate-hairy; styles 2–3 mm long, persistent.

Roadsides and disturbed sites; in southern Alaska; adventive from Europe.

BRASSICA L.

Plants glabrous or hirsute annuals from taproots; leaves alternate or basal, variously lobed to entire, the basal ones often lyrate-pinnatifid, reduced upwards and sessile to petiolate or auriculate-clasping; flowers in racemes, the pedicels not subtended by bracts; sepals 4, deciduous; pet-als 4, yellow; stamens 6, the staminal filaments lacking processes; style short, thick, the stigma bilobed; fruit a silique, several times longer than broad, linear, terete or nearly so, the valves 1–3-nerved, the apical portion produced into a stout beak; seeds several to many. *Note:* Several cultivated members of this genus are present in our region in addition to the weedy species distinguished below. They are: *B. caulorapa* Pasq. (kohlrabi); *B. napobrassica* Mill. (rutabaga); *B. oleracea* L. var. *botrytis* L. (cauliflower), var. *capitata* L. (cabbage), var. *gemifera* Zenker (brussels sprout), var. *italica* Plenck (broccoli); *B. rapa* L. (turnip).

1a. Upper cauline leaves auriculate-clasping, sessile. *B. campestris*

1b. Upper cauline leaves petiolate to sessile, not auriculate-clasping. (2)

2a. Valves of fruit (and often of pedicels and raceme rachis) hirsute with coarse, spreading, or descending hairs. *B. hirta*

2b. Valves of fruit, pedicels, and raceme rachis glabrous. (3)

3a. Plants glabrous or nearly so, glaucous; valves of fruit 1-nerved, the beak sterile, pedicels slender. *B. juncea*

3b. Plants hirsute with long, spreading hairs, especially near the base, green; valves of fruit 3-nerved, the beak often with 1 seed; pedicels thick. *B. kaber*

Brassica campestris L.
Field Mustard

Plants erect, glabrous or with very few hairs, the stems 2.5–10 dm tall or more, simple or branched; basal leaves lyrate-pinnatifid, 5–18 cm long, the terminal lobe mostly 2–5 cm broad, crenate-dentate; lower cauline leaves similar to the basal ones, reduced upwards, becoming auriculate-clasping and dentate to entire; pedicels 7–20 mm long, slender, ascending, glabrous; sepals 4.5–6 mm long, yellowish, glabrous; petals 6–10 mm long, yellow; siliques 30–70 mm long, 2.5–3.5 mm thick, the beak 10–15 mm long, 1-nerved, the valves 1-nerved, glabrous.

Fields, roadsides, gardens, and other disturbed sites; in much of temperate Alaska and Yukon (?); introduced from Europe (*B. rapa* authors; *B. napus* authors).

Brassica hirta Moench
White Mustard

Plants erect, pubescent with coarse, descending hairs, at least below, the stems 2–10 dm tall, usually branched; basal leaves lyrate-pinnatifid, mostly 5–15 cm long, the terminal lobe 3–10 cm broad, obscurely dentate; cauline leaves reduced upwards, becoming merely lobed, not auriculate; pedicels 5–10 mm long, slender, spreading, often hirsute with descending hairs; sepals 4–5 mm long, yellowish, glabrous; petals 7–10 mm long, yellow; siliques 30–50 mm long, 3–4 mm broad, the beak 8–12 mm long, the valves 3-nerved, hirsute with coarse hairs.

Weed of waste places; reported from near Dawson City, Yukon; to be expected in temperate portions of Alaska; introduced from Europe (*Sinapis alba* L.).

Brassica juncea (L.) Czern.
Indian Mustard

Plants erect, glabrous or nearly so, the stems 3–10 dm tall or more, usually branched; basal leaves lyrate-pinnatifid, 8–25 cm long, the terminal lobe 5–15 cm broad, dentate; cauline leaves reduced upwards, short-petiolate or sessile, not auriculate-clasping; pedicels 9–14 mm long, ascending, glabrous; sepals 4–6 mm long, yellowish, glabrous; petals 7–12 mm long, yellowish; siliques 20–50 mm long, 2–3 mm thick, the beak 6–12 mm long, the valves 1-nerved, glabrous.

Weed of cultivated land and other disturbed sites; in southern and southwestern Alaska (to be expected in the Yukon); introduced from Asia.

Brassica kaber (DC.) Wheeler
Charlock

Plants erect, pubescent with coarse, spreading hairs, at least below, the stems 3–10 dm tall or more, simple or branched; basal leaves lyrate-pinnatifid to merely dentate, 5–20 cm long, 3–10 cm broad; cauline leaves reduced upwards, short-petiolate or sessile, but not auriculate-clasping; pedicels 2–6 mm long, ascending, stout; sepals 4–5 mm long; petals 8–14 mm long, yellow; siliques 30–50 mm long, 2–3 mm thick, the beak 10–15 mm long, the valves 3-nerved, glabrous.

Weed of cultivated land and disturbed soils; in southern Alaska and Yukon; introduced from Europe (*Sinapis arvensis* L.).

BRAYA Stern. & Hoppe

Plants perennial (biennial), pubescent with forked or simple hairs, arising from taproots; leaves alternate or basal, simple, entire or dentate, tapering to the base or merely sessile, not clasping; flowers in elongate or subcapitate racemes, the pedicels not subtended by bracts; sepals 4, deciduous; petals 4, white to pink or pink purple; stamens 6, the filaments lacking processes; style short, the stigma slightly bilobed; fruit a silique, 3-many times longer than broad, slightly compressed parallel to the septum, the valves obscurely 1-nerved; seeds few to many.

1a. Sepals at least somewhat hairy; cauline leaves usually 3–8 or more; siliques 8–30 times longer than broad. *B. humilus*

1b. Sepals often glabrous; cauline leaves usually lacking, sometimes with a single, foliose bract subtending the inflorescence; siliques mostly 4–7 times longer than broad. (2)

2a. Longest styles often more than 1 mm

long, slender; plants 0.3–2 dm tall in fruit. *B. glabella*

2b. Longest styles 1 mm long or less, stout; plants (0.2)0.5–1.0(1.2) dm tall in fruit. *B. purpurascens*

Braya glabella Richards.
Smooth Braya

Plants perennial, 0.3–2 dm tall, from a branching caudex, acaulescent or nearly so, the scapes sparsely pubescent with forked hairs; leaves all basal or rarely 1–2 cauline ones present, 0.6–5 cm long, 0.1–0.6 cm broad, the blades linear to oblanceolate, entire to denticulate, glabrous above and below, ciliate with simple or forked hairs, at least on the slender petioles; racemes subcapitate in flower, much elongating in fruit; pedicels 1–11 mm long, ascending to erect, pubescent; sepals 1.8–3.5 mm long, greenish, glabrous or pubescent near the apex, tardily deciduous; petals 3–8 mm long, white or suffused with purple; siliques (6)8–16 mm long, 1.5–2.5 mm broad, pubescent to glabrate or glabrous, erect or ascending, straight or slightly curved, somewhat compressed, the styles variable in length, those on the lowermost fruits often very short, 0.5–1.5 mm long, conical to cylindrical, less than 0.5 mm in diameter; seeds in 2 rows, 1–1.2 mm long.

Arctic and alpine tundra; in western and northern Alaska, and less commonly in the interior; eastward to Hudson Bay (*B. bartlettiana* Jordal; *B. henryae* Raup; *B. pilosa* Hook.).

Braya humilis (C. A. Mey.) Robins. ex Gray
Low Braya

Plants perennial or biennial (annual), 0.5–3.5 dm tall, the stems 1–several from a taproot, rather densely pubescent with forked hairs; basal leaves 0.5–4(6) cm long, 0.1–1 cm broad, spatulate to oblanceolate, entire or toothed, ciliate, glabrous above and below or more commonly hairy; cauline leaves 3–8 or more, similar to the basal ones except reduced upwards; racemes subcapitate in flower, greatly elongated in fruit; pedicels 2–6 mm long, ascending, pubescent; sepals 1.5–2.5 mm long, greenish or pinkish, pubescent, deciduous; petals 2.8–5 mm long, white or tinged pink or purplish; siliques 8–25(30) mm long, 0.5–1 mm broad, pubescent, ascending, straight or curved, subcylindrical or slightly compressed, more or less contracted between the seeds, the styles 0.5–1 mm long; seeds in 1 row, 0.7–1 mm long.

Braya humilis (C. A. Mey.) Robins. (× 0.5).

Alpine slopes to open woodlands; from the Seward Peninsula and Kotzebue Sound eastward along the Brooks Range through eastern Alaska and southern to northern Yukon, and along the north coast of Alaska; east to Greenland and disjunctly in Colorado (*Torularia humilis* [C. A. Mey.] Schulz). Our material belongs to ssp. *arctica* (Bocher) Rollins (*B. humilis* ssp. *richardsonii* [Rydb.] Hultén).

Braya purpurascens (R. Br.) Bunge ex Ledeb.
Purplish Braya

Plants perennial, 0.2–1(1.2) dm tall, from a branching caudex, acaulescent or with a single foliose bract, the scapes rather densely pubescent with forked or simple hairs or rarely glabrate; leaves 0.7–3.6 cm long, 0.1–0.4 cm broad, linear to narrowly oblanceolate, entire to denticulate, sparsely hairy to glabrous or ciliate (especially along the petiole); pedicels 1–7 mm long, ascending, pubescent; sepals 1.5–2.5 mm long, purplish, glabrous, or rarely with a few hairs, tardily deciduous; petals 2.5–4 mm long, white or suffused with pink; siliques 5–10 mm long, 1.5–2.5 mm broad, pubescent or glabrous, erect or ascending, usually straight, somewhat compressed, the styles (0.4)0.5–1 mm long, stout; seeds in 2 rows, 0.9–1.1 mm long.

Alpine tundra or heathlands, often in gravel or on rock outcrops; from the Seward Peninsula eastward along the Brooks Range, and in northwestern Alaska and northern Yukon; then disjunctly to Mackenzie and Greenland; Eurasia. There is need for monographic study of *Braya*, and for numerous additional specimens to be collected.

CAKILE L.

Plants glabrous or nearly so, fleshy, annual, from taproots; leaves alternate, serrate to lobed, tapering to the base, not clasping; flowers in racemes, the pedicels not subtended by bracts; sepals 4, deciduous; pet-

als 4, white to purplish; stamens 6, the filaments lacking processes; styles obsolete, the stigma obscurely bilobed; fruit a silique, mostly 3–10 times longer than broad, 2-jointed, the lower joint more slender than the beaked, upper joint, terete, each joint usually 1-seeded or the lower one sterile, separating at maturity.

Cakile edentula (Bigel.) Hook.
Searocket

Stems decumbent to erect, glabrous, 1–4 dm tall; leaves 1.5–7 cm long, 0.3–2.5 cm broad, oblanceolate to oblong, serrate to lobed, glabrous; pedicels 1–6 mm long, ascending, stout; sepals 3–4 mm long, greenish or pinkish, pubescent apically; petals 5–8 mm long, white or purplish; siliques 12–20 mm long, 3–4.5 mm broad, erect, terete, glabrous, the upper segment tapering to a flattened beak.

Along the coast; on Kodiak Island and at Yakutat, and to be expected elsewhere; east to the Atlantic Coast and around the Great Lakes (*C. edentula* ssp. *californica* [Heller] Hultén).

CAMELINA Crantz

Plants pubescent with forked or stellate hairs, annual, from taproots; leaves alternate, simple, entire, auriculate-clasping basally; flowers in racemes, the pedicels not subtended by bracts; sepals 4, deciduous; petals 4, pale yellowish; stamens 6, the filaments lacking processes; styles slender, the stigma capitate; fruit a silicle, less than twice longer than broad, obovate in outline, somewhat compressed parallel to the septum, the valves 1-nerved; seeds several per locule.

Camelina sativa (L.) Crantz
False Flax

Plants erect, the stems 2.5–10 dm tall or more; leaves 2–8 cm long, 0.5–2 cm broad, lanceolate to elliptic, clasping basally, pubescent to glabrate; pedicels slender, 10–30 mm long, ascending; sepals 3–4 mm long, greenish, pubescent; petals 4–5 mm long, pale yellow; siliques 6–9 mm long, 4–6 mm broad, glabrous, the style 1–2 mm long, persistent.

Weed of cultivated land and other disturbed soils; in southern Alaska; introduced from Europe.

CAPSELLA Medic. Nom. Cons.

Plants stellate-pubescent and often with coarse simple hairs, annual, from taproots; leaves alternate or basal, simple, dentate or variously lobed or entire, the cauline ones auriculate-clasping; flowers in racemes, the pedicels not subtended by bracts; sepals 4, deciduous; petals 4, white; stamens 6, the filaments lacking processes; style short, the stigma capitate; fruit a silicle, less than twice longer than broad, cuneate-obcordate in outline, compressed at right angles to the septum, the valves reticulately veined, strongly keeled; seeds many per locule.

Capsella bursa-pastoris (L.) Medic
Shepherd's Purse

Stems 1–5 dm tall; basal leaves 3–20 cm long, 0.5–4 cm broad, oblanceolate, subentire, dentate, pinnately lobed or lyrate-pinnatifid; cauline leaves lanceolate to oblong or oblanceolate, serrate to dentate or entire, clasping basally; pedicels 5–15 mm long, slender, spreading; sepals 1.2–2.5 mm long, greenish or pinkish, glabrous; petals 2–4 mm long, white or pinkish; silicles 4–8 mm long, 3–5 mm wide, cuneate-obcordate, glabrous; style 0.5–1 mm long, persistent.

Weed of disturbed soils; in much of southern Alaska and southern Yukon; introduced from the Old World (*C. rubella* Reuter, a form with concave sides on the silicles).

CARDAMINE L.

Plants glabrous or sparsely pubescent with simple hairs, annual, biennial, or perennial, from taproots or rhizomes; leaves alternate or basal, simple to pinnately compound, petiolate, not clasping basally; flowers in racemes (often subcorymbose), the pedicels not subtended by bracts; sepals 4, deciduous; petals 4, white to pink or pink purple; stamens 6, the filaments lacking processes; style stout, the stigma obscurely bilobed; fruit a silique, several to many times longer than broad, slightly compressed parallel to the septum, the valves obscurely 1-nerved or nerveless; seeds several to many.

Schulz, O. E. 1903. Monographie der Gattung *Cardamine*. Engl. Bot. Jahresb. 32:280–623.

1a. Leaves all simple; plants caespitose, usually 7(14) cm tall or less. *C. bellidifolia*

1b. Leaves compound, or only the lowermost simple (see *C. scutata*); plants usually not caespitose, mostly more than 7 cm tall. (2)

2a. Leaflets of basal and cauline leaves linear to narrowly elliptic; flowers white; plants mostly north of the 66th parallel, except on the Seward Peninsula. *C. digitata*

2b. Leaflets, at least of basal leaves, oblanceolate or broader; flowers white or pink to pink purple; plants variously distributed. (3)

3a. Leaflets of cauline leaves 9–15(17), linear to narrowly oblong, 5 times longer than broad or more; flowers pink to pink purple, or less commonly white. *C. pratensis*

3b. Leaflets of cauline leaves 3–9, often less than 5 times longer than broad; flowers white or pink to pink purple. (4)

4a. Leaflets 3 (rarely 5) on basal and cauline leaves; petals 8–12 mm long; plants 2–8 dm tall, growing in extreme southeastern Alaska. *C. angulata*

4b. Leaflets 5 or more, at least on some leaves (3 in some *C. purpurea*); petals 3–8 mm long; plants mostly less than 3 dm tall, of various distribution. (5)

5a. Stems with spreading or descending hairs, at least above; flowers pink to pink purple or white, or varicolored. *C. purpurea*
5b. Stems glabrous; flowers white. (6)
6a. Leaves 1–2.5 cm long (rarely longer); petals 6–8 mm long; plants usually less than 1 dm tall, from west-central, north-central, and northeastern Alaska, northern Yukon, and adjacent Mackenzie. *C. microphylla*
6b. Leaves (1.2)2.5–10 cm long; petals (2)3–5(7) mm long; plants mostly more than 1 dm tall, variously distributed. (7)
7a. Plants from tap or fibrous roots, lacking rhizomes, of broad distribution in southern Alaska and Yukon. (8)
7b. Plants rhizomatous, of westernmost Aleutians or extreme southeastern Alaska. (9)
8a. Lateral leaflets, at least of the lower cauline leaves, obovate to oblanceolate; siliques usually more than 1 mm broad; plants of broad distribution in southern Alaska and Yukon. *C. oligosperma*
8b. Lateral leaflets of cauline leaves linear to oblong or oblanceolate; siliques usually less than 1 mm broad; plants of southeastern Alaska, southern Yukon, and adjacent British Columbia. *C. pensylvanica*
9a. Leaflets of lower leaves 7–9, the terminal one only slightly larger than the lateral ones; plants of southeastern Alaska. *C. occidentalis*
9b. Leaflets of lower leaves 3–5, or the leaves simple, or the terminal leaflet much larger than the others; distribution various. (10)
10a. Leaves of apical leaflet cordate to ovate or reniform; plants of southeastern Alaska. *C. breweri*

10b. Leaves or apical leaflet cuneate to oblanceolate; plants of westernmost Aleutians. *C. scutata*

Cardamine angulata Hook.
Seaside Bittercress

Plants rhizomatous, perennial, the stems 2–8 dm tall, pubescent to glabrate, at least below; leaves mainly cauline, 3 (rarely 5)-foliolate, the terminal leaflet 1.5–5 cm long, 1.2–3 cm broad, lobed or toothed, somewhat larger than the lateral leaflets, ciliate; pedicels 8–20 mm long, ascending, glabrous; sepals 2–3 mm long, glabrous, greenish; petals 8–12 mm long, white to pinkish; siliques 15–35 mm long, 1.5–2 mm broad, erect, glabrous, the styles 1.5–3 mm long.
Woods; in extreme southeastern Alaska; south to California.

Cardamine bellidifolia L.
Alpine Bittercress

Plants perennial, from a taproot and often with rhizomelike caudex branches, the stems usually less than 0.7(1.4) dm tall, glabrous; leaves mainly basal, all simple, 0.3–3 cm long, the blades 0.2–1.8 cm long, 0.1–1 cm broad, ovate to lanceolate or elliptic, usually entire (or rarely pinnatifid), glabrous, cauline leaves 1–3, much reduced; pedicels 1–10 mm long, ascending to erect, glabrous; sepals 1.8–3.5 mm long, glabrous, pinkish; petals 3.5–5 mm long, white; siliques 10–35 mm long, 1–2 mm broad, erect, glabrous, the styles 1–3 mm long.
Arctic and alpine tundra and heathlands; over most of Alaska and Yukon; east to the Atlantic and south to Oregon and New Hampshire; circumpolar (*C. bellidifolia* var. *beringensis* Porsild; *C. bellidifolia* var. *pinnatifida* Hultén). Our plants belong to var. *bellidifolia*.

Cardamine breweri Wats.

Plants perennial, rhizomatous, the stems 2–6 dm tall, glabrous or pubescent near the base; leaves mostly cauline, the basal ones simple, the blades 1–3 cm long, subcordate basally, reniform to oval, the cauline with 3–5(7) leaflets, the terminal leaflet cordate to ovate, glabrous; pedicels 5–20 mm long, ascending, glabrous; sepals 1.5–2.5 mm long, greenish to pinkish, glabrous; petals 3–7 mm long, white; siliques 18–30 mm long, 1–1.5 mm broad, glabrous, erect, the style 0.5–2 mm long.

Woods; in southeastern Alaska; southward to California.

Cardamine digitata Richards.
Richardson Bittercress

Plants perennial, rhizomatous, the stems (0.9)1–2(2.5) dm tall, decumbent at the base, glabrous; leaves mainly cauline (the lower ones arising directly from the rhizome), 1.2–7.5 cm long, the leaflets 5–9, linear to narrowly elliptic, entire or those of the lower leaves sometimes toothed or lobed, 0.6–4.5 cm long, 0.1–0.4 cm broad, glabrous; pedicels 4–15 mm long, ascending, glabrous; sepals 2–4.5 mm long, glabrous, greenish or pinkish; petals 5–10 mm long, white; siliques 14–30 mm long, 1–1.5 mm broad, erect, glabrous, the styles 1–2 mm long.

Arctic and alpine tundra and heathlands; from the Seward Peninsula northward and eastward through northern Alaska and northern Yukon; east to Hudson Bay; Asia (*C. richardsonii* Hultén; *C. hyperborea* Schultz).

Cardamine microphylla Adams

Plants perennial, rhizomatous, the stems 0.3–1.5 dm tall, decumbent at the base, glabrous; leaves mainly cauline (sometimes in a basal tuft), 0.8–2.5(5) cm long, the leaflets 3–5, the apical similar to the lateral, orbicular and somewhat lobed to lanceolate or lance-linear, 0.2–1.2 cm long, 0.1–0.9 cm broad, glabrous; pedicels 5–25 mm long, ascending, glabrous; sepals 2.5–3 mm long, glabrous, greenish or purplish; petals 6–8 mm long, white; siliques 25–35

mm long, about 1 mm broad, erect, glabrous, the style (1)1.5–2 mm long.

Arctic and alpine tundra and heathlands; from the Seward Peninsula northward and eastward to northeastern Alaska, northern Yukon, and adjacent Mackenzie; Asia (*C. blaisdellii* Eastw.).

Cardamine occidentalis (Wats.) Howell
Western Bittercress

Plants perennial, rhizomatous, the stems 2–4 dm tall, decumbent at the base, glabrous or pubescent; basal leaves with (3) 5–7 leaflets, the lateral leaflets 0.3–0.8 cm long, usually entire, the terminal leaflet 1–2.5 cm long, oval to cordate, often shallowly lobed; cauline leaves reduced upwards, mostly with 2–4 broadly lanceolate to linear leaflets, 1–2 cm long, the terminal leaflet 1–3 cm long, ovate to cuneate; pedicels 7–15 mm long; sepals 1.5–2.5 mm long, glabrous, greenish or pinkish; petals 3–5 mm long; siliques 15–30 mm long, 1–1.5 mm broad, erect, glabrous, the style 0.8–1.2 mm long.

Moist soils in woods; reported for southeastern Alaska; south to Oregon (*C. neglecta* Greene).

Cardamine oligosperma Nutt. ex T. & G.

Plants annual or biennial (short-lived perennial?), with taproots (sometimes rooting from the lower nodes); stems (0.4)1–4.5 dm tall, pubescent with spreading hairs to glabrate or glabrous; basal leaves with (1–3)7–11 oval to ovate leaflets, the lateral leaflets 0.3–2 cm long, 0.2–1.5 cm broad, entire or lobed, the terminal leaflet 0.3–2.3 cm long, 0.4–1.6 cm broad, reniform and entire to cuneate-ovate or orbicular and lobed, glabrous or pubescent; cauline leaves reduced upwards; pedicels 4–15 mm long, spreading-ascending to suberect, glabrous; sepals 1.5–2 mm long, glabrous, greenish or purplish; petals 3–5 mm long, white; siliques 15–30 mm long, 1.2–2 mm broad, glabrous, erect, the style 0.4–1 mm long.

Moist sites in tundra, heath, or woods; from the Seward Peninsula southward and eastward through most of southern Alaska and southern Yukon; southward to California; Asia (*C. umbellata* Greene). Specimens of *C. oligosperma* from Alaska and Yukon are assignable to var. *kamtschatica* (Reg.) Detl. In southeastern Alaska, this entity is apparently transitional with *C. pensylvanica*. The entire complex requires additional study.

Cardamine pensylvanica Muhl.

Plants annual or biennial, from taproots (sometimes rooting at the lower nodes), the stems mostly 1.5–3.5 dm tall, pubescent or glabrous; basal leaves with 7–11 oval to lanceolate or oblanceolate leaflets, the lateral leaflets 0.3–1.5 cm long, 0.2–1.2 cm broad, entire or lobed, the terminal leaflet orbicular to cuneate-oblanceolate, 0.4–2 cm long, 0.3–1.5 cm broad, glabrous or pubescent; cauline leaves reduced upwards, the lateral leaflets oblanceolate to linear; pedicels 3–10 mm long, spreading-ascending, glabrous; sepals 1.2–1.8 mm long, glabrous, pinkish; petals 2–3 mm long, white; siliques 15–30 mm long, 0.7–1 mm broad, erect, glabrous, the style 0.4–0.8 mm long.

Moist woods, lake shores, and seeps; in southeastern Alaska and southern Yukon; south to California and east to the Atlantic.

Cardamine pratensis L.
Cuckoo Flower

Plants perennial, rhizomatous, the stems 0.8–4.5 dm tall, glabrous; basal leaves with 9–17 orbicular to lanceolate leaflets, these 0.2–0.9 cm long, 0.9–1 cm broad, glabrous; cauline leaves with 9–15 linear to narrowly oblong leaflets, glabrous; pedicels 5–20 mm long, spreading-ascending, glabrous; sepals 3–4.5 mm long, glabrous, pinkish or greenish; petals 9–15 mm long, pink, pink purple, or less commonly white; siliques 14–30 mm long, 1–1.5 mm broad, erect, glabrous, the styles 1–1.5 mm long.

Tundra and heath to woodlands; over most of Alaska and Yukon; east to the Atlantic; circumpolar (*C. pratensis* var. *angustifolia* Hook.; *C. pratensis* ssp. *angustifolia* [Hook.] Schulz).

Cardamine purpurea Cham. & Schlecht.

Plants perennial, rhizomatous, the stems 0.5–1.7 dm tall, pubescent with spreading hairs, at least above; basal leaves with 3–5 leaflets on a broad rachis, the lateral leaflets obovate to oval, 0.2–0.6 cm long, 0.2–0.5 cm broad, entire; terminal leaflet 0.5–0.7(1.2) cm long, 0.6–0.9(1.7) cm broad, entire or lobed, often subcordate, pubescent or glabrous; cauline leaves with 3–5 oval to orbicular or less commonly lanceolate leaflets, or simple, pubescent to glabrous; sepals 1.5–3 mm long, pubescent or glabrous, greenish or purplish; petals 5–8 mm long, pink to pink purple or white, or varicolored; siliques 13–20 mm long, 1–1.5 mm broad, erect, glabrous, the styles 1–2.5 mm long.

Arctic and alpine tundra and heathlands; on islands of the Bering Sea and continental western Alaska, eastward throughout central Alaska to central and southwestern Yukon; Asia (*C. purpurea* var. *albiflos* Hultén).

Cardamine scutata Thunb.

Plants perennial, rhizomatous, the stems 0.7–3 dm tall, glabrous; basal leaves with 3(5) leaflets or simple, the lateral leaflets oblanceolate, 0.4–0.7 cm long, 0.1–0.3 mm broad, entire, the terminal leaflet obovate to oblanceolate, 1–1.5 cm long, 0.5–1 cm broad, entire or lobed, cuneate basally, glabrous or hairy; cauline leaves with 5–7 oblanceolate leaflets, the terminal one much the largest, glabrous or pubescent; pedicels 5–10 mm long, spreading-ascending, glabrous; sepals 1.2–1.5 mm long, glabrous, greenish or pinkish; petals 3–4 mm long, white; siliques 15–25 mm long, 1–1.5 mm broad, erect, glabrous.

Stream banks and shallow water; in westernmost Aleutian Islands (Attu); Asia (*C. regeliana* Miq.).

COCHLEARIA L.

Plants glabrous, annual to perennial, succulent, maritime herbs from taproots; leaves mainly basal, simple, entire or somewhat lobed, the cauline usually toothed, not auriculate-clasping; flowers in racemes, the pedicels not subtended by bracts; sepals 4, deciduous; petals 4, white; stamens 6, the filaments lacking processes; style short, the stigma capitate; fruit a silicle, less than 3 times longer than broad, lance-ovoid to elliptic or suborbicular in outline, slightly compressed at right angles to the septum, the valves 1-nerved (and often reticulate); seeds several to many per locule.

Cochlearia officinalis L.
Scurvy-grass

Stems decumbent to ascending or erect, 0.4–3.5 dm long; basal leaves 0.5–10 cm long, the blades 0.2–2 cm long, 0.2–1.8 cm broad, reniform to cordate, ovate, or oblong, entire to toothed; cauline leaves reduced upwards, becoming sessile or subsessile, the blades cuneate basally and at least somewhat toothed; racemes subcapitate in flower, elongating in fruit; pedicels 2–12 mm long, ascending; sepals 2–3 mm long, greenish or tinged with yellow or purple; petals 3–5 mm long, white; silicles 4–10 mm long, 2–5 mm broad, the style 0.2–0.5 mm long.

Along the coasts of Alaska and Yukon; south to Washington; circumpolar. Three poorly differentiated varieties are present in our region.

1a. Silicles 2–3.5 mm broad, lance-ovoid in outline, mostly (1.5)2–3 times longer than broad; plants of coastal and insular western and northern Alaska and northern Yukon (*C. officinalis* ssp. *arctica* [Schlecht.] Hultén).

C. officinalis var. *arctica* (Schlecht.) Gel. ex Anders. & Hessel.

1b. Silicles (3)4–5.5 mm broad, oval to elliptic, or less commonly lance-ovoid in outline, mostly 1–2(2.5) times longer than broad; plants of western and southern Alaska. (2)

2a. Silicles 7–10 mm long, lance-ovoid in outline; plants of Kodiak Island and Prince William Sound vicinity (*C. sessilifolia* Rollins). *C. officinalis* var. *sessilifolia* (Rollins) Welsh

2b. Silicles 4–7(11) mm long, oval to elliptic in outline; plants broadly distributed in western and southern Alaska (*C. officinalis* ssp. *oblongifolia* [DC.] Hultén). *C. officinalis* var. *oblongifolia* (DC.) Gel. ex Anders. & Hessel.

DESCURAINIA Webb & Berth. Nom. Cons.

Plants stellate-pubescent, stipitate-glandular, or glabrate annuals or biennials from taproots; leaves basal or cauline and alternate, 1–3 times pinnately compound or pinnatifid, not clasping basally; flowers in racemes, the pedicels not subtended by bracts; sepals 4, deciduous; petals 4, yellow to cream (or white); stamens 6, the filaments lacking processes; style short, the stigma capitate; fruit a silique, more than 5 times longer than broad, linear to oblong, terete or nearly so, the valves 1-nerved, glabrous; seeds several to many per locule.

Detling, L. E. 1939. *Descurainia* in North America. Am. Midl. Nat. 22:481–520.

1a. Longest siliques 7–12 mm long, or if longer, then shorter than the pedicels, or the seeds less than 20, or at least partially in 2 rows; partition veinless or 1-veined. (2)

1b. Longest siliques 10–33 mm long, mostly over 20-seeded, the seeds in 1 row; partition obscurely 2–3-veined. (3)

2a. Stems stipitate-glandular, at least above; siliques somewhat clavate; plants of southern Yukon and northern British Columbia. *D. pinnata*

2b. Stems not at all stipitate-glandular; siliques linear; plants of broad distribution in southern Alaska and Yukon. *D. richardsonii*

3a. Pedicels 7–12 mm long; stems not at all stipitate-glandular; leaf segments less than 1 mm broad; young fruits not conspicuously subumbellately clustered near the raceme apex. *D. sophia*

3b. Pedicels 1–7(9) mm long; stems sparsely to densely stipitate-glandular, at least above (rarely glandless); leaf segments more than 1 mm broad; young fruits tending to be subumbellately clustered near the raceme apex. *D. sophioides*

Descurainia pinnata (Walt.) Britt.
Western Tansy Mustard

Annual, the stems pubescent with stellate hairs and also stipitate-glandular, at least above, 1–5 dm tall, simple or freely branched; leaves mostly cauline, 3–12 cm long, once to twice pinnatifid, the segments linear to oblong, often toothed, the upper leaves usually once pinnatifid; pedicels 5–15 mm long, ascending-spreading; sepals 1–2 mm long, pubescent to glabrous, greenish or yellowish; petals 1.5–3 mm long, yellow to cream (or white); siliques 8–12 (15) mm long, 1–1.5 mm broad, the seeds at least partially in 2 rows, the style very short.

Waste places; in southern Yukon and northern British Columbia (to be expected in Alaska); widely distributed in the United States and southern Canada (*D. pinnata* ssp. *filipes* [Gray] Detl.). Our plants belong to var. *filipes* (Gray) Peck.

Descurainia richardsonii (Sweet) Schulz
Richardson Tansy Mustard

Annual or biennial, the stems pubescent with stellate hairs, not at all stipitate-glandular, 2–12 dm tall or more, simple or branched; leaves basal and cauline, 2–13 cm long, the lower ones 2–3 times pinnately compound, becoming once pinnatifid above, the segments oblong to lanceolate; pedicels 2–10 mm long, spreading to ascending; sepals 1–1.8 mm long, yellowish; petals 1–3 mm long, yellowish; siliques 7–11 mm long, 1–1.5 mm broad, erect or ascending and often appressed; seeds more or less in 1 row, the style 0.4–0.6 mm long.

Weed of waste places; in much of southern Alaska, from the Aleutians to the Panhandle, and in southern Yukon; distributed over much of North America. Specimens from our region belong to var. *richardsonii*.

Descurainia sophia (L.) Webb ex Prantl
Tansy Mustard

Annual (or biennial?), the stems with stellate hairs, not at all stipitate-glandular, 2–10 dm tall, simple or branched; leaves chiefly cauline, 2–9 cm long, the lower ones once to thrice pinnately compound, becoming once to twice pinnately compound or pinnatifid above, the segments linear, usually less than 1 mm broad; pedicels 7–12 mm long, ascending-spreading; sepals 1.5–2 mm long, yellowish or greenish; petals 0.8–1.5 mm long, yellowish; siliques 11–23(30) mm long, 0.7–1 mm broad, ascending to erect, the seeds in 1 row; style almost obsolete to 0.2 mm long.

Weed of waste places; in southern Alaska and disjunctly in western and northern Alaska, and southern Yukon; widely distributed in North America.

Descurainia sophioides (Fisch.) Schulz
Northern Tansy Mustard

Annual (or biennial?), the stems with stellate hairs (often sparse) usually stipitate-glandular, at least above, 1–15 dm tall; leaves mainly cauline, 2–12 cm long, the lower ones once to thrice pinnately compound, becoming once to twice pinnately compound above, the segments oblong to oblanceolate, usually 1–2 mm broad; pedicels 1–7(9) mm long, spreading-ascending, often curved; sepals 1.5–2 mm long, yellowish; petals 1.2–2 mm long, yellowish; siliques 16–33 mm long, 0.7–1 mm broad, spreading-ascending, the seeds in 1 row, the style almost obsolete, to 0.2 mm long.

Waste places; over much of Alaska and Yukon; east to Hudson Bay; Asia. This is the common tansy mustard of our region. The bulk of the specimens have at least some stipitate glands, at least in the inflorescence. There are some specimens which lack such glands. They may be distinguished from the similar *D. sophia* by their broader leaf segments, usually shorter pedicels, and by the tendency for the young fruits to be subumbellately disposed.

DRABA L.

Plants with stellate, forked, or simple hairs, or glabrate, annual, biennial, or perennial, from taproots and often with well-developed caudices; leaves all basal or some cauline, alternate, simple, entire or toothed, tapering to the base or rounded, not auriculate; flowers in racemes, the pedicels usually not subtended by bracts; sepals 4, deciduous; petals 4, white, yellow, or cream; stamens 6, the filaments lacking processes; style obsolete to prominent and slender, the stigma obscurely bilobed; fruit a silicle (or short silique), mostly 1–10 times longer than broad, oval to ovate, lanceolate, or linear, compressed parallel to the septum, plane or twisted, straight or curved, the valves obscurely 1-nerved or nerveless; seeds usually numerous.

Note: This is a particularly difficult genus. Due at least partially to hybridization, specific lines are not sharply drawn. Minute differences, primarily of pubescence type and position, have been utilized as diagnostic characters. Thus the determination of entities depends upon critical examination with the aid of strong magnification. Further work of a revisionary nature is indicated.

Fernald, M. L. 1934. *Draba* in temperate northeastern America. Rhodora 36: 241–61, 285–305, 314–44, 353–71, 392–404.

Hitchcock, C. L. 1941. A revision of the Drabas of western North America. Univ. Wash. Publ. Biol. 11:1–132.

Schulz, O. E. 1927. Cruciferae—*Draba* et *Erophila*. Pflanzenr. IV. 105. Heft 89 (2):1–396.

1a. Basal leaves fleshy, 1–22 cm long, 0.3–4 cm broad, cuneate-oblanceolate, commonly dentate; plants maritime, from islands of the Bering Sea, Aleutians, and southern and southeastern Alaska. *D. hyperborea*

1b. Basal leaves seldom fleshy, 0.3–5 cm long, 0.1–2 cm broad, oblanceolate to oblong or elliptic; plants maritime or not. (2)

2a. Plants with 1–many cauline leaves. (3)

2b. Plants scapose, lacking cauline leaves (or rarely with 1). (18)

3a. Plants annual. (4)

3b. Plants biennial or perennial. (6)

4a. Pedicels and rachis pubescent with usually branched hairs; plants of central Alaska and southern Yukon. *D. praealta*

4b. Pedicels and rachis glabrate; plants rather broadly distributed in Alaska and Yukon. (5)

5a. Pedicels usually over 1.5 times longer than the silicles; silicles obtuse to rounded apically. *D. nemorosa*

5b. Pedicels usually less than 1.5 times longer than the silicles; silicles acute apically. *D. stenoloba*

6a. Leaves glabrous, or pubescent with simple hairs only, merely ciliate (see also *D. crassifolia*). (7)

6b. Leaves pubescent, at least some of the hairs forked or stellate. (9)

7a. Petals white; sepals 1–1.5 mm long; styles to 0.2 mm long. *D. fladnizensis*

7b. Petals yellow; sepals 2–3.5 mm long; styles 0.3–0.7 mm long. (8)

8a. Plants glabrous; caudex branches 5 cm long or more; silicles 1–2 mm broad, glabrous; central western Yukon. *D. ogilviensis*

8b. Plants pubescent; caudex branches less than 5 cm long; silicles 2–4 mm broad, glabrous or hairy; broadly distributed. *D. alpina*

9a. Flowers yellow (occasionally fading whitish). (10)

9b. Flowers white (sometimes cream to yellowish in *D. glabella* and *D. longipes*). (12)

10a. Cauline leaves 1–2; plants usually less than 1 dm tall. *D. crassifolia*

10b. Cauline leaves 1–20; plants 1–4 dm tall. (11)

11a. Petals 4–6 mm long; silicles pubescent, usually twisted; plants perennial. *D. aurea*

11b. Petals 2.5–4 mm long; silicles glabrous (rarely puberulent), plane; plants biennial or short-lived perennial. *D. stenoloba*

12a. Plants biennial or short-lived perennial; styles 0.2 mm long or less; from interior Alaska and southern Yukon. *D. praealta*

12b. Plants perennial; styles 0.2–1 mm long; distribution various. (13)

13a. Pedicels usually much longer than the fruit, or if shorter, then the siliques glabrous or sparsely pubescent and the leaf margins ciliate with at least some simple hairs; cauline leaves 1–3(5). *D. longipes*

13b. Pedicels mostly subequal to the fruit or shorter; cauline leaves 1–many. (14)

14a. Cauline leaves 1 or 2(3); plants usually less than 1 dm tall; petals usually not notched apically. *D. nivalis*

14b. Cauline leaves usually more than 2; plants often over 1.5 dm tall; petals usually notched apically. (15)

15a. Lower part of stem with long, spreading, simple hairs, sometimes mixed with forked hairs; plants of coastal and insular southern and western Alaska (less commonly in interior southern Alaska and Yukon). *D. borealis*

15b. Lower part of stem with chiefly stellate hairs, rarely with a few long simple ones; plants of western, northern, and interior Alaska and southern to northern Yukon, rarely in coastal southern Alaska. (16)

16a. Silicles, pedicels, and rachis of inflorescence glabrous or glabrate. *D. glabella*

16b. Silicles, pedicels, and rachis of inflorescence sparsely to densely pubescent (silicles rarely glabrous in *D. lanceolata*). (17)

17a. Pedicels spreading-ascending; cauline leaves 2–5. *D. cinerea*

17b. Pedicels erect, usually appressed to the rachis; cauline leaves mostly 4–10. *D. lanceolata*

18a. Leaves long-pilose with simple or forked hairs; inflorescence 0.5–4 cm long; silicles obovate in outline, 4-seeded; plants of coastal and insular southern Alaska westward through the Aleutians. *D. aleutica*

18b. Leaves not long-pilose, or if so, then the silicles oblong to lanceolate in outline and with more than 4 seeds; inflorescence usually much surpassing the leaves; plants of various distribution. (19)

19a. Leaves glabrous above and below, or with simple hairs only, often ciliate (see also *D. alpina*). (20)

19b. Leaves pubescent on one or both surfaces, at least some of the hairs branched or stellate, sometimes ciliate. (21)

20a. Plants small, biennial or short-lived perennial; leaves not much imbricated; petals 2–3 mm long; sepals 1–

1.5 mm long; silicles 5–12 mm long.
D. crassifolia

20b. Plants caespitose perennials; leaves much imbricated; petals 3–5 mm long; sepals 2–2.5 mm .long; silicles 3–5 mm long. *D. densifolia*

21a. Flowers white. (22)
21b. Flowers yellow (sometimes fading whitish) or purplish. (25)

22a. Leaves with branched or stellate hairs only, not ciliate with long simple hairs (except occasionally near the base). *D. nivalis*
22b. Leaves ciliate with at least some long simple hairs. (23)

23a. Silicles distinctly pubescent at maturity. *D. pseudopilosa*
23b. Silicles glabrous at maturity, sometimes slightly pubescent when young. (24)

24a. Sepals 1.5–2.2 mm long; petals 2.5–4 mm long; lower leaf surface with at least some branched or stellate hairs; cilia of mixed, simple and forked hairs. *D. lactea*
24b. Sepals 1–1.5 mm long; petals 1.5–2.5 mm long; lower leaf surface with simple hairs; ciliate with simple hairs. *D. fladnizensis*

25a. Pubescence of long, simple or forked hairs, these sometimes mixed with stellate hairs. (26)
25b. Pubescence of stellate hairs only, lacking long, simple or forked hairs, or with a few long simple ones near the base of leaves or scape. (28)

26a. Leaves without long, simple or branched cilia, or only sparsely ciliate at the base, the lower surface uniformly pubescent with minute, stellate hairs. *D. eschscholtzii*
26b. Leaves conspicuously ciliate with long, simple or forked hairs, the lower surface not uniformly pubescent with minute, stellate hairs. (27)

27a. Leaves in dense, basal rosettes, commonly 5 mm long or less; petals less than 1 mm broad, yellowish or purplish. *D. stenopetala*
27b. Leaves in. dense to loose, basal rosettes, commonly over 5 mm long; petals 2–3(4) mm broad, oval to obovate, yellow. *D. alpina*

28a. Leaves grooved, about 1 mm broad; plants of southern Yukon. *D. oligosperma*
28b. Leaves flat, mostly more than 2 mm broad. (29)

29a. Silicles glabrous; plants of western and northern Alaska and northern Yukon. *D. caesia*
29b. Silicles pubescent (glabrous in some *D. incerta*); distribution various, the species rare or obscure. (30)

30a. Leaves 1–1.5 times longer than broad, less than 10 mm long; plants from Kenai and Seward peninsulas. *D. exalata*
30b. Leaves mostly 2–4 times longer than broad, often more than 10 mm long; plants from western, south-central, and southern Alaska and southern Yukon. (31)

31a. Styles 1–1.8 mm long; plants of western, central, and eastern Alaska and southern Yukon. *D. eschscholtzii*
31b. Styles 0.4–1(1.2) mm long; plants of south-central Alaska and southern Yukon. *D. incerta*

Draba aleutica Ekman
Aleutian Rockcress

Plants perennial, caespitose, matted; leaves all basal, oblanceolate to obovate, 0.3–1.6 cm long, 0.1–0.3 cm broad, at least some pilose with long, simple or forked hairs, often glabrous on at least one surface, ciliate with long simple hairs, persistent on the caudex branches; scapes 0.3–4 cm long, seldom much surpassing the leaves; racemes 2–6(10)-flowered; sepals about 2 mm long, glabrous; petals 2–3 mm long, yellowish; silicles 2–5 mm long, oval to obovate in outline, inflated, glabrous or nearly so; style 0.2–0.5 mm long; seeds 4.

Gravelly, alpine situations; from Seward west to the western Aleutian Islands; Asia.

Draba alpina L.
Alpine Rockcress

Perennial, caespitose, the caudex branches seldom more than 5 cm long, densely clothed with persistent leaves; leaves all basal (rarely 1 cauline), 0.5–2.5 cm long, 0.1–0.4 cm broad, oblanceolate to oblong or elliptic, the surfaces with few to many simple or forked hairs, long-ciliate with simple or forked hairs; midribs often conspicuous in age and persistent; scapes 2.5–13 cm tall, pubescent with at least some simple hairs among the branched or stellate ones, sometimes glabrous; racemes 2–7 (10)-flowered; pedicels 2–10 mm long, pubescent like the scapes or glabrous; sepals 2–3.5 mm long, sparsely pilose with simple (rarely forked) hairs; petals 4–5 mm long, yellow, emarginate; silicles 5–11 mm long, 2–3(4) mm broad, lance-ovate to elliptic, plane, pubescent or glabrous; style 0.3–0.7 mm long, seeds 2–10 or more.

Arctic and alpine tundra and heathlands; from the islands of the Bering Sea and Seward Peninsula northward and eastward through northern and central southern Alaska to western and northern Yukon; eastward to Labrador; circumpolar (*D. macrocarpa* Adams; *D. micropetala* Hook.; *D. oblongata* R. Br.; *D. pilosa* Adams ex DC.).

Draba aurea Vahl ex Hornem.
Golden Rockcress

Plants perennial, not caespitose, the caudex simple or branched; basal leaves 1–4 cm long, 0.2–1.3 cm broad, oblanceolate; cauline leaves mostly 3–20, ovate to lanceolate or oblanceolate, 0.5–3 cm long, 0.3–1.2 cm broad, entire or denticulate, pubescent with stalked, 4-rayed hairs and usually also with some simple and forked hairs; stems erect or decumbent at the base, mostly 1–4 dm tall, pubescent with at least some long simple hairs, and often with branched hairs as well; racemes several- to many-flowered, occasionally with solitary flowers or lateral racemes in the axils of the upper leaves; pedicels 3–15 mm long, ascending, often curved, pubescent; sepals 2–3 mm long, pilose; petals 4–6 mm long, yellow, rounded; silicles 8–17 mm long, 2–4 mm broad, lance-elliptic, usually twisted, pubescent; style 0.5–1.5 mm long.

Gravelly soils and rock outcrops, from low to high elevations; in southern Alaska from the Alaska Peninsula eastward, and in southern Yukon; south to Arizona and New Mexico and eastward to Labrador and Greenland (*D. luteola* Greene).

Draba borealis DC.
Northern Rockcress

Perennial, loosely caespitose, the caudex simple or branched; basal leaves 0.6–5 cm long, 0.3–1.8 cm broad, oblanceolate,

Draba borealis DC. (× 0.4).

toothed or entire; cauline leaves 3–15, ovate to lanceolate, elliptic, or oblong, 0.4–4 cm long, 0.2–1.8 cm broad, toothed or entire, pubescent with 4–6-rayed, stellate hairs and often with straight, simple or forked hairs as well; stems erect or decumbent at the base, 0.7–5.5 dm tall, pubescent with at least some long simple hairs and sometimes with branched hairs also; racemes several- to many-flowered, occasionally with solitary flowers in the upper leaf axils; pedicels ascending, 3–16 mm long, usually straight, pilose; sepals 2–3 mm long, pilose; petals 4–6 mm long, white (or cream); silicles 7–14 mm long, 2–4 mm broad, lanceolate to oblong, twisted or plane, pubescent or glabrous; style 0.3–1 mm long.

Alpine tundra, heath, and open woods, often on rock outcrops; in southern Alaska and Yukon and in coastal and insular southern and western Alaska; Asia. *D. borealis* is one of our most distinctive Drabas. It is similar in many respects to *D. glabella* with which it is apparently closely related, and from which it cannot always be distinguished. Two more or less well-defined but intergrading varieties are recognizable.

1a. Lower stems pubescent with mixed, long, simple and branched or stellate hairs; cauline leaves usually with 2–8 poorly developed lateral teeth; plants 0.7–3 dm tall, mostly from Kodiak Island westward along the Aleutians and northward to Point Hope. *D. borealis* var. *borealis*
1b. Lower stems pubescent with long simple hairs; cauline leaves usually with 8–18 well-developed, lateral teeth; plants 1.5–5.5 dm tall, in southern Alaska and Yukon, and from Kodiak Island eastward to the Panhandle and adjacent British Columbia (*D. maxima* Hultén). *D. borealis* var. *maxima* (Hultén) Welsh

Draba caesia Adams

Perennial, caespitose, the caudex branches short, clothed with persistent leaves; leaves all basal (or with 1–2 cauline), 0.3–1 cm long, 0.1–0.2 cm broad, oblanceolate to obovate, the surfaces with grayish, overlapping, stellate hairs or only sparsely hairy above, sparingly long-ciliate only near the leaf bases; midribs persistent; scapes 2–8 cm long, pubescent with soft, stellate hairs; racemes 5–15-flowered; pedicels 4–12 mm long, pubescent to glabrate; sepals 2–3 mm long, pubescent to glabrate; petals 3.5–5 mm long, yellow (cream), rounded; silicles 3–6(8) mm long, 1.5–2.5 mm broad, oblong to elliptic, plane, glabrous or pubescent only when young; style 0.4–0.8 mm long.

Rocky hillsides, in arctic and alpine sites; in western and northern Alaska and in continental southeastern Alaska and western and northern Yukon; Asia. This entity has received differing interpretations by various workers. More specimens must be collected before a real understanding of the relationships of this taxon will be possible (*D. chamissonis* G. Don; *D. palanderiana* Kjellm.).

Draba cinerea Adams

Perennial, caespitose, the caudex simple or branched; basal leaves 0.5–3 cm long, 0.2–0.6 cm broad, oblanceolate, entire or toothed; cauline leaves (0)2–5, lanceolate to ovate, entire or toothed, 0.3–1.5 cm long, 0.1–0.7 cm broad, pubescent with overlapping, stellate hairs, or glabrate; stems erect, 0.6–2 dm tall, pubescent with stellate hairs often mixed with forked or simple hairs; racemes several- to many-flowered, sometimes with solitary flowers in the upper leaf axils; pedicels ascending to spreading-ascending, not appressed to the rachis, 3–8 mm long or more, pubescent; sepals 1.8–2.2 mm long, pubescent; petals 3–4 mm long, white to cream, emarginate; silicles 6–8 mm long, 2–2.5 mm broad, lanceolate to oblong, plane or twisted, pubescent (often densely so); style 0.3–0.8 mm long.

Alpine and arctic tundra or heath; widely scattered in Alaska and Yukon; circumboreal. This taxon is apparently closely related to *D. lanceolata* with which it is often confused.

Draba crassifolia Grah.

Biennial or short-lived perennial, the caudex simple or branched; leaves all basal (rarely 1–2 cauline), 0.3–1.5 cm long, 0.1–0.3 cm broad, oblanceolate, glabrous on both surfaces, or with a few hairs on the upper surface, ciliate with simple hairs or the margin glabrous; midribs seldom persistent; scapes 2–8 cm tall, usually glabrous; racemes 3–12-flowered; pedicels 2–10 mm long, glabrous; sepals 1–1.5 mm long, glabrous; petals 2–3 mm long, yellow (fading white); silicles 5–12 mm long, 1.5–2 mm broad, lanceolate to oblong, plane, glabrous or pubescent; styles to 0.3 mm long.

Arctic and alpine tundra; in scattered localities in southeastern continental Alaska and southern Yukon; disjunctly eastward to Greenland and south to Nevada and Arizona; Europe.

Draba densifolia Nutt. ex T. & G.

Perennial, caespitose, the caudex usually branched; leaves all basal, 0.2–0.9 cm long, 0.5–3 mm broad, obovate to oblanceolate, glabrous above and below, or with a few hairs only, ciliate with simple hairs; midribs persistent; scapes 0.5–5 cm tall, pubescent with branched hairs; racemes 3–10-flowered; pedicels 2–5 mm long, pubescent; sepals 2–2.5 mm long, pilose; petals 3–5 mm long, yellow; silicles 3–5 mm long, 2–3.5 mm broad, ovate to elliptic, plane, pubescent or glabrous; styles 0.5–1 mm long.

Alpine tundra; known from widely spaced sites in central and eastern Alaska and disjunctly from southern Canada southward to California, Nevada, and Utah.

Draba eschscholtzii Pohle ex Busch
Eschscholtz Rockcress

Perennial, caespitose, the caudex branches seldom more than 5 cm long, with some persistent leaves; leaves all basal, 0.5–2 cm long, 0.2–0.4 cm broad, oblanceolate, the surfaces with few to many stellate hairs, sparsely long-ciliate with simple hairs, the midribs seldom long persistent; scapes 5–12 cm long, pubescent with at least some simple hairs among the branched or stellate ones; racemes 5–11-flowered; pedicels 5–9 mm long, puberulent to glabrate; sepals 2–3 mm long, pilose; petals about 6 mm long, yellow, emarginate; silicles pubescent(?), the styles 1–1.8 mm long.

Tundra or heath; in western and central to eastern Alaska and southern Yukon; south to British Columbia; Asia. *D. eschscholtzii* is apparently allied to *D. alpina*, under which it has been placed previously as a synonym. Possibly, it might best be included with that species at some infraspecific level.

Draba exalata Ekman

Perennial, caespitose, the caudex usually branched; leaves all basal, 0.4–1 cm long, 2.5–3.5 mm broad, oblanceolate to obovate, pubescent with overlapping, stellate hairs, and often with some simple hairs, sparsely long-ciliate with simple hairs near the base; midribs persistent; scapes 1–8 cm tall, pubescent with at least some long simple hairs among branched ones; racemes 2–12-flowered; pedicels 2–7 mm long, pilose; sepals 1.5–2 mm long, pilose; petals about 3–4 mm long, yellow, emarginate; silicles 3–12 mm long, 2–3.5 mm broad, ovate to lanceolate, plane or twisted, pubescent, at least when young, glabrate in age; style 0.8–1 mm long.

Alpine tundra; known from scattered sites, from the Seward Peninsula, Kenai Peninsula, and southeastern continental Alaska (and Yukon?); Asia. This is a poorly understood entity which might represent the northern extension of some Cordilleran species. Indeed, some of our materials have been ascribed tentatively to *D. ventosa* Gray.

Draba fladnizensis Wulf. ex Jacq.

Plants perennial, caespitose, the caudex simple or branched; leaves all basal (or with 1–2 cauline), 0.3–1 cm long, 0.1–0.2 cm broad, oblanceolate, more or less pubescent with simple or rarely with some branched hairs, strongly ciliate with long simple hairs; midribs persistent; scapes 1–8 cm tall, glabrous or pubescent near the base with simple or forked hairs; racemes 3–12-flowered; pedicels 2–10 mm long, glabrous; sepals 1–1.5 mm long, sparsely pilose; petals 1.5–2.5 mm long, white; silicles 3–6 mm long, 1.2–2 mm broad, glabrous or puberulent; styles lacking or to 0.2 mm long.

Arctic and alpine tundra; in western, northern, and central to eastern Alaska (and on Kodiak Island), and in central and northern Yukon and adjacent Mackenzie; southward to Colorado and Utah; Eurasia. This is a rarely collected entity with affinities to both *D. lactea* and *D. pseudopilosa*.

Draba glabella Pursh

Perennial, loosely caespitose, the caudex usually branched; basal leaves 0.5–4 cm long, 0.2–1 cm broad, oblanceolate, entire or toothed; cauline leaves 1–10, ovate to lanceolate, 0.8–3 cm long, 0.3–1.2 cm broad, toothed or entire, pubescent with 4–5-rayed, stellate hairs, the longer 2 rays often paralleling the leaf axis and again branched, or the upper and lower surface glabrate or glabrous, ciliate with mostly forked or branched hairs; stems erect, simple or branched, 0.8–4.2 dm tall, stellate-pubescent, rarely with a few simple hairs below; racemes several- to many-flowered, occasionally with solitary flowers in the upper leaf axils; pedicels ascending, 2–15 mm long, usually straight, not appressed to the rachis, sparsely pilose to glabrous; sepals 2–3 mm long, sparsely pilose; petals 3.5–5 mm long, white (rarely cream to yellow), emarginate; silicles 6–15 mm long, 1.5–3 mm broad, lanceolate to oblong, plane or rarely twisted, glabrous or puberulent, the style 0.2–0.8 mm long.

Tundra, heath, or less commonly in open woods; over much of Alaska (except for the southwestern portion) and most of the Yukon; east to Newfoundland; circumpolar (*D. hirta* authors, not L.). This entity is apparently related to *D. borealis*, with which it seems to hybridize in western Alaska.

Draba hyperborea (L.) Desv.

Perennial, caespitose, from a thick taproot, the caudex base roughened by persistent peglike leaf bases, sometimes elongated and rhizomatous; basal leaves 1–22 cm long, 0.3–4 cm broad, cuneate-oblanceolate, the petiole broadly winged, usually toothed, fleshy; cauline leaves 3–10, oblanceolate to elliptic or lanceolate, 0.9–9.5 cm long, 0.3–2 cm broad, fleshy, dentate or entire, pubescent with simple hairs or less commonly with forked or stellate hairs; racemes few- to many-flowered; pedicels 3–25 mm long, ascending, pubescent to glabrate; sepals 3–4 mm long; petals 4–6 mm long, yellow; silicles oval to lanceolate, 6–20 mm long, 4–7 mm broad, glabrous, plane, often inflated; styles 0.5–1.5 mm long.

Near the seacoast; on islands of the Bering Sea, Aleutian Islands, and in southern and southeastern Alaska; south to British Columbia; Asia.

Draba incerta Payson

Perennial, caespitose, the caudex usually much branched; leaves all basal (except on new shoots), 0.7–1.5 cm long (to 3 cm on new shoots), 0.1–0.4 cm broad, narrowly oblanceolate, pubescent with 4-rayed, stellate hairs, the hairs again branched, ciliate with branched or stellate hairs; midribs somewhat persistent as threadlike processes; scapes 2–20 cm tall, stellate-pubescent; racemes 3–15-flowered; pedicels 4–18 mm long, ascending, stellate-pubescent; sepals 2.5–3.5 mm long, sparsely pilose; petals 4–5.5 mm long, yellow; silicles 6–10 mm long, 2.5–3.5 mm broad, lanceolate to ovate, plane, pubescent to glabrate or glabrous; styles 0.4–1 mm long.

Rocky or gravelly slopes, evidently rare; in south-central Alaska (Turnagain Arm), southern Yukon, and northern British Columbia; south to Washington, Idaho, and Wyoming.

Draba lactea Adams

Perennial, caespitose, the caudex usually branched; leaves all basal (sometimes with 1 cauline leaf), 0.5–2 cm long, 0.1–0.5 cm broad, pubescent with at least some branched or stellate hairs, ciliate with simple or forked hairs; midribs prominent and persistent; scapes 0.5–11 cm tall, glabrous or pubescent with soft, stellate or simple hairs; racemes 3–12-flowered; pedicels 1–8 mm long, glabrous or rarely stellate-puberulent; sepals 1.5–2.2 mm long, sparsely pilose; petals 2.5–4 mm long, white; silicles 4–9 mm long, 2–3 mm broad, lanceolate to ovate, plane, glabrous; styles to 0.5 mm long.

Tundra, mostly near the coast; in northern, northwestern, and western Alaska, and disjunctly from alpine sites in southern Alaska and Yukon; east to Labrador; circumpolar. This species is closely related to *D. fladnizensis* and to *D. pseudopilosa*.

Draba lanceolata Royle

Perennial, loosely caespitose, the caudex simple or branched; basal leaves 0.5–4 cm long, 0.1–0.4 cm broad, oblanceolate, entire, pubescent with overlapping, stellate or many-branched hairs; midribs not or only somewhat persistent; cauline leaves several, lanceolate, commonly toothed; stems erect or decumbent at the base only, 0.5–3.5 dm tall, pubescent with soft, many-branched hairs; racemes several- to many-flowered, sometimes with solitary flowers in the upper leaf axils; pedicels 2–9 mm long, erect, usually appressed to the rachis, pubescent; sepals 1.5–2 mm long, sparsely pilose; petals 2.2–4 mm long, white; silicles 5–12 mm long, 1.5–2.5 mm broad, narrowly lanceolate to oblong, plane or more commonly twisted, densely soft pubescent (rarely glabrous), the style 0.2–0.8 mm long.

Tundra, heath, or open woods; in northern, central, and eastern Alaska and southern Yukon; south to Nevada and Utah and east to the Atlantic; Eurasia.

Draba longipes Raup

Perennial, loosely caespitose, the caudex usually branched; basal leaves 0.5–3 cm long, 0.2–1.2 cm broad, oblanceolate, entire or less commonly obscurely toothed; cauline leaves (0)1–3, lanceolate to ovate-lanceolate, 0.7–2 cm long, 0.3–1 cm broad, entire or sometimes toothed, pubescent with unequally 4-rayed, stellate hairs, rarely mixed with some 5-rayed or merely forked hairs, usually ciliate with simple, forked, or stellate hairs; stems often decumbent at the base, erect above, 0.7–2 (3.5) dm tall, pubescent with branched or stellate hairs, rarely with a few long, simple ones near the base; racemes 2–15-flowered; pedicels 3–15 mm long, pubescent (sometimes sparingly so); sepals 1.5–2.5 mm long, sparsely pilose; petals 3–5 mm long, white, cream, or yellow; silicles 6–12 mm long, 1–2.5 mm broad, lanceolate to oblong, plane, glabrous or nearly so, the style 0.5–1 mm long.

Arctic and alpine tundra and heathlands; in coastal and insular western and northern Alaska, and disjunctly in the Alaska Range, on the Kenai Peninsula, and in most of Yukon; south to British Columbia. *D. longipes* combines characteristics of *D. borealis* (pubescence type, except for simple hairs), *D. glabella* (nearly or quite glabrous silicles), and *D. stenoloba* (general habit). It differs from *D. borealis* in its longer styles and racemes with fewer flowers, from *D. glabella* in having chiefly 4-rayed hairs, and from *D. stenoloba* in style length and flower color.

Draba nemorosa L.

Annual, from a slender taproot, the stems simple or branched, 0.5–2.5 dm tall, pubescent with mixed forked and stellate hairs, or less commonly with some simple ones, or glabrate; leaves oblanceolate to lanceolate, ovate, or oblong, 0.5–3 cm long, 0.2–0.8 cm broad, entire or toothed, pubescent with branched or simple hairs; racemes few- to many-flowered; pedicels 0.5–2.5 cm long, spreading-ascending, glabrous; sepals 1–1.5 mm long, sparsely pilose to glabrous; petals 1.2–4 mm long, yellow to white; silicles 4–10 mm long, 1.5–3 mm broad, oblong to narrowly oblanceolate, plane, glabrous, the style obsolete.

A weedy species, usually growing in disturbed or open sites; known from widely scattered sites in southern Alaska and southern Yukon; in much of North America; Eurasia.

Draba nivalis Lilj.

Perennial, loosely to densely caespitose, the caudex usually branched; leaves all basal (or with 1–3 cauline), 0.5–1.5 cm long, 0.1–0.4 cm broad, pubescent with usually overlapping stellate hairs, marginal pubescence entirely stellate or with simple hairs only near the base; midribs prominent and persistent; scapes 3–12(20) cm tall, glabrous or pubescent with stellate hairs; racemes 3–12-flowered; pedicels 1–6(11) mm long, ascending to erect; sepals 1.5–2 mm long, glabrous or pubescent; petals 2.5–4 mm long, white; silicles 5–12 mm long, 1–2 mm broad, lance-linear to oblong or lance-oblong, plane or twisted, glabrous or pubescent, the style 0.2–0.5 mm long.

Arctic and alpine tundra and heathlands; over much of Alaska and Yukon; south through British Columbia to Utah and Colorado and east to Newfoundland; Eurasia. This species has been subject to numerous and often conflicting interpretations. As viewed herein, *D. nivalis* is a broadly ranging, polymorphic entity encompassing a broad range of morphologically differing phases.

1a. Cauline leaves 1–3. (2)

1b. Cauline leaves lacking or rarely 1. (3)

2a. Pedicels ascending or spreading-ascending; silicles oblong to lance-oblong; plants of south-central Alaska (*D. kamtschatica* [Ledeb.] Busch; *D. lonchocarpa* ssp. *kamtschatica* [Ledeb.] Calder & Taylor). *D. nivalis* var. *kamtschatica* (Ledeb.) Pohle

2b. Pedicels erect or steeply ascending; silicles linear-lanceolate; plants from Prince William Sound eastward through the Panhandle (perhaps not sufficiently distinct from var. *elongata*). *D. nivalis* var. *denudata* (Schultz) C. L. Hitchc.

3a. Leaves and stems pubescent with stellate or branched hairs only; silicles mostly 3–5 times longer than broad; plants broadly distributed. *D. nivalis* var. *nivalis*

3b. Leaves and/or stems usually with some simple hairs; silicles mostly 7–10 times longer than broad or more; plants of southeastern Alaska and southern Yukon (*D. lonchocarpa* Rydb.). *D. nivalis* var. *elongata* Wats.

Draba ogilviensis Hultén

Perennial, loosely caespitose, the caudex branches commonly more than 5 cm long, not obscured by persistent leaves; leaves cauline, 0.7–2 cm long, 0.2–0.7 cm broad, elliptic to oblanceolate, the surfaces glabrous, the margins glabrous; midribs neither persistent nor conspicuous; peduncles 2.5–6 cm long, glabrous; racemes 6–12-flowered; pedicels 2–10 mm long, glabrous; sepals 2–2.5 mm long, glabrous; petals 4–5 mm long, yellow, truncate to emarginate; silicles 4–6 mm long, 1–2 mm broad, glabrous; styles 0.4–0.7 mm long; seeds 5 or more.

Alpine tundra; in the Ogilvie Mountains, Yukon; endemic (*D. sibirica* authors, not Thell.).

Draba oligosperma Hook.

Perennial, densely caespitose, the caudex usually much branched; leaves all basal, 3–10 mm long, 0.7–1.5 mm broad, narrowly oblanceolate to oblong, pubescent with 4-rayed stellate hairs with the rays again branched, marginally pubescent with stellate hairs, moderately V-shaped in cross section; midribs prominent, often persistent; scapes 1–10 cm tall, glabrous or stellate-pubescent near the base; racemes 3–12-flowered; pedicels 2–10 mm long, glabrous, ascending; sepals 2–2.5 mm long, sparsely pilose; petals 3–4.5 mm long, yellow; silicles 3–7 mm long, 2–3 mm broad, ovate to lanceolate or oval, plane, glabrous or pubescent, the styles 0.2–1 mm long.

Dry hillsides; in southeastern, continental Alaska and southern Yukon; southward to California, Nevada, Utah, and Colorado.

Draba praealta Greene

Biennial or short-lived perennial (annual), from a slender taproot; basal leaves 0.5–3 cm long, 0.2–0.6 cm broad, oblanceolate; cauline leaves 1–6, lanceolate to ovate, 0.5–3 cm long, 0.2–1.4 cm broad, entire or toothed, pubescent with 4–7-rayed hairs, often mixed with forked or simple hairs; stems 1–3 dm tall, simple or branched, pubescent with simple, forked, or stellate hairs; racemes few- to many-flowered; pedicels 3–10 mm long, pubescent with forked or stellate hairs, ascending; sepals 1.5–2 mm long, pilose; petals 2–3.5 mm long, white; silicles 7–14 mm long, 1.5–2.5 mm broad, lanceolate to linear-lanceolate, pubescent, plane, the style 0.2 mm long or less.

Dry, open slopes; from central western to eastern Alaska and southern Yukon; east to Mackenzie and south to Oregon and Wyoming. D. praealta is allied with D. stenoloba, from which it can be distinguished by the usually whitish flowers and pubescent silicles.

Draba pseudopilosa Pohle

Plants perennial, densely caespitose, the caudex usually branched; leaves all basal, 0.5–1.5 cm long, 0.1–0.3 cm broad, narrowly oblanceolate, pubescent with long simple hairs often mixed with forked or stellate hairs, the marginal hairs conspicuous, simple or forked; midribs prominent and persistent; scapes 2–8 cm long, pubescent with forked, stellate, or simple hairs; racemes 3–10-flowered; pedicels 1–5 mm long, pubescent with simple or stellate hairs; sepals 1.5–2.2 mm long, sparsely pilose; petals 2–3.5 mm long, white; silicles 5–7 mm long, 2–3 mm broad, lanceolate to ovate, plane, pubescent (often sparsely so), the styles 0.1–0.6 mm long.

Arctic tundra; in northern, northwestern, and western Alaska, and islands of the Bering Sea, and disjunctly in the Alaska Range; Asia. This species has been treated as D. lactea var. pseudopilosa (Pohle) Schulz. Certainly D. pseudopilosa is closely related to D. lactea and might represent only a minor form of it. However, there does appear to be a tendency for the two types to remain distinct even when they are sympatric.

Draba stenoloba Ledeb.

Annual, biennial, or short-lived perennial, from a slender taproot; basal leaves 0.4–4 cm long, 0.2–0.8 cm broad, oblanceolate, toothed or entire; cauline leaves 1–8, ovate to lanceolate, toothed or entire, 0.6–1.7(3) cm long, 0.3–0.8(1.4) cm broad, pubescent with simple, forked, or branched hairs; stems 0.8–3 dm tall, simple or branched, pubescent below with stellate, forked, or simple hairs, commonly glabrous above; racemes 3–many-flowered; pedicels 4–15 mm long, commonly glabrous (seldom pubescent), ascending; sepals 1.5–2.5 mm long, pilose or glabrous; petals 2.5–4 mm long, yellow to cream; silicles 8–15 mm long, 1.5–2.5 mm broad, narrowly oblong to linear, plane, glabrous or rarely pubescent, the style to 0.1 mm long.

Alpine tundra and heathlands; widely distributed in southern Alaska and southern Yukon; south to California, Utah, and Colorado.

1a. Basal portion of stem with chiefly stellate or branched hairs; hairs of leaves mostly 4-rayed; our common variety. *D. stenoloba* var. *stenoloba*

1b. Basal portion of stem glabrous or with simple or forked hairs; pubescence of leaves simple or forked; plants of southern Yukon (*D. nitida* Greene). *D. stenoloba* var. *nana* (Schulz) C. L. Hitchc.

Draba stenopetala Trautv.

Perennial, caespitose, the caudex branches seldom more than 5 cm long, densely clothed with persistent leaves; leaves all basal, 0.4–0.7 cm long, 0.1–0.2 cm broad, oblanceolate to oblong or elliptic, the surfaces glabrous or more commonly with few to several simple, forked hairs, long-ciliate with simple hairs; midribs not conspicuous nor persistent in age; scapes 0.3–2.5 cm long, pubescent with forked or stellate hairs; racemes (1)2–3(4)-flowered; pedicels 2–9 mm long, pubescent like the scapes; sepals 2.5–3.5 mm long, pubescent with simple and forked hairs; petals 2–5 mm long, yellow or purplish, oblong to narrowly oblanceolate, entire; silicles 3–4 (5) mm long, 1–2 mm broad, lance-ovate to elliptic, pubescent or glabrous; styles 0.2–0.3 mm long; seeds 4–6.

Alpine tundra; in mountains of interior Alaska and southern Yukon; Asia (*D. stenopetala* var. *purpurea* Hultén).

ERYSIMUM L.

Plants pubescent with appressed, 2–3(4)-rayed hairs, annual to perennial, from taproots; leaves alternate or basal, simple, entire to dentate, tapering to the base, not clasping; flowers in racemes, the pedicels not subtended by bracts (occasionally bracteate in *E. pallasii*); sepals 4, deciduous; petals 4, yellow or pink to pink purple; stamens 6, the filaments lacking processes; style prominent, short to elongate, the stigma bilobed; fruit a silique, many times longer than broad, linear, compressed parallel to the partition or more commonly subterete, the valves 1–several-nerved; seeds many per locule.

Rossbach, G. B. 1958. The genus *Erysimum* (Cruciferae) in North America north of Mexico—a key to the species and varieties. Madroño 14:261–67.

1a. Petals bright pink to pink purple; siliques (25)30–100 mm long, (1.5) 2–3 mm broad, conspicuously flattened; plants usually less than 3 dm tall, from western, northwestern, and northern Alaska and northern Yukon and disjunctly in alpine sites elsewhere. *E. pallasii*

1b. Petals yellow; siliques 12–80 mm long, 1–1.5(2) mm broad, subterete or slightly flattened; plants often over 3 dm tall, variously distributed. (2)

2a. Plants annual, the basal rosettes lacking or withered and dead by flowering time; petals 3–5 mm long; siliques 12–27 mm long, about 1 mm thick; broadly distributed in Alaska and Yukon. *E. cheiranthoides*

2b. Plants biennial or short-lived perennials, the basal rosette usually evident at flowering time; petals (6)7–18(20) mm long; siliques mostly (10)25–80 mm long, often more than 1 mm broad; distribution various. (3)

3a. Petals (6)7–10(11) mm long; plants of broad distribution in eastern half of Alaska and in the Yukon. *E. inconspicuum*

3b. Petals 12–20 mm long; plants local in south-central and west-central Yukon and east-central and south-central Alaska. *E. asperum*

Erysimum asperum (Nutt.) DC.
Wallflower

Biennial or short-lived perennial, with simple or less commonly branched caudex, the stems (1)2–8 dm tall or more; leaves 3–12 cm long, 0.1–1 cm broad, linear to oblong, elliptic, or oblanceolate, entire or denticulate; racemes bractless; pedicels 5–15 mm long, ascending to spreading; sepals 7–10 mm long, yellowish; petals 12–20 mm long, bright yellow; siliques 30–80 mm long or more, 1–2 mm broad, subterete to flattened, ascending to spreading, the stylar beak 1–5 mm long; seeds 1.5–2.3 mm long.

Open woods and rock outcrops; in south-central and east-central Alaska and in west-central and south-central Yukon; disjunctly southward to California and eastward to Oklahoma, Kansas, and Minnesota.

1a. Stems 1–2(3.5) dm tall; leaves 1–2 mm broad; beak of silique 3–5 mm long; plants from west-central and south-central Yukon. *E. asperum* var. *angustatum* (Rydb.) B. Boi.

1b. Stems (1)2–8 dm tall; leaves mostly more than 2 mm broad; beak of silique 1–3 mm long; plants known from south-central Alaska and southern Yukon. *E. asperum* var. *asperum*

Erysimum chieranthoides L.
Wormseed

Annual, the stems simple or branched, 2–12 dm tall; leaves 2–8 cm long, 0.2–1.5 cm broad, linear to oblong, lanceolate, or oblanceolate, entire or denticulate; racemes bractless; pedicels 4–15 mm long, spreading or spreading-ascending; sepals 2–3 mm long, yellowish or greenish; petals 3–5 mm long, pale yellow; siliques 12–27 mm long, about 1 mm broad, subterete, ascending to erect, the style about 1 mm long; seeds 1–1.2 mm long.

Weedy species of disturbed sites; in much of Alaska and Yukon; south to California and east to the Atlantic; Eurasia (*E. chieranthoides* ssp. *altum* Ahti).

Erysimum inconspicuum (Wats.) MacM.
Prairie Violet

Biennial or short-lived perennial, with usu-

ally unbranched caudex, the stems 2–10 dm tall, simple or branched; leaves 1.5–8 cm long, 0.2–0.8 cm broad, linear to oblong, elliptic, lanceolate, or oblanceolate, usually entire; racemes bractless; pedicels 3–8 mm long, ascending; sepals 4–7 mm long, often purplish; petals (6)7–10(11) mm long, pale to bright yellow; siliques 15–50 mm long, 1–2 mm broad, subterete, ascending to erect, the stylar beak 1–1.5 mm long; seeds about 1.5 mm long.

Open woods, dry slopes, stream gravels, and roadsides; in the eastern half of Alaska and most of the Yukon; south to Oregon and Colorado and east to the Atlantic.

Erysimum pallasii (Pursh) Fern.
Pallas Wallflower

Perennial (or biennial), with branched or unbranched caudex, the stems 0.3–3.5 dm tall, simple; leaves 1.2–6 cm long, 0.2–0.7 cm broad, oblanceolate to oblong or linear, dentate to entire; racemes often with 1–few bracts, each subtending one of the lower flowers; pedicels 5–28 mm long, ascending to erect; sepals 5–8 mm long, often purplish; petals 2–3 mm broad, pink to pink purple; siliques (25)30–100 mm long, 1.5–3 mm broad, distinctly compressed parallel to the septum, ascending to erect, the stylar beak (1)1.5–3 mm long.

Arctic and alpine tundra and heath; in central, southwestern, southeastern (not Panhandle), and northern Alaska and western and northern Yukon; east to Greenland; Asia (*E. pallasii* var. *bracteosum* Rossb.).

EUTREMA R. Br.

Plants glabrous, perennial, from a thick taproot; leaves cauline and alternate or basal, simple, entire, tapering to the base, not clasping; flowers in racemes, the pedicels not subtended by bracts, or the lowermost sometimes bracteate; sepals 4, deciduous; petals 4, white; stamens 6, the filaments lacking processes; style short, stout, the stigma capitate; fruit a short silique, usually 5–10 times longer than broad, el-

Erysimum pallasii (Pursh) Fern. (× 0.5).

liptic to oblong, or narrowly oblanceolate, subterete, the septum obsolete, the valves 1–3(5)-nerved; seeds few to several.

Eutrema edwardsii R. Br.

Plants from a taproot usually 3–6 mm thick, the stems decumbent-ascending to erect, 1–3.5(4) dm tall, arising from a simple or branching caudex; basal leaves long-petiolate, 2–15 cm long, the blades oval to ovate, lanceolate, elliptic, or oblong, 0.6–4 cm long, 0.3–3 cm broad, entire; cauline leaves reduced upwards, becoming sessile or subsessile, lance-oblong to lanceolate, entire; racemes mostly 3–15-flowered; pedicels 3–14 mm long, ascending; sepals 2–2.5 mm long, purplish; petals 3–4.5 mm long, white or cream; siliques 9–20 mm long, 1.5–3 mm broad, ascending to erect, subterete, glabrous, the style to 0.5 mm long.

Arctic and alpine tundra and heathlands; in much of Alaska and Yukon north of the 62nd parallel; east to Labrador; circumpolar.

HESPERIS L.

Plants pubescent with simple and forked hairs, biennial or perennial; leaves alternate, simple, entire or denticulate, petiolate, subsessile, or sessile, not auriculate-clasping; flowers in racemes, the pedicels not subtended by bracts; sepals 4, deciduous; petals 4, pink to pink purple, or white; stamens 6, the filaments lacking processes; style cone-shaped, the stigma bilobed; fruit a silique, many times longer than broad, linear, subterete, the valves 1(3)-veined; seeds numerous.

Hesperis matronalis L.
Sweet-rocket

Stems 5–12 dm tall, simple, or less commonly branched, pubescent with coarse, spreading, simple or forked hairs; leaves 1.5–15 cm long, 0.5–4 cm broad, lanceolate to ovate, dentate, usually petiolate below, becoming smaller and sessile upwards, pubescent with simple and forked hairs; pedicels 3–15 mm long, ascending to spreading; sepals 5–7 mm long, pilose; petals 15–25 mm long, pink to pink purple

Eutrema edwardsii R. Br. (× 0.4).

or white; siliques 40–100 mm long, 1–2 mm broad, somewhat constricted between the seeds, the style 1.5–3 mm long.

Cultivated and escaping; in southern and southeastern Alaska; native to Eurasia.

IBERIS L.

Plants sparsely pubescent with simple hairs or glabrous, annual, from taproots; leaves alternate, simple, entire to pinnatifid, petio- late, not auriculate-clasping; flowers in ra- cemes, the pedicels not subtended by bracts; sepals 4, deciduous; petals 4, white or pink, the outer 2 much larger than the others; stamens 6, the filaments lacking processes; style slender, the stigma obscure- ly bilobed; fruit a silicle, less than twice longer than broad, oval to orbicular in outline, compressed at right angles to the partition, broadly wing-margined, the valves reticulately veined; seeds 1 per locule.

Iberis amara L.
Rocket Candytuft

Stems erect or ascending, 1–3 dm tall, simple or branched, minutely puberulent to glabrous; leaves 1–8 cm long, 0.2–1 cm broad, irregularly lobed to pinnatifid, petiolate, reduced upwards, glabrous or sparsely pubescent and ciliate along the petiole; pedicels 3–7 mm long, spreading- ascending; sepals 1.5–2.5 mm long, gla- brous; petals 3–5 mm long, white or pink; silicles 4–7 mm long, 4–9 mm broad, gla- brous, the style 1–2 mm long.

Cultivated ornamental; escaping and per- sisting in southeastern Alaska.

LEPIDIUM L.

Plants glabrous or puberulent with simple hairs, annual or biennial, from slender tap- roots; leaves alternate and basal, pinnati- fid to subentire or entire, tapering to the base, not clasping; flowers in racemes, the pedicels not subtended by bracts; sepals 4, deciduous; petals 4 or lacking, white; stamens 2, 4, or 6, the filaments lacking processes; style obsolete or very short; fruit a silicle, about as broad as long, oval to orbicular or obovate in outline, strongly compressed at right angles to the partition, the valves not veined; seeds 1 per locule.

Hitchcock, C. L. 1936. The genus *Lepi- dium* in the United States. Madroño 3: 265–320.

Mulligan, G. A. 1961. The genus *Lepi- dium* in Canada. Madroño 16:77–90.

1a. Styles 0.2–0.3 mm long; silicles 5–6 mm long, with margins somewhat winged and upturned; plants cultivated and escaping in west-central Yukon, and to be expected elsewhere. *L. sativum* L.

1b. Styles obsolete; silicles 2–4 mm long, the margins neither winged nor upturned; plants of various distribution. (2)

2a. Petals lacking or very small; silicles oblong-obovate to obovate, widest above the middle. *L. densiflorum*

2b. Petals usually conspicuous; silicles elliptic to oval, widest at or below the middle. *L. virginicum*

Lepidium densiflorum Schrad.
Common Peppergrass

Plants puberulent, the stems 1–5 dm tall, usually branched; basal leaves 1.5–8 cm long, 0.3–2 cm broad, entire to serrate or pinnatifid, usually not present at flowering; cauline leaves reduced upwards, becoming toothed or entire; pedicels 1–4 mm long, spreading; sepals 0.8–1.2 mm long, pubescent; petals lacking, or to 1 mm long, white; silicles oblong-obovate, 2–3.5 mm long, 1.8–2.8 mm broad, shallowly notched apically, glabrous or puberulent; style obsolete.

Weedy plants of disturbed soils; in much of southern Alaska and southern Yukon; widely distributed in North America; Europe. Two rather poorly differentiated varieties are present in our region.

1a. Silicles glabrous (*L. densiflorum* var. *bourgeanum* authors). *L. densiflorum* var. *macrocarpum* Mulligan

1b. Silicles ciliate. *L. densiflorum* var. *elongatum* (Rydb.) Thell.

Lepidium virginicum L.
Wild Peppergrass

Plants pubescent to glabrate, the stems 1.5–6 dm tall, usually branched; basal leaves 2–12 cm long, 0.5–3 cm broad, pinnatifid or merely toothed, usually absent at flowering; pedicels 2–6 mm long, spread-ing; sepals about 1 mm long; petals 1–3 mm long, white; silicles elliptic to orbicular, 2.5–4 mm long, usually glabrous, the style obsolete.

Waste places; rarely collected in southeastern Alaska.

LESQUERELLA Wats.

Plants pubescent with appressed, many-rayed, stellate hairs, perennial, from well-developed taproots; leaves alternate and basal, simple, entire, tapering to the base, not clasping; flowers in subcorymbose racemes, the pedicels not subtended by bracts; sepals 4, deciduous; petals 4, yellow; stamens 6, the filaments lacking processes; style slender, persistent, the stigma obscurely 2-lobed; fruit a silicle, less than twice longer than broad, 1-veined; seeds 2–several per locule.

Payson, E. B.   1921.   A monograph of the genus *Lesquerella*. Ann. Mo. Bot. Gard. 8:102–236.

Lesquerella arctica (Wormskj.) Wats.
Arctic Bladderpod

Plants decumbent to erect, the herbage stellate-pubescent, the stems 0.4–2(2.5) dm tall; leaves spatulate to oblanceolate, 0.5–8 cm long, 0.1–1.2 cm broad, stellate on both surfaces, the cauline ones much reduced; pedicels 3–20 mm long in fruit, ascending to erect; sepals 3–4.5 mm long, stellate-pubescent; petals 3.5–7(10) mm long, yellow; silicles 5–8 mm long, 4–5.5 (7) mm broad, glabrous or stellate-pubescent; styles 1–2(2.5) mm long.

Rocky hillsides, ridges, and gravelly stream banks; from the Seward Peninsula eastward along the Brooks Range and northward along the coast; in the southeastern quarter of Alaska proper and in northern and southern Yukon; eastward to Labrador and Greenland; Asia (*L. arctica* var. *scammanae* Rollins).

1a. Petals (6)7–10 mm long and about as broad; plants of the Ogilvie Mts., Yukon, and Richardson Mts., Mackenzie. *L. arctica* var. *calderi* (Mulligan & Porsild) Welsh stat. nov. (based on: *Lesquerella calderi* Mulligan & Porsild 1969. Can. Jour. Bot. 47:215).

1b. Petals 3.5–7 mm long and about half as broad; plants broadly distributed. *L. arctica* var. *arctica*

NESLIA Desv.

Plants pubescent with forked or stellate hairs, annual, from taproots; leaves alternate, simple, serrate to entire, auriculate-clasping basally; flowers in racemes, the pedicels not subtended by bracts; sepals 4, deciduous; petals 4, light yellow; stamens 6, the filaments lacking processes; styles slender, persistent or deciduous, the stigma capitate; fruit a silicle, about as broad as long or broader, orbicular, inflated, slightly compressed parallel to the septum, the valves reticulately pitted, indehiscent; seeds usually 1 per locule.

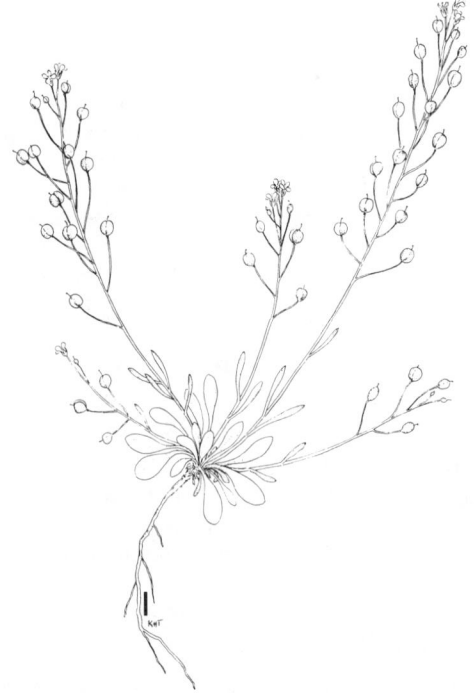

Lesquerella arctica (Wormskj.) Wats. (× 0.3).

Neslia paniculata (L.) Desv.
Ball Mustard

Plants erect, the stems 2–10 dm tall; leaves 1.5–7 cm long, 0.2–2 cm broad, lanceolate to elliptic, the cauline ones clasping basally, pubescent with stellate or branched hairs; pedicels slender, 3–25 mm long, ascending; sepals 1.2–2 mm long, greenish, glabrous or puberulent; petals 2–2.5 mm long, light yellow; siliques 2–2.5 mm long, 2.2–2.8 mm broad, glabrous, the style 0.7–1 mm long, persistent or deciduous.

Weed of cultivated land; in Alaska and Yukon (?); introduced from Europe.

PARRYA R. Br.

Plants pubescent with stipitate-glandular hairs (rarely glabrous), perennial from a simple or branched caudex and strong taproot; leaves chiefly basal, entire to dentate, lobed, or pinnatifid, tapering to a slender petiole, not auriculate-clasping; flowers in racemes, the pedicels not subtended by bracts; sepals 4, deciduous; petals 4, white, pink, or pink purple; stamens 6, the filaments lacking processes; styles stout, persistent, the stigma bilobed; fruit a silique, usually several to many times longer than broad, oblong in outline, constricted between the seeds, strongly compressed parallel to the septum, the valves with 1 main vein and reticulate lateral ones, stipitate-glandular to glabrous; seeds 1–several per locule.

Parrya nudicaulis (L.) Reg.

Caudex branches clothed with persistent leaf bases; stems 0.5–3 dm tall, erect or ascending, simple, pubescent with stipitate-glandular hairs; leaves 1.5–12 cm long, 0.4–3 cm broad, entire, dentate, lobed, or pinnatifid, stipitate-glandular to glabrous, petiolate; pedicels 3–50 mm long, ascending to erect, stipitate-glandular to glabrous; sepals 5–7.5 mm long, stipitate-glandular to glabrous, often tinged pinkish or purplish; petals 12–20 mm long, pink to pink purple or white; siliques 15–50 mm long, 3.5–7.5

Parrya nudicaulis (L.) Reg. (× 0.3).

mm broad, constricted between the seeds, the style 1.5–4 mm long.

Arctic and alpine tundra and heathlands; in most of Alaska and Yukon; east to the Mackenzie; Eurasia (*P. nudicaulis* ssp. *interior* Hultén; *P. nudicaulis* var. *grandiflora* Hultén; *P. nudicaulis* ssp. *septentrionalis* Hultén). The proposed infraspecific taxa can be recognized only arbitrarily, and it seems best to treat all of our material as belonging to a single, highly variable entity.

RAPHANUS L.

Plants sparsely pubescent with coarse, simple hairs, annual or biennial, from taproots; leaves alternate, lyrate-pinnatifid, petiolate to subsessile, not auriculate-clasping; flowers in racemes, the pedicels not subtended by bracts; sepals 4, deciduous; petals 4, pink, pink purple, or white; stamens 6, the

filaments lacking processes; styles slender, persistent, the stigma obscurely bilobed; fruit a silique, several to many times longer than broad, 2-jointed, the lower joint slender, stipelike and seedless, the upper joint several-seeded and terminating in a slender beak, indehiscent.

Raphanus sativus L.
Radish

Stems 3–10 dm tall, simple or branched; leaves 3–20 cm long, 1–8 cm broad, lyrate-pinnatifid, petiolate, reduced upwards; pedicels 8–27 mm long, ascending; sepals 7–11 mm long, glabrous; petals 13–20 mm long, pink, pink purple, or white; siliques 20–55 mm long, 5–8 mm broad, slightly if at all constricted between the seeds, not regularly breaking into 1-seeded segments, terete, slightly if at all grooved, glabrous.

Cultivated for the succulent, tuberous roots and occasionally escaping; introduced from Eurasia.

RORIPPA Scop.

Plants glabrous, or pubescent with simple hairs, annual, biennial, or perennial; leaves alternate, pinnately compound, pinnatifid, lobed, or entire, the cauline leaves often auriculate-clasping; flowers in racemes, the pedicels not subtended by bracts; sepals 4, deciduous, sometimes tardily so; petals 4, yellow or white; stamens 6, the filaments lacking processes; style short, stout, the stigma bilobed or capitate; fruit a silicle or short silique, from about as broad as long to several times longer than broad, terete or slightly compressed parallel to the septum, oval to oblong or clavate, the valves 1-nerved; seeds several to many.

1a. Petals white; plants perennial. (2)
1b. Petals yellow; plants annual or biennial. (3)

2a. Plants aquatic; leaves pinnately compound; roots fibrous. *R. nasturtium-aquaticum*
2b. Plants not aquatic; leaves simple; roots tuberous. *R. armoracia*

3a. Pedicels mostly subequal to the fruits or longer, usually over 4 mm long; fruit ovate to oblong or clavate in outline, 3–12 mm long and 2–3 mm broad; carpels 2–4; stems erect or ascending; our common *Rorippa. R. islandica*

3b. Pedicels mostly shorter than the fruits, usually less than 4 mm long; fruit oval to linear in outline, 2–16 mm long; carpels usually 2; stems decumbent to ascending or erect; plants of southeastern Alaska. (4)

4a. Fruit less than 4 times longer than broad, usually not curved. *R. obtusa*

4b. Fruit more than 5 times longer than broad, usually curved. *R. curvisiliqua*

Rorippa armoracia (L.) A. S. Hitchc.
Horseradish

Perennial, from stout, tuberous roots, the stems erect, 4–10 dm tall, glabrous; basal leaves 15–50 cm long, long-petiolate, the blades 5–15 cm broad, oblong to lanceolate, irregularly crenate-dentate to lobed, glabrous; cauline leaves reduced upwards, irregularly lobed or toothed, not auriculate-clasping; pedicels 3–10 mm long, ascending; sepals 2–3 mm long, glabrous; petals 5–6 mm long, white; fruit seldom if ever formed (reputedly a sterile triploid).

Cultivated for the root; in southern Alaska and Yukon, persisting; introduced from Europe (*Armoracia rusticana* [Lam.] Gaertn.).

Rorippa curvisiliqua (Hook.) Bessey ex Britt.

Annual or biennial, from slender taproots, the stems decumbent to ascending, 1–4 dm tall, glabrous or sparingly puberulent; leaves 2–7 cm long, 0.8–2 cm broad, oblong to oblanceolate, toothed, lobed, or pinnatifid, glabrous or puberulent; cauline leaves reduced upwards, auriculate-clasping basally; pedicels 1–5 mm long, spreading; sepals 1–2 mm long, glabrous; petals 1.5–2 mm long, yellow; fruit 6–12 mm long, 1–1.5 mm broad, curved, 2-carpelled; style to 1 mm long.

Disturbed, moist soil; in southern Alaska; southward to Baja California and eastward to Montana and Colorado.

Rorippa islandica (Oed.) Borbas
Marsh Yellowcress

Annual or biennial, from a slender taproot, the stems erect or ascending, 1.5–10 dm tall or more, glabrous or pubescent with coarse hairs; leaves 1.5–15 cm long, 0.5–4 cm broad, pinnatifid or toothed, the cauline ones auriculate-clasping basally, glabrous or pubescent; pedicels 3–8(10) mm long, spreading-ascending; sepals 1–2 mm long, glabrous, tardily deciduous, often yellowish or purplish; petals 2–3 mm long, yellowish; fruit 3–12 mm long, 2–3 mm broad, obovoid to cylindrical or clavate, 2–4 carpelled; styles to 1 mm long.

Disturbed soils along streams, roadsides, in woods, and on open slopes, in most of Alaska and Yukon; widely distributed in United States and Canada; circumboreal. The nomenclature of this species is tangled and further work is indicated (*R. palustris* [L.] Besser; *R. islandica* var. *fernaldiana* Butters & Abbe; *R. islandica* ssp. *fernaldiana* [Butters & Abbe] Hultén).

1a. Siliques usually oblong, 2-carpelled; plants widely distributed in Alaska and Yukon (*R. islandica* var. *microcarpa* [Regel] Fern.; *R. islandica* var. *williamsii* [Britt.] Hultén; *R. hispida* [Desv.] Britt.). *R. islandica* var. *hispida* (Desv.) Butters & Abbe

1b. Siliques clavate to ovoid or obovoid, often 4-carpelled. (2)

2a. Fruit less than twice longer than broad, ovoid to obovoid; plants of interior Alaska and Yukon, and of coastal western Alaska (*R. hispida* var. *barbariifolia* [DC.] Hultén). *R. islandica* var. *barbariifolia* (DC.) Welsh

2b. Fruit more than twice longer than broad, oblong to clavate; plants of coastal, southwestern, southern, and southeastern Alaska. *R. islandica* var. *occidentalis* (Wats.) Butters & Abbe

Rorippa nasturtium-aquaticum (L.) Schinz & Thell.

Watercress

Perennial, aquatic, the stems floating or decumbent, rooting at the nodes, succulent, 1–2 dm long or more, glabrous or sparsely pubescent; leaves pinnately compound, or simple only on juvenile growth, 3–15 cm long, petiolate, with 3–11 leaflets, not auriculate-clasping, glabrous; pedicels 6–15 mm long, spreading, puberulent; sepals 2.5–3.5 mm long, glabrous; petals 3–5 mm long, white; fruit 10–25 mm long, 2–3 mm broad, oblong, curved, 2-carpelled; style to 1 mm long.

In streams and springs or on mud; in interior (Manley Hot Springs) and southern (Matanuska) Alaska, to be expected elsewhere; introduced from Europe.

Rorippa obtusa (Nutt.) Britt.

Annual, from a taproot, the stems decumbent to erect, 1–4 dm long, glabrous; leaves 1.5–6 cm long, 0.5–2 cm broad, lobed to pinnatifid, the cauline ones auriculate-clasping, glabrous; pedicels 2–5 mm long, spreading to ascending; sepals 1–1.5 mm long, glabrous; petals 1–2 mm long, yellowish; fruit 3–8(10) mm long, 1.5–2.5 mm broad, oblong, straight or slightly curved, 2-carpelled; style to 1 mm long.

Moist soil; in southeastern Alaska; south to California and east to Michigan and Texas.

SISYMBRIUM L.

Plants pubescent with coarse, simple hairs, or glabrate, annual or biennial from taproots; leaves alternate, pinnatifid, petiolate or sessile, not auriculate-clasping; flowers in racemes, the pedicels not subtended by bracts; sepals 4, deciduous; petals 4, yellow; stamens 6, the filaments lacking processes; style short, stout, the stigma bilobed; fruit a silique, many times longer than broad, linear to linear-subulate, terete, the valves 1–3-veined, the apical portion produced into a short beak; seeds several to many per locule.

1a. Siliques 50–100 mm long, spreading, not appressed to the rachis; pedicels 4–10 mm long; petals 6 mm long or more. *S. altissimum*
1b. Siliques 8–15 mm long, erect and closely appressed to the rachis; pedicels 3 mm long or less; petals 4 mm long or less. *S. officinale*

Sisymbrium altissimum L.

Tumbling Mustard

Stems erect, 2.5–10 dm tall or more, simple or branched, pubescent with coarse, spreading hairs, especially below; leaves 3–20 cm long, 0.8–5 cm broad, lanceolate to oblanceolate in outline, lobed or pinnatifid below, pinnatifid into linear segments above; pedicels 4–10 mm long, spreading-ascending; sepals 3–4.5 mm long, glabrous; petals 6–8 mm long, yellow; siliques 50–100 mm long, 1–2 mm broad, linear, the stylar beak about 2 mm long.

Weed of disturbed soils; in southern Alaska and southern Yukon; introduced from Europe.

Sisymbrium officinale (L.) Scop.

Hedge Mustard

Stems erect, 2–8 dm tall or more, simple or branched, pubescent with coarse, spreading hairs; leaves 4–25 cm long, 1–6 cm broad, oblanceolate in outline, lyrate-pinnatifid, reduced upwards; pedicels 1–3 mm long, erect; sepals 1.5–2.5 mm long, glabrous; petals 3–4 mm long, yellowish; siliques 8–15 mm long, 1–1.5 mm broad, linear-subulate, the stylar beak about 2 mm long.

Weed of disturbed soils; in southeastern Alaska; introduced from Europe.

SMELOWSKIA C. A. Mey.

Plants pubescent with branched hairs, at least in part, perennial, arising from a simple or branched caudex and stout taproot; leaves alternate or chiefly basal, entire to palmately lobed or pinnatifid, tapering to the base, not clasping; flowers in subcorymbose to elongate racemes, the pedicels not subtended by bracts; sepals 4, deciduous or persistent; petals 4, white or purplish; stamens 6, the filaments lacking processes; fruit a silique (or silicle), 3–several times longer than broad, terete or variously compressed, the valves 1-nerved; seeds few to several per locule.

Drury, W. H. & R. C. Rollins. 1952. The North American representatives of *Smelowskia*. Rhodora 54:85–119.

1a. Caudex usually branched, less than 5 mm in diameter; stems mostly simple above the caudex; pedicels spreading to ascending. *S. calycina*

1b. Caudex usually simple, mostly 5 mm in diameter or more; stems branched; pedicels spreading to ascending or curved. (2)

2a. Basal leaves palmately 3–5-lobed; petals purple; sepals 2–3 mm long; siliques ovate to oblong or linear. *S. borealis*

2b. Basal leaves pinnately divided to the midrib, 7–9-lobed; petals white; sepals 1–1.5 mm long; siliques pear-shaped. *S. pyriformis*

Smelowskia borealis (Greene) Drury & Rollins

Caudex stout, 5 mm in diameter or more, simple, clothed with old leaf bases; stems 4–20(25) cm tall, branched from near the base, pubescent with branched and simple hairs; basal leaves 5–20 mm long, 5–12(15) mm broad, oblong to ovate or spatulate, palmately 3–7-lobed, densely pubescent; cauline leaves reduced upwards; pedicels 5–20 mm long, spreading or curved, pubescent like the stems; sepals 1.5–3 mm

long, pilose; petals 3–6 mm long, pink purple to purple; siliques 5–15(18) mm long, (2)3–6 mm broad, obovate to oblong, glabrous, flattened at right angles to the partition; styles 0.2–1.2 mm long.

Mountains of northern, central, and eastern Alaska and northern and western Yukon; east to Mackenzie; endemic. Several varieties are recognizable among our specimens (*Melanidion boreale* Greene; *Ermania borealis* [Greene] Hultén).

1a. Sepals caducous; styles 0.5 mm long or less; rachis and pedicels white-villous; plants from the Brooks Range. *S. borealis* var. *jordalii* Drury & Rollins

1b. Sepals persistent; styles 0.5 mm long or more; rachis and pedicels various; plants of other distribution. (2)

2a. Pubescence of herbage mostly of white-villous hairs; siliques oblong, the valves rigid; petals 4–6 mm long; plants of Mt. McKinley National Park. *S. borealis* var. *villosa* Drury & Rollins

2b. Pubescence of herbage not mostly of white-villous hairs; siliques ovate to obovate, the valves various; petals 3–4.5 mm long. (3)

3a. Siliques rigid, not inflated; plants of northern Yukon and Mackenzie. *S. borealis* var. *borealis*

3b. Siliques membranous, inflated; plants of Mt. McKinley National Park and vicinity. *S. borealis* var. *koliana* (Gambocz) Drury

Smelowskia calycina (Steph.) C. A. Mey.

Caudex usually branched, the branches mostly less than 5 mm in diameter, clothed with old leaf bases; stems 1.5–20 cm tall, simple, sparsely to densely pubescent with branched and simple hairs; basal leaves 5–45 mm long, 1–15 mm broad, linear to cuneate or obovate, entire to pinnately divided or tridentate, sparsely to densely pubescent; cauline leaves reduced upwards,

Smelowskia calycina (Steph.) C. A. Mey. (× 0.7).

often more narrowly lobed than the basal ones; pedicels 1–10 mm long, ascending to spreading, pubescent with usually simple hairs; sepals 2–3 mm long, pilose; petals 3–8 mm long, white to purplish; siliques 5–12 mm long, 1–2.5 mm broad, linear to oblong or elliptic, glabrous (rarely pubescent), terete or somewhat compressed parallel to the septum; styles 0.3–1.5 mm long.

Mountains; in western, northern, and northeastern Alaska and northern Yukon; east to Mackenzie and south to Washington, Nevada, Utah, and Colorado.

1a. Basal leaves pinnately lobed (some leaves occasionally almost entire); cauline leaves pinnately lobed; plants of northeastern Alaska and northern Yukon. S. *calycina* var. *media* Drury & Rollins

1b. Basal leaves entire or merely shallowly lobed at the tips; cauline leaves entire to pinnately lobed; plants of other distribution. (2)

2a. Blades of basal leaves obovate to oval; petioles shorter than the leaf blades; plants of northwestern and western Alaska (S. *calycina* ssp. *integrifolia* [Seem.] Hultén). S. *calycina* var. *integrifolia* (Seem.) Rollins.

2b. Blades of basal leaves linear to narrowly spatulate; petioles longer than the leaf blades; plants of north-central Alaska. S. *calycina* var. *porsildii* Drury & Rollins

Smelowskia pyriformis Drury & Rollins

Caudex stout, 5 mm in diameter or more, simple, clothed with old leaf bases; stems 5–14 mm long, branched from near the base, pubescent with branched and simple hairs; basal leaves 25–40 mm long, 5–12 mm broad, oblong to ovate in outline, pinnately divided into 7–9 segments, densely pubescent; cauline leaves reduced upwards; pedicels 3–7 mm long, spreading to ascending, pubescent like the stem; sepals 1.5–2 mm long, sparsely pilose; petals 2.8–

3.2 mm long, white or cream; siliques 5–8 mm long, 1.5–2 mm broad, pear shaped, glabrous, terete, the style to 0.5 mm long.

Rocky slopes; in the Farewell Mountains in the Kuskokwim River Drainage, Alaska; endemic.

SUBULARIA L.

Plants glabrous aquatic annuals with fibrous roots; leaves basal, simple, awl-shaped, tapering to the apex; flowers in racemes, the pedicels not subtended by bracts; sepals 4, deciduous; petals 4, white; stamens 6, the filaments not bearing processes; fruit a silicle, less than twice longer than broad, obovoid to ellipsoid, slightly compressed at right angles to the septum, the valves nerveless; seeds few to several per locule.

Subularia aquatica L.
Awlwort

Plants erect, the herbage glabrous, the stems 0.1–1 dm tall; leaves linear-subulate, 0.8–5 cm long, about 0.1 cm broad or less; pedicels ascending to erect, 1–2 mm long; sepals 0.5–0.8 mm long, glabrous; petals 0.2–1 mm long, white; silicles 2–3 mm long, 1.5–2 mm broad, glabrous, the style obsolete.

Shallow ponds; mostly in coastal and insular southern Alaska, but also in interior southern Alaska and Yukon; eastward to the Atlantic and south to California and Wyoming; Eurasia (S. *aquatica* ssp. *americana* Mulligan & Calder).

THLASPI L.

Plants glabrous, annual or perennial, from taproots; leaves alternate or chiefly basal, simple, entire to dentate or lobed, the cauline ones auriculate-clasping; flowers in racemes, the pedicels not subtended by bracts; sepals 4, deciduous; petals 4, white; stamens 6, the filaments lacking processes; style slender and conspicuous or obsolete, the stigma obscurely bilobed; fruit a silicle,

less than 3 times longer than broad, orbicular, obovate, oblanceolate or elliptic in outline, compressed at right angles to the septum, the valves reticulately veined and often wing-margined; seeds 2–several per locule.

Holmgren, P. K. 1971. A biosystematic study of North American *Thlaspi montanum* and its allies. Mem. N. Y. Bot. Gard. 21(2):1–106.

Payson, E. B. 1926. The genus *Thlaspi* in North America. Univ. Wyo. Publ. Bot. 1:145–86.

1a. Plants perennial, 0.5–1.5 dm tall; silicles not wing-margined; mostly of alpine or arctic sites. *T. arcticum*

1b. Plants annual, 1–5 dm tall; silicles conspicuously wing-margined; weedy species of disturbed soils. *T. arvense*

Thlaspi arcticum Porsild

Perennial, the stems 0.5–1.5 dm tall, simple, glabrous; basal leaves 0.8–2.5 cm long, 0.2–0.8 cm broad, oblanceolate to obovate, entire; cauline leaves reduced upwards; pedicels 2–8 mm long, spreading to ascending; sepals 2–3 mm long, glabrous; petals 4–5 mm long, white; silicles 5–8 mm long, 2–3 mm broad, not strongly compressed, the valves keeled but not wing-margined, obovoid to ellipsoid, glabrous; style 0.5–1 mm long.

Arctic and alpine tundra; known from widely disjunct localities in northeastern and southeastern Alaska, and in northern and southwestern Yukon; endemic.

Thlaspi arvense L.
Pennycress

Annual, the stems 1–5 dm tall or more, simple or branched, glabrous; basal leaves 1.5–7 cm long, 1–2 cm broad, oblanceolate, dentate to subentire; cauline leaves becoming sessile upwards; pedicels 3–15 mm long, spreading to ascending; sepals 1.5–2.5 mm long, glabrous; petals 3–4 mm long, white; silicles 9–18 mm long, 9–15 mm broad, strongly compressed, conspicuously wing-margined, oval to subcordate, emarginate; style almost obsolete.

Disturbed soils; in southern Alaska and Yukon; widely distributed in North America; Eurasia.

DIAPENSIACEAE
Diapensia Family

Plants low-growing evergreen shrubs (appearing herbaceous); leaves alternate, imbricated and rosettelike on the branches, simple; flowers perfect, regular, solitary, borne erect on subscapose peduncles; calyx 5-lobed, the sepals distinct or nearly so; corolla 5-lobed, the lobes shorter than or subequal to the tube; stamens 5, inserted on the corolla, alternate with the lobes; pistils 1, the ovary superior, 3-loculed; styles 1, the stigma 3-lobed; fruit a loculicidal capsule.

DIAPENSIA L.

Plants depressed, mat-forming; leaves spatulate to oblanceolate, persistent, usually curved and somewhat revolute, pale beneath; flowers showy; calyx immediately subtended by 3 bracts, persistent and investing the fruit; corolla campanulate, deciduous; staminal filaments short and broad, the anthers with a broad connective, opening throughout; capsules with numerous seeds.

Diapensia lapponica L.
Diapensia

Stems prostrate, much branched, forming a mat or tuft; leaves crowded, overlapping, spatulate to oblanceolate, mostly 5–12 mm long, entire, rounded apically, cuneate basally; flowering stems 0.5–4 cm long, leafless or with 1–few small leaves; flowers solitary, immediately subtended by 3 bracts; sepals 4–7 mm long, oblong to spatulate, becoming woody and enclosing the fruit at maturity; corolla white to cream, 9–12 mm long; capsules 4–6 mm long, ovoid.

Arctic and alpine tundra and heathlands; in most of Alaska and in western Yukon; circumboreal (*D. lapponica* ssp. *obovata* [F. Schmidt] Hultén; *D. lapponica* var. *rosea* Hultén). Our material is assignable to var. *obovata* F. Schmidt.

DROSERACEAE
Sundew Family

Perennial or biennial (annual?) herbs; leaves alternate, in basal rosettes, simple, the surfaces covered with viscid, stalked glands which trap insects; flowers perfect, regular, borne in racemes; sepals 4–5, distinct or connate basally, persistent; petals usually 5, distinct, usually persistent; stamens 4–10; pistils 1, the ovary superior, 1-loculed, 3–5-carpelled; styles 3–5, distinct, often forked; fruit a loculicidal capsule.

DROSERA L.

Plants acaulescent scapose herbs; leaves

Diapensia lapponica L. (× 0.5).

long-petioled, the blades orbicular or spatulate to linear or oblong; flowers 1–several, in 1-sided racemes (cymes); sepals 5, petals 5; stamens usually 5; capsules with numerous seeds.

1a. Leaves spatulate to linear or oblong, the blades more than thrice longer than broad. *D. anglica*

1b. Leaves oval to orbicular, the blades as broad as long or broader. *D. rotundifolia*

Drosera anglica Huds.
Long-leaf Sundew

Plants 0.5–1.7 cm tall; leaves 1.5–9 cm long, the blades spatulate to oblong or linear, 0.5–3 cm long, 0.1–0.4 cm broad; flowers mostly 2–7; sepals 3–5 mm long, connate below; petals 5–8 mm long, white; capsules 5–7 mm long.

Bogs and swamps; in interior and coastal southern Alaska (rarely northward) and in southern Yukon; south to California, Nevada, and Montana, and in eastern Canada; Eurasia.

Drosera rotundifolia L.
Round-leaf Sundew

Plants 0.5–1.8 dm tall; leaves 1–6 cm long, the blades oval to orbicular, 0.4–1 cm long, 0.3–1 cm broad; flowers mostly 2–8; sepals 2–5 mm long, connate basally; petals 4–6 mm long, white; capsules 5–7 mm long.

Bogs, swamps, and wet meadows; in much of Alaska south of the 67th parallel and in southern Yukon; east to the Atlantic and south to California; Eurasia.

ELAEAGNACEAE

Oleaster Family

Shrubs or small trees with stellate pubescence; leaves simple, alternate or opposite, entire; flowers perfect or imperfect, regular, borne in axillary clusters; perianth 4-lobed, in a single whorl, from the apex of the hypanthium; stamens 4 or 8; **pistils 1**, the ovary superior (though often **appearing** inferior), 1-loculed; styles and stigmas 1; fruit a dry, indehiscent achene enveloped by the fleshy, persistent perianth, hence drupelike.

1a. Leaves alternate; flowers perfect or polygamous; stamens 4. *Elaeagnus*

1b. Leaves opposite; flowers imperfect

Drosera rotundifolia L. (× 0.5).

(or rarely perfect); stamens 8. *Shepherdia*

ELAEAGNUS L.

Shrubs with ascending alternate branches; leaves alternate, silvery on both surfaces, sparsely to rather densely brownish-scaly; flowers axillary, perfect or polygamous; perianth tubular, investing the ovary and extending beyond as a tube usually longer than the spreading lobes, deciduous above the ovary, not glandular-thickened at the apex of the lobes; stamens 4, borne near the apex of the hypanthium; ovary wall bony, enclosed by the fleshy hypanthium base; fruit drupelike.

Elaeagnus commutata Bernh.
Silverberry

Shrubs mostly 1–2 m tall, the branchlets brownish-scaly; leaves 1.3–7 cm long, 0.5–3 cm broad, elliptic to oblanceolate or lanceolate, acute to obtuse or rounded apically, acute to obtuse basally, silvery on both sides and somewhat brownish-scaly; flowers 1–4 per axil, 10–15 mm long, the lobes yellowish; stamens 4, alternating with the perianth lobes; fruits 6–10 mm long, the hypanthium base mealy, silvery.

Sandy stream banks, lake shores, flood plains, road cuts, and in open woods or thickets; in the southeastern half of mainland Alaska and in most of the Yukon; east to Quebec and south to Idaho and Utah.

SHEPHERDIA Nutt. Nom. Cons.

Shrubs with spreading-ascending, opposite branches; leaves opposite, green above and pale beneath, sparsely to rather densely brownish-scaly; flowers axillary, imperfect (or rarely perfect); perianth short-tubular, investing the ovary in pistillate flowers, not deciduous above the ovary; staminate flowers with 8 stamens alternating with glandular thickenings at the base of the perianth lobes; ovary wall somewhat woody, enclosed by the juicy hypanthium; fruit drupelike.

Shepherdia canadensis (L.) Nutt.
Soapberry

Shrubs mostly 1–2 m tall, the branchlets brownish-scaly; leaves 0.5–8 cm long, 0.3–3 cm broad, ovate to lanceolate, rounded apically and basally, green above and pale

Elaeagnus commutata Bernh. (× 0.6).

beneath, sparsely to rather densely brownish-scaly; flowers 1–several per axil, 2–3 mm long, the lobes brownish; stamens 8; fruit 4–7 mm long, juicy, red, bitter.

Woods, thickets, and less commonly in heathlands; in much of Alaska (except for the southwestern portion) and most of the Yukon; south to Oregon, Utah, and Colorado and east to Newfoundland and New York.

EMPETRACEAE
Crowberry Family

Low, evergreen, heathlike shrubs; leaves alternate or whorled, simple, leathery and persistent; flowers perfect or imperfect, regular, mostly axillary; calyx of 3–6 sepals, in 2 whorls (the inner whorl sometimes interpreted as petals), subtended by 3 bracts; petals apparently lacking; stamens 2–4; pistils 1, the ovary superior, with 2–several carpels and locules; styles 1, the stigma lobed; fruit a fleshy, 2–several-seeded berry.

EMPETRUM L.

Stems prostrate to decumbent; leaves alternate, whorled, or subverticillate, not 4-ranked, overlapping, short-petiolate, with bulbous-decurrent bases on the stem; flowers inconspicuous, sessile in the axils; perianth purplish, 6-segmented, the segments distinct; staminal filaments filiform, glabrous, the anthers long-exserted, opening by slits, unawned; fruit black at maturity, each locule 1-seeded.

Empetrum nigrum L.
Crowberry

Stems 1–5 dm long, mat-forming, the branches ascending, hairy; leaves 3–7 mm long, spreading-ascending, oblong, the margin sharp to blunt, minutely glandular, alternate or whorled (especially on long shoots); flowers solitary, 2–3 mm long, immediately subtended by 3 bracts; fruit black, 6–9 mm broad, more or less edible.

In bogs and swamps from low to high elevations, in heathlands, and in tundra; in most of Alaska and Yukon; east to Newfoundland and south to California; circumboreal. The flowers of our plants are either perfect, monoecious, or dioecious and have received a variety of taxonomic treatments. In the species at large, two chromosome compliments are known (n=26, reported from North America and n=13, reported for the Old World). Evidently most Old World specimens have imperfect flowers and many of ours have perfect flowers. Our plants with perfect flowers are designated as var. *hermaphroditum* (Lge.) Sor. (*E. jamesii* Fern. & Wieg.; *E. nigrum* ssp. *hermaphroditum* [Lge.] Bocher). Those with imperfect flowers are designated as belonging to var. *nigrum*.

Empetrum nigrum L. (× 0.5).

ERICACEAE

Heath Family

Shrubs or subshrubs; leaves simple, alternate or opposite, often leathery and persistent; flowers perfect, regular or nearly so, axillary, in terminal clusters, or solitary; sepals mostly 4–5, distinct or more usually connate; petals mostly 4–5, connate or distinct, the corolla rotate to funnelform or urn shaped; stamens as many as the corolla lobes and alternate with them or twice as many, the anthers dehiscent by terminal pores or by longitudinal slits; pistils 1, the ovary superior or inferior, usually with 4–10 carpels and locules; styles 1, the stigma capitate or lobed; fruit a capsule or a berry.

1a. Ovary inferior or apparently so. (2)
1b. Ovary superior. (4)

2a. Plants evergreen; ovary in reality superior, but surrounded by the fleshy, purplish black or whitish calyx when mature. *Gaultheria*
2b. Plants mostly deciduous; ovary truly inferior; fruit red or blue. (3)

3a. Petals apparently distinct, sharply reflexed; plants slender, trailing evergreens. *Oxycoccus*
3b. Petals united into a distinct tube; plants ascending to erect, mostly deciduous (evergreen in V. *vitisidaea*). *Vaccinium*

4a. Leaves less than 6 mm long, heathlike, thick and imbricate, usually 4-ranked on the stem. *Cassiope*
4b. Leaves mostly with well-developed blades, if less than 6 mm long and heathlike, then not 4-ranked. (5)

5a. Corolla with petals distinct or nearly so, not at all tubular. (6)
5b. Corolla of united petals, urn shaped, bowl shaped, or campanulate. (7)

6a. Flowers in terminal corymbs; petals white, less than 8 mm long; leaves evergreen, villous-tomentose with yellowish, reddish, or brownish hairs, revolute. *Ledum*
6b. Flowers solitary, terminating lateral branchlets (rarely axillary); petals salmon or copper colored, 10–15 mm long; leaves deciduous, glabrous beneath, flat. *Cladothamnus*

7a. Corolla broadly funnelform to saucer shaped or rotate, not constricted at the apex. (8)
7b. Corolla campanulate to urn shaped, more or less constricted at the apex. (10)

8a. Leaf blades 2–8 mm long, distinctly opposite; corolla 4–5 mm long. *Loiseleuria*
8b. Leaf blades mostly over 10 mm long, alternate or opposite; corolla 6–28 mm long. (9)

9a. Leaves either brownish-scurfy or long-ciliate on the margin, alternate; corolla lobes longer than the tube. *Rhododendron*
9b. Leaves neither brownish-scrufy nor long-ciliate on the margin, opposite; corolla lobes shorter than the tube. *Kalmia*

10a. Pedicels with 2 bracts immediately below the flowers; leaves and calyces scurfy-pubescent. *Chamaedaphne*
10b. Pedicels without bracts subtending the flowers, but often bearing bracts at the base of the pedicel; leaves and calyces not scurfy-pubescent. (11)

11a. Leaves mostly 10 mm long or less, the bases bulbous-decurrent on the stem. *Phyllodoce*
11b. Leaves over 10 mm long, the bases not bulbous-decurrent on the stem. (12)

12a. Flowers 4-merous; plants ascending to erect, 5–15 dm tall or more, the young herbage stipitate-glandular; leaves deciduous. *Menziesia*
12b. Flowers 5-merous; plants decumbent to erect, mostly less than 3 dm tall

(the trailing stems often much longer); the young herbage not stipitate-glandular, or the leaves evergreen. (13)

13a. Plants maritime; anthers 4-awned; fruit a capsule embedded in a fleshy calyx. *Gaultheria*

13b. Plants usually not maritime; anthers 2-awned; fruit a capsule or a berry, the calyx not fleshy. (14)

14a. Leaves entire, the margins revolute, glaucous beneath; flowers in umbels, the pedicels 5–20 mm long, recurved apically (at least in flower); plants of bogs. *Andromeda*

14b. Leaves serrate, crenate, or entire, flat, not glaucous beneath; flowers in racemes, the pedicels 2–6 mm long, straight or curved; plants not of bogs. *Arctostaphylos*

ANDROMEDA L.

Small shrubs with creeping, hypogaeous rhizomelike stems and erect leafy branches; leaves alternate, evergreen, leathery, entire, revolute; flowers in terminal umbels, perfect, regular; calyx 5-lobed, connate at the base; corolla urn shaped, the 5 lobes much shorter than the tube, reflexed or spreading; stamens 10, included; filaments flattened, tapering to the apex, pubescent; anthers opening by terminal pores, with 2 slender awns; ovary superior, 5-carpelled, the stigma capitate; capsule 5-loculed, 5-valved, loculicidally dehiscent.

Andromeda polifolia L.
Bog Rosemary

Shrubs mostly 0.5–2.5 dm tall above ground; leaves (0.8)1–4 cm long, 0.2–0.6 cm broad, narrowly elliptic to oblong, entire, revolute, the lower surface glaucous; umbels (1)2–6-flowered, the pedicels 5–20 mm long, recurved apically; sepals 1–2 mm long, pink; corolla 5–7 mm long, pink; capsule 4–6 mm thick, glabrous.

Bogs, swamps, moist meadows, or wet tun-

dra; in most of Alaska (except for the Aleutians) and throughout the Yukon; east to Newfoundland and south to British Columbia, Alberta, and the Great Lakes; circumboreal (*A. polifolia* var. *concolor* B. Boi.; *A. polifolia* f. *concolor* [B. Boi.] B. Boi.).

ARCTOSTAPHYLOS Adans.

Prostrate to ascending shrubs with creeping, stolonlike stems and ascending leafy branches; leaves alternate, evergreen or dying at the end of one season, leathery or thin, entire or crenate-serrate, flat, not glaucous beneath; flowers in terminal racemes, perfect, regular; calyx 5-lobed, connate at the base; corolla urn shaped, the

Andromeda polifolia L. (× 0.7).

lobes much shorter than the tube, reflexed or spreading; stamens 10, included, the filaments flattened, tapering to the apex, pubescent; anthers opening by apparently terminal pores, with 2 recurved awns; ovary superior, usually 5-loculed; fruit a berry.

1a. Leaves crenate-serrate, becoming brightly colored in autumn, withering and persistent below leaves of the current year. *A. alpina*

1b. Leaves entire, not brightly colored in autumn, evergreen, leathery, not persisting as dead leaves below the current leaves. *A. uva-ursi*

Arctostaphylos alpina (L.) Spreng.
Alpine Bearberry

Prostrate shrub with subrhizomatous stems, mat-forming, the branches erect, short, the internodes short; leaves apparently in rosettes, the dead ones of previous seasons persistent below those of the current season, 1–9 cm long, 0.5–3 cm broad, oblanceolate to obovate, rounded to obtuse or rarely acute apically, cuneate to acute basally, crenate-serrate, the petiole short to moderately long; racemes few-flowered, the pedicels 3–6 mm long, straight, borne in the axils of ciliate bracts; sepals 1–2 mm long; corolla 4–6 mm long, white to greenish white (fading yellowish) or yellowish; berries red or black, 5–10 mm thick.

Arctic and alpine tundra, heath, or woods; in most of Alaska and throughout the Yukon; east to Labrador and south to British Columbia and New Hampshire; circumboreal. Our plants have been distinguished as belonging to two taxa, primarily on the basis of fruit color. Apparently there are no other really definitive features by which the proposed segregates can be distinguished when plants lack mature fruit. However, there do appear to be developmental trends which tend to set the entities apart, and it seems worthwhile to recognize them at some taxonomic rank.

1a. Fruit black; leaves mostly 1–5 cm long and 0.5–1.7 cm broad, the short

petioles often prominently ciliate; plants mostly of alpine sites. *A. alpina* var. *alpina*

1b. Fruit red; leaves mostly 2.7–9 cm long and 1.2–3 cm broad, the usually well-developed petioles often glabrous; plants mostly of lower elevations. (*A. erythrocarpa* Small; *A. rubra* [Rehd. & Wils.] Fern.). *A. alpina* var. *rubra* (Rehd. & Wils.) Bean

Arctostaphylos uva-ursi (L.) Spreng.
Bearberry, Kinnikinnick

Prostrate shrub with substoloniferous rooting stems, mat-forming, the branches ascending, the internodes usually apparent, puberulent and sometimes glandular; leaves not in rosettes, evergreen, not persistent after dying, mostly 1–3 cm long and 0.3–1.2 cm broad, oblanceolate to obovate, rounded to obtuse or rarely acute apically, cuneate to acute basally, entire, the petiole short; racemes few-flowered, the pedicels 2–5 mm long, straight or curved, borne in the axils of pubescent bracts; sepals 1–1.5 mm long; corolla 4–6 mm long, pink; berries red, 5–10 mm broad.

Exposed slopes and flats in open woods or heath; in much of Alaska (except for the western and northwestern portions) and most of the Yukon; east to the Atlantic and south to California, New Mexico, Illinois, and Georgia; Eurasia (*A. uva-ursi* var. *coactilis* Fern. & Macbr.; *A. uva-ursi* var. *adenotricha* Fern. & Macbr.; *A. uva-ursi* var. *pacifica* Hultén).

CASSIOPE D. Don

Shrubs with decumbent, rhizome- or stolon-like stems and ascending to erect branches; leaves opposite or alternate, 4-ranked, evergreen, leathery, overlapping, flat or concave above, convex or grooved below, sessile or short-petiolate; flowers solitary, arising from axillary or terminal buds, the pedicels short to long, recurved apically; sepals 5, nearly distinct; corolla campanulate, the 5 lobes shorter or longer than the tube, spreading; stamens 10, included, the filaments flat, glabrous; anthers opening by apparently terminal pores, with 2 recurved awns; ovary superior, 5-carpelled; capsule 5-valved, loculicidally dehiscent.

Good, R. D'O. 1926. The genera *Phyllodoce* and *Cassiope*. Lond. Jour. Bot. 64: 1–10.

1a. Pedicels subequal to the subtending leaves or shorter; leaves flat above, alternate, spreading; flowers terminal. *C. stelleriana*

1b. Pedicels much longer than the subtending leaves; leaves deeply concave above, opposite, appressed; flowers lateral. (2)

2a. Lower leaf surface grooved nearly the entire length, often pubescent with short, white hairs; plants of broad distribution, but seldom in coastal southern Alaska. *C. tetragona*

2b. Lower leaf surface convex, not longitudinally grooved, glabrous or pubescent near the base; plants of coastal and insular southern Alaska. (3)

3a. Main leaves 1.5–3 mm long, with scarious margins, often pubescent apically with long, brownish hairs and commonly pubescent within; plants widely distributed in coastal and insular southern Alaska. *C. lycopodioides*

3b. Main leaves 3–5 mm long, the margins not scarious or only slightly so, lacking long apical hairs; plants of southeastern Alaska. *C. mertensiana*

Cassiope lycopodioides (Pallas) D. Don
Club-moss Mountain-heather

Stems prostrate or decumbent, mat-forming, the branches 0.5–2 dm long; leaves opposite, appressed, 4-ranked, the main ones 1.5–3 mm long, scarious-margined, often ciliate and apically pubescent with long hairs, the concave upper surface often pubescent, the convex lower surface not grooved, sessile; flowers solitary, arising from axillary buds, the pedicels longer than the subtending leaves; sepals 2–2.5 mm long, scarious-margined; corolla 5–7 mm long, cream to white, the lobes shorter than the tube.

Alpine tundra to heath; in coastal and insular southern Alaska from the Aleutians to the vicinity of Juneau; Asia (*C. lycopodioides* ssp. *cristapilosa* Calder & Taylor).

Cassiope mertensiana (Bong.) D. Don
Mertens Mountain-heather

Stems mat-forming, the branches ascending, 0.5–3 dm long; leaves opposite, appressed, 4-ranked, the main ones 3–5 mm long, not scarious-margined, rarely ciliate, not apically pubescent with long hairs, the concave upper surface usually glabrous, the convex lower surface not grooved, sessile; flowers solitary, arising from axillary buds, the pedicels longer than the subtending leaves; sepals 2.5–3 mm long, scarious-margined; corolla 6–8 mm long, cream to white, the lobes shorter than the tube.

Alpine tundra and heath; in southeastern Alaska; south to California and east to Montana. *C. mertensiana* is similar to *C. lycopodioides*, with which it occurs in southeastern Alaska. The presence of thin scarious margins in some specimens of *C. mertensiana* suggests that there is some genetic exchange between the two species.

Cassiope stelleriana (Pallas) DC.
Alaska Moss Heath

Stems prostrate to decumbent, the branches 0.5–1.5 dm long, ascending to erect; leaves alternate, spreading, more or less 4-ranked, the main ones 2.5–4.5 mm long, not distinctly scarious-margined (except when young), not ciliate, the flat upper surface glabrous, the convex lower surface not grooved, petiolate, decurrent on the stem; flowers solitary, arising from terminal buds, the pedicels subequal to or shorter than the leaves; sepals 3–3.5 mm long, somewhat scarious-margined; corolla 5–7 mm long, white to pinkish, the lobes longer than the tube.

Alpine tundra and heath; mostly in coastal, southern and south-central Alaska and in southwestern Yukon; south to Washington; Asia.

Cassiope tetragona (L.) D. Don
Four-angle Mountain-heather

Stems spreading, the branches 0.5–3 dm long, ascending to erect; leaves opposite, appressed, 4-ranked, the main ones 2–6 mm long, not scarious-margined, ciliate but not apically pubescent with long hairs, usually pubescent and grooved on the outer (lower) surface, glabrous or glandular on the concave inner (upper) surface, sessile; flowers solitary, arising from axillary buds, the pedicels longer than the subtending leaves; sepals 2–2.5 mm long, whitish; corolla (4)5–8 mm long, cream to white, the lobes shorter than the tube.

Alpine and arctic tundra; in most of Alaska, except for the southern and southwestern coastal regions, and in most of Yukon; east to Labrador and south to New Hampshire and Washington; Eurasia. This is a remarkably pretty plant when in full flower, but its odor is offensive. The great bulk of our material belongs to var. *tetragona*. The var. *saximontana* enters southern Yukon and southeastern Alaska from the Rocky Mountains of British Columbia.

1a. Flowers 4–5(6) mm long; pedicels seldom over twice longer than the subtending leaves, not overtopping the branch on which it is borne (*C. tetragona* ssp. *saximontana* [Small] Porsild). *C. tetragona* var. *saximontana* (Small) C. L. Hitchc.

1b. Flowers 5–8 mm long; pedicels more than twice longer than the subtending leaves, overtopping the branch on which it is borne. *C. tetragona* var. *tetragona*

CHAMAEDAPHNE Moench

Erect or ascending shrubs, rooting along the lower stem; leaves alternate, evergreen, leathery, serrulate, flat or slightly revolute, scurfy-pubescent with collapsed globular hairs; flowers in terminal, leafy-bracted racemes, perfect, regular; calyx of 5 distinct sepals, immediately subtended by 2 bracts; corolla campanulate or shortly cylindrical, the lobes shorter than the tube; stamens 10, included, the filaments flattened, glabrous; anthers opening by apparently terminal pores at the ends of tubular beaks, not awned; ovary superior, usually 5-loculed; fruit a loculicidally dehiscent, 5-valved capsule.

Chamaedaphne calyculata (L.) Moench
Leatherleaf

Stems 2–12 dm tall, the branchlets pubescent with short, white hairs and sometimes scurfy as well; leaves 0.7–4 cm long, 0.3–1.5 cm broad, oblanceolate to oblong or elliptic, rounded to obtuse apically, acute basally, serrulate, flat or revolute, scurfy-pubescent; racemes poorly differentiated, the flowers borne in the axils of reduced foliage leaves (bracts), the pedicels 1–4 mm long, straight, bearing 2 bracts at the base of the calyx; sepals 2–3 mm long, scurfy-pubescent; corolla 4.5–6 mm long, white; capsules 3–4 mm broad.

Bogs, lake shores, wet meadows, and moist tundra; in much of Alaska south of the 68th parallel (occasional north of the Brooks Range), and in much of the Yukon; east to Newfoundland and south to British Columbia, Iowa, and Georgia; Eurasia (*C. calyculata* var. *nana* [Lodd.] Busch.).

CLADOTHAMNUS Bong.

Erect shrubs with ascending branchlets, puberulent in decurrent lines below the leaf bases; leaves alternate, deciduous, entire, flat, pale below; flowers solitary, terminating lateral branchlets, or rarely with a second flower borne laterally below the terminal one, perfect, regular; sepals 5, distinct or nearly so; corolla rotate, the petals distinct; stamens usually 10, the filaments flattened, glabrous; anthers dehiscing about half the length, unawned; ovary superior, 5-carpelled; fruit a 5-loculed, septicidally dehiscent capsule.

Cladothamnus pyrolaeflorus Bong.
Copper-bush

Stems 5–15 dm tall; leaves 1–5 cm long, 0.4–1 cm broad, oblanceolate to elliptic, obtuse and mucronate apically, acute basally, entire, flat, the lower surface pale, short-petioled; flowers showy; sepals 4–10 mm long, ciliate; petals 10–15 mm long, salmon or copper colored; capsules 5–9 mm broad.

Woods and thickets, in coastal regions; from Prince William Sound eastward through southeastern Alaska; south to Oregon.

GAULTHERIA L.

Prostrate to erect shrubs; leaves alternate, evergreen, leathery, serrate, flat; flowers in racemes, perfect, regular; calyx 5-lobed, united, enlarging and becoming fleshy at maturity; corolla urn shaped to subglobose, the lobes shorter than the tube; stamens usually 10(8), included, the filaments flattened, tapering to the apex, hairy or glabrous; anthers opening by terminal pores, 4-awned; ovary superior, usually 5-loculed; fruit a loculicidally dehiscent capsule, enclosed by the fleshy, expanded calyx.

1a. Calyx glabrous, white at maturity; plants nearly glabrous, known from the western Aleutians. *G. miqueliana*

1b. Calyx pubescent, purplish black at maturity; plants glandular-pubescent, at least in the inflorescence, from southeastern Alaska. *G. shallon*

Gaultheria miqueliana Takeda

Stems 1–3.5 dm long, procumbent, the branches ascending; leaves 1–3.5 cm long, 0.7–1.6 cm broad, broadly elliptic to oblong or obovate, obtuse apically, subacute to obtuse basally, serrate; racemes few-flowered, 2–6 cm long, the pedicels 5–7 mm long, recurved; calyx enlarging and surrounding the capsule, white at maturity, glabrous; corolla subglobose, 5–6 mm long, white; capsules about 6 mm broad.

Mountain slopes; in the western Aleutian Islands (Kiska); Asia.

Cladothamnus pyrolaeflorus Bong. (× 0.6).

Gaultheria shallon Pursh
Salal

Stems decumbent to ascending, 2–10 dm long; leaves 3–9 cm long and 1.5–6.5 cm broad, ovate to elliptic, acuminate apically, obtuse to rounded or subcordate basally, aristate-serrate; racemes few- to several-flowered, 4–10 cm long, the pedicels 3–10 mm long, recurved, bearing 1–2 bracts; calyx enlarging and surrounding the capsule, purplish black at maturity, glandular-hairy; corolla urn shaped, 7–9 mm long, pink, pubescent; capsules 5–10 mm broad.

Woods and meadows; in coastal and insular southeastern Alaska; south to California and east to Alberta.

KALMIA L.

Low shrubs with puberulent branches; leaves opposite, evergreen, leathery, decurrent, entire, revolute, glaucous beneath; flowers in terminal, leafy-bracted corymbs or solitary, perfect, regular; calyx 5-lobed, the segments almost distinct; corolla bowl shaped, the lobes shorter than the tube, the tube with 10 pouches in which the anthers are enclosed in bud; stamens usually 10, the filaments flattened, hairy below; anthers opening throughout, unawned; ovary superior, 5-loculed; fruit a septicidally dehiscent capsule.

Kalmia polifolia Wang.
Bog Laurel

Stems 1–3 dm tall; leaves opposite, (0.6) 0.8–3.5 cm long, (0.2)0.3–1.1 cm broad, lance-oblong to elliptic, entire, revolute, shining and green above or densely puberulent when young, grayish beneath, short-petioled; corymbs 1–12-flowered, the pedicels 1–3.5 cm long; sepals 2–3 mm long, ciliate; corolla 12–18 mm broad, pink; capsules 4–6 mm broad.

Bogs and wet meadows; in southeastern Alaska and southern Yukon; east to Newfoundland and south to California, Colorado, Minnesota, and New Jersey.

1a. Leaves mostly 1–2 cm long; plants usually less than 1 dm tall, growing in alpine sites (*K. polifolia* ssp. *microphylla* [Hook.] Calder & Taylor). *K. polifolia* var. *microphylla* (Hook.) Rehd.

1b. Leaves mostly 2–4 cm long; plants usually more than 2 dm tall, commonly growing at lower elevations (*K. occidentalis* Small). *K. polifolia* var. *polifolia*

LEDUM L.

Erect or spreading shrubs with reddish, villous-tomentose branchlets; leaves alternate, evergreen, leathery, entire, revolute, brownish (yellowish to reddish) woolly-tomentose below; flowers in terminal corymbs, perfect, regular; calyx small, the segments almost distinct; corolla rotate, the 5 petals distinct; stamens usually 5–10, the filaments filiform, usually hairy below; anthers opening by terminal pores, unarmed; ovary superior, 5-loculed; fruit a septicidally 5-valved capsule, opening from the base. *Note*: At least some species are poisonous to livestock.

1a. Pedicels pubescent with tangled, reddish hairs; leaves 0.3–1.5(3.8) cm long, 0.1–0.3(0.7) cm broad, linear to oblong. *L. decumbens*

1b. Pedicels pubescent with short, white hairs (rarely with some twisted, reddish hairs); leaves (1.2)2–4.5(6) cm long, 0.3–1.5 cm broad, oblong to elliptic or lance-oblong. *L. groenlandicum*

Ledum decumbens (Ait.) Lodd. ex Steud.
Northern Labrador-tea

Stems mostly 1–5 dm long, decumbent to erect, the branchlets reddish villous-tomentose; leaves 0.3–1.5(3.8) cm long, 0.1–0.3 (0.7) cm broad, linear to narrowly oblong, strongly revolute, reddish villous-tomentose beneath; petioles 0.5–2 mm long; corymbs several- to many-flowered, the pedicels pubescent with contorted reddish hairs, glandular, the longest pedicels 0.8–2.1

cm long in flower, somewhat longer and rather abruptly reflexed near the apex in fruit; calyx less than 1 mm long, ciliate with long, reddish hairs, glandular; corolla 10–12 mm broad, white; capsules 3–5 mm long, glandular.

Bogs, muskegs, and heathlands; over most of Alaska (except for the Aleutians and the Panhandle), and most of the Yukon; east to Newfoundland and Greenland, and south to British Columbia and Manitoba; Eurasia (*L. palustre* ssp. *decumbens* [Ait.] Hultén; *L. palustre* var. *decumbens* (Ait.). This species and the next are closely related to *L. palustre* L., and both have been treated at infraspecific level within that species. However, there are some rather distinctive characteristics which indicate a closer relationship between the allopatric *L. palustre* and *L. groenlandicum* than between the partially sympatric *L. palustre* and *L. decumbens*. Thus, a somewhat liberal treatment is indicated pending further study.

Ledum groenlandicum Oeder
Labrador-tea

Stems mostly 3–15 dm long, erect or spreading, the branchlets reddish (rarely whitish) villous-tomentose; leaves (1.2)2–4.5(6) cm long, 0.3–1.5 cm broad, oblong to elliptic or lance-oblong, strongly to slightly revolute, reddish or yellowish villous-tomentose beneath; petioles 1–5 mm long; corymbs many-flowered, the pedicels strigose with short, white hairs (rarely with a few contorted, reddish hairs), often glandular, the longest (0.6)1.2–2.5 cm long in flower, somewhat longer and gently curved well below the capsule in fruit; calyx less than 1 mm long, minutely white-ciliate, sometimes glandular; corolla 10–12 mm broad, white; capsules 4.5–6.5 mm long, glandular.

Bogs, muskegs, woods, and meadows; in the southeastern two-thirds of Alaska (roughly south of the 68th parallel and east of the 156th meridian) and most of

Ledum decumbens (Ait.) Lodd. (× 0.6).

Yukon; east to Greenland and south to New England and Oregon.

LOISELEURIA Desv.

Prostrate to ascending shrubs; leaves opposite, evergreen, leathery, entire, revolute, whitish beneath; flowers in terminal corymbs, perfect, regular; sepals 5, distinct or nearly so; corolla broadly campanulate, the lobes subequal to or longer than the tube; stamens 5, included, the filaments flattened, tapering to the apex, glabrous; anthers opening throughout, unawned; ovary superior, 2–5-loculed; fruit a 2–5-valved, septicidally dehiscent capsule.

Loiseleuria procumbens (L.) Desv.
Alpine Azalea

Stems 0.5–3 dm long, the branchlets glabrous or puberulent; leaves with broadened, partially clasping petioles 1–4 mm long, the blades 0.3–0.8 cm long, 0.1–0.3 cm broad, lance-oblong to oblong, revolute; corymbs (1)2–6-flowered, the pedicels 1–10 mm long, straight or curved; sepals 1.5–2.5 mm long, red, glabrous, persistent; corolla 4–5 mm long, pink; capsules ovoid, 2–3 mm broad, 3–5 mm long.

Arctic and alpine tundra and heathlands; in most of Alaska and Yukon; south to British Columbia and east to the Atlantic; Greenland, Eurasia.

MENZIESIA Smith

Erect shrubs, with puberulent and stipitate-glandular branches; leaves alternate, deciduous, serrulate, flat, pale beneath; flowers in terminal corymbs, from buds of the previous year (soon overtopped by current growth); calyx saucer shaped, indistinctly lobed; corolla urn shaped to campanulate, the 4 lobes shorter than the tube; stamens 8, the filaments flattened, tapering to the apex, pubescent; anthers opening by terminal pores, unawned; ovary superior, 4-loculed; fruit a 4-valved, septicidal capsule.

Menziesia ferruginea Smith
False Azalea

Stems 5–20 dm tall, the branchlets puberulent and stipitate-glandular; leaves with stipitate-glandular petioles 3–28 mm long, the blades 1–6 cm long, 0.3–2.2 cm broad,

elliptic to oblanceolate, serrulate, stipitate-glandular; corymbs (1)2–10-flowered, the pedicels 0.7–3.5 cm long, recurved in flower, erect in fruit, stipitate-glandular; calyx 1 mm long or less, stipitate-glandular ciliate; corolla 6–9 mm long, orange to pinkish orange; capsules lance-ovoid, 5–7 mm long.

Woods and thickets; in south-central and coastal southern Alaska, from the Alaska Peninsula eastward through the Panhandle; south to California and east to Alberta and Wyoming. Our plants belong to var. *ferruginea*.

OXYCOCCUS Hill

Prostrate, trailing shrub with filiform, rooting stems; leaves alternate, evergreen, leathery, entire, revolute, pale beneath; flowers solitary or in pairs from terminal or lateral buds, perfect, regular; calyx 4-lobed, united at the base; corolla reflexed, the 4 petals distinct; stamens 8, exserted, the filaments short, hairy; anthers opening by terminal pores at the ends of long tubular beaks; ovary inferior, usually 4-locular; fruit a several-seeded berry.

Oxycoccus microcarpus Turcz. ex Rupr.
Swamp Cranberry

Stems 1.5 dm long or more, 0.2–1 mm thick; leaves 2–8 mm long, 1–4 mm broad, lanceolate to ovate or elliptic, entire, revolute; flowers borne on slender pedicels 2–5 cm long, these with 2 small, subopposite bracts below the middle, recurved apically in flower, erect in fruit; sepals less than 1 mm long, pink; petals 5–7 mm long, pink, sharply reflexed (the flowers dodecatheon-like); berries 5–7 mm long.

Bogs and swamps; in much of Alaska and Yukon south of the 68th parallel (rarely somewhat farther north); east to the Atlantic and south to Oregon and Idaho; Eurasia. The bulk of our material belongs to var. *microcarpus*. However, in southeastern Alaska some of the plants have larger leaves (mostly 6–8 mm long and 2–4 mm broad), and puberulent pedicels. These have been designated as var. *intermedium* Gray (*O. quadripetalus* Gilib.; *O. palustris* Pers.).

PHYLLODOCE Salisb.

Ascending to erect, low shrubs; leaves alternate, not 4-ranked, evergreen, leathery, apparently revolute, overlapping, short-petiolate, with bulbous-decurrent bases on the stem; flowers apparently corymbose, borne singly in the axils of the apical leaves, the pedicels short to long, recurved apically in flower, erect in fruit, stipitate-glandular; sepals 5, distinct or nearly so; corolla urn shaped to campanulate, the 5 lobes shorter than the tube, included, the filaments flat, glabrous or pubescent; anthers opening by terminal slits, unawned; ovary superior, 5-carpelled; fruit a 5-valved, septicidally dehiscent capsule.

Stoker, F. 1939. The genus *Phyllodoce*. New Flora and Silva. 12:30–42.

1a. Flowers yellowish or greenish yellow. (2)
1b. Flowers pink, blue, or purple. (3)

2a. Corolla and stamens glabrous; leaf margins sharp, glandular-serrate; plants from the vicinity of Prince William Sound westward. *P. aleutica*
2b. Corolla stipitate-glandular externally, the staminal filaments pubescent; leaf margins blunt, irregularly serrate and often stipitate-glandular when young; plants from the vicinity of Prince William Sound eastward. *P. glanduliflora*

3a. Corolla blue to purple, urn shaped, usually stipitate-glandular externally; staminal filaments minutely puberulent; sepals stipitate-glandular; plants from central interior to western Alaska. *P. caerulea*
3b. Corolla pink, campanulate, glabrous externally; staminal filaments glabrous; sepals not stipitate-glandular; plants from southern Yukon, southeastern Alaska, and adjacent British Columbia. *P. empetriformis*

Phyllodoce aleutica (Spreng.) Heller
Aleutian Mountain-heather

Stems ascending to erect, 1–2.5 dm tall; leaves alternate, spreading, 4–12 mm long, the margin sharp, serrate (sometimes glandular); flowers few to several, the pedicels mostly less than 1 cm long in flower, longer in fruit; sepals 2.5–3.5 mm long, greenish, stipitate-glandular; corolla yellowish or greenish yellow, urn shaped, 6–8 mm long, glabrous externally; staminal filaments glabrous; capsules pubescent, 3–4 mm long, 2.5–3.5 mm broad.

Alpine tundra and heathlands; from the region of Prince William Sound west through the Aleutians, and in the interior as far north as Talkeetna Mountains; Asia.

Phyllodoce caerulea (L.) Bab.
Blue Mountain-heather

Stems ascending to erect, 0.8–2.5 dm tall;

Phyllodoce aleutica (Spreng.) Heller (× 0.6).

leaves alternate, spreading-ascending, 3–10 mm long, the margin sharp, glandular-serrate; flowers 2–several; pedicels often longer than 1 cm in flower; sepals 3–4 mm long, reddish, stipitate-glandular; corolla blue to purple, urn shaped, 7–9 mm long, stipitate-glandular to glabrous externally; staminal filaments minutely puberulent; capsules usually pubescent, 2.5–3.5 mm long and about as broad.

Alpine tundra and heathlands; from the upper Kuskokwim River Valley westward to the coast and northward to the Seward Peninsula, and on the westernmost Aleutian Islands; disjunctly eastward to the Atlantic; Greenland, Eurasia.

Phyllodoce empetriformis (J. E. Smith) D. Don
Pink Mountain-heather

Stems ascending, 0.5–2 dm tall; leaves alternate, spreading-ascending, 3–10 mm long, the margin sharp to blunt, glandular-serrate; flowers several to many, the pedicels usually more than 1 cm long in flower; sepals 2–3 mm long, reddish, lacking stipitate glands; corolla pink, campanulate, 5–9 mm long, glabrous externally; staminal filaments glabrous; capsules glandular-pubescent, 2.5–3.5 mm long, 3–4 mm broad.

Alpine tundra and heathlands; from southern Yukon southward through British Columbia and southeastern Alaska; south to California and east to Mackenzie, Alberta, and Montana.

Phyllodoce glanduliflora (Hook.) Coville
Yellow Mountain-heather

Stems 1–2 dm tall, erect or ascending; leaves alternate, spreading-ascending, 4–10 (12) mm long, the margin blunt, glandular-serrate (stipitate-glandular on young leaves); flowers several to many, the pedicels mostly longer than 1 cm in flower; sepals 4–5 mm long, greenish, stipitate-glandular; corolla yellowish or greenish yellow, urn shaped, 6–9 mm long, stipitate-

glandular externally; staminal filaments puberulent; capsules glandular-pubescent, 4–5 mm long and nearly as broad.

Alpine tundra and heathland in the coast ranges; from the vicinity of Prince William Sound eastward through the Alaskan Panhandle, and eastward to southeastern Yukon and western Mackenzie; south to Oregon and Wyoming (*P. aleutica* ssp. *glanduliflora* [Hook.] Hultén). This very pretty plant is known to hybridize with *P. empetriformis*. The hybrids are known as *P.* × *intermedia* (Hook.) Camp. *P. granduliflora* also apparently forms hybrids with *P. aleutica* in south-central Alaska.

RHODODENDRON L.

Low shrubs; leaves alternate, evergreen or deciduous, entire, flat or somewhat revolute; flowers terminal and solitary on short, leafy or bracteate branchlets, or in terminal umbels, perfect, irregular; calyx 5-lobed, the segments distinct or nearly so; corolla campanulate to subrotate, the lobes longer than the tube; stamens 5–10, exserted or included, the filaments flattened, densely hairy below; anthers opening by terminal pores, unarmed; ovary superior, 5-loculed; fruit a 5-valved, septicidally dehiscent capsule.

1a. Leaves deciduous, flat, margin and veins on lower surface long-hairy, not scurfy-glandular; flowers solitary, terminal, over 1.5 cm long; sepals foliose, 8 mm long or longer. *R. camtschaticum*

1b. Leaves evergreen or tardily deciduous, somewhat revolute, scurfy-glandular, lacking long hairs; flowers in terminal umbels, less than 1.5 cm long; sepals less than 2 mm long. *R. lapponicum*

Rhododendron camtschaticum Pallas
Kamchatka Rhododendron

Stems 0.5–3 dm long, decumbent to erect; leaves deciduous, 1–6 cm long, 0.4–2.2 cm broad, flat, spatulate to oblanceolate, rounded apically, cuneate to acute basally, marginally ciliate, pubescent on the veins on the lower surface with long hairs, subsessile; flowers solitary, terminal, large and showy, the pedicels 1–4 cm long, long-hairy or stipitate-glandular; sepals 8–19 mm long, stipitate-glandular ciliate; corolla reddish or pink or rarely white, 17–28 mm long; capsules 5–10 mm long, pubescent.

Arctic and alpine tundra and heathlands; from Prince William Sound westward to Kodiak, the Alaska Peninsula, and the Aleutians, and north to the Seward Peninsula (rarely in the Panhandle); Asia.

1a. Leaves distinctly stipitate-glandular; corolla glabrous externally; plants from the Seward Peninsula and lower Yukon River Valley (*Therorhodion glandulosum* Standl.; *R. camtschaticum* var. *glandulosum* [Standl.] B. Boi.). *R. camtschaticum* ssp. *glandulosum* (Standl.) Hultén

1b. Leaves lacking stipitate-glandular hairs, or sparsely stipitate-glandular;

corolla pubescent externally; plants from coastal and insular southern Alaska. *R. camtschaticum* ssp. *camtschaticum*

Rhododendron lapponicum (L.) Wahl.
Lapland Rosebay

Stems 0.5–2.5 dm long, prostrate to ascending; leaves evergreen or tardily deciduous, 0.5–1.7 cm long, 0.2–0.8 cm broad, somewhat revolute, elliptic, obtuse apically, obtuse to acute basally, scurfy-glandular on both surfaces, not ciliate, short-petiolate; flowers usually 2–4, in terminal umbels, the pedicels 0.3–1.4 cm long, lacking bracts, scurfy-glandular; sepals 0.5–1.5 mm long, ciliate, dorsally scurfy-glandular; corolla pink purple, 8–12 mm long; capsules scurfy-glandular, 5–6 mm long.

Arctic and alpine tundra and heathlands, or less commonly in woods or thickets; in most of Alaska (except for southern and southwestern portions) and most of the Yukon and adjacent British Columbia; east to the Atlantic; Greenland, Eurasia.

VACCINIUM L.

Decumbent-ascending to erect shrubs; leaves alternate, deciduous or evergreen, thin or leathery, entire to serrulate, flat or revolute, pale or green beneath; flowers solitary, axillary, or in terminal clusters, perfect, regular; calyx 4–6-lobed, united at the base; corolla urn shaped, depressed urn shaped, or campanulate, the 4–6 lobes shorter than the tube; stamens 8–12, included, the filaments glabrous or hairy; anthers opening by pores at the ends of tubular beaks, awned or unawned; ovary inferior, usually 4-locular; fruit a several-seeded berry.

Camp, W. H. 1942. A survey of the American species of *Vaccinium*, subgenus Euvaccinum. Brittonia 4:205–47.

———. 1945. The North American blueberries, with notes on other groups of Vacciniaceae. Brittonia 5:203–75.

Rhododendron camtschaticum Pallas (× 0.8).

1a. Leaves evergreen, slightly revolute, with dark glands on the lower surface; flowers borne in terminal, bracteate racemes; pedicels with bracts; fruit red. *V. vitis-idaea*

1b. Leaves deciduous (at least on mature plants), flat, glandular or glabrous or puberulent beneath; flowers axillary and solitary, or 1–few from buds of previous season; pedicels lacking bracts; fruit blue, blue black, black, or red. (2)

2a. Leaf margins serrulate throughout; leaf blades oblanceolate to elliptic, leathery; plants usually less than 3 dm tall. *V. caespitosum*

2b. Leaf margins entire or partially serrulate, not serrulate throughout (see *V. membranaceum*); leaf blades variously shaped. (3)

3a. Branchlets round in cross section; flowers 1–few from buds of the previous season; calyx deeply lobed, the lobes persistent on the fruit; plants widely distributed in Alaska and Yukon. *V. uliginosum*

3b. Branchlets slightly to conspicuously angled; flowers solitary in leaf axils of the current season; calyx shallowly lobed, the lobes often deciduous in fruit; plants mostly of southern Alaska. (4)

4a. Branchlets green, sharply angled; fruit red, the calyx lobes persistent; plants of southeastern Alaska. *V. parvifolium*

4b. Branchlets brownish or greenish brown, angled but seldom sharply so; fruit blue to blue black or black, the calyx lobes mostly deciduous; plants more widely distributed in southern Alaska. (5)

5a. Leaves sharply serrate nearly the full length; plants of northern British Columbia. *V. membranaceum*

5b. Leaves entire or inconspicuously serrulate below the middle; plants of coastal or near coastal Alaska. (6)

6a. Flowers appearing before the leaves; leaves glabrous, lacking glands on the lower veins; corolla pinkish, urn shaped, longer than broad, the style included; fruiting pedicels 2–8 mm long, often curved; plants from the Aleutians eastward through southeastern Alaska. *V. ovalifolium*

6b. Flowers appearing with the leaves; leaves glabrous or pubescent, often glandular on the lower veins; fruiting pedicels 6–14 mm long, straight or somewhat curved; corolla coppery pink, depressed urn shaped, often broader than long; plants from Prince William Sound eastward through southeastern Alaska. *V. alaskaense*

Vaccinium alaskaense Howell
Alaska Blueberry

Stems 5–10 dm tall or more, the branchlets angled; leaves deciduous, (1.2)2.4–6.6 cm long, (0.8)1.3–3.2 cm broad, lanceolate to lance-ovate, lance-elliptic or elliptic, the margin entire or serrulate only near the base or almost throughout, acute to obtuse apically, acute to obtuse basally, often pubescent and glandular on the lower veins; flowers solitary in the leaf axils, the pedicels 6–14 mm long in fruit, sometimes enlarged below the fruit; calyx shallowly lobed, the lobes deciduous in fruit; corolla coppery pink, depressed urn shaped, 4–6 mm long and often broader; styles often exserted; staminal filaments glabrous, the anthers awned; berries bluish black or black, 7–12 mm thick, edible.

Woods and thickets; in coastal and insular southern Alaska, from Prince William Sound eastward through the Panhandle; south to Oregon (*V. membranaceum* authors, not Dougl. ex Hook.). The Alaska blueberry is closely related to *V. ovalifolium*, and apparently is not always separable morphologically from that species.

Vaccinium caespitosum Michx.
Dwarf Blueberry

Stems mostly (0.5)1–3 dm tall, the branchlets somewhat angled, usually puberulent; leaves deciduous, somewhat leathery, 0.5–3.5 cm long, 0.2–1.5 cm broad, oblanceolate to obovate or elliptic, the margin serrulate throughout, obtuse to acute apically, acute basally, glabrous or puberulent above, often glandular below; flowers solitary in the leaf axils; calyx shallowly lobed, the lobes deciduous in fruit; corolla pink to whitish, urn shaped, 4–6 mm long, slender; staminal filaments glabrous, the anthers awned; berries blue to blue black, 5–8 mm thick, palatable.

Bogs, wet meadows, snow flushes, and tundra, from low to high elevations; in southern Alaska and southern Yukon; east to Newfoundland and New Hampshire and south to California and Idaho (*V. paludicola* Camp; *V. caespitosum* var. *paludicola* [Camp] Hultén).

Vaccinium membranaceum Dougl. ex Hook.
Mountain Huckleberry

Stems mostly (2)5–10 dm tall or more, the branchlets somewhat glabrous or puberulent; leaves deciduous, 1.5–5 cm long, (0.6) 1–3 cm broad, ovate or oblong to elliptic or obovate, the margin serrulate almost throughout, attenuate to acuminate apically, acute to rounded basally, glabrous above, often glandular below; flowers solitary in leaf axils; calyx shallowly lobed, the lobes deciduous in fruit; corolla yellowish pink, urn shaped, 4–6 mm long; staminal filaments glabrous, the anthers awned; berries purplish, 7–9 mm thick, palatable.

Woods; in northern British Columbia (to be sought in southeastern Yukon); southward to California, Idaho, and Montana.

Vaccinium ovalifolium Smith ex Rees
Early Blueberry

Stems 4–10 dm tall or more, the branchlets angled, often glabrous; leaves deciduous, 1.2–4.7 cm long, 0.7–2.9 cm broad, elliptic to oval or lance-oblong, entire to finely serrulate mostly in the lower half, obtuse to rounded or subcordate basally, glabrous, not glandular on the veins; flowers solitary in the leaf axils, the pedicels mostly 1–8 mm long in fruit, usually not enlarged below the fruit; calyx shallowly lobed, the lobes deciduous (rarely persistent) in fruit; corolla pinkish, urn shaped, 4–7 mm long, longer than broad; style usually included; staminal filaments glabrous, the anthers awned; berries bluish black, 6–10 mm thick, edible.

Woods and thickets or heathlands; in southern Alaska, from the Aleutians east through the Panhandle, and in interior Alaska as far north as Talkeetna; south to Oregon and disjunctly eastward to Newfoundland. *V. ovalifolium* should be studied in the field to determine its relationship to *V. alaskaense.*

Vaccinium parvifolium Smith ex Rees
Red Huckleberry

Stems mostly 5–15 dm tall or more, the branchlets angled, green, usually glabrous; leaves deciduous (except on juvenile growth), 0.9–3 cm long, 0.4–1.6 cm broad, elliptic to lance-elliptic or oblanceolate, entire, at least on adult growth, acute or more commonly obtuse to rounded apically, acute to rounded basally, glabrous or puberulent; flowers solitary in the leaf axils, the pedicels 2–8 mm long in fruit; calyx shallowly lobed, the lobes persistent in fruit; corolla yellowish pink, broadly urn shaped, 4–5 mm long; staminal filaments glabrous, the anthers awned; berries red, 7–10 mm thick, edible.

Woods and thickets; in coastal and insular southeastern Alaska; south to California.

Vaccinium uliginosum L.
Bog Blueberry

Stems 1–6 dm tall, the young branchlets puberulent, round in cross section; leaves deciduous, 0.4–2.8 cm long, 0.2–1.5 cm broad, oblanceolate to obovate or oval, entire, rounded to obtuse or acute apically, acute to obtuse basally, somewhat revolute and leathery, glabrous or puberulent; flowers 1–few from scaly buds of previous year, the pedicels short; calyx deeply lobed, the lobes persistent on the fruit, corolla pink, urn shaped, 3–4 mm long; staminal filaments glabrous, the anthers awned; berries blue black, 5–8 mm thick, edible.

Muskegs, swamps, woodlands, heath, or tundra; in most of Alaska and Yukon; south to California and east to the Atlantic; circumboreal (*V. uliginosum* var. *salicinum* [Cham. & Schlecht.] Hultén; *V. uliginosum* ssp. *microphyllum* Lange).

Vaccinium vitis-idaea L.
Mountain Cranberry

Stems decumbent-ascending, 0.5–2.5 dm tall, the branchlets terete or slightly angled, puberulent; leaves evergreen, leathery, 0.4–1.8 cm long, 0.2–0.9 cm broad, oblanceolate to obovate or elliptic, serrulate to entire along the slightly revolute margin, rounded to retuse apically, acute to obtuse basally, glabrous or ciliate, the lower surface with brownish glands; flowers 1–several in short, terminal racemes, the pedicels bearing 1–2 bracts below the flower; calyx deeply lobed, the lobes persistent in fruit; corolla pinkish, campanulate, 4–6 mm long; staminal filaments pubescent, the anthers awnless; berries red, 6–10 mm thick, edible.

Woods, thickets, heathlands, and tundra; over much of Alaska and Yukon; south to southern British Columbia and east to the Atlantic; Greenland, Eurasia (*V. minus* Lodd.). Our materials have been assigned to ssp. *minus* (Lodd.) Hultén.

EUPHORBIACEAE

Spurge Family

Annual or perennial herbs or woody shrubs; leaves simple alternate, entire or variously lobed; flowers imperfect, borne in cuplike involucres (cyathia), the staminate reduced to a solitary, pedicellate stamen, and the pistillate reduced to a solitary, pedicellate pistil; pistils 1, the ovary superior, 3-loculed, 3-carpellate; styles 3, often deeply lobed; fruit a capsule.

EUPHORBIA L.

Weedy annuals or cultivated shrubs; leaves petiolate, the upper ones sometimes becoming brightly colored and simulating a "flower;" cyathia solitary or in terminal cymes, often bearing glands along the margin; staminate flowers usually several, included in the cyathium; pistillate flowers solitary, often exserted from cyathium; capsules 3-seeded, usually reflexed.

1a. Plants annual, herbaceous; weeds of greenhouses and occasionally in gardens. *E. peplus*

1b. Plants perennial (under glass), woody; cultivated ornamentals of greenhouses and homes. *E. pulcherrima*

Euphorbia peplus L.
Spurge

Stems 1–5 dm tall, simple or freely branched, ascending or erect; leaves alternate, at least below, petiolate, the blades 1–3 cm long, entire; leaves of the inflorescence opposite, green; involucres 1–1.5 mm long, the glands bearing long, slender horns; capsules 2–3 mm long.

Introduced weed of greenhouses and gardens; in southeastern Alaska, and to be expected elsewhere; native to Europe.

Euphorbia pulcherrima Willd.
Poinsettia

Stems 6–20 dm tall or more, simple or branched; leaves alternate, petiolate, the blades 5–15 cm long, entire or more commonly lobed, those of the inflorescence becoming brightly colored (usually red, cream, or yellow) and simulating a "flower;" involucres 4–7 mm long, bearing a large, yellow gland on one side; capsules 4–6 mm long.

Cultivated greenhouse ornamental; native to Central America.

FUMARIACEAE
Fumitory Family

Annual, biennial, or perennial herbs with watery juice; leaves alternate or basal, compound; flowers perfect, irregular, usually borne in racemes or panicles; sepals 2, bractlike; petals 4, the outer two spreading at the apex and one or both saccate or spurred at the base, the inner pair not spurred, usually united at the apex; stamens 6, diadelphous in sets of 3 each; pistils 1, the ovary superior, 1-loculed, 2-carpelled; styles 1, the stigma usually 2-lobed; fruit a 2-valved, several-seeded capsule.

1a. Corolla with the two outer petals dissimilar, only 1 petal spurred or saccate at the base; plants with leafy stems or subscapose, indigenous, rarely if ever cultivated. *Corydalis*
1b. Corolla with the two outer petals alike, both petals spurred or saccate at the base; plants scapose, cultivated ornamentals. *Dicentra*

CORYDALIS Medic. Nom. Cons.

Annual, biennial, or perennial herbs from taproots or tuberous roots; leaves basal, and cauline, once to twice compound; flowers in terminal or axillary racemes or panicles, showy; sepals 2, scalelike, cauducous; petals 4, the outer pair dissimilar, only 1 petal spurred or saccate at the base, the inner petals connate at the apex; capsules slender, elongate.

Ownbey, G. B. 1947. Monograph of the North American species of *Corydalis*. Ann. Mo. Bot. Gard. 34:197–258.

1a. Plants subscapose perennials from deep-seated fleshy roots, the stems smaller below the ground level; flowers blue to blue purple, few to several in a compact raceme. *C. pauciflora*
1b. Plants caulescent annuals or biennials from taproots, the stem not extending below ground level; flowers yellow, or pinkish with yellow tips, few to many in contracted to lax racemes or panicles. (2)

2a. Flowers yellow; stems few to several, spreading. *C. aurea*
2b. Flowers pink with yellow tips; stems 1–few, erect. *C. sempervirens*

Corydalis aurea Willd.
Golden Corydalis

Annual or biennial, glaucous herbs from taproots, the stems (1) few to several, spreading, 1–6 dm long, often much branched; leaves from 1–4 times pinnately compound; racemes several-flowered, the pedicels longer than the spur or subequal to it; sepals 1–3 mm long, yellowish or whitish; corolla yellow, 12–16(18) mm long, the spur 3–6 mm long; capsules 1.5–2.5 cm long, constricted between the seeds, usually curved.

Disturbed soils, hillsides, roadsides, and open woods; in the southeastern quarter of Alaska and southern Yukon; east to the Atlantic and south to Mexico.

Corydalis pauciflora (Steph.) Pers.
Few-flowered Corydalis

Perennial, glaucous, subscapose herbs from deep-seated, tuberous roots, the stems usually solitary, erect, constricted below the ground level, mostly 0.5–1.5 dm tall above ground; leaves 2–5, clustered near the stem base, once-ternately compound, the leaflets lobed; racemes 2–6-flowered, the pedicels shorter than the spurs; sepals 1–2 mm long, purplish; corolla blue to blue purple, 15–22 mm long, the spur 6–10 mm long; capsules (0.8)1–2 cm long, slightly or not at all constricted between the seeds, usually straight.

Moist sites in arctic and alpine tundra and

heathlands; in most of Alaska (except for the Aleutians and the Alexander Archipelago) and most of the Yukon; east to Mackenzie and south to British Columbia; Asia (*C. pauciflora* var. *albiflora* Porsild).

Corydalis sempervirens (L.) Pers.
Pink Corydalis

Annual or biennial, glaucous herbs from taproots, the stems 1–few, usually erect, 2–10 dm tall, seldom much branched; leaves 3–5 times pinnately compound; racemes 3–10-flowered or more, solitary or several terminating the upper branches, the pedicels much longer than the spurs; sepals 3–4 mm long; corolla pink, yellow tipped,

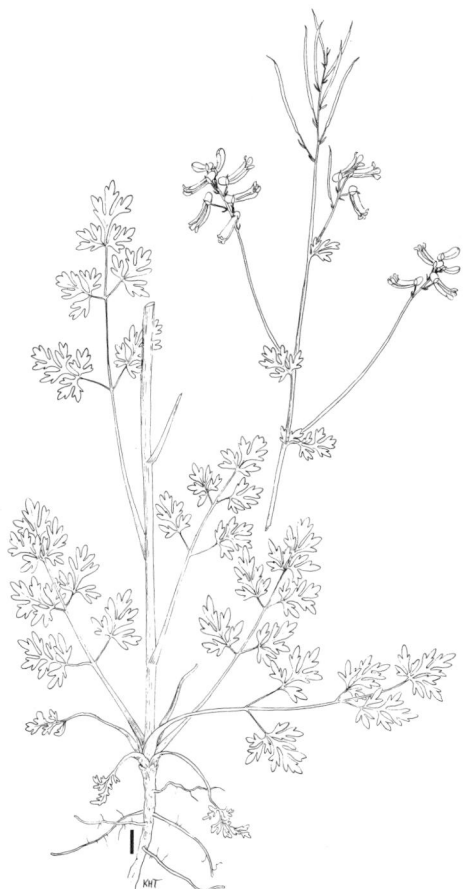

Corydalis sempervirens (L.) Pers. (× 0.3).

10–17 mm long, the spur 2–4 mm long; capsules 2–5 cm long, constricted between the seeds, straight or curved.

Open woods or shrublands, mostly in disturbed soils; in much of Alaska south of the Brooks Range and east of the 154th meridian (except for the Panhandle), and in southern Yukon; east to Newfoundland and south to Montana and Georgia.

DICENTRA Bernh. Nom. Cons.

Perennial, scapose, glaucous herbs from thick rhizomes; leaves basal or nearly so, 2–3 times compound; flowers in racemes or panicles, showy; sepals 2, somewhat persistent; petals 4, the outer pair alike, both spurred or saccate at the base, the inner petals connate at the apex; capsules elongate.

Dicentra eximia Torr.
Bleeding-heart

Scapose herbs from rhizomes, the scapes 1–5 dm long; leaves basal, long-petiolate, 2–3 times ternately compound, subequal to the scapes; flowers few to many, borne in panicles; sepals 5–7 mm long, lance-acuminate; corolla pink or pink purple, or the crests often white, 18–25 mm long, the spurs 1–2 mm long; capsules oblong to linear.

Cultivated ornamental; in southern Alaska; native to the eastern United States.

GENTIANACEAE
Gentian Family

Plants glabrous, annual, biennial, or perennial herbs; leaves opposite, or rarely alternate, simple, entire; flowers perfect, regular, solitary or in cymes; calyx of 4–5 sepals, often united; corolla of 4–5 united petals; stamens 4–5, alternate with the corolla lobes; pistils 1, the ovary superior, 1-loculed, 2-carpelled; styles 1, or obsolete, the stigma 2-lobed; fruit a capsule.

Gillett, J. M. 1963. The Gentians of Canada, Alaska, and Greenland. Can. Dept. Agr. Publ. 1180:1–99.

1a. Corolla rotate, the lobes equal to or longer than the tube. (2)
1b. Corolla tubular, the lobes usually shorter than the tube. (3)

2a. Plants perennial, rhizomatous; leaves mainly basal, the cauline ones reduced in size and number; stigmas not decurrent on the ovary. *Swertia*
2b. Plants annual, with taproots; leaves mainly cauline, not reduced in size or number upwards; stigmas decurrent along the sides of the ovary. *Lomatogonium*

3a. Corolla tube with folds between the lobes; nectary glands at the base of the ovary; cauline leaves mostly conspicuously connate-sheathing. *Gentiana*
3b. Corolla tube without folds between the lobes; nectary glands at the base of the corolla tube; cauline leaves distinct, or the lower ones slightly connate. *Gentianella*

GENTIANA L.

Plants perennial or annual (biennial?) herbs, the stems erect, ascending or prostrate; leaves all cauline or basal and cauline, the cauline ones opposite; flowers showy, large or small, solitary or in terminal cymes; calyx tubular, 4–5-toothed, or 2-lipped, the one lip 2-toothed, the other 3-toothed; corolla tubular or funnelform, the lobes shorter than the tube and bearing folds in the sinuses, persistent; stamens 5 (or 4), inserted on the corolla tube; nectary glands at the base of the ovary; stigmas 2; fruit a 2-valved, septicidal capsule.

1a. Plants annual (or biennial), from slender taproots. (2)
1b. Plants perennial, from rhizomes or caudices. (3)

2a. Plants prostrate to ascending, simple or branched from the base; flowers solitary, terminal. *G. prostrata*
2b. Plants erect or ascending, branched throughout or simple; flowers terminal and axillary, often more than 1. *G. douglasiana*

3a. Calyx 2-lipped, the one lip 2-toothed, the other 3-toothed; corolla usually bright blue, the lobes orbicular-acuminate; stems usually with more than 3 pairs of leaves below the bracts. *G. platypetala*
3b. Calyx subequally 5-toothed; corolla cream to yellowish green or blue, rarely white, the lobes merely acuminate or ovate; stems usually with 3 pairs of leaves below the bracts or fewer. (4)

4a. Flowers 3.5–5.3 cm long, cream to yellowish green with purple spots or stripes; leaves linear to narrowly oblong. *G. algida*
4b. Flowers 1.5–2 cm long, blue to yellowish green, rarely white; leaves oval to elliptic. *G. glauca*

Gentiana algida Pallas
Whitish Gentian

Plants perennial, the stems 0.8–2 dm tall, erect or ascending, simple; leaves chiefly basal, connate-sheathing, the cauline leaves usually in 2–3 pairs below the bracts (uppermost leaves), 2–12 cm long, 0.2–1 cm broad, linear to narrowly oblong; flowers solitary or in terminal cymes; calyx 5-toothed, 15–23 mm long, the teeth unequal, longer or shorter than the tube; corolla funnelform, cream to yellowish green, mottled or striped with purple, 35–53 mm long, the folds yellowish; pistil stipitate; capsule oblong-elliptic in outline.

Arctic and alpine tundra and heathlands; in much of Alaska south of the 66th parallel and in southwestern Yukon; disjunctly southward to Montana, Wyoming, and Colorado; Asia.

Gentiana douglasiana Bong.
Swamp Gentian

Plants annual, the stems 0.5–2.7 dm tall,

Gentiana algida Pallas (× 0.5).

often branched throughout or rarely simple, from a slender taproot; leaves basal and cauline, the lower ones forming a small rosette, the cauline ones 0.3–0.8 cm long and 0.2–0.7 cm broad, ovate to elliptic; flowers solitary or in cymes, and often axillary also; calyx 5-toothed, 4–7 mm long, the lobes shorter than the tube; corolla tubular-campanulate, 9–14 mm long, the tube yellowish green, the lobes whitish, spotted and streaked with purple, the folds whitish; pistil sessile; capsule oblong.

Bogs and wet meadows; from the Kenai Peninsula eastward through insular and coastal southeastern Alaska (rarely in south-central Alaska) to British Columbia and Washington.

Gentiana glauca Pallas
Glaucous Gentian

Plants perennial, the stems 0.2–1.6 dm tall, erect, simple, from a rhizome; leaves chief-

ly basal, the cauline leaves shortly connate, usually with 1–3 pairs below the bracts, elliptic to oval, 0.8–1.6 cm long, 0.5–0.9 cm broad; flowers in terminal cymes, enveloped by the upper leaves (bracts); calyx 5-toothed, 5–9 mm long, the lobes shorter than the tube, usually bluish; corolla tubular, blue to yellowish green, yellowish, or rarely white, 15–20 mm long, the folds pale; pistil stipitate; capsule lance-ovoid.

Arctic and alpine tundra and heathland; over much of Alaska (except for the Aleutians and the Alexander Archipelago) and most of the Yukon; south through British Columbia to Montana; Asia (*G. glauca* var. *paulensis* Kellogg; *G. glauca* f. *chlorantha* Jordal, a phase with yellow flowers).

Gentiana platypetala Griseb.

Plants perennial, the stems 1–3.5 dm tall, erect or ascending, simple from a rhizome; leaves cauline, slightly connate, ovate to elliptic or lanceolate, 1.5–3.2 cm long, 0.8–1.7(2.2) cm broad, becoming larger above; flowers solitary, sessile; calyx 2-lipped, one lip 2-toothed, the other lip 3-toothed; corolla funnelform, bright blue, spotted with green inside, 27–49 mm long; pistil sessile; capsule elliptic-oblong.

Alpine meadows; from Kodiak Island eastward to southeastern Alaska and adjacent British Columbia (*G. covillei* Nels. & Macbr.; *G. gormanii* How.).

Gentiana prostrata Haenke ex Jacq.
Moss Gentian

Plants annual, the stems 0.2–1.5(2.5) dm long, prostrate or ascending, branched from the base, from a slender taproot; leaves chiefly cauline, connate-sheathing, 0.2–1.5 cm long (including the sheath), 1–5 mm broad; flowers terminal, solitary; calyx 4–5-toothed, 8–16 mm long, the teeth shorter than the tube; corolla tubular, blue (or the tube whitish), 11–25 mm long; pistil stipitate; capsule lance-ovoid, long-stipitate.

Arctic and alpine tundra, heath, or woods and meadows; in much of Alaska (except for the Alexander Archipelago) and most of the Yukon; southward through British Columbia and Alberta to Montana and Wyoming.

GENTIANELLA Moench

Plants annual or biennial herbs, the stems erect or ascending; leaves cauline or basal and cauline, the cauline ones opposite, often bracteate, at least above; calyx tubular or funnelform, 4–5-toothed; corolla tubular to funnelform, 4–5-lobed, the lobes shorter than the tube, lacking folds in the sinuses, persistent; stamens 4–5, inserted on the corolla tube; nectary glands at the base of the corolla tube; stigmas 2; fruit a 2-valved, septicidal capsule.

1a. Flowers (17)20–50 mm long, blue, solitary or few, terminal on the main stem or on lateral branches, or with 1–2 axillary flowers. (2)

1b. Flowers (6)9–20(22) mm long, blue, greenish blue, yellowish, or white, usually in terminal cymes and with several axillary flowers. (3)

2a. Calyx lobes orbicular, rounded apically; corolla 17–28 mm long, with a fringed crown at the base of the lobes; plants of the westernmost Aleutians. G. auriculata

2b. Calyx lobes lance-acuminate; corolla 20–50 mm long, lacking a crown at the base of the lobes; plants broadly distributed, but not of the westernmost Aleutians. G. detonsa

3a. Sepals nearly distinct; pedicels longer than the next lower internode; each corolla lobe bearing 2 fringed scales. G. tenella

3b. Sepals connate for one-fourth to one-third their length; pedicels shorter than the next lower internode (except on some lateral branches); each corolla lobe with a continuous fringe, or naked at the base within. (4)

4a. Corolla mostly less than 15 mm long, the lobes fringed at the base within (at least some). G. amarella

4b. Corolla mostly more than 15 mm long, the lobes naked at the base within. G. propinqua

Gentianella amarella (L.) Borner
Northern Gentian

Plants annual, the stems 0.8–4.5(5) dm tall, erect or ascending, simple or more commonly branched, from a taproot; basal leaves oblanceolate to elliptic; cauline leaves ovate to lanceolate, 1–6 cm long, 0.3–1.5 cm broad, not connate; flowers several to many in cymes, or axillary, the pedicels shorter than the next lower internode; calyx 5–7 mm long, usually 5-lobed, the lobes longer than the tube; corolla tubular, 9–15 mm long, pink purple, bluish purple, pinkish, blue, or yellowish green, the lobes fringed within (at least some); pistil sessile; capsule cylindrical.

Open woods and meadows; in coastal or interior southern Alaska and southern Yukon; east to the Atlantic and south to Baja California, Arizona, New Mexico, North Dakota, and Vermont; circumboreal (*G. amarella* ssp. *acuta* [Michx.] Hultén; *G. acuta* var. *plebeja* [Cham.] Hultén; *G. amarella* ssp. *acuta* var. *plebeja* [Cham.] Hultén). Our materials belong to var. *acuta* (Michx.) Herder.

Gentianella auriculata (Pallas) Gillett
Eared Gentian

Plants annual, the stems 0.3–2.5 dm tall, erect or ascending, simple or branched above, from a taproot; basal leaves oblanceolate to elliptic; cauline leaves ovate to lance-elliptic, 1–4 cm long, 0.3–1.5 cm broad, not connate; flowers solitary or few in cymes, or axillary, the pedicels shorter or longer than the next lower internode; calyx 11–14 mm long, usually 5-lobed, the lobes orbicular, rounded apically, usually shorter than the tube; corolla tubular (17)

20–25(28) mm long, blue, the lobes fringed within; pistil sessile; capsule cylindrical.

Gravelly slopes and meadows; in the westernmost Aleutian Islands (Attu); Asia.

Gentianella detonsa (Rottb.) G. Don

Plants annual or biennial, the stems 1–5 dm tall, erect, simple or branched from near the base, from a taproot; basal leaves oblanceolate to spatulate; cauline leaves linear to lanceolate or oblanceolate, 1.5–5 cm long, 0.1–0.6 cm broad, not connate; flowers solitary and terminal, the pedicels usually much longer than the next lower internode; calyx 13–28 mm long, mostly 4-lobed, the lobes lance-attenuate, at least the longest subequal to the tube; corolla tubular, (20)25–50 mm long, blue, the lobes fringed marginally but not within; pistil short-stipitate; capsule cylindrical.

Arctic tundra, heathland, meadows, or open woods; from Kotzebue Sound disjunctly eastward to central eastern Alaska and southern Yukon; east to Newfoundland; Greenland, Iceland, Europe, Asia (*G. barbata* Froel.).

1a. Plants less than 2 dm tall, simple, or with branches from near the base; basal rosette poorly developed; known from western Alaska. *G. detonsa* ssp. *detonsa*

1b. Plants 2–5 dm tall, simple or with branches from lower or upper axils; basal rosette commonly well developed; known from east-central Alaska and southern Yukon. *G. detonsa* ssp. *yukonensis* Gillett

Gentianella propinqua (Richards.) Gillett

Plants annual, the stems 0.4–4 dm long, erect or ascending, simple or more commonly branched, from a taproot; basal leaves spatulate to elliptic; cauline leaves lanceolate to oblong, 0.5–3.5 cm long, 0.2–1 cm broad, not connate; flowers several to many in cymes, the terminal ones mostly larger, the pedicels shorter than the next lower internode; calyx 5–12 mm long, the lobes longer than the tube; corolla tubular, 10–20(22) mm long, blue to white, the lobes neither fringed marginally nor within; pistils short-stipitate; capsules cylindrical.

Arctic and alpine tundra, heath, and woods; in much of Alaska and Yukon; east to Newfoundland and south to British Columbia, Montana, and Quebec; Asia (*G. arctophila* Griseb.; *G. propinqua* ssp. *arctophila* [Griseb.] Hultén).

1a. Terminal flowers 8–13 mm long, subequal to the lateral ones, pale pink purple to white; plants from insular and coastal southern Alaska. *G. propinqua* var. *aleutica* (Cham. & Schlecht.) B. Boi.

1b. Terminal flowers 12–20(22) mm long, mostly larger than the lateral ones, usually blue; plants widespread in interior Alaska and Yukon, seldom in southern coastal regions. *G. propinqua* var. *propinqua*

Gentianella tenella (Rottb.) Borner

Plants annual, the stems 0.5–1.5(2) dm tall, ascending to erect, simple or more commonly branched, from taproots; basal leaves elliptic to obovate; cauline leaves lanceolate to elliptic, 4–10 mm long, 1–4 mm broad, not connate (except for the lower ones); flowers solitary, terminal or axillary, the pedicels longer than the next lower internode; calyx 4–10 mm long, blue to white, the lobes fringed within; pistils sessile; capsules lance-ovoid.

Heath or tundra; from islands of the Bering Sea disjunctly eastward along the mountains through eastern Alaska and southwestern Yukon, and northward along the coast to northwestern Alaska; eastward to Greenland and southward to California, Nevada, Arizona, and New Mexico; Eurasia. A phase from the Pribilof Islands with much contracted internodes and with 1–2 of the calyx lobes broadly ovate has been recognized as ssp. *pribilofii* Gillett.

The remainder of our material is designated as ssp. *tenella.*

LOMATOGONIUM A. Br.

Plants annual herbs, the stems erect or decumbent only at the base; leaves mainly cauline, opposite; flowers showy, moderately sized, borne in terminal cymes or in 1-flowered axillary cymes, or merely axillary; calyx (4)5-lobed, the lobes almost distinct; corolla rotate or nearly so, each segment with 2 scaly basal appendages, persistent; stamens (4)5, inserted at the base of the corolla; stigmas 2, decurrent over half the length of the ovary; fruit a 2-valved, septicidally dehiscent capsule.

Fernald, M. L. 1919. *Lomatogonium* the correct name for *Pleurogyne.* Rhodora 21:193.

Lomatogonium rotatum (L.) Fries
Marsh Felwort

Stems 0.5–5 dm tall, simple or branched from near the base; basal leaves elliptic to spatulate, early withering; cauline leaves ovate to linear-lanceolate, 0.5–2.5 cm long, 1–4 mm broad, not connate; sepals essentially distinct, 5–15 mm long, linear-lanceolate, acute; corolla rotate, 6-15 mm long, blue to pale blue or white; pistil sessile; capsule cylindrical.

Stream banks, lake shores, marshes, and bogs; in much of Alaska south and sometimes north of the 68th parallel and in much of the Yukon; east to the Atlantic and disjunctly southward to Idaho, Wyoming, and Colorado (*L. rotatum* var. *tenuifolium* Griseb.; *L. rotatum* ssp. *tenuifolium* [Griseb.] Porsild; *Pleurogyne rotata* [L.] Griseb.).

SWERTIA L.

Plants perennial, rhizomatous herbs, the stems erect; leaves basal and cauline, the cauline opposite or sometimes alternate; flowers showy, moderately sized, borne in terminal or axillary cymes or solitary; calyx (4)5-lobed, the lobes almost distinct; corolla rotate, each segment with 2 fringed basal glands; stamens (4)5, inserted near the base of the corolla; styles 1, short, the stigma capitate; fruit a 2-valved, septicidally dehiscent capsule.

Swertia perennis L. (× 0.4).

St. John, H. 1941. Revision of the genus *Swertia* (Gentianaceae) of the Americas and the reduction of *Frasera*. Am. Midl. Nat. 26:1–29.

Swertia perennis L.

Plants perennial, the stems 1–6 dm tall; basal leaves obovate or oblong-elliptic, petiolate, 4–20(25) cm long, 1–4.5 cm broad, cauline leaves short-petiolate to sessile, reduced in size upwards, clasping but not connate; sepals essentially distinct, 5–12 mm long, narrowly lanceolate, acuminate; corolla rotate, 10–16 mm long, dark bluish purple to whitish, often spotted; pistil sessile; capsule lance-ovoid.

Stream banks, moist meadows, and woods; in coastal and insular or less commonly interior southern Alaska (to be expected in southwestern Yukon); south to California, Utah, and Colorado; Eurasia.

GERANIACEAE
Geranium Family

Herbaceous annual or perennial plants; leaves alternate, or rarely opposite, compound or if simple then lobed or divided; flowers perfect, mostly regular, solitary, umbellate, cymose, or racemose; calyx 5-lobed; corolla of 5 distinct petals; stamens with anthers 5 or 10, more or less united by the filaments; pistils 1, the ovary superior, 5-loculed, 5-carpelled; styles 1; fruit capsular, separating at maturity into 5 1-seeded segments, coiling from the base.

1a. Flowers with a spur which is adnate to the pedicel; greenhouse or household ornamentals. *Pelargonium*
1b. Flowers lacking a spur; plants native or introduced but not as above. (2)

2a. Leaves palmately lobed or divided; stamens all bearing anthers; plants perennial or annual. *Geranium*
2b. Leaves pinnately dissected; only 5 stamens bearing anthers; plants annual. *Erodium*

ERODIUM L'Her.

Plants annual; leaves opposite, pinnately compound; flowers small, showy, borne in umbels on axillary peduncles; sepals 5, distinct or nearly so, persistent; petals 5, cauducous; staminal filaments 10, 5 short and sterile alternating with 5 longer, fertile ones; styles much longer than the ovary, spirally twisted at maturity.

Erodium cicutarium (L.) L'Her.
Alfilaria, Storksbill

Stems 0.5–4 dm long or more, decumbent to erect, spreading-hairy; leaves mainly basal, the cauline ones opposite, 2–30 cm long, the leaflets again lobed or cleft; flowers 6–12 mm broad; petals 3–5 mm long; petals pink; fruit 2–5 cm long.

Weed of waste places; introduced in southeastern Alaska, to be expected elsewhere; native to the Old World.

GERANIUM L.

Plants perennial or annual herbs; leaves alternate or opposite, or chiefly basal, palmately lobed or divided; flowers often large, showy, borne solitary or in umbels on axillary peduncles; sepals 5; petals 5, soon deciduous; stamens 10, usually all anther-bearing; styles much longer than the ovary, curved or coiling in fruit.

1a. Petals 5–12 mm long; plants annual or biennial from a slender taproot, introduced (except *G. bicknellii*). (2)
1b. Petals 12–20 mm long; plants perennial from a thick, caudexlike rhizome, indigenous (except *G. sanguineum*). (4)

2a. Leaves palmately compound; carpels glabrous; petals more than 7 mm long. *G. robertianum*
2b. Leaves palmately dissected; carpels pubescent; petals usually less than 7 mm long. (3)

3a. Fruiting pedicels usually twice longer

than the calyx or more; stylar beak usually more than 3 mm long; plants indigenous, widespread. *G. bicknellii*

3b. Fruiting pedicels usually less than 1.5 times longer than the calyx; stylar beak usually less than 3 mm long; plants introduced. *G. carolinianum*

4a. Stems with several petiolate leaves; peduncles usually with a single flower; pubescence of inflorescence not bulbous tipped; plants cultivated. *G. sanguineum*

4b. Stems with 1–2 petiolate leaves; peduncles usually with several flowers; pubescence of inflorescences with bulbous tips; plants indigenous. (5)

5a. Flowers blue to pink purple (rarely white); style branches 1–2 mm long; stylar beak 2.5–5 mm long in fruit, glabrous or merely strigose at the base; our common wild geranium. *G. erianthum*

5b. Flowers white, with pinkish or purplish veins; style branches 2.5–4 mm long; stylar beak 1–2 mm long in fruit, hairy; plants of southern Yukon. *G. richardsonii*

Geranium bicknellii Britt.
Bicknell Cranesbill

Plants annual or biennial, the stems 1.5–5 dm tall, erect or decumbent, spreading-hairy and often glandular, from a slender taproot; leaf blades cordate to rounded in outline, 1–3.5 cm long (from sinus to apex), the segments not distinct; flowers usually 2 per peduncle; fruiting pedicels usually 2–3 times longer than the calyx; sepals 4–8 mm long, aristate; petals pink, 3.5–7 mm long; stylar column 14–20 mm long (including the beak), pubescent, the beak 3–6 mm long, pubescent; style branches 1–1.5 mm long.

Disturbed soils along roadsides and in open woods; from central eastern Alaska eastward through southern Yukon; east to Newfoundland and south to Washington,

Utah, Michigan, and New York.

Geranium carolinianum L.
Carolina Cranesbill

Plants annual, the stems 1.5–5 dm tall, usually erect, retrorsely hairy and often glandular, from a slender taproot; leaf blades cordate to orbicular in outline, 1–4 cm long (from sinus to apex), the segments not distinct; flowers 2–more per peduncle, the peduncles often very short; fruiting pedicels usually less than 1.5 times longer than the calyx; sepals 5–8 mm long, aristate; petals pink, 5–7 mm long; stylar column 10–15 mm long (including the beak), pubescent, the beak 1–2 mm long, pubescent; style branches 1 mm long or less.

Introduced weed of waste places and gardens; uncommon in southeastern Alaska, to be expected elsewhere; widely distributed in North America.

Geranium erianthum DC.
Northern Geranium

Plants perennial, the stems 2–8 dm tall or more, erect or ascending, retrorsely hairy, from a thick, caudexlike rhizome; leaf blades orbicular in outline, 1.5–10 cm long (from sinus to apex), much broader than long, the segments not distinct; flowers 2–several per peduncle, the peduncles short to long; fruiting pedicels mostly once to twice longer than the calyx; sepals 8–12 mm long, pubescent with bulbous-capitate hairs, aristate; petals blue to pink purple (rarely white), 16–20 mm long, glabrous on the inner surface except at the base; stylar column 25–32 mm long (including the beak), pubescent, the beak 6–9 mm long, glabrous or merely strigose; style branches 1.5–2 mm long.

Woods, thickets, and meadows; in much of Alaska south of the 65th parallel, and in extreme southern Yukon; south to British Columbia; Asia. This is our common wild geranium. It is one of the most beautiful of the American species.

Geranium erianthum DC. (× 0.3).

Geranium richardsonii Fisch. & Trautv.
Richardson Geranium

Plants perennial, the stems 2–8 dm tall, erect, retrorsely hairy (often sparsely so), from a thick, caudexlike rhizome; leaf blades 2–8 cm long (from sinus to apex), much broader than long, the segments not distinct; flowers usually 2 per peduncle, the peduncles mostly long; fruiting pedicels usually 2–4 times longer than the calyx; sepals 6–11 mm long, aristate, with bulbous-capitate pubescence; petals white, with pinkish or purplish veins, 12–16 mm long, hairy on the inner surface; stylar column 16–22 mm long (including the beak), pubescent, the beak 2–4 mm long, pubescent, the style branches 2.5–4 mm long.

Meadows and woodlands; in southern Yukon; southward to California and New Mexico.

Geranium robertianum L.
Robert Geranium

Plants annual, the stems 1–6 dm tall, decumbent to erect, spreading-hairy with capitate hairs, from a slender taproot; leaf blades ovate or pentagonal in outline, 1.5–5 cm long (from sinus to apex), somewhat broader than long, the segments distinct; flowers usually 2 per peduncle, the peduncles moderately long; fruiting pedicels usually less than 1.5 times longer than the calyx (rarely much longer); sepals 6–9 mm long, aristate, pubescent with bulbous-capitate hairs; petals pink to reddish purple, 10–15 mm long; stylar column 13–20 mm long (including the beak), glabrous to the beak, the beak 4–6 mm long, pubescent or glabrate, the style branches about 1 mm long.

Cultivated ornamental; escaping in southeastern Alaska, to be expected elsewhere; native to eastern North America.

Geranium sanguineum L.

Plants perennial, the stems 1–4 dm tall or more, ascending to erect, spreading-hairy with noncapitate hairs, from a thick, caudexlike rhizome; leaf blades oval to pentagonal in outline, 1–3 cm long (from sinus to apex), somewhat broader than long, the segments not distinct; flowers 1 per peduncle, the peduncles moderately long; fruiting pedicels usually several times longer than the calyx; sepals 8–13 mm long, aristate, pubescent with noncapitate hairs; petals pink purple or white with pink or purple veins, 15–25 mm long; stylar column 20–25 mm long (including the beak), pubescent, the beak 2–4 mm long, pubescent, the style branches 2–4 mm long.

Introduced ornamental plant of gardens; escaping in southeastern Alaska; native to Eurasia.

PELARGONIUM L'Her.

Plants herbaceous, perennial (under glass) herbs; leaves alternate or opposite, usually palmately-veined and lobed; flowers small to large, showy, borne in umbels on axillary peduncles; sepals 5, produced into a spur adnate to the pedicel; petals 5, the upper 2 differing in size from the lower 3; stamens 10, usually only 5 with anthers; stylar column much longer than the ovary, usually coiling in fruit.

Pelargonium hortorum Bailey
Common Geranium

Stems mostly 2–5 dm long, somewhat succulent, pubescent; leaves alternate (rarely opposite), the blades 2.5–10 cm broad, orbicular to reniform, shallowly lobed and crenate; flowers several to numerous in involucrate umbels; petals pink, salmon, red, white, or variously spotted, 10–25 mm long.

The common geranium is widely cultivated as a household or greenhouse ornamental wherever domestic woman finds space and light.

HALORAGACEAE
Watermilfoil Family

Perennial, aquatic herbs with whorled, simple or dissected leaves; stems wholly or partially submerged or rarely emergent; flowers regular, minute, perfect or imperfect, sessile in the axils of the leaves, or in terminal spikes; sepals 2–4, or lacking; petals 2–4, distinct or lacking; stamens 1–8; pistils 1, the ovary inferior, 1–4-loculed; styles 1–4; fruit a nutlet or drupe.

1a. Submerged leaves similar to the emergent leaves, simple, entire; stems stout, erect; flowers often perfect, with 1 stamen. *Hippuris*
1b. Submerged leaves pinnatifid, the segments filiform; stems slender, usually not erect; flowers usually imperfect, with 4–8 stamens in staminate flowers. *Myriophyllum*

HIPPURIS L.

Plants aquatic or semiaquatic, rhizomatous herbs; leaves whorled, entire; flowers usually perfect (except in *H. montana*), sessile in the axils to the upper leaves; calyx adnate to the ovary, lobed or entire apically; petals lacking; stamens 1, the anther large; pistils 1-carpellate, 1-loculed; style 1; fruit nutlike, indehiscent, 1-seeded.

1a. Plants less than 1 dm tall; leaves 2–6 (10) mm long and 1 mm broad or less; stems less than 1 mm thick. *H. montana*
1b. Plants mostly 1–4 dm tall; leaves 6–30(50) mm long and 1–8 mm broad; stems more than 1 mm thick. (2)

2a. Leaves mostly 4–6(8) per whorl, 6–15 mm long and 2–8 mm broad, often obtuse. *H. tetraphylla*
2b. Leaves mostly (6)8–12 per whorl, 6–30(50) mm long and 1–2 mm broad, acute. *H. vulgaris*

Hippuris montana Ledeb. ex Reichenb.
Mountain Marestail

Stems 0.3–0.8 mm thick, mostly 1.5–8(10) cm tall; leaves 5–8 per whorl, 2–6(10) mm long, 0.5–1 mm broad; flowers mostly imperfect, the staminate ones below the pistillate; anthers sessile or on filaments to 0.5 mm long; fruit about 1 mm long.

Alpine sites on mossy banks, in wet rills, and in moist meadows; in southern Alaska, from the Aleutians to the Panhandle; south to Washington.

Hippuris tetraphylla L. f.
Four-leaf Marestail

Stems 1.5–4 mm thick, mostly 15–40 cm tall; leaves 4–6(8) per whorl, 6–15 mm long and 2–8 mm broad (at least the main ones), often obtuse; flowers mostly perfect; anthers subequal to the pistil in size, the filaments shorter than the anther; fruit about 2 mm long.

Shallow ponds or mud flats; from coastal northern Yukon westward around the coastal margin of Alaska to British Columbia; disjunctly eastward to the Atlantic; Eurasia (*H. maritima* Hook.; *H. vulgaris* var. *maritima* [Hook.] Macoun & Holm.).

Hippuris vulgaris L.
Common Marestail

Stems 1.5–5 mm thick, mostly 10–40 cm tall (at least as regards the aerial portion, the submerged portion often much longer); leaves (6)8–12 per whorl, 6–30(50) mm long, 1–2 mm broad, acute; flowers mostly perfect; anthers subequal to the pistil, the filaments to 1 mm long; fruit 2–2.5 mm long.

Shallow ponds, streams, or mud flats; in most of Alaska and Yukon; widely distributed in North America; circumboreal, also in South America.

MYRIOPHYLLUM L.

Plants aquatic, rhizomatous herbs; leaves whorled, pectinately dissected, the segments filiform; flowers imperfect, sessile in the axils of upper leaves; staminate flowers uppermost along the stem, each subtended by 2 bracts; sepals 4; petals 4, or lacking; stamens 4–8; pistillate flowers in the lower portion of the inflorescence; calyx adnate to the ovary, 4-toothed or entire; pistils 4-carpellate, 4-loculed; stigmas 4, plumose; fruit of 4, 1-seeded achenes or nutlets.

1a. Main leaves 3–12 mm long; fruit 1.5–2 mm long; at least the upper bracts and flowers alternate; plants evidently rare in our region. *M. alterniflorum*
1b. Main leaves 10–32 mm long; fruit 2–3 mm long; bracts and flowers opposite or whorled; plants common in our region. *M. spicatum*

Myriophyllum alterniflorum DC.

Stems to 10 dm long; leaves whorled, 3–10 (12) mm long; inflorescences immersed, 2–5 cm long; bracts mostly alternate (at

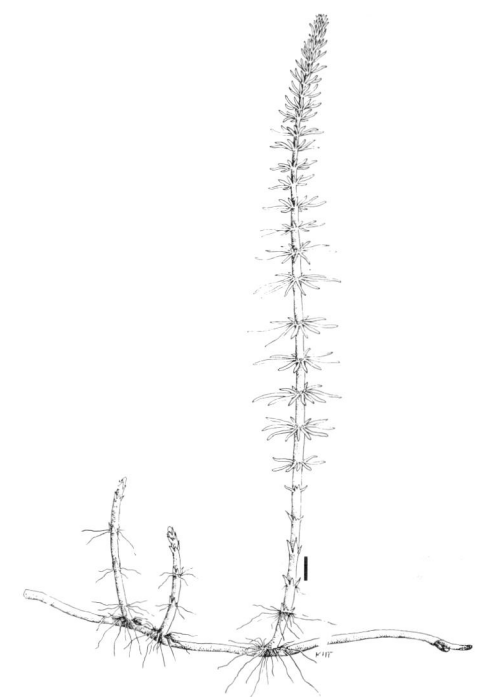

Hippuris vulgaris L. (× 0.3).

least those subtending the upper flowers), mostly shorter than the flowers, entire or the lowest ones pinnately dissected and exceeding the flowers; stamens 8; fruit 1.5–2 mm long.

Ponds; reported from widely disjunct sites in Alaska; eastward to Ontario and Maryland; Greenland, Europe.

Myriophyllum spicatum L.
Spike Watermilfoil

Stems 3–8 dm tall or more; leaves whorled, 10–32 mm long; inflorescences immersed or sometimes emergent; bracts opposite or whorled, from shorter than the flowers and entire to considerably longer and pinnately dissected; stamens 8; fruit 2–3 mm long.

Ponds and slow-moving streams; in much of Alaska and southern Yukon; widespread in North America; Eurasia.

1a. Bracts subtending flowers entire to serrulate, usually shorter than the fruits; plants of broad distribution in Alaska and southern Yukon, our common variety (*M. exalbescens* Fern.; *M. spicatum* ssp. *exalbescens* [Fern.] Hultén). *M. spicatum* var. *exalbescens* (Fern.) Jeps.

1b. Bracts subtending flowers pinnately dissected, usually longer than the fruits; plants from central and southeastern Alaska (*M. verticillatum* L.). *M. spicatum* var. *spicatum*

HYDROPHYLLACEAE
Waterleaf Family

Plants annual, biennial, or perennial herbs; leaves alternate or basal, or opposite (in *Nemophila*), pinnately or palmately lobed; flowers regular, perfect, arranged in cymes; calyx of 5 more or less united sepals; petals 5, united, the corolla rotate to campanulate; stamens 5, alternating with the corolla lobes; pistils 1, the ovary superior, 1 (or falsely 2)-loculed; styles 1, apically 2-cleft or entire; fruit a loculicidal capsule.

1a. Leaves opposite; plants annual, with spreading to erect stems; flowers commonly 15–25 mm broad, solitary in the leaf axils. *Nemophila*

1b. Leaves alternate or basal, or both, not opposite; plants annual, biennial, or perennial; flowers commonly less than 15 mm broad, few to numerous, in scorpioid or racemelike cymes. (2)

2a. Leaves pinnately veined and lobed; styles 2-cleft; pubescence mostly simple; inflorescence scorpioid. *Phacelia*

2b. Leaves palmately veined and lobed; styles 1, entire or nearly so; pubescence of multicellular hairs; inflorescence not scorpioid. *Romanzoffia*

NEMOPHILA Nutt. Nom. Cons.

Annual herbs from taproots; leaves opposite, pinnatifid; flowers solitary in the leaf axils; sepals distinct or nearly so, with an appendage at each sinus; corolla showy, saucer shaped; staminal filaments attached near the base of the corolla tube; styles 2-cleft; capsules loculicidal, with 2–several seeds in the single locule.

Nemophila menziesii H. & A.

Plants annual, the stems 1–3 dm tall, simple or more commonly branched, spreading to erect, strigose to retrorsely hairy; leaves mainly cauline, opposite, the blades once to twice pinnatifid, 2–4.5 cm long; flowers solitary; corolla white, sometimes suffused with purple or purplish-veined, deciduous, 6–12 mm long, 15–25 mm broad; stamens exserted, the anthers purplish black; capsules 3–6 mm long.

Cultivated ornamental; escaping and persisting in southeastern Alaska; introduced from western United States (it is a common component of "native wild flower" seed mixtures).

PHACELIA Juss.

Annual, biennial, or perennial herbs from taproots; leaves alternate, pinnately dissected or lobed; flowers borne in scorpioid cymes; sepals distinct or nearly so; corolla showy, rotate to campanulate; staminal filaments attached near the base of the corolla tube; styles 2-cleft; capsules loculicidal, with 2–several seeds in each of the 2 locules.

1a. Plants annual or biennial, the stem base lacking a thick cluster of old leaf bases; staminal filaments not or only slightly exserted from the corolla, glabrous; corolla glabrous within. *P. franklinii*

1b. Plants perennial from a taproot and caudex, the stem base with a thick cluster of old leaf bases; staminal filaments long-exserted from the corolla, glabrous; corolla hairy within. (2)

2a. Corollas cream to yellowish, often tinged with blue, 8–10 mm long and 10–14 mm broad, deciduous or tardily pubescent; herbage rather densely and softly villous. *P. mollis*

2b. Corolla purple or dark blue, 5–7 mm long and 7–10 mm broad, persistent; herbage variously pubescent, seldom softly villous. *P. sericea*

Phacelia franklinii (R. Br.) Gray

Plants annual or biennial, the stems 1.2–6 dm tall, often branched from near the base and sometimes from above, usually purplish, pubescent with spreading hairs, densely so in the inflorescence; leaves mainly cauline, only slightly reduced upwards, the blades once to twice pinnatifid, 1.5–9 cm long; cymes solitary, or several at the end of stems or branches; corolla purplish (fading cream), usually deciduous, 6–9 mm long, 8–12 mm broad, glabrous inside, hairy outside; stamens included or slightly exserted, the filaments hairy; capsules 6–10 mm long.

Disturbed soils along roads, in clearings, and on stream gravels; in central eastern Alaska, southern Yukon, and in southeastern Alaska (Skagway); south to Idaho and Wyoming and east to Lake Superior.

Phacelia mollis Macbr.

Plants perennial, the stems 1.5–5 dm tall, branching from the caudex, greenish, rather densely pubescent with soft, villous-spreading hairs, about equally hairy in the inflorescence; leaves mainly basal, the blades once-pinnatifid (the lobes sometimes again-toothed), 2.5–10 cm long, persisting, the cauline leaves much reduced upwards; cymes few to numerous in compact to interrupted inflorescences; corolla cream to yellowish, often tinged with blue, persisting or deciduous, 8–10 mm long, 10–14 mm broad, hairy inside and out; stamens long-exserted, the filaments glabrous; capsules 5–7 mm long.

Sandy or gravelly soils or rock outcrops, usually in open woods; in central eastern Alaska, southern Yukon, and in southeastern Alaska (Haines, Skagway); endemic. This is a very attractive plant. It should find use in rock garden plantings.

Phacelia sericea (Grah.) Gray

Plants perennial, the stems 1–5 dm tall, branching from the caudex, greenish, thinly strigose to densely sericeous or villous, densely villous in the inflorescence; leaves mainly basal, the blades once to twice pinnatifid, 1.5–8(10) cm long, persisting, the cauline leaves much reduced upwards; cymes several to many in compact to interrupted inflorescences; corolla purple or dark blue, persistent, mostly 5–7 mm long and 7–10 mm broad, hairy both outside and inside; stamens long-exserted, the filaments glabrous; capsules 5–9 mm long.

Open woods; in scattered locations from southeastern (Klukwan) and eastern (Eagle) Alaska, and reported from Yukon (?); south to California and east to Alberta and Colorado. This species is evidently rare in our region. It represents another of the tremendous disjunctions between the main range of the species and Alaska-Yukon.

ROMANZOFFIA Cham.

Plants perennial herbs; leaves primarily basal, the cauline ones alternate, palmately veined, toothed, or lobed; flowers borne in racemelike cymes; sepals distinct or nearly so; petals 5, united, the corolla campanulate; staminal filaments attached near the base of the corolla tube; styles entire or minutely 2-lobed; capsules loculicidal with numerous seeds in each of the 2 locules.

1a. Sepals glabrous or minutely stipitate-glandular, 2–3.5 mm long; capsules 5–6 mm long; plants from Kodiak Island and eastward. *R. sitchensis*
1b. Sepals rather densely villous with multicellular hairs, 4–6 mm long; capsules 6–7 mm long; plants from Prince William Sound and westward. *R. unalaschcensis*

Romanzoffia sitchensis Bong.
Mist-maid

Stems 0.4–2(3) dm long, sparsely villous

with multicellular hairs or sparsely stipitate-glandular; leaves long-petioled, the bases sheathing and long-ciliate, the blades reniform to orbicular, 0.3–2.5 cm long, 0.4–3.6 cm broad, with 3–9 ovate lobes; inflorescences longer than the leaves; pedicels minutely stipitate-glandular; sepals 2–3.5(4.5) mm long, minutely stipitate-glandular to glabrous; corolla cream to white, 6–10 mm long; capsules 5–6 mm long.

Moist sites on rock outcrops and along streams; in coastal and insular southern Alaska from Kodiak Island and vicinity, eastward to southeastern Alaska; southward to California and east to Alberta and Montana. This and the next species are frequently confused with species of *Saxifraga*.

Romanzoffia sitchensis Bong. (× 0.5).

Romanzoffia unalaschcensis Cham.

Stems 0.6–2 dm long, sparsely to densely villous with multicellular hairs; leaves long-petioled, the bases sheathing, ciliate and often dorsally villous, the blades reniform, 0.7–2.5 cm long, 0.9–3.5 cm broad, with 7–9 ovate lobes; inflorescences usually longer than the leaves; pedicels villous with multicellular hairs; sepals 4–6 mm long, sparsely to densely villous, ciliate; corolla cream to white, 8–10 mm long; capsules 6–7 mm long.

Moist banks and crevices of rock outcrops; Prince William Sound, Kodiak Island and vicinity westward along the Alaska Peninsula and Aleutian Islands; endemic (*R. unalaschcensis* var. *glabriuscula* Hultén).

LABIATAE

Mint Family

Plants annual or perennial, usually aromatic herbs, ordinarily with square or 4-angled stems; leaves simple, opposite or rarely whorled; flowers perfect, mostly irregular, borne in various types of cymose inflorescence; calyx of 5 united sepals, regular or irregular, usually 5-lobed, or the lobes obscure; corolla of 5 united petals, usually bilabiate, 5-lobed, or apparently 4-lobed by fusion of the upper 2 lobes; stamens 4, in 2 unequal pairs, or only 2 by abortion; pistils 1, the ovary superior, 2-carpelled, falsely 4-loculed and 4-lobed; styles 1, usually bifid apically; fruit a schizocarp, beaking at maturity into 4 1-seeded nutlets.

1a. Stamens with anthers 2 only (2 staminodes sometimes also present); corolla regular or nearly so. *Lycopus*

1b. Stamens with anthers 4; corolla 2-lipped (obscurely so in *Mentha*). (2)

2a. Calyx 2-lipped, the lips entire, the upper one bearing a cap or crest. *Scutellaria*

2b. Calyx not 2-lipped, or if so, then at least 1 lip toothed or lobed, no cap or crest present. (3)

3a. Corolla regular or nearly so, mostly 4-lobed, the upper lobe sometimes larger than the others and indistinctly 2-lobed; plants strongly aromatic. *Mentha*

3b. Corolla 2-lipped, the upper lip often 2-lobed and the lower lip 3-lobed; plants aromatic or scentless. (4)

4a. Calyx teeth 10, hooked at the apex; stems densely white-woolly. *Marrubium*

4b. Calyx teeth 5 or less, not hooked at the apex; stems variously pubescent or glabrous. (5)

5a. Inflorescence appearing axillary, the flowers or flower clusters subtended by more or less ordinary foliage leaves. (6)

5b. Inflorescence appearing terminal, the clusters mostly subtended by non-foliose bracts. (9)

6a. Lower lip of corolla constricted at the base of the enlarged central lobe, the lateral lobes each terminating in a short tooth. *Lamium*

6b. Lower corolla lip not constricted at the base of the central lobe, the lateral lobes not as above. (7)

7a. Flowers sessile, in compact clusters; plants erect, not rooting at the nodes. *Galeopsis*

7b. Flowers distinctly pediceled, solitary or in loose clusters; plants decumbent to prostrate, often rooting at the nodes. (8)

8a. Pedicels 5–15 mm long; flowers mostly solitary; corolla 7–10 mm long. *Satureja*

8b. Pedicels less than 5 mm long; flowers usually 2–3; corolla commonly 12–20 mm long. *Glecoma*

9a. Bracts of inflorescence spinose-toothed; upper calyx tooth ovate, twice as broad as the others. *Moldavica*

9b. Bracts of inflorescence, if present, not spinose-toothed; calyx various but not as above. (10)

10a. Flowers in clusters in the axils of upper leaves (at least some), sometimes in only 1 terminal, leaf-subtended cluster. *Lamium*

10b. Flowers in dense or somewhat interrupted spikes, if leafy at all, then only at the base. (11)

11a. Calyx distinctly 2-lipped; flowers in dense, interrupted spikes; bracts ovate to reniform, not at all leaflike. *Prunella*

11b. Calyx not 2-lipped, the teeth equal or nearly so; flowers in rather interrupted spikes; bracts when present leaflike, except smaller. (12)

12a. Upper (inner) pair of stamens longer than the lower pair; lower flower clusters distinctly pedunculate; stems branched; leaves petioled; plants aromatic. *Nepeta*

12b. Upper (inner) pair of stamens shorter than the lower pair; lower flower clusters sessile; stems usually unbranched; leaves sessile or subsessile; plants not aromatic. *Stachys*

GALEOPSIS L.

Plants annual, from taproots, the stems erect, retrorsely hispid; leaves petiolate, coarsely serrate; flowers sessile, numerous, borne in dense axillary clusters, subtended by foliage leaves; calyx with 5 spinescent teeth; corolla bilabiate, the upper lip entire, the lower lip 3-lobed, with a pair of prominent processes near the base; stamens with anthers 4, the lower pair longer; anther sacs transversely 2-valved, hairy.

Galeopsis tetrahit L.
Hemp Nettle

Plants 2–8 dm tall, the stems simple or branched above; leaf blades 2.5–8 cm long, 1–4.5 cm broad, ovate to lanceolate or elliptic, coarsely serrate, acute to obtuse or rounded basally, acute to acuminate apically, hairy to glabrate on both sides; flower clusters usually 2–6; calyx 8–15 mm long, the spinescent teeth subequal to the tube; corolla purplish or pinkish to white, 15–22 mm long, pubescent.

Weedy plants of disturbed sites, roadsides, gardens, and agricultural land; mostly in southern Alaska, but also in the interior (to be expected in the Yukon); introduced from Europe (*G. bifida* Boenn.; *G. tetrahit* ssp. *bifida* [Boenn.] Fries). Our plants belong to var. *bifida* (Boenn.) Lej. & Court.

GLECOMA L.

Plants perennial from stolons, the stems decumbent and rooting at the nodes, retrorsely scabrous to glabrous; leaves petiolate, crenate; pedicellate, few, borne in loose, axillary clusters, subtended by foliage leaves; flowers with 5 spinose teeth; corolla bilabiate, the upper lip 2-lobed, the lower lip 3-lobed (the central one largest); stamens with anthers 4, the upper pair longer; anther sacs opening by separate slits.

Glecoma hederacea L.
Ground-ivy

Plants with decumbent stems, the branches erect, 1–4 dm long; leaf blades reniform, 0.5–3 cm long (from sinus to apex), 0.7–4 cm broad, crenate, glandular-punctate and somewhat hairy below, hairy above; flower clusters usually 2–3; calyx 5–7 mm long, rough-hairy, the teeth much shorter than the tube; corolla bluish purple, mostly 12–20 mm long.

Weedy plant of moist, shaded sites, introduced as a cultivated ornamental; in southeastern Alaska, and to be expected elsewhere; native of Eurasia.

LAMIUM L.

Plants perennial from tap or fibrous roots, retrorsely hairy to glabrate; leaves petiolate, crenate-serrate, often doubly so; flowers sessile, borne in loose axillary clusters, subtended by foliage leaves, rarely terminal; calyx with 5 subequal teeth; corolla bilabiate, the upper lip entire, hooded, the lower lip very unequally 3-lobed, the central lobe broad and constricted at the base, the lateral lobes represented by slender teeth; stamens with anthers 4, the lower pair the longer; anthers pubescent.

1a. Terminal tooth of at least the upper leaves more than twice longer than broad, acute; flowers white. *L. album*

1b. Terminal teeth of upper leaves about as broad as long, obtuse; flowers purplish, pink, or white. *L. maculatum*

Lamium album L.
White Dead-nettle

Stems 2–5 dm tall, decumbent to ascending or erect; leaf blades green, ovate to lanceolate, 3–10 cm long, 2–5 cm broad, once or twice crenate-serrate, the apical tooth slender, acute, spreading-hairy; calyx 10–13 mm long, spreading-hairy, the teeth longer than the tube; corolla white, 25–30 mm long, the upper lip 7–12 mm long, villous.

Cultivated ornamental; escaping and persisting in southeastern Alaska; introduced from Eurasia.

Lamium maculatum L.
Spotted Dead-nettle

Stems 1.5–6 dm long, decumbent (sometimes rooting at the nodes) to ascending or erect; leaf blades with a whitish strip along the midvein, ovate to cordate, 2–8 cm long, 1.5–4 cm broad, once to twice crenate-serrate, the apical tooth about as broad as long, obtuse, spreading-hairy; calyx 8–10 mm long, the teeth subequal to the tube or shorter; corolla purplish, pink, or white, 20–25 mm long, the upper lip 7–12 mm long, villous.

Cultivated ornamental; persisting in more temperate portions of Alaska and Yukon (?); introduced from Europe.

LYCOPUS L.

Plants rhizomatous perennials, the stems scabrous to glabrate; leaves short-petiolate to nearly sessile, coarsely serrate; flowers sessile, borne in dense, axillary clusters subtended by foliage leaves; calyx 5-toothed; corolla regular or nearly so, 4-lobed, the upper lobe often broader than the others and usually emarginate; stamens with anthers 2, the upper pair represented by staminodes, or lacking; anthers glabrous.

1a. Calyx lobes lance-acuminate, 1–2 mm long; leaves subsessile to sessile. *L. asper*

1b. Calyx lobes broadly triangular, 1 mm long or less, obtuse or acute; leaves with short but distinct petioles. *L. uniflorus*

Lycopus asper Greene
Water Horehound

Plants 2–8 dm tall, the stems erect, simple or branched, sometimes stoloniferous from the base; leaves sessile to subsessile, the blades elliptic to oblong or oblanceolate, 2–8 cm long, 1–3 cm broad, coarsely serrate; calyx 2–3 mm long, the lobes lance-acuminate, scabrous; corolla white, 3–5 mm long; nutlets 1.6–2.1 mm long, 1.4–1.8 mm broad, truncate apically.

Marshy areas; known in Alaska from Circle; disjunctly southward to California and east to Saskatchewan (*L. lucidus* var. *americanus* Gray; *L. asper* ssp. *americanus* [Gray] Hultén).

Lycopus uniflorus Michx.
Northern Bugleweed

Plants 1–4 dm tall, the stems erect or ascending, usually simple, sometimes stoloniferous near the base; leaves short-petiolate, the blades lanceolate to oblong, 2–6 cm long, 0.6–2.5 cm broad, coarsely serrate; calyx 1–2 mm long, the lobes triangular, obtuse to acute, scabrous; corolla white or tinged pinkish, 2.5–3.5 mm long; nutlets

1.1–1.8 mm long, 0.8–1.2 mm broad, truncate and toothed apically.

Marshy areas; in central (Manley Hot Springs) and southeastern Alaska; east to Newfoundland and south to California, Nebraska, and North Carolina.

MARRUBIUM L.

Plants perennial from taproots, the stems woolly-tomentose; leaves petiolate, serrate-dentate; flowers sessile, borne in dense axillary clusters subtended by foliage leaves; calyx with 10 spinescent teeth, each hooked at the apex; corolla strongly bilabiate, the upper lip entire or emarginate, the lower lip 3-lobed, the central lobe often emarginate; stamens with anthers 4, the lower pair longer; anthers glabrous.

Marrubium vulgare L.
Horehound

Plants 3–10 dm tall, the stems frequently several from the root crown, basally decumbent to ascending or erect, woolly-tomentose; leaf blades broadly elliptic to ovate, 2–6 cm long, 1.5–5 cm broad, dentate-serrate, the alternate teeth often smaller, softly villous; calyx 5–8 mm long, villous, the hooked teeth from shorter than, to subequal to, the tube; corolla whitish, 8–11 mm long, pubescent.

Introduced weed; in southeastern Alaska (Juneau), and to be expected elsewhere; native to Eurasia.

MENTHA L.

Plants aromatic, rhizomatous perennials, the stems retrorsely scabrous to glabrate; leaves petiolate to sessile, serrate; flowers pedicellate, borne in axillary clusters, or the clusters aggregated into terminal, spikelike inflorescences; calyx with 5, nonspinescent teeth; corolla regular or nearly so, 4-lobed, the upper lobe often broader and emarginate to entire; stamens with anthers 4, usually exserted; anthers glabrous.

1a. Flower clusters axillary, subtended by foliage leaves; plants native. *M. arvensis*

1b. Flower clusters crowded in terminal, spikelike inflorescences, the individual clusters not subtended by foliage leaves; plants introduced. (2)

2a. Leaves distinctly petiolate; calyx 2.5–4 mm long; spikes 1 cm broad or more. *M. piperita*

2b. Leaves sessile or subsessile; calyx 1.5–2 mm long; spikes usually less than 1 cm thick. *M. spicata*

Mentha arvensis L.
Field Mint

Stems 1.5–8 dm tall, ascending to erect, simple or branched; leaves short-petiolate, the blades elliptic to lanceolate, ovate or oval, 1–8 cm long, 0.7–3 cm broad, once to twice serrate, acuminate, acute, obtuse, or rounded apically, usually glandular-punctate; calyx 2–3 mm long, pubescent and often glandular, the triangular teeth shorter than the tube; corolla purplish to pinkish or white, 4–6 mm long.

Seeps, springs, stream banks, and bogs; in much of Alaska south of the 66th parallel (except for the southwestern portion) and less commonly in southern Yukon; widely distributed in North America; circumboreal. Our plants belong to var. *glabrata* (Benth.) Fern. (*M. canadensis* L.; *M. arvensis* var. *villosa* [Benth.] S. R. Stewart).

Mentha piperita L.
Peppermint

Stems 3–10 dm tall or more, usually branched; leaves petiolate, the blades lance-ovate to elliptic, 1.5–6 cm long, 1–3 cm broad, serrate, glandular-punctate; calyx 2.5–4 mm long, the tube glandular, the lance-acuminate teeth hispid, shorter than the tube; corolla pink purple to white, 4–5 mm long.

Introduced aromatic herb; persisting in

southeastern Alaska, and to be expected elsewhere; native of Europe.

Mentha spicata L.
Spearmint

Stems 3–10 dm tall or more, usually branched; leaves sessile or subsessile, the blades lanceolate to lance-oblong, elliptic or ovate, 2–8 cm long, 0.6–2.5 cm broad, serrate, glandular-punctate; calyx 1.5–2 mm long, the tube glandular, the lance-acuminate teeth hispid, subequal to the tube; corolla pink to white, 3–4 mm long.

Introduced aromatic herb; persisting in more temperate regions of Alaska; native of Europe.

MOLDAVICA Adams.

Plants biennial or perennial (annual?) herbs from taproots, the stems ascending to erect, rough-hairy; leaves petiolate, toothed or cleft; flowers pedicellate, borne in clusters aggregated in dense, terminal, spikelike inflorescences, or sometimes with additional clusters in the axils of the upper leaves, the bracts spinose-toothed; calyx 5-toothed, the upper tooth ovate, twice as broad as the others; corolla bilabiate, the upper lip 2-lobed, the lower lip 3-lobed, the central lobe often notched; stamens with anthers 4, the upper pair the longer; anthers glabrous.

Moldavica parviflora (Nutt.) Britt
Dragonhead

Stems 1–7 dm tall, simple or branched; leaf blades lanceolate to lance-elliptic, 2–10 cm long, 0.4–2 cm broad, serrate to coarsely toothed or incised, rough-hairy; spikes 2–3.5 cm thick, 2–6 cm long; calyx 8–12 mm long, villous, the teeth shorter than the tube; corolla pinkish or purplish, 10–14 mm long, slightly surpassing the calyx.

Disturbed soils along roads and streams; in eastern Alaska and southern Yukon; east to Quebec and south to Oregon, Arizona,

Missouri, and New York (*Dracocephalum parviflorum* Nutt.).

NEPETA L.

Plants aromatic, perennial herbs from taproots, the stems ascending to erect, softly pubescent with short, retrorse hairs; leaves petiolate, toothed to cleft; flowers sessile or subsessile, borne in clusters aggregated into spicate or paniculate inflorescences, the bracts not spinose-toothed; calyx 5-toothed, the teeth subequal; corolla bilabiate, the upper lip entire or 2-lobed, the lower lip 3-lobed, the central lobe largest and often notched; stamens with anthers 4, the anthers glabrous.

Nepeta cataria L.
Catnip

Plants 3–10 dm tall or more, the stems usually branched; leaf blades cordate-ovate, 2–9 cm long, 0.8–6 cm broad, coarsely and bluntly serrate to somewhat incised, soft-hairy; spikes or panicles 1.5–3 cm broad, 1.5–6 cm long; calyx 5–6 mm long, villous, the lance-acuminate teeth shorter than or subequal to the tube; corolla whitish, often purple spotted, 7–10 mm long, hairy.

Introduced aromatic herb; in southeastern Alaska, and to be expected elsewhere; native to Europe.

PRUNELLA L.

Perennial herbs from taproots (or fibrous roots) and often with rhizomes, pubescent with multicellular hairs; leaves petiolate, crenate to entire; flowers short-pedicellate, borne in clusters aggregated into dense spikelike inflorescences, the bracts entire, ciliate; calyx 2-lipped, the upper lip shallowly 3-toothed, the lower lip deeply cleft into 2 narrow teeth; corolla bilabiate, the upper lip entire or nearly so, the lower lip 3-lobed; stamens with anthers 4, the filaments notched near the apex, the anthers glabrous.

Prunella vulgaris L.
Heal-all

Stems 0.6–5 dm long, ascending to erect, usually simple; leaf blades lance-ovate to oblong or elliptic, 2–9 cm long, 0.7–4 cm broad, minutely puberulent to glabrous; spikes 1–2 cm broad, 2–8 cm long; calyx 6–10 mm long, sparsely villous, purplish, the lower teeth subequal to the tube; corolla pink purple to pink or white, 12–18 mm long, glabrous.

Moist sites in woods, meadows, or on talus slopes; in most of coastal and insular southern Alaska (rarely in the interior), and to

Prunella vulgaris L. (× 0.3).

be sought in the Yukon; circumboreal (*P. vulgaris* ssp. *lanceolata* [Barton] Hultén; *P. vulgaris* var. *aleutica* Fern.; *P. vulgaris* ssp. *aleutica* [Fern.] Hultén). Our material belongs to var. *lanceolata* (Barton) Fern.

SATUREJA L.

Plants perennial rhizomatous herbs, the stems puberulent, prostrate and frequently rooting at the nodes, with ascending to erect branches; leaves short-petiolate to subsessile, irregularly toothed; flowers pedicellate, borne singly in the axils of foliage leaves; calyx regular, the 5 teeth subequal; corolla bilabiate, the upper lip emarginate, the lower lip 3-lobed; stamens with anthers 4, the lower pair the longer; anthers glabrous.

Satureja douglasii (Benth.) Briq.

Stems to 10 dm long or more, prostrate, with ascending to erect branches; leaves short-petiolate to subsessile, the blades ovate to oval, 1–3.5 cm long, 0.7–3 cm broad, irregularly crenate; pedicels 5–15 mm long; calyx 4–5 mm long, puberulent, the teeth subequal; corolla white or suffused with purple, 7–10 mm long, pubescent.

Introduced plants; in southeastern Alaska; native in western North America from British Columbia south to California and east to Idaho.

SCUTELLARIA L.

Plants perennial rhizomatous herbs, the stems retrorsely scabrous to glabrate; leaves short-petiolate to subsessile, irregularly toothed to subentire; flowers pedicellate, borne singly in the axils of foliage leaves; calyx 2-lipped, the lips entire, the upper one bearing a transverse cap or crest; corolla bilabiate, the upper lip entire, the lower lip 3-lobed, the lateral lobes partially connate with the upper lip; stamens with anthers 4, the lower pair the longer; anthers pubescent.

Scutellaria galericulata L.
Skullcap

Stems 1.5–9 dm tall, erect or ascending, simple or branched; leaves short-petiolate to subsessile, the blades lanceolate, 2–8 cm long, 0.4–3 cm broad, irregularly serrate, sparsely hispid; calyx 3.5–6 mm long, pubescent, the upper lobe distinct from the lower one; corolla pink purple to blue or white, 15–20 mm long, pubescent; nutlets yellowish, warty, about 1.5 mm long.

Margins of ponds, lakes, and streams, and in marshes and bogs; in the southeastern third of mainland Alaska and southern Yukon; widely distributed in North America; circumboreal (*S. galericulata* var. *pubescens* Benth.; *S. epilobiifolia* Hamilt.).

STACHYS L.

Plants rhizomatous perennial herbs, the stems retrorsely or spreading hairy; leaves short-petiolate to subsessile, or sessile, crenate-serrate; flowers subsessile, the clusters borne in open terminal spikelike inflorescences, or sometimes with additional clusters in the upper leaf axils; calyx with 5 subequal spinescent teeth; corolla bilabiate, the upper lip entire or emarginate, the lower lip 3-lobed, the central lobe the largest; stamens with anthers 4, the lower pair the longer; anthers glabrous.

1a. Leaves distinctly petiolate, ovate-lanceolate; plants of extreme southeastern Alaska. *S. mexicana*
1b. Leaves sessile or subsessile, lance-oblong to lance-elliptic; plants of broad distribution. *S. palustris*

Stachys mexicana Benth.
Ciliate Hedge Nettle

Stems 3–8 dm or more tall, erect, simple or branched; leaves distinctly petioled, the blades ovate-lanceolate, 2–12 cm long, 1.5–8 cm broad, subcordate basally, crenate-serrate, sparsely hispid; spikes 6–15 cm long, interrupted; calyx 5–9 mm long, pubescent with long hairs and often stipitate-glandular, the lobes shorter than the tube; corolla pink purple or pink, 12–18 mm long, hairy.

Reported from extreme southeastern Alaska (Annette Island); south to California (S. emersoni Piper).

Stachys palustris L.

Stems 2–7 dm or more tall, erect, simple or branched; leaves sessile or subsessile, the blades lance-oblong to lance-elliptic, 3–10 cm long, 1–3 cm broad, rounded to subcordate and somewhat clasping basally, crenate-serrate, hispid; spikes 7–25 cm long, interrupted; calyx 6–9 mm long, long-hairy and also stipitate-glandular, the lobes shorter than the tube; corolla purplish, white-spotted, 11–15 mm long, hairy.

Disturbed soils; in central, southern, and southeastern Alaska; widely distributed in North America; circumboreal. Our materials belong to var. *pilosa* (Nutt.) Fern. (S. palustris ssp. *pilosa* [Nutt.] Epling).

LEGUMINOSAE
Pea Family

Plants herbaceous, or woody (*Caragana*); leaves alternate, compound or rarely simple, mostly stipulate; inflorescence various, but usually racemose; flowers perfect, irregular, papilionaceous; calyx tubular, 4–5-lobed; corolla with 5 petals, the upper large and broad (banner), enclosing the lateral pair (wings) and the lower usually connate pair (keel) of petals in bud; keel petals united and enclosing the stamens and pistil; stamens usually 10, diadelphous (9 united and 1 free) or monadelphous; pistils 1, the ovary superior, 1-carpelled, 1–2-loculed; fruit a sessile to stipitate legume (pod) or loment.

1a. Leaves even-pinnate. (2)
1b. Leaves odd-pinnate, or palmately compound or simple. (5)

2a. Leaf rachis terminated by a bristle; plants woody shrubs. *Caragana*

2b. Leaf rachis terminated by a simple or branched tendril; plants herbaceous. (3)
3a. Styles strongly dilated; stipules foliaceous, larger than the leaflets; cultivated annuals. *Pisum*
3b. Styles not strongly dilated; stipules not foliaceous, or if so, then smaller than the leaflets and the plants perennial; indigenous or cultivated annuals or perennials. (4)
4a. Styles bearded down one side; plants generally coarse. *Lathyrus*
4b. Styles bearded at the apex; plants generally slender. *Vicia*
5a. Leaflets toothed (subentire in some *Trifolium*); leaves commonly 3-foliolate. (6)
5b. Leaflets (or leaves) entire; leaves pinnately or palmately 5–many-foliolate, or simple. (8)
6a. Leaves palmately trifoliolate (pinnately so in T. *dubium* and T. *procumbens;* 5-foliolate in T. *lupinaster*); corolla withering and persistent. *Trifolium*
6b. Leaves pinnately trifoliolate; corolla deciduous. (7)
7a. Leaflets toothed along the distal one-third or less (more in some M. *lupulina*); racemes compact; pods curved or twisted. *Medicago*
7b. Leaflets toothed along the distal one-half or more; racemes elongate; pods straight or nearly so. *Melilotus*
8a. Leaves palmately compound; stamens monadelphous. *Lupinus*
8b. Leaves pinnately compound or simple; stamens diadelphous. (9)
9a. Fruit a loment; wings shorter than the keel. *Hedysarum*
9b. Fruit a legume; wings mostly as long as or longer than the keel. (10)
10a. Keel with an apical, porrect beak; ventral suture of pods produced internally forming a partial or complete

partition, or the partition absent; plants mostly acaulescent (caulescent in some *O. deflexa*). *Oxytropis*

10b. Keel lacking a beak; sutures variously produced internally, or not at all produced; plants mostly caulescent. *Astragalus*

ASTRAGALUS L.

Plants herbaceous, perennial, caulescent; stipules connate and sheathing or free; leaves alternate, odd-pinnate, with 5–many leaflets, the leaflets entire; racemes axillary, pedunculate, elongate or spikelike and contracted or subcapitate; flowers showy, the keel lacking a beak; stamens diadelphous; calyx cylindrical to campanulate, 5-toothed; pods sessile or stipitate, of several shapes and textures, glabrous or pubescent, 1-loculed, partially 2-loculed, or completely 2-loculed, by intrusion of the dorsal (lower) suture, dehiscent or indehiscent.

Barneby, R. C. 1964. Atlas of North American *Astragalus*. Mem. N. Y. Bot. Gard. 13:1–1188.

Welsh, S. L. 1963. Legumes of Alaska: *Astragalus* L. Iowa State Jour. Sci. 37: 353–88.

1a. Calyx cylindrical, the tube more than 5 mm in length (occasionally only 4 mm in *A. polaris*). (2)
1b. Calyx campanulate, the tube less than 4(4.8) mm in length. (5)

2a. Plants with malpighian hairs (hairs double pointed, attached some distance from the ends); pods erect, strigose. *A. adsurgens*
2b. Plants with simple hairs (hairs attached at one end); pods spreading or reflexed, or if erect then villous. (3)

3a. Stipules foliose; flowers yellowish, the banner margins white; pods stipitate, pendulous. *A. umbellatus*
3b. Stipules not foliose; flowers usually some shade of pink purple (rarely white), the banner margin seldom if ever white; pods sessile or substipitate, erect or spreading. (4)

4a. Racemes with 1–3 (rarely 5) flowers; pods spreading, greatly inflated at maturity, minutely strigose. *A. polaris*
4b. Racemes several-flowered; pods erect, not inflated at maturity, villous. *A. agrestis*

5a. Stipules reflexed, foliose; plants glabrous or nearly so. *A. americanus*
5b. Stipules erect, not foliose; plants pubescent. (6)

6a. Racemes with 1–3 (rarely 5) flowers. (7)
6b. Racemes with 5–many flowers. (8)

7a. Pods and ovaries substipitate; pods greatly inflated at maturity. *A. polaris*
7b. Pods and ovaries distinctly stipitate; pods not greatly inflated, falcately curved at maturity. *A. nutzotinensis*

8a. Pods and ovaries glabrous (strigose in some *A. australis*). (9)
8b. Pods and ovaries pubescent. (10)

9a. Leaflets 9–15; pods (16)18–33 mm long, not especially flattened; plants broadly distributed in Alaska and Yukon. *A. australis*
9b. Leaflets 15–21; pods 12–18 mm long, strongly flattened; plants known from southern Yukon southward. *A. tenellus*

10a. Pods and ovaries sessile, subsessile, or substipitate (the stipe less than 2 mm long). (11)
10b. Pods and ovaries stipitate (stipe exceeding 2 mm in length). (13)

11a. Flowers 6–9 mm long; pods pilose, with either black or white hairs. *A. eucosmus*
11b. Flowers 8–15 mm long; pods merely strigose. (12)

12a. Flowers pink to purplish, 8–11(12) mm long; pods 6–9 mm long; plants prostrate. *A. bodinii*

12b.   Flowers yellowish, 12-15 mm long; pods 9–15 mm long; plants erect or ascending. *A. williamsii*

13a.   Leaflets 17–25(27); pods pendulous. *A. alpinus*

13b.   Leaflets 9–15; pods spreading to pendulous. (14)

14a.   Leaves densely gray villous to pilose beneath (or thinly hairy in southeastern Alaska); flowering inflorescence commonly 3 cm long or less; calyx teeth 2.5–4.8 mm long; plants of coastal southern Alaska. *A. harringtonii*

14b.   Leaves thinly strigulose to pilose beneath; flowering inflorescence commonly 3 cm long or more; calyx teeth 1–2(2.5) mm long; plants of interior Alaska. *A. robbinsii*

Astragalus adsurgens Pallas
Standing Milk-vetch

Stems 1–4 dm long, decumbent to erect, several to many from a caudex and a stout taproot, strigose with malpighian hairs; stipules 5–10 mm long, chartaceous, connate-clasping, the free ends lanceolate to triangular-subulate, strigose on the dorsal surface; leaves 6–12 cm long, the leaflets 11–17, oblong to elliptic, 15–30 mm long, 2–6 mm broad, acute to obtuse, strigose; peduncles 5–15 cm long, equaling or shorter than the subtending leaves; racemes 1–5 cm long, few- to many-flowered, subcapitate or slightly elongate; flowers 12–15 mm long, ascending, whitish or yellowish (pink purple in southern Yukon) or suffused with pink; calyx short-cylindrical, the tube 5–6 mm long, teeth 0.8–1.5(4) mm long; pods 6–10 mm long, substipitate (stipe to 1.5 mm long), sulcate on the lower (dorsal) suture, bilocular or nearly so, erect, strigose.

Sandy or gravelly soils, stream beds, moraines, and roadsides, or less commonly on grassy bluffs; in interior Alaska from the vicinity of McKinley National Park and Fairbanks southward and eastward to southern Yukon and British Columbia; widely distributed in the western plains and mountains of United States and Canada; Asia.

1a.   Calyx teeth 0.5–1.5(2) mm long; pods substipitate (the stipe to 1.5 mm long); plants of east-central Alaska and southern Yukon (*A. viciifolius* Hultén; *A. tananaicus* Hultén; *A. adsurgens* var. *tananaicus* [Hultén] Barneby). *A. adsurgens* ssp. *viciifolius* (Hultén) Welsh

1b.   Calyx teeth 2–4 mm long; pods sessile or nearly so; plants of southern Yukon (*A. striatus* Nutt.; *A. adsurgens* var. *robustior* Hook.). *A. adsurgens* ssp. *robustior* (Hook.) Welsh

Astragalus agrestis Dougl. ex G. Don
Field Milk-vetch

Stems 1–2.5(4) dm long, few to several from a rhizome, decumbent to erect, strigose with simple hairs; stipules 4–11 mm long, connate-clasping below; leaves 4–7 (10) cm long; leaflets 11–25, 6–15 mm long, 2–4 mm broad, narrowly elliptic, obtuse; peduncles 4–15 cm long; racemes 1–4 cm long, subcapitate, several-flowered; flowers 17–24 mm long, purplish, ascending; calyx cylindrical, villous with dark and light hairs; tube 6–7 mm long; teeth 2.8–5.5 mm long, half as long as the tube; pods 8–10 mm long, woolly-villous, erect, subsessile, 2-loculed by intrusion of the dorsal (lower) suture.

Usually in damp open sites in grasslands or woods; known in our region only from southern Yukon where it is locally abundant (to be sought in east-central Alaska); eastward to Ontario and south to Washington, Utah, New Mexico, and Iowa; Asia (*A. hypoglottis* authors, not L.; *A. dasyglottis* Fisch. ex DC., not Pallas; *A. tarletonis* Rydb.; *A. goniatus* Nutt.).

Astragalus alpinus L.
Alpine Milk-vetch

Stems 0.5–3.5 dm long, few to several from

Astragalus alpinus L. (× 0.25).

a rhizome, decumbent to erect, strigulose; stipules 3–10 mm long, the lower ones connate-clasping, the upper ones nearly free, ovate to triangular, herbaceous, strigulose to glabrous on the dorsal surface; leaves 5–15 cm long; leaflets 17–25(27), 6–20 mm long, 2–10 mm wide, ovate to elliptic or oblong, the apex retuse to rounded, strigulose above and below with simple hairs; peduncles 3–15 cm long; racemes 1–6 cm long in flower (longer in fruit), subcapitate to elongate, 2–many-flowered, the flowers erect at first, later spreading and finally reflexed in age; flowers 9–15 mm long, light to dark pink purple, often fading to yellow on drying; calyx campanulate, strigulose with black hairs, the tube 2–3.5 mm long, the teeth 1–3.2 mm long; pods 10–17 mm long, stipitate, the stipe 2–5 mm long, strigose with black hairs, the dorsal suture sulcate, the body oblong-lanceolate.

Arctic and alpine tundra, heath, and woods, river beds, terraces, or moraines; in most of Alaska (except for the Aleutians) and most of the Yukon; eastward to Newfoundland and south to Utah, Colorado, South Dakota, and Vermont; Eurasia (A. *alpinus*

ssp. *arcticus* [Bunge] Hultén; A. *alpinus* ssp. *alaskanus* Hultén). The present writer has been unable to satisfactorily place our specimens into the various named segregates, and it seems best to treat them as belonging to the highly variable var. *alpinus.*

Astragalus americanus (Hook.) M. E. Jones
American Milk-vetch

Stems 2.5–9(10) dm tall, few to several from a caudex, ascending to erect, sparsely villous with simple hairs; stipules 10–22 mm long, ovate to lanceolate, distinct, the lower ones reflexed, turning brown in age, glabrous to sparsely villous with simple hairs; leaves 9–17 cm long; leaflets 9–15, oblong to lanceolate, 1.5–6 cm long, 0.5–2 cm broad, obtuse, sparsely villous below, glabrous above; peduncles 5–16 cm long; racemes 1.2–9 cm long, several- to many-flowered; flowers 8–13 mm long, whitish, the keel tip purplish or pinkish; calyx campanulate, oblique, the tube 3.5–5 mm long, glabrous or nearly so, the teeth 0.2–0.8 mm long, villous with black or white hairs; pods 22–32 mm long (including the 5–9 mm stipe), ovoid-inflated, 1-loculed, pendulous, glabrous, the stipe exceeding the calyx.

Gravelly or sandy soils in woods or thickets; in the southeastern quarter of mainland Alaska and in southern Yukon; south to British Columbia, Colorado, South Dakota, and Saskatchewan and east to Ontario and Quebec.

Astragalus australis (L.) Lam.
Southern Milk-vetch

Stems 1–5 dm long, few to several from a branching caudex, decumbent to ascending or erect, strigose to villous with simple hairs, or glabrate; stipules 3–11 mm long, the lower ones connate-clasping, those above nearly free; leaves 2–10 cm long, short-petiolate; leaflets (5)9–15, linear to oblong, elliptic or lanceolate, 7–25 mm long, (1)2–8 mm wide, variously pubescent to glabrous, acute or obtuse apically; peduncles 3–14 cm long; racemes 2–15 cm long, several- to many-flowered; flowers 8–13 mm long, yellowish white (often tinged with pink), the keel purple tipped; calyx campanulate, strigulose with black hairs; tube 3–4.5 mm long, the teeth 1.5–4 mm long; pods (16)18–25 mm long (including the stipe 5–10 mm long), the ventral suture curved, the dorsal suture straight, 1-loculed, dehiscent, glabrous or sometimes strigose to pilose.

Arctic and alpine tundra to heathlands and woods, often on sandy or gravelly soils; from the Seward Peninsula northward and eastward through interior Alaska and in northern and southern Yukon; east to Quebec and south to Nevada, Utah, Colorado, and South Dakota; Eurasia (*A. aboriginorum* Richards.; *A. glabriusculus* [Hook.] Gray; *A. richardsonii* Sheld.; *A. aboriginorum* var. *richardsonii* [Sheld.] B. Boi.; *A. scrupulicola* Fern. & Weath.; *A. linearis* [Rydb.] Porsild; *A. lepagei* Hultén; *A. aboriginorum* var. *lepagei* [Hultén] B. Boi.; *A. aboriginorum* var. *major* Gray; *A. aboriginorum* var. *muriei* Hultén; *A. forwoodii* Wats.). This species shows a very wide range of variation in our region. Major variable features include pubescence of herbage, size of pods, length of peduncles, and pubescence of pods. Plants with various combinations of these characteristics have formed the basis of several different species or varieties. The nature of these variations is not understood. Some populations demonstrate several possible character combinations on which various taxa have been based, thus it does not seem practical to recognize all of the recombination types as taxonomic units.

Astragalus bodinii Sheld.
Bodin Milk-vetch

Plants prostrate or sprawling, the stems 2–6 dm long or more, very slender, arising from a caudex and a slender taproot, matforming, strigose; stipules 2–8 mm long, ovate, all connate-clasping; leaves 1–9 cm long, the leaflets 7–17, lanceolate to lance-oblong, elliptic or ovate, 2–15 mm long, obtuse to retuse or emarginate apically; peduncles 3–12 cm long; racemes 2–9 cm long, few- to several-flowered; flowers 8–11 mm long, pink to purplish; calyx campanulate, black-strigose, the tube 2.5–3.5 (4) mm long, the teeth 1.2–2.4 mm long; pods 6–10 mm long, substipitate, ovoid to oblong-ovoid, black-strigose.

In gravelly or sandy soils, stream terraces, flood plains, roadsides, heath, and woodlands; from west-central, central, and east-central Alaska and southern Yukon; disjunctly east to Newfoundland and south to Utah, Colorado, and Nebraska (*A. yukonis* Jones; *A. bodinii* var. *yukonis* [Jones] B. Boi.; *A. stragulus* Fern.; *A. prebblei* [Rydb.] Tidestr.; *Homalobus retusus* Rydb.).

Astragalus eucosmus Robins.
Elegant Milk-vetch

Stems 1.5–7 dm long, erect or ascending, few to several from a branching caudex and taproot, minutely strigose with simple hairs; stipules 3–9 mm long, connate-clasping below, free above; leaves 3–11 cm long, the leaflets 9–15, elliptic to lance-elliptic or oblong, 10–22(30) mm long, 4–7 mm broad, obtuse, glabrous or glabrate above, strigose beneath; peduncles 8–18 cm long; racemes 1–22 cm long, contracted in flower, usually elongating in fruit, several- to many-flowered; flowers 6–8 mm long, purplish or whitish; calyx campanulate, the tube 2.5–3.5 mm long, the teeth 0.9–1.6 mm long, narrowly lanceolate; pods 8–12 mm long, spreading or reflexed, 1-loculed, laterally compressed, the ventral suture prominent, the dorsal one intruded as a thin, partial partition, substipitate, black- or white-pilose.

In sandy or gravelly soils, or less commonly in clay, in heathlands, woods, or thickets, and less commonly in tundra; from the Seward Peninsula northward and eastward through most of interior Alaska and Yukon; east to Newfoundland and Maine and south to Utah and Colorado (*Atelophragma atratum* Rydb.; *Astragalus sealei* Lepage; *A. eucosmus* ssp. *sealei* [Lepage] Hultén).

Astragalus harringtonii (Rydb.) Hultén
Harrington Milk-vetch

Stems 1.5–5 dm long, few to several from a caudex and taproot, ascending to erect, strigose with simple hairs; stipules 5–8 mm long, connate below, distinct above; leaves 3–11 cm long, the leaflets (9)11–15, lanceolate to elliptic or oblong, 10–25 mm long, 3–10 mm broad, gray-villous to pilose beneath, often black-hairy along the lower midvein (or less commonly only thinly hairy); peduncles 3–21 cm long, strigulose to strigose; racemes 2–8(10) cm long, subcapitate in flower, elongating in fruit; several- to many-flowered; flowers 9–12 mm long, yellowish white (sometimes fading bluish), spreading; calyx campanulate, the tube 3.5–4.3 mm long, the teeth 2.5–4.8 mm long, linear; pods 15–22 mm long (including the naked stipe 2–6 mm long), spreading to pendulous, 1-loculed, the dorsal suture produced internally as a partial partition.

Terraces and floodplains, in woods or thickets, usually in gravelly soil; in coastal southern Alaska from Kodiak to Glacier Bay; endemic (*A. robbinsii* var. *harringtonii* [Rydb.] Barn.; *A. robbinsii* ssp. *harringtonii* [Rydb.] Hultén). For discussion of the relationship of this entity see *A. robbinsii*.

Astragalus nutzotinensis Rousseau
Nutzotin Milk-vetch

Stems prostrate, 1–4(5) dm long, few to several from a caudex and taproot, minutely strigose with white and black simple hairs; stipules 3–6 mm long, connate-clasping; leaves 1–6 cm long, the leaflets 9–15, elliptic to oblong, 1.5–7 mm long, 1–3 mm broad, obtuse or retuse; peduncles 3–10 cm long; racemes 1–4-flowered, mostly 1–3 cm long in fruit; flowers 12–18 mm long, cream, suffused with pink or purplish;

calyx campanulate, black-strigose to glabrate, the tube 3–4.5 mm long, the teeth 1–2.5 mm long, triangular-acuminate to linear; pods 3–5 cm long, curved into a semicircle, 1-loculed, minutely strigose, the stipe 4–10 mm long, articulated to the deciduous pod.

Gravelly stream channels, floodplains, or on rock outcrops; in the Brooks Range and in the southeastern third of mainland Alaska, in southwestern Yukon, and adjacent British Columbia; endemic (*Gynophoraria falcata* Rydb.; *Astragalus falciferus* Hultén, not Lam.; *A. gynophoraria* Tidestr.).

Astragalus polaris Seem. ex Hook.
Polar Milk-vetch

Stems prostrate, 0.5–2 dm long or more, few to several from shallow rhizomes attached to a caudex and taproot, mat-forming, minutely strigose; stipules 2–5 mm long, connate-clasping; leaves 1–6 cm long, the leaflets 9–17, elliptic to oblong, 2–6 mm long, 1–3 mm broad, emarginate, retuse, or obtuse; peduncles 2–7 cm long; racemes 1–3(5)-flowered; flowers 11–16 mm long, pink purple (rarely white); calyx campanulate or short-cylindrical, black-strigose, the tube 3.5–6 mm long, the teeth 1–2.4 mm long, triangular to lanceolate; pods 18–43 mm long, ovoid-inflated at maturity, 1-loculed, articulated to a short stipe (to 1 mm long), deciduous, minutely black-strigose, both sutures prominent.

Gravelly soils in arctic and alpine tundra to heathlands; in coastal northwestern Alaska, eastward along the Brooks Range, and disjunctly in the Alaska Range and along the Alaska Peninsula; Asia (*A. amblyodon* [Rydb.] Hultén; *Homalobus amblydon* Rydb.; *A. atlasovii* Kom.).

Astragalus robbinsii (Oakes) Gray
Robbins Milk-vetch

Plants (1)3–7 dm tall, the stems ascending to erect, few to several from a caudex and a taproot, minutely strigose with simple hairs, or rarely glabrous; stipules 5–15 mm long, the lower ones connate; leaves (3)7–12 cm long; leaflets 9–15, lance-elliptic to oblong, (8)20–50(60) mm long, 4–16 mm broad, rounded or retuse apically; peduncles (4)9–23 cm long, minutely strigose; racemes 2–20 cm long, several- to many-flowered; flowers 7–11 mm long, yellowish to whitish (sometimes fading bluish); calyx campanulate, strigose with black hairs, the tube 3.5–4.2 mm long, the teeth 1.5–2(2.5) mm long, lance-linear; pods 19–26(28) mm long, (including a stipe 3–6 mm long), spreading to pendulous, lance-ovoid, the dorsal suture intruded as a partial partition 0.5–1.2 mm broad, minutely black-strigose to glabrous.

Sandy or gravelly soils in woods, along rivers and streams, and on talus slopes; disjunct in the interior of Alaska from Mt. McKinley National Park and along the Yukon River from near Dawson City, Yukon, westward in the central Yukon Valley, southward to south-central Yukon and Mt. Roberts in southeastern Alaska (depauperate *A. harringtonii?*), and from the Tetsa River Valley in northern British Columbia; disjunctly in southern British Columbia, Oregon, and Colorado, and eastward to Vermont (*A. collieri* [Rydb.] Porsild; *A. macounii* Rydb.). Our material is highly variable, with each local population differing in some manner from the others; it belongs to var. *minor* (Hook.) Barneby. Specimens from Glacier Bay have been assigned to this variety chiefly on the basis of the thin pubescence on the lower leaflet surface. This difference does not seem to be correlated with other features of var. *minor*, and the pubescence differs only in degree from more typical *A. harringtonii* with which all other characteristics agree. Thus, the Glacier Bay materials are herein assigned to *A. harringtonii*.

Astragalus tenellus Pursh
Pulse Milk-vetch

Stems 2–5.5 dm tall, usually several from a caudex and taproot, decumbent to erect,

strigose with simple hairs; stipules 2–6 mm long, connate-clasping, often blackening on drying; leaves 5–8 cm long, the petioles short or lacking; leaflets 15–21, linear-oblong to narrowly lanceolate or elliptic, 8–20 mm long, 2–4 mm wide, rounded apically; peduncles 1–4 cm long, occasionally arising in pairs; racemes 3–10 cm long, several- to many-flowered; flowers 7–9 mm long, ochroleucous, the keel purple tipped; calyx campanulate, strigose, the tube 2–2.5 mm long, the teeth 0.9–1.7 mm long; pods 12–18 mm long (including the stipe which is 1–5 mm long), laterally compressed and nearly flat, 1-loculed, glabrous, both sutures prominent.

Sandy and gravelly bluffs and road cuts, usually in woods; in southern Yukon; east to Manitoba and south to Nevada, Utah, New Mexico, Nebraska, and Minnesota.

Astragalus umbellatus Bunge
Tundra Milk-vetch

Stems solitary or few from ascending, rhizomelike caudex branches, 0.5–3 dm tall, erect, villous; stipules 7–15 mm long, foliose, ovate-lanceolate; leaves 4–12 cm long; leaflets 7–13, elliptic to oblong or lance-elliptic, 10–30 mm long, 4–14 mm broad, obtuse to rounded; peduncles 3–9 cm long; racemes 1–5 cm long, few- to several-flowered; flowers 13–19 mm long, yellowish (sometimes shaded with purple), the banner bordered with white; calyx short-cylindrical, sparsely black-villous, the tube 6–8 mm long, the teeth 0.6–0.8 mm long, broadly triangular; pods 20–26 mm long (including the stipe 5–7 mm long), ovoid-inflated, 1-loculed, reflexed, black-pilose.

Arctic and alpine tundra, heath, and woods; through most of continental and coastal western and northern Alaska and most of the Yukon; eastward to Mackenzie and in northern British Columbia; Asia.

Astragalus williamsii Rydb.
Williams Milk-vetch

Stems 4–7 dm tall, erect, few to several

from a branching caudex, minutely strigose with simple hairs; stipules 7–11 mm long, the lower ones connate; leaves 5–10 cm long, short-petiolate; leaflets 9–17, narrowly oblong to linear or elliptic, 18–34 mm long, 3–10 mm wide, rounded to retuse apically; peduncles 8–16 cm long, strigose; racemes 1–13 cm long, several- to many-flowered; flowers 12–15 mm long, yellowish, the keel purple tipped; calyx campanulate, black-strigose, the tube 3.5–4.5(4.8) mm long, the teeth 1.5–2.5 mm long, linear-lanceolate; pods 9–15 mm long, partially 2-loculed by intrusion of the dorsal suture, erect, substipitate, cross-reticulate, minutely strigose.

Gravelly or sandy soil along stream courses, lake shores, and on shaly bluffs, mostly in woods; in central and east-central Alaska and in southern Yukon; endemic (*A. gormanii* Wight ex Jones).

CARAGANA Lam.

Small to large deciduous shrubs; stipules small and deciduous or persistent as slender spines; leaves even-pinnate, the rachis extended as a bristle or spine, the leaflets oblanceolate, oblong or oval, entire; flowers solitary, yellow, showy; calyx campanulate to turbinate, the teeth nearly obsolete; stamens diadelphous; pods subcylindrical, 1-loculed, straight, glabrous, the valves coiling upon dehiscence.

Komarov, V. L. 1908. Monograph of *Caragana*. Acta Horti Petropolitani 29: 179–399.

Caragana arborescens Lam.
Siberian Pea-tree

Plants 1–3(6) m tall, with several to many branches from the ground level; leaves 4–10 cm long, the leaflets 8–12, lance-oblong to elliptic or oval, 12–25 mm long, 5–15 mm broad, cuspidate apically, villous above and below, becoming glabrate in age; stipules narrow, often persisting as spines; bracts reduced to rudiments at

the juncture of the peduncle and pedicel; flowers 17–23 mm long, borne singly on peduncles 12–35 mm long, few to several from each bud; pedicels 5–15 mm long; calyx broadly campanulate, pubescent, the tube 4.5–7.5 mm long, the teeth small or nearly obsolete, the margin villous; pods 3.5–5.5 cm long, sessile.

The Siberian pea-tree is cultivated in the more temperate regions of Alaska and Yukon where it evidently fares well. It is a handsome ornamental and should be more widely planted than at present. The golden pea-tree, *C. aurantiaca* Koehne, is cultivated to a limited extent. It is a low shrub to 1.2 m tall, has only four leaflets per leaf, and has decidedly spiny stipules. Other species of *Caragana* might be expected to do well in our region.

Caragana arborescens Lam. (× 0.4).

HEDYSARUM L.

Plants herbaceous, perennial, caulescent; stipules connate and sheathing or free; leaves alternate, odd-pinnate, with 5–many leaflets, the leaflets entire, minutely glandular-dotted; racemes axillary, pedunculate, elongate or contracted; flowers showy, the keel with a straightish leading edge, lacking a beak, longer than the wings; stamens diadelphous; calyx campanulate, 5-toothed; loments 5, constricted between the seeds into indehiscent segments, breaking transversely.

Northstrom, T. E., and S. L. Welsh. 1970. Revision of the *Hedysarum boreale* complex. Great Basin Nat. 30:109–30.

Rollins, R. C. 1940. Studies in the Genus *Hedysarum* in North America. Rhodora 42: 217–39.

1a. Leaflets thin, with prominent lateral veins; upper pair of calyx lobes distinctly shorter than the lower three; wing auricles linear, subequal to the claw. *H. alpinum*

1b. Leaflets thick, the lateral veins usually not evident; upper pair of calyx lobes subequal to the lower three; wing auricles broad, shorter than the claw. *H. boreale*

Hedysarum alpinum L.
Alpine Sweet-vetch

Plants 1–9(11) dm tall, arising from a caudex and taproot, the stems erect or decumbent-ascending, strigulose with simple hairs; stipules 5–25 mm long, connate below, brownish; leaves 4–16 cm long, short-petiolate; leaflets 9–23, lanceolate to oblong, elliptic, or lance-elliptic, 10–50 mm long, 3–15 mm broad, rounded to acute apically, the apex cuspidate; peduncles 6–15 cm long; racemes 1.5–16(18) cm long, several- to many-flowered; flowers 10–19 mm long, pink to pink purple (rarely white), the keel longer than the wings or banner; calyx campanulate, oblique, puberulent, the tube 1.5–2.5 mm long, the

teeth 0.7–2.5 mm long, the upper pair shorter than the lower three; loments usually 1–4-segmented, the segments 5–9 mm long, 3–7 mm broad, reticulately veined, glabrous.

Arctic and alpine tundra to heath and woods, in sandy or gravelly soils, or less commonly in clay, and on rocky outcrops; almost cosmopolitan in Alaska (except for southwestern and southeastern portions) and most of the Yukon; east to Newfoundland, Maine, and Vermont and south to British Columbia, Wyoming, and South Dakota; Eurasia (*Astragalus hedysaroides* L.; *H. hedysaroides* [L.] Schinz & Thell.; *H. obscurum* L.; *H. arcticum* Fedtsch.; *H. truncatum* Eastw.; *H. americanum* [Michx.] Britt.; *H. auriculatum* Eastw.). Alpine sweet-vetch is a beautiful plant that occurs in our region in several forms. Perhaps the most familiar form is the tall woodland type which has numerous flowers borne in long racemes. These have been designated as var. *americanum* Michx. ex Pursh, but they are indistinguishable from typical *H. alpinum*. At higher elevations, higher latitudes, and in other less hospitable sites the tall form intergrades with populations of smaller, more decumbent plants with fewer flowers in shorter racemes. In some plants of this smaller phase, the flowers average larger than in the tall phase, especially in plants from the Seward Peninsula. These latter plants have been designated either as var. *alpinum* or var. *grandiflorum* Rollins (or as *H. hedysaroides*). They appear to intergrade with the tall phase in every way.

Hedysarum boreale Nutt.
Northern Sweet-vetch

Stems 1–5(6) dm long, arising from a caudex and taproot, decumbent to ascending, strigulose with simple hairs; stipules (2.5) 5–10 mm long, shortly connate to distinct, usually straw-colored; leaves 3–12(14) cm long, short-petiolate; leaflets 9–15, oblong to elliptic or lance-elliptic, 10–40 mm long, 3–14 mm broad, rounded to truncate or rarely acute apically, the apex cuspidate; peduncles 3–14 cm long; racemes 2–15 cm long, few- to many-flowered; flowers (14) 18–22 mm long, pink to red purple (rarely white), the keel longer than the wings or banner; calyx campanulate, strigulose, the tube 2–3 mm long, the teeth 2.5–5 mm long, subequal; loments usually 1–4(6)-segmented, the segments 5–8 mm long, 4–7 mm broad, reticulately veined, pubescent.

In gravelly or sandy soil in stream channels, on terraces, and along roadsides, or less commonly in tundra or heathlands and open woods; in most of Alaska (except for southwestern, southern, and southeastern portions) and most of the Yukon; east to Newfoundland and south to Oregon, Arizona, and New Mexico (*H. mackenzii* Richards.; *H. boreale* var. *mackenzii* [Richards.] C. L. Hitchc.; *H. mackenzii* f. *proliferum* Dore). Our material is assignable to ssp. *mackenzii* (Richards.) Welsh.

LATHYRUS L.

Plants herbaceous, annual, from a taproot, or perennial from a rhizome, stipules prominent, semihastate, semisagittate or ovate, usually smaller than the leaflets; stems erect to clambering or prostrate, angled or winged; leaves alternate, even-pinnate, the terminal extension of the rachis either bristlelike or forming prehensile tendrils; racemes axillary, pedunculate, few- to several-flowered; flowers showy; calyx obliquely campanulate, 5-toothed; stamens diadelphous; styles laterally compressed, bearded along the ventral edge; pods 1-loculed, oblong to linear, several-seeded, the valves coiling upon dehiscence.

Hitchcock, C. L. 1952. A revision of North American species of *Lathyrus*. Univ. Wash. Publ. Bio. 15:1–104.

1a. Leaflets 2; plants annual, cultivated ornamentals. *L. odoratus*
1b. Leaflets 4 or more; plants perennial, indigenous, rarely if ever cultivated. (2)

2a. Stipules foliaceous, subequal in size to the leaflets or larger; plants of strands and beaches, rarely some distance from the coast. *L. japonicus*
2b. Stipules much smaller than the leaflets; plants of coastal woods or tidal flats, rarely growing on strands or beaches. (3)

3a. Stems winged; flowers pink to pink purple; leaflets narrowly lanceolate to linear; plants of broad distribution in coastal regions from Norton Sound south and east, and in interior Alaska. *L. palustris*
3b. Stems merely angled; flowers cream to yellowish, sometimes suffused with pink; leaflets lance-ovate to broadly elliptic; plants from extreme southeastern Alaska and adjacent British Columbia. *L. ochroleucus*

Lathyrus japonicus Willd.
Beach-pea

Plants perennial, from rhizomes, the stems 0.5–10 dm long or more, decumbent or prostrate to clambering, angled but not winged; stipules foliaceous, subequal to the leaflets or larger; leaflets 6–12, broadly elliptic to oblanceolate, 1–6 cm long, 0.5–3 cm broad; tendrils simple to branched, slightly to strongly prehensile; flowers (2)3–6(8), pink purple, (17)20–26(29) mm long; calyx 10–14 mm long, the lower teeth subequal to or longer than the tube; pods 3–6 cm long, usually pubescent.

Beaches, strands, cliff bases, or rarely in open woods near the coast; in coastal and insular Alaska, from Pt. Lay south along the Chukchi Sea and Bering Strait to the Aleutians, eastward to southeastern Alaska, and on the arctic coast of Yukon; south to California, and also along the Atlantic Coast and westward to Minnesota and Hudson Bay; Eurasia. Two rather weak and mostly sympatric varieties are recognized.

1a. Leaflets mostly over 3 cm long; tendrils either branched or over 3 cm long; plants most common along the southern coast (*L. maritimus* [L.] Fries; *L. japonicus* var. *pellitus* Fern.; *L. japonicus* var. *aleuticus* [Greene] Fern.). *L. japonicus* var. *glaber* (Ser.) Fern.

1b. Leaflets mostly less than 3 cm long; tendrils unbranched or rarely so, usually less than 3 cm long; plants most common along the western coast and on islands of the Bering Sea (*L. maritimus* var. *pubescens* Hartm.; *L. maritimus* ssp. *pubescens* [Hartm.] Reg.). *L. japonicus* var. *japonicus*

Lathyrus ochroleucus Hook.

Plants perennial, from rhizomes, the stems 3–10 dm tall, clambering, angled, but not winged; stipules narrow to subfoliaceous, much smaller than the leaflets; leaflets (4)6–8, lance-ovate to broadly elliptic, 2–6 cm long, 1–3(3.5) cm broad; tendrils usually branched and prehensile; flowers (5) 6–12, cream to yellowish, 12–16 mm long; calyx 8–10 mm long, the lower teeth subequal to the tube; pods 3–7 cm long, glabrous.

Woods and thickets; known in our region only from extreme southeastern Alaska (Hyder) and adjacent British Columbia (to north of Liard Hot Springs), and to be sought in southeastern Yukon (*L. venosus* authors, not Muhl. ex. Willd.).

Lathyrus odoratus L.
Sweet-pea

Plants annual, from taproots, the stems 3–15 dm long or more, clambering, winged, pubescent; stipules narrowly lanceolate, smaller than the leaflets; leaflets 2, oval to oblong, 2–5 cm long; tendrils branched, prehensile; racemes 2–5-flowered, pink, white, lavender, or red, 25–30 mm long, fragrant; pods 3–6 cm long, pubescent.

Cultivated ornamentals; in more temperate portions of Alaska and Yukon; Europe.

Lathyrus palustris L.
Wild-pea

Plants perennial, from rhizomes, the stems 3–10 dm long or more, clambering, winged (at least above); stipules narrow, much smaller than the leaflets; leaflets 4–8, lance-oblong to lance-linear, elliptic or linear, 1.5–6 cm long, 0.1–1.2 cm broad; tendrils usually branched and prehensile; flowers 2–8, pink to pink purple (rarely white), 14–20 mm long; calyx 8–12 mm long, the lower three teeth subequal to the tube; pods 3–6 cm long, pubescent to glabrate.

Meadows, tidal flats, beaches, or woods and lake shores; most common in southern and southeastern Alaska, but also in widely scattered sites in interior Alaska as far north as the 66th parallel, and in southwestern Yukon; south to California, and from the Atlantic Coast west to the Dakotas; Eurasia (*L. palustris* ssp. *pilosus* [Cham.] Hultén.). Our material has pubescent calyces and leaves, and belongs to a phase of the species passing under the name of var. *pilosus* (Cham.) Ledeb.

LUPINUS L.

Plants herbaceous, perennial, the stems solitary or several from a caudex; stipules adnate to the petioles; leaves cauline and alternate, or chiefly basal, palmately compound, with 5–11 leaflets; leaflets oblanceolate, entire; racemes terminal, pedunculate, several- to many-flowered; flowers showy; calyx 2-lipped, the lower lip of 3 connate lobes, the upper lip of 2 more or less connate lobes; stamens monadelphous, the anthers of 2 kinds, 5 large alternating with 5 small ones; pods flattened, oblong, 1-loculed, both sutures usually prominent, the valves coiling upon dehiscence.

Dunn, D. B. 1965. The interrelationships of the Alaskan Lupines. Madroño 18:1–17.

————. 1966. *Lupinus*, pp. 10–89. *In*: Dunn, D. B. and J. M. Gillett. The Lupines of Canada and Alaska. Can. Dept. Agr. Mono. 2:1–89.

1a. Leaflets pilose with silvery hairs on both surfaces; plants appearing silvery pubescent, often less than 3 dm tall. (2)

1b. Leaflets glabrous or thinly pubescent, at least above; plants appearing green, various in height. (3)

2a. Plants with keel conspicuously ciliate along the upper edge, known from extreme southeastern Alaska. *L. lepidus*

2b. Plants with keel glabrous or only slightly ciliate along the upper edge, in east-central Alaska, southwestern Yukon, and northern British Columbia. *L. kuschei*

3a. Longest petioles cauline, 2–10 cm long; stems with usually 5 or more leaves above the base; lower lip of calyx broad, boat shaped; plants of coastal and insular southern Alaska, rarely some distance from the coast. *L. nootkatensis*

3b. Longest petioles mostly basal (see couplet 5), 9–50 cm long; stems with mostly 2–5 leaves above the base; lower lip of calyx slender lanceolate; plants of various distribution. (4)

4a. Stems slender, mostly 3–4(5) mm in diameter, usually 1–3 dm tall; racemes mostly less than 20 cm long; plants of interior Alaska and Yukon, or alpine in coastal, southern regions. *L. arcticus*

4b. Stems mostly more robust, (4)5–10 mm in diameter, often more than 3 dm tall; racemes commonly over 20 cm long; plants of southern Alaska, or rarely in the interior. (5)

5a. Petioles of basal leaves often much longer than the cauline ones; flowers uniformly lupine blue (pink in cultivated forms); plants mostly of southern Alaska, often near habitations, less commonly in the interior. *L. polyphyllus*

5b. Petioles of basal leaves subequal to the cauline ones; flowers lupine blue, pale blue, pinkish, pink, or white (often varicolored within the same raceme) in populations; plants common on the Kenai Peninsula and less common at other sites where *L. polyphyllus* contacts *L. nootkatensis. L. × pseudopolyphyllus*

Lupinus arcticus Wats.
Arctic Lupine

Stems 1.3–4.5(6) dm tall, ascending to erect, few to several from the caudex, pubescence appressed to spreading; petioles of basal leaves mostly much longer than the upper cauline ones, mostly 6–21 cm

Lupinus arcticus Wats. (× 0.3).

long; leaflets 5–9, narrowly oblanceolate to narrowly elliptic, 1.3–9 cm long, 0.4–1.9 cm broad, acute apically, glabrous above, thinly strigose below; peduncles 4–10 cm long; racemes 4–14 cm long; pedicels 3–9 mm long; calyx strigose to villous, the lower lip 6–11 mm long, slender, lanceolate, the upper lip somewhat gibbous basally, 5–8 mm long; flowers 15–20 mm long, lupine blue (rarely white), the banner orbicular, glabrous, reflexed slightly below the midpoint, the keel glabrous or sparsely ciliate along the upper margin; pods 2–4 cm long, silky-pilose, the seeds 5–8.

Arctic and alpine tundra, heath, and woods; over most of Alaska (except for coastal and insular southern Alaska) and throughout the Yukon; east to Mackenzie (*L. borealis* Heller; *L. donnellyensis* C. P. Sm.; *L. gakonensis* C. P. Sm.; *L. multicaulis* C. P. Sm.; *L. multifolius* C. P. Sm.; *L. nootkatensis* var. *kjellmanii* Ostenf.; *L. polyphyllus* ssp. *arcticus* [Wats.] Phillips; *L. toklatensis* A. Nels.; *L. yukonensis* Greene). Our abundant material belongs to ssp. *arcticus*. However, a second entity, ssp. *subalpinus* (Piper & Rob.) Dunn, approaches extreme southeastern Alaska, where it should be expected in alpine sites. The leaves of ssp. *subalpinus* are mostly cauline with the longest petioles 4–8 cm long.

Lupinus kuschei Eastw.
Yukon Lupine

Stems 1.5–5(6) dm long, decumbent to erect, few to several from the caudex, pubescence silvery-silky; petioles of lower leaves 4–15 cm long, mostly 2–3 times longer than the upper, cauline leaves; leaflets 5–9, oblanceolate to elliptic, 1.5–7 cm long, 0.3–1.1 cm broad, acute apically, silvery-silky below and above, although often less hairy above; peduncles 2.5–7 (13) cm long; racemes 3–10(12) cm long; pedicels 2–5(7) mm long; calyx silky-villous, the lower lip 5–7 mm long, slender, the upper lip somewhat gibbous basally, 4–6 mm long; flowers 10–13 mm long, lu-

pine blue, the banner suborbicular, glabrous, reflexed at or below the midpoint, the keel glabrous or with a few cilia along the upper edge; pods 2–3 cm long, silky-pilose, the seeds 4–6.

Sandy alluvium, sand dunes, or roadsides and open woods; in east-central Alaska, southwestern Yukon, and northern British Columbia (*L. jacob-andersonii* C. P. Sm.; *L. porsildianus* C. P. Sm.; *L. sericeus* var. *kuschei* [Eastw.] B. Boi.).

Lupinus lepidus Pursh

Stems 1–3.5 dm long, decumbent to erect, pubescent with spreading hairs; petioles of basal leaves 2–10 cm long or more, somewhat longer than the upper ones; leaflets 5–9, oblanceolate, 1–4 cm long, 0.2–0.8 cm broad, sericeous to villous on both surfaces; peduncles 2–7 cm long; racemes 4–15 cm long; pedicels 2–5 mm long; calyx villous, the lower lip 4–6 mm long, broad, the upper lip 4–5 mm long, not gibbous basally; flowers 10–13 mm long, lupine blue, lavender, or white, the banner suborbicular, glabrous, reflexed at or below the midpoint, the keel finely ciliate along the upper margin; pods 3–4 cm long, strigose, the seeds 2–4.

Hillsides and open woods; in extreme southeastern Alaska (where probably introduced); disjunctly from southern British Columbia south to Oregon (*L. sellulus* authors, not Kellogg).

Lupinus nootkatensis Donn ex Sims
Nootka Lupine

Stems mostly 3–12 dm long or more, erect or ascending, few to several from the caudex, often fistulose, shaggy-villous to strigose; petioles all cauline, 2–10 cm long, seldom twice as long as the leaflets; leaflets 5–9, oblanceolate to elliptic, 1–7 cm long, 0.3–2 cm broad, obtuse to rounded apically, hairy below, glabrous above (except near the margins); peduncles mostly 4–9 cm long; racemes 5–35 cm long; pedicels 6–14 mm long; calyx villous, the lower lip 6–12 mm long, broad, boat shaped, often reflexed, the upper lip 6–8 mm long, not gibbous basally; flowers 16–21 mm long, lupine blue (rarely white), the banner orbicular, glabrous, reflexed below the midpoint, the keel ciliate along the upper margin; pods 3–6 cm long, silky-hairy, the seeds 7–11.

Meadows, open woods, thickets, and roadsides; from near sea level to a few hundred feet in elevation, mostly in insular and coastal southern and southwestern Alaska, but extending into the interior as far as Cantwell and southwestern Yukon (*L. kiskensis* C. P. Sm.; *L. nootkatensis* var. *ethel-looffiae* C. P. Sm.; *L. nootkatensis* var. *henry-looffii* C. P. Sm.; *L. nootkatensis* var. *perlanatus* C. P. Sm.; *L. nootkatensis* f. *leucanthus* Lepage; *L. perennis* ssp. *nootkatensis* [Donn ex Sims] Phillips; *L. nootkatensis* β *glaber* Hook.; *L. nootkatensis* var. *unalaskensis* Wats.; *L. arboreus* var. *fruticosus* [Sims] Wats.). Two noteworthy extremes are recognizable within the Nootka lupine.

1a. Pubescence spreading, long shaggy-villous; bracts subtending pedicels long and filamentous; stems generally fistulose, mostly 10–15 mm in diameter (passing into the next). *L. nootkatensis* var. *nootkatensis*

1b. Pubescence appressed or short-villous; bracts lance-subulate; stems sometimes fistulose, mostly less than 9 mm in diameter; plants growing with the above variety. *L. nootkatensis* var. *fruticosus* Sims

Lupinus polyphyllus Lindl.
Large-leaf Lupine

Stems mostly 4–10 dm tall, rarely more, erect or ascending, few to several from the caudex, glabrous or appressed-pubescent to spreading-hairy; petioles of basal leaves (7)10–30 cm long or more, mostly much longer than the upper cauline ones; leaflets (8)9–17, oblanceolate to elliptic,

3–12 cm long, 0.6–2.7 cm broad, sparsely strigose below, glabrous above; peduncles 4–10(14) cm long; racemes 6–25(40) cm long; pedicels 5–10 mm long; calyx villous, sometimes sparsely so, the lower lip 4–6 mm long, slender, somewhat reflexed, upper lip subequal to the lower, slightly gibbous basally; flowers 13–16 mm long, lupine blue (rarely pink), the banner suborbicular, reflexed at the midpoint, glabrous, the keel glabrous; pods 3–5 cm long, villous, the seeds 6–10.

Open to rather dense woods or along roadsides, often near habitations, introduced and cultivated as ornamentals; escaping and locally well established (or native?) in numerous sites from the vicinity of Fairbanks to southern Alaska where especially common in the Anchorage vicinity (*L. matanuskensis* C. P. Sm.).

Lupinus × pseudopolyphyllus C. P. Sm.
Kenai Lupine

Stems 2.5–6 dm tall, ascending to erect, few to several from the caudex, pubescence appressed to spreading; petioles of basal leaves from shorter than, to longer than, the upper cauline leaves, 4–13 cm long; leaflets 9–13, oblanceolate, 2–9 cm long, 0.6–1.3 cm broad, rounded to obtuse or acute apically, minutely strigose to strigulose below, glabrous above or nearly so; peduncles 2–11 cm long; racemes 7–20 cm long; pedicels 6–10 mm long; calyx villous, the lower lip 3–6 mm long, somewhat thickened, the upper lip 3–5 mm long, slightly gibbous basally; flowers 13–17 mm long, lupine blue, light blue, pink, or white (often varicolored in the same raceme), the banner orbicular, reflexed at or below the midpoint, the keel glabrous or ciliate along the upper margin; pods mostly 2–5 cm long, villous.

Roadsides, in burned over woodland, and near habitations; mostly in the Kenai Peninsula (the type from Moose Pass), especially abundant from Soldatna at mile 95 northward to mile 63 on the Sterling Highway where populations of plants occur along both sides of the highway, the racemes beautifully varicolored, and less commonly in other sites where both *L. nootkatensis* and *L. polyphyllus* occur. In most features, the Kenai lupine is intermediate to the Nootka and Large-leaf lupines. Apparently, the Kenai lupine has resulted from hybridization and introgression of *L. polyphyllus* with both varieties of *L. nootkatensis*. The presence of numerous specimens trending towards the parental types indicates a free exchange of genes and a normally segregating series of highly heterozygous populations. Several phases are sufficiently striking that they might be investigated for their horticultural possibilities (*L. stationis* C. P. Sm.).

MEDICAGO L.

Plants herbaceous, perennial or annual, caulescent; stipules distinct, not sheathing; leaves alternate, pinnately trifoliolate, the leaflets apically dentate to prominently apiculate; racemes axillary, pedunculate, mostly spike or headlike; flowers showy, the keel lacking a beak; stamens diadelphous; calyx campanulate to short cylindrical, 5-toothed; pods sessile or nearly so, curved to coiled at maturity, reticulately veined, sometimes armed with prickles, 1-loculed, glabrous to pubescent, indehiscent, 1–several-seeded.

1a. Flowers 2–3 mm long; inflorescence less than 10 mm long in flower; pods coiled through a single spiral, 1-seeded, unarmed; plants prostrate to decumbent (rarely erect). *M. lupulina*

1b. Flowers 4–10 mm long; inflorescence more than 10 mm long (including flower length), or the pods differing in several ways from above; plants various. (2)

2a. Flowers 4–5 mm long, yellow, 2–5 per inflorescence, the flower cluster less than 10 mm long; pods armed with prickles, several-seeded; plants annual. *M. hispida*

2b. Flowers 6–10 mm long, yellow, or blue to pink or white, 6–many on at least some inflorescences, the flower cluster over 10 mm long; pods unarmed, several-seeded; plants annual to perennial. (3)

3a. Flowers yellow (sometimes tinged with violet); pods merely curved. *M. falcata*

3b. Flowers blue, pink, or white; pods spirally coiled. *M. sativa*

Medicago falcata L.
Yellow Alfalfa

Plants annual to perennial, the stems 4–10 dm long or more, ascending to erect (rarely decumbent), strigulose; stipules entire, 4–12 mm long, persistent, conspicuously veined; leaves short-petiolate, the leaflets linear, oblong, oblanceolate, or elliptic, 0.6–2 cm long, 0.1–0.6 mm broad, obscurely few-toothed, tridentate, or merely long-apiculate apically, pubescent; peduncles subequal to the subtending leaves; racemes (2)6–20-flowered, usually 10–20 mm long, hemispheric; flowers yellow (sometimes violet tinged), 6–9 mm long; calyx campanulate, the tube 1–2 mm long, the teeth 1.5–3 mm long, lance-subulate; pods 6–10 mm long, merely curved, unarmed, several-seeded.

Cultivated forage plant; escaping and persisting where it behaves as an annual in much of our region (including southern Yukon). It is often established as a roadside weed; introduced from the Old World.

Medicago hispida Gaertn.
Bur Clover

Plants annual, the stems prostrate to erect, 1–4 dm long; stipules deeply divided into few to several, long teeth, mostly 3–7 mm long; leaves short-petiolate, the leaflets cuneate to obovate or obcordate, 1–2.5 cm long, 0.6–1.8 cm broad, toothed in the apical one-third (rarely more), pubescent to glabrous; peduncles mostly much shorter than the subtending leaves; racemes 2–5-flowered, less than 10 mm long; flowers yellow, 4–5 mm long; calyx campanulate, the tube 1–1.5 mm long, the lance-subulate teeth 1–2 mm long; pods spirally coiled, armed with spines, several-seeded.

An occasional weed; in southeastern Alaska, and to be expected elsewhere; introduced from Europe.

Medicago lupulina L.
Black Medick, Hop Clover

Plants annual, the stems prostrate to erect, 1–3(4) dm long; stipules entire or nearly so, 3–6 mm long, persistent; leaves short-petiolate, the leaflets cuneate to obcordate, 0.4–1.5 cm long, 0.2–1.2 cm broad, toothed in the apical one-third (rarely more), pubescent to glabrous; peduncles mostly equaling or longer than the subtending leaves; racemes mostly 6–25-flowered, less than 10 mm long in flower (to 15 mm long in fruit); flowers yellow, 2–3 mm long; calyx campanulate, about 1 mm long; pods spiral (through about 1 coil), unarmed, 1-seeded.

Occasional weed; in widely scattered locations in Alaska and to be expected in the Yukon; introduced from Europe.

Medicago sativa L.
Alfalfa

Plants annual to perennial, the stems 4–10 dm long or more, ascending to erect (rarely decumbent), strigulose; stipules entire, 4–12 mm long, persistent; leaves short-petiolate, the leaflets elliptic-oblanceolate, 0.8–4 cm long, 0.2–1.5 cm broad, apically few-toothed, pubescent; peduncles subequal to or longer than the subtending leaves; racemes mostly 6–25-flowered, hemispheric to subcylindrical, 10–20 mm long or more; flowers blue to pink, pink purple, or white, 6–10 mm long; calyx campanulate to short-cylindrical, the tube 1.5–2.5 mm long, the teeth 2–4 mm long, lance-subulate; pods spirally coiled, unarmed, several-seeded.

Introduced forage plant, escaping and sometimes persistent, often annual or short-lived perennial where it survives at all under our climatic conditions; introduced from Europe.

MELILOTUS Mill.

Plants herbaceous, annual or biennial, caulescent; stipules distinct, not sheathing; leaves alternate, short-petiolate, pinnately trifoliolate, the leaflets dentate-serrate through half or more of the length; racemes axillary, pedunculate, elongate; flowers small, but showy; stamens diadelphous; calyx campanulate, 5-toothed; pods sessile, straight, ovoid, reticulately veined, or cross-ribbed, unarmed, 1-loculed, glabrous, usually indehiscent, mostly 1–2-seeded.

1a. Flowers white, 3–5 mm long; pods reticulately veined. *M. alba*
1b. Flowers yellow, 5–7 mm long; pods cross-ribbed. *M. officinalis*

Melilotus alba Descr. ex Lam.
White Sweet Clover

Plants annual or less commonly biennial, the stems mostly 5–15 dm tall or more, erect, strigulose; stipules entire or hastately lobed, mostly 5–10 mm long, persistent; leaflets obovate to elliptic or oblanceolate, 1–4 cm long or more, pubescent to glabrous; peduncles shorter to longer than the subtending leaves; racemes mostly 20–50-flowered or more, 5–20 cm long; flowers white, 4–5 mm long; calyx campanulate, the tube 1.2–1.8 mm long, the teeth 1–1.5 mm long, acuminate; pods 3–6 mm long, reticulately veined or ridged, 1–2-seeded.

Introduced forage plant, locally established as a roadside weed where it occurs mostly as an annual; in southern Alaska and southern Yukon; introduced from Europe.

Melilotus officinalis (L.) Lam.
Yellow Sweet Clover

Plants annual or less commonly biennial,

the stems mostly 5–15 dm tall or more, erect, strigulose; stipules entire or with 1–3 basal teeth, mostly 5–10 mm long, persistent; leaflets cuneate to elliptic or oblanceolate, 1–4 cm long, 0.3–1.2 cm broad, usually toothed or obscurely sinuate from near the base, pubescent to glabrous; peduncles shorter or longer than the subtending leaves; racemes mostly 20–50-flowered or more, 3–15 cm long; flowers yellow, (4.5)5–7 mm long; calyx campanulate, the tube 1–1.8 mm long, the teeth 1–1.5 mm long, acuminate; pods 3–5 mm long, cross-ribbed, 1–2-seeded.

Forage plant, escaping and locally established (often with *M. alba*) as a roadside weed which survives mostly as an annual; in southern Alaska and southern Yukon; introduced from Europe.

OXYTROPIS DC.

Plants herbaceous, perennial, acaulescent to short-caulescent; stipules mostly connate, adnate to the petiole; leaves alternate, odd-pinnate, with 5–many leaflets or with 1–3 leaflets (in *O. mertensiana*), the leaflets entire; racemes axillary, pedunculate or scapose, 1–many-flowered, subcapitate to elongate; flowers showy, the keel tip produced into a porrect beak; stamens diadelphous; calyx cylindrical to campanulate, 5-toothed; pods sessile to stipitate, straight, erect or reflexed, 1-loculed or partially 2-loculed by intrusion of the ventral (upper) suture, dehiscent.

Barneby, R. C. 1952. A revision of the North American species of *Oxytropis* DC. Proc. Cal. Acad. Sci. IV. 27:177–312.

Welsh, S. L. 1967. Legumes of Alaska II: *Oxytropis* DC. Iowa State Jour. Sci. 41:277–303.

1a. Base of plant sheathed with reddish brown or purplish brown stipules. (2)
1b. Base of plant not reddish or purplish brown, the stipules light tan or grayish (or some dark brown to black, or purplish only in *O. kobukensis*). (3)

2a. Flowers yellow or yellowish; racemes with more than 5 flowers; pods erect, not over 3 times longer than broad. *O. maydelliana*
2b. Flowers bright pink to pink purple; inflorescence typically 2–3-flowered; pods spreading (reclining on the ground at maturity), usually at least 3 times longer than broad. *O. kokrinensis*

3a. Leaves simple or trifoliolate (rarely 5-foliolate), the leaflets decurrent or obscurely articulated with the rachis. *O. mertensiana*
3b. Leaves 5–many-foliate, the leaflets jointed to the rachis. (4)

4a. Racemes 1–5-flowered (sometimes more in *O. arctica*). (5)
4b. Racemes 5–many-flowered. (8)

5a. Flowers usually more than 15 mm long; leaflets 11–many, scattered or pseudoverticillate; stipular margins beset with clavate processes mixed with cilia. *O. arctica*
5b. Flowers usually less than 15(20) mm long; leaflets (1)3–13, scattered or opposite, but not pseudoverticillate; stipular margins with cilia only (except in some *O. nigrescens*). (6)

6a. Flowers and pods elevated above the ground; pods erect, black-pilose (rarely white) or glabrous; stipules usually prominent, glabrous or sparsely pilose dorsally, straw-colored. *O. scammaniana*
6b. Flowers and pods spreading, the peduncle weak and at length reclining; pods spreading to ascending, glabrous to pilosulose; stipules various but usually not conspicuous. (7)

7a. Pods short, twice longer than broad or less, glabrous to strigose, 1-loculed; flowers 1(2) per peduncle. *O. huddelsonii*
7b. Pods elongate, usually over 3 times longer than broad; pilosulose (rarely

glabrous), 2-loculed or nearly so; flowers (1)2–3 per peduncle. *O. nigrescens*

8a. Plants caulescent or subacaulescent; stipules shortly adnate to the petiole base; pods pendulous; flowers usually less than 10(12) mm long. *O. deflexa*

8b. Plants acaulescent; stipules adnate to the petiole base through half their length or more; pods erect or spreading; flowers usually more than 10 mm long. (9)

9a. Plants glandular-viscid (at least to some degree); bracts glabrous dorsally. *O. viscida*

9b. Plants glandular; bracts pilose dorsally. (10)

10a. Flowers white to yellowish (the keel tip maculate or immaculate). (11)

10b. Flowers pinkish or purplish. (13)

11a. Flowers 12–17(18) mm long; stipular margins beset with clavate processes as well as cilia; plants of broad distribution, possibly our most common species. *O. campestris*

11b. Flowers 18–22 mm long; stipular margins beset with clavate processes or lacking them; plants of northern and northwestern Alaska, or from southern Yukon. (12)

12a. Leaflets 9–15, thin, green; flowers 5–9, borne in a short to subcapitate raceme; plants of northern and northwestern Alaska. *O. arctica*

12b. Leaflets 11–19(25), thick, sericeous; flowers (8)9–15, borne in subcapitate to elongate racemes; plants of southern Yukon and northern British Columbia. *O. sericea*

13a. Flowers usually over 16 mm long; stipular margins with clavate processes as well as cilia; plants of northern or western Alaska, or local elsewhere. (14)

13b. Flowers 16(18) mm long or less; stipular margins beset with clavate pro-

cesses or lacking them; plants occurring as scattered individuals in Alaska, and as populations in the Yukon, Mackenzie, and northern British Columbia. (15)

14a. Stipules firm, purplish; leaflets not fascicled; plants of western Alaska. *O. kobukensis*

14b. Stipules fragile, grayish or yellowish; leaflets all, or at least some, fascicled. *O. arctica*

15a. Leaves with all or most of the leaflets fascicled; calyx tube about one-half as long as the banner; stipules lacking marginal, clavate processes; plants of eastern Alaska and southern Yukon. *O. splendens*

15b. Leaves with several or none of the leaflets fascicled; calyx tube less than half as long as the banner; stipules with or without marginal, clavate processes; plants from scattered sites in Alaska, in southeastern and northern Yukon, Mackenzie, and northern British Columbia. *O. campestris*

Oxytropis arctica R. Br.
Arctic Oxytrope

Plants caespitose, acaulescent from branching caudex; leaves 4-21 cm long, the leaflets 9–many, lanceolate to elliptic or oblong, 4–40 mm long, 2–5 mm broad, opposite, scattered, or some fasciculate, variably pilose above and below with simple hairs; stipules 10–20 mm long, connate, pilose dorsally, becoming glabrate in age, adnate to the petiole, the margins ciliate and beset with clavate processes; scapes 6–21 cm long, strigulose to spreading villous; racemes 2–many-flowered; bracts mostly longer than the pedicels, pilose dorsally; flowers (14)16–22 mm long, pink purple, pink, bluish, or rarely white; calyx cylindrical, villous with light and dark hairs, the tube 5–8.7 mm long, the teeth 1.5–6 mm long, linear-lanceolate to triangular; pods 10–25 mm long, ovoid-acuminate, bilocular or nearly so.

Sand and gravel bars, terraces, flood plains, sand dunes, rocky ridges and roadsides, from arctic and alpine tundra to heath and woodlands; mostly north of the 66th parallel in both Alaska and Yukon; east to the Canadian Arctic Archipelago and Coronation Gulf, but also known from disjunct localities south of that parallel. Three poorly understood, but rather striking varieties are present among our materials.

1a.   Flowers 10–many; plants mostly more than 15 cm tall; leaflets often pseudofasciculate and numerous; disjunct from Shaktolik, Anaktuvuk, Umiat, and Northway vicinity (*O. roaldi* Ostenf., in part). *O. arctica* var. *koyukukensis* (Porsild) Welsh

1b.   Flowers 2–8(9); plants mostly less than 15(22) cm tall; leaflets various in disposition and number; plants mostly north of the 66th parallel. (2)

2a.   Leaflets 11–many, narrowly lanceolate to lance-elliptic; flowers pink purple to bluish, or rarely white; plants widely distributed (*O. coronaminis* Fern.; *O. roaldi* Ostenf., in part). *O. arctica* var. *arctica*

2b.   Leaflets 9–15, broadly elliptic to oblong or lance-elliptic; flowers white (fading cream); plants from the vicinity of Kotzebue and scattered sites eastward. *O. arctica* var. *barnebyana* Welsh

Oxytropis campestris (L.) DC.
Field Oxytrope

Plants caespitose, acaulescent from branching caudex; leaves (2)4–30 cm long, the leaflets 11–45, lance-elliptic to oblong or narrowly lanceolate, 3–40 mm long, 2–11 mm wide, opposite, scattered, or fasciculate, pilose above and below; stipules ovate to lanceolate, acuminate, connate, adnate to the petioles, straw-colored to black, pilose dorsally, becoming glabrate, the margins ciliate and usually beset with clavate processes; scapes 3.5–36 cm long; racemes 2–26 cm long, becoming lax; bracts narrowly lanceolate, longer than the pedicels, villous dorsally; flowers 10–17(18) mm long, whitish, yellowish, pink, purplish tinged, or purplish, the keel tip maculate or immaculate; calyx cylindrical, villous with dark and light hairs, the tube (4) 4.5–8 mm long, the teeth 1–3 mm long, triangular to linear; pods 8–18 mm long, erect, sessile or subsessile, pilose, partially bilocular by intrusion of the ventral suture.

Gravel bars, terraces, rock outcrops, and roadsides, most commonly in woods and heath, but also in open alpine meadows; widely distributed in Alaska and Yukon south of the 68th parallel; east to the Atlantic and south to Washington, Utah,

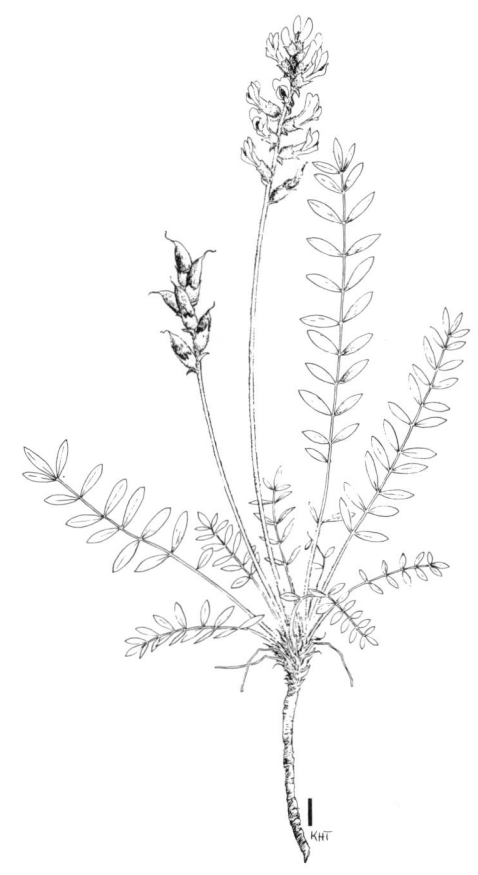

Oxytropis campestris (L.) DC. (× 0.4).

Colorado, Wisconsin, and Maine; circumboreal. The species is composed of a large number of more or less distinctive races. Of these, three occur in the region from northern British Columbia through the Yukon and Alaska.

1a. Plants usually less than 10-flowered; flowers 14 mm long or less; leaflets 9–17, merely scattered or opposite; restricted to alpine situations from north of the Brooks Range eastward to Mackenzie, and south to Juneau (*O. jordalii* Porsild; *O. campestris* ssp. *jordalii* [Porsild] Hultén). *O. campestris* var. *jordalii* (Porsild) Welsh

1b. Plants 10–many-flowered; flowers mostly 12–17 mm long; leaflets 19–45, scattered, opposite, or fasciculate; seldom in alpine sites but sometimes so. (2)

2a. Flowers uniformly pink purple, pink, or bluish; plants of northern British Columbia (from Sikanni Chief River Valley north to Racing River Valley), and disjunctly in extreme southeastern Yukon, east to Yellowknife. *O. campestris* var. *davisii* Welsh

2b. Flowers white, yellowish, or rarely purplish in some individuals; plants of southeastern Alaska and southern Yukon northward and westward into Alaska proper (*O. alaskana* A. Nels.; *O. hyperborea* Porsild; *O. campestris* var. or ssp. *gracilis* authors). *O. campestris* var. *varians* (Rydb.) Barneby

Oxytropis deflexa (Pallas) DC.
Deflexed Oxytrope

Plants 0.8–5 dm tall, subacaulescent or short-caulescent, decumbent or ascending to erect, variously pubescent, the stems few to several from a caudex; leaves 4–22 cm long, the leaflets (9)17–41, ovate to narrowly lanceolate or oblong, 4–20 mm long, 2–8 mm broad, sparsely to densely pilose; stipules 7–20 mm long, connate,

slightly adnate to the petiole; peduncles 6–28 cm long; bracts lanceolate, usually not exceeding the calyx; racemes 1–18 cm long, few- to many-flowered, elongating in fruit or rarely compact; flowers 6–10 mm long, whitish, pinkish, bluish, or bright pink purple; calyx campanulate, the tube 2–4 mm long, the teeth 1–4 mm long, lanceolate to linear; pods 8–25 mm long (including a short stipe to 3 mm long), pendulous, straight or curved, oblong-ellipsoid, 1-loculed, the ventral suture sulcate, strigulose.

Woodlands, thickets, heath, or rarely in tundra, mostly in gravelly or sandy soils along drainages; in much of Alaska (except for the western and southwestern portions) and most of the Yukon; east to Newfoundland and south to California, Nevada, Utah, New Mexico, North Dakota, and Quebec. Two varieties are widely scattered and locally abundant in our region. A third, var. *deflexa*, has been reported from the Yukon.

1a. Flowers whitish, pinkish, bluish, or less commonly pink purple, 10–30 or more per raceme; racemes subcylindrical, 10–13(14) mm broad in flower (when pressed), much elongating in fruit (16–18 cm long); herbage usually conspicuously long-villous; plants commonly of open sites (*O. retrorsa* Fern.; *O. deflexa* var. *parviflora* B. Boi.). *O. deflexa* var. *sericea* T. & G.

1b. Flowers pink purple, pink, or less commonly whitish, 2–15(20) per raceme; racemes hemispheric (or subcapitate) to shortly subcylindrical, (14)15–20 mm broad in flower (when pressed), not much elongating in fruit (1–10 cm long); herbage variously pubescent; plants usually in woods or thickets. (2)

2a. Racemes usually 7–10(15)-flowered; herbage sparingly pilose, green; plants locally abundant (*O. deflexa* var. *capitata* B. Boi.). *O. deflexa* var. *foliolosa* (Hook.) Barneby

2b. Racemes usually 10–20-flowered; herbage copiously and loosely villous-pilose, sericeous; plants evidently rare. *O. deflexa* var. *deflexa*

Oxytropis huddelsonii Porsild
Huddelson Oxytrope

Plants pulvinate-caespitose, the caudex branches prostrate; leaves 1.5–5 cm long, the rachis obscurely purple, white-pilose, the leaflets 7–13, lanceolate to elliptic or oblong, 3–6 mm long, 1–2 mm broad, involute, pilose above, sparsely so below; stipules straw-colored and firm, glabrous or sparsely pilose dorsally, the free part short, obtuse-deltoid, long-ciliate, lacking clavate processes; peduncles 1–4 cm long, 1–2-flowered; flowers pink purple, 11–15 mm long; calyx campanulate, appressed-strigose, the tube 4–6 mm long, the teeth 1.2–2.2 mm long, triangular; pods 10–18 (23) mm long, 1-loculed, sessile or nearly so, glabrous or minutely strigose.

Ridge tops and frost boils, in alpine tundra, heath, or less commonly in woods; in interior eastern Alaska and southern Yukon; endemic. This is a very pretty dwarf oxytrope which resembles *O. nigrescens* and occurs in habitats similar to that of *O. nigrescens*. It has long been overlooked.

Oxytropis kobukensis Welsh
Kobuk Locoweed

Perennial herb to 17 cm tall, acaulescent, arising from a branching caudex, the branches ascending to erect, 2–6 cm long; leaves 3–10 cm long; leaflets 13–17, lanceolate to lance-oblong, 6–16 mm long, 2–3.5 mm wide, rounded at the base, the apex obtuse to acute, strigose to pilose below, pilose to glabrate above, the margin involute; stipules persistent, purplish, the margins scarious, ciliate, beset with clavate processes, adnate to the petiole, pilose dorsally, somewhat glabrate on the free ends, long attenuate; scapes 7–11 cm long, strigose with appressed and spreading hairs; racemes 5–6-flowered (or more?); bracts lanceolate, pilose dorsally, ciliate and beset with clavate processes on the margin; flowers 16–18 mm long, purplish; calyx short-cylindrical to campanulate, purplish, minutely strigulose with light and dark hairs, the tube 6 mm long, the teeth 1.5–2 mm long, linear-lanceolate, the sinuses broad; immature legume 17 mm long (including the style), black-pilose, bilocular or nearly so.

Sand dunes; Kobuk River, across from the mouth of the Hunt River; endemic.

Oxytropis kokrinensis Porsild
Kokrines Oxytrope

Plants caespitose, subcaulescent from a caudex, the caudex branches clothed with persistent, purplish brown stipules and petioles; leaves 1–5 cm long, the leaflets 7–9, elliptic to lanceolate, 4–6 mm long, 1–2 mm broad, revolute or folded; peduncles slightly if at all exceeding the leaves, 1–2-flowered; flowers 12–15 mm long, purplish; calyx campanulate, the tube 4–4.5 mm long, the teeth 2–2.5 mm long; pods 20–35 mm long (including a stipe 1.5–4 mm long), bilocular or nearly so, pilose.

Ridge tops; in the Kokrines Mountains and western Brooks Range of Alaska; endemic. Rare in collections.

Oxytropis maydelliana Trautv.
Maydell Oxytrope

Plants caespitose, acaulescent from a branching caudex, the branches clothed with persistent, reddish or purplish brown stipules and petioles; leaves 3–14 cm long, the leaflets 11–21, ovate to lanceolate, elliptic or oblong, 4–17 mm long, 2–6 mm broad, pilose to glabrous above, pilose on the midrib or sometimes generally below, acute or obtuse; stipules 12–20 mm long, becoming brown in age, adnate to the petiole, connate, the free ends caudate-acuminate, pilose to glabrous dorsally, ciliate; scapes 4–15 cm long, erect or ascending; bracts lanceolate, villous to sparsely pilose dorsally, ciliate; racemes subcapitate, mostly 1–4 cm long and 6–12-flowered (rarely more); flowers erect or ascending, 13–17 mm long, yellow to whitish yellow, the keel tip immaculate; calyx short-cylindrical, villous with light and dark hairs, the tube 5–6 mm long, the teeth 1.5–3.2 mm long; pods 15–21 mm long, ovoid-ellipsoid, sulcate ventrally, incompletely bilocular, pilose.

Arctic and alpine tundra, and heathlands; in most of Alaska (except for the Aleutians and Alexander Archipelago) and much of the Yukon; south to British Columbia and east to Quebec; Asia.

Oxytropis mertensiana Turcz.
Mertens Oxytrope

Plants acaulescent, from a branching, subterranean caudex, the elongated branches covered with persistent, pale stipules; leaves 1–7 cm long, the leaflets 1–3(5), mostly continuous with the rachis, elliptic to oblong, 7–25 mm long, 2–5 mm broad, glabrous below, sparingly pubescent above; scapes 3–8 cm long, bearing 1–2 flowers; flowers 12–16 mm long, pink purple; calyx campanulate, densely black-pilose, the tube 4.8–6.2 mm long, the teeth 2.1–4.1 mm long, lance-acuminate; pods 13–20 mm long, erect and borne aloft, pilose with black hairs, subunilocular, stipitate, the stipe 1.5–2 mm long.

Moist, alpine situations; mostly in northern and western Alaska, but also in central and eastern Alaska; Asia.

Oxytropis nigrescens (Pallas) Fisch. ex DC.
Blackish Oxytrope

Plants pulvinately caespitose to loosely matted, the caudex branches erect or ascending to prostrate-spreading; leaves 0.5–5 cm long; leaflets 5–15, elliptic to ovate, acute or obtuse, 2–10 mm long, 1–2 mm broad, sparsely and loosely pilose to densely villous or rarely glabrous; stipules 5–14 mm long, whitish or with pale, herbaceous

tips, more or less pilose or rarely glabrous dorsally, frequently long-ciliate and with clavate processes; peduncles 0.5–5 cm long, bearing 1–4 flowers, the flowers 12–20 mm long, bright pink purple or blue purple to white; calyx campanulate, usually black-pilose, the tube 3–6 mm long, the teeth 1.8–4 mm long, triangular to lanceolate; pods 18–38 mm long, subsessile or with a stipe 4–5 mm long, oblong-ellipsoid, bilocular or nearly so, short-pilosulose.

Arctic and alpine tundra and heathlands; in most of Alaska and Yukon; east to Baffin Island and south to British Columbia; Asia.

1a. Pods long-stipitate, the stipe 4–5 mm long, subequal to the calyx tube; Ogilvie Mts., Yukon. *O. nigrescens* var. *lonchopoda* Barneby
1b. Pods sessile or subsessile, the stipe 2 mm long or less. (2)
2a. Leaflets 9–15, variously pubescent to glabrous, greenish or silvery-canescent; plants widely distributed, our common blackish oxytrope (*O. pygmaea* [Pallas] Fern.; *O. glaberrima* Hultén; *Aragallus bryophilus* Greene; *O. nigrescens* var. *bryophila* [Greene] Lepage; *O. nigrescens* ssp. *bryophila* [Greene] Hultén; *Astragalus pygmaeus* Pallas; *O. nigrescens* var. *pygmaea* [Pallas] Cham.; *O. nigrescens* ssp. *pygmaea* [Pallas] Hultén). *O. nigrescens* var. *nigrescens*
2b. Leaflets 5–11, densely silky-canescent; plants alpine in southern Yukon, northern British Columbia, and coastal from the Mackenzie Delta eastward (*O. arctobia* Bunge; *O. arctica* var. *uniflora* Hook.; *O. nigrescens* ssp. *arctobia* [Bunge] Hultén). *O. nigrescens* var. *uniflora* (Hook.) Barneby

Oxytropis scammaniana Hultén
Scamman Oxytrope

Plants caespitose, low-growing, from a usually subterranean caudex, the frequently elongate branches clothed with persistent, straw-colored stipules; leaves 2–9 cm long, the leaflets 9–13, lanceolate to elliptic, 4–13 mm long, 1–4 mm broad, sparingly pilose (or glabrous), jointed to the rachis; scapes 2–8 cm long, bearing 1–5 (usually 2–3) flowers, the flowers purplish (rarely white), 12–17 mm long; calyx campanulate, the tube 4.5–6 mm long, the teeth 1.5–4 (5.5) mm long, linear-lanceolate; pods 11–18 mm long, subunilocular, oblong-ellipsoid, erect and borne aloft, black-pilose (rarely glabrous).

Arctic and alpine tundra and heathlands; in interior Alaska from the Brooks Range and Alaska Range east to southern Yukon and rarely in southeastern Alaska; endemic. A glabrous phase is known from Mt. Fairplay.

Oxytropis sericea Nutt. ex T. & G.
Silvery Oxytrope

Plants caespitose, acaulescent, from a branching caudex; leaves 3.5–30 cm long, the leaflets 11–21(25), ovate to elliptic or lanceolate, 7–30 mm long, 4–9 mm broad, pilose above and below, often densely so; stipules adnate to the petioles, villous to glabrate dorsally, ciliate; bracts narrowly lanceolate, shorter than or equaling the calyx tube, villous dorsally; scapes 7–15 (20) cm long; racemes 2–10 cm long, subcapitate to elongate, mostly 5–15-flowered; flowers 18–25 mm long, whitish or yellowish, the keel tip maculate or immaculate; calyx cylindrical, pilose with light and dark hairs, the tube 7.5–9.5 mm long, the teeth 2–3.5 mm long, narrowly triangular, usually dark hairy; pods 11–25 mm long, bilocular or nearly so, erect, shortly pubescent.

Gravelly or sandy bluffs, roadsides, and on stream gravels; in southern Yukon and northern British Columbia; southward to Nevada, Utah, New Mexico, Oklahoma, and Saskatchewan (*O. campestris* var. *spicata* Hook.; *O. spicata* [Hook.] Standl.). Our material has been assigned to var. *spicata* (Hook.) Barneby.

Oxytropis splendens Dougl. ex Hook.
Showy Oxytrope

Plants caespitose, acaulescent from a branching caudex; leaves 3–28 cm long, the leaflets verticillate on the rachis in 7–15 fascicles (32–70 leaflets), or rarely only a few verticillate, 4–25 mm long, 2–6 mm broad, narrowly lanceolate, villous throughout; stipules membranous, long-pilose dorsally, the hairs obscuring the margins, the free ends triangular to acuminate, adnate to the petioles; peduncles 9–29 cm long; bracts narrowly lanceolate, villous dorsally; racemes 2–10 cm long, little elongating in fruit, several- to many-flowered; flowers 12–15 mm long, pink purple; calyx cylindrical, long-villous, the tube 5–6.5 mm long, the teeth 2–4 mm long, lance-linear; pods 10–17 mm long, ovoid to ovoid-oblong, sulcate on both sutures, villous, subunilocular.

Sandy bluffs and open woodlands; in central eastern Alaska (Chitna Road, Little Salmon, Porcupine River) where evidently rare, through southern Yukon, and disjunctly to Ontario and south to Minnesota, North Dakota, New Mexico, Montana, and British Columbia.

Oxytropis viscida Nutt. ex T. & G.
Viscid Oxytrope

Plants caespitose, acaulescent from a branching caudex; leaves 2–25 cm long, the leaflets 19–51, lance-oblong to lanceolate, ovate or elliptic, 4–22 mm long, 1.5–6 mm broad, pilose to glabrous and often glandular (sometimes verrucose) above and below, acute or obtuse; stipules 8–22 mm long, membranous, pale, adnate to the petiole, the free ends acuminate, pilose to glabrous dorsally and often glandular, ciliate; scapes 5–27 cm long; bracts lanceolate to lance-linear, glandular or glabrous dorsally, ciliate; racemes 6–many-flowered, 2–18 cm long, seldom much elongated in fruit; flowers 11–21 mm long, pink purple, yellowish, or white, the keel tip maculate or immaculate; calyx cylindrical, villous with light and dark hairs, the tube 4.5–8.6

mm long, the teeth 2–6.5 mm long, narrowly lanceolate, glandular and villous; pods 12–17(30) mm long, including the 4–5 mm long beak, chartaceous, partially 2-loculed, strigulose to pilose.

Sandy or gravelly soil, in tundra, heath, or woods, or on rock outcrops; in most of Alaska and the Yukon; east to Quebec and south to Minnesota, Colorado, Utah, Nevada, and California; Asia. *O. viscida* is extremely polymorphic throughout its range. This fact is indicated by the numerous epithets which have been applied to various portions of the complex; e.g. *Aragallus viscidulus* Rydb.; *O. glutinosa* Porsild; *O. sheldonensis* Porsild; *O. verruculosa* Porsild; and *O. viscidula* ssp. *sulphurea* Porsild. The complex is in need of further work. The nomenclature of the taxon is confused. Almost surely some Old World name has precedence over *O. viscida*. Any name change must await typification of the epithets involved.

PISUM L.

Plants herbaceous, annual, from a taproot; stipules prominent, larger than the leaflets, semisagittate to ovate; stems clambering, not winged; leaves alternate, even-pinnate, the terminal extension of the rachis forming prehensile tendrils; racemes axillary, pedunculate, few-flowered; stamens diadelphous; style laterally compressed, bearded along the ventral edge; pods 1-loculed, oblong in outline, several-seeded, the valves coiling upon dehiscence.

Pisum sativum L.
Pea

Plants 2–20 dm tall or more, sprawling or clambering, the stems merely angled, glabrous; stipules foliaceous, larger than the leaflets; leaflets usually 4–6, elliptic to oblong-lanceolate, 2–6 cm long, 1.5–4 cm broad; tendrils with 1–3 pairs of lateral branches, prehensile; flowers 1–3, white (red or bicolored in some forms), 18–30 mm long; calyx 12–18 mm long, the teeth

longer than the tube; pods mostly 5–10 cm long, glabrous.

Cultivated garden or forage plants, grown for the edible immature seeds, or sown with other plants (especially barley) for cattle feed; in the more temperate regions of Alaska and Yukon; Eurasia. Several horticultural forms are grown.

TRIFOLIUM L.

Plants herbaceous, annual, biennial, or perennial, caulescent; stipules connate and sheathing or distinct; leaves alternate, short- to long-petiolate, palmately 3–5-foliolate, the leaflets toothed through half or more of the length or subentire; racemes axillary or terminal, pedunculate or subsessile, subcapitate, subtended by an involucre of scarious, connate bracts, or the involucre lacking; flowers showy; stamens diadelphous; calyx campanulate, 5-toothed; pods sessile or subsessile, straight, unarmed, 1-loculed, indehiscent, 1–few-seeded, enclosed within the calyx and persistent corolla.

1a. Leaflets 3–7-times longer than broad, often more than 3 per leaf. *T. lupinaster*

1b. Leaflets 2.5 times longer than broad or less, rarely more than 3 per leaf. (2)

2a. Peduncles shorter than the subtending leaves, mostly obsolete; calyx teeth strongly hairy; flowers red to pink; plants perennial, locally common. *T. pratense*

2b. Peduncles usually longer than the subtending leaves, never obsolete; calyx teeth glabrous, or if hairy then the plants annual; flowers of various colors. (3)

3a. Terminal leaflet much longer stalked than the lateral ones; leaflets often less than 1 cm long; plants low annuals. (4)

3b. Terminal leaflet sessile or subsessile, the stalk about as long as those of the lateral leaflets; leaflets mostly more than 1 cm long; plants annual, biennial, or perennial. (5)

4a. Heads usually with 3–12 flowers; flowers 2.5–3.5 mm long; petioles mostly shorter than the leaflets, even on the lower leaves. *T. dubium*

4b. Heads usually with 20–40 flowers; flowers 3.5–6 mm long; petioles often longer than the leaflets, at least on lower leaves. *T. procumbens*

5a. Heads of flowers not subtended by an involucre; plants biennial, perennial, or annual. (6)

5b. Heads of flowers subtended by an involucre; plants annual (perennial in *T. wormskioldii*). (8)

6a. Plants annual; flowers yellow; uncommon or rare in our region. *T. agrarium*

6b. Plants perennial (biennial?); flowers white to pink or red. (7)

7a. Plants stoloniferous, glabrous, the stems rooting at the nodes; leaves all long-petioled; flowers usually white (often tinged pink). *T. repens*

7b. Plants not stoloniferous, hairy to glabrate, the stems not rooting at the nodes; leaves with progressively shorter petioles upwards; flowers pink to reddish or white. *T. hybridum*

8a. Plants perennial, often rhizomatous; calyx glabrous. *T. wormskjoldii*

8b. Plants annual, from taproots; calyx hairy or glabrous. (9)

9a. Involucre villous, somewhat cup shaped. *T. microcephalum*

9b. Involucre glabrous, saucer shaped. *T. variegatum*

Trifolium agrarium L.

Plants annual, the stems 1–5 dm long, 1–several from a taproot, erect or ascending, hairy to glabrous; stipules entire, 8–20 mm long, persistent; petioles shorter than the leaflets; leaves palmately trifoliolate, the leaflets oblong to oblanceolate, 0.8–3 cm long, 0.3–2 cm broad, serrulate, pubescent to glabrous; peduncles longer than the subtending leaves; heads mostly 30–50-flowered or more, 8–20 mm long in flowers or fruit, lacking an involucre; flowers yellow, 5–7 mm long, soon reflexed; calyx 2–4 mm long; pods straight, ovoid, usually 1-seeded, enclosed by the persistent corolla.

Weedy plant; reported from the Anchorage vicinity and to be expected elsewhere; introduced from Europe (*T. aureum* Poll.).

Trifolium dubium Sibth.
Suckling Clover

Plants annual, the stems 1–5 cm long, 1–several from a taproot, prostrate, decumbent or erect, sparsely pubescent; stipules entire to toothed, 3–5 mm long, persistent; petioles shorter than the leaflets; leaves pinnately trifoliolate, the leaflets cuneate to obovate or obcordate, 0.4–1.5 cm long, 0.3–0.9 cm broad, usually denticulate, glabrous; peduncles longer than the subtending leaves; heads mostly 3–12(20)-flowered, less than 10 mm long in flower or fruit, lacking an involucre; flowers yellow, 2.5–3.5 mm long, soon reflexed; calyx 1–2 mm long; pods straight, ovoid, 1–2-seeded, enclosed by the persistent corolla.

Weed of disturbed soil; uncommon in southeastern Alaska and to be expected elsewhere; introduced from Europe. This species is sometimes confused with *Medicago lupulina*, but the round stems, reflexed flowers, and persistent perianth are diagnostic.

Trifolium hybridum L.
Alsike Clover

Plants perennial, the stems 1.5–8 dm tall or more, 1–several from a taproot, ascending to erect, usually not rooting at the nodes, sparsely pubescent; stipules entire, 5–30 mm long, persistent; petioles longer than the leaflets below, becoming shorter upwards; leaves palmately trifoliolate, the leaflets obovate or ovate to elliptic or lance-elliptic, 0.7–3.1 cm long, 0.6–2.5 cm broad, serrulate; peduncles usually much longer than the subtending leaves; heads many-flowered, 15–25 mm long in flower, lacking an involucre; flowers pink to reddish or white, 5–9(10) mm long, soon reflexed; calyx 3–4.5 mm long, hairy at the sinuses; pods 2–3-seeded.

Introduced forage plants, escaped from cultivation and established in disturbed sites; in more temperate parts of Alaska and Yukon; introduced from Europe.

Trifolium lupinaster L.

Plants perennial, the stems mostly 4–6 dm tall, 1–several from a taproot, erect or ascending, not rooting at the nodes, strigulose; stipules entire, 9–17 mm long, persistent; petioles entirely adnate to the stipules, much shorter than the leaflets; leaves palmately 3–5-foliolate, the leaflets oblong-elliptic, 15–40 mm long, 2.5–5 mm broad, serrulate; peduncles usually shorter than the subtending leaves; heads many-flowered, 15–25 mm long in flower, lacking an involucre; flowers pink, 15–18 mm long, not reflexed; calyx 8–10 mm long, hairy; pods 1–3-seeded.

Introduced (or possibly indigenous?) plant; known in Alaska from between Rampart and Tanana, and at Fairbanks; Eurasia.

Trifolium microcephalum Pursh.
Small-head Clover

Plants annual, the stems 0.8–5 dm long or more, 1–several from a taproot, prostrate to ascending or erect, not rooting at the nodes, pubescent; stipules entire to denticulate, 5–12 mm long, persistent; petioles mostly longer than the leaflets; leaves palmately trifoliolate, the leaflets obovate to obcordate, 0.4–1.8 cm long, 0.2–1.2 mm broad, serrulate; peduncles from shorter to longer than the subtending leaves; heads 10–many-flowered, 5–10 mm long, subtended by a pubescent involucre; flowers white to pinkish, 4–6 mm long, not reflexed; calyx 2.5–3.5 mm long, pubescent; pods 1–2-seeded.

Introduced weedy plants; known from interior Alaska (Manley Hot Springs), where it was collected many years ago, and to be expected elsewhere in our region; indigenous from British Columbia east to Montana and south to California and Arizona.

Trifolium pratense L.
Red Clover

Plants perennial, the stems 3–10 dm tall or more, 1–several from a taproot, erect or ascending, pubescent; stipules entire, 12–30 mm long, persistent, conspicuously veined; petioles longer than the leaflets, at least on the lower leaves; leaves palmately trifoliolate, the leaflets lance-elliptic to elliptic, 1.2–4 cm long, 0.5–2.2 cm broad, serrulate to subentire; peduncles shorter than the subtending leaves or obsolete; heads many-flowered, 20–35 mm long, usually subtended by 2 foliage leaves and their stipules; flowers red to pink, 12–20 mm long, not reflexed; calyx 10–14 mm long, hairy; pods 1–2-seeded.

Introduced forage plant, escaped and established in disturbed sites; in much of temperate Alaska and Yukon; introduced from Europe.

Trifolium procumbens L.
Hop Clover

Plants annual, the stems 1–6 dm long, 1–several from a taproot, prostrate to ascending, pubescent; stipules entire, 3–9 mm long, persistent; petioles often longer than the leaflets, at least below; leaves pinnately trifoliolate, the leaflets cuneate to obovate, 0.8–1.6 cm long, 0.8–1.2 cm broad, serrulate; peduncles shorter to longer than the subtending leaves; heads 20–40-flowered, 7–14 mm long, lacking an involucre; flowers yellow, 3.5–6 mm long, soon reflexed; calyx 1.5–2 mm long, glabrous to pubescent; pods 1–2-seeded.

nodes, glabrous or glabrate; stipules entire, 3–13 mm long; petioles mostly much longer than the leaflets; leaves palmately trifoliolate, the leaflets 0.7–2.7 cm long, 0.6–2.3 cm broad, serrulate; peduncles usually longer than the subtending leaves; heads many-flowered, 20–35 mm long in flower, lacking an involucre; flowers white or suffused with pink, 6–9(11) mm long, soon reflexed; calyx 4–6 mm long, glabrous or nearly so; pods 2–3-seeded.

Introduced forage plant, escaping and established in disturbed sites; in much of Alaska and Yukon; native to Europe.

Trifolium variegatum Nutt.
White-tip Clover

Plants annual, the stems 1–4 dm long, 1–several from a taproot, prostrate to ascending or erect, glabrous; stipules dentate, 3–5 mm long, persistent; petioles shorter to longer than the leaflets; leaves palmately trifoliolate, the leaflets obovate, 0.4–1.2 cm long, 0.2–0.8 cm broad, serrate; peduncles longer than the subtending leaves; heads 3–many-flowered, 8–15 mm long, subtended by a glabrous involucre; flowers purplish, often whitish tipped, 6–10 mm long; calyx 4–7 mm long, glabrous; pods 1–2-seeded.

Introduced weedy plant; known from western Alaska (St. Michael), and to be expected elsewhere; indigenous from British Columbia east to Montana and south to California and Utah.

Trifolium wormskjoldii Lehm.
Coast Clover

Plants perennial, the stems 1–6 dm long, arising from taproots and often with rhizomes, glabrous; stipules dentate, 10–30 mm long; petioles mostly longer than the leaflets; leaves palmately trifoliolate, the leaflets oblong-obovate, 1–3 cm long, 0.3–1 cm broad, serrulate; peduncles longer than the subtending leaves; heads few- to many-flowered, 12–25 mm long, subtended by a glabrous involucre; flowers reddish to

Trifolium pratense L. (× 0.4).

Introduced weedy plant of disturbed soils; known from southeastern Alaska (Juneau), and to be expected elsewhere; native to Europe (*T. campestre* Schreb.).

Trifolium repens L.
White Clover

Plants perennial (or biennial?), the stems 1–5 dm long, decumbent and rooting at the

purplish, 10–16 mm long, not reflexed; calyx 6–10 mm long, glabrous; pods 1–4-seeded.

This species has been reported from extreme southeastern Alaska (Loring), where it might occur naturally; indigenous from British Columbia south to California and east to Idaho and Colorado (*T. fimbriatum* Lindl.).

VICIA L.

Plants herbaceous, perennial or annual, from taproots or rhizomes; stipules prominent, semisaggitate or semihastate, toothed or entire, smaller than the leaflets; stems sprawling or clambering, usually angled; leaves alternate, even-pinnate, the terminal extension of the rachis forming prehensile tendrils; racemes axillary, pedunculate or subsessile, 1–several-flowered; flowers showy; calyx obliquely campanulate, 5-toothed; stamens diadelphous; style filiform, with a tuft of hair around the tip; pods 1-loculed, oblong to linear, several-seeded, the valves coiling upon dehiscence.

Herman, F. J. 1960. Vetches of the United States—native, naturalized, and cultivated. U.S.D.A. Agr. Handb. 168: 1–84.

1a. Flowers 1–2(3) borne subsessile in the leaf axils. V. *angustifolia*

1b. Flowers usually 2–many, borne in pedunculate racemes. (2)

2a. Racemes 2–12-flowered, shorter than the subtending leaves; plants perennial, indigenous, mostly in coastal southeastern Alaska. (3)

2b. Racemes mostly more than 12-flowered, at least some equaling to surpassing the subtending leaves; plants adventive in agricultural areas of Alaska and Yukon. (4)

3a. Leaflets 20–30; stems somewhat swollen; flowers yellowish, tinged with red or purple (sometimes fading purplish). V. *gigantea*

3b. Leaflets (6)8–12; stems not swollen; flowers pink to pink purple. V. *americana*

4a. Plants strigose to strigulose, perennial; flowers mostly 12–15 mm long, the lower calyx teeth subequal to the tube. V. *cracca*

4b. Plants spreading-hairy, annual or biennial; flowers mostly over 15 mm long, the lower calyx teeth often longer than the tube. V. *villosa*

Vicia americana Muhl. ex Willd.
American Vetch

Perennial, with stems mostly 2–8 dm tall, pubescent; stipules dentate, 3–11 mm long; leaflets (6)8–12, oblong to elliptic or linear, 0.5–3.5 cm long, 0.2–1.1 cm broad, acute to truncate apically; flowers pink to pink purple, 13–21 mm long, borne in 2–6 (10)-flowered pedunculate racemes; calyx 4–8 mm long, the teeth about half as long as the tube; pods 3–4 cm long.

Open woods, thickets, and meadows; mostly in southeastern Alaska and adjacent British Columbia, to be expected in southeastern Yukon; east to Ontario and south to Virginia, Missouri, Oklahoma, Texas, New Mexico, and California (V. *americana* var. *truncata* [Nutt.] Brew.). Our material belongs to var. *americana*.

Vicia angustifolia (L.) Reich.
Common Vetch

Perennial, the stems 3–8 cm long, spreading-hairy; stipules toothed, 3–10 mm long; leaflets 8–16, linear to oblanceolate, 1.5–3 cm long, 0.1–0.5 cm broad, acute; flowers pink purple to blue purple, 15–25 mm long, 1–2(3) borne subsessile in the leaf axils; calyx 7–11 mm long, the lower teeth equaling or exceeding the tube; pods 3–7 cm long.

Introduced forage plant, possibly not persisting; native to Europe (V. *sativa* L. var. *angustifolia* L.).

Vicia cracca L.
Bird Vetch

Perennial, with stems 5–20 dm long, pubescent; stipules entire or toothed, 7–15 mm long; leaflets 16–22, narrowly oblong to lance-elliptic, 0.7–3 cm long, 0.2–0.5 cm broad, acute; flowers pink purple or blue purple, 10–12 mm long, usually 20-many, borne in pedunculate racemes; calyx 3–5 mm long, the lower teeth subequal to the tube or shorter; pods 2–3 cm long.

Introduced forage plant, established along roadsides in agricultural regions; Alaska and Yukon; native to Europe.

Vicia gigantea Hook.
Giant Vetch

Perennial, the stems mostly 5–20 dm long, pubescent to glabrate, somewhat swollen; stipules toothed, 6–25 mm long; leaflets 20–30, oblong to lance-oblong or lanceolate, 1.5–6 cm long, 0.3–1.2 cm broad, acute apically; flowers yellowish, suffused with red or purple (or orange), 12–18 mm long, 5–12, borne in pedunculate racemes; calyx 5–8 mm long, the lower teeth subequal to the tube or shorter; pods 3–5 cm long.

Beaches and meadows; in coastal southeastern Alaska (also reported from the Cook Inlet); south to California.

Vicia villosa Roth
Hairy Vetch

Annual or biennial, the stems 5–20 dm long, spreading-hairy; stipules toothed or entire, 5–15 mm long; leaflets 12–18, oblong to narrowly lanceolate, 1–3 cm long, 0.3–0.6 cm broad, acute to obtuse; flowers pink purple or reddish purple, 15–18 mm long, mostly 20–many, in pedunculate racemes; calyx 4–9 mm long, the lower teeth equaling or longer than the tube; pods 2–3 cm long.

Introduced forage plant; in southern Alaska, doubtfully persisting but to be expected in agricultural regions; native to Europe.

LENTIBULARIACEAE
Bladderwort Family

Herbaceous perennials, of aquatic or merely wet sites; leaves alternate or basal; simple, or the submerged ones finely divided and bearing insectiverous bladders; flowers perfect, irregular, solitary or in scapose racemes; calyx 2–5-lobed; corolla of 5 united petals, bilabiate, the lower lip saccate or spurred at the base; stamens 2, sometimes with 2 staminodes also; pistils 1, the ovary superior, unilocular, bicarpellate; styles 1 or lacking, the stigma 2-lobed; fruit a capsule.

1a. Plants of wet places; leaves simple, entire, in a basal rosette; flowers purplish. *Pinguicula*
1b. Plants aquatic, growing in water; leaves dissected, at least the submerged ones; flowers yellowish. *Utricularia*

PINGUICULA L.

Perennial herbs with fibrous roots; leaves simple, entire, all basal; flowers showy, borne singly on naked scapes; calyx 5-lobed, irregular, the lower lip 2-lobed, the upper one 3-lobed; corolla bilabiate, the upper lip 2-lobed, the lower one 3-lobed, and produced into a basal spur; fruit a 2-valved capsule.

1a. Flowers 6–10 mm long; leaves 0.4–1.5 cm long, 0.2–0.7 cm broad; scapes villous with capitate hairs. *P. villosa*
1b. Flowers 12–25 mm long; leaves 1.2–4.5(5) cm long, 0.8–2.4 cm broad; scapes glabrous to minutely pubescent with capitate hairs. *P. vulgaris*

Pinguicula villosa L.
Hairy Butterwort

Plants 0.2–1.2 dm tall; leaves 0.4–1.5 cm long, 0.2–0.7 cm broad, the margins revolute, glabrous below, villous above; scapes villous with capitate hairs, at least above; flowers nodding, lavender blue; sepals (0.8)

1–2 mm long; corolla 6–10 mm long, including the spur; capsules 3–5 mm long, erect.

In bogs and swamps, usually in moss; through most of Alaska and the Yukon; east to Labrador; Eurasia.

Pinguicula vulgaris L.
Bog-violet, Common Butterwort

Plants 0.3–1.6 dm tall; leaves 1.2–4.5(5) cm long, 0.8–2.4 cm broad, the margins revolute, usually glabrous below, viscid-glandular or apparently glabrous above; scapes glabrous to minutely pubescent with capitate hairs; flowers nodding, lavender blue; sepals 3–5 mm long; corolla 12–25 mm long, including the spur; capsules 4–6 mm long, erect.

Moist sites, bare soil, beaches, meadows, and mossy seeps; in most of Alaska and less commonly in the Yukon; eastward to Newfoundland and south to Oregon, Montana, Michigan, and New York.

1a. Lower corolla lobes oblong-obovate; plants of coastal southern Alaska, and less commonly in southwestern Yukon (*P. macroceras* Link.; *P. vulgaris* ssp. *macroceras* [Link] Calder & Taylor). *P. vulgaris* var. *macroceras* (Link) Herder

1b. Lower corolla lobes oblong; plants widely distributed in interior Alaska and Yukon, and less commonly in coastal and southern portions (*P. microceras* Cham.; *P. vulgaris* var. *microceras* [Cham.] Casper; *P. arctica* Eastw.). *P. vulgaris* var. *vulgaris*

UTRICULARIA L.

Perennial herbs without roots, or apparently so; leaves all submersed and finely dissected, or sometimes emergent and with leaves somewhat reduced, alternate, bearing bladderlike floats which serve to trap small animals (or the bladders borne on separate branches); flowers showy, usually few to several in a raceme; calyx 2-lobed, the lobes entire; corolla bilabiate, the upper lip entire or obscurely 2-lobed, the lower one entire or 3-lobed, produced into a basal spur; fruit a capsule.

1a. Leaf segments terete; bladders not on separate branches; flowers 12 mm long or more. *U. vulgaris*

1b. Leaf segments flat; bladders borne on separate branches or the flowers less than 10 mm long. (2)

2a. Bladders borne on separate branches; flowers 8–12 mm long. *U. intermedia*

2b. Bladders borne on the leaves; flowers 5–8 mm long. *U. minor*

Utricularia intermedia Hayne ex Schrad.
Flat-leaf Bladderwort

Plants of shallow water, the stems growing along the bottom; leaves alternate, mostly 0.5–1.5 cm long, dissected into flat, linear segments; bladders 2–4 mm broad, borne on separate, leafless branches; winter buds 5–7 mm long; scapes 5–20 cm long, emergent; flowers 2–5, yellow, 8–12 mm long, the spur about as long as the lower lip; fruiting pedicels erect.

Ponds and lakes; in much of Alaska (except for the extreme northern and southwestern portions), and in the Yukon; widely distributed in North America; Eurasia. The closely related *U. ochroleuca* R. Hartman has been reported from northern British Columbia (Liard Hot Springs), and might occur in the Yukon or in Alaska. It may be distinguished by having many or all of the pedicels or bladders subtended by dissected leafy bracts.

Utricularia minor L.
Lesser Bladderwort

Plants of shallow water, the stems very slender, growing along the bottom or floating; leaves alternate, mostly 0.4–0.9 cm long, divided into flat, linear segments; bladders 1–2 mm broad, borne on the leaves; winter buds 2–5 mm long; scapes 5–12 (15) cm long, emergent; flowers 2–9, yellow, 5–8 mm long, the spur short and sac-like, or lacking; fruiting pedicels recurved.

In shallow ponds; in the southeastern two-thirds of Alaska and southern Yukon; wide-ly distributed in North America; circum-boreal.

Utricularia vulgaris L.
Common Bladderwort

Plants of shallow to deep water (rarely terrestrial), the stems coarse, usually float-ing; leaves alternate, 1–4 cm long, divided into numerous terete, filiform segments; bladders 1–3 mm broad, borne on the leaves; winter buds 8–15 mm long; scapes mostly 8–25 cm long, emergent; flowers 5–15 (rarely more), yellow, 12–18 mm long, the spur well developed, curved, shorter than the lower lip; fruiting pedicels re-curved.

Utricularia vulgaris L. (× 0.3).

Ponds and lakes, or less commonly in bogs; in most of Alaska and the Yukon, south of the 70th parallel and less commonly northward, broadly distributed in North America; circumboreal (*U. macrorhiza* Le Conte; *U. vulgaris* var. *americana* Gray). Our materials belong to ssp. *macrorhiza* (Le Conte) Clausen.

LINACEAE
Flax Family

Perennial herbs; leaves alternate, simple; flowers perfect, regular, borne in racemes or paniculate cymes; sepals 5, persistent, distinct or connate at the base; petals 5, distinct, usually blue, soon falling; stamens 5, the filaments united at their bases; pistils 1, the ovary superior, 5-carpelled; styles 5, usually distinct; fruit a 10-loculed capsule, with 1–2 seeds per locule.

LINUM L.

Plants arising from a stout taproot, the stems usually several, ascending to erect; leaves entire, linear to oblong; flowers showy; staminal filaments forming a mem-branous sheath around the ovary, alternat-ing with small toothlike processes; capsule septicidal, falsely 10-loculed, the seeds flat-tened, slightly mucilaginous when wetted.

Linum perenne L.
Wild Flax

Plants mostly 1.5–5 dm tall; leaves linear to narrowly oblong, 1–3 cm long, 0.1–0.3 cm broad, 1-veined, acute to acuminate; sepals 3–7 mm long, entire; petals blue, 10–20 mm long; styles usually 5, longer than or equaling the stamens; capsules de-pressed globose, 5–7 mm long.

Bluffs along major rivers, hillsides, sandy flats, gravelly stream banks, and open meadows; in much of the eastern half of continental Alaska and much of the Yukon; throughout western North America; Eur-asia (*L. lewisii* Pursh; *L. perenne* ssp. *lewisii* [Pursh] Hultén). I have been unable to detect any substantial difference between North American and Old World specimens.

Linum perenne L. (× 0.3).

LYTHRACEAE
Loosestrife Family

Herbs; leaves opposite or alternate, estipulate, simple; flowers perfect, solitary in axils, regular, not showy; sepals and petals distinct, 4–6, borne on the margin of a hypanthium; stamens 4–12, inserted on the hypanthium tube; pistils 1, superior; styles 1, the stigma 2-lobed; fruit a capsule.

LYTHRUM L.

Slender plants with angled stems; flowers inconspicuous, in the axils of foliose bracts; hypanthium cylindrical, 8–12-ribbed; sepals 4–6, with as many intervening appendages; ovary 2-loculed; capsule 2-valved.

Lythrum hyssopifolia L.

Annual or perennial, with stems simple or branched, 1–3 dm tall or more; leaves linear to oblong, 5–15 mm long and 2–4 mm broad, sessile, obtuse, glabrous; flowers solitary, sessile, white or purplish; petals 1–2 mm long; stamens included in the hypanthium, this about 3–4 mm long.

Gardens; Anchorage (Strutz 1810, BRY) and to be expected in other disturbed sites in southern Alaska; southward from Washington to California, and on the Atlantic Coast; Europe.

MALVACEAE
Mallow Family

Annual or biennial herbs; leaves alternate, simple; flowers perfect, regular; sepals 5, distinct or connate basally; petals 5, adnate to the base of the staminal sheath; stamens numerous, the filaments connate into a sheath surrounding the styles; pistils 1, the ovary superior, with 5–several carpels; styles equal in number to the carpels, the stigmas elongate; fruit a schizocarp.

MALVA L.

Plants with strong taproots; leaves with long petioles, the blades shallowly lobed, palmately veined; flowers in axillary clusters; calyx subtended by 3 bractlets, expanding and surrounding the fruit at maturity; petals usually emarginate; carpels and styles usually 10–15.

Malva neglecta Wallr.
Cheese-weed

Stems mostly 1–5 dm long, prostrate to decumbent, radiating from the taproot; petioles 3–15 cm long or more; leaf blades reniform, 1–4 cm long, 1.5–6 cm broad, shallowly 5-lobed; corolla white to bluish or pinkish, the petals mostly 6–10 mm long; fruit clusters 4–10 mm broad, separating into 1-seeded segments at maturity.

Occasional weed of greenhouses; this is

potentially a troublesome weed of gardens; adventive from Eurasia.

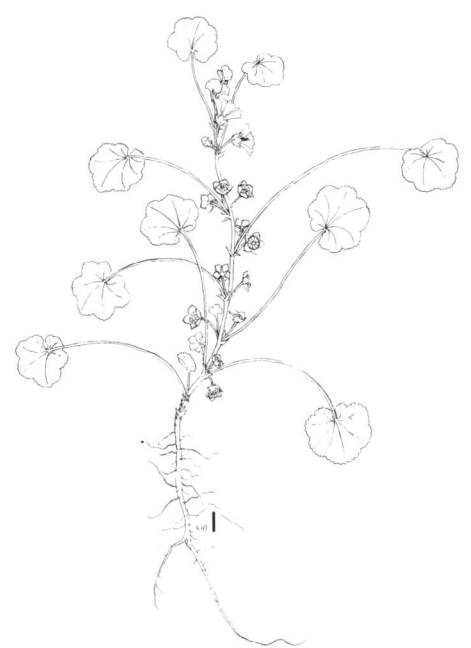

Malva neglecta Wallr. (× 0.3).

MENYANTHACEAE
Buckbean Family

Plants aquatic or semiaquatic, glabrous herbs with thick rhizomes; leaves alternate, long-petiolate, simple or trifoliolate, the petioles sheathing basally; flowers perfect, regular, arranged in paniculate cymes or racemes; sepals 5, adnate to the lower portion of the ovary; petals 5, united, the corolla rotate to funnelform; stamens 5, alternate with the corolla lobes; pistils 1, the ovary partially inferior, 1-loculed; styles long or wanting, the stigma simple or 2-lobed; fruit a septicidal capsule, tardily dehiscent to indehiscent; seeds smooth and shining.

Gillett, J. M. 1963. Menyanthaceae. pp. 82–90. In: The Gentians of Canada, Alaska, and Greenland. Can. Dept. Agr. Publ. 1180:1–99.

1a. Leaves simple, the blades reniform; corolla lobes each with an undulate flange on the midvein. *Fauria*
1b. Leaves trifoliolate; corolla lobes bearded. *Menyanthes*

FAURIA Franchet

Semiaquatic herbs; leaves long-petiolate, the blades inconspicuously bilobed, the margin crenate or bicrenate; flowers few to several, borne in paniculate cymes on long, naked scapes; calyx 5-lobed, the tube adnate to the ovary, the lobes distinct; corolla 5-lobed, the lobes spreading, each with an undulate flange along the midvein; capsules 1-loculed, dehiscent.

Fauria crista-galli (Menzies) Makino
Deer Cabbage

Rhizomes covered with old leaf bases; petioles 4–30 cm long; leaf blades reniform, 2–7 cm long (from sinus to apex), 5–14 cm broad; scapes 10–30(50) cm long; sepals lanceolate, 3–5 mm long, spreading, the adnate calyx tube conic; corolla white, rotate, the tube 2–4 mm long, the lobes 5–6 mm long, the midvein and margins with undulate flanges; capsules elongated at maturity, the free portion longer than the adnate lower part.

In bogs, swamps, wet meadows, and seeps; in coastal and insular southern and southeastern Alaska; south to Washington; Japan (*Menyanthes crista-galli* Menzies; *Nephrophyllidium crista-galli* [Menzies] Gilg). The flowers have a rank odor.

MENYANTHES L.

Aquatic to semiaquatic herbs; leaves long-petiolate, the blades trifoliolate, the leaflets entire or nearly so; flowers few to several, borne in racemes, or less commonly in panicles, on long naked scapes; calyx 5-lobed, the tube adnate to the ovary, the lobes distinct; corolla 5-lobed, the lobes spreading, each with numerous scaly hairs on the inner surface; capsules 1-loculed, indehiscent to tardily dehiscent.

Menyanthes trifoliata L. (× 0.3).

Menyanthes trifoliata L.
Buckbean

Rhizomes covered with membranous leaf bases; petioles 5–30 cm long, the sheathing stipules subequal in length to the free petioles; leaflets elliptic to oblanceolate or obovate, 2–9(12) cm long, 0.9–5 cm broad; scapes 5–30 cm long; sepals 2–5 mm long, oblong, spreading, the adnate calyx tube short-conic; corolla white to pink, funnelform, the tube 5–8 mm long, the lobes 5–8 mm long, usually purplish tinged at the apex, clothed with scaly hairs; capsules ellipsoid, almost entirely superior.

In ponds, bogs, swamps, wet meadows, and seeps; in most of Alaska and Yukon; south to California, Utah, and Colorado and east to the Atlantic; circumboreal.

MYRICACEAE
Bayberry Family

Plants dioecious or monoecious shrubs; leaves alternate, simple, deciduous; flowers imperfect, arranged in stiff, axillary spikes, the perianth lacking; stamens mostly 3–12; pistils 1, the ovary superior, 1-loculed; styles 2, distinct or nearly so; fruit a 1-seeded, wax-covered drupe.

MYRICA L.

Stems dark, the branches ascending; leaves clustered near the ends of the branches; spikes appearing before the leaves; flowers inconspicuous.

Myrica gale L. (× 0.5).

Myrica gale L.
Sweet Gale

Plants deciduous, 5–15 dm tall (rarely more); leaves oblanceolate, 1–6 cm long, 0.3–2 cm broad, rounded to obtuse apically, cuneate basally, coarsely few-toothed near the apex, or rarely entire, pubescent to glabrate or glabrous on both sides, dotted with yellow, resinous glands; staminate spikes mostly 1–2 cm long, the stamens mostly 3–5; pistillate spikes mostly 1 cm long or less; pistils fused to the lateral bracts, dotted with yellow glands; drupes 2.5–3.5 mm long.

Marshlands, bogs, and along streams; in much of Alaska south of the 67th parallel, but more common in coastal swamps in southern and southeastern Alaska, and in west-central Yukon; east to Newfoundland and south to Oregon, Michigan, and Virginia; Eurasia (*M. gale* var. *tomentosa* C. DC. ex DC.).

NYMPHAEACEAE
Waterlily Family

Perennial, aquatic plants, from submerged rhizomes; leaves alternate, simple, long-petioled, the blades floating on the surface or less commonly elevated above, peltate or cordate; flowers perfect, regular, solitary, long-peduncled; perianth segments distinct; sepals mostly 3–5, green or brightly colored; petals 3–many, usually showy (scalelike in *Nuphar*); stamens 3–6, or indefinite; pistils 3–several, distinct, or solitary; ovary superior, with few to several locules and carpels; fruit a follicle or a leathery berry.

1a. Leaves peltate; sepals 3; petals 3; carpels separate. *Brasenia*
1b. Leaves with petiole laterally attached to the blade; sepals 4–several; petals numerous; carpels united. (2)

2a. Flowers white; leaves mostly less than 10 cm long from sinus to apex. *Nymphaea*

2b. Flowers yellow; leaves mostly over 12 cm long from sinus to apex. *Nuphar*

BRASENIA Schreb.

Herbs with elongate, slender stems; leaves alternate, arising from near the stem apex, the blades peltate; flowers long-pedunculate, solitary, arising from the leaf axils, purplish; sepals 3 (rarely 4); petals 3 (rarely 4); stamens 12–30, the filaments filiform; pistils 5–several, each 1-carpelled and 1-loculed; fruit follicular, 1–2-seeded, indehiscent.

Brasenia schreberi Gmel.
Watershield

Plants with a gelatinous sheath (except for the upper surface of the leaves), arising from a slender rhizome; leaf blades elliptic, floating on the surface of the water, mostly 3–12 cm long, the petioles to 30 cm long or more; sepals and petals purplish, oblong, 10–15 mm long; stamens purplish; fruit 5–8 mm long.

In ponds; known in Alaska from the extreme southeastern portion (Gravina Island); southward to California, then disjunctly eastward to eastern United States and southeastern Canada; Asia.

NUPHAR J. E. Smith Nom. Cons.

Herbs with thick rhizomes; leaves alternate (appearing spirally arranged), arising from the rhizome, the blades laterally attached; flowers long-pedunculate, solitary, arising from the rhizomes, yellowish; sepals 5–several, yellowish to greenish or reddish, petaloid; petals 10–20, small and inconspicuous; stamens numerous, the filaments flattened; pistils 1, the ovary with several carpels and several locules; stigma broad, forming a circular disk; fruit a capsule, many-seeded, indehiscent.

Beal, E. O. 1956. Taxonomic revision of the Genus *Nuphar* Sm. of North America and Europe. Jour. Elisha Mitchell Sci. Soc. 72(2):317–46.

1a. Petioles compressed; sepals commonly 6 or fewer; generally 3.5 cm long or less; plants rare, reported for central Yukon. *N. variegatum*

1b. Petioles round; sepals commonly 7 or more, mostly over 3.5 cm long; plants abundant locally in much of Alaska and Yukon. *N. polysepalum*

Nuphar polysepalum Engelm.
Yellow Pondlily, Spatterdock

Plants coarse herbs; leaves with petioles to 10 dm long or more, the blades floating to emergent (rarely submersed), ovate, sagittately lobed basally, 8–25 cm long from sinus to apex, 10–23 cm broad, leathery; sepals usually 7–12, yellow (tinged with green or red), 3–6 cm long, 2–5 cm broad; petals yellowish to purple, about as long as the stamens; fruit ovoid, mostly 4–6 cm long.

In ponds; through much of Alaska and Yukon south of the 68th parallel (rarely northward); south to California, Utah, Colorado, and South Dakota and east to the Mackenzie (*N. luteum* ssp. *polysepalum* [Engelm.] Beal).

Nuphar variegatum Engelm. ex Clint.

Plants coarse herbs; leaves with petioles to 10 dm long or more, the blades floating (rarely emergent or submersed), ovate, sagittately lobed basally, mostly 10–25 cm long from sinus to apex, 10–20 cm broad, leathery; sepals usually 6, yellow or greenish, sometimes tinged red, 2–3.5 cm long, 1–2.5 cm broad; petals yellow or greenish, sometimes reddish tinged apically, about as long as the stamens; fruit ovoid, 3–5 cm long.

In ponds; central Yukon (Calder & Gillett 4056, 3 mi. N.E. of Mayo US); from the Mackenzie eastward to the Atlantic and south to British Columbia, Idaho, Montana, Nebraska, and New Jersey (*N. luteum* ssp. *variegatum* [Engelm.] Beal).

NYMPHAEA L. Nom. Cons.

Herbs with thick rhizomes; leaves alternate (appearing spirally arranged), arising from the rhizome, the blade laterally attached; flowers long-pedunculate, solitary, arising from the rhizome; sepals usually 4, greenish; petals numerous, white, conspicuous; stamens numerous, transitional from the petals, the filaments flattened; pistils 1, the ovary with several carpels and several locules; stigmas several to many, broad, petaloid; fruit a capsule, many-seeded, tardily dehiscent.

Nymphaea tetragona Georgi
Pygmy Waterlily

Plants slender herbs; leaves with petioles to 5 dm long or more, the blades floating, ovate, sagittately lobed basally, 2.5–6 cm long from sinus to apex, 2.5–8 cm broad, leathery; sepals 5, 1.8–2.5 cm long; petals white, subequal to the sepals, longer than the stamens; capsules 1–2 cm long, ovoid.

Lakes and ponds; in south-central to east-central Alaska and western Yukon, and dis-

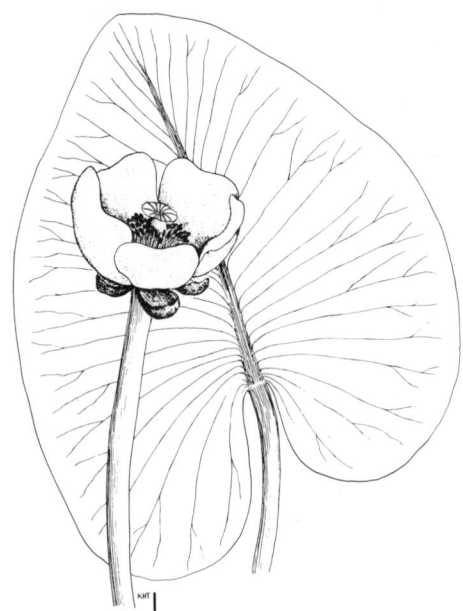

Nuphar polysepalum Engelm. (× 0.25).

junctly in southeastern Alaska; south to Washington and Idaho and eastward to Maine and Quebec; Eurasia (*N. tetragona* var. *leibergii* [Morong] Schuster). Our material has been assigned to ssp. *leibergii* [Morong] Porsild.

OLEACEAE
Olive Family

Shrubs or trees; leaves opposite (rarely alternate), simple; flowers perfect or imperfect, regular; calyx usually 4-lobed, sometimes obsolete; corolla of 4 or more united petals; stamens usually 2; pistils 1, the ovary superior, 2-loculed; fruit a capsule.

SYRINGA L.

Shrubs or small trees; leaves pinnately veined, petiolate, entire; flowers numerous, borne in terminal panicles; calyx campanulate, the teeth much shorter than the calyx tube, persistent; corolla funnelform, the tube slender, the lobes spreading; stamens inserted on the corolla tube; fruit a loculicidal capsule; seeds winged, 2 per locule.

1a. Leaves subcordate to truncate basally, ovate, acuminate to acute apically. S. *vulgaris*
1b. Leaves obtuse to acute basally, not at all subcordate to truncate, elliptic to oblong or lanceolate, acuminate (rarely acute) apically. S. *villosa*

Syringa villosa Vahl
Lilac

Shrubs to 3 m tall, the branches stout; leaves elliptic to oblong, 4–12 cm long, obtuse to acute basally, acuminate (rarely acute) apically, green above, pale beneath, pubescent on the midrib below; flowers lilac to white, in panicles 10–18 cm long; corolla 15–20 mm long; capsules 10–15 mm long.

Cultivated ornamentals; in southern Alaska and in the Yukon (?); Asia. This is a very attractive and hardy plant. It should be more widely used in the future.

Syringa vulgaris L.
Common Lilac

Shrubs or small trees, to 4 m tall or more, the branches stout; leaves ovate, 5–10 cm long, subcordate to truncate basally, acuminate to acute apically, green above, glaucous beneath, glabrous; flowers lilac to white, in panicles 10–20 cm long; corolla 12–18 mm long; fruit 10–15 mm long.

Occasionally cultivated ornamental; in southern Alaska and in the Yukon (?); Asia.

Syringa villosa Vahl (× 0.3).

ONAGRACEAE
Evening-Primrose Family

Herbaceous annual or perennial plants; leaves alternate or opposite, simple; flowers perfect, regular (irregular in *Circaea*), arranged in racemes; hypanthium adnate to the ovary and usually prolonged beyond as a tube; sepals 4 or 2; petals distinct, 4 or 2, inserted on the apex of the hypanthium; stamens as many, or twice as many, as the petals; pistils 1; ovary inferior, usually 4-loculed; styles 1, the stigma capitate, 4-lobed, or discoid; fruit a capsule or a nut.

   1a. Sepals, petals, and stamens 2; fruit indehiscent, usually with hooked hairs. *Circaea*

   1b. Sepals and petals 4; stamens 8; fruit dehiscent, lacking hooked hairs. *Epilobium*

CIRCAEA L.

Perennial herbs from rhizomes, stolons, or tubers; leaves opposite, petioled; flowers irregular, inconspicuous, borne in terminal racemes; hypanthium prolonged somewhat beyond the ovary; sepals 2, reflexed; petals 2, emarginate; stamens 2, alternate with the petals; ovary 1–2-loculed, the ovules 1 per locule; fruit indehiscent, pubescent with hooked or straight hairs.

Munz, P. A. 1965. *Circaea.* N. Am. Fl. II. 5:24–25.

Circaea alpina L.
Enchanter's Nightshade

Plants mostly 0.5–2.5 dm tall, the stems simple or more commonly branching from near the base; leaves petiolate, the blades ovate, 1–6 cm long, 0.8–4 cm broad, cordate to truncate or rounded basally, acuminate to acute apically, irregularly serrate-dentate; racemes several-flowered, the pedicels subtended by minute bracts; sepals 1–2 mm long, white to pinkish; petals 1–2 mm long, white to pinkish, notched to near the middle; fruit oblong to obovoid, 1-loculed, 1.5–2 mm long, pubescent with hooked or straight bristles.

Moist sites, mostly in woods; in much of coastal southern Alaska, and less commonly in the interior (to be sought in the Yukon); east to Newfoundland and Labrador and

south to California, Arizona, New Mexico, Iowa, Tennessee, and North Carolina; Eurasia. Our materials befong to var. *alpina.*

EPILOBIUM L.

Perennial herbs (though often flowering the first year), from rhizomes, stolons, bulblike offsets (turions), taproots or fibrous roots, the stems decumbent to ascending or erect; leaves all opposite or the upper ones alternate, or all alternate (rarely whorled), simple, entire to toothed; flowers perfect, regular or nearly so, solitary in leaf axils or in terminal racemes; hypanthium short or lacking; sepals 4; petals 4, entire or emarginate; stamens 8; stigmas capitate, cylindrical, or 4-lobed; fruit a loculicidal, 4-carpellate, 4-loculed, elongate capsule; seeds with a tuft of hair.

Note: Important diagnostic features are present on the plant parts at or below ground level. These must be collected in order to properly identify that entity to which a particular specimen belongs.

Hitchcock, C. L. 1961. *Epilobium.* pp. 473–85. *In*: C. L. Hitchcock, et al. Vascular plants of the Pacific Northwest. Univ. Wash. Publ. Bio. 17:1–615.

Munz, P. A. 1965. *Epilobium.* N. Am. Fl. II. 5:198–225.

   1a. Petals (8)10–30 mm long, rounded apically, spreading; hypanthium not extending beyond the ovary (the calyx cleft to the top of the ovary); Section Chamaenerion, the fireweeds. (2)

   1b. Petals 2–10 mm long (longer in *E. luteum* and *E.* × *treleasianum*), emarginate apically, ascending; hypanthium extending beyond the ovary (the calyx not cleft to the top of the ovary); Section Epilobium, the willow-herbs. (3)

   2a. Plants erect; styles pubescent at the base; floral bracts much reduced; herbage glabrous or nearly so; inflorescence usually with 15–more flowers. *E. angustifolium*

2b. Plants decumbent; styles glabrous; floral bracts similar to the leaves, though smaller; herbage usually minutely pubescent; inflorescence usually with fewer than 10 flowers. *E. latifolium*

3a. Petals yellow, 12–19 mm long; plants 1.3–8 dm tall, occurring in coastal southern Alaska. *E. luteum*

3b. Petals pinkish to purplish or white, mostly less than 12 mm long; plants with distribution various. (4)

4a. Petals pink to violet, (9)10–12 mm long; stigma 4-lobed; plants of the Aleutians and southwestern Alaska. *E.* × *treleasianum*

4b. Petals pinkish to purplish or white, 2–10 mm long; stigma neither lobed nor cleft; plants of broad distribution. (5)

5a. Leaves linear to narrowly lanceolate, mostly 8(12) mm broad or less, entire or nearly so; plants often grayish-hairy. *E. palustre*

5b. Leaves lanceolate to oblong or ovate-lanceolate, often more than 8 mm broad, often toothed; plants variously pubescent to glabrous. (6)

6a. Plants producing bulblike offsets (turions) at the base of the stem, the fleshy overlapping scales often persistent on the base of the current stem, either robust herbs with mostly simple stems, or low, usually much-branched herbs. (7)

6b. Plants arising from short to elongate rhizomes, or from taproots or fibrous roots, not producing turions; stems simple or branched. (8)

7a. Petals (3)5–10 mm long; stems coarse, simple, or with erect branches near the apex, 3–10 dm tall; leaves 2.5–10 cm long. *E. glandulosum*

7b. Petals 3–4 mm long; stems slender, usually branched from near the base,

1–2 dm tall; leaves 0.5–2.5 cm long. *E. leptocarpum*

8a. Stems often more than 3 dm tall; rhizomes short or lacking; leaves often denticulate; hair on seeds white (sometimes dingy); seeds minutely roughened, striate. *E. adenocaulon*

8b. Stems seldom to 3 dm tall; rhizomes mostly well developed; leaves sometimes entire; hair on seeds dingy; seeds smooth to somewhat roughened. *E. alpinum*

Epilobium adenocaulon Hausskn.
Northern Willow-herb

Plants 1–8(10) dm tall, arising from tap- or fibrous roots, ultimately shortly rhizomatous, perennating from somewhat fleshy rosettes, lacking turions; stems often purplish, usually erect, glabrous or glabrate below, usually glandular above, the pubescence decurrent in lines below the leaf bases or broadly scattered, usually simple below and branched above; leaves opposite (rarely alternate above), lanceolate to ovate, (1.5)2–9 cm long, 0.6–3 cm broad, rounded basally, tapering to a blunt apex, serrulate, pubescent to glabrous, short-petiolate to subsessile; sepals 2–4 mm long; petals 3–8 mm long, white to pink or reddish; capsules 3.5–8 cm long, glabrous to pubescent or glandular-pubescent; seeds 0.6–1 mm long, minutely roughened in parallel lines, the hair white or sometimes dingy.

Moist soil, along streams and roadsides, in meadows and waste places; in most of Alaska and Yukon, south of the 68th parallel (except for the southwestern portion of Alaska); widely distributed in North America; adventive in Europe (*E. boreale* Hausskn.; *E. ciliatum* authors, not [?] Raf.). Our specimens demonstrate considerable variation. Some of them approach the morphologically similar (and closely allied) *E. glandulosum*, with which this entity has previously been combined (as *E. glandulosum* var. *adenocaulon* [Hausskn.]

Fern.). The lack of true turions (a feature usually lacking in herbarium specimens) seems to be the singly most definitive feature by which *E. adenocaulon* can be distinguished. Two varieties are distinguished among our materials.

1a. Petals 3–5 mm long, white, pinkish, or reddish; our common form. *E. adenocaulon* var. *adenocaulon*

1b. Petals mostly 5–8 mm long, bright pink to reddish; uncommon in our region. *E. adenocaulon* var. *occidentale* Trel.

Epilobium alpinum L.
Alpine Willow-herb

Plants (0.2)0.4–3(5) dm tall, arising from elongate rhizomes and stolons, lacking turions; stems sometimes purplish, basally decumbent to ascending or erect, glabrous to glabrate below, pubescent in lines below the leaf bases, or sometimes almost entirely glabrous, simple to branched; leaves opposite, or the upper ones alternate, narrowly oblong to ovate, 1–4.5 cm long, 0.2–2 cm broad, acute to obtuse basally, the apex blunt, entire to serrulate, sessile to short-petioled, glabrous to minutely pubescent; sepals 2–5 mm long; petals 3–8 mm long, white to bright pink; capsules 2–7 cm long, glabrous to puberulent or glandular; seeds 1–2 mm long, smooth to somewhat roughened, the hair dingy to white.

Moist soil, near seeps and springs, along streams, in meadows, and bogs; mostly in southern Alaska, but also in interior and western Alaska and southern Yukon; widely distributed in North America; Eurasia. Here the assemblage is treated as a single species consisting of several varieties; however, it is often regarded as a series of separate species. The entities appear to be based on the extreme forms of a series of morphologically intergrading populations whose relationships are poorly understood. The following somewhat mechanical key will serve to distinguish most specimens.

1a. Plants 2–4 cm tall; mature capsules longer than the stems; leaves 0.5–2 cm long, lance-oblong, thickish, longer than the internodes; mostly of southern and southwestern Alaska (*E. sertulatum* Hausskn.). *E. alpinum* var. *sertulatum* (Hausskn.) Welsh

1b. Plants 5–50 cm tall; mature capsules shorter than the stems; leaves various, but if 0.5–2 cm long, then often shorter than the internodes. (2)

2a. Petals white to pinkish, 2–4(5) mm long; seeds smooth; plants 1–4 dm tall, of south-central and southeastern Alaska and southern Yukon (*E. lactiflorum* Hausskn.) *E. alpinum* var. *lactiflorum* (Hausskn.) C. L. Hitchc.

2b. Petals mostly pink or over 5 mm long; plants variable in size. (3)

3a. Stems mostly 5–15(22) cm tall, often curved; leaves mostly 1–2(2.5) cm long, obscurely petioled; petals 3–6 mm long. (4)

3b. Stems mostly 10–40 cm tall, straight or curved; leaves 1.5–4 cm long, sessile to subsessile; petals 5–8 mm long. (5)

4a. Capsules broader toward the apex, mostly 1.5–2 mm broad; inflorescence often erect in bud; leaves broadly ovate, more or less serrulate; seeds roughened, 1.5–2 mm long; plants of southeastern Alaska (*E. clavatum* Trel.). *E. alpinum* var. *clavatum* (Trel.) C. L. Hitchc.

4b. Capsules linear, about 1 mm broad; inflorescence nodding in bud; leaves lanceolate to elliptic or oblong; seeds smooth, about 1 mm long; plants of most of Alaska and Yukon south of the 66th parallel (*E. anagallidifolium* Lam.; *E. pseudo-scaposum* Hausskn.; *E. anagallidifolium* var. *pseudo-scaposum* [Hausskn.] Hultén). *E. alpinum* var. *alpinum*

5a. Base of stem with several pairs of broad, withered leaves; seeds smooth; leaves often sharply toothed; plants of coastal and insular southern Alaska

(*E. behringianum* Hausskn.). *E. alpinum* var. *behringianum* (Hausskn.) Welsh

5b. Base of stem with leaves inconspicuous or lacking; seeds more or less roughened; leaves sharply serrate to subentire; plants of most of Alaska and Yukon south of the 66th parallel (*E. hornemanni* Reichb.). *E. alpinum* var. *nutans* (Hornem.) Hook.

Epilobium angustifolium L.
Fireweed

Plants mostly (3)5–20 dm tall, arising from rhizomelike roots which produce buds freely; stems often purplish, at least above, basally decumbent to erect, usually simple (less commonly much branched), glabrous below, commonly puberulent above; leaves alternate, lanceolate to elliptic or linear, 5–20 cm long, 0.3–4 cm broad, entire or nearly so, sessile or subsessile, glabrous or pubescent only along the lower midvein; sepals 8–17 mm long, puberulent; petals 8–20 mm long, pink purple, pink, or rarely white; styles longer than the stamens, pubescent near the base, the stigmas 4-lobed; capsules 3–9 cm long, pubescent; seeds 1–1.5 mm long, smooth, the hair dingy.

Arctic and alpine tundra, heathlands, woods, meadows, and roadsides; through nearly all of Alaska and Yukon; eastward to the Atlantic and south to California, Arizona, New Mexico, South Dakota, Minnesota, Indiana, and Virginia; circumboreal. This is one of the most beautiful of all the boreal plants. This is the territorial flower of the Yukon. Roadsides blaze with color from the flowers of this remarkable plant from mid- to late summer.

1a. Leaves long-attenuate apically, the lower midribs always glabrous, often less than 2 cm broad and less than 10 cm long (pollen commonly less than 85 μ broad); plants broadly distributed in Alaska and Yukon (*E. angustifolium* var. *intermedium* [Wormskj.] Fern.; *E. angustifolium* f.

albiflorum [Dumort.] Hausskn., a phase with pale petals and whitish sepals; *E. angustifolium* f. *spectabile* [Simmons] Fern., a phase with pale petals and red sepals. *E. angustifolium* var. *angustifolium*

1b. Leaves blunt to acute apically, rarely long-attenuate, the lower midribs glabrous to pubescent, mostly over 2 cm broad and more than 10 cm long pollen commonly over 85 μ broad); plants of coastal and insular southern Alaska, rarely interior (*E. angustifolium* var. *platyphyllum* [Daniels] Fern.; *E. angustifolium* ssp. *circumvagum* Mosq.; *E. angustifolium* ssp. *macrophyllum* [Hausskn.] Hultén.) *E. angustifolium* var. *abbreviatum* (Lunell) Munz

Epilobium glandulosum Lehm.
Glandular Willow-herb

Plants 3–10 dm tall, arising from tap- or fibrous roots, or from short to elongate rhizomes, perennating from turions; stems sometimes purplish, erect, simple or branched above, glabrous below, often glandular above, the pubescence often decurrent in lines below the leaf bases; leaves opposite (sometimes alternate above), narrowly lanceolate to ovate-lanceolate, 2.5–10 cm long, 0.5–4 cm broad, rounded to subcordate basally, acuminate to acute (the apex usually blunt) apically, serrulate, glabrous or pubescent, sessile and somewhat clasping to short-petiolate; sepals 3–5 mm long, glandular-pubescent; petals 3–10 mm long, pink to purplish; capsules 3–7 cm long, usually glandular-puberulent; seeds 1–1.8 mm long, raggedly roughened in parallel lines, the hairs dingy or white.

Moist sites, in woods, thickets, meadows, roadsides, and along beaches; mostly in coastal and insular southern Alaska, but also known from scattered locations in interior Alaska and Yukon; east to Labrador and south to California and Colorado; Asia. This entity is perplexingly difficult to distinguish from *E. adenocaulon*, but the largely coastal distribution, along with the more robust nature and conspicuously roughened seeds, seems sufficient to set these plants apart, even though turions are rarely collected (*E. saximontanum* Hausskn.).

Epilobium latifolium L.
Dwarf Fireweed

Plants mostly 1–4(7) dm long, arising from a caudex, lacking rhizomes; stems occasionally purplish, decumbent to ascending, glabrous below, puberulent above; leaves opposite (or whorled) below, usually alternate above, lanceolate to elliptic or ovate, 1.5–8 cm long, 0.5–3 cm broad, entire or denticulate, sessile or nearly so, glaucous, usually pubescent; sepals 10–18 mm long, puberulent; petals 15–30 mm long, bright pink to pink purple or rarely white; styles shorter than the stamens, glabrous, the stigma 4-lobed; capsules 3–9(10) cm long, glabrate to pubescent, usually purplish; seeds (1)1.5–2 mm long, smooth, the hair dingy.

Gravelly flats, stream banks, roadsides, talus slopes, or on rocky outcrops; in most of Alaska and Yukon; east to Labrador and Newfoundland and south to California, Utah, Colorado, South Dakota, and Quebec; Greenland, Iceland, Asia (*E. latifolium* f. *leucanthum* [Ulke] Fern., a phase with sepals and petals whitish; *E. latifolium* f. *munzii* Lepage, a phase with colored sepals and whitish petals).

Epilobium leptocarpum Hausskn.

Plants mostly 0.5–2 cm tall, arising from taproots, perennating from turions; stems often reddish, erect or decumbent at the base, simple or much branched, glabrous except for decurrent lines of pubescence from the leaf bases or pubescent throughout; leaves mostly opposite (rarely alternate), lanceolate to lance-oblong or elliptic, 0.5–2.5 cm long, 0.2–1 cm broad, obtuse to acute basally, tapering to a blunt apex, serrulate, glabrous or pubescent, short-

Epilobium latifolium L. (× 0.4).

petiolate; sepals 1–2 mm long, glabrous or nearly so; petals 2–3 mm long, whitish to pink; capsules 1–4 cm long, glabrous or nearly so; seeds 0.8–1 mm long, roughened in parallel lines, the hairs dingy.

Moist sites; mostly in coastal and insular

southern Alaska, but less commonly some distance from the coast (to be expected in the Yukon). Our material has been treated as *E. glandulosum* var. *macounii* (Trel.) C. L. Hitchc., or as *E. leptocarpum* var. *macounii* Trel. It seems to be quite distinct from the plants herein considered as belonging to *E. glandulosum*.

Epilobium luteum Pursh
Yellow Willow-herb

Plants 1.3–8 dm tall, arising from rhizomes, perennating from turions; stems greenish, erect, simple, or branched above, glabrous or with decurrent lines of pubescence from the leaf bases; leaves opposite, lanceolate to elliptic, 2–8 cm long, 0.6–3 cm broad, acute to obtuse or ·rounded basally, acute to acuminate apically, denticulate, glabrous (pubescent basally), sessile or subsessile; sepals 10–15 mm long, glabrate; petals 12–19 mm long, yellow, the margin undulate; styles glabrous, longer than the petals; capsules 4–7 cm long, glandular-puberulent; seeds 1–2 mm long, smooth, the hairs dingy.

Moist sites along streams, springs, and lakes; in insular and coastal southern Alaska; south to northern California and east to Alberta. This distinctive species flowers from late July to early September. It is seldom collected in fruit.

Epilobium palustre L.
Swamp Willow-herb

Plants mostly 1–4(8) dm tall, arising from slender rhizomes (or taproots), which frequently produce loose turions; stems sometimes reddish, erect, simple or branched, glabrous to variously pubescent; leaves mostly opposite, linear to narrowly oblong or narrowly lanceolate, 1–6 cm long, 0.1–0.8(1.2) cm broad, acute basally, the apex blunt, entire to minutely denticulate, glabrous or nearly so, short-petiolate to subsessile; sepals 1.5–2.5 mm long, puberulent; petals 3–5 mm long, white to pinkish; capsules 3–7 cm long, pubescent to glabrate or glabrous; seeds 1–2 mm long, smooth or nearly so, the hair dingy.

Moist soil, in bogs, roadside ditches, along streams, and in meadows; in most of Alaska and Yukon; eastward to the Atlantic and south to Washington, Colorado, and New York; circumboreal. Several minor, intergrading varieties are present among our materials. No single characteristic appears to be definitive in separating the various entities and the segregation is more or less arbitrary.

1a. Main leaves 0.5–1.2 cm broad; fruiting pedicels shorter than the subtending leaves; plants broadly distributed in Alaska and Yukon (*E. leptophyllum* Raf.). *E. palustre* var. *palustre*
1b. Main leaves 0.1–0.5 cm broad; fruiting pedicels equal to the subtending leaves or longer. (2)
2a. Leaves 2 mm broad or less, or less than 1.5 cm long; fruiting capsules glabrous or glabrate; plants ordinarily less than 2 dm tall, scattered in northern and south-central to eastern Alaska and much of the Yukon (*E. davuricum* Fisch. ex Hornem.). *E. palustre* var. *davuricum* (Fisch.) Welsh
2b. Leaves more than 2 mm broad (at least some), and commonly more than 1.5 cm long; fruiting capsules usually pubescent, rarely glabrate (*E. palustre* var. *grammadophyllum* Hausskn.); our most common variety. *E. palustre* var. *lapponicum* Wahl.

Epilobium × treleasianum Levl.
Trelease Willow-herb

Plants mostly 2–3 dm tall, arising from rhizomes, perennating from turions (?); stems greenish, erect, simple, glabrous or with decurrent lines of pubescence from the leaf bases; leaves opposite or nearly so, oblong-ovate, 1.5–5 cm long, obtuse to acute apically, denticulate, glabrous or nearly so, subsessile; sepals 7–8 mm long, glabrate; petals (9)10–12 mm long, pink to violet; stigma distinctly 4-lobed, shorter than the petals; capsules 2.5–3.5 cm long, pubescent; seeds smooth.

Moist sites; known in our region from the Alaska Peninsula, and from the easternmost Aleutians. This entity is poorly understood. It has some characteristics in common with *E. luteum,* and may be a hybrid involving that species as one parent. The name has also been cited as a synonym of *E. alpinum* var. *nutans.* Further investigation and more specimens are required to allow for adequate interpretation of *E.* × *treleasianum.*

OROBANCHACEAE
Broomrape Family

Herbaceous root parasites, without green foliage; stems erect, usually yellowish or purplish; leaves reduced to alternate scales; flowers perfect, irregular; calyx regularly to irregularly 4–5-toothed or -cleft, or almost truncate; corolla 5-lobed, more or less bilabiate; stamens 4, didynamous; pistils 1, the ovary superior, 1-loculed, mostly 2(3)-carpellate; styles 1, the stigma terminal, discoid or lobed; fruit a loculicidal capsule.

1a. Plants glabrous (the bracts and corolla lobes sometimes ciliate), thick, fleshy, brownish red or yellowish red to purplish; inflorescence compact, many-flowered, the flowers subsessile. *Boschniakia*

1b. Plants glandular-pubescent, occasionally somewhat thickened and fleshy, usually whitish or yellowish; inflorescence 1–few-flowered, the flowers long-pedicellate. *Orobanche*

BOSCHNIAKIA C. A. Mey.

Plants erect, arising from thickened, cormous bases, attached to the root of the host; leaves numerous, alternate or apparently spirally arranged; inflorescence many-flowered, the raceme spicate; flowers subsessile, subtended by broad bracts; calyx irregularly lobed or toothed; corolla bilabiate, the upper lip curved, usually shallowly cleft, the lower lip greatly shortened, 3-lobed;

stamens with flattened filaments, the anthers rounded apically; ovary 2(3)-carpelled; capsules irregularly dehiscent, the seeds numerous.

Boschniakia rossica (Cham. & Schlecht.) Fedtsch.
Ground-cone, Poque

Plants 1–4 dm long, from a cormous base 1.5–2.5 cm in diameter, the stems 4–15 mm in diameter at base of inflorescence; scale leaves triangular to ovate, short-acuminate to acute, yellowish to purplish, entire to minutely shredded or short-ciliate, 3–10 mm long; inflorescence 6–25 cm long, many-flowered; bracts ovate, acuminate to acute or obtuse, minutely shredded to ciliate; calyx 3–6 mm long, irregularly lobed, the lobes short-ciliate; corolla purplish, mostly 8–13 mm long, the lobes ciliate; anther sacs only slightly exceeding the connective, glabrous or nearly so.

Thickets, woodlands, and heath to tundra, most commonly found on *Alnus* species, but also reported on the roots of *Betula, Vaccinium, Picea, Salix,* and *Chamaedaphne;* widely distributed in Alaska and Yukon; east to Mackenzie and south to British Columbia; Asia. *B. hookeri* Walp. has been reported from northern British Columbia, and might be expected in southeastern Alaska. It differs from *B. rossica* in having entire, nonciliate bracts, decidedly pubescent filaments, and anthers much exceeding the connective.

OROBANCHE L.

Plants erect, the bases usually only slightly thickened; leaves few to several, alternate; inflorescence few-flowered, corymbose; flowers borne on long pedicels, subtended by elongate bracts; calyx usually regularly 4–5-lobed; corolla bilabiate, the upper lip 2-lobed, the lower lip about as long as the upper one, 3-lobed; stamens with anthers pointed apically; ovary 2-carpelled; capsules regularly dehiscent, the seeds numerous.

Boschniakia rossica (Cham. & Schlecht.) Fedtsch. (× 0.7).

Achey, D. M. 1933. A revision of the section Gymnocaulis of the genus *Orobanche*. Bull. Torrey Club 60:441–51.

1a. Pedicels 4–10 per stem, subequal to the stem. *O. fasciculata*

1b. Pedicels 1–3 per stem, longer than the stem. *O. uniflora*

Orobanche fasciculata Nutt.

Cluster Cancer-root

Plants mostly 0.6–2 dm long, the stems mostly 3–12 cm long, solitary or several, yellowish or purplish, glandular-pubescent; scale leaves few to several; pedicels 4–10, mostly 2–10 cm long, subequal in length to the stem or shorter; calyx 5-lobed, about 10 mm long; corolla yellowish to purplish, 12–25 mm long, the lobes not ciliate; anthers glabrous or hairy; capsule 10–12 mm long.

Plants parasitic on species of *Artemisia;* in southern Yukon (the most common host is *A. frigida*) and to be sought in eastern Alaska; south to California and east to Michigan and Indiana.

Orobanche uniflora L.

One-flower Cancer-root

Plants mostly 0.5–1.5 dm tall, the stems mostly 1–4 cm long, solitary or several, yellowish or whitish, glandular-pubescent; scale leaves few; pedicels 1–3 per stem, mostly 3–10 cm long, exceeding the stem in length; calyx 5-lobed, 6–10 mm long; corolla yellowish to purplish, 15–30 mm long, the lobes ciliate; anthers glabrous or hairy.

Plants parasitic on species of Crassulaceae, Saxifragaceae, and Compositae; known in our region from Kodiak and from the Shumagin Islands; eastward to Newfoundland and south to Florida and California.

PAPAVERACEAE

Poppy Family

Perennial or annual herbaceous plants, usually with milky, whitish or yellowish juice; leaves all basal or alternate, mostly lobed or divided; flowers perfect, regular, solitary or in clusters; sepals 2–3, caducous; petals 4–6, rarely more, brightly colored; stamens numerous, distinct; pistils 1, the ovary superior, 1-loculed, with 2–many carpels; stigmas as many as the carpels; fruit a poricidal capsule.

PAPAVER L.

Herbage hispid-hairy to glabrate (rarely glabrous); leaves all basal, or cauline and alternate, merely lobed to once or twice pinnatifid; flowers solitary on scapes or on axillary peduncles, nodding in bud, erect or spreading in flower and fruit; capsules oblong to elliptic or cup shaped; stigmas radiating from the apex.

Gjaerevoll, O. 1963. *Papaver*. pp. 39–43. *In*: O. Gjaerevoll. Botanical investigations in central Alaska, especially in White Mts. Part II. Dicotyledones: Salicaceae-Umbelliferae. Kgl. Norske Skrift. 1963 (4): 1–97.

Löve, D. 1956. Papaveraceae. pp. 173-88. *In*: D. Löve and N. J. Freedman. A plant collection from SW Yukon. Bot. Notis. 109:153–211.

1a. Plants with elongate leafy stems, the leaves alternate, cultivated annuals (rarely escaping). *P. rhoeas*
1b. Plants with leaves all basal, perennial, indigenous or cultivated. (2)

2a. Leaves 3-lobed, the blades glabrous or nearly so, 3 cm long or less; scapes 10 cm long or less; plants endemic to the Seward Peninsula. *P. walpolei*
2b. Leaves usually 5–several-lobed, the blades at least somewhat hairy (except in cultivated forms of *P. nudicaule*); scapes often more than 10 cm long; plants of broad distribution. (3)

3a. Flowers white to pale pink; scapes decumbent; plants of gravelly soils and rock outcrops, in the Cook Inlet vicinity, Kenai Peninsula, St. Elias Mts., and northern British Columbia. *P. alboroseum*
3b. Flowers usually yellow (occasionally red, apricot, or white in indigenous

forms, or rarely pink, or multicolored in cultivated forms); scapes erect, decumbent only at the base (except in some *P. radicatum*); plants of tundra, heathlands, or woods over much of our region. (4)

4a. Caudex branches clothed with conspicuous, elongate, straw-colored (rarely brownish) leaf bases, the clustered bases often over 4 cm long; plants of the Alaska Range, Aleutian Islands, islands of the Bering Sea, western Alaska, and less commonly elsewhere. *P. alaskanum*
4b. Caudex branches clothed with brownish to straw-colored leaf bases, the cluster of bases seldom to 4 cm long; plants broadly distributed. (5)

5a. Populations with flowers polychrome, the plants cultivated or growing near habitations. *P. nudicaule*
5b. Populations with flowers yellow, rarely red, white, or pink, the plants indigenous, rarely cultivated. (6)

6a. Capsules 2–4 times longer than broad, less than 6(8) mm broad; plants often over 2.5 dm tall. *P. macounii*
6b. Capsules mostly 1.5–2.5 times longer than broad, often over 6 mm broad; plants seldom over 2.5 dm tall. *P. radicatum*

Papaver alaskanum Hultén
Alaska Poppy

Plants 1–2.6 dm tall; caudex conspicuously clothed with straw-colored (rarely somewhat brownish) leaf bases; leaves 3–13 cm long, once (less commonly twice) pinnatifid, with 5–11 main lobes, pubescent with bent coarse whitish spreading hairs; scapes erect, pubescent like the leaves, the pubescence brownish below the flower and on the sepals; buds nodding, 10–20 mm long; petals yellow, apricot, or white, 13–27 mm long; capsules mostly 7–12 mm long and 5–6 mm broad, brown hispid.

Arctic and alpine tundra and heathlands; from the Kenai Peninsula westward through the Aleutians, islands of the Bering Sea, western Alaska, and less commonly in northern Alaska and southwestern Yukon; endemic (*P. alaskanum* var. *latilobum* Hultén, a form with broad leaf lobes; *P. alaskanum* var. *macranthum* Hultén, a form with large flowers; *P. denalii* Gjaerevoll; *P. mcconellii* authors, not Hultén). Perhaps it should best be treated as *P. radicatum* ssp. *alaskanum* (Hultén) J. P. Anders.

Papaver alboroseum Hultén
Pale Poppy

Plants 0.6–1.8 dm tall; caudex clothed with brownish leaf bases; leaves 2–8 cm long, twice pinnatifid, with 7–11 main lobes (the lower two pairs of lobes again lobed), pubescent with bent coarse whitish to pinkish spreading hairs; scapes decumbent, pubescent with ascending-spreading, whitish to brownish hairs, the pubescence pinkish to brownish below the flower, brownish to blackish on the sepals; buds nodding, 8–14 mm long; petals white to pale pink, often with a basal, yellowish spot, 12–20 mm long; capsules 10–15 mm long, 6–11 mm broad, hispid with yellowish to brownish hairs.

Gravelly soils and rock outcrops; in the vicinity of Cook Inlet and on the Kenai Peninsula and in northern British Columbia and southwestern Yukon; Kamchatka. The pale poppy is closely related to *P. radicatum* and might best be regarded at some infraspecific level. However, the combination of decumbent scapes with small pale flowers on a population basis appears to set this pretty poppy apart from all others in Alaska. Reports of this entity from northern Alaska all belong to other taxa.

Papaver macounii Greene
Macoun Poppy

Plants 0.9–4.3 dm tall; caudex clothed with brownish (straw-colored when young) leaf bases; leaves 2–15 cm long, once (or less commonly twice) pinnatifid with 5–9 main lobes, pubescent with bent coarse whitish to yellowish or brownish spreading hairs; scapes erect, pubescent (often sparsely so) with appressed-ascending to spreading, usually brownish hairs, the pubescence usually darker and more dense below the flowers, brownish to blackish on the sepals; buds nodding, 8–16 mm long; petals yellow (rarely red or white), 11–30 mm long; capsules 14–22 mm long, 3–8 mm broad, pubescent with stiff brown to whitish hairs.

Arctic and alpine tundra or heathlands; in most of Alaska and Yukon north of the 62nd parallel; east to the Mackenzie; Asia (*P. hultenii* Knaben; *P. hultenii* var. *salmonicolor* Hultén; *P. keelei* Porsild; *P. scammanianum* D. Löve; *P. nudicaule* ssp. *radicatum* var. *kamtschaticum* [Reg.] Fedde, at least in part; *P. macounii* var. *discolor* Hultén).

Papaver nudicaule L.
Iceland Poppy

Plants 1.2–5 dm tall; caudex clothed with brownish (straw-colored when young) leaf bases; leaves 4–21 cm long, once or twice pinnatifid, with 5–11 main lobes, glabrous or pubescent (often sparsely so) with bent coarse whitish spreading hairs; scapes erect, pubescent with ascending or spreading, whitish to brownish hairs, the pubescence often darker and more dense below the flowers, reddish brown to blackish on the sepals; buds nodding, 9–16 mm long; petals yellow, red, orange, white, or pink, 20–35 mm long; capsules mostly 9–12 mm long, 6–12 mm broad, pubescent with stiff brown (white on old capsules) hairs.

Gardens, roadsides, and other disturbed sites, usually near dwellings; in widely scattered locations in Alaska and Yukon. *P. nudicaule* is cultivated in the more temperate portions of Alaska and Yukon, where it is often broadcast along roadsides near residences.

Papaver radicatum Rottb.
Arctic Poppy

Plants 0.6–2.5 dm tall; caudex clothed with

brownish (straw-colored when young) leaf bases; leaves 1.5–11 cm long, once, or more commonly twice, pinnatifid, with (3)5–9 main lobes, pubescent with bent coarse whitish or yellowish spreading hairs; scapes erect, pubescent with spreading or ascending-spreading, yellowish or brownish to black hairs, the pubescence usually darker and more dense below the flowers, brownish or blackish on the sepals; buds nodding, 9–13 mm long; petals yellow

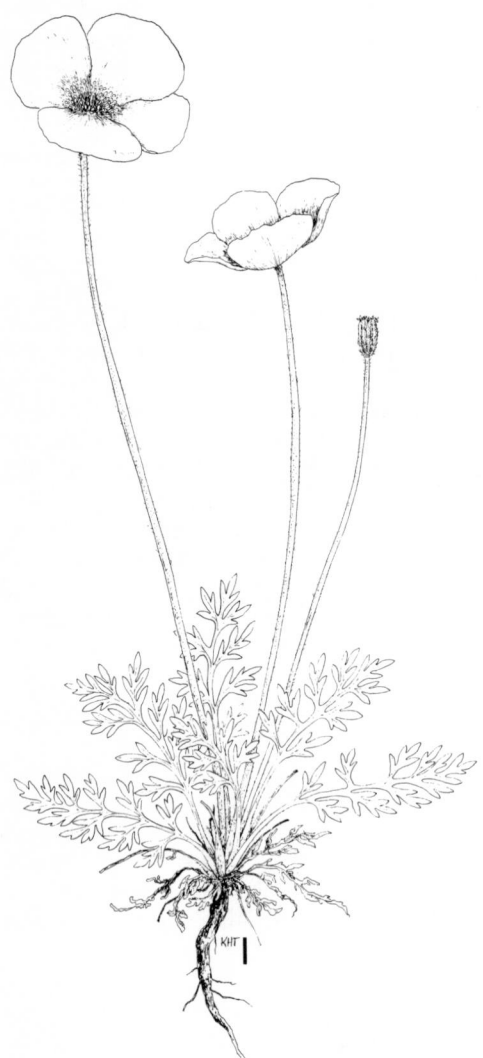

Papaver radicatum Rottb. (× 0.4).

(rarely red, white, or pinkish), 10–25 mm long; capsules 10–18 mm long, 5–9 mm broad, pubescent with stiff brownish (rarely whitish) hairs, occasionally wrapped in the corolla which usually dries bluish.

Arctic and alpine tundra or heathlands; in most of coastal western and northern Alaska and less commonly in interior Alaska and widely scattered in the Yukon, mostly in alpine sites; circumboreal (*P. freedmanianum* D. Löve; *P. kluanensis* D. Löve; *P. lapponicum* [Tolm.] Nordh.; *P. radicatum* ssp. *lapponicum* [Tolm.] Knaben; *P. lapponicum* ssp. *porsildii* Knaben; *P. mcconellii* Hultén; *P. nigro-flavum* D. Löve; *P. nudicaule* ssp. *radicatum* [Rottb.] Fedde; *P. radicatum* ssp. *occidentale* Lundstr.). *P. radicatum* is made up of a series of populations which vary from each other in minor ways. The entire complex is in serious need of a monographic study before segregation of infraspecific taxa is attempted.

Papaver rhoeas L.
Corn Poppy

Plants annual, 3–9 dm tall; stems elongate, leafy; leaves alternate, simple, merely toothed or irregularly once to twice pinnatifid, pubescent to glabrate; peduncles 7–30 cm long, pubescent with spreading yellowish hairs; buds nodding, 10–15 mm long, pubescent with spreading yellowish hairs; petals purple, scarlet, white, or bicolored, 20–40 mm long; capsules mostly 10–15 mm long, glabrous.

Cultivated ornamental; in the temperate regions of Alaska and Yukon (?), escaping and persisting; Eurasia.

Papaver walpolei Porsild
Walpole Poppy

Plants 0.4–1.1 dm tall; caudex clothed with brownish leaf bases; leaves 1–2(3) cm long, once pinnatifid (rarely simple), 3-lobed, sparsely pubescent with bent coarse whitish hairs, or glabrous; scapes erect, pubescent with spreading hairs, or glabrate, brownish-hairy below the flowers and on the sepals; buds nodding, 5–7 mm long; petals yellow (or white with a yellow spot), 9–20 mm long; capsules 8–10 mm long, 3–4 mm broad, sparsely pubescent with stiff whitish or brownish hairs.

Tundra to heathlands, often in gravel or on rocky outcrops; in the Seward Peninsula and Kuskokwim Delta region (Goodnews Bay), and easternmost Siberia; endemic (*P. walpolei* var. *sulphureo-macalatum* Hultén). This is our most distinctive indigenous poppy.

PLANTAGINACEAE
Plantain Family

Annual or perennial, acaulescent or short-stemmed herbs; leaves all basal or nearly so; flowers small, perfect or imperfect, regular, borne in bracteate spikes, or in heads; sepals 4; corolla scarious, 4-lobed; stamens 4, alternate with the corolla lobes, or only 2; pistils 1, the ovary superior, with 1–4 locules, the carpels 2; styles 1; fruit a circumscissile capsule (indehiscent in *P. macrocarpa*).

PLANTAGO L.

Plants from taproots; leaves simple, entire or variously lobed; flowers several to many, inconspicuous, each subtended by a bract; calyx sometimes somewhat irregular; corolla scarious or membranous, persistent; stamens included or exserted; fruit included in the calyx or exserted.

1a. Bracts of the inflorescence long-exserted, commonly more than twice longer than the flowers; plants annual or short-lived perennial. *P. aristata*
1b. Bracts of the inflorescence not at all exserted, shorter than the flowers; plants perennial or biennial. (2)

2a. Leaves linear to narrowly oblong or narrowly elliptic, 2–8 mm broad; tube of corolla pubescent; plants maritime. *P. maritima*

2b. Leaves oblong to elliptic, lanceolate, or ovate, commonly at least some of them more than 8 mm broad; tube of corolla glabrous; plants mostly not maritime (see *P. macrocarpa*). (3)

3a. Leaves with well-defined blade and petiole, the blade ovate to broadly elliptic, mostly 1–3 times longer than broad; plants mostly weeds of waste places. *P. major*
3b. Leaves with blades tapering to the petioles, the blades narrowly elliptic to lanceolate, oblanceolate, or oblong, 3–several times longer than broad; plants introduced weeds, or nonweedy and indigenous. (4)

4a. Outer pair of sepals (those next to the bract) connate, appearing as a solitary, 2-veined, apically notched or entire sepal; plants adventive weeds. *P. lanceolata*
4b. Outer pair of sepals distinct; plants indigenous, not weedy. (5)

5a. Leaves conspicuously membranous-sheathing basally; plants essentially glabrous (except on the scapes and spike); capsules indehiscent, 5–7 mm long. *P. macrocarpa*
5b. Leaves not membranous-sheathing basally; plants distinctly pubescent on the scapes or leaves, or brown-woolly at the base; capsules circumscissile, 2–4 mm long. (6)

6a. Plants white-pubescent on scapes or leaves or both, not brown-woolly at the base. *P. canescens*
6b. Plants white-pubescent to glabrous on leaves and scapes, distinctly brown-woolly at the base. *P. eriopoda*

Plantago aristata Michx.
Large-bracted Plantain

Plants annual or short-lived perennial, not woolly at the base, 0.5–3.5 dm tall or taller; leaves linear to narrowly oblong, 5–18 cm long, 0.1–0.7 cm broad, membranous-sheathing basally, 3-veined, entire; inflorescence dense, conspicuously hairy, 2–10 cm long; bracts linear, the lower ones 3–several times longer than the flowers; scapes villous; corolla lobes 1–2 mm long, spreading; stamens 4; capsule circumscissile, the seeds 2.

Introduced weed of waste places; reported from the Yukon (Dawson), and to be expected in Alaska; indigenous from Illinois to Texas and Louisiana, and widely naturalized elsewhere.

Plantago canescens Adams

Plants perennial, not woolly at the base, 0.5–3 dm tall; leaves narrowly elliptic to oblong or linear, 5–25 cm long, 0.3–2.5 cm broad, somewhat expanded but not membranous basally, 3–5-veined, entire to irregularly denticulate; inflorescence dense or the lower flowers scattered, 1.5–8 cm long, inconspicuously hairy (on the rachis); bracts ovate, glabrous dorsally, ciliate, shorter than the flowers; scapes villous; corolla lobes (1)1.5–2 mm long, spreading; stamens 4; capsules circumscissile, the seeds 4.

Riverbanks, roadsides, sandy terraces, and on rocky outcrops; in most of Alaska east of the 152nd meridian and in much of the Yukon; eastward to the Mackenzie and south to Alberta; Asia (*P. septata* Morris ex Rydb.).

Plantago eriopoda Torr.
Saline Plantain

Plants perennial, brown-woolly at the base, 1.5–4 dm tall; leaves elliptic to oblong, 5–25 cm long, 0.6–4 cm broad, somewhat expanded but not membranous basally, 3–5-veined, entire; inflorescence dense to loose, 5–20 cm long, inconspicuously villous to glabrate (on the rachis); bracts ovate to elliptic, shorter than the flowers, glabrous dorsally, short-ciliate; scapes villous to glabrate; corolla lobes 1–1.5 mm long, spreading or reflexed; stamens 4; capsules 3–4

mm long, circumscissile, the seeds usually 4.

Meadows and dry hillsides; in south- and east-central Alaska and southern Yukon and in adjacent northern British Columbia; east to the Mackenzie, south to Mexico, and in eastern North America.

Plantago lanceolata L.
Ribgrass, Buckhorn

Plants perennial, not woolly at the base, 1.5–5 dm tall; leaves elliptic to narrowly lanceolate or oblanceolate, 5–30 cm long, 0.3–4 cm broad, somewhat expanded but not membranous at the base, 3–several-veined, entire to obscurely denticulate; inflorescence dense, 1–8 cm long; bracts ovate, the apex sometimes acuminate and slightly surpassing the flowers, dorsally pubescent to glabrous, ciliate; scapes strigose; corolla lobes 2–2.5 mm long, spreading; stamens 4; capsules 2–4 mm long, circumscissile, the seeds usually 2.

Weeds of disturbed soils; mostly in coastal southern Alaska, but to be expected elsewhere; adventive from Eurasia, and widely naturalized.

Plantago macrocarpa Cham. & Schlecht.
Seashore Plantain

Plants perennial, essentially glabrous, except in the scape and inflorescence, 0.5–4 dm tall; leaves elliptic to lanceolate, 7–40 cm long, 0.3–3 cm broad, conspicuously expanded and membranous-sheathing basally, 3–several-veined, entire; inflorescence dense or interrupted, 1–7 cm long; bracts ovate to elliptic, shorter than the flowers, glabrous dorsally, not ciliate; scapes villous to glabrate; corolla lobes 1.5–2 mm long, spreading; stamens 4; capsules indehiscent, 5–7 mm long, the seeds 2.

Moist beaches, meadows, and salty marshes; in coastal southern Alaska, from the Aleutians to the Panhandle; southward to Oregon.

Plantago major L.
Common Plantain

Plants perennial, not woolly at the base, mostly 1–5 dm tall; leaves ovate to lanceolate or broadly elliptic, acute to cordate basally, short- to long-petiolate, the blades 3–15 cm long, 2–12 cm broad, expanded and often somewhat membranous basally, 5–several-veined, denticulate to entire; inflorescence dense to lax (especially below), 3–25 cm long, essentially glabrous; bracts ovate, shorter than the flowers, glabrous, not ciliate; corolla lobes spreading to reflexed, about 1 mm long; stamens 4; capsules 3–4 mm long, circumscissile, the seeds several to many.

Adventive weeds of waste places, or indigenous nonweedy plants; widely distributed in Alaska and in the Yukon; cosmopolitan (*P. major* var. *pilgeri* Fern.).

Plantago maritima L.
Goose-tongue

Plants perennial, not woolly at the base, 0.5–2.5 dm tall; leaves linear to narrowly elliptic or narrowly lanceolate, 4–20(25) cm long, 0.2–0.8 cm broad, expanded and membranous-sheathing basally, inconspicuously veined, entire or obscurely denticulate; inflorescence dense or lax, 2–10 cm long, glabrous (except on the corolla and rachis); bracts lanceolate, acute, shorter than the flowers, glabrous dorsally, minutely ciliate; corolla lobes 1–1.5 mm long, spreading; stamens 4; capsules 3–4 mm long, circumscissile; seeds 2–4.

Coastal marshes; in western, southern, and southeastern Alaska; widely distributed along coasts (sometimes in the interior) of North America and Eurasia. Our material has been assigned to ssp. *juncoides* (Lam.) Hultén.

PLUMBAGINACEAE
Plumbago Family

Perennial herbs; leaves simple, all basal; flowers perfect, regular, borne in headlike

Plantago maritima L. (× 0.6).

cymules; calyx 5-lobed, scarious; corolla tubular and 5-lobed, or the petals distinct; stamens 5, opposite the corolla lobes; pistils 1, the ovary superior, 1-loculed, 5-car-

pelled, 5-seeded; styles 5, the stigmas filiform; fruit a utricle, often enclosed within
the calyx.

ARMERIA Willd.

Plants acaulescent, scapose, from a taproot
and branching caudex; leaves arising from
the apex of the caudex, linear, entire;
flowers several to many, borne in scarious-
bracted, headlike clusters, subtended by
an involucre of broad bracts, the outer
bracts connate and forming a tube around
the scape; calyx of 5 united sepals, funnelform, scarious, 10-veined; petals 5, united
or distinct, scarious; styles united basally.

Lawrence, G. H. M. 1940. Armerias native and cultivated. Gentes Herb. 4:398–
418.

Armeria maritima (Mull.) Willd.
Sea-pink, Thrift

Plants 0.5–3 dm tall; leaves 1.5–10 cm long,
0.1–0.2 cm broad, linear to narrowly lanceolate, glabrous or ciliate; scapes 1–several; headlike clusters of flowers 1.5–3 cm
broad; pedicels short; flowers in clusters of
3, subtended by paired transparent bracts;
calyx 4–7 mm long, pubescent at the base
and also along the ribs; corolla included
within the calyx, pinkish.

River banks, spits, and tundra to heathlands; in coastal southwestern (rarely
southeastern), western, and northern Alaska (and in the Alaska Range); eastward
to Newfoundland and south to California
and Colorado; circumpolar. Two rather
poorly defined varieties are recognized
among our materials.

1a. Outer involucral bracts (not the con-
nate ones) usually one-half as long
as the inner ones, or less; leaves flat
(*A. maritima* ssp. *sibirica* Turcz. ex
DC.). *A. maritima* var. *sibirica*
(Turcz.) Lawr.

1b. Outer involucral bracts more than
half as long as the inner ones; leaves
flat, recurved, or slightly contorted

and canaliculate (*A. maritima* f. *arctica* Cham.; *A. maritima* ssp. *arctica*
[Cham.] Hultén). *A. maritima* var.
purpurea (Koch) Lawr.

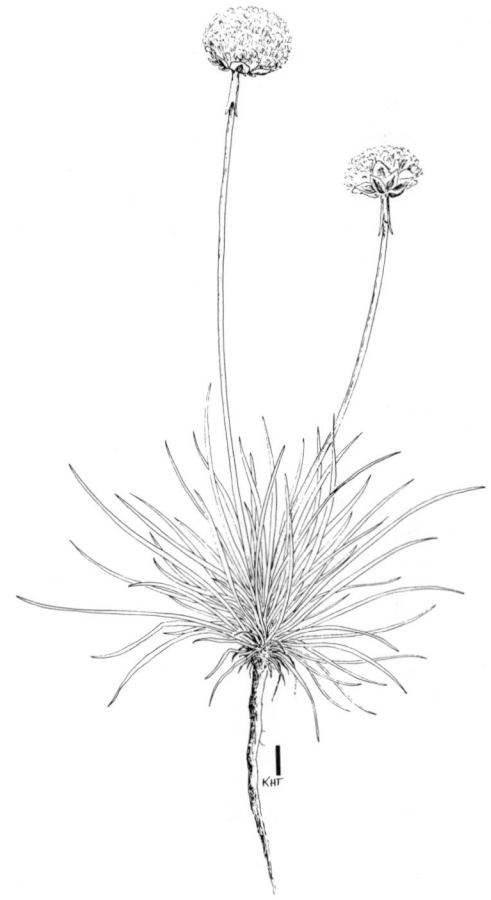

Armeria maritima (Mull.) Willd. (× 0.5).

POLEMONIACEAE
Phlox Family

Annual, biennial, or perennial herbs; leaves
simple and entire, or variously pinnatifid to
compound; flowers perfect, regular, solitary, or variously clustered; calyx 5-lobed;
corolla salverform to rotate, 5-lobed; stamens 5, alternate with the corolla lobes;
pistils 1; ovary superior, 3-loculed; styles
1; stigmas 3 (rarely only 2); fruit a loculicidal capsule.

1a. Leaves entire, partly or wholly opposite. (2)

1b. Leaves variously lobed, divided, or compound, all alternate (entire and sometimes with the lowermost leaves opposite in *Collomia*). (3)

2a. Plants depressed-caespitose perennials; leaves all or nearly all opposite. *Phlox*

2b. Plants erect annuals; leaves alternate above. *Microsteris*

3a. Leaves simple, entire; calyx enlarged in fruit; stamens unequally inserted on the corolla tube. *Collomia*

3b. Leaves pinnatifid or compound; calyx not, or little, enlarging in fruit; stamens equally inserted on the corolla tube. (4)

4a. Leaves pinnately compound (at least the lower leaflets distinct); calyx green or variously tinged; corolla campanulate to rotate-funnelform. *Polemonium*

4b. Leaves bipinnatifid (at least the lower ones); calyx more or less scarious in the sinuses; corolla salverform or trumpet shaped. *Gilia*

COLLOMIA Nutt.

Plants annual, from slender taproots; leaves alternate (or the lower ones opposite), entire; flowers few to several, in terminal headlike long-bracteate clusters; calyx of 5 united sepals, the tube scarious, enlarging in fruit, the lobes greenish; corolla of 5 united petals, tubular-funnelform, the limb 5-lobed; stamens 5, unequally inserted on the corolla tube; capsules enclosed by the calyx; seeds 1–2 per locule, mucilaginous when moistened.

Collomia linearis Nutt.

Plants erect, 0.5–4 dm tall, the stems simple or more commonly branching from most of the nodes; herbage subglabrous to pubescent below, becoming glandular above; leaves all alternate, or sometimes the lower ones opposite, sessile or subsessile, entire, 1–9 cm long, 0.2–1.2 cm broad, lanceolate to oblong or linear; calyx 4–6 mm long in flower, longer in fruit, the triangular lobes glandular-pubescent; corolla 8–14 mm long, pink or white to bluish, exserted from the calyx.

Weedy plant of disturbed soils, mostly along roadsides; in eastern Alaska and southern Yukon; eastward to the Atlantic and south to California, New Mexico, Nebraska, and Wisconsin.

GILIA R. & P.

Plants annual or biennial (?), from slender taproots; leaves all alternate, at least the lower ones bipinnatifid; flowers several to many in terminal bractless cymose heads; calyx of 5 united sepals, the tube scarious (and often brightly colored) at least in the sinuses, enlarging and investing the fruit; corolla of 5 united petals, tubular-salverform, the limb 5-lobed; stamens 5, inserted at the apex of the corolla tube; capsules enclosed by the calyx; seeds 1–3 per locule.

Grant, V. 1950. Genetic and taxonomic studies in *Gilia*. I. *Gilia capitata*. El Aliso 2:239–316.

Gilia capitata Sims

Plants erect, 2–8 dm tall, the stems simple or branching from the upper nodes; herbage glabrous to stipitate-glandular; leaves bipinnatifid, at least the lower ones, becoming smaller upwards, 2–10 cm long, the lobes linear; calyx 4–5 mm long, the lanceolate lobes greenish, pubescent to glandular; corolla 6–10 mm long, bluish, exserted from the calyx.

Plants cultivated, escaping and persisting (?) (the seeds are common in mixtures of wildflower seeds); in southwestern and southeastern Alaska and western Yukon, and to be expected elsewhere; native from southern British Columbia and Idaho southward to California (*G. achilleaefolia* Benth.).

MICROSTERIS Greene

Plants annual, from slender taproots; leaves alternate above, opposite below, entire; flowers paired or solitary, axillary, borne near the apex of the stem; calyx of 5 united sepals, the tube scarious (at least in the sinuses), ultimately ruptured by the fruit, the lobes greenish; corolla of 5 united petals, salverform; stamens 5, unequally inserted on the corolla tube; capsules rupturing the calyx; seeds 1 per locule, mucilaginous when moistened.

Microsteris gracilis (Hook.) Greene

Plants erect, 0.3–2.5 dm tall, the stems simple or branched; herbage puberulent to glandular, at least above; leaves linear to lance-elliptic, 1–5 cm long, 3–8 mm broad; calyx 7–10 mm long, somewhat enlarged in fruit, glandular-pubescent; corolla 8–14 mm long, the tube whitish, the limb purplish to violet, exserted from the calyx.

Weedy species of disturbed soils, mostly along roadsides; adventive in southeastern Alaska and in southern Yukon (to be expected elsewhere); native from British Columbia and Montana south to California and Utah; South America.

PHLOX L.

Plants perennial, from taproots; leaves opposite, entire, spinulose-tipped; flowers solitary or in terminal cymes; calyx of 5 united sepals, the tube scarious (at least in the sinuses), ultimately ruptured by the fruit, the lobes greenish; corolla of 5 united petals, salverform, the limb 5-lobed; stamens 5, unequally inserted on the corolla tube; capsule rupturing the calyx; seeds 1–2 per locule, not at all mucilaginous when moistened.

Wherry, E. T. 1955. The genus *Phlox*. Morris Arboretum Monog. 3:1–174.

1a. Corolla tube 4–6 mm long; leaves 3–5 mm long, cobwebby-tomentose (at least basally); plants of central

Yukon and Porcupine valleys. *P. hoodii*

1b. Corolla tube 7–12 mm long; leaves mostly 6–22 mm long, not cobwebby-tomentose (except in var. *richardsonii*); plants of broader distribution. *P. sibirica*

Phlox hoodii Richards.
Moss Phlox

Plants pulvinate, forming a mat or low cushion; leaves firm, mostly 3–5 mm long and 0.4–1 mm broad, cobwebby-tomentose (at least basally); flowers white to pink, solitary, the pedicels 0.5–1 mm long; calyx 3–6 mm long, tomentose, the lobes subulate; corolla tube 4–6 mm long, the lobes 2.5–4 mm long.

Dry open slopes; in central and southwestern (and northern ?) Yukon, and in central eastern Alaska; southward to California, Nevada, Utah, and Colorado.

Phlox sibirica L.
Siberian Phlox

Plants pulvinate, forming low mats or cushions; leaves mostly 8–22 mm long, 1–2.5 mm broad, ciliate (cobwebby in var. *richardsonii*); flowers solitary (sometimes 2–3), pink to lilac or white; calyx 5–12 mm long, pubescent with mixed glandular and glandless multicellular hairs, the lobes cuspidate; corolla tube 6–12 mm long, the lobes 6–14 mm long.

Arctic and alpine tundra, often in gravel or on rocky outcrops; in western Alaska from Norton Sound northward to the vicinity of Point Hope, eastward to the central Yukon River Valley and in northern Yukon and adjacent Mackenzie, less commonly in southwestern Alaska and southwestern Yukon; Asia. Our specimens belong to a series of apparently intergrading populations which have been treated previously as belonging to one or more specific taxa. All are closely related and fit readily into ssp. *sibirica* in a broad sense.

1a. Pedicels mostly 5–20(35) mm long; corolla lobes mostly 8–14 mm long; plants widely distributed in northern and western Alaska and northern Yukon (*P. borealis* Wherry; *P. sibirica* ssp. *borealis* [Wherry] Shetler). *P. sibirica* var. *borealis* (Wherry) B. Boi.

1b. Pedicels mostly less than 5 mm long; corolla lobes 6–9(10) mm long; plants of restricted distribution. (2)

2a. Sepals mostly 5–7 mm long; leaves cobwebby-tomentose, at least near the base; plants of northeastern Alaska, northern Yukon (?), and eastward (*P. richardsonii* Hook.; *P. sibirica* ssp. *richardsonii* [Hook.] Hultén). *P. sibirica* var. *richardsonii* (Hook.) Welsh

2b. Sepals 7–10 mm long; leaves not cobwebby-pubescent, usually ciliate with coarse hairs; plants from the Brooks Range (*P. alaskensis* Jordal; *P. richardsonii* var. *alaskana* [Jordal] Wherry). *P. sibirica* var. *alaskana* (Jordal) B. Boi.

POLEMONIUM L.

Plants perennial, from rhizomes; leaves alternate, compound or deeply pinnatifid (the apical leaflets may not be distinct); flowers showy, borne in terminal or axillary cymes; calyx of 5 united sepals, green throughout, enlarging in fruit; corolla of 5 united petals, rotate-campanulate to funnelform, the limb 5-lobed; stamens 5, equally inserted on the corolla tube; capsules enclosed by the calyx; seeds 1–10 per locule.

Davidson, J. F. 1950. The genus *Polemonium*. Univ. Calif. Pub. Bot. 23:329–82.

Wherry, E. T. 1952. The genus *Polemonium* in North America. Am. Midl. Nat. 27:741–60.

1a. Leaflets glabrous or nearly so; corolla lobes tapering apically, acute or abruptly obtuse, often ciliate; cauline leaves 1–several; plants often over 3 dm tall. *P. caeruleum*

1b. Leaflets distinctly pubescent (at least when young); corolla lobes rounded apically (not at all tapering), rarely ciliate; cauline leaves 1–few; plants 3 dm tall or less. (2)

2a. Pedicels shorter than the length of the calyx; calyx 5–10 mm long; corolla 15–20 mm long. *P. boreale*

2b. Pedicels exceeding (or about equal) to the length of the calyx; calyx 4–6(7) mm long; corolla 8–12(13) mm long. *P. pulcherrimum*

Polemonium boreale Adams
Northern Jacobs-ladder

Plants 0.8–3 dm tall, the stems solitary or few together from the rhizome apex, erect or ascending, spreading-hairy to glandular-pubescent (especially above); leaves primarily basal, pinnately compound, the leaflets 13–23, mostly 4–12 mm long and 1–5 mm broad, elliptic to oblong or oval, acute to obtuse apically, pubescent at least when young; pedicels shorter than the calyx; calyx campanulate, 5–10 mm long, the lobes lanceolate to oblong, about twice longer than broad; corolla violet to blue (rarely white), 15–20 mm long, the lobes only slightly longer than the tube.

Arctic and alpine tundra and heathlands; widely distributed in Alaska and in the Yukon; circumboreal (*P. boreale* var. *villosissimum* Hultén; *P. boreale* ssp. *richardsonii* [Grah.] J. P. Anders.; *P. boreale* ssp. *macranthum* [Cham.] Hultén; *P. richardsonii* Grah.; *P. humile* var. *macranthum* Cham.).

Polemonium caeruleum L.
Blue Jacobs-ladder

Plants 2–10 dm tall, the stems solitary, erect (decumbent at the base), glandular-pubescent, at least above; leaves pinnately compound, reduced in size upwards, the leaflets 19–27, mostly (4)10–25 mm long

Polemonium caeruleum L. (× 0.3).

and 3–10 mm broad, lanceolate to elliptic, acute to acuminate apically, glabrous or nearly so; pedicels shorter than the calyx; calyx campanulate, the lobes lanceolate, about twice longer than broad; corolla blue to purple or white, 10–20 mm long, the lobes about twice longer than the tube.

Tundra, heath, or woodlands; widely distributed in Alaska and in the Yukon; in much of North America; Eurasia. Two poorly defined subspecies have been reported.

1a. Flowers mostly more than 15 mm long (from base of calyx to tip of petals); anthers reaching almost to the base of the style branches, or surpassing them; this is the truly common, tall jacobs-ladder in Alaska and Yukon (*P. acutiflorum* Willd. ex R. & S.). *P. caeruleum* ssp. *villosum* (Rud.) Brand

1b. Flowers mostly less than 15 mm long; anthers not reaching to the vicinity of the style branches; plants uncommon in east-central Alaska and in the Yukon (*P. occidentale* Greene). *P. caeruleum* ssp. *amygdalinum* (Wherry) Munz

Polemonium pulcherrimum Hook.
Pretty Jacobs-ladder

Plants 1–3 dm tall, the stems few to several from the rhizome apex, forming hemispheric clumps, erect, ascending, or decumbent, spreading-hairy to glandular; leaves primarily basal, pinnately compound; leaflets 11–23(37), 2–15 mm long, 1–5 mm broad, ovate to lanceolate or oval, acute to obtuse apically; pedicels exceeding the calyx (at least in full flower); calyx campanulate, 4–6(7) mm long, the lobes lance-oblong, once to twice longer than broad; corolla blue to violet or white, 8–12 (13) mm long, the lobes about equal to the yellowish tube.

Open woods and meadows, often in gravelly soil; mostly south of the 68th parallel in Alaska and Yukon; eastward to Mackenzie and south to California and Colorado (*P. lindleyi* Wherry; *P. pulcherrimum* var. *lindleyi* [Wherry] J. P. Anders.; *P. fasciculatum* Eastw.; *P. rotatum* Eastw.).

POLYGONACEAE
Buckwheat Family

Herbaceous annual, biennial, or perennial plants; leaves alternate, basal, or opposite; stipules sheathing (ocrea) or lacking; flowers perfect or rarely imperfect, regular, arranged in axillary or terminal racemes, panicles, spikes, or umbels; perianth with 2–6 segments; stamens 6–9; pistils 1, the ovary superior, 1-loculed, 2–4-carpelled; styles 1, the stigmas 2–4; fruit an achene.

1a. Inflorescence umbellate, subtended by an involucre of few to several elliptic to oblanceolate, foliose bracts; plants either diminutive, glabrous annuals or scapose grayish-tomentose perennials. (2)
1b. Inflorescence a spike, raceme, or panicle, not subtended by an involucre; plants annual or perennial, usually glabrous. (3)

2a. Plants diminutive annuals; leaves cauline; stipules sheathing. *Koenigia*
2b. Plants scapose perennials; leaves all basal; stipules lacking. *Eriogonum*

3a. Leaf blades reniform; perianth 4-segmented; pistil 2-carpelled. *Oxyria*
3b. Leaf blades not reniform, the base sometimes cordate or hastate; perianth with 5 or more lobes; pistil mostly 3-carpelled. (4)

4a. Perianth 5-segmented, erect in fruit, about equal in size, not much enlarging in fruit. (5)
4b. Perianth 6-segmented, the inner ones erect and enlarged in fruit, or the achenes winged, the outer ones reflexed and usually smaller. (6)

5a. Leaf blades ovate-cordate, acuminate at the apex; achenes much exceeding the perianth at maturity; cultivated annuals. *Fagopyrum*
5b. Leaf blades variously shaped, but if ovate-cordate and acuminate at the apex, then the plants twining; achenes little if at all exceeding the perianth at maturity; indigenous or adventive, usually not cultivated. *Polygonum*

6a. Stipular sheaths prominent; stamens usually 9; leaf blades ovate to orbicular; plants cultivated and persisting. *Rheum*
6b. Stipular sheaths evanescent; stamens commonly 6; leaf blades mostly elliptic to lanceolate; plants indigenous or adventive, not cultivated. *Rumex*

ERIOGONUM Michx.

Perennial herbs; leaves simple, basal, arising from the apex of a stout caudex, tomentose; stipules lacking; inflorescence umbellate; flowers borne on slender pedicels inserted at the base of a cup-shaped involucre; perianth of 6 petaloid segments, tomentose; stamens 9; pistil 3-carpelled, the ovary 1-loculed, 1-ovuled; styles 3; fruit a 3-angled achene.

Eriogonum flavum Nutt.
Wild Buckwheat

Scapose, caespitose herbs from a stout taproot and spreading caudex; herbage grayish-tomentose; scapes 1.5–10 cm long, leafless; leaves crowded at the apex of elongate caudex branches clothed with persistent leaves, 1–3.5 cm long, 0.4–1.0 cm broad, narrowly elliptic, tapering to petioles up to 2 cm long; involucres campanulate (4)6–8 mm long, deeply 4-lobed; perianth 3–5 mm long, yellow, silky-pubescent externally, borne on a slender, stipelike base; staminal filaments hairy basally; achenes sparsely pubescent.

Basaltic greenstone slopes at Mission Bluff, near Eagle, Alaska, far north of the normal range of the species which is known from southern British Columbia and Alberta south to Oregon and along the Rocky Mountains to Colorado and extreme northern Arizona and adjacent southern Utah. Our materials have been designated as var. *aquilinum* Reveal in Hultén.

FAGOPYRUM Gaertn.

Annual herbs from taproots; leaves simple, alternate, petiolate, the blades ovate-cordate, acuminate; stipules sheathing; flowers borne in axillary or terminal racemes, not subtended by an involucre; perianth of 5 petaloid segments, glabrous; stamens 8(5 and 3); pistil 3-carpelled, the ovary

1-loculed, 1-ovuled; styles 3; fruit a 3-angled achene.

Fagopyrum esculentum Moench
Common Buckwheat

Plants 3–10 dm tall, or more; stems erect or sprawling, not twining; leaves ovate-cordate to ovate-hastate, the blades mostly 3–10 cm long, 1.5–8 cm broad; racemes axillary or terminal, several- to many-flowered, compact; perianth 1.5–3 mm long, white, glabrous; staminal filaments glabrous, 5 alternating with the perianth segments, 3 connivent around the styles; achenes glabrous, conspicuously longer than the persistent perianth.

Cultivated; occasionally escaping in southern Alaska; Eurasia. This is the buckwheat of commerce.

KOENIGIA L.

Diminutive annual herbs from slender taproots; leaves simple, alternate or opposite; stipules sheathing; flowers few to several, borne in terminal or subterminal umbellate clusters, subtended by 2–4 foliose involucral bracts; perianth of 3 subpetaloid segments, glabrous; stamens usually 3(2–4); pistils 2–3-carpelled, the ovary 1-loculed, 1-ovuled; styles short, 2–3; fruit a 3-angled or lenticular achene.

Koenigia islandica L.
Koenigia

Plants 2–15 cm long; stems very slender, simple or branched; leaves 2–9 mm long, 1.5–3 mm broad, oblong, lanceolate or oblanceolate; umbels with few to several flowers; perianth about 1(1.4) mm long, greenish, whitish, or reddish, glabrous; staminal filaments glabrous, usually 3, alternating with the perianth segments; achenes glabrous, enclosed by the perianth.

Moist, bare soil; known from widely scattered sites in Alaska and southern Yukon; southward to Colorado; circumpolar. The species is easily overlooked.

OXYRIA Hill

Perennial subrhizomatous herbs, from long taproots; leaves simple, alternate, or mostly basal; stipules sheathing; flowers usually numerous, borne in panicles, not subtended by an involucre; perianth of 4 sepaloid segments, glabrous; stamens 6; pistil 2-carpelled, the ovary 1-loculed, 1-ovuled; styles 2, short, the stigmas fringed; fruit a flattened, wing-margined achene.

Oxyria digynia (L.) Hill
Mountain Sorrel

Plants 0.5–6 dm tall, the herbage often reddish tinged; steps usually simple, the

Oxyria digynia (L.) Hill (× 0.4).

juice acrid; leaves mostly basal, the petioles mostly 1–10(25) cm long, the blades 0.5–5 cm long, 0.6–6 cm broad, reniform to cordate; panicles 2–20 cm long; perianth 1.5–2.5 mm long, the 2 segments at the edges of the achene more slender than those at the sides; achenes flattened, 3–6 mm broad, prominently winged.

Moist sites in tundra, heathlands, or woods, often in gravelly or rocky places; in most of Alaska and Yukon; east to Labrador, south to California, Arizona, and New Mexico; circumpolar (*Rumex digynus* L.).

POLYGONUM L.

Plants annual, biennial, or perennial herbs, from taproots or rhizomes; leaves alternate, cauline or basal; stipules sheathing; flowers single or clustered in leaf axils or in axillary or terminal, spikelike racemes or panicles, not subtended by a regular involucre; perianth of 5 petaloid (or sepaloid) segments; stamens 8(5 and 3), rarely lacking; pistils usually 3-carpelled, the ovary 1-loculed, 1-ovuled; styles 2–3, often very short; fruit a lens-shaped or 3-angled achene.

1a. Plants with twining stems; leaves ovate-cordate, acuminate. *P. convolvulus*
1b. Plants with stems various but not twining; leaves variously shaped, but rarely both ovate-cordate and acuminate. (2)

2a. Stems erect, from an expanded to somewhat bulbous caudex; leaves mostly basal; flowers borne in terminal spicate racemes. (3)
2b. Stems usually not from an expanded caudex, prostrate to erect; leaves mostly cauline; flowers axillary, or in axillary and terminal racemes, or in panicles. (4)

3a. Racemes slender, the lower flowers at least replaced by bulblets; flowers usually greenish or whitish (less commonly pink). *P. viviparum*

3b. Racemes thick, lacking bulblets; flowers usually bright pink to rose. *P. bistorta*

4a. Plants often over 8 dm tall; flowers in open terminal or axillary panicles. (5)
4b. Plants usually less than 8 dm tall; flowers axillary, or in axillary and/or terminal, spikelike racemes. (6)

5a. Leaves cordate-ovate; plants cultivated, persistent and escaping; in coastal southeastern Alaska (*P. zuccarinii* Small). *P. cuspidatum* Sieb. & Zucc.
5b. Leaves lanceolate to lance-ovate; plants indigenous, rarely cultivated. *P. alpinum*

6a. Leaves not jointed at the base; flowers in terminal and/or axillary spikes or racemes. (7)
6b. Leaves with a hingelike joint at the point of attachment of leaf base with the sheath; flowers in small axillary clusters, or solitary. (11)

7a. Inflorescences all terminal, usually solitary; plants perennial, aquatic or semiaquatic; flowers bright pink. *P. amphibium*
7b. Inflorescences not all terminal, at least some axillary; plants mostly annual, seldom aquatic (see *P. hydropiperoides*); flowers pink, green, or white. (8)

8a. Stipular sheaths lacking marginal bristles (or merely short-ciliate); veins of the outer pair of perianth segments branched and recurved at the tip. *P. lapathifolium*
8b. Stipular sheaths with well-developed marginal bristles; veins of the outer pair of perianth segments not branched and recurved at the tip. (9)

9a. Plants perennial, from rhizomes, growing in or near water; spikes slender, mostly less than 5 mm broad. *P. hydropiperoides*

9b. Plants annual, from taproots, growing in moist sites, but not aquatic; spikes slender to thick. (10)

10a. Mature perianth glandular-punctate, greenish to white; spikes slender, arching, interrupted near the base. *P. hydropiper*

10b. Mature perianth not glandular-punctate, pink to purplish; spikes dense, erect or nearly so, not interrupted. *P. persicaria*

11a. Leaves ovate to elliptic or obovate, pointed at both ends, cuspidate, about half as broad as long; achenes almost black, lustrous; plants usually less than 1 dm tall. *P. minimum*

11b. Leaves variously shaped, usually more than twice longer than broad, or obtuse to rounded apically; achenes various; plants often over 1 dm tall. (12)

12a. Plants maritime, growing on strands; perianth 3–4 mm long; achenes highly lustrous; plants prostrate. *P. fowleri*

12b. Plants seldom maritime in distribution; perianth often less than 3 mm long, or if longer, then the achenes dull or sublustrous, or the plants erect. (13)

13a. Plants ascending to erect; achenes highly lustrous. (14)

13b. Plants prostrate, or less commonly ascending or erect (especially when crowded); achenes dull or sublustrous, often striate. (15)

14a. Pedicels included within the sheaths; leaves linear-oblong, the larger ones 1–5(8) mm broad; achenes about 2 mm long. *P. prolificum*

14b. Pedicels longer than the sheaths; leaves narrowly lanceolate to oblanceolate or linear; achenes 2.2–4 mm long. *P. ramosissimum*

15a. Inner perianth segments distinctly narrower than the outer ones; achenes usually exserted from the perianth,

straw-colored to brownish. *P. caurianum*

15b. Inner perianth segments not conspicuously narrower than the outer ones; achenes usually included within the perianth, light to dark brown. (16)

16a. Leaves crowded, elliptic to obovate, rounded apically; perianth 3.4–4 mm long in fruit, strongly constricted below the apex. *P. achoreum*

16b. Leaves not especially crowded, linear to oblong or elliptic, acute to rounded apically; perianth 2–3(4) mm long in fruit, not especially constricted below the apex. (17)

17a. Leaves of branches subequal in size to those of the main stems; fruiting perianth divided about half its length; achenes with 2 sides convex and 1 side concave. *P. arenastrum*

17b. Leaves of branches conspicuously smaller than those of the main stems; fruiting perianth divided to well below the middle; fruit with 3 concave sides. (18)

18a. Leaves of main stems with well-developed petioles above the joint; achenes 3.5–4 mm long. *P. boreale*

18b. Leaves of main stems lacking petioles (or with very short ones) above the joint; achenes mostly 2.5–3.5 mm long. *P. aviculare*

Polygonum achoreum Blake

Plants annual, prostrate to ascending, the stems striate, 1–6 dm long; leaves crowded, oval to obovate or elliptic, 10–24(35) mm long, 3–10(14) mm broad, somewhat smaller upwards but not conspicuously smaller on lateral branches, rounded apically, the short petiole jointed basally; stipules shredded, 4–10 mm long; flowers 1–4, axillary; pedicels included; perianth 2–3.5 mm long, becoming 3.4–4 mm long in fruit, united about half the length, 5-lobed, greenish, with white to pink edges, the outer ones usually broader than the inner; styles 3; achenes 3-angled, yellowish brown.

Weedy species of disturbed soils; in central and eastern Alaska; widely distributed in North America; adventive in Europe. This entity is closely allied to *P. aviculare*.

Polygonum alpinum All.
Alaska Wild Rhubarb

Plants perennial, erect, the stems striate, fistulose, 3–20 dm tall or more, pubescent, especially at the nodes, or glabrous; leaves with petioles not jointed basally, the blades 4–20 cm long and 2–10 cm broad or more, lanceolate to lance-ovate, long-acuminate apically, truncate to cordate basally, ciliate, pubescent to glabrous below, glabrous above; stipules brownish, not shredded-ciliate, glabrous to long-hairy; flowers numerous, borne in open panicles; perianth 2–3.5 mm long, united near the base, 5-

Polygonum alpinum All. (× 0.27).

lobed, white, cream, or greenish, the lobes subequal; styles 3; achenes 3-angled, falsely winged, lustrous, whitish to yellowish brown, the seed lance-ovoid, terete, rough.

Arctic or alpine tundra, heathlands, and woods; from the Bering Sea eastward to the Yukon and Mackenzie, mostly between the 62nd and 68th parallels; southward to Washington and California; Eurasia (*P. alpinum* var. *lapathifolium* Cham. & Schlecht; *P. alaskanum* [Small] Wight ex Harshb. var. *glabrescens* Hultén; *P. phytolaccaefolium* Meissn. ex Small; *P. alpinum* var. *alaskanum* Small). Specimens from Alaska are very similar to the Eurasian *P. alpinum* ssp. *alpinum*, differing mainly in leaf size and in achene characteristics. In our materials, the achenes are falsely winged due to inflation of the pericarp. The roughened seeds are lance-ovoid and terete in cross section. Thus it seems best to recognize our entity as ssp. *alaskanum* (Small) Welsh.

Polygonum amphibium L.
Water Smartweed

Plants perennial, rhizomatous to stoloniferous, the stems prostrate to ascending, floating when aquatic, 5–10 dm long or more; leaves with petioles not jointed basally, the blades 5–16 cm long, 1–5 cm broad, narrowly elliptic to oblong or lanceolate, acuminate to acute apically, acute to subcordate basally, ciliate or not, pubescent to glabrous below, glabrous above; stipules not shredded, sometimes flaring and greenish at the apex, ciliate with stout bristles or smooth, glabrous to pubescent on the sheath; peduncles glabrous; flowers in 1–2 terminal compact spikelike panicles, these 1–3(4) cm long; perianth 4–5 mm long, bright pink, the segments distinct to below the middle, the lobes subequal; styles 2; achenes lens shaped, dark brown.

Ponds, streams, and mud banks; in central western to east-central and southeastern Alaska and southern Yukon; widely distributed in North America; Eurasia and elsewhere (*P. amphibium* ssp. *laevimarginatum* Hultén).

Polygonum arenastrum Jordal ex Bor.

Plants annual, prostrate to ascending, the stems striate and usually triangular (at least when young) 1–5 dm long; leaves scattered or somewhat crowded, elliptic to oblong, lanceolate or oblanceolate, 7–20 mm long, 2–6 mm broad, becoming smaller upwards, but not conspicuously smaller on lateral branches, rounded to obtuse apically, the short to elongate petiole jointed basally; stipules shredded, 4–6 mm long; flowers 1–5, axillary; pedicels included; perianth 1.8–3 mm long, united about half the length, 5-lobed, the lobes greenish, with white to pink edges, the outer lobes usually broader than the inner; styles 3; achenes 3-angled, brown.

Weedy species of disturbed soils; in much of southern Alaska and southern Yukon; widely distributed in North America and Eurasia. *P. arenastrum* is closely allied to *P. aviculare* from which it differs principally in the more nearly uniform size of the leaves and in the less deeply divided perianth.

Polygonum aviculare L.
Knotweed

Plants annual, prostrate to ascending or erect, the stems striate, terete to triangular, 1–10 dm long; leaves usually not crowded, oblong to elliptic or oblanceolate, 5–40 mm long, 2–10 mm broad, smaller on the branchlets than on the main stem, acute to obtuse (rarely rounded), the blades sessile or short-petiolate above the basal joint; stipules shredded, 3–6 mm long; flowers 1–5, axillary; pedicels included or shortly exserted; perianth 2–3 mm long, united about one-third the length, 5-lobed, the lobes greenish with white or pink edges, the outer lobes only slightly broader than the inner; styles 3; achenes 3-angled, brown.

A weed of waste places; in southern Alaska and southern Yukon; widely distributed

in most continents (*P. buxiforme* Small; *P. neglectum* Besser; *P. heterophyllum* Lindm.).

Polygonum bistorta L.
Meadow Bistort

Plants perennial, erect, from expanded, bulblike bases, the stems 0.8–4(5) dm tall; basal leaves well developed, 3–22 cm long, the blades lanceolate to elliptic, 2–12 cm long, 0.5–3 cm broad, abruptly acute to rounded apically, cuneate to acute basally; petioles mostly well developed, not jointed; cauline leaves reduced upwards; stipules 1–8 cm long, brownish, often flaring and membranous apically, the upper ones often bladeless; flowers numerous, borne in terminal spikelike racemes 1.5–6 cm long; perianth 4–6 mm long, connate only at the base, 5-lobed, the lobes bright pink, about equal in size; stamens exserted; styles 3, exserted; achenes 3-angled, brown, lustrous.

Arctic and alpine tundra and heathlands; in most of Alaska (except for coastal southern portions) and western and northern Yukon; east to Mackenzie; Eurasia (*P. plumosum* Small). Our material belongs to ssp. *plumosum* (Small) Hultén. This is one of the truly attractive flowers of the tundra.

Polygonum boreale (Lange) Small
Northern Knotweed

Plants annual, prostrate to ascending or erect, the stems striate, terete to triangular, (0.2)1–10 dm long; leaves usually not crowded, oblanceolate to oblong or elliptic, 10–55 mm long, 2–10(14) mm broad, smaller on the branchlets than on the main stem, acute to obtuse or rounded, distinctly petioled above the basal joint; stipules shredded, 3–8 mm long; flowers 1–5, axillary; pedicels mostly included; perianth 2.8–4 mm long, united about one-third the length, 5-lobed, greenish with white or pink edges, the lobes subequal; styles 3; achenes 3-angled, brown.

Weedy species of waste places; in southern Alaska and Yukon (?); closely related to

P. aviculare, and sometimes included within that entity.

Polygonum caurianum Robins.
Alaska Knotweed

Plants annual, prostrate to ascending, the stems striate and often triangular, 0.5–5 dm long; leaves crowded or not crowded, oblong to oblanceolate or elliptic, 3–20 mm long, 1–5 mm broad, somewhat smaller on the branchlets than on the main stem, rounded apically, the blades sessile to shortly petiolate above the basal joint; stipules shredded, 2–5 mm long; flowers 1–5, axillary; pedicels included; perianth 1.5–2 mm long, the 5 segments greenish with white to red margins, the outer ones somewhat broader than the inner; styles 3; achenes 3-angled (at least apically), straw-colored to brownish.

Waste places, on beaches, dunes, roadsides and gravel bars; in much of Alaska south of the Brooks Range, and in southern Yukon; Asia. *P. caurianum* seems to be indigenous to our region.

Polygonum convolvulus L.
Black Bindweed

Plants annual, erect (when young), soon prostrate or twining, the stems 1–10 dm long or more; leaves with long petioles, not jointed basally, the blades 1.5–8 cm long (from sinus to apex), 0.7–5 cm broad, sagittate-ovate, acuminate; stipules 2–5 mm long, shredded and soon deciduous; flowers few to several, borne in axillary or terminal racemes (or both); perianth 3–4.5 mm long, greenish, 5-lobed, the outer lobes keeled; styles 3-cleft; achenes 3-angled, black.

Disturbed sites; in widely scattered locations in Alaska and southern Yukon; widely ranging in North America; adventive from Europe.

Polygonum fowleri Robins.
Fowler Knotweed

Plants annual (biennial?), prostrate (erect when young), the stems striate, terete, 1–5 dm long or longer; leaves usually not crowded, elliptic to oblanceolate, acute to rounded apically, 1–6(7) cm long, 0.3–1.5 (2.4) cm broad, gradually reduced upwards, the blades subsessile to distinctly petiolate, jointed basally; stipules shredded, 3–8 mm long; flowers 1–4, in axillary clusters; pedicels included; perianth 3–4.5 mm long, united about one-third the length, 5-lobed, the lobes greenish with pink margins, about equal in size; styles 3; achenes 3-angled, yellowish brown, exserted.

Coastal strands; in southern Alaska from Kodiak Island eastward to the Panhandle; south to California; also in coastal regions of eastern North America (*P. buxifolium* Nutt. ex Bong.).

Polygonum hydropiper L.
Water-pepper

Plants annual (or sometimes perennial?), the stems occasionally rooting at the nodes, 3–8 dm tall; leaves with short petioles or subsessile, not jointed at the base, the blades 3–10 cm long, 0.5–3 cm broad, lanceolate to elliptic, acute to acuminate apically, acute to cuneate basally, sparsely strigose to glabrous, ciliate; stipules 8–15 mm long, not shredded, strigose to glabrous, ciliate with long bristles; flowers several to many, borne in terminal and usually also lateral, spikelike, interrupted racemes 2–8 cm long; perianth 2.5–4 mm long, glandular-dotted, united about one-third the length, usually 4-lobed, the lobes greenish with white or pink margins; styles 2–3, distinct; achenes lens shaped or 3-angled, brown.

Weedy species of moist, disturbed sites; reported from the vicinity of Juneau, where probably adventive from Europe. The herbage has a peppery flavor.

Polygonum hydropiperoides Michx.
Mild Water-pepper

Plants perennial, with elongate rhizomes or stolons, the stems decumbent, 2–10 dm long or more; leaves with short petioles or subsessile, not jointed at the base, the blades 3–14 cm long, 0.8–4 cm broad, lanceolate to oblong or elliptic, acuminate apically, acute to cuneate basally, glabrous to pubescent, ciliate; stipules not shredded, 9–20 mm long, strigose, ciliate apically with long bristles; flowers few to many, borne in terminal and usually lateral, spikelike racemes; perianth 2.5–3 mm long, not glandular dotted, united about one-third the length, 5-lobed, the lobes greenish to white or pink; styles 3; achenes 3-angled, black, lustrous.

Moist sites; reported from Circle Hot Springs, Alaska, where probably adventive; widely distributed in North America, Mexico, and South America.

Polygonum lapathifolium L.
Willow Weed

Plants annual, the stems erect or prostrate (rarely rooting at the nodes), 1–9 dm long; leaves petiolate to subsessile, not jointed at the base, the blades 2–20 cm long, 0.6–7 cm broad, lanceolate to oblong or elliptic, acuminate to acute (abruptly rounded) apically, acute to cuneate basally, glabrous to pubescent, ciliate or glabrous marginally; stipules 5–20 mm long, not shredded, glabrous to pubescent, sparsely short-ciliate to glabrous apically; flowers several to many, borne in spikelike racemes, often aggregated in panicles, the peduncles often stipitate-glandular (or sessile); perianth 2–3 mm long, not glandular-dotted, united only near the base, 4–5-lobed, the lobes greenish, white, or pink, strongly veined, the veins branched apically and the ends recurved; styles 2; achenes lens shaped (rarely 3-angled), brown, lustrous.

Weedy plants of disturbed sites, roadsides, gardens, abandoned fields; adventive (or indigenous, in part?) in widely scattered locations in southern Alaska (and to be sought in the Yukon); throughout much of North America; Eurasia (*P. nodosum* Pers.; *P. scabrum* Moench; *P. pensylvanica*, au-

thors, not L.). Two rather distinctive types occur in Alaska.

1a. Leaves lanceolate, acute to shortly cuneate basally; plants usually erect; in interior and coastal southeastern Alaska. *P. lapathifolium* var. *lapathifolium*

1b. Leaves elliptic to lance-elliptic, long-cuneate basally; plants usually prostrate to decumbent, in interior and coastal, southern Alaska (*P. oneillii* Brenckle; *P. pensylvanicum* ssp. *oneillii* [Brenckle] Hultén). *P. lapathifolium* var. *prostratum* Wimm.

Polygonum minimum Wats.
Least Knotweed

Plants annual, ascending to erect, the stems not conspicuously striate, terete, or triangular, 0.5–2.5 dm long; leaves crowded only near the stem tips, oblong to elliptic, ovate, or obovate, 5–15 mm long, 2.5–8 mm broad, somewhat smaller above, acute and mucronate apically, acute basally, the blades sessile at the basal joint; stipules shredded, 2–4 mm long; flowers 1–4, axillary; pedicels included; perianth 1.5–2 mm long, united about one-third the length, 5-lobed, the lobes greenish with white or pink edges, subequal; stigmas 3; achenes 3-angled, black, lustrous.

Weed of waste places; known from the vicinity of Haines, Alaska, where probably adventive from the range of the entity in British Columbia and south to California, Nevada, Utah, and Colorado.

Polygonum persicaria L.
Lady's-thumb

Plants annual, the stems erect or ascending, 3–10 dm tall; leaves petiolate to subsessile, not jointed at the base; blade 3–15 cm long, 0.8–4 cm broad, lanceolate to elliptic or oblong, acuminate apically, acute to cuneate basally, with a purplish spot near the center, usually glabrous, ciliate; stipules 5–15 mm long, not shredded, usually pubescent, long-ciliate apically; flowers several to many, borne in terminal and usually axillary racemes; perianth 1.5–3 mm long, not glandular-dotted, united only near the base, 5-lobed, the lobes pinkish, not strongly veined nor with vein ends recurved; styles 2–3; achenes lens shaped to 3-angled, black, lustrous.

Weeds of disturbed sites; known from disjunct localities in central and southern Alaska and in southern Yukon, adventive; widely distributed in North America and Eurasia.

Polygonum prolificum (Small) Robins.
Proliferous Knotweed

Plants annual, the stems erect or ascending, striate and somewhat 3-angled, 1–8 dm tall; leaves scattered to somewhat crowded, linear-oblong, 5–30 mm long, 1–8 mm broad, smaller on the branchlets than on the main stems, acute to rounded, the blades short-petioled above the basal joint; stipules shredded, 3–6 mm long; pedicels included; perianth 1.2–2 mm long, united about one-third the length, 5-lobed, the lobes green, with white to pink margins, subequal; styles 3; achenes 3-angled, brownish, sublustrous.

Weedy plant; reported from the Yukon (Whitehorse), and southeastern Alaska (Skagway); widely distributed in North America. This entity is sometimes treated as a variety of *P. ramosissimum*.

Polygonum ramosissimum Michx.
Bushy Knotweed

Plants annual, erect or ascending, the stems striate and somewhat 3-angled, 1–10 dm tall; leaves not conspicuously crowded, linear-oblong to lance-elliptic, 10–50 mm long, 2–6 mm broad, gradually reduced upwards, usually acute, the blades short-petiolate above the joint; stipules shredded, 5–10 mm long; pedicels exserted; perianth about 3 mm long, united about one-third the length, 5-lobed, the lobes greenish with yellow margins, the outer lobes broader than the inner; stigmas 3; achenes 3-angled, brown to nearly black, lustrous.

Weed of disturbed sites; reported from southeastern Alaska (Tlehini); widely distributed in North America.

Polygonum viviparum L.
Alpine Bistort

Plants perennial, erect, from short, expanded bases, the stems somewhat striate, 0.7–4.5(5.5) dm tall; basal leaves well developed, 3–25 cm long, the blades narrowly oblong to oblong, elliptic, lanceolate, or oval, 2–13 cm long, 0.3–2.5 cm broad, attenuate to acute apically, cuneate to subcordate basally, petioles well developed, not jointed; cauline leaves reduced upwards; stipules 1–6 cm long, not shredded, often flaring and brownish apically, the upper ones seldom bladeless; flowers several to many, borne in terminal spikelike racemes 2–12 cm long, at least the lower ones replaced by bulblets; perianth 2–3.5 mm long, connate only near the base, 5-lobed, the lobes greenish basally, white (cream) to pink apically, about equal; stamens often vestigial, when fertile then included or shortly exserted; styles 3, exserted; achenes 3-angled, pale brownish, lustrous, seldom developing.

Arctic and alpine tundra, heathlands, and woods; in most of Alaska and Yukon; east to Newfoundland, south to Oregon, Idaho, New Mexico, Minnesota, New Hampshire, and Maine; circumpolar (*P. macounii* Small; *Bistorta littoralis* Greene; *B. ophioglossa* Greene; *P. fugax* Small; *P. viviparum* var. *macounii* [Small] Hultén).

RHEUM L.

Plants perennial, from stout, tuberous roots; leaves alternate, cauline or mostly basal; stipules sheathing, prominent; flowers numerous, borne in terminal or axillary panicles, not subtended by a regular involucre; perianth of 6 petaloid segments, open and spreading; stamens mostly 9 (rarely 6);

pistils 3-carpelled, the ovary 1-loculed, 1-ovuled; styles 3; fruit a strongly winged achene.

Rheum rhaponticum L.
Rhubarb

Plants erect, the stems striate, fistulose, 3–20 dm tall or more, glabrous or nearly so; leaves mostly basal, the edible petioles thick, the blades 2–5 dm long, cordate-ovate, entire but undulate; stipules long-sheathing; flowers numerous, in compound branching panicles; perianth greenish white; achenes 6–12 mm long, winged.

Cultivated throughout the more temperate parts of Alaska and Yukon, where it persists and escapes sparingly; native to Siberia. Rhubarb is grown as a fruit substitute and is often used in mixtures with native fruits to make delightful jams, jellies, and beverages.

RUMEX L.

Plants annual, biennial, or perennial, from stout taproots or rhizomes; leaves alternate, basal or mostly cauline, gradually reduced upwards; stipules sheathing; flowers borne in panicles, not subtended by a regular involucre; perianth of 6 (rarely 4), petaloid or sepaloid segments, the inner 3 segments enlarging in fruit and forming the "wings" or valves which enclose the achene, the midveins of the valves sometimes thickened and forming grainlike tuberosities on the segments; stamens usually 6; pistils 3-carpelled, the ovary 1-loculed, 1-ovuled; styles 3; fruit a 3-angled achene.

Rechinger, K. H. Jr. 1937. The North American Species of *Rumex*. Field Mus. Pub. Bot. 17:1–151.

1a. Flowers mostly imperfect; leaves sagittate or hastate basally, or linear to narrowly oblong; plants often less than 3 dm tall. (2)
1b. Flowers mostly perfect; leaves cordate, truncate, rounded, or cuneate basally, usually not linear to narrow-

ly oblong; plants often more than 3 dm tall. (4)

2a. Leaves primarily basal, linear to narrowly oblong, only some of them hastately lobed, or all entire; inner perianth segments not conspicuously enlarged in fruit. *R. graminifolius*
2b. Leaves variously disposed, but seldom primarily basal, mostly hastately or sagittately lobed, or oblong to lanceolate; inner perianth segments expanding or not. (3)

3a. Leaves sagittate (rarely somewhat hastate), the basal lobes broadly to narrowly triangular; plants from taproots; inner perianth segments conspicuously enlarged; fruiting pedicels jointed near the middle. *R. acetosa*
3b. Leaves hastate or rarely entire, the basal lobes linear to oblong, lanceolate, or triangular; plants from rhizomes; inner perianth segments not conspicuously enlarged; fruiting pedicels jointed at the base of the perianth. *R. acetosella*

4a. Stems decumbent to ascending, with axillary branches at most or all of the nodes; leaves usually pale, mostly cauline, not much reduced in size upwards. *R. salicifolius*
4b. Stems erect or nearly so, usually lacking axillary branches below the inflorescence; leaves usually bright green, often mostly basal and conspicuously reduced in size upwards. (5)

5a. Inner perianth segments with distinct spreading or curved elongate teeth. (6)
5b. Inner perianth segments entire or minutely toothed. (7)

6a. Plants perennial; pedicels jointed well above the base; inner perianth segments often lacking distinctive tuberosities. *R. obtusifolius*
6b. Plants annual; pedicels jointed at or very near the base; inner perianth

segments usually bearing distinctive tuberosities. *R. maritimus*

7a. Well-developed tuberosities present on 1, 2, or all 3 of the inner perianth segments in fruit. *R. crispus*

7b. Well-developed tuberosities lacking on all of the inner perianth segments in fruit. (8)

8a. Cauline leaves usually much reduced in size and number; herbage (including the inflorescence) often purplish tinged; plants broadly distributed in the interior, and in western and northern coastal regions, less common in southern coastal regions. *R. arcticus*

8b. Cauline leaves smaller above, but not conspicuously so; herbage and inflorescence usually green (rarely purplish on *R. occidentalis*); plants mostly of coastal southern Alaska, rarely in the interior. (9)

9a. Inner perianth segments reniform, broadly rounded apically, as broad as long or broader; pedicels conspicuously jointed near the middle; weedy plants, usually growing around habitations. *R. longifolius*

9b. Inner perianth segments ovate to ovate-orbicular, tapering to the apex, often longer than broad; fruiting pedicels obscurely or not at all jointed near the middle. *R. occidentalis*

Rumex acetosa L.
Green Sorrel

Plants perennial, dioecious, erect from a taproot, the stems 3–10 dm tall or more, not branching below the inflorescence; basal leaves long-petiolate, the blades oblong to lanceolate or ovate, 2–10 cm long, 0.8–3 cm broad, sagittately lobed basally, obtuse to rounded apically; flowers numerous, imperfect, borne in leafless panicles, usually purplish tinged; fruiting pedicels jointed near the middle; perianth 0.5–1.5 mm long in flower, the outer ones soon reflexed; inner segments much enlarged and enclosing the achene in fruit, 3–4 mm long;

cordate to orbicular, entire or nearly so, each with a small, basal tuberosity; achenes 2–2.5 mm long, dark brown, lustrous.

Moist sites, in tundra and heath or woods; mostly in western Alaska, but also in the interior and in coastal southern Alaska, and in southwestern Yukon; widely distributed in North America; Eurasia. Two more or less distinctive entities have been distinguished among our materials. Presumably, ssp. *alpestris* is indigenous (at least in part), while ssp. *acetosa* is adventive.

1a. Stipular sheaths strongly shredded apically. *R. acetosa* ssp. *acetosa*

1b. Stipular sheaths not shredded apically (fragile and soon deciduous by fragmentation). *R. acetosa* ssp. *alpestris* (Scop.) A. Löve

Rumex acetosella L.
Sheep Sorrel

Plants perennial, dioecious, erect from slender rhizomes, the stems 1–6 dm tall, usually not branching below the inflorescence; basal leaves long-petiolate, the cauline ones becoming short-petiolate to subsessile, the blades oblong to ovate, linear, lanceolate, or elliptic, 1–8 cm long, 0.2–2.5 cm broad, hastately lobed basally, attenuate, acute, or obtuse apically; flowers numerous, imperfect, borne in leafless panicles, often purplish tinged; fruiting pedicels jointed at the base of the flower; perianth segments 0.5–1.8 mm long in flower, the outer segments not reflexed; inner segments enlarging sufficiently to invest the achene, 1–2 mm long, ovate, entire, lacking tuberosities; achenes 1–2 mm long, yellowish brown, lustrous, sometimes adherent to the inner perianth segments.

Weedy species of disturbed sites, roadsides, abandoned fields, beaches and gardens; in interior and southern Alaska, from the Aleutians eastward to the Panhandle; widely distributed in North America; introduced from Eurasia (*R. angiocarpus* Murb.; *R. acetosella* ssp. *angiocarpus* [Murb.] Murb.).

Rumex arcticus Trautv. ex Middend.
Arctic Dock

Plants perennial, erect from a taproot or sometimes from a short thick rhizome, the stems 1–12 dm tall or more; basal leaves long-petiolate, the blades lanceolate to oblong, linear-oblong, or ovate, 6–23 cm long, 1–5 cm broad, subcordate, truncate, obtuse or acute basally, acuminate to rounded apically, at least somewhat undulate; cauline leaves much reduced upwards, short-petiolate to subsessile; flowers numerous, perfect, borne in panicles which are leafy-bracted at least near the base, often purplish tinged; fruiting pedicels indistinctly or not at all jointed; perianth 2–3 mm long in flower, the outer segments not reflexed, the inner segments much enlarged in fruit, 4–7 mm long, ovate to lance-ovate, irregularly denticulate to subentire, lacking tuberosities; achenes 3–4 mm long, brown, lustrous.

Arctic and alpine tundra to heath and woods; in most of Alaska (except for the southeastern portion and the Aleutians) and in most of the Yukon; Asia (*R. arcticus* var. *perlatus* Hultén). *R. arcticus* shows considerable variation in stature, shape of basal leaves, and shape of fruiting perianth segments. Numerous additional fruiting specimens are required along with mature sterile basal rosettes before infraspecific segregation is attempted. This entity is closely allied to *R. occidentalis*.

Rumex crispus L.
Curled Dock

Plants perennial, erect from taproots, the stems 3–10 dm tall or more; basal leaves long-petiolate, the blades oblong-lanceolate to elliptic, 8–30(40) cm long, acute to rounded basally, acuminate to acute apically, undulate-crisped (the margin appearing irregularly lobed due to numerous overlapping folds in pressed specimens); cauline leaves somewhat smaller upwards, short-petiolate; flowers numerous, perfect, borne in panicles with large leafy bracts to midlength or above, usually greenish; fruiting pedicels jointed above the base; perianth 1.5–2 mm long, the outer segments not reflexed; inner segments much enlarged in fruit, 3–5 mm long, cordate to deltoid or ovate, denticulate to entire, usually each (though sometimes only 1–2) bearing a reticulately patterned tuberosity almost half as long as the segment; achenes 2–3 mm long, brown, lustrous.

Weedy plants of disturbed sites, old fields, and roadsides; in widely disjunct regions in southern Alaska and southern Yukon; adventive from Europe.

Rumex graminifolius Lamb.
Grass-leaved Sorrel

Plants perennial, dioecious, erect from a caudex and taproot, the stems 0.5–2.5 dm tall; leaves primarily basal, long-petiolate, the blades oblong to linear, 1–6 cm long, 0.2–0.5 cm broad, entire, or less commonly hastately lobed near the base; flowers numerous, imperfect, borne in leafless (or occasionally leafy bracted) panicles, often purplish tinged; fruiting pedicels jointed near the base of the flower; perianth 1–2 mm long in flower, the outer segments not reflexed; inner segments not much expanded in fruit, enclosing the achene, 1.5–2 mm long, ovate, entire, lacking tuberosities; achenes 1–1.5 mm long, yellowish brown, not sharply angled, lustrous.

Sandy areas, dunes, beaches, river banks, and lake shores; in southwestern, western, and northwestern Alaska; Eurasia.

Rumex longifolius DC. ex Lam. & DC.
Garden Dock

Plants perennial (biennial?), erect from taproots, the stems 2–15 dm tall; basal leaves long-petiolate, the blades lanceolate to oblong-lanceolate, mostly 10–50 cm long and 2–10 cm broad, subcordate to truncate or acute basally, acute to acuminate apically, flat to conspicuously undulate; cauline leaves only slightly smaller upwards, short-petiolate; flowers numerous, perfect, borne in panicles with large leafy bracts to the middle or above, usually greenish; fruiting pedicels pointed near the middle or slightly below; perianth 1.5–2 mm long, the outer segments not reflexed; inner segments much enlarged in fruit, 4–7 mm long (mostly broader than long), reniform-ovate, broadly rounded to obtuse apically, denticulate to entire, lacking well-developed tuberosities; achenes 2.5–3 mm long, brown, lustrous.

Weedy species of vacant lots, abandoned fields, and roadsides; in western, central, and coastal southern Alaska (to be expected in the Yukon); adventive from Europe (*R. domesticus* Hartm.).

Rumex maritimus L.
Golden Dock

Plants annual (or biennial?), erect from taproots, the stems 1–8 dm tall; basal leaves reduced; cauline leaves well developed but reduced in size upwards, short-petiolate, the blades oblong to lanceolate, 4–15 cm long, 1–4 cm broad, rounded to subcordate or less commonly acute basally, acute to acuminate or obtuse apically, undulate to plane; flowers numerous, borne in compact axillary clusters, the inflorescence leafy throughout, often over half the total plant height, greenish; pedicels jointed near or at the base; perianth 1–2 mm long in flower, the outer ones not reflexed; inner segments much enlarged in fruit, 3–7 mm long (including the acuminate apex), ovate, with 2–4 slender teeth per segment, each tooth 2–5 mm long, each segment usually with a well-developed tuberosity about half as long as the segment; achenes 1.5–2 mm long, brown, lustrous.

Moist places on lake shores, in marshes, and along streams; disjunct in east-central and southeastern Alaska and southern Yukon; widely distributed in North and South America; Europe. Two entities are sometimes recognized among materials of *R. maritimus*.

1a. Principal cauline leaves cuneate basally, distinctly pubescent with flattened hairs along the petiole and midvein. *R. maritimus* var. *maritimus*

1b. Principal cauline leaves subcordate to truncate basally, indistinctly pubescent (*R. fueginus* Phill.; *R. maritimus* ssp. *fueginus* [Phill.] Hultén). *R. maritimus* var. *fuegina* (Phill.) Dusen.

Rumex obtusifolius L.
Bitter Dock

Plants perennial, erect from taproots, the stems 5–12 dm tall or more, usually unbranched below the inflorescence; basal leaves long-petioled, the blades ovate to oblong or lanceolate, mostly 10–40 cm long, 5–15 cm broad, cordate to truncate basally, obtuse to acute (or acuminate) apically, undulate; cauline leaves like the basal ones, somewhat smaller and with shorter petioles upward; flowers numerous, perfect, borne in panicles with leafy bracts to the middle or above, usually greenish; perianth segments 2–3 mm long, the outer ones not reflexed; inner segments much enlarged in fruit, 3.5–5 mm long, ovate, with 4–6 teeth per segment, each tooth 0.5–2 mm long, at least some segments bearing a prominent tuberosity; achenes 1.5–2 mm long, brown, lustrous.

Weed of disturbed soils; in coastal and insular southern Alaska; widely distributed in North America; adventive from Eurasia. Our materials have been assigned to ssp. *agrestis* (Fries) Danser.

Rumex occidentalis Wats.
Western Dock

Plants perennial, erect from taproots, the stems 5–20 dm tall, usually unbranched below the inflorescence, often reddish tinged; basal leaves long-petioled, the blades oblong to ovate or oblong-lanceolate, 0.6–4 dm long, 3–15 cm broad, deeply cordate to truncate basally, rounded to obtuse or acute apically, usually more or less undulate-crisped; cauline leaves reduced upwards; flowers numerous, perfect, borne in panicles with leafy bracts only near the base, greenish; fruiting pedicels obscurely jointed near or below the middle; perianth segments 2–4 mm long, the outer ones not reflexed; inner segments greatly enlarged in fruit, 4–10 mm long, ovate to oval (mostly longer than broad), denticulate to entire, lacking tuberosities; achenes 3–4 mm long, brown, lustrous.

Moist sites, marshes, beaches, stream banks, and tidal flats; in coastal and insular southern Alaska and less commonly in interior Alaska and Yukon; east to Quebec and south to California, Nevada, New Mexico, and South Dakota (*R. fenestratus* Greene; *R. fenestratus* ssp. *puberulus* Hultén). This taxon is apparently closely related to *R. arcticus*, and it is not always possible to segregate all specimens consistently.

Rumex salicifolius Weinm.
Beach Dock

Plants perennial from taproots, the stems decumbent to ascending (rarely erect), 2–6 dm tall, branching from the lower nodes; leaves mostly cauline, short-petiolate, not much reduced upwards, the blades 5–20 cm long, 0.3–3 cm broad, narrowly lanceolate to oblong or linear, acute to rounded basally, acuminate apically, plane to undulate, not crisped; flowers numerous, perfect, borne in panicles with leafy bracts mostly near the base, usually greenish; fruiting pedicels jointed near the base; perianth segments 1–2 mm long, the outer ones not reflexed; inner segments much enlarged in fruit, 2–4 mm long, ovate to deltoid, entire to denticulate, tuberosities usually well developed on all segments (lacking in var. *montigenitus*); achenes 1.5–2.5 mm long, brown, lustrous.

Disturbed soils, beaches, abandoned fields, roadsides, and river banks; in much of Alaska and Yukon south of the 66th parallel; east to Quebec and south to New York, Texas, and California; Asia.

1a. Inner perianth segments with tuberosities usually well developed on all of them in fruit; plants broadly distributed in Alaska and Yukon (*R. sibiricus* Hultén, a phase from the interior with usually narrower leaves; *R. transitorius* Rech. f., a coastal phase with broader leaves). *R. salicifolius* ssp. *salicifolius* var. *salicifolius*

1b. Inner perianth segments lacking tuberosities; plants known from the Yukon (*R. utahensis* Rech. f.; *R. mexicanus* Meisn.). *R. salicifolius* ssp. *triangulivalvis* Danser var. *montigenitus* Jeps.

PORTULACACEAE
Purslane Family

Perennial or annual, more or less succulent herbs; leaves alternate, opposite, or basal, entire; flowers perfect, regular (at least as regards the corolla), arranged in cymes or racemes; calyx of 2 distinct or partly united sepals; petals mostly 5(3–16), often more or less united at the base; stamens few to many, opposite the petals when of the same number; pistils 1; ovary superior, 1-loculed, the carpels 2–3; styles and stigmas 2–4; fruit a 2–3-valved capsule.

1a. Plants from fibrous roots, annuals, or slenderly rhizomatous to stoloniferous perennials; flowers seldom or not at all corymbose; ovules 1–3(6 in some species). *Montia*

1b. Plants from distinctly thickened taproots or corms; flowers usually corymbose; ovules 6–40, occasionally only 3. (2)

2a. Petals and stamens commonly more than 5; capsules opening by a circumscissile lid. *Lewisia*

2b. Petals and stamens commonly 5; capsules opening by 3 valves. *Claytonia*

CLAYTONIA L.

Plants perennial, from distinctly thickened taproots or corms; basal leaves 1–many or lacking; cauline leaves 2(3), opposite, bractlike, subtending the inflorescence; flowers showy, (1)2–several, borne in terminal corymbose racemes; sepals 2, persistent; petals 5(6), distinct or basally united; stamens 5(6), opposite the petals and adnate to them at the base; styles 1, deeply 3-cleft; capsules 3-valved, usually 6-ovuled; seeds 2–6.

1a. Plants from a short thick deep-seated corm; basal leaves 1–few, or lacking. *C. tuberosa*

1b. Plants from a fleshy thickened taproot; basal leaves several to many, with membranous sheathing bases. (2)

2a. Basal leaves linear to narrowly oblong-lanceolate, 4–13 cm long; plants of western, northern, and central Alaska. *C. acutifolia*

2b. Basal leaves oblanceolate to spatulate, mostly 2–4 cm long; plants of western Aleutians and Pribilofs. *C. arctica*

Claytonia acutifolia Pallas ex Roem. & Schult.
Bering Sea Spring-beauty

Plants 5–15 cm tall, from taproots 4–20 mm in diameter; flowering stems 1–several from the root crown; basal leaves linear to narrowly oblong-lanceolate, 2–13 cm long, 0.1–0.9 cm broad; cauline leaves lanceolate to linear, 1–3 cm long; inflorescence corymbose; flowers mostly 2–5, the pedicels 6–20 mm long in flower, subtended by lanceolate bracts; sepals 2, 4–12 mm long, ovate to lanceolate, obtuse or acute; petals white to pink, 12–18 mm long, persistent and investing the capsule, united basally; capsule 3–5 mm long, ovoid; seeds 2–6.

Moist sites, in tundra or heathlands; in islands of the Bering Sea, Seward Peninsula, Yukon River Delta, and eastward to Mt. McKinley National Park; Asia (*C. eschscholtzii* Cham.; *C. acutifolia* var. *graminifolia* [Hultén] B. Boi). Our materials are known as ssp. *graminifolia* Hultén.

Claytonia arctica Adams
Arctic Spring-beauty

Plants 3–6 cm tall or more, from a thick, fleshy taproot to 10 mm in diameter; flowering stems 1–few from the root apex; basal leaves 2–4 cm long or longer, membranous-sheathing basally, the blades oblanceolate to spatulate, 4–6 mm broad; cauline leaves 2, opposite, broadly elliptic, ovate, or obovate, twice longer than broad or less; inflorescence nodding, racemose; flowers 2–5; pedicels 5–10 mm long in flower, not subtended by bracts; sepals 2, 4–5 mm long, ovate to elliptic, obtuse to rounded; petals 5, white, 9–12 mm long, persistent, distinct; stamens 5; ovules 6.

Mountain slopes and ridges, in alpine tundra; in the Aleutian Islands and in the Pribilofs; Asia.

Claytonia tuberosa Pallas ex Roem. & Schult.
Tuberous Spring-beauty

Plants mostly 6–15 cm tall (above ground level), from short, thick corms 4–15 cm below ground level; stems mostly solitary, usually whitish and very slender below ground; basal leaves 1–2 (usually lacking in pressed specimens), arising from the corm, the subterranean petiole whitish, very slender, the blades elliptic to oblong, 2–6 cm long, 0.4–0.8 cm broad; cauline leaves narrowly elliptic to oblanceolate or narrowly oblong, 1.5–6 cm long; inflorescence corymbose; flowers 3–8; pedicels 5–28 mm long, erect or ascending in flower, recurved in bud and in fruit, subtended by small, ovate bracts; sepals 2, ovate to orbicular, 4–7 mm long, acute to obtuse; petals white (rarely pink?), 9–15 mm long, somewhat persistent, basally united; capsules 4–6 mm long; seeds 2–6, reddish brown, about 2 mm long.

Moist sites, in tundra or heathlands; islands of the Bering Sea and Seward Peninsula;

eastward to the central Yukon River Valley of eastern Alaska and southern Yukon, and disjunctly in northeastern Alaska; Asia (*C. czukczorum* Volk.; *C. tuberosa* var. *czukczorum* [Volk.] Hultén).

LEWISIA Pursh

Plants perennial, more or less succulent herbs from fleshy roots; basal leaves few to numerous; cauline leaves 2, bractlike; flowers showy, 1 per peduncle; sepals 2, persistent; petals 5–9, distinct; stamens 4–90 or more; styles 1, deeply 3–6-cleft; capsules opening by a circumscissile lid; seeds 15 or more.

Lewisia pygmaea (Gray) Robins.

Plants mostly 2–6 cm tall, from thick tap or fascicled roots; stems solitary or 2–several; basal leaves few to numerous, linear to oblanceolate, 2–6 cm long, 0.1–0.3 cm broad; cauline leaves bractlike, opposite, borne at or below the middle of the stem; flowers solitary; pedicels 10–25 mm long or more; sepals 2, ovate, 3–6 mm long, acute to rounded; petals 5–9, white to pink, 6–10 mm long; stamens 4–10 or more; style branches 3–6; capsules 3–6 mm long; seeds 15 or more, dark brown, 1–1.2 mm long, shiny.

Alpine tundra; vicinity of Kluane Lake, Yukon; disjunctly southward from Washington to California and westward to Montana and New Mexico (*Talinum pygmaeum* Gray).

MONTIA L.

Plants annual to perennial, somewhat succulent herbs from fibrous roots, or with rhizomes or stolons; cauline leaves 2 and opposite, or 2–several and opposite or alternate; basal leaves well developed and petiolate, or lacking; flowers showy or inconspicuous, usually several (rarely 1) in axillary or terminal racemes; sepals 2, persistent; petals usually 5(2–6), distinct or basally united; stamens 2–5, opposite the petals and usually adnate to them at the base; styles deeply 3-cleft; capsules 3-valved; seeds 1–3(6 in some species).

1a. Cauline leaves alternate. (2)
1b. Cauline leaves opposite. (3)

2a. Basal leaves linear-oblanceolate; cauline leaves usually 2–3(5); plants of central Yukon. *M. bostockii*
2b. Basal leaves oval or obovate or broadly oblanceolate; cauline leaves usually more than 3; plants of coastal southeastern Alaska. *M. parvifolia*

3a. Plants with more than a single pair of opposite cauline leaves. (4)
3b. Plants with a single pair of opposite cauline leaves. (5)

4a. Corolla 5–8 mm long; plants perennial; leaves mostly over 20 mm long. *M. chamissoi*
4b. Corolla 1.5–2 mm long; plants annual; leaves mostly less than 15 mm long. *M. fontana*

5a. Petals usually less than 5 mm long; cauline leaves connate around the stem, forming a disklike structure. *M. perfoliata*
5b. Petals generally more than 5 mm long; cauline leaves distinct or nearly so, not at all disklike. (6)

6a. Cauline leaves narrowly lanceolate; basal leaves with conspicuously sheathing membranous bases and narrowly oblong-spatulate to linear blades. *M. scammaniana*
6b. Cauline leaves broadly ovate to ovate-lanceolate; basal leaves not both membranous-sheathing and with narrowly oblong to linear blades. (7)

7a. Plants from fibrous roots; racemes elongate, usually with many flowers; pedicels subtended by bracts. *M. sibirica*
7b. Plants from rhizomes (and occasionally with stolons); racemes 1–5(8)-flowered; pedicels not subtended by bracts. *M. sarmentosa*

Montia bostockii (Porsild) Welsh

Plants perennial, 6–12 cm tall or taller, from elongate rhizomes and with slender stolons; flowering stems solitary from the nodes of the rhizomes; basal leaves 10–30 mm long, 1–3 mm broad, linear-oblanceolate, not sheathing; cauline leaves commonly 2–3(5), alternate, linear or narrowly oblong, several times longer than broad; inflorescence somewhat corymbose; flowers 2–6; pedicels 5–25 mm long, erect in flower, nodding in bud, at least the lowermost one subtended by a bract; sepals 2, 3.5–6 mm long; petals whitish with pink veins, 10–12 mm long; capsules not seen.

Moist places near springs and on alpine slopes; in southwestern Yukon and east-central Alaska; endemic (*Claytonia bostockii* Porsild), possibly *vassilievii*. The *Claytonia vassilievii* Kuzen, (Flora USSR 6:880. 1936) description is similar to *M. bostockii*, but no transfer is proposed.

Montia chamissoi (Ledeb.) Robins. & Fern
Toad-lily

Plants perennial, 5–25 cm tall, from slender rhizomes and stolons (which produce bulbletlike offsets); flowering stems often branched; basal leaves reduced or lacking; cauline leaves opposite, 4–several, 1–5 cm long, 0.2–1.8 cm broad, oblanceolate to elliptic, tapering to the petiole; inflorescence erect, racemose, terminal and axillary; flowers few to several, sometimes replaced by bulblets; pedicels 8–30 mm long in flower, nodding in bud, a single bract at the base of the lowest pedicel; sepals 2, 2–3 mm long; petals 5, white (or pinkish), 5–8 mm long; stamens commonly 5; capsules 2–3 mm long; seeds 1–3.

Marshes, seeps, springs, or streams; in coastal and insular southern Alaska and less commonly in interior southern Alaska; east to Manitoba and southward to California, New Mexico, and Iowa.

Montia fontana L.
Blinks, Water Chickweed

Plants annual, slender, decumbent to ascending, from taproots, or rooting at the nodes; stems 3–25 cm long, often much-branched; basal leaves reduced or lacking; cauline leaves opposite, 4–several, 0.3–1.5 cm long, 0.1–0.5 cm broad, oblanceolate to oblong or elliptic, subsessile or short-petiolate; inflorescence nodding, racemose, terminal; pedicels 3–10 mm long, sometimes longer, nodding in bud and in fruit; flowers 2–7; sepals 2, 1–1.5 mm long, broader than long; petals white, 1–2 mm long, united about half the length; stamens usually 3; capsules 1–1.5 mm long; seeds usually 1–2, black.

Wet soil, or in shallow water; in coastal and insular western and southern Alaska and northern Yukon; circumboreal. Two remarkably similar varieties occur in our region.

1a. Seeds reticulately patterned, smoothish, lustrous, 1–1.2 mm long. *M. fontana* var. *lamprosperma* (Cham.) Fenzl.

1b. Seeds minutely roughened, sublustrous, 0.7–0.9 mm long (*M. hallii* [Gray] Greene). *M. fontana* var. *tenerrima* (Gray) Fern & Wieg.

Montia parvifolia (Moq.) Greene

Plants perennial, from elongate rhizomes or fibrous roots and with elongate stolons; flowering stems erect or ascending, 10–30 cm tall; basal leaves 1–6 cm long, petiolate, somewhat sheathing basally, the blades oval to orbicular, obovate, or oblanceolate, 7–15 mm long, 4–13 mm broad; cauline leaves several, alternate, mostly smaller and narrower than the basal leaves; inflorescence racemose, erect, terminal; flowers 3–8; pedicels 4–15 mm long, nodding in bud, the lower one subtended by a bract; sepals 2, 2–3 mm long, rounded; petals 5, white to pink, 7–14 mm long; stamens 5; capsule 2–4 mm long; seeds commonly 2, black, lustrous.

Moist, rocky outcrops, and on beaches; in southeastern Alaska; southward to California and east to Montana and Utah. Two intergrading taxa are present within our range.

1a. Basal leaf blades lance-ovate to spatulate, mostly less than 5 mm broad, longer than broad. *M. parvifolia* var. *parvifolia*

1b. Basal leaf blades oval to orbicular, mostly more than 5 mm long, about as broad as long or broader (*Claytonia parvifolia* ssp. *flagellaris* [Bong.] Hultén). *M. parvifolia* var. *flagellaris* (Bong.) C. L. Hitchc.

Montia perfoliata (Donn) Howell
Miner's-lettuce

Plants annual, from slender taproots; flowering stems few to several from the root crown, 5–20 cm tall; basal leaves 2–10 cm long, somewhat sheathing basally, the blades linear to spatulate or broader; cauline leaves 2, opposite, usually connate around the stem and forming a disklike structure 0.8–3 cm broad; inflorescence racemose, terminal and sometimes also axillary; flowers mostly 3–8; pedicels 2–10 mm long or longer, nodding, only the lowermost subtended by a bract; sepals 2, 1.5–3 mm long, rounded; petals 5, white or pinkish, 3–5 mm long; stamens 5; capsule 2–4 mm long; seeds usually 3, black, lustrous.

Adventive in the Aleutian Islands (Unalaska); indigenous from British Columbia eastward to South Dakota and south to Arizona and California.

Montia sarmentosa (C. A. Mey.) Robins.
Alaska Spring-beauty

Plants perennial, from elongate rhizomes 1–2 mm in diameter, and with stolons; flowering stems erect, 5–23 cm tall, 1–few from the apex of the rhizome; basal leaves

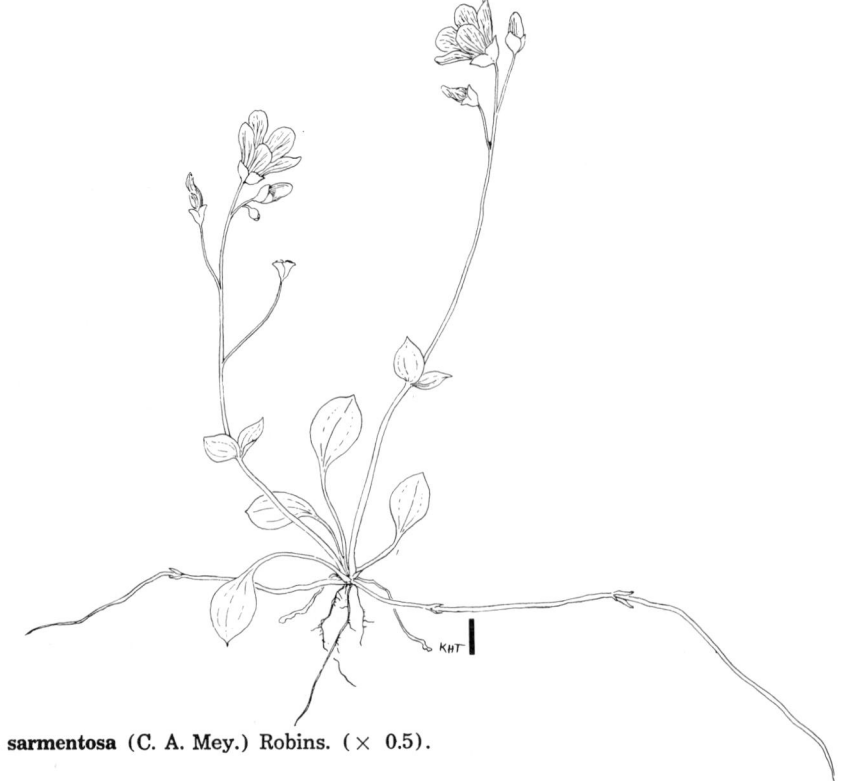

Montia sarmentosa (C. A. Mey.) Robins. (× 0.5).

1.5–11 cm long, sheathing basally, the blades broadly elliptic to ovate or spatulate, 10–30 mm long, 3–22 mm broad, tapering to the petiole; cauline leaves 2, opposite, ovate to ovate-lanceolate, twice longer than broad or less; inflorescence nodding, racemose, terminal; flowers 2–8; pedicels 8–25 mm long, nodding in bud, not subtended by bracts; sepals 2, 4–6 mm long, oval to ovate, rounded; petals 5, pink (fading white), 10–14 mm long, persistent, distinct or nearly so; stamens 5; capsules 3–5 mm long; seeds 2–6, black, lustrous.

Arctic and alpine tundra and heathlands (rarely in woods); in most of Alaska (except for the central to western Aleutians, southeastern and extreme northern Alaska) and in southern Yukon; Asia (*Claytonia sarmentosa* C. A. Mey.).

Montia scammaniana (Hultén) Welsh
Scamman Spring-beauty

Plants perennial, from elongate rhizomes 1.5–3 mm in diameter; flowering stems erect, 4–10 cm tall, few to several from the apex of the rhizome; basal leaves 1–10 cm long, conspicuously membranous-sheathing basally, the blades linear to oblong-spatulate, 1.5–7 mm broad, tapering to the petiole; cauline leaves 2, opposite, broadly lanceolate to elliptic, mostly 2–4 times longer than broad; inflorescence erect, racemose, terminal; flowers 1–2; pedicels 7–35 mm long in flower, nodding in bud, not subtended by bracts; sepals 2, 4–7 mm long, oval to orbicular, rounded; petals 5, pink (fading white), 10–18 mm long, persistent, distinct or nearly so; stamens 5; capsules 5–7 mm long; seeds 2–6.

Alpine tundra; from the Alaska Peninsula and Kuskokwim drainage eastward along the Alaska Range to the central Yukon River Valley of east-central Alaska and west-central Yukon; endemic.

Montia sibirica (L.) Howell
Siberian Spring-beauty

Plants annual, from fibrous roots or from a slender taproot, or short-lived perennials (rarely rhizomatous); flowering stems 15–50 cm long, decumbent to erect, few to several from the root crown; basal leaves 3–20(30) cm long, long-petiolate, not conspicuously sheathing basally, the blades lanceolate to ovate, orbicular, or elliptic, 0.7–6 cm long, 0.3–6 cm broad; cauline leaves 2, opposite, ovate to obovate or broadly lanceolate, 1.4–6 cm long, 1–5 cm broad, less than twice longer than broad, not connate; inflorescence erect or spreading, racemose, terminal and axillary; flowers many; pedicels 1–5 cm long, nodding in bud, spreading in flower, subtended by bracts; sepals 2, 2–5 mm long, oval, broadly rounded; petals 5, white to pink, 6–12 mm long, basally united; stamens 3–5; capsules 3–5 mm long; seeds 1–3, black, lustrous.

Moist thickets, woods, and beaches; in coastal and insular southern Alaska, from the Aleutians through the Panhandle; south to California, Utah, and Montana (*Claytonia sibirica* L.).

PRIMULACEAE
Primrose Family

Annual or perennial herbs; leaves simple, alternate, opposite, or whorled; flowers perfect, regular, variously arranged, terminal or axillary; sepals commonly 5, more or less united; petals 5, united, or lacking (in *Glaux*); stamens 5, opposite the corolla lobes; pistils 1; ovary superior, 1-loculed, 5-carpelled; styles 1; stigma capitate; fruit a capsule.

1a. Leaves cauline; flowers axillary or in axillary inflorescences. (2)

1b. Leaves all basal; flowers terminal or in terminal inflorescences. (4)

2a. Corolla lacking; plants of saline soils. *Glaux*

2b. Corolla present; plants usually not of saline soils. (3)

3a. Flowers sulfur yellow, in dense

axillary clusters; plants tall, semi-aquatic. *Lysimachia*

3b. Flowers white or pinkish, solitary on long peduncles; plants of various habitats, seldom semiaquatic. *Trientalis*

4a. Corolla lobes distinctly reflexed; flowers nodding. *Dodecatheon*

4b. Corolla lobes spreading to erect, not reflexed; flowers erect or spreading. (5)

5a. Plants densely caespitose perennials with persistent, densely imbricate, narrow leaves; peduncles solitary, each with a single flower; corolla pink. *Douglasia*

5b. Plants various, but seldom densely caespitose; peduncles 1–more, each with 1–few flowers; corolla white to pink. (6)

6a. Flowers constricted at the throat, white and borne in umbels (rarely solitary), or sometimes fading pinkish; corolla tube usually shorter than the calyx. *Androsace*

6b. Flowers open at the throat, white, or more commonly pink, the inflorescence umbellate, commonly with 2–more flowers; corolla tube usually longer than the calyx. *Primula*

ANDROSACE L.

Annual or perennial scapose herbs; leaves in basal rosettes, simple, usually persistent; herbage pubescent with stellate, forked, or simple hairs, or glabrous; flowers in umbels of 2–20 or more, or rarely solitary, borne on peduncles that arise from the basal rosette; calyx 5-lobed, the tube nearly equal to the lobes; corolla tubular, contracted and sometimes with 5 crests at the throat, 5-lobed; stamens 5, the filaments adnate to the corolla; capsules subglobose.

Robbins, G. T. 1944. North American species of *Androsace*. Am. Midl. Nat. 32:137–63.

1a. Pedicels mostly 1–3 cm long or more (if shorter, then the peduncles glabrous), the flowers mostly 4–more in an open umbel; peduncles 1–8 or more per rosette. *A. septentrionalis*

1b. Pedicels usually less than 1 cm long, or only 1–3 flowers per inflorescence, or peduncles only 1 per rosette. (2)

2a. Peduncles 1 per rosette, pubescent with long, simple hairs; corolla large and showy, the lobes longer than the calyx; flowers 2–several. *A. chamaejasme*

2b. Peduncles usually several per rosette, pubescent with stellate hairs or glabrous; flowers usually 1(2–3). *A. alaskana*

Androsace alaskana Cov. & Standl. ex Hultén

Plants annual or biennial from a taproot, 3–15 cm tall; leaves numerous in a dense basal rosette, 10–30 mm long, apically 3-lobed, ciliate, glabrous below, pubescent or glabrous above; peduncles 4–20 or more per rosette, glabrous or stellate-hairy; flowers 1–2(3) immediately subtended by a lanceolate bract; calyx 4.5–5.5 mm long, the teeth 1.5–3 mm long; corolla white, 4–5 mm long, the lobes 1–1.5 mm long; capsule 5–6 mm long.

Rocky alpine sites; in southwestern and south-central Alaska and in southwestern Yukon; endemic.

Androsace chamaejasme Host

Plants mat-forming perennials, with a caudex and prostrate stems, each with a terminal rosette, the flowering stems 1.2–15 cm tall; leaves several to many, 3–16 mm long, entire, ciliate, pubescent to glabrate or glabrous on both sides; peduncles 1 per rosette, pubescent with long, simple hairs; flowers 2–several on pedicels 1–12 mm long, the umbels subtended by few to several saccate bracts; calyx 2–3 mm long, the teeth about equal to the tube; corolla white to cream (fading pinkish), with a yellowish center, the tube subequal to the calyx, the lobes 2–5 mm long; capsules 2–3 mm long.

Arctic and alpine tundra, heathlands, or less commonly in woods; in most of Alaska (except for the southeastern portion) and in northern and western Yukon; east to Mackenzie and south to Montana; Eurasia (*A. chamaejasme* var. *arctica* Kunth; *A. chamaejasme* var. *andersonii* Hultén; *A. chamaejasme* ssp. *andersonii* [Hultén] Hultén; *A. lehmanniana* Spreng.). Our material belongs to ssp. *lehmanniana* (Spreng.) Hultén.

Androsace septentrionalis L.

Plants annual (or biennial?) from a taproot; leaves numerous in a dense basal rosette, 5–50 mm long, usually with 5–more apical teeth, sparsely to densely pubescent with simple or forked hairs; peduncles 1–many per rosette, glabrous or pubescent with stellate hairs; flowers mostly several per inflorescence, the pedicels (0.3)1–4 cm long or more; calyx 2.5–4 mm long, the lobes shorter than the tube; corolla white, the tube subequal to the calyx, the lobes 0.5–1 mm long; capsules 2–4 mm long.

Gravelly or sandy soil, along streams, roadsides, and ridge tops; in most of Alaska and Yukon; eastward to Labrador and south to California, Arizona, and New Mexico; circumboreal.

DODECATHEON L.

Perennial, subrhizomatous, scapose herbs; leaves all basal, petiolate, the blades broad; herbage glabrous or glandular-pubescent; flowers (1)2–several, borne in terminal umbels, nodding in anthesis, erect in fruit; calyx cup shaped, 4–5-lobed; corolla 4–5-lobed; the lobes much longer than the tube, reflexed; stamens 4–5, the filaments short and distinct or connate, adnate to the tube; anthers connivent around the style; anther sacs with a prominent connective that is dilated at the base; capsules elongate.

Beamish, K. I. 1955. Studies in the genus *Dodecatheon* of Northwestern America. Bull. Torrey Club 82:357–66.

Thompson, H. J. 1953. The biosystematics of *Dodecatheon*. Contr. Dudley Herb. 4:73–154.

1a. Staminal filaments 1 mm long or more, united into a yellow tube. **D.** *pulchellum*

1b. Staminal filaments less than 1 mm long, distinct or connate by a narrow, purplish membrane. (2)

2a. Roots brownish; rhizome commonly vertical; stigma slightly or not at all enlarged; base of connective narrower than the base of the anther; plants in interior Alaska and Yukon. *D. frigidum*

2b. Roots whitish; rhizome commonly horizontal; stigma often twice the diameter of the style; base of connective as broad as the anther or nearly so; plants of coastal, southeastern Alaska. *D. jeffreyi*

Dodecatheon frigidum Cham. & Schlecht.
Northern Shooting-star

Plants 0.6–4 dm tall, the roots brownish; leaves 2–15 cm long, 0.8–3.5 cm broad, irregularly crenate to subentire; scape 5–32 cm long; umbels 2–7-flowered, the pedicels 0.5–4.5(5) cm long; calyx tube 1.5–2.5 mm long, the lobes 2.5–3.5 mm long; corolla magenta to lavender, the lobes 8–15 mm long; staminal filaments less than 1 mm long, distinct or narrowly connate; anthers 4–5 mm long, the connective black, smooth; stigma not conspicuously enlarged; capsule 7–12 mm long, cylindrical.

Moist tundra, heathlands, or woods; mostly in the interior, but also along the coasts in western and northern Alaska, and broadly distributed in the Yukon and adjacent Mackenzie; south to British Columbia; Asia.

Dodecatheon frigidum Cham. & Schlecht.
(× 0.6).

Dodecatheon jeffreyi Van Houtte
Jeffrey Shooting-star

Plants 1.8–6 dm tall, the roots pale; leaves 4–30(50) cm long, 1–5 cm broad, irregu-

larly crenate to entire; scapes 15–69 cm long; umbels 3–15-flowered, the pedicels 0.6–9 cm long; calyx tube 2–5 mm long, the lobes 5–10 mm long; corolla magenta to lavender or white, the lobes 10–25 mm long; staminal filaments short (to 1.5 mm long) or obsolete, distinct or narrowly connate; anthers 7–10 mm long, the connective maroon to black (rarely yellow), rough; stigma conspicuously enlarged; capsule 7–12 mm long, ovoid.

Moist meadows; in coastal, south-central and southeastern Alaska; south to California, Idaho, and Montana (*D. viviparum* Greene). Our specimens belong to ssp. *jeffreyi*.

Dodecatheon pulchellum (Raf.) Merrill
Pretty Shooting-star

Plants 1.2–6 dm tall, the roots pale; leaves 3–30 cm long, 1–6 cm broad, usually entire; scapes 7–50 cm long; umbels 3–20-flowered; pedicels 1–7 cm long; calyx tube 2–3.5 mm long, the lobes 2.5–6 mm long; corolla magenta to lavender (white), 9–21 mm long; staminal filaments 0.5–3.5 mm long, united into a distinctive tube; anthers 3–8 mm long, the connective maroon to black (rarely yellow), smooth; stigma not conspicuously enlarged; capsules 7–17 mm long, cylindrical to ovoid.

Moist meadows; in coastal southern and southeastern Alaska, and less commonly in moist situations in the interior of Alaska and Yukon; south to Mexico, and disjunctly in the eastern United States (*D. macrocarpum* authors, not [Gray] Kunth; *D. pauciflorum* authors not [Durand] Greene; *D. superbum* Pennell & Stair). There is much confusion regarding the proper epithet for this distinctively beautiful shooting-star. The problem is nomenclatural and not easily soluble. The coastal phase of our material is assignable to var. *alaskanum* (Hultén) B. Boi. (*D. pulchellum* ssp. *alaskanum* [Hultén] Hultén; *D. pulchellum* ssp. *superbum* [Pennell & Stair] Hultén). Specimens from the interior tend to have shorter anthers and represent var. *pulchellum* (*P. pulchellum* ssp. *pauciflorum* [Greene] Hultén).

DOUGLASIA Lindl. Nom. Cons.

Perennial, caespitose, cushion plants, with successively dichotomously branched stems from a taproot; leaves in imbricate rosettes, persistent, alternate; herbage pubescent with forked or simple hairs; flowers solitary (rarely 2), borne on peduncles which arise from the apex of the terminal rosettes; calyx 5-lobed, the tube nearly equal to the lobes; corolla funnelform, with 5 crests in the throat, 5-lobed; stamens 5, the filaments adnate to the corolla tube; capsule subglobose.

Constance, L. 1938. A revision of the genus *Douglasia* Am. Midl. Nat. 19:249–59.

1a. Leaves pubescent on the dorsal (lower) surface, not distinctly ciliate on the margin or ciliate with forked hairs; plants of high mountains in interior Alaska and southern Yukon. *D. gormanii*
1b. Leaves glabrous on the dorsal (lower) surface, ciliate with simple hairs on the margin; plants of western, northern, and northeastern Alaska and northern Yukon, and less commonly in the interior. (2)

2a. Leaves ascending, glabrous above and below; plants of south-central and east-central Alaska and northern Yukon. *D. arctica*
2b. Leaves spreading, glabrous below, hairy above; plants of the Brooks Range, northwest coast, and Seward Peninsula and in northern Yukon. *D. ochotensis*

Douglasia arctica Hook.

Plants depressed-caespitose or in rounded hemispheric clumps to 2 dm broad, 2-5(10) cm tall; leaves closely imbricated, ascending 4–8 mm long, 1–2 mm wide, narrowly oblanceolate, glabrous above and below, ciliate; peduncles pubescent with stellate, forked, or simple hairs, or merely scabrous, solitary from the terminal rosette, mostly 5–80 mm long; calyx cup shaped, the tube 2–3(4) mm long, the lanceolate to ovate lobes 1–3 mm long; flowers solitary, rose pink; corolla tube 4.5–6 mm long, the lobes 2.5–6 mm long; capsules 5–6-valved, shorter than or subequal to the calyx.

Arctic and alpine tundra; in south-central and east-central Alaska and northern Yukon; east to Mackenzie; endemic.

Douglasia gormanii Constance

Plants caespitose, 2–5 cm tall; leaves closely imbricate, 4–10 mm long, 1–2 mm wide, oblanceolate, pubescent on the lower surface and margin with stellate hairs; peduncles stellate-pubescent (sometimes sparsely so), solitary from the terminal rosettes, 1–35 mm long; calyx cup shaped, the tube 2–3 mm long, the ovate to triangular lobes 1.5–2 mm long; flowers solitary, rose pink; corolla tube 3–4 mm long, the lobes 1–2 mm long.

Mountain slopes, in alpine tundra; from west-central to east-central Alaska and

Douglasia arctica Hook. (× 0.75).

southern Yukon; endemic. This entity may not be sufficiently distinct from *D. arctica*. Much more material is needed.

Douglasia ochotensis (Willd.) Hultén

Plants mat-forming perennials with a caudex; leaves closely imbricate, 1–7 mm long, 1–2 mm wide, oblanceolate, glabrous below, pubescent above with simple hairs, ciliate; peduncles pubescent with simple hairs and often glandular, solitary from the terminal rosettes, 1–15 mm long; flowers solitary, closely subtended by a single bract, or the bract lacking; calyx 2–4 mm long, the teeth about equal to the tube; corolla pink (rarely white), the tube subequal to the calyx, the lobes (2.5)3–6 mm long; capsules 2.5–3 mm long.

Arctic and alpine tundra; from the Seward Peninsula northward to the northwest coast and eastward along the Brooks Range to northeastern Alaska and northern Yukon; eastward to the Mackenzie; Asia (*Androsace ochotensis* Willd. ex Roem. & Schult.).

GLAUX L.

Plants succulent perennial herbs from short rhizomes with fibrous or tuberous roots; leaves opposite below, subopposite or alternate above, entire, sessile; herbage glabrous; flowers solitary, sessile or subsessile in the axils near the middle of the stem; calyx cup shaped, the 5 petaloid lobes equaling or longer than the tube; corolla lacking; stamens 5, alternate with the calyx lobes; capsule subglobose, few-seeded.

Glaux maritima L.
Sea Milkwort

Plants 3–25(30) cm tall; leaves 3–20(25) mm long, oval to narrowly oblong, jointed to the stem; calyx 3–5 mm long, the lobes white or pinkish; stamens subequal to the calyx lobes, inserted at the base of the ovary; capsules 2–3 mm long.

Saline soils, along beaches, in seeps, and in meadows; in coastal south-central and southeastern Alaska and in southern Yukon; broadly distributed in North America; circumboreal (*G. maritima* var. *obtusifolia* Fern.).

LYSIMACHIA L.

Perennial, rhizomatous herbs; leaves opposite or whorled, large, minutely spotted with red, sessile or subsessile; flowers few to many, pedicellate, borne in axillary racemes near the middle of the stem; calyx usually with 5 more or less united sepals; corolla usually with 5 petals which are united at the base; stamens usually 5, attached at the base of the ovary, exserted; capsule subglobose.

Ray, J. D. Jr. 1956. The genus *Lysimachia* in the New World. Ill. Bio. Monogra. 24(3–4):1–160.

Lysimachia thyrsiflora L.
Tufted Loosestrife

Plants 2–8 dm tall; stems erect, simple; leaves scalelike below, enlarged above, 3–16 cm long, 0.5–6 cm broad, lanceolate, elliptic, or oblanceolate; flowers in dense pedunculate racemes; pedicels 1–4 mm long; calyx glandular-spotted, 5(7)-lobed, the lobes 1.8–3.5 mm long; corolla yellow, purple-spotted, 5(7)-lobed, the lobes 3–7 mm long; ovary dark, glandular; style 4–6 mm long; capsule about 2.5 mm broad, few-seeded.

Moist sites, in swamps, along streams, lake shores, and ponds; in coastal southern Alaska from the Aleutians to the Panhandle, and also in scattered sites in interior southwestern to east-central Alaska and southern Yukon; widespread in North America; Eurasia.

PRIMULA L.

Perennial, scapose herbs; leaves all basal, simple, not conspicuously imbricate nor persistent; herbage glabrous, glandular, or mealy; flowers 1–several, in involucrate umbels; calyx tubular, 5-lobed; corolla tubular, not conspicuously contracted, the crests absent or reduced, 5-lobed, the lobes usually emarginate; stamens 5, the filaments adnate to the corolla; capsule cylindrical.

Fernald, M. L. 1928. *Primula* Section Farinosae in America. Rhodora 30:59–77, 85–104.

Porsild, A. E. 1965. The North American races of *Primula tschuktschorum* Kjellm. of the sect. Novales. pp. 84–90. *In*: Some new or critical vascular plants of Alaska and Yukon. Can. Field-Nat. 79:79–90.

Smith, W. W. and H. R. Fletcher. 1942. The genus *Primula*: Section Nivales. Trans. Roy. Soc. Edinburgh 60(2):536–627.

————. 1943. The genus *Primula*: Section Farinosae. Trans. Roy. Soc. Edinburgh 61(1):1–69.

————. 1948. The genus *Primula*: Sections Cuneifolia, Floribundae, Parryi, and Auricula. Trans. Roy. Soc. Edinburgh 61 (3):631–86.

1a. Corolla lobes rounded apically (emarginate in var. *beringensis*); leaves mostly 3–20 cm long, the blade tapering to the base of the leaf, not distinctly petiolate. *P. tschuktschorum*

1b. Corolla lobes emarginate apically; leaves mostly less than 3 cm long, often conspicuously petiolate. (2)

2a. Leaves with 5–11 conspicuous apical teeth; flowers 10-20 mm broad; scapes often less than 3 cm long (much longer in some). *P. cuneifolia*

2b. Leaves various, but if apically 5–11-toothed, then the flowers 10 mm broad or less. (3)

3a. Bracts subtending umbels oblong to oblanceolate, rounded or obtuse (rarely acute) apically. *P. sibirica*

3b. Bracts subtending umbels lanceolate to lance-subulate, acute or acuminate apically. (4)

4a. Corollas 10–20 mm broad. (5)

4b. Corollas 5–10 mm broad (sometimes broader in *P. matsumurae*). (6)

5a. Bracts subtending umbels produced below the point of insertion into a saclike pouch; leaves mostly with conspicuous petioles. *P. borealis*

5b. Bracts slightly or not at all saclike below the point of insertion; leaves merely cuneate at the base, or with a broad subpetiolar base. *P. mistassinica*

6a. Leaves entire, undulate, or slightly dentate, with slender petioles; mature capsules 2–3 times longer than the calyx, 7–13 mm long. *P. egaliksensis*

6b. Leaves often dentate, or at least crenate, the blade often cuneate almost to the base of the leaf; mature capsule twice as long as the calyx or less, 2–5 mm long. (7)

7a. Leaves mostly distinctly toothed; corollas white. *P. tenuis*

7b. Leaves not, or indistinctly and irregularly, toothed; corollas mostly pink to violet (seldom white). (8)

8a. Leaves green beneath, seldom slightly mealy; calyx not mealy, 3.8–6 mm long. *P. stricta*

8b. Leaves mealy below; calyx mealy, 4–6 mm long or longer. (9)

9a. Leaves densely yellowish-mealy below; calyx 4–6 mm long; plants from the region of the Bering Sea. *P. matsumurae*

9b. Leaves usually whitish- to (less commonly) yellowish-mealy below; calyx (5.5)6–10 mm long; plants from interior eastern Alaska and southern Yukon. *P. incana*

Primula borealis Duby
Northern Primrose

Plants (1.5)3–17 cm tall; leaves 0.5–5 cm long, 0.1–1 cm broad, not mealy, the petioles mostly longer than the blades, the blades cuneate-obovate to spatulate; bracts subtending umbels lance-subulate, 2–6 mm long, somewhat saccate at the base; umbels 1–12-flowered; pedicels 2–25 mm long; calyx 3–6 mm long, the teeth equaling, to much shorter than, the tube; corolla pink to violet, the tube equaling, or somewhat longer than, the calyx, 10–20 mm broad; capsule cylindrical, somewhat exserted from the calyx; seeds smooth.

Open arctic or alpine tundra, meadows, or heathlands; in near coastal to coastal northern and western Alaska and northern Yukon; east to Mackenzie; Asia (*P. parvifolia* Duby).

Primula cuneifolia Ledeb.
Wedge-leaf Primrose

Plants 1.5–20 cm tall; leaves 0.5–8 cm long, 0.5–2 cm broad, oblanceolate to obovate, cuneate, tapering to the petiole base, with 5–11 apical teeth; bracts lance-subulate, not saccate at the base; umbels 1–9-flowered; pedicels 1–25 mm long; calyx 3–6 mm long, the teeth longer than the tube; corolla pink to rose or white, the tube slightly longer than the calyx, 10–20(24) mm broad; capsule subglobose, shorter than the calyx; seeds angled.

Arctic and alpine tundra or heathlands; from Kotzebue Sound southward to the Aleutians and eastward to the Panhandle, and less commonly some distance in the interior; Asia. Two completely intergrading varieties have been recognized in this species. The extreme forms are strikingly different and seem to warrant some taxonomic recognition.

1a. Plants low, seldom over 6 cm tall; leaves not conspicuously petiolate; distribution broad (*P. saxifragifolia* Lehm.; *P. cuneifolia* ssp. *saxifragifolia* [Lehm.] Smith & Forrest). *P. cuneifolia* var. *saxifragifolia* (Lehm.) Pax ex Engler
1b. Plants tall, often over 10 cm; leaves conspicuously petiolate; distribution mostly insular. *P. cuneifolia* var. *cuneifolia*

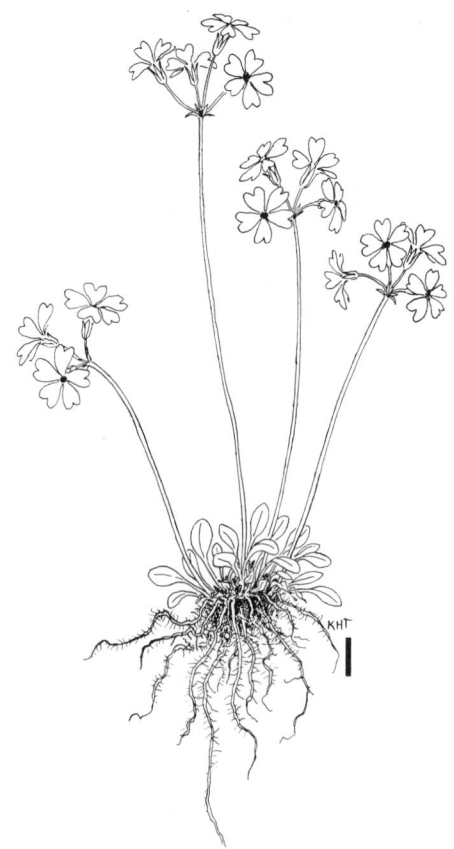

Primula borealis Duby (× 0.5).

Primula egaliksensis Wormskj. ex Hornem.
Greenland Primrose

Plants 1–27 cm tall; leaves 0.5–5 cm long, 0.2–1.5 cm wide, ovate, oblong, obovate, or spatulate, entire or undulate, the petiole slender; bracts lanceolate to lance-subulate, 2–7 mm long, saccate at the base; umbels 1–9-flowered; pedicels 2–30 mm long; calyx 3.5–6 mm long, the teeth shorter than the tube; corolla white or violet, the tube longer than the calyx, 5–9 mm broad, the lobes shorter than the tube; capsule at maturity 2–3 times longer than the calyx; seeds smooth.

Moist soils, stream banks, marshes, and rock outcrops; disjunct at widely separated locations almost throughout Alaska and the Yukon; eastward to Labrador and Greenland and south to British Columbia.

Primula incana M. E. Jones
Silvery Primrose

Plants 5–40 cm tall; leaves 1.5–8 cm long, 0.5–2 cm broad, oblanceolate to spatulate, shallowly denticulate, usually mealy below, the petioles broad; bracts lanceolate to linear, 3–10 mm long, swollen at the base; umbels 2–15-flowered, the pedicels 2–25 mm long; calyx (5.5)6–10 mm long, mealy, the teeth shorter than the tube; corolla lilac, the tube somewhat longer than the calyx, 6–10 mm broad, the lobes shorter than the tube; capsule only slightly longer than the calyx, the seeds roughened.

Moist sites in meadows and saline pans; in interior eastern Alaska and southern Yukon; southward to Utah and Colorado and eastward to Hudson Bay.

Primula matsumurae Petitm.

Plants 5–15 cm tall; leaves 1.5–10 cm long and 0.3–2 cm broad, ovate or obovate to spatulate, obscurely denticulate to crenate, usually yellowish-mealy below, the petioles broad; bracts lance-subulate, 3–10 mm long, swollen at the base; umbels (1)2–12-flowered, the pedicels 3–30 mm long; calyx 4–6 mm long, yellowish-mealy, the teeth about as long as the tube; corolla violet, the tube slightly exceeding the calyx, 10–12 mm broad, the lobes about equal to the tube; capsule about twice longer than the calyx.

Moist sites; in southwestern Alaska adjacent to the Bering Sea and on Nunivak Island; Asia (*P. ajanensis* E. Busch). This is a poorly known entity. Further material is necessary for a proper evaluation of its relationships. It is considered by some to be best treated as *P. modesta* Biss. & Moore var. *matsumurae* (Petitm.) Takeda.

Primula mistassinica Michx.
Mistassini Primrose

Plants 3–25 cm tall; leaves 0.5–7 cm long,

0.2–1.6 cm broad, oblanceolate to spatulate, often dentate, usually not mealy, but rarely so, the petioles broad; bracts linear-subulate, 2–6 mm long, usually not saccate at the base; umbels 1–10-flowered; pedicels 2–35 mm long; calyx 3–6 mm long, usually not mealy, the lobes about as long as the tube; corolla pink to violet (rarely white), the tube longer than the calyx, (8)10–20 mm broad, the lobes equaling or longer than the tube; capsules 1–2 times longer than the calyx, the seeds smooth or nearly so.

Moist sites; in central eastern Alaska (?) and western and southeastern Yukon, where evidently rare; eastward to Newfoundland and south to British Columbia, Alberta, and New Hampshire.

Primula sibirica Jacq.
Siberian Primrose

Plants 4–28 cm tall; leaves 1–7 cm long, 0.3–1.3 cm wide, orbicular to ovate, elliptic or oblanceolate, entire to irregularly dentate, the blade usually shorter than the long, narrow petiole, not mealy; bracts oblong to obovate, 2–11 mm long, obtuse, rounded, or rarely acute apically, conspicuously saccate basally; umbels 1–8-flowered; pedicels 3–50 mm long; calyx 5–8 mm long, the teeth shorter than the tube, not mealy; corolla pink purple (rarely white), the tube longer than the calyx, 10–18 mm broad, the lobes shorter or longer than the tube; capsules from slightly exserted to twice as long as the calyx.

Moist sites, in tundra, heath, or woods; disjunct from the northwestern arctic coast southward to the Yukon River Delta, and in south-central and east-central Alaska and southern Yukon; Eurasia.

Primula stricta Hornem.

Plants 1.5–30 cm tall; leaves 0.5–4 cm long, 0.2–1.5 cm broad, oblanceolate, entire to irregularly dentate, not or only slightly mealy below, broadly petioled; bracts lance-subulate, 3–8 mm long, saccate or swollen at the base; umbel 2–8-flowered; pedicels 3–20 mm long or more; calyx 3.8–6 mm long, the lobes about half as long as the tube, not mealy; corolla pink purple, the tube longer than the calyx, 5–8 mm broad, the lobes equaling or shorter than the tube; capsules only slightly longer than the calyx, the seeds angled.

Moist sites; in western, northern, and southeastern Alaska and southern Yukon; eastward to Labrador, Greenland, Iceland, and Europe. This entity is poorly understood in Alaska; further work is indicated.

Primula tenuis Small
Slender Primrose

Plants 3–15 cm tall; leaves 0.4–3 cm long, 0.1–0.7 cm broad, obovate to spatulate, dentate, the upper leaves at least with narrow petioles, not mealy; bracts lance- to linear-subulate, 2–5 mm long, not saccate basally; umbels 1–9-flowered; pedicels 3–20 mm long; calyx 3–5 mm long, the lobes about equal to the tube; corolla white, the tube subequal to the calyx, (6)7–10 mm broad, the lobes about as long as the tube; capsule about twice as long as the calyx tube, the seeds smooth.

Moist meadows and heathlands; from the Seward Peninsula southward to the Yukon River Delta; Asia (*P. parvifolia* authors, not Duby).

Primula tschuktschorum Kjellm.
Chukch Primrose

Plants 3–25(40) cm tall; leaves 1.5–18 cm long, 0.5–3 cm broad, oblanceolate to linear, entire or dentate, the petioles obsolete or broadly winged, mealy or not; bracts lanceolate, 4–10 mm long, acute to rounded apically, not saccate basally (but sometimes swollen); umbels 1–15-flowered; pedicels 3–40 mm long; calyx 3.5–9 mm long, the teeth longer than the tube; corolla pink purple to rose, the tube longer than the calyx, 12–20 mm broad, the lobes rounded, or less commonly emarginate or

cleft; capsules slightly longer than the calyx to twice as long, the seeds rough.

Moist meadows, marshes, and heathlands; in the Aleutian Islands, islands of the Bering Sea, coastal western and southern Alaska, and less commonly disjunct in interior Alaska and western to northern Yukon; Asia (*P. nivalis* authors, not Pallas). This most beautiful and largest of our primulas appears to be the most variable. Various proposals have been made to segregate the materials into infraspecific taxa (e.g., Porsild, 1965; Fernald, 1928). In specimens from the islands of the Bering Sea, the leaves tend to be slender and entire or nearly so. At least some of the leaves on plants of the mainland tend to be dentate, often conspicuously so. Flowers vary in number from 1–several. The number appears to be subject to ecological control and not related especially to vegetative features. Corolla lobes are usually rounded, but in some plants from the Bering Sea islands, the lobes are emarginate to cleft. There is apparent intergradation of each of the types into the others.

1a. Leaves dentate, at least some, narrowly to broadly oblanceolate; plants mostly from the continental regions of Alaska and Yukon (*P. eximia* Greene; *P. macounii* Greene; *P. tschuktschorum* ssp. *eximia* [Greene] Porsild; *P. tschuktschorum* ssp. *cairnesiana* Porsild). *P. tschuktschorum* var. *arctica* (Koidz.) Fern.

1b. Leaves usually entire or nearly so, linear to narrowly oblanceolate; plants mostly insular, less commonly in mainland Alaska (*P. tschuktschorum* var. *beringensis* Porsild). *P. tschuktschorum* var. *tschuktschorum*

TRIENTALIS L.

Perennial herbs from short rhizomes and tubers; stems erect, simple; leaves simple, entire, alternate below (and reduced in size), crowded or whorled above; flowers solitary, axillary, on slender pedicels; calyx with 5–9 sepals united at the base, the lobes linear to linear-lanceolate; corolla with 5–9 petals united at the base, showy; stamens 5–9, the filaments connate at the base, inserted at the base of the ovary; capsules subglobose to depressed globose.

1a. Plants with small foliage leaves below the terminal cluster; flowers white (often tinged with pink); our common starflower. *T. europaea*

1b. Plants with leaves below the terminal cluster reduced to small, scaly bracts; flowers pinkish; reported only from the Yukon, evidently rare. *T. latifolia*

Trientalis europaea L.
Arctic Starflower

Plants mostly 5–35 cm tall, arising from slender rhizomes and inconspicuously thickened, horizontal tubers; leaves in the terminal whorl, 1–8 cm long, 0.5–3 cm broad, the blades broadly elliptic, oblanceolate or obovate, subsessile to petiolate; corolla white, 12–25 mm broad.

Moist woods, thickets, meadows, or tundra; in most of Alaska and Yukon south of the 67th parallel. Our specimens have been variously interpreted. They have been treated as belonging to *T. arctica* Fisch ex Hook., or segregated into ssp. *arctica* (Fisch.) Hultén, and ssp. *europaea*. A comparison of European and American specimens has not demonstrated consistent objective differences. Thus a conservative treatment is suggested.

Trientalis latifolia Hook.
Broad-leaf Starflower

Plants mostly 10–25 cm tall, from distinctly thickened, vertical tubers; foliage leaves all in a terminal cluster, 3–10 cm long, 1.5–4 cm broad, ovate to elliptic or obovate, short-petiolate to subsessile; corolla pink, 8–17 mm broad.

This entity has been reported from a single location in the Yukon (Keno). It is often

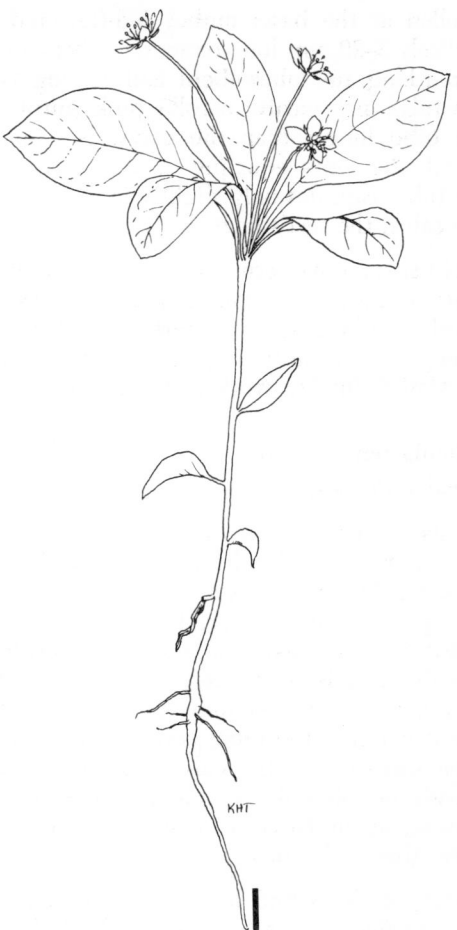

Trientalis europaea L. (× 0.6).

treated at infraspecific level within *T. europaea* (as var. *latifolia* [Hook.] Torr., or as *T. borealis* Raf. ssp. *latifolia* [Hook.] Hultén).

PYROLACEAE
Wintergreen Family

Suffrutescent or herbaceous perennials; leaves simple, alternate, opposite, or appearing whorled, evergreen or much reduced and lacking chlorophyll; flowers usually perfect, regular or irregular; calyx with 4–5 more or less distinct sepals; corolla with 4–5 more or less distinct petals (united in *Pterospora*); stamens twice as many as the petals, the anthers pendulous, opening by apparently terminal pores or slits, or the anthers erect, sometimes bearing 2 awn-like appendages; pistils 1; ovary superior, 4–5-loculed; styles 1; fruit a capsule.

1a. Plants lacking chlorophyll; leaves reduced and scalelike, purplish, yellowish, reddish, white, or pink, not green. (2)

1b. Plants with chlorophyll; leaves not reduced to scales, evergreen. (4)

2a. Flowers solitary at the apex of the stem; plants usually white or pink when fresh (drying black). *Monotropa*

2b. Flowers few to many, in contracted, to greatly elongate, racemes; plants reddish, brownish, purplish, or yellowish when fresh (drying black). (3)

3a. Corolla of separate petals; anthers lacking awns; flowers in few-flowered racemes. *Hypopitys*

3b. Corolla of united petals; anthers 2-awned; flowers in several- to many-flowered racemes; reported from the vicinity of Paxon, Alaska, but the report requires confirmation; the species is distributed widely in North America. *Pterospora andromeda* Nutt.

4a. Flowers solitary, the petals rotate or nearly so. *Moneses*

4b. Flowers few to several, the petals concave. (5)

5a. Stems leafy, though short, the leaves apparently whorled; flowers corymbose; staminal filaments dilated near the base; styles very short or lacking. *Chimaphila*

5b. Stems leafy at the base only; flowers in elongated racemes; filaments not especially dilated at the base; styles in most species over 2 mm long. *Pyrola*

CHIMAPHILA Pursh

Rhizomatous subshrubs; leaves with chloro-phyll, leathery, persistent, in one or more whorls along the stem; flowers one to few, borne in terminal corymbs, the peduncle leafless; flowers usually 5, persistent; petals usually 5, distinct, spreading, concave; stamens 10, the filaments conspicuously expanded near the base, the anthers awnless, opening by means of apparently terminal pores; ovary superior, 5-loculed, the stigma sessile or subsessile; fruit a loculicidal capsule.

Chimaphila umbellata (L.) Bart.
Pipsissewa, Prince's Pine

Plants 1–3 dm tall; leaves 2–7 cm long, 0.5–2.5 cm broad, oblanceolate to elliptic, abruptly acute apically, cuneate-acute basally, sharply serrate, usually 3–5 per whorl, the petioles 3–8 cm long; peduncles 4–10 cm long; flowers mostly 3–9; petals 4–7 mm long, pink; capsules 5–7 mm broad.

Coastal woodlands; in southeastern Alaska; southward to California and Mexico, east to New Mexico and Colorado, and in the eastern United States; Eurasia (*C. occidentalis* Rydb.; *C. umbellata* ssp. *occidentalis* [Rydb.] Hultén). Our materials are referable to var. *occidentalis* (Rydb.) Blake.

HYPOPITYS Hill

Yellowish or reddish to pink, succulent, herbaceous saprophytes; leaves lacking chlorophyll, scalelike, alternate along the length of the stem; flowers few to several, in terminal racemes; sepals usually 4 (rarely 3 or 5); petals usually 4 (rarely 3 or 5), saclike at the base; stamens 6–10 (twice as many as the petals), the filaments hairy, linear, the anthers erect, unawned, opening by a slit; ovary superior, 4–5-loculed, the stigma elevated on a hairy, elongated style; fruit a loculicidal capsule.

Hypopitys monotropa Crantz
Pinesap

Plants (0.5)1–3 dm tall; leaves entire or somewhat fringed; racemes usually nodding in flower and becoming erect in fruit; sepals 4–8 mm long, ciliate; petals 10–14 mm long, erect, pubescent; stamens shorter than the petals; capsule 5–8 mm broad.

Coastal woodlands; in southeastern Alaska; southward to California and eastward to the Atlantic; Eurasia (*H. latisquama* Rydb.; *Monotropa latisquama* [Rydb.] Hultén; *M. lanuginosa* Michx.; *M. hypopitys* ssp. *lanuginosa* [Michx.] Hara).

MONESES Salisb.
Rhizomatous herbs; leaves with chorophyll, leathery, persistent, mainly basal, but sometimes opposite or in whorls of 3 on a short, vegetative stem; flowers solitary, nodding, borne on a long peduncle; sepals usually 5, persistent; petals usually 5, distinct, spreading; stamens usually 10, the filaments tapering to the apex, the anthers awnless, nodding, opening by means of apparently terminal pores; ovary superior, 5-loculed, the stigma borne on an elongate, glabrous style; fruit a loculicidal capsule.

Moneses uniflora L.
Single Delight, Wax-flower

Plants 4–17 cm tall; leaves (including petioles) 0.8–4 cm long, 0.6–2 cm broad, serrate to crenate-serrate; peduncles 3–15 cm long, usually with 1–2 bracts along its length; flowers 1.3–2.5 cm broad, white to cream; sepals 1.5–2.5 mm long, ciliate; petals 7–11 mm long, spreading; style 2–4 mm long; capsule 5–8 mm broad.

Moist woods and thickets; almost throughout Alaska (except for the southwestern portion) and much of the Yukon south of the 68th parallel; widely distributed in North America; Eurasia (*M. reticulata* Nutt.; *M. uniflora* var. *reticulata* [Nutt.] Blake).

MONOTROPA L.
White to pinkish, succulent, herbaceous saprophytes; leaves lacking chlorophyll, scalelike, alternate along the length of the stem; flowers solitary, terminal, at first nodding, finally erect; sepals (in reality probably lacking) of 1–4 bractlike scales; petals usually 5, saclike at the base; stamens usually 10, the filaments hairy, linear, the anthers unawned, erect, opening by a slit; ovary superior, 5-loculed, the stigma elevated on a short, glabrous style; fruit a loculicidal capsule.

Copeland, H. F. 1941. Further studies on Monotropoideae. Madroño 6:104–8.

Monotropa uniflora L.
Indian-pipe

Plants 0.5–2.5 dm tall; leaves entire or somewhat erose; petals 15–20 mm long, erect, ciliate and pubescent within, waxy white (drying black); stamens shorter than the petals; capsules 5–7 mm broad.

Coastal woodlands; in extreme southeastern Alaska; southward to California and east to the Atlantic; Asia.

PYROLA L.
Rhizomatous herbs; leaves with chlorophyll, leathery, persistent, all basal, or rarely lacking and the plants then partially or completely saprophytic; flowers regular to irregular, in terminal racemes; sepals 5, united at the base; petals 5, distinct, usually concave, deciduous; stamens 10, the filaments tapering to the apex, the anthers unawned, pendulous, opening by means of apparently terminal pores; ovary superior, 5-loculed, the stigma borne on an elongate, straight or curved style; fruit a loculicidal capsule.

Copeland, H. F. 1947. Observations on structure and classification of the Pyroleae. Madroño 9:65–102.

1a. Styles straight or nearly so; pores of anthers sessile; stigma usually much broader than the style. (2)
1b. Styles bent or curved; pores of anthers usually borne on short tubes (sessile or nearly so in *P. grandiflora*); stigmas only slightly broader than the styles. (3)

2a. Styles 2 mm long or less, not (or seldom) exserted from the flower; flowers not secund; petals pinkish to cream. *P. minor*

2b. Styles over 2 mm long, exserted from the flower; flowers secund; petals greenish white. *P. secunda*

3a. Flowers pink to purplish; sepals longer than broad. *P. asarifolia*

3b. Flowers pale, greenish yellow, or creamy white to pinkish; sepals various. (4)

4a. Sepals broader than long; flowers greenish yellow. *P. virens*

4b. Sepals longer than broad; flowers creamy white to pinkish. *P. grandiflora*

Pyrola asarifolia Michx.
Liverleaf Wintergreen

Plants 1.3–4 dm tall; leaves basal, the blades 2–8 cm long, oval, rotund, elliptic, or obovate, subcordate to rounded or somewhat acute basally, rounded to obtuse or emarginate at the apex, entire to serrulate, green above, usually purplish beneath; petioles 2–9 cm long; racemes mostly 8–20-flowered; pedicels 3–8 mm long; sepals 1.5–4 mm long; petals pink to purplish, 5–7 mm long; anthers pink, the pores on short tubes; style curved, with a flaring collar below the stigma.

Woods and thickets to alpine meadows; in most of Alaska and Yukon south of the 68th parallel; east to Newfoundland and south to California, New Mexico, South Dakota, and New England; Asia (*P. bracteata* Hook.; *P. incarnata* Fisch. ex DC.; *P. asarifolia* var. *incarnata* [Fisch.] Fern.; *P. rotundifolia* var. *purpurea* Bunge). Our materials belong to var. *purpurea* (Bunge) Fern.

Pyrola grandiflora Radius
Large-flower Wintergreen

Plants 0.6–2.5 dm tall; leaves basal, the blades 1–4.5 cm long, 1–5 cm broad, oval,

Pyrola grandiflora Radius (× 0.6).

ovate, elliptic, or rotund, subcordate, truncate, rounded or acute basally, obtuse to rounded apically, subentire to crenate or undulate, green above, often reddish beneath, the petioles 1–8 cm long; racemes 4–11-flowered; pedicels 2–8 mm long; sepals 2–4 mm long; petals creamy white suffused with pink, 7–12 mm long; anthers yellow to pink, the pores on very short tubes; style curved, with a collar below the stigma.

Arctic and alpine tundra, heathlands, and woods; almost throughout Alaska (except for the southwestern and southeastern portions) and nearly all of the Yukon; south to northern British Columbia and east to Greenland; circumpolar (*P. gormanii* Rydb.; *P. grandiflora* var. *gormanii* [Rydb.] Porsild; *P. canadensis* Andres; *P. grandiflora* var. *canadensis* [Andres] Porsild).

Pyrola minor L.
Lesser Wintergreen

Plants 0.6–2.4 dm tall; leaves basal, the blades 1–3 cm long, 1.4–2.5 cm broad, oval, elliptic or ovate, obtuse to rounded or subcordate basally, obtuse to rounded apically, crenate to subentire; petioles 1–3 cm long; racemes mostly 5–13-flowered; pedicels 2–3 mm long; sepals 1–1.5 mm long; petals pale pink to cream, 3.5–4.5 mm long; anthers with the pores sessile; style straight, with a more or less distinctive collar below the stigma.

Meadows, heathlands, or woods; in much of Alaska and Yukon south of the 66th parallel; eastward to Greenland and south to California and Colorado; circumboreal.

Pyrola secunda L.
One-sided Wintergreen

Plants 0.6–1.8(2.1) dm tall; leaves basal (though sometimes with a naked stem below the leaves), the blades 1.5–4 cm long, 1–2.5(3) cm broad, ovate, oval, elliptic, orbicular, obtuse to rounded basally, obtuse to rounded apically, crenate-serrate, usually green on both surfaces but paler beneath, the petioles 1–2 cm long; racemes mostly 4–15-flowered, the flowers secund; pedicels 2–3 mm long; sepals 0.5–1(1.5) mm long; petals greenish white, 4–6 mm long; anthers with pores sessile; style straight, exserted from the flower, lacking a collar.

Arctic and alpine tundra, heathlands, and woods; in most of Alaska and Yukon, lacking in the Aleutians; broadly distributed in North America; Eurasia. Our plants have been segregated into two minor phases, which may represent nothing more than ecological variants.

1a. Leaves elliptic to ovate, acute to obtuse; racemes mostly 8–15-flowered; plants of woodlands at low elevations, mostly in southern Alaska and southern Yukon. *P. secunda* var. *secunda*

1b. Leaves oval, obtuse to rounded apically; racemes mostly 4–10-flowered; plants of woods, heath, or tundra, often at middle elevations and at higher latitudes (*P. secunda* ssp. *obtusata* [Turcz.] Hultén). *P. secunda* var. *obtusata* Turcz.

Pyrola virens Schweigg. ex Schweigg. & Koerte

Plants 0.9–2.5 dm tall; leaves basal, the blades 0.6–3.5 cm long and 0.5–3 cm broad, broadly elliptic, oval or obovate, obtuse to rounded basally, rounded apically, crenate to serrate, usually green on both surfaces, the petioles 0.8–6 cm long; racemes mostly 2–9-flowered; pedicels 3–8 mm long; sepals 0.5–1.5 mm long; petals greenish yellow, 5–7 mm long; anthers yellowish, the pores on elongate tubes; style curved, with a flaring collar below the stigma.

Woodlands; in southern Alaska, mostly east of the 168th meridian, south of the 64th parallel, and in southern Yukon; broadly distributed in North America; Eurasia (*P. chlorantha* Sw.).

RANUNCULACEAE
Buttercup Family

Plants herbaceous; leaves alternate, opposite, or basal, simple, deeply divided, or variously compound; flowers perfect or rarely imperfect, regular, or irregular, lacking a hypanthium; sepals 3–many, often petaloid; petals 3–many, or lacking; stamens many; pistils 1–many, superior; fruit a follicle, achene, or a berry.

1a. Flowers distinctly irregular, mostly dark blue to purple. (2)
1b. Flowers regular, seldom dark blue to purple. (3)

2a. Upper sepal spurred at the base; petals usually 4. *Delphinium*
2b. Upper sepal not spurred, but hooded at the apex; petals usually 2. *Aconitum*

3a. Petals conspicuously spurred. *Aquilegia*
3b. Petals not at all spurred. (4)

4a. Perianth segments all alike in color and texture (arbitrarily called sepals). (5)
4b. Perianth segments of 2 distinctive types (apparently sepals and petals), although the sepals may be caducous. (7)

5a. Leaves simple, the blades reniform to cordate. *Caltha*
5b. Leaves compound, or if simple, then deeply lobed or dissected. (6)

6a. Stem leaves opposite or whorled; flowers often showy. *Anemone*
6b. Stem leaves alternate or flowers inconspicuous. *Thalictrum*

7a. Flowers in terminal racemes; pistils 1; fruit a fleshy red or white berry. *Actaea*
7b. Flowers not in terminal racemes; pistils few to numerous; fruit an achene or follicle. (8)

8a. Leaves 1–3 times ternately parted or dissected; fruit of distinctly stipitate follicles. *Coptis*
8b. Leaves various, but usually not as above; fruit of sessile or subsessile follicles or achenes. (9)

9a. Petals reduced to small, linear organs, much smaller than the petaloid sepals; fruit a several-seeded follicle. *Trollius*
9b. Petals mostly as broad as, or broader than the sepals; fruit a 1-seeded achene. *Ranunculus*

ACONITUM L.

Perennial herbs from tuberous roots; stems simple, erect; leaves alternate, palmately divided; flowers perfect, irregular, large and showy, borne singly or in terminal racemes; sepals 5, petaloid, the upper one large, helmetlike, the lateral ones oval, broader than the lower two; petals 2, enclosed within the helmetlike hood, highly modified (with a slender, filamentous claw, a pendulous blade, and a saclike or curved spur), sometimes 3 additional, scalelike petals present; stamens numerous, the filaments flattened; pistils 3–5, distinct; fruit a follicle.

1a. Pubescence of inflorescence of 2 types, spreading and recurved; stems usually conspicuously contracted immediately below the ground level, mostly less than 5 mm in diameter; plants widespread. *A. delphinifolium*
1b. Pubescence of inflorescence entirely of recurved hair; stems usually not contracted at ground level, often over 5 mm in diameter; plants of the Aleutian Islands. *A. maximum*

Aconitum delphinifolium DC.
Monkshood

Plants 0.8–12.5 dm tall, the stems glabrous below, pubescent above with spreading and recurved hairs, usually contracted at or below the ground level, mostly less than 5 mm in diameter; leaves mostly cauline, 5–25 cm long, the blade 5-lobed, the sinuses lateral to the terminal lobe extending almost or quite to the base, the terminal lobe 1.5–13 cm long; sepals blue purple, greenish purple, yellowish, or rarely white; hood 12–22 mm high; only the upper two petals usually present, the spur straight or curved; follicles glabrous to sparsely hairy, 15–20 mm long.

Arctic and alpine tundra to heathlands and woods; throughout most of Alaska and Yukon; eastward to Mackenzie and south to British Columbia; Asia. This species is tremendously variable. There are several intergrading types which differ in such characteristics as plant height, leaf size and lobing, and in flower size. Some of these undoubtedly have a genetic basis, and some probably should be recognized in taxonomic rank, but the paucity of definitive criteria seems to indicate a conservative treatment and the necessity for further work. Reports of *A. columbianum* Nutt. ex T. & G. from Alaska probably belong here. The southern limits of *A. delphinifolium* are not known, but the species appears to be distinct from *A. columbianum* which has spreading hairy upper stems.

1a. Stems 1–12.5 dm tall; spur of petals curved; plants of broad distribution (*A. chamissonianum* Reichb.; *A. delphinifolium* ssp. *chamissonianum* [Reichb.] Hultén; *A. delphinifolium* var. *chamissonianum* (Reichb.) B. Boi.; *A. delphinifolium* var. *albiflorum* Porsild). *A. delphinifolium* var. *delphinifolium*

1b. Stems 0.8–2.5(5) dm tall; spur of petals short, not or only slightly curved; plants of extreme western and northwestern Alaska and islands of the Bering Sea, less commonly in interior Alaska and Yukon (*A. nivatum* A. Nels.; *A. paradoxum* Reichb.; *A. delphinifolium* ssp. *paradoxum* [Reichb.] Hultén). *A. delphinifolium* var. *paradoxum* (Reichb.) Welsh

Aconitum maximum Pallas ex DC.
Kamchatka Aconite

Plants 5–10.5 dm tall, the stems glabrous below, pubescent above with recurved hairs, usually not contracted at the ground level, mostly more than 5 mm in diameter; leaves mostly 10–20 cm long, the blade 5-lobed, the sinuses lateral to the apical lobe sometimes not reaching the base, the terminal lobe 5–10 cm long; sepals blue purple (rarely white); hood 20–25 mm high, only the upper two petals usually present, the spur curved; follicles glabrous to sparsely hairy, 14–18 mm long.

Beaches to heathlands and meadows; from the tip of the Alaska Peninsula westward through Aleutians; Asia.

ACTAEA L.

Perennial rhizomatous herbs, the stems simple or more usually branched, erect; leaves all cauline, alternate, twice ternately or twice ternate-pinnately compound; flowers perfect, regular, small, borne in terminal racemes; sepals 3–5, caducous; petals 5–10 (rarely lacking); stamens numerous; pistils 1; fruit a red or white berry.

Actaea rubra (Ait.) Willd.
Baneberry

Plants 3–10 dm tall; stems glabrate to glabrous; leaves all cauline, the segments broad, ovate to lanceolate, oblong, or obovate, 2–10 cm long, sharply toothed and often lobed; sepals 2–3 mm long, whitish or purplish; petals white 2–3.5 mm long; stamens longer than the petals; berries 5–10 mm long.

Woods, stream banks, beaches, and open slopes; almost throughout Alaska and Yukon south of the 66th parallel; southward to California and east to the Atlantic (*A. arguta* Nutt. ex T. & G.; *A. spicata* var. *arguta* [Nutt.] Torr.; *A. spicata* var. *rubra* Ait.; *A. rubra* ssp. *arguta* [Nutt.] Hultén). White-fruited specimens belong to f. *neglecta* (Gillman) Robins. (*A. eburnea* Rydb.).

ANEMONE L.

Perennial herbs from rhizomes or caudices; stems with usually a single whorl of 3 leaves below the inflorescence; basal leaves long-petioled, the blades simple and palmately lobed or ternately compound and often deeply dissected; flowers perfect, regular, showy, borne 1 per peduncle, the peduncles solitary or 2-several in umbellate cymes; sepals 4–10, petaloid, variously colored; petals lacking; stamens numerous; pistils numerous; fruit an achene.

1a. Sepals 2–5 cm long, blue to purplish (rarely white); styles plumose, 2–4 cm long in fruit. A. patens

1b. Sepals 0.5–2(2.5) cm long, white, yellow, purplish, or bicolored; styles not plumose, less than 1 cm long in fruit. (2)

2a. Ovaries and achenes glabrous. (3)

2b. Ovaries and achenes hairy, often densely so. (4)

3a. Flowers bright yellow; plants with elongate rhizomes; leaves simple, merely lobed or incised. A. richardsonii

3b. Flowers white to cream (often tinged purplish on the outer surface, sometimes drying yellowish); plants from caudices; leaves palmately compound. A. narcissiflora

4a. Leaflets of basal leaves merely crenate or lobed, or cleft to near the middle; flowers white, solitary, usually large. (5)

4b. Leaflets of basal leaves dissected into narrowly oblong to lanceolate lobes; flowers purplish, maroon, yellow, or white, large and showy or inconspicuous. (6)

5a. Cauline leaves cleft to near the middle; basal leaves usually several; plants widespread in Alaska and Yukon. A. parviflora

5b. Cauline leaves commonly crenate to shallowly lobed; basal leaves 1–few; plants reported from northern British Columbia. A. deltoidea

6a. Plants 3–25 cm tall; sepals (8)11–21 mm long; styles 1.5–3.5 mm long; terminal leaflet of basal leaves 0.9–

2.6 cm long; flowers bluish, purplish, or white. A. drummondii

6b. Plants (10)20–50 cm tall; sepals 6–12 mm long; styles 0.7–1.2 mm long; terminal leaflet of basal leaves 2.5–6 cm long; styles 0.7–1.2 mm long; flowers purple, reddish, yellow, or white (often colored differently on the two surfaces). A. multifida

Anemone deltoidea Hook.

Plants from elongate rhizomes, 1–3 dm tall, the herbage hairy to glabrous; basal leaves 2–15 cm long, 3-foliolate, the leaflets crenate to lobed or rarely incised to near the middle, the segments broad, the terminal leaflet 2–5 cm long; cauline leaves 2–7 cm long, in one whorl, glabrous to hairy; peduncles 1 per inflorescence, commonly 5–12 cm long; flowers solitary, large and showy; sepals white, 15–25 mm long, glabrous; cluster of achenes ovoid, about 1 cm long; achenes 2.5–4 mm long, hairy below the middle, the styles 1.5–3 mm long.

Open woods; reported from northern British Columbia; southward to Washington and California.

Anemone drummondii Wats.
Drummond Anemone

Plants from caudices, 0.3–2.5 dm tall, the herbage long-hairy to glabrate or glabrous; basal leaves 2–13 cm long, 3-foliolate, the leaflets dissected to below the middle into narrowly oblong or lanceolate lobes, the terminal leaflet 0.9–2.6 cm long; cauline leaves 0.9–3.2 cm long, in 1 whorl, silky-hairy to villous; peduncles solitary, 1–13 cm long; flowers solitary, showy; sepals bluish, purplish, or rarely white (often bicolored), (8)11–21 mm long, silky-hairy to villous on the outer surface; cluster of achenes subglobose, 8–17 mm long; achenes silky-hairy, 2.5–4 mm long, the styles 1.5–3.5 mm long.

Arctic and alpine tundra to heathlands; disjunct in western, northern, and interior Alaska and western Yukon; southward to

California, Montana, and Alberta. Apparently there are no major morphological differences between *A. drummondii* and *A. multiceps* (Greene) Standl. When a series of specimens are examined, the styles and staminal filaments show every possible combination of coloration, and the pubescence of the achenes becomes more pronounced as the fruit matures. This accounts for previous collections wherein the flowering specimens were designated as *A. multiceps* and fruiting specimens as *A. drummondii*.

Anemone multifida Poir. ex Lam
Cut-leaf Anemone

Plants from caudices, 1.5 dm tall, the herbage pubescent with long, spreading hairs; basal leaves 6–22 cm long, 3-foliolate, the leaflets dissected to below the middle into narrowly oblong or lanceolate lobes, the terminal leaflet 2.5–6 cm long; cauline leaves 2.5–6.5 cm long, in 1–2 whorls, silky-hairy to glabrate; peduncles 1–3, 5–22 cm long; flowers solitary, often showy; sepals purple, reddish, yellow, or white (often bicolored), 6–12 mm long, silky-hairy on the outer surface; cluster of achenes ovoid to subglobose, 1–1.6 cm long; achenes 3–4 mm long; silky-hairy, the styles 0.7–1.2 mm long.

Open sunny places in woods and meadows; in the eastern half of Alaska and southern Yukon; eastward to Newfoundland and south to California, New Mexico, and New York. Our material belongs to var. *multifida*.

Anemone narcissiflora L.

Plants from caudices, 0.5–6 dm tall, the herbage glabrous or more commonly spreading-hairy; basal leaves 5–32 cm long, 3-foliolate, the leaflets incised to below the middle, the terminal leaflet 1.5–8 cm long; cauline leaves 1.5–5.5 cm long, in 1 whorl, silky-hairy to glabrate; peduncles 1–5 per inflorescence, 1–15 cm long; flowers solitary, showy; sepals white to cream,

often bluish-tinged on the outer surface, 12–20 mm long, glabrous; achenes 6–9 mm long, glabrous, the styles 0.8–1 mm long.

Arctic and alpine tundra, heathlands, and woods; through most of Alaska and Yukon; eastward to Mackenzie and south to British Columbia; Asia.

1a. Plants robust (15)25–60 cm tall, conspicuously long-hairy; leaf blades reniform to orbicular in outline; Aleutian Islands (*A. narcissiflora* ssp. *villosissima* [DC.] Hultén). *A. narcissiflora* var. *villosissima* DC.

1b. Plants slender, mostly less than 30 cm tall, variously pubescent to glabrate; leaf blades oval to ovate (rarely reniform) in outline; broadly dis-

Anemone narcissiflora L. (× 0.3).

tributed (*A. narcissiflora* ssp. *alaskana* Hultén; *A. narcissiflora* ssp. *interior* Hultén; *A. sibirica* L.; *A. narcissiflora* ssp. *sibirica* [L.] Hultén; *A. narcissiflora* var. *uniflora* Eastw.). *A. narcissiflora* var. *monantha* Schlecht.

Anemone parviflora Michx.
Northern Anemone

Plants from elongate rhizomes, 0.4–3.2 dm tall, the herbage spreading-villous to glabrate or glabrous; basal leaves 1.5–17 cm long, 3-foliolate, the leaflets merely lobed or incised to near the middle, the segments broad, the terminal leaflet 0.5–3 cm long; cauline leaves 0.7–2.8 cm long, in 1 whorl, silky-hairy to glabrous; peduncles 1 per inflorescence, 2–26 cm long; flowers solitary, large and showy; sepals white to cream, often tinged bluish externally, 9–20 mm long, silky-hairy externally; cluster of achenes ovoid, 0.8–1.3 cm long; achenes 2–3 mm long, woolly-hairy, the styles 1.2–2 mm long.

Arctic and alpine tundra to heathlands and woods; through most of Alaska and Yukon; eastward to the Atlantic and south to Oregon, Idaho, and Colorado; Asia (*A. parviflora* var. *grandiflora* Ulbr.).

Anemone patens L.
Pasque-flower, Wild Crocus

Plants from caudices, 1–5.5 dm tall, the herbage more or less grayish-villous; basal leaves 8–35 cm long, 3-foliolate, the leaflets dissected into narrowly oblong or lanceolate segments, the terminal leaflet 2.5–6 cm long; cauline leaves 3–6 cm long, in 1 whorl, silky-hairy; peduncles 1 per inflorescence, 2.5–35 cm long; flowers solitary, large and showy; sepals 2–4.5 cm long, silky-hairy externally; achenes 4–6 mm long, silky-hairy, the elongate, plumed style 2–3.5 cm long.

Arctic and alpine tundra to heathlands and woods; in central, interior and northern Alaska, and eastward through most of the Yukon to the Mackenzie; southward to Washington, Utah, Texas, and east to Illinois (*Pulsatilla patens* [L.] Mill. ssp. *multifida* [Pritzel] Zamels). This is one of the very earliest plants to flower in our region (sometimes flowering again in late summer). Our plants are assignable to var. *multifida* Pritzel.

Anemone richardsonii Hook.
Yellow Anemone

Plants from elongate rhizomes, 0.5–3 dm tall, the herbage more or less villous; basal leaves 3–18 cm long, simple, the blade palmately lobed, 1–4.5 cm long; cauline leaves 1.1–4 cm long in a single whorl; peduncles 1 per inflorescence, 3–20 cm long; flowers solitary, showy; sepals yellow, 8–14(15) mm long, sometimes bluish tinged externally; achenes 3–4 mm long, glabrous, the styles 4–6 mm long, hooked at the apex.

Arctic and alpine tundra to heath and woods; throughout Alaska and the Yukon; south to British Columbia and east to Labrador; Asia.

AQUILEGIA L.

Perennial herbs from taproots; stems simple, erect; leaves alternate, or mainly basal, the cauline ones much reduced, bi- or ternately compound; flowers regular, showy, solitary, or 2–several in bracteate long-pedicellate terminal racemes; sepals 5, petaloid, not spurred; petals 5, the lower part forming a spur with a bulbous, glandular tip; stamens numerous, the inner ones often sterile (staminodia); pistils mostly 5, distinct; fruit a follicle.

Munz, P. A. 1946. *Aquilegia.* The cultivated and wild Columbines. Gentes Herb. 7:1–150.

Payson, E. B. 1918. The North American species of *Aquilegia.* Contr. U. S. Nat. Herb. 20:133–57.

1a. Spur of petal shorter than the blade, hooked; spurs and sepals bluish. *A. brevistyla*

1b. Spur of petal longer than the blade, straight or nearly so; spurs and sepals red, reddish, or rarely yellowish. *A. formosa*

Aquilegia brevistyla Hook.
Small-flower Columbine

Plants 2.5–7 dm tall; stems glabrous below, sparsely spreading-hairy above; leaves mainly basal, 5–20 cm long, (once) twice ternately compound, the blades glabrous to villous, often bluish tinged, paler and glaucous below; flowers usually 2–3(5), spreading to nodding; sepals 12–15 mm long, bluish; petals (including spur) 14–18 mm long, the hooked spur bluish, the blade white to cream, often tinged bluish externally; follicles usually 5, pubescent, 17–25 mm long, the styles less than 5 mm long.

Gravelly or shaly soil, in open woods, and along streams; in the southeastern quarter of continental Alaska and southern Yukon; eastward to the Mackenzie and south to Alberta, South Dakota, and Minnesota.

Aquilegia formosa Fisch. ex DC.
Western Columbine

Plants 3–10.5 dm tall; stems glabrous to the inflorescence; leaves mainly basal, 10–30 cm long, twice ternately compound, the blades glabrous to villous, green above, paler and glaucous beneath; flowers usually 2–4, nodding; sepals 14–27 mm long, red, reddish, or rarely yellowish; petals (including the spur) 15–20 mm long, the straight spur red, reddish, or rarely yellowish, the blade yellowish; follicles usually 5, pubescent, 15–25 mm long; styles more than 10 mm long.

Heathlands to open woods; in southern and southeastern Alaska, and in southern Yukon (from the vicinity of Dawson City southward); south to Baja California and Utah. This is a very handsome plant. It is easily grown from seed and should be utilized as a garden ornamental throughout southern Alaska. Numerous forms of exotic columbines are cultivated successfully in that region.

CALTHA L.

Perennial subrhizomatous or stoloniferous herbs with fibrous roots; stems prostrate to ascending or erect; leaves basal or alternate, the blades simple, reniform to cordate or oval-cordate; flowers perfect, regular, showy, solitary, or sometimes 2–3; sepals 5–12, petaloid; petals lacking; stamens numerous, rarely petaloid and resulting in "double flowers," pistils 5–several; fruit a stipitate follicle.

1a. Plants semiaquatic or aquatic, with 2–more cauline leaves; stems decumbent to creeping or sometimes erect to ascending; sepals yellow, or if white, then 6 mm long or less. (2)

1b. Plants of various habitats, but seldom semiaquatic; cauline leaves (0)1 (rarely 2); stems erect; sepals white, over 7 mm long. (3)

2a. Sepals yellow, 6–20 mm long; plants semiaquatic; follicles straight, 5–14 mm long. *C. palustris*

2b. Sepals white, 4–5 mm long; plants aquatic or semiaquatic; follicles curved apically, 4–6 mm long. *C. natans*

3a. Leaf blades oval-cordate, from slightly broader than long to longer than broad, the sinus apparent, the auricles not overlapping. *C. leptosepala*

3b. Leaf blades reniform, distinctly broader than long, the sinus not apparent, the auricles overlapping. *C. biflora*

Caltha biflora DC.
Broad-leaf Marsh-marigold

Plants erect, 1.5–3.5 dm tall; basal leaves 12–30 cm long, the petioles 10–25 cm long, the blades 2.5–7 cm long (from sinus to apex) and 4–12.5 cm broad, reniform, crenate to dentate; stems usually with a single leaf subtending the inflorescence; peduncles usually 2, 3–12 cm long (or longer), 1-flowered; sepals whitish, 7-17 mm long; follicles short-stipitate, 12–16 mm long.

Moist woods and meadows; in southeastern Alaska; southward to California and eastward to Utah and Colorado. Our materials belong to var. *biflora*.

Caltha leptosepala DC.
Mountain Marsh-marigold

Plants erect, 0.5–4 dm tall; basal leaves 4–22 cm long, the petioles 2–20 cm long, the blades 1.5–2.5 cm long (from sinus to apex) and 3–4.5 cm broad, oval-cordate, crenate; stems usually with a single leaf (rarely 2); peduncles 1–2, 4–20 cm long, 1-flowered; sepals whitish, often tinged purplish externally, 12–27 mm long; follicles subsessile, 12–15 mm long.

Moist heath and alpine meadows, often along streams; in coastal and insular southwestern, southern, and southeastern Alaska, and less commonly some distance in the interior in southern Alaska and southern Yukon; southward to Oregon, Nevada, Utah, and Colorado. Our plants belong to var. *leptosepala*.

Caltha natans Pallas
Floating Marsh-marigold

Plants prostrate, mostly 3–10 dm long; leaves 2–26 cm long, the petioles 2.5–22 cm long, the blades 1.5–2.5 cm long (from sinus to apex) and 3–4.5 cm broad, floating on water or reclining on mud, reniform, crenate; stems leafy, stoloniferous, rooting at the nodes; peduncles 1–2, 1–6 cm long, 1-flowered; sepals white (often tinged with pink), 4–5 mm long; follicles sessile or subsessile, 4–6 mm long, straight.

Ponds and mud banks; disjunct from the Seward Peninsula and Alaska Peninsula eastward through southern and central to eastern Alaska and west central Yukon; eastward to Hudson Bay and south to Alberta, Minnesota, and Michigan; Asia.

Caltha palustris L.
Yellow Marsh-marigold

Plants erect or ascending, at length reclining and rooting at the nodes, 1–6 dm long or more; leaves 3–40 cm long, the petioles 2.5–3.2 cm long, the blades 0.6–7 cm long (from sinus to apex) and 0.6–13 cm broad, seldom floating on water, reniform to ovalcordate, crenate; stems leafy, sometimes stoloniferous; peduncles 1–3, 1–7 cm long, 1-flowered; sepals yellow (often purplish externally), 6–20 mm long; follicles sessile or subsessile, mostly 5–14 mm long, the style bent.

Moist situations; at widely scattered locations in Alaska and Yukon; southward to Oregon and eastward to Labrador and south to Nebraska and South Carolina; Eurasia. Two intergrading morphological races of *C. palustris* are present in Alaska.

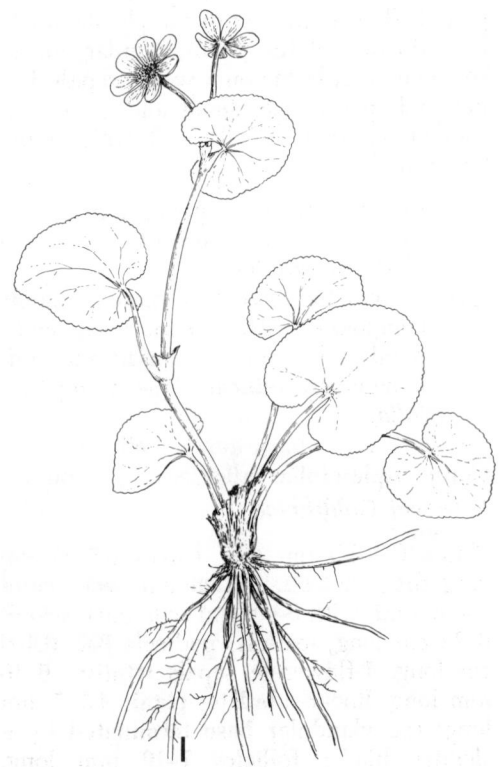

Caltha palustris L. (× 0.3).

1a. Plants 10–25(40) cm long; sepals 7–13 mm long; follicles 5–10 mm long; leaf blades 0.6–3.5 cm long and 0.6–5 cm broad; northern, western, and interior Alaska and western to northern Yukon (*C. arctica* R. Br.; *C. palustris* ssp. *arctica* [R. Br.] Hultén). *C. palustris* var. *arctica* (R. Br.) Huth.

1b. Plants (13)20–60 cm long or more; sepals 11–20 mm long; follicles 8–14 mm long; leaf blades 2–7 cm long and 2.8–13 cm broad; coastal and insular southern Alaska and disjunct in interior, western, and northern Alaska (*C. asarifolia* DC.; *C. palustris* ssp. *asarifolia* [DC.] Hultén). *C. palustris* var. *asarifolia* (DC.) Huth.

COPTIS Salisb.

Perennial, rhizomatous, scapose herbs with fibrous roots; leaves all basal, 3-foliolate or biternately to ternate-pinnately compound; flowers perfect or merely staminate by reduction of the pistils, regular, inconspicuous, 1–3, borne on a scape; sepals 5–8, petaloid; petals 3–8, small and glandular; stamens 12–many; pistils 5–10; fruit a stipitate follicle.

1a. Leaves 3-foliolate; petals represented by a flattened nectary, lacking a blade. *C. trifolia*

1b. Leaves biternate to ternate-pinnately compound, with 5 or more segments; petals represented by a nectary terminated by a linear blade. *C. asplenifolia*

Coptis asplenifolia Salisb.
Fern-leaf Goldthread

Plants 0.9–3.5 dm tall; leaves 3.5–17 cm long, the petioles 1.5–11 cm long, the blades compound, with 5–several segments; scapes 6–22 cm long, leafless; pedicels 1–2, 0.4–9 cm long, 1-flowered; sepals whitish, 6–10 mm long, linear-subulate; petals 4.5–7 mm long, the glandular base terminated by a slender blade; follicles 7–10 mm long, spreading, borne on stipes 6–8 mm long.

Moist woods and bogs; in coastal, southern and southeastern Alaska; southward to southern British Columbia.

Coptis trifolia (L.) Salisb.
Trifoliate Goldthread

Plants 0.5–1.4 dm tall; leaves 1.5–11 cm long, the petioles 1–9 cm long, the blades 3-foliolate; scapes 3.5–10.5 cm long; pedicels 1, 0.8–3.5 cm long, 1-flowered; sepals whitish, often tinged with pink, 5–10 mm long, with an expanded, glandular, bladeless apex; follicles 5–10 mm long, spreading, borne on stipes 5–8 mm long.

Alpine tundra, moist meadows, heath, muskegs, or woods; in coastal and insular west-central, southwestern, southern, and southeastern Alaska; southward to southern British Columbia and disjunctly eastward to Labrador; Greenland; Asia.

DELPHINIUM L.

Perennial herbs with fibrous roots or taproots; stems simple, erect; leaves alternate, the blades palmately divided; flowers perfect, irregular, large and showy, borne in terminal racemes, subtended by a pair of bracts; sepals 5, petaloid, the upper one produced into a prominent spur, the lateral ones often shorter than the lower two; petals 4, in 2 pair, the upper ones spurred and clawless, the lower ones clawed and with expanded blades; stamens numerous; pistils 3–5; fruit a follicle.

Ewan, J. 1945. A synopsis of the North American species of *Delphinium*. Univ. Colo. Stud. Ser. D. 2:55–244.

1a. Stems at least somewhat hairy, almost or quite to the base, 0.7–5.5 cm tall; lateral sepals 12–20 mm long; racemes 2–10-flowered. *D. brachycentrum*

1b. Stems glabrous below the inflorescence, (3)5–15 dm tall or more; lateral sepals various; racemes 7–many-flowered. (2)

2a. Plants often over 15 dm tall; flowers

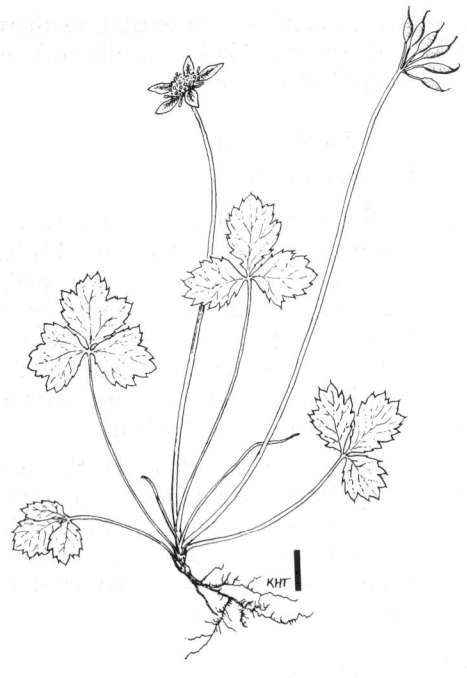

Coptis trifolia (L.) Salisb. (× 0.6).

white, pink, blue, or blue purple, often double; cultivated ornamentals. *D. elatum*

2b. Plants seldom over 15 dm tall; flowers mostly blue purple (rarely white), not double (or only rarely so); indigenous, rarely cultivated. *D. glaucum*

Delphinium brachycentrum Ledeb.
Northern Larkspur

Plants 0.7–5.5 dm tall, the stems at least somewhat hairy almost or quite to the base; leaves 4–17 cm long, the petioles 2.5–14 cm long, spreading-hairy, the blades 1.5–6.5 cm long, with 5–7 main lobes, pubescent above and below, at least on the veins, ciliate; racemes 4–18 cm long, 2–10-flowered, the pedicels 1–6 cm long; sepals bluish or purplish (rarely white), the upper one (including spur) 2.2–3.5 cm long, the lateral ones 1.2–2 cm long; follicles pubescent.

Arctic and alpine tundra to heathlands; disjunct in west-central, northwestern, northeastern and east-central Alaska, and west-central and northern Yukon; Asia (*D. blaisdellii* Eastw.; *D. chamissonis* Pritz.; *D. ruthae* A. Nels; *D. nutans* A. Nels.).

Delphinium elatum L.
Candle Larkspur

Plants mostly 5–25 dm tall, the stems usually glabrous and glaucous; leaves more than 20 cm long, glabrous, (3)5–7-lobed; racemes often more than 3 dm long, many-flowered; sepals blue, blue purple, pink or white, often double and very large; follicles glabrous, 12–20 mm long.

Cultivated ornamental; throughout the warmer portions of Alaska and Yukon, where the plants grow to be very large; native to Siberia.

Delphinium glaucum Wats.
Glaucous Larkspur

Plants (3)5–15 dm tall or more, the stems glabrous and glaucous below the inflorescence; leaves 3–23 cm long, the petioles 1.5–19 cm long, glabrous or pubescent, the blades 2–8 cm long, with 5–7 main lobes, glabrous to pubescent above and below, ciliate; racemes 4–50 cm long, mostly 7–many-flowered, the pedicels 1–2 cm long; sepals bluish or purplish (rarely white), the upper one (including spur) 1.2–1.9 cm long, the lateral ones (4)7–12 mm long; follicles glabrous or pubescent, 4–14 mm long.

Open woods, thickets, stream banks, and meadows; in most of Alaska (except for the northwestern and southwestern portions) and most of the Yukon; southward to California and Nevada (*D. brownii* Rydb.; *D. alatum* A. Nels.; *D. hookeri* A. Nels.).

RANUNCULUS L.

Perennial, biennial, or annual, aquatic or terrestrial herbs, with fibrous to tuberous roots; stems erect, ascending, or rarely prostrate and stoloniferous; leaves basal or cauline and alternate, simple or compound; flowers perfect, regular, solitary or few to several in cymes; sepals 3–5, herbaceous or petaloid, usually deciduous; petals 5–16, rarely more; stamens 10–numerous, sometimes the outer ones petaloid; pistils 5–many; fruit an achene.

Benson, L. 1948. A treatise on the North American Ranunculi. Am. Midl. Nat. 40:1–264.

Padmore, P. A. 1957. The varieties of *Ranunculus flammula* L. and the status of *R. scoticus* E. S. Marshall and of *R. reptans* L. Watsonia 4:19–27.

1a. Sepals 3–4; flowering stems arising from rhizomes. (2)
1b. Sepals 5–6; flowering stems arising from fibrous roots, rarely from rhizomes. (3)

2a. Petals white, (5)7–10; rhizomes and scapes 2–6 mm in diameter; leaf blades entire or 3-lobed; plants aquatic or semiaquatic. *R. pallasii*
2b. Petals yellow, 5–6; rhizomes and scapes 1–2 mm in diameter; leaf blades 3-lobed, the lobes again toothed or lobed; plants terrestrial, but often in moist sites. *R. lapponicus*

3a. Petals white; plants aquatic; submersed leaves divided into linear-filiform segments. *R. aquatilis*
3b. Petals yellow, red, or rarely white; plants terrestrial or aquatic; leaves variable, but not with linear-filiform segments. (4)

4a. Petals reddish or white; sepals pubescent with long, brown hairs, persistent; leaves 3-parted, the lobes again 3–5-lobed; plants from islands of the Bering Sea, Seward Peninsula, and rarely elsewhere. *R. glacialis*
4b. Petals yellow; sepals variously pubescent to glabrous, deciduous in early to late anthesis (persistent in *R. kamtschaticus*); leaves various but

usually not as above; distribution various. (5)

5a. Leaves entire or merely crenate-serrate, the blades simple, not dissected; plants with conspicuous, well-developed, strawberrylike stolons. (6)
5b. Leaves deeply lobed or compound (at least some); plants sometimes stoloniferous, but more commonly with erect stems. (7)

6a. Leaves all entire, the blades narrowly elliptic to lanceolate or oblong to linear, acute at both ends; achenes less than 50, not longitudinally ribbed. *R. flammula*
6b. Leaves crenate-serrate (at least some), the blades oval, cordate basally, rounded to obtuse apically; achenes more than 50, longitudinally ribbed. *R. cymbalaria*

7a. Plants aquatic or semiaquatic; leaves often dimorphic, at least some finely dissected; stems often prostrate and rooting at the nodes (erect in *R. sceleratus*); achenes glabrous, beakless or the beak less than 0.5 mm long. (8)
7b. Plants neither aquatic nor semiaquatic and with finely dissected leaves, though sometimes growing in wet places; stems mostly erect, seldom rooting at the nodes; achenes either hairy or with a beak more than 0.5 mm long. (10)

8a. Plants annual, erect, not rooting at the nodes; achenes almost beakless, the beak less than 0.2 mm long. *R. sceleratus*
8b. Plants perennial; stems usually prostrate and rooting at the nodes or floating; achenes with the beak more than 0.2 mm long (shorter in some *R. hyperboreus*). (9)

9a. Leaf blades with 3–5 broad entire lobes, mostly 3–8(9) mm long; beak of achene less than 0.2 mm long. *R. hyperboreus*
9b. Leaf blades dissected into numerous narrow lobes, mostly 10–30 mm long;

beak of achenes over 0.3 mm long. *R. gmelinii*

10a. Petals 7–16; plants glabrous; cauline leaves 1, or more commonly lacking. (11)

10b. Petals usually 5; plants pubescent or glabrous; cauline leaves usually more than 1. (12)

11a. Leaf blades 3–5-lobed, the lobes crenately toothed or lobed; sepals yellow, deciduous; plants of coastal south-central to southeastern Alaska. *R. cooleyae*

11b. Leaf blades slightly lobed to subentire; sepals greenish, tinged with red or purple, persistent; plants of southwestern and western Alaska. *R. kamtschaticus*

12a. Sepals with conspicuous reddish brown or blackish hair externally. (13)

12b. Sepals with yellowish or whitish hair, or glabrous externally. (14)

13a. Receptacle glabrous. *R. nivalis*

13b. Receptacle with brown hair. *R. sulphureus*

14a. Leaf blades less than 3 cm long; plants rarely more than 2 dm tall; stems below the inflorescence glabrous or nearly so (the pedicels sometimes hairy). (15)

14b. Leaf blades more than 3 cm long, or plants more than 2 dm tall, or stems below the inflorescence long-hairy, or all of these. (20)

15a. Petals 1–3.5(4) mm long; plants mostly 7 cm tall or less. *R. pygmaeus*

15b. Petals 4–18 mm long; plants often more than 7 cm tall. (16)

16a. Pedicels glabrous (sparsely hairy in some *R. eschscholtzii*). (17)

16b. Pedicels hairy. (18)

17a. Blades of lowermost leaves deeply cordate; petals subequal to or slightly longer than the sepals; beaks of

achenes curved, to 0.5 mm long. *R. verecundus*

17b. Blades of lowermost leaves truncate to rounded or cordate; petals distinctly longer than the sepals; beaks of achenes straight, 0.8–1 mm long. *R. eschscholtzii*

18a. Upper cauline leaves with (3)5–7 linear lobes; petals (6)8–10 mm long. *R. pedatifidus*

18b. Upper cauline leaves merely lobed, or with 3 oblong lobes; petals 4–6 mm long. (19)

19a. Blades of basal leaves deeply cordate; achenes about 2.5 mm long. *R. gelidus*

19b. Blades of basal leaves cuneate to obtuse or rounded; achenes about 1.5 mm long. *R. sabinii*

20a. Upper cauline leaves divided into 3–5 narrowly oblong to linear lobes; blades of basal leaves merely crenatedentate. *R. abortivus*

20b. Upper cauline leaves not as above, the lobes oblanceolate or broader; blades of basal leaves deeply incised or compound. (21)

21a. Stems decumbent, often rooting at the nodes and more or less stoloniferous. *R. repens*

21b. Stems erect or ascending, usually neither rooting at the nodes nor stoloniferous. (22)

22a. Flowers mostly less than 15(20) mm broad. (23)

22b. Flowers (12)15–30 mm broad. (25)

23a. Sepals about twice longer than the petals; cluster of achenes 11–17 mm long. *R. pensylvanicus*

23b. Sepals subequal to the petals or shorter than them; cluster of achenes 3–12 mm long. (24)

24a. Cluster of achenes 3–5(7) mm long; beak of achenes hooked; receptacle glabrous. *R. uncinatus*

24b. Cluster of achenes 7–12 mm long;

beak of achenes straight; receptacle hairy. *R. macounii*

25a. Blades of basal leaves pinnately compound with 3–7 leaflets, the leaflets all stalked. (26)

25b. Blades of basal leaves palmately lobed or palmately compound with 3 sessile or short-stalked leaflets. (27)

26a. Beak of achene 1.5–2 mm long; blades of basal leaves 3-foliolate. *R. pacificus*

26b. Beak of achene 2.5–4 mm long; blades of basal leaves 3–7-foliolate. *R. orthorhynchus*

27a. Sepals reflexed as the flower opens. *R. occidentalis*

27b. Sepals spreading as the flowers open, not reflexed. (28)

28a. Achene beaks 0.3–0.6 mm long; blades of basal leaves (2.8)4–8 cm long; sepals 4–7(8) mm long. *R. acris*

28b. Achene beaks 1–2 mm long; blades of basal leaves 1.3–3 cm long; sepals 8–10 mm long. *R. turneri*

Ranunculus abortivus L.
Smooth-leaf Crowfoot

Plants biennial or short-lived perennials; stems erect, 1.5–5 dm tall, not rooting at the nodes, glabrous to hairy; basal leaves long-petioled, the blades 1–5 cm long, reniform to oval, crenate-serrate to shallowly lobed; cauline leaves transitional to bracts, alternate; bracts deeply divided into 3–5 narrowly oblong to lanceolate lobes, sessile; pedicels 0.7–10 cm long, glabrous or hairy; sepals 5, 2.5–4 mm long, spreading to reflexed, usually purplish tinged, glabrous, deciduous; petals 5, yellow, 2–3.5 mm long; achenes 1–1.5 mm long, 10–50 in an ovoid cluster 3–6 mm long, glabrous, the beak 0.1–0.2 mm long.

Weedy (probably adventive) plants of moist, disturbed sites; disjunct in south-central and southeastern Alaska and southern Yukon; eastward to Labrador and south to Washington, Idaho, Texas, and Florida.

Ranunculus acris L.
Tall Buttercup

Plants perennial; stems erect, 4–10 dm tall, not rooting at the nodes, pubescence spreading or appressed-ascending, or rarely glabrous; basal leaves long-petioled, the blades (2.8)4–8 cm long, deeply 3-parted and again lobed, cordate basally; cauline leaves alternate, petioled; bracts with 3 linear segments, sessile; pedicels 1–12 cm long; sepals 5, 4–7 mm long, spreading, greenish, pubescent externally with pale hairs; petals 5, yellow, 8–14 mm long; achenes 2–2.5 mm long, 25–40 in a subglobose cluster about 6 mm long, glabrous, the beak 0.3–0.6 mm long, curved apically.

Moist meadows, roadsides, and muskegs; from the Aleutian Islands eastward to southeastern Alaska; widely distributed in North America; Eurasia. Apparently hybridization occurs between *R. acris* and *R. uncinatus.*

1a. Hair of stems and petioles spreading; plants mostly of southern and southeastern Alaska, where adventive. *R. acris* var. *acris*

1b. Hair of stems and petioles appressed-ascending; plants mostly of the Aleutians, where possibly indigenous (*R. grandis* Honda var. *austrokurilensis* [Tatew.] Hara; *R. acris* var. *austrokurilensis* Tatew.). *R. acris* var. *frigidus* Reg.

Ranunculus aquatilis L.
Water Crowfoot

Plants aquatic, perennial; stems floating or submerged, 2–20 dm long, rooting at the lower nodes, glabrous or sparsely short-hairy; leaves all cauline, alternate, the blades (1)2–4(5) cm long, dissected into filiform segments, the upper ones sometimes floating and simple, with the blades palmately 3–5-lobed; pedicels 1–5(9) cm long; sepals 5, 2–3 mm long, spreading to reflexed, greenish, glabrous; petals 5, white,

4-8(10) mm long; achenes 1-2.5 mm long, 10-20 in a globose cluster, glabrous or pubescent, the beak 0.1-0.3 mm long.

Ponds and streams; disjunct through most of Alaska and Yukon; broadly distributed in North America; circumboreal. Three closely related and intergrading morphological forms are recognized at intraspecific level.

1a. Leaves dimorphic, the submerged ones dissected into filiform segments, the floating ones broadly 3-5-lobed (*R. grayanus* Freyn; *Batrachium grayanum* [Freyn] Rydb.; *R. trichophyllus* var. *hispidulus* [E. Drew] W. Drew.). *R. aquatilis* var. *hispidulus* E. Drew

1b. Leaves all alike, all submerged and dissected into filiform segments. (2)

2a. Stamens 5-8; petals 4-6 mm long; achenes about 1 mm long, glabrous; stems 1-2.5 mm in diameter (*R. confervoides* [Fries] Fries). *R. aquatilis* var. *eradicatus* Laestad.

2b. Stamens 10-25; petals 4-8 mm long; achenes 1-1.5 mm long, glabrous or glabrate; stems 0.4-1(2) mm in diameter (*R. trichophyllus* Chaix. ex Vill.). *R. aquatilis* var. *capillaceus* (Thuill.) DC.

Ranunculus cooleyae Vasey & Rose
Cooley Buttercup

Plants perennial; stems erect, 0.5-3 dm long, not rooting at the nodes, glabrous; leaves all basal, the petioles 2-8 cm long, the blades simple, 1-2.5 cm long, 3-5-lobed and then crenate-dentate, orbicular-reniform, cordate basally; bracts lanceolate or 3-lobed, or lacking; pedicels 2-20 cm long; sepals 5, 7-9 mm long, yellow, glabrous; petals 7-16, yellow, 7-10(15) mm long; achenes 30-70 or more in a hemispheric cluster, about 2.5 mm long, glabrous, the beak 1-1.5 mm long, curved.

Alpine meadows and slopes in snow flushes; in south-central and southeastern Alaska; southward to Washington.

Ranunculus cymbalaria Pursh

Plants perennial; vegetative stems produced into strawberrylike stolons, rooting at the nodes, glabrous to pubescent; flowering stems 0.2-3 dm tall; basal leaves simple, the petioles 1-15 cm long, the blades 0.5-4 cm long, cordate, ovate, or reniform, crenate; pedicels 2-10 cm long; sepals 5, 2-4 mm long, glabrous, deciduous; petals 5-more, yellow, (2.5)3-8 mm long; achenes about 1.5 mm long, 25-200 in a cylindrical cluster, glabrous, the beak about 0.3 mm long, straight.

Muddy sites along streams, pools, or in marshlands; disjunct from the Seward Peninsula and Kotzebue Sound eastward through central and southern Alaska to southern Yukon; broadly distributed in North and South America; Eurasia. (*R. cymbalaria* var. *alpina* Hook.). Our material belongs to var. *cymbalaria*.

Ranunculus eschscholtzii Schlecht.
Eschscholtz Buttercup

Plants perennial; stems erect, 0.5-2.5(3) dm tall, not rooting at the nodes, glabrous; basal leaves simple, the petioles 2-11 cm long, the blades 1-3 cm long, reniform to oval in outline, 3-cleft and again lobed; cauline leaves alternate, 1-3, or lacking; bracts usually with 3 entire lobes, sessile; pedicels 2-12 cm long; sepals 5, 4-8 mm long, yellowish, glabrous or pubescent with yellowish or brownish hairs, deciduous; petals 5, yellow, 7-12 mm long; achenes 1.3-1.8 mm long, 20-50 or more in an elongate cluster, glabrous, the beak 0.8-1 mm long.

Alpine tundra; in meadows and talus slopes, in southwestern, southern, and southeastern Alaska, and southern Yukon; southward to California, Nevada, Utah, and Colorado. This species is closely related to *R. nivalis*. Our material belongs to var. *eschscholtzii*.

Ranunculus flammula L.

Creeping Spearwort

Plants perennial; vegetative stems produced into strawberrylike stolons, often rooting at the nodes, glabrous or nearly so; flowering stems 0.5–1 dm tall; leaves simple, alternate (appearing basal), the petioles 2–13 cm long, the blades linear to narrowly elliptic, lanceolate, or oblanceolate, acute at both ends, entire; pedicels 2–10 cm long; sepals 5, 2–5 mm long, glabrous or pubescent, greenish, deciduous; petals 5, yellow, 2–8 mm long; achenes 1.2–1.8 mm long, 5–30 in a subglobose cluster, glabrous, the beak 0.1–0.7 mm long.

Shallow ponds, wet soil, or muddy banks; disjunct from Kotzebue Sound eastward and southward through most of Alaska and southern Yukon; eastward to Newfoundland and south to California, Arizona, New Mexico, and New England; circumboreal. Two more or less distinct varieties occur in our region.

1a. Leaf blades 1.5–7 mm broad, markedly broader than the petioles; plants uncommon in our region (*R. filiformis* var. *ovalis* Bigel.). *R. flammula* var. *ovalis* (Bigel.) Benson
1b. Leaf blades 0.5–1.5 mm broad, not markedly broader than the petioles; plants common in our region (*R. flammula* var. *intermedius* Hook.; *R. filiformis* Michx.). *R. flammula* var. *filiformis* (Michx.) Hook.

Ranunculus gelidus Kar. & Kir.

Plants perennial; stems erect, 0.2–1 dm tall, not rooting at the nodes, glabrous or sparsely pubescent; basal leaves simple, the petioles 1.5–6 cm long, the blades 0.5–2.5 cm long, cordate to reniform, 3-parted and again cleft or lobed; cauline leaves alternate, sessile or short-petioled; pedicels 1.5–5 cm long, sepals 5, 2.5–5 mm long, greenish, pubescent with pale hairs, deciduous; petals 5, yellow, 4–5 mm long; achenes about 2.5 mm long, 50–80 in a cylindrical cluster, glabrous, the beak 0.6–0.7 mm long.

Stream banks and hillsides; in tundra or heathlands; disjunct from northern to south-central and southwestern Alaska, and northern to southern Yukon; southward to Montana and Colorado (*R. grayi* Britt.; *R. gelidus* ssp. *grayi* [Britt.] Hultén; *R. gelidus* var. *shumaginensis* Hultén).

Ranunculus glacialis L.

Glacier Buttercup

Plants perennial; stems erect, 1–3 dm tall, not rooting at the nodes, pubescent with reddish brown hairs, at least above; basal leaves compound, the petioles 1–8 cm long, the blades 0.8–3.5 cm long, 3-foliolate, the leaflets divided into narrow segments; cauline leaves alternate, 2–3, reduced upwards; pedicels (1)1.5–8 cm long; sepals 5, 9–13 mm long, reddish, straw-colored, or white, 10–15 mm long; achenes 2–3 mm long, 30–50 in a hemispheric cluster, glabrous, the beak 2–3 mm long.

Wet soil, on grassy slopes and ridges, in tundra; in the Seward Peninsula and islands of the Bering Sea, and rarely in disjunct sites in northwestern and east-central Alaska; Eurasia (*R. chamissonis* Schlecht.; *R. glacialis* ssp. *chamissonis* [Schlecht.] Hultén). Our material belongs to var. *chamissonis* (Schlecht.) Benson.

Ranunculus gmelinii DC.

Plants perennial; stems prostrate or floating, 1–5 dm long, rooting at the nodes, glabrous or pubescent; leaves mostly cauline, alternate, the petioles 1–4 cm long, the blades 3-parted and again parted, at least some with very narrow segments, oval, deeply cordate basally; pedicels 1–5 cm long; sepals 5, 2.5–6 mm long, greenish, glabrous or pubescent, deciduous; petals 5, yellow, 4–8 mm long; achenes 1–1.5 mm long, 50–70 in an ovoid cluster, glabrous, the beak 0.6–0.8 mm long.

Mud flats, shallow ponds, and marshes; in most of Alaska (except for coastal and insular southern Alaska) and most of the

Yukon; widely distributed in North America; Asia.

1a. Plants sparsely to rather densely pubescent; stems 0.5–1 mm in diameter; sepals 2.5–3 mm long (*R. yukonensis* Britt.; *R. purshii* ssp. *yukonensis* [Britt.] Porsild; *R.* gmelinii var. yukonensis [Britt.] Benson). *R. gmelinii* var. *gmelinii*

1b. Plants glabrous or nearly so; stems 1.5–3 mm in diameter; sepals 4–6 mm long (*R. purshii* Richards.; *R. gmelinii* var. *purshii* [Richards.] Hara; *R. gmelinii* ssp. *purshii* [Richards.] Hultén). *R. gmelinii* var. *hookeri* (G. Don) Benson

Ranunculus hyperboreus Rottb.
Arctic Buttercup

Plants perennial; stems prostrate to erect (sometimes floating), 1–5 dm long, rooting at the nodes, glabrous; leaves mostly cauline, alternate, the petioles 1.5–5 cm long, the blades simple, 3–8(10) mm long, with 3–5 entire lobes, truncate to rounded basally; pedicels 0.5–2.5(4) cm long; sepals 5, 2–3 mm long, greenish, glabrous, deciduous; petals 5, yellow, 2–4 mm long; achenes 0.6–0.8 mm long, 10–20 in a globose cluster, glabrous, the beak about 0.1 mm long.

Mud flats or shallow ponds; widely distributed in Alaska and Yukon; eastward to Labrador and south to Montana and Quebec; circumpolar (*R. hyperboreus* ssp. *arnellii* Scheutz.).

Ranunculus kamtschaticus DC.
Kamchatka Buttercup

Plants perennial; stems erect, 0.2–1 dm tall, not rooting at the nodes, glabrous; leaves all basal, simple, the petioles about 1 cm long, the blades slightly lobed apically or subentire, orbicular to ovate or oblong, 0.8–2.2 cm long, 0.6–1.2 cm wide, truncate or tapering basally; sepals 4, 4–10 mm long, greenish, tinged with red or purple, gla-

brous, persistent; petals 7–16, yellow, 5–15 mm long; achenes 1.5–2 mm long, about 20 in a hemispheric cluster, glabrous, the beak 0.7–2 mm long.

Wet situations, meadows, and tundra; in the Seward Peninsula, Aleutian Islands, and Shumagin Islands; Asia (*Ficaria glacialis* Fisch.; *Oxygraphis glacialis* [Fisch.] Bunge).

Ranunculus lapponicus L.
Lapland Buttercup

Plants perennial from elongate slender rhizomes; flowering stems erect, 0.5–2.5 dm tall, glabrous; basal leaves arising directly from the rhizome, simple, the petioles 4–15 cm long, the blades 1–4 cm long, deeply 3-parted, the lobes again broadly lobed, reniform, cordate basally; cauline leaves solitary or lacking; sepals 3(4), 5–7 mm long, greenish, glabrous, reflexed, deciduous; petals 5(6), yellow, 4–6(7) mm long; achenes 3.5–5 mm long, 10–20 in a subglobose cluster, glabrous, the beak 1.5–2 mm long, curved.

Swamps, muskegs, moist woods, heathlands, or tundra; in most of Alaska (except for coastal and insular southwestern and southeastern portions) and Yukon; eastward to Labrador and south to British Columbia, Alberta, Minnesota, Ontario, and Maine; Eurasia. This species resembles *Anemone richardsonii* (please compare).

Ranunculus macounii Britt.
Macoun Buttercup

Plants perennial; stems 2–10 dm tall, erect to decumbent, sometimes rooting at the lower nodes, pubescent with long spreading hairs, or rarely glabrous; basal leaves simple or pinnately 3–5-foliolate, the petioles 5–20 cm long, long-hairy (or glabrous), the blades 3–8 cm long; cauline leaves alternate, similar to the basal; pedicels 0.3–10 cm long; sepals 5, 4–7 mm long, yellowish (often tinged with purple), pubescent or glabrous, deciduous; petals 5, yellow, 3–8(10) mm long; achenes 2–3 mm long, 20–50 in an elongate cluster, glabrous, the beak 1–1.2 mm long, straight.

Moist meadows, beaches, and along streams; in the southeastern quarter of continental Alaska; in coastal and insular Alaska from Kodiak eastward through the Panhandle, and in southern Yukon; eastward to Newfoundland and south to California, Arizona, New Mexico, Nebraska, and Michigan. Our material belongs to var. *macounii*.

Ranunculus nivalis L.
Snow Buttercup

Plants perennial; stems erect or ascending, 0.4–3 dm tall, not rooting at the nodes, hairy to glabrous; basal leaves simple, the petioles 0.5–10 cm long, the blades 0.6–2 cm long, 3-lobed, the lobes entire or again lobed, truncate to cordate basally; cauline leaves 2–3, alternate; bracts 3–5-lobed, sessile; pedicels 1–12(19) cm long, brown-hairy or rarely glabrous; sepals 5, 6–8 mm long, brown-hairy, deciduous; petals 5, yellow, 8–11 mm long; achenes about 1.5 mm long, mostly 40–50 in an elongate cluster, glabrous, the beak 1–2 mm long, straight.

Moist meadows, bogs, and stream banks, in arctic and alpine tundra; in most of Alaska (except for the southwestern portion) and most of the Yukon; east to Labrador and south to British Columbia; circumboreal.

Ranunculus occidentalis Nutt. ex T. & G.
Western Buttercup

Plants perennial; stems erect, 0.5–7 dm tall, not rooting at the nodes, pubescent with spreading hairs or glabrous; basal leaves simple (or sometimes pinnately 3-foliolate), the petioles 2–17 cm long, the blades 1–5 cm long, 3-parted, the lobes again lobed; cauline leaves alternate; bracts with 3 oblong lobes, sessile; pedicels 0.3–18 cm long; sepals 5, 4–8 mm long, greenish, re-

Ranunculus nivalis L. (× 0.4).

flexed, pubescent, deciduous; petals 5, rarely more, yellow, 6–12(18) mm long; achenes 2.5–3.5 mm long, 8–15 in a hemispheric cluster, glabrous, the beak 0.5–2 mm long, slightly curved.

Moist meadows, heathlands, and tundra; in southwestern, south-central, and southeastern Alaska and southern Yukon; south to California.

1a. Stems mostly 2–3 mm in diameter, rather sparsely pubescent to glabrous; plants of south-central and southeastern Alaska. *R. occidentalis* var. *brevistylus* Greene

1b. Stems mostly 3–5 mm in diameter,

densely pubescent to glabrous; plants of southwestern Alaska and the Aleutians (*R. occidentalis* ssp. *insularis* Hultén). *R. occidentalis* var. *nelsonii* (DC.) Benson

Ranunculus orthorhynchus Hook.
Straight-beak Buttercup

Plants perennial; stems decumbent to erect, 2–10 dm tall, not rooting at the nodes, densely to sparsely long-hairy; basal leaves pinnately 3–7-foliolate, the petioles 3–25 cm long, the blades 3–8 cm long; cauline leaves alternate; bracts deeply lobed to simple, sessile or short-petioled; pedicels 1.5–15 cm long; sepals 5, 6–10 mm long, greenish, tinged with purple, reflexed, glabrous, deciduous; petals 5, yellow, mostly 10–12 mm long; achenes 3–4 mm long, mostly 10–20 in a hemispheric cluster, glabrous, the beak 2.5–4 mm long.

Moist open sites in southeastern Alaska; southward to California and Utah. (*R. orthorhynchus* ssp. *alaschensis* [Benson] Hultén). Our plants belong to var. *alaschensis* Benson.

Ranunculus pacificus (Hultén) Benson
Pacific Buttercup

Perennial herbs; stems decumbent to erect, 2–5 dm tall, not rooting at the nodes, sparsely appressed-pubescent to glabrous; basal leaves pinnately 3-foliolate, the petioles 0.5–2 dm long, appressed-pubescent, the blades 2–6 cm long, the leaflets 3-lobed and again toothed; cauline leaves alternate; sepals 5, 7–8 mm long, purplish tinged, reflexed, pubescent, deciduous; petals 5, yellow (tinged purplish externally), 8–13 mm long; achenes to 4 mm long, about 25 in a subglobose cluster, glabrous, the beak about 2 mm long, hooked apically.

Moist sites in meadows and along streams; in southeastern Alaska; endemic(?). It seems probable that *R. pacificus* might be better considered at some infraspecific level within *R. macounii*.

Ranunculus pallasii Schlecht.
Pallas Buttercup

Plants perennial from elongate slender rhizomes; flowering stems erect, 0.5–3 dm tall, glabrous; basal leaves arising directly from the rhizome, simple, the petioles 5–26 cm long, the blades 1–5 cm long, entire and narrowly oblong to linear, or 3-lobed; cauline leaves alternate; pedicels 1–20 cm long; sepals 3–4, greenish (often purplish tinged), 8–10 mm long, glabrous, deciduous; petals 5–10, white, 6–13 mm long; achenes about 4–6 mm long, 12–25 in a hemispheric cluster, glabrous, the beak about 1–2 mm long, straight.

Shallow water or mud of ponds and swamps; in coastal or near coastal and insular western and northern Alaska and northern Yukon; eastward to Labrador; Asia.

Ranunculus pedatifidus J. E. Smith ex Rees
Northern Buttercup

Perennial herbs; stems erect, 0.8–4 dm tall, pubescent to glabrous, not rooting at the nodes; basal leaves simple, the petioles 3–10 cm long, the blades 1–3.5 cm long, parted into 3 main lobes, the terminal one often entire, the lateral ones again dissected into 3–more lobes, cordate basally; cauline leaves alternate; bracts divided into 3–7 linear lobes, sessile; pedicels 1–15 cm long; sepals 5, greenish (sometimes tinged purplish, 4–6 mm long, spreading, pubescent, deciduous; petals 5, yellow, (6)8–10 mm long; achenes about 2 mm long, 25–70 in an elongate cluster, glabrous, the beak 0.6–1 mm long.

Meadows, in arctic or alpine tundra, in west-central, northern, south-central, and eastern Alaska and western to northern Yukon and adjacent Mackenzie (*R. affinis* R. Br.; *R. pedatifidus* ssp. *affinis* [R. Br.] Hultén; *R. verticillatus* Eastw.; *R. eastwoodianus* Benson). Our materials belong to var. *affinis* (R. Br.) Benson.

Ranunculus pensylvanicus L. f.
Bristly Buttercup

Plants annual or short-lived perennials; stems erect, 3–10 dm tall, pubescent with long spreading hairs, not rooting at the nodes; basal leaves pinnately 3-foliolate, the petioles 3–20 cm long, spreading-hairy, the leaflets short-stalked (rarely sessile), 3–7 cm long, the middle leaflet 3-parted, the lateral ones 2-parted; cauline leaves alternate; bracts similar to the cauline leaves; pedicels 0.3–6 cm long; sepals 5, 4–5 mm long, yellowish, reflexed, pubescent, deciduous; petals 5, yellow, 2–3 mm long; achenes about 2.5 mm long, 60–80 in an elongate cluster, glabrous, the beak 0.6–0.9 mm long, straight.

Weedy plant of moist, disturbed sites; adventive in southwestern, south-central, east-central, and southeastern Alaska (to be expected in Yukon); widely distributed in North America.

Ranunculus pygmaeus Wahl.
Pygmy Buttercup

Plants perennial; stems erect or ascending, 0.1–1.2(1.8) dm tall, sparsely pubescent to glabrous, not rooting at the nodes; basal leaves simple or compound, the petioles 0.2–4 cm long, the blades 0.3–1.2 mm long, 3-parted or 3-foliolate, the central lobe entire or 3-lobed, the lateral ones 2–4-lobed, subcordate, truncate, or rounded basally; cauline leaves alternate; bracts 3-lobed, sessile; pedicels 0.5–9 cm long, pubescent; sepals 5, 2–3.5 mm long, greenish (often tinged purplish), sparsely pubescent, deciduous; petals 5, yellow, 1.5–4 mm long; achenes about 1 mm long, mostly 40–50 in an elongate cluster, glabrous, the beak 0.3–0.7 mm long, straight.

Moist meadows and beaches, in arctic and alpine tundra; in much of Alaska (except for the central, coastal and insular southern portions) and most of the Yukon; eastward to Labrador and south to British Columbia, Montana, Wyoming, and Colorado; circumpolar.

Ranunculus repens L.
Creeping Buttercup

Plants perennial; stems usually decumbent and rooting at the nodes, or less commonly erect, to 10 dm long, pubescent with long spreading hairs or glabrate; basal leaves pinnately (or obscurely palmately) compound, the petioles 4–40 cm long, the blades 3-foliolate, 3–10 cm long, the main segments stalked, and usually deeply 3-lobed and again toothed; cauline leaves alternate; bracts 5-lobed to simple; pedicels 1.5–15 cm long; sepals 5, 5–7 mm long, greenish, pubescent, spreading, deciduous; petals 5–more, yellow, 6–17 mm long; achenes 2–3 mm long, 10–50 in a subglobose cluster, glabrous, the beak 0.7–1.2 mm long, curved.

Weedy species of open sites, along roads, ditches, and in vacant lots; in widely scattered locations in much of Alaska south of the 65th parallel, and in southern Yukon; probably adventive in our region, a cosmopolitan weed; native to Europe.

Ranunculus sabinii R. Br.
Sabine Buttercup

Plants perennial; stems erect, 0.3–1.5 dm tall, not rooting at the nodes, sparsely pubescent to glabrous; basal leaves simple, the petioles 1.5–5 cm long, the blades 0.8–3 cm long, 3-lobed, the central lobe entire, the lateral ones entire or again 2-lobed, cuneate to rounded basally; cauline leaves alternate; bracts 3-lobed, sessile; pedicels 1–9 cm long, sparsely pubescent; sepals 5, 4–7 mm long, yellowish (often tinged purplish), spreading, pubescent with pale hairs, deciduous; petals 5, yellow, 5–8 mm long; achenes about 1 mm long, mostly 50–80 in a cylindrical cluster, glabrous, the beak about 0.5 mm long, straight.

Moist tundra; in western and northern Alaska and northern Yukon; eastward to Greenland; Asia (R. pygmaeus ssp. sabinii [R. Br.] Hultén). This is a poorly understood and rarely collected entity with close affinities to *R. pygmaeus;* it may not be altogether distinct.

Ranunculus sceleratus L.
Celery-leaf Crowfoot

Plants annual; stems erect, 1–4(5) dm tall, not rooting at the nodes, sparsely long-hairy or glabrous; basal leaves simple, the petioles 2–12 cm long, the blades 1–5 cm long, deeply 3-parted, the lobes again lobed, subcordate to cordate basally; cauline leaves alternate; pedicels 0.3–3.5 cm long; sepals 5, 2–4.5 mm long, greenish, spreading, glabrous to sparsely pubescent, ultimately deciduous; petals 5, yellow, 2–5 mm long; achenes 0.8–1 mm long, 40–numerous in a cylindrical cluster, glabrous, the beak about 0.1 mm long.

Shallow ponds, mud flats, stream banks, swamps, and muskegs; in west-central, east-central, and southeastern continental Alaska and southern Yukon; widely distributed in North America; circumboreal (*R. sceleratus* ssp. *multifidus* [Nutt.] Hultén). Our material is assignable to var. *multifidus* Nutt.

Ranunculus sulphureus Soland. ex Phipps
Sulphur Buttercup

Plants perennial; stems erect, 0.3–3(4) dm tall, not rooting at the nodes, glabrous to pubescent with brown hairs; basal leaves simple, the petioles 2–16 cm long, the blades 0.6–4 cm long, crenately 3–9-lobed, the central lobe largest, cuneate to truncate basally; cauline leaves 1–3, alternate; bracts 3–5-lobed, the lobes narrowly oblong to lanceolate, sessile; pedicels 0.3–19 cm long, sparsely to rather densely brown-hairy sepals 5, 6–8 mm long, brown-hairy, deciduous; petals 5, yellow, 8–12 mm long; achenes 1.8–2 mm long, mostly 50–90 in an elongate cluster, the beak 0.9–1 mm long, straight.

Moist tundra or heathlands along streams and in bogs; in the Aleutians, islands of the Bering Sea, west-central, northern, east-central, and south-central Alaska, and northern to southwestern Yukon; eastward to Greenland; circumpolar.

1a. Basal leaf blades merely crenately lobed; plants broadly distributed. *R. sulphureus* var. *sulphureus*
1b. Basal leaf blades conspicuously 3-lobed; plants of the Aleutian and Pribilof Islands (*R. nivalis* var. *intercedens* [Hultén] B. Boi.). *R. sulphureus* var. *intercedens* Hultén

Ranunculus turneri Greene
Turner Buttercup

Plants perennial; stems erect, 1–5 dm tall or more(?), not rooting at the nodes, pubescent with long, spreading hairs; basal leaves simple, the petioles 5–10(14) cm long, the blades 1.3–3(4.5) cm long, 3-parted and again cleft or toothed, cordate to truncate basally; cauline leaves alternate; bracts simple or with 3 narrowly lanceolate segments (these sometimes again dissected), sessile or shortly petiolate; pedicels 2–14(21) cm long; sepals 5, (6)8–10 mm long, greenish, spreading, pubescent with yellowish hairs, deciduous; petals 5(6), yellow, 8–15 mm long; achenes about 2.5–3.5 mm long, 20–25 in a subglobose cluster, glabrous, the beak 1–2 mm long, curved.

Moist sites; disjunct from islands of the Bering Sea to northeastern Alaska and northern Yukon; east to the Mackenzie (*R. occidentalis* ssp. *turneri* [Greene] J. P. Anders.; *R. occidentalis* var. *turneri* [Greene] Benson).

Ranunculus uncinatus D. Don

Plants annual or perennial; stems erect, 3–10 dm tall, not rooting at the nodes, glabrous or pubescent; basal leaves simple, the petioles 5–25 cm long, the blades 2–9 cm long, 3-parted, the segments lobed and toothed, cordate basally; cauline leaves alternate; pedicels 1–5 cm long; sepals 5, about 3 mm long, greenish, reflexed, pubescent, deciduous; petals 5, yellow, 2.5–7 mm long; achenes 1.8–3 mm long, 5–30 in a hemispheric cluster, glabrous or hairy, the beak 1–2(2.5) mm long, curved.

Moist soil, in woods, thickets, meadows, beaches, and along streams; in coastal southwestern, south-central, and southeastern Alaska. Two varieties are present among our materials. They are not geographically distinct (*R. bongardii* Greene; *R. tenellus* Nutt.).

1a. Achenes glabrous; plants sparsely hairy to glabrous. *R. uncinatus* var. *uncinatus*
1b. Achenes pubescent; plants (glabrous) sparsely to densely hairy. *R. uncinatus* var. *parviflorus* (Torr.) Benson

Ranunculus verecundus Robins. ex Piper

Plants perennial; stems ascending to erect, 0.7–2 dm tall, not rooting at the nodes, glabrous to sparsely pubescent above; basal leaves simple, the petioles 1–5 cm long, the blades 0.7–2 cm long, 3-lobed, the lobes again lobed, cordate basally; cauline leaves 1–3, alternate; pedicels 1–10 cm long, glabrous; sepals 5, 3.5–5 mm long, yellowish (tinged purplish), spreading, pubescent, deciduous; petals 5, yellow, 3–5 mm long; achenes 1–1.5 mm long, 25–70 in an elongate cluster, glabrous, the beak about 0.5 mm long, curved.

Reported from Russell Fjord and Egg Island, in coastal southeastern Alaska; southward to Oregon and east to Montana and Alberta.

THALICTRUM L.

Perennial glabrous rhizomatous herbs, the stems simple or branched above, erect; leaves alternate or basal, bi- or triternately compound; flowers perfect or imperfect, regular, inconspicuous, borne in racemes or panicles; sepals 4–5, green or purplish, caducous; petals lacking; stamens 8–numerous; pistils 2–several; fruit a sessile or stipitate achene.

Boivin, B. 1944. American Thalictra and their Old World allies. Rhodora 46:337–77; 391–445; 453–87.

———. 1948. Key to the Canadian species of Thalictra. Can. Field-Nat. 63: 169–70.

1a. Plants with perfect flowers, broadly distributed in Alaska and Yukon; cauline leaves lacking or solitary, or the anthers less than 1 mm long. (2)
1b. Plants dioecious (perfect in *T. minus*), occurring in the Aleutian Islands and in southern Yukon and southeastern Alaska, rarely in the interior; cauline leaves several and the anthers more than 1 mm long. (3)

2a. Cauline leaves lacking or stems with a single one near the base; flowers in racemes; staminal filaments linear; stems mostly 2.5 cm tall or less. *T. alpinum*
2b. Cauline leaves more than 1; flowers in panicles; staminal filaments broadened near the apex; stems mostly 3 dm tall or more. *T. sparsiflorum*

3a. Plants with perfect flowers, occurring in the Aleutians and in near central to southwestern Alaska; achenes sessile or nearly so. *T. minus*
3b. Plants dioecious, occurring in southern Yukon and southeastern Alaska; achenes with a stipe 0.4–1.2 mm long. *T. occidentale*

Thalictrum alpinum L.
Arctic Meadowrue

Plants 0.3–2.5(3) dm tall; leaves all basal or the stem with a single leaf near the base, 2–8(13) cm long, biternate, the segments 3–8 mm long; racemes elongate; pedicels recurved in fruit; flowers perfect; sepals purplish tinged; staminal filaments linear, or slightly expanded apically; anthers 1.5–3 mm long; achenes 2–3.5 mm long, subsessile.

Arctic and alpine tundra to heathlands; through most of Alaska and the Yukon;

Thalictrum alpinum L. (× 0.7).

eastward to Newfoundland and south to California, Nevada, Utah, and New Mexico; circumboreal. Our material belongs to var. *alpinum*.

Thalictrum minus L.

Plants 3–10 dm tall; leaves mostly cauline, 10–25 cm long, 2–4 times ternate, the segments 10–30 mm long; panicles rather large; flowers perfect; sepals yellowish green; staminal filaments slightly expanded apically; anthers 2–3 mm long; achenes about 3 mm long, sessile.

Meadows; in the eastern Aleutians and southwestern Alaska; Eurasia (*T. kemense* Fries; *T. minus* ssp. *kemense* [Fries] Hultén; *T. hultenii* B. Boi.). Our material belongs to var. *stipellatum* (C. A. Mey.) Tamura.

Thalictrum occidentale Gray
Western Meadowrue

Plants 3–10 dm tall; leaves mostly cauline, 10–25 cm long or longer, 3–4 times ternate, the segments 10–30 mm long; panicles rather large; flowers imperfect (the plants dioecious); sepals greenish or purplish; staminal filaments linear; anthers 1.5–4 mm long; achenes 4–10 mm long, the stipe 0.4–1.2 mm long.

Open woods and meadows; in southern Yukon and southeastern Alaska; southward to California, Utah, and Colorado (*T. breitungii* B. Boi.).

Thalictrum sparsiflorum Turcz. ex Fisch. & Mey.
Few-flower Meadowrue

Plants (2)3–10 dm tall; leaves mostly cauline, 6–20 cm long, 2–3 times ternate, the segments 5–20 mm long; panicles rather small; flowers perfect; sepals whitish or greenish, often purplish tinged; staminal filaments broadened upwards; anthers 0.6–0.8 mm long; achenes 4–6 mm long, the stipe 0.8–1.5 mm long.

Meadows, thickets, and woods; in much of

Alaska and Yukon south of the 66th parallel (rarely northward); eastward to Manitoba and south to California, Utah, and Colorado; Asia (*T. richardsonii* Gray). Our material is assigned to var. *richardsonii* (Gray) B. Boi.

TROLLIUS L.

Perennial, rhizomatous herbs; stems erect; leaves basal and cauline, alternate, palmately lobed to compound; flowers perfect, regular, solitary or few; sepals 5–several, petaloid; petals 5–10, linear, glandular (sometimes regarded as staminodia); stamens numerous; pistils few to numerous; fruit of few- to several-seeded follicles.

1a. Basal and lower cauline leaves long-petioled; plants indigenous to the Aleutians (Kiska). *T. riederianus*
1b. Basal (if present) and lower cauline leaves short-petioled; plants cultivated ornamentals. (2)

2a. Petals about as long as the stamens, much shorter than the petaloid sepals. *T. hybridus*
2b. Petals much longer than the stamens, about as long as the petaloid sepals. *T. chinensis*

Trollius chinensis Bunge

Plants glabrous, the stems erect, mostly 5–15 dm tall, arising from stout rhizomes; basal leaves small or lacking, the lower cauline ones short-petioled, becoming sessile upwards, the terminal segment of the blade 6–14 cm long or more, pinnately lobed, the lobes toothed; flowers few to several, each terminating a branch of the upper stem; sepals 10–15, yellow to orange, 25–35 mm long; petals 25–35 mm long, yellow to orange, much longer than the stamens.

Cultivated ornamentals; in southern Alaska; introduced from China.

Trollius hybridus Hort.

Plants glabrous, the stems erect, mostly 5–10 dm tall, arising from a stout rhizome; basal leaves short-petioled, the lower cauline ones short-petioled, becoming sessile upwards, the terminal segment of the blade 2.5–6 cm long, subpalmately lobed, the lobes toothed; flowers few to several, each terminating a branch of the upper stem; sepals 5–10, yellow, 15–20 mm long; petals 10–12 mm long, yellow, subequal to the stamens.

Cultivated ornamentals; in southern Alaska; introduced from the Old World.

Trollius riederianus Fisch. & Mey.

Plants glabrous or nearly so, the stems 1.5–4.5(6) dm tall, arising from a stout rhizome with a thatch of old leaf bases; basal and lower cauline leaves with petioles mostly 10–40 cm long, the terminal segment of the blade 3.5–6.5 cm long, subpalmately lobed, the lobes toothed or incised; sepals 5–7, yellow, 15–27 mm long; petals 9–11 mm long, yellow, subequal to the stamens; follicles 7–9 mm long.

In lush glades; on Kiska, Aleutian Islands, Alaska; Asia.

ROSACEAE
Rose Family

Herbs, shrubs, or trees; leaves usually alternate and stipulate, simple, variously lobed, or variously compound; flowers perfect or imperfect, regular, often showy; sepals and petals distinct, 5 each (8–10 in *Dryas*), borne on the margin of a hypanthium (petals lacking in *Alchemilla* and *Sanguisorba*); stamens few to numerous; pistils 1–several, superior, inferior, or half-inferior; styles 1–several; fruit a drupe, pome, aggregate, accessory, achene, or hip.

1a. Plants herbaceous. (2)
1b. Plants woody, shrubs or trees. (12)

2a. Petals lacking; flowers small, numerous, borne in dense spikes or cymes; leaves pinnately compound or merely palmately lobed. (3)

2b. Petals present; plants of various habit; flowers not at once small, numerous, and borne in a dense spike, or if so, then the leaves ternately dissected. (4)

3a. Leaves palmately lobed; flowers in corymbose cymes. *Alchemilla*

3b. Leaves pinnately compound; flowers borne in dense spikes. *Sanguisorba*

4a. Leaves bi- or triternately dissected into linear segments; petals white. (5)

4b. Leaves various, but not bi- or triternately dissected into linear segments; petals white, yellow, or pink. (6)

5a. Flowers in cymes; stems erect from a taproot. *Chamaerhodos*

5b. Flowers racemose; stems prostrate from rhizomes or stolons. *Luetkea*

6a. Leaves ternate to triternate, the segments broad; plants mostly 1–2 m tall, dioecious; flowers small (petals about 1 mm long), numerous, borne in panicles to 5 dm long. *Aruncus*

6b. Leaves various, but not as above; plants mostly less than 0.5 m tall, mostly perfect; flowers larger, not in panicles. (7)

7a. Flowers solitary on scapose peduncles; leaves simple, entire to crenate; fruit of plumose achenes; sepals and petals 8–10 each. *Dryas*

7b. Flowers usually more than 1; leaves compound or lobed, rarely simple; fruit not of plumose achenes; sepals and petals usually 5 each. (8)

8a. Leaflets tridentate apically (entire along the sides); stamens 5; plants prostrate or mat-forming. *Sibbaldia*

8b. Leaflets variously toothed or lobed, but not regularly tridentate apically, or the leaves simple; stamens 10 or more; plants usually not prostrate or mat-forming. (9)

9a. Calyx lacking bractlets between the sepals; fruit an aggregate. *Rubus*

9b. Calyx with bractlets alternating with the sepals; fruit an achene. (10)

10a. Leaves trifoliolate; plants with well-developed stolons; flowers white; receptacle ripening into an accessory fruit. *Fragaria*

10b. Leaves mostly with more than 3 leaflets, but if 3-foliolate or rarely simple, then lacking stolons; flowers mostly yellow (rarely white); receptacle not ripening. (11)

11a. Leaves palmately lobed or compound, or pinnately lobed or compound, not lyrate-pinnatifid; styles at maturity not elongate and conspicuous. *Potentilla*

11b. Leaves pinnately lobed or compound, or more usually lyrate-pinnatifid, the terminal segment often greatly enlarged, or rarely simple; styles at maturity elongate and conspicuous. *Geum*

12a. Leaves compound. (13)

12b. Leaves simple. (16)

13a. Stems usually with prickles; pistils several, separate, enclosed in a hollow receptacle, which at maturity ripens into a hip. *Rosa*

13b. Stems armed or unarmed; pistils various, but not as above; fruit not a hip. (14)

14a. Flowers in corymbs; ovary inferior, the carpels united; fruit a pome. *Sorbus*

14b. Flowers not in corymbs; ovary superior; fruit an achene or aggregate. (15)

15a. Calyx with bractlets alternating with the sepals; fruit of dry achenes; stems unarmed. *Potentilla*

15b. Calyx lacking bractlets; fruit fleshy (an aggregate); stems sometimes armed. *Rubus*

16a. Plants armed with thorns. *Crataegus*

16b. Plants unarmed, lacking thorns. (17)

17a. Ovary inferior (fused to the calyx tube), with 2–5 united carpels. (18)

17b. Ovary superior, the carpels 1–several, separate. (19)

18a. Flowers in corymbs; fruit with 2–5 locules, each with 1–2 seeds. *Malus*

18b. Flowers in racemes; fruit appearing to have 10 locules, each with a single seed. *Amelanchier*

19a. Plants with prostrate branches; flowers solitary on scapose peduncles; fruit of plumose achenes; sepals and petals 8–10 each. *Dryas*

19b. Plants various, but not as above; flowers 1–several, not on scapose peduncles; fruit not of plumose achenes. (20)

20a. Leaves palmately lobed. (21)

20b. Leaves serrate or serrulate, not palmately lobed. (22)

21a. Flowers large, solitary, or in few-flowered cymes; fruit a fleshy aggregate (raspberry). *Rubus*

21b. Flowers small, in terminal corymbs; fruit a cluster of papery carpels. *Physocarpus*

22a. Leaves with glands on petioles; carpels 1; fruit a drupe. *Prunus*

22b. Leaves lacking glands on petioles; carpels usually 5; fruit a follicle. *Spiraea*

ALCHEMILLA L.

Perennial herbs; leaves alternate, palmately lobed; stipules well developed, connate-sheathing; flowers perfect, borne in compound cymes; hypanthium saucer to cup shaped; sepals 4, sometimes alternating with 4 sepallike bracteoles; petals none; stamens 4, alternate with the sepals; pistil commonly 1; fruit an achene.

Alchemilla vulgaris L.
Lady's Mantle

Plants erect or ascending, from a rhizome; stems 1.5–5 dm tall, simple or branched; leaf blades 2.5–10 cm wide, reniform, 3–9-lobed, the lobes serrate; flowers green,

3–5 mm wide, borne in terminal corymbose cymes; hypanthium with a wide glandular disk; sepals to about 1.5 mm long, alternating with smaller bractlets; stamens 4; pistils usually 1.

Known in Alaska only from Dutch Harbor, Unalaska Island (Strutz 1731, BRY), where it has been introduced and is now locally abundant; eastern North America, Eurasia.

AMELANCHIER Medic.

Shrubs or small trees with unarmed branches; leaves alternate, simple, not lobed; stipules linear, caducous; flowers perfect, regular, borne in racemes; hypanthium short, with a glandular disk on the inner surface; sepals 5, persistent; petals 5; stamens usually 10 or more; pistils 1, the ovary inferior, usually 5-loculed (appearing as 10); styles usually 5, the stigmas capitate; fruit a reddish to purplish, often glaucous pome.

Amelanchier alnifolia (Nutt.) Nutt.
Northern Serviceberry

Plants low shrubs to small trees, mostly 0.3–5 m tall; leaves petiolate, the blades mostly 20–50 mm long and 15–40 mm broad, oval to oblong, acute to rounded or subcordate basally, rounded to truncate apically, serrate near the apex, glabrous to hairy; flowers in short racemes; sepals 1.5–3 mm long; petals 5–14 mm long, white to pinkish; fruit purplish, subglobose, 6–10 mm long, palatable.

Open woods, often on dry slopes; in central interior, southern, and southeastern Alaska, and southern Yukon; eastward to Hudson Bay and Saskatchewan and southward to California, Arizona, New Mexico, and Nebraska (A. *florida* Lindl.; A. *gormani* Greene).

ARUNCUS L.

Perennial dioecious herbs; leaves alternate, ternate to triternate, the segments broad; stipules lacking; flowers imperfect, regular, borne in panicles; hypanthium saucer shaped, with an internal glandular disk; sepals 5; petals 5; stamens (or staminodes in the pistillate flowers) 15–20; pistils 2–5 (vestigial in staminate flowers), distinct, the ovaries superior, each 1-loculed; styles 1 per pistil, the stigma capitate; fruit a few-seeded follicle.

Fernald, M. L. 1936. Memoranda on *Aruncus*. Rhodora 38:179–82; 237.

Aruncus sylvester Kostel.
Goatsbeard

Plants erect or ascending, mostly (0.3) 1–2 m tall; leaves mostly 10–50 cm long, smaller

Aruncus sylvester Kostel. (× 0.3).

and with fewer segments above, the segments sharply serrate, long-acuminate, pubescent to glabrous; panicles mostly 10–50 cm long, the spreading to pendulous branchlets spikelike, with numerous short-pedicellate flowers; sepals less than 0.5 mm long; petals about 1 mm long; follicles 2–3 mm long, glabrous, reflexed.

Moist woods; from Attu Island and the Shumagin Islands eastward along the coast to the Panhandle; southward to California; Eurasia (*Spiraea aruncus* L.; *S. acuminata* Dougl.; *Aruncus acuminatus* [Dougl.] Rydb.; *A. kamchaticus* Rydb.).

CHAMAERHODOS Bunge

Plants perennial (biennial?) herbs; leaves alternate or basal, bi- or triternately divided, the segments narrow; stipules foliose, narrowly oblong, simple or divided, persistent; flowers perfect, regular, borne in bracteate, corymbose cymes; hypanthium cup shaped, long-hairy within; sepals 5; petals 5; stamens 5, borne at the base of the petals; pistils 5–15, distinct, the ovaries superior, each 1-loculed; styles 1 per pistil, the stigma capitate; fruit an achene.

Chamaerhodos erecta (L.) Bunge ex Ledeb.
American Chamaerhodos

Plants erect, mostly 10–30 cm tall, from a taproot and a basal rosette, the stems freely branched above the base; leaves mostly 15–40 mm long, the ultimate segments linear to oblong, sparsely long-hairy; cymes making up one-quarter to one-half or more of the plant height; flowers short-pedicellate, inconspicuous; sepals 1.2–2.5 mm long, triangular, sparsely hirsute; petals white, equaling or slightly longer than the sepals; achenes 1.2–1.5 mm long, glabrous, grayish.

Limestone outcrops, gravelly lake shores, dry slopes, and sandy flats; in south-central Alaska (upper Matanuska Valley) and southern Yukon; eastward to Michigan and south to Utah, Colorado, and North Dakota; Asia (*C. erecta* var. *nuttallii* T. & G.;

C. erecta ssp. *nuttallii* [T. & G.] Hultén; *C. nuttallii* [T. & G.] Rydb.; *Sibbaldia erecta* var. *parviflora* Nutt.). Our material belongs to var. *parviflora* (Nutt.) C. L. Hitchc.

Chamaerhodos erecta (L.) Bunge (× 0.4).

CRATAEGUS L.

Shrubs or small trees, the branches armed with stout thorns; leaves alternate, simple, toothed to lobed; stipules broad, toothed, deciduous; flowers perfect, regular, borne in corymbs; hypanthium short, lined with a glandular disk; sepals 5, persistent; petals 5; stamens 10–20; pistils 1, the ovary inferior, usually 5-loculed; styles usually 5, the stigmas capitate; fruit a blackish pome.

Crataegus douglasii Lindl.
Black Hawthorn

Plants mostly 1–4 m tall, the thorns 1–3 cm long; leaves petiolate, the blades mostly 3–8 cm long and 1.5–5 cm broad, elliptical to oval, obovate, or oblong, cuneate basally, acute to acuminate apically, serrate and often lobed, pubescent to glabrous; flowers showy; sepals 1.5–2.5 mm long, triangular, reflexed; petals 5–7 mm long, white; fruit blackish when ripe, subglobose, 8–10 mm long, edible.

Woods; in coastal south-central and extreme southeastern Alaska; southward to California and east to Ontario, Michigan, and New Mexico. Our material belongs to var. *douglasii*.

DRYAS L.

Shrubs or subshrubs, with stoloniferous branches; leaves alternate, simple, crenate to entire, sometimes incised at the base, evergreen; stipules narrowly lanceolate, adnate to the petioles, persistent; flowers perfect (rarely imperfect), regular, solitary; hypanthium saucer shaped, with an internal glandular disk; sepals 8–10, persistent; petals 8–10; stamens numerous; pistils numerous, distinct, the ovaries superior, each 1-loculed; styles 1 per pistil, much elongating in fruit; fruit an achene with a long plumose style.

Hultén, E. 1959. Studies in the genus *Dryas*. Sv. Bot. Tidskr. 53:507–42.

Porsild, A. E. 1947. The genus *Dryas* in North America. Can. Field-Nat. 6:175–92.

1a. Sepals ovate; petals yellow; flowers never opening flat; staminal filaments pubescent near the base; scape with 1–4 bracts; leaves tapering to the base. *D. drummondii*

1b. Sepals narrowly oblong or lanceolate; petals white; flowers opening flat; staminal filaments glabrous; scape bractless or with a solitary bract; leaves usually rounded to subcordate at the base. (2)

2a. Leaf blades entire, or crenate only in the lower portion (rarely almost all crenate), neither glandular (or uncommonly so) nor with hairy processes on the lower midveins or petioles. *D. integrifolia*

2b. Leaf blades crenate throughout, often with glandular or hairy processes on the lower midveins or petioles. *D. octopetala*

Dryas drummondii Richards. ex Hook.
Yellow Dryas

Petioles hairy, often with stalked reddish glands; leaf blades mostly 0.9–5 cm long and 0.6–2 cm broad, elliptic to oblong or oblanceolate, obtuse to rounded apically, tapering (and often lobed or incised) at the base, crenate, green and glabrous or pubescent above, densely tomentose below, often somewhat revolute; scapes 4–30 cm long, tomentose, sparsely stipitate-glandular; sepals 4–8(10) mm long, ovate, stipitate-glandular; petals yellow, 8–12 mm long; staminal filaments hairy; styles plumose in fruit, to 4 cm long.

Gravelly soils of alluvial fans, terraces, and floodplains; in northeastern, south-central, and southeastern Alaska and southern Yukon; disjunctly eastward to Newfoundland and south to Oregon and Montana (*D. tomentosa* Farr.; *D. drummondii* var. *tomentosa* [Farr.] L. O. Williams; *D. drummondii* var. *eglandulosa* Porsild). In full fruit, when in morning or evening light, this species presents a remarkably beautiful sight.

Dryas integrifolia Vahl

Petioles sparsely to densely tomentose, lacking glands (rarely with them) and processes; leaf blades mostly 0.8–4.5 cm long and 0.3–1.2 cm broad, narrowly oblong to lanceolate or ovate, obtuse to rounded apically, rounded to subcordate (often obliquely so) at the base, entire or crenate along the lower portion (rarely crenate throughout), green and glabrous to sparsely tomentose above, densely tomentose below, often somewhat revolute; scapes 1.5–23 cm long, tomentose, often sparsely stipitate-glandular; sepals 5–10 mm long, narrowly oblong to lanceolate, stipitate-glandular; petals white (fading yellowish), 8–15 mm long; staminal filaments glabrous; styles plumose in fruit, to 4 cm long.

Arctic and alpine tundra to heathlands and woods; in most of Alaska (except for the southwestern portion) and most of the Yukon; eastward to Greenland and south to Montana and New Hampshire; Asia.

1a. Leaf blades mostly 0.7–1.5(1.9) cm long and 0.3–0.7 cm broad; scapes mostly 1.5–14 cm long; plants common in arctic and alpine tundra or heathlands (*D. chamissonis* Spreng. ex Juz.; *D. integrifolia* var. *canescens* Simm.; *D. integrifolia* f. *canescens* [Simm.] Fern.). *D. integrifolia* var. *integrifolia*

1b. Leaf blades mostly 1.5–4.5 cm long and 0.6–1.2 cm broad; scapes commonly (3)9–23 cm long; plants most common in woods or heathlands (*D. crenulata* Juz.; *D. integrifolia* var. *subintegrifolia* Hultén; *D. integrifolia* f. *sylvatica* [Hultén] Porsild; *D. integrifolia* ssp. *sylvatica* [Hultén] Hultén; *D. sylvatica* [Hultén] Porsild). *D. integrifolia* var. *sylvatica* Hultén

Dryas octopetala L.

Petioles sparsely to densely tomentose, often stipitate-glandular, or with white hairy processes; leaf blades mostly 1–4 cm long and 0.3–1.5 cm broad, elliptic to oblong, oblanceolate, or lanceolate, obtuse to rounded apically, truncate to rounded or subcordate basally, crenate, green and glabrous to pubescent above, tomentose (rarely glabrate below), mostly with stipitate glands or with hairy processes on the midrib below, often revolute; scapes 1–18 cm long, tomentose, often somewhat stipitate-glandular; sepals 6–10 mm long, narrowly oblong to lanceolate, stipitate-glandular; petals white (fading yellowish) or rarely yellowish, 9–17 mm long; staminal filaments glabrous; styles plumose, in fruit to 4 cm long.

Arctic and alpine tundra and heath or woodlands; through most of Alaska (except for the southwestern portion) and most of the Yukon; eastward to Newfoundland and south to Oregon, Utah, and Colorado (*D. octopetala* var. *luteola* Hultén, a form with yellowish petals). Two main types of processes are present to a greater or lesser extent (sometimes absent) on the midveins of the lower leaf surface and also on the petioles. These must be viewed under high magnification (20–30 x) for accurate determination. On the basis of process type, our material has been segregated into various taxa.

1a. Leaves commonly bearing short to elongate hairy processes on the lower midveins or petioles (often both); plants broadly distributed in Alaska and Yukon (*D. punctata* Juz.; *D. octopetala* var. *viscida* Hultén; *D. octopetala* var. *argentea* Blytt; *D. octopetala* f. *argentea* [Blytt] Hultén). *D. octopetala* var. *octopetala*

1b. Leaves commonly bearing sessile to elongate, often reddish, capitate glands on the lower midveins or petioles (often both); plants with about the same range as var. *octopetala* (*D. alaskensis* Porsild; *D. octopetala* ssp. *alaskensis* [Porsild] Hultén; *D. kamtschatica* Juz.). *D. octopetala* var. *kamtschatica* (Juz.) Hultén

FRAGARIA L.

Plants perennial stoloniferous herbs from short rhizomes, with fibrous roots; leaves basal, long-petiolate, 3-foliolate; stipules lanceolate, adnate to the petioles, persistent; flowers perfect, regular, borne in bracteate cymes; hypanthium saucer shaped, short; sepals 5, alternating with 5 sepallike bracteoles; petals 5; stamens about 20; pistils numerous, borne on a hemispherical receptacle, distinct, the ovaries superior, each 1-loculed; styles 1 per pistil, attached laterally; fruit an achene, borne on the fleshy receptacle which ripens to form the edible "accessory fruit."

Rydberg, P. A. 1908. *Fragaria. In*: Rosaceae, by P. A. Rydberg. N. Am. Fl. 22: 239–533.

1a. Pubescence of petioles and scapes of long spreading to descending hairs; leaves thick, with conspicuous elevated veins on the lower surface; plants maritime or cultivated. *F. chiloensis*

1b. Pubescence of petioles and scapes appressed to ascending; leaves not especially thick, the veins of the lower surface not conspicuously elevated; plants mostly of the interior, seldom cultivated. *F. virginiana*

Fragaria chiloensis (L.) Duchesne
Beach Strawberry

Petioles 2–20 cm long or more, often reddish tinged, silky-villous with long yellowish, broadly spreading to descending hairs; leaflets stalked, 1.5–4 cm long, thick, coarsely serrate, the upper surface green, shiny, glabrate or glabrous, the lower surface pale, silky-villous (glabrate in age) and often somewhat tomentose, the veins conspicuously elevated; scapes 3–15 cm long, villous with long yellowish broadly spreading to descending hairs; sepals 5–10 mm long, enlarging somewhat in fruit; petals white, mostly 6–12(15) mm long; accessory fruit mostly at least 1.5 cm in diameter; achenes 1.5–2 mm long.

Woods, meadows, and gravelly beaches; along coastal southern Alaska, from the Aleutians through the southeast; southward to California; South America (*F. vesca* var. *chiloensis* L.; *F. chiloensis* ssp. *pacifica* Staudt.) This species has served as parental stock for most of the cultivated strawberries, which are designated as var. *ananassa* Bailey. Several different horticultural strains are cultivated in the more temperate parts of Alaska and Yukon. The fruits often grow to a remarkable size.

Fragaria virginiana Duchesne
Wild Strawberry

Petioles 2–17 cm long, usually not reddish tinged, sparsely villous with inconspicuous appressed to ascending hairs; leaflets stalked, 1.5–9.5 cm long, not especially thick, coarsely serrate, the upper surface green or bluish green, glabrous, the lower surface glaucous, silky-villous to glabrate, not at all tomentose, the veins not conspicuously elevated; scapes 3–15 cm long, sparsely villous with appressed to ascending hairs; sepals mostly 4–6 mm long, somewhat enlarging in fruit; petals white, commonly 6–8(12) mm long; accessory fruit usually less than 1.5 cm broad; achenes 1–1.5 mm long.

Open woods in dry to moist sites; in central to east-central Alaska and southern Yukon; eastward to the Atlantic and south to California, Colorado, and Georgia (*F. yukonensis* Rydb.; *F. virginiana* ssp. *glauca* [Wats.] Staudt.). Our material belongs to var. *glauca* Wats.

GEUM L.

Plants perennial, more or less rhizomatous herbs; leaves basal, or some cauline, pinnatifid to pinnately compound, the terminal lobe often larger than the lateral (lyrate); stipules lanceolate or broader, adnate to the petioles, persistent; flowers perfect, regular, solitary or few to several borne in bracteate cymes; hypanthium saucer to cup shaped; sepals 5, alternating with sepallike bracteoles; petals 5; stamens numerous; pistils numerous, the ovaries superior, each 1-loculed; styles 1 per pistil, often elongate and persistent on the achene; fruit a hispid achene.

1a.   Leaves glabrous; petals white (fading yellow); basal leaves mostly 1.5–3 cm long; plants of the Aleutians. *G. pentapetalum*

1b.   Leaves at least somewhat hairy; petals yellow (rarely whitish); basal leaves mostly 3–30 cm long; plants of various distribution. (2)

2a.   Leaves with the terminal lobe or leaflet much expanded (lateral lobes smaller and sometimes lacking), mostly more than 2 cm broad. (3)

2b.   Leaves with the terminal lobe not expanded, the lateral lobes or leaflets similar in size or shape, or larger, often less than 1 cm broad (except in *G. aleppicum*). (4)

3a.   Terminal lobes of leaves broadly rounded, reniform to orbicular, not or only slightly lobed; bracteoles 2–6 mm long, persistent; styles not abruptly bent near the apex; plants mostly less than 3 dm tall. *G. calthifolium*

3b.   Terminal lobes of leaves ovate to orbicular, or less commonly reniform, often conspicuously lobed; bracteoles 0.5–3 mm long, deciduous; styles with an abrupt S-shaped bend near the apex, the apical segment finally deciduous; plants often over 3 dm tall. *G. macrophyllum*

4a.   Plants commonly 3–8 dm tall; cauline leaves 3–5, well developed, the lowermost with conspicuous petioles; plants uncommon in south-central Alaska and southern Yukon. *G. aleppicum*

4b.   Plants 0.3–2.6 dm tall (or if taller, then distribution usually different); cauline leaves 1–3, rarely well developed, sessile or short-petiolate. (5)

5a. Plants conspicuously yellowish-pilose, the hairs 2–3 mm long or more; petals 12–18(20) mm long. *G. glaciale*

5b. Plants inconspicuously and often sparsely strigulose, the hairs less than 1 mm long; petals 8–12(15) mm long. (6)

6a. Styles glabrous except at the base; basal leaves mostly 3–12 cm long; plants 0.3–2.6 dm tall, widely distributed in Alaska and Yukon. *G. rossii*

6b. Styles stiffly hairy except at the apex; basal leaves 8–16 cm long; plants 2–4 dm tall, known only from the Alaska Peninsula and the Aleutian Islands. *G. × macranthum*

Geum aleppicum Jacq.

Plants mostly 3–8 dm tall, the stems erect or ascending, pubescent with spreading hairs, with 3–5 well-developed leaves below the inflorescence (the lowermost cauline leaves conspicuously petiolate); basal leaves 8–25 cm long, long-petiolate, the leaflets 7–15 or more, the apical lobe (leaflet) 2–7 cm long and 1.5–7.5 cm broad, cuneate to obtuse basally, acute to acuminate apically, dentate and conspicuously lobed or incised, glabrate to glabrous above, hairy along the veins beneath; sepals 5–8 mm long, reflexed in fruit; bracteoles 1.5–3 mm long, narrowly lanceolate; petals yellow, 5–9 mm long; styles elongate, with a distinct S-shaped bend near the tip, commonly hairy above the bend, at length descending or reflexed and the apical sections deciduous.

Meadows, open woods, and thickets; in south-central Alaska and southern Yukon; eastward to Newfoundland and south to California, New Mexico, Illinois, and New Jersey; Eurasia (*G. strictum* Ait.; *G. aleppicum* var. *strictum* [Ait.] Fern.; *G. aleppicum* ssp. *strictum* [Ait.] R. T. Clausen).

Geum calthifolium Menzies ex Rees
Caltha-leaf Avens

Plants mostly 1–3 dm tall, the stems erect, pubescent with spreading hairs, often with 1–3 sessile leaves below the inflorescence; basal leaves 4–25 cm long, long-petiolate, simple or with 1–few tiny leaflets below the greatly enlarged terminal leaflet, the apical leaflet 1.5–6 cm long (from apex of rachis) and 3–10 cm broad, reniform to orbicular, cordate basally, rounded apically, dentate, but not or only slightly lobed, glabrate to glabrous above, hairy along the veins below; sepals 8–14 mm long; bracteoles 2–6 mm long, lanceolate, simple or cleft; petals yellow, 8–12 mm long; styles

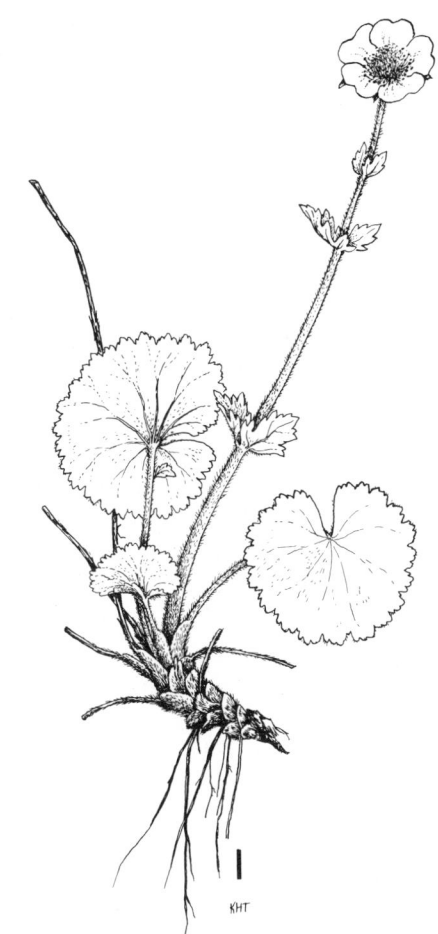

Geum calthifolium Menzies (× 0.4).

elongate, stiffly hairy (except for the apical part), erect.

Moist sites in meadows or heathlands, often in alpine regions; in coastal and insular southern Alaska, from the Aleutians eastward through the Panhandle and southward to British Columbia; Asia.

Geum glaciale Adams ex Fisch.
Glacier Avens

Plants 0.3–2.5 dm tall, the stems erect, pubescent with spreading, yellowish hairs, the hairs mostly 2–3 mm long or more, often with 1–4 greatly reduced leaves below the inflorescence; basal leaves 3–10 cm long, sessile or short-petiolate, pinnatifid, with mostly 17–23 entire or 2-toothed lobes, the apical lobe similar to the lateral ones except smaller, glabrous to glabrate above, pilose below along the veins, ciliate with hairs 2–3 mm long or longer; sepals 7–12 (15) mm long, ovate-lanceolate; bracteoles 5–8 mm long, lanceolate; petals yellow, 12–18(20) mm long; styles elongate, stiffly hairy (except for the apical part), erect.

Damp slopes, in arctic or alpine tundra or heathlands; from the Seward Peninsula northward to the northwest coast, then east along the Brooks Range to northeastern Alaska, northern Yukon and adjacent Mackenzie, and less commonly in central eastern Alaska and on St. Lawrence Island; Asia. This is a most beautiful species.

Geum × macranthum (Kearney) B. Boi.

Plants mostly 2–4 dm tall, the stems erect, pubescent with short, spreading hairs, with 1–2 small leaves below the inflorescence; basal leaves 8–16 cm long, pinnatifid, with mostly 11–19 several-toothed lobes, the apical lobe like the larger lateral ones, except often more deeply cleft, somewhat pubescent above and below, especially along the veins, ciliate with short, stiff hairs; sepals 7–10 mm long, ovate-lanceolate; bracteoles 4–8 mm long, lanceolate; petals yellow, 8–12(15) mm long; styles

stiffly hairy, except for the apical portion, erect.

Alpine meadows; in the Aleutian Islands and the Alaska Peninsula. *G. × macranthum* is made up of a series of robust, large-flowered plants whose vegetative aspect resembles that of *G. rossii*. However, the plants tend to average larger than in that entity. There is also a tendency for the terminal lobe of the leaf to be more coherent than in *G. rossii*. In that regard and in the stiffly hairy styles, *G. × macranthum* resembles *G. calthifolium*. The pollen of both *G. rossii* and *G. calthifolium* is readily stainable and fully rounded. That of *G. × macranthum* is collapsed, clumped, and not stainable. In addition, fully formed fruit of *G. × macranthum* is unknown. It appears that *G. × macranthum* is of hybrid origin, having resulted from hybridization between *G. rossii* and *G. calthifolium*.

Geum macrophyllum Willd.
Large-leaf Avens

Plants 2.8–10 dm tall, the stems erect, pubescent with spreading hairs, often with 2–4(5) subsessile to short-petiolate leaves below the inflorescence; basal leaves 12–30 cm long, long-petiolate, the leaflets 9–25 or more, the apical lobe (leaflet) 2.5–10 cm long (from apex of rachis) and 3–12 cm broad, acute to truncate or cordate basally, rounded to obtuse apically, dentate and often conspicuously lobed, glabrate to glabrous above, hairy along the veins beneath; sepals 4–5 mm long, reflexed in fruit; bracteoles 0.5–3 mm long, linear to lanceolate, or lacking; petals yellow, 4–7 mm long; styles elongate, with a distinctive S-shaped bend near the tip, hairy above the bend or glabrous, often glandular below the bend, at length descending or reflexed, and the apical sections deciduous.

Woods, beaches, wet meadows, and roadsides, commonly in disturbed sites; through much of Alaska from the Brooks Range southward (except for western Alaska) and in southern Yukon; eastward to Newfoundland, and south to Mexico; Asia.

1a. Terminal segments of basal leaves only shallowly lobed; those of the cauline leaves usually not deeply cleft; peduncles lacking glandular hairs or nearly so; plants of coastal and insular southern Alaska. *G. macrophyllum* var. *macrophyllum*

1b. Terminal segments of the basal leaves conspicuously lobed, those of cauline leaves deeply cleft or lobed; peduncles often glandular; plants mostly of interior Alaska and Yukon (*G. perincisum* Rydb.; *G. macrophyllum* ssp. *perincisum* [Rydb.] Hultén). *G. macrophyllum* var. *perincisum* (Rydb.) Raup

Geum pentapetalum (L.) Makino
Low Avens

Plants mostly 0.4–0.8 dm tall, from elongate, subrhizomatous caudex, the flowering stems erect, pubescent with short spreading hairs, often with 1–2 reduced leaves below the usually solitary flower; basal leaves 1.5–3 cm long, short-petiolate, pinnately compound; leaflets 3–11, few- to several-toothed; apical leaflet 0.8–1 cm long and 0.3–0.5 cm broad, cuneate, glabrous; sepals 5–7 mm long, lanceolate; bracteoles 3–4 mm long, lance-oblong; petals white (fading yellowish), 7–9 mm long; styles elongate, stiffly hairy except for the apical part, erect.

Moist sites; in the middle Aleutian Islands; Asia.

Geum rossii (R. Br.) Ser. ex DC.
Ross Avens

Plants 0.3–2.6 dm tall, the stems erect, often with 1–3 greatly reduced leaves below the inflorescence; pubescent with spreading hairs, the hairs mostly less than 1 mm long; basal leaves 3–12 cm long, short-petiolate, pinnatifid, with 15–21 entire to several-toothed or lobed lateral divisions, the apical lobe similar to the lateral ones, except smaller, glabrous to pubescent along the veins above, pubescent along the veins below, ciliate with hairs less than 1 mm long; sepals 5–10 mm long, ovate-lanceolate; bracteoles 3–7 mm long, lanceolate; petals yellow, 9–12 mm long; styles elongate, glabrous except at the base, erect.

Arctic and alpine tundra, in gravelly or peaty soils; from the Seward Peninsula disjunctly eastward along the Brooks Range to northern Yukon, and from the Aleutians and islands of the Bering Sea through southwestern, south-central, and east-central Alaska to central Yukon; east to Melville Island and south to Oregon, Nevada, Arizona, and New Mexico; Asia.

LUETKEA Bong.

Plants perennial herbs or subshrubs; leaves alternate, biternate, the segments narrow; stipules lacking; flowers perfect, regular, borne in bracteate racemes; hypanthium cup shaped, with an internal glandular disk; sepals 5; petals 5; stamens about 20; pistils 3–6, distinct, the ovaries superior, each 1-loculed; styles 1 per pistil, the stigmas capitate; fruit a stipitate, several-seeded follicle.

Luetkea pectinata (Pursh) Kuntze
Luetkea

Plants prostrate from rhizomes and stolons, mat-forming; flowering stems erect or ascending, 0.3–1.5 dm tall; leaves mostly 7–20 mm long, the ultimate segments linear-lanceolate to oblong, glabrous; racemes several-flowered; flowers pedicellate, moderately conspicuous; sepals 1.5–2.5 mm long, triangular, glabrous; petals white, 2.5–4 mm long; follicles mostly 4–5 mm long.

Alpine or subalpine meadows or stream gravels; from southwestern to south-central and southeastern Alaska and southern Yukon; southward to California, Idaho, and Montana (*Saxifraga pectinata* Pursh).

MALUS Mill.

Shrubs or trees, with unarmed branches; leaves alternate, simple, rarely somewhat lobed; stipules linear, caducous; flowers perfect, regular, borne in corymbose racemes; hypanthium short, lined with glandular disk; sepals 5; petals 5; stamens 15–many; pistils 1, the ovary inferior, usually 3–5-loculed; styles usually 3–5, the stigmas capitate; fruit a pome.

1a. Petals mostly 8–14 mm long; styles and carpels usually 3, the styles glabrous; plants indigenous to coastal and insular southern and southeastern Alaska. *M. fusca*

1b. Petals mostly 15–25 mm long; styles and carpels usually 4–5, the styles often hairy; plants cultivated. (2)

2a. Leaves sharply serrate; sepals deciduous. *M. floribunda*

2b. Leaves crenate-serrate; sepals persistent. *M. sylvestris*

Malus floribunda Sieb.
Showy Crab-apple

Small trees, mostly less than 5 m tall; leaves with long petioles, the blades mostly 4–8 cm long and 2–3.5 cm broad, elliptic to ovate, sharply serrate, pubescent to glabrous; flowers showy, bright pink to white, borne on long pedicels; sepals 8–12 mm long, pubescent to glabrate, narrowly oblong, acuminate, deciduous; petals mostly 15–20 mm long; styles usually 4, hairy at the base; fruit globose, about 1 cm long.

Sparingly cultivated ornamental tree; in southern and southeastern Alaska.

Malus fusca (Raf.) Schneid.
Western Crab-apple

Shrubs or small trees, mostly 2–5 m tall; leaves with long petioles, the blades 3–8.5 cm long and 1.2–5.5 cm broad, ovate to oblong or lanceolate, sharply serrate and often irregularly lobed; pubescent to glabrous; flowers showy, white, borne on long pedicels; sepals 3–5 mm long, narrowly triangular, acuminate, tardily deciduous; petals 8–14 mm long; styles usually 3, glabrous; fruit ovoid, 1–1.5 cm long.

Coastal and insular woodlands; from Prince William Sound and Kenai Peninsula eastward through the Panhandle; southward to California (*Pyrus fusca* Raf.; *Pyrus diversifolia* Bong.; *M. diversifolia* [Bong.] Roem.).

Malus sylvestris Mill
Apple

Small to moderate trees, to 10 m tall; leaves with long petioles, the blades mostly 3–10 cm long and 1.5–5.5 cm broad, elliptic to oblong, oval, or ovate, crenate-serrate, not lobed, pubescent to glabrate or glabrous; flowers showy, white suffused with pink; sepals mostly 6–10 mm long, lanceolate, persistent; petals mostly 15–25 mm long; styles usually 5, hairy at the base; fruit shaped variously, usually over 2 cm long.

Sparingly cultivated in southern and southeastern Alaska. Usually the plantings represent one or more of the several ornamental horticultural varieties.

PHYSOCARPUS Maxim. Nom. Cons.

Shrubs with unarmed branches; leaves alternate, simple, palmately 3-lobed; stipules narrowly lanceolate, deciduous; flowers perfect, regular, borne in terminal corymbs; hypanthium cup shaped, lined with a glandular disk; sepals 5; petals 5; stamens 20, or more; pistils 4–5, distinct or nearly so, the ovaries superior, each 1-loculed; styles 1 per pistil, the stigmas capitate; fruit a several-seeded, 2-valved capsule.

Physocarpus capitatus (Pursh) Kuntze
Pacific Ninebark

Plants 1–3(5) m tall, the branches with shreddy, exfoliating bark; leaves with long petioles, the blades 3–10 cm long and 4–12 cm broad, 3-lobed, ovate to orbicular in outline, truncate to subcordate basally, doubly serrate, stellate pubescent to glabrous; flowers showy; sepals 3–5 mm long, lanceolate, stellate-pubescent; petals white, mostly 4–6 mm long; capsules mostly 7–10 mm long, glabrous.

Woodlands; in coastal and insular southeastern Alaska; southward to California and Idaho (*Spiraea capitata* Pursh).

POTENTILLA L.

Plants perennial or annual herbs with rhizomes, stolons, or taproots; leaves basal or cauline and alternate, short- or long-petiolate, pinnately or palmately 3–several-foliolate; stipules lanceolate, persistent; flowers perfect (rarely imperfect), regular, solitary, or borne in bracteate cymes; hypanthium saucer to cup shaped, lined with a glandular disk; sepals 5, alternating with 5 sepallike bracteoles; petals 5; stamens mostly 10–30; pistils numerous, distinct, the ovaries superior, each 1-loculed; styles 1 per pistil, attached laterally or apically, jointed to the achene, usually deciduous; fruit an achene; receptacle not fleshy at maturity.

Hiitonen, I. 1949. Über die Ostfennoskandischen Formen und Bastarde der Kollektivart *Potentilla nivea* L. nebst Erörterung einiger anderen Arten der Nivea-Gruppe. Arch. Soc. Zoo. Bot. Fenn. 'Vanamo' 2:23–33.

Rydberg, P. A. 1908. Rosaceae. N. Am. Fl. 23:239–388.

Rousi, A. 1965. Biosystematic studies on the species aggregate *Potentilla anserina* L. Ann. Bot. Fenn. 2:47–112.

1a. Flowers reddish purple to dark red or maroon; petals less than half as long as the sepals. *P. palustris*
1b. Flowers white, cream, or yellow, not at all red or purple. (2)
2a. Plants woody shrubs; ovaries and achenes hairy. *P. fruticosa*
2b. Plants herbaceous; ovaries and achenes glabrous. (3)
3a. Plants strongly stoloniferous; flowers solitary on naked scapes. *P. anserina*
3b. Plants lacking stolons; flowers one to several on leafy stems. (4)
4a. Leaves pinnately compound with 5 or more leaflets. (5)
4b. Leaves palmately compound with 5 or more leaflets, or 3-foliolate. (9)
5a. Stems, petioles, and leaflets distinctly glandular-villous, the leaflets green on both sides, mostly once to twice longer than broad. *P. arguta*
5b. Stems, petioles, and leaflets variously pubescent or glabrous, but not glandular-villous, the leaflets often white-tomentose below, mostly more than twice longer than broad. (6)
6a. Leaves green above and below, or white-tomentose only on the lower surface. (7)
6b. Leaves silky-pilose on both surfaces. (8)
7a. Leaves green or equally grayish-hairy on both surfaces, the 5–7 leaflets irregularly toothed apically, cuneate and entire in the lower portion. *P. diversifolia*
7b. Leaves green above and below, or white-tomentose below, the 5–11 leaflets toothed or lobed to near the base. *P. pensylvanica*
8a. Leaves with 7–15 leaflets; plants mostly more than 1.5 dm tall, occurring in east-central Alaska. *P. hippiana*
8b. Leaves with (3)5–7 leaflets; plants mostly less than 1.5 dm tall, mostly of northern Alaska and Yukon. *P. pulchella*
9a. Leaves palmately compound, with 5 or more leaflets (at least some). (10)
9b. Leaves 3-foliolate. (13)
10a. Leaves green or equally grayish-hairy on both surfaces, the leaflets irregularly toothed apically, cuneate and entire in the lower portion (sometimes entire throughout). (11)

10b. Leaves green above, often tomentose below, the leaflets regularly or irregularly toothed to near the base. (12)

11a. Leaflets linear to oblong-oblanceolate, 0.8 cm broad to less, toothed only at the very apex; plants rare, north of the Brooks Range. *P. stipularis*

11b. Leaflets oblanceolate to oblong, 0.7–2 cm broad, toothed for some distance below the apex; plants of south-central Alaska and southern Yukon. *P. diversifolia*

12a. Plants mostly more than 3 dm tall; leaflets incised about one-half the distance to the midrib, always 5 or more on the basal leaves; distributed mostly in coastal and insular southern Alaska, less commonly in interior Alaska and Yukon. *P. gracilis*

12b. Plants mostly less than 3(3.5) dm tall; leaflets incised almost to the midrib, usually only 3 on at least some basal leaves; distributed mostly in interior to northern Alaska and Yukon, rarely coastal. *P. hookeriana*

13a. Plants annual or biennial, from taproots; leaves mostly cauline, green on both surfaces, never white-hairy or tomentose below. (14)

13b. Plants perennial from caudices or rhizomes; leaves mostly basal, variously pubescent, but most species with a white tomentum on the lower surface. (15)

14a. Stems and lower leaf surfaces with stiff hairs; stamens 15–20; plants of broad distribution in Alaska and Yukon. *P. norvegica*

14b. Stems and lower leaf surfaces with soft, slender, often glandular hairs; stamens 10–15; plants known from the vicinity of Dawson City, Yukon. *P. biennis*

15a. Leaflets not whitish- or grayish-tomentose on either side, variously pubescent to glabrous on both surfaces. (16)

15b. Leaflets tomentose on one or both surfaces, the upper surface mostly sharply contrasting in color with the lower. (18)

16a. Plants mostly 1.5–4 cm tall; leaves (0.5)1.5–2 cm long, the terminal leaflet 2.5–7 mm long and 2.5–5 mm broad; sepals 1.8–2.5 mm long. *P. elegans*

16b. Plants mostly 3–20 cm tall; leaves 2–10 cm long, the terminal leaflet 7–22 mm long or more and 5–17 mm broad; sepals 3.5–6.5 mm long. (17)

17a. Lateral leaflets usually deeply incised into 2 narrowly oblong lobes, and the terminal into 3 narrow lobes, usually glaucous below, revolute. *P. biflora*

17b. Lateral leaflets and the terminal ones 3–11-toothed, not at all glaucous below, usually not revolute. *P. hyparctica*

18a. Leaflets white-villous or tomentose on both surfaces, often 5 on at least some leaves. *P. pulchella*

18b. Leaflets variously pubescent, but evidently green above (rarely tomentose above in *P. nivea*), tomentose beneath. (19)

19a. Petioles and at least the lower portions of stems cobwebby-tomentose, not at all pilose or villous with long hairs. *P. nivea*

19b. Petioles and stems pilose or villous with long hairs (sometimes tomentose beneath the longer outer hairs). (20)

20a. Bracteoles of the calyx mostly 1.5–2 times longer than broad, ovate to lanceolate; plants mostly long-villous throughout, usually coastal. *P. villosa*

20b. Bracteoles of the calyx commonly 2–4 times longer than broad, narrowly oblong to lanceolate; plants variable in pubescence, usually not long-villous throughout, mostly in the interior, rarely coastal. (21)

21a. Plants often less than 1.5 dm tall; stems commonly somewhat tomentose beneath the longer outer hairs; flowers often over 1.5 cm broad. *P. uniflora*

21b. Plants often more than 1.5 dm tall; stems not at all tomentose beneath the longer outer hairs; flowers mostly less than 1.5 cm broad. *P. hookeriana*

Potentilla anserina L.
Common Silverweed

Perennial herbs with long, strawberrylike stolons; leaves 0.2–3.5 dm long, pinnately compound with 5–17(25) main leaflets interspersed by smaller ones, the terminal leaflet 0.6–5.5 cm long and 0.3–2.6 cm broad, oval to oblong, coarsely serrate, green and glabrous to pubescent above, pale and villous to tomentose (rarely glabrous) beneath; scapes 3–28 cm long, glabrous to spreading-villous, leafless; sepals ovate, 3–8(10) mm long, pubescent or glabrous, erect and enlarging in fruit; petals yellow, 7.5–16 mm long; achenes 1.5–2 mm long.

Beaches, tidal flats, river banks, lake shores, and meadows; along the coast and river systems of Alaska and Yukon; circumboreal. Three more or less poorly defined infraspecific taxa are known from our region.

1a. Leaflets with silvery, silky hairs over the tomentum on the lower surface; bracteoles commonly lobed; scapes and rachis of leaves often conspicuously spreading-hairy; plants primarily of interior Alaska and Yukon (*P. yukonensis* Hultén; *P. egedii* ssp. *yukonensis* [Hultén] Hultén). *P. anserina* var. *anserina*

1b. Leaflets with silvery, silky hairs only along the veins, the tomentum distinctly visible; bracteoles not or only rarely lobed; scape and leaf rachis glabrous to sparsely spreading-hairy; plants primarily of coastal Alaska. (2)

2a. Apical leaflet less than twice longer than broad, mostly with 4–6 rounded teeth along each side; plants primarily from western Alaska (*P. egedii* Wormskj.; *P. anserina* ssp. *egedii* [Wormskj.] Hiit.; *P. egedii* var. *groenlandica* [Tratt.] Polunin). *P. anserina* var. *groenlandica* Tratt.

2b. Apical leaflet often more than twice longer than broad, mostly with 6–10 sharp teeth along each side; plants from southwestern, southern and southeastern Alaska, rarely elsewhere (*P. pacifica* Howell; *P. anserina* ssp. *pacifica* [Howell] Rousi; *P. egedii* var. *grandis* [T. & G.] Hara; *P. egedii* ssp. *grandis* [T. & G.] Hultén). *P. anserina* var. *grandis* T. & G.

Potentilla arguta Pursh
Glandular Cinquefoil

Perennial, glandular-pubescent herbs, 2.5–8 dm tall, from a caudex; basal leaves 0.6–3 dm long, pinnately compound with 5–11 leaflets, the terminal one 1.5–9 cm long and 1.2–4 cm broad, oval, elliptic or obovate, doubly dentate to somewhat lobed, green and glandular-pubescent or sparsely hirsute to glabrate on both surfaces; flowers several to many, showy; sepals 4–8 mm long, lance-ovate, longer in fruit; bracteoles 2–6 mm long, oblong to narrowly lanceolate; petals yellow to cream or white, mostly 5–8 mm long; receptacle sparsely hairy; achenes 1–1.5 mm long.

Dry to wet woods or bogs; in south-central Alaska and southern Yukon; southward to Oregon, Nevada, Arizona, and New Mexico and eastward to Pennsylvania and New Jersey (*P. convallaria* Rydb.; *P. arguta* ssp. *convallaria* [Rydb.] Keck; *P. arguta* var. *convallaria* [Rydb.] Wolf).

Potentilla biennis Greene

Annual or biennial herbs, mostly 1–6 dm tall, from taproots; leaves mostly cauline, palmately 3(4–5)-foliolate, the terminal leaflet 1–5 cm long and 1–3 cm broad, obovate to oblanceolate, closely crenate-serrate, pubescent with spreading to appressed hairs and multicellular, glandular ones; flowers several, inconspicuous; sepals ovate-triangular, mostly 2–4 mm long; bracteoles ovate-lanceolate to oblong, 2–3 mm long; petals yellow, about 1.5–3 mm long; achenes numerous, about 1 mm long.

Weedy species; known in our region from the vicinity of Dawson City, Yukon; widely distributed in western North America. The occurrence of *P. biennis* in the Yukon may have been due to introduction with forage or food grains during the era of the Gold Rush at the turn of the century. It has not been collected in recent times.

Potentilla biflora Willd. ex Schlecht.
Two-flower Cinquefoil

Perennial herbs, 4–15 cm tall, from a caudex; basal leaves mostly 0.2–0.7 dm long, pinnately 3-foliolate, the lateral leaflets commonly deeply cleft into 2 narrowly oblong lobes, the terminal leaflet commonly with 3 narrowly oblong lobes, 0.8–1.7 cm long, oval in general outline, green and sparsely hairy to glabrous above, pale and glaucous to glabrous or less commonly hairy below, the margin revolute; flowers one to several, showy; sepals 3.5–6 mm long, ovate-lanceolate; bracteoles 4–6.5 mm long, lanceolate to oblong; petals yellow, 6–10 mm long, emarginate; receptacle hairy; achenes numerous, 1.5–2 mm long.

Dry to moist hillsides, in gravelly soils or in humus, in tundra and heathlands; in coastal, southwestern Alaska, northward to the northwest coast, eastward along the Brooks Range to northeastern Alaska and northern Yukon, and in south- to east-central Alaska and western Yukon; eastward to the Mackenzie and south to British Columbia; Asia.

Potentilla diversifolia Lehm.

Perennial, nonglandular herbs, mostly 1.5–4.2 dm tall, from a caudex; basal leaves 0.3–2 dm long, palmately, subpalmately, or pin-

nately compound, with 5–7 leaflets, the terminal one 1.5–4.5 cm long and 0.7–2 cm broad, oblanceolate to oblong, coarsely toothed apically, entire and cuneate basally, green and glabrate to glabrous above, green (though often paler) and usually strigose below; flowers few to several, showy; sepals 4–7 mm long, lanceolate; bracteoles 2.5–5 mm long, narrowly lanceolate; petals yellow, 6–11 mm long, emarginate; receptacle hairy; achenes numerous, 1–1.5 mm long.

Woods, heathlands, and alpine meadows; in south-central Alaska and southern Yukon; eastward to Greenland and south to California, Nevada, Utah, New Mexico, and Saskatchewan (*P. diversifolia* var. *genuina* Wolf; *P. diversifolia* var. *glaucophylla* Lehm.). Our material belongs to var. *diversifolia*.

Potentilla elegans Cham. & Schlecht.
Elegant Cinquefoil

Dwarf perennial herbs, mostly 1.5–4 cm tall, from a caudex; leaves (0.5)1.5–2 cm long, palmately 3-foliolate, the terminal leaflet 2.5–7 mm long and 2.5–5 mm broad, obovate, apically toothed, green and sparsely hairy to glabrate on both surfaces, the lateral leaflets mostly 5-toothed; flowers solitary, inconspicuous; sepals 1.8–2.5 mm long, lanceolate; bracteoles 1.5–2 mm long; petals pale yellow to white, 2–2.8(3.5) mm long; receptacle hairy; achenes several, 1–1.5 mm long.

Alpine regions, often on rocky outcrops; known from disjunct areas from the Seward Peninsula, Brooks Range, Alaska Range, and east-central Alaska and eastern to northern Yukon; east to Mackenzie; Asia.

Potentilla fruticosa L.

Shrubby Cinquefoil, Yellow Rose, Tundra Rose

Shrubs to 1 m tall or more; bark shreddy; leaves 1–5 cm long, pinnately 3–7-foliolate, the terminal leaflet 0.5–2.5 cm long and

cept for southwestern and southeastern portions) and most of the Yukon; east to Newfoundland and south to California, Utah, New Mexico, Iowa, and New Jersey; Eurasia. *P. fruticosa* is a beautiful plant which lends itself to ornamental plantings.

Potentilla gracilis Dougl. ex Hook.
Slender Cinquefoil

Perennial, nonglandular herbs, 2.8–10 dm tall, from a caudex; basal leaves 0.5–4 dm long, palmately (rarely subpinnately) compound with 5–9 leaflets, the terminal one 3–12 cm long and 1.3–3.5 cm broad, obovate to oblanceolate, coarsely toothed to near the base, the sinuses usually about one-half the distance to the midrib, green and hairy to glabrous above, usually paler and more densely hairy beneath; flowers several to many, showy; sepals 4.5–7 mm long, ovate-lanceolate; bracteoles 2.5–5 mm long, lanceolate; petals yellow, 7–10 mm long, emarginate; receptacle glabrous or nearly so; achenes numerous, 1–1.5 mm long.

Woodlands, thickets, and meadows; mostly in coastal southern Alaska, but also along major drainage systems in eastern Alaska and southern Yukon; eastward to Saskatchewan and south to Baja California, Arizona, New Mexico, and Nebraska (*P. alaskana* Rydb.; *P. flabelliformis* Lehm.). *P. gracilis* is usually distinctive in our region. However, along Turnagain Arm, near Anchorage, there exists a series of populations which are apparently intermediate to *P. hookeriana,* and the 5-foliolate phase of the taxon in that region may represent the influence of *P. gracilis.*

Potentilla hippiana Lehm.
Woolly Cinquefoil

Perennial, silky-villous herb, mostly 3–5 dm tall, from a caudex; basal leaves pinnately 5–7-foliolate, the leaflets white or grayish-silky on both surfaces, mostly 2–5 cm long, coarsely toothed; flowers several to many,

Potentilla fruticosa L. (× 0.7).

0.2–1 cm broad, oblong to elliptic, entire, green and sparsely hairy to glabrate above, grayish and silvery hairy below, somewhat revolute; flowers 1–several, conspicuous; sepals 3.5–9 mm long, ovate-lanceolate; bracteoles 4–13 mm long, lanceolate to elliptic; petals yellow, 6–14 mm long, rounded; receptacle hairy; achenes 1.5–2 mm long, white-hairy.

Woods, heathlands, and tundra, in dry to wet habitats; through most of Alaska (ex-

showy; sepals 5–7 mm long, lanceolate; bracteoles lanceolate, shorter than the sepals; petals yellow, 6–8 mm long, emarginate; achenes numerous.

Porsild (Rhodora 41:246. 1939) reports this species from Telegraph Creek, British Columbia, and from the area between Summit and McCarty in the Alaska Range district. The plant has not been collected in recent years.

Potentilla hookeriana Lehm.
Hooker Cinquefoil

Perennial, nonglandular herbs, 8–25(50) cm tall, from a caudex, the stems and petioles with long spreading hairs, not at all tomentose; basal leaves 0.2–1.3 dm long, palmately (rarely subpinnately) compound with 3 (less commonly 5) leaflets, the terminal one 0.8–3.5 cm long, and 0.4–1.5 cm broad, elliptic or oblong to oblanceolate or obovate, coarsely toothed to near the base, the sinuses often extending to near the midrib, green and strigose above, pale and tomentose below; flowers few to several, small but showy; sepals 3–6 mm long, ovate-lanceolate; bracteoles 2–5 mm long, oblong to lanceolate; petals yellow, 4–6(7) mm long, emarginate; receptacle sparsely hairy; achenes several, 1–1.5 mm long.

Gravelly soil or rocky outcrops, in open woodlands to tundra; in most of Alaska (except for southwestern and southern coastal regions) and most of the Yukon; eastward to Mackenzie and southward to British Columbia, Alberta, and Saskatchewan; Asia (*P. chamissonis* Hultén; *P. hookeriana* ssp. *chamissonis* [Hultén] Hultén; *P. furcata* Porsild; *P. hookeriana* var. *furcata* [Porsild] Hultén; *P. nivea* ssp. *chamissonis* [Hultén] Hiit.; *P. nivea* ssp. *hookeriana* [Lehm.] Hiit.; *P. nivea* var. *hookeriana* [Lehm.] Wolf). *P. hookeriana* is closely allied to *P. nivea*. Perhaps it would be best treated as part of an expanded *P. nivea*. As considered herein, it includes *P. rubricaulis* authors, not Lehm.

Potentilla hyparctica Malte
Arctic Cinquefoil

Perennial, nonglandular herbs, 0.3–2(3) dm tall, from a caudex, the stems and petioles spreading long-hairy, not at all tomentose; basal leaves 0.2–1 dm long, palmately 3-foliolate, the terminal leaflet 0.7–2.2 cm long or more, 0.5–1.7 cm broad, obovate to oval, coarsely 3–11-toothed to near the base, entire and cuneate in the lower part, green and strigose to glabrous above, somewhat paler and silky-pilose below, especially along the veins, or rarely almost glabrous; flowers one to few, showy; sepals 4–7 mm long, lanceolate; bracteoles 4–6.5 mm long, lanceolate to oblong or oblanceolate; petals yellow, 6–8 mm long, emarginate; receptacle sparsely hairy; achenes several, 1–1.5 mm long.

Coastal beaches to arctic and alpine tundra; in most of Alaska (except for the central valleys) and southern Yukon; eastward to Labrador and south to British Columbia; circumpolar (*P. emarginata* Pursh, not Desf.; *P. nana* Willd.; *P. hyparctica* ssp. *nana* [Willd.] Hultén).

Potentilla nivea L.
Snow Cinquefoil

Perennial, nonglandular herb, 0.7–3 dm tall, from a caudex, the stems and petioles distinctly cobwebby-tomentose; basal leaves 0.2–0.9 dm long, palmately compound with 3 (rarely 5) leaflets, the terminal one 0.8–2.7 cm long, 0.6–1.7 cm broad, obovate to elliptic or oblanceolate, coarsely toothed to near the base, green and silky-pubescent, or rarely whitish-tomentose above, whitish-tomentose below; flowers 1–several, showy; sepals 2.5–5 mm long, lanceolate; bracteoles 1.8–4.5 mm long, oblong to lanceolate; petals yellow, 4–7 mm long, emarginate; receptacle sparsely hairy; achenes several, 1–1.5 mm long.

Meadows, open gravelly or sandy slopes, and rock outcrops, in arctic or alpine tundra; mostly in the Brooks Range or on the north slope, but also in the Alaska and St.

Elias Ranges in southern Alaska and Yukon and in northern Yukon; eastward to Labrador and south to Nevada, Utah, and Colorado; circumboreal (*P. nivea* ssp. *fallax* Porsild). This entity is a near ally of both *P. hookeriana* and *P. uniflora*. The phase with tomentose upper leaf surface has been designated as var. *tomentosa* Nilsson-Ehle ex Hultén.

Potentilla norvegica L.
Rough Cinquefoil

Annual or biennial (short-lived perennial?) herbs, mostly 1–10 dm tall, from a taproot, the stems and petioles sparsely stiff-hairy; leaves mostly cauline, palmately (or subpinnately) compound with 3 (rarely 5) leaflets, the terminal one 1–8 cm long, 0.6–2.7 cm broad, obovate to oblanceolate, coarsely toothed to near the base, or entire and cuneate in the lower part, green and sparsely stiff-hairy to glabrous above, paler and stiff-hairy beneath, especially along the veins; flowers several to many, inconspicuous; sepals 4–6 mm long, ovate-lanceolate; enlarging in fruit; bracteoles 3.5–6 mm long, oblong to elliptic or lanceolate; petals yellow or whitish, 2.5–3.6 mm long, emarginate; receptacle glabrous; achenes numerous, about 1 mm long.

Disturbed soils along roads, old fields, stream banks, and meadows; in much of Alaska south of the Brooks Range, except for the western and southwestern portions, and most of the Yukon; widely distributed in the Northern Hemisphere (*P. monspeliensis* L.; *P. hirsuta* Michx.; *P. norvegica* ssp. *hirsuta* [Michx.] Hyl.; *P. norvegica* var. *hirsuta* [Michx.] Lehm.; *P. norvegica* var. *labradorica* [Lehm.] Fern.; *P. norvegica* ssp. *monspeliensis* [L.] Aschers. & Graebn.).

Potentilla palustris (L.) Scop.
Marsh Cinquefoil

Perennial, rhizomatous, palustrine herbs, the stems prostrate to ascending, purplish, glabrous to hairy, mostly 1–10 dm long; leaves mostly cauline, the lower ones the largest, mostly 0.5–2 dm long, pinnately (3)5–7-foliolate, the terminal leaflet 1.8–8.5 cm long and 0.7–3.2 cm broad, oblanceolate to elliptic, coarsely serrate (sometimes doubly so), green and hairy to glabrous above, paler and hairy beneath; flowers few to several, showy; sepals 7–15 mm long, ovate-lanceolate, enlarging in fruit, purplish; bracteoles 3.5–11 mm long, lanceolate to linear; petals purplish or reddish purple, 2.5–5 mm long; receptacle glabrous; achenes numerous, 1–1.5 mm long.

Wet meadows, bogs, muskegs, stream banks, and lake shores; throughout Alaska and most of the Yukon; eastward to Labrador and south to California, Wyoming, and Ohio; circumboreal. The individual flowers with their intense purplish color are remarkably pretty.

Potentilla pensylvanica L.
Pennsylvania Cinquefoil

Perennial, nonglandular to glandular herbs, 0.7–8 dm tall, from a caudex; basal leaves 0.4–2.5 dm long, pinnately compound with 5–11 leaflets, the terminal one 1.5–6 cm long, 0.7–2.5 cm broad, oblong to elliptic, obovate, or oblanceolate in outline, coarsely toothed to narrowly lobed, the sinuses shallow or extending almost or quite to the midrib, green and somewhat hairy to glabrous above, whitish-tomentose to strigulose or glabrate and greenish beneath; flowers few to many, showy; sepals 3.5–7.5 mm long, ovate-lanceolate; bracteoles 3–6 mm long, narrowly lanceolate; petals yellow, 5–8 mm long, emarginate; receptacle glabrous; achenes numerous, 1–1.5 mm long.

Open woods, meadows, heathlands, and tundra; in the southeastern third of Alaska and southern Yukon, eastward to Labrador and south to Nevada, New Mexico, Kansas, Minnesota, and New Hampshire; Eurasia. Our specimens are separable into two loosely coherent and completely intergrading varieties.

1a. Plants mostly erect or ascending, usu- ally not with reddish stems; leaflets coarsely toothed to lobed, the sinuses usually not extending completely to the midrib, often green on both sur- faces, though paler below (*P. glab- rella* Rydb.; *P. pectinata* Raf.; *P. pen- sylvanica* var. *strigosa* Pursh; *P. pen- sylvanica* var. *glabrata* Wats.). *P. pensylvanica* var. *pensylvanica*

1b. Plants decumbent to ascending, the stems often reddish; leaflets narrow- ly lobed, the sinus extending nearly or quite to the midrib, green above, and usually grayish-tomentose be- neath (*P. multifida* L., at least in part; *P. virgulata* A. Nels.). *P. pen- sylvanica* var. *virgulata* (A. Nels.) Wolf

Potentilla pulchella R. Br. ex Ross
Pretty Cinquefoil

Perennial, nonglandular herbs, 0.2–1.5 dm long, from a caudex; basal leaves 0.2–0.6 dm long, pinnately compound with (3)5 leaflets, the terminal one 0.8–1.5 cm long, 0.5–1 cm broad, obovate in outline, lobed, the sinus extending almost or quite to the midrib, silvery-villous on both surfaces, or less hairy and greenish above; flowers 1– few, showy; sepals 3–5 mm long, ovate; bracteoles 2–4 mm long, oblong; petals yellow, 3.5–5 mm long; receptacle glabrous or nearly so; achenes numerous, about 1 mm long.

Sandy or gravelly soils, or less commonly in clay, on spits, islands, and less commonly stream banks; in coastal or near coastal, west-central and northern Alaska and northern Yukon; eastward to Greenland; circumpolar. *P. pulchella* might be treated best at some infraspecific level within *P. pensylvanica*. More material and further study is necessary before one can determine the affinities of this entity.

Potentilla stipularis L.

Perennial, nonglandular herbs, mostly 1.5– 3.5 dm tall, from a caudex; basal leaves 0.4–2 dm long, palmately compound, with 7–9 leaflets, the terminal one 1–3.5 cm long, 0.2–0.8 cm broad, linear to oblong- oblanceolate, coarsely few-toothed at the apex, entire along the sides, cuneate ba- sally, green above and below; flowers few to several, showy; sepals 4–7 mm long, lance-ovate; bracteoles 3–5 mm long, lance- linear; petals yellow, 4–8(9) mm long, emarginate; receptacle hairy; achenes nu- merous, about 1.5 mm long.

Meadows and river banks; in north-cen- tral Alaska (Umiat vicinity); Asia.

Potentilla uniflora Ledeb.
One-flower Cinquefoil

Perennial, nonglandular herbs, 0.4–1.5(2.3) dm tall, from a caudex, the stems and peti- oles with long spreading hairs (sometimes with a tomentum beneath the longer hairs); basal leaves 0.1–0.8 dm long, palmately 3 (rarely 5)-foliolate, the terminal leaflet 0.9– 2.2 cm long, 0.5–1.7 cm broad, obovate, coarsely toothed (at least apically), green and sparsely to rather densely hairy above, white-tomentose beneath; flowers 1–few, conspicuous; sepals 3.8–6 mm long, ovate- lanceolate; bracteoles 3–6 mm long, nar- rowly lanceolate to elliptic or oblong; pet- als yellow, 6–11 mm long, emarginate; re- ceptacle pubescent; achenes several to many, 1–1.5 mm long.

Rocky outcrops and gravelly soils, in arc- tic and alpine tundra to heathlands and woods; in coastal, western and northern Alaska, and almost throughout mountainous interior Alaska and southern to northern Yukon; east to Hudson Bay and south to British Columbia and Colorado; Asia (*P. ledebouriana* Porsild; *P. vahliana* Lehm.). *P. uniflora* as apparently not always dis- tinguishable from either *P. hookeriana* or *P. villosa* and might be best regarded at some level within *P. nivea*.

Potentilla villosa Pallas ex Pursh
Villous Cinquefoil

Perennial, nonglandular herbs, 0.5–2.8 dm

tall, from a caudex, the stems and petioles with long spreading hairs; basal leaves 2–12 cm long, palmately 3-foliolate, the terminal leaflet 1.4–4 cm long, 1–3 cm broad, obovate, coarsely toothed apically, entire and cuneate at the base, green and silky-pilose above, white or grayish-tomentose below; flowers 1–several, showy; sepals 4–7 mm long, ovate to lanceolate; bracteoles 4–6 mm long, ovate to broadly lanceolate; petals 8–12 mm long, emarginate; receptacle pubescent; achenes several to many, 1–1.5 mm long.

Rocky outcrops, gravelly spits, and in meadows; mostly in coastal and insular western and southern Alaska, but also in interior central to southern Alaska and southern Yukon; southward to Washington and Alberta; Asia (*P. villosa* var. *unifoliolosa* Hultén; *P. nivea* var. *villosa* f. *unifoliolosa* (Hultén) B. Boi.).

PRUNUS L.

Trees or shrubs with unarmed branches; leaves alternate, simple, not lobed, the petiole or base of leaf blades usually with conspicuous, elevated glands; stipules linear, caducous; flowers perfect, regular, borne in racemes, or in umbels; hypanthium cup shaped, lined with a glandular disk; sepals 5, deciduous; petals 5; stamens usually 20 or more; pistils 1, the ovary superior, 1-loculed; styles 1, terminal, the stigma capitate; fruit a blackish to reddish drupe.

1a. Flowers several to many, borne in racemes; plants indigenous or introduced. (2)
1b. Flowers solitary, or several and borne in umbellate clusters; plants introduced. (3)
2a. Hypanthium pubescent within; plants large shrubs or small trees, cultivated in southern Alaska. *P. padus*
2b. Hypanthium glabrous within; plants small to large shrubs, indigenous in northern British Columbia. *P. virginiana*

3a. Leaves pubescent beneath, at least along the veins; fruit yellowish, dark red, or purplish. *P. avium*
3b. Leaves glabrous; fruit usually bright red. *P. cerasus*

Prunus avium L.
Sweet Cherry

Trees, mostly 5–10 m tall, with smooth, reddish or purplish brown bark marked by conspicuous horizontal lenticles; leaves long-petiolate, the blades mostly 5–12 cm long, oblong to ovate or obovate, serrate; flowers large and showy, usually few to several, borne in umbels; calyx lobes entire or nearly so; petals white, mostly 10–15 mm long; fruit subglobose, about 1.5–2.5 cm long.

Sparingly cultivated fruit plant; in southeastern Alaska; native to Eurasia.

Prunus cerasus L.
Sour Cherry

Small trees or shrubs, mostly less than 4 m tall, with grayish bark; leaves long-petiolate, the blades mostly 5–10 cm long, ovate to obovate, doubly serrate; flowers showy, usually 1–few in umbels; calyx lobes crenate; petals white, mostly 8–12 mm long; fruit subglobose, mostly 12–18 mm long.

Sparingly cultivated fruit plant; in southeastern Alaska; native to Eurasia.

Prunus padus L.
European Bird Cherry

Shrubs or small trees, mostly 2–5 m tall, with purplish gray to greenish bark; leaves long-petiolate, the blades 1.5–10 cm long, elliptic to obovate, sharply serrate; flowers showy, commonly numerous, borne in elongate terminal racemes; petals white to cream, mostly 4–6 mm long, much exceeding the stamens; fruit ovoid, 5–8 mm long, astringent.

Commonly cultivated ornamental; in southern Alaska; introduced from Eurasia.

Prunus virginiana L.
Chokecherry

Shrubs or small trees, mostly 1–4 m tall, with purplish gray bark; leaves long-petiolate, the blades 2–10 cm long, elliptic to ovate or oblong, sharply serrate; flowers small but showy, several to many, borne in elongate, terminal racemes; sepals glandular; petals white to cream, mostly 4–6 mm long, subequal to the stamens; fruit ovoid, 5–10 mm long, astringent.

Woods and thickets at Liard Hot Springs, British Columbia; eastward to Newfoundland and south to California, New Mexico, Kansas, Missouri, and North Carolina. Our material is referable to var. *melanocarpa* (A. Nels.) Sarg.

ROSA L.

Shrubs with the branches usually armed with prickles or spines; leaves alternate, pinnately compound, the leaflets serrate to doubly serrate; stipules foliose, adnate to the petioles, persistent; flowers perfect, regular, borne singly or few to several in cymes, large and showy; hypanthium ellipsoid to subglobose, constricted at the top; sepals 5; petals 5 (or many in cultivated forms); stamens numerous; pistils several to many, superior, 1-loculed; styles 1 per pistil, the stigma capitate; fruit of hairy achenes, enclosed by the fleshy hypanthium which ripens to form the rose "hip."

Lewis, W. H. 1959. A monograph of the genus *Rosa* in North America. I. *R. acicularis*. Brittonia 11:1–24.

1a. Stems densely pubescent, armed with prickles and often with stipitate glands; plants cultivated in southern Alaska. *R. rugosa*

1b. Stems not pubescent, armed or unarmed, lacking stipitate glands; plants indigenous, rare in cultivation. (2)

2a. Stems armed with slender prickles along the internodes; infrastipular spines, if any, not much different from the others; our most common rose. *R. acicularis*

2b. Stems with or without prickles along the internodes; infrastipular spines often well developed and different from the internodal prickles (when present). (3)

3a. Sepals 1.5–4 cm long; petals (2.2) 2.5–4 cm long; hips 1–2 cm thick at maturity; plants common in coastal southern and southeastern Alaska. *R. nutkana*

3b. Sepals 1–1.5 cm long; petals 1.5–2.5 cm long; hips 0.8–1 cm thick at maturity; plants uncommon in interior Alaska and Yukon. *R. woodsii*

Rosa acicularis Lindl.
Prickly Rose

Shrubs, mostly 3–12 dm tall, the stems armed with few to many prickles along the internodes, the infrastipular spines lacking, or if present, then similar to the internodal prickles; leaves 3–15 cm long, 3–7-foliolate, the terminal leaflet 1.5–6 cm long, 1–3.5 cm broad, somewhat pubescent to glabrous on both surfaces, serrate to doubly serrate; flowers solitary (rarely 2–3); sepals 1.5–4 cm long; petals pink, 2–3 cm long; hips ellipsoid, mostly 1–2 cm long and wide, purplish or red.

Rosa acicularis Lindl. (× 0.32).

Woods, thickets, and meadows; almost throughout Alaska (except for northern, southwestern, and southeastern portions) and most of the Yukon; eastward to the Atlantic and southward to Idaho and New Mexico; Eurasia. This is our most common indigenous rose. Two infraspecific taxa are represented within our region.

1a. Pedicels glandular; leaflets commonly 5 per leaf, singly serrate; plants uncommon in Alaska. *R. acicularis* var. *acicularis*

1b. Pedicels rarely glandular; leaflets 5–7, mostly doubly serrate; plants common in Alaska (*R. acicularis* var. *cucurbiformis* Raup; *R. acicularis* ssp. *sayi* [Schw.] Lewis). *R. acicularis* var. *bourgeauiana* Crepin

Rosa nutkana Presl
Nootka Rose

Shrubs, mostly 5–20 dm tall, the stems armed with distinctive, infrastipular spines (or rarely unarmed), the internodal prickles lacking, or few in number and different from the infrastipular spines; leaves 6–13 cm long, with 5–7(9) leaflets, the terminal one 2–7 cm long and 1.2–3.5 cm broad, pubescent to glabrous below, once or twice serrate; flowers solitary (rarely 2–3); sepals (1.5)2–4 cm long; petals pink, 2.2–4 cm long; hips subglobose to depressed-globose, 1–2 cm long at maturity, purplish.

Coastal woods and thickets; from the Aleutians eastward through southern and southeastern Alaska; southward to California, Utah, and Colorado.

Rosa rugosa Thunb.
Sitka Rose

Shrubs, mostly 5–15 dm tall, the stems armed with many internodal prickles, and often stipitate-glandular as well, lacking infrastipular spines, densely hairy; leaves mostly 10–15 cm long, with 7–11 leaflets, the terminal one 2–5.5 cm long, 1–3.5 cm broad, pubescent below, green and wrinkled above, once to twice serrate; flowers solitary (sometimes 2–5); sepals 2.5–3.5 cm long; petals usually pink, 3.5–5 cm long; hips subglobose, about 1 cm long.

Cultivated ornamental; in coastal southern Alaska; native to Asia.

Rosa woodsii Lindl.
Woods Rose

Shrubs, mostly 5–20 dm tall, the stems unarmed or more commonly armed with infrastipular spines and often with internodal prickles as well; leaves 3–10 cm long, with 5–9 leaflets, the terminal one 1–3(5) cm long, 0.7–1.5(2.5) cm long, glabrous (in our material), once serrate to partially twice serrate; flowers solitary, or 2–several in cymes; sepals mostly 1–1.5 cm long; petals pink, 1.5–2.5 cm long; hips globose to ellipsoid, 6–15 mm long.

Woodlands; disjunct in east-central Alaska and southern Yukon; eastward to Hudson Bay and south to California, Texas, Missouri, and Wisconsin.

RUBUS L.

Shrubs or semiwoody herbs, with armed or unarmed branches; leaves alternate, palmately or pinnately compound, or simple and palmately veined or lobed; stipules various in shape, usually persistent; flowers perfect or imperfect, regular, solitary, or few to several in cymes; hypanthium short, saucerlike, lined with a glandular disk; sepals 5–7 or more, lacking bracteoles; petals the same number as the sepals (rarely more); stamens 15 to numerous, linear to clavate; pistils several to many, the ovaries superior, each 1-loculed; styles 1 per pistil, the stigma capitate; fruit of separate drupelets, or the drupelets coherent and free of the receptacle, hence an "aggregate" fruit.

Bailey, L. H. 1941. Species Batorum. The genus *Rubus* in North America. Gentes Herb. 5:1–932.

Fernald, M. L. 1919. *Rubus idaeus* and some of its variations in North America. Rhodora 21:89–98.

1a. Leaves simple (at least some), broadly and shallowly to deeply 3–5-lobed. (2)

1b. Leaves compound, with 3–5 leaflets. (4)

2a. Plants distinctly woody, with large, maplelike leaves, the lobes acute or acuminate, usually over 4 dm tall. *R. parviflorus*

2b. Plants herbaceous, or only semi-woody, the leaves with broadly rounded lobes; stems trailing, usually less than 4 dm tall. (3)

3a. Flowers white, imperfect, the plants dioecious; fruit yellowish when ripe, the flavor fetid. *R. chamaemorus*

3b. Flowers pink, perfect; fruit reddish or purplish at maturity, not fetid. *R. stellatus*

4a. Stems erect or arching, woody, seldom less than 5 dm tall. (5)

4b. Stems low, often trailing, herbaceous or only slightly woody, seldom as much as 5 dm tall. (7)

5a. Petals red or pink, 1.5–2.5 cm long; fruit yellowish to red, mostly 12–20 mm broad, sour and soapy; stems unarmed or armed mostly near the base. *R. spectabilis*

5b. Petals white (rarely pink), 0.6–1 cm long; fruit red to reddish purple or black, often less than 12 mm broad; stems armed with spines or prickles. (6)

6a. Stems armed with hooked spines; fruit reddish purple to black, agreeably flavored; plants uncommon, in coastal southeastern Alaska. *R. leucodermis*

6b. Stems armed with bristly, straight prickles; fruit red, agreeably flavored; our common red raspberry. *R. idaeus*

7a. Petals white, mostly 6–9 mm long. (8)

7b. Petals pink, mostly 10–15 mm long or longer. (9)

8a. Leaves 3-foliolate; upright stems more than 1 dm tall. *R. pubescens*

8b. Leaves apparently 5-foliolate (by lobing of the lower pair of leaflets); upright stems less than 0.5 dm tall. *R. pedatus*

9a. Plants (16)20–52 cm tall; glandular, pricklelike processes often present along the stems, petioles, and veins on the lower surface of the leaflets *R. × alaskensis*

9b. Plants 3–30 cm tall; lacking glandular pricklelike processes (though sometimes glandular along the pedicels and on the calyx). *R. arcticus*

Rubus × alaskensis Bailey
Alaska Bramble

Plants semiwoody, (1.6)2–5.2 dm tall, the stems, petioles, and veins on the lower leaf surfaces often with distinctive glandular, pricklelike processes (often sparse); stipules lanceolate; leaves 4–15 cm long, palmately or pinnately 3-foliolate (rarely some leaves simple), the terminal leaflet 2.7–8 cm long, 1.8–5.2 cm broad, glabrous to sparsely pubescent above, paler and usually hairy below, at least along the veins; flowers perfect, showy, 1–3; sepals 8–17 mm long, lance-attenuate, glandular; petals pink, 10–14 mm long; fruit unknown.

Woods and thickets; known mostly from the region along the Susitna and Matanuska valleys. As herein interpreted, *R. × alaskensis* represents a series of hybrid derivatives from initial hybridization of *R. arcticus* or *R. stellatus* with *R. idaeus*. Specimens from the Susitna and Matanuska valleys are intermediate between those species in such features as plant size, flower size, and in the presence of glandular, pricklelike processes. It is probable that other species are involved as parents for some part of the specimens interpreted as belonging to *R. × alaskensis*. This is true for the majority of specimens from other parts of our range, where putative parents include such species as *R. parviflorus* and *R. stellatus*. Additional specimens and population studies are required.

Rubus arcticus L.
Nagoon Berry, Kneshenaka

Plants herbaceous (rarely somewhat woody), 0.3–3 dm tall, the stems, petioles, and lower leaf surfaces variously pubescent but lacking glandular pricklelike processes; stipules lanceolate to ovate; leaves 2–14 cm long, palmately or pinnately 3-foliolate, the terminal leaflet 1–6.5 cm long, 0.9–5 cm broad, glabrous to pubescent and green on both surfaces, paler and usually hairy below; flowers perfect, showy, usually solitary (rarely 2); sepals 4–13 mm long, narrowly lanceolate, often somewhat expanded apically, variously pubescent and glandular to almost glabrous and glandless; petals pink, 8–17 mm long, 3–7 mm broad; staminal filaments flattened; seeds 2–3 mm long, the drupelets red to purplish, several to many.

Arctic and alpine tundra, heath, and woodlands; in most of continental Alaska (except for the northwestern portion) and most of the Yukon; eastward to Newfoundland and south to Montana, Wyoming, and Colorado; Eurasia (*R. acaulis* Michx.; *R. arcticus* var. *pentaphylloides* Hultén; *R. arcticus* ssp. *acaulis* [Michx.] Focke; *R. arcticus* ssp. *stellatus* var. *acaulis* [Michx.] B. Boi.). Attempts have been made to segregate *R. arcticus* as herein interpreted into two species or subspecies. The criteria which form the basis of such segregation include petal width, sepal form, pubescence of sepals, flower number, and plant height. There is little apparent correlation of any of these features, which occur in more or less haphazard recombinations throughout the range of the complex. Thus it seems best to recognize only a single polymorphic species for our region.

Rubus chamaemorus L.
Cloudberry

Plants herbaceous, 0.2–3 dm tall, the stems, petioles, and veins on the lower leaf surfaces variously pubescent and often glandular or glabrate; stipules ovate; leaves 3–11 cm long, simple, the blades palmately veined and lobed, 1.5–6.5 cm long, 2–12 cm broad, green and glabrous to sparsely pubescent above, paler and more pubescent below; flowers imperfect, showy, solitary; sepals 2.5–16 mm long, lanceolate or ovate, pubescent to glabrate; petals white, 6–14 mm long; staminal filaments linear; seeds 3–4.2 mm long, the drupelets yellowish, several to many, soapy.

Tundra to heath or woods, often in moist sites; throughout Alaska and Yukon; eastward to Newfoundland and south to British Columbia and New Hampshire; circumboreal.

Rubus idaeus L.
Raspberry

Plants woody shrubs, (2)5–15 dm tall or more, the stems, petioles, and veins on lower leaf surfaces with glandular, pricklelike processes or prickles or both; stipules linear; leaves 5–20 cm long, pinnately compound, with 3–5 leaflets, the terminal leaflet 3–10 cm long, 1.5–7.5 cm broad, green and glabrous to hairy above, paler and grayish-hairy to glabrate and greenish below; flowers perfect, not conspicuous, 1–4; sepals 4–12 mm long, lanceolate; petals white, 4–7 mm long; staminal filaments slender, often somewhat clavate; seeds 2–2.5 mm long, the drupelets red, several to many, the flavor agreeable.

Woods and thickets; in much of Alaska south of the 68th parallel (except for the western and southwestern portions); in central to southern Yukon; eastward to the Atlantic and south to California, Mexico, Iowa, and North Carolina; Eurasia (*R. strigosus* Michx.; *R. idaeus* var. *strigosus* [Michx.] Maxim.; *R. sachalinensis* Lev.; *R. idaeus* ssp. *sachalinensis* [Lev.] Focke; *R.*

subarcticus Rydb.; *R. idaeus* var. *canadensis* Richards.; *R. melanolasius* Dieck). Our material belongs to ssp. *melanolasius* (Dieck) Focke. The cultivated raspberry (ssp. *idaeus*) is grown in southern Alaska where it spreads vigorously. It can be distinguished by the lack of glands in the inflorescence.

Rubus leucodermis Dougl. ex T. & G.
Black Raspberry

Plants woody shrubs, 1–3 m tall, the stems and petioles armed with hooked spines; stipules linear; leaves pinnately 3–5-foliolate, the leaflets 1.5–9 cm long, green and glabrous to sparsely hairy above, white-tomentose beneath; flowers perfect, usually 2–10; sepals 6–12 mm long, narrowly lance-acuminate; petals white, 5–10 mm long; staminal filaments slender; drupelets several to many, reddish purple to black, agreeably flavored.

Woods and thickets; in coastal and insular south-central and southeastern Alaska; southward to California, Nevada, and Utah.

Rubus parviflorus Nutt.
Thimbleberry

Plants woody shrubs, mostly 4–15 dm tall, the stems, petioles, and veins on the lower leaf surface with stipitate glands; stipules linear to lanceolate; leaves 8–25 cm long, simple, the blades 5–15 cm long, 8–20 cm broad, cordate at the base, the lobes acuminate to acute, glabrous or somewhat hairy; flowers perfect, large and showy, 2–9 in terminal corymbs; sepals 10–25 mm long, lance-attenuate into a long appendage; petals white, 10–25 mm long; staminal filaments linear; seeds 1.5–2.5 mm long, the drupelets numerous, red, pubescent, hardly edible.

Woods; in southeastern Alaska; eastward to the Great Lakes and south to California, New Mexico, Mexico, and the Dakotas (*R. nutkanus* Moc. ex DC.; *R. parviflorus* var. *grandiflora* Farw.). Our materials belong to var. *parviflorus*.

Rubus pedatus J. E. Smith
Five-leaf Bramble

Plants herbaceous, with trailing stoloniferous stems, the leaves and flowering stems mostly less than 1 dm tall; stems, petioles, and veins on lower leaf surfaces pubescent and sometimes glandular; stipules ovate; leaves 2–9 cm long, palmately 3-foliolate, but appearing 5-foliolate by incision of the lateral leaflets, the terminal leaflet 0.9–5 cm long, 0.7–3 cm broad, variously pubescent, greenish on both surfaces; flowers perfect, small, 1–2; sepals 4–11 mm long, lanceolate; petals white, 6–9(10) mm long; staminal filaments linear; seeds 4–6 mm long, drupelets 1–few, red, the flavor agreeable.

Woods and heathlands; in southern Alaska, from the Shumagin Islands eastward through the Panhandle; southward to Oregon, Idaho, and Montana; Asia.

Rubus pubescens Raf.
Dwarf Blackberry

Plants herbaceous, mostly 1.5–5 dm tall, the stems, petioles, and veins on the lower leaf surfaces pubescent; stipules lanceolate; leaves 5–15 cm long, palmately 3-foliolate, the terminal leaflet 4–8 cm long, 2.5–4.5 cm broad, variously pubescent, pale below, green above; flowers perfect, small, 1 or 2, sepals 3–6 mm long, lanceolate; petals white, 4–8 mm long; staminal filaments flattened: seeds 2.5–3.5 mm long, the drupelets several, red.

Woods; in southeastern Yukon and adjacent British Columbia; eastward to Newfoundland and south to Washington, Colorado, Iowa, and Indiana.

Rubus spectabilis Pursh
Salmonberry

Plants woody shrubs, mostly 10–30 dm tall, the stems, petioles, and veins of lower leaf surfaces pubescent and armed with prickles, or unarmed; stipules linear; leaves 7–22 cm long, pinnately 3 or rarely 5-foliolate, the apical leaflet (including petiolule) 4–15 cm long, 2.5–7.5 cm broad, pubescent to glabrous, paler below; flowers showy, perfect, mostly 1–2; sepals 9–20 mm long, ovate-lanceolate, acuminate; petals pink to red, 15–25 mm long; staminal filaments somewhat flattened; seeds about 3 mm long, the drupelets numerous, yellowish to dark red, insipid.

Woods and thickets; in coastal and insular southern Alaska from the Aleutians through the Panhandle; southward to California; Japan. The flowers and fruit of this plant are beautiful, but the flavor of the fruit is poor. It is used to make jellies of excellent quality, however.

Rubus stellatus J. E. Smith
Nagoon Berry

Plants herbaceous, 0.3–2.8 dm tall, the stems, petioles, and lower leaf surfaces pubescent; stipules ovate to lanceolate; leaves 2–12 cm long, palmately 3-lobed (rarely some 3-foliolate), the blade 1.5–4.5(5.5) cm long, 2–7(9) cm broad, cordate to truncate basally, the lobes rounded to obtuse apically, glabrous to pubescent on both surfaces, paler beneath; flowers perfect, showy, usually solitary (rarely 2); sepals 7–13 mm long, narrowly lanceolate, variously pubescent and often glandular; petals pink, 8–19 mm long, 3–7 mm broad; staminal filaments flattened; seeds 2–3 mm long, the drupelets red to purplish, several to many, agreeably flavored.

Tundra, heathlands, and woods; in west-central to southwestern, south-central, and southeastern Alaska (mostly near the coast, but also in the interior); in southern Yukon and adjacent British Columbia; Asia (*R. arcticus* var. *stellatus* [J. E. Smith] B. Boi.).

SANGUISORBA L.

Plants perennial, rhizomatous herbs; leaves basal and alternate, pinnately compound; stipules adnate to the petioles, persistent; flowers perfect (rarely imperfect), regular, numerous in short to elongate, dense spikes; hypanthium subglobose, constricted near the apex; sepals 4, petaloid; petals lacking; stamens usually 4; pistils 1–3, the ovary superior, 1-loculed; styles 1 per pistil, the stigma capitate, fringed; fruit an achene, enclosed by the usually 4-angled to 4-winged hypanthium.

1a. Flowers greenish white, the whitish stamens long-exserted, flattened, and clavate. S. *stipulata*
1b. Flowers reddish purple, the purplish stamens various, but not as above. (2)

2a. Staminal filaments linear, not expanded upwards, about equaling the sepals. S. *officinalis*
2b. Staminal filaments flattened and broadened upwards, much longer than the sepals. S. *menziesii*

Sanguisorba menziesii Rydb.
Menzies Burnet

Plants mostly 3–5 dm tall, basal leaves 1.5–3.5 dm long, with 9–15, ovate to oblong, coarsely serrate leaflets; cauline leaves small, 1–2; spikes 1.5–7 cm long; flowers reddish purple; sepals 2–3 mm long; stamens 4, the filaments flattened and broadened upwards, usually long-exserted, about twice longer than the sepals; hypanthium pubescent, winged in fruit.

Woods and meadows; from the vicinity of Cook Inlet and disjunctly in southeastern Alaska; southward to Washington. S. *menziesii* is morphologically intermediate between S. *officinalis* and S. *stipulata*. It might have originated by hybridization of those two entities. It is largely sympatric with S. *stipulata,* but allopatric with S. *officinalis*. Further work is necessary to determine the true nature of this entity.

Sanguisorba officinalis L.
Official Burnet

Plants mostly 2–9 dm tall; basal leaves 1–3 dm long, with 7–15 ovate to oblong or lanceolate, coarsely serrate leaflets; cauline leaves small, 1–2; spikes 1–3 cm long; flowers reddish purple; sepals 2–3 mm long; stamens 4, subequal to the sepals or somewhat longer, the filaments linear; hypanthium pubescent, winged in fruit.

Tundra, heath, and woods; from the Seward Peninsula and Point Hope eastward through central interior Alaska to western Yukon and southeastern Alaska; southward to California; Eurasia.

Sanguisorba stipulata Raf.
Sitka Burnet

Plants mostly 2.5–8 dm tall; basal leaves 1–6.5 dm long, with 7–17 ovate to oblong, coarsely serrate leaflets; cauline leaves small, 1–3; spikes 2.5–12.5 cm long; flowers greenish white; sepals 2.5–3 mm long; stamens 4, the filaments much longer than the sepals, flattened and expanded upwards, white; hypanthium glabrous, narrowly winged.

Woods and thickets; in most of Alaska south of the 65th parallel and in southern Yukon; southward to Oregon and Idaho; Asia (S. *sitchensis* C. A. Mey.; S. *canadensis* var. *sitchensis* [C. A. Mey.] Koidz.; S. *canadensis* var. *latifolia* Hook.; S. *canadensis* ssp. *latifolia* [Hook.] Calder & Taylor; S. *latifolia* [Hook.] Coville).

SIBBALDIA L.

Plants perennial herbs, from a caudex; leaves basal or cauline and alternate, long-petioled, palmately 3-foliolate; stipules adnate to the petioles, persistent, lanceolate; flowers perfect, regular, borne in leafy-bracted cymes; hypanthium short, saucer shaped, lined with a glandular disk; sepals 5, alternating with 5 sepallike bracteoles; petals 5; stamens mostly 5; pistils 5–20, distinct, the ovaries superior, each 1-loculed; styles 1 per pistil, the stigmas capitate; fruit an achene.

Sibbaldia procumbens L.
Sibbaldia

Plants low, mat-forming, the flowering stems 0.4–1.4 dm tall; leaves 2–12 cm long, the 3 leaflets oblanceolate to obovate, 3-rarely 5-toothed apically, the terminal leaflet 11–32 mm long, 7–18 mm broad, stiff-hairy on both surfaces; flowers inconspicuous; sepals 2.5–5 mm long; petals pale yellow, 1.5–3 mm long; achenes stipitate, about 1 mm long.

Alpine tundra to heath and open woods; in coastal and insular southern Alaska, and in the southeastern quarter of continental Alaska and southern Yukon (rarely elsewhere); east to Newfoundland and south to California, Utah, Colorado, Quebec, and New Hampshire; circumboreal.

SORBUS L.

Shrubs or small trees with unarmed branches; leaves alternate, pinnately compound; stipules persistent or deciduous; flowers perfect, regular, borne in corymbose cymes; hypanthium short, lined with a glandular disk; sepals 5, persistent; petals 5, cream to white; stamens 15–20; pistils 1, the ovary inferior, 2–5-loculed; styles 2–5, the stigmas capitate; fruit a pome.

Jones, G. N. 1939. A synopsis of the North American species of *Sorbus*. Jour. Arnold Arb. 20:1–43.

1a. Leaflets 7–11; flowers more than 10 mm broad; plants of the western Aleutian Islands. S. *sambucifolia*

1b. Leaflets 7–15; flowers less than 10 (12) mm broad (rarely more in S. *sitchensis*); plants not of the Aleutian Islands. (2)

2a. Winter buds densely white-villous; inflorescence usually whitish-pilose at flowering time; leaflets 11–15; plants cultivated and escaping. S. *aucuparia*

2b. Winter buds glabrous or sparsely pilose with whitish or reddish hairs; leaflets 7–11(15); plants indigenous. (3)

3a. Winter buds sparsely pilose with whitish hairs and more or less glutinous, or glabrous; pedicels sparsely whitish-pilose; leaflets acuminate to acute apically. *S. scopulina*

3b. Winter buds pilose with reddish hairs; pedicels pilose with reddish hairs; leaflets rounded to acute apically. *S. sitchensis*

Sorbus aucuparia L.
European Mountain-ash

Trees, mostly 3–6 m tall, with grayish or yellowish green smooth bark; winter buds densely white-villous; leaflets 11–15, 3–5 cm long, 1–1.8 cm broad, the margins coarsely serrate except at the base; petioles and branches of the inflorescence white-hairy at least in flower; stipules persistent; flowers 8–9 mm broad; sepals triangular; petals white to cream, orbicular, 3–4 mm long; fruit 9–11 mm long, scarlet, drying purplish.

Cultivated ornamental trees; in southern and southeastern Alaska, where it has escaped around the major communities; introduced from Europe.

Sorbus sambucifolia (Cham. & Schlecht.) Roemer
Elder-leaf Mountain-ash

Shrubs, 1–4(8) m tall, with gray smooth bark; winter buds glutinous and glossy, somewhat reddish-hairy; leaflets 7–11, 2–7 cm long, 1–2.5 cm broad, broadest near the base, serrate almost to the base; petioles glabrous; branches of inflorescence reddish-hairy or glabrate; stipules deciduous; flowers 10–15 mm broad; sepals triangular; petals white to cream, oval, 4–8 mm long; fruit (8)10–15 mm long at maturity.

Hillsides; in the westernmost Aleutian Islands; Asia (*Pyrus sambucifolia* Cham. & Schlecht.).

Sorbus scopulina Greene
Western Mountain-ash

Shrubs, 1–4 m tall, with grayish red or yellowish bark; winter buds glutinous and glossy, white-hairy to glabrous, the twigs white-hairy; leaflets 7–13, 2–9 cm long, 0.7–3 cm broad, sharply serrate almost to the base; branches of inflorescence sparsely to rather densely pubescent with white hairs; stipules persistent or tardily deciduous; flowers 6–12 mm broad; sepals triangular; petals white to cream, oval, 4–6 mm long; fruit 5–10 mm long, scarlet to orange, drying purplish.

Woods, thickets, and alpine tundra; in much of Alaska south of the 65th parallel (except for the Aleutians) and in southern Yukon; southward to California and east to Alberta, the Dakotas, and New Mexico (*S. alaskana* G. N. Jones; *S. andersonii* G. N. Jones).

Sorbus sitchensis Roemer
Sitka Mountain-ash

Shrubs, 1–4 m tall, with grayish red bark; winter buds and young twigs pubescent with reddish hairs; leaflets 9–11, 2.5–8 cm long, 1–3 cm broad, sharply serrate to the middle or below; petioles and branches of inflorescence pubescent to glabrous in age; stipules deciduous; flowers 6–9(12) mm broad; sepals triangular; petals white to cream, oval, 4–6 mm long; fruit 8–10 mm long, red, drying purplish.

Woods and thickets; in coastal and insular southern Alaska, from Kodiak Island and the Alaska Peninsula eastward through the Panhandle; southward to California, Idaho, and Montana.

SPIRAEA L.

Shrubs with unarmed branches; leaves alternate, simple, toothed and sometimes lobed; stipules obsolete; flowers perfect, regular, borne in terminal corymbs or panicles; hypanthium cup shaped; sepals 5, persistent; petals 5; stamens 25 or more; pistils 3–7 (usually 5), distinct, the ovaries superior, each 1-loculed; styles 1 per pistil, the stigma capitate; fruit a few-seeded follicle.

1a. Flowers white, borne in flattopped or hemispheric corymbs; plants of broad distribution. *S. beauverdiana*

1b. Flowers bright pink, borne in elongate panicles; plants of southeastern Alaska. *S. douglasii*

Spiraea beauverdiana Schneid.
Beauverd Spiraea

Shrubs 2–10 dm tall or more, with pubescent reddish branchlets, the bark on older stems shredding; leaves 0.9–6 cm long and 0.5–3.5 cm broad, oblong to elliptic or lanceolate, subentire to once or twice serrate, green and glabrous to glabrate above, paler and glabrous or pubescent along the veins beneath; flowers few to many in flattopped or hemispheric corymbs; sepals 1–1.5 mm long, triangular; petals white, 1.5–2.5 mm long; follicles 2.5–3.5 mm long, pubescent.

Tundra, heathlands, and woods; in most of Alaska (except for the northwestern, southwestern, and extreme southeastern portions) and much of the Yukon and adjacent Mackenzie; northern British Columbia; Asia (*S. beauverdiana* var. *stevenii* Schneid.; *S. stevenii* [Schneid]. Rydb.).

Spiraea douglasii Hook.
Menzies Spiraea

Shrubs 5–20 dm tall or more, with pubescent reddish branchlets; leaves 2.5–10 cm long, 0.8–2.5 cm broad (or broader), oblong to elliptic, ovate, or obovate, serrate about halfway to the base, green and glabrous to minutely hairy above, paler and glabrous to pubescent beneath; flowers numerous, borne in elongate panicles; sepals about 1 mm long, triangular; petals bright pink, 1.5–2 mm long; follicles 2.5–3 mm long, glabrous except along the suture.

Swamps, lake shores, and stream banks; in southeastern Alaska; southward to California and Idaho (*S. menziesii* Hook; *S. douglasii* ssp. *menziesii* [Hook.] Calder & Taylor). Our material belongs to var. *menziesii* (Hook.) Presl.

RUBIACEAE
Madder Family

Plants annual or perennial herbs; leaves opposite or appearing whorled, simple, entire; flowers perfect or imperfect, regular; calyx 4–5-lobed, or obsolete; corolla of united petals, rotate, 4–5-lobed; stamens as many as the corolla lobes and alternate with them; pistils 1, the ovary inferior, 2-carpellate, usually 2-loculed; styles 1, the stigmas capitate; fruit separating at maturity into two 1-seeded schizocarps.

GALIUM L.

Plants with square stems and whorled leaves; flowers perfect, borne in cymes or cymose panicles, or solitary; calyx lobes obsolete; corolla rotate, mostly 4-lobed; stamens mostly 4; ovary pubescent or glabrous; stigmas capitate; fruit dry at maturity, indehiscent.

1a. Plants with erect stems; leaves with 3 main veins. (2)

1b. Plants with stems ultimately sprawling; leaves with 1 main vein. (3)

2a. Flowers in terminal panicles; leaves more than 4 times longer than broad; plants of diverse habitats and broad distribution, but seldom in damp woods and thickets. *G. boreale*

2b. Flowers in axillary cymes; leaves seldom more than twice longer than broad; plants of damp woods or thickets, in coastal and insular southern Alaska. *G. kamtschaticum*

3a. Ovaries and fruits glabrous. (4)

3b. Ovaries and fruits pubescent. (5)

4a. Flowers several to many, in a dichotomous cyme; pedicels horizontally spreading; plants of southern Yukon. *G. palustre*

4b. Flowers 1–2(3), on curved pedicels; plants broadly distributed. *G. trifidum*

5a. Plants perennial, from rhizomes;

leaves with marginal cilia pointing toward the apex. *G. triflorum*

5b. Plants annual, from taproots; leaves with marginal cilia spreading or reflexed. *G. aparine*

Galium aparine L.
Cleavers

Herbs annual, from taproots, the stems sprawling, mostly 3–10 dm long, retrorsely bristly along the angles; leaves mostly 1.5–7.5 cm long, 0.2–0.8 cm broad, linear to oblong, borne in whorls of 6–8, retrorsely bristly along the margin, 1-veined, cuspidate; flowers mostly 3–5, borne in cymes on axillary peduncles; corolla whitish or greenish, 1–2 mm broad, 4-lobed; fruit mostly 2–4.5 mm long, pubescent with hooked bristles.

Maritime plants; along beaches in southern Alaska; from the Aleutians eastward through the Panhandle; widely distributed in North America; circumboreal.

Galium boreale L.
Northern Bedstraw

Herbs perennial, from rhizomes, the stems mostly 2–8(10) dm tall, erect or ascending, rough to smooth on the angles, not retrorsley bristly; leaves mostly 1–5(6.5) cm long, 0.2–1.2 cm broad, narrowly lanceolate to oblong or linear, borne in whorls of 4, glabrous or roughened marginally and on the veins beneath, 3-veined, rounded apically; flowers several to numerous, borne in terminal, cymose panicles; corolla white to cream, 3–7 mm broad, 4-lobed; fruit 1.5–2 mm long, pubescent with straight or curled hairs, rarely glabrous.

Tundra, heathlands, or open woods, on slopes and roadsides; in most of Alaska south of the Brooks Range (except for the Aleutians and much of the Panhandle) and central to southern Yukon; widely distributed in North America; circumboreal. This species is very showy and fragrant in full flower.

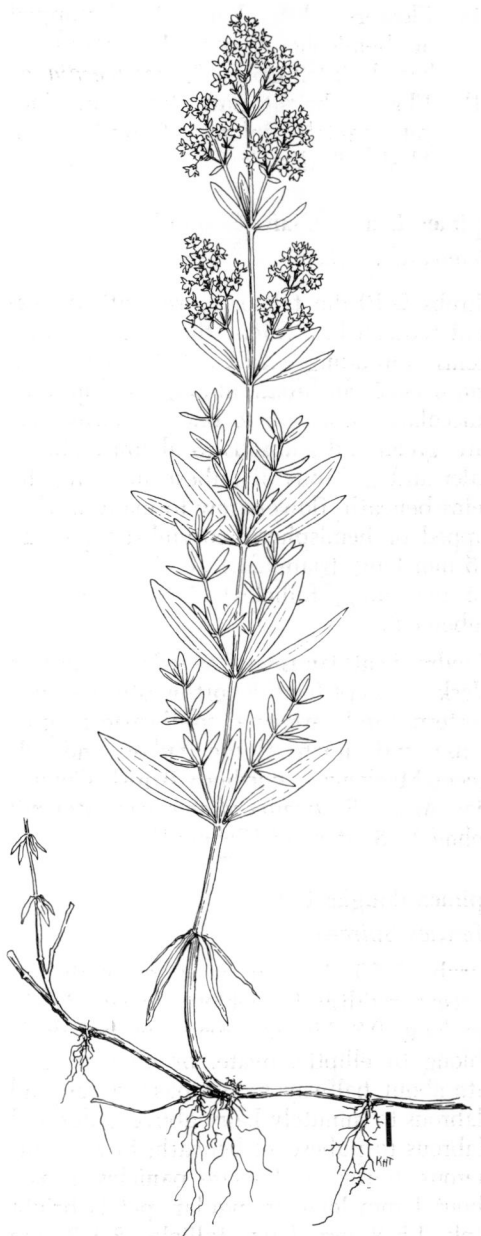

Galium boreale L. (× 0.5).

Galium kamtschaticum Stellar ex Schult. & Schult.
Northern Wild-licorice

Herbs perennial from rhizomes, the stems erect, mostly 1–3 dm tall, sparingly hirsute with spreading hairs or glabrous; leaves mostly 1–2.5 cm long, 0.6–1.5 cm broad, oval to broadly elliptic or obovate, borne in whorls of 4, the marginal hairs pointed forward, 3-veined, mucronate apically; flowers mostly 1–3 (rarely to 6), borne in cymes on axillary peduncles; corolla whitish to greenish, 2.5–3.5 mm broad, 4-lobed; fruit mostly about 1.5 mm long, pubescent with hooked bristles.

Moist woods or thickets; in coastal and insular southern Alaska, from the Aleutians through the Panhandle; southward to Washington and also in eastern Canada and northeastern United States; Asia.

Galium palustre L.

Herbs perennial from rhizomes, the stems sprawling, mostly 2–5 dm long, more or less minutely retrorsely bristly on the angles; leaves mostly 0.4–1.2 cm long, 0.1–0.2 mm broad, narrowly elliptic to spatulate, borne in whorls of 4–6, retrorsely bristly along the margins, 1-veined, rounded to acutish apically; flowers several to many, borne in terminal cymes; corollas whitish, 2.5–3 mm broad; 4-lobed; fruit about 1.5 mm long, glabrous.

Moist sites; reported for south-central Yukon, where probably adventive; eastern North America; Eurasia.

Galium trifidum L.
Small Bedstraw

Herbs perennial from rhizomes, the stems sprawling, mostly (0.5)1–6 dm long, minutely retrorsely bristly along the angles; leaves mostly 0.3–2 cm long, 0.1–0.3 cm broad, oblong to linear or elliptic, borne in whorls of 4(5–6), retrorsely bristly along the margin, 1-veined, rounded apically; flowers 1–3, borne in axillary or terminal cymes; corolla whitish, about 1.5 mm broad, 3–4-lobed; fruit about 1 mm long, glabrous.

Moist areas, in bogs, swamps, margins of ponds, and along streams; in much of

Alaska and Yukon south of the 68th parallel (except for portions of central western Alaska); eastward to the Atlantic and south to California, Texas, and Georgia; circumboreal.

1a. Leaves 4 per whorl, usually less than 12 mm long and 2 mm broad; plants mostly of interior Alaska and Yukon (*G. brandegei* Gray; *G. trifidum* var. *pusillum* Gray). *G. trifidum* var. *trifidum*

1b. Leaves 5 or 6 per whorl (at least some), mostly more than 12 mm long and over 2 mm broad; plants of coastal and insular southern Alaska (*G. columbianum* Rydb.; *G. trifidum* ssp. *columbianum* [Rydb.] Hultén; *G. trifidum* ssp. *pacificum* [Wieg.] Piper). *G. trifidum* var. *pacificum* Wieg.

Galium triflorum Michx.
Sweet-scented Bedstraw

Herbs perennial from rhizomes, the stems sprawling, mostly 2–10 dm long or more, retrorsely bristly to smooth on the margins; leaves mostly 1–5 cm long, 0.3–1.2 mm broad, elliptic to lanceolate, oblanceolate, or oblong, borne in whorls of 5–6(4 on some branchlets), the marginal hairs pointed forward, 1-veined, cuspidate; flowers mostly 3 (rarely more), borne in axillary cymes; corolla whitish, 2–3 mm broad, 4-lobed; fruit about 1.5–2 mm long, pubescent with hooked prickles.

Woods; in central, interior to coastal and insular southern Alaska and southern Yukon; widely distributed in North America; circumboreal. The herbage is vanilla-scented in drying.

SALICACEAE
Willow Family

Plants dioecious trees and shrubs; leaves simple, deciduous, alternate, stipitate; flowers in aments (catkins), solitary in the axils of scalelike bractlets; perianth lacking; stamens 1–many; pistils 1, the ovary superior, solitary, 1-loculed, 2–4-carpelled; stigmas as many as the carpels, sometimes cleft; fruit a 2–4-valved capsule; seeds numerous, bearing an apical tuft of hair.

1a. Visible bud scales several; bractlets of catkins incised; flowers borne on broad or cup-shaped disks; catkins pendulous; pith conspicuously to obscurely 3–5-angled; plants trees. *Populus*

1b. Visible bud scales 1; bracts of catkins entire or merely dentate; flowers with 1–2 basal glands, lacking a disk; catkins various, but often upright; pith usually terete in cross section; plants mostly shrubs. *Salix*

POPULUS L.

Plants small to large trees, with smooth whitish bark or with dark deeply furrowed bark at maturity; twigs of 2 types (short, lateral spurs, with contiguous nodes or with the internodes somewhat elongated, and long usually terminal shoots with elongate internodes); buds with several scales visible, often coated with resinous varnish; leaves dimorphic, those of long shoots usually more variable (and often larger) than those of the spur shoots, oval to cordate, ovate, lanceolate, or elliptic, acute to acuminate apically, crenate-serrate, lighter on the lower surface than above; catkins pendulous, appearing before the leaves; scales incised or fringed; pistils subtended by a disk; stamens 4–many; capsules 2–4-valved; seeds with long, copious hair.

1a. Petioles strongly laterally flattened; buds not distinctly resin-coated; bark smooth, white or greenish (rarely furrowed). *P. tremuloides*

1b. Petioles round in cross section or minutely channeled above; buds distinctly resin-coated; bark becoming rough, furrowed, and grayish with age. (2)

2a. Capsules glabrous, with 2 (rarely 3–4) carpels; plants mostly of coastal, western and interior Alaska and Yukon. *P. balsamifera*

2b. Capsules pubescent (rarely glabrous?), with 3 (rarely 2–4) carpels; plants of coastal southern to southeastern Alaska (rarely elsewhere). *P. trichocarpa*

Populus balsamifera L.
Balsam Poplar

Trees of small to moderate size, mostly less than 10(15) m tall; bark smooth and greenish when young, becoming furrowed and gray in age; trunks mostly less than 6 dm in diameter; leaves long-petiolate, the petioles terete or channeled above, the blades of mature leaves on spur shoots ovate to lanceolate or elliptic, subcordate to truncate, obtuse or acute basally, mostly 4–9 (12) cm long, 2–5(7) cm broad; pistillate catkins mostly 6–16(21) cm long; mature capsules 4–7(9) mm long, glabrous, with 2 (rarely 3–4) valves.

Populus balsamifera L. (× 0.40).

Stream sides, margins of lakes and seeps, and in mixed woods; in most of continental Alaska and Yukon, and less commonly in coastal southern Alaska; eastward to Newfoundland and southward to Alberta, Michigan, and New York (*P. tacamahacca* Mill.)

Populus tremuloides Michx.

Aspen

Trees slender, small to moderate in size, mostly less than 10 m tall; bark smooth, white or greenish (rarely furrowed and grayish in age); trunks mostly less than 3 dm in diameter; leaves long-petiolate, the petioles laterally flattened, trembling in the slightest breeze; blades of mature leaves on lateral branches oval to ovate, subcordate to truncate or obtuse at the base, mostly 1–7.5 cm long, 1–6 cm broad, often infected by "leaf miners"; pistillate catkins 5–10 mm long; mature capsules 4–6 mm long, on pedicels 1–2 mm long, 2-valved.

Woods to heathlands; in much of continental Alaska south of the Brooks Range (except for western and southwestern portions), and most of the Yukon; eastward to the Atlantic and south to Mexico. Aspen is often a major component of the boreal forest.

Populus trichocarpa T. & G.

Black Cottonwood

Trees of moderate to large size (our largest deciduous tree), often over 15(30) m tall; bark smooth and greenish when young, becoming deeply furrowed and grayish in age; trunks often much over 6 dm (to 30 dm) in diameter; leaves long-petiolate, the petioles terete or channeled above; blades of mature leaves on spur shoots cordate to ovate or lanceolate, cordate to truncate or obtuse basally, mostly 8–10 cm long, 4–8 cm broad; pistillate catkins mostly 12–21 cm long; mature capsules 5–8 mm long, pubescent (rarely glabrous?), with 3 (rarely 2 or 4) valves.

Moist sites along streams, seeps, and ponds; in south-central to southeastern Alaska; southward to Baja California, Utah, Wyoming, Montana, and Alberta. In southern Alaska, the range of *P. trichocarpa* overlaps that of *P. balsamifera*. These species are remarkably similar to one another. Indeed, vegetative specimens apparently lack definitive distinguishing characteristics, and have been separated arbitrarily on a geographical basis. The primary differential features emphasized in the key are not always definitive—as the noted exceptions prove. Further work is needed, especially in the Cook Inlet-Kenai Peninsula region. Brayshaw (Can. Field-Nat. 79:91–95. 1965) has reviewed this complex in Alberta where *P. balsamifera* and *P. trichocarpa* intergrade, and recognizes this entity as *P. balsamifera* ssp. *trichocarpa* (T. & G.) Brayshaw.

SALIX L.

Contributed by George W. Argus

Plants dioecious, dwarf or trailing to erect shrubs or occasionally trees; buds with a single visible scale; leaves alternate, simple, usually stipulate, more or less dimorphic; catkins mostly erect or spreading (rarely pendulous), developing before, with, or after the leaves, sessile or on leafy peduncles; bracts entire to slightly toothed, 1 per flower; pistillate flowers with a single pistil, borne sessile or on a stipe (pedicel), subtended by 1(2) stalked glands, lacking a disk; staminate flowers with 2 (1 or 3–5) stamens, with 1–2 basal glands; capsules 2-valved, the single style sometimes bifid, the 2 stigmas usually bifid or horseshoe shaped; seeds several, with long, copious hairs.

Argus, G. W. 1965. The taxonomy of the *Salix glauca* complex in North America. Contr. Gray Herb. 196:1–142.

———. 1973. The genus *Salix* in Alaska and the Yukon. Nat. Mus. Canad. Publ. Bot. 2:1–279.

Hultén, E. 1967. Comments on the flora of Alaska and Yukon. Ark. Bot. II 7:147.

Raup, H. M. 1959. The willows of boreal western America. Contr. Gray Herb. 185: 1–95.

Skvortsov, A. K. 1966. Salicaceae. In: A. I. Tolmatchev, Flora Arctic U.S.S.R.

1a. Plants dwarf or prostrate trailing shrubs, under 2 dm tall. KEY I. p. 403
1b. Plants erect shrubs exceeding 2 dm tall, or trees. KEY II. .... p. 404

KEY I. PLANTS DWARF OR PROS-TRATE SHRUBS, UNDER 2 DM TALL.

1a. Pistillate and staminate flowers with 2 nectaries, one on either side of the stipe; leaves prominently reticulate and pale beneath; aments borne on prominent, subterminal, floriferous branchlets. S. reticulata
1b. Pistillate flowers with 1 nectary between the stipe and the ament axis; aments borne on lateral, floriferous branchlets. (2)

2a. Pistils pubescent, sometimes only on the beak (see also S. stolonifera). (3)
2b. Pistils glabrous. (11)

3a. Leaves distinctly and minutely serrulate. S. chamissonis
3b. Leaves entire or toothed only on the lower half. (4)

4a. Leaves green or pale green beneath, not glaucous. (5)
4b. Leaves glaucous beneath. (6)

5a. Branches clothed with persistent, skeletonized leaves; leaf margins usually ciliate; nectaries usually shorter than the stipes. S. phlebophylla
5b. Branches without persistent, skeletonized leaves; leaf margins rarely ciliate; nectaries 2–5 times as long as the stipes. S. polaris

6a. Leaves 0.9–1.5 cm long, margins prominently ciliate; aments globose. S. ovalifolia
6b. Leaves longer than 1.5 cm, the margins not ciliate (except in some S. arctica); aments cylindrical. (7)

7a. Styles 0.1–0.5 mm long. (8)
7b. Styles longer than 0.5 mm long. (9)

8a. Leaves obovate to elliptic, 1.7–2.7 times as long as wide, glabrous above, the margins distinctly toothed on lower half; pistils sparsely pubescent with rusty-colored hairs. S. fuscescens
8b. Leaves narrowly elliptic to narrowly obovate, 2–3.7 times as long as wide, pubescent on both sides, the margins entire; pistils densely pubescent with white hairs. S. brachycarpa

9a. Pistils sparsely pubescent with crinkly, refractive hairs; nectaries shorter than the stipes; branchlets slender and trailing, glabrous; leaves glabrous. S. arctophila
9b. Pistils sparsely or densely pubescent with nonrefractive hairs; nectaries equal to or longer than the stipes; branchlets and leaves various. (10)

10a. Pistils densely pubescent; leaves dark green and usually glossy above, cuneate to rounded at base; branchlets trailing to erect. S. arctica
10b. Pistils sparsely pubescent on the beak, or glabrous; leaves pale green and dull above, cuneate at base; branchlets trailing and rooting. S. sphenophylla

11a. Leaves green (not glaucous) beneath. (12)
11b. Leaves glaucous beneath. (14)

12a. Plants decumbent or trailing forest shrubs; leaves narrowly elliptic to narrowly obovate, 2–5 cm long, the margins crenate to crenate-serrulate. S. myrtillifolia
12b. Plants dwarf, sometimes trailing, arctic shrubs; leaves circular to obovate or narrowly elliptic, 0.5–1.4 cm long, the margins entire or toothed only at the base. (13)

13a. Leaves circular or sometimes narrowly elliptic, not prominently reticulate, the margins entire and ciliate; branchlets erect, not trailing; styles 0.5–1 mm long; plants widely distributed. *S. rotundifolia*

13b. Leaves subcircular, prominently reticulate, the margins glandular-toothed on the lower half, not ciliate; branchlets more or less trailing; styles 0.2–0.4 mm long; plants known only from St. Paul Island. *S. numularia*

14a. Branchlets usually densely woolly; leaves lemon green, leathery, obovate to narrowly obovate and tapering to a short petiole; petioles to 0.3 mm long; bracts tawny, the apex often retuse; pistils commonly brick red. *S. setchelliana*

14b. Branchlets glabrous to sparsely pubescent; leaves thin, elliptic to subcircular; petioles 2–20 mm long; bracts brown to blackish; pistils reddish, purplish, or greenish. (15)

15a. Pistillate nectaries shorter than the stipes; leaf margins distinctly toothed in the lower half; petioles 2–5.6 mm long. *S. fuscescens*

15b. Pistillate nectaries longer than the stipes; leaf margins usually entire; petioles usually 4–20 mm long. (16)

16a. Branches short and erect, sometimes trailing, often glaucous; plants often rhizomatous; styles 0.8–1.6 mm long. *S. stolonifera*

16b. Branches long and trailing, not glaucous; styles 0.2–0.8 mm long. *S. ovalifolia*

KEY II. PLANTS ERECT SHRUBS EXCEEDING 2 DM TALL, OR TREES.

1a. Plants flowering before the leaves appear (precocious). (2)

1b. Plants flowering as the leaves appear (coetaneous) or after (serotinous). (10)

2a. Pistils glabrous. (3)

2b. Pistils pubescent. (6)

3a. Stipules absent; branchlets brittle and with persistent, long-villous hairs at the base; plants of coastal southern Alaska. *S. hookeriana*

3b. Stipules present, often persistent; branchlets wiry, without long villous hairs at the base; plants not of coastal southern Alaska. (4)

4a. Stipules persistent for several years, linear to ovate, the apex attenuate; styles longer than 1.2 mm; nectaries 2–3 times as long as the stipes. *S. lanata*

4b. Stipules not persistent for more than 1 year, elliptic to broadly ovate, rounded apically; styles shorter than 1.2 mm; nectaries shorter than the stipes. (5)

5a. Aments precocious, on floriferous branchlets to 0.5 cm long; styles 0.8–1.2 mm long; branchlets sparsely pubescent; leaves elliptic or obovate. *S. monticola*

5b. Aments subprecocious, on floriferous branchlets 0.3–1.3 cm long; styles 0.5–0.75 mm long; branchlets glabrescent; leaves narrowly oblong to narrowly obovate. *S. rigida*

6a. Leaves densely white-lanate beneath, bright green above; stipes to 0.4 mm long. *S. alaxensis*

6b. Leaves sericeous or densely villous to sparsely pubescent, glabrescent or glabrous beneath; stipes 0.2–2 mm long. (7)

7a. Branchlets with thick glaucescence. *S. drummondiana*

7b. Branchlets not glaucous (rarely thinly so in *S. planifolia*). (8)

8a. Buds and stipules oily; stipules broadly ovate, the margins prominently glandular; leaves white or gray sericeous-lanate beneath. *S. barrattiana*

8b. Buds and stipules not oily; stipules ovate to linear; leaves glabrous, glabrate, or sericeous beneath. (9)

9a. Branchlets velutinous; styles 0.2–0.5 mm long; stipes 0.8–2 mm long. *S. scouleriana*

9b. Branchlets pubescent to villous or glabrous; styles 0.5–1.8 mm long; stipes 0.2–0.8 mm long. *S. planifolia*

10a. Pistils glabrous. (11)

10b. Pistils pubescent. (23)

11a. Plants introduced species, cultivated in some southern centers. (12)

11b. Plants indigenous species. (13)

12a. Leaves broadly ovate to narrowly elliptic, green or pale beneath; branches not pendulous; staminate flowers with 5 stamens. *S. pentandra*

12b. Leaves narrowly ovate, glaucous beneath; branchlets pendulous; staminate flowers with 2 stamens. *S. babylonica*

13a. Leaves green or pale beneath, not glaucous. (14)

13b. Leaves glaucous beneath. (17)

14a. Leaves linear, 7–18 times as long as wide, the margins distantly denticulate; aments often branched; bracts deciduous after flowering; stipes pubescent. *S. interior*

14b. Leaves not linear, only 2–5 times as long as wide, the margins serrulate or crenate; aments unbranched; bracts persistent; stipes glabrous. (15)

15a. Leaves coarsely villous on both sides, the margins glandular-serrulate or partly entire. *S. commutata*

15b. Leaves glabrous or glabrescent, the margins crenate to crenate-serrulate. (16)

16a. Shrubs decumbent, 0.1–0.9 m tall; stipules minute or 1–2 mm long; styles 0.3–0.5 mm long. *S. myrtillifolia*

16b. Shrubs erect, 0.6–4 m tall; stipules 0.5–7 mm long; styles 0.5–0.9 mm long. *S. novae-angliae*

17a. Immature leaves membranous and translucent, glabrate and green on both sides, becoming glaucous beneath; leaves and buds with persistent balsamlike fragrance. *S. pyrifolia*

17b. Immature leaves thickened and opaque, glabrous to pubescent, lacking a balsamlike fragrance. (18)

18a. Petioles glandular near the leaf base; stamens 5; leaf apex acuminate to caudate. *S. lasiandra*

18b. Petioles not glandular; stamens 2; leaf apex acute to rounded. (19)

19a. Styles 0.1–0.4 mm long. (20)

19b. Styles 0.5–2 mm long. (21)

20a. Stipes (1.6)2–3.2 mm long, glabrous; leaves leathery, glabrous; styles 0.1–0.2 mm long. *S. pedicellaris*

20b. Stipes 0.4–1.2 mm long, pubescent; leaves thin, glabrescent, with rust-colored hairs persistent along the midrib; styles 0.2–0.4 mm long. *S. hastata*

21a. Stipules absent; leaves pubescent beneath, at least on the midrib; branchlets brittle, with long, persistent, villous hairs at the base; styles red in life. *S. hookeriana*

21b. Stipules present; leaves glabrous beneath; branchlets wiry, lacking persistent hairs at the base; styles greenish. (22)

22a. Leaves elliptic or obovate, the immature ones green and opaque; petiole green; branchlets densely to sparsely villous; styles 0.6–1.6 mm long. *S. barclayi*

22b. Leaves narrowly oblong to narrowly obovate, the immature ones reddish and translucent; petioles reddish; branchlets glabrescent; styles 0.5–0.75 mm long. *S. rigida*

23a. Stipes 2.8–4.8 mm long, about 10 times as long as the nectaries. *S. bebbiana*

23b. Stipes to 2 mm long, 0.5–2 times as long as the nectaries. (24)

24a. Leaves sericeous beneath, the margins glandular-serrulate to distantly so. (25)

24b. Leaves densely pubescent to glabrescent to glabrous beneath, not sericeous. (26)

25a. Leaves narrowly lanceolate, 5–7 times as long as wide, sericeous beneath with short white or rusty-colored hairs oriented toward the apex, the margins prominently glandular-serrulate; styles 0.3–0.5 mm long. S. *arbusculoides*

25b. Leaves narrowly elliptic to obovate, 2.5–3 times as long as wide, appearing satiny beneath with matted, sericeous hairs, the margins distantly and inconspicuously glandular-serrulate to glandular-crenate; styles 0.5–0.8 mm long. S. *sitchensis*

26a. Leaves pale green and glabrescent beneath, the margins prominently glandular-dotted to glandular-serrulate; immature leaves sericeous with a mixture of white and rusty-colored hairs; stipes 0.8–2 mm long. S. *maccalliana*

26b. Leaves glaucous and/or variously pubescent, the margins entire; stipes 0.1–1.5 mm long. (27)

27a. Leaves densely dull white lanate-floccose beneath, and floccose to glabrescent above, narrowly elliptic to narrowly ovate, 3.5–7 times as long as wide; styles red. S. *candida*

27b. Leaves not or only slightly pubescent above, obovate to broadly or narrowly elliptic, 2–4 times as long as wide; styles yellow green. (28)

28a. Leaves with rusty-colored hairs sparsely distributed on both sides, especially on immature leaves. S. *athabascensis*

28b. Leaves without rusty-colored hairs. (29)

29a. Petioles 3–15 mm long, yellowish; stipes 0.5–2 mm long. S. *glauca*

29b. Petioles 1–3 mm long, often reddish; stipes to 0.5 mm long. S. *brachycarpa*

Salix alaxensis (Anderss.) Coville
Alaska Willow

Shrubs or small trees, mostly 0.3–8(10) m tall; branches dark brown to chestnut brown; branchlets densely villous-lanate or glabrate to glabrous; leaves narrowly ovate, to oblong or obovate to narrowly obovate, the largest mature leaves (3.5)5–11 cm long and (1)1.5–4 cm wide, acute to rounded apically, narrowly cuneate basally, the margins entire; petioles (3)5–15(20) mm long, yellowish; stipules 5–15(25) mm long, linear to lanceolate; aments precocious or subprecocious; staminate aments 3–3.5(4) cm long; pistillate aments (3) 6–15 cm long; pistils about 1.5 mm long, green, sparsely pubescent; capsules 4–5(8) mm long; styles 1.3–2.8 mm long; nectaries 1–2 times as long as the stipes; bracts 1.5–2.5 mm long, dark brown to black.

Gravel bars and river terraces, margins of streams and lakes in arctic and boreal alpine meadows; throughout Alaska and Yukon (except most of the Aleutian Islands, some Bering Sea islands, and southeastern Alaska); south to British Columbia and Alberta; Asia, from Yenisei River to Chukotsk Peninsula. Two varieties are recognized.

1a. Plants shrubs 0.3–4 m tall; branchlets and branches not evidently glaucous, densely white or yellowish villous-lanate; buds usually large and petiole base inflated; bract acute apically or obtuse (S. *speciosa* var. *alaxensis* Anderss.). S. *alaxensis* var. *alaxensis*

1b. Plants tall shrubs or trees 2–10 m tall; branchlets and branches glaucous, sometimes sparsely pubescent but soon glabrate; buds often small; petiole base not inflated; bract apex sometimes obtuse (S. *longistylis* Rydb.; S. *alaxensis* ssp. *longistylis* [Rydb.] Hultén). S. *alaxensis* var. *longistylis* (Rydb.) Schneid.

Salix arbusculoides Anderss.

Shrubs 1–4 m tall or trees 5–6 m tall; branches slender and glossy; branchlets sparsely velutinous; largest mature leaves (3.8)6–7(8) cm long, (0.7)1–1.5(2) cm wide, narrowly acute to obtuse apically, acute basally, glossy and glabrous above when mature; petiole (3)5–8(11) mm long; stipules 0.9–2.4 mm long, the margins glandular (a white secretion often accumulates and dries at the base); aments coetaneous or subprecocious, the staminate 1.8–2.5 cm long, the pistillate 2–7.5 cm long, usually loosely flowered, borne on floriferous branchlets to 5 mm long; pistils 3–4.5 mm long, sericeous with white and rusty-colored hairs; capsules 4–7 mm long; styles to 0.8 mm long; stipes 0.6–0.9 mm long, velutinous; nectaries equal to or slightly exceeding the stipes.

Stream banks, openings in forests, muskegs and in willow thickets at the edge of arctic and alpine tundra; common throughout interior Alaska and the Yukon (not reaching the Pacific Coast); eastward to Hudson Bay and south to northeastern British Columbia and Alberta.

Salix arctica Pallas
Arctic Willow

Dwarf shrubs, usually prostrate or trailing but sometimes up to 3–5 dm tall in protected habitats; branchlets yellow green to chestnut brown, glossy, glabrous or sparsely pubescent with straggly hairs, sometimes glaucous; leaves narrowly obovate to elliptic (Arctic Ocean populations), suborbicular to broadly elliptic or narrowly obovate (Bering Sea populations), obovate to narrowly obovate or broadly elliptic (Pacific Coast and interior populations), the largest mature leaves 1.9–7.6 cm long (median 4.5, Arctic Ocean and interior), 2.5–8.5 cm long (median 5, Bering Sea and Pacific Coast), and 0.7–3.4 cm wide (median 1.8, Arctic Ocean and interior), and 1.9–6 cm wide (median 3, Bering Sea and Pacific Coast), acute to rounded apically, the margins entire, glabrescent beneath, the apex usually "bearded" with long, straight hairs; petioles (3)9–15(35) mm long; aments coetaneous, the staminate 2.5–

5 cm long, the pistillate (1.5)3–9(12) cm long (1.5–5 cm long in Arctic Ocean populations), the floriferous branchlets (1)3–8 (12) cm long; capsules 5.6–10 mm long, reddish or tawny, sparsely pubescent, glossy; styles 0.6–2.2 mm long, red in life; stipes 0.2–1.6 mm long; nectaries usually 1.5–4 times as long as the stipes; bracts broadly oblong, uniformly brown or bicolored.

In a wide variety of tundra sites (arctic: meadows, sandy tundra, beach ridges, *Empetrum*-lichen heath, snowbed vegetation, and polygonal tundra—alpine: glacial moraine, talus slopes, tundra, and subalpine shrubby tundra); throughout Alaska and Yukon; circumpolar (*S. torulosa* Trautv.; *S. arctica* ssp. *torulosa* [Trautv.] Hultén; *S. crassijulis* Trautv.; *S. arctica* ssp. *crassijulis* [Trautv.] Skvortz.; *S. pallasii* Anderss.). *S. arctica* is a complex circumpolar taxon which has been subdivided into numerous indistinct infraspecific taxa. It requires monographic study.

Salix arctophila Cock. ex Heller

Dwarf shrubs with trailing branches; branchlets usually yellow green; leaves broadly elliptic to obovate, the largest mature leaves 1.7–3.7 cm long, 0.7–1.6 cm broad, obtuse or rounded apically, the margin entire and minutely glandular or minutely glandular-serrulate, glabrous, glossy and yellow green above, glabrous and plane beneath; petioles 3–13 mm long; stipules absent, or 0.4–4 mm long; aments coetaneous, the staminate about 2–5 cm long, the pistillate 2.1–6 cm long, the floriferous branchlets erect, 1.5–5.3 cm long; capsules 5–6 mm long; styles 0.6–1 mm long; stipes 0.8–1.2 mm long; bracts broadly oblong with rounded apex, 0.8–1.6 mm long, distinct, usually purplish red or black.

Wet arctic tundra; along the Arctic Coast from Bullen, Alaska, to King Point, Yukon, and eastern Brooks Range to the Mackenzie-Yukon River divide; eastern American Arctic south to northeastern Saskatchewan and Manitoba.

Salix athabascensis Raup
Athabasca Willow

Shrubs 0.6–2.3 m tall; branchlets reddish brown, glossy, densely or sparsely pubescent, becoming glabrescent; leaves narrowly elliptic to elliptic or narrowly obovate, the largest mature leaves 1.7–5 cm long and 0.8–1.8 cm wide, the apex acute, the base round to acute, the margins entire and often glandular on lower portion of blade; petioles 3–10 mm long; stipules minute, 0.2–0.5 mm long, aments coetaneous, the staminate 0.5–0.9 cm long, the scales with 2 nectaries, dorsal and ventral, the pistillate 1.2–3.7 cm long, loosely flowered, the floriferous branchlets 0.2–1.5 cm long; pistils 1.8–2 mm long, greenish or tawny, densely sericeous with white or sometimes with white and rusty-colored hairs; styles 0.5–1 mm long; stipes 0.8–1.2 mm long, pubescent; nectaries 1, ventral, usually 0.5 times as long as the stipes.

Fens, muskegs, and bogs; in central Alaska along the Tanana River, and in southern Yukon; eastward to the Northwest Territories and Hudson Bay and south to British Columbia and Alberta. S. *athabascensis* hybridizes with S. *pedicellaris.*

Salix babylonica L.
Weeping Willow

Trees to 10 m tall; branchlets long, pendulous, often yellowish, glabrous or glabrescent; leaves lanceolate to lance-linear, commonly 6–16 cm long and 1–3 cm broad, long-acuminate apically, obtuse to acute basally, the margin serrulate, green above, grayish beneath, glabrous; petioles 3–5 mm long; stipules rarely developed; aments coetaneous, the staminate 2–5 cm long, the pistillate 1.5–3 cm long, the floriferous branchlets pendulous; ovary sessile or nearly so, glabrous; style very short; stipes obsolete or nearly so; nectaries 1, much longer than the stipes.

Introduced trees; known to be cultivated in Petersburg and Wrangell in southeastern Alaska; Europe.

Salix barclayi Anderss.
Barclay Willow

Shrubs (prostrate–0.7) 1–3(5) m tall; branchlets yellow green, glossy, densely to sparsely villous, glabrescent; leaves elliptic or obovate, the largest mature leaves 3.3–7(9.9) cm long, 1.2–3.5(4.8) cm wide, the apex broad, the tip acuminate or acute, the base commonly rounded, the margin glandular-serrulate, the upper side of mature leaves glabrescent, the indumentum may persist along the midrib, green in life, drying dark brown or blackish along the veins; petioles 3–14(20) mm long; stipules ovate or narrowly elliptic, often glandular-dotted below; aments coetaneous, the staminate 2.5–4 cm long, the pistillate (2.5) 4–7(8) cm long, the floriferous branchlets (0.5)1–2(3) cm long; pistils green and glabrous; capsules 5–6.5 mm long; styles 0.6–1.6 mm long; stipes 0.4–1.4 mm long; nectaries about 0.5 times as long as the stipes; bracts narrowly oblong, with acute to attenuate apex.

Glacial moraines, lake and river shores, on subalpine and alpine slopes, and occasionally in muskegs, fens, and forests; in southern coastal Alaska from the eastern Aleutian Islands to Hyder, in the Alaska Range and adjacent Tanana River Valley and southern Yukon; Northwest Territories and south to British Columbia, Washington, and Alberta.

Salix barrattiana Hook.
Barratt Willow

Low, often depressed, alpine shrubs, 0.3–1 m tall; branches gnarled, reddish brown and coarsely villous; leaves crowded on branchlets, elliptic to obovate or narrowly obovate, the largest mature leaves 3.7–9.5 cm long and 1–2.9 cm wide, the margin entire or glandular to very finely glandular-serrulate, the upper side of mature leaves sparsely pubescent and glossy; petioles 4–15 mm long; aments precocious, the staminate 3–5 cm long, the pistillate 4.5–11 cm long; pistils densely gray-white sericeous; capsules 4.5–6 mm long; styles 0.6–

1.6 mm long; nectaries longer than the stipes; bracts narrowly oblong, black to dark brown.

Alpine, in river bottoms, gravel stream channels, hillsides, meadows, and wet tundra; in eastern Brooks Range and Alaska Range, and mountains of southern Yukon and adjacent Northwest Territories; south to southwestern Alberta, British Columbia, and northwestern Montana.

Salix bebbiana Sarg.
Bebb Willow

Shrubs to small trees 0.5–10 m tall; branches divaricate, reddish brown, pubescent to glabrescent; leaves elliptic to obovate or narrowly ovate, the largest mature leaves 2.6–6(7.2) cm long, 1–2(3) cm wide; base round to obtuse, the margin entire to crenate, the upper side of mature leaves pubescent or glabrescent, the lower side sericeous-tomentose or glabrescent, glaucous and rugose; petioles 2–9(12) mm long; stipules deciduous; aments coetaneous or subprecocious, the staminate 0.6–1.5 cm long, the pistillate 2.8–5 cm long; floriferous branchlets 3–15 mm long; pistils 5.5–7 mm long, long beaked, greenish, sericeous; styles 0.1–0.4 mm long; bracts narrowly oblong, tawny.

Woods along rivers and on uplands, wet lowland thickets, muskegs, prairie margins, dry south-facing slopes, and in disturbed areas such as roadsides and burns; in central Alaska, eastern Alaska Peninsula, Kodiak Island, and Kenai Peninsula (absent from Pacific Coast of Alaska from Prince William Sound to southeastern Alaska), and southern (to northern) half of Yukon and adjacent Northwest Territories; transcontinental in boreal and temperate Canada and south in cordillera to Arizona and New Mexico and east across the northern United States; Eurasia, Kola Peninsula to Chukotsk Peninsula.

Salix brachycarpa Nutt.

Low shrubs, 0.3–0.9 (rarely 2–3) m tall; branchlets usually densely white or gray villous-lanate; leaves obovate to broadly or narrowly elliptic to narrowly obovate, the largest mature leaves (1.2)2.3–3(4) cm long, 0.6–1.6 cm wide, the base rounded or cuneate, the margin entire; petioles (0.5)1–3(4) mm long; stipules 0.5–1.5(4) mm long, often obscured by the dense branchlet pubescence; aments coetaneous, the staminate 0.6–3.7 cm long; floriferous branchlets rarely more than 10 mm long, with 2 nectaries; pistillate aments 1.5–5 cm long; floriferous branchlets 3–20 mm long, pistils densely white-lanate; stipes usually absent, of 0.2–0.5 mm long; nectaries 1, ventral; bracts elliptic to broadly so, tawny to blackish.

1a. Staminate and pistillate aments globose or short-cylindrical, densely flowered; pistillate aments 1.5–2 cm long; styles 0.5–0.8 mm long; branchlets thick; leaves pubescent on both sides with densely matted, grayish white hairs. On alpine slopes, unstable limestone scree, boreal fens, margins of alkaline or marly ponds, and fens, unstable gravel margins of streams and lakes; in southeastern Yukon and adjacent British Columbia and Northwest Territories; south to Utah and Colorado and east to Hudson Bay, James Bay, Ungava, and the Gaspé Peninsula, and across southern Alberta and Saskatchewan. *S. brachycarpa* ssp. *brachycarpa*

1b. Staminate and pistillate aments narrowly cylindrical, loosely flowered; pistillate aments 2–5 cm long; styles 0.2–0.5 mm long; branchlets thin and flexible; leaves sparsely pubescent to glabrescent above, appressed-lanate or villous-lanate beneath. Arctic tundra, *Salix* thickets on stream margins and in sandy blowouts, dry alpine slopes, unstable limestone talus, glacial outwash and subalpine shrubby tundra, and in early successional stages on boreal alluvial deposits and margins of semisaline prairies; in central Alaska, Alaska Range, arctic Alaska, and the Alaska Peninsula, and throughout Yukon and adjacent British Columbia; east to Hudson Bay (*S. glauca* ssp. *niphoclada* [Rydb.] Wiggins; *S. niphoclada* var. *mexiae* [Ball] Hultén; *S. muriei* Hultén). *S. brachycarpa* ssp. *niphoclada* (Rydb.) Argus

Salix candida Fluegge ex Willd.

Shrubs 0.3–3 m tall; branchlets densely lanate to floccose, rarely sparsely pubescent; largest mature leaves 5–8 cm long, 0.8–1.8 cm wide, the apex acute, acute basally, the margin subentire; petioles 3–9 mm long; stipules narrowly ovate, 2–3 mm long, lanate; aments coetaneous, the staminate 1–1.5 cm long, subsessile, the pistillate 2.2–3(5) cm long, the floriferous branchlets 2–7 mm long; pistils 4–6 mm long, densely dull white lanate; styles 0.3–1 mm long; nectaries 0.5–1 times as long as the stipes; bracts narrowly oblong, 1.2–1.5 mm long, pale to dark brown.

Occasional in wet, usually alkaline fens and thickets at edges of ponds and on river terraces; in central Alaska from Fort Yukon to the Tanana River southeast of Fairbanks, and eastward to Old Crow Flats, to Mayo, Whitehorse, and Watson Lake, Yukon; south and east through the boreal forest and northern prairies of Canada and northern United States. This species reaches the northwestern end of its range in central Alaska where its occurrence is sporadic.

Salix chamissonis Anderss.
Chamisso Willow

Prostrate, trailing shrubs, with aments and some vegetative branchlets arising at right angles to the branch, the branchlets yellow green, glabrous or sparsely pubescent; leaves obovate to elliptic-obovate, the largest mature leaves 3–5 cm long, 1.7–3 cm wide, the apex obtuse to round or retuse, the base cuneate, the upper side of mature leaves glabrous and semiglossy, the lower side glabrous and glaucous; petioles 5–13 mm long; stipules narrowly elliptic, 2.8–9 mm long; aments coetaneous, the staminate 2.2–3.2 cm long, the pistillate 3–6 cm long, the floriferous branchlets 1.5–4 cm long; pistils 2.4–4.8 mm long, greenish red with reddish sutures, variously pubescent with flat, crinkled, refractive hairs; styles 0.8–1.2 mm long; stipes 0.2–0.4 mm long; nectaries equal to or slightly exceeding the stipes; bracts ovate, black to dark brown.

Wide ranging but infrequent in arctic and alpine tundra; in Attu Island, St. Lawrence Island, Seward Peninsula, Arctic Slope, and Brooks Range south to Wiseman and Twelvemile Summit and eastward to Richardson Mountains, Yukon, and adjacent Mackenzie. *S. chamissonis* forms relatively large disjunct populations on Twelvemile Summit and Eagle Summit region, but is not known elsewhere in central Alaska. In Asia, the species extends from eastern Siberia from the northern shores of the Sea of Okhotsk, Kamchatka Peninsula, Commander Islands, and the Chukotsk Peninsula.

Salix commutata Bebb

Low shrubs 0.2–1(2) m tall; branchlets densely white-lanate to sparsely villous; leaves elliptic to broadly or narrowly so, the largest mature leaves (2.8)3.5–5.5(10) cm long, and 1.3–3.4(4.4) cm wide, the apex acute, the base round; petioles 1.5–7 (10) mm long; stipules half-ovate, (0.8) 1–6(9) mm long; aments coetaneous, the staminate 1.5–3.5 cm long; pistils pyriform, the capsules 4.4–6.4 mm long, reddish, greenish, or tawny; styles 0.6–1.2 mm long; stipes 0.3–1.2 mm long; nectaries 0.5–0.7 times as long as the stipe; bracts narrowly oblong.

An early species on glacial moraine and rocky slopes occurring in alpine tundra (up to 1550 m), gravelly benches along rivers and in *Salix* thickets in wet fens; in coastal southern Alaska from Unalaska, Aleutian

Islands, Kodiak Island, Kenai and central Alaska Range south to the northern end of southeastern Alaska and southern Yukon (especially in the mountains near the Northwest Territories boundary); east to Northwest Territories and south to British Columbia, California, Utah, Montana, and Wyoming.

Salix drummondiana Barr.
Drummond Willow

Shrubs 1–4 m tall; branches dark reddish brown, glabrous or glabrescent, brittle; branchlets glabrous or rarely sparsely pubescent, glaucous; leaves elliptic, elliptic-obovate to narrowly elliptic-obovate, the largest mature leaves 4.2–8.5 cm long, 1.1–2.6 cm wide, the apex acute, the base acute, the margin entire, distantly glandular-dotted or glandular-crenate, the upper side of mature leaves dark green, sparsely pubescent, the lower side densely to sparsely sericeous with white or white and rusty-colored hairs, pale, not glaucous; petioles 2–12 mm long; stipules (0.2)1.2–3.2(7.2) mm long; aments precocious, the staminate 2–2.7 cm long, with 1 ventral nectary, the pistillate 2.4–8 cm long; capsules 2–5.6 mm long; styles (0.4)0.7–1.3 mm long; stipes 0.6–1.4 mm long; nectaries 2, ventral and dorsal, 0.5–2 times as long as the stipes; bracts ovate, brown.

Margins of streams and rivers, and in subalpine scrub; in southeastern Yukon; southward to California, Nevada, Utah, Montana, and Alberta (*S. subcoerulea* Piper).

Salix fuscescens Anderss.
Brownish Willow

Low trailing shrubs, with branches spreading from a central caudex and rooting at the nodes, and with aments arising at right angles; branchlets reddish or yellowish brown; leaves obovate to elliptic, the largest mature leaves (1.4)1.7–2.7(3.7) cm long, 0.7–2.1 cm wide, the margin entire or partly glandular-serrulate; leaves glabrous, glossy and bright green above, glaucous or pale green beneath; stipules absent or minute; aments coetaneous, the staminate 0.8–1.3 cm long, the pistillate 1.5–6 cm long, the floriferous branchlets 1.5–4 cm long; pistils about 5 mm long, usually sparsely pubescent with short, rust-colored hairs or densely white-sericeous; styles 0.1–0.4 mm long; stipes 0.8–2.5 mm long; bracts oblong, bicolored.

Trailing in moss, in tundra, meadows, and muskegs; almost throughout Alaska (except for the Aleutian Islands and Alexander Archipelago) and in northern half of Yukon; eastward in the Northwest Territories from the Mackenzie Delta region south to northern Manitoba; Asia, from the Lena River to Chukotsk Peninsula, Kamchatka, and region around Okhotsk (*S. arbutifolia* authors, not Pallas).

Salix glauca L.
Diamond Willow, Glaucous Willow

Erect shrubs 0.3–5 m tall; twigs reddish brown to grayish, variously pubescent to glabrate, sometimes glaucous; leaves most-

Salix glauca L. (× 0.25).

ly 2–10 cm long, 0.4–3 cm broad, elliptic to oblong, oblanceolate, or lanceolate, acute, attenuate, obtuse, or rounded apically, attenuate to acute, obtuse, or rounded to subcordate basally, entire or rarely minutely glandular-serrulate on the lower part, villous pubescent above and below, or glabrate to glabrous above and sometimes below in age, the petioles mostly (2) 4–15 mm long; catkins appearing with the leaves, the staminate 1.6–2.5 cm long, with 2 nectaries, dorsal and ventral, the pistillate mostly 2–7.5 cm long, with 1 nectary, ventral, usually shorter than the stipes; pistils pubescent (sometimes sparsely so), 4–8 mm long, the stipes (0.5)1–2 mm long; bracts yellowish to brownish (rarely blackish), ovate to narrowly elliptic, mostly short-hairy.

This entity is widely distributed in a variety of habitats in our vicinity. Three poorly defined phases, which may be treated as varieties occur in our region (*S. glauca* ssp. *acutifolia*, ssp. *callicarpaea*, and ssp. *desertorum* in sense of Hultén, 1968).

1a. Shrubs (0.3)0.9–1.2(3) m tall; leaves 2.4–4(5) cm long, elliptic to obovate, 1.6–3 times as long as wide; stipules minute, 2–6 mm long, generally inconspicuous; pistillate aments 2–4 cm long; along rivers and creeks, in openings in spruce woods, and forming thickets on subalpine slopes; in the Rocky Mountains from northern British Columbia and adjacent Yukon; south to Utah and New Mexico and east to Hudson Bay. Rocky Mountain Phase. *S. glauca* var. *villosa* (D. Don) Anderss.

1b. Shrubs prostrate, or to 4.5(5) m tall; leaves 4–10 cm long, obovate to narrowly obovate, 2.8–4 times as long as wide; stipules 4–10(17) mm long, prominent; pistillate aments 3.5–7 cm long. (2)

2a. Shrubs 0.9–2.1(5) m tall; leaves dark green and glabrescent above, villous-lanate or sometimes glabrescent beneath; petioles 3–16 mm long; stipules very prominent; branchlets villous to pubescent; pistillate aments stout, long-cylindrical; bracts light brown to tawny; forests and muskegs, subarctic willow thickets and alpine tundra; in central Alaska and Alaska Range east to the Yukon; east to Great Bear and Great Slave lakes and south to northern British Columbia. Western Phase. *S. glauca* var. *acutifolia* (Hook.) Schneid.

2b. Shrubs prostrate, 0.3–0.9 m tall; leaves light green above, pubescent on both sides, becoming glabrescent above, never villous-lanate beneath; petioles 2–10 mm long; stipules variable; branchlets densely villous, the internodes usually short; pistillate aments shorter, narrowly cylindrical; bracts often dark brown; arctic tundra, commonly forming thickets on creeks and rivers and on old beaches; in western Alaska Peninsula eastward to Kodiak and Lake Iliamna, north along the Bering Sea coast to the Seward Peninsula, and east along the Arctic coast and along the Brooks Range to the Mackenzie River Delta. Beringia Phase. *S. glauca* var. *glauca*

Salix hastata L.

Shrubs (0.2)0.7–1(3) m tall; branchlets reddish brown, commonly white-villous; leaves elliptic to obovate, the largest mature leaves 2.5–6.8 cm long and 0.1–3.2 cm wide, acute to rounded basally, the margin entire or indistinctly and irregularly glandular-serrulate; petioles 1.5–6(9) mm long; stipules 1–4(6) mm long; aments coetaneous, the staminate 1.3–2.5 cm long, the pistillate 2.5–7 cm long, often loosely flowered, the floriferous branchlets 5–12 mm long; pistils 2–4 mm long, glabrous, green, reddish below; nectaries about 0.5 times as long as stipes; bracts narrowly oblong, light brown or bicolored.

Willow thickets; in sandy, arctic tundra and stream banks and meadows; in northern Alaska, in mountains of central Alaska and Alaska Range, and east to Dawson, Yukon; northwestern Northwest Territories (*S. farrae* Ball var. *walpolei* Coville & Ball).

Salix hookeriana Barr.
Hooker Willow

Shrubs 0.6–1.5 m tall; branches thick; branchlets usually densely white villous-lanate; leaves broadly elliptic to broadly obovate, the largest mature leaves 3.6–7 (10.3) cm long and 1.9–6.3 cm broad, rounded to obtuse apically, the base rounded, the margin distantly and irregularly glandular-crenate, the upper side of mature leaves sparsely pubescent, yellow green and glossy, the lower side sparsely villous (especially the midrib) to glabrescent, the petiole 4–12 mm long; aments coetaneous or subprecocious, the staminate 2.5–5.5 cm long, usually very thick, the pistillate 2.2–14 cm long, capsules at base of ament usually reflexed, the floriferous branchlets 1.3 cm long; pistils 3.6–6.4 mm long, green, glabrous or partially to completely lanate; capsules 8–10 mm long; styles 1.1–2 mm long; stipes 0.6–1.8 mm long; nectaries about 0.5–0.7 times as long as stipes; bracts broadly oblong, brown or blackish.

Stabilized sand dunes, meadows, and beach ridges to willow thickets; from Yakutat Bay region, Childs Glacier, Middleton Island, and Kodiak (?); southward to British Columbia, Washington, Oregon, and California (*S. amplifolia* Cov.).

Salix interior Rowlee
Interior Willow

Shrubs 0.5–4 m tall, colonial, with shoots arising from the roots; branches grayish; branchlets reddish brown, sparsely sericeous, becoming glabrescent; largest mature leaves 4–12.8 cm long, 0.3–1 cm wide, acute apically, narrowly cuneate basally, the upper side of mature leaves glabrous and green, the lower side sparsely sericeous to glabrescent; petiole 0.8–5 mm long;

stipules minute, glandular lobes; aments coetaneous, the staminate 2–5 cm long, often branched and bearing lateral, secondary aments, the pistillate 2.5–6 cm long, unbranched; floriferous branchlets 1.3–5.5 cm long; pistils long-beaked, 2–4.5 mm long, the capsules 5–8 mm long; styles to 0.1 mm long; stipes 0.6–0.8 mm long; nectaries 2, dorsal and ventral, 2–3 times as long as stipes; bracts oblong, tawny.

Boreal, on river banks and alluvial deposits; along the Yukon River and its tributaries west to west-central Alaska and in southern Yukon; south to California and east across Canada to eastern United States.

Salix lanata L.
Lanate Willow

Shrubs 0.6–3(7) m tall; branches pubescent with persistent, coarse, spreading indumentum; leaves elliptic, narrowly ovate, or broadly obovate, the largest mature leaves (3)3.8–6.8(8) cm long, (1.1)1.5–4(7.3) cm wide, acute apically, the margin entire or glandular-serrulate, the upper side of mature leaves glabrous or sparsely villous, the lower side glabrous or sparsely villous, glaucous; petioles 5–18 mm long; stipules 6–17(25) mm long, the margin glandular-serrulate or prominently and irregularly toothed; aments precocious, the staminate 2.5–4.5 cm long, the pistillate 4.5–9.5 cm long; pistils 2.5–3.6 mm long, green, the capsules 4.5–7.2 mm long; styles 1.2–1.6 (3.2) mm long; stipes 0.2–0.5 mm long; bracts narrowly obovate, dark brown.

Arctic and alpine thickets, on sands or gravels, in spruce scrub on mountain slopes, and in wet meadows; almost throughout Alaska and Yukon, except for the Aleutian Islands and the Pacific coastal regions; south to northern British Columbia and east to Baffin Island (*S. richardsonii* Hook.; *S. lanata* ssp. *richardsonii* [Hook.] Skvortz.).

Salix lasiandra Benth.

Shrubs or small trees 1–7(11) m tall;

branchlets tawny, lanate to sparsely pubescent; leaves narrowly to broadly ovate, the largest mature ones 6.7–14.2 cm long, 1.3–3 cm wide, obtuse to rounded basally, the margin glandular serrate-crenate, the upper side glabrescent, green and glossy when mature, the lower side sparsely pubescent, sometimes with rust-colored hairs, becoming glabrescent, thinly glaucous or pale; petiole 1.3–3 cm long; stipules semiovate, 1.4–5.6 mm long; aments coetaneous, the staminate 2–3.5 cm long, with 2 nectaries, dorsal and ventral, the pistillate 2.5–5 cm long, the floriferous branchlets 1.3–3.5 cm long; pistils 3–4.8 mm long; styles 0.4–0.8 mm long; stipes 0.9–1.2 mm long; nectaries 1, ventral, about 0.2–0.5 times as long as stipes; bracts narrowly oblong, tawny, deciduous after flowering.

River banks, alluvial deposits, and wet meadows; along the Yukon River and its tributaries and in south-central and southeastern Alaska; eastward to Saskatchewan and south to California and New Mexico (*S. lasiandra* var. *lancifolia* [Anderss.] Bebb).

Salix maccalliana Rowlee
MacCall Willow

Shrubs 0.9–3.5 m tall; branches dark reddish brown, glabrous and glossy; leaves coriaceous, narrowly elliptic to oblong, the largest mature leaves 5.2–7 cm long and 0.8–2 cm wide, acute apically, the upper side of mature ones glabrescent, the midrib sometimes remaining puberulent, glossy; petioles 5–10 mm long; stipules usually small, glandular lobes 0.2–0.5(2) mm long; aments coetaneous, the staminate 1.8–2.7 cm long; nectaries cuplike and surrounding the stamens; pistillate aments 2–6 cm long; floriferous branchlets 1–2.8 cm long; pistils 6–8 mm long, densely sericeous, often with white and rusty-colored hairs, tawny or green; styles 0.8–1.2 mm long; nectaries 1, ventral, about 0.5 times as long as the stipes; bracts narrowly oblong, tawny.

Muskegs, fens, and river banks; known in our area only from Watson Lake, Yukon; southward to British Columbia and east to Alberta and Quebec.

Salix monticola Bebb

Shrubs 1–4 m tall; branchlets yellow green, glabrous or sparsely pubescent; largest mature leaves 4.1–6.9(8) cm long and 2–3.3 (4) cm wide, broad and abruptly acuminate to acute apically, obtuse to rounded or subcordate basally, the margin glandular crenate-serrulate, the upper side of mature leaves green and glabrous, the lower side glabrescent and glaucous; petioles 6–12(20) mm long; stipules ovate, 1–12 mm long; aments precocious; the staminate 2.5–3 cm long, the pistillate 4–9 cm long, usually sessile; pistils green and glabrous, the capsules 5.5–6.5 mm long; styles 0.8–1.2 mm long; stipes 0.9–2.2 mm long; bracts narrowly oblong, dark brown to blackish.

Fens and drainages in forests, muskegs, and moist woods; in central Alaska, eastern Alaska Range, and eastward to southern Yukon; east to the Northwest Territories, Saskatchewan, South Dakota, and south to New Mexico (*S. padophylla* Rydb.; *S. pseudomonticola* Ball).

Salix myrtillifolia Anderss.

Low shrubs 1–9 dm tall, usually decumbent or trailing and rooting along the stem; branches grayish brown; branchlets greenish brown to dark reddish brown, glossy, sparsely pubescent; leaves narrowly elliptic to narrowly obovate, the largest mature leaves 2–5.1 cm long, 0.9–2.2 cm wide, the apex broad, the sides forming an acute angle but the tip round, rounded to cuneate basally, the margin glandular-crenate, glossy, glabrous, and green when mature, sometimes blackening on drying; petiole 1.5–8 mm long; stipules narrowly elliptic, 0.2–1.8(3) mm long; aments coetaneous, the staminate 1.3–3.5 cm long, the pistillate 1.3–4.2 cm long, the floriferous branchlets

3–10 mm long; pistils 3–3.5 mm long; styles 0.3–0.5 mm long; stipes 0.6–1.6 mm long; nectaries 0.1–0.5 times as long as stipes; bracts narrowly oblong, dark brown or bicolored.

Fens, muskegs, lake margins, river banks, and subalpine spruce thickets; in the eastern half of Alaska (rarely westward) and southern Yukon; east to Newfoundland and south to British Columbia and Alberta.

Salix novae-angliae Anderss.
New England Willow

Erect shrubs (0.2)0.6–4 m tall, the branchlets usually densely white-villous, becoming sparsely pubescent in age or glabrous; leaves narrowly elliptic to narrowly obovate, the largest mature leaves 3.2–6.8 cm long, 1–2.7 cm wide, rounded basally, the margin usually glandular-crenate, the upper side glabrescent when mature, the midrib remaining pubescent with white or sometimes rust-colored hairs, glossy, the lower side usually glabrescent, glossy; petioles 2.5–7 mm long; stipules 0.5–7 mm long; aments coetaneous, the staminate 1.5–2 cm long, the pistillate (1.5)2.2–4.5 cm long, the floriferous branchlets 0.5–1.8 cm long; pistils green, glabrous, the capsules 4.4–6.4 mm long; stipes 0.8–1.4 mm long; nectaries about one-fourth as long as stipes; bracts oblong, bicolored.

Shores of lakes, streams, and rivers, prairie margins, and muskegs; in eastern central Alaska and southern Yukon; south to California and east to Saskatchewan (S. *pseudocordata* Rydb.; S. *myrtillifolia* var. *pseudo-myrsinites* Ball ex Hultén).

Salix numularia Anderss.

Dwarf, trailing shrubs less than 2 dm tall, the branches slender, arising from a stout caudex; branchlets yellow brown and sparsely pubescent; leaves subcircular, the largest mature leaves about 1–1.2 cm long, 0.8–1 cm wide, rounded to retuse apically, rounded to subcordate basally, the margin entire or with 3–4 pairs of glands or

glandular teeth on lower half, glabrous and glossy on both sides when mature, with 4–5 pairs of secondary veins prominent, persistent for 2–3 years or more commonly deciduous each year; petioles 1.5–2 mm long; stipules minute; aments probably coetaneous or serotinus, the staminate unknown in North America, the pistillate about 3–5 mm long, 4–5-flowered, the floriferous branchlets about 1.5 mm long; capsules about 3.5 mm long; styles 0.2–0.4 mm long; stipes about 0.4 mm long; nectaries about twice as long as stipes; bracts obovate, pale brown.

Arctic tundra; known in our region only from St. Paul Island, but widely distributed in northern U.S.S.R. from the Kola Peninsula to Chukotsk Peninsula.

Salix ovalifolia Trautv.
Round-leaf Willow

Dwarf shrubs usually less than 2 dm tall, the branches long, slender and trailing; branchlets yellow or greenish brown, glabrous or sparsely pubescent; leaves obovate to elliptic, the largest mature leaves 0.8–4.6 cm long, 0.4–2.2 cm wide, acute to rounded apically and basally, the margin entire, often ciliate, the upper side glabrescent when mature and with prominently reticulate venation, the lower side sparsely villous to glabrescent, glaucous; petioles 1.1–16 mm long; stipules consisting of minute glandular lobes; aments coetaneous, the staminate 0.4–1.8 cm long, the pistillate 0.7–5 cm long, the floriferous branchlets 0.5–3 cm long; pistils 2.5–9.6 mm long, dark purple or reddish, glabrous and glaucous or pubescent, the capsules 4–10 mm long; styles 0.2–0.8 mm long; stipes 0.2–1.4 mm long; nectaries 1–3 times as long as stipes; bracts narrowly oblong, dark brown to blackish.

Beach ridges, sand spits, meadows, saline marshes, and occasionally in upland tundra; in arctic coast of Alaska and Yukon; east to the mouth of the Mackenzie River, Northwest Territories, and from the Aleutian Islands eastward to Kodiak Island. Four intergrading varieties can be recognized.

1a. Pistils usually pubescent and not glaucous (15 percent of plants are glabrous); leaves 0.8–1.4 cm long, the margins ciliate; aments globose; capsules 4.3–5.2 mm long; stipes 0.2–0.8 mm long. S. *ovalifolia* var. *glacialis* (Anderss.) Argus

1b. Pistils usually glabrous and glaucous (about 10 percent have some pubescence); leaves 1.3–4.6 cm long, the margins sometimes ciliate; aments cylindrical; capsules 5.2–10 mm long; stipes 0.2–1.4 mm long. (2)

2a. Leaves subcircular, the largest 1–1.5 times as long as wide. S. *ovalifolia* var. *cyclophylla* (Rydb.) Ball

2b. Leaves obovate, elliptic or narrowly so, 1.5–3.5 times as long as wide. (3)

3a. Leaves obovate, elliptic or broadly so, 2.5–4.6 cm long; pistillate aments 0.9–4.5 cm long (median 1.5 cm); pistils 2.5–4 mm long. S. *ovalifolia* var. *ovalifolia*

3b. Leaves often narrowly elliptic, 2.5–4.6 cm long; pistillate aments 2.2–5 cm long; pistils 5.2–9.6 mm long. S. *ovalifolia* var. *arctolitoralis* (Hultén) Argus

Salix pedicellaris Pursh

Shrubs 2–15 dm tall; branchlets commonly reddish brown, minutely puberulent; leaves narrowly elliptic, narrowly rectangular or narrowly obovate, the largest mature leaves 2.3–5.3 cm long and 0.7–1.3 cm wide, obtuse to rounded apically, acute to obtuse basally, the margin entire, the upper side glabrous at maturity, dull, usually glaucescent and with raised reticulate venation, the lower side glabrous and glaucous; petioles 3–8 mm long; stipules reduced to minute glands; aments coetaneous, the staminate 0.9–1.5 cm long, the pistillate 1.3–3 cm long, the floriferous branchlets 1.5–3 cm long; pistils glabrous and often glau-

cous, the capsules 5.6–6.4 mm long; nectaries 0.2–0.5 times as long as stipes; bracts ovate to narrowly oblong, tawny.

Fens and muskegs; in southern and northern Yukon; eastward from the Northwest Territories to Newfoundland and northeastern United States and south to Washington. S. *pedicellaris* hybridizes with S. *athabascensis*.

Salix pentandra L.

Introduced shrubs or small trees up to 7 m tall; branchlets glabrous and glossy; leaves broadly ovate to narrowly elliptic, the largest mature ones (3.5)7–8.5(11) cm long (excluding apex), (1.5)2.5–3(4.3) cm wide, the apex acuminate on later leaves, 7–12 mm long, the base rounded, the margin glandular-serrulate, coriaceous at maturity, the upper side dark green, the lower side green or pale, not glaucous; petioles 4–10 mm long, glandular at the upper end; stipules minute, deciduous; aments coetaneous, the staminate 2–6 cm long, the staminate flowers with 5 stamens, the pistillate aments 3.5–6 cm long, on floriferous branchlets 1.5–4 cm long; capsules 5–6 mm long, glabrous; styles about 1 mm long; stipes 0.5–1 mm long; nectaries cuplike, with lobes dorsally and ventrally, about half as long as stipe; bracts narrowly elliptic, 2–3 mm long, pale yellow, deciduous after flowering.

Cultivated ornamental; known from Petersburg in southeastern Alaska; native to Eurasia.

Salix phlebophylla Anderss.

Dwarf shrub forming compact mats; branches thick and clothed with persistent, skeletonized leaves, reddish brown; branchlets glabrous, not glaucous; leaves narrowly to broadly obovate or elliptic, the largest mature leaves 0.7–1.5 cm long, 0.3–1.1 cm wide, obtuse apically, cuneate basally, the margin entire, the upper side glabrous and glossy at maturity, the lower side pubescent, glossy, the veins prominent on both sides; petioles 1.2–3.2(4.8) mm long; stipules minute; aments coetaneous, the staminate 1.3–2.5 cm long, the nectaries 1–2, the pistillate 1.6–2.5 cm long, bearing more than 25 flowers, on floriferous branchlets 0.8–2.4 cm long; pistils about 1.8 mm long, sericeous, occasionally pubescent on beak only or entirely glabrous, not glaucous; capsules 2.9–4.8 mm long; styles 0.3–1 mm long; stipes 0.4–1.4 mm long; bracts broadly oblong, dark brown to black.

Arctic and alpine tundra; on St. Lawrence Island; in northern Alaska and Yukon; on the Alaska Peninsula and mountains of central Alaska and in the Alaska Range; eastward to northern Mackenzie; Asia. Apparently, S. *phlebophylla* forms hybrids with S. *rotundifolia*.

Salix planifolia Pursh

Shrubs prostrate, or to 1–4 m tall; branchlets brownish, glossy, glabrous or sparsely pubescent to densely whitish-gray villous; leaves elliptic to narrowly elliptic or rhombic, the largest mature leaves (2.2)3.5–6 (7.5) cm long and (0.8)1–2(2.6) cm wide, acute apically and basally, the margin entire to glandular crenate, the upper side glabrescent and glossy when mature, the lower side sparsely pubescent with white or white and rust-colored hairs, or glabrous and glaucous; petioles 3–10(15) mm long; aments precocious, the staminate 1.5–5 cm long, the pistillate 1.2–6 cm long; pistils 2–2.8 mm long, tawny, densely sericeous; capsules 5.5–6 mm long; nectaries equal to or up to 3 times as long as stipes; bracts oblong, black to dark brown.

Three infraspecific taxa are recognizable within S. *planifolia* in our region.

1a. Stipules narrowly elliptic, not persistent for more than one year, 0.8–2.8 mm long; in fens, margins of streams and lakes, muskegs and forests, in southern Yukon and adjacent Northwest Territories and British Columbia; east to Newfoundland and northeastern United States, and south to California (S. *phylicifolia* ssp. *planifolia* [Pursh] Hiit.). S. *planifolia* ssp. *planifolia*

1b. Stipules linear, often persistent for 2–4 years, 3.5–14(32) mm long; in most of Alaska and Yukon, and eastward to the Mackenzie; Asia. (2)

2a. Branchlets glabrous or sparsely pubescent; in arctic and alpine tundra, and subarctic woods (S. *pulchra* Cham.). S. *planifolia* ssp. *pulchra* (Cham.) Argus var. *pulchra*

2b. Branchlets densely whitish-gray villous; in boreal forests and muskegs (S. *pulchra* var. *yukonensis* Schneid.). S. *planifolia* ssp. *pulchra* var. *yukonensis* (Schneid.) Argus

Salix polaris Wahl.
Polar Willow

Dwarf shrubs often partly subterranean, with reddish brown branches, glaucous; leaves obovate to narrowly elliptic, the largest mature leaves 1–2.8 cm long and 0.8–1.8 cm wide, rounded to retuse or obtuse apically, cuneate to rounded basally, entire, glabrous and glossy; petioles 2–10 mm long; stipules absent or minute; aments coetaneous, the staminate 1.5–1.8 cm long, the pistillate 1.5–3.5 cm long, on floriferous branchlets 1.2–2 cm long; pistils reddish and glossy, variously pubescent, the capsules 4.8–6.4 mm long; styles 0.7–1.6 mm long; stipes 0.2–0.7 mm long; bracts oblong to broadly obovate, brownish.

Arctic and alpine tundra; in the Bering Sea region, northward to Point Hope, eastward along the Brooks Range to northern Yukon and Mackenzie, and in central Alaska, Alaska Range, southeastward to the St. Elias Mountains, Yukon, and eastward to northern British Columbia and southern Yukon; Eurasia.

Salix pyrifolia Anderss.

Shrubs 1–3 m tall; branchlets glabrous, shiny, dark reddish brown, rarely greenish, drying black; buds and foliage with a balsamlike fragrance; leaves narrowly elliptic, narrowly ovate to ovate, the largest mature leaves 3–6(8.5) cm long, 2–4 cm wide, and 1.6–2.5 times as long as wide, acute apically, cordate to rounded basally, margins glandular-serrulate on immature leaves, becoming coarsely serrate or crenate in age; immature leaves membranaceous and translucent, thinly pubescent or glabrescent, green on both sides or thinly glaucous beneath; mature leaves subcoriaceous, opaque, the lower surface reticulately veined and glaucous; petioles 7–15 mm long, pubescent, sometimes glandular at the distal end; stipules small, caducous; aments coetaneous, on short leafy floriferous branchlets, the staminate 2–5 cm long, on floriferous branchlets 5–7 mm long; stamens 2, the filaments glabrous or pubescent at base; pistillate aments loosely flowered, 2.5–9 cm long, on floriferous branchlets 0.5–3 cm long; pistils glabrous, the capsules 5–9 mm long, spreading or reflexed; styles 0.5–1 mm long; nectary 1, ventral; bracts oblong, 1.5 mm long, tawny and pilose.

Muskegs, fens, wet lake and slough margins; in the Yukon (known only from Palmer Lake on the western flanks of the Mackenzie Mountains) and disjunctly in the western Northwest Territories, northeastern British Columbia, and eastward to Newfoundland and south to New York (*S. balsamifera* Barr. ex Anderss.).

Salix reticulata L.
Netted Willow

Branchlets greenish brown, glabrous; leaves elliptic-circular to oblong, the largest mature leaves 1.2–6.6 cm long, 0.8–5 cm wide, subentire, the upper side of mature leaves dark green and more or less glossy, glabrous, the lower side pale green and not glaucous to conspicuously so, sparsely pubescent; petioles (3)10–25(30–46) mm long; stipules minute; aments coetaneous, the staminate 1.1–5.2 cm long, the nectaries 2–3, more or less surrounding

the stamens, the pistillate 1–6 cm long, on floriferous branchlets 1.2–4.5 cm long, as long as the vegetative branchlets; pistils densely sericeous with white or mixed white and rust-colored hairs, the capsules about 4.5–5 mm long; styles 0.2–0.8 mm long; stipes lacking or 0.4–0.8 mm long; nectaries 2, ventral, equaling or to twice as long as stipes; bracts oblong to obovate.

Arctic and alpine tundra, rarely in boreal forest and muskegs; throughout Alaska and Yukon; south to British Columbia and east to Newfoundland and Greenland; circumpolar. Two subspecies are recognized.

1a. Pistils densely sericeous; stipes sericeous; bract apex rounded or retuse; plants broadly distributed (*S. reticulata* ssp. *orbicularis* [Anderss.] Flod.). *S. reticulata* ssp. *reticulata*

1b. Pistils glabrous and glaucous or with patchy pubescence; stipes glabrous; bract apex retuse; plants known only from Mt. Gastineau near Juneau, and the Queen Charlotte Islands, British Columbia. *S. reticulata* ssp. *glabellicarpa* Argus

Salix rigida Muhl.

Shrubs 0.3–3 m tall; branchlets reddish brown to yellow green; leaves 5–10.5 cm long, 1.2–2 cm wide, serrulate, the upper side of mature leaves glabrescent, the midrib remaining velutinous; staminate aments 1.5–2.5 cm long; pistillate aments 3–5 cm long; pistils reddish or greenish, the capsules 4–5 mm long; stipes 1–2 mm long; nectaries 1, one-fourth to one-half as long as stipes; bracts narrowly elliptic, tawny to dark brown.

Sand bars and mud flats along rivers; infrequent in southern Yukon and adjacent British Columbia; south to California and Arizona and east to Newfoundland and Virginia.

Salix rotundifolia Trautv.

Dwarf, largely subterranean shrubs about 0.2–0.3 dm tall; branches yellow brown and glabrous; branchlets bearing 2–3 leaves; leaves circular, elliptic, or narrowly elliptic, the largest mature leaves 0.4–1.4 cm long, 0.2–0.9(1.1) cm wide, rounded apically, entire, sometimes ciliate, the upper side of mature leaves glossy and glabrous, the primary veins usually prominently raised, the lower side glossy, glabrous and green, not glaucous, the venation prominent; aments coetaneous or serotinous, the staminate 0.3–1 cm long, the pistillate 0.7–2 cm long, on floriferous branchlets 0.7–2.5 cm long; pistils glabrous and reddish brown, the capsules 4–7.2 mm long; stipes 0.4–0.8 mm long; nectaries equal to or up to thrice as long as stipes; bracts broadly obovate, brown or bicolored.

Arctic and alpine tundra; throughout Alaska and northern to southwestern Yukon; disjunctly southward to northwestern Wyoming and adjacent Montana. Two subspecies are recognized; ssp. *rotundifolia* apparently hybridizes with *S. phlebophylla.*

1a. Pistillate aments 4–12(16)-flowered; leaves 0.5–1.4 cm long, 0.9–1.3 times as long as wide, prominently reticulate above; petioles 1.4–3 mm long. *S. rotundifolia* ssp. *rotundifolia*

1b. Pistillate aments 2–4(9)-flowered; leaves 0.4–0.6 cm long, 1.7–2.7 times as long as wide, the venation on upper side less prominent; petioles 0.8–1.6 mm long (*S. dodgeana* Rydb.). *S. rotundifolia* ssp. *dodgeana* (Rydb.) Argus

Salix scouleriana Barr.
Scouler Willow

Shrubs 2–7 m tall or trees 10–20 m tall; branches reddish brown, glossy; leaves obovate, elliptic to narrowly elliptic, the largest mature leaves 5–8 cm long, (1.3)2–3 cm wide, usually entire, the upper side of mature leaves pubescent, becoming glabrate, the lower side sericeous, with appressed white and/or rusty colored hairs, sometimes sparsely to densely white-villous,

glaucous; petioles 5–10(19) mm long; stipules usually 0.8–3.5 mm long; aments precocious, the staminate 1.5–4 cm long, the pistillate 1.5–5 cm long, on floriferous branchlets 0.2–15 mm long; pistils gray green, densely sericeous, the capsules 4.5–11 mm long; nectaries one-third to one-half as long as stipes; bracts narrowly elliptic, dark brown, sometimes bicolored.

Dry forests, muskegs, mature forests, willow thickets, meadows, and disturbed sites; in central Alaska, Kenai Peninsula, and Alaska Peninsula, and in southeastern Alaska and southern Yukon; south to California and New Mexico and east to Manitoba and South Dakota.

Salix setchelliana Ball
Setchell Willow

Prostrate or semiprostrate shrubs to 2.5 dm tall; branches gray brown, the epidermis loose and leathery; branchlets reddish; largest mature leaves coriaceous, narrowly obovate, 2.5–6.6 cm long and 1–2 cm wide, entire or glandular-serrulate, the upper side lemon green, glabrous and with impressed venation when mature, the lower side glabrous, glaucescent or pale yellow green; petioles lacking, or to 3 mm long, the branchlet below the petiole glabrous and appearing to be part of it; stipules absent or minute; aments coetaneous, the staminate 1.3–2 cm long, on floriferous branchlets 0.7–1.3 cm long and with 2 nectaries, one dorsal and one ventral, the pistillate 4–20-flowered, 1.5–2.5 cm long, on floriferous branchlets 1–2 cm long; pistils 3.5–4.8 mm long, brick red, glabrous, the capsules 3.6–10 mm long; styles 0.3–0.4 mm long; stipes obsolete or to 0.6 mm long; nectaries 1, ventral, equal to or to one-third longer than the stipe; bracts broadly obovate, 2–3.6 mm long, papery.

Pioneer plant on sandy beaches, sandy gravel margins of glacial rivers and on glacial moraines; in the Alaska Range, in the upper Tanana, Matanuska, and Chitina River valleys, and in southwestern Yukon and adjacent Alaska; endemic (*S. aliena* Flod.).

Salix sitchensis Sanson in Bong.
Sitka Willow

Shrubs 0.5–3.8 m tall; branches dark brown; branchlets densely sericeous, becoming sparsely pubescent; leaves narrowly elliptic, narrowly obovate, or obovate, the largest (3.1)4.5–8.5(12) cm long, (1.3) 1.9–3.2(4.8) cm wide, the apex round with a pointed tip, the base cuneate, the margin distantly and inconspicuously glandular-serrulate, the upper side sparsely sericeous or villous, becoming glabrate at maturity, bright green and dull, the lower side densely to sparsely sericeous, appearing satiny, the epidermis glossy; petioles (3)4–10(16) mm long; stipules 0.4–1.5 mm long; aments usually coetaneous, the staminate 2.2–3 cm long, the staminate flowers with 1 stamen; pistillate aments (1.5)3–6.5(9.5) cm long, on floriferous branchlets 1–2 cm long; pistils 2.4–3.6 mm long, densely white-sericeous, the capsules 2.6–5.6 mm long; styles 0.5–0.8 mm long; stigmas 0.1–0.2 mm long; stipes 0.8–1.4 mm long; nectary equal to or one-half as long as stipe; bracts narrowly oblong to obovate, 1.5–2.4 mm long, usually bicolored.

Gravel bars, thickets, and openings in forests; in coastal and insular southern Alaska from Kodiak Island eastward through southeastern Alaska and adjacent British Columbia; south to Washington and California.

Salix sphenophylla Skvortz.
Wedge-leaf Willow

Dwarf shrubs; branches trailing and rooting; leaves elliptic or narrowly so, 1.9–6 cm long and 1–2(2.6) cm wide, entire, glabrous or sparsely pubescent beneath; aments coetaneous, the staminate about 2.7–3 cm long, the pistillate 2–4.5(6) cm long, on floriferous branchlets 2.4–4.5(5) cm long; pistils greenish, the capsules abous 5–6 mm long, glabrous or sparsely pubescent; styles 0.6–1.8 mm long; nectaries 1, equal to or to twice as long as stipes; bracts narrowly elliptic, dark brown to blackish.

Arctic tundra, marshes, and rocky slopes; in western arctic Alaska, on Little Diomede Island, Seward Peninsula, and on the eastern arctic slope on Barter Island and Nuvagapak Point; Asia, from Lake Baikal and Lena River to Kamchatka and Chukotsk peninsulas.

Salix stolonifera Cov.
Stoloniferous Willow

Dwarf shrubs, with trailing or subrhizomatous branches; branchlets greenish brown, glabrous and glossy, sometimes glaucous; leaves broadly obovate or broadly elliptic, the largest mature leaves 1.6–4.2 cm long, 1.2–3(3.8) cm wide, rounded, obtuse, or retuse apically, entire or irregularly glandular or glandular-toothed in the lower half, the upper side of mature leaves generally glabrous and glossy, the lower side sparsely pubescent to glabrescent or glabrous, glaucous; petioles 3–20 mm long; stipules 0.2–1.2 mm long; aments coetaneous, the staminate 1.2–2 cm long, the pistillate aments 1.5–3.7(6) cm long, on floriferous branchlets 0.8–6 cm long; pistils 4–5.6 mm long, greenish to reddish brown, usually glabrous, glossy; styles (0.6)0.8–1.6 mm long; stipes 0.2–0.8 mm long; nectary longer than the stipe; bracts broadly oblong, brown.

Alpine tundra, slide rock, moraine, and sandy lake margins to sea level; from King Cove, Alaska Peninsula to Kodiak Island, Kenai Peninsula and southeastern Alaska and adjacent British Columbia.

SANTALACEAE
Sandalwood Family

Perennial, parasitic herbs; leaves alternate, short-petiolate to nearly or quite sessile, entire, simple, green; flowers perfect, at least some; sepals 5 (rarely 3, 6–7); petals lacking; stamens as many as the sepals and opposite them; pistils 1; ovary inferior, 1-loculed; styles 1; fruit drupaceous.

1a. Flowers in terminal cymes; fruit purplish or greenish at maturity. *Comandra*

1b. Flowers axillary, solitary or in cymules; fruit scarlet at maturity. *Geocaulon*

COMANDRA Nutt.

Plants glabrous rhizomatous root parasites; leaves nearly or quite sessile; flowers several to many in terminal, corymbose cymes; sepals white to pink (fading yellowish), hairy on the ventral surface, the hairs adherent to the anthers; style filiform; ovary free from the floral tube at the apex; fruit a leathery 1-seeded drupe.

Piehl, M. A. 1965. The natural history and taxonomy of *Comandra* (Santalaceae). Mem. Torrey Bot. Club 22(1): 1–97.

Comandra umbellata (L.) Nutt.
Bastard Toadflax

Plants mostly 10–25 cm tall; stems ascending to erect from a whitish to bluish rhizome, glabrous; leaves mostly 10–40 mm long, lanceolate to elliptic, ovate, or oblanceolate; sepals oblong to lanceolate, 2–5 mm long; fruit 4–9 mm in diameter, purplish or greenish at maturity.

Reported from west-central Yukon, where evidently rare, and from northern British Columbia and Alberta; southward and eastward through most of southern Canada and United States; Europe (*C. pallida* DC.). Our material belongs to ssp. *pallida* (DC.) Piehl.

GEOCAULON Fern.

Plants glabrous rhizomatous root parasites; leaves short-petiolate; flowers (1)2–4, in peduncled cymules in the axils of the middle to subterminal leaves, perfect or the outer flowers imperfect; sepals green to reddish purple, not hairy on the ventral surface; style short, conic; ovary completely inferior; fruit a fleshy drupe.

Geocaulon lividum (Richards.) Fern.

Plants 10–30 cm tall, the stems ascending to erect from a brown to reddish rhizome, glabrous; leaves mostly 15–40 mm long, ovate to oblong, elliptic, or oblanceolate; sepals triangular, about 1.5 mm long; fruit 5–10 mm in diameter, scarlet.

Woods to alpine or arctic tundra; in the southeastern third of continental Alaska and in southwestern Alaska (less commonly in the Panhandle), and in southern Yukon; south to Washington and Idaho and east to Michigan, New York, and Newfoundland.

SAXIFRAGACEAE
Saxifrage Family

Perennial herbs or shrubs; leaves simple, alternate, opposite, or basal; flowers usually perfect, regular or irregular, the hypanthium well developed to obsolete, often with a central, glandular disk; sepals 4–5, distinct or united; petals 4–5, distinct, or sometimes lacking; stamens usually as many as the petals, or twice as many (only 3 in *Tolmiea*), rarely numerous; pistils 1, rarely 2; ovary inferior to superior; fruit a capsule, follicle, or berry.

1a. Plants woody shrubs; fruit an inferior berry. *Ribes*

1b. Plants herbaceous, not woody (at least not above the base); fruit a capsule or follicle. (2)

2a. Petals lacking; sepals 4; stamens 4–8; flowers in the axils of the upper leaves, inconspicuous. *Chrysosplenium*

2b. Petals usually present; sepals 5; stamens (3)5–10; flowers often showy. (3)

3a. Pistil with 4 carpels; stamens 5, alternating with staminodia. *Parnassia*

3b. Pistil usually with 2 carpels, rarely more; stamens (3)5–10; staminodia lacking. (4)

4a. Ovary with 2 locules, rarely with more; placentae axile. (5)

4b. Ovary with a single locule; placentae 2, rarely 3, parietal to basal. (7)

5a. Stamens 5; plants (1)2–8 dm tall; petals about 1 cm long; leaves conspicuously stipitate-glandular ciliate. *Boykinia*

5b. Stamens 8–10; plants mostly less than 3 dm tall; petals mostly less than 1 cm long; leaves various, but not conspicuously stipitate-glandular. (6)

6a. Leaves leathery, broadly elliptic to obovate, crenate; carpels nearly distinct (at least in flower). *Leptarrhena*

6b. Leaves leathery or not, but if so, not broadly elliptic to obovate; carpels united for some distance above the base. *Saxifraga*

7a. Petals with pinnate lateral lobes; flowers in racemes. (8)

7b. Petals entire; flowers in panicles or racemes. (9)

8a. Calyx distinctly united, the hypanthium appearing tubular, mostly over 6 mm long; plants over 30 cm tall. *Tellima*

8b. Calyx saucer shaped, much less than 6 mm long; plants often less than 30 cm tall. *Mitella*

9a. Stamens 10; carpels unequal in size, one about twice as large as the other; sepals white to pink. *Tiarella*

9b. Stamens 3–5; carpels essentially equal in size; sepals green or greenish purple. (10)

10a. Flowers in panicles; petals white; stamens 5. *Heuchera*

10b. Flowers in racemes; petals brownish purple; stamens 3. *Tolmiea*

BOYKINIA Nutt. Nom. Cons.

Perennial, robust, glandular-pubescent, often brownish-hairy herbs from short rhizomes; leaves chiefly basal, the alternate cauline ones much reduced upwards; leaf blades reniform to orbicular, dentate, and lobed to shallowly incised; stipules clasping the stem; flowers regular or nearly so, perfect, borne in terminal cymose panicles; hypanthium short-campanulate; sepals 5, equal; stamens 5, opposite the sepals; petals 5, entire; pistil 2-carpellate, 2-loculed; ovary one-third to one-half inferior; placentation axile; styles 2, beaklike; fruit a capsule, the seeds numerous.

Boykinia richardsonii (Hook.) Gray
Richardson Boykinia

Plants (1)2–8 dm tall, the stem base clothed with persistent, brownish leaf bases and stipules; stems erect, simple, conspicuously glandular; basal leaves long-petiolate, the blades 2–9(14) cm long from sinus to apex, 3–10(24) cm broad, conspicuously glandular-ciliate; stem leaves becoming sessile or subsessile and finally merging with the bracts upwards; flowers large and showy; sepals ovate-acuminate, often purplish; petals white to pink, with darker veins, 10–15 mm long, abruptly acuminate to acute apically; stamens short; capsules 9–12 mm long.

Alpine and arctic tundra, heathlands, and less commonly in open woods; through most of northern, west-central, and central Alaska and west-central to northern Yukon; endemic (*Saxifraga richardsonii* Hook.; *Therofon richardsonii* [Hook.] Kuntze).

CHRYSOSPLENIUM L.

Small, perennial, glabrous or sparsely villous herbs from rhizomes or stolons; leaves alternate or basal, the blades reniform to orbicular, crenate-dentate; stipules lacking; flowers regular, perfect, solitary, or borne in few-flowered terminal, or apparently axillary, bracteate cymes; hypanthium short-campanulate to saucer shaped; sepals usually 4; petals lacking; stamens 4 or 8, arising from an epigynous disk, the fila-

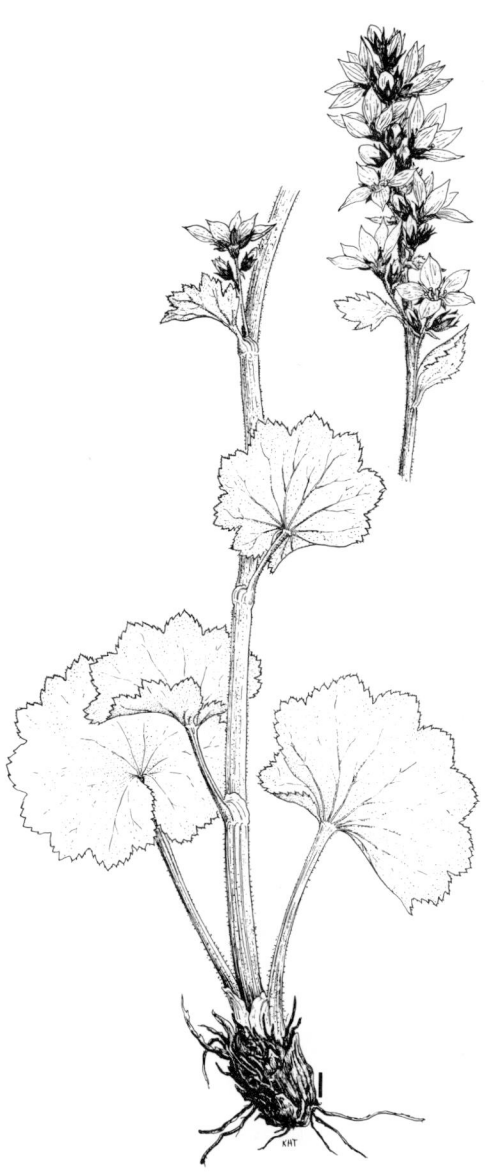

Boykinia richardsonii (Hook.) Gray (× 0.3).

ments very short; pistil 2-carpellate, 1-loculed; ovary about one-half inferior; placentae 2, parietal; styles 2; fruit a capsule, tardily and irregularly dehiscent, seeds several to many.

1a. Stamens 4; leaves thin, the veins often apparent under magnification, glabrous or sparsely villous, especially along the petiole base. *C. tetrandrum*

1b. Stamens 8; leaves thick, the veins not apparent under magnification, sparsely to densely villous, especially along the petiole base. *C. wrightii*

Chrysosplenium tetrandrum (Lund) Fries
Northern Water-carpet

Plants 0.1–1.5 dm tall, from short to elongate, subrhizomatous stolons; stems ascending to erect, simple or branched, glabrous or rarely somewhat villous; leaves long- to short-petiolate (upwards, the blades 3–10 mm long, 5–20 mm broad, with mostly 3–5 broad rounded teeth; bracts sessile or subsessile, involucrate; flowers inconspicuous; sepals usually 4, broadly rounded, green or tinged with purple; stamens mostly 4, opposite the sepals; capsules mostly 3–4 mm long.

Moist sites near pools, along streams, and in seeps; throughout Alaska (except for the western Aleutians and southern Panhandle) and in most of the Yukon; eastward to Labrador and south to Washington and Colorado; circumpolar.

Chrysosplenium wrightii Franch & Sav.
Wright Water-carpet

Plants 0.2–1 dm tall, from short, often thickened rhizomes or elongate, subrhizomatous stolons; stems erect, simple or branched above, sparsely villous; leaves long- to short-petiolate (upwards), the blades 4–9 mm long, 2–14 mm broad, mostly with 5–9 broad rounded teeth; bracts sessile or subsessile, involucrate; flowers inconspicuous; sepals 4, broadly rounded, often purplish; stamens 8; capsules 3–4 mm long.

Moist sites on tundra, on gravelly ridges, in mud, or sometimes emergent aquatics; in northern, western, and southwestern Alaska, and from the Alaska Range through east-central Alaska to southwestern Yukon, and in northern Yukon; Asia (*C. beringianum* Rose).

HEUCHERA L.

Perennial, slender, glabrous to glandular-pubescent herbs from short rhizomes; principal leaves basal, but often with 1–3 alternate cauline leaves; leaf blades ovate-orbicular to orbicular, cordate basally, palmately lobed, and once to twice serrate-dentate; stipules membranous, clasping the stem; flowers regular or nearly so, perfect, borne in open cymose panicles; hypanthium very short; sepals 5; petals 5, entire; stamens 5, opposite the calyx lobes; pistils 2-carpellate, 1-loculed; ovary one-half inferior; placentae 2, parietal; styles 2; fruit a capsule, the seeds numerous.

Rosendahl, C. O., F. K. Butters, and O. Lakela. 1936. A monograph of the genus *Heuchera*. Minn. Stud. Pl. Sci. 2: 1–180.

Heuchera glabra Willd. ex R. & S.
Alpine Heuchera

Plants 1.5–6 dm tall, the stem base clothed with persistent brown leaf bases and stipules; basal leaves long-petiolate, the blades (1)5–12 cm long from sinus to apex, 1.5–14 cm broad, glabrous or minutely glandular-puberulent below; stem leaves (0)1–3, like the basal ones but becoming short-petiolate to sessile and with smaller blades; flowers small, showy; sepals triangular, 0.5–1 mm long, green; petals white, longer than the sepals, 2–4 mm long, acute; stamens exserted; capsules 5–6 mm long.

Stream banks, moist cliffs, and slopes, mostly in woodlands; in or near coastal and insular southern Alaska, from the Alaska Peninsula eastward through the Panhandle; southward to Oregon.

LEPTARRHENA R. Br.

Perennial, slender, glabrous to glandular-pubescent herbs from elongate rhizomes; principal leaves basal, the 1–3 cauline

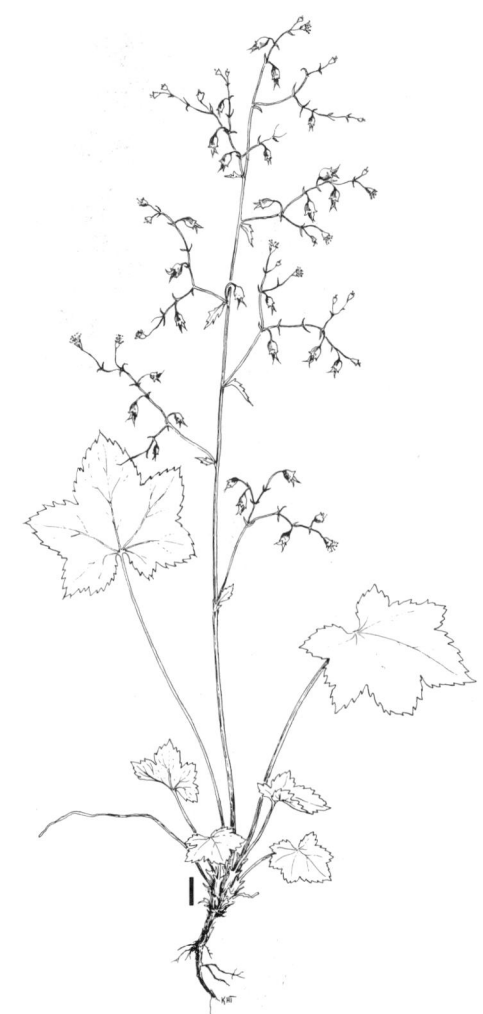

Heuchera glabra Willd. (× 0.4).

leaves greatly reduced; leaf blades oval, elliptic, oblong, or obovate, acute or cuneate basally, once to twice crenate-serrate, not lobed; stipules narrow, membranous; flowers regular or nearly so, perfect, borne in rather compact cymose panicles which expand in fruit; hypanthium obsolete; sepals 5; petals 5, entire; stamens 10; pistils 2-carpellate, the carpels connate only at the base in flower, each 1-loculed; ovary inferior at the base only; placentae basal or nearly so; styles 2, short; fruit consisting of 2 follicles, the seeds numerous.

Leptarrhena pyrolifolia (D. Don) R. Br. ex DC.

Leatherleaf Saxifrage

Plants 1.5–4.5 dm tall, the stem base clothed with old leaf bases and often with dead and dying leaves of previous years; basal leaves short-petiolate, the blades 2–10 cm long, 1–5 cm broad, glabrous; stem leaves sessile, much reduced, clasping; flowers small, inconspicuous; sepals ovate, rounded, 0.5–1.2 mm long; petals white or whitish, 1–2 mm long, rounded to obtuse; stamens about as long as the petals; fruit 5–7 mm long, often brightly colored with red or purple.

Moist alpine meadows to heathlands; in or near coastal and insular southern Alaska and southern Yukon; south to Oregon, Idaho, and Montana (*Saxifraga pyrolifolia* D. Don). The brightly colored fruiting inflorescence is very showy, accounting for the numerous fruiting specimens in collections. The plant should be introduced into cultivation.

MITELLA L.

Perennial slender glandular-pubescent to glabrate herbs from rhizomes, and sometimes also stoloniferous; principal leaves basal, the cauline leaves lacking or with 1–3 greatly reduced ones; the blades cordate-ovate to orbicular or reniform, cordate basally, once to twice crenate or crenate-serrate; stipules membranous, clasping the stem; flowers regular, perfect, borne in racemes; hypanthium very short; sepals 5; petals 5, pinnately lobed, the lobes linear; stamens 10, or 5 and opposite or alternate with the petals, arising from an epigynous disk; pistil 2-carpellate, 1-loculed; ovary less than one-half to almost completely inferior; placentae 2, parietal; styles 2; fruit a capsule, the seeds numerous.

1a. Stamens 10; ovary less than one-half inferior. *M. nuda*

1b. Stamens 5; ovary at least one-half inferior. (2)

2a. Stamens opposite the sepals; petals white or purplish; plants of southeastern Alaska. *M. trifida*

2b. Stamens alternate with the sepals; petals yellowish green; plants of broad distribution. *M. pentandra*

Mitella nuda L.

Stoloniferous Mitrewort

Plants 0.7–2 dm tall, arising from rhizomes and often stoloniferous as well; aerial stems finely glandular-pubescent, scapose, lacking cauline leaves or with one near the base; basal leaves long-petiolate, the blades 1–3 cm long, 1.5–4.5 cm broad, cordate to reniform, once to twice crenate; flowers small, inconspicuous; sepals triangular to ovate, acute to obtuse, 0.8–1.5 mm long; petals greenish yellow, (2)3–4 mm long, with usually 4 pairs of lateral lobes; stamens 10; ovary about one-third inferior; capsules 2–3 mm long.

Moist woods, along streams, and in marshes; in southeastern Alaska and southern Yukon; eastward to Newfoundland and south to Washington, Montana, Minnesota, and Pennsylvania; Asia.

Mitella pentandra Hook. ex Curtis

Alpine Mitrewort

Plants 1.5–6 dm tall, arising from rhizomes, and rarely producing stolons; aerial stems glandular-pubescent to glabrous, scapose, or occasionally with 1–2 reduced cauline leaves; basal leaves long-petiolate, the blades 2.5–5.5(6) cm long from sinus to apex, 1.5–6(7) cm wide, cordate-ovate to orbicular, palmately lobed and once to twice crenate-serrate; flowers small, inconspicuous; sepals broadly triangular, obtuse, 0.8–1.2 mm long; petals greenish yellow, 1.5–3 mm long, with 2–5 pairs of lateral lobes; stamens 5; ovary at least one-half inferior; capsules 2–3 mm long.

Moist woods, along streams, and in marshes; in southwestern, south-central, and southeastern Alaska and southern Yukon; southward to California and Colorado.

Mitella trifida Grah.

Plants 1.2–4 dm tall, arising from short rhizomes; aerial stems glandular-pubescent, scapose or with 1–2 membranous bracts; basal leaves long-petiolate, the blades 1.2–4(6) cm long from sinus to apex and about as broad, cordate-ovate to orbicular, palmately lobed and once to twice crenate; flowers small, inconspicuous; sepals triangular-ovate, obtuse, 1–1.5 mm long; petals white or purplish, 1–2 mm long, simple or 3-lobed; stamens 5; ovary about one-half inferior; capsules 2–3 mm long.

Moist woods; in coastal southeastern Alaska (Chilcat, Williams 1475, BRY); southward to California and Montana.

PARNASSIA L.

Perennial slender glabrous herbs from short rhizomes; leaves chiefly or entirely basal, the cauline leaves represented by a single, sessile bract or lacking, the blades reniform to cordate, ovate, or elliptic, entire; stipules membranous; flowers regular, perfect, solitary; hypanthium short or obsolete; sepals 5; stamens with anthers 5, opposite the sepals, alternating with broad, dissected, glandular-tipped staminodia; petals 5, entire, or fimbriate-pectinate on the lower half; pistil 4-carpellate, 1-loculed; ovary superior to about one-fourth inferior; placentae 4, parietal; styles obsolete; fruit a capsule, the seeds numerous.

1a. Petals fimbriate-pectinate along the margins of the lower half. *P. fimbriata*
1b. Petals entire, not fimbriate along the margins. (2)
2a. Petals subequal to the sepals, mostly 3–5-veined; stems bractless, or the bract near the base of the scape; staminodia mostly 4–6-lobed. *P. kotzebuei*
2b. Petals mostly much longer than the sepals, mostly 7–9-veined; stems usually with a bract above the base of the scape; staminodia with 5–many segments. *P. palustris*

Parnassia fimbriata Konig
Fringed Grass-of-Parnassus

Plants 1.5–4(5) dm tall; flowering stems solitary or clustered, the bract from below to above the middle of the scape; leaves long-petioled, the petioles 3–12 cm long, the blades 10–25 mm long, 15–50 mm broad, reniform to cordate; flowers showy; sepals 4–7 mm long, 5–7-veined, entire or fimbriate; petals white, 8–12 mm long, 5–7-veined, fimbriate along the lower margins; staminodia mostly with 5–9 segments; stamens about as long as the sepals; capsules 9–12 mm long.

Alpine meadows to woodlands; in south-central to southeastern Alaska and southern Yukon; southward to California and Colorado.

Parnassia kotzebuei Cham. ex Spreng.
Kotzebue Grass-of-Parnassus

Plants 0.3–2.2(3.5) dm tall; flowering stems solitary or clustered, the bract lacking or near the base of scape; leaves short-petiolate to subsessile, the blades 0.5–2.5 cm long, 0.3–1.5 cm broad, ovate to elliptic or orbicular; petioles (0)0.5–5 cm long; flowers inconspicuous, not especially showy; sepals 4–7 mm long, 3–7-veined; petals white, 3.5–7.5 mm long, 3–5-veined, entire; staminodia mostly 4–6-segmented; stamens about as long as the sepals; capsules 8–10 mm long.

Tundra to heath and woodlands; almost throughout Alaska and Yukon; eastward to Labrador and south to Washington, Nevada, and Wyoming; Greenland, Asia.

Parnassia palustris L.
Northern Grass-of-Parnassus

Plants 1–4.5 dm tall; flowering stems solitary or clustered, the bract mostly below the middle of the scape, rarely above the middle; leaves subsessile to short-petiolate; petioles (0)0.5–10 cm long, the blades 0.5–3 cm long, 0.5–2 cm broad, often smaller than the bract, ovate to cordate-ovate or elliptic; flowers showy; sepals 5–11 mm long, 5–9-veined; petals white, 8–15 mm long, 7–9-veined; staminodia mostly 7–many-segmented; stamens often much shorter than the sepals; capsules 8–12 mm long.

Tundra to woods; throughout most of Alaska and Yukon; southward to California, Utah, and Colorado, and eastward through Canada to Newfoundland; Eurasia (*P. palustris* ssp. *neogaea* [Fern.] Hultén). Most of our plants belong to var. *neogaea* Fern. The var. *montanensis* (Fern. & Rydb.) C. L. Hitchc. has been reported from the Yukon. It differs from var. *neogaea* in having petals less than 1.5 times as long as the sepals and fewer segments on the staminodia (7–9 as compared to 7–many in var. *neogaea*).

RIBES L.

Sprawling, ascending, or erect shrubs; stems unarmed, or with nodal spines or internodal bristles or both; leaves alternate, simple, mostly 3–7-lobed, palmately veined; flowers regular or nearly so, perfect, arranged in bracteate racemes; hypanthium short to somewhat elongate; sepals 5; petals 5, smaller than the sepals, entire; stamens 5, alternate with the petals; pistil 2-carpellate, 1-loculed; ovary inferior; placentae 2, parietal; styles 2, partially connate to distinct; stigmas capitate; fruit a berry, glabrous or variously armed with bristles, spines, or glands.

Berger, A. 1924. A taxonomic review of Currants and Gooseberries. N. Y. State Agr. Expt. Sta. Tech. Bull. 109:1–118.

1a. Stems armed with spines or bristles. (2)
1b. Stems unarmed, lacking both spines and bristles. (3)

2a. Racemes 1–3-flowered; pedicels not jointed below the ovary; flowers yellowish or greenish, rarely suffused with pink; sepals narrowly oblong, 3–5 mm long, subequal to the sub-cylindrical hypanthium. *R. oxyacanthoides*
2b. Racemes (3)5–several-flowered; pedicels jointed below the ovary; flowers usually pink to reddish or purplish; sepals broadly ovate, 2–3 mm long, mostly much longer than the saucer shaped hypanthium. *R. lacustre*

3a. Ovary with sessile yellow crystalline glands; herbage usually similarly glandular, at least when young. (4)
3b. Ovary lacking glands, or with dark stipitate glands; herbage lacking glands, or if glandular, then the fruit with stipitate glands or the fruit red at maturity. (5)

4a. Flowers white; racemes mostly 5–12-flowered, 2–6(10) cm long; cylindrical hypanthium and sepals villous; ovary glabrous except for the sessile yellow glands. *R. hudsonianum*
4b. Flowers greenish or purplish; racemes mostly 15–50-flowered, 8–30 cm long; saucer-shaped hypanthium, sepals, and ovary often strigose to short-villous in addition to being glandular. *R. bracteosum*

5a. Ovary glabrous; fruit glabrous, red when ripe. (6)
5b. Ovary glandular with stipitate glands or rarely strigulose; fruit stipitate-glandular or strigulose, dark purple, black, or dark red when ripe. (7)

6a. Flowers yellowish to greenish, or the petals sometimes tinged with red; anther sacs separated by a wide connective; plants known in our region only in cultivation. *R. sativum*
6b. Flowers purplish or reddish purple, the petals reddish purple; anther sacs touching at the apex; plants indigenous, rarely (or not at all) cultivated. *R. triste*

7a. Flowers greenish white to pinkish; petals longer than broad; hypanthium saucer shaped; fruit dark red; plants mostly of interior Alaska and Yukon. *R. glandulosum*

7b. Flowers reddish or purplish, rarely greenish; petals as broad as long or broader; hypanthium shallowly cup shaped; fruit purple or black; plants mostly of coastal regions. *R. laxiflorum*

Ribes bracteosum Dougl. ex Hook.
Blue Currant

Shrubs erect, unarmed, 1–2(3) m tall, glandular throughout with sessile yellow crystalline glands; leaf blades 4–20 cm long, 4–25 cm broad, cordate basally, 5–7-lobed; petioles long; racemes erect to ascending, 15–50-flowered, 8–30 cm long; sepals 3–4 mm long, ovate, longer than broad, purplish to greenish, or rarely white, prominently 7–9-veined; hypanthium saucer shaped to shallowly cup shaped; petals white, about 1 mm long; berries subglobose to ovoid, 8–10 mm long, black, glaucous, glandular with subsessile

Ribes bracteosum Dougl. (× 0.25).

yellow crystalline glands, with peculiar flavor and odor.

Woods and thickets; in coastal south-central to southeastern Alaska; southward to California. Putative hybrids are known from the vicinity of Juneau involving *R. bracteosum* and *R. laxiflorum*. The problem requires additional work.

Ribes glandulosum Grauer
Skunk Currant

Reclining or sprawling shrubs, unarmed, to 1 m tall; leaf blades 1.5–6 cm long, 3–8 cm wide, cordate basally, 5–7-lobed; petioles long; racemes erect or ascending, 6–15-flowered, 3–6 cm long; pedicels and rachis usually stipitate-glandular; sepals 2–2.5 mm long, obovate to oblong, longer than wide, white, or suffused with pink, 3–5-veined; hypanthium saucer shaped, less than 1 mm long; petals white or pink (red), 1–1.5 mm long, cuneate-obovate; berries subglobose, dark red, stipitate-glandular, 6–8 mm in diameter, with a disagreeable odor and flavor.

Tundra to woodlands; mostly in interior southern Alaska and southern Yukon, less commonly in coastal southwestern Alaska; eastward to Newfoundland and southward to British Columbia, Michigan, and North Carolina (*R. prostratum* L'Her.).

Ribes hudsonianum Richards.
Northern Black Currant

Shrubs erect, unarmed, 0.5–2 m tall, glandular throughout with sessile yellow crystalline glands; leaf blades 2–8 cm long, 3–11 cm broad, cordate basally, 3–5-lobed; petioles long; racemes erect to spreading or rarely deflexed, mostly 6–15-flowered, 3–8 cm long; pedicels and rachis more or less pubescent, but not stipitate-glandular; sepals 3–5 mm long, oblong to lanceolate, whitish, mostly 3-veined; hypanthium saucer shaped to shallowly cup shaped, 1–1.5 mm long; petals white, 1–1.5 mm long, cuneate-obovate; berries subglobose,

5–12 mm long, black, glaucous, somewhat glandular with sessile yellow glands, with a disagreeable flavor.

Woodlands and thickets; in the southeastern three-quarters of continental Alaska (less commonly in coastal, southern regions) and southern Yukon; east to Hudson Bay and south to California, Utah, and Minnesota.

Ribes lacustre (Pers.) Poir. ex Lam.
Swamp Gooseberry

Shrubs erect to spreading, armed with nodal spines and often densely bristly along the internodes, mostly 1–2 m tall, stipitate-glandular throughout, often sparsely so; leaf blades 2–6.5 cm long, 2–6 cm broad, cordate basally, 3–5-lobed; petioles long; racemes spreading to pendulous, mostly 6–20-flowered, 3–9 cm long; pedicels and rachis more or less pubescent and often stipitate-glandular; sepals 2–3 mm long, broadly oblong, about as broad as long, usually pink to reddish or purplish, obscurely 7–9-veined; hypanthium saucer shaped, about 1 mm long; petals pink, 1–1.5 mm long, cuneate-obovate, berries globose, 5–9 mm in diameter, black, stipitate-glandular, palatable.

Woods and thickets; mostly in coastal and insular south-central and southeastern Alaska, but also in central to eastern interior Alaska and southern Yukon; eastward to Newfoundland and south to California, Colorado, Michigan, and Pennsylvania (R. oxyacanthoides var. lacustre Pers.).

Ribes laxiflorum Pursh
Trailing Black Currant

Shrubs, decumbent (often rooting along the stem) to erect, unarmed, to 1 m long or more, stipitate-glandular to glabrous; leaf blades 2.5–9 cm long, 3–12 cm broad, cordate basally, 5 (or obscurely 7)-lobed; petioles long; racemes ascending to erect, 6–18-flowered, 4–10 cm long; pedicels and rachis stipitate-glandular; sepals 2.5–3.5

mm long, broadly ovate, longer than wide, reddish to purplish, 3–5-veined; hypanthium shallowly cup shaped, about 1 mm long; petals reddish to purplish, 1–1.5 mm long, as broad as long or broader, cuneate-reniform; berries subglobose, 6–10 mm long, purplish black, stipitate-glandular or merely strigulose, the flavor disagreeable.

Woods and thickets; in or near coastal south-central to southeastern Alaska and southern Yukon; southward to California, Idaho, and Alberta. This entity is closely allied to the interior species, R. glandulosum.

Ribes oxyacanthoides L.
Northern Gooseberry

Shrubs sprawling to ascending or erect, armed with nodal spines and usually with internodal bristles, mostly 0.5–1.5 m tall, and the herbage with very short stalked yellowish glands; leaf blades 0.5–4 cm long, 0.6–4.5 cm broad, truncate to subcordate basally, 3–5-lobed; petioles long; racemes spreading, 1–3-flowered, to 1.5 cm long; pedicels and rachis glabrous; sepals 3–5 mm long, narrowly oblong, much longer than broad, greenish yellow to whitish, obscurely 7-veined; hypanthium subcylindrical, 3–4 mm long; petals colored like the sepals, 2–3 mm long, oblanceolate; berries subglobose, 9–12 mm long, bluish purple, glabrous, palatable.

Woods; in south-central Alaska and the southern two-thirds of the Yukon; south to British Columbia and east to Hudson Bay.

Ribes sativum (Reichb.) Syme
Red Currant

Shrubs erect, unarmed, to 1.5 m tall, the herbage pubescent and often somewhat stipitate-glandular; leaf blades 3.5–7 cm long, 4–8 cm broad, cordate basally, 3–5-lobed; petioles long; racemes ascending to pendulous, 6–20-flowered, 3–7 cm long; pedicels and rachis glabrous to somewhat glandular; sepals about 2 mm long, broadly ovate, broader than long, greenish to pinkish; hypanthium saucer shaped, to 1 mm long; petals yellowish to reddish, to 1 mm long, cuneate-obovate; berries globose, 6–10 mm long, red, glabrous.

Sparingly cultivated; in southern and southeastern Alaska; native to western Europe. *R. sativum* is apparently closely related to the indigenous *R. triste*.

Ribes triste Pallas
American Red Currant

Shrubs decumbent to erect, to 1.5 m tall, the herbage pubescent and often somewhat stipitate-glandular; leaf blades 2–9 cm long, 2.5–12 cm broad, cordate basally, 3–5-lobed; petioles long; racemes pendulous, 6–15-flowered, 3–8 cm long; pedicels and rachis glabrous to pubescent, and often somewhat stipitate-glandular; sepals 1.5–2 mm long, very broadly elliptic, as broad as long, reddish or purplish, obscurely veined; hypanthium saucer shaped, to 1 mm long; petals reddish or purplish, to 1 mm long, cuneate; berries globose, 6–10 mm in diameter, red, glabrous.

Woods and thickets; through much of Alaska (except for extreme northern, southwestern, and southeastern portions) and most of the Yukon; eastward to Newfoundland and south to Oregon, South Dakota, and Virginia; Asia.

SAXIFRAGA L.

Perennial slender tufted, solitary or mat-forming, glabrous to pubescent herbs with rhizomes, stolons, or fibrous roots; leaves alternate, basal, or opposite, the blades variously shaped, dentate, lobed, or entire; stipules lacking or present, obscure to conspicuous; flowers regular or nearly so, perfect (rarely imperfect), solitary, racemose, or in cymose panicles, sometimes replaced by bulblets; hypanthium obsolete or well developed; sepals 5; petals 5, entire, variously colored, often spotted; stamens 10; pistils 2(3–5)-carpellate, with 1 locule per carpel, or 1-loculed; ovary almost entirely inferior to almost completely superior; placentation axile or basal; styles 2 or more; fruit a capsule or a follicle (when the carpels are almost distinct), the seeds numerous.

Calder, J. A., and D. B. O. Savile. 1959. Studies in Saxifragaceae–II. *Saxifraga* Sect. Trachyphyllum in North America. Brittonia 11:228–49.

———. 1960. Studies in Saxifragaceae–III. *Saxifraga odontoloma* and *Lyallii*, and North American subspecies of *S. punctata*. Can. Jour. Bot. 38:409–35.

Engler, A., and E. Irmscher. 1916–18. Saxifragaceae–*Saxifraga* I, II. Das Pflanzenreich IV. 117 (Heft 67, 69):1–709.

Johnson, A. M. 1923. A revision of the North American species of the Section Boraphila Engler of the genus *Saxifraga*. Minn Stud. Pl. Sci. 4:1–190.

1a. Plants not scapose; cauline leaves present below the inflorescence, or the inflorescence reduced to a solitary flower. (2)
1b. Plants scapose; cauline leaves lacking below the branches of the inflorescence (the branches sometimes subtended by reduced, leaflike bracts); inflorescence with few to many flowers. (16)

2a. Leaves toothed or lobed (at least some). (3)
2b. Leaves all entire (though coarsely ciliate in some). (8)

3a. Basal leaves leathery, tridentate, the teeth spinulose-tipped. *S. tricuspidata*
3b. Basal leaves not leathery, with 3–7 teeth or lobes, these not spinulose-tipped. (4)

4a. Bulblets present in the axils of the upper leaves; inflorescence with only the uppermost flower (rarely 2) developing. *S. cernua*
4b. Bulblets lacking in the axils of the uppermost leaves; inflorescence with 1–several flowers. (5)

5a. Petals 9–13 mm long; uppermost lobed cauline leaves with narrow, lanceolate to oblong lobes. *S. exilis*

5b. Petals 3–8 mm long; lobes (if any are lobed) of uppermost lobed cauline leaves ovate, or if lanceolate to oblong, then similar to those of the basal leaves. (6)

6a. Leaves reniform, with 3–7 ovate to ovate-lanceolate lobes. *S. rivularis*

6b. Leaves cuneate, 3-toothed, or with 3–7 triangular to lanceolate or oblong lobes. (7)

7a. Leaves merely 3-toothed; staminal filaments subequal to the sepals; plants of southern and southeastern Alaska. *S. adscendens*

7b. Leaves with 3–7 distinct lobes; staminal filaments longer or shorter than the sepals; plants of broad distribution. *S. caespitosa*

8a. Leaves opposite; petals pink purple to blue purple, or rarely white. *S. oppositifolia*

8b. Leaves alternate; petals white, yellow, reddish purple, or suffused with pink. (9)

9a. Leaves marginally setose-ciliate. (10)
9b. Leaves glabrous, not ciliate. (12)

10a. Flowers solitary, borne sessile on the basal rosette, or on a stem to 1 cm long; petals minute, to 1 mm long. *S. eschscholtzii*

10b. Flowers 1–several, borne on stems usually over 3 cm long; petals 3–10 mm long. (11)

11a. Plants bearing elongate, slender stolons; petals often bright yellow, about as broad as long. *S. flagellaris*

11b. Plants lacking stolons; petals pale yellow to white, much longer than broad. *S. bronchialis*

12a. Petals white or greenish yellow; flowering stems subscapose, or less than 2 cm long. (13)

12b. Petals yellow; flowering stems with several leaves (subscapose in *S. serpyllifolia*). (14)

13a. Flowering stems less than 2 cm tall; staminal filaments subulate, not petaloid; plants of western Aleutians. *S. aleutica*

13b. Flowering stems mostly more than 2 cm tall; staminal filaments clavate, petaloid; plants of southeastern Alaska. *S. tolmiei*

14a. Flowering stems subscapose; cauline leaves (0)1–3 above the basal rosette; petals 4–7 mm long, 5(7)-veined. *S. serpyllifolia*

14b. Flowering stems distinctly leafy; cauline leaves mostly more than 3 above the basal rosette; petals 7–15 mm long, or if less than 7 mm long, then only 3-veined. (15)

15a. Petals 7–15 mm long, 5–7-veined; flowers solitary (rarely more); basal leaves mostly over 12 mm long, petiolate; plants of broad distribution. *S. hirculus*

15b. Petals 3–7 mm long, 3-veined; flowers often more than 1; basal leaves mostly less than 10 mm long, sessile or subsessile; plants of eastern Yukon. *S. aizoides*

16a. Leaves entire. (17)
16b. Leaves lobed or toothed or both (at least some). (18)

17a. Staminal filaments clavate, subpetaloid; plants less than 10 cm tall; leaves mostly less than 10 mm long. *S. tolmiei*

17b. Staminal filaments subulate, not petaloid; plants mostly 20–30 cm tall; leaves mostly over 20 mm long; reported from Buckland River, Bering Strait District (Porsild 1518), but the report requires verification. *S. integrifolia* Hook.

18a. Leaves with 3–7 lanceolate to oblong lobes; petals variously colored, but often pink or pink purple; plants introduced, cultivated. *S. rosacea*

18b. Leaves various, toothed or lobed, but when lobed the segments ovate or broader; petals white, yellowish, or reddish purple, sometimes suffused with pink; plants indigenous, rarely cultivated. (19)

19a. Bulblets replacing flowers on much of the inflorescence, with only the terminal flower (or the terminal one on each branch of inflorescence) developing. S. *foliolosa*

19b. Bulblets lacking, or only a few present; lateral as well as terminal flowers developing on each branch of the inflorescence (if rarely as above, then the rachis conspicuously flexuous, and plants of southern and southeastern coastal regions). (20)

20a. Leaf blades distinctly cordate at the base, or broader than long, or both. (21)

20b. Leaf blades never cordate, always cuneate or acute at the base, longer than broad (except in S. *lyallii*). (24)

21a. Leaf blades broadly lobed, each lobe with 3-more teeth. S. *mertensiana*

21b. Leaf blades toothed or lobed but not both. (22)

22a. Flowers pale yellow, borne in slender elongate racemelike panicles (rarely somewhat expanded in fruit); leaves irregularly dentate to lobed, the alternate teeth usually smaller. S. *spicata*

22b. Flowers white, often suffused with pink, borne in hemispherical to conic, or corymbose panicles; leaves lobed (broadly dentate), the lobes not conspicuously of 2 sizes. (23)

23a. Petals 4–5 mm long, indistinctly 3–5-veined; staminal filaments subulate; plants from Seward Peninsula and islands of the Bering Sea. S. *nudicaulis*

23b. Petals (1.8)2.5–4(4.5) mm long, 1-veined; staminal filaments obscurely

to distinctly clavate; plants of broad distribution. S. *punctata*

24a. Flowers numerous, borne in dense clusters along an elongate, spikelike panicle; petals reddish purple at flowering; leaf blades elliptic, mostly 2–3 times as long as broad, often over 5 cm long. S. *hieracifolia*

24b. Flowers few to several or many, in open panicles (compact in S. *nivalis*); petals white to pink at flowering (becoming reddish purple in fruit in S. *davurica*); leaf blades mostly not as above. (25)

25a. Leaves with reddish glandular tomentum on the lower surface, the blades with mostly 15–30 teeth; inflorescence corymbose, open; plants of southeastern Alaska. S. *occidentalis*

25b. Leaves pubescent or glabrous on the lower surface, lacking a reddish glandular tomentum (except in S. *nivalis*), the blades often with less than 15 teeth; plants of broad distribution. (26)

26a. Staminal filaments clavate; capsules and ovaries various, but if reddish purple, then the branches of the inflorescence with stipitate glands, or the flowers more than 10. (27)

26b. Staminal filaments subulate to linear, but if clavate and the capsules reddish purple, then branches of the inflorescence not stipitate-glandular and the flowers less than 10. (28)

27a. Leaves crenate-dentate, the alternate teeth often differing in size, the blades mostly longer than broad, usually distinctly pubescent, often purplish tinged on the lower surface. S. *reflexa*

27b. Leaves acutely to acuminately dentate or lobed, the alternate teeth or lobes not differing in size, the blades mostly as broad as long or broader, not at all or only slightly purplish tinged beneath, usually glabrous or glabrate. S. *lyallii*

28a. Sepals not reflexed during or following flowering; flowers in more or less compact spikelike panicles, the lower 1–2 branches sometimes elongating. *S. nivalis*

28b. Sepals reflexed during flowering and in fruit; flowers mostly in open panicles, usually with more than 2 elongate branches. (29)

29a. Flowers of inflorescence often more than 10, sometimes some replaced by bulblets; petals white; carpels 2–3, sometimes turning dark reddish purple following flowering. *S. ferruginea*

29b. Flowers of inflorescence less than 10, not replaced by bulblets; petals white to pink, often turning reddish purple as the fruit matures; carpels 2–5, dark reddish purple following flowering. *S. davurica*

Saxifraga adscendens L.
Wedge-leaf Saxifrage

Plants 0.3–1 dm tall, from a taproot and a dense, imbricate, basal rosette; stems simple or freely branched, conspicuously glandular-pubescent; leaves 0.5–1.5 cm long, entire to apically 3 or rarely 5-toothed, obovate, cuneate, the basal leaves persistent and often reddish or brownish, the 3–several cauline leaves alternate, often entire; flowers small, more or less conspicuous; sepals often reddish purple, ovate, 1–2.5 mm long in flower, enlarging in fruit, glabrate to glandular dorsally, stipitate-glandular ciliate, erect; petals white, 2–6 mm long; stamens subequal to the sepals, the filaments subulate; ovary almost completely inferior; capsule 3–5 mm long.

Rock crevices and glacial moraines; in western and central to south-central and southeastern Alaska (where evidently rare) and southern Yukon; east to Mackenzie and south to Oregon, Utah, and Colorado; Europe (*Ponista oregonensis* Raf.; *S. adscendens* ssp. *oregonensis* [Raf.] Bacigalupi). Our material belongs to var. *oregonensis* (Raf.) Breit.

Saxifraga aizoides L.
Golden Saxifrage

Plants 0.5–1.4 dm tall; flowering stems erect or ascending, more or less glandular-pubescent, at least above, arising from leafy prostrate vegetative stems; leaves 0.5–1 cm long, entire, narrowly oblong, the basal leaves usually smaller than the several alternate cauline ones, glabrous or ciliate; sepals greenish, ovate to lanceolate, 3–4 mm long, glabrous, not ciliate, spreading; petals yellow, sometimes spotted orange, 4–6 mm long, 3-veined; stamens subequal to the petals, the filaments subulate; ovary about one-third inferior; capsules 4–6 mm long.

Along streams, on moist cliffs, and on talus slopes; in eastern Yukon; eastward to Labrador and southward to British Columbia, Alberta, Michigan, and New York; Europe.

Saxifraga aleutica Hultén
Aleutian Saxifrage

Plants 0.1–0.3 dm tall, mat-forming, arising from a taproot; flowering stems 0.5–2(2.5) cm long, glandular-pubescent, arising from the apex of the densely leafy vegetative stems; leaves 0.2–0.5 cm long, entire, fleshy, oblong to oblanceolate, not ciliate, the cauline leaves alternate, (0)1 or more; flowers not showy; sepals greenish, oblong to lanceolate, 2–3 mm long, glabrous, the margin glandular or smooth, spreading; petals greenish yellow, 2.5–3.5 mm long, indistinctly 3-veined; stamens subequal to the petals, the filaments subulate; ovary only slightly inferior; capsules 3.5–5 mm long.

Gravelly soil; on mountain tops in the western Aleutian Islands; endemic.

Saxifraga bronchialis L.
Spotted Saxifrage

Plants (0.1)0.3–1.5 dm tall, arising from a caudex with a taproot; caudex branches prostrate to ascending, clothed with densely imbricate leaves; flowering stems glandular-pubescent, with few to several, alternate leaves; leaves 0.3–1.5 cm long, 0.1–0.3 cm broad, entire, leathery, elliptic to oblong or spatulate, setose-ciliate and spinulose-tipped; flowers showy, yellowish or whitish; sepals greenish, ovate, 1–2.5 mm long, erect to spreading, glabrous, not ciliate; petals whitish to yellowish, yellow- to red-spotted, 3.5–8 mm long, indistinctly 3-veined; stamens shorter to longer than the petals, the filaments subulate; ovary only slightly inferior; capsules 6–10 mm long.

Cliffs, talus slopes, gravelly flats, from tundra to woodlands; in much of Alaska (except for the central valleys) and northern to west central and southern Yukon; southward to Oregon, Idaho, and New Mexico; Asia. Our material has been treated as belonging to two or more infraspecific taxa, and sometimes as separate species.

1a. Leaves spatulate, the apex obtusely pointed; flowering stems 1–4 cm tall; plants of the Aleutian Islands and less commonly in southern Alaska (S. cherlerioides D. Don; S. bronchialis ssp. cherlerioides [D. Don] Hultén). S. bronchialis var. cherlerioides (D. Don) Engler

1b. Leaves elliptic to oblong, the apex acutely pointed; flowering stems 3–15 cm tall; plants of wide distribution (Leptasea funstonii Small; S. bronchialis ssp. funstonii [Small] Hultén). S. bronchialis var. purpureo-maculata Hultén

Saxifraga caespitosa L.
Tufted Saxifrage

Plants 0.1–1.6 dm tall, arising from a caudex with a taproot; caudex branches prostrate to ascending, clothed with densely imbricate leaves; flowering stems glandular-pubescent, often densely so, with 1–several alternate leaves; basal leaves 0.5–2 cm long, 0.2–0.7 cm broad, with 3–5(7) apical triangular to lanceolate or narrowly oblong lobes, cuneate, glandular-ciliate, not spinulose-tipped; cauline leaves often entire; flowers moderately showy, white to yellowish; sepals often purplish, ovate, 1–3 mm long, erect, glandular-pubescent and ciliate; petals white to yellowish, 3.5–5.5 mm long, 3-veined; stamens shorter than the petals, longer than the sepals, the filaments subulate; ovary almost completely inferior; capsule 6–10 mm long.

Meadows in tundra, gravelly ridges, cliffs, and talus slopes, often on limestone; in much of Alaska (except for the central valleys) and southern to northern Yukon; eastward to Labrador and south to Oregon, Nevada, Utah, and New Mexico; circumpolar. Alaskan materials of S. caespitosa have been treated as belonging to ssp. sileneflora (Sternb.) Hultén and ssp. uniflora (R. Br.) Porsild. Both of the subspecies supposedly differ from typical Old World specimens and they are said to differ from each other on the basis of flower size, flower number, and sepal color. It has not been possible to segregate our specimens into one or the other of the proposed entities, except on a completely arbitrary basis.

Saxifraga cernua L.
Nodding Saxifrage

Plants 0.8–2.5(3) dm tall, arising from fibrous roots; stems sparsely to densely glandular-villous, with several alternate leaves; basal leaves long-petiolate, bearing bulblets in the axils, the blades (3)5–7-lobed, reniform, 0.4–1.5 cm long, 0.6–2.5(3) cm broad; cauline leaves becoming smaller, fewer lobed, and shorter petioled upwards, at least the upper ones bearing usually purplish bulblets in the axils; flowers white, showy, solitary (rarely 2) at the apex of the inflorescence, the others replaced by bulblets; sepals green to dark reddish purple, ovate to lanceolate, 2.5–3.5 mm long, erect, sparsely to densely glandular, somewhat ciliate; petals white, 8–14 mm long, 3–5-veined; stamens longer than the sepals, much shorter than the petals; ovary slightly if at all inferior; capsules rarely developing.

Arctic and alpine tundra, mostly in wet situations, or on cliffs and talus slopes; in northern, western, and southwestern to southeastern continental Alaska and northern and southern Yukon; eastward to Labrador and south to Utah and New Mexico; circumpolar.

Saxifraga davurica Willd.

Plants 0.3–1.5 dm tall, arising from a caudex with a taproot; caudex branches often rooting and subrhizomatous; flowering stems conspicuously villous to glabrate or glabrous, leafless (rarely with 1 cauline leaf); leaves 1.5–4.5 cm long, 0.5–1.3 cm broad, with 5–11 teeth, the blades cuneate-oblanceolate to spatulate or wedge shaped, glabrous, ciliate; flowers moderately showy; sepals purple or greenish purple, lanceolate, 2–3 mm long, reflexed, glabrous or somewhat ciliate; petals pink to maroon, 3–4 mm long, only the midvein apparent; stamens about equaling the sepals, the filaments subulate to linear or obscurely clavate; ovary only slightly inferior; capsule 6–10 mm long, dark reddish purple, often more than 2-carpellate.

Arctic and alpine tundra and heathlands; in northern, western, southwestern and in the southeastern quarter of mainland Alaska and southern to northern Yukon; east to Mackenzie; Asia. Specimens treated herein as belonging to *S. davurica* have been interpreted differently by previous authors. Much confusion is evident concerning the typification of the various entities. Attempts to follow the various monographic treatments (Calder and Savile, 1959, 1960; Engler and Irmscher, 1916–18; Johnson, 1923) have resulted in complete chaos. It is only on an arbitrary basis that the two infraspecific taxa cited below are listed with *S. davurica*. Both entities have distinctly reflexed sepals, a characteristic not attributed to *S. davurica* by Engler and Irmscher (1916–18), but definitely a characteristic of *S. lyallii*. Both of the varieties

have characteristics in common with *S. lyallii* and *S. reflexa*, and might be best included within one of those taxa. However, until the true nature of *S. davurica* can be determined, both are best retained under the older epithet.

1a. Stems distinctly villous with multicellular hairs; leaves mostly 2–3 times as long as broad (from base of petiole to apex); plants from extreme western Alaska, from Cape Lisburne southward, islands of the Bering Sea, and the Aleutians (*S. unalaschcensis* Sternb.; *S. flabellifolia* T. & G.). *S. davurica* var. *unalaschcensis* (Sternb.) Engler

1b. Stems glabrous to glabrate, the bracts sometimes somewhat villous at the base; leaves mostly over 3 times as long as broad (from base of petiole to apex); plants from west-central, northern, and interior central to eastern Alaska and in the Yukon (*S. davurica* f. *grandipetala* Engler & Irmscher; *S. grandipetala* [Engler & Irmscher] Los.; *S. davurica* ssp. *grandipetala* [Engler & Irmscher] Hultén). *S. davurica* var. *grandipetala* (Engler & Irmscher) B. Boi.

Saxifraga eschscholtzii Sternb.
Ciliate Saxifrage

Plants 0.1–0.3 dm tall, mat-forming, the prostrate caudex branches rooting, subrhizomatous; flowering stems obsolete or to 1 cm long, leafless, minutely glandular to glabrous, borne at the end of the densely leafy caudex branches; leaves 1–3 mm long, entire, oblong to obovate, setose-ciliate, alternate, densely imbricate; flowers inconspicuous, imperfect, the sexes on separate plants; sepals greenish to yellowish or reddish purple, about 1 mm long, spreading to reflexed, setose-ciliate; petals less than 1 mm long, caducous, pink to white, 1-veined; stamens shorter than the sepals in pistillate flowers (the pistil well developed), longer than the sepals in staminate flowers (the pistil poorly developed), the filaments in both subulate; ovary only slightly inferior; capsules 2–4 mm long.

Rocky slopes, cliffs, and gravelly spits; from the Seward Peninsula northward and eastward along the Brooks Range to northeastern Alaska and northern Yukon, and from the Alaska Peninsula and Kodiak eastward along the Alaska Range, rarely in southeastern Alaska; Asia (*S. fimbriata* D. Don).

Saxifraga exilis Steph.

Plants 0.6–1.9 dm tall, arising from fibrous roots; stems glandular-villous, with mostly 1–4 alternate cauline leaves; basal leaves short-petiolate, bearing bulblets in the axils, the blades 5–7-lobed, reniform to orbicular, cordate or hastate basally, 0.4–1 cm long, 0.6–1.8 cm broad; cauline leaves becoming smaller and the lobes more slender, fewer, and shorter petioled upwards, not bearing bulblets; flowers showy, 1–6, not replaced by bulblets; sepals green to purplish, ovate to lanceolate, 2–3.5(4) mm long, erect, sparsely glandular-pubescent to glabrate, somewhat ciliate; petals white, 9–13 mm long, 3(5)-veined; stamens mostly longer than the sepals, the filaments linear; ovary one-fourth to one-third inferior; capsules 9–12 mm long.

Tundra to heathlands and woods; in central western to northwestern, northeastern, and east-central Alaska and northern to central Yukon; east to Mackenzie; Asia (*S. radiata* Small).

Saxifraga ferruginea Grah.
Alaska Saxifrage

Plants 0.6–3.5 dm tall, arising from a basal rosette with fibrous roots; flowering stems conspicuously villous with long, spreading, glandular hairs, leafless (though the lower bracts of the inflorescence are often foliose); leaves 1.5–5.5(10) cm long, 0.3–1.7(3) cm broad, with 5–15 forward pointing teeth (rarely lobed), the alternate teeth occasionally differing in size, the blades cuneate-oblanceolate to spatulate, glabrous to glandular-villous above and below, stipitate-glandular ciliate (often conspicuously so); flowers showy, occasionally replaced by bulblets; sepals green to purplish, 1.2–3 mm long, lanceolate to oblong, reflexed; petals white, 3.2–5.5 mm long, distinctly stalked, auriculate to hastately lobed at the base, indistinctly 1–3-veined; stamens longer than the sepals, the filaments subulate; ovary only slightly inferior; capsules 3–5 mm long.

Woods, wet meadows, and rock outcrops, from sea level to alpine; in coastal and insular southern Alaska from the Alaska Peninsula through the Panhandle; southward to California, Idaho, and Montana (*S. ferruginea* var. *newcombei* Small). The var. *macounii* Engler & Irmscher has a portion of the flowers replaced by bulblets or by foliose plantlets. The nature of these viviparous plants is not understood. They may represent the response of a variable genotype to differences in the environment. Very robust plants with exceptionally large leaves are known. They probably represent nothing more than plants from favorable habitats.

Saxifraga flagellaris Willd. ex Sternb.
Flagellate Saxifrage

Plants 0.2–1.4 dm tall; flowering stems erect, stipitate-glandular villous, arising from a basal rosette with fibrous roots, with few to several cauline leaves; basal leaves 0.5–1.5 cm long, 0.1–0.4 cm broad, entire, cuneate-oblanceolate to spatulate, setose-ciliate; cauline leaves similar to the basal ones except becoming glandular-ciliate and smaller above; flowers 1–few, showy; sepals greenish to reddish purple, oblong to lanceolate or ovate, 2.5–4.5 mm long, glandular-villous, stipitate-glandular ciliate; petals bright yellow, 6–12 mm long, 7–9-veined; stamens longer than the sepals, the filaments subulate; ovary only slightly to one-fourth inferior; capsules 8–10 mm long.

Arctic and alpine tundra; in much of Alaska (except for the central valleys, the Aleutians, and the Alexander Archipelago) and western Yukon; east to Mackenzie and south to Arizona and New Mexico; circumpolar.

1a. Plants with dark reddish purple calyx lobes and glands, the upper leaves and stems often similarly colored; ovary about one-fourth inferior; mostly north of the Brooks Range (*S. platysepala* [Trautv.] Tolm.; *S. flagellaris* ssp. *platysepala* [Trautv.] Porsild). *S. flagellaris* var. *platysepala* Trautv.

1b. Plants with green to yellowish calyx lobes and glands, the leaves and stems not dark reddish purple; ovary only slightly inferior; widely distributed (*S. setigera* Pursh; *S. flagellaris* ssp. *setigera* [Pursh] Tolm.; *S. flagellaris* var. *stenosepala* Trautv.). *S. flagellaris* var. *flagellaris*

Saxifraga foliolosa R. Br.
Foliose Saxifrage

Plants 0.4–2.8 dm tall, arising from a basal rosette with fibrous roots; flowering stems sparsely villous to glabrate, the hairs gland tipped, leafless, the bracts small; leaves 0.6–2.2 cm long, 0.2–0.9 cm broad, with 3–5(7) forward pointing teeth (rarely some entire), the blades cuneate-oblanceolate to spatulate, sparsely glandular-ciliate to completely glabrous; flowers showy, mostly replaced by bulblets; sepals green to reddish purple, 1–2 mm long, oblong to ovate, spreading to reflexed; petals white, 3–5 mm long, distinctly stalked, auriculate to hastately lobed at the base, indistinctly 3-veined; stamens longer than the sepals, the filaments subulate; ovary only slightly inferior; capsules 3–5 mm long, seldom developing.

Tundra, often in wet sites; in western, northern, and south-central to east-central Alaska and southwestern Yukon; eastward to Labrador and south to British Columbia; circumpolar (*S. foliolosa* var.

multiflora Hultén). *S. foliolosa* is apparently closely allied to the more southern *S. ferruginea*. The geographic ranges of these two taxa approach one another in south-central and southwestern Alaska, but in those regions *S. ferruginea* seldom produces bulblets in the inflorescence. The more bulbiferous phase of that species (the var. *macounii*) occurs in southeastern Alaska.

Saxifraga hieracifolia Waldst. & Kit.
Hawkweed-leaf Saxifrage

Plants 0.4–8.5(10) dm tall, arising from a basal rosette with short, stout rhizome; flowering stems sparsely to densely glandular-pilose, leafless; leaves 2–12(18) cm long, 0.4–4 cm broad, 5–19-toothed to nearly or quite entire, the blades elliptic to oblanceolate, cuneate at the base, sparsely pubescent to glabrous above and below, glandular-ciliate; flowers showy, in dense clusters along the spikelike panicle; sepals green to reddish purple, 1.5–3 mm long, ovate to triangular, spreading; petals reddish purple, 2–3 mm long, elliptic to lanceolate, indistinctly 1-veined; stamens shorter than the sepals, the filaments subulate to linear, the pollen orange to whitish; ovary about one-fourth inferior; capsules 5–6 mm long.

Tundra to heathlands and open woods; in most of Alaska (except for the coastal and insular southern portions) and western and northern Yukon; east to Greenland; circumpolar.

Saxifraga hirculus L.
Yellow Marsh Saxifrage

Plants 0.5–2.2 dm tall; flowering stems erect, brownish or yellowish tomentose, at least above, arising from a basal rosette with fibrous roots, with usually several cauline leaves; basal leaves 1.2–3.5 cm long, 0.1–0.3(0.4) cm broad, entire, linear-oblanceolate to spatulate, glabrous or glabrate; cauline leaves similar to the basal only smaller above; flowers 1–few, showy; sepals greenish to reddish, oblong to lanceolate, 2.5–5.5 mm long, spreading, glandular-villous to glabrate, ciliate; petals bright yellow, 7–15 mm long, 5–7-veined; stamens longer than the sepals, the filaments subulate; ovary only slightly inferior; capsules 7–12 mm long.

Tundra to heathlands, commonly in wet meadows; in most of Alaska (except for the southeastern portion) and western to northern Yukon; east to Hudson Bay; circumpolar (S. propinqua R. Br.; S. hirculus var. propinqua [R. Br.] Simm.).

Saxifraga lyallii Engler
Red-stem Saxifrage

Plants 0.7–3 dm tall; arising from a basal rosette and short to elongate rhizome; flowering stems conspicuously villous to glabrate, leafless; leaves 0.8–8 cm long, 0.3–5 cm broad, with 7–17 teeth or lobes, the blades cuneate-oblanceolate to spatulate or fan shaped, glabrous to pilose, somewhat ciliate; flowers showy, few to several in usually open panicles; sepals purplish to green, lanceolate, 1.7–4 mm long, reflexed, glabrous or somewhat ciliate; petals white to yellowish, 2.4–5 mm long, only the midvein distinct; stamens from shorter to longer than the sepals, the filaments clavate, somewhat petaloid; ovary only slightly inferior; capsules 6–11 mm long.

Wet places, along streams and seeps, in tundra, heathlands, or woods; in southwestern, south-central, and southeastern Alaska and southern Yukon, and less commonly in north-central Alaska; east to Mackenzie and south to Washington, Idaho, and Montana (S. lyallii spp. hultenii [Calder & Savile] Calder & Taylor). Our material belongs to var. hultenii Calder & Savile.

Saxifraga mertensiana Bong.
Wood Saxifrage

Plants 1.5–3.5 dm tall with basal leaves from short rhizomes; flowering stems villous with spreading glandular hairs, leafless; leaves 5–19 cm long, 2–8 cm broad, with 13–17 lobes, at least the main lobes with (2)3–5 teeth, the blades cordate at the base, oval to orbicular or reniform, sparsely pilose to glabrate or glabrous on both surfaces, sparsely ciliate; flowers showy, few to many, the lateral ones often replaced by bulblets; sepals green or purplish, 1.5–3 mm long, lanceolate, reflexed; petals white, 2.5–3.5 mm long, indistinctly 1-veined; stamens longer than the sepals, the filaments clavate; ovary only slightly inferior; capsule 4–6 mm long.

Wet sites, from woodlands to alpine; in coastal and insular southern Alaska from Kodiak eastward through the Panhandle; south to California, Idaho, and Montana.

Saxifraga nivalis L.
Alpine Saxifrage

Plants 0.5–2.2 dm tall, arising from a basal rosette with a short stout rhizome; flowering stems glandular-villous, leafless; leaves 1.2–4.5 cm long, 0.4–1.6(2) cm broad, 9–21-toothed, the blades oblanceolate to spatulate, elliptic or ovate, cuneate at the base, sparsely villous (rarely somewhat brownish-tomentose) to glabrous, ciliate; flowers showy, in 1–3 dense clusters, the inflorescence spikelike, or the lower 1–2 branches elongating; sepals green to purplish, 1–3 mm long, ovate to triangular, erect; petals white to pink (rarely reddish purple), ovate to lanceolate, indistinctly 3-veined; stamens shorter to longer than the sepals, the filaments subulate; ovary one-third to one-half inferior; capsules 5–7 mm long.

Tundra to heathlands and woods; in most of Alaska (except for the central valleys, Aleutians, and southern Panhandle) and most of the Yukon; east to Labrador and south to British Columbia and Alberta; circumpolar (S. nivalis var. tenuis Wahl.; S. nivalis var. rufopilosa Hultén).

Saxifraga nudicaulis D. Don
Naked-stem Saxifrage

Plants 0.5–1.6 dm tall, from short to elongate, slender rhizomes; flowering stems minutely stipitate-glandular to glabrate, leafless; leaves 1.6–4.5 cm long, 0.9–2.4 cm broad, with (3)5–7(9) lobes or teeth, the blades cordate (at least some) to abruptly acute basally, reniform to orbicular, glandular-villous to glabrous above and below, somewhat ciliate, especially along the petioles; flowers showy, few to several in a contracted to open, corymbose panicle; sepals green to reddish or purplish, 1.2–2.3 mm long, ovate to triangular, erect, glabrous, not ciliate; petals white to yellowish or pinkish, 4–5 mm long, elliptic, indistinctly 3–5-veined; stamens longer than the sepals, the filaments subulate; ovary one-fourth to one-third inferior; capsules 4–6.5 mm long.

Moist sites, in tundra or heathlands; in the lower Yukon River Delta and Seward Peninsula, and islands of the Bering Sea; Asia. When the plants grow in deep moss, the stems are stoloniferous. The basal internodes are often much elongated and roots develop at the nodes. Thus the stems appear to have cauline leaves. There is a distinct scape above the uppermost cluster of leaves.

Saxifraga occidentalis Wats.
Rusty Saxifrage

Plants 1–2.5 dm tall, arising from a basal rosette with a short, stout rhizome; flowering stems glandular-villous, leafless; leaves 3–8 cm long, 1–3 cm broad, with mostly 15–30 teeth, the blades ovate to elliptic, cuneate at the base, at least the younger reddish glandular-tomentose on the lower surface, ciliate; flowers showy, several to many, in an open corymbose panicle; sepals green to purplish, 1–2.5 mm long, oblong to lance-ovate, spreading; petals white to pink, ovate to oblong, indistinctly 1-veined; stamens longer than the sepals, clavate; ovary one-fourth inferior or less; capsules 3–6 mm long.

Woods; in southeastern Alaska, (where evidently rare) and adjacent British Columbia; southward to Washington and

Oregon (*Micranthes rufidula* Small; *S. rufidula* [Small] Macoun). Our material belongs to var. *rufidula* (Small) C. L. Hitchc.

Saxifraga oppositifolia L.
Purple Mountain Saxifrage

Plants 0.2–0.7 dm tall, mat-forming, the prostrate caudex branches rooting, subrhizomatous; flowering stems obsolete or to 5 cm tall, sparsely villous to glabrate, borne at the ends of densely leafy caudex branches, with one to few pairs of cauline leaves; leaves 0.1–0.4 cm long, connate-clasping, entire, setose-ciliate (rarely glabrous), opposite (the upper cauline ones sometimes alternate), densely imbricate; flowers conspicuous, solitary; sepals green to purplish, 2–3.5 mm long, erect, setose-ciliate; petals bright pink purple to blue purple or rarely white, 6–9 mm long, 5–7-veined; stamens longer than the sepals, the filaments subulate; ovary about one-fourth inferior; capsule 6–9 mm long.

Tundra to heathlands, on gravelly slopes and cliffs; in most of Alaska (except for the central valleys) and most of the Yukon; east to Labrador and south to Oregon, Idaho, and Wyoming; circumpolar (*S. oppostifolia* subvar. *smalliana* Engler & Irmscher; *S. oppositifolia* ssp. *smalliana* [Engler & Irmscher] Hultén).

Saxifraga punctata L.
Brook Saxifrage

Plants 0.3–6 dm tall, with basal leaves from short to elongate rhizomes; flowering stems villous to glandular-villous, especially in the inflorescence, leafless; leaves 2–12.5 cm long, 1–6 cm broad, with (7)9–19 teeth or lobes (these not further toothed), the blades cordate at the base (at least some), orbicular to reniform, pilose, often glandular, to glabrous above and below, usually ciliate; flowers showy, few to several in contracted to open, hemispherical to conic panicles; sepals green to purplish, 1–2 mm long, ovate to lance-ovate, reflexed, often ciliate; petals white to pink or rose, (1.8) 2.5–4(4.5) mm long, 1-veined; stamens subequal to the sepals or longer, the filaments obscurely to distinctly clavate; ovary only slightly inferior; capsules 3–6 mm long.

Tundra to heathlands or woods; in most of Alaska (except for the central valleys) and most of the Yukon; eastward to Northwest Territories and south to British Columbia and Alberta; Asia. *S. punctata* is tremendously variable in Alaska and the Yukon. This variation has been summarized by Hultén (1960), and by Calder and Savile (1960). The correct name for this species might well be *S. nelsoniana* D. Don. Hultén (Bot. Not. 126:494–95. 1973) transfers all entities to that species at subspecific level. Transfers at varietal level have not been made, nor are they intended or implied herein. Explicit typification of *S. punctata* and of *S. nelsoniana* should be completed before wholesale transfers are made. The various segregate taxa are based on minor, and often subjective characteristics. It seems, nevertheless, that the various entities should be recognized in some infraspecific rank.

1a. Principal hairs of the inflorescence short, distinctly glandular; plants of western, northern, south-central, and east-central Alaska, and on islands of the Bering Sea and west-central to northern Yukon; Asia (*S. nelsoniana* D. Don; *S. punctata* ssp. *nelsoniana* [D. Don] Hultén). *S. punctata* var. *nelsoniana* (D. Don) Macoun

1b. Principal hairs of the inflorescence elongate (multicellular), usually not conspicuously gland tipped; plants mostly of southern Alaska and southern Yukon. (2)

2a. Leaves fleshy, thick, often somewhat hairy; petals sometimes pinkish; plants of the Shumagin and Aleutian Islands (*S. punctata* ssp. *insularis* Hultén). *S. punctata* var. *insularis* (Hultén) B. Boi.

2b. Leaves not fleshy, drying thin, usually glabrous except for being marginally ciliate; petals white; plants mostly of other distribution. (3)

3a. Principal leaves 1–2(3.8) cm wide, glabrous to sparsely pubescent above and below; plants of central to eastern Alaska and southern Yukon (*S. punctata* ssp. *porsildiana* Calder & Savile). *S. punctata* var. *porsildiana* (Calder & Savile) B. Boi.

3b. Principal leaves 2.3–6 cm wide, glabrous to somewhat ciliate; widely distributed in southern Alaska, from the Alaska Peninsula eastward through the Panhandle, and in southwestern Yukon (*S. punctata* ssp. *carlottae* Calder & Savile; *S. punctata* ssp. *pacifica* Hultén). *S. punctata* var. *pacifica* (Hultén) Welsh

Saxifraga reflexa Hook.
Yukon Saxifrage

Plants 0.9–6 dm tall, arising from a caudex with a short stout rhizome; flowering stems sparsely to densely glandular-villous, leafless; leaves 2–8.5 cm long, 0.8–2.5 cm broad, with 9–21 crenate-serrate teeth, the blades orbicular to ovate, lanceolate, oblanceolate, or spatulate, cuneate to acute at the base, sparsely villous on both surfaces, often purplish below; flowers showy, mostly 10–numerous in an open cymose panicle; sepals green, rarely purplish, 1.2–3 mm long, ovate to lanceolate, reflexed, minutely ciliate; petals white, 2–3 mm long, distinctly 1-veined; stamens longer than the sepals, the filaments distinctly clavate, often petaloid; ovary only slightly inferior; capsule follicular, the carpels almost distinct, 3–6 mm long, sometimes dark reddish purple.

Tundra, heathlands, and open woods; in much of continental and coastal western and northwestern Alaska and southern to northern Yukon; eastward to the Mackenzie; endemic. This species exhibits a large amount of variation. Small specimens from northwestern Alaska approach S. *davurica*.

Saxifraga rivularis L.
Brook Saxifrage

Plants 0.2–2.5(3) dm tall, arising from fibrous roots; stems sparsely to densely villous to tomentose, with 1–few alternate cauline leaves; basal leaves short- to long-petiolate, often with bulblets in the axils, the blades 3–9-lobed, reniform to orbicular, 0.3–1.2(2.7) cm long, 0.4–2.3(4.3) cm broad; cauline leaves somewhat smaller, fewer lobed, and shorter petioled above, not bearing bulblets; flowers showy to inconspicuous, 1–few at the apex of the inflorescence, not replaced by bulblets; sepals green to reddish purple, ovate to lanceolate or lance-oblong, 1–4.5 mm long, sparsely glandular-pubescent; petals white to pink, 3–5 mm long, 1–3-veined; stamens shorter to longer than the sepals, the filaments subulate; ovary one-third to one-half inferior; capsules 4–9 mm long.

Gravelly soil and cliffs, often in tundra; in much of Alaska (except for the central valleys) and most of the Yukon; east to Labrador and south to Colorado and New Hampshire; circumpolar. S. *rivularis* consists of two more or less distinct types which overlap in range and in morphological characteristics. They have been recognized previously as two species or as entities within an expanded S. *rivularis*.

1a. Flowers mostly distinctly pedicellate; stems below the inflorescence mostly less than 0.5 mm broad, glandular-villous; leaves usually less than 10 mm broad; capsules 4–7 mm long; plants widely distributed (S. *flexuosa* Sternb.; S. *rivularis* ssp. *flexuosa* [Sternb.] Gjaerevoll; S. *rivularis* var. *flexuosa* [Sternb.] Engler & Irmscher). S. *rivularis* var. *rivularis*
1b. Flowers mostly lacking distinct pedicels; stems below the inflorescence often more than 0.5 mm broad, glandular-villous to tomentose; leaves

often more than 10 mm broad; capsules 7–9 mm long; plants mostly of western Alaska and the Aleutians (S. *bracteata* D. Don; S. *laurentiana* Ser.). S. *rivularis* var. *laurentiana* (Ser.) Engler

Saxifraga rosacea Moench

Plants 1.5–3 dm tall, arising from a caudex and a taproot; caudex branches ascending, clothed apically with imbricate leaves; flowering stems glandular-pubescent, usually with at least 1 cauline leaf; basal leaves 1.2–2 cm long, 0.5–1 cm broad, with 3–7 apical, oblong to lanceolate lobes, cuneate basally, glandular-ciliate, somewhat mucronate, but not spinulose-tipped; cauline leaves lobed; flowers showy, several in a corymbose cyme; sepals purplish, 2–4 mm long, erect, glandular-pubescent to glabrous dorsally, glandular-ciliate; petals red to crimson or pink purple, 11–15 mm long, 3–5-veined; stamens longer than the sepals, the filaments subulate; ovary more than one-half inferior; capsule 10–14 mm long.

Cultivated ornamental; in southern and southeastern Alaska; native to Europe. S. *rosacea* is related to the indigenous S. *caespitosa*, and is sometimes regarded as a subspecies of it.

Saxifraga serpyllifolia Pursh
Thyme-leaf Saxifrage

Plants 0.2–8 dm tall; flowering stems erect, stipitate-glandular, not at all tomentose, arising from long subrhizomatous caudex branches; caudex branches densely clothed with imbricate leaves; basal leaves 0.3–0.9 cm long, 0.1–0.2 cm broad, entire, linear-oblanceolate to spatulate, glabrous; cauline leaves alternate, similar to the basal, or lacking; flowers showy, solitary (rarely 2); sepals green to reddish, oblong to lanceolate, 1.5–3 mm long, spreading to reflexed, glabrous, not ciliate; petals yellow (rarely purplish), 4–7 mm long, 5(7)-veined; stamens mostly longer than the sepals, the filaments subulate; ovary only slightly inferior; capsules 5–7 mm long.

Moist sandy or gravelly soils, or on rock outcrops, commonly in arctic or alpine tundra; in much of Alaska (except for the central valleys, coastal south-central, and southeastern portions) and western to northern Yukon; south to British Columbia; Asia (*S. serpyllifolia* var. *purpurea* Hultén; *S. serpyllifolia* f. *purpurea* [Hultén] B. Boi.).

Saxifraga spicata D. Don
Spiked Saxifrage

Plants 1.5–7 dm tall, with basal leaves, from short stout rhizomes; flowering stems glandular-villous, leafless; leaves 7–26 cm long, 3.5–8 cm broad, with 21–45 teeth (the alternate teeth often differing in size), the blades cordate, orbicular, or reniform, cordate at the base, pilose to glabrate on both surfaces, ciliate; flowers showy, several to numerous, borne in slender compact panicles (becoming somewhat open in fruit); sepals green, 1.5–2.5 mm long, lanceolate, reflexed (at least in fruit); petals yellowish, 3–4.5 mm long, obscurely 1(3)-veined; stamens longer than the petals, the filaments linear to obscurely clavate; ovary only slightly inferior.

Moist sites in tundra or heathlands; in west-central to southwestern, south-central and central to eastern Alaska and west-central Yukon; endemic.

Saxifraga tolmiei T. & G.
Tolmie Saxifrage

Plants 0.2–0.8 dm tall; flowering stems erect, stipitate-glandular, not at all tomentose, arising from long stoloniferous caudex branches; caudex with alternate imbricate to distant leaves; basal leaves 0.2–0.6 cm long, 0.1–0.2 cm broad, entire, oblanceolate to spatulate, glabrous, usually with 1–2 cilia near the base; cauline leaves alternate, 1–few, similar to the basal ones; flowers inconspicuous, 1–3 (rarely more); sepals

green to purplish, lanceolate, 1.5–2.5 mm long, spreading, glabrous; petals white, 4–5 mm long, obscurely 1-veined; stamens mostly longer than the calyx lobes, the filaments clavate, subpetaloid; ovary only slightly inferior; capsules 7–12 mm long.

Rock crevices, talus slopes, and open meadows, mostly in alpine tundra; in southeastern Alaska; southward to California.

Saxifraga tricuspidata Rottb.
Three-tooth Saxifrage

Plants 0.3–2.5 dm tall, arising from a substoloniferous caudex; caudex branches prostrate to ascending, clothed with densely imbricate leaves; flowering stems minutely glandular-pubescent to glabrous, with few to several alternate leaves; leaves 0.6–1.9 cm long, 0.1–0.7 cm broad, apically tridentate, leathery, linear to oblong or somewhat cuneate, setose-ciliate, spinulose-tipped, cauline leaves like the basal, except usually entire; flowers showy, few to several in a corymbose cyme; sepals green (often tinged reddish), ovate to lanceolate, 1.5–3 mm long, glandular-ciliate, spreading; petals white or pale yellow, 4–7 mm long, indistinctly 3-veined; stamens mostly longer than the sepals, the filaments subulate; ovary only slightly inferior; capsules 4–8 mm long.

Cliffs, gravelly slopes and flats, in tundra to heathlands or woods; in much of Alaska (except for the west-central to southwestern portions) and almost throughout the Yukon; eastward to Greenland and south to British Columbia, Alberta, Saskatchewan, Michigan, and Labrador.

TELLIMA R. Br.

Perennial, robust, glandular-villous to hirsute herbs from rhizomes; principal leaves basal, the cauline ones usually 2–3, somewhat smaller and with shorter petioles; leaf blades cordate-ovate to orbicular or reniform, cordate basally, palmately lobed, the lobes mostly doubly crenate-serrate; stipules membranous, clasping the stem; flowers regular, perfect, borne in racemes; hypanthium cup shaped; sepals 5; petals 5, pinnately lobed, the lobes subulate; stamens 10, arising from the hypanthium; pistil 2-carpellate, 1-loculed; ovary about one-third inferior; placentae 2, parietal; styles 2; fruit a capsule, the seeds numerous.

Tellima grandiflora (Pursh) Dougl. ex Lindl.
Fringe-cups

Plants 3–10 dm tall, the stems erect, decumbent at the base; basal leaves long-petiolate, the blades 2.5–12 cm long from sinus to apex, 3–14 cm broad, sparsely hirsute on both surfaces; flowers inconspicuous, 10–35; sepals triangular to lanceolate, 2–4 mm long, arising from the apex of the cup-shaped hypanthium, glandular-ciliate; petals greenish to yellowish (rarely reddish), 4–6 mm long, with 2–3 pairs of lateral lobes (these sometimes further lobed); stamens shorter than the sepals; capsules 6–9 mm long.

Moist places in thickets and woods; in coastal and insular southern Alaska from the Alaska Peninsula eastward through the Panhandle; south to California and Idaho.

TIARELLA L.

Perennial slender glandular-pubescent herbs from rhizomes; principal leaves basal, the cauline ones reduced in size and with shorter petioles; leaves simple, palmately lobed or palmately trifoliolate, the blades ovate-cordate to orbicular in outline, cordate basally, once to twice crenate-serrate; stipules membranous to herbaceous, somewhat clasping; flowers irregular, perfect, borne in cymose panicles; hypanthium cup shaped; sepals 5, white to pink; petals 5, entire; stamens 10; pistil 2-carpellate, 1-loculed; ovary superior; placentae 2, parietal; styles 2; fruit a capsule, the valves very unequal in size, the seeds few.

Lakela, O. 1937. A monograph of the genus *Tiarella* in North America. Am. Jour. Bot. 24:344–51.

1a. Leaves trifoliolate. *T. trifoliata*
1b. Leaves simple, the blades merely 3–5-lobed. *T. unifoliata*

Tiarella trifoliata L.
Trifoliate Foamflower

Plants 1.5–5 dm tall, the stems erect or ascending, the base not conspicuously clothed with leaf bases; basal leaves long-petiolate, the blades 1.5–8 cm long, 2.5–12 cm broad, trifoliolate, the leaflets stalked, sparsely hispid on both surfaces; flowers moderately showy, several to many; sepals 1.5–2 mm long, white to pink, stipitate-glandular; petals white, linear to clavate, resembling the filaments, 2.5–4 mm long; stamens longer than the sepals; capsules with the valves 3–5 and 5–10 mm long.

Moist woods; in coastal and insular south-central to southeastern Alaska; south to Oregon, Idaho, and Montana.

Tiarella unifoliata Hook.
Unifoliate Foamflower

Plants 2–5 dm tall, the stems erect or ascending, the base not conspicuously clothed with leaf bases; basal leaves long-petiolate, the blades 2.5–6 cm long, 3.5–9 cm broad, simple, sparsely hispid on both surfaces; flowers moderately showy, several to many; sepals 1.5–2.5 mm long, white to pink, stipitate-glandular; petals white, linear to clavate, resembling the filaments, 2.5–5 mm long; stamens mostly longer than the sepals; capsules with valves 3–5 and 5–10 mm long.

Moist sites in woods and thickets; in coastal and insular southeastern Alaska; southward to California, Idaho, Montana, and Alberta (*T. trifoliata* var. *unifoliata* [Hook.] Kurtz; *T. trifoliata* ssp. *unifoliata* [Hook.] Kern).

TOLMIEA T. & G. Nom. Cons.

Perennial, slender, glandular-hispid herbs from rhizomes; principal leaves basal, the cauline ones reduced in size and with shorter petioles; leaves simple, palmately lobed, the blades cordate-ovate to orbicular, cordate basally, mostly twice crenate-dentate; stipules green, clasping; flowers irregular, perfect, borne in racemes; hypanthium tubular; sepals 5, 3 large and 2 small, deeply cleft between the two smaller ones; petals 4, linear-subulate, entire; stamens 3, opposite the larger calyx lobes; pistils 2-carpellate, 1-loculed; placentae 2, parietal; styles 2; fruit a capsule.

Tolmiea menziesii (Pursh) T. & G.
Youth-on-Age

Plants 4–7 dm tall, the stems erect or ascending, the base not conspicuously clothed with leaf bases; basal leaves long-petiolate, the blades 1.5–10 cm long, 1.2–9 cm broad, sparsely hispid on both surfaces; flowers inconspicuous, several to many; sepals greenish purple, 3–5 mm long, oblong to lanceolate, glandular-pubescent; petals brownish purple, 6–10 mm long; stamens of 2 lengths, the lower ones longer than the upper; capsules 9–14 mm long, the valves equal.

Moist sites in woods; in southeastern Alaska; south to California (*Tiarella menziesii* Pursh).

SCROPHULARIACEAE
Figwort Family

Annual, biennial, or perennial herbs; stems usually terete; leaves opposite, alternate, or rarely whorled, simple, pinnatifid, or pinnately compound; flowers perfect, usually more or less irregular (regular or nearly so in *Veronica* and *Limosella*); calyx mostly with 5(4) more or less united sepals; corolla of 5 united petals, usually 2-lipped; stamens commonly 4, or the fifth present and sterile (rarely fertile), or less commonly only 2; pistils 1; ovary superior, 2-loculed; styles 1; stigmas entire, or 2-lobed; fruit a capsule.

1a. Anther-bearing stamens normally 2. (2)
1b. Anther-bearing stamens normally 4. (4)
2a. Leaves opposite (rarely alternate), all cauline, the bracts sometimes foliose and alternate. *Veronica*
2b. Leaves alternate, chiefly basal, the cauline much reduced in size upwards. (3)
3a. Basal leaves cordate to ovate or orbicular, as broad as long or broader than long, hairy, especially along the margins; sepals 4, distinct or nearly so. *Synthyris*
3b. Basal leaves ovate to lance-oblong or elliptic (rarely reniform to orbicular), mostly longer than broad, glabrous; sepals all connate, appearing spathelike. *Lagotis*
4a. Corolla with a distinct spur at the base. *Linaria*

Tiarella unifoliata Hook. (× 0.25).

4b. Corolla without a spur at the base. (5)

5a. Staminal filaments 5, or the fifth represented by a gland at the base of the corolla, 4 anther-bearing, the fifth sterile and lacking an anther; leaves opposite. (6)

5b. Staminal filaments 4, all anther-bearing; leaves alternate, basal, whorled, or opposite. (7)

6a. Plants annual; corolla 7 mm long or less, the middle lobe of the lower lip deeply concave and enclosing the stamens; sterile stamen reduced, minute and glandlike. *Collinsia*

6b. Plants perennial; corolla over 7 mm long, the middle lobe of the lower lip not enclosing the stamens; sterile stamen elongate, bearded. *Penstemon*

7a. Leaves opposite. (8)

7b. Leaves alternate or basal (whorled in some *Pedicularis* species). (10)

8a. Corolla lobes of upper lip spreading, not developed into a hood or a beak. *Mimulus*

8b. Corolla lobes of upper lip inconspicuous, developed into a hood which arches over the stamens. (9)

9a. Leaves lanceolate, serrate, hispid-pubescent; calyx somewhat inflated in flower, much so in fruit. *Rhinanthus*

9b. Leaves ovate to orbicular, coarsely crenate-dentate to lobed, not rough-hairy; calyx not much inflated. *Euphrasia*

10a. Leaves all basal; corolla small and inconspicuous, essentially regular; plants diminutive aquatics, growing in mud. *Limosella*

10b. Leaves cauline, at least some (except in some *Pedicularis*); corolla conspicuous, irregular; plants not aquatic, although sometimes growing in mud. (11)

11a. Plants biennial, cultivated and escaping; flowers open at the throat, spotted within, subtended by short, entire bracts. *Digitalis*

11b. Plants perennial (sometimes biennial or annual), not or only rarely cultivated; flowers neither open at the throat nor spotted within; bracts mostly toothed, lobed, or dissected, often subequal to the calyx. (12)

12a. Leaves toothed or dissected, or rarely pinnately compound, mostly basal, the cauline ones often poorly developed or lacking; pollen sacs similar. *Pedicularis*

12b. Leaves entire, or pinnately cleft, the cauline ones well developed, the basal ones reduced in size, or lacking; pollen sacs unequal. (13)

13a. Plants annual; lower lip of corolla subequal to the upper; bracts green. *Orthocarpus*

13b. Plants perennial or biennial (rarely annual); lower lip of corolla much shorter than the upper; bracts mostly brightly colored. *Castilleja*

CASTILLEJA Mutis ex L. f.

Caulescent, erect or ascending, annual or perennial herbs, often more or less parasitic on roots of other plants; leaves alternate, sessile, or entire to lobed or dissected; flowers irregular, borne in short to elongate, bracted, terminal spikes or spicate racemes, the bracts brightly colored; calyx tubular, 4-lobed, more deeply cleft above and below than on the sides, the lobes connate in lateral pairs, expanding (although ultimately rupturing) and investing the fruit; corolla elongate, but often almost completely enclosed by the calyx, strongly 2-lipped, the upper lip (galea) beaklike, its lobes united to the apex and enclosing the anthers, the lower lip shorter than the upper, 3-toothed, or the teeth almost lacking; stamens 4, in 2 pairs, the anther sacs unequal; stigmas capitate, entire or 2-lobed; capsule loculicidal, the seeds numerous.

Pennell, F. W. 1934. *Castilleja* in Alaska

and northwestern Canada. Proc. Acad. Nat. Sci. Phila. 86:517–40.

1a. Lower lip of corolla less than one-fifth as long as the galea; plants mostly of coastal, southern, and southeastern Alaska (less commonly in southern Yukon). (2)

1b. Lower lip of corolla at least one-fourth as long as the galea; plants of coastal, western, and northern Alaska and Yukon, common in the interior, and less common in southern and southeastern coastal areas. (3)

2a. Bracts of inflorescence yellow to yellow orange; calyx lobes and bracts obtuse to rounded apically. *C. unalaschcensis*

2b. Bracts of inflorescence red to crimson or red orange (rarely yellowish); calyx lobes and bracts obtuse to acute apically. *C. miniata*

3a. Corolla 13–16(18) mm long in flower, to 21 mm long in fruit; plants 3–5 dm tall, biennial or short-lived perennials, with ordinarily (1)3–5(8) branches in the inflorescence; plants of interior, eastern Alaska. *C. annua*

3b. Corolla 17–25(30) mm long in flower, to 35 mm long in fruit, or if less than 16 mm long, then the plants seldom over 3 dm tall and usually of different distribution. (4)

4a. Bracts of inflorescence and calyx lobes predominantly pinkish, lavender, or violet purple. (5)

4b. Bracts of inflorescence and calyx lobes predominantly whitish, cream, yellowish, or greenish. (7)

5a. Leaves with 1–3 pairs of prominent lobes; corolla mostly 12–17 mm long; plants of southeastern Alaska. *C. parviflora*

5b. Leaves entire, or the upper ones rarely with 1 pair of lateral lobes; corollas mostly more than 17 mm long; plants of interior Alaska and Yukon. (6)

6a. Plants commonly (2)2.5–4.5 dm tall, often branched above; bracts and sepals not uniformly pinkish throughout; plants from Mt. McKinley National Park eastward. *C. raupii*

6b. Plants commonly less than 2.5 dm tall, seldom if ever branched above; bracts and sepals uniformly pinkish or purplish; plants from Mt. McKinley National Park westward. *C. elegans*

7a. Leaves all or nearly all with a pair of lateral lobes; plants 0.7–1.5(2) dm tall; inflorescence commonly one-fourth to one-half the plant height. *C. hyperborea*

7b. Leaves entire, or only the upper ones lobed; plants often over 2 dm tall; inflorescence commonly less than one-fourth (to one-third) the plant height. (8)

8a. Lateral calyx lobes distinct for 3–8 mm from the apex; our most common paintbrush. *C. pallida*

8b. Lateral calyx lobes distinct for 1–3 mm from the apex; plants of eastern Alaska and southern Yukon. (9)

9a. Bracts and calyces yellowish; plants of southern Yukon and southeastern continental Alaska. *C. yukonis*

9b. Bracts and calyces purplish, at least the lower ones; plants of eastern Alaska and southern Yukon. *C. raupii*

Castilleja annua Pennell
Alaska Indian Paintbrush

Plants biennial or short-lived perennials, 2.5–5 dm tall; stems simple or with 3–8 ascending branches near the top, finely appressed-pubescent below, becoming villous with white hairs in the inflorescence; leaves narrowly lanceolate, 1.5–7(8.5) cm long, entire, pubescent to glabrate; bracts lance-ovate, entire or with 1–3 pairs of lateral lobes, purplish, greenish, or yellowish throughout or greenish yellow apically; calyx 12–13 mm long, the lateral lobes distinct for 3–4 mm, purplish or yellowish;

corolla 13–16(18) mm long in flower to 21 mm long in fruit, the galea 5–6 mm long, green, the lower lip 3–4 mm long, the lobes distinct; capsules 6–8 mm long.

Woods, bluffs, terraces, bars, and roadsides; in central eastern Alaska; endemic. The epithet *annua* is a misnomer since the entity is clearly not annual. The extreme forms approach if not pass into *C. raupii* on the one hand and into *C. pallida* on the other.

Castilleja elegans Malte
Elegant Indian Paintbrush

Plants perennial, 0.5–2.5 dm tall; stems simple, moderately to sparingly short-hairy below, villous with whitish or yellowish hairs in the inflorescence; leaves narrowly lanceolate, 2–6 cm long, entire or the upper with a pair of lateral lobes, pubescent; bracts lanceolate to ovate, entire or with 1–3 pairs of lateral lobes, purplish or pink purple throughout; calyx 15–19 mm long, the lateral lobes distinct for 2.5–4 mm, purplish throughout; corolla 20–24 mm long, the galea 4–6 mm long, green or purplish, the lower lip 3–4 mm long, the lobes distinct; capsule 6–10 mm long.

Tundra, moraines, talus slopes, and fell fields; in west-central to northwestern Alaska, eastward along the Brooks Range to northern Yukon, and in central, eastern to southwestern, continental Alaska; east to Hudson Bay; Asia.

Castilleja hyperborea Pennell
Northern Indian Paintbrush

Plants perennial, 0.7–1.6 dm tall; stems simple, sparsely to densely villous, the hairs longer and more numerous in the inflorescence; leaves lance-linear, attenuate, 1.5–4 cm long, with one (rarely 2) pair of lateral lobes, or the lowermost entire; bracts lanceolate to lance-oblong, often lobed, yellowish; calyx 10–20 mm long, the lateral lobes distinct 0.5–1.5 mm from the apex, yellowish; corolla 10–22 mm long,

the galea 5–8 mm long, green, the lower lip 3–5 mm long, the lobes distinct; capsule 9–13 mm long.

Arctic and alpine tundra to open slopes and flats in boreal forest; in central western Alaska, disjunctly along the Brooks Range to northern Yukon and Mackenzie, and in southeastern continental Alaska and southwestern Yukon and adjacent Mackenzie; endemic (*C. villosissima* Pennell).

Castilleja miniata Dougl. ex Hook.

Plants perennial, 2–8 dm tall; stems simple or branched, glabrous or short-hairy to viscid-villous, especially above; leaves lance-linear to lanceolate, entire, but sometimes the upper ones lobed; bracts lanceolate to oblong-ovate, simple or toothed, red or scarlet to crimson or rarely yellowish; calyx 15–30 mm long, the lateral lobes distinct for 3–9 mm, red to scarlet; corolla 20–40 mm long, the galea 9–12 mm long, green, the lower lip 1–1.5 mm long, the teeth connate or nearly so; capsule 9–11 mm long.

Coastal southern Alaska, from the Alaska Peninsula eastward through the Panhandle, and in the interior in the Copper River Drainage; south to California, Arizona, and New Mexico (*C. hyetophila* Pennell and *C. chrymactis* Pennell are both distinguished on characters lacking in constancy in Alaska specimens and present in specimens from the continental range of the species.)

Castilleja pallida (L.) Spreng.
Pale Indian Paintbrush

Plants perennial, 1–5.5 dm tall; stems simple or branched, glabrate to short-hairy to densely villous, especially in the inflorescence; leaves 3–9 cm long, linear to lance-linear or lanceolate, usually caudate-acuminate, simple or uncommonly the upper ones toothed; bracts lanceolate to ovate, entire or toothed, greenish, yellowish, or cream; calyx (12)14–23 mm long, the lateral lobes distinct for (1.5)3–7 mm, colored like the bracts; corolla (14)16–26 mm long, the galea 3.5–8 mm long, green, the lower lip mostly 2.5–3 mm long, the teeth distinct; capsule 7–10 mm long.

Woods, thickets, bars, terraces, meadows, and tundra; in most of Alaska north of the 62nd parallel, and from southwestern to northern Yukon; Asia (*C. caudata* [Pennell] Rebr; *C. pallida* var. *caudata* [Pennell] B. Boi.; *C. pallida* ssp. *auricoma* Pennell; *C. pallida* var. *auricoma* [Pennell] B. Boi.). This entity exhibits wide variation in eastern Alaska where it meets both *C. annua* and *C. raupii*. Intermediates occur which possibly represent hybrids. Further work is indicated. Specimens from northwestern Alaska are more uniform and may warrant taxonomic recognition. Our materials belong to ssp. *caudata* Pennell.

Castilleja parviflora Bong.

Plants perennial, 1–4(5) dm tall, mostly much blackened on drying, glabrate to sparsely villous; leaves 1.5–5 cm long, lanceolate, all except the lowermost ones with 1–2 pairs of lateral lobes; bracts lanceolate to ovate, 3–5-lobed (or rarely entire), rose pink to crimson; calyx 10–15 (20) mm long, the lateral lobes distinct for 3–6 mm, colored like the bracts; corolla 12–17(20) mm long, the galea 3–7 mm long, green, the lower lip 1.5–3 mm long, the teeth distinct; capsules 8–11 mm long.

Alpine meadows; in coastal and insular, south-central to southeastern Alaska and adjacent British Columbia; south to Oregon (*C. henryae* Pennell).

Castilleja raupii Pennell
Raup Indian Paintbrush

Plants perennial, 2–4(6) dm tall; stems simple or branched above, sparsely pubescent below, becoming villous upwards; leaves 2–8(13) cm long, linear to lance-linear, narrowly attenuate; bracts lanceolate to oblong, entire, or more commonly with

a pair of lateral lobes, rose pink to violet purple (rarely greenish); calyx 13–20(25) mm long, the lateral lobes distinct for 0.5–3.5(7) mm, purplish (or rarely greenish); corolla 15–20(26) mm long, the galea 4–6 mm long, green, the anterior lip 2–3 mm long, purplish, the teeth distinct; capsule 8–11 mm long.

Woods, terraces, bars, and slopes; in central eastern Alaska and west-central and northern Yukon; eastward to the Mackenzie and in northern British Columbia. *C. raupii* appears to be an eastern vicariad of *C. elegans*, and possibly should be placed at infraspecific level within that entity.

Castilleja unalaschcensis (Cham. & Schlecht.) Malte
Unalaska Indian Paintbrush

Plants perennial, 2–8 dm tall; stems simple or sometimes branched above, glabrate to sparsely villous below, the hairs more numerous in the inflorescence; leaves 3.5–10 cm long, lance-linear to lanceolate, acuminate; bracts lanceolate to ovate or cuneate, entire or with 2 broad, lateral lobes, greenish to bright yellow; calyx 15–24 mm long, the lateral lobes distinct for 5–10 mm, greenish to bright yellow; corolla 18–28 mm long, the galea 6–11 mm long, green, the lower lip 1–2 mm long, the teeth connate or nearly so; capsule 8–12 mm long.

Tundra to heathlands, open woods, and tidal flats; in coastal and insular southern Alaska from the Aleutians eastward to the Panhandle and in southern Yukon and adjacent British Columbia; endemic (*C. pallida* var. *unalaschcensis* Cham. & Schlecht.; *C. unalaschcensis* ssp. *transnivalis* Pennell).

Castilleja yukonis Pennell
Yukon Indian Paintbrush

Plants perennial, 1.5–3.5(4.2) dm tall, simple or rarely branched above, finely pubescent below, villous with whitish or yellowish hairs above; leaves lance-linear, 2–6 cm long, attenuate, entire or the uppermost

with a pair of lateral lobes; bracts lanceolate, entire or with 1–3 pairs of lateral
lobes; bracts lanceolate, entire or with 1–3
pairs of lateral lobes, yellow or yellowish;
calyx 13–18 mm long, the lateral lobes
distinct for 1–2(4) mm, yellowish; corolla
12–20(28) mm long, the galea 6–8 mm
long, yellowish or greenish, the lower lip
3–5 mm long, the teeth distinct; capsule
8–10 mm long.

Lake shores, bars, terraces, and open slopes;
in southwestern Yukon and adjacent continental Alaska; endemic (*C. muelleri* Pennell).

COLLINSIA Nutt.

Caulescent, erect or ascending, annual
herbs; leaves opposite or whorled, simple,
entire or toothed; flowers irregular, 1–several, in axils of upper leaves (bracts);
calyx of 5 united subequal sepals; corolla
2-lipped, the upper lip 3-lobed, the shorter
central lobe deeply concave and enclosing
the style and stamens; stamens 5, 4 in 2
pairs, the fifth stamen (staminode) represented by a gland near the base of the corolla, the anther cells confluent at the tip;
stigma capitate or somewhat 2-lobed; capsule dehiscing by 4 valves; seeds usually
several in each locule.

Newsom, V. M. 1929. A revision of the
genus *Collinsia* (Scrophulariaceae). Bot.
Gaz. 87:260–301.

Collinsia parviflora Lindl.
Blue-eyed Mary

Plants mostly 0.5–4 dm tall; stems simple
or branched, minutely pubescent to glabrous; leaves mostly 1–5 cm long, entire to
serrulate, oblong to lance-linear, or the
lower ovate to spatulate; flowers long pedicelled, solitary in the lower axils, but commonly 2–5 in the upper ones; calyx 3–6
mm long, the lobes longer than the tube;
corolla white to blue, 4–7 mm long; capsule 3–4 mm long, normally 4-seeded.

Shaded areas along roadways; in southeastern Alaska, southern Yukon, and ad-

Castilleja unalaschcensis (Cham. & Schlecht.)
Malte (× 0.5).

jacent British Columbia; east to Ontario and south to California, Arizona, Colorado, and Michigan; possibly adventive in our region.

DIGITALIS L.

Caulescent, erect, biennial herbs; leaves alternate, simple, entire to dentate; flowers irregular, borne in elongate, bracteate, often one-sided racemes; calyx of 5 more or less united sepals; corolla declined, tubular-campanulate, open and spotted within, somewhat 2-lipped, the 5 lobes much reduced; stamens 4, in 2 pairs, the anther sacs equal, confluent at the apex, included within the corolla, but not enclosed by either lip; stigmas 2; capsule septicidally dehiscent, the seeds numerous.

Digitalis purpurea L.
Common Foxglove

Plants (3)5–15 dm tall; stems puberulent to glabrate below, becoming glandular-viscid above; leaves wrinkled, the basal ones petiolate, 15–50 cm long, crenate-serrate, the cauline ones reduced upwards; bracts entire; calyx foliaceous, 10–18 mm long; corolla 4–8 cm long, pink purple, or variously colored, the lower side often lighter, spotted or mottled within; capsule ovoid, longer than the calyx; seeds minute.

Cultivated ornamental; in more temperate regions of Alaska, sparingly escaping in southeastern Alaska; native of Europe.

EUPHRASIA L.

Caulescent erect annual herbs, reputedly parasites on roots of various plants; leaves opposite, simple, toothed; flowers irregular, borne in prominently leafy-bracted terminal spikes; calyx 4-lobed, more deeply cleft above and below than laterally; corolla 2-lipped, the 2 lobes of the upper lip inconspicuous, developed into a hood which arches over the stamens and stigma, the lower lip 3-lobed; stamens 4, in 2 pairs, the anther sacs equal, but unequally spurred at the base; stigma capitate, or somewhat

2-lobed; fruit a loculicidal capsule, the seeds numerous.

Fernald, M. L., and K. M. Wiegand. 1915. The genus *Euphrasia* in North America. Rhodora 17:181–201.

Euphrasia arctica Lange ex Rostrup
Arctic Eyebright

Plants 0.2–3 dm tall; stems simple, or less commonly branched, puberulent, often glandular in the inflorescence; leaves mostly 1–4 pairs above the cotyledons, 5–12 mm long, prominently toothed; bracts often larger than the leaves, crenate-dentate to dentate, the dentations often spinulose-tipped; flowers small, inconspicuous; calyx 3–4 mm long, the teeth triangular to lanceolate, the lateral sinuses shallower than the upper or the lower; corolla 4–5.5 mm long, whitish; capsule 3.5–5 mm long, pubescent to glabrate.

Moist sites, in bogs, seeps, and stream banks; in much of Alaska south of the 65th parallel and in southern Yukon; eastward to Newfoundland and south to British Columbia and the Great Lakes.

1a. Lateral calyx teeth triangular, the upper and lower sinuses often more than twice as deep as the lateral ones; plants of coastal and insular southern Alaska from the Aleutians eastward to the Panhandle (*E. officinalis* var. *mollis* Ledeb.; *E. mollis* [Ledeb.] Wettst.). *E. arctica* var. *mollis* (Ledeb.) Welsh

1b. Lateral calyx teeth lanceolate, the upper and lower sinuses less than twice as deep as the lateral ones; plants of continental Alaska and Yukon, or less commonly in coastal sites (*E. disjuncta* Fern. & Wieg.; *E. subarctica* Raup; *E. pennellii* var. *incana* Callen). *E. arctica* var. *disjuncta* (Fern. & Wieg.) Cronq.

LAGOTIS Gaertn.

Caulescent, decumbent to erect, glabrous, perennial herbs from short to elongate rhizomes; leaves simple, basal and cauline, alternate or the lower opposite or subopposite; flowers irregular, borne in leafy-bracted terminal spikes; calyx 2-lobed, spathelike, the lower sinus extending to the base or nearly so, the upper sinus shallow or lacking, membranous, the margin glandular-ciliate; corolla 2-lipped, the upper consisting of a single entire or crenulate lobe, the lower with 2 entire or emarginate lobes; stamens 2, the filaments adnate to the corolla tube, the shortly stalked to sessile anthers appearing to arise from the basal edges of the upper corolla lobe; anther sacs equal, confluent at the apex and dehiscing so as to open out flat; stigma capitate; fruit a 2-lobed capsule, with 1 seed in each of the 2 locules.

Lagotis glauca Gaertn.

Plants 0.7–3.8 dm tall; stems glabrous, simple; leaves glabrous, the blades of the basal ones mostly lanceolate to elliptic or oblanceolate, occasionally ovate to reniform, 4–26 cm long, the cauline ones ovate to lanceolate, elliptic, or oblanceolate, merging upward with the bracts, the margin broadly crenate, crenate-serrate, serrate, or rarely somewhat lobed; bracts entire, or the lower ones toothed, usually suffused with purple or blue, the margin often membranous; calyx spathelike, membranous except for the 2 greenish veins, mostly 6–11 mm long; corolla violet purple to blue, 9–14 mm long, curved, enlarging and enclosing the capsule at maturity; capsule ovoid, laterally flattened; seeds 2–3 mm long.

Tundra and heathlands; over much of Alaska (except for the southeastern portion) and western Yukon; east to Mackenzie; Asia (*Gymnandra minor* Willd.; *L. glauca* ssp. *minor* [Willd.] Hultén; *G. stelleri* Cham. & Schlecht.; *L. glauca* var. *stelleri* [Cham. & Schlecht.] Trautv.). *L. glauca* is highly variable throughout our area. The characters appear to be largely haphazard, and appear not to be correlated with other features (e.g. leaf shape does not seem to be associated with any particular type of

margin, nor with any special plant habit). Thus, attempts to segregate our plants into infraspecific categories seem impractical due to the large number of intermediate forms encountered.

LIMOSELLA L.

Acaulescent glabrous perennial scapose herbs with fibrous roots; leaves simple, entire, long-petiolate; flowers borne singly on elongate scapes; calyx 5-lobed; corolla regular or nearly so, with 5 spreading lobes; stamens 4, the anther sacs equal, confluent; stigma capitate; fruit a septicidal, incompletely 2-loculed capsule; seeds numerous.

Limosella aquatica L.
Mudwort

Plants 0.2–0.8 dm tall; stems not elongating; leaves 2–8 cm long, the blades narrowly elliptic to oblong, entire; scapes slender, 0.8–3 cm long; calyx 2–3 mm long; corolla tube nearly equaling the calyx, the lobes shorter than the tube; capsule 2–3.5 mm long.

Mud banks and shallow water; in widely disjunct localities in the southern half of Alaska and southern Yukon; eastward to Newfoundland and south to California; Eurasia.

LINARIA Mill.

Caulescent erect glabrous perennial herbs from elongate rhizomes; leaves alternate or the basal ones opposite, simple, entire; flowers irregular, conspicuously spurred at the base, borne in terminal racemes; calyx of 5 more or less distinct sepals; corolla 2-lipped, the upper lip 2-lobed, the lower 3-lobed, closed at the throat; stamens 4, in 2 pairs, the anther sacs equal, confluent at the apex; stigma capitate, hairy; capsule dehiscent by pores or slits below the apex; seeds numerous.

Linaria vulgaris Hill
Butter-and-Eggs

Plants 1–6 dm tall; stems glabrous and often glaucous; leaves linear to narrowly lanceolate or narrowly oblong, 2–10 cm long; bracts inconspicuous; calyx 2–6 mm

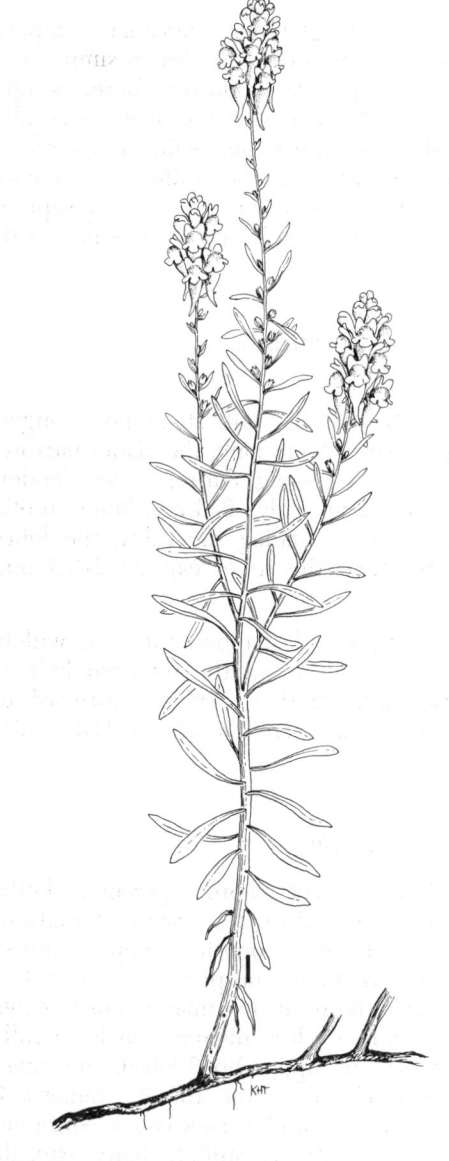

Linaria vulgaris Hill (× 0.4).

long; corolla 20–35 mm long including the spur, bright yellow, bearded and bright orange at the throat; capsule 8–10 mm long.

Cultivated ornamental, occasionally escaping from cultivation and persisting in warmer parts of Alaska; native to Europe.

MIMULUS L.

Caulescent annuals with fibrous roots, or perennials with rhizomes or stolons or both; leaves opposite, simple, usually toothed; flowers irregular, in the axils of the upper leaves (bracts); calyx 5-angled, of 5 united sepals, the teeth equal or unequal; corolla only slightly to strongly 2-lipped, yellow or pink purple, the upper lip 2-lobed, the lower 3-lobed; stamens 4, in 2 pairs, the anther sacs equal, confluent at the apex; stigmas flattened and flaring apically; capsule loculicidal, the seeds numerous.

Grant, A. 1924. A monograph of the genus *Mimulus*. Ann. Mo. Bot. Gard. 11: 99–388.

1a. Corolla yellow, often marked with red or purple; calyx teeth unequal, the upper one larger than the lateral ones. *M. guttatus*
1b. Corolla pink purple, marked with yellow; calyx teeth subequal; plants of extreme southeastern Alaska. *M. lewisii*

Mimulus guttatus DC.
Yellow Monkey-flower

Plants 1–7 dm tall or more, either perennials with rhizomes and often stolons as well or rarely annuals with fibrous roots; stems simple or branched, glabrous or pubescent, somewhat orbicular to obovate, palmately veined, the lower ones petiolate, the upper ones sessile or subsessile; flowers long-pedicelled, large and showy; calyx mostly 10–20 mm long, expanding at maturity, the upper tooth largest; corolla 20–40 mm long, strongly 2-lipped, often purple-mottled or -spotted in the throat, not spurred; capsule 10–20 mm long, many-seeded.

Moist places, along seeps, springs, and streams; in much of Alaska south of the 65th parallel, but most common in coastal and insular southern Alaska and in southern Yukon; southward to California, Utah, and Colorado; introduced in Europe (*M. langsdorfii* Donn ex Sims; *M. tilingii* authors, not Reg.).

Mimulus lewisii Pursh
Lewis Monkey-flower

Plants 3–10 dm tall, from stout rhizomes; stems simple, viscid-villous; leaves 2–10 cm long, dentate to entire, lanceolate to obovate or elliptic, palmately veined, sessile or subsessile; flowers long-pedicelled, large and showy; calyx 12–25 mm long, the teeth subequal; corolla 30–50 mm long, evidently 2-lipped, pink purple, marked with yellow, not spurred; capsule 10–20 mm long, many seeded.

Moist sites, in woods; in extreme southeastern Alaska; southward to California, Utah, and Wyoming.

ORTHOCARPUS Nutt.

Caulescent erect annual herbs; leaves alternate, sessile, entire to dissected; flowers irregular, borne in short to elongate bracted terminal spikes or spicate racemes, the bracts herbaceous or brightly colored; calyx tubular, 4-lobed, more deeply cleft above and below, the lobes connate in lateral pairs; corolla elongate, strongly 2-lipped, the upper lip (galea) beaklike, its lobes united to the apex and enclosing the anthers, the lower lip saclike, subequal to the galea, with 3 slender teeth; stamens 4, in 2 pairs, the anther sacs unequally attached, one below the other along the filament; stigma capitate; capsule loculicidal, the seeds numerous.

Keck, D. D. 1927. A revision of the genus *Orthocarpus*. Proc. Cal. Acad. Sci. IV. 16:517–71.

Orthocarpus hispidus Benth.
Lesser Paintbrush

Plants 1–4 dm tall; stems simple or less commonly branched, with spreading hairs throughout; leaves linear, entire, or broader and pinnately 3–5-lobed, pubescent with long, spreading hairs; bracts with 3–5 lobes, green or suffused with purple, but not showy; calyx 7–10 mm long, the lateral lobes distinct for 2–3 mm, herbaceous; corolla whitish to yellowish, 12–20 mm long, the galea 3–5 mm long, the saclike lower lip 3-lobed, the teeth small and inconspicuous; capsule 6–8 mm long.

Disturbed sites at Skagway in southeastern Alaska; possibly introduced; southward from British Columbia to California, Nevada, and Idaho.

PEDICULARIS L.

Caulescent, subacaulescent, or acaulescent, perennial, biennial, or annual herbs, reputedly more or less parasitic on the roots of other plants; leaves alternate, basal, or whorled, pinnatifid, pinnately dissected, or pinnately compound; flowers irregular, borne in terminal bracted spikes or spike-like racemes, the bracts not usually brightly colored; calyx tubular, 5(4 or 2)-lobed, cleft on the lower side and sometimes on the upper; corolla strongly 2-lipped, the lower lip 3-lobed, the upper lip (galea) modified into a hood or beak which encloses the stamens; stamens 4, in 2 pairs, the anther sacs parallel and equal; stigmas capitate; capsule loculicidal, the seeds several.

1a. Flowers cream to yellow, sometimes suffused with pink, red orange, or red purple. (2)
1b. Flowers rose pink to pink purple or purplish red, often variously marked with other colors. (5)
2a. Plants annual or biennial (short-lived perennial), the stems branching; bracts within the inflorescence subequal to the flowers or much longer. *P. labradorica*

2b. Plants perennial, the stems simple or rarely branching from the base; bracts within the inflorescence mostly much shorter than the flowers. (3)

3a. Flowers cream-colored (often suffused with pale pink or red purple), 25–40 mm long; racemes usually capitate, few-flowered. *P. capitata*

3b. Flowers pale to bright yellow, 12–20 (23) mm long; racemes usually elongate, mostly several- to many-flowered. (4)

4a. Leaves dissected to the midrib, or nearly so; galea not prolonged into a beak; calyx 5-lobed. *P. oederi*

4b. Leaves merely lobed, the blade broadly continuous along both sides of the midrib; galea prolonged into a conical beak; calyx 2-lobed. *P. lapponica*

5a. Stems with one or more sets of whorled leaves. (6)

5b. Stems with leaves alternate (some rarely subopposite), or the leaves all basal (and still alternate). (7)

6a. Flowers 18–25 mm long, the galea prolonged into a beak; plants glabrous, known from the Aleutian Islands and southwestern Alaska. *P. chamissonis*

6b. Flowers 11–16(18) mm long, the galea not prolonged into a beak; plants pubescent to glabrate, widely distributed. *P. verticillata*

7a. Galea prolonged apically into a prominent beak, the beak 2 mm long or more; plants of southeastern Alaska or southern Yukon. (8)

7b. Galea not prolonged into a beak, blunt apically or with a pair of subterminal teeth; plants widely distributed. (9)

8a. Calyx 7.5–10 mm long; galea 6–8 mm long, the beak conical, 2–4 mm long; plants of southeastern Alaska. *P. ornithorhyncha*

8b. Calyx 5–7 mm long; galea (6)8–18 mm long, prolonged into a recurved beak 6–16 mm long; plants of southern Yukon. *P. groenlandica*

9a. Plants annual or biennial (short-lived perennial?); stems often branched above the base, or if simple then the inflorescence often much elongated. (10)

9b. Plants perennial; stems simple or branched from the caudex; inflorescence seldom much elongated. (11)

10a. Stems 0.5–2 dm tall; inflorescence compact, subcapitate, or occasionally with a few flowers in the axils of the lowermost bracts; galea with a pair of subapical teeth; plants of interior northern and coastal western to southwestern Alaska. *P. pennellii*

10b. Stems (1.5)2–7 dm tall; inflorescence with several apparent internodes below the few-flowered, subcapitate apex; galea with or without subapical teeth; plants of the southeastern quarter of Alaska, including the Panhandle, and southern Yukon. *P. parviflora*

11a. Galea lacking subterminal teeth; inflorescence often densely tomentose. *P. lanata*

11b. Galea with a pair of subterminal teeth; inflorescence tomentose to glabrate. (12)

12a. Stems subscapose, with usually (0) 1–4 leaves (more in var. *gymnocephala*); flowers arranged in a spiral; staminal filaments glabrous. *P. sudetica*

12b. Stems subscapose to leafy, with usually 1–8 leaves; flowers not conspicuously spirally arranged; staminal filaments villous with multicellular hairs or glabrate. *P. langsdorfii*

Pedicularis capitata Adams
Capitate Lousewort

Plants perennial, 0.5–1.6 dm tall; stems simple, sparsely villous, leafless or with

1 or 2 leaves; leaves alternate, long-petiolate, pinnately once-compound, the leaflets lobed or dissected; bracts similar to the leaves or less dissected; inflorescence capitate, 1–8-flowered; calyx 5-toothed, 9–16 mm long, the teeth 4–10 mm long, somewhat crenate apically; corolla 25–40 mm long, cream, often suffused with pink or red purple, the galea strongly arched, 15–25 mm long, bearing a pair of subapical teeth (or the teeth obsolete) but not beaked; staminal filaments glabrous; capsule 12–15 mm long.

Moist or dry tundra or heathlands; in most of Alaska and Yukon; eastward to Greenland and south to British Columbia; Asia (*P. nelsonii* R. Br.).

Pedicularis chamissonis Steven
Chamisso Lousewort

Plants perennial, 1–6 dm tall; stems simple or branched from the caudex, glabrous or sparsely villous in the inflorescence, with usually 1–3 whorls of leaves below the bracts; leaves short-petiolate to subsessile, deeply pinnatifid, the lobes toothed to incised; bracts similar to the leaves but much reduced; inflorescence capitate to elongate, many-flowered; calyx 5-toothed, 5–9 mm long, the teeth 1–2 mm long, entire; corolla 18–25 mm long, rose pink, the galea strongly arched, 6–10 mm long, prolonged apically into a conical beak at least 2 mm long, lacking subapical teeth; staminal filaments sparsely villous; capsules 8–12 mm long.

Moist marshy areas and stream sides; in the Alaska Peninsula and Aleutian Islands; Asia.

Pedicularis groenlandica Retz.
Elephant's Head

Plants perennial, 1.5–6 dm tall, glabrous or nearly so; leaves alternate, both cauline and basal, the basal ones long-petiolate, becoming subsessile upwards, pinnately

parted, the lobes toothed or again parted and the ultimate lobes toothed; bracts similar to the leaves, but much reduced; inflorescence many-flowered, elongate, the lower internodes commonly apparent; calyx 5–7 mm long, 5-toothed, the teeth 0.8–1.5 mm long, ciliate-fringed; corolla (10)12–25 mm long, purplish or pinkish, the galea strongly attenuate, curved (resembling an elephant trunk), (6)8–18 mm long; staminal filaments glabrous; capsules 8–14 mm long.

Swamps and wet meadows to open woods; in southern Yukon; eastward to Labrador and Greenland and south to California and New Mexico.

Pedicularis labradorica Wirsing
Labrador Lousewort

Plants annual or biennial (short-lived perennial?); stems simple or more commonly branched along the stem as well as from the base, sparsely to densely villous, the hair commonly in lines below the leaf bases; leaves alternate, with short- to long-winged petioles, lobed to pinnatifid, the lobes toothed to incised; bracts similar to the leaves but reduced and less deeply lobed upwards; inflorescence capitate to elongate, often with several apparent internodes, several- to many-flowered; calyx 4–6 mm long, 2–3 (rarely obscurely 5)-lobed, the lateral teeth connate and appearing as a single tooth (0.5–1 mm long) near the upper edge of the tube, the upper tooth present or absent; corolla 13–17 mm long, yellow, often marked with orange or red orange, the galea slightly arched, 4–7 mm long, somewhat beaked and subapically toothed; staminal filaments villous or glabrous; capsule 10–13 mm long.

Dry to moist tundra to heath and woodlands; in most of Alaska (except for the Aleutians and most of the Panhandle) and most of the Yukon; eastward to Labrador and south to British Columbia; Asia (*P. labradorica* var. *sulphurea* Hultén).

Pedicularis lanata Cham. & Schlecht
Kane Lousewort

Plants perennial, 0.5–2.5(4) dm tall; stems simple, villous to densely tomentose, especially in the inflorescence; leaves alternate or basal, long- to short-petiolate, pinnatifid to pinnately compound, the lobes toothed to pinnatifid; bracts much reduced upwards; inflorescence many-flowered, elongate, but the internodes seldom apparent; calyx 4–5 mm long, 5-toothed, the teeth 0.5–1 mm long; corolla 15–20(25) mm long, rose pink to lavender, the galea slightly arched, 3–5(8) mm long, neither toothed nor beaked (or vestigial teeth rarely apparent); staminal filaments villous; capsule 8–13 mm long.

Wet to dry tundra or heathlands; throughout Alaska and Yukon; eastward to Greenland and south to British Columbia; Asia (*P. kanei* Durand).

Pedicularis langsdorfii Fisch. ex Steven
Langsdorf Lousewort

Plants perennial, 0.5–2.3 dm tall; stems simple, tomentose to glabrate; leaves alternate or basal, with long- to short-winged petioles, lobed to pinnatifid, the lobes toothed to entire; bracts similar to the leaves, but less complex and reduced upwards; inflorescence several- to many-flowered, capitate to elongate, a few internodes sometimes visible; calyx 7–10 mm long, 5-toothed, the teeth 2–3 mm long, entire or serrate; corolla 20–25 mm long, rose pink to pink purple, the galea 8–12 mm long, moderately arched, with a pair of subapical teeth, not beaked; staminal filaments villous to glabrate; capsule 12–15 mm long.

Dry to moist tundra and heathlands; in most of Alaska (except for the central valleys and the Panhandle) and southern to northern Yukon; eastward to the Canadian Arctic Archipelago and south to British Columbia; Asia (*P. arctica* R. Br.; *P. langsdorfii* ssp. *arctica* [R. Br.] Pennell; *P. hians* Eastw.).

Pedicularis lapponica L.
Lapland Lousewort

Plants perennial, 0.8–2.5 dm tall; stems simple, puberulent, arising from elongate rhizomes; leaves alternate, subopposite or basal, petiolate below, becoming sessile above, pinnately lobed to pinnatifid, the

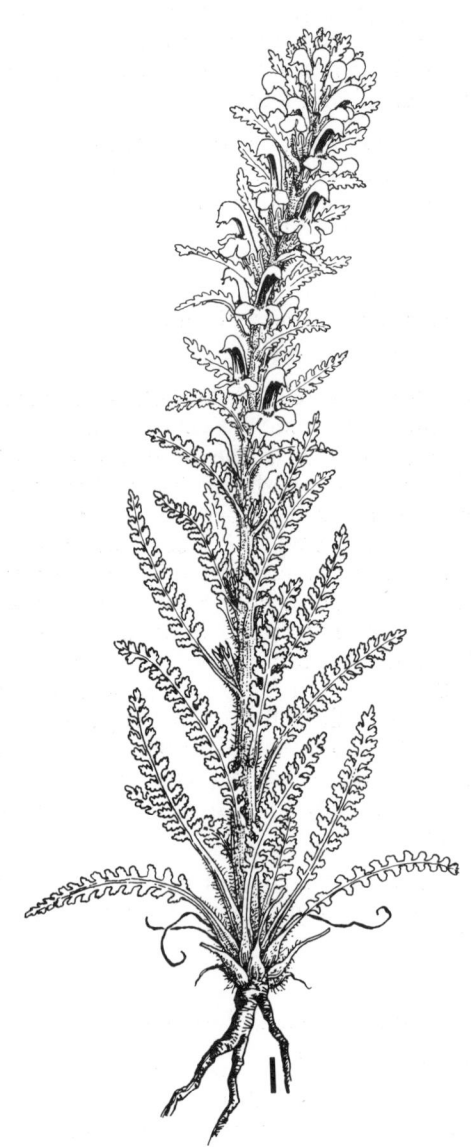

Pedicularis langsdorfii Fisch. (× 0.5).

lobes toothed; bracts similar to the leaves but smaller; inflorescence few- to several-flowered, the lower internodes apparent; calyx 3–4.5 mm long, 2-lobed (by connation of the paired lateral lobes), the lobes 0.2–0.5 mm long; corolla 11–15 mm long, pale yellow, the galea only slightly arched, 7–9 mm long, both subapically toothed and prolonged into a conical beak, the beak 1.5–2.5 mm long; staminal filaments glabrous; capsule 6–9 mm long.

Arctic and alpine tundra; in northwestern to northeastern Alaska and central to northern Yukon; east to Labrador and Greenland; circumpolar.

Pedicularis oederi Vahl ex Hornem.
Oeder Lousewort

Plants perennial, 0.4–2.5 dm tall; stems simple, villous-tomentose to glabrate; leaves alternate or basal (sometimes all basal), petiolate below and above, pinnatifid, the lobes toothed or incised; bracts similar to the leaves but becoming smaller and only lobed or toothed above; inflorescence several- to many-flowered, capitate, or more commonly elongate, the lower internodes often apparent; calyx 7–11 mm long, 5-toothed, the teeth 2–3 mm long, sometimes serrate or crenate apically; corolla 17–25 mm long, yellow, sometimes marked with orange purple, the galea slightly to moderately arched, 5–9 mm long, neither subapically toothed nor beaked (or only vestigial teeth present); staminal filaments villous; capsule 11–16 mm long.

Moist to dry tundra and heathlands; in much of Alaska and western to northern Yukon; southward to Alberta, Montana, and Wyoming; Eurasia (*P. flammea* authors, not L.).

Pedicularis ornithorhyncha Benth. ex Hook.

Plants perennial, 0.5–3 dm tall; stems simple, glabrate to glabrous or somewhat villous in the inflorescence; leaves alternate or mostly basal, long-petiolate, pinnatifid to pinnately compound, the lobes deeply cleft and often again toothed; bracts similar to the leaves but often much reduced; inflorescence few- to several-flowered, capitate or less often elongate, the lower internodes often apparent; calyx 7.5–10 mm long, 5-toothed, the teeth 1.5–3 mm long, often serrate or dentate apically; corolla 10–17 mm long, purplish, the galea strongly arched, 6–8 mm long, prolonged into a conical beak 2–4 mm long; staminal filaments villous; capsules 10–15 mm long.

Alpine meadows; in coastal and insular southeastern Alaska; southward to Washington.

Pedicularis parviflora Smith ex Rees

Plants annual or biennial (short-lived perennial?), (1.5)2–7 dm tall; stems simple or more commonly branched, puberulent to glabrate or glabrous; leaves alternate, mostly cauline, sessile or subsessile, pinnately lobed to pinnatifid, the lobes crenate to serrate or incised; bracts similar to the leaves but much reduced above; inflorescence several- to many-flowered, elongate, with several apparent internodes; calyx 4–6 mm long in anthesis, expanding and partially investing the capsule in fruit, cleft above and below, the 2 lateral lobes irregularly toothed; corolla 11–14 mm long, purple or bicolored, the galea straight to slightly arched, 4–5 mm long, neither beaked nor with subapical teeth (or the teeth more or less well developed), with a pair of processes near the middle of the lower edge; staminal filaments villous; capsules 8–17 mm long.

Wet meadows, muskegs, marshes, and bogs; in south-central to west-central and southeastern Alaska and southwestern Yukon; eastward to Hudson Bay and south to British Columbia and Saskatchewan. Our materials are divisible into two closely allied infraspecific taxa.

1a. Bracts subtending flowers with 3–4 well-developed lateral segments;

plants of coastal or near coastal habitats. *P. parviflora* var. *parviflora*

1b. Bracts subtending flowers with 1–2 poorly developed segments (rarely more but none really elongate); plants from interior Alaska and Yukon, or less commonly in coastal regions. *P. parviflora* var. *macrodonta* (Richards.) Welsh comb. nov. (based on: *Pedicularis macrodonta* [as *macrodontis*] Richards. Bot. Append. Frankl. Jour. 742. 1823)

Pedicularis pennellii Hultén
Pennell Lousewort

Plants annual or biennial (short-lived perennial?), 0.5–2 dm tall; stems simple, or more usually branched, glabrous or sometimes sparsely and minutely puberulent; leaves alternate, mostly cauline, sessile or subsessile, pinnately lobed to pinnatifid, the lobes toothed to incised; bracts similar to the leaves, reduced upwards; inflorescence several-flowered, subcapitate or with a few lower internodes apparent; calyx 5–6 mm long, enlarging in fruit, cleft above and below, the 2 lateral lobes irregularly toothed; corolla 12–17 mm long, purplish or bicolored, the galea slightly arched, 4–6.5 mm long, not beaked, bearing a pair of minute, subapical teeth as well as a pair of processes near the middle of the lower edge; staminal filaments villous; capsule 8–12 mm long.

Moist tundra and heathlands; in north-central to northwestern, west-central, and southwestern Alaska and islands of the Bering Sea; Asia (*P. parviflora* ssp. *pennellii* [Hultén] Hultén). The affinities of *P. pennellii* lie with both *P. parviflora* and *P. palustris* L. This complex is in need of monographic revision.

Pedicularis sudetica Willd.

Plants perennial, 0.4–5 dm tall; stems simple, villous-tomentose to glabrate or glabrous; leaves cauline and alternate or all basal, long-petiolate, pinnately lobed or pinnatifid, the lobes toothed or incised and the ultimate divisions again toothed; bracts variable, usually resembling the leaves, always reduced upwards; inflorescence several- to many-flowered, capitate to elongate, the lower internodes usually not apparent at anthesis, the flowers spirally arranged; calyx 8–12 mm long, 5-toothed, the teeth 3–5 mm long, these entire to variously crenate to dentate apically; corolla 15–24 mm long, rose pink, pink purple, or bicolored (the galea dark purple and the lower lip pink, often spotted), the galea 6–13 mm long, moderately arched, not (or only slightly) beaked, subapically toothed; staminal filaments glabrous; capsule 9–14 mm long.

Moist to dry tundra, heathlands, and woods; through most of Alaska and Yukon; east to Baffin Island and south to Colorado; Eurasia. In Alaska and Yukon there occur a series of populations of *P. sudetica* which vary in several morphological features. Most of the variations can be summarized as follows; presence or absence of cauline leaves, nature of the cauline leaves when present, size and shape of bracts, plant size, pubescence or the lack of it, size of the inflorescence at anthesis, flower size, thickness of the galea, and the shape of the lower corolla lobes. The range of habitats is likewise broader than for most other louseworts in Alaska. The species simulates *P. langsdorfii* on the one hand and *P. lanata* on the other. It is unified by the conspicuous spiral arrangement of the flowers in the inflorescence and by the glabrous staminal filaments. The infraspecific status of *P. sudetica* has been reviewed by Hultén (1961. Two *Pedicularis* species from northwestern America, *P. albertae* n. sp. and *P. sudetica* sens. lat. Svensk Bot. Tidssk. 55:193–204). I have not been able to segregate specimens into all of the infraspecific taxa recognized by Hultén. Thus it seems best to give a more conservative treatment.

1a. Flowers bicolored, the galea dark purplish and the lower lip variously pink to nearly white and often purple- or red-spotted; plants of northern and western Alaska and northern Yukon (*P. sudetica* ssp. *albolabiata* Hultén). *P. sudetica* var. *bicolor* Walpers

1b. Flowers uniformly colored (pink or pink purple), or not dark purplish, the lower lip sometimes pink. (2)

2a. Plants tall, often with several well-developed cauline leaves; inflorescence often elongate at anthesis; plants widely distributed in interior Alaska, east to Hudson Bay, and south to British Columbia (*P. sudetica* ssp. *interioroides* Hultén; *P. sudetica* var. *interior* Hultén; *P. sudetica* ssp. *interior* [Hultén] Hultén). *P. sudetica* var. *gymnocephala* Trautv.

2b. Plants short, the cauline leaves reduced or lacking; inflorescence mostly capitate at anthesis; plants of western, southwestern, and southeastern Alaska (*P. sudetica* ssp. *pacifica* Hultén). *P. sudetica* var. *pacifica* (Hultén) Welsh

Pedicularis verticillata L.
Whorled Lousewort

Plants perennial (short-lived?), 0.8–4 dm tall; stems simple or branched from the caudex, sparsely villous to glabrate, villous to tomentose in the inflorescence, with usually 1–2 whorls of leaves below the bracts; leaves short- to long-petiolate, deeply pinnatifid, the lobes toothed to incised; bracts similar to the leaves, but much reduced above; inflorescence capitate to elongate (especially so at maturity), many-flowered; calyx 3–7 mm long, 5-toothed, the teeth 0.5–1 mm long, entire; corolla 11–16(18) mm long, rose pink, the galea slightly arched, 5–8 mm long, neither subapically toothed nor beaked; staminal filaments sparsely villous; capsules 10–15 mm long.

Moist to dry tundra, heathlands, and woods; through most of Alaska and western Yukon; Eurasia.

PENSTEMON Mitchell

Caulescent perennial herbs or subshrubs from a caudex; leaves opposite, simple, entire or toothed; flowers irregular, borne in verticels or in paniculate clusters; calyx 5-lobed; corolla showy, 2-lipped, the upper lip 2-lobed, the lower one 3-lobed; stamens 5, the four with anthers in 2 pairs, the fifth (staminode) lacking an anther but the filament well developed and bearded apically, the anther sacs equal, confluent at the apex; stigma capitate; capsule septicidal; seeds numerous.

1a. Leaves entire; flowers 6–10 mm long; plants glabrous. *P. procerus*

1b. Leaves toothed or entire; flowers 12–25 mm long; plants pubescent, at least in the inflorescence. (2)

2a. Leaves of stem strongly serrate-dentate, ovate to lanceolate; plants of extreme southeastern Alaska. *P. serrulatus*

2b. Leaves of stem irregularly serrate to entire, narrowly lanceolate to oblong; plants of eastern Alaska and southern Yukon. *P. gormanii*

Penstemon gormanii Greene
Gorman Beardtongue

Plants herbaceous, 0.6–5 dm tall; stems tufted, ascending to erect, glabrous to glabrate below, pubescent in the inflorescence; leaves entire or irregularly serrate, the lower ones petiolate and cuneate-oblanceolate to oblong, the upper ones sessile or subsessile, narrowly lanceolate to oblong; bracts similar to the upper leaves; calyx 7–11 mm long, the lobes almost distinct; corolla 19–30 mm long, white, pink, or purplish; anthers glabrous, horseshoe shaped, opening throughout; staminode bearded; capsules 6–10 mm long.

Dry, steep, sandy or gravelly slopes along river canyons, bluffs, and road cuts; in central eastern Alaska and southern Yukon; east to the Mackenzie and south to northern British Columbia; endemic. The closely related *P. eriantherus* Pursh occurs in the high plains of the United States and adjacent Canada.

Penstemon procerus Dougl. ex R. Grah.

Plants herbaceous, 0.5–4 dm tall; stems tufted, ascending to erect, glabrous throughout; leaves entire, the basal ones petiolate and oblanceolate to elliptic, the upper ones sessile or nearly so and narrowly lanceolate; bracts similar to the upper leaves; calyx 4–6 mm long; corolla blue purple (sometimes pink), 6–10 mm long; anthers glabrous, opening throughout and laying out flat at maturity; staminode bearded; capsules 4–5 mm long.

Dry sandy banks, meadows, and open woods; in southern Yukon; eastward to Saskatchewan and south to California, Utah, and Colorado. Our plants belong to var. *procerus.*

Penstemon serrulatus Menzies ex Rees

Plants woody below, 2–7 dm tall; stems puberulent to glabrate below, conspicuously pubescent in the inflorescence; leaves serrate-dentate, all cauline, sessile or nearly so, ovate to lanceolate; bracts similar to the upper leaves; calyx 5–9 mm long; corolla blue to purple, 13–25 mm long; anthers glabrous, opening only at the confluent apex; staminode bearded, capsule 5–8 mm long.

Moist sites; in extreme southeastern Alaska (Hyder); southward to Oregon (*P. diffusus* Dougl. ex Lindl.).

RHINANTHUS L.

Caulescent erect annual herbs; leaves opposite, simple, toothed; flowers irregular, borne in leafy-bracted terminal spikes; calyx 4(5)-lobed, somewhat inflated at anthesis, conspicuously so in fruit; corolla 2-lipped, the 2 lobes of the upper lip inconspicuous or obsolete, developed into a hood which arches over the stamens, the lower lip 3-lobed; stamens 4, in 2 pairs, the anther sacs equal, not spurred; stigma capitate; fruit a loculicidal capsule, the seeds numerous.

Rhinanthus crista-galli L.
Rattlebox

Plants 0.8–7.5 dm tall; stems simple or branched, puberulent; leaves mostly 2–6(9) pairs above the cotyledons, 1.5–6.5 cm long, prominently toothed, rough-hairy; bracts similar to the leaves, the teeth often spinu-

Penstemon gormanii Greene (× 0.4).

lose-tipped; flowers showy; calyx 7–10 mm long in anthesis, to 17 mm long and much inflated in fruit; corolla 9–15 mm long, yellow, the galea with a pair of broad, subapical teeth; capsules laterally flattened, orbicular, 5–12 mm long, completely enclosed by the calyx.

Meadows, open woods, and roadsides; in much of Alaska and Yukon south of the 62nd parallel; eastward to Greenland and New York and south to Oregon and Colorado; circumboreal. During the past few decades, Alaskan plants of *Rhinanthus* have been recognized under a series of epithets: *R. arcticus* (Stern.) Pennell; *R. borealis* (Stern.) Boiss.; *R. kyrollae* Chab.; *R. groenlandicus* Chab.; *R. minor* L.; *R. minor* ssp. *groenlandicus* (Chab.) Neum.; and *R. minor* ssp. *borealis* (Stern.) A. Löve. The problems of recognition of numerous taxa within this genus are similar to those of the related genus, *Euphrasia*. The plants are circumboreal in distribution and numerous minor variations occur among the disjunct populations. These minor and often ill-defined variations form the bases of the numerous epithets. It seems best to recognize our plants as belonging to a single polymorphic species.

SYNTHYRIS Benth.

Caulescent, decumbent to erect, villous perennial herbs from short fibrous-rooted rhizomes; leaves basal and cauline (or the cauline lacking), the basal ones long-petiolate, the cauline ones sessile or subsessile and alternate or subopposite; flowers irregular, borne in bracted, terminal, spike-like racemes; calyx 4-lobed, the lobes distinct or nearly so; corolla unequally 4-lobed, the upper lobe (lip) wider and often longer than the lower 3; stamens 2, on long filaments, the anther sacs equal, confluent at the apex, horseshoe shaped; stigma capitate; capsule loculicidal, the seeds few to several.

Pennell, F. W. 1933. A revision of *Synthyris* and *Besseya*. Proc. Acad. Nat. Sci. Phila. 85:77–106.

Synthyris borealis Pennell
Kittentails

Plants 0.5–1.4 dm tall, villous throughout with multicellular hairs; stems simple, decumbent to erect; leaves simple, the blades of the basal leaves ovate to oval or orbicular, flattened, 4–6 mm long.

Alpine tundra; in the Alaska Range from Mt. McKinley National Park east to east-central Alaska and west-central to southwestern Yukon and Richardson Mountains, Northwest Territories; endemic. The materials from Northwest Territories (Calder 34014, 8 July 1962, US) are sparingly villous to glabrous on lower stem portion, leaf blades, and petioles.

VERONICA L.

Caulescent, erect, decumbent, or procumbent annuals with taproots, or perennials with stolons or rhizomes; leaves simple, entire, or toothed, opposite or the upper ones alternate; flowers essentially regular, borne in racemes or solitary in the upper axils; calyx 4(5)-lobed, the lobes distinct or nearly so; corolla irregularly 4-lobed, the upper lobe largest; stamens 2, the anther sacs equal; stigma capitate; fruit a septicidal capsule, the seeds few to several.

Pennell, F. W. 1921. *"Veronica"* in North and South America. Rhodora 23:1–22, 29–41.

1a. Flowers in axillary racemes, the main stem not terminating in an inflorescence; leaves opposite throughout. (2)
1b. Flowers in terminal racemes, the main stem terminating in an inflorescence, or the flowers axillary and solitary, or less commonly with axillary racemes also. (6)

2a. Plants distinctly pubescent; leaves ovate to lance-ovate, 1–3 times as long as broad. (3)
2b. Plants essentially glabrous; leaves lanceolate to ovate, or lance-linear to linear, mostly more than 3 times as long as wide. (4)

3a. Racemes 3–7-flowered; corolla 5–6 mm long; stems 4–7 cm long; plants indigenous to the western Aleutians. V. grandiflora

3b. Racemes 10–20-flowered; corolla 8–9 mm long; stems 10–30 cm long; plants adventive weeds. V. chamaedrys

4a. Leaves all with short petioles, the blades lanceolate to lance-ovate; plants broadly distributed in southern Alaska and Yukon. V. americana

4b. Leaves sessile, at least the middle and upper ones, the blades lance-linear to linear, or if lanceolate to lance-ovate, then known only from southeastern Alaska. (5)

5a. Leaves lance-linear to linear, mostly 4–10 times longer than wide or more; plants of central eastern Alaska and southern Yukon. V. scutellata

5b. Leaves lanceolate to elliptic, mostly 1.5–3 times longer than wide; plants known from southeastern Alaska. V. anagallis-aquatica

6a. Plants indigenous perennials. (7)

6b. Plants adventive annuals. (8)

7a. Stems tending to creep and to root at several nodes, and to produce lateral branches; main leaves mostly less than 15(29) mm long, glabrous or nearly so, entire or inconspicuously toothed; capsules wider than long; plants mostly of wet places. V. serpyllifolia

7b. Stems erect, or merely decumbent at the base, rooting mostly at the base, seldom branched; main leaves mostly more than 15 mm long, sparsely to densely villous, often distinctly toothed; capsules longer than wide; plants of various habitats. V. wormskjoldii

8a. Corollas 6–11 mm wide; fruiting pedicels 12–35 mm long. V. persica

8b. Corollas 2–3.5 mm wide; fruiting pedicels 3–5 mm long. (9)

9a. Leaves ovate to elliptic, mostly 1–2 times longer than wide; corolla violet blue. V. arvensis

9b. Leaves narrowly oblong to oblanceolate, mostly more than 3 times longer than wide; corolla white. V. peregrina

Veronica americana Schwein. ex DC.
Speedwell

Perennial herbs, 0.7–10 dm long; stems decumbent at the base and rooting at the nodes or from rhizomes, simple or branched, glabrous; leaves opposite, all short-petioled, obscurely toothed to serrate, lanceolate to elliptic or oblong, the main leaves 2–9 cm long, 2–4 times as long as broad; racemes all axillary, mostly 10–25-flowered; calyx 2.5–4 mm long; corolla 5–9 mm broad, blue; fruiting pedicels 5–16 mm long; capsules about as wide as long, slightly notched; seeds numerous.

Marshes, seeps, springs, and along streams; in much of Alaska and Yukon south of the 65th parallel; eastward to Newfoundland and south to California, Utah, Colorado, and Virginia.

Veronica anagallis-aquatica L.
Water Speedwell

Perennial (or biennial?) herbs, mostly 2–10 dm tall; stems more or less erect from a decumbent to trailing and rooting base, simple or branched, glabrous or slightly puberulent, especially above; leaves opposite, sessile and more or less clasping or the lower ones narrowed to a petiolate base, lanceolate to elliptic, the main leaves 2–10 cm long, 1.5–3 times as long as broad; racemes all axillary, many-flowered; calyx 2.5–4 mm long; corolla 4–7 mm broad, blue; fruiting pedicels 3–8 mm long; capsules about as wide as long, slightly notched; seeds numerous.

Along streams and ponds or in shallow water; in extreme southeastern Alaska; widely introduced and established in North America; native to Europe.

Veronica arvensis L.
Corn Speedwell

Annual herbs, 0.5–3 dm tall; stems prostrate to erect, simple or branched, hairy; leaves opposite, short-petioled to sessile, crenate-serrate, ovate to elliptic, mostly 0.5–1.5(2) cm long, 1–2 times as long as wide; racemes terminal, but often also axillary, or appearing so; calyx 2.5–4 mm long; corolla bluish violet, 2–4 mm broad; fruiting pedicels 0.5–2 mm long; capsules about as long as wide, heart shaped; seeds several.

Adventive weed of cultivated and disturbed soil; introduced in southeastern Alaska and southern Yukon; native to Eurasia.

Veronica chamaedrys L.
Germander Speedwell

Perennial herbs, 1–3 dm tall; stems erect or ascending, from elongate rhizomes, villous, mostly simple; leaves opposite, short-petiolate to sessile, crenate-serrate to somewhat incised, ovate, mostly 1.5–3.5 cm long, 1–2 times as long as broad; racemes all axillary, several-flowered; calyx 3–5 mm long; corolla blue, 9–12 mm broad; capsules broader than long, broadly heart shaped, several-seeded.

Adventive weed of lawns and disturbed soils; in southeastern Alaska (not collected in recent years); native to Europe.

Veronica grandiflora Gaertn.
Large-flower Speedwell

Perennial herbs, 0.5–1.5 dm tall; stems decumbent to erect, from elongate rhizomes (or stolons?), sparsely to densely villous, simple; leaves opposite, short-petioled to sessile, crenate-serrate to subentire, broadly elliptic to oblanceolate, sparsely villous, mostly 1–3 cm long, 1.5–3 times as long as broad; racemes all axillary, 3–8-flowered; calyx 3–4 mm long; corolla violet blue, 10–16 mm broad; capsules broader than long, or about as broad as long; seeds several (?).

Moist to dry banks and slopes; in the middle to western Aleutians; Asia.

Veronica peregrina L.
Neckweed

Annual herbs, 0.5–5 dm tall; stems erect or ascending, simple or more usually branched, glabrous or more commonly glandular-pubescent; leaves opposite, sessile, toothed or entire, oblong or narrowly oblong to oblanceolate, 1–3 cm long, more than 3 times as long as broad; racemes few- to many-flowered, terminal, but some often axillary, or the flowers appearing to be borne singly in the axils of leaves (bracts); calyx 3.5–5 mm long; corolla whitish, about 2 mm broad; fruiting pedicels 1–2 mm long; capsule about as broad as long, heart shaped; seeds numerous.

A weed of damp soil along roadsides, in pastures, and in greenhouses; in south-central to east-central and southeastern Alaska and southern Yukon; widespread in North and South America; introduced from Europe (V. peregrina ssp. xalapensis [H. B. K.] Pennell). Our materials contain both the typical glabrous var. peregrina (somewhat rare?) and the common, glandular-pubescent var. xalapensis (H.B.K.) St. John & Warren.

Veronica persica Poir. ex Lam.
Persian Speedwell

Annual herbs, 0.5–4 dm tall; stems erect, ascending, or decumbent, simple, or more commonly branched, villous; leaves opposite, short-petioled, serrate to crenate-serrate, the blades ovate to orbicular, 0.5–1.5 cm long and almost as broad to broader; racemes few- to several-flowered, terminal, but the bracts resembling foliage leaves and subtending the long-pedicellate flowers; calyx 4–6.5 mm long; corolla blue, 5–11 mm broad; fruiting pedicels 12–35(40) mm long; capsules much broader than long, broadly heart shaped; seeds several.

A weed of lawns and roadsides; in southeastern Alaska where evidently rare and possibly not persisting; widely distributed in North America; introduced from Europe.

Veronica scutellata L.
Skullcap Speedwell

Perennial glabrous herbs, 1–6 dm tall; stems from rhizomes, simple, erect or ascending, or branched and the lower branches decumbent; leaves opposite, sessile, entire or irregularly toothed, narrowly lanceolate to narrowly elliptic or oblong, 1.5–10 cm long or more, more than 4 times as long as broad; racemes all axillary, few- to several-flowered; calyx 1–2.5 mm long; corolla bluish, 6–9 mm wide; fruiting pedicels 5–15 mm long; capsules broader than long, notched; seeds several.

Moist places, bogs, swamps, and lake shores; in central eastern Alaska and southern Yukon; eastward to Newfoundland and south to California, Colorado, and New York; Eurasia.

Veronica serpyllifolia L.
Thyme-leaf Speedwell

Perennial herbs, (0.6)1.5–4 dm tall; stems from rhizomes or stolons or both, decumbent and creeping at the base, simple or branched from near the base, sparsely to rather densely short-villous; leaves opposite, short-petioled to sessile or subsessile, entire or inconspicuously toothed, glabrous or nearly so, the blade lance-ovate to ovate or oval, 8–15(29) mm long, mostly 1–2 times as long as broad; racemes terminal on the main stem, or terminal on lateral stems, several- to many-flowered; calyx 2–4.5 mm long; corolla blue to white, 4–8 mm broad; fruiting pedicels 4–10 mm long; capsules broader than long, notched; seeds numerous.

Moist soil, in seeps, swamps, or along streams; widely distributed in coastal and insular southern Alaska and rarely in southern Yukon; widely distributed in temperate North America, Eurasia, and South America.

1a. Pedicels and rachis of inflorescence puberulent with incurved hairs; corolla pale blue to whitish with violet veins, 2–4 mm broad; plants of south-central to southeastern Alaska and southern Yukon. *V. serpyllifolia* var. *serpyllifolia*

1b. Pedicels and rachis of inflorescence with spreading, multicellular, often glandular-tipped hairs; corolla blue violet, mostly more than 4 mm broad; plants widely distributed in southern Alaska (*V. tenella* All.; *V. humifusa* Dickson; *V. serpyllifolia* ssp. *humifusa* [Dickson] Syme). *V. serpyllifolia* var. *humifusa* (Dickson) Vahl

Veronica wormskjoldii Roem. & Schult.
Alpine Speedwell

Perennial herbs, 0.3–4 dm tall; stems from rhizomes, erect or decumbent at the base, simple, spreading-villous with multicellular hairs; leaves opposite or rarely alternate, sessile or subsessile, crenate-serrate to serrate, subentire or entire, villous to glabrate or rarely glabrous, (0.8)1.7–4 cm long, mostly 1–3 times as long as broad; racemes terminal, few- to several-flowered; calyx 2.5–7 mm long, the pubescence moniliform the terminal cell either expanded and often glandular or conic or laterally collapsed; corolla light blue to violet purple, 5–11 mm broad; fruiting pedicels 4–15 mm long; capsules longer than broad, pubescent like the sepals; seeds numerous.

Moist to dry meadows, in tundra and heathlands; in much of Alaska and Yukon south of the 65th parallel, and rarely northward; eastward to Labrador and Greenland and south to California, New Mexico, and New Hampshire; Asia. The similarities between *V. stelleri* and *V. wormskjoldii* have long been recognized, and in their ultimate expression are readily separable. The presence of intermediates complicates the situation. The primary basis upon which the two entities have been segregated rests on the presence or absence of glandular (or apparently glandular) hairs among the numerous glandless, multicellular hairs of the calyx and fruit. There is a gradation exhibited in those plants which have glandular hairs from numerous hairs with glands to few (or none?). Secondary characteristics such as dentation of leaves and leaf length/width ratio similarly fail as definitive features, either singly or in combination where the two basic types overlap on a geographic basis. The *stelleri-wormskjoldii* complex is only a portion of a much broader problem, that of the V. *alpina* complex. Thus it seems best to reduce our materials to a single polymorphic species.

1a. Main leaves 1–2 times as long as broad, often sharply serrate; pubescence of calyx and capsule lacking glandular hairs (use 30 ×); plants of insular and coastal southern Alaska (*V. stelleri* Pallas ex Schrad. & Link). V. *wormskjoldii* var. *stelleri* (Pallas) Welsh

1b. Main leaves 2–3 times as long as broad, crenate-serrate to subentire; pubescence of calyx and capsule mostly with at least some glandular hairs (use 30 ×); plants more widely distributed (*V. nutans* Bong.; V. *wormskjoldii* var. *nutans* [Bong.] Pennell; V. *alpina* var. *unalaschcensis* Cham. & Schlecht.; V. *alpina* var. *wormskjoldii* [Roem. & Schult.] Hook.; V. *alpina* var. *alterniflora* Fern.; V. *wormskjoldii* ssp. *alterniflora* [Fern.] Pennell; V. *stelleri* var. *glabrescens* Hultén). V. *wormskjoldii* var. *wormskjoldii*

SOLANACEAE

Potato Family

Plants herbaceous; leaves alternate, rarely opposite, mostly simple, but often pinnately divided or compound; flowers perfect, usually regular, arranged in cymes; calyx usually of 5 more or less united sepals; corolla of 5 united petals, rotate, campanulate, funnelform, or salverform; stamens as many as the corolla lobes and alternate with them

or fewer; pistils 1; ovary superior, 2-loculed (more by intrusion of placental tissue in cultivated tomato); styles 1; stigmas 2-lobed; fruit a berry or a capsule.

1a. Corolla funnelform to salverform; fruit a capsule; leaves simple, entire. *Petunia*

1b. Corolla rotate or rotate-campanulate; fruit a berry; leaves mostly pinnately parted or compound (simple but not entire in some *Solanum* species). (2)

2a. Corolla white, pink, or blue purple; anthers opening by a terminal pore or slit; plants bearing underground tubers, or arising from taproots. *Solanum*

2b. Corolla yellow; anthers opening by a longitudinal slit from base to apex; plants from taproots, not tuberous. *Lycopersicon*

LYCOPERSICON Mill.

Caulescent, viscid-pubescent, decumbent to ascending, annual herbs (at least appearing so in cultivation); leaves alternate, pinnately compound or pinnatifid; flowers small, yellow; calyx 5-lobed, the lobes lanceolate; corolla rotate, 5-lobed (rarely with more lobes), regular; stamens 5, connivent around the style, the anthers projected into sterile tips, dehiscent from top to base; fruit a 2–few-loculed berry.

Lycopersicon esculentum Mill.
Tomato

Plants 5–15 dm long (or longer in some forms), strongly scented; stems sprawling to erect; leaves pinnatifid to pinnately compound, glandular-pubescent; flowers 1–2 mm broad, nodding, in clusters of 5–7; fruit red or yellow at maturity, of various sizes, colors and shapes.

Cultivated food plant, especially under glass, but also grown in outside plantings in some parts of Alaska for at least a part of the season; rarely escaping but not persisting; native to western South America.

PETUNIA Juss.

Caulescent, viscid-pubescent, decumbent to ascending, annual herbs from taproots; leaves alternate, or the upper opposite, simple, entire; flowers large and showy, white, pink, purple, or varicolored; calyx 5-lobed, the lobes oblong to linear; corolla funnelform to salverform, 5-lobed, regular or nearly so; stamens 5, in 2 pairs, the fifth reduced or obsolete; fruit a 2-valved capsule.

Petunia hybrida Vilm.
Common Petunia

Plants sprawling, the stems 2–6 dm long; leaves ovate to ovate-lanceolate, short-petioled; flowers mostly 5–8 cm long and broad, variously colored, funnelform; fruit a capsule, the seeds minute.

Cultivated ornamental, as a border plant or grown inside as potted plants. Materials placed within *P. hybrida* are cultigens which have apparently originated from hybridization of *P. axillaris* BSP and *P. violacea* Vilm.

SOLANUM L.

Caulescent, pubescent to glabrous herbs, with prostrate to decumbent or ascending stems; leaves alternate, pinnately compound, pinnatifid, or simple and merely dentate or lobed; flowers small, white or pink to purplish; calyx 5-lobed, the lobes oblong to lanceolate; corolla rotate to rotate-campanulate; stamens 5, connivent around the style, the anthers lacking sterile tips, opening by a terminal pore or slit; fruit usually a 2-loculed berry.

Stebbins, G. L., and E. F. Paddock. 1949. The *Solanum nigrum* complex in Pacific North America. Madroño 10:70–81.

1a. Plants from rhizomes, these producing large, succulent tubers at their apices; cultivated. *S. tuberosum*
1b. Plants from taproots, not producing tubers; adventive weeds. (2)

Petunia hybrida Vilm. (× 0.4).

2a. Stems and leaves glabrate, puberulent, or strigose; fruit black when ripe; calyx not expanding and investing the lower part of the fruit when ripe. *S. nigrum*
2b. Stem and leaves conspicuously villous; fruit greenish or yellowish when ripe; calyx expanding at maturity and enveloping the lower part of the fruit. *S. sarachoides*

Solanum nigrum L.
Black Nightshade

Plants annual, glabrate, puberulent, or strigose, from taproots, the prostrate to ascending stems 1.5–3 dm long; leaves ovate to deltoid, irregularly toothed to subentire; peduncles few-flowered; calyx 2–3 mm long; corolla white to bluish, mostly 5–10 mm broad; fruit mostly 5–8 mm in diameter, black, subglobose.

A weed of restricted distribution in central and southeastern Alaska, but to be expected in all agricultural areas and greenhouses; widely distributed in northern and southern hemispheres. The plants and immature berries are poisonous.

Solanum sarachoides Sendt. ex Mart.
Nightshade

Plants annual, conspicuously villous, the hairs often gland-tipped; stems prostrate, 1.5–3 dm long; leaves ovate to deltoid or ovate-lanceolate, irregularly toothed; calyx expanding at maturity, 4–9 mm long; enveloping the lower portion of the fruit; flowers white, mostly 4–7 mm across; fruit mostly 5–8 mm in diameter, greenish- to yellowish-globose.

A weed of cultivated land and greenhouses; in southern Alaska.

Solanum tuberosum L.
Potato

Plants grown as annuals, producing rhizomes, these bearing tubers; stems decumbent to erect, mostly less than 10 dm long; leaves pinnately compound, 10–25 cm long, with mostly 7–9 main leaflets (other smaller ones between), these ovate to ovate-lanceolate; flowers white to pink or purplish, 2–3 cm broad; fruit seldom produced, mostly 1–2 cm in diameter, greenish or yellowish.

Cultivated food plant; in the agricultural regions of southern Alaska.

UMBELLIFERAE
Carrot Family

Herbaceous, acaulescent or caulescent, annual, biennial, or perennial plants; leaves alternate (rarely opposite) or basal, compound or rarely simple, usually much incised or divided and with sheathing petioles; flowers small, regular, perfect, in compound umbels; rays (stems of umbellets) sometimes subtended by bracts forming an involucre; umbellets often subtended by bractlets, forming an involucel; calyx teeth small or obsolete; petals 5; stamens 5, inserted on an epigynous disk; pistils 1; ovary inferior, 2-loculed; styles 2, sometimes swollen at the base and forming a stylopodium; fruit a schizocarp.

Mathias, M. E., and L. Constance. 1944–45. Umbelliferae. N. Am. Fl. 28B:43–297.

1a. Leaves simple, entire. (2)
1b. Leaves compound, at least some, or deeply dissected. (3)

2a. Plants aquatic; leaves reduced to long narrow phyllodes, bladeless; reported from southeastern Alaska, but the report requires confirmation; it is otherwise known from Vancouver Island, British Columbia to California. *Lilaeopsis occidentalis* Coult. & Rose
2b. Plants terrestrial; leaves well developed, the blades expanded. *Bupleurum*

3a. Leaves dissected into small and narrow segments, distinct leaflets not apparent. (4)
3b. Leaves not dissected into small and narrow segments, at least some with distinct leaflets. (6)

4a. Plants biennial (cultivated as annuals), taprooted; introduced, widely cultivated. *Daucus*
4b. Plants perennial, with or without a taproot. (5)

5a. Leaves ovate to triangular in outline, the segments acute; cauline leaves well developed; fruit 4–8 mm long. *Conioselinum*

5b. Leaves ovate in outline, the segments obtusish; cauline leaves poorly developed; fruit 2.5–3 mm long; reported from Old Man Creek, Central Yukon River District, Alaska, but more material is necessary for positive identification of this Asiatic species. *Cnidium ajanense* (Reg. & Tiling) Drude

6a. Leaves trifoliolate, the terminal leaflet mostly 1–3 dm long and broad at maturity; marginal flowers of umbel enlarged, the outer corolla lobes often 2-cleft. *Heracleum*

6b. Leaves with more than 3 leaflets, these usually much less than 5 cm broad; marginal flowers neither enlarged nor with corolla lobes 2-cleft. (7)

7a. Plants acaulescent, or nearly so. (8)

7b. Plants with well-developed, leafy stems. (9)

8a. Involucres and involucels both conspicuously developed; flowers not sessile; leaves glabrous to hispidulose beneath, not tomentose; plants not of coastal southern Alaska. *Podistera*

8b. Involucres lacking; involucels present; flowers sessile; leaves tomentose beneath; plants of coastal and insular southern Alaska. *Glehnia*

9a. Leaves palmately once-compound, or palmately cleft; plants of northern British Columbia. *Sanicula*

9b. Leaves ternately or pinnately once to several times compound; plants of various distribution. (10)

10a. Flowers yellow; introduced biennials. *Pastinaca*

10b. Flowers white, pinkish, or greenish, not yellow; indigenous perennials. (11)

11a. Leaves ternately once, twice, or more times compound; leaflets mostly ovate. (12)

11b. Leaves pinnately once to twice compound; leaflets ovate to lanceolate or linear. (14)

12a. Plants maritime; leaflets mostly 9, glabrous or coriaceous. *Ligusticum*

12b. Plants maritime or not; leaflets more than 9, or if 9 or fewer, not both glabrous and coriaceous. (13)

13a. Stems slender, mostly less than 6 mm in diameter; leaves mostly once or twice compound; ovary and fruit armed with bristles or prickles. *Osmorhiza*

13b. Stems thick, mostly over 10 mm in diameter; leaves often thrice compound; ovary and fruit unarmed. *Angelica*

14a. Plants decumbent to ascending, often rooting at the nodes; leaflets ovate to lanceolate; styles 1–3 mm long, persistent on the fruit. *Oenanthe*

14b. Plants erect, not or only seldom rooting at the nodes; leaflets lanceolate to linear; styles less than 1 mm long, not conspicuous in fruit. (15)

15a. Stem base thickened, hollow, with well-developed partitions; leaves mostly more than once compound. *Cicuta*

15b. Stem base not thickened, lacking transverse partitions; leaves once compound. *Sium*

ANGELICA L.

Caulescent perennials from taproots, these with internal, transverse partitions at the apex of the root crown; leaves ternate-pinnately once to thrice compound, the 12–50 leaflets broad and distinct, serrate to crenate-serrate; petioles sheathing basally, the cauline sheaths often inflated and bladeless; umbels compound; involucre lacking or less commonly present; involucels of numerous, narrow bractlets, the rays few to numerous; pedicels slender; flowers white or pink; calyx teeth minute or obsolete; styles short to long, the stylopodium low-conic; fruit oblong-oval to orbicular, strongly dorsally flattened, the dorsal ribs filiform to winged, the lateral ones broadly winged.

1a. Leaf rachis flexuous; leaflets lanceolate to narrowly ovate, mostly 2–3 times as long as broad, reflexed, sharply serrate, the serrations spinulose apically. *A. genuflexa*
1b. Leaf rachis straight; leaflets ovate to lanceolate, mostly 1–2 times as long as broad, not reflexed, bluntly serrate to crenate-serrate, the serrations at most bluntly cuspidate. *A. lucida*

Angelica genuflexa Nutt.

Plants mostly 10–15 dm tall, the stem leafy; leaves 10–60 cm long, ternate-pinnately twice compound, the rachis bent at the point of insertion of the pinnae, the pinnae reflexed; leaflets lanceolate to narrowly ovate or elliptic, mostly 2–3 times as long as broad, 2.5–10 cm long, 0.8–5 cm broad, serrate, the serrations spinulose apically; rays 20–45, unequal, 2.5–8 cm long; involucre mostly lacking, rarely with a few slender bracts present; involucel present, the bractlets narrow; flowers white or pinkish; ovaries minutely pubescent; fruit glabrous, 3–4 mm long, the dorsal wings narrow, the lateral ones about as wide as the body.

Moist sites; in or near coastal and insular southern Alaska, from Bristol Bay and the Alaska Peninsula eastward through the Panhandle; south to California and east to Alberta.

Angelica lucida L.

Plants mostly less than 10 dm tall, rarely more; leaves 10–60 cm long, ternate-pinnately twice to thrice compound, the rachis not bent at the point of insertion of the pinnae, the pinnae ascending; leaflets ovate to lanceolate, mostly 1–2 times as long as broad, 2–7 cm long, 1.5–5 cm broad, serrate or crenate-serrate, the serrations at most bluntly cuspidate; rays 20–45, unequal, 1.5–15 cm long; involucre lacking; involucel well-developed, of long slender bractlets; flowers white; ovaries pubescent or glabrous; fruit glabrous, 4–7 mm long, the wings all about alike.

Moist sites; in central to western, southwestern, south-central, and southeastern Alaska and less commonly in southern Yukon; southward to California, and on the east coast from Labrador to New York; Asia (*Archangelica gmelinii* DC.; *Coelopleurum gmelinii* [DC.] Ledeb.; *C. lucidum* [L.] Fern.).

BUPLEURUM L.

Caulescent perennials from a branching caudex and taproot, the caudex sheathed with dark brown leaf bases, the dead leaves often coiled upon drying; leaves simple, the basal ones petiolate, the blades entire, parallel-veined, the cauline ones sessile and clasping; umbels compound; involucre of conspicuous foliose bracts; involucel of foliose, often connate, bractlets; rays few to several; pedicels short; flowers yellow, greenish, or purplish; calyx teeth lacking; styles short, the stylopodium low-conic; fruit oblong to orbicular, somewhat laterally flattened, glabrous or roughened; ribs filiform.

Bupleurum triradiatum Adams
Thorough-wort

Plants 1.5–5(7) dm tall, with few to several, usually simple stems; leaves simple, basal and cauline, 2–25 cm long, narrowly oblong to linear, with 3–5 prominent, parallel veins; rays 1–8(14), 0.5–5 cm long; involucre of 1–several unequal, lanceolate to ovate bracts; involucel of 5–8 ovate to lanceolate bractlets; flowers yellow, greenish, or purplish; ovaries glabrous; fruit glabrous, 3–4 mm long, the ribs raised but wingless.

Dry, rocky slopes, terraces, flood plains, in grasslands and woods; in much of Alaska (except for the coastal and insular portions) and in western Yukon; south to Montana, Idaho, and Wyoming (*B. americanum* Coult. & Rose; *B. ranunculoides* var. *arcticum* Reg.). Our material belongs to ssp. *arcticum* (Reg.) Hultén.

CICUTA L.

Caulescent, robust to slender perennials from taproots, these with well-developed internal transverse partitions and an orange yellow resin within the apex of the root crown; leaves once to thrice pinnately (rarely ternate-pinnately) compound, the 7–45 leaflets linear to lanceolate, serrate to somewhat incised; petioles sheathing basally, the cauline sheaths sometimes bladeless, but little inflated; involucre lacking or with a few slender bracts; involucel of several narrow bractlets or lacking; rays several to numerous; pedicels slender; flowers white or greenish; calyx teeth evident; styles short, the stylopodium low-conic; fruit oval to ovoid or orbicular, laterally flattened, the ribs usually prominent but not winged. *Note*: Plants of this genus are noted for being poisonous to man and to livestock. All parts are poisonous, but the roots and young plants are especially so.

1a. Leaflets mostly 3–4 times as long as broad, lanceolate to narrowly oblong or elliptic; fruit from as broad as long to longer than broad. *C. douglasii*

1b. Leaflets mostly more than 5 times as long as broad, linear to narrowly lanceolate; fruit broader than long. *C. mackenzieana*

Cicuta douglasii (DC.) Coult. & Rose
Water Hemlock

Plants 5–20 dm tall, robust, from a taproot or cluster of tuberous roots; leaves once to thrice pinnately compound, the pinnae ascending; leaflets mostly 3–4 times as long as broad, 3–10 cm long, 0.6–3 cm broad, lanceolate to narrowly oblong or elliptic, serrate to somewhat incised, the main lateral veins directed toward the sinuses of the marginal serrations; rays 12–38, unequal, 3–10 cm long; involucre lack-

ing or with a few slender bracts; involucel of several lanceolate to linear bractlets; flowers white to greenish; ovaries glabrous; fruit glabrous, 2–4 mm long, ovate to orbicular.

Marshes and along streams; in central to east-central and coastal southern Alaska from Kodiak Island eastward; southward to California and east to Alberta, Colorado, and New Mexico (*Sium douglasii* DC.; *C. occidentalis* Greene). Fruit of two rather distinct types are present among the broad leafletted forms of water hemlock in Alaska. Those from the interior lack the commissural constriction of fruits from coastal plants (which resemble those of *C. mackenzieana*), and are longer than broad. These long-fruited plants were distinguished by Anderson as *C. maculata* L. However, the limited number of samples with fruit make it difficult to evaluate the importance of the differences. Thus until additional specimens having fruit are available from the interior, it seems best to include them along with the coastal plants within *C. douglasii*.

Cicuta mackenzieana Raup
Mackenzie Water Hemlock

Plants 3–10 dm tall, slender to robust, from a cluster of fibrous or tuberous roots; leaves once to twice pinnately compound, the pinnae ascending, the leaflets mostly more than 5 times as long as broad, 2–10 cm long, 0.2–2 cm broad, linear to narrowly lanceolate, serrate to somewhat incised or rarely almost entire, the veins mostly not conspicuous (when apparent then similar to the above species); rays 14–33, unequal, 1.5–7 cm long; involucre lacking; involucel of a few narrow bractlets; flowers white to cream or pink; ovaries glabrous; fruit glabrous, 1.5–2.2 mm long, orbicular.

Marshes, lake and pond margins, and along streams; in central western and southwestern Alaska, eastward to central Yukon; east to Hudson Bay.

Cicuta mackenzieana Raup (× 0.3).

CONIOSELINUM Hoffm.

Caulescent, robust or slender perennials, from taproots or from a cluster of tuberous roots, these lacking conspicuous transverse partitions within the apex of the root crown; leaves ternate-pinnately several times compound, the leaflets lobed or dissected; petioles sheathing basally, the cauline sheaths with reduced blades; involucre of narrow or foliaceous bracts or lacking; involucel of few to several linear to broad bractlets; rays several to many; pedicels short; flowers white; calyx teeth lacking; styles short, the stylopodium conic; fruit oblong-oval to oval, flattened dorsal-

ly, the dorsal ribs low or narrowly winged, the lateral ribs more broadly thin winged.

1a. Bractlets of the involucel and involucre (when present) scarious, abruptly contracted apically into a narrowly aristate tip; dorsal ribs of the fruit almost as broadly winged as the lateral. *C. cnidiifolium*
1b. Bractlets of the involucel and involucre (when present) not scarious, linear to oblong throughout the entire length; dorsal ribs of the fruit more narrowly winged than the lateral. *C. chinense*

Conioselinum chinense (L.) BSP
Western Hemlock-parsley

Plants 1.5–10 dm tall, robust; leaves ovate to deltoid in general outline, ternate-pinnately dissected, the secondary pinnae lobed to incised, or rarely pinnatifid, the ultimate segments often over 4 mm broad; rays 8–28, unequal, 1–5 cm long; involucre with 1–few linear bracts or lacking; involucel of several, linear, nonscarious bractlets; flowers white; ovaries glabrous; fruit glabrous, 5–8 mm long, oblong, wings of dorsal ribs narrower than the lateral.

Gravelly slopes, beaches, and marshes; in or near coastal and insular western, southwestern, south-central, and southeastern Alaska; south to California and disjunctly in eastern North America; Asia (*Athamantha chinensis* L.; *Selinum benthami* Wats.; *Conioselinum benthami* [Wats.] Fern.; *S. pacificum* Wats.; *C. pacificum* [Wats.] Coult. & Rose; *Ligusticum gmelini* Cham. & Schlecht.; *C. gmelini* [Cham. & Schlecht.] Coult. & Rose, not [Bray] Steud.).

Conioselinum cnidiifolium (Turcz.) Porsild

Plants 2.5–9 dm tall, robust; leaves ovate to deltoid in general outline, ternate-pinnately dissected, the secondary pinnae pinnatifid, the ultimate segments seldom over 3 mm broad; rays 6–28, unequal, 1–3(20) cm long; involucre of 1–several, scarious (or rarely pinnatifid) bracts or lacking;

involucel of several narrowly oblong scarious bractlets which are abruptly narrowed apically to an aristate tip; flowers white; ovaries glabrous; fruit glabrous, 4–5 mm long, oblong-oval, wings of dorsal ribs almost as broad as the lateral.

Dry open hillsides, flats, terraces, and open woods; in west-central to northwestern Alaska and eastward through northern and central Alaska to northern Yukon and central to southwestern Yukon; east to Mackenzie; Asia (*Selinum cnidiifolium* Turcz.; *Cnidium cnidiifolium* [Turcz.] Schischk.; *S. dawsonii* Coult. & Rose; *Conioselinum dawsonii* [Coult. & Rose] Coult. & Rose).

DAUCUS L.

Caulescent biennial (cultivated as annual) pubescent herbs from taproots, these lacking internal transverse partitions; leaves pinnately several times compound, the ultimate divisions small and narrow; petioles sheathing basally; umbels compound; involucre of numerous dissected or entire bracts, or wanting; involucel of numerous toothed or entire bractlets, or wanting; rays few to numerous, spreading or at length incurved; pedicels slender; flowers white or the center flower in each umbellet purple; calyx teeth lacking or evident; styles short, the stylopodium conic; fruit oblong to ovoid, dorsally flattened, evidently ribbed, with stout prickles along alternate ribs, the intermediate ribs merely bristly or hairy.

Daucus carota L.
Carrot

Plants 1.5–12 dm tall when in flower, producing only a basal rosette of leaves the first year (rarely flowering the first season); leaves oblong in general outline, several times pinnately compound, the ultimate divisions linear to lanceolate, glabrous to pubescent; involucre of slender pinnately divided (rarely entire) scarious-margined bracts; involucel of slender entire or pinnate ciliate bractlets; rays numerous, unequal, 3–7 cm long; fruit ovoid, 3–4 mm long.

Cultivated food plant; regularly grown in the agricultural areas of Alaska and Yukon. The cultivated carrot belongs to var. *sativa* DC.

GLEHNIA Schmidt

Subacaulescent, low pubescent perennials from taproots; leaves imperfectly once to twice ternate or ternate-pinnately compound, the (3)9–21 leaflets broad and distinct, crenate-dentate; petioles sheathing basally; umbels compound; involucre of few narrow bracts or wanting; involucel of several narrow bractlets; rays few to several; flowers white; calyx teeth inconspicuous; styles short, the stylopodium lacking; fruit ovate-oblong to subglobose, somewhat dorsally flattened, pubescent to glabrate, the ribs all broadly corky winged.

Mathias, M. E. 1928. Studies in Umbelliferae. I. Ann. Mo. Bot. Gard. 15:91–103.

Glehnia leiocarpa Mathias

Plants mostly 1 dm tall or less, acaulescent, or with some elongate internodes usually buried in sand; leaves ovate in general outline, spreading to prostrate, the leaflets oblong-obovate or cuneate, 1–6 cm long, 1–3 cm broad, tomentose beneath, glabrous above; rays 5–13, 1.5–5 cm long, tomentose; bracts and bractlets both tomentose; fruit 4–13 mm long, glabrous or hairy at the apex.

Sandy beaches; in coastal and insular southern Alaska, from Kodiak Island eastward to the vicinity of Yakutat; disjunctly southward to British Columbia and California (*G. littoralis* ssp. *leiocarpa* [Mathias] Hultén).

HERACLEUM L.

Caulescent robust perennial or biennial herbs from taproots, these lacking transverse partitions within the root crown;

leaves ternately or pinnately compound, the (1)3 leaflets broad, coarsely serrate, and variously lobed; petioles sheathing basally, the cauline sheaths often much expanded; umbels compound; involucre lacking or present as a few deciduous bracts; involucel of numerous narrow bractlets or lacking; rays numerous; pedicels slender; flowers white, those of the margin of the inflorescence enlarged and often 2-cleft; calyx teeth minute or lacking; styles short, the stylopodium conic; fruit orbicular to obovate or elliptic, strongly dorsally flattened, the dorsal ribs filiform, the lateral, broadly thin winged.

Heracleum lanatum Michx.
Cow Parsnip

Plants 4.5–20 dm tall or more; leaves orbicular to reniform in general outline, ternately compound, the 3 leaflets ovate to orbicular, 1–4 dm long and wide, tomentose to villous below, the sheaths often woolly-villous at the juncture with the stem; rays 15–45, unequal, 3–15 cm long; involucre of slender bracts or lacking; involucel of slender tomentose bractlets; flowers white; fruit obovate, 7–12 mm long, somewhat pubescent.

Stream banks, moist slopes, and woods; in west-central to southwestern Alaska and east to central eastern and southeastern Alaska and southern Yukon; east to Newfoundland and south to California, Arizona, and Georgia; Asia.

LIGUSTICUM L.

Caulescent slender to robust glabrous perennials from taproots, lacking transverse partitions within the root crown; leaves ternately once to twice compound, the usually 9 leaflets broad and distinct, coarsely serrate and variously lobed or incised; petioles sheathing basally, the cauline sheaths not much expanded; involucre of few to several linear to lanceolate or rarely pinnate bracts; involucel of narrowly lanceolate to linear, somewhat scarious

bractlets; rays few to several, subequal; pedicels slender; flowers white to pinkish; calyx teeth minute to evident; styles short, the stylopodium low-conic; fruit ovoid to oblong, somewhat laterally flattened or terete, glabrous, the ribs narrowly winged.

Ligusticum scoticum L.
Hultén Sea-lovage

Plants 1–8 dm tall; leaves orbicular to reniform in general outline, mostly twice ternate, the leaflets obovate, obtuse apically, often cuneate basally, 2–6 cm long, 1–3 cm broad, glabrous; rays 5–16, subequal, 1–4 cm long; involucre of 2–4 narrowly lanceolate bracts; involucel of several narrow bractlets; flowers white or pinkish; fruit oblong, 7–8 mm long, terete, glabrous.

Beaches and strands along the coast; in western, southwestern, south-central, and southeastern Alaska; south to British Columbia, and in eastern North America; Europe, Asia (*L. hulténii* Fern.; *L. scoticum* var. *hulténii* [Fern.] B. Boi.). Our materials belong to ssp. *hulténii* (Fern.) Calder & Taylor.

OENANTHE L.

Caulescent slender to robust glabrous perennial herbs from fibrous or tuberous roots, often rooting at the lower nodes; leaves pinnately twice to thrice compound, the 25–71 leaflets (or ultimate segments) serrate to incised; petioles sheathing basally, the cauline sheaths not expanded; involucre lacking, or with 1–few narrow bracts; rays numerous, subequal; flowers white; calyx teeth lanceolate, persistent in fruit; styles erect, persistent in fruit, the stylopodium conic; fruit oblong, terete or somewhat laterally flattened, glabrous, the ribs low and not winged.

Oenanthe sarmentosa Presl ex DC.
Water Parsley

Plants 5–15 dm tall, decumbent to ascending; leaves oblong to ovate in general outline, twice or imperfectly thrice pinnate, the leaflets ovate to lanceolate, acute apically, acute to rounded basally, toothed to cleft, 1–6 cm long, 1–5 cm broad, glabrous; involucre of 1–few linear bracts or lacking; involucel of several evident narrow bractlets; rays 5–21, subequal, 1.5–3 cm long; flowers white; calyx teeth lanceolate; styles 1–3 mm long; fruit oblong, 2.5–3.5 mm long, the ribs broader than the intervals.

Low wet thickets along streams and in marshes; in coastal and insular southern Alaska from Prince William Sound eastward through the Panhandle; southward to California.

OSMORHIZA Raf.

Caulescent slender pubescent to glabrate perennial herbs from fascicled roots; leaves thin and membranous, ternate to ternate-pinnate, the usually 9 leaflets lanceolate to orbicular, coarsely serrate to lobed; petioles sheathing basally; umbels compound; involucre lacking or of 1–few bracts; involucel of several bractlets or lacking; rays few; pedicels slender; flowers white, purple, or greenish yellow; calyx teeth lacking; styles slender to obsolete, the stylopodium depressed-conic to conic; fruit linear-oblong to clavate, obtuse, tapering to a beak or constricted at the apex, caudate at the base, bristly-hispid.

Constance, L., and R. Shan. 1948. The genus *Osmorhiza* (Umbelliferae). Univ. Calif. Publ. Bot. 23:111–56.

1a. Fruit (or ovaries) rounded to acute at the apex, neither constricted below the apex nor tapering into a beaklike tip; stylopodium low-conic, as wide as high or wider. *O. depauperata*
1b. Fruit (or ovaries) tapering at the apex, or constricted below the tip into a distinct beaklike apical portion; stylopodium low-conic to conic. (2)

2a. Stylopodium more or less conic, mostly as high as broad; flowers whitish or greenish white; fruit 12–20 mm long. *O. chilensis*
2b. Stylopodium depressed-conic, mostly broader than high; flowers purplish or greenish white; fruit 10–13 mm long. *O. purpurea*

Osmorhiza chilensis Hook. & Arn.

Chile Sweet-cicely

Plants 3–10 dm tall, slender, more or less pubescent; leaves orbicular in general outline, biternate, the leaflets ovate-lanceolate to orbicular, 2–9 cm long, 1–5 cm broad, coarsely crenate-serrate, incised or lobed, hairy especially along the veins or glabrate in age; involucre usually lacking; involucel lacking; rays 3–8, spreading to ascending, 2–12 cm long; pedicels 5–30 mm long; flowers greenish white; fruit linear-oblong, 12–20 mm long, the apex tapering into a beak, caudate and hispid basally.

Woods and thickets; in coastal and insular southern Alaska from the eastern Aleutians eastward to southeastern Alaska; disjunctly east to the Atlantic and south to California, Arizona, Colorado, and South Dakota; South America.

Osmorhiza depauperata Phil.

Plants 1.5–7 dm tall, slender, more or less pubescent; leaves orbicular in general outline, biternate or ternate-pinnate, the leaflets lanceolate to ovate, 1.5–7 cm long, 1–4 cm broad, coarsely crenate-serrate, incised, or lobed, hairy to glabrate; involucre lacking or consisting of a solitary foliaceous bract; involucel lacking; rays 2–6, widely divergent, 2–7 cm long; pedicels 10–30 mm long; flowers greenish white; fruit clavate, 10–15 mm long, obtuse or abruptly acute apically, caudate and hispid basally.

Woods; in coastal and insular south-central and southeastern Alaska; eastward to Newfoundland and south to California, New Mexico, South Dakota, and Vermont; South

America (*Washingtonia obtusa* Coult. & Rose; *O. obtusa* [Coult. & Rose] Fern.).

Osmorhiza purpurea (Coult. & Rose) Suksd.
Sitka Sweet-cicely

Plants 2–7 dm tall, slender, pubescent or glabrous; leaves deltoid to orbicular in outline, once to thrice ternate, the leaflets lanceolate to ovate, 1–8 cm long, 0.5–4 cm broad, coarsely crenate-serrate, incised or lobed, hairy to glabrate; involucre lacking; involucel lacking; rays 2–6, spreading to ascending, 2.5–7 cm long; pedicels 5–20 mm long; flowers purplish or greenish white; fruit linear-fusiform, 10–13 mm long, constricted below the shortly beaked apex, caudate basally.

Woods; in coastal and insular southern Alaska from Kodiak Island eastward through the Panhandle; southward to California and east to Montana and Idaho (*Washingtonia purpurea* Coult. & Rose).

PASTINACA L.

Caulescent biennial glabrous to pubescent robust herbs from taproots; leaves pinnately once (or imperfectly twice) compound, the 7–21 leaflets broad, toothed to pinnatifid; petioles sheathing basally; umbels compound; involucre and involucel usually lacking; rays few to numerous; pedicels slender; flowers yellow; calyx teeth small or lacking; styles short, the stylopodium low-conic; fruit oval to obovate, dorsally flattened, glabrous, the dorsal ribs filiform, the lateral ones broadly thin winged.

Pastinaca sativa L.
Parsnip

Plants 3–15 dm tall; leaves oblong to ovate in general outline; leaflets oblong to ovate, 5–10 cm long, 2.5–8 cm broad, coarsely toothed and lobed or divided, puberulent to glabrate; rays 15–33, unequal, 2–10 cm long; flowers yellow; fruit oval to elliptic, 5–6 mm long, 4–5 mm broad.

Cultivated food plant; in the agricultural

regions of Alaska, where it sometimes escapes and persists (as at Manley Hot Springs); native to Europe.

PODISTERA Wats.

Acaulescent low glabrous to scabrous scapose perennials from taproots, the caudex branches clothed with old leaf bases; leaves pinnately once to twice compound or the lower pinnae ternate; petioles sheathing basally; umbels compound; involucres of 1–several bracts; involucel of several linear to lanceolate bractlets; rays few; pedicels short, stout; flowers white, yellow, or purplish; calyx teeth conspicuous; styles short, spreading, the stylopodium conic; fruit oblong to ovoid, somewhat laterally flattened, glabrous, the ribs prominent, not winged.

1a. Flowers purplish or yellow; leaflets lobed or dissected; plants of broad distribution. *P. macounii*

Pastinaca sativa L. (× 0.25).

1b. Flowers white; leaflets entire; plants of west-central Yukon and east-central Alaska (?). *P. yukonensis*

Podistera macounii (Coult. & Rose) Mathias & Const.

Plants 0.4–2.3 dm tall; leaves oblong in general outline, 3–12 cm long, the leaflets ovate to orbicular, 0.4–1.5 cm long, 0.3–1.5 cm broad, coarsely lobed or incised; peduncles 6–15 cm long, shorter to much longer than the leaves; involucre of several linear-lanceolate, often pinnatifid bracts; involucel similar to the bracts; rays 3–12; flowers purplish or yellow; fruit oval, 4–5 mm long, 2–3 mm broad, the ribs prominent.

Alpine tundra and heathlands; in much of continental Alaska south of the 66th parallel, in coastal west-central regions, and in northern Yukon and adjacent Mackenzie; Asia (*Ligusticum macounii* Coult. & Rose; *L. mutellinoides* authors, not [Crantz.] Willar [?]).

Podistera yukonensis Mathias & Const.

Plants 1–1.5 dm tall, acaulescent, caespitose; leaves oval to narrowly oblong in general outline, 3–6 cm long, the leaflets orbicular to narrowly lanceolate, 0.5–1.5 cm long, 0.2–0.5 mm broad, entire; peduncles 10–12 cm long, slightly exceeding the leaves; involucre of 1 or more small linear entire bracts; involucel of few to several linear-acuminate, entire, purplish bractlets, connate at the base; rays 4–8, 5–10 mm long; pedicels 1.5–3 mm long; flowers white; fruit oval-oblong, about 3 mm long and 1.5 mm broad, the ribs prominent.

Talus slopes; in west-central Yukon and east-central Alaska (?); endemic.

SANICULA L.

Caulescent slender erect glabrous or glabrate perennial herbs from fibrous roots; leaves palmately 5–7-parted, the divisions toothed and often lobed; petioles sheathing basally; umbels compound; involucre foliaceous; involucel of slender bractlets; rays few, unequal; pedicels lacking (at least on fertile flowers); flowers greenish white, perfect or staminate, the staminate predominantly pedicellate; sepals prominent, connate, persistent; styles elongate, persistent, the stylopodium lacking; fruit ovoid, somewhat laterally flattened, densely covered with uncinate bristles, the ribs obsolete.

Shan, R., and L. Constance. 1951. The genus *Sanicula* (Umbelliferae) in the Old World and the New. Univ. Calif. Publ. Bot. 25:1–78.

Sanicula marilandica L.
Black Snakeroot

Plants 4–12 dm tall; leaves reniform to suborbicular in general outline, palmately 5–7-parted, the primary divisions oval to cuneate-obovate, doubly serrate to dentate-serrate with spinulose teeth, deeply incised apically; involucre of a few leaflike bracts; involucel of minute narrow bractlets; rays 1–4; flowers of 2 types (fertile and sterile) in the same umbellet or the sterile in separate umbellets, greenish white; calyx deeply cleft; styles long, usually recurved; fruit ovoid, 4–6 mm long, 3–5 mm wide, covered with numerous uncinate prickles.

Along streams and hot pools; at Liard Hot Springs, British Columbia; eastward to Newfoundland and south to Idaho, Wyoming, and Florida.

SIUM L.

Caulescent stout glabrous perennials from tuberous roots, these lacking transverse partitions within the root crown; leaves once pinnately compound, the 7–13 leaflets lanceolate to linear, serrate or incised; petioles sheathing basally; involucre of subfoliaceous, entire, or incised bracts; involucel of conspicuous narrow bractlets; rays few to several; flowers white; calyx teeth minute or lacking; styles short, the stylopodium depressed-conic to conic; fruit oval to orbicular, somewhat laterally flattened, glabrous, the ribs prominent, but not winged.

Sium suave Walt.
Hemlock Water-parsnip

Plants 5–12 dm tall; leaves oblong to ovate in general outline, pinnately once compound or rarely simple, the leaflets 1–9 cm long, 0.3–1.5 cm broad, the main lateral veins not directed toward the sinuses of the marginal serrations; rays 10–30, subequal, 1.5–3 cm long; involucre of 6–10 lanceolate or linear bracts; involucel of 4–8 linear-lanceolate bractlets; calyx teeth minute; fruit glabrous, 2–3 mm long, oval to orbicular.

Marshes and ponds; along the Yukon River Valley from central eastern to central western Alaska; northern British Columbia, eastward to the Atlantic, and south to California and Virginia.

URTICACEAE
Nettle Family

Plants perennial or annual herbs; leaves opposite or alternate, simple, with or without stinging hairs; plants monoecious or dioecious; flowers imperfect, inconspicuous, arranged in spicate cymes or in small cymose clusters; perianth present or absent, when present, of 4–5 more or less distinct segments; staminate flowers with 3–5 stamens; pistils 1; ovary superior or inferior, 1-loculed; styles 1; stigmas 1; fruit an achene.

1a. Leaves alternate, entire; plants unarmed. *Parietaria*
1b. Leaves opposite, toothed; plants armed with stinging hairs. *Urtica*

PARIETARIA L.
Plants annual, herbaceous, monoecious or rarely perfect, lacking stinging hairs, but often otherwise pubescent; leaves simple, alternate, entire, lacking stipules; flowers in axillary clusters, the perianth 4-lobed; stamens 4; ovary superior, the stigma tufted; fruit an ovoid achene.

Parietaria pensylvanica Muhl. ex Willd.

Plants erect or ascending, the stem simple or branched, 0.5–5 dm tall; leaves 1.5–9 cm long, the petiole 0.5–2 cm long, the blade 1–7 cm long and 1–2.5 cm broad, lanceolate to ovate or elliptic, strigulose to glabrate above and below, ciliate; flowers inconspicuous, in axillary clusters; perianth segments 1–2 mm long, brownish; achenes smooth and shining, 1–1.2 mm long.

Woods; at Liard Hot Springs, British Columbia; eastward to the Atlantic Coast and south to California, Mexico, and Alabama.

URTICA L.

Plants perennial or annual, herbaceous, monoecious to dioecious, armed with stinging hairs (in addition to other types of pubescence); leaves simple, opposite, toothed, stipulate; flowers in axillary spicate cymes, the perianth 4-lobed; staminate flowers with 4 stamens and a rudimentary pistil; ovary superior, the stigma tufted; fruit a lens-shaped achene.

Fernald, M. L. 1926. *Urtica gracilis* and some related North American species. Rhodora 28:191–99.

Selander, S. 1947. *Urtica gracilis* Ait. in Fennoscandia. Svensk. Bot. Tidskr. 41: 264–82.

1a. Plants rhizomatous perennials; stipules more than 5 mm long; plants of broad distribution in southern Alaska and southern Yukon. *U. dioica*
1b. Plants taprooted annuals; stipules less than 5 mm long; plants known from the western Aleutians. *U. urens*

Urtica dioica L.
Stinging Nettle

Plants perennial, rhizomatous, the stems 6–20 dm tall, armed with stinging hairs

Urtica dioica L. (× 0.25).

(and otherwise hairy to glabrous); leaves 4–15 cm long, the petioles 1–6 cm long, the blades 3–15 cm long, 1–8 cm wide, lanceolate to ovate, coarsely serrate, acuminate apically, from cordate to truncate or acute at the base; stipules 5–15 mm long; flowers inconspicuous, the perianth 1–2 mm long, greenish.

Meadows, thickets, open woods, and stream banks; in much of Alaska south of the 65th parallel (except for the Aleutians) and in southern Yukon; widely distributed in North America; Eurasia.

1a. Leaves ovate, the blades mostly less than twice as long as broad, mostly cordate to truncate at the base; plants of coastal southern Alaska from Kodiak Island eastward through the Panhandle (*U. lyallii* Wats.). *U. dioica* ssp. *gracilis* (Ait.) Selander var. *lyallii* (Wats.) C. L. Hitchc.

1b. Leaves lanceolate (rarely broader), the blades usually 2–3 times as long

as broad, acute to rounded at the base; plants of interior southwestern to east-central Alaska and southern Yukon, and less commonly in coastal southern Alaska (*U. gracilis* Ait.). *U. dioica* ssp. *gracilis* (Ait.) Selander var. *gracilis*

Urtica urens L.
Dog Nettle

Plants annual, taprooted, the stems 1–5 dm tall, armed with stinging hairs and sparsely to moderately pubescent; leaves 2–7 cm long, the petioles 0.5–3 cm long, the blades 1–4 cm long, 0.5–3 cm wide, elliptic to ovate or obovate, coarsely toothed, obtuse to broadly cuneate basally; stipules 2–3 mm long; flowers inconspicuous, in clusters much shorter than the leaves, the perianth 1–2 mm long, greenish.

Weedy species; reported from the western Aleutians (Attu), but to be expected elsewhere; widely distributed in North America and Eurasia (where probably native).

VALERIANACEAE
Valerian Family

Perennial, frequently odoriferous herbs; leaves basal or opposite, simple or pinnately divided; flowers more or less irregular, perfect or polygamous, in corymbose, paniculate, or capitate cymes; calyx vestigial or present as a ring, or 2–4-toothed; corolla tubular, usually 5-lobed; stamens 2–4, alternating with the corolla lobes; pistils 1; ovary inferior, 1–3-loculed, 3-carpelled; styles 1; stigma simple or 2–3-lobed; fruit an achene.

Meyer, F. G. 1951. *Valeriana* in North America and the West Indies (Valerianaceae). Ann. Mo. Bot. Gard. 38:377–503.

VALERIANA L.

Plants with rhizomes, the stems leafy, hollow; flowers whitish or pink, in compact cymes; calyx limb at first inrolled, developing at maturity into 5–20 plumose bristles; corolla funnelform to rotate, often saccate at the base; stamens mostly 3; achenes flattened, 1-nerved on one side, 3-nerved on the other.

1a. Corolla 2–3.5 mm long; main cauline leaves with usually 3 or more pairs of lateral lobes (or leaflets). *V. dioica*

1b. Corolla 4–8 mm long; main cauline leaves with usually 1–2 pairs of lateral lobes (or leaflets), or the leaves simple. (2)

2a. Main cauline leaves simple or with 1 pair of lateral lobes; plants (0.6) 1.5–3(6.5) dm tall; leaves at second node above base sessile or with petiole to 1.5 cm long. *V. capitata*

2b. Main cauline leaves with 1–2 pairs of lateral leaflets; plants (3)4–7(10) dm tall; leaves at second node above base with petioles 1.5–10 cm long. *V. sitchensis*

Valeriana capitata Pallas ex Link
Capitate Valerian

Plants 0.6–6.5 (usually 1.5–3) dm tall; stems glabrous or somewhat hairy below, with usually 2–3 nodes below bracts of inflorescence; leaves 3–5.5 cm long, 1.5 cm wide, those of the basal node usually simple, those of the second node usually with 1 pair of lateral lobes, the terminal lobe lanceolate to ovate or suborbicular, 0.8–3 cm wide, 1.5–6 cm long, sessile or with petioles to 1.5 cm long; inflorescence 1.5–3.5 cm broad at flowering time, expanding in fruit; corolla pink to white, 4–6(7) mm long, glabrous externally; achenes 3–4 mm long.

In tundra, heath, and less commonly in open woods; in most of Alaska (except for the southern portion of the Panhandle) and most of the Yukon; Asia (*V. bracteosa* Britt.; *V. capitata* var. *bracteosa* [Britt.] Hultén.).

Valeriana dioica L.

Plants 1–4(5) dm tall; stems glabrous or nearly so, usually with 2–4 nodes below the bracts of inflorescence; leaves 3–20 cm long, those of the basal node well developed, mostly simple, those of the second node with mostly 3–7 pairs of lateral lobes, the terminal lobe ovate to oblong or nar-

Valeriana capitata Pallas (× 0.6).

rower, 1.5–5 cm long, 0.7–2.5 cm wide, sessile or the petiole to 1.5 cm long; inflorescence 1.5–3(5) cm wide at flowering, elongating in fruit; corolla white, 2–3.5 mm long, glabrous; achenes 3–5 mm long.

Open meadows, woodlands, and rocky slopes; in southern Yukon; eastward to Newfoundland and south to Washington, Idaho, and Wyoming; Europe (*V. sylvatica* Richards., not Schmidt; *V. dioica* ssp. *sylvatica* [Richards.] F. G. Meyer). Our materials belong to var. *sylvatica* (Richards.) Wats.

Valeriana sitchensis Bong.
Sitka Valerian

Plants 4–10(12) dm tall; stems pubescent or glabrous, usually with (3)4–5 nodes below bracts of inflorescence; leaves 3–25 cm long, those of the basal node simple or with 3–5 lobes, those of the second node with (1)2 pairs of lateral leaflets (or lobes), the terminal lobe 3–10 cm long and 2–7(8) cm wide, broadly lanceolate to ovate or orbicular, the petiole 1.5–10 cm long; inflorescence 2.5–8 cm wide at flowering time, expanding in fruit; corolla white or pinkish tinged, glabrous or pubescent externally, 5–7 mm long; achenes 3–6 mm long.

Moist meadows, heathlands, or woods; in south-central to southeastern Alaska and southern Yukon; southward to Oregon, Idaho, and Montana.

VIOLACEAE
Violet Family

Perennial herbs; leaves basal or cauline and alternate, simple; flowers perfect, irregular, sometimes cleistogamous, solitary; sepals 5, distinct or nearly so; petals 5, the lowermost spurred; stamens 5; pistils 1; ovary superior, 1-loculed, 3–5-carpelled; styles 1; stigma usually lobed; fruit a loculicidal capsule.

Baker, M. S. 1935. Studies in western violets—I. Madroño 3:51–56.

Brainerd, E. 1921. Violets of North America. Bull. Vt. Agr. Expt. Sta. 224: 1–250.

VIOLA L.

Plants with rhizomes or stolons; erect aerial stems present or lacking; leaves with stipules and petioles, the blades ovate to orbicular or reniform; flowers solitary on axillary peduncles, the early ones openflowered, the later ones not opening (cleistogamous); sepals persistent; petals variously colored; stamens with short filaments, the connective extended beyond the apex of the anther as a broad appendage; capsule explosively dehiscent.

1a. Petals white, at least on the inner surface (yellowish at the base in *V. canadensis*), though often purple veined. (2)
1b. Petals purplish, violet, lavender, pink, or yellow. (3)

2a. Flowers axillary on leafy elongate aerial stems. *V. canadensis*
2b. Flowers pedunculate from the main rhizome; leafy elongate stems lacking. *V. renifolia*

3a. Petals yellow. (4)
3b. Petals purplish, violet, lavender, or pink, not yellow. (6)

4a. Aerial stems stoloniferous, rooting at the nodes; flowers from near the base of a leafy stem; plants from extreme southeastern Alaska. *V. sempervirens*
4b. Aerial stems not stoloniferous; flowers from near the tip of an erect leafy stem; plants of broad distribution. (5)

5a. Leaf blades 1–2.5(3) cm broad, reniform, the apex rounded or emarginate (rarely acute). *V. biflora*
5b. Leaf blades 3–6 cm broad, orbicular to cordate, the apex acuminate to acute or with a broad blunt cusp. *V. glabella*

6a. Plants with slender elongate stolons,

the leaves arising from the stolons.
V. *palustris*

6b. Plants without stolons, the leaves arising from erect stems or from the apex of a thickened rhizome. (7)

7a. Head of style bearded; plants usually with elongate aerial stems, usually hairy; spur slender, to half as long as the blade of the lowest petal. V. *adunca*

7b. Head of style glabrous (except in V. *langsdorffii*); plants acaulescent or short caulescent, glabrous or minutely pubescent on the upper surface; spur thick, much less than half as long as the blade of the lowest petal. (8)

8a. Leaves pubescent on the upper surface; petals glabrous. V. *selkirkii*

8b. Leaves glabrous or glabrate on the upper surface; petals variously bearded. (9)

9a. Plants acaulescent; petals all bearded. V. *nephrophylla*

9b. Plants with short but conspicuous stems, at least one internode usually apparent; only lateral petals bearded. V. *langsdorffii*

Viola adunca Smith ex Rees

Plants with short to elongate rhizomes, usually stemless early in the season, but aerial stems develop as the season advances; stipules 5–10 mm long, entire or toothed; petioles 2–10 cm long, glabrous or hairy; leaf blades cordate to ovate or reniform, crenate, 1–3 cm broad, 1–3 cm long, mostly longer than broad, hairy or glabrous; peduncles 2–10 cm long, with 2 bracts borne above the middle; flowers 5–15 mm long, the spur to half as long as the lowest petal; petals blue to violet, the lateral pair bearded; style bearded.

Moist sites in woods, thickets, and meadows; in coastal and insular southern Alaska from Kodiak Island eastward through the Panhandle, and in southern Yukon; eastward to Newfoundland and south to California, New Mexico, and the Great Lakes.

Viola biflora L.
Two-flower Violet

Plants with short to elongate rhizomes; aerial stems well developed, with usually 1–3 elongate internodes; stipules 2–5 mm long, entire; petioles 0.5–10 cm long, glabrous; leaf blades orbicular to reniform, crenate to serrate, 0.9–2.7 cm broad, 0.6–2 cm long, broader than long, hairy above and along the veins below; peduncles of open flowers 2–8 cm long, those of cleistogamous flowers often less than 1 cm long, the minute bracts above the middle; flowers 9–12 mm long, the spur 1–2 mm long; petals yellow, with purple lines beardless; style glabrous.

Moist tundra and woods; in islands of the Bering Sea, southern Alaska, and west-central Yukon; Eurasia.

Viola canadensis L.
Canada Violet

Plants with short to elongate rhizomes and often with stolons; aerial stems 10–40 cm tall, usually with 1–3 elongate internodes; stipules 8–20 mm long, entire; petioles 2–25 cm long, glabrous to puberulent; leaf blades cordate, acute to acuminate, puberulent to glabrous, 3–10 cm long, 2–10 cm broad; peduncles 2–5 cm long, arising from the upper axils, the bracts near the middle; flowers 10–15 mm long, the spur short; petals white, yellow at the base, the lower ones with purple lines, the lateral pair bearded, purplish tinged outside; style bearded.

Woods; at Liard Hot Springs, British Columbia; eastward to the Atlantic and south to Oregon, Arizona, and New Mexico.

Viola glabella Nutt. ex T. & G.
Stream Violet

Plants with thickened, short to elongate rhizomes; aerial stems 5–30 cm tall, with 1–3 elongate internodes; stipules 5–10 mm long, entire or glandular-toothed; petioles of basal leaves 2–20 cm long, those of upper leaves frequently less than 1 cm long; leaf blades reniform to ovate or cordate, abruptly acute, crenate to serrate, mostly 2–4 cm long from sinus to apex, 2–5 cm broad, puberulent to glabrate, ciliate; peduncles 1.5–4 cm long, from the upper axils, the bracts near the middle or above; flowers 7–16 mm long, the spur short and blunt; petals yellow on both sides, the lower ones with purple lines, the lateral pair bearded; style bearded.

Moist sites in woods to alpine tundra; in coastal and insular southern Alaska from Kodiak and the Alaska Peninsula eastward through the Panhandle; southward to Oregon and Montana. A related species, V. orbiculata Geyer ex Hook., is known from British Columbia and might be expected to occur in southeastern Alaska. V. orbiculata is distinguished from V. glabella by the stem which has elongate upper internodes and which produces flowers along much of its length.

Viola langsdorffii (Reg.) Fisch. ex DC.
Alaska Violet

Plants with thickened, short to elongate rhizomes; aerial stems poorly developed but usually with at least one internode readily apparent; stipules 5–17 mm long, entire or toothed; petioles 2–22 cm long, glabrous; leaf blades reniform to orbicular or cordate-ovate, rounded or obtusely cuspidate apically, crenate to serrate, (1.1)2–5 cm long from sinus to apex, (1)2–5 cm broad, glabrous or minutely puberulent above, especially along the veins, not ciliate; peduncles 3–22 cm long, the bracts commonly near the middle but occasionally much above or below the middle; flowers 16–25 mm long, the spur 2–4 mm long, thick and blunt; petals violet, the lower ones white at the base, the lateral pair bearded; style bearded or glabrous.

Moist to dry tundra, heathlands, and woods; in coastal and insular west-central, southwestern, south-central, and southeastern Alaska and in interior south-central Alaska and southern Yukon; southward to California; Asia.

Viola nephrophylla Greene

Plants with thickened, elongate rhizomes, acaulescent, the leaves arising from the rhizome apex; stipules 5–12 mm long, entire; petioles 3–25 cm long, glabrous; leaf blades ovate to cordate, rounded to obtusely cuspidate apically, crenate-dentate, 2–7 cm broad, 2–7 cm long (from sinus to apex), glabrous to minutely puberulent along the veins of the upper surface; peduncles 5–25 cm long, mostly surpassing the leaves, the bracts mostly near the middle; flowers 10–22 mm long, the spur 2–5 mm long; petals bluish violet, the lower three whitish at the base and bearded, the upper pair bearded, or less commonly glabrous; style glabrous.

Moist places, in woods, meadows, and along streams; Liard Hot Springs, British Columbia; east to Newfoundland and south to California, Utah, Colorado, and Wisconsin.

Viola langsdorffii (Reg.) Fisch. (× 0.4).

Viola palustris L.
Marsh Violet

Plants with elongate rhizomes and creeping stolons; elongate aerial stems lacking; stipules 4–9 mm long, entire or glandular-toothed; petioles 2–10 cm long, glabrous; leaf blades reniform to cordate-ovate, crenate to serrate, rounded to obtuse apically, 0.5–3.5 cm long from sinus to apex, 0.5–4 cm broad, glabrous, not ciliate; peduncles 2–15 cm long, often longer than the leaves, the bracts from well below to well above the middle; flowers 9–16 mm long, the spur 2–3 mm long; petals lavender to almost white, the lower 3 petals with purple lines, the lateral pair sparsely bearded to glabrous; style glabrous.

Arctic and alpine tundra, heathlands, and woods; in most of Alaska and Yukon; eastward to Hudson Bay and Labrador and south to California; Eurasia (*V. epipsila* Ledeb.; *V. repens* Turcz.; *V. palustris* ssp. *repens* [Turcz.] Becker).

Viola renifolia Gray
White Violet

Plants with short to elongate somewhat thickened rhizomes and occasionally with stolons as well, acaulescent or nearly so, the leaves arising from the rhizome apex or from a very short stem; stipules 3–10 mm long, entire or glandular-toothed; petioles 1.5–10 cm long, pubescent to glabrous; leaf blades reniform to orbicular or cordate-ovate, crenate to serrate, rounded or obtusely cuspidate apically, 2–5 cm broad, 1.5–3.5 cm long, glabrous to pubescent; peduncles 1.5–8 cm long, often shorter than the leaves, the bracts usually above the middle; flowers 8–10 mm long, the spur very short; petals white, the lower 3 with purple veins, all glabrous; style glabrous.

Moist woods; in or near coastal south-central and southeastern Alaska and southern Yukon; east to the Atlantic and south to Washington, Colorado, Michigan, and Pennsylvania (*V. brainerdii* Greene). Our material is referable to var. *brainerdii* (Greene) Fern.

Viola selkirkii Pursh
Selkirk Violet

Plants with short to elongate somewhat thickened rhizomes, acaulescent, the leaves arising from the apex of the rhizome; stipules 6–15 mm long, glandular-toothed; petioles 1.5–5 cm long, glabrous to pubescent; leaf blades cordate-ovate, crenate-serrate, rounded or more often acute apically, 1.2–3 cm long from sinus to apex, 1–3 cm broad, pubescent above, especially along the veins, not ciliate; peduncles 2–6 cm long, the bracts mostly above the middle; flowers 8–13 mm long, the spur 3–5 mm long; petals pale violet, all glabrous.

Moist sites in thickets, woods, and fens; in southwestern, south-central and southeastern Alaska, and southern Yukon; disjunctly eastward to the Atlantic; Eurasia.

Viola sempervirens Greene

Plants with rhizomes and with stoloniferous aerial stems, short-caulescent, with 2–3 or more elongate internodes; stipules 2–8 mm long, entire or toothed; petioles 2–10 cm long; leaf blades cordate-ovate to nearly reniform, 1–3 cm long from sinus to apex, 1–3 cm broad, about as broad as long, serrulate to crenulate, usually puberulent; peduncles of open flowers 3–10 cm long, the bracts usually above the middle; flowers 6–14 mm long, the spur short; petals yellow, the lower 3 with purple lines, the lateral pair bearded; style bearded.

Woods; in extreme southeastern Alaska; southward to California.

VISCACEAE
Mistletoe Family

Parasitic plants, attached to host plants by specialized roots; stems swollen-jointed; leaves opposite, scalelike; flowers unisexual, the plants dioecious or monoecious; perianth 2–5-lobed; stamens as many as the perianth lobes; pistils 1, the ovary inferior, 1-loculed, apparently 1-carpelled; styles and stigmas 1; fruit a 1-seeded (rarely 2-seeded) mucilaginous berry.

ARCEUTHOBIUM Bieb. Nom. Cons.

Small glabrous yellowish or greenish brown shrubs parasitic on the branches of conifers, with fragile jointed angled stems; leaves reduced to small opposite scales; flowers imperfect; staminate flowers with 2–5 greenish segments, the stamens adnate to the segments; pistillate flowers with the ovary inferior, the perianth segments usually 2; berries greenish or bluish, ovoid.

Arceuthobium campylopodum Engelm. ex Gray
Hemlock Dwarf Mistletoe

Stems 2–8 cm long, the segments mostly 1–2.5 mm thick; staminate flowers 2–3 mm broad, borne on short lateral branches; pistillate flowers 2–several, short-pedicellate; fruit maturing the second season.

Our materials are parasitic on western hemlock (*Tsuga heterophylla*), and have been assigned to *A. tsugense* (Rosend.) G. N. Jones; *A. campylopodum* f. *tsugensis* (Rosend.) Gill (see: Trans. Conn. Acad. Sci. 32:111–245. 1935). This form is common in the vicinity of Juneau and near Sitka. It probably occurs throughout southeastern Alaska, but is usually high up in the host and is seldom noticed.

Subdivision PTEROPSIDA
Class ANGIOSPERMAE
Subclass MONOCOTYLEDONEAE

1a. Plants small, free-floating aquatics with leaflike stems. LEMNACEAE (*Lemna*) .... p. 622
1b. Plants with stems and leaves, terrestrial or aquatic, but not free-floating. (2)

2a. Flowers borne on a spadix (a fleshy spike) subtended by a spathe (an enlarged, often showy bract). ARACEAE .... p. 486
2b. Flowers variously arranged, but not on a spadix; spathaceous bracts lacking, or if present, then 2–more and seldom showy. (3)

3a. Perianth well developed, at least the inner segments petaloid in color and texture. (4)
3b. Perianth lacking or reduced and inconspicuous, its parts often of bristles or scales, not petallike in texture or color. (7)

4a. Stamens more than 6; pistils numerous, or the upper flowers staminate only. ALISMACEAE (*Sagittaria*) p. 485
4b. Stamens 1–4 or 6; pistils 1; plants of various habitat. (5)

5a. Ovary superior; stamens 6 or 4. LILIACEAE .... p. 623
5b. Ovary inferior; stamens 3, 2, or 1. (6)

6a. Fertile stamens 3; flowers regular or nearly so. IRIDACEAE .... p. 606
6b. Fertile stamens 1–2; flowers irregular. ORCHIDACEAE .... p. 632

7a. Flowers sessile in the axils of chaffy or husklike scales; leaves with sheathing bases; perianth usually much reduced. (8)
7b. Flowers not both sessile and in the axils of chaffy bracts; leaves often without sheathing bases, or if sheathing basally, then perianth well developed. (9)

8a. Leaves 2-ranked on the stem; stems mostly hollow and terete; anthers versatile; flowers subtended by 2 bracts (palea and lemma). GRAMINEAE .... p. 540
8b. Leaves 3-ranked; stems mostly solid, triangular in cross section (or sometimes subterete); anthers basifixed; flowers subtended by a single bract. CYPERACEAE .... p. 487

9a. Plants terrestrial, or growing in shallow water, usually leaves and flowers both emergent. (10)
9b. Plants floating or submersed aquatics, usually not raised above the surface of the water. (14)

10a. Inflorescence a spike or spicate raceme, either dense (cattail) or elongate. (11)
10b. Inflorescence of subglobose heads or of racemes or otherwise, but not in spikes. (12)

11a. Inflorescence a double spike, the staminate above and the pistillate below; plants usually over 10 dm tall. TYPHACEAE (*Typha*) .... p. 651
11b. Inflorescence a simple spicate raceme, the sexes intermingled or the flowers perfect; plants usually under 6 dm tall. JUNCAGINACEAE (*Triglochin*) ...... p. 620

12a. Flowers imperfect, the lower heads pistillate; perianth of chaffy scales. SPARGANIACEAE (*Sparganium*) p. 650
12b. Flowers perfect, usually not in heads (although they may appear headlike in some species); perianth 6-parted, in 2 whorls. (13)

13a. Pistils 3 per flower, divergent; fruit a follicle. JUNCAGINACEAE (*Scheuchzeria*) .... p. 619
13b. Pistils 1 per flower; fruit a capsule. JUNCACEAE .... p. 608

14a. Flowers clustered in the leaf axils; pistils mostly 3–5. ZANNICHELLIACEAE (*Zannichellia*) .... p. 652
14b. Flowers in spikes, umbels, or heads; pistils various. (15)

15a. Plants strictly maritime; flowers in flattened spikes. ZOSTERACEAE .... p. 653
15b. Plants not maritime (although sometimes growing in brackish water); flowers variously arranged, but not in flattened spikes. (16)

16a. Flowers in few-rayed umbels; leaves

seldom over 1 mm wide. RUPPIACEAE (*Ruppia*) .... p. 649
16b. Flowers in spikes or heads; leaves mostly over 2 mm wide. (17)

17a. Flowers perfect, usually in elongate spikes; pistils 4 per flower. POTAMOGETONACEAE (*Potamogeton*) .... p. 643
17b. Flowers monoecious, in heads (pistillate below, staminate above); pistils 1 per flower. SPARGANIACEAE (*Sparganium*) .... p. 650

ALISMACEAE
Water-plantain Family

Perennial herbs aquatic or amphibious; leaves all basal, sheathing at the base, emersed or emergent; plants monoecious; flowers imperfect, arranged in racemes or panicles in whorls of 3, the upper ones staminate, the lower pistillate; sepals 3, reflexed; petals 3, white, stamens numerous; pistils numerous, the ovaries superior, each 1-loculed; styles simple; fruit of achenes.

SAGITTARIA L.

Perennial, scapose, aquatic or semiaquatic, rhizomatous herbs; leaves with sheathing bases and sagittate blades, or bladeless in submersed plants; flowers borne in whorls in a simple raceme, the lower ones usually pistillate and the upper staminate; sepals 3, persistent; petals white, exceeding the sepals; stamens numerous; pistils numerous; fruit a laterally flattened, winged achene.

Sagittaria cuneata Sheld.
Arrowhead

Plants 1.5–4 dm tall, arising from fibrous roots and elongate slender rhizomes; leaves all basal, long-petioled, the blades sagittate or lanceolate, 5–10 cm long or more (from petiole insertion to apex) and 1–10 cm broad; scapes 0.5–2.5 cm long; flowers in whorls of 3, the upper staminate (or perfect), the lower pistillate; sepals 5–8 mm

Sagittaria cuneata Sheld. (× 0.3).

long, ovate, acuminate apically; petals 8–10 mm long, white; stamens numerous; achenes 2–3 mm long.

Ponds and lake shores; in central eastern Alaska and southern Yukon; eastward to Quebec and south to California, New Mexico, Indiana, and New Jersey.

ARACEAE
Arum Family

Perennial herbs from thickened rhizomes or from elongate rhizomatous stolons; leaves petiolate, the bases sheathing, netted-veined; inflorescence a fleshy spike (spadix) which is subtended by a large,

often showy bract (spathe); flowers perfect or the uppermost staminate; perianth 4-lobed, or absent; stamens 4–7; pistils 1, superior, the ovary 1–2-loculed; fruit a berry.

1a. Flowers lacking a perianth; spathe white; leaf blades cordate, shorter than the petioles. *Calla*
1b. Flowers with a perianth; spathe yellow; leaf blades ovate to broadly oblong or elliptic, mostly much longer than the petioles. *Lysichitum*

CALLA L.

Stoloniferous (subrhizomatous) herbs with acrid juice; leaves long-petiolate, sheathing at the base, the blades cordate; inflorescence borne on an elongate scape, the spathe small but longer than the spadix; spadix subsessile, with flowers perfect or the uppermost staminate only; perianth lacking; stamens mostly 5–7; ovary 1-loculed, with 6–9 ovules; fruit a berry.

Calla palustris L.
Water Arum

Plants mostly 2–4.5 dm tall; leaf blades 6–15 cm long, 4–12 cm wide, broadly cordate; petioles mostly 10–35 cm long, with long membranous sheathing stipules; scapes as long as the petioles or longer; spathe white to cream, ovate-lanceolate, 4–7 cm long, 3–4(5) cm wide; spadix 2–3 cm long, becoming ovoid-ellipsoid in fruit; berries red.

Shallow water, in ponds, bogs, and streams; in southwestern to central, east-central, and south-central Alaska, and central western to northern Yukon; east to the Atlantic; circumboreal.

LYSICHITUM Schott

Herbaceous plants from thick rhizomes; roots fleshy; sap acrid and ill scented; leaves petiolate, sheathing at the base, large, ovate to broadly elliptic; spathe sheathing, the spadix long-peduncled;

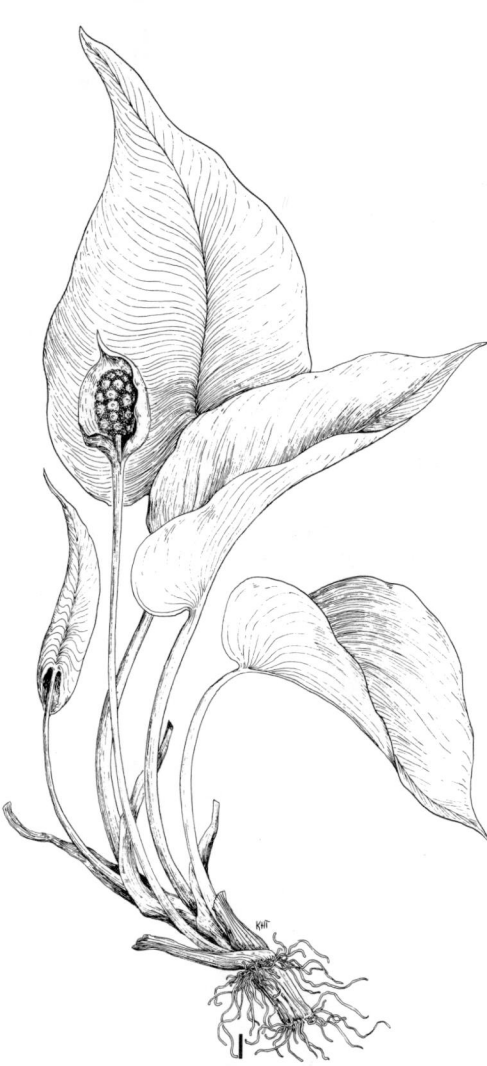

Lysichitum americanum Hultén & St. John
(× 0.24).

ovate-lanceolate, the blade 10–30 cm long;
spadix 7–12 cm long, on a thick peduncle
3–5 dm long; fruit greenish to reddish.

Moist sites in woods; in coastal and insular
southern Alaska from Kodiak Island eastward through the Panhandle; southward
to California and Montana.

CYPERACEAE
Sedge Family

Perennial grasslike plants; stems (culms)
usually solid and triangular; leaves 3-
ranked, with closed sheaths; flowers perfect or imperfect, each subtended by 1 or
sometimes 2 scales, arranged in spikes;
perianth composed of bristles or absent;
stamens 1–3; pistils 1, the ovary superior,
1-loculed, surrounded by a saclike organ
(perigynium) or naked, the ovules 1; stigmas 2–3; fruit an achene.

Calla palustris L. (× 0.3).

spadix with perfect flowers; perianth 4-
lobed; stamens 4; ovary 2-loculed, 2-ovuled;
fruit a berry.

Lysichitum americanum Hultén & St. John
Yellow Skunk-cabbage

Plants mostly 3–15 dm tall; leaf blades
ovate to broadly elliptic, 30–130 cm long,
10–70 cm wide; petioles much shorter than
the blades, estipulate; spathe yellowish,

1a. Flowers all unisexual; fruit naked or enclosed. (2)

1b. Flowers perfect, or perfect and staminate; fruit naked. (3)

2a. Ovary and fruit enclosed in a perigynium except for the apex. *Carex*

2b. Ovary and fruit without a closed perigynium, the enclosing scale split to the base. *Kobresia*

3a. Perianth bristles many, long-silky; inflorescence often appearing like a tuft of cotton. *Eriophorum*

3b. Perianth bristles few to several; inflorescence not at all like a tuft of cotton (except in *Scirpus hudsonianus*). (4)

4a. Fertile flowers or achenes solitary (rarely 2) in each spikelet, the lower scales of the spikelet empty. *Rhynchospora*

4b. Fertile flowers or achenes few to several in each spikelet (rarely solitary), the lower scales usually fertile. (5)

5a. Involucral leaves lacking; style base persistent as a tubercle at the achene apex; spikes solitary. *Eleocharis*

5b. Involucral leaves present (1 or more); style base deciduous, the tubercle lacking; spikes mostly more than 1, though solitary in some. *Scirpus*

CAREX L.

Plants monoecious or seldom dioecious perennials, rhizomatous and sod-forming or caespitose; stems (culms) triangular to terete, solid; leaves 3-ranked, the sheaths closed, the lower leaves bladeless or nearly so (aphyllopodic) or with well-developed blades (phyllopodic); flowers imperfect, borne in 1 or more spikes; spikes subtended by a bract longer, equaling, or shorter than the inflorescence or the bract lacking, sessile or pedunculate or sometimes borne axillary near the stem base, pistillate or staminate or with staminate above and pistillate below (androgynous) or with pistillate above and staminate below (gynaecandrous); flowers solitary in the axils of scales, the pistillate with 2–3 stigmas and the ovary surrounded by a saclike organ (perigynium), the staminate with 3 stamens; perianth lacking; achenes 3-angled or lens shaped.

Krechetovich, V. I. 1935. *Carex* L. pp. 111–464. *In*: Komarov, V. L. Flora of the U.S.S.R. Bot. Inst. Akad. Nauk U.S.S.R. 3:1–512.

Mackenzie, K. K. 1931–35. *Carex* L. Fl. N. Am. 18:9–478.

1a. Spikes 1 per culm. KEY I. .... p. 488

1b. Spikes 2 or more per culm (sometimes densely clustered and appearing as 1). (2)

2a. Lateral spikes sessile, monoecious (pistillate in some) or plants dioecious (see *C. macrocephala*). KEY II. .... p. 490

2b. Lateral spikes borne on short to elongate peduncles (subsessile or sessile in some, see Key VI), commonly pistillate though sometimes monoecious, the terminal ones commonly staminate or sometimes monoecious. (3)

3a. Stigmas 2. KEY III. .... p. 492

3b. Stigmas 3. (4)

4a. Perigynia pubescent to long-hairy. KEY IV. .... p. 493

4b. Perigynia glabrous, or scabrous only along the margins or on the beak. (5)

5a. Perigynia distinctly beaked, bearing well-developed teeth at the mouth or the leaf sheaths septate-nodulose or both. KEY V. .... p. 494

5b. Perigyna beaked or beakless, the mouth entire or minutely toothed or with soft blunt teeth. KEY VI. p. 495

KEY I. SPIKES 1 PER CULM.

1a. Stigmas 2; achenes lens shaped. (2)

1b. Stigmas 3; achenes 3-angled. (7)

2a. Spikes entirely staminate or pistillate, rarely androgynous; plants sod-forming, with well-developed rhizomes; plants broadly distributed south of the 68th parallel. *C. dioica*

2b. Spikes gynaecandrous or androgynous; plants clump-forming; rhizomes lacking or short. (3)

3a. Spikes gynaecandrous, the staminate flowers at the base; culms usually curved or drooping; plants of the Arctic Coast. *C. ursina*

3b. Spikes androgynous, the pistillate flowers at the base; culms straight or nearly so. (4)

4a. Perygnia spreading at right angles to the rachis or nearly so, sessile, the margins entire. *C. capitata*

4b. Perigynia erect or ascending to spreading-ascending, stipitate or substipitate, the margins entire or serrulate. (5)

5a. Perigynia distinctly stipitate; pistillate scales early deciduous; plants of southern and western Alaska. *C. pyrenaica*

5b. Perigynia substipitate; pistillate scales persistent; plants of western, northern, and eastern Alaska and northern Yukon. (6)

6a. Beak of perigynium serrulate; plants of northern and eastern Alaska and most of the Yukon. *C. nardina*

6b. Beak of perigynium smooth; plants of western Alaska. *C. jacobi-peteri*

7a. Plants dioecious; perigynia puberulent. *C. scirpoidea*

7b. Plants monoecious (rarely dioecious in *C. anthoxanthea*); perigynia glabrous (puberulent in some *C. filifolia*). (8)

8a. Perigynia strongly reflexed, narrowly lanceolate, pale yellowish green, sessile or subsessile; scales early deciduous. (9)

8b. Perigynia not reflexed, or if so then strongly stipitate and ovate to ovate-lanceolate and not pale yellowish

green; scales persistent or early deciduous. (10)

9a. Perigynia 3–5 mm long; rachilla axis of the flower projecting out of the perigynium, the stigmas appearing lateral. *C. microglochin*

9b. Perigynia 6–7 mm long; rachilla lacking, the stigmas terminal on the exserted style. *C. pauciflora*

10a. Pistillate scales early deciduous; perigynia stipitate, the lower ones often reflexed. (11)

10b. Pistillate scales persistent; perigynia sessile to subsessile or substipitate, not reflexed. (12)

11a. Plants rhizomatous; leaves 1.5–2 mm broad, flat or channeled basally; staminate flowers occupying one-third to one-half of the spike. *C. nigricans*

11b. Plants clump-forming; leaves 1.5 mm broad or less, channeled throughout; staminate flowers occupying less than one-third of the spike. *C. pyrenaica*

12a. Perigynia puberulent; pistillate scales truncate, with broad hyaline margins clasping around the subcylindrical spike; leaves filiform; plants of the southeastern quarter of Alaska and southern Yukon. *C. filifolia*

12b. Perigynia glabrous; pistillate scales usually not as above; leaves filiform or flat; distribution various. (13)

13a. Perigynia tapering to a long slender beak. (14)

13b. Perigynia beakless, or abruptly contracted into a short to elongate distinct beak. (15)

14a. Plants with well-developed rhizomes; leaves flat, 1–2 mm broad; perigynia glabrous. *C. anthoxanthea*

14b. Plants clump-forming; leaves channeled, less than 1 mm broad; perigynia obscurely serrulate along the margin. *C. circinata*

15a. Perigynia beakless, many-veined, green, much surpassing the greenish to pale brownish scales. *C. leptalea*

15b. Perigynia beaked, few- to many-veined, whitish to chestnut brown, obscured by the scales or only slightly surpassing them. (16)

16a. Pistillate scales acute to obtuse; perigynia dull, whitish to yellowish brown, obscurely veined, the beak 0.1–0.3 mm long, the mouth neither flaring nor hyaline. *C. rupestris*

16b. Pistillate scales obtuse or acutish; perigynia shining, chestnut brown, distinctly several-veined, the beak 0.5–0.8 mm long with a flaring hyaline mouth. *C. obtusata*

KEY II. LATERAL SPIKES SESSILE, MONOECIOUS, RARELY OTHERWISE.

1a. Plants dioecious (rarely monoecious); stigmas 3; pistillate spikes aggregated into heads 3–6 cm long and 2–4 cm thick; staminate heads to 4 cm long and 1 cm broad; plants of coastal southern Alaska. *C. macrocephala*

1b. Plants monoecious; stigmas 2; spikes various, but if aggregated into heads then less than 3 cm long and 2 cm thick (except in *C. stipata*). (2)

2a. Spikes androgynous, the staminate flowers at the apex, the pistillate below. (3)

2b. Spikes gynaecandrous, the pistillate flowers at the apex, the staminate below (the lateral spikes pistillate in some). (9)

3a. Inflorescence 2–5 cm long; plants clump-forming, often over 4.5 dm tall; perigynia tapering to slender flattened distinctly serrulate beaks. (4)

3b. Inflorescence 0.7–2.5 cm long (longer in some *C. praegracilis*); plants with elongate rhizomes, seldom over 4 dm tall; perigynia with shorter, only slightly flattened, inconspicuously serrulate or entire beaks. (5)

4a. Perigynia 4–6 mm long; leaves 4–8 mm broad; plants of coastal southern Alaska. *C. stipata*

4b. Perigynia 2.5–3 mm long; leaves 1.5–2.5 mm broad; plants of interior Alaska and Yukon and less commonly in coastal southern Alaska. *C. diandra*

5a. Rhizomes with internodes (3)4–7 cm long, giving rise to 1–2 upright leafy culms per node; plants of bogs and lake shores in interior, northern, and western Alaska. *C. chordorrhiza*

5b. Rhizomes with internodes less than 2 cm long, giving rise to 1–more leafy culms per node; plants of moist maritime sites (seldom interior) or of dry interior sites. (6)

6a. Perigynia 1–4(5) per spike, conspicuously several-nerved, abruptly short-beaked; spikes separated by 3–12 mm (at least the lower). *C. disperma*

6b. Perigynia usually more than 4 per spike; spikes closely aggregated into a compact head or if not, then perigynia not as above. (7)

7a. Culms sharply angled, the edges minutely toothed; inflorescence slender, the lowermost spike often 4–8 mm below the next higher. *C. praegracilis*

7b. Culms with rounded angles, these smooth; inflorescence compact, the lowermost spike closely subtending the next higher. (8)

8a. Perigynia 2.8–3.7 mm long, distinctly stipitate; spikes aggregated into an ovoid head; plants of moist maritime or, less commonly, interior sites. *C. maritima*

8b. Perigynia 2.5–3.3 mm long, substipitate; spikes aggregated into an oblong or narrowly lance-ovoid head; plants of dry hillsides and grasslands. *C. stenophylla*

9a. Margins of perigynia winged or sharp-edged, the beaks attenuate, usually minutely serrulate. (10)

9b. Margins of perigynia rounded or merely acute, not winged; beaks short, abruptly acuminate or sometimes attenuate, glabrous or minutely serrulate. (19)

10a. Bracts much surpassing the inflorescence. (11)

10b. Bracts shorter than or rarely subequal to the inflorescence (in some with 1 bract longer). (12)

11a. Bracts 2–4 mm broad basally, commonly all surpassing the inflorescence; plants of eastern Alaska and southern Yukon. *C. sychnocephala*

11b. Bracts seldom over 1.5 mm broad basally, commonly with only the lower ones much exceeding the inflorescence; plants mostly of coastal southern and southeastern Alaska. *C. athrostachya*

12a. Inflorescence usually 2–5 times longer than broad, at least the lower spikes separated by 3–12 mm or more (see also *C. phaeocephala*). (13)

12b. Inflorescence usually 1–2 times longer than broad, the spikes separated by less than 3 mm or only the lowermost separated by more than 3 mm. (15)

13a. Beak of perigynium merely sharp-edged, not winged; lateral spikes pistillate; plants of southern Yukon. *C. deweyana*

13b. Beak of perigynium winged; lateral spikes gynaecandrous; plants of various distribution. (14)

14a. Perigynia 3.5–5.2 mm long; plants widely distributed in the eastern third of Alaska and southern Yukon. *C. praticola*

14b. Perigynia 5–7 mm long; plants of the Yukon River Valley in southwestern Yukon. *C. petasata*

15a. Beak of perigynium entire at the apex or nearly so. (16)

15b. Beak of perigynium serrulate to the apex. (18)

16a. Perigynia concealed by the pistillate scales or nearly so; lowermost spikes often separated by more than 3 mm; plants 1–2 dm tall, of southeastern Alaska. *C. phaeocephala*

16b. Perigynia not concealed by the pistillate scales (at least when mature); lowermost spikes separated by less than 3 mm; plants commonly over 2 dm tall, of southern Alaska and Yukon. (17)

17a. Heads of spikes ovate to orbicular in outline, often more than 10 mm broad; perigynia inconspicuously hyaline at tip. *C. macloviana*

17b. Heads of spikes lanceolate to oblong in outline, 10 mm broad or less; perigynia not hyaline at the tip. *C. preslii*

18a. Perigynia 3–3.5 mm long, 1.5–2 mm broad, ovate. *C. bebbii*

18b. Perigynia 3.5–4 mm long, 0.8–1.2 mm broad, narrowly lanceolate. *C. crawfordii*

19a. Perigynia tapering to slender more or less flattened elongate beaks; plants clump-forming, the rhizomes seldom well developed. (20)

19b. Perigynia beakless, or if abruptly short-beaked, then the beak conic. (26)

20a. Pistillate scales brown to chestnut brown. (21)

20b. Pistillate scales pale, hyaline or straw-colored or pale brownish near the center. (22)

21a. Terminal spikelets only gynaecandrous, the others lacking empty staminate scales at the base. *C. glareosa*

21b. Terminal and lateral spikelets gynaecandrous, commonly all with empty staminate scales at the base. *C. lachenalii*

22a. Beaks of perigynia with a few weak serrulations, the body broadest near the middle; perigynia ascending; plants of near coastal southern and southeastern Alaska. *C. laeviculmis*

22b. Beaks of perigynia distinctly serrulate, the body broadest below the middle; perigynia ascending or spreading; distribution various. (23)

23a. Beaks of perigynia sharply bidentate; perigynia spreading or spreading-ascending. (24)

23b. Beaks of perigynia not or only obscurely bidentate; perigynia ascending or spreading. (25)

24a. Pistillate scales yellowish brown; perigynia 2.5–3.7 mm long; plants known only from Unalaska. *C. echinata*

24b. Pistillate scales greenish or brownish; perigynia (3)3.5–4.5 mm long; plants of broad distribution. *C. phyllomanica*

25a. Perigynia ascending, 1–1.5 mm broad, 2–3 mm long; spikes 5–10 or more; plants of southeastern quarter of Alaska and southern Yukon. *C. arcta*

25b. Perigynia spreading, (1.2)1.5–2 mm broad, 2.2–3.3 mm long; spikes usually 2–4(6); plants of southern Yukon and northern British Columbia. *C. interior*

26a. Spikes 2–3(4), closely aggregated, forming ovoid or oblique heads, or the lowermost spikes separated by 2–4 mm. (27)

26b. Spikes (2)3–6 or more, at least the lowermost separated by 3–15 mm or more. (29)

27a. Plants 0.2–1 dm tall, densely tufted; culms seldom much surpassing the leaves; spikes 1 or rarely with a small second one at its base. *C. ursina*

27b. Plants 1–4 dm tall or more, densely or loosely tufted; culms commonly surpassing the leaves; spikes 2–3(4). (28)

28a. Plants rhizomatous, forming small tufts in loose clumps. *C. tenuiflora*

28b. Plants not rhizomatous, forming large to small clumps. *C. heleonastes*

29a. Spikes 2–4(5), each with (1)2–10 perigynia; plants caespitose. *C. loliacea*

29b. Spikes 3–6 or more, each usually with 10 or more perigynia; plants clump-forming, seldom rhizomatous. (30)

30a. Margins of perigynia smooth near the beak. (31)

30b. Margins of perigynia minutely serrulate, at least below the beak. (34)

31a. Perigynia membranous or papery, whitish to straw-colored; scales much shorter than the perigynia. (32)

31b. Perigynia rigid, leathery, yellowish to brownish; scales equaling or shorter than the perigynia. (33)

32a. Perigynia 1.5–1.8(2) mm long; scales brownish, obtuse or nearly so. *C. bonanzensis*

32b. Perigynia (1.8)2–3 mm long; scales yellowish green to whitish, acute. *C. canescens*

33a. Terminal spike club shaped, the lower two-thirds clothed with staminate scales; perigynia 3–3.5 mm long; leaves 2–3 mm broad. *C. mackenziei*

33b. Terminal spike oblong to ovoid, with few staminate scales basally; perigynia 2.5–3 mm long; leaves 1.5–2 mm broad. *C. heleonastes*

34a. Perigynia loosely spreading to ascending, with a distinct short minutely bidentate beak. *C. brunnescens*

34b. Perigynia compactly appressed-ascending, the very short beaks not bidentate. (35)

35a. Perigynia 1.5–1.8(2) mm long; scales brownish, obtuse or nearly so. *C. bonanzensis*

35b. Perigynia 2–3 mm long; scales straw-colored to yellowish brown, acute. *C. canescens*

KEY III. LATERAL SPIKES PEDUNCULATE; STIGMAS 2.

1a. Bracts subtending lowermost spike of inflorescence long-sheathing; perigynia beakless or nearly so. *C. bicolor*

1b. Bracts subtending lowermost spike of inflorescence sheathless or very short-sheathing; perigynia with at least a very short beak present. (2)

2a. Perigynia with beaks (0.4)0.5 mm long or longer, dark brown to blackish (rarely greenish). *C. saxatilis*

2b. Perigynia with beaks 0.4 mm long or less, green to whitish or rarely mottled brownish. (3)

3a. Bract of lowermost spike shorter than the inflorescence. (4)

3b. Bract of lowermost spike surpassing the inflorescence or subequal to it. (5)

4a. Perigynia ascending to spreading-ascending, overlapping but not densely aggregated; plants common in much of our region. *C. bigelowii*

4b. Perigynia spreading, densely compacted in the spike; plants reported for west-central and southwestern Yukon. *C. scopulorum*

5a. Spikes long-pedunculate, at least the lowermost pendulous; plants of coastal southern and southwestern Alaska. *C. lyngbyei*

5b. Spikes short to moderately long-pedunculate, not, or seldom, any of them pendulous (except in *C. sitchensis*); plants of various distribution. (6)

6a. Uppermost spike gynaecandrous, the lateral spikes congested. (7)

6b. Uppermost spike staminate (rarely with some pistillate flowers in *C. lenticularis*), the lateral spikes not congested. (8)

7a. Perigynia distinctly stipitate; plants of coastal southern Alaska. *C. enanderi*

7b. Perigynia sessile or substipitate; plants of interior Alaska and Yukon. *C. eleusinoides*

8a. Perigynia conspicuously 5–7-nerved, commonly surpassing the scales. *C. lenticularis*

8b. Perigynia nerveless or nearly so, commonly surpassed by the scales. (9)

9a. Culms mostly 0.5–2.5(3.5) dm tall; lowest bract commonly less than 10 cm long. *C. subspathacea*

9b. Culms mostly 2.5–6 dm tall or more; lowest bract commonly more than 10 cm long. (10)

10a. Plants tufted, with rhizomes lacking or very short; lowermost peduncle seldom less than 3 cm long, the spike tending to nod. *C. sitchensis*

10b. Plants not tufted, with well-developed horizontal rhizomes; peduncles all less than 3 cm long, the spikes erect. *C. aquatilis*

KEY IV. LATERAL SPIKES PEDUNCULATE; STIGMAS 3; PERIGYNIA HAIRY.

1a. Bracts subtending inflorescence tubular-sheathing. (2)

1b. Bracts subtending inflorescence sheathless, or at most auriculate or open-sheathing, or if sheathing the blade well developed. (3)

2a. Lowermost bract 0.4–1 cm long; perigynia 2.5–3.5 mm long; peduncles 0.3–0.8 cm long. *C. concinna*

2b. Lowermost bract 1.5–10 cm long; perigynia 3.5–6.3 mm long; peduncles 1.2–4.5 cm long. *C. petricosa*

3a. Pistillate spikes cylindrical, many-flowered; plants 2.5 dm tall or more. (4)

3b. Pistillate spikes ovoid to orbicular, 3–12-flowered; plants commonly less than 2.5 dm tall. (6)

4a. Perigynia pubescent only marginally and along the beak, the beak indistinctly bidentate; lowermost bract seldom more than 5 cm long. *C. parryana*

4b. Perigynia pubescent throughout, distinctly bidentate; lowermost bract commonly over 5 cm long. (5)

5a. Leaf blades strongly involute; perigynia 1.4–1.7 mm broad, ovate. *C.
lasiocarpa*

5b. Leaf blades flat; perigynia 1.7–2 mm
broad, elliptic to obovate. *C. lanuginosa*

6a. Plants lacking short fertile culms near
the base, all or nearly all inflorescences elevated above the leaves. *C.
peckii*

6b. Plants with both short and elongate
fertile culms, at least some culms
very short and much surpassed by
the foliage. (7)

7a. Beak of perigynium 1–1.5 mm long,
one-half to two-thirds as long as the
body. *C. rossii*

7b. Beak of perigynium 0.4–0.5 mm long,
less than one-half as long as the body.
C. deflexa

KEY V. LATERAL SPIKES PEDUNCU-
LATE; STIGMAS 3; PERIGYNIA GLA-
BROUS, THE BEAK TOOTHED, OR
THE LEAF SHEATHS SEPTATE-NODU-
LOSE, OR BOTH.

1a. Teeth of beak of perigynia 1–2 mm
long; perigynia lanceolate, gradually
tapering to the beak; scales ciliate,
awned apically, much shorter than the
perigynia; leaf sheaths softly long-
hairy. *C. atherodes*

1b. Teeth of perigynia 0.3–0.8 mm long;
perigynia various; scales awned or
not, various in length; leaf sheaths
not long-hairy. (2)

2a. Pistillate scales acuminate to attenuate, subequal to the ascending perigynia; lower sheaths purplish. *C.
spectabilis*

2b. Pistillate scales obtuse to acute, or
less commonly acuminate to attenuate, various in length, but if subequal
to the perigynia, then the perigynia
spreading; lower sheaths not or seldom purplish. (3)

3a. Beak of perigynium 2–3 mm long,

about as long as the body; at least the
lowermost perigynia reflexed. *C. flava*

3b. Beak of perigynium 0.2–1.5(1.8) mm
long, much shorter than body; perigynia spreading to spreading-ascending. (4)

4a. Leaf sheaths not septate-nodulose;
spikes 1–2 times longer than broad.
(5)

4b. Leaf sheaths with white transverse
septae; spikes commonly more than
twice longer than broad. (6)

5a. Lowermost bract longer than the inflorescence; perigynia 2–3 mm long,
yellowish green; pistillate scales
straw-colored to pale brownish;
spikes less than 1 cm long. *C. oederi*

5b. Lowermost bract shorter than the inflorescence; perigynia 4–5 mm long,
straw-colored to brownish; pistillate
scales brown; spikes commonly more
than 1 cm long; plants local near Carcross, Yukon. *C. sabulosa*

6a. Lowermost bract commonly more
than 15 cm long; pistillate scales and
perigynia straw-colored to pale
brownish or brown; staminate spikes
commonly 2–4 (rarely to 6). (7)

6b. Lowermost bract commonly less than
15 cm long; pistillate scales dark
brown to purplish black; staminate
spikes 1–2(3). (8)

7a. Perigynia spreading to spreading-ascending, diverging about 90 degrees
from the spike axis, abruptly contracted into a beak 0.7–2 mm long.
C. rhynchophysa

7b. Perigynia ascending to spreading-ascending, commonly diverging less
than 90 degrees from the spike axis,
tapering to a beak 1–2 mm long. *C.
rostrata*

8a. Leaf blades involute, 1–3 mm broad;
perigynia merely dark brown apically.
C. rotundata

8b. Leaf blades flat, 3–4.5 mm broad
or more; perigynia purplish black or
less commonly straw-colored. (9)

9a. Pistillate spikes with filiform peduncles 1–3 cm long or more, commonly nodding. *C. saxatilis*

9b. Pistillate spikes subsessile or with peduncles to 1 cm long or more, stiff, coarse, erect. *C. membranacea*

KEY VI. LATERAL SPIKES PEDUNCULATE; STIGMAS 3; PERIGYNIA GLABROUS, THE BEAK TRUNCATE, OBLIQUE, OR VARIOUSLY MINUTELY BIDENTATE, BUT NOT DISTINCTLY TOOTHED.

1a. Pistillate scales awned, the awn short-ciliate. (2)

1b. Pistillate scales lacking ciliate awns, the apex rounded, obtuse, acute, or acuminate, or with a short glabrous awn (the awn ciliate in some *C. buxbaumii*). (4)

2a. Terminal spike staminate; awns of pistillate scales commonly 3–6 mm long or more. *C. macrochaeta*

2b. Terminal spike monoecious; awns of pistillate scales 1–3(4) mm long. (3)

3a. Lowermost spikes with peduncles 1–6 mm long; pistillate scales tapering to the awn. *C. buxbaumii*

3b. Lowermost spikes with peduncles 6–15 mm long or more; pistillate scales abruptly contracted to the awn. *C. gmelinii*

4a. Lowermost bract of inflorescence lacking a blade, or the blade narrowly subulate and less than 10 mm long (to 20 mm long in some *C. supina*). (5)

4b. Lowermost bract of inflorescence with more or less well-developed blade, the blades variously shaped but commonly over 10 mm long (shorter in some *C. media*). (8)

5a. Lateral spikes pendulous on capillary peduncles; perigynia 3–4 mm long; leaves 2 mm broad or more. *C. pluriflora*

5b. Lateral spikes erect or ascending on short thick or capillary peduncles; perigynia 1.5–3 mm long; leaves 1 mm broad or less. (6)

6a. Perigynia 2.5–3.5 mm long, yellowish brown, glabrous, smooth or serrulate along the beak. *C. supina*

6b. Perigynia 1.5–2.5 mm long, greenish to brown to chestnut brown, smooth. (7)

7a. Lowermost bract with a closed sheath 4–6 mm long, bladeless or nearly so; pistillate scales hyaline, straw-colored. *C. eburnea*

7b. Lowermost bract with a closed sheath 1–2 mm long or the sheath open, the blade 2–10 mm long; pistillate scales chestnut brown. *C. glacialis*

8a. Terminal spike gynaecandrous, the pistillate flowers above the staminate. (9)

8b. Terminal spike staminate, all flowers staminate or rarely some basal flowers pistillate. (21)

9a. Inflorescence more or less nodding; lower spikes borne on peduncles 1–3 cm long or more (sometimes shorter in *C. atrata* and *C. atratiformis*). (10)

9b. Inflorescence erect or nearly so; lower spikes subsessile or borne on peduncles less than 1 cm long. (15)

10a. Spikes (4)6–8, cylindrical, all or nearly all gynaecandrous, 2–4 cm long; perigynia 4–5 mm long; lowermost bract surpassing the inflorescence; plants of coastal southern Alaska. *C. mertensii*

10b. Spikes 2–5(6), cylindrical to lance-ovoid, only the terminal ones gynaecandrous, 0.8–2(2.3) cm long; perigynia 2–3 mm long or if 4–5 mm long then the lowermost bract shorter than the inflorescence. (11)

11a. Lowermost bract with a closed cylindrical sheath; perigynia serrulate to scabrous marginally. (12)

11b. Lowermost bract not sheathing, merely auriculate; perigynia glabrous or serrulate marginally. (13)

12a. Pistillate scales purplish black; perigynia 3.5–5 mm long. *C. misandra*

12b. Pistillate scales straw-colored to brownish; perigynia 2.5–3 mm long. *C. capillaris*

13a. Perigynia brownish throughout, not conspicuously contrasting with the dark scales; spikes often more than twice longer than broad. *C. atratiformis*

13b. Perigynia straw-colored to greenish or brownish except apically, contrasting strongly with the dark scales; spikes various. (14)

14a. Beak of perigynium 0.3–0.5 mm long; perigynia 3–4 mm long. *C. atrata*

14b. Beak of perigynium to 0.1 mm long; perigynia 2.2–3 mm long. *C. magellanica*

15a. Pistillate scales acuminate or shortly awned apically (especially the lowermost ones); perigynia several-veined. *C. buxbaumii*

15b. Pistillate scales rounded, obtuse, or acute, not at all awned; perigynia veinless or several-veined. (16)

16a. Perigynia beakless, merely with a dark notch at the summit. *C. adelostoma*

16b. Perigynia distinctly beaked. (17)

17a. Perigynia ciliate along the upper third and along the beak; spikes cylindrical, 3 times longer than broad or more; inflorescence 3–6 cm long. *C. parryana*

17b. Perigynia glabrous or ciliate only along the beak; spikes ovoid or short-cylindrical, 1–2 times longer than broad; inflorescence less than 3 cm long. (18)

18a. Perigynia rather gradually tapering to a sparingly ciliate beak; pistillate scales much shorter than the perigynia. *C. media*

18b. Perigynia abruptly contracted to a short glabrous beak; pistillate scales subequal to the perigynia (shorter in some *C. holostoma*). (19)

19a. Perigynia conspicuously several-veined, straw-colored or purplish-spotted; plants of coastal southern Alaska. *C. enanderi*

19b. Perigynia not veined or only inconspicuously so, brownish or purplish black; plants of interior Alaska. (20)

20a. Perigynia 1.5–2.5 mm long; pistillate scales with a very narrow hyaline margin less than 0.2 mm broad. *C. holostoma*

20b. Perigynia 2.5–3 mm long; pistillate scales with a broad hyaline margin (0.1)0.2–0.5 mm broad. *C. albo-nigra*

21a. Lowermost bract sheathless or the sheath less than 3 mm long or the sheath open. (22)

21b. Lowermost bract with a closed cylindrical sheath 3 mm long or more. (31)

22a. Lateral spikes subsessile or with peduncles less than 0.6 cm long; perigynia minutely granular. *C. holostoma*

22b. Lateral spikes on peduncles more than 0.6 cm long, at least some; perigynia smooth, not granular. (23)

23a. Style thick, rigid, long-exserted from the beak; leaves 1–2(3) mm broad. *C. stylosa*

23b. Style short, thick or slender, included within the beak or only slightly exserted; leaves often over 2 mm broad (except in *C. limosa*). (24)

24a. Leaves 1–2 mm broad, channeled; pistillate scales brown. *C. limosa*

24b. Leaves 2–6 mm broad; pistillate scales purplish black to brown (brown to chestnut in *C. magellanica*). (25)

25a. Perigynia ciliate marginally. *C. atrofusca*

25b. Perigynia glabrous marginally. (26)

26a. Lateral spikes erect or ascending on rather stiff peduncles. *C. nesophila*

26b. Lateral spikes pendulous to widely spreading or rarely some ascending, borne on filiform peduncles. (27)

27a. Lowermost bract leaflike, longer than the inflorescence; pistillate scales brown to chestnut. *C. magellanica*

27b. Lowermost bract various, but generally shorter than the inflorescence; pistillate scales purplish brown to black. (28)

28a. Perigynia beakless; spikes loosely 5–20 (25)-flowered. (29)

28b. Perigynia with beaks 0.2–0.5 mm long; spikes densely (15)20–40-flowered. (30)

29a. Perigynia commonly 5–10 per spike; leaves flat to channeled or involute. *C. rariflora*

29b. Perigynia commonly 10–20 per spike; leaves flat with revolute margins. *C. pluriflora*

30a. Pistillate scales acute to acuminate, the midvein not or seldom reaching the apex; plants common in much of Alaska and Yukon. *C. podocarpa*

30b. Pistillate scales acuminate, the midvein reaching the apex and often produced as a short awn; plants uncommon in coastal and insular southern Alaska. *C. spectabilis*

31a. Leaves filiform, less than 1 mm broad; inflorescences less than 3(4.5) cm long; plants commonly 1.2 dm tall or less. *C. williamsii*

31b. Leaves 1–4 mm broad, flat, channeled, or involute; inflorescences commonly over 3 cm long; plants mostly more than 15 cm tall. (32)

32a. Pistillate scales straw-colored, tinged with brown, the margins hyaline, broad; spikes 1.5–2.5 mm thick; perigynia straw-colored to yellowish brown, gradually tapering to a hyaline beak. *C. capillaris*

32b. Pistillate scales purplish black to brown, the margins various but seldom broadly hyaline; spikes more than 2.5 mm thick; perigynia various but not as above. (33)

33a. Lateral spikes erect or nearly so. (34)

33b. Lateral spikes ascending to spreading or pendulous. (35)

34a. Perigynia beakless or nearly so, minutely granular; spikes compact. *C. livida*

34b. Perigynia distinctly beaked, smooth; spikes loosely flowered. *C. vaginata*

35a. Beak of perigynium 0.7–1 mm long, glabrous, not hyaline apically. *C. saxatilis*

35b. Beak of perigynium less than 0.7 mm long, or if more, then the apex hyaline and the margin ciliate. (36)

36a. Scales and perigynia purplish black or brown; perigynia short-ciliate marginally along the beak, the apex hyaline. (37)

36b. Scales and perigynia of contrasting colors, not as above; perigynia glabrous, the apex not hyaline. (38)

37a. Pistillate scales brownish purple, hyaline margined, erose, abruptly acuminate. *C. petricosa*

37b. Pistillate scales purplish black, hyaline only at the apex, entire, acute to obtuse. *C. atrofusca*

38a. Sheath of lowermost bract cylindrical, 3–20 mm long or more; pistillate scales with broad green or yellowish midribs and broad hyaline or brownish margins. *C. laxa*

38b. Sheath of lowermost bract not or seldom sheathing for more than 8 mm; pistillate scales purplish black, or if brown and with broad midribs, then the margins not hyaline. (39)

39a. Pistillate spikes mostly more than 10-flowered; leaves flat with revolute margins. *C. pluriflora*

39b. Pistillate spikes 10-flowered or less;

leaves flat to channeled or involute. *C. rariflora*

Carex adelostoma Krecz.

Key VI type; rhizomatous, 1.5–3.5 dm tall; leaf sheaths purplish red, the blades on the lower sheaths obsolete, becoming elongate upwards, 2–3 mm broad, flat; lowermost bract 0.3–0.5 cm long, the blade subulate or lacking; inflorescence 3–5 cm long, erect to nodding; spikes 3–4, the terminal gynaecandrous (sometimes staminate), the lateral pistillate, sessile, the lowermost separated from the upper ones; pistillate scales about as long as the perigynia, ovate, acuminate, brown with a green median stripe; perigynia 2.5–3 mm long, about 1.5 mm broad, ellipsoid, rounded basally and apically, obscurely veined, grayish green, glabrous, the beak obsolete, stigmas 3.

Moist sites; in southeastern quarter of Alaska; disjunctly eastward to Labrador; circumboreal.

Carex albo-nigra Mack.

Key VI type; caespitose, 1–3 dm tall; leaf sheaths greenish to straw-colored, the blades on lower sheaths well developed, 0.2–0.5 cm broad, flat; lowermost bract 2–4 cm long, the blade leaflike; inflorescence 2–5 cm long, 0.8–1.5 cm broad, erect or nodding; spikes usually 3, the terminal gynaecandrous, the lateral ones pistillate, congested and subsessile above, the lowermost separate and short-pedunculate; pistillate scales subequal to the perigynia, broadly ovate, obtuse to acutish, purplish black with white-hyaline margins and apex; perigynia 2.5–4 mm long, 1.4–2.3 mm broad, obovate to elliptic in outline, substipitate, abruptly short beaked, entire to ciliate; stigmas 3.

Dry to moist mountain slopes and summits; in scattered localities in eastern Alaska and southern Yukon (where evidently rare); southward to Colorado, Arizona, and California.

Carex anthoxanthea Presl

Key I type; rhizomatous, 0.4–3.5(4) dm tall; leaf sheaths greenish to whitish, the blades of lower sheaths short or lacking, elongating upwards, 1–2 mm broad, flat or somewhat channeled; lowermost bract 0.5–1.5 cm long, the blade (when present) linear-subulate; spike solitary, 0.7–3 cm long, 0.2–0.3 cm broad, androgynous (sometimes unisexual); pistillate scales subequal to or longer than the perigynia, lanceolate to elliptic, dark brown with green to straw-colored centers; perigynia 3.8–4.2 mm long, 1.3–1.6 mm broad, narrowly elliptic in outline, tapering to both ends, veined dorsally and ventrally, yellowish green to straw-colored, substipitate, glabrous; stigmas 3.

Bogs and wet meadows; in coastal and insular southern and southwestern Alaska, and less commonly in south-central Alaska and southern Yukon; south to British Columbia; Siberia.

Carex aquatilis Wahl.
Water Sedge

Key III type; rhizomatous, 1.5–8(10) dm tall; leaf sheaths mostly reddish to purplish black or brown, the blades on lower sheaths lacking, elongating above, 2–5 mm broad, flat or channeled at the base; lowermost bract 7–35 cm long, the blade leaflike; inflorescence 5–25 cm long, erect or nodding; spikes 3–6, the terminal 1–3 staminate, the lateral pistillate or the middle androgynous, sessile or the lower on peduncles to 2 cm long; pistillate scales shorter than or subequal to the perigynia, ovate-lanceolate, acute to obtuse, blackish, with straw-colored to greenish median; perigynia (2)2.5–3 mm long, 1.2–1.8 mm broad, appressed-ascending, obovate, glabrous, substipitate, veinless, greenish, the beak 0.1–0.3 mm long, not toothed; stigmas 2.

Bogs, meadows, stream banks, and shallow ponds; in most of Alaska and Yukon; eastward to Newfoundland and south to California, Nevada, Utah, and Colorado; circumboreal. Two more or less distinctive subspecies are present in our region.

1a. Staminate spikes 1; plants mostly less than 2 dm tall, in much of Alaska north of the 66th parallel, and in scattered localities in southeastern Alaska and in much of the Yukon (*C. stans* Drej.). *C. aquatilis* ssp. *stans* (Drej.) Hultén

1b. Staminate spikes 2–3; plants mostly more than 2 dm tall; in most of Alaska (except for the southern portion of the Panhandle) and Yukon. *C. aquatilis* ssp. *aquatilis*

Carex arcta Boott

Key II type; caespitose, 1.5–8 dm tall; leaf sheaths light brown to straw-colored, the blades of lower sheaths well developed, 2–4 mm broad, flat; lowermost bract 0.3–1.5 cm long, the blade (when present) linear-subulate, inflorescence 1.5–3 cm long, 3–7 mm broad, erect; spikes 5–10 or more, all gynaecandrous, sessile, the lower separated by 2–6 mm; pistillate scales shorter than or subequal to the perigynia, ovate-lanceolate, acuminate to obtusish, hyaline, with green midvein; perigynia 2–3 mm long, 1–1.5 mm broad, lanceolate, appressed-ascending, not wing-margined, ciliate, strongly veined dorsally, substipitate, straw-colored to greenish, tapering to an elongate serrulate flattened beak 0.4–0.7 mm long, bidentate; stigmas 2.

Wet meadows and bogs; in scattered localities in the southeastern quarter of Alaska and southern Yukon; southward to California and Montana and disjunctly eastward to Labrador and New England.

Carex atherodes Spreng.

Key V type; caespitose, 3–10 dm tall or more; leaf sheaths purplish tinged to straw-colored, septate-nodulose, spreading-hairy, the blades of lower sheaths short or lacking, elongating upwards, 3–12 mm broad, flat; lowermost bract commonly 20–30(40) cm long, leaflike; inflorescence commonly 15–25 cm long, erect; spikes 2–7, the upper 1–3 staminate, those below the staminate often androgynous, the lower ones pistillate, pedunculate, separated by 4–10 cm or more; pistillate scales shorter than the perigynia, lance-elliptic, awned, the awns ciliate basally, reddish brown, the margins hyaline, the median greenish to straw-colored; perigynia 6.5–10 mm long, 1–2 mm broad, narrowly lanceolate, short-stipitate, prominently veined dorsally and ventrally, straw-colored, glabrous, tapering to a slender bidentate beak, the teeth 1.2–2 mm long; stigmas 3.

Wet meadows, bogs, and streamsides; in central eastern Alaska and southern Yukon; east to Hudson Bay and Maine and south to Utah, Colorado, Nebraska, and Missouri; Eurasia.

Carex athrostachya Olney

Key II type; caespitose, commonly 2–6 dm tall; leaf sheaths brownish to straw-colored, the blades of the lower sheaths short or obsolete, elongating upwards, 1–4 mm broad, flat; lowermost bract 1–12 cm long, the blade linear; inflorescence 1–2.5 cm long, erect; spikes 4–10 or more, all gynaecandrous, sessile, the lower separated by 1–3(4) mm; pistillate scales about equaling the perigynia, elliptic to lanceolate, acute to acuminate; brownish, the margins white-hyaline, the median greenish to straw-colored; perigynia 3–4.5 mm long, 1–1.5 mm broad, lanceolate, appressed, wing-margined, ciliate, substipitate, straw-colored to brownish, tapering to an elongate flattened beak about 1 mm long, bidentate; stigmas 2.

Dry to wet meadows; in coastal southern and southeastern Alaska, and less commonly in interior Alaska; disjunctly southward to California, Utah, and Colorado.

Carex atrata L.
Black-scale Sedge

Key VI type; caespitose, commonly 2–8 dm

tall; leaf sheaths chestnut brown to reddish purple (at least the lowermost), the blades of the lower sheaths lacking, elongating upwards, 2–6 mm broad, flat or the margins revolute; lowermost bract 1–6 cm long, the blade leaflike; inflorescence 2–6 cm long, more or less nodding; spikes 3–5(7), the terminal gynaecandrous, the lateral pistillate or some gynaecandrous, the lowermost with peduncles 0.5–4 cm long; pistillate scales shorter than to longer than the perigynia, lance-elliptic, acuminate to acute, purplish brown to dark brown, with narrow, hyaline margins; perigynia 3–4 mm long, 1.2–2.5 mm broad, elliptic to obovate, glabrous, substipitate, veinless, straw-colored to yellowish brown, abruptly beaked, the purplish beak 0.3–0.5 mm long, bidentate; stigmas 3.

Alpine meadows, marshy slopes and flats, heathlands, and open, moist woods; in the southeastern third of Alaska and southern Yukon; east to Labrador and south to Nevada, Utah, and Colorado; circumboreal (*C. atrosquama* Mack.; *C. atrata* ssp. *atrosquama* [Mack.] Hultén).

Carex atratiformis Britt.
Black Sedge

Key VI type; caespitose, commonly 2–8 dm tall; leaf sheaths chestnut brown to reddish purple (at least the lowermost), the blades of the lower sheaths lacking, elongating upwards, 2–6 mm broad, flat with revolute margins; lowermost bract 2.5–11 cm long, the blade leaflike; inflorescence 2–9.5 cm long (from base of lowermost, often sterile bract), more or less nodding; spikes 3–5(6), the terminal gynaecandrous, the lateral pistillate or some gynaecandrous, the lowermost with peduncles 0.5–4 cm long; pistillate scales shorter than to subequal to the perigynia, lance-elliptic, acute to acuminate, purplish brown to dark brown, with narrow hyaline margins; perigynia (2.5)3–4 mm long, 1.5–2 mm broad, elliptic to obovate, glabrous, substipitate, veinless, purplish to greenish brown, not much lighter than the scales, abruptly beaked, the beak 0.3–0.5 mm long, bidentate; stigmas 3.

Streamsides, muskegs, wet to dry meadows, open heath, and woodlands; in the southeastern quarter of Alaska and southern Yukon; east to Labrador and south to British Columbia, Michigan, and Maine (*C. raymondii* Calder; *C. atratiformis* ssp. *raymondii* [Calder] Porsild). This entity is closely allied to *C. atrata* and possibly would be best included within that species at some taxonomic rank.

Carex atrofusca Schkuhr.
Dark-brown Sedge

Key VI type; rhizomatous (often shortly so and more or less caespitose), 1–5 dm tall; leaf sheaths straw-colored to brownish, the blades of lower sheaths well developed, elongating upwards, 2–4 mm broad, flat or the margins revolute; lowermost bract 1.5–7 cm long, tubular-sheathing basally, the sheath (0.2)0.6–3.8 cm long, the blade leaflike or subulate; inflorescence 3–12.5 cm long, more or less nodding, the terminal staminate (rarely gynaecandrous?), the lateral pistillate, with slender spreading to pendulous peduncles 0.5–4.5(9) cm long; pistillate scales shorter than or subequal to the perigynia, lance-elliptic, acute to obtuse, purplish to brownish black, with narrow white-hyaline margins; perigynia 4–5 mm long, 1.7–2 mm broad, elliptic to ovate, glabrous, ciliate, subsessile, veinless, purplish black or straw-colored basally, the beak 0.2–0.5 mm long, more or less bidentate, hyaline apically; stigmas 3.

Bogs and wet meadows and heathlands, and less commonly in shallow ponds; in much of Alaska and Yukon north of the 62nd parallel; east to Labrador; circumboreal (*C. ustulata* Holm; *C. koraginensis* Meinsch.).

Carex bebbii Olney
Bebb Sedge

Key II type; caespitose, 2–8 dm tall; leaf sheaths straw-colored to brownish, the blades of the lower sheaths short to well developed, elongating upwards, 0.2–0.4 cm broad, flat; lowermost bract 0.3–1 cm long, the blade linear-subulate; inflorescence 1–2.5 cm long, erect; spikes 3–10 or more, all gynaecandrous, sessile and contracted or the lowermost separated by 1–3 mm; pistillate scales about equaling the perigynia, lanceolate, acute to attenuate, yellowish to brownish with hyaline margins and greenish to straw-colored median; perigynia 3–3.5 mm long, 1.5–2 mm broad, ovate, appressed-ascending, wing-margined, ciliate, subsessile, straw-colored to greenish, tapering to an elongate flattened beak 0.7–1 mm long, bidentate; stigmas 2.

Wet meadows, bogs, and muskegs; in central Alaska and northern British Columbia (to be expected in the Yukon); eastward to Nova Scotia.

Carex bicolor All.
Two-color Sedge

Key III type; rhizomatous (often shortly so and more or less caespitose), 0.3–3(4) dm tall; leaf sheaths yellowish brown to straw-colored, the blades of the lower sheaths short to elongate, elongating upwards, 1–3 mm broad, flat or channeled below; lowermost bract (0.8)1.5–9 cm long, the blade leaflike or linear-subulate, tubular-sheathing basally; inflorescence 1–6 cm long, erect or nodding; spikes 2–6, the terminal gynaecandrous or staminate, the lateral pistillate, short-pedunculate and clustered, or the lower ones with peduncles 0.5–3.5 cm long or more; pistillate scales subequal to or shorter than the perigynia, ovate, obtuse or abruptly acuminate, purplish black to reddish or yellowish brown, the margins narrowly hyaline, the median greenish to straw-colored; perigynia 2–3 mm long (longer in diseased specimens), 1.2–1.7 mm broad, obovate, spreading to ascending, veined, glabrous, short-stipitate, greenish or whitish to yellowish or brown, granular-roughened, beakless or nearly so; stigmas 2.

Muskegs, fens, stream banks, lake shores, and seeps; in much of Alaska (except for the western portion) and much of the Yukon; east to Labrador and south to California, New Mexico, Nebraska, Indiana, and Pennsylvania. Our materials, as herein interpreted, include elements recognized by other authors as belonging to three species. The morphological variation, though large, is not greater than allowed in other portions of the genus. Thus it seems best to treat all parts as belonging to a single polymorphic species.

1a. Sheath of lower bract terminated by short yellowish to brownish or blackish auricles, the blade shorter than or subequal to the inflorescence. *C. bicolor* var. *bicolor*

1b. Sheath of lower bract lacking auricles, the blade surpassing the inflorescence (*C. aurea* Nutt.; *C. garberi* Fern.; *C. garberi* var. *bifaria* Fern.). *C. bicolor* var. *androgyna* (Olney) Welsh, comb. nov. (based on *Carex aurea* var. *androgyna* Olney ex Wats. Bot. King Expl. 371. 1871)

Carex bigelowii Torr.
Bigelow Sedge

Key III type; rhizomatous, 1–6 dm tall; leaf sheaths brown to chestnut or purplish brown, the blades of the lower sheaths short to elongate, elongating upwards, 1–4 mm broad, flat or less commonly channeled below, the margins revolute; lowermost bract 0.4–5 cm long, the blade linear-subulate to leaflike, sheathless; inflorescence 1–7.5 cm long, erect; spikes 2–5, the terminal staminate, the lateral pistillate or the upper androgynous; pistillate scales subequal to the perigynia, elliptic to oblong or oblanceolate, acute to obtuse, purplish to brownish black with hyaline margins and lighter midrib; perigynia (1.7)2.5–3.5 mm long, 1.2–2 mm broad, spreading-ascending, elliptic, glabrous, substipitate, veinless, greenish or straw-colored to purplish black, the beak 0.1–0.3 mm long, entire apically; stigmas 2.

Bogs, muskegs, heathlands, tundra, open woods, lake shores, and meadows; in most of Alaska and Yukon (except for the southern portions); east to Nova Scotia and south to Utah, Colorado, and New Hampshire; circumboreal (*C. concolor* authors, not R. Br.?; *C. consimilis* Holm; *C. cyclocarpa* Holm; *C. lugens* Holm; *C. yukonensis* Holm). This is one of the most common sedges in our region. It displays considerable more or less continuous variation, with one extreme having narrow leaves and small perigynia. This phase has been segregated as *C. lugens,* but since only arbitrary separation seems possible, it seems best to include it as a portion of the total *bigelowii* complex.

Carex bonanzensis Britt. ex Britt. & Rydb.
Yukon Sedge

Key II type; caespitose, 2–8 dm tall; leaf sheaths brownish to straw-colored, the blades of the lower sheaths short, much elongating upwards, 1–3 mm broad, flat; lowermost bract obsolete or to 0.5 mm long, the blade (when present) subulate; inflorescence 1–5 cm long, 0.3–0.5 cm broad, erect; spikes 3–7, all gynaecandrous, sessile, the lower separated by 3–16 mm; pistillate scales shorter than the perigynia, obovate to broadly elliptic, obtuse or abruptly acute, yellowish brown with hyaline margins and paler midrib; perigynia 1.5–1.8 mm long, 1–1.2 mm broad, ovate to elliptic, spreading to ascending, not winged, glabrous, veined dorsally (and obscurely so ventrally), substipitate, straw-colored to brownish, abruptly short beaked, the beak 0.1–0.2 mm long, entire apically; stigmas 2.

Muskegs, open woods, seeps, margins of ponds and lakes; in the southeastern quarter of continental Alaska and in west-central Yukon; disjunctly eastward to Mackenzie; Asia.

Carex brunnescens (Pers.) Poir
Brownish Sedge

Key II type; caespitose, 2–7 dm tall; leaf sheaths brownish to straw-colored, the blades of lower sheaths short, elongating upwards, 1–3 mm broad, flat; lowermost bract 0.2–1.8 cm long, the blade linear-subulate; inflorescence 1.5–5.5 cm long, 0.4–0.8 mm broad, erect; spikes 3–7 or more, all gynaecandrous, sessile, the lower separated by 4–24 mm; pistillate scales commonly shorter than the perigynia, lanceolate to elliptic, rounded to obtuse, acute, or acuminate, straw-colored or hyaline to brownish, the median sometimes greenish to straw-colored; perigynia 2–2.5 mm long, 1–1.5 mm broad, ovate, spreading to spreading-ascending, wingless, minutely scabrous along the margin, substipitate, lightly veined, greenish to straw-colored or brownish, the beak 0.3–0.5 mm long, conic or slightly flattened, scabrous; stigmas 2.

Bogs, muskegs, open woods, and subalpine meadows; in interior northwestern to southern Alaska and southern Yukon; east to Nova Scotia and south to Oregon, Utah, Colorado, Minnesota, and New York (*C. brunnescens* ssp. *alaskana* Kalela; *C. brunnescens* ssp. *pacifica* Kalela).

Carex buxbaumii Wahl.
Buxbaum Sedge

Key VI type; rhizomatous (often shortly so and more or less caespitose), 2.5–7 dm tall or more; leaf sheaths purplish red to brownish purple, the blades of lower sheaths short or lacking, elongating upwards, 1–3 mm broad, flat or the margins revolute; lowermost bract 1.5–9 cm long or more, sheathless, the blade leaflike; inflorescence 3.5–9 cm long, erect or nearly so; spikes 2–5, the terminal gynaecandrous, the lateral pistillate, the upper more or less congested and subsessile to sessile, the lower separated by 10–36 mm or more and borne on peduncles 1–6 mm long, erect; pistillate scales longer than the perigynia, ovate-lanceolate, awned to acuminate or acute, purplish brown or black with straw-colored to purplish-mottled or greenish midvein; perigynia 2.5–4 mm long, 1.5–2 mm broad, elliptic to obovate, spreading-ascending, wingless, glabrous, substipitate, lightly veined, greenish to straw-colored, the beak 0.2–0.3 mm long, minutely bidentate; stigmas 3.

Bogs, wet meadows, and open woods; in southern Alaska and Yukon; east to Newfoundland and south to California, Utah, Colorado, Missouri, and Georgia; circumboreal.

Carex canescens L.
Silvery Sedge

Key II type; caespitose, 1.5–8 dm tall; leaf sheaths brownish, the blades of lower sheaths short or obsolete, elongating upwards, 1–3 mm broad, flat; lowermost bract 0.3–4.5 cm long, the blade linear to subulate; inflorescence 1.5–8.5 cm long, erect; spikes 3–8, all gynaecandrous, sessile, the lower separated by 4–30 mm; pistillate scales shorter than or subequal to the perigynia, ovate to elliptic, acute to obtuse, hyaline to straw-colored, with greenish to straw-colored median; perigynia (1.8)2–3 mm long, 1–1.5 mm broad, ovate, appressed-ascending, wingless, minutely scabrous along the margin, substipitate, lightly veined, greenish to straw-colored, the beak 0.1–0.3 mm long, entire or minutely bidentate, conic or slightly flattened; stigmas 2.

Bogs, wet meadows, heathlands, open woods, and tundra; in most of Alaska and Yukon south of the 68th parallel; east to Newfoundland and south to California, Arizona, Colorado, Indiana, and Virginia; circumboreal, also in South America, Australia (C. lapponica Lang).

Carex capillaris L.
Hairlike Sedge

Key VI type; caespitose, 0.3–5 dm tall; leaf sheaths brownish to straw-colored, the blades of the lower sheaths short to well developed, 1–3 mm broad, flat; lowermost bract 2–6.5(8) cm long, long tubular-sheathing, the blade leaflike; inflorescence 2–14 cm long, nodding to erect; spikes 2–5, the terminal staminate or gynaecandrous, the lateral pistillate, spreading to pendulous, at least the lower ones with slender peduncles 1.5–5 cm long or more; pistillate scales shorter than the perigynia, ovate to obovate, obtuse or acute, chestnut brown with broad hyaline margins, the median green to straw-colored or hyaline throughout; perigynia (1.8)2.4–3 mm long, 0.8–1 mm broad, lanceolate to ovate, ascending, wingless, glabrous, short-stipitate, veinless, greenish brown to greenish, the beak 0.3–0.6 mm long with white-hyaline apex; stigmas 3.

Muskegs, seeps, roadsides, and stream, lake, and pond margins, in tundra, heathlands, and open woods; in much of Alaska (except for the southwestern portion) and throughout the Yukon; east to the Atlantic and south to Utah, Colorado, and New Hampshire; Eurasia (C. krausei Boeck.; C. capillaris var. krausei [Boeck.] Kurtz; C. capillaris ssp. krausei [Boeck.] Böcher; C. krausei ssp. porsildiana Löve & Löve).

Carex capitata L.
Capitate Sedge

Key I type; caespitose, 0.5–4.5 dm tall; leaf sheaths chestnut to purplish brown, the blades of lower sheaths short or obsolete, elongating upwards, 1 mm broad or less, involute; lowermost bract scalelike, bladeless; spike solitary, 0.4–1 cm long, 0.3–0.8 cm broad, androgynous; pistillate scales shorter than the perigynia, ovate to elliptic, obtuse to abruptly acuminate, brown with margins hyaline, the median brown to straw-colored; perigynia 2–3.8 mm long, 1.5–2.4 mm broad, ovate, spreading to spreading-ascending or the lowermost descending, veinless, wingless, glabrous, sessile, greenish or brownish near the apex, the beak 0.4–0.8 mm long, brownish with white-hyaline apex; stigmas 2.

Muskegs and stream sides, in tundra or

open woods; in much of Alaska (except for the southern and southwestern portions), and throughout the Yukon; east to the Atlantic and south to California, Colorado, and New Hampshire; Eurasia, South America.

Carex chordorrhiza Ehrh. ex L. f.
Creeping Sedge

Key II type; rhizomatous or stoloniferous or both, the fertile culms 0.8–3 dm long; leaf sheaths purplish or reddish brown or straw-colored, the blades of lower sheaths lacking, elongating upwards, 1–3 mm broad, flat or some folded; lowermost bract scalelike, bladeless; inflorescence 0.8–1.3 cm long, erect; spikes 3–5, all androgynous, closely clustered and appearing as a single spike; pistillate scales commonly longer than the perigynia, ovate to lance-attenuate, acute to acuminate, brown with broad hyaline margins, the median brown to straw-colored; perigynia (2.3)2.5–3.5(4) mm long, (1.6)1.8–2.2 mm broad, ovate to elliptic, appressed-ascending, veined on both sides, wingless, substipitate, glabrous, obscurely serrulate, yellowish brown or the base straw-colored, the beak 0.3–0.5 mm long, hyaline tipped; stigmas 2.

Wet meadows, lake shores, and bogs, in tundra, heathlands, and open woods; in much of Alaska (except for the southwestern and southeastern portions) and in northern Yukon; east to Newfoundland and south to British Columbia, Iowa, Indiana, and New York; Eurasia.

Carex circinata C. A. Mey.
Coiled Sedge

Key I type; caespitose, 0.5–2.5 dm tall; leaf sheaths light brown, the blades of lower sheaths short, elongating upwards, less than 1 mm broad, involute, usually curved; lowermost bract 0.4–0.8 cm long, the blades (when present) subulate; spikes solitary, 1.5–2.7 cm long, 0.3–0.5 cm broad, androgynous; pistillate scales from shorter to longer than the perigynia, lanceolate to elliptic, obtuse to acute or acuminate, brownish with hyaline margins and straw-colored midvein; perigynia 4.5–6 mm long, 1–1.5 mm broad, erect-ascending, narrowly lanceolate, tapering at both ends, veined dorsally and ventrally, substipitate, wingless, glabrous on both sides, serrulate marginally, straw-colored, tapering to a slender reddish beak 1–2 mm long, white-hyaline apically; stigmas 3.

Ridge tops, rock crevices, and meadows; in coastal southern Alaska from the Aleutians eastward to the Panhandle, and in southwestern Yukon; south to Washington.

Carex concinna R. Br. ex Richards.
Low Northern Sedge

Key IV type; caespitose, 0.4–2.6 dm tall; leaf sheaths brown to hyaline or straw-colored, the blades of lower sheaths short to elongate, elongating upwards, 1–3.2 mm broad, commonly much shorter than the culms, flat or slightly revolute apically and involute basally; lowermost bract 0.4–1 cm long, with tubular-sheathing base 1–4 mm long, the blade linear-subulate; inflorescence 1–3 cm long, erect or nodding; spikes 2–4, the terminal staminate, the lateral pistillate, erect or ascending, the lower ones with peduncles 0.3–0.8 cm long, the upper more or less densely clustered below the terminal spike; pistillate scales shorter than the perigynia, ovate, rounded to obtuse, brown with broad hyaline margins and lighter midrib; perigynia 2.5–3.5 mm long, 1–1.5 mm broad, elliptic to obovate, spreading-ascending, wingless, moderately to densely hairy, substipitate, obscurely veined, whitish to greenish or straw-colored (or sometimes brownish), the beak 2–4 mm long, brown, the apex hyaline; stigmas 3.

Dry gravelly or sandy slopes or flats, or in bogs and muskegs and in open woods; in much of the eastern half of continental Alaska and most of the Yukon; east to Newfoundland and south to Colorado, South Dakota, and Wisconsin.

most bract 0.8–3.5 cm long, the blade linear-subulate; inflorescence 1–3.5 cm long, erect or nearly so; spikes 3–10 or more, all gynaecandrous, sessile, densely aggregated or the lowermost separated by 1–3 mm; pistillate scales shorter than the perigynia, narrowly lance-attenuate, brownish with more or less hyaline margin and pale midvein; perigynia (3)3.5–4 mm long, 0.5–1.2 mm broad, narrowly lanceolate, appressed-ascending, obscurely veined, wing-margined, substipitate, glabrous on both sides, serrulate marginally, tapering to a slender flattened bidentate beak 0.8–1.2 mm long; stigmas 2.

Moist sites in open woods; in the southeastern quarter of Alaska and southern Yukon; east to Newfoundland and south to Washington, Idaho, Minnesota and New Jersey.

Carex deflexa Hornem.
Northern Sedge

Key IV type; caespitose, 0.4–2.5 dm tall; leaf sheaths purplish brown or reddish brown, the blades of lower sheaths short or lacking to well developed, elongating upwards, 1–2(3) mm broad, flat or channeled near the base; lowermost bract 0.5–4 cm long, sheathless, merely auricled basally, the blade leaflike; inflorescence 0.4–2.5 cm long, erect or nearly so; spikes (1)2–5, the terminal staminate, the lateral pistillate, erect or ascending, the lower ones with peduncles 0.1–1 cm long, the upper more or less densely clustered below the terminal spike; pistillate scales shorter than the perigynia, lanceolate, acute to acuminate, purplish brown or reddish brown with hyaline margins and greenish to straw-colored median; perigynia 2.2–3 mm long, 1–1.2 mm broad, obovate, spreading-ascending, wingless, moderately to densely hairy, stipitate, veinless, greenish to straw-colored, the beak 0.4–0.7 mm long, greenish, ciliate, bidentate; stigmas 3.

Open woods, in sandy or gravelly soil; in central eastern Alaska and southern Yukon; east to Newfoundland and south to British

Carex concinna R. Br. (× 0.6).

Carex crawfordii Fern.
Crawford Sedge

Key II type; caespitose, 0.8–7 dm tall; leaf sheaths pale brown to straw-colored, the blades of lower sheaths short or lacking, elongating upwards, 0.7–3 mm broad, flat or channeled (especially below); lower-

Columbia, Minnesota, and New York. This entity is closely related to *C. rossii* Boott.

Carex deweyana Schwein.
Dewey Sedge

Key II type; caespitose, 2–10 dm tall or more; leaf sheaths brownish to straw-colored, the blades of lower sheaths short, elongating upwards, 1–4 mm broad, flat; lowermost bract 1–4 cm long, the blade linear-subulate; inflorescence 1.5–5 cm long, erect or nodding; spikes 2–4, the terminal gynaecandrous, the lateral pistillate, all sessile, the lowermost separated by 8–25 mm; pistillate scales shorter than the perigynia, lanceolate, shortly awned or cuspidate, hyaline except for the green midrib or else tinged brownish; perigynia 4–5 mm long, 1.2–1.8 mm broad, lanceolate, appressed-ascending, veinless or obscurely veined dorsally, wing-margined only along the beak, subsessile, glabrous, tapering to a slender flattened serrulate bidentate beak 1–2 mm long; stigmas 2.

Open woods; in east-central Yukon (where possibly introduced); east to Nova Scotia and south to Idaho, Colorado, South Dakota, Iowa, and Pennsylvania.

Carex diandra Schrank

Key II type; caespitose, 2–9 dm tall or more; leaf sheaths brown to dark brown, the blades of lower sheaths short or lacking, elongating upwards, 1–3 mm broad, flat or more or less channeled; lowermost bract 0.5–1.8 cm long, the blade linear-subulate; inflorescence 1.3–4 cm long, erect; spikes 5–many borne in simple or compound spikes, all androgynous, all sessile, the lowermost separated by 1–11 mm; pistillate scales shorter than the perigynia, ovate, attenuate to acute or shortly awned, brown with margins hyaline and midrib greenish to brownish; perigynia 2–2.8 mm long, 1–1.4 mm broad, lance-ovate, spreading-ascending, veined dorsally, wing-margined only along the beak, shortly stipitate, glabrous, brownish, tapering to a

slender flattened serrulate bidentate beak 0.8–1.5 mm long; stigmas 2.

Lake and pond margins, muskegs, wet meadows, and bogs; in central and southeastern continental Alaska and southern to northern Yukon; east to Nova Scotia and south to California, Colorado, Nebraska, Iowa, and New Jersey.

Carex dioica L.
Northern Bog Sedge

Key I type; rhizomatous, dioecious or rarely monoecious, 0.4–2 dm tall or more; leaf sheaths straw-colored to brownish, the blades of lower sheaths short or lacking, elongating upwards, less than 1 mm broad, involute; lowermost bract not developed; spikes solitary, the pistillate 6–10 mm long, 4–8 mm broad, the staminate 5–10 mm long, 1–2 mm broad, the androgynous like the pistillate; pistillate scales from shorter to longer than the perigynia, ovate, acute to obtuse, brownish with hyaline margins and paler midvein; perigynia 2.5–4 mm long, 1.5–2 mm broad, spreading-ascending, elliptic to lanceolate or ovate, rounded basally, sessile, veined dorsally and ventrally, wingless, glabrous, obscurely serrulate near the apex, brownish, abruptly beaked, the beak 0.3–0.5 mm long, bidentate, white-hyaline apically; stigmas 2.

Streamsides, terraces, open woods, bogs, and wet meadows; in most of Alaska south of the 68th parallel (except for the southern Panhandle region) and in southern Yukon; east to Newfoundland and south to Colorado, Minnesota, and New York; Eurasia (*C. gynocrates* Wormsk.; *C. alascana* Boeck.). Our material belongs to ssp. *gynocrates* (Wormsk.) Hultén.

Carex disperma Dewey
Soft-leaved Sedge

Key II type; rhizomatous (but still caespitose), 0.6–5 dm tall; leaf sheaths straw-colored to light brown, the blades of lower sheaths short or lacking, elongating upwards, mostly 1–2 mm broad, flat; lowermost bract 0.3–0.8 cm long, the blade (when present) subulate; inflorescence 1–3.5 cm long, erect or nearly so; spikes 2–4, all androgynous, sessile, the lowermost separated by 3–12 mm or more; pistillate scales from shorter to longer than the perigynia, ovate, acute to obtuse, hyaline, the midrib greenish; perigynia 1–4(5) per spike, 2–3 mm long, 1.2–1.5 mm broad, elliptic, spreading-ascending, veined dorsally and ventrally, wingless, short-stipitate, glabrous, greenish to brownish and shining at maturity, abruptly short beaked, the beak 0.1–0.3 mm long, minutely bidentate at the hyaline apex; stigmas 2.

Muskegs, stream banks, lake margins, and in woods; in the southeastern two-thirds of Alaska and southern Yukon; east to Nova Scotia and south to California, New Mexico, South Dakota, Minnesota, Indiana, and New Jersey; Eurasia.

Carex eburnea Boott

Key VI type; caespitose (but also rhizomatous), 1–3 dm tall; leaf sheaths brownish, the blades of lower sheaths short or lacking, elongating upwards, linear-filiform, less than 1 mm broad, involute; lowermost bract mostly 0.4–0.6 cm long, tubularsheathing, bladeless or nearly so; inflorescence 1–3 cm long, erect or nearly so; spikes 2–5, the terminal staminate, the lateral pistillate, erect or ascending, the lower on slender peduncles 0.6–1.5 cm long; pistillate scales shorter than the perigynia, ovate to obovate, obtuse to acute, hyaline to yellowish green or brownish with hyaline margins and greenish midrib; perigynia 1.7–2 mm long, 0.7–1 mm broad, elliptic, appressed-ascending, wingless, glabrous, subsessile, finely veined on both surfaces, greenish to brown, shining, the beak 0.2–0.3 mm long, hyaline at the oblique apex; stigmas 3.

Dry hillsides and rock outcrops; in central and southeastern mainland Alaska and southern Yukon; east to Newfoundland and south to Texas, Arkansas, Alabama, and Virginia.

Carex echinata Murr.
Prickly Sedge

Key II type; caespitose, 1–2.5 dm tall; leaf sheaths straw-colored to brownish, blades of lower sheaths obsolete, elongating upwards, 0.7–2 mm broad, flat or channeled; lowermost bract 0.2–1.3 cm long, scalelike or the blade linear-subulate; inflorescence 0.6–2.5 cm long, erect; spikes 2–5, all gynaecandrous or the lateral pistillate, sessile, the lower separated by 1–10 mm, spreading-ascending; pistillate scales shorter than the perigynia, ovate, acute or obtuse, brownish with translucent hyaline margins and lighter median; perigynia 2.5–3.7 mm long, 1–1.5 mm broad, lanceolate, spreading, wingless (merely sharp edged), glabrous, sessile, veined on both sides, straw-colored to brownish, tapering to a flattened, serrulate beak, the beak 0.8–1.5 mm long, bidentate; stigmas 2.

Bogs; in Unalaska, Alaska; disjunctly east to northeastern North America, Iceland, and Europe (*C. stellata* Good.; *C. muricata* authors, not L.?).

Carex eleusinoides Turcz.

Key II type; caespitose, 1.5–3 dm tall; leaf sheaths purplish brown, the blades of lower sheaths lacking, developed upwards, 1–3 mm broad, flat or channeled; lowermost bract 2–5(6.5) cm long, the blade leaflike; inflorescence 2–4 cm long, erect or nearly so; spikes 2–4, the terminal gynaecandrous, the lateral pistillate (or sometimes gynaecandrous), all more or less clustered or the lowermost separated by 4–15(21) mm and with peduncles 2–18 mm long, erect or nodding; pistillate scales commonly shorter than the perigynia, lanceolate to elliptic, obtuse to acute, blackish with pale midrib; perigynia 1.7–2.5 mm long, 1–1.2 mm broad, ovate to obovate, ascending, glabrous, substipitate, obscurely veined, straw-colored, the apex purplish, the beak 0.1–0.3 mm long, slightly emarginate; stigmas 2.

Gravelly or sandy terraces or bars; in western, central, and east-central Alaska and southern to northern Yukon and adjacent Mackenzie; Asia (*C. kokrinensis* Porsild).

Carex enanderi Hultén
Enander Sedge

Key VI type; rhizomatous or stoloniferous, or both (and loosely caespitose), 1.3–4 dm tall; leaf sheaths brownish to chestnut brown, the blades of lower sheaths short or lacking, elongating upwards, 1–3 mm broad, flat to channeled throughout; lowermost bract 1.5–7.5 cm long, the blade leaflike; inflorescence 1.5–4 cm long, erect or nearly so; spikes 2–5, the terminal gynaecandrous, the lateral pistillate, appressed-ascending to erect, all clustered and subsessile or less commonly the lower separated by 12–25 mm and with peduncles 0.6–1.2 cm long; pistillate scales slightly shorter than the perigynia, lanceolate to obovate, rounded to obtuse or emarginate, purplish black with straw-colored to greenish median; perigynia 2.5–3.5 mm long, 1.2–1.6 mm broad, ovate to elliptic, spreading-ascending, wingless, glabrous, with a stipe 0.3–0.6 mm long, many-veined on both faces, straw-colored or purple spotted or purplish apically, abruptly beaked, the beak 0.2–0.4 mm long, truncate; stigmas 2.

Wet swampy or boggy sites; in coastal and insular southern Alaska; southward to British Columbia and Alberta. This entity is apparently closely related to *C. lenticularis* Michx.

Carex filifolia Nutt.
Thread-leaf Sedge

Key I type; caespitose in dense clumps 1–3 dm tall; leaf sheaths brown, the blades of lower sheaths short to elongate, elongating upwards, less than 1 mm broad, involute; lowermost bract not developed; spikes solitary, androgynous, 8–25 mm long, 2–6 mm broad; pistillate scales longer than or subequal to the perigynia, oval, truncate

to rounded or abruptly short-acuminate, brownish with broad hyaline margins and straw-colored midvein; perigynia 3–4 mm long, 1.5–2.5 mm broad, spreading-ascending, ovate to elliptic, rounded or tapering basally, veinless, sessile, wingless, obscurely short-hairy, abruptly short beaked, the beak 0.2–0.4 mm long, hyaline at the truncate apex; stigmas 3.

Dry open sandy or shaly slopes; in southeastern quarter of Alaska and southern Yukon; east to Saskatchewan and south to California, Arizona, New Mexico, and Texas.

Carex flava L.
Yellow Sedge

Key V type; caespitose, 1–5 dm tall or more; leaf sheaths straw-colored to greenish, the blades of the basal leaves short or lacking, elongating upwards, 2–4 mm broad, flat; lowermost bract commonly 3–10 cm long, short-sheathing, spreading, the blade leaflike; inflorescence 2.5–6 cm long, erect; spikes 2–6, the terminal staminate, the lateral pistillate, the upper commonly clustered, the lower separated by 5–25 mm or more and borne on peduncles 5–15 mm long or more; pistillate scales much shorter than the perigynia, lanceolate, acute to obtuse or cuspidate, brownish and more or less hyaline, the midrib greenish; perigynia 4–6 mm long, 1.5–2.5 mm broad, the body lanceolate to oblanceolate or obovate, spreading or reflexed, wingless, glabrous, sessile, prominently veined, yellowish green or greenish, the beak 2–3 mm long, obscurely serrulate, bidentate; stigmas 3.

Moist sites; in south-central and southeastern Alaska and northern British Columbia; east to Newfoundland and south to Montana, Minnesota, Indiana, and Pennsylvania; Eurasia.

Carex glacialis Mack.
Glacier Sedge

Key VI type; caespitose, 0.4–1(1.5) dm

tall; leaf sheaths reddish to purplish brown to straw-colored, the blades of the lower sheaths short or lacking, elongating upwards, less than 1 mm broad, involute or channeled; lowermost bract 0.2–1 cm long, shortly tubular-sheathing, the blade (when present) subulate; inflorescence 0.7–1.2 cm long, erect; spikes 2–4 (appearing as 1), the terminal staminate, the lateral pistillate, clustered and subsessile or the lower separated by 2–5 mm and with peduncles 1–2 mm long; pistillate scales about as long as perigynia, ovate to elliptic, obtuse to acute, chestnut brown with broad hyaline margins and greenish to straw-colored median; perigynia 2–2.5 mm long, 0.8–1 mm broad, obovate to oblanceolate, spreading-ascending, wingless, glabrous, substipitate, veinless, yellowish green to straw-colored or chestnut brown apically, the beak 0.3–0.5 mm long, the apex hyaline, truncate; stigmas 3.

Dry, gravelly or sandy sites in arctic or alpine tundra; in much of Alaska (except for the southwestern quarter) and southern Yukon; east to Greenland and southward to Alberta and Newfoundland; Eurasia.

Carex glareosa Wahl.
Clustered Sedge

Key II type; rhizomatous (but loosely caespitose), 0.8–4 dm tall; leaf sheaths brownish to straw-colored, the blades of lower sheaths short or lacking, commonly 1–3 mm broad or less, flat to channeled; lowermost bract 0.3–1 cm long, the blade (when present) subulate; inflorescence 1–4 cm long, erect or nearly so; spikes 2–4, the terminal gynaecandrous, the lateral pistillate, sessile, clustered or the lower separated by 2–8 mm; pistillate scales about equaling the perigynia, ovate to obovate, obtuse, chestnut brown with hyaline margins and yellowish brown median; perigynia 2–3.5 mm long, 1–1.5 mm broad, obovate to oblanceolate, appressed-ascending, wingless, glabrous, short-stipitate, veined on both sides, straw-colored, tapering to a more or less flattened beak, the

beak 0.3–0.5 mm long, obliquely cleft; stigmas 2.

Tidal flats and brackish marshlands; throughout coastal and insular Alaska and Yukon; east to Newfoundland; circumpolar (*C. marina* Dewey; *C. bipartita* Bellardi var. *glareosa* [Wahl.] Polunin; *C. glareosa* var. *amphigena* Fern.; *C. pribylovensis* Macoun).

Carex gmelinii H. & A.
Gmelin Sedge

Key VI type; caespitose (1)2–9 dm tall; leaf sheaths dark brown, blades of lower sheaths short or lacking, elongating upwards, 2–5 mm broad (rarely less), flat, the margins revolute or sometimes channeled below; lowermost bract 2–8 cm long, sheathless, the blade leaflike; inflorescence 2–8 cm long, erect to nodding; spikes 3–6, the terminal gynaecandrous, the lateral pistillate, the upper more or less congested and commonly short-pedunculate, the lower separated by 10–35 mm and with peduncles 6–15 mm long or more, ascending to erect; pistillate scales mostly longer than the perigynia, elliptic to ovate, with ciliate awns 1–3(4) mm long or some merely cuspidate, purplish black with narrow hyaline margins and lighter median; perigynia 4–5 mm long, 1.8–2.2 mm broad, lance-elliptic to ovate, appressed-ascending, wingless, glabrous, short-stipitate, veined on both sides (often lightly so), yellowish brown, the beak 0.2–0.3 mm long, purple tipped, minutely bidentate; stigmas 3.

Sea beaches, meadows, and sea cliffs; in coastal, western, southwestern, southern, and southeastern Alaska; south to British Columbia; Asia.

Carex heleonastes Ehrh.
Hudson Bay Sedge

Key II type; caespitose, 1–4 dm tall; leaf sheaths brownish to straw-colored, the blades of lower sheaths short or lacking, elongating upwards, 1–2 mm broad, flat or channeled; lowermost bract 0.3–0.6 cm long, bladeless or the blade subulate; inflorescence 0.7–3 cm long, erect; spikes 2–5, all gynaecandrous, sessile, more or less clustered, or the lower separated by 3–12 mm; pistillate scales shorter than to equaling the perigynia, elliptic to ovate, acute to obtusish, yellow brown with broad hyaline margins and brownish to straw-colored median; perigynia 2–3 mm long, 1–1.5 mm broad, lance-ovate, spreading-ascending, wingless, glabrous, substipitate to subsessile, veined on both sides, straw-colored to brownish, tapering to a more or less conical beak 0.3–0.5 mm long, the beak serrulate or smooth, obliquely cleft; stigmas 2.

Muskegs, bogs, and seeps; in widely disjunct sites in much of mainland Alaska and Yukon; east to Greenland; Eurasia (*C. amblyorhyncha* Krecz.; *C. neurochlaena* Holm; *C. heleonastes* ssp. *neurochlaena* (Holm) Böcher).

Carex holostoma Drej.

Key VI type; caespitose (loosely so and more or less rhizomatous), 1–2.5 dm tall; leaf sheaths purplish or reddish, blades of lower sheaths short, elongating upwards, 1–2 mm broad, flat, the margins revolute; lowermost bract 1–4 cm long, the blade leaflike; inflorescence 1.5–3 cm long, erect; spikes 2–4, the terminal staminate, the lateral pistillate, clustered, or the lower separated by 3–8 mm and with peduncles 2–6 mm long; pistillate scales about as long as the perigynia, ovate to obovate, obtuse or acute, purplish black with lighter midvein; perigynia 2–2.5 mm long, 1.2–1.6 mm broad, obovate, appressed-ascending, wingless, glabrous, sessile, obscurely veined, straw-colored or the apex purplish black, the beak about 0.1 mm long, entire or nearly so; stigmas 3.

Moist sites; in widely disjunct localities in Alaska; east to Greenland; circumpolar. This species is evidently rare in our region.

Carex interior Bailey

Key II type; caespitose, 1.5–6 dm tall or more; leaf sheaths dull brown to straw-colored, the blades of lower sheaths short or lacking, elongating upwards, 1–3 mm broad, flat or channeled; lowermost bract 0.2–0.8 mm long, the blade subulate; inflorescence 1–3.5 cm long, erect; spikes (2)3–4, the terminal gynaecandrous, the lateral pistillate or gynaecandrous, all sessile, the lower separated by 2–8 mm; pistillate scales shorter than the perigynia, ovate to lanceolate, obtuse to acutish, brownish with broad hyaline margins and greenish midvein; perigynia 2.2–3.3 mm long, 1.2–2 mm broad, lanceolate, spreading, wingless (but sharp edged), veined dorsally, substipitate, greenish to brownish, tapering to an elongate serrulate flattened beak 0.5–0.8 mm long, the beak sharp edged, more or less bidentate; stigmas 2.

Wet meadows, seeps, and bogs; in southern Yukon and northern British Columbia; east to Newfoundland and south to Mexico, Kansas, Indiana, and Pennsylvania.

Carex jacobi-peteri Hultén
Anderson Sedge

Key I type; caespitose, 0.3–1 dm tall; leaf sheaths brown, persistent, the blades of lower sheaths well developed, 0.5–1.5 mm broad, flat; lowermost bract scalelike; spikes solitary, androgynous, 4–11 mm long, 3–8 mm broad; pistillate scales subequal to the perigynia, ovate to elliptic, acute to acuminate, brown with hyaline margins (?) and lighter median; perigynia 2.5 mm long, 1.2 mm broad, ovate, substipitate, wingless, glabrous, obscurely veined, brownish, tapering to a smooth beak about 0.2 mm long, truncate, hyaline apically; stigmas 2.

Arctic tundra; in the Seward Peninsula of Alaska (Tin City); endemic. This entity is apparently most closely related to *C. nardina* Fries and might be best considered within it. More materials are necessary before final decisions can be made.

Carex lachenalii Schkuhr.

Hare'sfoot Sedge

Key II type; caespitose (often shortly rhizomatous), 1–4 dm tall; leaf sheaths brownish to straw-colored, the blades of lower sheaths lacking, developed upwards, 1–3 mm broad, flat or channeled; lowermost bract scalelike, bladeless (or the blade 0.2–2 cm long and subulate); inflorescence 1–3.5 cm long, erect; spikes 2–5, all gynaecandrous or some pistillate only, clustered, or the lower separated by 2–16 mm, sessile or the lower with peduncles 2–3 mm long, erect-ascending; pistillate scales shorter than the perigynia, brownish, ovate to elliptic, obtusish, with hyaline margins and paler midrib; perigynia 2–4 mm long, 1–1.5 mm broad, lanceolate to elliptic, ascending, veined on both faces, wingless, stipitate, brownish or greenish, tapering to a more or less flattened beak, the beak 0.2–0.6 mm long, obliquely cleft; stigmas 2.

Bogs, stream banks, shallow ponds, wet meadows, snow flushes, and open woods; in most of Alaska and Yukon; circumpolar (*C. bipartita* All.; *C. tripartita* All.). *C. lachenalii* is closely allied to *C. glareosa* Wahl.

Carex laeviculmis Meinsch.

Smooth-stem Sedge

Key II type; caespitose, 1.5–9 dm tall; leaf sheaths straw-colored to light brownish, the blades of lower sheaths short or lacking, developed upwards, 1–2 mm broad, flat; lowermost bract 0.5–3.5 cm long, the blade linear to subulate; inflorescence 1.5–6.5 cm long, erect; spikes 3–8, all gynaecandrous or the lower sometimes pistillate, sessile, separated by 3–30 mm, erect-ascending; pistillate scales shorter than the perigynia, oval to ovate, obtuse to acutish, yellowish brown with broad hyaline margins and greenish to brownish median; perigynia 2.2–3.5 mm long, 1–1.5 mm broad, lanceolate to ovate, ascending, wingless, glabrous, short-stipitate, veined on both faces, brownish to greenish, tapering to a more or less flattened serrulate

beak, the beak 0.4–0.8 mm long, obliquely cleft; stigmas 2.

Open gravelly slopes, muskegs, and wet woods; in near coastal and insular southern and southeastern Alaska; south to California and Idaho.

Carex lanuginosa Michx.

Key IV type; rhizomatous, but loosely caespitose, 2–10 dm tall; leaf sheath purplish to brownish, obscurely septate-nodulose, the blades of lower sheaths short to elongate or lacking, developed upwards, 2–5 mm broad, flat; lowermost bract 5–15 cm long or more, sheathless or tubular-sheathing, the blade leaflike; inflorescence 6–18 cm long or more, erect or nodding; spikes 3–6, the upper 1–3 staminate, the lower pistillate, separated by 25–100 mm or more, and borne on peduncles 5–70 mm long, erect; pistillate scales shorter than the perigynia, lance-acuminate, more or less awned, reddish brown with hyaline margins and greenish to brownish median; perigynia 2.5–4 mm long, 1.7–2 mm broad, elliptic to obovate, ascending, wingless, densely hairy, subsessile, veined on both faces, brownish, rather abruptly beaked, the beak 0.8–1.2 mm long, sharply bidentate; stigmas 3.

Wet meadows; in south-central Alaska; disjunctly southward to Mexico and east to the Atlantic. This entity is closely related to *C. lasiocarpa* Ehrh., and it is sometimes included within that species as var. *lanuginosa* (Michx.) Kuk.

Carex lasiocarpa Ehrh.

Key IV type; rhizomatous, but loosely caespitose, 3–10 dm tall or more; leaf sheaths purplish, obscurely septate-nodulose, the blades of lower sheaths short to elongate, developed upwards, 1–2 mm broad, involute; lowermost bract commonly 6–15 cm long or more, sheathless or short-sheathing, the blade involute; inflorescence commonly 8–15 cm long or more, erect or nearly so; spikes 3–6, the upper 1–3 staminate, the lower ones pistillate, separated by 30–150 mm or more, with peduncles 0.2-3 cm long, erect; pistillate scales lance-acuminate, acuminate and more or less awned, brown with hyaline margins and greenish to straw-colored median; perigynia 2.8–4 mm long, 1.4–1.7 mm broad, ovate, ascending, wingless, densely hairy, sessile, obscurely veined, brownish, rather abruptly beaked, the beak 0.6–1 mm long, sharply bidentate; stigmas 3.

Bogs, stream banks, and shallow ponds; in south-central Alaska and southwestern Yukon; east to Nova Scotia and south to Washington, Idaho, Iowa, Ohio, and New Jersey (*C. lasiocarpa* ssp. *americana* [Fern.] Hultén). Our material belongs to var. *americana* Fern.

Carex laxa Wahl.
Weak Sedge

Key VI type; rhizomatous, mostly 1.5–3 dm tall; leaf sheaths brownish, the blades of lower sheaths short to elongate, developed upwards, 1–2 mm broad, flat; lowermost bract 1.5–10 cm long, the sheath 8–20 mm long or more, the blade leaflike; inflorescence 4–10 mm long, more or less nodding; spikes 2–4, the terminal staminate, the lateral pistillate, separated by 20–50 mm or more, and with filiform peduncles 1.5–4.5 cm long, ascending to spreading or nodding; pistillate scales longer than or subequal to the perigynia, elliptic to lanceolate, obtuse to acute, brown with broad greenish or yellowish median; perigynia 2.5–3 mm long, 1–1.5 mm broad, elliptic to lanceolate, ascending, wingless, glabrous, subsessile, obscurely veined on both faces, greenish to straw-colored; beak 0.1–0.2 mm long, truncate; stigmas 3.

Swamps and muskegs; in disjunct sites in southeastern and northeastern continental Alaska; east to the Mackenzie; Eurasia. This species is evidently rare in Alaska.

Carex lenticularis Michx.

Key III type; caespitose (the rhizomes short to elongate), 1–6 dm tall; leaf sheaths straw-colored to reddish brown, the blades of lower sheaths short or lacking, elongating upwards, 1–3 mm broad, flat or channeled; lowermost bract 2–15(20) cm long, sheathless, the blade leaflike; inflorescence 3–14 cm long, erect or nearly so; spikes 3–6, the terminal staminate or rarely gynaecandrous, the lateral pistillate, all congested or the lower separated by 6–30 mm or more, with peduncles 3–15(30) mm long; pistillate scales commonly shorter than the perigynia, elliptic to oblong, obtuse or acutish, purplish brown to purplish black with hyaline margins and straw-colored to greenish median; perigynia 2–3 mm long, 1.2–1.5 mm broad, ovate to elliptic, spreading-ascending, wingless, glabrous, with a stipe 0.3–0.6 mm long, veined on both faces, greenish to straw-colored or brownish-mottled, abruptly beaked, the beak 0.1–0.3 mm long, truncate; stigmas 2.

Muskegs, wet meadows, gravel bars, and lake shores; in south-central and coastal southern Alaska; south to California, Utah, and Colorado (*C. hindsii* Clarke; *C. kelloggii* Boott).

Carex leptalea Wahl.
Bristle-stalk Sedge

Key I type; rhizomatous and more or less caespitose; leaf sheaths brown to yellowish, the blades of lower sheaths lacking, developed upwards to 1 mm broad, flat or channeled; lowermost bract scalelike; spikes solitary, 0.5–1.5 cm long, 0.2–0.3 cm broad, androgynous; pistillate scales much shorter than the perigynia, elliptic to ovate, obtuse, or abruptly acuminate, brownish to hyaline with a greenish median; perigynia 2–4 mm long, 1–1.5 mm broad, lanceolate to oblanceolate, appressed-ascending, veined on both sides, wingless, glabrous, substipitate, greenish to straw-colored, beakless, the apex truncate or emarginate; stigmas 3.

Mossy bogs, lake shores, and muskegs; in the southeastern two-thirds of Alaska and southern Yukon; east to Nova Scotia and south to California, Colorado, Missouri, and North Carolina.

Carex limosa L.
Shore Sedge

Key VI type; rhizomatous, 2–6 dm tall; leaf sheaths purplish to brown, the blades of lower sheaths short or lacking, developed upwards, 1–2 mm broad, channeled; lowermost bract 2–7 cm long, erect to nodding; spikes 2–4, the terminal staminate, the lateral pistillate, separated by 25–50 mm or more, with filiform peduncles 1.5–5 cm long, pendulous; pistillate scales subequal to the perigynia, ovate to elliptic, acute to cuspidate, brownish with lighter base and greenish to straw-colored median; perigynia 3–4 mm long, 1.8–2.2 mm broad, ovate to obovate, ascending, wingless, glabrous, short-stipitate, veined on both sides, greenish to straw-colored; beak 0.1–0.2 mm long, truncate; stigmas 3.

Bogs, muskegs, and shallow ponds; in much of continental Alaska and southern to northern Yukon; east to Nova Scotia and south to California, Montana, Iowa, Indiana, and Delaware; circumboreal.

Carex livida (Wahl.) Willd.
Livid Sedge

Key VI type; rhizomatous, seldom somewhat caespitose, 0.7–5 dm tall; leaf sheaths light brownish, the blades of basal sheaths well developed, 1–3 mm broad, channeled; lowermost bract 2–9.5 cm long, the closed sheath 3–10 mm long or more, the blade leaflike; inflorescence 2–8 cm long, erect or nearly so; spikes 2–4, the terminal staminate, the lateral pistillate, the lower separated by 7–30 mm, with peduncles 0.3–2.7 cm long, erect; pistillate scales from shorter to longer than the perigynia, lance-ovate to elliptic, obtuse to acute, brownish to purplish brown with hyaline margins and greenish to straw-colored median; perigynia 3–4.5 mm long, 1.5–2.5 mm broad, elliptic, ascending, wingless, glabrous,

stipitate, veinless, greenish to whitish or straw-colored, the beak about 0.1 mm long, truncate; stigmas 3.

Bogs and muskegs; in southern, southeastern, and northeastern Alaska and southern Yukon; east to Newfoundland and south to California, Idaho, Minnesota, Michigan, and New York.

Carex loliacea L.

Key II type; caespitose, 1–4 dm tall; leaf sheaths brownish, the blades of lower sheaths short or lacking, elongating upwards, 1–2 mm broad, flat; lowermost bract 2–10 mm long, bladeless or the blade narrowly subulate; inflorescence 0.5–3.5 cm long, erect; spikes 2–5, all gynaecandrous, sessile, the lower separated by 1–18 mm, erect or spreading; pistillate scales shorter than the perigynia, ovate, obtuse or acutish, white-hyaline with greenish median; perigynia 2.5–3 mm long, 1.1–1.4 mm broad, elliptic to lanceolate, spreading to ascending, wingless, glabrous, short-stipitate, veined on both sides, greenish to straw-colored, essentially beakless; stigmas 2.

Muskegs, open woods, and seeps; in the southeastern three-quarters of mainland Alaska and southern Yukon; east to Quebec and south to British Columbia and Alberta; Eurasia.

Carex lyngbyei Hornem.
Lyngbye Sedge

Key III type; rhizomatous and caespitose, 2–9 dm tall; leaf sheaths purplish brown, the blades of lower sheaths lacking, developed upwards, 3–8 mm broad or more, flat with revolute margins; lowermost bract 8–30 cm long, sheathless, the blade leaflike; inflorescence 7–25 cm long, erect to nodding; spikes 4–6, the upper 1–3 staminate, the lower pistillate (or some androgynous), separated by 15–105 mm or more, with peduncles 1.5–9 cm long, at least the lowermost commonly pendulous; pistillate scales longer than the perigynia, lance-attenuate,

attenuate to acute, purplish brown to purplish black with lighter median; perigynia 2.5–3.5 mm long, 1.5–2.5 mm broad, ovate to oval, ascending, wingless, glabrous, substipitate, obscurely veined on both sides, greenish to straw-colored, the beak 0.1–0.2 mm long, truncate; stigmas 2.

Wet meadows, tidal flats, and bogs; in near coastal to coastal and insular western, southwestern, southern, and southeastern Alaska and southwestern Yukon; south to California and disjunctly to the Atlantic coast; Greenland, Iceland, Asia (C. *cryptocarpa* C. A. Mey.; C. *lyngbyei* ssp. *cryptocarpa* [C. A. Mey.] Hultén).

Carex mackenziei Krecz.
Mackenzie Sedge

Key II type; caespitose, 1–4 dm tall; leaf sheaths straw-colored to brownish, blades of lower sheaths short, elongating upwards, 1–3 mm broad, flat; lowermost bract 0.3–0.9 cm long, scalelike or with a short, subulate blade; inflorescence 1.5–4.5 cm long, erect; spikes 2–6, all gynaecandrous, sessile, erect or ascending, the terminal club shaped, with the lower one-half to two-thirds staminate; pistillate scales from subequal to longer than the perigynia, ovate to oval, obtuse to rounded, brown or yellowish brown with hyaline margins and green to straw-colored median; perigynia 2.5–3.5 mm long, 1.2–1.6 mm broad, elliptic to ovate, appressed-ascending, wingless, glabrous, substipitate, veined on both sides, straw-colored to greenish, the beak 0.1–0.5 mm long, serrulate; stigmas 2.

Shallow ponds, bogs, and muskegs; in coastal western, southern, and southeastern Alaska; disjunctly east to the Atlantic; Eurasia. C. *mackenziei* is a close relative of both C. *glareosa* and C. *lachenalii*. The three might best be considered as portions of a single species.

Carex macloviana d'Urville
Thick-head Sedge

Key II type; caespitose, 1.5–5 dm tall; leaf sheaths brownish to yellowish, the blades of lowermost sheaths lacking or short, developed upwards, 1–4.5 mm broad, flat; lowermost bract 0.4–3 cm long, sheathless, the blade subulate to leaflike; inflorescence 1–2 cm long, 0.7–2 cm broad, erect; spikes 3–10, gynaecandrous, sessile, congested and headlike or the lowermost separated by 1–3 mm; pistillate scales shorter than the perigynia, lanceolate, attenuate to acute, brown with hyaline margins and commonly paler median; perigynia 3–4.5 mm long, 1.2–2 mm broad, lanceolate, appressed-ascending, winged, glabrous, subsessile, finely veined on both sides, greenish to brown apically and marginally, straw-colored below, the beak 1–2 mm long, flattened, serrulate, hyaline apically, minutely bidentate; stigmas 2.

Bogs, open woods, heathlands, open slopes, flats, and meadows; in southern Alaska, from Unalaska east through the Panhandle, and in the southeastern quarter of mainland Alaska and southern to northern Yukon; east to Labrador and south to California, Utah, and Colorado; Greenland (C. *pachystachya* Cham.; C. *macloviana* ssp. *pachystachya* [Cham.] Hultén; C. *microptera* Mack.). C. *macloviana* approaches C. *phaeocephala* in some respects and might be confused in some portions of its range with that entity.

Carex macrocephala Willd.
Large-head Sedge

Key II type; dioecous (rarely monoecious), rhizomatous, 1–4 dm tall; leaf sheaths brownish, sometimes obscured by old leaves, the blades of basal sheaths short or lacking, elongating upwards, 4–10 mm broad, channeled; lowermost bract scalelike or to 6 cm long and the blade leaflike; inflorescence mostly 4–6(7) cm long and 1–4 cm broad, erect, the pistillate inflorescence with several to many spikes, all sessile, compactly clustered; staminate inflorescences with numerous spikes, all sessile; pistillate scales shorter than the perigynia, lanceolate, acuminate to awned, chestnut brown with hyaline margins and green midvein or with broad straw-colored median; perigynia 10–12(13) mm long, (3) 4–6 mm broad, lanceolate, ascending, winged, spinulose and serrulate, glabrous, substipitate, strongly veined on both sides, brownish, the beak 4–6 mm long, serrulate, sharply bidentate; stigmas 3.

Sandy seashores; in coastal southern Alaska from the Alaska Peninsula eastward through the Panhandle; south to Oregon; Asia. Because of the size of the spikes this is the most singular species of *Carex* in our flora.

Carex macrochaeta C. A. Mey.
Long-awn Sedge

Key III type; rhizomatous, 1–9 dm tall; leaf sheaths purplish brown (at least below), blades of lower sheaths short or lacking, elongating upwards, 2–5 mm broad, flat with revolute margins; lowermost bract 2.5–16 cm long, tubular-sheathing, the sheath 3–10 mm long or more, the blade leaflike; inflorescence 3–15 cm long, more or less nodding; spikes 2–5, the terminal staminate, the lateral pistillate, separated by 12–70 mm and with slender peduncles 1–8 cm long, erect or more commonly spreading to nodding; pistillate scales longer than the perigynia, oblong to obovate or lanceolate, awned (the awn often over 3 mm long), purplish black to black with more or less hyaline margins and straw-colored to greenish median; perigynia 3–5 mm long, 1–2 mm broad, lanceolate to elliptic, appressed-ascending, wingless, glabrous, substipitate, finely veined, straw-colored, or greenish, sometimes purplish spotted, the beak 0.1–0.3 mm long, the apex truncate or nearly so; stigmas 3.

Meadows, sandy beaches, bogs, and heathlands; in coastal western, southwestern, southern, and southeastern Alaska and less commonly in interior southern Alaska and Yukon; south to Oregon; Asia.

Carex magellanica Lam.
Bog Sedge

Key VI type; rhizomatous and more or less caespitose, 1–8 dm tall; leaf sheaths brownish, blades of lower sheaths lacking, developed upwards, 2–4 mm broad, flat; lowermost bract 2–11 cm long, sheathless, the blade leaflike; inflorescence 3–12 cm long, more or less nodding; spikes 2–6, the terminal staminate, sometimes gynaecandrous, the lateral pistillate or gynaecandrous, separated by 9–40 mm and with peduncles 1.2–5 cm long, more or less pendulous; pistillate scales longer than the perigynia, lanceolate, attenuate to shortly awned, brown to chestnut with greenish to straw-colored median; perigynia 2.2–3 mm long, 1.3–1.8 mm broad, obovate to elliptic, ascending, wingless, glabrous, short-stipitate, obscurely veined, greenish to brown, the beak about 0.1 mm long, truncate to obscurely bidentate; stigmas 3.

Bogs, muskegs, and lake shores; in the southeastern third of Alaska and disjunctly in the Brooks Range; east to Newfoundland and south to Utah, Colorado, Minnesota, and Pennsylvania; circumboreal (*C. magellanica* ssp. *irrigua* [Wahl.] Hultén; *C. paupercula* Michx.).

Carex maritima Gunn.
Curved Sedge

Key II type; rhizomatous, 0.5–3 dm tall; leaf sheaths brownish to hyaline, the blades of lower sheaths short or lacking, developed upwards, commonly 0.5–2 mm broad, flat or involute; lowermost bract scalelike, inflorescence 0.7–1.5 cm long, 0.5–1.4 cm broad, erect; spikes 3–5 or more, androgynous, sessile, spreading or spreading-ascending, borne in a compact head and appearing as a single spike; pistillate scales shorter than the perigynia, ovate to lanceolate, acute to obtuse, brownish with broad hyaline margins and lighter median; perigynia 2.8–3.7 mm long, 1–1.6 mm broad, lanceolate to elliptic, ascending, wingless, glabrous, short-stipitate, veined on both sides, brown to straw-colored, tapering to a slender serrulate to smooth beak, the beak 0.5–1 mm long, obliquely cleft; stigmas 2.

Stream banks, lake shores, bogs, and spits, in arctic and alpine tundra, heathlands, and open woods; in northern, western, and southeastern Alaska and in northern and southwestern Yukon; east to Newfoundland; circumpolar (*C. incurva* Lightf.; *C. incurviformis* Mack.).

Carex media R. Br. ex Richards.

Key VI type; caespitose (and more or less rhizomatous), 1–9 dm tall; leaf sheaths purplish brown to brownish, the blades of lower sheaths short or lacking, elongating upwards, 1–3 mm broad, flat, the margins revolute; lowermost bract 0.5–4.5 cm long, sheathless, subulate to leaflike; inflorescence 1–3 cm long, erect; spikes usually 3(2–4), the terminal gynaecandrous, the lateral pistillate, clustered or the lower separated by 2–11 mm and with peduncles 0.1–0.7 cm long, erect or ascending; pistillate scales shorter than the perigynia, ovate, obtuse to acute, purplish brown to purplish black; perigynia 2–3(3.5) mm long, 0.7–1.2 mm broad, elliptic to oblanceolate, ascending, wingless, glabrous, substipitate, veinless, greenish to straw-colored or brownish, the beak 0.2–0.5 mm long, minutely bidentate, often brownish; stigmas 3.

Stream banks, muskegs, open woods, shallow ponds, meadows, and bogs; in most of continental Alaska (except for northern and southwestern portions) and southern Yukon; east to Newfoundland and south to Utah, Colorado, Minnesota, and Michigan; Eurasia (*C. norvegica* Retz ssp. *inferalpina* [Wahl.] Hultén; *C. angarae* Steud.; *C. vahlii* authors, not Schkuhr.).

Carex membranacea Hook.
Fragile Sedge

Key V type; rhizomatous, 1.5–7 dm tall; lower leaf sheaths purplish to brownish, septate-nodulose, the blades of lowermost sheaths well developed, 2–6 mm broad, flat with revolute margins, sometimes channeled apically and basally; lowermost bract 2.5–15 cm long, sheathless or the sheath 1–15 cm long or more, the blade leaflike; inflorescence 4–14 cm long, erect or nearly so; spikes 2–5, the upper 1–3(4) staminate, the lower 1–3 pistillate, 1.5–5.5 cm long, separated by 10–110 mm and with stiff, erect peduncles 0.2–9.5 cm long; pistillate scales from shorter to longer than the perigynia, ovate to lanceolate, rounded to obtuse or acute, brownish to purplish black with more or less white-hyaline margins and brownish to purplish black median; perigynia 3–4.7 mm long, 1.5–2.5 mm broad, elliptic to obovate, spreading to descending, inflated, membranous, wingless, glabrous, short-stipitate, obscurely veined, purplish black or the base straw-colored or rarely straw-colored to greenish throughout, abruptly beaked, the beak 0.3–0.7 mm long, bidentate to emarginate; stigmas 3.

Wet meadows, muskegs, bogs, lake shores, and shallow ponds; in most of Alaska, except for the southern coastal portion, and in most of the Yukon; east to Labrador and south to British Columbia; Asia. *C. membranacea* is allied to both *C. saxatilis* and *C. rostrata*. Apparent intermediates are difficult to place with certainty.

Carex mertensii Prescott ex Bong.
Mertens Sedge

Key VI type; caespitose from stout vertical rhizomes, 3–12 dm tall; leaf sheaths purplish to brownish purple, the blades of lower sheaths short or lacking, elongating upwards, 3–10 mm broad, flat with revolute margins; lowermost bract 8–27 cm long, sheathless or short-sheathing, the blade leaflike; inflorescence 5.5–18 cm long, more or less nodding; spikes (4)5–8, all gynaecandrous, the terminal conspicuously so, the lower separated by 15–50 mm or more, and with slender peduncles 2–5 cm long; pistillate scales equaling or shorter than the perigynia, lance-elliptic, acute to shortly awned, purplish brown with narrow hyaline margins and lighter median; perigynia 3.4–5 mm long, 2.5–3.5 mm broad, ovate to oval, appressed-ascending, flattened, membranous, wingless, glabrous, sessile, 2-veined, whitish with greenish to straw-colored apex, the beak 0.2–0.5 mm long, emarginate, whitish to purplish tipped; stigmas 3.

Open slopes, in woods, meadows, and bogs; in coastal (rarely interior) southern and southeastern Alaska; south to California and east to Montana. This is one of our most distinctive sedges.

Carex microglochin Wahl.

Key I type; rhizomatous, 0.6–2(2.5) dm tall; leaf sheaths brownish, the blades of lower sheaths short, elongating upwards, less than 1 mm broad, involute; lowermost bract scalelike, deciduous; spikes solitary, androgynous, 0.5–1.3 cm long, 0.5–0.8 cm broad; pistillate scales shorter than the perigynia, ovate to lance-elliptic, obtuse to acutish, brown with hyaline margins and lighter median; perigynia 3.5–4.2 mm long (exclusive of the protruding rachilla 0.5–2 mm long), 0.5–0.8 mm broad, lance-linear, abruptly reflexed by a basal bend, sessile, wingless, glabrous, tapering to the apex, several-veined, brownish to straw-colored, the apex hyaline; stigmas 3, appearing lateral on the rachilla.

Marshy heath, lake shores, and stream banks; in western and northern Alaska, in the Alaska Range, and in northern and southern Yukon; east to Labrador and Greenland; Eurasia.

Carex misandra R. Br.
Short-leaf Sedge

Key VI type; caespitose, 0.8–3.5 dm tall; leaf sheaths brownish, the blades of basal sheaths well developed, 1–3 mm broad, flat or channeled; lowermost bract 1.2–5 cm long, tubular sheathing, the sheath 0.7–2.5 cm long, the blade leaflike to subulate; inflorescence 4–14 cm long, nodding; spikes 2–4, the terminal gynaecandrous, the lateral pistillate, the lower separated by 18–80 mm, with slender drooping peduncles 1.5–8 cm long; pistillate scales shorter than the perigynia, ovate to lanceolate or elliptic, obtuse to acuminate, purplish black with hyaline margins, the median light to dark; perigynia (3.5)4–5 mm long, 1–1.3 mm broad, lance-attenuate, appressed-ascending, wingless, glabrous, short-stipitate, veinless, purplish black or the base straw-colored, tapering to a flattened serrulate beak, the beak blending with the body, white-hyaline apically; stigmas 3.

Dry sphagnum slopes, meadows, heathlands, and beaches; in western, northern, and south-central to southeastern Alaska and much of the Yukon; east to Labrador; circumboreal. *C. misandra* is allied to *C. atrofusca*.

Carex nardina Fries

Key I type; caespitose, 0.3–2 dm tall; leaf sheaths brown, persistent, the blades of lower sheaths well developed, less than 1 mm broad, filiform, lowermost bract scalelike, 0.3–0.5 cm long, bladeless or with a short, subulate blade; spikes solitary, androgynous, 6–13 mm long, 3–5 mm broad; pistillate scales commonly longer than the perigynia, ovate to oval, rounded to obtuse, dark brown with broad hyaline margins and lighter median; perigynia 3–4.5 mm long, 1.2–1.8 mm broad, elliptic to lance-ovate, short-stipitate, wingless, glabrous, obscurely veined, brown to straw-colored, tapering to a serrate beak, the beak 0.2–0.5 mm long, emarginate to truncate, hyaline apically; stigmas 2.

Ridges, talus slopes, sandy terraces, and meadows; in western, northern, and southeastern Alaska and most of the Yukon; east to Labrador and south to Washington and Colorado (*C. hepburnii* Boott).

Carex nesophila Holm
Bering Sea Sedge

Key VI type; rhizomatous, 1–5.5 dm tall;

leaf sheaths chestnut to brownish to straw-colored, the blades of lower sheaths lacking, elongating upwards, 2–4.5 cm long, flat with revolute margins; lowermost bract 1.8–8 cm long, sheathless or seldom with closed sheaths 0.2–0.6 cm long, the blade leaflike; inflorescence 2–7(9) cm long, more or less nodding; spikes 2–4, the terminal 1 (rarely 2) staminate, the lateral pistillate, the lower separated by 6–30 mm, with erect to ascending peduncles 0.3–1.8 (3) cm long; pistillate scales from shorter than to longer than the perigynia, ovate to lanceolate, acute to obtuse, purplish brown to purplish black, the margins hyaline, the midvein brown to straw-colored; perigynia 2.4–3.5 mm long, (1)1.2–1.8 mm broad, obovate to ovate, appressed-ascending, obscurely veined on both sides, wingless, substipitate, glabrous, purplish brown or brown only apically or greenish to straw-colored throughout, the beak 0.1–0.3 mm long, truncate to emarginate, not hyaline apically; stigmas 3.

Ridge tops and slopes; in alpine tundra or heathlands; in western, southwestern, and south-central Alaska; Asia. *C. nesophila* is apparently closely allied to *C. podocarpa* and all specimens cannot be placed with one or the other entity with certainty.

Carex nigricans C. A. Mey.
Blackish Sedge

Key I type; rhizomatous, 0.5–4.5 dm tall; leaf sheaths brownish to straw-colored, the blades of lowermost sheaths short or lacking, elongating upwards, 1–2.5 mm broad, flat or channeled; lowermost bract scalelike; spikes solitary, 0.8–1.7 cm long, 0.5–0.9 cm broad, androgynous; pistillate scales deciduous, about equaling the perigynia, lanceolate to ovate, obtuse to acute or attenuate, brown with lighter midvein; perigynia 3.5–4.8 mm long, 1–1.8 mm broad, lanceolate, ascending to spreading, veinless, wingless, glabrous, with stipes 0.6–1.2 mm long, brownish, the beak 0.3–0.6 mm long, oblique apically, hyaline; stigmas 3.

Alpine meadows and mountain slopes; in coastal and insular southern Alaska from the Aleutians through the Panhandle; south to California and Colorado.

Carex obtusata Lilj.

Key I type; rhizomatous, 0.4–2.5 dm tall; leaf sheaths purplish brown, the blades of lower sheaths lacking, developed upwards, 0.5–2 mm broad, flat or channeled; lowermost bract scalelike; spikes solitary, 0.5–1.2 cm long, 0.2–0.6 cm broad, androgynous; pistillate scales from shorter to longer than the perigynia, ovate to obovate, acuminate to obtuse, brown with hyaline margins and lighter midvein; perigynia 3–4 mm long, 1.5–2 mm broad, obovate, spreading-ascending, obscurely veined, wingless, glabrous, subsessile, chestnut brown to blackish, shining, the beak 0.5–0.8 mm long, with a flaring hyaline apex; stigmas 3.

Dry bluffs, shale outcrops, and hilltops; in northern, central, and east-central Alaska and in northern and southwestern Yukon; east to the Mackenzie and south to New Mexico; Eurasia.

Carex oederi Retz.
Oeder Sedge

Key V type; caespitose, 0.5–5 dm tall or more; leaf sheaths straw-colored, the blades of lower sheaths well developed, 1–4 mm broad, flat or channeled basally; lowermost bract 3–15 cm long, sheathless or with closed sheaths 1–10 mm long, the blades leaflike; inflorescence 1.2–6 cm long, erect; spikes 2–5, the terminal staminate, the lateral pistillate, all clustered, or the lower separated by 5–35 mm or more, and with peduncles 0.1–2 cm long or more, ascending to spreading; pistillate scales shorter than the perigynia, ovate, obtuse to acute, brownish to yellowish with hyaline margins and lighter median; perigynia 2–3 mm long, 1–1.5 mm broad, obovate to elliptic, spreading, wingless, glabrous, substipitate, veined on both sides, green to yellowish, the beak 0.5–0.8 mm long, bidentate.

Marshes, muskegs, and shallow ponds; in east-central and southeastern Alaska and southern Yukon; east to the Atlantic and south to California, Utah, New Mexico, Indiana, and New Jersey; Eurasia (*C. viridula* Michx.). Our plant belongs to ssp. *viridula* (Michx.) Hultén.

Carex parryana Dewey
Parry Sedge

Key VI type; rhizomatous (and loosely caespitose), 1.5–6 dm tall; leaf sheaths brown to pale brownish, the blades of lower sheaths obsolete, developed upwards, 2–3 mm broad, with revolute margins; lowermost bract 0.9–5.5 cm long, sheathless, the blade linear-subulate; inflorescence 1.5–6 cm long, erect; spikes 2–5, the terminal gynaecandrous, the lateral pistillate, erect, the lower separated by 5–30 mm, with peduncles 0.1–0.7 cm long; pistillate scales subequal to the perigynia or shorter, ovate to oval, obtuse to mucronate, dark brown with hyaline margins and lighter midvein; perigynia 2–2.5 mm long, 1–1.5 mm broad, obovate, appressed-ascending, wingless, ciliate, substipitate, veinless, straw-colored or purplish apically, the beak 0.1–0.2 mm long, emarginate, ciliate; stigmas 3.

Tidal flats, marshes, and wet meadows; in central to southern Alaska and southern Yukon; east to Hudson Bay and south to Alberta.

Carex pauciflora Lightf.

Key I type; rhizomatous, 1–3(4) dm tall; leaf sheaths brownish, the blades of lower sheaths obsolete, developed upwards, 0.5–1.8 mm broad, flat or more commonly involute or channeled; lowermost bract scalelike, deciduous; spikes solitary, androgynous, 0.5–1.5 cm long, 0.5–1.2 cm broad; pistillate scales subequal to or shorter than the perigynia, lance-attenuate, obtuse, pale brownish with hyaline margins and light median; perigynia 6–7 mm long, 1–1.5 mm broad (the rachilla not exserted), lance-

linear, abruptly reflexed by a basal bend, sessile, wingless, glabrous, tapering to the apex, several-veined, greenish to straw-colored or brownish, the apex hyaline to brownish; stigmas 3.

Marshes, bogs, and muskegs; in southern Alaska, mostly near the coast from Kodiak Island and the Alaska Peninsula eastward through the Alexander Archipelago; east to Newfoundland and south to Washington, Minnesota, and Connecticut; Eurasia.

Carex peckii Howe ex Peck
Peck Sedge

Key IV type; caespitose (and more or less rhizomatous), 1–5 dm tall or more; leaf sheaths purplish to brownish, the blades of lowermost sheaths short to elongate, developed upwards, 1–2 mm broad, flat; lowermost bract 0.4–1.2 cm long, sheathless, the blade subulate; inflorescence 0.3–2.5 cm long, erect; spikes 2–5, the terminal staminate, the lateral pistillate, all clustered or the lower separated by 2–7 mm, subsessile, erect or ascending; pistillate scales shorter than the perigynia, ovate, obtuse to acuminate, brownish with broad hyaline margins and greenish median; perigynia 3–3.5 mm long, 1–1.5 mm broad, obovate, ascending, wingless, short-hairy overall, short-stipitate, veinless, greenish, the beak 0.4–0.6 mm long, hyaline at the oblique apex; stigmas 3.

Open woods; reported from central Alaska and west central Yukon; disjunctly eastward to Quebec and south to British Columbia, South Dakota, and Michigan.

Carex petasata Dewey

Key II type; caespitose, 2–8 dm tall; leaf sheaths brownish to straw-colored, the blades of lower sheaths short, elongating upwards, 1.5–4 mm broad, flat or more or less channeled; lowermost bract 0.3–2.5 cm long, the blade (when present) linear-subulate or seldom leaflike; inflorescence 2–5 cm long, erect to nodding; spikes (1)

2–6, all gynaecandrous, tapering to a slender, staminate base, sessile, more or less clustered, or the lower separated by 4–10 mm, ascending; pistillate scales from shorter to longer than the perigynia, lanceolate, acute, yellowish brown with broad hyaline margins and greenish to straw-colored median; perigynia 5–7 mm long, 1.7–2.2 mm broad, lanceolate, appressed-ascending, wing-margined and serrulate, glabrous, substipitate, veined on both sides, greenish to brownish, tapering to a slender flattened beak, the beak 1–2 mm long, serrulate, obliquely cleft; styles 2.

Meadows and open woods; reported from along the Yukon River Valley in southwestern Yukon; south to Oregon, Nevada, Utah, and Colorado. *C. petasata* is similar to *C. praticola* and might be confused with that entity.

Carex petricosa Dewey

Key IV or VI type; caespitose (and often shortly rhizomatous), 1.4–6 dm tall; leaf sheaths brown, the blades of basal sheaths short, elongating upwards, 0.5–3 mm broad, flat or channeled; lowermost bract 1.5–10 cm long with closed sheath 6–30 mm long or more, the blade leaflike; inflorescence 4–13 cm long, more or less nodding; spikes 3–7, the terminal 1–3 staminate or androgynous, the lateral 1–4 pistillate, the lower separated by 20–80 mm, and with filiform ascending to pendulous peduncles 1.7–9 cm long; pistillate scales commonly shorter than the perigynia, lanceolate to elliptic, acute to acuminate or short awned, brown to purplish brown with hyaline margins and light median; perigynia 3.5–6.3 mm long, 1–2 mm broad, lanceolate to lance-ovate, appressed-ascending, wingless, sparsely to moderately short-hairy, or merely ciliate marginally, substipitate, obscurely veined, greenish to straw-colored or brownish apically, the beak 0.4–0.8 mm long, white-hyaline apically; stigmas 3.

Dry, gravelly slopes and rock outcrops; in much of continental Alaska from the Seward Peninsula eastward along the Brooks and Alaska ranges and through the Yukon Valley and in much of the Yukon; east to Labrador and south to Alberta; Asia (*C. franklinii* Boott ex Hook.).

Carex phaeocephala Piper

Key II type; caespitose, 0.8–2(2.5) dm tall; leaf sheaths brown, the blades of lower leaves lacking or short, developed upwards, 1–3 mm broad, flat or channeled to involute; lowermost bract 0.5–2 cm long, scalelike or the blade subulate; inflorescence 1.5–3.5 cm long, 0.8–1.7 cm broad (pressed), erect or nearly so; spikes 2–6, all gynaecandrous, all clustered or the lower separated by 2–11 cm, all sessile or the lower with peduncles to 3 mm long; pistillate scales subequal to and more or less obscuring the perigynia, lanceolate, acute to obtuse, brown with white-hyaline margins and lighter median; perigynia 4–6 mm long, 1.3–2.5 mm broad, lanceolate, appressed-ascending, winged, glabrous except for serrulate margin, substipitate, lightly veined dorsally, straw-colored to brownish, tapering to a flattened beak, the beak 0.8–1.2 mm long, obliquely cleft, the apex hyaline; stigmas 2.

Wet meadows; in south-central and southeastern Alaska and southern Yukon; east to the Mackenzie and south to California, Utah, and Colorado.

Carex phyllomanica Boott
Stellate Sedge

Key II type; caespitose, 1.4–6.5 dm tall; leaf sheaths straw-colored to brownish, blades of lower sheaths obsolete, elongating upwards, 1–3.5 mm broad, flat or channeled; lowermost bract 0.2–1.5 cm long, scalelike or the blade linear-subulate; inflorescence 1.5–2.5 cm long, erect; spikes 2–4(5), all gynaecandrous, sessile, the lower separated by 3–6 mm, ascending-spreading; pistillate scales shorter than the perigynia, ovate, acute to obtusish, greenish or brownish with translucent hyaline margins and lighter median; perigynia 3–4.5 mm long, 1.2–1.7 mm broad, lanceolate, spreading, wingless (merely sharp edged), glabrous, sessile, veined on both sides, straw-colored to brownish, tapering to a flattened serrulate beak, the beak 1–1.5 mm long, bidentate; stigmas 2.

Bogs and muskegs; in coastal and insular southern and southeastern Alaska and less commonly in interior southern Alaska; south to California.

Carex pluriflora Hultén
Many-flower Sedge

Key VI type; rhizomatous, 1–4.5 dm tall; leaf sheaths purplish black to brown or tan, blades of lower sheaths obsolete, elongating upwards, 1.2–4 mm broad, flat with revolute margins; lowermost bract 0.5–2.5 (4.5) cm long, sheathless or short-sheathing, the blade linear-subulate to leaflike; inflorescence 3–7.5 cm long, nodding (at least the lower spikes); spikes 2–3, the terminal staminate, the lateral pistillate, with usually 10 or more fertile scales, the lower separated by 8–35(45) mm and with filiform, arched peduncles 0.7–3 cm long; pistillate scales commonly longer than the perigynia, ovate to elliptic, acute to cuspidate, purplish black with lighter median; perigynia 3.5–4.5 mm long, 1.7–2.3 mm broad, ovate, ascending, wingless, glabrous, substipitate, veined on both sides, greenish to brown, beakless, the apex truncate; stigmas 3.

Bogs, muskegs, and wet meadows; in coastal and insular southern Alaska and less commonly some distance from the coast in south-central Alaska; southward to Washington (*C. stygia* authors, not Fries; *C. rariflora* ssp. *stygia* Hultén; *C. limosa* var. *stygia* Vasey). This species is apparently closely related to *C. rariflora*.

Carex podocarpa R. Br. ex Richards.
Short-stalk Sedge

Key VI type; rhizomatous, 0.7–7 dm tall; leaf sheaths brown to purplish black, blades of lower sheaths obsolete, elongating upwards, 2–6.5 mm broad, flat, with revolute margins; lowermost bract 2.5–9 cm long, sheathless, the blade leaflike; inflorescence 2.5–9 cm long, more or less nodding; spikes 2–5, the terminal 1–2 staminate, the lateral pistillate (rarely some androgynous), the lower separated by 8–30 mm and with capillary arched peduncles 1–4 cm long; pistillate scales longer than the perigynia, lanceolate, acute to attenuate, purplish brown to purplish black; perigynia 3.3–4.5 mm long, 1.3–2.2 mm broad, lanceolate to ovate or elliptic, appressed-ascending, wingless, glabrous, substipitate, veinless, greenish to straw-colored, often purplish spotted, the beak 0.2–0.5 mm long, truncate to emarginate; stigmas 3.

Heathlands, alpine tundra, open woods, and meadows; in most of Alaska (except for the northern and southern coastal portions) and most of the Yukon; east to Mackenzie and south to Oregon and Wyoming; Asia (*C. microchaeta* Holm).

Carex praegracilis Boott

Key II type; rhizomatous, 2–4 dm tall (seldom more); leaf sheaths purplish black to gray or straw-colored, blades of lower sheaths obsolete, developed upwards, 1–4 mm broad, flat or channeled; lowermost bract 0.7–1.8 cm long, sheathless, bladeless or the blade lance-subulate; inflorescence 1.5–4.5 cm long, erect; spikes 5–10 or more, androgynous or the upper staminate, sessile, all clustered or the lower separated by 2–8 mm, ascending; pistillate scales shorter than or equaling the perigynia, ovate to lanceolate, acute to acuminate or shortly awned, brownish to straw-colored with broad hyaline margins and lighter median; perigynia 3.5–4.5 mm long, 1.2–1.6 mm broad, lanceolate, appressed-ascending, wingless, sharp edged, glabrous except for the serrulate margin, veined on both faces, short-stipitate, straw-colored or greenish to brownish, tapering to a flattened beak, the beak 0.8–1.2 mm long, serrulate, obliquely cleft; stigmas 2.

Meadows and grasslands; in east-central Alaska and southern Yukon; south to Mexico, and east to Manitoba and Iowa; South America.

Carex praticola Rydb.
Meadow Sedge

Key II type; caespitose, (1.5)2–8(10) dm tall; leaf sheaths brownish to straw-colored, the blades of lower sheaths short, elongating upwards, 2–5 mm broad, flat; lowermost bract 0.3–5.5 cm long, the blade (when present) linear-subulate; inflorescence 2–7 cm long, erect to nodding; spikes 3–8(10), all gynaecandrous, tapering to a slender staminate base, sessile, clustered, or the lower separated by 5–20 mm or more, ascending; pistillate scales shorter than to subequal to the perigynia, lanceolate, acuminate to acute, yellowish brown except for the green to brown median and hyaline margins; perigynia 3.5–5.2 mm long, 1.3–1.8 mm broad, lanceolate to lance-ovate, appressed-ascending, wing-margined and serrulate, glabrous, substipitate, veined on both sides or the ventral veins suppressed, greenish to brownish, tapering to a slender flattened beak, the beak 1–2 mm long, serrulate, obliquely cleft; styles 2.

Open woods, muskegs, bluffs, floodplains, and meadows; in much of Alaska and Yukon south of the 66th parallel (except for the southwestern portion of Alaska); east to Newfoundland and south to California, Colorado, and Quebec (*C. aenea* Fern.; *C. foenea* authors, not Willd.?).

Carex preslii Steud.

Key II type; caespitose, 2–7 dm tall; leaf sheaths blackish, the blades of lower sheaths short or obsolete, developed upwards, 1–4 mm broad, flat; lowermost bract scalelike; inflorescence 1–2 cm long, erect; spikes 3–8, gynaecandrous, sessile, clustered or the lower separated by 1–3 mm, spreading to ascending; pistillate scales shorter than to subequal to the perigynia, lanceolate to ovate, acute to acuminate, brown with hyaline margins and green to straw-colored median; perigynia 3.5–4.2 mm long, 1.5–2.2 mm broad, lance-ovate, ascending, wing-margined, obscurely veined, substipitate, greenish to brownish, tapering to an elongate flattened beak 0.8–1.5 mm long, the beak serrulate, obliquely cleft, more or less bidentate; stigmas 2.

Wet meadows; in coastal southern Alaska from the Kenai Peninsula east to Yakutat; disjunctly south to California and Montana.

Carex pyrenaica Wahl.

Key I type; caespitose, commonly 1–3 dm tall; leaf sheaths brownish to chestnut, blades of lower sheaths obsolete, developed upwards, 0.5–1.5 mm broad, channeled; lowermost bract scalelike; spikes solitary, androgynous, 0.8–2.1 cm long, 0.4–0.7 cm broad, erect; pistillate scales shorter than to subequal to the perigynia, lanceolate, attenuate to acute, brown to chestnut with hyaline margins and lighter median; perigynia 2.8–4.1 mm long, 0.9–1.3 mm broad, lanceolate, ascending, wingless, veinless, stipitate (the stipe 0.4–0.8 mm long), brownish to straw-colored, tapering to an elongate glabrous beak, the beak 0.2–0.6 mm long, oblique, hyaline apically; stigmas 3.

Alpine meadows; in central and south-central to southeastern and southwestern Alaska and southern Yukon; east to Mackenzie and south to Oregon and Colorado (*C. micropoda* C. A. Mey.). Our material belongs to ssp. *micropoda* (C. A. Mey.) Hultén.

Carex rariflora (Wahl.) Smith

Key VI type; rhizomatous, 0.8–2.5 dm tall; leaf sheaths brown to gray or straw-colored, the blades of lower sheaths short or obsolete, developed upwards, 1.5–2.5 mm broad, flat or channeled to involute; lowermost bract 0.5–3 cm long, sheathless or short-sheathing, the blade when present subulate to leaflike; inflorescence 2–5.5 cm long, nodding (at least the lower spikes); spikes 2–4, the terminal staminate, the lateral pistillate, with usually 2–10 fertile scales, the lower separated by 6–25 mm and with filiform, more or less arched peduncles 0.5–2 cm long; pistillate scales from shorter to longer than the perigynia, elliptic to ovate, obtuse to acute or shortly awned, purplish black with lighter median; perigynia 3.5–4.5 mm long, 1.4–2 mm broad, ovate to elliptic, ascending, wingless, glabrous, substipitate, veined on both sides, greenish to straw-colored, beakless, the apex truncate; stigmas 3.

Shallow ponds, lake shores, meadows, tundra, heath, and muskegs; in northern, western, central, and south-central Alaska and in northern Yukon; eastward to Newfoundland and south to Maine; circumboreal (*C. limosa* var. *rariflora* Wahl.). *C. rariflora* is closely allied to *C. pluriflora* with which it is partially sympatric.

Carex rhynchophysa C. A. Mey.

Key V type; rhizomatous, 5–10 dm tall or more; leaf sheaths purplish to brownish purple or straw-colored, septate-nodulose, glabrous, blades of lower sheaths well developed, 3–13 mm broad, flat with revolute margins, sometimes channeled apically or basally; lowermost bract 12–35 cm long or more, sheathless or with a sheath 1–15 mm long or more, the blade leaflike; inflorescence 10–35 cm long, erect or nearly so; spikes 4–9, the upper 1–6 staminate, the lower 1–4 pistillate (or some androgynous or rarely gynaecandrous), (2)3–9 cm long, separated by 30–100 mm or more and with stiff, erect peduncles 0.3–10 cm long (rarely longer); pistillate scales shorter to longer than the perigynia, lanceolate, acute to attenuate, straw-colored to purplish black with dark to hyaline margins and dark to light median; perigynia 4.8–7.5 mm long, 2.2–3 mm broad, ovoid- to obovoid-inflated, spreading-ascending to spreading, membranous, wingless, glabrous, short-stipitate, obscurely veined, straw-colored to brownish or greenish, more or less abruptly beaked, the beak 0.7–2 cm long, bidentate, the teeth 0.3–1 mm long; stigmas 3.

Bogs, muskegs, heathland and tundra, stream banks, lake and stream shores, and mud flats; in southwestern, south-central, and east-central Alaska and southern Yukon; Eurasia. This entity is apparently closely related to both *C. membranacea* and *C. rostrata*, from which all specimens cannot be segregated with certainty.

Carex rossii Boott
Ross Sedge

Key IV type; caespitose, 0.6–2.8 dm tall; leaf sheaths purplish brown to reddish brown, the blades of the lower sheaths short to well developed, 0.4–2 mm broad, flat; lowermost bract 1–7.5 cm long, sheathless, merely auricled basally, the blade leaflike; inflorescence 1–3 cm long, erect or nearly so; spikes (1)2–5, the terminal staminate, the lateral pistillate, erect or ascending, the lower ones with peduncles 0.1–1.2 cm long, the upper more or less clustered below the terminal spike; pistillate scales commonly shorter than the perigynia, ovate to lanceolate, acute to acuminate or shortly awned, brownish with broad hyaline margins and greenish median or less commonly hyaline throughout; perigynia 3.5–4.2 mm long, 1.2–1.6 mm broad, elliptic to obovate, ascending, wingless, short-hairy to glabrate, stipitate, veinless, greenish to straw-colored, the beak 0.9–1.3 mm long, greenish, the apex hyaline, bidentate; stigmas 3.

Gravelly flats and slopes, stream banks, open woods, and roadsides usually in dryish sites; in the southeastern quarter of mainland Alaska and southern Yukon; east to the Great Lakes and south to California and Colorado.

Carex rostrata Stokes
Beaked Sedge

Key V type; rhizomatous, 4–10 dm tall or more; leaf sheaths purplish brown to brown, straw-colored, or hyaline, septate-nodulose, glabrous, blades of lower sheaths commonly well developed, 3–12 mm broad, flat with revolute margins or more or less channeled; lowermost bract 15–60 cm long or more, sheathless or with a sheath 1–35 mm long or more, the blade leaflike; inflorescence 14–35 cm long, erect or nearly so; spikes 4–9, the upper 2–6 staminate, the lower 1–4 pistillate (or some androgynous), 2.5–9.5 cm long, separated by 35–105 mm or more, and with stiff, erect peduncles 0.3–4 cm long or more; pistillate scales shorter to longer than the perigynia, lanceolate to lance-oblong, acute to attenuate, brown to brownish with hyaline margins and lighter median; perigynia 4.2–6.5 mm long, 1.7–2.8 mm broad, ovoid-inflated, ascending to ascending-spreading, membranous, wingless, glabrous, short-stipitate, obscurely veined, straw-colored to brownish or greenish, the body tapering to the beak, the beak 1–2 mm long, bidentate, the teeth 0.2–0.8 mm long; stigmas 3.

Stream banks, marshes, lake and pond shores, muskegs, and in shallow ponds; in the southeastern three-quarters of Alaska and southern Yukon; east to the Atlantic and south to California, New Mexico, South Dakota, Indiana, and Delaware; circumboreal (*C. utriculata* Boott).

Carex rotundata Wahl.

Key V type; rhizomatous, 1–4.5 dm tall; leaf sheaths brown to straw-colored, septate-nodulose, blades of lower sheaths well developed, 0.5–2.5 mm broad, involute; lowermost bract 1–12 cm long, sheathless or with a sheath 1–10 mm long, the blade leaflike; inflorescence 2.5–9 cm long, erect; spikes 2–4, the terminal 1–2 staminate, the lower 1–2 pistillate, 0.6–2 cm long, separated by 7–55 mm and subsessile or with stiff erect peduncles 0.1–1.2 cm long; pistillate scales commonly shorter than the perigynia, ovate to elliptic, acute to obtuse, purplish brown to purplish black with hyaline apex and paler median; perigynia 3–4 mm long, 1.8–2.4 mm broad, obovate, spreading, inflated, membranous,

wingless, glabrous, short-stipitate, veinless, purplish black to straw-colored, abruptly beaked, the beak 0.3–1 mm long, emarginate to bidentate; stigmas 3.

Muskegs, pond and lake shores, bogs, meadows, and shallow ponds; in most of continental Alaska (except for the southeastern portion) and in northern Yukon (?); east to Hudson Bay; Eurasia.

Carex rupestris All.
Rock Sedge

Key I type; rhizomatous, 0.5–1.5 dm tall; leaf sheaths brown to blackish, blades of lower sheaths commonly well developed, 0.5–2.5 mm broad, flat to channeled or involute; lowermost bract scalelike; spikes solitary, 0.6–2.5 cm long, 0.2–0.4 mm broad, androgynous; pistillate scales from shorter to longer than the perigynia, ovate to oval, acute to obtuse, yellowish brown with broad hyaline margins and brown to pale median; perigynia 3–4 mm long, 1–1.5 mm broad, obovate to oblanceolate, ascending, obscurely veined, wingless, glabrous (or serrulate along the margin), substipitate, brown to straw-colored, dullish, the beak 0.1–0.3 mm long, neither flaring nor hyaline apically; stigmas 3.

Rocky outcrops, mountain slopes, and meadows; in western, northern, east-central, and south-central Alaska and most of the Yukon; east to Newfoundland and south to Colorado; circumpolar.

Carex sabulosa Turcz.

Key V type; rhizomatous, 1.5–3.5(4) dm tall; leaf sheaths brown, not septate-nodulose, the blades of lower sheaths short to well developed, 0.7–3.5 mm broad, flat with revolute margins to channeled or involate, all aggregated or the lower separated less, the blade subulate; inflorescence 2–5 cm long, erect or nearly so; spikes 3–5, the terminal gynaecandrous, the lateral pistillate, all aggregated or the lower separated by 1–27 mm and sessile or with stiff erect peduncles 0.1–0.7 cm long; pistillate scales commonly longer than the perigynia, ovate to lanceolate, acute to attenuate or acuminate, purplish brown to purplish black with narrow hyaline margins and lighter median; perigynia 4–5.1 mm long, 1.9–2.5 mm broad, obovate, spreading-ascending, wingless, glabrous, short-stipitate, veinless or nearly so, straw-colored to greenish or brownish mottled, the beak 0.6–1 mm long, sharply bidentate; stigmas 3.

Sand dunes; known in North America only from the vicinity of Carcross, Yukon; Asia (*C. leiophylla* Mack.).

Carex saxatilis L.

Key III or VI type; rhizomatous, 1.5–10 dm tall; leaf sheaths chestnut brown to purplish or reddish, blades of lower leaves obsolete, developed upwards, 2–4.5(6.4) mm broad, flat with revolute margins or channeled below; lowermost bract 2–21 cm long, sheathless or rarely with a cylindrical sheath 1–3 mm long or more, the blade leaflike; inflorescence 5–16 cm long, erect to more or less nodding; spikes 2–4, the terminal 1–2(3) staminate, the lower 1–3 pistillate, separated by 20–75 mm or more and with slender more or less arched peduncles 0.6–7.5 cm long; pistillate scales shorter than the perigynia, lanceolate to ovate, acute to acuminate or obtusish, purplish brown or purplish black, the margins narrowly to broadly hyaline, the midrib lighter and commonly with a broadly hyaline apex; perigynia 3.5–5.7 mm long, 1.5–2.5 cm broad, ovate to lanceolate or lance-oblong, ascending to spreading-ascending, wingless, glabrous, subsessile, veinless or nearly so, membranous, purplish brown to purplish black or yellowish to greenish and often purplish mottled, shining at maturity, the beak 0.4–0.7 mm long, truncate to emarginate or shortly bidentate; stigmas 2 (rarely 3).

Shallow ponds, muskegs, stream banks, floodplains, heathlands, and tundra; throughout Alaska and Yukon; south to Utah and Colorado; circumpolar (*C. ambusta* Boott; *C. physocarpa* Presl; *C. procerula* Krecz.; *C. pulla* var. *laxa* Trautv.). Our material belongs to ssp. *laxa* (Trautv.) Kalela.

Carex scirpoidea Michx.

Key I type; rhizomatous, dioecious, 1–5 dm tall; leaf sheaths purplish brown to purplish black or purplish, blades of lower sheaths lacking, developed upwards, 1–3 mm broad, flat or channeled; lowermost bract 0.5–1.5 cm long or more, sheathless, the blade subulate to leaflike; spikes solitary, erect, the pistillate 1.3–3.7 cm long, 0.2–0.4(0.5) cm broad, the staminate 1.0–2.5 cm long, 0.4–0.6 cm broad; pistillate scales shorter to longer than the perigynia, ovate to lanceolate, obtuse to acute, purplish brown with more or less hyaline ciliate margins and light to dark median; staminate scales brown, ciliate; perigynia (2.4)3–4 mm long, (0.9)1.1–1.7 mm broad, obovate to elliptic, appressed-ascending, veinless, wingless, more or less densely hairy, substipitate, purplish brown to straw-colored, the beak 0.2–0.5 mm long, truncate to emarginate, hyaline apically; stigmas 3.

Muskegs, talus slopes, rock outcrops, heathlands, woods, gravelly spits, hilltops, and tundra; in most of Alaska (except for the southwestern portion) and throughout the Yukon; east to Newfoundland and south to British Columbia, Colorado, Michigan, and New Hampshire; Greenland, Europe, Asia (*C. stenochlaena* Holm).

Carex scopulorum Holm

Key III type; rhizomatous, 1–4.5 dm tall; leaf sheaths purplish brown or reddish brown, the blades of lower sheaths lacking or present, developed upwards, 2–6 mm broad, flat with revolute margins; lowermost bract 0.8–6 cm long, the blade leaflike; inflorescence 2.5–6 cm long, erect or nearly so; spikes 2–6, the terminal staminate (rarely androgynous), the lateral pistillate (or the upper androgynous), subsessile or the lower with peduncles to 1.5

cm long; pistillate scales shorter to longer than the perigynia, ovate to obovate, acute to obtuse, black to purplish black, often with more or less hyaline margins and lighter median; perigynia 2.5–3.5 mm long, 1.5–2 mm broad, spreading, obovate, glabrous, substipitate, veinless, greenish or the apex blackish or blackish throughout, the beak 0.2–0.5 mm long, truncate apically; stigmas 2.

Wet meadows, open woods, and stream banks; in east-central Alaska and western Yukon (where evidently rare); southward to California, Nevada, Utah, and Colorado.

Carex sitchensis Prescott
Sitka Sedge

Key III type; rhizomatous (and more or less caespitose), commonly 4–10 dm tall; leaf sheaths reddish brown, the blades of lowermost sheaths obsolete, developed upwards, 2.5–9 mm broad, flat with revolute margins; lowermost bract 15–45 cm long or more, sheathless, the blade leaflike; inflorescence 7–36 cm long, erect or more or less nodding; spikes 4–8, the terminal 1–4 staminate, the lower 2–5 pistillate (or some androgynous), separated by (10)50–150 mm, and with erect to spreading or arched peduncles 0.6–8 cm long; pistillate scales longer than the perigynia, lanceolate to lance-attenuate, acute to attenuate, purplish brown to brownish, the margins often hyaline and the median straw-colored; perigynia 2.2–3 mm long, 1.2–1.7 mm broad, ovate to oval, appressed-ascending, wingless, glabrous, veinless or nearly so, substipitate, greenish to straw-colored, the beak 0.1–0.3 mm long, truncate; stigmas 2.

Bogs and marshes; in coastal and insular south-central Alaska from the Kenai Peninsula eastward through the Panhandle (rarely some distance into the interior); south to California and Idaho.

Carex spectabilis Dewey
Showy Sedge

Key VI type; rhizomatous (and more or less caespitose), 1.5–5 dm tall or more; leaf sheaths purplish to brownish, the blades of lower sheaths obsolete, developed upwards, 2–7 mm broad, flat with revolute margins; lowermost bract 7–15 cm long, sheathless, the blade leaflike; inflorescence 7–14 cm long, erect or nearly so; spikes 3–5, the terminal 1–2 staminate, the lower 2–4 pistillate, separated by 8–80 mm and with peduncles 0.2–0.5 cm long, erect to spreading; pistillate scales commonly longer than the perigynia, lanceolate, acute to shortly awned, purplish brown to purplish black with more or less hyaline margins and lighter median; perigynia 3–4 mm long, 1.5–2 mm broad, appressed-ascending, oval to elliptic, wingless, glabrous (or the margin sometimes serrulate), substipitate, veinless, straw-colored to greenish or blotched purplish, the beak 0.1–0.5 mm long; stigmas 3.

Marshes; in coastal and insular southwestern to southeastern Alaska and reported for the Yukon; south to California and Montana. This is a poorly understood entity which requires additional study.

Carex stenophylla Wahl.

Key II type; rhizomatous, 0.3–1.5 dm tall; leaf sheaths brown to gray brown, the blades of lower sheaths commonly well developed, 0.3–1.2 mm broad, channeled or more or less flattened; lowermost bract 0.5–1 cm long, scalelike or with a short subulate blade; inflorescence 0.8–1.2 cm long, 0.3–0.7 cm broad, erect; spikes 3–5, androgynous, sessile, ascending, borne in a compact head and often appearing as a single spike or the lower separated by 1–4 mm; pistillate scales shorter to longer than the perigynia, ovate, acute to obtuse, brownish with broad hyaline margins and lighter median; perigynia 2.5–3.3 mm long, 1.2–1.5 mm broad, lanceolate, ascending, wingless, glabrous, substipitate, veined above or veinless on both faces, brown to straw-colored, tapering to a slender serrulate beak, the beak 0.4–0.8 mm long, obliquely cleft; stigmas 2.

Steep grassy slopes, river terraces, and sandy flats; in central and east-central Alaska and southern Yukon; disjunctly south to Oregon, Utah, New Mexico, and Iowa; Eurasia (*C. eleocharis* Bailey). Our material belongs to ssp. *eleocharis* (Bailey) Hultén.

Carex stipata Muhl.

Key II type; caespitose, 2.5–7 dm long or more; leaf sheaths brown to straw-colored, blades of basal sheaths obsolete, developed upwards, mostly 3–11(12) mm broad, flat; lowermost bract 0.7–2 cm long, scalelike or the blade linear-subulate; inflorescence 1.5–5.5 cm long, erect; spikes several to many in simple or compound clusters, androgynous, sessile, ascending, all aggregated or the lower separated by 1–4 mm; pistillate scales commonly shorter than the perigynia, lanceolate to ovate, acute to attenuate or shortly awned, brownish to yellowish brown and more or less hyaline, the median straw-colored to greenish; perigynia 4–5 mm long, 1–2 mm broad, lance-attenuate, ascending, wingless, glabrous, short-stipitate, veined on both sides, brownish to straw-colored, tapering to a slender serrulate beak 2–2.5 mm long, the beak bidentate; stigmas 2.

Streamsides, wet meadows, and muskegs; in coastal south-central to southeastern Alaska; east to the Atlantic and south to California, Utah, New Mexico, Kansas, and North Carolina; Asia.

Carex stylosa C. A. Mey.
Variegated Sedge

Key VI type; caespitose, 2–6 dm tall; leaf sheaths brown, blades of lower sheaths short or lacking, developed upwards, 0.8–3 mm broad, flat with revolute margins or channeled below; lowermost bract 1.2–4 cm long, sheathless, the blade leaflike; inflorescence 3–6 cm long, erect or more or less nodding; spikes 2–4, the terminal staminate, the lateral pistillate, with usually 10–more fertile scales, the lower sepa-rated by 15–30 mm and with stiff erect to ascending peduncles 1–3 cm long; pistillate scales from shorter to longer than the perigynia, ovate, obtuse to acute, purplish black with hyaline margins and lighter median; perigynia 2.5–3.5 mm long, 1–1.5 mm broad, ovate to elliptic, ascending, wingless, glabrous, short-stipitate, veinless, straw-colored to brownish or tinged purplish, the beak 0.1–0.3 mm long, truncate; stigmas 3.

Bogs and muskegs; mostly in coastal southern Alaska but also reported from interior Alaska and Yukon; disjunctly from Quebec to Newfoundland and south to Washington; Greenland, Asia.

Carex subspathacea Wormskj.
Hoppner Sedge

Key III type; rhizomatous, 0.3–2.5 dm tall; leaf sheaths brown, blades of lower sheaths short or lacking, developed upwards, 0.6–3 mm broad, flat or more or less involute; lowermost bract (1.5)2.5–10 cm long, sheathless, the blade leaflike; inflorescence 2–8 cm long, erect or nearly so; spikes 2–4(5), the terminal staminate, the lateral pistillate, the lower separated by 5–30 mm, the peduncles 0.1–0.5 cm long; pistillate scales from shorter to longer than the perigynia, ovate, obtuse to acute or shortly awned, purplish brown to black with paler median; perigynia 2.4–3.5 mm long, 1.2–2 mm broad, appressed-ascending, ovate to elliptic, wingless, glabrous, subsessile, veinless, greenish to straw-colored, the beak 0.1–0.2 mm long, truncate, seldom serrate; stigmas 2.

Saline marshes and tidal flats; in coastal northern, western, and south-central Alaska; circumpolar (*C. ramenskii* Kom.).

Carex supina Willd.

Key IV type; caespitose and more or less rhizomatous, 0.4–2.5 dm tall; leaf sheaths brown to chestnut, blades of lower sheaths short to elongate, developed upwards, 0.5–

1.8 mm broad, flat or more or less channeled; lowermost bract 0.4–1.2(2) cm long, not sheathing basally, the blade subulate to linear; inflorescence 0.8–2.3 cm long, erect; spikes 2–4, the terminal staminate, the lateral pistillate, all clustered, or the lower separated by 1–10 mm, subsessile, spreading to ascending; pistillate scales subequal to or shorter than the perigynia, broadly ovate, acute to acuminate or shortly awned (sometimes obtuse), brownish with broad hyaline margins and pale to brownish median; perigynia 2.5–3.5 mm long, 1.4–1.8 mm broad, ovate to elliptic, spreading, wingless, glabrous or hairy along the beak, subsessile, obscurely veined, yellowish green to brownish, the beak 0.4–0.8 mm long, hyaline at the oblique apex; stigmas 3.

Steep south-facing slopes with *Artemisia frigida,* on rock outcrops, and in grasslands; in north-central and central to eastern Alaska and southern to northern Yukon; eastward to Greenland; Eurasia (*C. spaniocarpa* Steud.). Our material belongs to ssp. *spaniocarpa* (Steud.) Hultén.

Carex sychnocephala Carey

Key II type; caespitose, 1–6 dm tall; leaf sheaths brownish or grayish, blades of lower sheaths short or lacking, elongating upwards, 1–3 mm broad, flat or channeled; lowermost bract 8–16 cm long, the blade leaflike; inflorescence 1.5–4 cm long, erect; spikes 5–10 or more, all gynaecandrous, sessile, the lower separated by 2–12 mm; pistillate scales commonly shorter than the perigynia, lanceolate, attenuate to shortly awned, whitish to greenish with hyaline margins and whitish to greenish median; perigynia 4.8–6.5 mm long, 0.6–1.2 mm broad, lance-attenuate, appressed-ascending, more or less winged, ciliate, short-stipitate, straw-colored or greenish, tapering to an elongate flattened beak 2–3 mm long or more, bidentate; stigmas 2.

Stream banks and meadows; in eastern Alaska and southern Yukon; east to Ontario and New York and south to Montana and Iowa.

Carex tenuiflora Wahl.

Key II type; caespitose or with long vertical rhizomes, 1.2–3.5 dm tall; leaf sheaths brownish, blades of lower sheaths well developed, 0.5–1.5 mm broad, flat or channeled; lowermost bract 0.1–0.5 cm long, scalelike or less commonly with a short subulate blade; inflorescence 0.7–1.2 cm long, erect; spikes 2–4, all gynaecandrous, sessile, the lower separated by 1–4 mm, spreading; pistillate scales shorter than or subequal to the perigynia, ovate to elliptic, white-hyaline with greenish median; perigynia 2–3 mm long, 1–1.3 mm broad, elliptic to ovate, spreading-ascending, wingless, glabrous, subsessile, obscurely veined, greenish to straw-colored, beakless or the beak 0.1–0.2 mm long; stigmas 2.

Lake shores, muskegs, and bogs; in most of Alaska (except for portions of the northwest, southwest, and southeast) and in much of the Yukon; east to Newfoundland and south to Saskatchewan, Minnesota, and Maine.

Carex ursina Dewey

Key I (II) type; caespitose, 0.1–0.6 dm tall; leaf sheaths brownish, blades of lower sheaths short or obsolete, developed upwards, 0.3–1.2 mm broad, flat or more commonly channeled or involute; lowermost bract scalelike; inflorescence 5–7 mm long, 2–4 mm broad, erect; spikes solitary (rarely 2), gynaecandrous; pistillate scales commonly shorter than the perigynia, ovate to oval, obtuse, brown, the margins narrowly to broadly hyaline; perigynia 1.4–1.8 mm long, 1–1.2 mm broad, ovate to obovate or elliptic, ascending, obscurely veined, wingless, glabrous, short-stipitate, straw-colored, the beak to 0.2 mm long or beakless, truncate; stigmas 2.

Arctic tundra; in coastal northern Alaska and northern Yukon; circumpolar.

Carex vaginata Tausch.

Sheathed Sedge

Key VI type; rhizomatous, 1–4 dm tall; leaf sheaths brownish, blades of basal sheaths short, elongating upwards, 1–3 mm broad, flat; lowermost sheaths 1.5–5.5 cm long, the closed sheath 10–25 mm long, the blade leaflike; inflorescence 4–20 cm long or more, erect or nearly so; spikes 2–4, the terminal staminate, the lateral pistillate, the lower separated by 2.5–14 cm and with erect peduncles 1.5–6.5 cm long, erect; pistillate scales shorter than the perigynia, ovate, obtuse to acuminate, brown with narrow hyaline margins and broad greenish median; perigynia 3.5–4.5 mm long, 1.5–2 mm broad, ovate to elliptic, ascending, wingless, glabrous, substipitate, veined, greenish to brownish, the beak 0.6–1 mm long, oblique, hyaline or brown apically; stigmas 3.

Muskegs and bogs; in much of Alaska (except for southwestern, south-central, and southeastern portions) and in much of the Yukon; circumboreal (*C. saltuensis* Bailey).

Carex williamsii Britt.

Williams Sedge

Key VI type; caespitose, 0.3–1.5 dm tall; leaf sheaths purplish brown, blades of lower sheaths well developed, 0.3–1 mm broad, channeled; lowermost bract 0.7–3.5 cm long, the closed cylindrical sheath 3–12 mm long, the blade linear; inflorescence 1.5–4.5 cm long, erect or more or less nodding; spikes 2–5, the terminal staminate, the lateral pistillate, ascending to spreading, separated by 3–40 mm and with ascending peduncles 0.5–2 cm long; pistillate scales shorter than the perigynia, yellowish brown with hyaline margins and lighter median; perigynia 2.5–3.5 mm long, 0.7–1 mm broad, lanceolate, ascending, wingless, glabrous, short-stipitate, several-veined, greenish to brownish, the beak 0.3–0.8 mm long, the apex hyaline; stigmas 3.

Lake shores and bogs; in northern and central to eastern Alaska and southern to northern Yukon; east to Labrador; Asia.

ELEOCHARIS R. Br.

Plants perfect, perennial, rhizomatous or caespitose; stems (culms) triangular to terete, solid; leaves 3-ranked, the sheaths closed, bladeless; flowers perfect, borne in a solitary, apical spike, few to many, each subtended by a single scale, the scales spirally arranged in the spike; perianth of 3–12 bristles, these rough or barbed downward; stamens 3; pistils terminated by a 2–3-cleft style, the style base bulbous, persistent as a tubercle on the apex of the achene; achenes 3-angled or lens shaped.

Svenson, H. K. 1957. *Eleocharis*. N. Am. Fl. 18(9):509–40.

1a. Stems very slender, less than 0.5 mm broad, commonly less than 1 dm tall. (2)

1b. Stems commonly 0.6 mm broad or more, usually over 1 dm tall. (3)

2a. Scales of spike mostly 1.5–2.5 mm long; tubercle of achene conic with a flangelike base; plants broadly distributed, mostly in interior Alaska and Yukon. *E. acicularis*

2b. Scales of spike about 1 mm long; tubercle of achene flat, with a central process; plants of coastal and insular southwestern Alaska. *E. nitida*

3a. Base of style continuous with the achene, not developed into a definite tubercle; lowermost scales about half as long as the spikelet; plants mostly of interior Alaska and Yukon. *E. quinqueflora*

3b. Base of style enlarged, definitely tuberculate; lowermost scales much shorter than half the length of the spikelet; plants of various distribution. (4)

4a. Scales at base of spikelet paired (rarely 3), clasping about halfway around the stem; plants common in central and eastern interior to coastal southern Alaska and Yukon. *E. palustris*

4b. Scales at base of spikelet solitary, almost or quite encircling the stem; plants uncommon in much of Alaska and Yukon south of the 68th parallel. (5)

5a. Tubercle subequal to the achene in size; plants of coastal and insular western, southwestern, south-central and southeastern Alaska. *E. kamtschatica*

5b. Tubercle substantially smaller than the achene; plants of central to eastern interior to south-central Alaska and southern Yukon. *E. uniglumis*

Eleocharis acicularis (L.) Roem. & Schult.
Needle Spikerush

Plants caespitose and with slender rhizomes and stolons, 0.1–1 dm tall (rarely more), the stems 0.5 mm or less in diameter; leaf sheaths membranous, hyaline, bladeless; spikes solitary, 2.5–6.5 mm long, 0.9–2 mm broad, flattened; scales 1.5–2.5 mm long, reddish brown to brownish with hyaline margins and lighter to greenish median; achenes 0.7–1 mm long, white, transversely ridged, the tubercle conic with a flange-like base; perianth bristles 3–4.

Mud banks, along streams, lakes, or ponds, or in muskegs or roadside ditches; in much of continental Alaska and southern Yukon; broadly distributed in North America; circumboreal (*Scirpus acicularis* L.).

Eleocharis kamtschatica (C. A. Mey.) Kom.
Kamchatka Spikerush

Plants caespitose and more or less rhizomatous, mostly 1–4 dm tall, the stems 0.8–1.5 mm in diameter; leaf sheaths purplish below, somewhat flaring upwards, bladeless; spikes solitary, 0.5–1.4 cm long, 0.3–0.7 mm broad, flattened, subtended by a single bract almost or quite encircling the stem; scales 3.5–5.5 mm long, purplish black to purplish brown with hyaline margins and

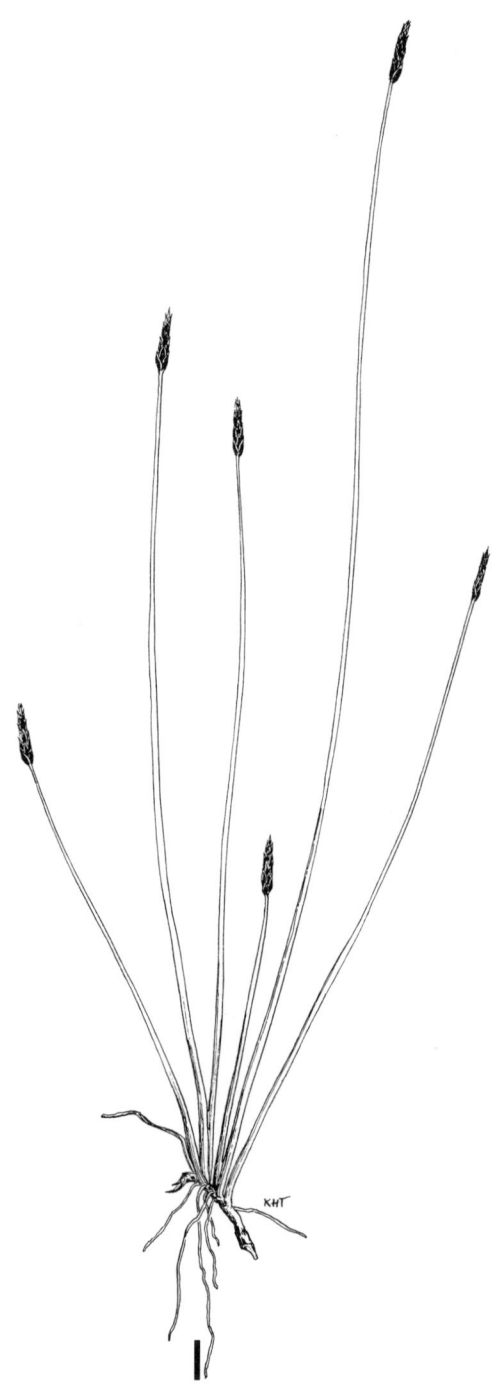

Eleocharis kamtschatica (C. A. Mey.) Kom.
(× 0.6).

lighter median; achenes 1.3–1.8 mm long, punctate in longitudinal rows, the tubercle slightly smaller than the achene and about as broad; perianth bristles 4–6.

Tidal flats and saline meadows; in coastal west-central, south-central, and southeastern Alaska; southward to British Columbia and disjunctly eastward to Labrador; Asia (*Scirpus kamtschaticus* C. A. Mey.).

Eleocharis nitida Fern.
Slender Spikerush

Plants rhizomatous and loosely caespitose, 0.2–1 dm tall, the stems 0.4 mm in diameter or less; leaf sheaths greenish to yellowish, bladeless; spikes solitary, 1.5–4.5 mm long, 1.5–3 mm broad, flattened; scales about 1 mm long, purplish brown with hyaline margins; achenes 0.7–0.8 mm long, pale yellowish, minutely wrinkled, the tubercle flat; perianth bristles lacking.

Moist soils; in the Shumagin and Kodiak islands; eastward to northeastern North America (*Scirpus nitidus* [Fern.] Hultén).

Eleocharis palustris (L.) Roem. & Schult.
Creeping Spikerush

Plants caespitose and rhizomatous to stoloniferous, mostly 1–7 dm tall, the stems 0.8–2.9 mm in diameter; leaf sheaths purplish or reddish below, somewhat flaring upwards, bladeless; spikes solitary, 0.5–2 cm long, 0.2–0.5 cm broad, terete or nearly so, subtended by 2(3) bracts encircling about half of the stem; scales mostly 3–4 mm long, reddish brown or purplish brown to brown or less commonly purplish black with hyaline margins and lighter median; achenes 1.2–1.8 mm long, light yellowish to brown, minutely wrinkled, the tubercle bulbous, longer than broad, much smaller than the achene; perianth bristles 4.

Muskegs, bogs, river and stream banks, and pond margins; in much of interior Alaska south of the Brooks Range and southward to the coast, from the Alaska Peninsula eastward, and in southern Yukon; widely distributed in North America; circumboreal, Southern Hemisphere (*Scirpus palustris* L.; *E. macrostachya* Britt.)

Eleocharis quinqueflora (Hartm.) Schwarz

Plants caespitose and rhizomatous to stoloniferous, mostly 0.6–3.5 dm tall, the stems 0.5–1 mm in diameter; leaf sheaths brownish to straw-colored below, cylindrical, not flaring upwards, bladeless; spikes solitary, 0.4–0.9 cm long, 0.2–0.4 cm broad, flattened, subtended by 2 bracts encircling about half the stem and about half as long as the spike or more; scales mostly 3–5 mm long, brown to purplish brown with hyaline margins and lighter median; achenes 1.8–2.2 mm long, straw-colored, minutely reticulate, the tubercle not or poorly developed; perianth bristles 4–6.

Moist sites in marshes and muskegs; in interior central to eastern Alaska and southern Yukon; eastward to Newfoundland and south to California, Nevada, Utah, Colorado, Iowa, Indiana, and New York (*Scirpus pauciflorus* Lightf.; *E. pauciflora* [Lightf.] Link; *S. quinqueflora* Hartm.).

Eleocharis uniglumis (Link.) Schult.

Plants caespitose and more or less rhizomatous and stoloniferous, mostly 1–2 dm tall, the stems 0.6–1.2 mm in diameter; leaf sheaths reddish to purplish below, slightly flaring upwards, bladeless; spikes solitary, 0.5–1.3 cm long, 0.2–0.6 cm broad, subterete, subtended by a single bract almost or quite encircling the stem; scales 2–5 mm long, brown to purplish brown with hyaline margins and lighter median; achenes 1.2–1.8 mm long, yellowish to brown, smoothish; tubercle bulbous, much smaller and narrower than the achene; perianth bristles variable.

Beaches and riverbanks; in the southeastern quarter of Alaska, including the Panhandle, and southern Yukon; east to Newfoundland and south to Oregon, Wyoming, and North Dakota (*Scirpus uniglumis* Link; *E. halophila* Fern. & Brack., at least in part).

ERIOPHORUM L.

Plants perennial, rhizomatous, or caespitose; stems (culms) triangular to terete, solid; leaves 3-ranked, the sheaths closed, the blades developed or obsolete; flowers perfect, borne in 1 or more spikes, numerous, each subtended by a single scale; perianth of numerous silky bristles, elongate at maturity; stamens 1–3; pistils terminated by 3-cleft styles, the style base not enlarged; achenes compressed, 3-angled.

Raymond, M. 1954. What is *Eriophorum chamissonis* C. A. Meyer? Svensk. Bot. Tidskr. 48(1):65–82.

1a. Spikes 2–several (rarely some solitary); inflorescence subtended by 1–more leafy bracts. (2)
1b. Spikes solitary, not subtended by leafy bracts. (4)

2a. Leaves 1–2 mm broad, channeled throughout; peduncles minutely short-hairy overall; plants uncommon in northwestern and south-central to southeastern Alaska. *E. gracile*
2b. Leaves commonly (1.5)2–4 mm broad or more, at least some, flat below the middle; peduncles various. (3)

3a. Anthers 1–1.3 mm long; midrib of scale prominent at the tip; peduncles minutely hairy throughout; plants uncommon. *E. viridi-carinatum*
3b. Anthers 2–5 mm long; midrib of scale not prominent at the tip; peduncles glabrous or merely serrate marginally; plants common and widespread. *E. angustifolium*

4a. Stems solitary or few, arising from rhizomes; sterile basal scales of spikelet commonly less than 7. (5)
4b. Stems several to numerous in a tuft, not rhizomatous; sterile basal scales of spikelet commonly more than 7. (6)

5a. Scales lance-attenuate (at least the middle and upper ones); anthers 0.5–1.5 mm long; spikes mostly less than 2.5 cm long; bristles white. *E. scheuchzeri*
5b. Scales ovate to lanceolate; anthers 1.5–3(3.2) mm long; fruiting spikes mostly over 2.5 cm long; bristles cinnamon brown or white. *E. chamissonis*

6a. Empty scales at base of spikelet reflexed; uppermost cauline leaf sheath commonly much expanded upwards; achenes 1.9–2.8(3.5) mm long. *E. vaginatum*
6b. Empty scales at base of spikelet ascending to spreading, not reflexed; uppermost cauline leaf sheath not or only slightly expanded upwards; achenes 2–2.3 mm long. (7)

7a. Plants commonly less than 2.2 dm tall; cauline leaf sheaths borne below middle of stem. *E. callitrix*
7b. Plants commonly 3 dm tall or more; cauline leaf sheaths commonly borne above middle of stem (at least some). *E. brachyantherum*

Eriophorum angustifolium Honck.
Tall Cottongrass

Plants rhizomatous, the culms commonly 1–9 dm tall, clothed basally with brownish to purplish septate-nodulose persistent sheaths; leaves usually flat below the middle, triangular channeled or folded above the middle; cauline leaves commonly 2–more, the blades 2–6 mm broad; involucral leaves 2–3, purplish basally; spikes 2–10, at least some usually pendulous (the peduncles glabrous or serrulate marginally), in flower ovoid and 1–2 cm long, in fruit 2–4 cm long; scales grayish, ovate to lanceolate, appressed-ascending, the midvein not reaching the apex, acutish; anthers 2.5–5 mm long; achenes 2.5–3.5 mm long; bristles numerous, white to cream.

Muskegs, bogs, meadows, and shallow ponds, in arctic and alpine tundra, heathlands, and open woods; almost throughout Alaska and Yukon; east to the Atlantic and

Eriophorum angustifolium Honck. (× 0.4).

gion bear serrations more often than those from the interior, but the presence or lack of them does not appear to be correlated with other features (*E. angustifolium* var. *major* Schulz; *E. subarcticum* Vassiljev; *E. angustifolium* ssp. *subarcticum* [Vassiljev] Hultén; *E. angustifolium* ssp. *scabriusculum* Hultén; *E. angustifolium* var. *triste* T. Fries; *E. angustifolium* ssp. *triste* [T. Fries] Hultén; *E. angustifolium* var. *coloratum* Hultén; *E. angustifolium* var. *giganteum* Hultén). This is our most common many-spikeleted cottongrass.

Eriophorum brachyantherum Trautv. & Mey.

Plants caespitose, the culms (2)2.5–7 dm tall, clothed basally with brownish septate-nodulose persistent sheaths; leaves linear, channeled throughout, 0.4–1.2 mm broad; cauline leaves commonly 2–more, at least 1 commonly above the middle of the stem, bladeless or the lower ones with short blades less than 2 mm broad, the sheath not or only slightly expanded upwards; involucral leaves lacking; spikes solitary, erect in flower, obovoid and 1–2 cm long, in fruit obovoid to subglobose and 2–3.5 cm long; lowermost scale (spathe) 0.5–1 cm long; sterile scales commonly 7–more; scales grayish, lance-attenuate, appressed-ascending, the midvein not reaching the apex; anthers 1–2 mm long; achenes 1.8–2.4 mm long; bristles numerous, creamy white.

Bogs, shallow ponds, moist meadows, in arctic and alpine tundra, heathlands, and open woods; in most of Alaska, except for the Aleutians, Kodiak, and the Panhandle, and in most of Yukon; east to Newfoundland and south to British Columbia and Ontario; Eurasia (*E. vaginatum* var. *opacum* Bjornstr.; *E. opacum* [Bjornstr.] Fern.).

Eriophorum callitrix Cham.
Arctic Cottongrass

Plants caespitose, the culms 0.8–2(2.2) dm long, clothed basally with brownish sep-

south to Oregon, Colorado, Iowa, Indiana, and Massachusets; circumpolar. This entity has been segregated by previous authors into 2–more infraspecific taxa. The basis for this segregation has been primarily the presence of small sharp serrations on the peduncles of some plants. Specimens from the southern coastal retate-nodulose persistent sheaths; leaves linear, channeled throughout, 0.4–1.2 mm broad; cauline leaves usually solitary (or lacking) and placed below the middle of the stem, bladeless or with a short blade, the sheath commonly expanded upwards; involucral leaves lacking; spikes solitary, erect in flower, obovoid, commonly 1–1.5 cm long, in fruit obovoid to subglobose, commonly 2–2.5 cm long; lowermost scale (spathe) 0.5–0.8 cm long; sterile scales commonly 7–more; scales blackish (or yellowish), lance-attenuate, appressed-ascending, the midvein not reaching the apex; anthers 0.6–1 mm long; achenes 1.8–2.1 mm long; bristles numerous, shiny, white.

Rocky outcrops, bogs, and peaty slopes, in arctic and alpine tundra and heathlands; from the islands of the Bering Sea northward to the Seward Peninsula and Point Hope, east along the Brooks Range to the Richardson Mts., Yukon, and in the mountains of central eastern Alaska and southwestern Yukon; east to Newfoundland; Eurasia (*E. callitrix* var. *moravium* Raymond; *E. callitrix* var. *pallidus* Hultén). The similarities between *E. callitrix* and *E. brachyantherum* make precise determination of all specimens exceedingly difficult. The disposition of the cauline leaves appears to be definitive in most cases.

Eriophorum chamissonis C. A. Mey.
Russett Cottongrass

Plants rhizomatous, the culms mostly 2–7.5 dm tall, clothed basally with brown to purplish brown septate-nodulose persistent sheaths; leaves linear-filiform, channeled throughout, 0.4–1.3 mm broad; cauline leaves usually 1–3, placed below the middle of the stem (rarely 1 above), the sheaths not or only somewhat expanded upwards; involucral leaves lacking; spikes solitary, erect, in flower oblong-cylindrical, 1.5–2 cm long, in fruit obovoid to subglobose, 2–4 cm long; lowermost scale (spathe) 0.4–2 cm long; sterile scales mostly less than 7; scales blackish to grayish, ovate to ovate-lanceolate, appressed-ascending, obtusish to acutish, the midvein not reaching the apex; anthers 1.5–3(3.2) mm long; achenes 1.7–2.7 mm long, smooth or ciliate-serrulate along the margin near apex; bristles numerous, cinnamon to white.

Bogs, lake shores, muskegs, wet meadows, and stream banks, in alpine and arctic tundra, heathlands, and open woods; in most of Alaska (except the central valley region) and from central and northern Yukon; east to Newfoundland and south to Utah and Colorado; Eurasia. Problems of nomenclature often are more difficult than taxonomic ones. This is the case with *E. chamissonis*. The original description is based upon discordant materials, but under the International Code of Botanical Nomenclature (1972) it is apparent that the portion of the specimens cited (the Chamisso collection) from Unalaska adequately typify this entity. And, if one is unable to distinguish *E. russeolum* Fries from *E. chamissonis*, then the name for the entire complex of forms must be maintained as *E. chamissonis*. Two more or less distinctive infraspecific entities are recognizable among our materials.

1a. Perianth bristles cinnamon-colored; culms commonly over 1.5 mm broad near the base; plants mostly of coastal and insular western and southern Alaska (*E. rufescens* Anders.; *E. russeolum* ssp. *rufescens* [Anders.] Hyl.; *E. medium* Anders.; *E. russeolum* var. *majus* Somm.). *E. chamissonis* var. *chamissonis*

1b. Perianth bristles white to tawny; culms commonly less than 1.5 mm broad; plants mostly of northern, western, and south-central Alaska (less commonly in coastal southern Alaska) and northern Yukon (*E. russeolum* var. *albidum* Nyl.; *E. chamissonis* f. *albidum* [Nyl.] Fern; *E. russeolum* f. *leucothrix* Blomgr.; *E. chamissonis* var. *leucothrix* [Blomgr.] Hultén). *E. chamissonis* var. *albidum* (Nyl.) Fern.

Eriophorum gracile W. D. J. Koch
Slender Cottongrass

Plants rhizomatous, the culms (1.5)2–6 dm tall, clothed basally with brownish to reddish brown septate-nodulose persistent sheaths, commonly lacking young basal leaves at flowering time; leaves commonly channeled throughout; cauline leaves 2–3, the blades 0.8–1.5(2) mm broad, purplish to brownish basally; involucral leaves solitary, not purplish basally; spikes 2–8, at least some pendulous (the peduncles minutely short hairy), in flower ovoid and 0.5–1 cm long, in fruit 1.5–2(2.5) cm long; scales grayish or blackish, lance-ovate to lanceolate, appressed-ascending, the midvein not reaching the apex, obtuse (except the uppermost); anthers 1–2 mm long; achenes 1.5–2 mm long; bristles numerous, white.

Peaty soil in muskegs or marshes; in south-central and less commonly in northwestern and southeastern Alaska; east to Newfoundland and south to California, Nebraska, Illinois, and Delaware; Eurasia.

Eriophorum scheuchzeri Hoppe
White Cottongrass

Plants rhizomatous, the culms 0.5–7 dm tall, clothed basally with brownish to reddish septate-nodulose persistent sheaths; leaves commonly channeled throughout, 0.5–1.3 mm broad; cauline leaves usually 1(2) placed below the middle of the stem, the sheath slightly expanded upwards; involucral leaves lacking; spikes solitary, erect, in flower subglobose to oblong-cylindrical, 0.8–2 cm long, in fruit obovoid to subglobose, 1.8–4 cm long; lowermost scale (spathe) 0.4–1 cm long; sterile scales mostly less than 7; scales blackish to grayish, ovate-lanceolate, appressed-ascending, attenuate, the midvein not reaching the apex; anthers 0.5–1.5 mm long; achenes 1.7–2.3 mm long, not ciliate-serrulate; bristles numerous, white.

Muskegs, wet meadows, river banks, lake and pond shores, shallow pools, and road-sides; in most of Alaska and Yukon; east to the Atlantic; circumpolar (*E. capitatum* Torr.; *E. altaicum* authors, not Meinsch.?; *E. scheuchzeri* var. *tenuifolium* Ohwi). *E. scheuchzeri* is closely allied to *E. chamissonis*. Intermediates do occur, and therefore it is not possible to place all specimens accurately.

Eriophorum vaginatum L.
Tussock Cottongrass

Plants caespitose, tussock-forming, the culms 1–6 dm tall, clothed basally with a thatch of brownish septate-nodulose persistent sheaths; leaves channeled, 0.5–1.2 mm broad; cauline leaves 1–3, the sheath conspicuously expanded upwards; involucral leaves lacking; spikes solitary, erect, in flower obovoid to subglobose, 1–2 cm long, in fruit subglobose to depressed-globose, 1.7–4 cm long; lowermost scale (spathe) 0.5–1 cm long; sterile scales mostly 10–more; scales grayish to blackish, ovate-lanceolate to lance-acuminate, acuminate apically, at least the lower finally reflexed, the midvein not reaching the apex; anthers 1–2 mm long; achenes 1.9–2.8(3.5) mm long, not ciliate-serrulate; bristles numerous, white.

Arctic and alpine tundra, heathlands, and less commonly in open woods, in wet meadows, swales, and marshy ground; in most of continental Alaska and western Yukon; east to the Atlantic; Eurasia. The present writer has been unable to segregate *E. vaginatum* into the proposed infraspecific entities. Apparently the characteristics chosen as diagnostic are entirely subjective and their application arbitrary. Thus it seems best to recognize only one highly variable entity (*E. spissum* Fern.; *E. vaginatum* ssp. *spissum* [Fern.] Hultén).

Eriophorum viridi-carinatum (Engelm.) Fern.

Plants rhizomatous, mostly 2–7 dm tall, clothed basally with brownish septate-nodulose persistent sheaths; leaves flat except at very tip; cauline leaves 2–more, the blades 2–6 mm broad; involucral leaves 2–3(4), green to merely brownish basally; spikes 2–10 or more, at least some pendulous (the peduncles minutely hairy throughout), in flower oblong-ovoid, 0.6–1 cm long, in fruit obovoid and 1.5–3 cm long; scales grayish to greenish, lance-attenuate, appressed-ascending, the conspicuous midrib extending to the tip, obtusish; anthers 1–1.3 mm long; achenes 3–4 mm long; bristles numerous, creamy white to tawny.

Wet meadows, marshes, and bogs; in south-central and southeastern Alaska and southern Yukon; disjunctly eastward to the Atlantic (*E. latifolium* var. *viridi-carinatum* Engelm.). This species is uncommon to rare in our region.

KOBRESIA Willd.

Plants monoecious caespitose perennials; stems (culms) triangular, solid; leaves 3-ranked, with blades well developed, the sheaths closed; flowers unisexual, borne in 1–few-flowered spikelets aggregated in spikes or panicles; flowers solitary in the axils of scales, the lower flower usually pistillate, the upper staminate, the pistillate flowers with 2–3 stigmas and the ovary surrounded by a spathelike glume (homologous with the perigynium in *Carex*), the staminate flowers with usually 3 stamens; perianth lacking; achenes 3-angled.

1a. Spikelets arranged in spikelike panicles, numerous; inflorescence 5–12 mm broad. *K. simpliciuscula*
1b. Spikelets arranged in compact spikes, several to many; inflorescence 2–5(7) mm broad. (2)

2a. Scales subtending spikelets 4.5–6 mm long; inflorescence more than 3 mm broad. *K. sibirica*
2b. Scales subtending spikelets 2.5–4.5 mm long; inflorescence 3 mm broad or less. *K. myosuroides*

Kobresia myosuroides (Vill.) Fiori & Paol.

Plants caespitose, commonly 0.4–3 dm tall; leaf sheaths brownish to dark brown, per-

Kobresia myosuroides (Vill.) Fiori & Paol. (× 0.9).

sistent, the blades well developed, 0.2–0.6 mm broad, linear-filiform; lowermost bract scalelike or with a short blade; spikes solitary, ·0.8–2 cm long, 0.2–0.3 cm broad, erect; spikelets few to numerous, commonly 2-flowered, erect-ascending; glumes subtending spikelets 2.5–3.5 mm long, brown with broad hyaline margins; achenes 1.5–2 mm long, compressed 3-angled; stigmas 2–3.

Spits, rocky ridges, stream banks and stream beds, rock outcrops, and meadows; in arctic and alpine tundra and heathlands; in western, northern, and central to southeastern continental Alaska and most of the Yukon; east to Labrador and south to Utah and Colorado; circumpolar (*Carex myosuroides* Vill.; *C. bellardii* All.; *K. bellardii* [All.] Koch).

Kobresia sibirica Turcz.

Plants caespitose, commonly 0.4–3 dm tall; leaf sheaths brownish to brown, persistent, the blades well developed, 0.5–1.2 mm broad, linear to filiform; lowermost bract scalelike; spikes solitary, 1.4–1.9 cm long, 0.4–0.7 cm broad, erect; spikelets several to numerous, commonly 2–4-flowered, erect or ascending; glumes subtending spikelets 3.5–6 mm long, brown with narrow hyaline margins; achenes 2.5–3.5 mm long, 3-angled; stigmas 3.

Bogs to dry sphagnum mats; in west-central to northwestern, north-central and northeastern Alaska and northern Yukon and less commonly in central Alaska and southwestern Yukon; east to the Northwest Territories and disjunctly southward to Colorado; Asia (*K. arctica* Porsild; *K. hyperborea* Porsild; *K. macrocarpa* Clokey).

Kobresia simpliciuscula (Wahl.) Mack.

Plants caespitose, commonly 0.7–3.5 dm tall; leaf sheaths brown to yellowish brown, persistent, the blades well developed, 0.4–1.5 mm broad, linear; lowermost bract scalelike or with a linear-subulate blade; spikes apparently solitary or obviously

paniculate, 1.2–3 cm long, 0.4–1.1 cm broad; spikelets numerous, commonly 2-flowered, erect-ascending, 2.3–4.5 mm long, brown to brownish with broad hyaline margins; achenes 2.3–2.8 mm long, compressed 3-angled; stigmas 3.

Bog margins, heathlands, meadows, pond shores, rocky outcrops, and sandy beaches; in west-central to northwestern Alaska, eastward along the Brooks Range to northern Yukon, in south-central to eastern Alaska, and in southern Yukon; south to British Columbia and Alberta (*Carex simpliciuscula* Wahl.).

RHYNCHOSPORA Vahl

Plants perfect or monoecious, caespitose to shortly rhizomatous perennials; leaves 3-ranked, with blades short to elongate, the sheaths closed (often ruptured); spikelets aggregated in dense, headlike clusters; flowers solitary in axils of scales, few to several per spikelet, the lower perfect, the upper imperfect; ovary lacking an enclosing scale; stigmas 2; stamens commonly 3; perianth of 9–15 bristles; achenes lens shaped.

Rhynchospora alba (L.) Vahl
White Beaked-rush

Plants caespitose, the rhizomes short, commonly 1.5–3.5(5) dm tall; leaf sheaths straw-colored, more or less persistent, the blades 0.3–1 mm broad, linear; lowermost bract 0.5–6 cm long, sheathless or with a closed cylindrical sheath 1–15 mm long; spikelets aggregated into 1–4 dense headlike clusters; scales ovate to lance-ovate, whitish; bristles retrorsely barbed, about as long as the brownish lens-shaped tuberculate achene; achenes 1.5–2.5 mm long, topped by an acuminate tubercle about one-third to one-half as long as achene.

Bogs; in southwestern and southeastern Alaska; disjunctly eastward to Newfoundland and south to California, Idaho, Illinois, Ohio, and North Carolina (*Schoenus albus* L.).

SCIRPUS L.

Plants perfect, caespitose or rhizomatous perennials; leaves 3-ranked, the blades well developed to much reduced, the sheaths closed (often ruptured); spikelets 1–many in an umbel, panicle, or headlike cluster (when more than one); flowers solitary in axils of scales, few to many per spikelet, all perfect; ovary lacking an enclosing scale; stigmas 2–3; stamens 3; achenes lens shaped.

Beetle, A. A. 1947. *Scirpus*. N. Am. Fl. 18(8):481–504.

1a. Bracts subtending inflorescence 2–more, leaflike; culms leafy. (2)
1b. Bracts subtending inflorescence solitary or lacking or if more than 1 then not leaflike; culms with leaves greatly reduced or leafy only toward the base. (3)

2a. Spikelets 10–20 mm long, densely clustered or shortly umbellate, few to many. S. *paludosus*
2b. Spikelets 3–6 mm long, in open paniculate inflorescences, numerous. S. *microcarpus*

3a. Inflorescence lacking a persistent bract, the lowermost scale merely longer than the others, commonly early deciduous; spikelets solitary. (4)
3b. Inflorescence subtended by a persistent terete or compressed bract (rarely lacking in S. *rufus*); spikelets commonly more than 1 (solitary in S. *subterminalis*). (6)

4a. Perianth bristles mostly 10–25 mm long; culms 3-angled, the angles scabrous. S. *hudsonianus*
4b. Perianth bristles 3–5 mm long or lacking; culms terete, smooth. (5)

5a. Lowermost scale with scabrous awn; spikelets less than 4 mm long, ovate; plants rare in southern Yukon. S. *pumilus*
5b. Lowermost scale with smooth awn; spikelets often over 4 mm long, lance-ovate; plants common, widely distributed. S. *caespitosus*

6a. Spikelets several, 2-ranked in a terminal spike; perianth bristles lacking; plants of south-central Alaska. S. *rufus*
6b. Spikelets solitary or few to numerous in compact to open umbels or panicles; plants of various distribution. (7)

7a. Culms 0.6–2 cm in diameter at the base; spikelets numerous, borne in an open panicle (at least when mature). S. *validus*
7b. Culms commonly less than 0.6 cm in diameter at the base; spikelets solitary or few to several borne in headlike clusters. (8)

8a. Spikelets solitary; culms terete; plants slender aquatics, known from extreme southeastern Alaska. S. *subterminalis*
8b. Spikelets (1)2–4 or more; culms sharply triangular; plants slender to robust, growing in moist sites, known from central eastern Alaska. S. *americanus*

Scirpus americanus Pers.

Three-square

Plants rhizomatous, the culms sharply triangular, 3–10 dm tall or more, less than 6 mm broad at the upper sheath; leaf sheaths straw-colored to brownish, septate-nodulose, the blades of lower sheaths short, elongated upwards; leaves all near the base of the culm; spikelets 5–20 mm long, (1)2–5 or more borne in a headlike cluster appearing as if lateral, subtended by a solitary bract 3–10 cm long; scales ovate, brown, shortly awned; bristles 2–6, retrorsely barbed; achenes smooth, 2.5–3 mm long.

Wet areas; known in our region only from Circle Hot Springs, Alaska; widely distributed in North America and South America.

Scirpus caespitosus L.
Tufted Clubrush

Plants densely caespitose (though sometimes with rhizomes connecting the tufts), the culms terete or nearly so, 0.5–4 dm long, 1.5 mm in diameter; leaf sheaths straw-colored to brownish, not septate-nodulose, the blades lacking or the upper ones with blades to 8 mm long; leaves all near the base of the culm; spikelets solitary, 3.5–6 mm long, terminal, subtended by a solitary scale about as long as the spikelet; scales ovate, yellowish brown, the lower 2–3 bluntly awn tipped; bristles 6, smooth, about twice as long as the achene; achenes 1.5–2 mm long.

Muskegs, pond margins, bogs, meadows, in arctic to alpine tundra, heathlands, and woods; in most of Alaska (except for the central valleys) and southern to northern Yukon; east to Newfoundland and south to Washington, Colorado, Minnesota, and New York; circumboreal (*Trichophorum caespitosum* [L.] Hartm.; *Baeothryon caespitosum* [L.] Dietr.).

Scirpus hudsonianus (Michx.) Fern.

Plants rhizomatous, the culms in rows from the shortened rhizomes, sharply triangular, 0.8–4 dm tall, 0.5–1 mm in diameter; leaf sheaths straw-colored to brownish, not septate-nodulose, bladeless or the uppermost with short blunt blades; leaves all near the base of culm; spikelets solitary, 4–8 mm long (excluding bristles), terminal, subtended by a solitary scale about half as long as the spikelet; scales lanceolate to lance-ovate, the lowermost with short blunt awn tip; bristles 6, smooth, exserted 10–20 mm or more; achenes 1.3–1.6 mm long.

Bogs and muskegs in open woods and meadows; in north-central to south-central and southeastern quarter of continental Alaska and southern Yukon; east to Newfoundland and south to Montana, Minnesota, and New York; circumboreal (*Eriophorum alpinum* L.; *E. hudsonianum* Michx.; *Trichophorum alpinum* [L.] Pers.).

Scirpus microcarpus Presl
Small-fruit Bullrush

Plants with a rhizomatous caudex, the culms triangular, 4–15 dm tall, 4–13 mm in diameter near the base; leaf sheaths purplish to brownish tinged or straw-colored to greenish, septate-nodulose, the blades all well developed, 0.7–2 cm broad; leaves distributed along most of the stem; spikelets numerous, 3–6 mm long, borne in an open paniculate cluster; inflorescence subtended by 2–3 or more leaflike bracts; scales ovate, suffused with purplish black; bristles 4, about as long as the achene, retrorsely barbed; achenes 1–1.5 mm long.

Swamps and riverbanks; disjunctly in central and south-central continental Alaska and in coastal and insular southern Alaska from Kodiak Island east through the Panhandle; east to Newfoundland and south to California, Nevada, Colorado, Minnesota, and New York; Asia (*S. rubrotinctus* Fern.; *S. microcarpus* var. *rubrotinctus* [Fern.] Jones).

Scirpus paludosus A. Nels.
Bayonet-grass

Plants rhizomatous, the culms triangular, 3–10 dm tall or more, 5–20 mm in diameter at base; leaf sheaths brownish, septate-nodulose, the blades well developed, 5–10 mm broad or more; leaves distributed along much of the stem; spikelets 3–10 or more in a compact headlike cluster subtended by 2–3 leaflike bracts; scales ovate, brownish to brown, 2-toothed and shortly awned apically; bristles 1–3, much shorter than the achene, caducous; achenes 2.5–4 mm long.

Wet beaches and tidal flats; in south-central and southeastern Alaska (where evidently rare); east to Quebec and south to Mexico, Texas, Minnesota, and New Jersey.

Scirpus pumilus Vahl

Plants in tufts from slender rhizomes, the culms terete or nearly so, 0.5–1.5 dm tall, to 1 mm in diameter; leaf sheaths blackish to brownish, not septate-nodulose, the upper 1–3 with blades to 3 cm long; leaves all near the culm base; spikelets solitary, 2–3 mm long, terminal, subtended by a soltiary scale; scales ovate, brown, awnless; bristles lacking; achenes 1.2–1.5 mm long.

Bogs and lake shores; in southwestern Yukon and to be sought elsewhere; east to Quebec and south to California and Colorado; Eurasia (*Trichophorum pumilum* [Vahl] Schinz & Thell.; *S. rollandii* Fern.; *S. pumilis* var. *rollandii* [Fern.] Hultén).

Scirpus rufus (Huds.) Schrad.
Red Clubrush

Plants rhizomatous, the culms terete or nearly so, 0.5–3 dm tall or more, to 2.5 (3) mm in diameter near the base; leaf sheaths dark brown, obscurely septate-nodulose, the upper 1–3 with erect blunt blades to 15 cm long or more; leaves all near the culm base; spikelets few to several, 6–8 mm long, 2-ranked in a more or less compact spike 1–2 cm long, the spike subtended by a single scale- to leaflike bract; scales lance-oblong to lance-attenuate, brown, awnless; bristles lacking; achenes 4–5 mm long.

Tidal flats; in south-central Alaska (Matanuska); disjunctly east to Newfoundland; Eurasia (*Schoenus rufus* Huds.; *Blysmus rufus* [Huds.] Link).

Scirpus subterminalis Torr.
Swaying Rush

Plants rhizomatous, the culms terete, 3–10 dm tall; leaf sheaths straw-colored to whitish, at least the upper with blades to 25 cm long or more and 0.5–1 mm broad; leaves all near the culm base; spikelets solitary, 6–12 mm long, terminal, subtended by an erect, leaflike bract to 3 cm long; scales lance-ovate, pale brown, acute; bristles about 6, retrorsely barbed, about as long as the achene; achenes 2.5–3.5 mm long.

Aquatic sites, in ponds and bogs; in extreme southeastern Alaska; disjunctly to Newfoundland and south to Washington, Idaho, Michigan, and Georgia.

Scirpus validus Vahl
Great Bulrush

Plants coarsely rhizomatous, the culms terete, 5–15 dm tall or more, commonly 6–20 mm in diameter near the base; leaf sheaths brown to straw-colored or mottled brownish, usually all bladeless; leaves all near the culm base; spikelets several to numerous, 5–12 mm long, in a compound compact to open paniculate cluster subtended by 1 (rarely more) reduced leaflike bract to 7 cm long; scales ovate, brown to purplish brown, shortly awned; achenes 1.5–2.5 mm long.

Tidal flats, muskegs, fens, ponds, and lakes; in the southeastern third (including Panhandle) of Alaska and southern Yukon; east to the Atlantic and south to California, Mexico, Oklahoma, Missouri, and Georgia.

GRAMINEAE
Grass Family

Herbaceous annual or perennial plants; stems (culms) usually hollow in the internodes; leaves 2-ranked, divided into a basal sheath and a terminal blade, a ligule often present at juncture of blade and sheath; inflorescence a spike, raceme, or panicle; flowers perfect or imperfect, borne in florets consisting of a flower and 2 bracts (glumes); perianth reduced, consisting of 2 fleshy lodicules; stamens usually 3, the anthers versatile; pistils 1; ovary superior, 1-loculed, 1-ovuled; stigmas 2, plumose; fruit a caryopsis.

1a. Inflorescence a spike; spikelets sessile. KEY I. .... p. 541
1b. Inflorescence a panicle (this sometimes dense and spikelike); spikelets pedicellate. (2)
2a. Spikelets awned; awns arising from glumes or lemmas or both. KEY II. .... p. 541

2b. Spikelets awnless. (3)

3a. Spikelets with a single floret. KEY III: .... p. 542

3b. Spikelets with 2–several florets. (4)

4a. Glumes equaling or longer than a spikelet. KEY IV. .... p. 543

4b. Glumes much shorter than the spikelet. KEY V. .... p. 543

KEY I. INFLORESCENCE A SPIKE; SPIKELETS SESSILE.

1a. Spikelets arranged along one side of the rachis (though often in 2 rows). *Beckmannia*

1b. Spikelets arranged on opposing sides of the rachis. (2)

2a. Spikelets averaging 2–3 per node of the rachis. (3)

2b. Spikelets usually 1 at each node of the rachis (sometimes with 2 per node near the base). (4)

3a. Spikelets 1-flowered, 3 per node, the 2 lateral ones on short pedicels reduced to awns (fertile in *H. vulgare*). *Hordeum*

3b. Spikelets 2–6 flowered, usually 2–3 per joint, all alike. *Elymus*

4a. Spikelets placed edgewise on the rachis; glumes narrow, the upper one (on the side toward the rachis) lacking, except in the terminal one. *Lolium*

4b. Spikelets placed broadside to the rachis; glumes broad, both present. (5)

5a. Glumes narrowly subulate, 1-nerved; spikelets 2-flowered. *Secale*

5b. Glumes usually with 3–more nerves; spikelets often more than 2-flowered. (6)

6a. Plants annual, cultivated and escaping; lemmas toothed and often awned. *Triticum*

6b. Plants perennial, indigenous or less commonly cultivated; lemmas not toothed, awnless or awned. *Agropyron*

KEY II. SPIKELETS AWNED, AWNS ARISING FROM GLUMES OR LEMMAS OR BOTH.

1a. Spikelets with a single floret. (2)

1b. Spikelets with 2–several florets. (9)

2a. Lemmas hardened, much firmer than the glumes, awned from the apex. (3)

2b. Lemmas membranous, of about the same texture as the glumes, or not as firm as the glumes, variously awned or awnless. (4)

3a. Awns 0.5–1 mm long, caducous; spikelets 3–4 mm long. *Oryzopsis*

3b. Awns 15–50 mm long or more, persistent; spikelets 5–15 mm long (excluding awns). *Stipa*

4a. Panicle very dense, spikelike, cylindrical, ovoid, or globose. (5)

4b. Panicle open or contracted, but not forming dense heads. (7)

5a. Panicle thick, softly hairy; glumes with slender awns to 6 mm long or more. *Polypogon*

5b. Panicle cylindrical to globose; glumes awnless or with awns to 2 mm long. (6)

6a. Glumes awned, stiffly hairy on the keels; lemmas awnless. *Phleum*

6b. Glumes blunt, awnless, softly hairy on the keels; lemmas awned. *Alopecurus*

7a. Florets with conspicuous, long hairs at base; palea present, the rachilla prolonged behind it. *Calamagrostis*

7b. Florets naked at the base, or with short hairs only; palea lacking in some species. (8)

8a. Lemmas awned from the tip; palea well developed, about as long as the lemma. *Muhlenbergia*

8b. Lemmas awned from the back; palea often much reduced or lacking. *Agrostis*

9a. Spikelets with 2 hairy rudimentary lemmas below the single perfect flower; lemmas hard and shiny, pale yellowish. (10)

9b. Spikelets without rudimentary lemmas below; imperfect florets, when present, borne above the perfect ones. (11)

10a. Lower florets staminate; spikelets brownish, shining. *Hierochloe*

10b. Lower floret sterile; spikelets greenish or yellowish. *Anthoxanthum*

11a. Glumes shorter than the first floret; lemmas awned from the tip or from the bifid apex. (12)

11b. Glumes (at least one) as long as the lowest floret, often as long as the entire spikelet; lemmas variously awned. (14)

12a. Callus of florets strongly bearded; lemmas 7–13-nerved, awned from a deeply bifid apex. *Schizachne*

12b. Callus of florets not bearded; lemmas 5–9(11)-nerved, awned from the tip or from between the teeth of a bifid apex. (13)

13a. Lemmas awned from the tip. *Festuca*

13b. Lemmas awned from below the tip, between the teeth of a bifid apex. *Bromus*

14a. Spikelets with 2 florets, one perfect, the other staminate. (15)

14b. Spikelets with 2–more florets, all alike except the upper ones reduced. (16)

15a. Lower floret staminate, awned, the awn twisted, geniculate, exserted; upper floret perfect, awnless. *Arrhenatherum*

15b. Lower floret perfect, awnless; upper floret staminate, shortly awned from below the top. *Holcus*

16a. Ligules a fringe of hairs; lemmas bifid at the apex, awned from between the lobes. *Danthonia*

16b. Ligules various, but not a fringe of hairs; lemmas awned from the back, toothed apically. (17)

17a. Spikelets (10)12–20 mm long or more; glumes often surpassing the florets. (18)

17b. Spikelets 8 mm long or less; glumes shorter to longer than the florets. (19)

18a. Glumes more than 15 mm long, 7–9-nerved; spikelets pendulous; plants annual. *Avena*

18b. Glumes less than 15 mm long, 3–5-nerved; spikelets erect; plants perennial. *Helictotrichon*

19a. Lemmas keeled, awned from above the middle. (20)

19b. Lemmas convex, awned from above or below the middle. (21)

20a. Lemmas with a straight awn from the apex; rachilla joints glabrous or only short-hairy; panicle contracted. *Koeleria*

20b. Lemma with a bent awn borne from along the back; rachilla joints long-hairy; panicle open or contracted. *Trisetum*

21a. Plants annual, slender, delicate; leaves less than 1 mm broad; glumes 3 mm long or less. *Aira*

21b. Plants perennial, the leaves commonly more than 1 mm broad, or if annual then glumes more than 5 mm long. *Deschampsia*

KEY III. SPIKELETS AWNLESS, WITH A SINGLE FLORET.

1a. Spikelets with 2 slender hairy rudimentary lemmas below the single perfect floret; lemmas hard and shiny, pale yellowish. *Phalaris*

1b. Spikelets without rudimentary lemmas below; imperfect florets, if any, above the perfect ones. (2)

2a. Panicle very dense and spikelike, cylindrical, ovoid, or globose. (3)

2b. Panicle open or contracted, but not forming dense heads. (4)

3a. Glumes with short sharp points, stiffly hairy on the keels; lemmas awn-less. *Phleum*

3b. Glumes blunt, soft hairy on the keels; lemmas with short awns. *Alopecurus*

4a. Florets on short but distinct pedicels; rachilla prolonged behind the palea; tall drooping plants of damp or wet shady places. *Cinna*

4b. Florets sessile; rachilla not prolonged behind the palea (except in some *Agrostis*). (5)

5a. Plants usually 0.4–1.5(2.3) dm tall, rarely over 1.5 dm; glumes usually much shorter than the lemma. *Phippsia*

5b. Plants 2–6 dm tall or more; glumes longer or shorter than the lemma. (6)

6a. Glumes longer than the lemma; spike-lets mostly less than 3 mm long. *Agrostis*

6b. Glumes shorter than the lemma; spikelets variable in length, shorter or longer than 3 mm. (7)

7a. Plants densely tufted, to 3(4) dm tall; leaves 1 mm broad or less; spike-lets less than 3 mm long. *Muhlenbergia*

7b. Plants loosely tufted, mostly 3–15 dm tall; leaves more than 1 mm broad; spikelets over 3 mm long. *Arctagrostis*

KEY IV. SPIKELETS AWNLESS, WITH 2–SEVERAL FLORETS; GLUMES EQUALING OR LONGER THAN THE SPIKELET.

1a. Spikelets with 2 staminate lemmas below the single perfect floret; slender plants of wet soil. *Hierochloe*

1b. Spikelets without staminate lemmas below; imperfect flowers, if any,. above the perfect ones. (2)

2a. Spikelets terete or essentially so; lemmas not keeled, rounded on the back. *Dupontia*

2b. Spikelets compressed, not terete; lemmas at least somewhat keeled. (3)

3a. Spikelets falling with glumes intact; second glume much wider than the first. *Sphenopholis*

3b. Spikelets falling apart at maturity, leaving glumes attached to plant; second glume similar to the first. *Koeleria*

KEY V. SPIKELETS AWNLESS, WITH 2–SEVERAL FLORETS; GLUMES SHORTER THAN THE SPIKELET.

1a. Callus of florets bearded or pilose (sparsely so in *Arctophila*). (2)

1b. Callus of florets not bearded or pi-lose (cobwebby in some species of *Poa*). (3)

2a. Ligules lacerate; panicle branches often twisted and drooping. *Arctophila*

2b. Ligules not lacerate; panicle branches rigid, spreading, not twisted. *Scolochloa*

3a. Spikelets nearly sessile, in dense 1-sided clusters at the ends of the few panicle branches. *Dactylis*

3b. Spikelets variously arranged, but not in 1-sided clusters. (4)

4a. Glumes papery; lemmas firm, strong-ly nerved, scarious-margined; spike-lets tawny or purplish, usually not green. *Melica*

4b. Glumes not papery; lemmas and spikelets various, but not as above, or seldom so. (5)

5a. Nerves of lemma converging at the summit. (6)

5b. Nerves of lemma parallel or nearly so, not converging at the summit. (8)

6a. Lemmas keeled or rarely rounded on the back, awnless; leaf tips boat shaped. *Poa*

6b. Lemmas rounded on the back, often shortly awned; leaf tips not or only slightly boat shaped. (7)

7a. Lemmas acute or more or less obtuse or awned, the awn inserted near the apex; leaf sheaths open. *Festuca*

7b. Lemmas acute to obtuse, usually awned from below the tip (usually between the teeth of a bifid apex); leaf sheaths closed (usually to near the top). *Bromus*

8a. Florets 2 per spikelet; lemmas 3-nerved. *Catabrosa*

8b. Florets 3–several per spikelet; lemmas 5–9-nerved. (9)

9a. Nerves of lemmas prominent; glumes much shorter than the adjacent lemma. (10)

9b. Nerves of lemmas inconspicuous; glumes much shorter or nearly as long as the adjacent lemma. (11)

10a. Leaf sheaths closed (usually to near the top); plants often aquatic or semiaquatic. *Glyceria*

10b. Leaf sheaths open to the base; plants of moist places but seldom or uncommonly aquatic. *Torreyochloa*

11a. Glumes longer than half the length of the adjacent lemma; plants rare or uncommon. *Colpodium*

11b. Glumes shorter than half the length of the adjacent lemma; plants common. *Puccinellia*

AGROPYRON Gaertn.

Perennial caespitose or rhizomatous plants; culms hollow; leaves with open sheaths, the blades flat or involute, the auricles mostly well developed, the ligules membranous or membranous-ciliate; spikelets borne in a terminal spike, usually sessile and solitary (rarely 2–3), arranged flatwise to the continuous rachis, 4–12-flowered, disarticulating above the glumes; glumes (1)3–9-nerved, awnless or short-awned, the awn straight or curved; palea subequal to the lemma; lodicules 2; stamens 3.

Bowden, W. M. 1965. Cytotaxonomy of the species and interspecific hybrids of the genus *Agropyron* in Canada and neighboring areas. Canad. Jour. Bot. 43: 1421–48.

Hitchcock, C. L. 1969. *Agropyron.* pp. 447–61 *In*: Hitchcock, C. L., et al., Vascular Plants of the Pacific Northwest. 1: 1–914. Seattle: University of Washington Press.

Melderis, A. 1968. *Agropyron.* pp. 181–90. *In*: Hultén, E., Flora of Alaska and Neighboring Territories. Stanford: Stanford University Press.

Note: The genus *Agropyron* has received a variety of treatments by modern authors. The complex assemblages treated here as *A. caninum* and *A. boreale* have been considered as belonging to several species previously, or the various entities have been placed at infraspecific level in one or more of the component entities. Because of the possibility of misinterpretation of the proper position of each taxon, it is apparent that all types must be reexamined and the complexes subjected to a monographic study. Only then will a definitive treatment be possible. The following outline is tentative at best.

1a. Spikes pectinate, the spikelets crowded, at least some spikelets to 4 times as long as the internodes of the rachis, strongly divergent. *A. cristatum*

1b. Spikes not pectinate, the spikelets widely spaced to crowded, rarely to 3 times as long as the internodes of the rachis, not strongly divergent. (2)

2a. Rhizomes present, usually well developed; plants sod-forming, not bunch grasses. (3)

2b. Rhizomes lacking or poorly developed; plants forming dense tufts, bunch grasses. (6)

3a. Lemmas and often glumes as well pubescent with stiff divergent hairs; plants of southern Yukon and east-central Alaska. *A. yukonense*

3b. Lemmas and glumes glabrous or merely puberulent; plants of various distribution. (4)

4a. Leaf blades flat, 5–10 mm broad; lowermost sheaths commonly stiffly hairy; plants introduced, mostly in disturbed sites. *A. repens*

4b. Leaf blades involute, or if flat then seldom to 5 mm broad; lowermost sheaths commonly glabrous or merely puberulent. (5)

5a. Glumes lance-subulate, the hyaline margin not continuing to the tip; anthers mostly 2.5–4 mm long; spikelets usually much longer than internodes of the rachis. *A. smithii*

5b. Glumes elliptic to lance-elliptic, the hyaline margin continuing to the tip or nearly so; anthers mostly 4.5–5.5 mm long; spikelets usually shorter than internodes of the rachis. *A. spicatum*

6a. Glumes about three-quarters the length of the spikelet or subequal to the spikelet. *A. caninum*

6b. Glumes about half the length of the spikelet or less. (7)

7a. Anthers mostly 4.5–5.5 mm long; awns of lemmas (when present) curved or twisted, to 1 cm long or more; leaf blades commonly involute; spikelets (excluding awns) usually shorter than internodes of the rachis. *A. spicatum*

7b. Anthers mostly 1–2 mm long; awns lacking or if present then straight and seldom more than 0.5 cm long; leaf blades commonly flat; spikelets (excluding awns when present) much longer than internodes of the rachis. (8)

8a. Glumes scabrous, usually tapering from the middle to the tip, the margin seldom broadened upwards, the lemmas strigulose to strigose. *A. macrourum*

8b. Glumes and lemmas strigulose or scabrous to glabrous, the margin of glumes often broadened above the middle and more or less abruptly contracted apically. *A. boreale*

Agropyron boreale (Turcz.) Drobov
Northern Wheatgrass

Plants tufted, lacking rhizomes, mostly 1.5–5.5 dm tall, the nodes of culms almost always hairy; lower leaf sheaths usually glabrous; ligules entire or erose-ciliate, less than 0.5 mm long; blades usually flat, mostly 1.5–6 mm broad, glabrous to scabrous above; spikes 4–9 cm long, more or less compact; spikelets usually much longer than the rachis internodes, 12–16(20) mm long (excluding awns), 3–5(6)-flowered, the rachilla rarely exposed; glumes one-third to one-half as long as the spikelet, lance-elliptic to elliptic or oblanceolate, the hyaline margin usually expanded toward the tip and abruptly contracted apically, strongly 3–4(5)-ribbed, acute or slender-awned, glabrous or strigose to strigulose; lemmas glabrous or more commonly puberulent to strigulose, awnless or with straight awns 2–6 mm long; anthers 1–2 mm long.

Woods, river banks, rock outcrops, gravel bars, and squirrel mounds; in much of continental Alaska and Yukon, especially in the more mountainous districts; eastward to Ungava and Greenland; circumboreal (*Triticum boreale* Turcz.; *Roegneria borealis* [Turcz.] Nevski). *A. boreale* combines characteristics of some phases of *A. caninum* (the broadened hyaline margin) and of *A. macrourum* (the glumes only one-third to one-half as long as the spikelets). The relationship of *A. boreale* is apparently with those taxa. The final disposition of *A. boreale* and its proposed segregates must await revisionary study. Three more or less well-defined varieties occur in Alaska and the Yukon.

1a. Lemmas and glumes strigulose, the glumes sometimes sparsely so; plants commonly 1.5–3 dm tall (*A. boreale* ssp. *hyperarcticum* [Polunin] Melderis; *A. latiglume* var. *pilosiglume* Hultén). *A. boreale* var. *hyperarcticum* (Polunin) Welsh comb. nov. (based on: *A. violaceum* var. *hyperarcticum* Polunin Bull. Nat. Mus. Canad. 92 [Bio. Ser. 24]:95. 1940.).

1b. Lemmas and glumes glabrous or the lemmas hairy only in upper portion; plants of various heights. (2)

2a. Lemmas more or less hairy, especially toward the apex; plants commonly 2.5–5 dm tall (*A. boreale* ssp. *alaskanum* [Scribn. & Merr.] Melderis). *A. boreale* var. *alaskanum* (Scribn. & Merr.) Welsh comb. nov. (based on: *Agropyron alaskanum* Scribn. & Merr. Contr. U. S. Nat. Herb. 13: 85. 1910).

2b. Lemmas glabrous; plants of various size. *A. boreale* var. *boreale*

Agropyron caninum (L.) Beauv.

Plants tufted, lacking (or seldom with short) rhizomes, mostly 2.5–9(12) dm tall, the nodes of culms glabrous or less commonly puberulent; ligules entire or erose-ciliate, to 0.8 mm long; blades usually flat or less commonly more or less involute, mostly 1.5–6 mm broad, scabrous, glabrous, or less commonly puberulent above; spikes 4–21 cm long, more or less compact; spikelets usually much longer than the rachis internodes, 9–18(22) mm long (excluding awns), 3–6-flowered, the rachilla rarely exposed; glumes three-fourths to about as long as the spikelet, lanceolate to oblanceolate, hyaline margined, 3–7-ribbed, acute to slender awned; lemmas glabrous or scaberulous, or less commonly puberulent, awnless to awned, the awn (1–3)4–20(25) mm long, straight or curved; anthers 1–2 mm long.

Open slopes, meadows, woodlands, gravel bars, beaches, and talus slopes; in most of Alaska (except for the Aleutian Islands) and most of the Yukon; eastward to the Atlantic Ocean and southward to Mexico; Eurasia. *A. caninum* is represented in our area by a vast series of forms which differ by perceptible but highly variable morphological features. Previous workers have treated many of the variants at the level of species, but despite the presence of cleistogamy and its consequent production of uniform progeny, the relationships of the entities involved are apparently close and it seems best to include all taxa as portions of a single complex species. *A. caninum* is known to hybridize with *Hordeum jubatum*, resulting in a plant known as *Agrohordeum* × *macounii* (Vasey) Lepage. The entire complex, including such closely related species as *A. boreale* and *A. macrourum*, should be treated monographically.

The following infraspecific taxa are recognizable among our materials.

1a. Lemmas strigulose, bearing curved awns 5–16 mm long; leaf blades with mixed short and long hairs on the upper surface; plants of south-central Alaska. *A. caninum* var. *mitchellii* Welsh var. nov. *A* var. *canino* differt lemmibus strigulosis aristis curvatis 5–16 mm longis. (Holotype S. L. Welsh 4616 BRY, Eklutna Lake, in *Betula, Populus, Salix* woodland, 12 July 1965).

1b. Lemmas glabrous, scabrous, or sparingly strigulose, awnless, or if awned, the awns straight; leaf blades variously pubescent or glabrous, but rarely or not as above; plants of various distribution. (2)

2a. Lemmas with at least some awns (6) 8–20 mm long or more. (3)

2b. Lemmas awnless or with awns to about 5 mm long. (4)

3a. Spikelets mostly 1–2(2.5) times as long as the rachis internodes; plants mostly from the southeastern quarter of continental Alaska and southern Yukon (*A. caninum* ssp. *majus* var. *unilaterale* [Vasey] C. L. Hitchc.; *A. violacescens* Beal; *A. trachycaulum* var. *unilaterale* [Vasey] Malte). *A. caninum* var. *unilaterale* Vasey

3b. Spikelets mostly 2.5–3 times as long as the rachis internodes; plants from northern and east-central Alaska and southern Yukon (*A. violaceum* var. *andinum* Scribn. & Smith; *A. andinum* [Scribn. & Smith] Rydb.; *Triticum subsecundum* Link; *A. subsecundum* [Link] A. S. Hitchc.; *A. caninum* ssp. *majus* var. *andinum* [Scribn. & Smith] C. L. Hitchc.). *A. caninum* var. *andinum* (Scribn. & Smith) Pease & Moore

4a. Scarious margin of glumes broader toward the tip, often asymmetrical; inflorescence often purplish; plants of broad distribution in continental and coastal south-central Alaska and southern Yukon, transitional to and passing into the following variety (*A. violaceum* [Hornem.] Lang; *Triticum violaceum* Hornem.; *A. violaceum* var. *latiglume* Scribn. & Smith; *T. pauciflorum* Schwein.; *A. pauciflorum* [Schwein.] A. S. Hitchc., not Schur.; *A. trachycaulum* [Link] Malte; *T. trachycaulum* Link; *A. caninum* ssp. *major* var. *latiglume* [Scribn. & Smith] C. L. Hitchc.). *A. caninum* var. *latiglume* (Scribn. & Smith) Pease & Moore

4b. Scarious margin of glumes not or only seldom broader toward the tip, usually symmetrical; inflorescence commonly green; plants in most of Alaska south of the Brooks Range (excluding the Aleutians) and in much of southern Yukon (*A. violaceum* var. *majus* Vasey; *A. trachycaulum* var. *majus* [Vasey] Fern.; *A. caninum* ssp. *majus* [Vasey] C. L. Hitchc.; *A. teslinense* Porsild & Senn; *A. pauciflorum* ssp. *teslinense* [Porsild & Senn] Melderis; *A. tenerum* var. *novae-angliae* Scribn.; *A. trachycaulum* var. *novae-angliae* [Scribn.] Fern.; *A. pauciflorum* ssp. *novae-angliae* [Scribn.] Melderis). (*Note*: the correct name at varietal level is subject to interpretation and until dem-onstrated otherwise the correct name appears to be based on *Triticum biflorum* var. *hornemannii* Koch). *A. caninum* var. *hornemannii* (Koch) Pease & Moore

Agropyron cristatum (L.) Gaertn.
Crested Wheatgrass

Plants tufted, lacking rhizomes, mostly 3–8 dm tall, the nodes of culms glabrous; lower leaf sheaths glabrous or puberulent; ligules erose-ciliate, less than 1 mm long; blades flat, mostly 2–5 mm broad, commonly pubescent at least above; spikes 1.5–6(10) cm long, compact; spikelets closely imbricated, many times longer than the internodes, (4)5–12 mm long (excluding awns), 5–8-flowered, the rachilla not exposed; glumes about three-fourths as long as the spikelet, lance-acuminate, hyaline margined, 1(2)-ribbed, shortly awned; lemmas glabrous, with awns 2–4 mm long; anthers 2.5–4 mm long or more.

Openings in cleared woodlands, cultivated and escaping; in the southeastern quarter of Alaska and southern Yukon; widely cultivated in North America; introduced from Eurasia (*Bromus cristatus* L.; *Triticum pectinatum* Bieb.; *A. pectiniforme* Roem. & Schult.; *T. sibiricum* Willd.; *A. sibiricum* [Willd.] Beauv.; *T. desertorum* Fisch.; *A. desertorum* [Fisch.] Schult.).

Agropyron macrourum (Turcz.) Drobov

Plants tufted, lacking (or seldom with short) rhizomes, mostly (2)3.5–10 dm tall, the nodes of culms glabrous; lower leaf sheaths glabrous or less commonly puberulent; ligules entire or more commonly erose-ciliate, less than 1 mm long; blades flat, (1.5)3–8(9) mm broad, scabrous above; spikes 4.5–15(20) cm long, compact to loose; spikelets from slightly longer to much longer than the rachis internodes, 10–23(26) mm long (excluding short awn tips), 3–9(10)-flowered; glumes 5–10 mm long, mostly less than one-half as long as the spikelet, lanceolate to oblanceolate, the hyaline margin often broader apically and sometimes toothed, 3–5-ribbed, scabrous on the ribs, acute or awn tipped; lemmas shortly pilose, the hairs spreading-ascending, awnless or with awns to 7 mm long; anthers 1–2 mm long.

Open slopes, gravel bars, lake shores, and terraces in tundra and woodlands; in much of the northern three-fourths of Alaska and most of the Yukon; eastward to the Mackenzie River and southward to British Columbia (?). (*Triticum macrourum* Turcz.; *A. sericeum* A. S. Hitchc.). *A. macrourum* is closely allied to *A. caninum* and apparent intermediates exist (Welsh & Rigby 11437; Welsh & Moore 7372, BRY). The nature of these morphological intermediates is not known, but *A. macrourum* is known to form intergeneric hybrids with *Elymus sibiricus* (*Agroelymus* × *palmerensis* Lepage) and with both *Hordeum brachyantherum* (*Agroelymus* × *jordalii* Melderis) and *Hordeum jubatum* (*Agrohordeum* × *pilosilemma* Mitchell & Hodgson).

Agropyron repens (L.) Beauv.
Quackgrass

Plants not strongly tufted, sod-forming, with elongate rhizomes, mostly 4–12 dm tall, the nodes of culms glabrous; lower leaf sheaths usually stiffly spreading-hairy; ligules erose-ciliate, less than 0.5 mm long; blades flat, 3–10 mm broad, usually scabrous at least above; spikes 6–15 cm long, more or less compact; spikelets commonly much longer than the internodes, 10–20 mm long (excluding the awns), 3–6(8)-flowered; glumes mostly one-half to three-fourths as long as the spikelet, lance-attenuate, hyaline margined, 3–7-ribbed, acute or awn tipped; lemmas glabrous or scabrous, awnless or with a straight awn to 10 mm long; anthers 3.5–5 mm long.

A weedy plant of cultivated land and disturbed sites such as roadsides; in widely scattered sites in Alaska and to be expected in the Yukon; introduced from Eurasia (*Triticum repens* L.)

Agropyron smithii Rydb.
Western Wheatgrass

Plants sod-forming, with elongate rhizomes, mostly 3–8 dm tall, the nodes of culms glabrous; lower leaf sheaths glabrous; ligules 1–4 mm broad, often glaucous; spikes 4–15 cm long, more or less compact; spikelets from shorter to much longer than the internodes, 8–20 mm long (excluding awns), 3–10-flowered; glumes one-third to three-fourths as long as the spikelet, stiffly lance-subulate, narrowly if at all hyaline margined, 3–5-ribbed, acuminate or awn tipped; lemmas glabrous or scaberulous, awn tipped or the awn to 5 mm long; anthers 3–4.5 mm long.

Disturbed soils along rights-of-way and cultivated lands; in south-central Alaska and to be expected elsewhere; possibly introduced; widely distributed in North America (*A. glaucum* var. *occidentale* Scribn.; *A. occidentale* [Scribn.] Scribn.).

Agropyron spicatum (Pursh) Scribn. & Smith
Bluebunch Wheatgrass

Plants tufted, lacking (or seldom with short) rhizomes, mostly 4–10 dm tall, the nodes of culms glabrous; lower leaf sheaths glabrous or minutely retrorsely puberulent; ligules erose-ciliate, less than 1 mm long; blades flat to involute, 1–4 mm broad, puberulent to pilose above; spikes 4–17(20) cm long, loose; spikelets shorter or slightly longer than the internodes, 11–22 mm long (excluding awns), 5–8-flowered; glumes one-third to one-half as long as the spikelet, lance-oblong, hyaline margined, 3–7-ribbed, acute to obtuse; lemmas glabrous or short-puberulent, awnless or awned, the awn usually divergent and 10–20 mm long; anthers 4–6 mm long.

Steep south-facing open slopes, river banks, and limestone outcrops; in central eastern Alaska and southern Yukon; disjunctly southward to the western United States and Canada (*Festuca spicata* Pursh; *Triticum divergens* Nees; *A. divergens* [Nees] Vasey;

A. *vaseyi* Scribn. & Smith; A. *divergens* var. *inerme* Scribn. & Smith; A. *spicatum* var. *inerme* [Scribn. & Smith] Heller; A. *inerme* [Scribn. & Smith] Rydb.).

Agropyron yukonense Scribn. & Merr.
Yukon Wheatgrass

Plants sod-forming, rhizomatous, mostly 2–5 dm tall, the nodes of culms glabrous; lower leaf sheaths puberulent; ligules entire or erose-ciliate, scarcely 0.5 mm long; leaf blades flat or more or less involute, 1–4 mm broad, scabrous or puberulent above; spikes 4–10 cm long, more or less compact; spikelets much longer than the internodes, 9–12(15) mm long, 3–5(6)-flowered; glumes about one-half as long as the spikelet, lanceolate, hyaline margined, 2–4-ribbed, pilose; lemmas pilose, awnless or less commonly with awns to 10 mm long; anthers 3–5 mm long.

Meadows and open sandy or gravelly sites; in east-central Alaska and southern Yukon and in extreme northern British Columbia, endemic. A. *yukonense* forms hybrids with other entities within *Agropyron* and possibly with taxa of closely related genera as well.

AGROSTIS L.

Perennial tufted, rhizomatous, or stoloniferous plants; culms hollow; leaves with open sheaths, the blades flat, folded, or involute, lacking auricles, the ligules membranous, more or less puberulent; spikelets borne in open to contracted panicles, 1-flowered, articulate above the glumes; glumes 1-nerved, acute to acuminate or aristate, awnless; palea lacking, rudimentary, or well developed and about equaling the lemma; lemma (3)5-nerved, the midnerve sometimes produced into a straight to geniculate and sometimes twisted awn; rachilla usually lacking but sometimes prolonged as a bristle behind the palea; lodicules 2; stamens 3.

Beetle, A. A. 1945. A new section Micro-

phyllae in *Agrostis*. Bull. Torrey Club. 72:541–49.

Hitchcock, A. S. 1905. North American species of *Agrostis*. Bull. U. S. Bur. Pl. Ind. 68:1–68.

1a. Palea well developed, usually at least half as long as the lemma. (2)
1b. Palea lacking, or if present never as much as half the length of the lemma. (5)

2a. Anthers 0.5–0.6 mm long; plants usually less than 3 dm tall. A. *thurberiana*
2b. Anthers mostly 1–1.4(1.8) mm long; plants often over 3 dm tall. (3)

3a. Rachilla prolonged behind the palea as a short stub or bristle 0.5–1 mm long; plants uncommon, in coastal and insular southern Alaska from the easternmost Aleutians eastward. A. *aequivalvis*
3b. Rachilla obsolete, not evident as a stub or bristle; plants with various distribution. (4)

4a. Panicle branches not spikelet-bearing at the base; ligules mostly 1–3 mm long; plants mostly 3–5 dm tall, known from coastal and insular south-central and southeastern Alaska. A. *tenuis*
4b. Panicle branches usually spikelet-bearing at the base; ligules mostly 3–6 mm long; plants mostly 5–10 dm tall or more, known from the southern half of Alaska except for the Aleutians. A. *alba*

5a. Lemmas with exserted geniculate awns. A. *borealis*
5b. Lemmas awnless or with included or only slightly exserted, usually straight awns. (6)

6a. Panicle dense, interrupted, at least some of the lower branches spikelet-bearing from the base; plants of coastal and insular southern Alaska. A. *exarata*
6b. Panicle loose, open, sometimes dif-

fuse, none of the branches spikelet-bearing from the base. (7)

7a. Anthers 0.5–0.8 mm long or more; awns lacking or very small; plants of coastal and insular southern Alaska, seldom interior. *A. alaskana*

7b. Anthers less than 0.5(0.6) mm long; awns lacking or present; plants of various distribution. (8)

8a. Panicle diffuse; lemmas shorter than the glumes; plants widespread in Alaska and Yukon, mostly south of the Arctic Circle. *A. scabra*

8b. Panicle open but not diffuse; lemmas about as long as the glumes; plants evidently rare in central eastern Alaska and west-central Yukon. *A. clavata*

Agrostis aequivalvis (Trin.) Trin.
Northern Bentgrass

Plants perennial, rhizomatous, mostly 3–8 dm tall, forming small tufts; ligules 1–5 mm long; leaf blades flat to involute, 1–3 mm broad, minutely scabrous; panicle loose, 5–15 cm long or more, purplish or brownish, with spreading or ascending branches; glumes subequal, 3.1–4.5 mm long; lemma shorter than or subequal to the glumes, awnless, slightly or not at all hairy on the callus; palea subequal to the lemma; rachilla joint 0.6–1.1 mm long; anthers 1–1.4 mm long.

Bogs and lake margins; in coastal and insular southern Alaska from the easternmost Aleutians eastward; southward to Washington (*A. canina* var. *aequivalvis* Trin. in Bong.; *Deyeuxia aequivalvis* [Trin.] Benth. ex Vasey; *Podagrostis aequivalvis* [Trin.] Scribn. & Merr.).

Agrostis alaskana Hultén
Alaska Bentgrass

Plants perennial, shortly rhizomatous, mostly (0.5)1–8 dm tall, forming small to large tufts; ligules 2–6 mm long; leaf blades flat to involute, 0.5–3 mm wide, minutely sca-brous; panicle loose, 2–16 cm long or more, purplish or greenish purple, with ascending or spreading branches; glumes slightly unequal, the first 3.2–4.5 mm long; lemma shorter than the glumes, awnless or with a straight awn shorter than the spikelet or less commonly exserted, not or slightly hairy on the callus; palea lacking or very small; rachilla joint obsolete; anthers 0.5–0.8 mm long.

Bogs, lake shores, and sea beaches; in coastal and insular southern Alaska from the westernmost Aleutians eastward through the Panhandle (*A. aenea* [Trin.] Trin.; *A. melaleuca* [Trin.] A. S. Hitchc.). *A. alaskana* as interpreted here does not seem to represent a unified entity. Rather it seems to be an assemblage of forms intermediate between *A. borealis* on the one hand and *A. scabra* on the other, held together by the questionable feature of somewhat larger anthers.

Agrostis alba L.
Creeping Bentgrass; Redtop

Plants perennial, rhizomatous or sometimes stoloniferous or both, mostly 5–10 dm tall or more, ligule (1.5)2–6.5 mm long; leaf blades flat or folded, 1–10 mm broad, scaberulous, panicle loose or more or less contracted, 5–25 cm long or more, the branches ascending or appressed, bearing spikelets almost or quite to the base; glumes subequal, 2.0–3.1 mm long; lemma shorter than the glumes, awnless or sometimes awned, the awn straight, not or only slightly exserted, minutely hairy on the callus; palea well developed, usually more than one-half as long as the lemma; rachilla joint obsolete; anthers 1–1.3 mm long.

Roadsides and disturbed soils; in much of southern Alaska and Yukon (and to be expected elsewhere); widespread in the agricultural regions of the world; introduced. There are three more or less intergrading varieties recognizable among our materials. These entities have been treated at species level but a conservative treatment is indicated.

1a. Panicle contracted, commonly less than 15 mm broad when pressed; plants commonly stoloniferous, and often shortly rhizomatous (*A. palustris* Huds.; *A. stolonifera* var. *palustris* [Huds.] Farw.; *A. stolonifera* var. *compacta* Hartm.). *A. alba* var. *palustris* (Huds.) Pers.

1b. Panicle only somewhat or not at all contracted, commonly more than 15 mm broad when pressed; plants stoloniferous or rhizomatous. (2)

2a. Leaf blades often over 4 mm broad; plants with stolons (seldom if ever also rhizomatous), intergrading with var. *alba* (*A. stolonifera* L.). *A. alba* var. *stolonifera* (L.) J. E. Smith

2b. Leaf blades commonly less than 4 mm broad; plants with rhizomes (seldom stoloniferous) (*A. gigantea* Roth.; *A. alba* var. *gigantea* [Roth] Griseb.). *A. alba* var. *alba*

Agrostis borealis Hartm.
Red Bentgrass

Plants perennial, tufted, (0.5)1–3.5(4.0) dm tall; ligules 1–3 mm long; leaf blades flat to involute, mostly (0.5)1–3 mm broad, minutely scabrous; panicles loose, with spreading or ascending branches, 2–12 cm long, purplish; glumes unequal to sub-equal, the first (2)2.5–3.5 mm long; lemma shorter than the glumes, awned, the awn curved and often twisted, exserted from the spikelet, the callus hairy; palea very small; rachilla joint obsolete; anthers 0.6–1.1(1.5) mm long.

Stream banks, bars, and open slopes, mostly in arctic and alpine tundra; in most of Alaska south of the Brooks Range and most of the Yukon; southward to Utah and Colorado and eastward to Newfoundland and the northeastern United States; circumboreal (*A. trinii* Turcz.). *A. trinii* is here treated as a synonym of *A. borealis* because of lack of consistent characteristics for separation. In our specimens neither anther size nor the nature of the panicle is definitive.

Agrostis clavata Trin. in Spreng.

Plants perennial, shortly rhizomatous, 3–7 dm tall, tufted; ligules 1–3 mm long; leaf blades flat, commonly less than 1.5 mm broad, scaberulous; panicles loose, the branches erect, 7–25 cm long or more, purplish or straw-colored; glumes unequal, the first 1.7–2.5 mm long; lemmas slightly shorter than the glumes; palea to about one-third as long as the lemma; rachilla joint obsolete; anthers 0.3–0.4 mm long.

Wet meadows; in our region known only from east-central Alaska and from west-central Yukon; Eurasia. Our materials were at first identified with *A. idahoensis* Nash, but seem to be adequately distinct from that entity. The specimens are not unquestionably distinguishable from *A. scabra*, however.

Agrostis exarata Trin.
Spike Redtop

Plants perennial, rhizomatous or rhizomes not developed, sometimes stoloniferous, 1–10 dm tall or more, tufted; ligules 2–10 mm long or more; leaf blades flat, (1)2–10 mm broad, scabrous; panicles contracted, commonly interrupted, 5–25 cm long or more, the branches ascending to erect, green to purplish; glumes unequal, the first commonly 2–3.5 mm long; lemmas shorter than the glumes, awnless or rarely awned; palea to about one-third as long as the lemma; callus slightly hairy; rachilla joint obsolete; anthers 0.4–0.7 mm long.

Sea beaches, meadows, and open woods; in coastal and insular southern Alaska from the westernmost Aleutians eastward through the Panhandle; southward to California, Mexico, Texas, and Nebraska. Two subspecies have been treated within our materials.

1a. Glumes long-aristate, commonly more than 3 mm long, the awnlike tip to 1 mm long (*A. exarata* var. *purpurascens* Hultén; *A. scouleri* Trin.). *A. exarata* ssp. *exarata*

1b. Glumes acute but not aristate, mostly less than 3 mm long (*A. exarata* var. *minor* Hook.). *A. exarata* ssp. *minor* (Hook.) C. L. Hitchc.

Agrostis scabra Willd.
Ticklegrass

Plants perennial, shortly rhizomatous (rarely stoloniferous) or merely tufted, mostly 1.5–8 dm tall; ligules 2–4 mm long, rarely longer; leaf blades flat or folded, 0.5–2(3) mm broad, scaberulous; panicles diffuse, open, 6–30 cm long, the branches spreading to erect, purplish; glumes unequal, the first 2.1–3.0 mm long; lemmas shorter than the glumes, awnless or awned, the awn straight or nearly so, this usually included; anthers 0.3–0.6 mm long.

Roadsides, woods, muskegs, pond margins, and disturbed soils; in much of Alaska and Yukon south of the Arctic Circle and less commonly northward (except for the Aleutians); widespread in North America (*A. hiemalis* var. *scabra* [Willd.] Blomquist; *A. geminata* Trin.; *A. hiemalis* var. *geminata* [Trin.] A. S. Hitchc.; *A. hiemalis* var. *nutkaensis* [Kunth] Scribn. & Merr.; *A. scabra* var. *aristata* Hultén). This is apparently our most common bentgrass species. It produces pleasing pinkish or purplish filmy margins along roadsides in midsummer.

Agrostis tenuis Sibth.
Colonial Bentgrass

Plants perennial, more or less stoloniferous and often rhizomatous, mostly 3–6 dm tall; ligules 0.5–2(3) mm long; leaf blades flat, 2–5 mm broad, scabrous; panicle open (or less commonly contracted?), 4–15 cm long, purplish, the branches spreading-ascending (or erect?); glumes subequal or slightly unequal, awnless or with a short straight or bent awn, this usually included; anthers 1–1.3 mm long.

Fields, roadsides, beaches, and lawns; in coastal and insular south-central and south-eastern Alaska, introduced; widely distributed in temperate regions of the world.

Agrostis thurberiana A. S. Hitchc.
Thurber Bentgrass

Plants perennial, tufted, shortly rhizomatous, mostly 1–3 dm tall; ligules 1–3 mm long; leaf blades flat, 0.5–2(3) mm broad; panicle loose, narrowing in age, 2–10 cm long, the branches ascending, purple to greenish; glumes about equal, the first 2–2.5 mm long; lemma about equaling the shorter glume, awnless, minutely hairy on the callus; rachilla joint 0.1–0.3 mm long; anthers 0.5–0.6 mm long.

Wet meadows, mostly in alpine sites; in coastal southern Alaska from the easternmost Aleutians eastward to southeastern Alaska (*Podagrostis thurberiana* [A. S. Hitchc.] Hultén).

AIRA L.

Annual herbs with hollow culms; leaves with sheaths open, the blades filiform, involute, lacking auricles, the ligules membranous, usually puberulent; spikelets borne in open panicles, 2-flowered, articulate above the glumes; glumes 1-nerved, awnless, longer than the uppermost floret; lemmas 3–5-nerved, rounded on the back, awned from below the middle, the awn twisted and geniculate, usually exserted; palea about equaling the lemma; rachilla lacking; lodicules 2; stamens 3.

Aira caryophyllea L.
Silver Hairgrass

Stems slender, solitary or tufted, 0.5–3 dm tall; leaves 0.3–0.7 mm broad, scaberulous; ligules 1–3 mm long; panicle open, brownish or purplish, 2–6 cm long; glumes 2.5–3 mm long; lemmas awned, the awn 2–3 mm long; anthers 0.2–0.3 mm long.

Roadsides; in our region known only from southwestern Yukon (Anderson and Brown 10338 B, ISC), to be expected elsewhere; widely distributed in North America; South America.

ALOPECURUS L.

Perennial (or winter annual in some) herbs with hollow culms; leaves with sheaths open, the blades usually flat, lacking auricles, the ligules membranous, usually puberulent; spikelets borne in cylindrical spikelike panicles, 1-flowered, articulate below the glumes, strongly flattened; glumes subequal, more or less connate up to half their length; lemmas indistinctly 3–5-nerved, strongly flattened, awned from the back, the awn straight to geniculate and twisted, shorter or longer than the glumes; palea lacking; rachilla lacking; lodicules absent; stamens 3.

1a. Glumes usually more than 4 mm long, not silky-woolly; plants commonly (3) 5–12 dm tall; introduced forage grass; escaping. A. pratensis

1b. Glumes usually less than 4 mm long (more in some A. alpinus); plants mostly 1–5(9) dm tall; indigenous or introduced species. (2)

2a. Glumes densely woolly-silky over entire surface; anthers 2–2.3 mm long; plants of broad distribution in Alaska and Yukon, except for coastal southeastern portion. A. alpinus

2b. Glumes variously hairy but pubescence mainly on the nerves and keel; anthers often less than 2 mm long; distribution various. (3)

3a. Awn usually less than 1.5 mm longer than the glume, inserted near the middle of the lemma; anthers less than 1 mm long. A. aequalis

3b. Awn usually more than 1.5 mm longer than the glume, inserted in the lower third of the lemma; anthers more than 1 mm long. A. geniculatus

Alopecurus aequalis Sobol.
Short-awn Foxtail

Plants perennial (or sometimes flowering first year), tufted, mostly 1.0–3.5(9) dm tall, erect or decumbent and rooting at the nodes; ligules 3–5 mm long or more; leaf blades usually flat, 1–5 mm broad, scabrous (at least below); panicle 1.4–7.5 cm long, pale green; glumes commonly 1.9–2.4 mm long, villous-hairy on the nerves, more or less silky over the back, connate only near the base; lemma subequal to the glumes, the awn inserted near the middle and subequal to the glumes or somewhat longer; anthers less than 1 mm long.

Muskegs, river banks, roadsides, in shallow ponds, and along streams; in much of Alaska and Yukon south of the Arctic Circle and rarely northward; widespread in North America; circumboreal.

Alopecurus alpinus J. E. Smith
Mountain Foxtail

Plants perennial, shortly rhizomatous and sometimes stoloniferous, tufted, 1–8 dm tall; ligules 1–3 mm long; leaf blades usually flat, 1–7 mm broad, scabrous; panicle from almost as broad as long to much longer

Alopecurus alpinus J. E. Smith (× 0.3).

than broad, 1–4.5 cm long; glumes usually 3–4(6.5) mm long, silky-woolly throughout, about one-third connate; lemma slightly shorter than the glumes, awned from below the middle or awnless, the awn slightly to strongly geniculate, scarcely to much longer than the glumes; anthers 2–2.3 mm long.

Moist tundra, river banks, spits, shallow ponds, and bogs; in most of Alaska (except for the southeastern portion) and most of the Yukon; eastward to Newfoundland and south to Utah and Colorado; circumpolar. Three phases have been recognized within our region, the *alpinus, glaucus,* and *stejnegeri* phases. Only the *stejnegeri* phase appears to be sufficiently distinctive to warrant treatment apart from the main body, since the *glaucus* phase is connected by a series of intermediates with *alpinus* in a strict sense.

1a. Panicles only slightly longer than broad, commonly more than 12 mm broad; glumes mostly 5–6.5 mm long; leaf sheaths often much inflated; plants usually of islands of the Bering Sea and Aleutians, seldom in continental Alaska (*A. stejnegeri* Vasey; *A. alpinus* ssp. *stejnegeri* [Vasey] Hultén). *A. alpinus* var. *stejnegeri* (Vasey) Hultén

1b. Panicles conspicuously longer than broad or less than 12 mm broad or both; glumes mostly 3–4 mm long; leaf sheaths not or scarcely inflated; plants widespread in Alaska and Yukon (*A. occidentalis* Scribn. & Tweedy; *A. glaucus* Less.; *A. alpinus* ssp. *glaucus* [Less.] Hultén). *A. alpinus* var. *alpinus*

Alopecurus geniculatus L.
Water Foxtail

Plants perennial, more or less stoloniferous, tufted, 2–5 cm tall; ligules 2–5 mm long; leaf blades flat, 2–5 mm broad, scabrous (at least above); panicle 2–6(7) cm long, often purplish; glumes (2)2.5–3.5(4.2) mm long, hairy on nerves and keels and silky over the back, connate only at the base; lemma about equaling the glumes, the awn inserted near the base and surpassing the glumes by 1.5–4 mm or more; anthers more than 1 mm long.

Wet places, often in shallow water; known in Alaska from south-central and southeastern regions, where probably introduced, and to be expected elsewhere; widely distributed in temperate regions of the earth.

Alopecurus pratensis L.
Meadow Foxtail

Plants perennial, more or less stoloniferous, tufted; ligules mostly 2–6 mm long; leaf blades flat, 2–8(10) mm broad, scabrous; panicle 3–10 cm long, straw-colored or with a purplish cast; glumes mostly 4–6 mm long, long-hairy on nerves and keel, connate in the lower one-fifth or one-fourth; lemma subequal to the glumes, awned from near the base, this surpassing the glumes by 2–4 mm or more; anthers 2.3–5 mm long.

Roadsides; introduced in southern Yukon and in southeastern Alaska; widespread in temperate regions of the earth.

ANTHOXANTHUM L.

Perennial herbs with hollow culms; leaves with sheaths open, the blades flat, often with short rounded auricles, the ligules membranous; spikelets borne in congested panicles, 3-flowered (the lower 2 with empty lemmas), articulate above the glumes; glumes unequal, the first 1-nerved, the second 3-nerved and about twice as long as the first; sterile lemmas hairy, brownish, compressed, awned from the back; fertile lemma unawned; palea present; lodicules absent; stamens 2.

Anthoxanthum odoratum L.
Sweet Vernalgrass

Plants tufted, 2.5–7 dm tall; ligules 1–3 mm long; leaf blades flat, 1–5(7) mm broad, commonly more or less pilose, scabrous; panicle narrow, congested, 2–5(9) cm long, yellowish brown; glumes mostly 4–5 and 8–10 mm long respectively; sterile lemmas awned, the awns mostly 3–10 mm long; palea about equaling the lemmas; anthers 4–5 mm long.

Roadsides; in insular southern Alaska and to be expected elsewhere, introduced; widespread in North America; Eurasia.

ARCTAGROSTIS Griseb.

Perennial rhizomatous more or less tufted plants; culms hollow; leaves with open sheaths, the blades flat or more or less channeled or involute, lacking auricles, the ligules membranous to firm, more or less puberulent; spikelets borne in open to more or less contracted panicles, 1-flowered, articulate above the glumes; glumes 1-nerved, acute to acuminate, awnless; palea subequal to the lemma; lemma 3-nerved, unawned or the midvein sometimes awn tipped; rachilla commonly produced as a short bristle behind the palea; lodicules 2; stamens 2–3.

Arctagrostis latifolia (R. Br.) Griseb.
Polargrass

Plants perennial, rhizomatous, mostly 2.5–14.5 dm tall, forming tufts, or solitary; ligules 3.5–6.5(9.5) mm long, lacerate, often suffused red or purplish at base; leaf blades 2–10(15) mm broad, scabrous; panicle contracted to loose and open, 5–30(43) cm long, yellowish or green or more commonly purplish, with ascending to suberect or less commonly spreading branches; glumes unequal, the first 1.7–4.7 mm long; lemma longer than the glumes, awnless, 3.0–6.0 mm long, callus not long-hairy; palea subequal to the lemma; rachilla joint to about 1 mm long; anthers mostly 1.5–3 mm long.

One of our most common grasses, in tundra, heathland, and woods; in most of Alaska and Yukon; circumpolar (*Colpodium latifolium* R. Br.; *Vilfa arundinacea* Trin.; *A.*

poaeoides Nash; *A. latifolia* var. *arundinacea* [Trin.] Griseb.; *A. latifolia* ssp. *nahanniensis* Porsild). *A. latifolia* is a highly polymorphic species. The various phases have been treated at specific or infraspecific rank on the basis of characters whose application seems wholly arbitrary. Anther size, spikelet size, and spikelet color and texture, either separately or in combination, do not seem to be definitive.

ARCTOPHILA Rupr.

Perennial rhizomatous aquatic herbs; culms hollow; leaves with sheaths open only near the apex, the blades flat, the auricles lacking, the ligules membranous, lacerate and more or less puberulent; spikelets borne in open panicles, 2–5(7)-flowered, articulate above the glumes; glumes 1- or obscurely 3-nerved, shorter than the spikelet, obtuse, awnless; palea shorter than the lemma; lemmas obscurely 3(5)-nerved, the nerve glabrous, unawned; callus with a tuft of short stiff hairs; lodicules 2; stamens 3.

Arctophila fulva (Trin.) Anderss.
Pendant Grass

Plants mostly 2–8(9) dm tall, commonly solitary, 3–8 mm broad, smooth; panicle open, the branches recurved or drooping, 5–21(30) cm long, greenish purple; spikelets 3.5–7 mm long; glumes 3.0–3.7 mm long, obtuse, with broad hyaline margins; lemmas 2.7–4 mm long, with texture like the glumes.

Lake and pond margins and stream banks; in most of Alaska and Yukon (Aleutians and most of the Alexander Archipelago excepted); circumpolar (*Poa fulva* Trin.; *Colpodium fulvum* [Trin.] Griseb.). This is a very common species in arctic portions of Alaska and Yukon.

ARRHENATHERUM Beauv.

Perennial herbs with hollow culms; leaf sheaths open, the blades flat, lacking auricles, the ligules membranous; panicles 2-flowered, the lower usually staminate, the upper perfect, articulate above the glumes, the rachilla produced as a bristle behind the upper palea; glumes membranous, unequal, the first 1-nerved, the second 3-nerved, about equaling the second floret; lemmas 7-nerved, the callus bearded, the first with a stout twisted bent awn inserted near the middle, the second with a straight subterminal awn; lodicules 2; stamens 3.

Arrhenatherum elatius (L.) Presl
Tall Oatgrass

Plants tufted, 8–12 dm tall or more, often rooting at the lowermost nodes; ligules 1–3 mm long; leaf blades 2–8 mm long, scabrous (at least below); panicle 8–25 cm long, narrow, shining; glumes unequal, the first mostly 5–7 long, the second 7–10 mm long; awn of lower lemma 8–15 mm long or more, that of the second commonly less than 10 mm long or rarely lacking; paleas subequal to the lemmas.

Introduced forage grass; known from southeastern Alaska (Petersburg, J. P. Anderson 5643, ISC); widespread in North America; introduced from Europe.

AVENA L.

Annual herbs with hollow culms; leaf sheaths open, the blades flat, lacking auricles, the ligules membranous, 2–3 flowered, articulate above the glumes, usually pendulous; glumes about equal, commonly surpassing the florets, the first 7-nerved, the second 9-nerved; lemmas 7-nerved, with a stout twisted bent awn inserted near the middle or the awn sometimes reduced or lacking; rachilla produced behind the upper floret; palea shorter than the lemma; lodicules 2; stamens 3.

1a. Florets usually 2, the upper floret awnless. *A. sativa*
1b. Florets usually 3, the first and second usually awned. *A. fatua*

Avena fatua L.
Wild Oats

Plants 0.5–10 dm tall or more, tufted or solitary; leaf blades 3–10 mm broad or more, scabrous; ligules 3–6 mm long, ciliate-erose; spikelets commonly 20–30 mm long, usually with 3 florets, the rachilla readily disarticulating between the florets; lemmas hardened at base, usually with densely bearded callus; awned on first 2 florets, with twisted strongly bent awns; anthers 4–5 mm long.

Introduced weedy annual; known in our area from southeastern Alaska (Juneau, J. P. Anderson 2A-442 ISC); widespread in temperate regions of the earth. This plant is distributed in seed grains and should be expected elsewhere.

Avena sativa L.
Oats

Plants 0.5–10 dm tall or more, tufted or solitary; leaf blades 3–10 mm broad or more, scabrous (at least above); ligules 2–4 mm long, ciliate-erose; spikelets commonly over 20 mm long, usually with 2 florets, the rachilla not readily disarticulating between the florets; lemmas hardened to above middle, usually with naked or sparsely bearded callus; awned on first floret, or awns lacking, when present then straight or nearly so; anthers 3–4 mm long.

Introduced grain plant; in southern Alaska and Yukon where it is often grown as fodder in oat-pea (*Pisum sativum*) mixtures; occasionally escaping but not persisting; widespread in temperate regions of the earth.

BECKMANNIA Host

Annual herbs with hollow culms; leaf sheaths open, the blades flat, lacking auricles, the ligules membranous, more or less puberulent; spikelets borne closely aggregated in 2 rows on one side of the axis in numerous appressed to ascending or spreading racemose-paniculate spikes, 1-flowered, articulate below the glumes, the glumes about equal, about as long as the floret, 3-nerved, acute; lemmas 5-nerved, awnless; palea about equaling the lemma; rachilla a mere stub; lodicules 2; stamens 3.

Beckmannia syzigachne (Steud.) Fern.
Sloughgrass

Plants mostly 2.5–10 dm tall, tufted or less commonly solitary; leaf blades 2–10 mm broad, scabrous; ligules 4–10 mm long, lacerate, often folded back; panicle narrow, more or less congested, 6–26(30) cm long; spikelets 2–3.2 mm long and about or not quite as broad; anthers 1–1.4 mm long.

Shallow ponds, streams, and muskegs; in much of continental Alaska and Yukon south of the arctic slope; widespread in North America; circumboreal (*B. erucaeformis* var. *baicalensis* Kuzn.; *B. erucaeformis* ssp. *baicalensis* [Kuzn.] Hultén; *B. syzigachne* ssp. *baicalensis* [Kuzn.] Koyama & Kawano). Our material differs from the Eurasian *B. erucaeformis* (L.) Host in having only one floret per spikelet.

BROMUS L.

Annual or perennial tufted or rhizomatous herbs; culms hollow; leaf sheaths closed to near the top, auricles present or absent, the ligules membranous, more or less puberulent; spikelets borne in open or congested panicles, several-flowered, articulate above the glumes; glumes unequal, shorter than the spikelet, the first 1–3-nerved, the second 3–5(7)-nerved, rarely awn tipped; lemmas 5–9-nerved, usually bifid, awnless or awned from between the teeth of the bifid apex; palea shorter than the lemma; lodicules 2; stamens 3.

Harlan, J. R. 1945. Cleistogamy and chasmogamy in *Bromus carinatus* Hook. & Arn. Am. Jour. Bot. 32:66–72.

Mitchell, Wm. W. 1967. Taxonomic synopsis of *Bromus* section Bromopsis (Gramineae) in Alaska. Can. Jour. Bot. 45: 1309–13.

1a. Lemmas narrowly lanceolate, with a hard sharp callus, the lateral teeth acuminate, very thin; introduced weedy annual. *B. tectorum*
1b. Lemmas broad, rounded or tapering to the apex, the callus neither sharp nor prolonged; the lateral teeth short, acute; introduced or indigenous annuals or perennials. (2)
2a. Plants annual, introduced weedy species. (3)
2b. Plants perennial, indigenous or adventive. (6)
3a. Lemmas awnless or merely awn tipped, inflated; spikelets 9–15-flowered, 15–30 mm long. *B. brizaeformis*
3b. Lemmas with awns well developed, not inflated; spikelets various. (4)
4a. Lemmas with margins inrolled, the rachilla joints thus apparent, awnless or awned; awn flexuous; leaf sheaths glabrous or only the lowermost sparsely hairy. *B. secalinus*
4b. Lemmas not as above, the rachilla thus obscured; awns usually straight; leaf sheaths hairy. (5)
5a. Spikelets pubescent; nerves of lemma prominent; inflorescence congested and erect; lemmas 6–8 mm long. *B. mollis*
5b. Spikelets glabrous or scabrous; nerves of lemma not prominent; inflorescence rather open, the branches ascending to spreading; lemmas mostly more than 8 mm long. *B. commutatus*
6a. First glume 3–5-nerved; plants indigenous in coastal and insular southern Alaska or adventive in interior sites. (7)
6b. First glume 1-nerved; plants indigenous or introduced, with various distribution. (8)
7a. Ligules mostly more than 4 mm long; spikelets mostly 1–2 at tips of branches; leaves often more than 10 mm broad; plants of coastal southern Alaska. *B. sitchensis*
7b. Ligules mostly less than 4 mm long; spikelets several per branch; leaves

mostly less than 10 mm broad; plants introduced in widely scattered interior sites. *B. carinatus*

8a. Anthers over 3.4 mm long; panicles erect to nodding; plants rhizomatous or tufted. (9)

8b. Anthers less than 3.2 mm long; panicles drooping; plants tufted, not rhizomatous. (10)

9a. Lemmas glabrous to finely appressed-puberulent across the lower portion, awnless or with awn to 1.5 mm long; glumes glabrous; plants strongly rhizomatous; introduced forage grass. *B. inermis*

9b. Lemmas hairy throughout or along margin and keel only, with awns (1) 1.5–6 mm long; glumes pubescent to glabrous; plants shortly rhizomatous or tufted; indigenous. *B. pumpellianus*

10a. Glumes hairy; lemmas pubescent throughout; nodes retrorsely pubescent; plants generally tall and robust, with 5–9 nodes. *B. pacificus*

10b. Glumes glabrous; lemmas with conspicuous fringe of hairs along margin, glabrous or pubescent across the back; nodes pubescent or glabrous; plants medium to tall, with 3–7 nodes. (11)

11a. Nodes and sheaths pubescent; lemmas hairy along the marginal nerve only; first glume 4.5–7.5 mm long; uppermost ligule 0.4–1 mm long. *B. ciliatus*

11b. Nodes and sheaths glabrous; lemmas (at least the upper) pubescent throughout; first glume 7.5–12.5 mm long; uppermost ligule 0.8–2.0 mm long. *B. richardsonii*

Bromus brizaeformis Fisch. & Mey.
Rattlesnake Chess

Plants annual, 2–6 dm tall; leaf sheaths commonly densely pilose, the auricles lacking, the ligules to 1.5 mm long, the blades commonly 2–5 mm broad, pilose or short-pubescent; panicle 3–10 cm long or more, the branches usually curved to reflexed, mostly with a single spikelet; spikelets strongly flattened, oblong-ovate in outline, mostly 15–30 mm long, glabrous, mostly with 9–15 flowers; glumes unequal, the first 3–5-nerved and 4–6 mm long; lemmas inflated, 9–11 mm long, awnless or with an awn tip to 1 mm long; palea shorter than the lemma; anthers 0.7–0.8 mm long.

Disturbed sites near habitations; in widely scattered sites in Alaska and to be expected in the Yukon; introduced. This striking weedy species occurs widely in North America and in Europe.

Bromus carinatus Hook. & Arn.
California Brome

Plants perennial, tufted, mostly 3–10 dm tall or more; leaf sheaths glabrous to pilose or canescent-puberulent, the auricles tiny, usually present, the ligules mostly 1–3 mm long, the blades commonly 3–10 mm broad, glabrous to scabrous or pilose; panicle 5–20 cm long or more, the branches erect or ascending or less commonly spreading to reflexed; spikelets strongly compressed, narrowly lanceolate in outline, mostly 20–35 mm long, mostly with 5–10 flowers; glumes unequal, the first usually 3-nerved and 5–9 mm long; lemmas lanceolate, pubescent, awned, the awn 3–10 mm long, straight; palea about as long as the lemma; anthers 4–5 mm long (at least the exserted ones).

Introduced species; in widely scattered sites in Alaska and Yukon; possibly introduced as a forage plant or as a waif in hay or grain mixtures (*B. marginatus* Nees in Steud.). This species is a valuable forage plant in the western states.

Bromus ciliatus L.
Fringed Brome

Plants perennial, tufted, mostly 4.5–12 dm tall; leaf sheaths hairy, at least at the base, the auricles lacking the ligules 0.4–1 mm long, the blades 4–12 mm broad, generally pilose on the upper surface; panicle 8–18 cm long, distinctly nodding; spikelets only slightly compressed, lance-oblong in outline, 14–23 mm long, with 5–9 flowers; glumes unequal, the first 1(3)-nerved and 4.5–7.5 mm long; lemmas lanceolate, with a fringe of hairs along the margin, glabrous or scabrous across the back, with awn 1–4 mm long; palea shorter than the lemma; anthers 1–1.6(2) mm long.

Open woods, heathlands, and meadows; in central and southern Alaska and southern Yukon; eastward to Newfoundland and south to Mexico, Texas, and New Jersey.

Bromus commutatus Schrad.
Meadow Brome

Plants annual, tufted or less commonly solitary, mostly 2–10 dm tall; leaf sheaths softly hairy with retrorse or spreading hairs, the auricles lacking, the ligules mostly 0.5–2 mm long, the blades 2–5(10) mm broad, hairy; panicle 7–15 cm long or more, open, the branches ascending to spreading; spikelets compressed, lanceolate to lance-oblong in outline, 13–20 mm long, with 5–9 flowers; glumes unequal, the first 3(5)-nerved and 4–7 mm long; lemmas elliptic to oblong, glabrous, the nerves inconspicuous, the awn 4-10 mm long, straight to curved; palea shorter than the lemma; anthers 1–1.5(2) mm long.

Introduced weedy grass of disturbed soils; in widely separated locations in southern Alaska and to be sought elsewhere; widespread in North America; Europe.

Bromus inermis Leyss.
Smooth Brome

Plants perennial, rhizomatous, the rhizomes extensive, erect, 4–16.5 dm tall; leaf sheaths glabrous, auricles present on at least some leaves, to 1 mm long, the ligules 0.5–3 mm long, the blades 3–10 mm wide, usually glabrous; panicle 5–24(27) cm long, the branches ascending to erect; spikelets nearly cylindrical, 15–30 mm long or more, with 5–13 flowers; glumes unequal, the

first 1(3)-nerved and 4–8 mm long, glabrous; lemmas oblong-elliptic, glabrous or finely puberulent across the back or along nerves at base, awnless or with awns to 1(2) mm long; anthers 4–6 mm long.

Introduced forage and revegetation grass; in much of Alaska south of the Brooks Range and in southern Yukon; widespread in the north temperate regions of the earth. Smooth brome is confined in our region to areas of disturbance, particularly around settlements, and is spreading along the road systems. *B. inermis* hybridizes with *B. pumpellianus* along lines of contact. (q.v.).

Bromus mollis L.
Soft Chess

Plants annual, tufted or less commonly solitary, 2–7 dm tall (rarely more); leaf sheaths long-pilose, the hairs more or less retrorse, lacking auricles, the ligules 0.5–1.5 mm long, the blades 1.5–4 mm broad, usually hairy; panicle 3–10 cm long, the branches ascending to erect; spikelets compressed, 10–20 mm long, with 5–9 flowers; glumes unequal, the first 3–5-nerved and 4–6 mm long; lemmas broadly oblong-lanceolate, the nerves prominent, usually softly hairy, the awn straight or nearly so, 6–10 mm long; anthers 0.5–3 mm long.

Introduced weedy grass of widely disjunct localities; in much of Alaska south of the Yukon River and in southern Yukon (*B. hordeaceus* auth. not L.; *B. racemosus* L.); widespread in North America; introduced from Europe. The writer follows Hitchcock (Fl. Pac. N. W. 1:511. 1969) in the interpretation of nomenclature presented herein.

Bromus pacificus Shear
Pacific Brome

Plants perennial, tufted, 6–17 dm tall, retrorsely pubescent at the nodes; leaf sheaths pilose, the hairs more or less retrorse, lacking auricles, the ligules 1.8–3.2 (5) mm long, the blades 6–16 mm wide, pilose on upper surface; panicle 13–22(30) cm long, open, with spreading to drooping branches; spikelets somewhat compressed, 20–28(35) mm long, with 5–11 flowers; glumes unequal, pubescent, the first 1 (rarely 3)-nerved and 6–8.5 mm long; lemmas lance-elliptic, sericeous throughout, the awns 3.5–6 mm long; anthers 2–3 mm long.

Moist woods and beaches; mainly in insular southeastern Alaska and also disjunctly to Cook Inlet; southward to Washington and Oregon.

Bromus pumpellianus Scribn.
Arctic Brome

Plants perennial, tufted or rhizomatous, the rhizomes seldom extensive, erect or decumbent at base, 3–15 dm tall; leaf sheaths glabrous to densely hairy, auricles present at least on some leaves, the ligules 0.5–4 mm long, the blades 2.5–12 mm wide, glabrous to densely hairy; panicle 4–33 cm long, the branches erect to spreading; spikelets moderately compressed, 12–60 mm long, with (5)7–15 flowers; glumes unequal, the first usually 1-nerved, 4.5–13.5 mm long, glabrous to densely hairy; lemmas lance-oblong to oblong-elliptic, sparsely to densely hairy along the margin and keel or throughout, with awns 1–6 mm long; anthers 3.5–6.8 mm long.

Indigenous, in meadows, open woods, slopes, gravel bars, rock outcrops and stream banks; in most of continental Alaska and Yukon; southward to Colorado and Wyoming and eastward to the Mackenzie; Asia. This entity is known to form hybrids with *B. inermis* Leyss. Two more or less distinctive subspecies occur within the included area.

1a. Panicles strict, erect to open and arching; plants rhizomatous, spreading, with erect stems; nodes generally pubescent; ligules of uppermost leaves 0.5–2.5(3.5) mm long; plants widespread in Alaska and Yukon (*B. inermis* ssp. *pumpellianus* [Scribn.] Wagnon; *B. inermis* var. *purpurascens* [Hook.] Wagnon; *B. purgans*

var. *purpurascens* Hook.; *B. arcticus* Shear; *B. inermis* ssp. *pumpellianus* var. *arcticus* [Shear] Porsild; *B. pumpellianus* var. *villosissimus* Hultén). *B. pumpellianus* ssp. *pumpellianus*

1b. Panicles generally open, nodding to erect; plants tufted to shortly rhizomatous, with stems ascending; nodes glabrous or pubescent; ligules of uppermost leaves 1.5–4 mm long; plants known only from west-central Alaska, along the Yukon River from Ruby to Kaltag. *B. pumpellianus* ssp. *dicksonii* Mitchell & Wilton

Bromus richardsonii Link
Richardson Brome

Plants perennial, tufted, mostly 6.5–14.5 dm tall; leaf sheaths glabrous except for a tuft of auricular hairs, the auricles lacking, the ligules 0.4–2.0 mm long, the blades 4.5–12 mm wide, glabrous; panicle 10–20 cm long, distinctly nodding; spikelets only slightly compressed, lance-oblong to oblong-elliptic in outline, 16–26 mm long, with (3)5–7 flowers; glumes unequal, the first 1(3)-nerved and 7.5–12.5 mm long; lemmas lanceolate to oblong-elliptic with a fringe of hairs along the margin and short hairs across the back of at least the uppermost, with awns 2–5 mm long; palea slightly shorter than the lemma; anthers 1–1.2 mm long.

In mixed forb and grass communities of the upper alder zone, above tree line; in the Matanuska Valley, north and east of Palmer; in western North America; Asia. The known range of this entity seems likely to be enlarged as southern Alaska is more thoroughly examined. Most previous treatments have placed *B. richardsonii* with *B. ciliatus*, a species which it resembles and with which it forms sterile hybrids.

Bromus secalinus L.
Ryebrome

Plants annual, tufted, mostly 3–8 dm tall; leaf sheaths glabrous to finely, retrorsely pubescent, the auricles lacking, the ligules 1.5–3 mm long, the blades 2–5 mm broad or more, pilose to glabrous; panicle 5–20 cm long, loose, usually nodding; spikelets strongly compressed, narrowly lanceolate to ovate-lanceolate in outline, 10–20 mm long, mostly with 5–9 flowers; glumes unequal, the first 3–5-nerved and 4–6 mm long; lemmas ovate to elliptic, the margin inrolled exposing the rachilla, with flexuous awns mostly 3–6 mm long or the awn rarely reduced; palea equal to or only slightly shorter than the lemma; anthers about 1 mm long.

Weedy plants; reported from scattered localities in southern Alaska and southern Yukon; widely distributed in North America; Eurasia.

Bromus sitchensis Trin.
Alaska Brome

Plants perennial, tufted, nonrhizomatous, 5–15 dm tall or more; leaf sheaths glabrous or pilose, lacking auricles, the ligules 2–8 mm long, the blades 5–15 mm broad, glabrous or pilose; panicle 10–35 cm long, the branches erect to spreading or drooping; spikelets strongly compressed, narrowly lance-elliptic to oblong in outline, (25) 35–52 mm long, mostly with 5–11 flowers; glumes unequal, the first 3–5-nerved and 7.5–14 mm long; lemmas lanceolate, glabrous or puberulent, keeled on the back, with straight or moderately bent awn to 10 mm long or more; anthers to 6 mm long.

River banks, beaches, and mountain slopes; in coastal and insular southern Alaska from the westernmost Aleutians eastward through the Panhandle; southward to Oregon. Our materials are separable into two more or less distinct varieties.

1a. Panicle with branches erect or stiffly ascending; plants mostly from the Aleutians (*B. aleutensis* Trin.). *B. sitchensis* var. *aleutensis* (Trin.) Hultén

1b. Panicle with branches spreading or drooping; plants mostly of the Panhandle. *B. sitchensis* var. *sitchensis*

Bromus tectorum L.
Cheatgrass

Plants annual, tufted, mostly 2–8 dm tall or rarely more; leaf sheaths pilose, lacking auricles, the ligules 1–3 mm long, the blades mostly 2–4 mm wide, puberulent with short straight hairs; panicle 3–15 cm long or more, drooping or nodding; spikelets tapering from the base, to 20 mm long or more, with 3–6 flowers; glumes unequal, the first 1-nerved and mostly 5–7 mm long; lemmas glabrous to villous, narrowly lance-elliptic, with straight or slightly curved awns to 15 mm long; anthers 0.5–0.7 mm long.

Disturbed soils, mostly along roads and near habitations; at widely disjunct localities in southern Alaska and southern Yukon; widely distributed in North America; Eurasia.

CALAMAGROSTIS Adans.

Perennial tufted or rhizomatous herbs; culms hollow; leaves with open sheaths, the blades flat to involute, lacking auricles, the ligules membranous, more or less puberulent; spikelets borne in open to contracted panicles, 1-flowered, articulate above the glumes; glumes about equal, the first 1-nerved, the second 3-nerved, awnless; palea well developed, from about one-half to quite as long as the lemma; lemma usually 5-nerved, the midnerve produced into a straight to geniculate awn; rachilla prolonged behind the palea, bearing an apical tuft of long hairs; callus hairs long, white; lodicules 2; stamens 3.

Note: The taxa within this genus belong to widespread, complex species groups. Their interpretation has varied greatly depending on the overview of the worker and on the material at his disposal. Definitive criteria are not clear cut and defini-

tions of taxa are apt to be arbitrary. Therefore attempts at identification are fraught
with difficulties of subjective interpretation
of characteristics present in individual
specimens. Not all specimens will fit a
category easily and much further work is
necessary.

Inman, D. L. 1922. *Calamagrostis canadensis* and some related species. Rhodora 24:142–44.

Kawano, S. 1965. *Calamagrostis purpurascens* R. Br. and its identity. Acta
Phytotax. Geobot. 21:73–89.

Stebbins, G. L., Jr. 1930. A revision of
some North American species of *Calamagrostis*. Rhodora 32:35–57.

1a. Awn geniculate, exserted beyond
the spikelet tip by more than 1 mm;
plant a bunch grass, usually of dryish, open slopes. *C. purpurascens*

1b. Awn straight, curved, or sometimes
geniculate, included within the spikelet or seldom surpassing the glumes
by less than 1 mm; plants various.
(2)

2a. Plants mostly 1–3 dm tall; inflorescence compact, slender, mostly 3–5
cm long; spikelets 4.5–5 mm long;
plants known only from coastal and
near coastal northern Alaska and Yukon. *C. holmii*

2b. Plants commonly more than 3 dm
tall; inflorescence open or if compact
then over 5 cm long and spikelets
usually less than 4.5 mm long; plants
of various distribution, usually not
as above. (3)

3a. Callus hairs rarely more than half as
long as the lemmas; awn commonly
curved, often projecting sidewise
from the spikelet; plants of coastal
and insular southern Alaska (rarely
in the interior?). *C. nutkaensis*

3b. Callus hairs usually at least three-
fourths as long as the lemma; awn
straight or curved, rarely projecting

sidewise from the spikelet; plants of
various distribution. (4)

4a. Plants low-growing, to 4.5 dm tall;
inflorescence open; occurring in
brackish marshes, meadows, and on
sea cliffs, in coastal and insular Alaska, from Prince William Sound westward and northward around the margin of Alaska and eastward through
coastal Yukon. *C. deschampsioides*

4b. Plants often much more than 4.5 dm
tall, or the inflorescence contracted;
widely distributed in Alaska and Yukon, but seldom in brackish marshes
in coastal regions. (5)

5a. Panicle almost always open, when
pressed usually more than 2 cm
broad; callus hairs from almost as
long as the lemma to longer than it;
possibly our most abundant and most
widespread *Calamagrostis* species. *C.
canadensis*

5b. Panicle relatively contracted, when
pressed usually 2 cm broad or less
(rarely much over 2 cm wide); callus hairs commonly about three-
fourths as long as the lemma. (6)

6a. Panicle contracted, mostly 4–10 cm
long; glumes 2.8–4 mm long. *C. neglecta*

6b. Panicle contracted to loose, mostly
10 cm long or more, or if less then
panicle branches short, ascending,
and stiff, or the glumes mostly more
than 4 mm long. (7)

7a. Spikelets mostly 4.5–5.5 mm long,
often dark purple in color, the panicle
branches slender, often coiled. *C.
lapponica*

7b. Spikelets mostly 3–4 mm long, often
dull brownish purple, the panicle
branches thickish and stiff, not or
seldom coiled. *C. inexpansa*

Calamagrostis canadensis (Michx.) Beauv.
Bluejoint

Plants strongly rhizomatous, 4.5–20 dm tall;
stems smooth; sheath glabrous to scabrous;

ligules 2.5–12 mm long; leaf blades involute or more commonly flat and 3–8 mm broad, scabrous on both surfaces; panicle usually open, commonly more than 2 cm broad (when pressed), 6–23(25) cm long; glumes usually purplish but sometimes greenish or straw-colored, mostly 3.2–5.8 mm long, subequal to unequal, usually scabrous; lemma distinctly to barely shorter than the glumes, awned from the median one-third to the apical one-third of the lemma, the awn straight or seldom curved, shorter to slightly longer than the glumes; callus hairs at least three-fourths as long and often slightly surpassing the lemma; anthers 1–1.8 mm long.

Meadows, open woods, hillsides, river banks and bars, beaches, and in muskegs; in practically all of Alaska and Yukon; eastward and southward through much of North America; Asia. This is likely our most common and most widely distributed species of *Calamagrostis*, and indeed of all of our grasses. It is also one of our most handsome grasses. Within our region two more or less distinct varieties are commonly recognized. Further segregation might be possible. More work is indicated. The nomenclature of varietal names is only tentative.

1a. Glumes mostly more than 4.5 mm long, long-acuminate, rather strongly scabrous (*Deyeuxia scabra* Kunth; *C. canadensis* var. *scabra* [Kunth] A. S. Hitchc.; *Arundo langsdorffii* Link; *C. langsdorffii* [Link] Trin.; *C. canadensis* ssp. *langsdorffii* [Link] Hultén). *C. canadensis* var. *langsdorffii* (Link) Inman

1b. Glumes mostly less than 4.5 mm long, abruptly acute to shortly acuminate, scabrous to glabrous (*Deyeuxia macouniana* Vasey; *C. macouniana* [Vasey] Vasey; *C. canadensis* var. *macouniana* [Vasey] Stebbins; *C. pallida* Vasey & Scribn.; *C. canadensis* var. *pallida* [Vasey & Scribn.] Stebbins; *C. canadensis* var. *acuminata* Vasey ex Shear & Rydb.; *C. alaskana*

Kearney; *C. atropurpurea* Nash; *C. scribneri* var. *imberbis* Stebbins; *C. canadensis* var. *imberbis* [Stebbins] C. L. Hitchc.). *C. canadensis* var. *canadensis*

Calamagrostis deschampsioides Trin.

Plants rhizomatous and more or less stolon-stems smooth; leaf sheaths glabrous; ligules 1–2.5 mm long; leaf blades involute, glabrous below, scabrous on upper surface; panicle usually open, commonly 3–10 cm long, relatively few-flowered; glumes 3–4.5 mm long, brownish purple to straw-colored, subequal, glabrous; lemmas subequal to the glumes or distinctly shorter, awned in the median one-third, the awn curved to decidedly geniculate, shorter to longer than the glumes; callus hairs about half as long as the lemma or slightly more than half; anthers 1.5–2 mm long.

Brackish marshes and meadows and on sea cliffs; in coastal and insular Alaska from the Prince William Sound region westward and northward around the margin of Alaska and northern Yukon; eastward to Labrador; circumpolar.

Calamagrostis holmii Lange in Holm
Holm Reedgrass

Plants rhizomatous and more or less stoloniferous, mostly 1.5–2 dm tall, decumbent at the base; stems smooth, or scabrous below the inflorescence; leaf sheaths smooth, glabrous; ligules 1.5–3 mm long; leaf blades flat or more commonly more or less involute, 2–3 mm broad, glabrous below, scabrous on upper surface; panicle contracted to slightly loose, 3–5 cm long, purplish; glumes 4–5 mm long, unequal, glabrous; lemmas shorter than the glumes, awned in the median one-third, the awn straight or curved, shorter than or subequal to the glumes; callus hairs about three fourths as long as the lemma; anthers 1.5–2 mm long.

Maritime tundra; islands of the Bering Sea, coastal northern Alaska, and northern

Yukon; Asia. Our material has been identified with *C. kolymaensis* Kom. by some workers. Recent investigators indicate that *C. kolymaensis* is conspecific with *C. holmii*. The former entity has been distinguished on the basis of position of awn insertion in the upper one-third of the lemma, the latter in the lower one-fourth. Our material has also been treated as *C. neglecta* var. *borealis* (Laest.) Kearney. Position of awn insertion does not appear to be a reliable character in distinguishing most taxa in *Calamagrostis*, varying as it does not only from specimen to specimen, but even in different spikelets from the same inflorescence. Since the larger spikelets seem to distinguish these dwarf plants from similar short plants in *C. neglecta*, it seems best to treat our material as *C. holmii*, which has priority over *C. kolymaensis*, until such time as a definitive monograph is available.

Calamagrostis inexpansa Gray
Northern Reedgrass

Plants rhizomatous, tufted, commonly 4–8.5 dm tall; stems scabrous, at least below the inflorescence; leaf sheaths smooth or minutely scaberulous; ligules 1–2 mm long; leaf blades flat or more or less involute, 2–4 mm long, scabrous on both surfaces and along the margins; panicle contracted and more or less verticillate, mostly 9–18 cm long, purplish brown; glumes subequal, 3.2–4 mm long, scabrous at least on the keel; lemmas subequal to the glumes, awned from the median one-third, the awn straight, about equaling or slightly longer than the lemma; callus hairs about two-thirds the length of the lemma; anthers 1.5–1.7 mm long.

Muskegs and meadows; in much of Alaska and at least southern Yukon; eastward to Newfoundland and southward to California, New Mexico, Missouri, and Maine; Asia. Alaskan materials of *C. inexpansa* have been shuffled for years between *C. canadensis* on the one hand and *C. neglecta* or *C. lapponica* on the other. *C. in-*expansa seems to be defined from *C. canadensis* by the compact racemes, from *C. lapponica* by the smaller spikelets, and from *C. neglecta* by the harshly rough-scabrous herbage and stiffish inflorescence branchlets.

Calamagrostis lapponica (Wahl.) Hartm.
Lapland Reedgrass

Plants tufted, shortly rhizomatous, commonly 2.7–8 dm tall; stems smooth or less commonly scabrous below the inflorescence; leaf sheaths smooth; ligules 2–4 mm long; leaf blades flat and 2–4 mm long or more commonly involute, smooth to scabrous; panicle contracted to slightly loose, 7–15(20) cm long, purplish to dark purplish; glumes subequal or slightly unequal, 4.5–5.5 mm long, scabrous at least on the keel; lemmas much shorter than the glumes, awned from the median one-third to lower one-third, the awn straight or somewhat curved, shorter than or subequal to the lemma; callus hairs about three-fourths to fully as long as the lemma; anthers 1.5–2.1 mm long.

River banks, heathlands, muskegs, meadows, open woods, and tundra; in much of continental Alaska and Yukon; eastward to Labrador and southward to British Columbia; Eurasia (*Arundo lapponica* Wahl.).

Calamagrostis neglecta (Ehrh.) Gaertn., Mey., Schreb.
Narrow Reedgrass

Plants rhizomatous, more or less tufted, (1.5)2.5–5 dm tall; stems smooth; leaf sheaths smooth; ligules 0.7–3.5 mm long; leaf blades flat, or more commonly involute, 1–3(4) mm long, scabrous above or smooth on both sides; panicle contracted, 3–10 cm long, purplish or sometimes greenish to straw-colored; glumes subequal, 2.8–4 mm long, glabrous or sometimes scabrous on the keel; lemma slightly to distinctly shorter than the glumes, awned from the median one-third to the lower one-third, the awn straight or curved, from shorter than to surpassing the lemma; anthers 1.3–1.8 mm long.

Gravel bars, meadows, stream banks, and pond margins; in much of Alaska north of the 62nd parallel and in most of the Yukon; eastward to Greenland and south to Oregon, Utah, Colorado, and North Dakota; circumboreal (*Arundo neglecta* Ehrh.).

Calamagrostis nutkaensis (Presl) Steud.
Pacific Reedgrass

Plants strongly tufted, (2.5)3–9 dm tall or more, shortly rhizomatous; stems smooth or scabrous below the inflorescence only; leaf sheaths smooth; ligules mostly 2–4 mm long; leaf blades flat and 4–8 mm broad or more, or more or less involute, often scabrous on one or both sides; panicle narrow, more or less loose, mostly 9–20 cm long, greenish or purplish; glumes unequal, very narrow, (4.5)5–8.3 mm long, glabrous or scabrous on the keel; lemma distinctly shorter than the longer glume, awned from the median one-third, the awn straight or geniculate and exserted from the side of the spikelet, seldom vestigial; anthers 1.8–3 mm long.

Beaches, stream banks, mountain slopes, and meadows; in coastal and insular southern Alaska from the Aleutians eastward through the Panhandle; south to California (*Deyeuxia nutkaensis* Presl; *C. aleutica* Trin. in Bong.).

Calamagrostis purpurascens R. Br. in Richards.
Purple Reedgrass

Plants strongly tufted and often shortly rhizomatous, mostly 2.5–10 dm tall; stems smooth or scabrous to puberulent below the inflorescence; leaf sheaths smooth to scabrous; ligules 1–6 mm long; leaf blades flat or involute, 2–4 mm broad, scabrous at least on upper surface; panicle congested to loose, 5–17 cm long; glumes unequal, (4.5)5–7.8 mm long, scabrous over the back or at least on the keel; lemma distinctly shorter than the glume, awned from the basal one-third, the awn twisted and geniculate, exceeding the spikelet by 1.5–2 mm or more; anthers 2–3 mm long.

Open slopes and ridge tops; in much of Alaska and most of the Yukon; eastward to Greenland and south to California, Utah, Colorado, and South Dakota; Asia (*C. yukonensis* Nash; *C. purpurascens* var. *sylvatica* Thurb.; *C. arctica* Vasey; *C. purpurascens* ssp. *arctica* [Vasey] Hultén; *C. purpurascens* var. *arctica* [Vasey] Kearney).

CATABROSA Beauv.

Perennial rhizomatous aquatic or semiaquatic herbs; culms hollow; leaf sheaths at least partially closed near the base; leaf blades flat, lacking auricles, the ligules membranous; spikelets borne in open panicles, mostly 2-flowered, articulate above the glumes; glumes very unequal, nerveless, shorter than the first floret, obtuse to acute, awnless; palea subequal to the lemma; lemmas 3-nerved, nerves more or less parallel, not converging apically, unawned; callus not hairy; lodicules 2; stamens 3.

Catabrosa aquatica (L.) Beauv.
Brookgrass

Plants 1–5 dm tall, commonly decumbent and rooting at the nodes; ligules 2–8 mm long, erose to almost entire, not or seldom bent back, minutely scabrous but not puberulent; leaf blades 2–10 mm broad or more, smooth; panicles open, commonly 5–20 cm long, the branches spreading to ascending, purplish to straw-colored; spikelets usually 2-flowered, rarely 1-flowered, mostly 3–4 mm long; glumes unequal, the first less than (0.5)1 mm long, the second 1–2 mm long; lemmas 2.5–3 mm long.

Marshes, shallow ponds and lakes; in Alaska known only from the Alaska Peninsula, to be expected elsewhere; widely distributed in North America; Eurasia (*Aira aquatica* L.).

CINNA L.

Perennial shortly rhizomatous tufted herbs; culms hollow; leaf sheaths open; leaf blades flat, lacking auricles, the ligules membranous; spikelets borne in open panicles, 1-flowered, articulate below the glumes; glumes subequal or somewhat unequal, 1-nerved, about as long as the floret, very acute, awnless; palea subequal to the lemma; lemmas 3-nerved, awnless or usually with a short subterminal awn; callus not hairy; floret shortly stipitate, the rachilla prolonged behind the palea; lodicules 2; stamens (1)3.

Cinna latifolia (Trev.) Griseb. in Ledeb.
Woodreed

Plants tufted, shortly rhizomatous, 8–12 dm tall or more; stems smooth; sheaths glabrous or scaberulous; ligules 3–8 mm long; leaf blades 7–15 mm broad, scabrous on both sides; panicle open, the branches spreading to pendulous, 15–30 cm long, greenish or purplish; glumes mostly 3–4 mm long, usually scabrous at least on the keel; lemma subequal or slightly longer than the glumes, awnless or with a straight, subterminal awn to about 1 mm long; anthers 0.5–1 mm long.

Stream banks, talus slopes, meadows, and moist woods; in coastal and insular southern Alaska, from the Alaska Peninsula eastward, in northern British Columbia (Liard Hot Springs), and to be sought in southern Yukon; widespread in North America; Eurasia (*Agrostis latifolia* Trev. ex Goepp.).

COLPODIUM Trin.

Perennial shortly rhizomatous tufted herbs; culms hollow; leaves with open sheaths, the blades flat or folded to involute, lacking auricles, the ligules membranous, entire; spikelets born in open or contracted panicles, mostly with 2–4 flowers, articulate above the glumes; glumes unequal, the first 1-nerved, the second 3-nerved, shorter than the spikelet, obtuse, awnless; palea subequal or somewhat shorter than the lemma; lemmas 3–5-nerved, unawned; callus not hairy; lodicules 2; stamens 3.

1a. Panicle contracted, rather dense, 2–4 cm long; plants mostly 0.5–1.5 dm tall; known from southwestern Yukon. *C. vahlianum*

1b. Panicle open, 4–9 cm long; plants mostly 3–5 dm tall; known from Seward Peninsula, northwest coast, and north slope of the Brooks Range. *C. wrightii*

Colpodium vahlianum (Liebm.) Nevski

Plants mostly 0.5–1.5 dm tall, tufted and shortly rhizomatous (?); ligules 1.5–2(4) mm long; leaf blades flat or folded, 1–2 mm broad, smooth; panicle contracted, 2–4 cm long, the branches appressed-ascending, purplish; spikelets 4–5 mm long, 2–4-flowered; glumes unequal, scarcely shorter than the lower florets, the first 1-nerved and 2.5–3 mm long, the second 3-nerved and 3–3.5 mm long; lemmas 3–4 mm long, 3-nerved.

Stony tundra; in our region known only from southwestern Yukon; disjunctly eastward to Greenland; circumpolar (*Poa vahliana* Liebm.).

Colpodium wrightii Scribn. & Merr.

Plants mostly 3–5 dm tall, tufted (rhizomatous?); ligules mostly 1–2(3) mm long; leaf blades involute, less than 1 mm broad, smooth; panicle open, 4–9 cm long, the branches spreading to ascending or deflexed, purplish; spikelets 6–8 mm long, with 3–4 flowers; glumes unequal, slightly more than half as long as the lower florets, the first 1-nerved and 1.5–2.5 mm long, the second 3-nerved and 2.5–3.5 mm long; lemmas 4.5–5 mm long, 5-nerved.

Wet meadows; in west-central and northwestern Alaska; Siberia.

DACTYLIS L.

Perennial tufted herbs; culms hollow; leaf sheaths open, the blades flat, lacking auricles, the ligules membranous; spikelets borne subsessile in dense, 1-sided clusters in dense panicles, 3 (rarely 5)-flowered, articulate above the glumes; glumes unalike, the first mostly 2-nerved, the second 1-nerved, each with a soft awnlike tip; palea subequal to the lemma; lemma 5-nerved, the obscure nerves convergent apically, usually awn tipped; lodicules 2; stamens 3.

Dactylis glomerata L.
Orchard Grass

Plants tufted, mostly 5–10 dm tall or more; ligules 3–7(9) mm long, more or less pubescent; leaf blades flat, mostly 3–10 mm wide, usually scabrous; panicle 3–10 cm long or more, the branches stiffly ascending to erect or less commonly spreading or even deflexed; spikelets 5–9 mm long, strongly compressed; glumes 4–6 mm long, awn tipped; lemmas mostly 5–8 mm long, the awn tip to 1 mm long; anthers 3–4 mm long.

Disturbed sites, mostly near habitations; in widely scattered localities in coastal and insular southern Alaska and to be expected elsewhere; widespread in temperate regions of the earth. Orchard grass is cultivated as a forage grass and in stabilization of cultivated lands.

DANTHONIA Lam. & DC.

Perennial tufted herbs; culms hollow; leaf sheaths open, blades flat or more or less involute, lacking auricles, the ligules consisting of a fringe of short hairs; spikelets borne in compact to elongate panicles, several-flowered, articulate above the glumes; glumes often as long as the spikelets (exclusive of the awns), mostly 3–5-nerved, unawned; palea shorter than the lemma; lemmas strongly bifid, with a flattened twisted geniculate awn from below the lobes; lodicules 2; stamens 3.

1a. Lemmas pilose over the back and along the margins, commonly 4–5 mm long; plants known from southeastern Alaska. *D. spicata*

1b. Lemmas glabrous over the back, pilose only along the margins, mostly 7–10 mm long; plants known from south-central and southeastern Alaska and southern Yukon. *D. intermedia*

Danthonia intermedia Vasey
Timber Oatgrass

Plants tufted, commonly 0.5–2.5 dm tall or more; ligules very short, with a fringe of short hairs; leaf blades flat or involute, commonly 1–3 mm broad, pilose to glabrous; panicle spicate, mostly 3–7 cm long, often with 2–5 spikelets, purplish to straw-colored; glumes subequal, 7.5–12 mm long, obscurely 3–5-nerved; lemmas mostly 7–10 mm long, the callus bearded, margins hairy, glabrous on the back, the teeth 1–2 mm long; awn to 10 mm long; anthers 3–4 mm long.

Meadows; in south-central and continental southeastern Alaska and southern Yukon; eastward to Newfoundland and south to California, Utah, Colorado, and Michigan.

Danthonia spicata (L.) Beauv. ex Roem. & Schult.
Poverty Oatgrass

Plants tufted, commonly 2–7 dm tall; ligule very short, with a fringe of short hairs; leaf blades usually involute, commonly 0.5–2 mm broad, glabrous or pilose; panicle slender, open, mostly 2–5 cm long, often with 3–5 spikelets, green; glumes mostly 9–12 mm long, subequal, obscurely 3–5-nerved; lemmas 4–5 mm long, the callus lightly bearded, margins and back hairy, the teeth 0.5–2 mm long, awn to 8 mm long; anthers 2–3 mm long.

Known in Alaska from extreme southern portion of the Panhandle and to be expected elsewhere; widely distributed in North America (*Avena spicata* L.).

DESCHAMPSIA Beauv.

Annual or perennial caespitose herbs; culms hollow; leaves with open sheaths, the blades flat to involute, the ligules membranous, pubescent; spikelets borne in open to contracted panicles, 2–3-flowered, articulate above the glumes; glumes 1–3-nerved, acute to acuminate or attenuate, awnless; palea shorter than the lemma; lemmas obscurely 5-nerved, awned from about or below the middle, the awn twisted and more or less geniculate, the callus bearded; rachilla prolonged behind the uppermost floret; lodicules 2; stamens 3.

Kawano, S. 1963. Cytogeography and evolution of the *Deschampsia caespitosa* complex. Can. Jour. Bot. 41:719–42.

Lawrence, Wm. E. 1945. Some ecotypic relations of *Deschampsia caespitosa*. Am. Jour. Bot. 32:298–314.

1a. Plants slender annuals with few spikelets borne on slender ascending to erect branches; leaves with blades filiform, rarely to 1 mm broad. *D. danthonioides*

1b. Plants perennial with spikelets mostly numerous and with panicle branches various; leaf blades various, often more than 1.5 mm broad. (2)

2a. Panicle narrow, the branches appressed-ascending; glumes usually equaling or exceeding the uppermost floret. *D. elongata*

2b. Panicle usually open, or if narrow, then the glumes shorter than the uppermost floret. (3)

3a. Awn of lemma geniculate, twisted below, much longer than the spikelet; leaf blades filiform; plants uncommon or rare, introduced. *D. flexuosa*

3b. Awn of lemma straight or more or less bent and twisted, seldom longer than the glumes; leaf blades flat or folded; plants indigenous, widespread. (4)

4a. Leaf blades flat, (2)3–6 mm wide; ligules mostly 1–3.5 mm long; spikelets purplish or green becoming purplish, with awns mostly much less

than 3 mm long. *D. atropurpurea*

4b. Leaf blades folded or involute, or less commonly flat, rarely to 3 mm broad; ligules over 3.5 mm long (at least some); spikelets tawny to straw-colored or purplish, shining. (5)

5a. Glumes 4.8–7.3 mm long; anthers (1.5)1.9–2.2 mm long; plants of coastal and insular southern and southwestern Alaska and islands of the Bering Sea. *D. beringensis*

5b. Glumes 3.0–4.5(5.2) mm long; anthers (1.2)1.4–1.9 mm long; plants of broad distribution in Alaska (including coastal southern portions) and Yukon. *D. caespitosa*

Deschampsia atropurpurea (Wahl.) Scheele
Mountain Hairgrass

Plants tufted, perennial, 1–6.5 dm tall; ligules obtuse to truncate, erose to subentire, 1–3.5 mm long; leaf blades flat, (2) 3–6 mm broad, pubescent to glabrous; panicles opening with age, 3–12 cm long, the branches often long and curved; spikelets green becoming purplish, occasionally 3-flowered; glumes subequal, 4–6.3 mm long, the first 1-nerved, the second 3-nerved, longer than the uppermost floret; lemmas obscurely 3-nerved with callus hairs about half as long, awned from above the middle, the awn twisted and more or less geniculate or straight, to 2.5 mm long; anthers 0.8–1.2 mm long.

Meadows, heathlands, and open woods; in much of southern Alaska and in southern Yukon; eastward to Newfoundland and south to California, Colorado, and New Hampshire (*Aira atropurpurea* Wahl.; *Avena atropurpurea* [Wahl.] Link; *Vahlodea atropurpurea* [Wahl.] Fries; *Aira latifolia* Hook.; *D. latifolia* [Hook.] *Vasey*; *V. latifolia* [Hook.] Hultén; *V. atropurpurea* ssp. *latifolia* [Hook.] Porsild; *D. atropurpurea* var. *paramushirensis* Kudo; *V. atropurpurea* ssp. *paramushirensis* [Kudo] Hultén). Our materials are variable in pubescence, the chief feature used previously to separate them into two infra-specific taxa. The segregation is not an "either-or" case, but one of transition from hairy to not hairy, hardly a sufficient basis for taxa at any rank. Our specimens belong to the var. *latifolia* (Hook.) Scribn. ex Macoun.

Deschampsia beringensis Hultén
Bering Hairgrass

Plants tufted, perennial, 2.5–14 dm tall; ligules 3.8–7 mm long, acute; leaf blades involute or folded, or more commonly flat, mostly 1.8–3(4) mm broad, scabrous at least on upper side; panicles open, 9–33 (40) cm long, the branches spreading to spreading-ascending, not markedly curved; spikelets tawny to straw-colored or pur-plish, occasionally 3-flowered; glumes sub-equal, 4.8–7.3 mm long, the first 1–3-nerved, the second 3-nerved, from longer than to shorter than the uppermost floret; lemmas obscurely 3(5)-nerved, with callus hairs about one-third as long, awned from the lower one-third, the awn straight or curved, to 5 mm long or more; anthers (1.5)1.9–2.2 mm long.

Pond and lake margins, slopes, beach-fronts, meadows, tidal flats, and wet beaches; in coastal, insular southwestern, and southern Alaska and islands of the Bering Sea; southward to California; Asia (*D. caespitosa* ssp. *beringensis* [Hultén] Lawrence; *D. beringensis* var. *atkensis* Hul-tén; *D. caespitosa* var. *arctica* Vasey). *D. beringensis* is a phase of the *D. caespitosa* complex and might best be treated at vari-etal level (var. *arctica*) or subspecific level (ssp. *beringensis*) within that entity. Glume size and anther length appear to be the best diagnostic features, and even these features are not absolute. They form a con-tinuum with those of *D. caespitosa*. The averages seem to differ sufficiently to al-low segregation of most specimens, but some specimens from the interior, placed herein in *D. caespitosa*, are more similar to *D. beringensis* than would be allowed for most taxa treated as species.

Deschampsia caespitosa (L.) Beauv.
Tufted Hairgrass

Plants tufted, perennial, (0.8)2–12 dm tall; ligules (1)2.2–6 mm long, acute; leaf blades involute or folded or sometimes flat and 1–2(3) mm broad, scabrous at least on up-per side; panicles open or more or less con-tracted, 4–26 cm long, the branches spread-ing or ascending to almost erect, not markedly curved; spikelets tawny to straw-colored or purplish, occasionally 3-flow-ered; glumes subequal or somewhat un-equal, 3.0–4.5(5.2) mm long, the first 1-nerved, the second 3-nerved; lemmas ob-scurely 5-nerved, with callus hairs about one-fourth as long, awned from near the base, the awn straight or slightly curved, to 4 mm long; anthers (1.2)1.4–1.9 mm long.

Bogs, shallow ponds, river banks, spits, bars, and beaches; in practically all of Alas-ka and Yukon; eastward to Greenland and south to California, Arizona, Michigan and Wisconsin; Mexico; Eurasia (*Aira caespi-tosa* L.; *D. alpicola* Rydb.; *A. alpicola* [Rydb.] Rydb.; *D. glauca* Hartm.; *D. caes-pitosa* var. *glauca* [Hartm.] Sam.; *D. caes-pitosa* ssp. *orientalis* Hultén; *D. brevifolia* R. Br.; *D. caespitosa* var. *brevifolia* [R. Br.] Trautv.; *A. arctica* Spreng.; *D. arctica* [Spreng.] Ostenf.; *D. brevifolia* var. *pu-mila* Trin.; *D. pumila* [Trin.] Ostenf.). *D. caespitosa* is an extremely diverse taxon, varying in plant height, panicle size and compactness, spikelet size, and anther size. Attempts at segregation of the specimens into subordinate taxa are extremely diffi-cult, especially since many infraspecific entities have been recognized both in North America and in the Old World. Until the complex receives proper monographic study, it seems best to treat all of our speci-mens as portions of a single polymorphic entity.

Deschampsia danthonioides (Trin.) Munro ex Benth.
Annual Hairgrass

Plants simple or somewhat tufted annuals,

0.5–4 dm tall (rarely more); ligules 1–5 mm long, acute; leaf blades involute, to 1 mm broad, basal or nearly so, glabrous; panicles open though sometimes narrow, mostly 5–25 cm long, the branches ascending to nearly erect; spikelets greenish to straw-colored, 2-flowered; glumes unequal, 3-nerved, 5–8 mm long; lemmas obscurely 5-nerved, with callus hairs about one-fourth as long, awned from near the middle, the awn geniculate, commonly 3–7 mm long.

Disturbed sites, mostly near habitations; in widely scattered sites in Alaska and southern Yukon, where apparently introduced; disjunctly in the western states (*Aira danthonioides* Trin.).

Deschampsia elongata (Hook.) Munro ex Benth.
Slender Hairgrass

Plants tufted, perennial, mostly 3–8 dm tall; tall; ligules 3–8 mm long, acute; leaf blades flat or folded, mostly less than 1.5 mm broad, glabrous or scabrous; panicles narrow, 5–25 cm long or more, the branches ascending to erect; glumes subequal, 3-nerved, 3–5.5 mm long; lemmas obscurely 5-nerved, with callus hairs about half as long, awned from near the middle, the awn straight or slightly curved, mostly 3–4 mm long.

Disturbed sites, mostly near habitations; at widely scattered sites in Alaska and southern Yukon, where apparently introduced; widespread in western states (*Aira elongata* Hook.).

Deschampsia flexuosa (L.) Trin.
Wavy Hairgrass

Plants tufted, perennial, mostly 3–8 dm tall; ligules 1.5–3.6 mm long, obtuse to acute; leaf blades involute, to 1 mm broad, scabrous; panicle loose, open, mostly 5–15 cm long, the branches spreading to ascending; spikelets purplish to tawny, 2-flowered; glumes subequal or slightly unequal,

1-nerved, 4–5 mm long; lemmas obscurely 5-nerved, with callus hairs about one-fifth as long, awned from near the base, the awn twisted, geniculate, 4–7 mm long.

Disturbed sites; known from westernmost Aleutians and from the Panhandle, where probably introduced, and to be expected elsewhere; widely distributed in eastern North America; Eurasia (*Aira flexuosa* L.).

DUPONTIA R. Br.

Perennial rhizomatous herbs; culms hollow; leaves with sheaths partially closed at least below the middle, the blades flat or more or less involute, the ligules membranous, glabrous; spikelets borne in open or narrow panicles, 2–3-flowered, articulate above the glumes; glumes slightly unequal, the first 1-nerved, the second 3-nerved, attenuate to obtuse, awnless; palea slightly shorter than the lemma; lemmas apparently 1-nerved, awnless (or merely awn tipped), the callus with short stiff hairs; rachilla prolonged behind the upper floret; lodicules 2; stamens 3.

Dupontia fisheri R. Br.
Dupontia

Plants rhizomatous, mostly 1.5–4 dm tall; ligules 1–2.5 mm long, truncate, erose; leaf blades flat to folded or involute, mostly (1)2–5 mm broad, glabrous or scabrous on upper surface; panicle narrow or open, 5–14 cm long, the branches appressed-ascending to spreading or the lower descending; spikelets purplish to straw-colored, usually 2-flowered; glumes slightly unequal, 5.5–8.3 mm long; lemmas rounded, glabrous or puberulent, callus hairs about one-fifth as long, awnless or minutely awn tipped; anthers 1.7–3.2 mm long.

Wet polygonal ground, meadows and bluffs, in maritime and low alpine tundra; in western and northern Alaska and northern Yukon; circumpolar (*D. psilosantha* Rupr.; *D. fisheri* ssp. *psilosantha* [Rupr.] Hultén). The segregation of *D. fisheri* into two subspecies has been based largely on the nature of the inflorescence, and indeed the extreme forms appear to differ sufficiently to allow adequate separation. However, intermediates linking the two types are abundant and other characteristics do not seem to be correlated. Therefore a conservative treatment is indicated.

ELYMUS L.

Perennial tufted or rhizomatous herbs; culms hollow; leaf sheaths open, the blades flat, often with well-developed auricles, the ligules membranous; spikelets borne in a solitary (rarely branched) terminal spike or spikelike inflorescence, 2–many-flowered, articulate above the glumes, borne flatwise to the rachis, usually 2 per node and both sessile, but sometimes one spikelet pedicellate, and some nodes with only 1, or rarely with 2 or more spikelets, or the inflorescence sometimes branched; glumes subequal, 1–5-nerved, acute; palea almost as long as the body of the lemma; lemmas faintly nerved, awnless or awned; lodicules 2; stamens 3.

Bowden, W. M. 1957. Cytotaxonomy of section Psammelymus of the genus *Elymus*. Can. Jour. Bot. 35:951–93.

———. 1964. Cytotaxonomy of the species and interspecific hybrids of the genus *Elymus* in Canada, and neighboring areas. Can. Jour. Bot. 42:547–601.

Lepage, E. 1957. × *Elymordeum,* gen. hybr. nov. Nat. Can. 84:97.

1a. Plants with well-developed rhizomes; lemmas awnless or merely awn tipped. (2)
1b. Plants tufted, lacking rhizomes; lemmas with awns well developed, or if lemma merely awn tipped then glumes setaceous and spikes very dense. (3)
2a. Glumes narrow, subulate, 1–3-nerved; anthers mostly 4–5 mm long; plants of interior Alaska and Yukon except

occasionally in northern coastal sector. *E. innovatus*
2b. Glumes lanceolate to very narrowly so, broadened some distance above the base, 3–5-nerved; anthers often more than 5 mm long; plants of coastal and insular Alaska and Yukon. *E. mollis*
3a. Glumes subulate; spikes compact; lemmas merely awn tipped; introduced forage grass in southern Yukon. *E. junceus*
3b. Glumes lanceolate, broadened some distance above the base; spikes not especially compact; lemmas with well-developed awns; indigenous or introduced plants of southern Alaska and southern Yukon. (4)
4a. Margin of lemma ciliate with long hairs, the cilia several times as long as any dorsal pubescence; plants of coastal and insular southern Alaska. *E. hirsutus*
4b. Margin of lemma not ciliate, or if so, then hairs not longer than any dorsal pubescence; plant distribution various (see below). (5)
5a. Spike stiff and erect, not or seldom proliferating branches; indigenous plants of coastal and insular southeastern Alaska and southern Yukon. *E. glaucus*
5b. Spike arching to pendulous, often proliferating branches and becoming paniculate; adventive plants of the Cook Inlet vicinity, McKinley National Park, and from near Delta Junction. *E. sibiricus*

Elymus glaucus Buckl.
Western Ryegrass

Plants tufted, not rhizomatous, mostly (5) 8–15 dm tall; ligules 0.5–1.5 mm long; auricles more or less developed; leaf blades flat, 5–12 mm broad, glabrous or scabrous to pilose; spikes erect, stiff, 5–13 cm long or more; spikelets mostly 2 per node and usually overlapping, with 3–5 flowers; glumes narrowly lanceolate, broadest some distance above the base, mostly about as long as the spikelet, 2–5-nerved, acuminate to shortly awned; lemmas mostly 8–12 mm long, not especially ciliate on margins, acuminate, merely awn tipped or with a slender straight to curved awn mostly 1–3 cm long; anthers 2.5–3 mm long.

Open woods and meadows; in coastal and insular southeastern Alaska, southern Yukon, and adjacent British Columbia; eastward to Ontario and south to California, Colorado, Iowa, and Indiana.

1a. Lemmas awnless or merely awn tipped, the awn usually less than 5 mm long (*E. virescens* Piper; *E. glaucus* ssp. *virescens* [Piper] Gould; *E. glaucus* var. *virescens* [Piper] Bowden; *E. howellii* Scribn. & Merr.). *E. glaucus* var. *breviaristatus* Davy in Jeps.

1b. Lemmas with awns 10–30 mm long. *E. glaucus* var. *glaucus*

Elymus hirsutus Presl
Northern Ryegrass

Plants tufted, not rhizomatous, mostly 5–12(15) dm tall; ligules 0.5–1 mm long; auricles lacking or small; leaf blades flat, 4–15(17) mm broad, glabrous or more commonly rough-hairy; spikes erect to nodding or drooping, mostly 6–15 cm long; spikelets mostly 2 per node, overlapping but not especially compact, with 3–5-flowers; glumes narrowly lanceolate, broadest some distance above the base, the body shorter than the spikelet, 3–5-nerved, produced apically into a conspicuous awn; lemmas mostly 7–10 mm long (excluding awn), strongly long-ciliate on margins, the awn 5–25 mm long; anthers 2–2.5 mm long.

Open woods and meadows; in coastal and insular southern Alaska from the westernmost Aleutians eastward through the Panhandle; southward to Oregon (*E. borealis* Scribn.).

Elymus innovatus Beal
Downy Ryegrass

Plants rhizomatous and more or less tufted, mostly 2–9 dm tall; ligules 0.5–1 mm long; auricles often well developed; leaf blades 2–7 mm broad, flat to involute, scabrous at least on the upper surface; spikes erect, mostly 4–11 cm long; spikelets mostly 2 per node, overlapping and usually compact, with 3–5 flowers; glumes subulate, broadest at the base, (0)1–3-nerved, much shorter than the spikelet, awn tipped, pubescent; lemmas mostly 7–9 mm long, villous to appressed-hairy, awn tipped or with an awn to 6 mm long; anthers 4–5 mm long.

Floodplains, open woods, slopes, tundra, bars, stream beds, and meadows; in much of interior, eastern two-thirds of Alaska (including the arctic slope), and most of the Yukon; eastward to Hudson Bay and south to Montana, Wyoming, and South Dakota (*E. innovatus* ssp. *velutinus* Bowden; *E. innovatus* var. *velutinus* [Bowden] Hultén).

Elymus junceus Fisch.
Russian Wild Rye

Plants tufted, not rhizomatous, mostly 3–10 dm tall (or more?); ligules to 1 mm long; auricles well developed; leaf blades flat or the margins involute, 2–5 mm broad, scabrous at least on upper surface; spikes erect, (3)5–13 cm long; spikelets usually 2–3 per node, overlapping and usually very compact with 3–5-flowers; glumes subulate, broadest at the base, (0)1–3-nerved, about half as long as the spikelet, awn tipped, hairy; lemmas mostly 6–10 mm long, pubescent over the back with short hairs, awn tipped, this to about 2 mm long; anthers 3–4.5 mm long.

Disturbed sites, mostly near habitations; in southern Yukon (Alaska Hwy mi 1019, M V Guttman, 25 July 1960 ISC; do mi 812, Welsh & Moore 7635, 2 July 1968, BRY, ISC) (*Psathyrostachys juncea* [Fisch.] Nevski). This is an introduced forage grass and is to be expected else-where in the region.

Elymus mollis Trin.
Dunegrass

Plants rhizomatous and more or less tufted, mostly 1.5–15 dm tall (or more?); ligules to 1 mm long; auricles lacking or present; leaf blades 3–18 mm broad, flat or more or less involute, scabrous at least above; spikes erect or nearly so, 5–25(30) cm long; spikelets mostly 2 per node, overlapping (except occasionally the lowermost sepa-rate), with (3)4–6 flowers; glumes lanceo-late, broader above the base, mostly 3–5-nerved, about equaling the spikelet, awn tipped, hairy; lemmas mostly 10–20 mm long, hairy, acuminate to mucronate, but scarcely awned; anthers 4.8–9 mm long.

Spits, sea beaches, tidal flats, sea cliffs, lake shores, and less commonly on dune sands; in most of coastal and insular Alas-ka and northern Yukon (*E. arenarius* ssp. *mollis* [Trin.] Hultén; *E. arenarius* ssp. *mollis* var. *villosissimus* [Scribn.] Hultén; *E. villosissimus* Scribn.; *E. arenarius* ssp. *villosissimus* [Scribn.] A. Löve; *E. arenari-us* var. *villosus* H. E. Meyer). *Elymus mollis* is a portion of the *E. arenarius* com-plex and is often placed within that entity. It is herein maintained as separate, pri-marily for convenience of citation. *E. mollis* is known to form hybrids with species in related genera. Among those which have received official recognition are: *Elyhor-deum* × *littorale* Hodgson & Mitchell, a hybrid between *E. mollis* and *Hordeum brachyantherum*; *Elyhordeum* × *dutilly-anum* Lepage, a hybrid between *E. mollis* and *Hordeum jubatum*; *Elymus* × *aleuti-cus* (Hultén) Bowden, a putative hybrid between *E. mollis* and *E. hirsutus*.

Elymus sibiricus L.
Siberian Wild Rye

Plants tufted, not rhizomatous, mostly 5–12 dm tall; ligules very short, mostly less than 0.5 mm long; auricles poorly devel-oped; leaf blades flat, 3–16 mm broad, sca-brous on both sides, sometimes hairy on the upper surface; spikes drooping, (10) 15–30 cm long, often proliferating branches and forming panicles with spreading-as-cending branches; spikelets mostly 2 per node, overlapping but loose, with 3–7 flow-ers; glumes narrowly lanceolate, broader above the base, 3-nerved, much shorter than the spikelet, attenuate into slender awns to 3 mm long or merely awn tipped, scabrous; lemmas 8–12 mm long (exclud-ing awn), scabrous, prolonged into an awn mostly 1–2.8 cm long, the awn spread-ing; anthers 1.0–1.8 mm long.

Cultivated land and roadsides; in south-central Alaska (Delta Jct. and Matanuska Valley vicinities) and to be expected else-where; introduced weedy species; Asia. A hybrid involving *E. sibiricus* and *Hordeum jubatum* has been identified. It is known

Elymus sibiricus L. (× 0.25).

as *Elyhordeum* × *arcuatum* Mitchell & Hodgson. *E. sibiricus* is also known to hybridize with *Agropyron macrourum* (q.v.).

FESTUCA L.

Annual or perennial, tufted to rhizomatous herbs; culms hollow; leaf sheaths open; ligule membranous; leaf blades involute, folded or flat, the auricles lacking or present; spikelets borne in open to contracted panicles, usually 2–10-flowered (rarely more), articulate above the glumes; glumes more or less unequal, the first 1-nerved, the second 3-nerved, awnless or merely awn tipped; palea about equaling the lemmas; lemmas mostly 5-nerved, acute to obtuse, rarely bifid, awnless or more commonly with well-developed awn; lodicules 2; stamens 1–3.

1a. Plants annual; glumes very short, the first to 2.5 mm, the second to about 5 mm long; introduced weed of disturbed sites in central western Yukon and central Alaska. *F. megalura*

1b. Plants perennial; glumes various, but seldom with the first only half as long as the second; indigenous or introduced species of various distribution. (2)

2a. Leaf blades flat, mostly more than 3 mm broad. (3)

2b. Leaf blades mostly folded or involute, usually much less than 3 mm broad. (4)

3a. Auricles well developed, ciliate margined; lemmas awnless or merely awn tipped; introduced plant of disturbed soils, in southern Alaska and Yukon. *F. arundinacea*

3b. Auricles lacking; lemmas with well-developed awns mostly 5–15 mm long; indigenous plants of coastal and insular southeastern Alaska. *F. subulata*

4a. Panicle open, the branches ascending to spreading or deflexed; first glume (3.4)4–6 mm long; anthers 3–4 mm long. *F. altaica*

4b. Panicle contracted or less commonly open and with branches ascending to spreading; first glume usually less than 4 mm long or stamens less than 3 mm long or both. (5)

5a. Innovations (sterile branches) extravaginal (i.e., breaking through the base of the sheath); basal sheaths thin, strongly nerved, ultimately shredding into filiform fibers (the veins); plants often shortly rhizomatous. *F. rubra*

5b. Innovations (sterile branches) intravaginal (i.e., included within the sheath); basal sheaths not especially thin, not or seldom shredded into fibers; plants ordinarily densely tufted, not rhizomatous. (6)

6a. Stems puberulent; anthers very short, mostly 0.3–0.5(0.7) mm long; plants mostly of arctic and arctic-alpine sites. *F. baffinensis*

6b. Stems glabrous; anthers mostly 0.5–1.5(3) mm long; plants variously distributed. *F. ovina*

Festuca altaica Trin. in Ledeb.

Plants perennial, strongly tufted, 2–9(10) dm tall; innovations intravaginal; stem bases clothed with conspicuous persistent sheaths; ligules less than 0.5 mm long; leaf blades folded, rarely flat, commonly less than 2 mm wide, scabrous; panicle more or less open, 7–15(18) cm long, the branches ascending to spreading or drooping; spikelets 7–13(15) mm long, 2–5(6)-flowered; glumes unequal, the first 3.7–7 mm long, the second 4.6–8.1 mm long, lanceolate, usually acute to attenuate; lemmas glabrous to scabrous or puberulent, acuminate or short-awned, purplish to green; anthers (2.5)3–4 mm long.

Open woods, heathlands, rock outcrops, mountainsides, stream banks, and gravel bars; in most of Alaska (except for Aleutians and Panhandle) and most of Yukon; eastward to Mackenzie and southward to British Columbia; Asia.

Festuca arundinacea Schreb.

Plants perennial, strongly tufted and usually shortly rhizomatous, mostly 5–10 dm tall or more; innovations intravaginal; stem bases clothed with persistent sheaths; ligules less than 0.5 mm long; leaf blades flat or somewhat involute, mostly 3–10 mm wide, scabrous above; panicle slender, 13–35 cm long, often somewhat nodding; spikelets 8–15 mm long; glumes lanceolate, the first 3–5 mm long, the second 5–7 mm long; lemmas glabrous acute or minutely awn tipped, green or less commonly purplish; anthers 3–4 mm long.

Disturbed sites; in widely scattered localities in southern Alaska and in southern Yukon, where introduced; widespread in North America; Eurasia.

Festuca baffinensis Polunin
Baffin Fescue

Plants perennial, mostly 0.5–1.5(2) dm tall, tufted; innovations intravaginal; stems puberulent with short curved hairs, at least below the inflorescence, stem bases clothed with persistent sheaths; ligules less than 0.5 mm long; leaf blades folded, less than 1 mm broad, more or less scabrous; panicle compact, mostly 2–4 cm long, commonly secund; spikelets mostly 7–11 mm long, 3–5-flowered; glumes lanceolate, the first 2.5–3.5 mm long, the second 3.3–4.6 mm long; lemmas glabrous or scabrous on the midvein, awned, the awn to 3 mm long or rarely more; anthers 0.2–0.5(0.7) mm long.

Spits, rock outcrops, ridges, in arctic or alpine tundra; in northern Alaska and Yukon and less commonly elsewhere; eastward to Greenland and south to Montana. *F. baffinensis* is very closely allied to *F. ovina* var. *brevifolia*, and might be placed at some level within *F. ovina*. It is distinguished mainly on the basis of hairy stems and anthers which seem to average smaller.

Festuca megalura Nutt.

Plants annual, simple or tufted, mostly 1–5 dm tall; stem bases not especially clothed with persistent sheaths; ligules to 0.6 mm long; leaf blades folded, less than 1 mm broad, glabrous below, scabrous on the inrolled upper surface; panicle very narrow, mostly 5–15 cm long or more; spikelets mostly 7–10 mm long (excluding awns), usually 3–6-flowered; glumes lance-subulate, the first usually less than 2 mm long, the second about twice as long; lemmas puberulent to scabrous, with an awn to 20 mm long or more; anthers 0.5–1 mm long.

Weedy plant of disturbed soils; in central Alaska and central western Yukon; disjunctly southward from British Columbia to California and Arizona (*Vulpia megalura* [Nutt.] Rydb.).

Festuca ovina L.
Sheep Fescue

Plants perennial, densely tufted, (0.3)0.7–3.5(4.5) dm tall; stem bases clothed with persistent sheaths, these not or seldom shredding; ligules less than 0.5 mm long; leaf blades folded to involute, less than 1 mm broad, glabrous to scabrous; panicle usually compact, mostly 1.5–9 cm long, often secund; spikelets 4.8–10(11) mm long (including awns), usually 3–4(6)-flowered; glumes lanceolate, the first 1.6–3.0(3.3) mm long, the second 2.5–4.2(4.6) mm long; lemmas glabrous to scabrous, awned, the awn to 3 mm long; anthers 0.5–2.1(3) mm long.

Arctic and alpine tundra, ridge tops, cliffs, woods, and gravel bars; in most of Alaska and Yukon; eastward to Newfoundland and south to California and New Mexico; Eurasia, South America. *F. ovina* is an extremely variable complex of forms with circumboreal representation. Viviparous forms of the species are known. The variability is accounted for in part by the presence of a wide series of polyploid levels, which are not accompanied by distinctive diagnostic features which will allow segregation of any but the most distinctive forms. The following key is tentative at best. A thorough monographic treatment is indicated.

1a. Anthers mostly 2 mm long or more; plants low, mostly of arctic-alpine sites along the arctic slope. *F. ovina* var. *alaskana* (Holm.) Welsh stat. nov. (based on: *F. ovina* ssp. *alaskana* Holm. Bot. Not. 117:115. 1964)

1b. Anthers usually less than 1.5(1.8) mm long; plants and distribution various. (2)

2a. Anthers mostly 0.5–1 mm long; plants commonly arctic or alpine in distribution (*F. brachyphylla* Schult.; *F. ovina* ssp. *brachyphylla* [Schult.] Piper; *F. brevifolia* R. Br. not Muhl.). *F. ovina* var. *brevifolia* (R. Br.) Wats.

2b. Anthers mostly 1–1.5(1.8) mm long; plants commonly of low elevation sites (*F. saximontana* Rydb.; *F. ovina* ssp. *saximontana* [Rydb.] St.-Yves; *F. brachyphylla* ssp. *saximontana* [Rydb.] Hultén). *F. ovina* var. *rydbergii* St.-Yves

Festuca rubra L.
Red Fescue

Plants perennial, loosely to rather densely tufted, (1.2)2–10 dm tall; stem bases clothed with persistent sheaths, these shredding into coarse fibers; ligules less than 0.5 mm long; leaf blades folded to involute (rarely flat), mostly less than 1.5 mm wide, glabrous or puberulent above; panicles compact or less commonly open, mostly 2.5–11 cm long, often secund; spikelets 8.5–17 mm long (including awns), usually 4–7-flowered, purple to green; glumes lanceolate, the first 2–4.6 mm long, the second 4–7 mm long; lemmas glabrous and sometimes glaucous to scabrous or villous, awned, the awns to 3 mm long; anthers 2–4 mm long.

Tidal flats, beaches, cliffs, muskegs, ridge tops, terraces, and bars; in practically all of Alaska and Yukon; widespread in North America; Eurasia (*F. richardsonii* Hook.; *F. rubra* ssp. *richardsonii* [Hook.] Hultén; *F.*

aucta Krecz. & Bobr.; *F. rubra* ssp. *aucta* [Krecz. & Bobr.] Hultén; *F. rubra* var. *lanuginosa* Mert.). This is a complex variable entity with a series of polyploid levels, some of which are probably worthy of taxonomic recognition; much work is needed to elucidate the nature and extent of taxonomic subunits.

Festuca subulata Trin. in Bong.
Bearded Fescue

Plants perennial, tufted, mostly 4.5–12 dm tall; innovations intravaginal; stem bases not especially clothed with persistent sheaths; ligules less than 1 mm long; leaf blades flat, 3–10 cm wide, glabrous or more or less scabrous on both sides; panicle open, 10–30 cm long or more, more or less drooping; spikelets 7–10 mm long (excluding awns), 3–6-flowered; glumes narrowly lanceolate, the first 2.5–4 mm long, the second 4.5–5.5 mm long, attenuate; lemmas minutely scabrous to smooth, awned, the awns to 15 mm long; anthers 2–3 mm long.

Stream banks, open woods, and rock outcrops; in coastal and insular southeastern Alaska; southward to California (*F. jonesii* Scribn.).

GLYCERIA R. Br. Nom. Cons.
Perennial rhizomatous or stoloniferous herbs; culms hollow; leaf blades flat or less commonly folded; spikelets borne in slender to open panicles, articulate above the glumes, mostly 5–12-flowered (or more); glume more or less unequal, 1-nerved, shorter than the lowest lemma, acute to obtuse, awnless; palea subequal to or slightly surpassing the lemma; lemmas 5–9-nerved, the nerves more or less parallel, not convergent at the apex, awnless; callus not hairy; lodicules 2; stamens 2–3.

Church, G. L. 1949. A cytotaxonomic study of *Glyceria* and *Puccinellia*. Am. Jour. Bot. 36:155–65.

1a. Spikelets nearly terete, 10 mm long or more, linear to narrowly oblong in outline, several times longer than broad. (2)

1b. Spikelets flattened, mostly much less than 10 mm long, ovate to oblong, mostly less than 3 times longer than broad. (3)

2a. Lemma minutely scabrous only on the nerves, if at all, not hairy or scabrous between the nerves; lower surface of blades minutely papillate; plants of rather broad distribution in the southern half of continental Alaska and in the Panhandle and in southern Yukon. *G. borealis*

2b. Lemma minutely scabrous on nerves and in internerve areas; lower surface of blades merely scabrous, not papillate; plants of the Panhandle. *G. leptostachya*

3a. Leaf sheaths smooth; inflorescence very large, mostly more than 20 cm long; plants stout. *G. grandis*

3b. Leaf sheaths retrorsely scabrous; inflorescence moderate, mostly less than 20 cm long; plants of moderate size. (4)

4a. First glume very short, less than 0.8 mm long; plants uncommon, in central and south-central Alaska and southeastern Yukon. *G. striata*

4b. First glume longer, mostly 1.5–2 mm long; plants rather common, in central and east-central Alaska and southern Yukon. *G. pulchella*

Glyceria borealis (Nash) Batchelder
Northern Mannagrass

Plants rhizomatous, rooting at the lower nodes, mostly 6–12 dm tall; leaf sheaths glabrous, open near the top; ligules 4–10 mm long; leaf blades flat or folded, 2–6 mm broad, the ventral surface minutely papillate, not scabrous; panicle mostly 15–45 cm long, narrow, the branches ascending to erect; spikelets narrowly oblong to linear, nearly terete, 10–13 mm long or more, 6–11-flowered; first glume 1.5–2.5 mm long, the second 3–3.5 mm long; lemma rather prominently 7-nerved, scabrous on the nerves.

Ponds and lake margins, bogs, swamps, and wet meadows; in much of the southern half of continental Alaska, in the Panhandle, and in southern Yukon; eastward to Newfoundland and south to California, Arizona, New Mexico, and Pennsylvania (*Panicularia borealis* Nash).

Glyceria grandis Wats. ex Gray
American Mannagrass

Plants rhizomatous, often rooting at the lower nodes, mostly 10–15 dm tall; leaf sheaths glabrous, closed nearly or quite to the apex; ligules (2.5)4–8 mm long; leaf blades flat, 3–10 mm broad or more, smooth on both surfaces or minutely scabrous above; panicle mostly 20–35 cm long, open, the branches spreading or spreading-ascending; spikelets more or less flattened, oblong in outline, 4.5–9 mm long, 4–7-flowered; first glume 1.5–2(2.3) mm long, the second 2.2–3 mm long; lemma rather prominently 7-nerved, minutely scabrous on the nerves.

Ponds, stream margins, bogs, and wet meadows; in the southern half of continental Alaska (and at Skagway) and in southern Yukon; eastward to Newfoundland and south to Oregon, Nevada, Arizona, Illinois, and Virginia (*G. maxima* ssp. *grandis* [Wats.] Hultén; *G. hulteniana* A. Löve). Our material belongs to the *G. maxima* complex, but differs in having consistently smaller glumes and lemmas.

Glyceria leptostachya Buckl.
Davy Mannagrass

Plants rhizomatous, mostly 7–12 dm tall or more; leaf sheaths retrorsely scabrous, closed except near the top; ligules 6–10 mm long; leaf bases flat, 3–9 mm broad, scabrous on both sides; panicle 20–60 cm long, loose, the branches ascending to appressed; spikelets narrowly oblong to linear, nearly terete, 11–18 mm long, 8–15-flowered; first glume 1–1.5 mm long, the second 1.5–2.6 mm long; lemmas prominently 7-nerved, scabrous on the nerves and in the inter-nerve areas.

Pond and lake margins; known only from southeastern Alaska (Wrangell, J. P. Anderson 5639, ISC); southward to California.

Glyceria pulchella (Nash) Schum.

Plants rhizomatous, often rooting at the lower nodes, 4–10 dm tall; leaf sheaths retrorsely scabrous, open at least near the top; ligules (1.5)2–4 mm long; leaf blades flat, 2–5 mm wide, more or less scabrous on both sides; panicle 15–25 cm long or more, loose, with the branches ascending to spreading-ascending; spikelets ovate to oblong in outline, compressed, 3.5–6 mm long, 3–6-flowered; first glume 0.8–1.5 mm long, the second 1.5–2 mm long; lemmas prominently 7-nerved, scabrous on the nerves.

Marshes, muskegs, ponds and stream margins; in central to east-central Alaska and southern Yukon; eastward to Northwest Territories (*Panicularia pulchella* Nash).

Glyceria striata (Lam.) A. S. Hitchc.
Fowl Mannagrass

Plants rhizomatous and usually tufted, mostly 2.5–9 dm tall; leaf sheaths retrorsely scabrous, usually closed to the top; ligules 1.5–4 mm long; leaf blades flat or folded, (1)2–6 mm broad, scabrous on one or both surfaces; panicle 7–20 cm long, open, the branches spreading or ascending; spikelets ovate to elliptic or obovate in outline, 3–4.1 mm long, 3–7-flowered; first glume 0.6–1.0 mm long, the second 0.9–1.3 mm long; lemmas prominently 7-nerved, glabrous or minutely roughened on the nerves.

Tidal flats, marshes, and fens; in central and south-central Alaska and in northern British Columbia (Liard Hot Springs); eastward to Newfoundland and south to California, Arizona, Texas, and Florida (*Poa striata* Lam.; *Panicularia nervata* var. *stricta* Scribn.; *G. striata* ssp. *stricta*

[Scribn.] Hultén). Our material is referable to var. *stricta* (Scribn.) Fern.

HELICTOTRICHON Besser

Plants perennial, tufted; culms hollow; leaf sheaths open; ligules membranous; leaf blades flat or folded, nonauriculate; spikelets borne in contracted panicles, 3–6-flowered, articulate above the glumes; glumes somewhat unequal, the first 3-nerved, awnless; palea shorter than the lemma; lemmas 5–7-nerved, the tip entire or bilobed, awned from near the middle, the awn twisted and geniculate, surpassing the spikelet; callus bearded; lodicules 2; stamens 3.

Helictotrichon hookeri (Scribn.) Henrard

Plants tufted, mostly 2.5–4.5 dm tall; leaf blades 1–4 mm broad, the margins thickened and usually whitish; ligules 1–3 mm long, erose; spikelets mostly 12–16 mm long, usually with 4–6 flowers, the rachilla disarticulating between the florets; lemma more or less hardened at the base, with a densely bearded callus, awned, the awn to 15 mm long, flattened, twisted, geniculate; anthers 4–5 mm long.

Dry grassy slopes; in southern Yukon; eastward and southward to New Mexico and Minnesota (*Avena pratensis* var. *americana* Scribn.; *Avena hookeri* Scribn. in Hack.; *Avena americana* [Scribn.] Scribn.). This entity is known from Duke River, Alaska Highway mi 1098 (Anderson & Brown 10068, ISC).

HIEROCHLOE R. Br. Nom. Cons.

Plants perennial, rhizomatous; culms hollow; leaf sheaths open; ligules membranous, often fringed; leaf blades flat to involute, nonauriculate; spikelets borne in open or contracted panicles, 3-flowered, the lower two staminate, the upper one perfect, articulate above the glumes; glumes subequal, 3-nerved, awnless, glabrous; palea somewhat shorter than the lemma; lemmas mostly 5-nerved, the lower two commonly more pubescent and of different texture than the upper one, awnless or awned, the awn usually borne between the lobes of the bifid apex; lodicules 2; stamens 2–3.

1a. Lower staminate lemma bearing a short awn tip; upper staminate lemma with a long geniculate awn which surpasses the spikelet. *H. alpina*
1b. Lower and upper staminate lemmas awnless or merely awn tipped. (2)

2a. Panicle loose to open, the branches ascending to spreading; ligules mostly 2–6 mm long. *H. odorata*
2b. Panicle narrow, the branches appressed-ascending; ligules mostly 0.5–1.5 mm long. *H. pauciflora*

Hierochloe alpina (Swartz) Roem. & Schult.
Alpine Holygrass

Plants tufted, shortly rhizomatous, 1.5–6 dm tall; basal sheaths purplish; ligules 0.6–1 mm long, with a conspicuous terminal fringe of hairs; leaf blades involute or upper ones greatly shortened, flat or nearly so and 1–3 mm broad; panicle more or less contracted, (1.5)2–5(5.5) cm long, the branches ascending to appressed-ascending; spikelets tawny or green with purplish margins, 5.3–7 mm long (excluding the awn); glumes about as long as the spikelet; lemmas of staminate florets pubescent, awned, the awn of the first very short, that of the second twisted and geniculate, inserted near the middle of the lemma, exserted from the spikelets for 2–4 mm or more; fertile lemma pubescent toward the apex, unawned.

Arctic and alpine tundra and heathlands, and open woods; in practically all of Alaska and Yukon; eastward to Newfoundland and south to British Columbia, Montana, New York, and Vermont; Eurasia (*Holcus alpinus* Swartz in Willd.). Viviparous specimens occur rarely.

Hierochloe odorata (L.) Beauv.
Vanilla Grass

Plants solitary or tufted, rhizomatous, 2.3–

Hierochloe alpina (Swartz) Roem. & Schult. (× 0.7).

6.7 dm tall; basal sheaths pale purplish to straw-colored; ligules 2.5–6 mm long, inconspicuously ciliate; leaf blades flat to folded, mostly 2–5 mm broad; panicle loose to open, 3–13.5 cm long, the branches ascending to spreading; spikelets tawny to straw-colored, 3.8–6 mm long; glumes commonly surpassing florets; lemmas of staminate florets strongly pubescent, awnless; fertile lemma pubescent toward the apex.

Moist meadows, marshes, tidal flats, stream banks, and less commonly on dryish sites in arctic and alpine tundra; over much of Alaska and Yukon; widely distributed in North America; Eurasia (*Holcus odoratus* L.; *Hierochloe arctica* Presl).

Hierochloe pauciflora R. Br.
Arctic Holygrass

Plants rhizomatous, not or seldom tufted, 0.5–2.5 dm tall; basal sheaths straw-colored, seldom very purplish; ligules 0.5–1.5 mm long, inconspicuously short-ciliate; leaf blades involute or the greatly shortened upper ones merely channeled, commonly less than 1 mm broad; panicle contracted, 1.2–3.3 cm long, the branches appressed-ascending; spikelets tawny to straw-colored, 3.3–5 mm long; glumes somewhat shorter than to equaling the florets; lemmas of staminate florets puberulent, awnless; fertile lemma pubescent towards the apex.

Spits, ponds, bogs, and moist tundra; in or near coastal and insular western (except for the Aleutians) and northern Alaska and northern Yukon; eastward to the Canadian Arctic Archipelago; Asia.

HOLCUS L. Nom. Cons.

Perennial, tufted; culms hollow; leaf sheaths open; ligules membranous, puberulent; leaf blades flat, nonauriculate; spikelets borne in somewhat congested panicles, 2-flowered, the lower one perfect, the upper staminate, articulate above the glumes; glumes subequal, longer than the florets, strongly keeled, the first 1-nerved, the second 3-nerved, awnless or shortly awned; palea subequal to the lemma; lemmas obscurely 5-nerved, shining, the lower one awnless, the upper with a short awn from near the top; lodicules 2; stamens 3.

Holcus lanatus L.
Velvet Grass

Plants tufted, velvety-pubescent, mostly 4–10 dm tall; ligules 1–2 mm long, strongly pubescent; leaf blades flat, mostly 3–8 mm wide; panicle 5–15 cm long, usually purplish tinged, the branches ascending to appressed; spikelets 3–4.5 mm long; glumes hairy, especially on the nerves, the first much narrower than the second; lemmas shining, the awn of the upper one strongly hooked, often exserted laterally from the spikelet.

Disturbed soils, mostly near habitations; in coastal and insular southeastern Alaska where introduced; widespread in North America; Europe, Asia.

HORDEUM L.

Plants annual or perennial, tufted; culms hollow; leaf sheaths open; ligules membranous; leaf blades flat, auriculate or auricles not developed; spikelets borne in a terminal spike, usually 3(2) per node, mostly 1(2)-flowered; central spikelet of each cluster perfect, the floret sessile or stalked, the lemma usually long-awned; lateral spikelets usually shortly stalked, the florets sessile or stalked, usually staminate or rudimentary; glumes narrow and awnlike; lemma 5-nerved, usually awned.

Bowden, W. M. 1962. Cytotaxonomy of the native and adventive species of *Hordeum, Eremopyrum, Secale, Sitanion,* and *Triticum* in Canada. Can. Jour. Bot. 40:1675–711.

1a. Auricles well developed on most leaves, these mostly more than 1 mm long; leaf blades mostly 5–15 mm broad; rachis of spike not disarticulating; cultivated and escaping. *H. vulgare*

1b. Auricles lacking, or if present less than 1 mm long; leaf blades mostly 3–8 mm broad; rachis of spike disarticulating; adventive or indigenous weedy species. (2)

2a.   Glumes mostly 3.5–9.8 cm long, the spike including awns usually about as thick as long. *H. jubatum*

2b.   Glumes mostly 0.5–3.5 cm long, the spike including awns mostly several times longer than thick. (3)

3a.   Glumes mostly 1.5–3.5 cm long; plants relatively rare, almost or quite sterile. *H.* × *caespitosum*

3b.   Glumes mostly 0.5–1.5 cm long; plants relatively common. *H. brachyantherum*

Hordeum brachyantherum Nevski
Meadow Barley

Plants tufted, perennial, 1.5–8.5 dm tall; leaf blades flat, mostly 2–6(7) mm broad, scabrous, lacking auricles; ligule less than 0.5 mm long; spike erect or nearly so, 4–10 cm long, several times longer than thick, easily shattering; central spikelet sessile, the pedicel of lateral spikelets about 1 mm long; glumes all slender, awnlike, mostly 7–12(15) mm long; central floret usually with an awn surpassing those of the glumes; lateral florets usually reduced, often awnlike, sometimes staminate.

Open woods, tidal flats, beaches, bluffs, and roadsides; in or near coastal and insular southern Alaska and less commonly in central interior and coastal western Alaska and southern Yukon; southward to California and New Mexico (*H. boreale* Scribn. & Smith, not Gaud.; *H. nodosum* var. *boreale* [Scribn. & Smith] A. S. Hitchc.; *H. jubatum* ssp. *breviaristatum* Bowden; *H. nodosum* sensu authors, not L.). *H. brachyantherum* forms intergeneric hybrids with *Elymus glaucus* (*Elyhordeum* × *stebbinsianum* Bowden) and with other species as well.

Hordeum × caespitosum (Scribn.) Mitchell & Wilton
Bobtail Barley

Plants tufted short-lived perennials, 1.5–8 dm tall; leaf blades flat, 2–6 mm broad, scabrous, lacking auricles; ligules less than 0.5 mm long; spike more or less nodding, mostly 4–10 cm long (including awns), commonly 2–3 times longer than thick, easily shattering; central spikelet sessile or subsessile, the pedicel of lateral spikelet commonly less than 1 mm long; glumes all slender, awnlike, mostly 15–30 mm long; central floret with awn often exceeded by the glumes; lateral florets usually reduced, often awnlike.

Roadsides, disturbed soils, mostly near habitations; in coastal and insular southern Alaska, and less commonly elsewhere in the range of *H. jubatum* and *H. brachyantherum* (*H. caespitosum* Scribn.; *H. jubatum* var. *caespitosum* [Scribn.] A. S. Hitchc.; *H. jubatum* ssp. × *intermedium* Bowden). Mitchell and Wilton (Madroño 17:269–80. 1964) have presented evidence supporting the hybrid nature of *H.* × *caespitosum*. It is intermediate in every way between *H. jubatum* and *H. brachyantherum* and additionally essentially sterile.

Hordeum jubatum L.
Foxtail Barley

Plants tufted, perennial, mostly 2.5–6.5 dm tall; leaf blades flat, mostly 1–4 mm broad, scabrous, lacking auricles or auricles present on some leaves; ligules less than 0.6 mm long; spike more or less nodding, 5–15 cm long (including awns), commonly almost as long as thick, easily shattering; central spikelet sessile, the pedicel of lateral spikelets mostly 0.8–1.5 mm long; glumes all slender, awnlike, mostly 30–100 mm long; central floret with awn often surpassing the glumes; lateral spikelets usually reduced, often awnlike.

Muskegs, disturbed sites, beaches, roadsides, tidal flats, terraces, and river banks; in much of Alaska and Yukon south of the 68th parallel; widespread in North America; Eurasia. *Hordeum jubatum* is interfertile with numerous species in related genera, forming hybrids with *Agropyron* caninum (*Agrohordeum* × *macounii* [Vasey] Lepage); *Agropyron macrourum* (*Agrohordeum* × *pilosilemma* Mitchell & Hodgson); *Elymus mollis* (*Elyhordeum* × *dutillyanum* Lepage); and *Elymus sibiricus* (*Elyhordeum* × *arcuatum* Mitchell & Hodgson).

Hordeum vulgare L.
Barley

Plants tufted, annual, commonly 8–12 dm tall or more; leaf blades flat, mostly 6–16 mm broad, scabrous or more or less smooth at least beneath, with auricles well developed; ligules 0.5–1 mm long; spike erect to more or less nodding, 5–22 cm long (including awns, if present), usually 2–several times longer than broad, not shattering; spikelets sessile, all three fertile; glumes narrow, not subulate, usually shortly awned; lemmas with awns mostly 8–15 cm long, much surpassing the glumes, or awnless and often 3-lobed at tip.

Cultivated forage and grain (?) plant; in agricultural regions of southern Alaska and Yukon, where commonly grown with peas (*Pisum sativum*), and occasional as a waif along roadsides and near habitations; widely cultivated in temperate regions of the earth.

KOELERIA Pers.

Plants perennial, tufted or rhizomatous; culms hollow; ligules membranous, leaf sheaths open; leaf blades folded, involute, or sometimes flat, nonauriculate; spikelets borne in congested spikelike panicles, 2–4-flowered, articulate above the glumes; glumes unequal, the first 1-nerved, much narrower than the second, the second 3–rarely 5-nerved and about as long as the first floret, awnless; palea shorter than the lemma; lemmas 5-nerved, awnless or shortly awned from a minutely bifid apex, keeled; lodicules 2; stamens 3.

1a. Stems pubescent, often densely so; rhizomes commonly well developed; plants of arctic northern Alaska and Yukon and less commonly in arctic-alpine sites elsewhere. *K. asiatica*

1b. Stems glabrous except below the inflorescence; rhizomes not developed; plants of southern Yukon, mostly at lower elevations. *K. nitida*

Koeleria asiatica Domin

Plants loosely tufted, usually with elongate rhizomes, mostly 1.4–3 dm tall; stems and leaf sheaths villosulous with descending hairs to glabrous; ligules mostly 0.5–1 mm long, ciliate; leaf blades commonly folded and less than 2.5 mm wide, puberulent to glabrous or scabrous; panicle 2–4 cm long (or more?), contracted; spikelets 2-flowered (seldom more); glumes awnless, merely attenuate, the first 2.7–4.1 mm long, the second 3.4–4.8 mm long, glabrous or nearly so; lemmas 3.7–5.5 mm long, awnless or merely awn tipped.

Gravel bars, stream banks, bluffs, and open slopes; in coastal or near coastal arctic northern Alaska, Yukon, and Northwest Territories, and in arctic alpine southwestern Yukon; Asia (*K. cairnesiana* Hultén). This entity seems to fit well into the complex of forms treated in a broad sense as *K. nitida* (see below) and should probably be treated within that entity.

Koeleria nitida Nutt.
Junegrass

Plants densely tufted, not or seldom rhizomatous, mostly 2–6 dm tall; stems and leaf sheaths glabrous to variously hairy; ligules 0.5–2 mm long, ciliate; leaf blades commonly folded and less than 3 mm wide, glabrous to puberulent or villous; panicle mostly 4–10 cm long, contracted; spikelets 2-flowered (seldom more); glumes awnless, usually attenuate but sometimes awn tipped, the first 3–5 mm long, awnless or awn tipped.

Open hillsides; in southern Yukon; southward to Mexico and Texas and eastward to Maine; (*K. cristata* Pers., illegitimate; *K. gracilis* Pers., illegitimate; *K. mac-*

rantha [Ledeb.] Spreng.?; *Aira macrantha* Ledeb.?). The problem of nomenclature of this entity has been reviewed by Shinners (Rhodora 58:93-96. 1956), and if our material is the same as that on which *Aira macrantha* was based, then the correct name might well be *K. macrantha* (Ledeb.) Spreng., as was proposed by Shinners. Certainly neither *K. cristata* nor *K. gracilis* are legitimate.

LOLIUM L.

Tufted annual, biennial, or perennial herbs; culms hollow; leaf sheaths open; ligules membranous; leaf blades flat to folded, with auricles often well developed; spikelets sessile, borne edgewise in a spike on the opposite side of a continuous rachis; first glume lacking on all but terminal spikelet, the second glume (on edge away from rachis) well developed; glumes 5–9-nerved, awnless; palea subequal to the lemma; lemmas 5-nerved, awnless or awned; lodicules 2; stamens 3.

1a. Glume usually longer than the spikelet; plants annual. *L. temulentum*
1b. Glume commonly much shorter than the spikelet; plants biennial or perennial (possibly annual). (2)

2a. Lemmas awned (at least some); plants biennial (annual). *L. multiflorum*
2b. Lemmas awnless; plants perennial. *L. perenne*

Lolium multiflorum Lam.

Italian Ryegrass

Plants perennial (or biennial, rarely annual), 3–10 dm tall or more; ligules 0.5–1.5 mm long; leaf blades flat or somewhat involute, 3–8 mm broad, scabrous (at least above); spikes mostly 10–25 cm long or more; spikelets 12–25 mm long, 8–15-flowered; glumes (6.8)8–12 mm long, often somewhat longer than the adjacent lemma, much shorter than the spikelet; lemmas minutely bifid, with a short subterminal awn (at least some).

Weedy species of disturbed sites, sometimes included in grass seed mixtures for lawns; known from widely disjunct localities in southern Alaska and Yukon; widespread in temperate regions of the earth.

Lolium perenne L.

Perennial Ryegrass

Plants perennial, 3–8 dm tall; ligules 0.5–1.5 mm long; leaf blades flat to folded or involute, mostly 2–5 mm broad, scabrous (at least above); spikes mostly (7)10–25 cm long; spikelets 12–21 mm long, 6–12-flowered; glumes 7–10 mm long, usually longer than the adjacent lemma but shorter than the spikelet; lemmas acute, awnless.

Weedy species; known from widely separated localities in Alaska and to be expected in the Yukon; widespread in temperate regions of the earth. Cultivated both as a forage grass and in lawns in some places; ours evidently occurring as waifs.

Lolium temulentum L.

Darnel

Plants annual, 3–9 dm tall; ligules 1–2 mm long; leaf blades flat, 3–10 mm broad, scabrous or smooth; spikes 10–25 cm long (or more); spikelets mostly 10–20 mm long, 5–10-flowered; glumes surpassing the florets; lemmas 6–8 mm long with a subterminal awn or awnless.

Disturbed sites, usually near habitations; in southern Alaska and southern Yukon; widespread in temperate regions of the earth.

MELICA L.

Plants perennial, rhizomatous and more or less caespitose; culms hollow; leaf sheaths closed; ligules membranous; leaf blades flat, lacking auricles; spikelets in open panicles, 2–5(7)-flowered, the upper 2–4 florets represented by empty lemmas; glumes shorter than the first lemma, unequal, the first 3-nerved, the second 5-nerved; lemmas rounded on the back, awnless, with usually 7 nonconvergent nerves; palea somewhat shorter than the lemma; callus not bearded; lodicules 2, fused; stamens 3.

Boyle, W. S. 1945. A cytotaxonomic study of the North American species of *Melica*. Madroño 8:1–26.

Melica subulata (Griseb.) Scribn.
Alaska Oniongrass

Plants tufted, mostly 3–8(10) dm tall, bulbous based atop short to elongate rhizomes; leaf sheath closed nearly or quite to the top (sometimes split); ligules 1.5–5 mm long, glabrous; leaf blades flat, mostly 1–5 (7) mm broad, usually scabrous, at least on upper surface; panicle 10–20 cm long, the branches ascending to erect or sometimes spreading; spikelets 12–20 mm long, 2–5-flowered; glumes acute to acuminate, the first about 5–6.5 mm long, the second mostly 7.5–9 mm long; lemmas slenderly acuminate, unawned, more or less hairy on margins and nerves near the base.

Meadows and woods; in insular southern (Unalaska) and southeastern Alaska; southward to California, Idaho, and Wyoming (*Bromus subulatus* Griseb. in Ledeb.).

MUHLENBERGIA Schreb.

Plants perennial, rhizomatous; culms solid or hollow; leaf sheaths open; ligules membranous; leaf blades flat or more or less involute, lacking auricles; spikelets borne in contracted spikelike panicles, 1-flowered, articulate above the glumes; glumes shorter than or about equaling or surpassing the lemma, the first 1-nerved, the second 1–3-nerved, awnless or awned, lemma 3-nerved, acute to attenuate or awned, sometimes hairy at the base; palea subequal to the lemma; lodicules 2, stamens 3.

Pohl, R. W. 1969. *Muhlenbergia*, Subgenus Muhlenbergia (Gramineae) in North America. Am. Midl. Nat. 82:512–42.

1a. Lemmas neither awned nor hairy at the base; culms solid; leaves involute, very narrow. *M. richardsonis*

1b. Lemmas awn tipped or awned, longhairy at the base; culms hollow; leaves flat or with margins involute or revolute. (2)

2a. Glumes distinctly awned; ligules 0.2–0.6 mm long; anthers 0.8–1.5 mm long. *M. glomerata*

2b. Glumes merely awn tipped or attenuate; ligules 0.5–1 mm long; anthers 0.3–0.5 mm long. *M. mexicana*

Muhlenbergia glomerata (Willd.) Trin.
Muhlygrass

Plants strongly rhizomatous, 3–6 dm tall or more; leaf blades flat or the margins more or less revolute or involute, 1–5 mm wide; ligules 0.2–0.6 mm long, somewhat ciliate; panicle dense, often more or less interrupted, 2–8(11) cm long; spikelets 3.5–8 mm long; glumes awned, both exceeding the awnless floret; lemma 2.3–3 mm long, rarely awn tipped, hairy on the callus and along the margins; anthers 0.8–1.5 mm long.

Moist sites; in southeastern Yukon and northern British Columbia (Liard Hot Springs); eastward to Newfoundland and south to Oregon, Utah, Colorado, Nebraska, Iowa, and West Virginia (*Polypogon glomeratus* Willd.; *Dactylogramma cinnoides* Link; *M. glomerata* var. *cinnoides* [Link] Herm.).

Muhlenbergia mexicana (L.) Trin.

Plants strongly rhizomatous, 2.5–6 dm tall or more; leaf blades flat or more or less revolute, 2–6 mm wide; ligule 0.5–1 mm long; panicle slender or sometimes with branches ascending, mostly 5–15 cm long; spikelets 1.7–4.4 mm long; glumes narrow, unawned or merely awn tipped, shorter than or equaling the floret; lemma 1.5–3.4 mm long, long-hairy on the callus, awnless or awn tipped; anthers 0.3–0.5 mm long.

Moist sites; known in northern British Columbia (Liard Hot Springs, Anderson & Brown 10363, ISC); east to Quebec and south to California, Arizona, Texas, Missouri, and Virginia (*Agrostis mexicana* L.). Our materials apparently belong to f. *mexicana*.

Muhlenbergia richardsonis (Trin.) Rydb.

Plants strongly rhizomatous, mostly 0.5–3 (4) dm tall; leaf blades involute, mostly less than 1(1.5) mm wide; ligule mostly 1–2 mm long; panicle narrow, spikelike, 2–7 cm long; glumes broad, unawned, much shorter than the floret; lemma 2–3 mm long, not long-hairy, the callus glabrous, awnless or very minutely awn tipped; anthers about 1.5 mm long.

Dry grasslands; in southern Yukon; southward to Mexico, New Mexico, and Nebraska (*Vilfa richardsonis* Trin.; *Vilfa squarrosa* Trin.; *M. squarrosa* [Trin.] Rydb.).

ORYZOPSIS Michx.

Plants perennial, caespitose; culms usually hollow; leaf sheaths open; ligules membranous; leaf blades flat to involute, sometimes reduced; spikelets borne in more or less congested panicles, 1-flowered, articulate above the glumes; glumes more or less unequal or subequal, usually 3–5-nerved; palea subequal to the lemma; lemma about equaling the glumes, of different texture than the glumes, hairy, awned from near or at the apex, the awn deciduous, straight or curved; callus usually more or less hairy; lodicules 2–3; stamens 3.

1a. Lemma 6–8 mm long, shining, puberulent above, long-pilose basally; leaf blades flat, 3–6 mm broad. *O. asperifolia*
1b. Lemma 3–4 mm long, dull, puberulent almost throughout, the basal hairs not especially longer; leaf blades involute. *O. pungens*

Oryzopsis asperifolia Michx.

Plants tufted, 2–5 dm tall; ligules about 0.5 mm long, ciliolate; leaf blades of fertile stem usually greatly reduced or lacking, those of sterile branches (innovations) flat, 3–6 mm wide; panicle reduced to a nearly simple raceme, 4–8 cm long; glumes 7-nerved, 6–7 mm long, somewhat unequal; lemma subequal to the glumes, appressed-puberulent, the hairs of the callus much longer, awned from the tip, the awn 5–10 mm long.

Dry grassy slopes; in southern Yukon (Watson Lake, Anderson & Brown 9923B, ISC); eastward to Newfoundland and south to Washington, Utah, New Mexico, South Dakota, Indiana, and West Virginia.

Oryzopsis pungens (Torr.) A. S. Hitchc. *Mountain Rice*

Plants tufted, 1.5–4.5 dm tall; ligules 1.5–2.5 mm long, ciliolate; leaf blades of fertile stems reduced upwards, these and those of the sterile branches involute, mostly less than 1 mm broad; panicle reduced to a nearly simple raceme, (3)4–8 cm long; glumes 5-nerved, 3–4 mm long, subequal; lemma about as long as the glumes, rather densely and uniformly puberulent, the hairs of the callus not much longer, awned from near the tip, the caducous awn to about 2 mm long.

Sandy soils in open woods; in southern Yukon (Watson Lake, Anderson & Brown 9923A, ISC; Alaska Highway, mi 839, Welsh & Moore 7653, ISC, BRY); eastward to Labrador and south to Colorado, Illinois, and New York (*Milium pungens* Torr.).

PHALARIS L.

Plants annual or perennial; culms hollow; leaf sheaths open; leaf blades flat, auriculate or nonauriculate; spikelets in congested, often spikelike panicles, articulate above the glumes, usually 3-flowered, the uppermost floret perfect, the lower ones sterile and often greatly reduced; glumes subequal, compressed, usually 3-nerved; sterile lemmas much smaller than the fertile one, sometimes lacking; fertile lemma hardened, hairy, shorter than the glumes; palea subequal to the lemma; lodicules 2; stamens 3.

Anderson, D. E. 1961. Taxonomy and distribution of the genus *Phalaris*. Iowa State Jour. Sci. 36:1–96.

1a. Plants perennial, rhizomatous, semi-aquatic or palustrine; panicle usually more than 8 cm long. *P. arundinacea*

1b. Plants annual, tufted, lacking rhizomes, waifs of disturbed soils; panicle mostly less than 5 cm long. (2)

2a. Sterile lemmas 2; fertile lemma 4–6 mm long; wing of glumes usually entire. *P. canariensis*

2b. Sterile lemmas 1; fertile lemma 2.7–4 mm long; wing of glumes usually toothed or erose. *P. minor*

Phalaris arundinacea L.

Reed Canary Grass

Plants perennial, strongly rhizomatous, mostly 8–16.5(20) dm tall; ligules 4–9 mm long, puberulent; leaf blades flat, 6–18 mm broad, minutely scabrous; panicles (8)11–20(30) cm long, compact or with branches more or less spreading; glumes subequal, (3.5)4.0–5.6(7) mm long, more or less acute, usually wingless; fertile lemma 2.7–4.5 mm long, shining at maturity; sterile lemmas 2, 1.2–2 mm long, subulate, pubescent.

Stream banks, margins of springs, and wet meadows; in central, south-central, and southeastern Alaska, southern Yukon, and northern British Columbia; widespread in North America; Eurasia. This grass is possibly indigenous to our region, but it has been used as a forage grass and some of our specimens might therefore be adventive.

Phalaris canariensis L.

Canary Grass

Plants annual, tufted, 3–8(10) dm tall; ligules 3–8 mm long, glabrous; leaf blades 3–8 mm wide, scabrous; panicles 1.5–4 cm long, very compact; glumes subequal, 6–8 (10) mm long, abruptly acute, the keel broadly winged, the wing to about 1 mm wide and entire or nearly so; fertile lemma 4.8–5.5(6.8) mm long, finely pubescent; sterile lemmas 2, 2.5–3(4) mm long, broad, finely pubescent.

Adventive weedy plant of disturbed sites, mostly near habitations; in central, south-central, and southeastern Alaska and to be expected through most of our region. This grass is used in birdseed mixtures and is about as widely distributed as is the practice of maintaining birds as household pets.

Phalaris minor Retz

Plants annual, tufted, 2–8 dm tall or more; ligules mostly 1.8–3.5 mm long, glabrous; leaf blades 2–6 mm wide or more, scabrous; panicles (1)1.5–5(6) cm long, compact; glumes subequal, 4–6(6.5) mm long, acute, the keel winged, the wing much less than 1 mm wide, more or less toothed or sometimes entire; fertile lemma 2.7–4 mm long, finely pubescent; sterile lemmas 1, (0.2) 1–1.8 mm long, subulate, pubescent.

Weedy adventive species of disturbed sites; in Alaska reported for Manley Hot Springs; widespread in both northern and southern hemispheres.

PHIPPSIA R. Br.

Plants perennial, tufted; culms hollow; leaf sheaths closed usually to above the middle; ligules membranous; leaf blades flat to folded, with a keel-shaped apex, nonauriculate; spikelets borne in congested or open panicles, 1 (rarely 2)-flowered, articulate above the glumes; glumes unequal, apparently nerveless, awnless, glabrous, or sometimes lacking; palea somewhat shorter than the lemma; lemmas 3-nerved, the nerves not convergent at the summit, awnless; lodicules 2; stamens 2–3.

1a. Panicle contracted, mostly 0.8–4 cm long; lemmas glabrous or sparingly hairy; our common *Phippsia*. *P. algida*

1b. Panicle open, mostly 3–8 cm long; lemmas copiously white-hairy at least below; reported from St. Lawrence Island, Bering Sea. *P. concinna* (Fries) Lindeb.

Phippsia algida (Soland.) R. Br.
Phippsia

Plants perennial, densely tufted, the stems erect or the marginal ones spreading and flexuous, mostly 0.4–1.5(2.3) dm tall; ligules 1–1.5 mm long, glabrous; leaf blades flat and 1–2.5 mm wide or folded and less than 1 mm wide, glabrous, prowlike apically, arising from inflated sheaths; panicles congested, mostly 0.8–4 cm long; spikelets 1.2–1.7 mm long; glumes unequal, much shorter than the lemma and palea, membranous, sometimes lacking; lemmas glabrous or sometimes sparingly hairy on the nerves; anthers 0.3–0.6 mm long.

Sandy beaches, peaty soil, bare ground, drained lake beds, and coastal tundra; in coastal and insular western (Bering Sea), west-central, and northern Alaska and Yukon (?) and in alpine sites in south-central and southeastern Alaska; disjunctly southward to Colorado; circumpolar (*Agrostis algida* Soland.).

PHLEUM L.

Plants perennial, tufted; culms hollow; leaf sheaths open; ligules membranous, glabrous; leaf blades flat, often with small auricles; spikelets borne in cylindrical spikelike panicles, 1-flowered, articulate above the glumes; glumes equal, 3-nerved, strongly flattened and keeled, ciliate on the keel, awn tipped; palea subequal to the lemma; lemma 5-nerved, membranous, usually pubescent, awnless; lodicules 2; stamens 3.

1a. Culms usually bulbous based; panicles (at least some) commonly more than 4.5 cm long and less than 1 cm broad; anthers usually over 1.6 mm long; introduced plants of disturbed soils near habitations or highways. *P. pratense*

1b. Culms not bulbous based; panicles rarely more than 4.5 cm long but often over 1 cm thick (when pressed); anthers usually less than 1.4(1.5) mm long; indigenous plants, rarely or not in disturbed soils. *P. alpinum*

Phleum alpinum L.
Alpine Timothy

Plants tufted, 1.2–6.5 dm tall, the stems sometimes rooting at the lower nodes; ligules 1–4 mm long; leaf blades flat or more or less folded to involute, 2–9 mm wide, more or less scabrous at least on the margins; panicle 1–4.5(7) cm long, often more than 1 cm broad (when pressed); glumes ciliate on the keel and usually puberulent on the sides, awned, the awn to 3 mm long; lemma puberulent; anthers 1–1.6 mm long.

Thickets, stream banks, meadows, open woods; in much of the southern half of Alaska and Yukon, rarely in the Brooks Range; widely distributed in North America; Eurasia (*P. pratense* var. *alpinum* [L.] Celak.; *P. haenkeanum* Presl; *P. alpinum* var. *americanum* Fourn.; *P. commutatum* var. *americanum* [Fourn.] Hultén). The suggestion that the name of our materials should be *P. commutatum* Gand., and that *P. alpinum* L. is restricted to the "Alps," is difficult to follow or believe. The specimen in the Linnaean herbarium designated *P. alpinum*, and marked with the annotation "Lappo" is certainly equivalent to our materials and indeed the phrase name in Species Plantarum is attributed to Fl. Lap. among others, despite the habitat reference as "In Alpibus." It seems clear that whatever else might be said, the nature of our materials and the application of the name *alpinum* to them is correct.

Phleum pratense L.
Timothy

Plants tufted, 3–15 dm tall, more or less bulbous based; ligules 2–4 mm long; leaf blades flat, 2–8 mm wide, scabrous at least along the margins; panicle (1.5)3–10 cm long or more, less than 1 cm thick (when pressed); glumes ciliate on the keel and usually puberulent on the sides, awned, the awn to 2 mm long; lemmas puberulent; anthers 1.6–2 mm long.

Roadsides, old fields, homesites, and other disturbed soils, usually near habitations; in much of southern Alaska and southern Yukon; widespread in the temperate regions of the earth. Timothy is included in forage plantings and should be expected to maintain itself in our region.

POA L.

Annual or perennial tufted stoloniferous or rhizomatous plants; culms hollow; leaf sheaths usually partially closed; ligules membranous; leaf blades flat to folded or involute, with tips usually prowlike, lacking auricles; spikelets in contracted to open panicles, 2–7-flowered, articulate above the glumes, the flowers perfect or imperfect; glumes somewhat unequal to equal, commonly 1–3-nerved, mostly shorter than the first lemma but sometimes longer, unawned; palea usually shorter than the lemma; lemmas with 5 obscure to prominent converging nerves, often with a tuft of cobwebby crumpled hairs at the base, and usually hairy on at least the keel and marginal nerves; lodicules 2; stamens 3.

This is a very difficult genus from a taxonomic standpoint. Part of this is due not only to a lack of consistent, easily distinguished diagnostic features, but also to the presence of apomyxis within some species, wherein seed is produced by asexual means, and by means of vivipary, where bulblets replace reproductive structures in the spikelets. Other considerations involve those dealing with aneuploid and polyploid series, in which consistent morphological criteria are lacking or else merely serve to connect broad series of types into a more or less continuous variation.

The treatment below represents an attempt to segregate our materials into those most recognizable entities. The earliest name has been applied, but the application of some of the names to those morphological units might be in error; much additional work is necessary.

1a. Stems stout, mostly 3–6 mm thick near the base; spikelets mostly 6.5–11.5 mm long, ascending-erect, in stiff compact panicles; glumes almost as long as spikelet or surpassing the florets; plants of sea beaches and tidal flats. *P. eminens*

1b. Stems various, seldom as much as 3 mm thick; spikelets variously arranged, sometimes 6.5–10 mm long, but then seldom in stiff compact panicles; glumes usually shorter than the spikelet; plants of various distribution, seldom in tidal flats. (2)

2a. Plants annual, the remains of old stems never present; anthers less than 1 mm long; second glume broadened near midlength; lemmas not webbed at base. *P. annua*

2b. Plants perennial, although sometimes flowering the first season, usually with remains of previous season's stems; anthers usually more than 1 mm long, but if less, then second glume not broadened near midlength and the lemmas webbed at base. (3)

3a. Anthers less than 1 mm long. (4)

3b. Anthers 1–2.5 mm long or more. (6)

4a. Leaves in dense basal tufts; cauline leaves lacking or the solitary one placed well below the middle; plants mostly of arctic or arctic-alpine distribution. *P. abbreviata*

4b. Leaves not especially in basal tufts, some of them usually cauline and placed well above the stem base; plants of various distribution. (5)

5a. Spikelets green, seldom slightly purplish; nerves of palea mostly villous at least below; leaf sheaths conspicuous, hyaline; plants mostly of coastal and insular sites (rarely inland), from

Prince William Sound westward. *P. brachyanthera*

5b. Spikelets purplish, seldom greenish; nerves of palea scabrous-ciliate, not villous; leaf sheaths usually opaque, not especially conspicuous; distribution various, sometimes coastal. *P. leptocoma*

6a. Stems strongly flattened, 2-edged; lemmas only slightly, if at all webbed at base; plants strongly rhizomatous, introduced to southern Alaska and the Yukon. *P. compressa*

6b. Stems terete or nearly so, not 2-edged; lemmas variously webbed or not. (7)

7a. Spikelets only slightly compressed to almost terete, in flower more than twice (sometimes several times) as long as broad; lemmas rounded on the back or only slightly keeled; plants of continental, south-central, and southeastern Alaska and southern Yukon. *P. scabrella*

7b. Spikelets compressed, in flower often less than twice as long as broad; lemmas more or less strongly keeled (except in *P. stenantha*, which has more or less pilose keel and marginal nerves and anthers less than 2 mm long); distribution various. (8)

8a. Plants tufted; rhizomes lacking (stolons sometimes present). (9)

8b. Plants with rhizomes present, these frequently well developed (see also *P. palustris* and *P. trivialis*). (16)

9a. Lemmas with long tangled cobwebby hairs at the base. (10)

9b. Lemmas not cobwebby at the base or if rarely with a few long folded hairs, then inflorescence very narrow and erect; plants seldom if ever in coastal southern Alaska (see *P. glauca*). (12)

10a. Leaf sheaths retrorsely scabrous; lemmas glabrous except for basal tuft of cobwebby hairs or pubescent only on the keel; plants mostly of coastal southern Alaska. *P. trivialis*

10b. Leaf sheaths smooth or scabrous, not retrorsely so; lemmas pubescent on marginal nerves and on the keel nerve. (11)

11a. Ligules less than 1 mm long, at least some; second glume usually about as long or even longer than the first lemma; introduced, mostly in coastal southern Alaska. *P. nemoralis*

11b. Ligules more than 1 mm long, at least some; second glume usually shorter than (sometimes equaling) the first floret; introduced, often in moist sites, in southern Alaska. *P. palustris*

12a. Lemmas glabrous or merely scabrous, not villous or pilose on the keel and marginal nerves; plants of central eastern Alaska and west-central Yukon. *P. leibergii*

12b. Lemmas pubescent on the keel and marginal nerves and sometimes also on the internerves. (13)

13a. Spikelets at flowering time about two-thirds as broad as long, subcordate; leaf blades flat, usually more than 2 mm wide; panicle pyramidal, about as long as broad; plants widespread. *P. alpina*

13b. Spikelets at flowering usually less than two-thirds as broad as long, usually not subcordate; leaf blades often folded or involute and less than 2 mm broad; panicle various. (14)

14a. Panicle open, more or less pyramidal; spikelets usually dark purplish; pubescence of marginal nerves usually softly long-villous; internerves usually pubescent towards the base. *P. arctica*

14b. Panicle more or less contracted, the branches ascending to appressed or uncommonly more or less open; spikelets green or purplish, seldom dark purple; pubescence of marginal nerves pilose, not especially soft; internerves pubescent or glabrous. (15)

15a. Lemmas hairy between the keel and marginal nerves; plants mostly of coastal southern Alaska (less commonly inland). *P. stenantha*

15b. Lemmas hairy only on the keel and marginal nerves, the internerves glabrous or rarely puberulent; plants widespread in Alaska and Yukon. *P. glauca*

16a. Spikelets mostly (5.7)6.0–10.4 mm long; plants mostly of coastal and insular southern, and less commonly, interior Alaska and Yukon(?). *P. macrocalyx*

16b. Spikelets commonly less than 5.5 mm long; distribution various. (17)

17a. Stems and or leaf sheaths retrorsely scabrous. (18)

17b. Stems and leaf sheaths glabrous, or if scabrous, not retrorsely so (except rarely in some *P. pratensis,* q.v.). (19)

18a. Ligules 1–2 mm long; lemmas pubescent on the keel and marginal nerves; plants reported for southernmost Alexander Archipelago. *P. laxiflora*

18b. Ligules often more than 3 mm long; lemmas not hairy on marginal nerves; plants rather widely distributed, mostly in insular and coastal southern Alaska, rarely elsewhere. *P. trivialis*

19a. Lemma glabrous or scabrous between the keel nerve and the marginal nerve, not truly hairy on the internerve area. *P. pratensis*

19b. Lemma with pilose or lanate hairs between the keel nerve and the marginal nerve, at least near the base. (20)

20a. Stem base clothed with prominent cylinder of hyaline sheaths; plants mostly of coastal or near coastal sites in western Alaska. *P. malacantha*

20b. Stem base clothed with rather inconspicuous mostly opaque sheaths; plants widespread. *P. arctica*

Poa abbreviata R. Br.
Low Speargrass

Plants perennial, densely tufted, lacking rhizomes, mostly 0.5–1.5 dm tall; leaf sheaths glabrous, often forming a persistent thatch; stems smooth; ligules 1.2–2 mm long, mostly acute; leaf blades folded or involute, 1 mm wide or less; cauline leaves lacking or the uppermost one near the stem base; panicle compact to somewhat loose, 1–2.5 cm long, the branches ascending to appressed; spikelets compressed (4)4.5–5.8 mm long, with 2 or 3(4) flowers, purple; glumes unequal, shorter than the adjacent lemma; lemmas villous on the keel and marginal nerves and densely puberulent to glabrous on the internerves, not or only sparingly webbed at the base; anthers 0.6–0.8 mm long.

Arctic and alpine tundra; in northern and central southern Alaska, westernmost Aleutians (?), and northern Yukon; eastward through the Canadian Arctic Archipelago; circumpolar (*P. pseudoabbreviata* Roshev.; *P. jordalii* Porsild [?]).

Poa alpina L.
Alpine Bluegrass

Plants perennial, densely tufted, lacking rhizomes, mostly 1–4 dm tall (to 6 dm in viviparous forms); leaf sheaths glabrous, sometimes persistent with the blades and forming a conspicuous thatch; stems smooth; ligules 2–4.4 mm long, lacerate to erose or entire, glabrous; leaf blades flat, less commonly folded, (1)2–4(5) mm broad, chiefly basal; cauline leaves 1–2 (very rarely 3), the uppermost placed near the middle of the stem or below in mature plants; panicle compact to more or less open, pyramidal, 1.5–6(8) cm long (longer in some viviparous specimens), the branches finally spreading; spikelets compressed, 4.5–7.7 mm long, with mostly 3–6 flowers, in flower about two-thirds as broad, green to purple; glumes unequal, shorter than the adjacent lemmas; lemmas villous on keel and marginal nerves, more

Poa alpina L. (× 0.5).

or less hairy on the internerve, not webbed at the base; anthers 1.5–2 mm long.

Heathlands, thickets, muskegs, stream banks, lake shores, open woods, alpine and arctic tundra; in most of Alaska and Yukon, (except for the Aleutians, where rare, and much of southeastern Alaska); eastward to the Atlantic and south to Oregon, Utah, Colorado, and Michigan; circumboreal *P. alpina* f. *vivipara* L.; *P. alpina* var. *vivipara* [L.] Willd.)

Poa annua L.
Annual Bluegrass

Plants annual, tufted, sometimes stoloniferous, mostly 0.4–2(3.5) dm tall; leaf sheaths glabrous, hyaline; stems smooth; ligules 0.7–3.2 mm long, erose to entire, glabrous; leaf blades flat or folded, 1–3.5 mm broad; cauline leaves usually 1–2, the uppermost commonly well below the panicle; panicles open, pyramidal, 1.6–5(8) cm long, the branches spreading; spikelets compressed, 3.8–6.5 mm long, 3–6-flowered, purple or purplish to green; glumes unequal, the second abruptly widened at or near the middle, shorter than the adjacent lemma; lemmas more or less villous on the keel and marginal nerves, not webbed at the base; anthers 0.5–1(1.1) mm long.

Roadsides, campgrounds, roadways, gardens, mostly in disturbed soils; in much of southern Alaska and southern Yukon; widespread in North America; circumboreal.

Poa arctica R. Br.
Arctic Bluegrass

Plants perennial, tufted or sod-forming, rhizomes more or less well developed, sometimes stoloniferous, mostly 1–4(5) dm tall; leaf sheaths glabrous, usually opaque; stems smooth; ligules 1.5–3.5 mm long, erose to lacerate, glabrous; leaf blades folded or involute, sometimes flat, mostly 1–2(3) mm broad; cauline leaves 1–2, the uppermost commonly well below the panicle; panicles open, pyramidal, 4–10(12) cm long (not including pendulous lower branches), the branches spreading or the lowermost sometimes drooping; spikelets compressed, 4.5–6(7.5) mm long, 2–6-flowered, dark purple or greenish purple; glumes unequal, shorter than the adjacent lemmas; lemmas villous on the keel and marginal nerves and densely hairy to subglabrous on the internerve, webbed or not webbed at the base; anthers 1.3–2.2 mm long.

Heathlands, tundra, and woods, on ridges, slopes, rock outcrops, stream banks, gravel bars, lake shores, meadows, and spits; almost throughout Alaska and Yukon; eastward to the Atlantic and southward to Utah and Colorado; circumboreal (*P. williamsii* Nash; *P. arctica* ssp. *williamsii* [Nash] Hultén; *P. brintnellii* Raup; *P. arctica* ssp. *longiculmis* Hultén; *P. cenisea* f. *caespitans* Simm.; *P. arctica* ssp. *caespitans* [Simm.] Nannf.; *P. lanata* authors, pro parte). *Poa arctica* is a variable entity; however, the variation does not seem to be correlated with either geographic or habitat differences. Thus, it seems best to treat our materials as portions of a single polymorphic entity. Indeed, *P. arctica* appears to share characteristics not only with *P. alpina* and *P. glauca,* but with *P. pratensis* as well.

Poa brachyanthera Hultén

Plants perennial (sometimes flowering the first year), tufted, often stoloniferous, sometimes viviparous, mostly 0.5–2 dm tall; leaf sheaths hyaline, glabrous, often forming a conspicuous thatch; stems smooth; ligules 1.5–2.5 mm long, erose, glabrous; leaf blades flat, 0.5–3 mm broad; cauline leaves 1–4, the uppermost commonly above the middle of the stem; panicles open, pyramidal, 1–4(6.5) cm long, the branches spreading; spikelets compressed, 4–6.8 mm long, 2–5-flowered, green; glumes unequal, the second considerably wider above the middle, shorter than the adjacent lemma; lemmas villous on the keel and marginal nerve, glabrous or puberulent on the internerve, not webbed at base; anthers 0.6–1 mm long.

Marshes, gravelly slopes, and roadsides; in or near coastal and insular southwestern Alaska; endemic.

Poa compressa L.
Canada Bluegrass

Plants perennial, rhizomatous and sometimes stoloniferous, mostly 2–5(6) dm tall; leaf sheaths smooth, opaque, sometimes reddish; stems smooth, flattened, 2-edged; ligules 0.5–2 mm, long-ciliate, pubescent; leaf blades flat to folded, 1–4 mm broad; cauline leaves 3–4, the uppermost placed well above the middle; panicles rather compact, (1)2–9 cm long or more, the short branches ascending to spreading; spikelets compressed, 3.5–6.5 mm long, 2–6-flowered, green to purplish; glumes subequal, shorter than the adjacent lemma; lemmas usually hairy on keel and marginal nerves, glabrous on the internerves, not or rarely webbed at base; anthers 1–1.5 mm long.

Beaches, roadsides, disturbed soils; at widely scattered sites in southern Alaska and southern Yukon; widely distributed in temperate regions of the earth; introduced.

Poa eminens Presl
Large-flower Speargrass

Plants perennial, strongly rhizomatous, mostly (2.5)5–13 dm tall; leaf sheaths smooth, opaque; stems smooth, terete, commonly 3–6 mm thick near the base; ligules 1.2–3 mm long, ciliate, glabrous; leaf blades flat, 4–11 mm broad; cauline leaves 2–4, the uppermost placed near the middle or above; panicles compact, (7)9–24 cm long, the stiff branches appressed-erect or appressed-ascending; spikelets compressed, (6.5)7.0–11.5 mm long, 2–6-flowered, pale green maturing purplish or brownish; glumes subequal to somewhat unequal, about equaling to surpassing the adjacent florets and often surpassing all of the florets; lemmas pilose on keel and marginal nerves, and more or less puberulent to long-hairy on the internerves, not webbed at base; anthers (1.6)2.3–4 mm long.

Sea beaches and tidal flats; throughout coastal and insular southern and western Alaska; southward to British Columbia and disjunctly to coastal Labrador; Asia (*P. glumaris* Trin.; *Glyceria glumaris* [Trin.] Griseb. in Ledeb.; *P. trinii* Scribn. & Merr.). This is perhaps our most striking and most easily recognizable species of *Poa*.

Poa glauca Vahl
Glaucous Bluegrass

Plants perennial, strongly tufted, not rhizomatous or stoloniferous, mostly 1–5.5 (rarely to 7.5) dm tall; leaf sheaths smooth or antrorsely scabrous, opaque, those above the basal tuft often purplish; stems smooth or more or less antrorsely scabrous, terete; ligules 1.2–3.8 mm long, more or less lacerate (often folded back), glabrous; leaf blades folded or involute or that of uppermost leaf flat and 1–2 mm wide; cauline leaves 1 or more, usually 2, the upper one often placed at or near the middle of the stem; panicles compact with appressed-ascending branches or less commonly open and with branches spreading-ascending to ascending, (2)3–10(12) cm long; spikelets more or less compressed, 3.2–7.2 mm long, 2–6-flowered, green to purplish or dark purple; glumes subequal or somewhat unequal, shorter than the adjacent lemma; lemmas villous on the keel and marginal nerves, usually glabrous on the internerves, though rarely sparingly puberulent, not webbed at the base, or sometimes with a few folded hairs; anthers 1.4–2 mm long.

Dry gravelly slopes, open woods, gravel bars, stream banks, rock stripes, dunes and rock outcrops; in most of Alaska (except for the Aleutians and the Panhandle) and most of the Yukon; eastward to the Atlantic and southward to Colorado, Minnesota, and New Hampshire; circumpolar (*P. ammophila* Porsild; *P. rupicola* Nash). This is one of our most abundant and most variable species of *Poa*. On the one hand it approaches (if not passes into) *P. arctica* and on the other it apparently intergrades, at least morphologically, with *P. pratensis*

and to a lesser extent with *P. alpina*. The narrow inflorescence, few-leaved stem, and compactly tufted habit seem to define this entity.

Poa laxiflora Buckl.
Loose-flower Bluegrass

Plants perennial, rhizomatous, mostly 8–12 dm tall; leaf sheaths strongly retrorsely scabrous, opaque; stems retrorsely scabrous, terete; ligules 1–3(4) mm long, erose, puberulent; leaf blades flat, 3–5 mm broad, scabrous on both sides; cauline leaves usually at least 3, the uppermost usually well above the middle; panicles open, mostly 12–20 cm long, the branches spreading to reflexed; spikelets compressed, mostly 5–6 mm long, 2–4-flowered, usually green; glumes unequal, shorter than the adjacent lemmas; lemmas silky-hairy on keel and marginal nerves and strongly webbed at base; anthers mostly 1–1.2 mm long.

Woods; in Alaska known only from the southernmost Panhandle; southward to Oregon (*P. leptocoma* ssp. *elatior* Scribn. & Merr., the type from Cape Fox, Alaska).

Poa leibergii Scribn.
Lieberg Bluegrass

Plants perennial, tufted, not rhizomatous, mostly 0.5–3 dm tall; leaf sheaths glabrous, opaque; stems smooth, terete; ligules 1–3 mm long, entire or erose and more or less lacerate, glabrous; leaf blades flat to folded or involute, mostly less than 1 mm broad; cauline leaves usually only one, placed variously; panicles 2–8 cm long, 2–6-flowered, purplish; glumes unequal, usually shorter than the adjacent lemma; lemmas smooth to scabrous, not villous or pilose on the marginal and keel nerves, not webbed at the base; anthers 2.5–3.2 mm long.

Mountain slopes; in central eastern Alaska and west-central Yukon; disjunctly southward from Washington and Oregon (*P. vaseyochloa* Scribn.; *P. pulchella* Vasey, not Salisb.).

Poa leptocoma Trin.

Bog Bluegrass

Plants perennial, tufted, not rhizomatous, mostly (0.3)0.5–4(5) dm tall; leaf sheaths smooth or minutely scabrous, opaque, sometimes the lower ones purplish; ligules 1.3–3 mm long, entire or lanceolate, glabrous; leaf blades flat or folded, 1–4 mm broad; cauline leaves 1–2, the uppermost variously placed; panicle compact, becoming open at maturity, (2)4–13(15) cm long, the branches slender, ascending to spreading or sometimes drooping; spikelets compressed, 4.3–7.7 mm long, 2–5-flowered, purple or green turning purple; glumes unequal, shorter than the adjacent lemmas; lemmas villous on the keel and marginal nerves, glabrous on the internerves, webbed at the base; anthers 0.5–0.9 mm long.

Alpine tundra, often at high elevations (to 1700 m); in much of Alaska (except for the Aleutians) and most of the Yukon; eastward to the Mackenzie and southward to California, Nevada, Utah, and New Mexico; Asia (*P. paucispicula* Scribn. & Merr.; *P. leptocoma* var. *paucispicula* [Scribn. & Merr.] C. L. Hitchc.; *P. glacilis* Scribn. & Merr., not Stapf.; *P. merrilliana* A. S. Hitchc.). The type of *P. leptocoma* was collected at Sitka (D. Mertens), that of *P. paucispicula* came from Hidden Glacier, Yakutat Bay and that of *P. merrillana* from Hubbard Glacier, Yakutat Bay. The similarity of the three has been recognized for some time and it seems inevitable that they should be placed together.

Poa macrocalyx Trautv. & Mey.

Large-glume Bluegrass

Plants perennial, rhizomatous and more or less stoloniferous, sometimes also tufted, mostly (1.5)2.2–7.5 dm tall; leaf sheaths smooth, often membranous and more or less hyaline, sometimes reddish above the base; stems smooth, terete; ligules 2.5–5.5 mm long, erose and often lacerate, glabrous; leaf blades flat or folded, (1)2–7 mm broad; cauline leaves (1)2–3, the up-

permost placed from below to above the middle; panicle rather compact to open, (2.5)4–12(14) cm long, the branches ascending to spreading; spikelets compressed, mostly (5.7)6.0–11.2 mm long, 2–6-flowered, green to purplish; glumes unequal to subequal, shorter than the adjacent lemma; lemmas silky-hairy on keel and marginal nerves, glabrous to puberulent, scabrous, or pilose on the internerve, webbed at base; anthers 1.7–2.5 mm long.

Sandy beaches, marshes, slopes, sea cliffs, and mountains; mostly in or near coastal and insular Alaska, from the northern portion of the Panhandle westward to the westernmost Aleutians and along coastal western Alaska (rarely elsewhere); Asia (*P. hispidula* Vasey; *P. turneri* Scribn.; *P. lanata* Scribn. & Merr.; *P. norbergii* Hultén; *P. hispidula* var. *aleutica* Hultén; *P. hispidula* var. *vivipara* Hultén). The materials herein interpreted as *P. macrocalyx* are composed of a series of intergrading morphological units which are separable only on arbitrary bases. All of the names cited as synonyms are based on type specimens collected along the coast of Alaska, undoubtedly on morphologically distinctive materials. These are, however, connected by a great many intermediate specimens and it seems best to treat all of our materials as belonging to a single polymorphic species. Our specimens seem to fit the descriptions of *P. macrocalyx* and that name, apparently the earliest available, is taken up. It seems that most of the continental records previously attributed to *P. lanata* belong to other species, especially to *P. arctica*.

Poa malacantha Kom.

Plants perennial, more or less tufted and shortly rhizomatous, mostly 1–3.5 dm tall; leaf sheaths smooth, hyaline, conspicuously persistent; ligules 2–4 mm long, lacerate; leaf blades flat or folded, 2–4 mm broad; cauline leaves mostly 1–2; panicles more or less open, pyramidal, 4–10 cm long, the branches ascending to spreading; spikelets compressed, 5–8 mm long, 2–6-flowered, glumes subequal, shorter than the adjacent lemma; lemmas silky-hairy on the keel and marginal nerves, puberulent on the internerves, webbed at the base; anthers more than 1 mm long.

Meadows and mountain slopes; in or near coastal southwestern, western, and northwestern Alaska; Asia (*P. komarovi* Roshev.). The above description is abstracted from that given in Flora of the USSR (2:333, 334. 1934). I have found no specimens that fit the description of this entity, nor, in fact, any clear diagnostic features. Rather, it seems probably that this taxon represents no more than a portion of the variation in *P. macrocalyx* as herein interpreted.

Poa nemoralis L.
Wood Bluegrass

Plants perennial, tufted, sometimes stoloniferous; mostly 3–10 dm tall; leaf sheaths smooth, opaque; ligules 0.3–0.6 mm long, ciliate, puberulent; leaf blades flat or less commonly involute, 1–3 mm broad; cauline leaves, usually 3–4, the uppermost variously placed; panicles narrow, mostly 6–20 cm long, erect to drooping, the branches ascending; spikelets compressed, 3–4.5 mm long, 2–5-flowered, green to purplish; glumes subequal to somewhat unequal, the second about equaling the first lemma; lemmas more or less hairy on keel and marginal nerves, glabrous on the internerves, scantily webbed at base; anthers 1–1.2 mm long.

Moist sites such as tidal flats, stream banks, and meadows; known in Alaska (where introduced) only from the coastal and insular southern portion; widely distributed in much of North America; Eurasia.

Poa palustris L.

Plants perennial, tufted, more or less decumbent at the base and often rooting at the nodes, mostly 3–10 dm tall or more; leaf sheaths glabrous or scabrous, opaque; ligules 2–5 mm long, more or less lacerate, ciliate, puberulent; leaf blades flat or folded, 1–3 mm broad; cauline leaves usually 3–4, the uppermost usually placed above the middle; panicles open, mostly 8–25 cm long or more, the branches ascending to spreading; spikelets compressed, 3.7–5 mm long, 2–4(5)-flowered, greenish to purplish; glumes somewhat unequal, the second only slightly shorter or sometimes equaling the first lemma; lemmas more or less hairy on keel and marginal nerves, glabrous on the internerves, webbed at the base; anthers 1–1.3 mm long.

Wet meadows, lake shores, sea beaches, and marshes; in central and east-central Alaska and in coastal and insular southern Alaska and southern Yukon; introduced from Europe, escaping and persisting; widespread in the north temperate regions. The specimen collected at Teikel Road House (Richardson Highway mile 52, J. P. Anderson 2762, ISC) and identified with *P. occidentalis* Vasey belongs to *P. palustris*.

Poa pratensis L.
Bluegrass

Plants perennial, rhizomatous, sod-forming (seldom tufted), mostly (1)2.5–10 dm tall; leaf sheaths glabrous or rarely retrorsely or antrorsely scabrous, opaque, often reddish above the base; ligules 1–3 mm long, erose, ciliate, puberulent; leaf blades flat or folded, 2–5 mm broad; cauline leaves mostly 2–3(4), the uppermost variously placed; panicle open or more or less compact, mostly (3)5–15(27) cm long, the branches appressed to ascending or spreading; spikelets compressed, (3)4.0–6.5 mm long, 2–6-flowered, green to purplish; glumes more or less unequal in length and in shape, shorter than the adjacent lemma, the second usually shorter than the first lemma; lemmas more or less silky-hairy on keel and marginal nerves, glabrous or rarely scabrous on the internerves, strongly webbed at the base; anthers 1.0–1.9 mm long.

River banks, roadsides, lawns, muskegs, spits, bars, talus slopes, meadows, and disturbed sites; in most of Alaska and the Yukon and to be expected practically anywhere; widespread in temperate regions of the earth (*P. eyerdamii* Hultén; *P. subcoerulea* J. E. Smith; *P. pratensis* var. *subcoerulea* [J. E. Smith] J. E. Smith; *P. pratensis* ssp. *subcoerulea* [J. E. Smith] Hiit.; *P. irrigata* Lindm.; *P. pratensis* var. *alpigena* Fries; *P. alpigena* [Fries] Lindem.; *P. pratensis* ssp. *alpigena* [Fries] Hiit.; *P. angustifolia* L.; *P. pratensis* ssp. *angustifolia* [L.] Lindb.; *P. pratensis* var. *angustifolia* [L.] Gaudin). From distributional and other data, there seems to be little doubt that *P. pratensis* in our region is both indigenous and adventive. The introduced phases have been segregated as *P. pratensis*, *P. angustifolia*, and *P. subcoerulea;* the indigenous as *P. alpigena*. That there is a lack of definitive criteria for segregation of these types seems clear, especially when one tries to separate our specimens by single features or even by means of combinations of characteristics. Further work might demonstrate a correlation of some characters, but for the present it seems best to treat all of our materials as portions of an apomictic polymorphic series whose chromosome number varies through a vast sequence of aneuploids. Viviparous plants are known also.

Poa scabrella (Thurb.) Benth.

Plants perennial, tufted, mostly 2–7.5(10) dm tall; leaf sheaths glabrous (or scabrous?), opaque; ligules lacerate to entire, mostly (1)3–4.5 mm long, glabrous or puberulent; leaf blades folded to involute or flat, 1–3 mm broad; cauline leaves usually 2–3, the uppermost often below the middle of the stem; panicles narrow, 5–12 cm long or more, the branches ascending to appressed; spikelets terete or only slightly compressed, 4.8–10.3 mm long, mostly 3–5-flowered, striped purplish; glumes unequal, much shorter than the adjacent lemma; lemmas more rounded than keeled, usually scabrous or merely puberulent, not long-hairy on median or marginal nerves, not webbed at base; anthers (1.8)2–3 mm long (in lower florets at least).

Steep dry slopes and dune sands in open woods; in south-central and southeastern continental Alaska and southern Yukon; southward to California and Colorado and eastward to Minnesota (*Atropis scabella* Thurb. in Wats.; *Glyceria canbyi* Scribn.; *P. canbyi* [Scribn.] Howell). Also included herein are those reports of *P. nevadensis* Vasey and *P. ampla* Merrill, both near congeners of *P. scabrella*.

Poa stenantha Trin.

Plants perennial, tufted, mostly 2–7 dm tall; leaf sheaths smooth, opaque; ligules 1–3.5 mm long, acute, usually lacerate, puberulent to glabrous; leaf blades involute to flat, 0.5–2 mm broad; cauline leaves commonly 1–3, the uppermost placed near the middle; panicles narrow to somewhat open, (5)7–15 cm long, the branches ascending to appressed or less commonly spreading-ascending; spikelets more or less compressed, keeled, (5.4)6–10 mm long, 3–5-flowered, striped purplish; glumes unequal, much shorter than the adjacent lemmas; lemmas keeled, more or less long-pilose along keel and marginal nerves and puberulent to glabrous on the internerves, not webbed at base; anthers 1.2–2 mm long.

Thickets, marshes, roadsides, sea cliffs, and rocky outcrops; in coastal and insular southern Alaska and less commonly some distance in the interior; southward to Oregon, Idaho, and Montana (*P. acutiglumis* Scribn.). This entity resembles *P. scabrella*, but the longer hairs on keel and marginal nerves and definitely keeled lemmas appear to be definitive. Viviparous specimens are known.

Poa trivialis L.
Rough Bluegrass

Plants perennial, tufted, decumbent and rooting at the base, mostly 3–10 dm tall;

leaf sheaths retrorsely scabrous, opaque; ligules entire or nearly so, 3–5 mm long or more, puberulent; leaf blades flat, 1.5–4 mm broad; panicles open, 8–15 cm long, the branches ascending to spreading; spikelets compressed, 2.5–4 mm long, commonly 2–3-flowered, usually green or greenish purple; glumes unequal, shorter than the adjacent lemmas; lemmas pilose on the keel nerve, the marginal nerve and internerves glabrous, webbed at the base.

Introduced; escaping and persisting; in southern Alaska and Yukon; widely distributed in temperate regions of the earth.

POLYPOGON Desf.

Annual solitary or tufted herbs; culms hollow; leaf sheaths open; ligules membranous; leaf blades flat, lacking auricles; spikelets borne in dense panicles, 1-flowered, articulate below the glumes; glumes keeled, equal, awned from between short lobes of a minutely bifid apex; palea subequal to the lemma; lemma shorter than the glumes, awned, the awn straight and surpassing the glumes; lodicules 2; stamens 1–3.

Polypogon monspeliensis (L.) Desf.
Rabbitfoot

Plants mostly 0.5–6(7) dm tall, often decumbent and rooting at the lower nodes; ligules 3–8 mm long or more, puberulent; leaf blades flat, 3–7 mm broad (or more); panicle dense, spikelike, mostly 3–7 cm long; glumes puberulent, awned, the awn 5–10 mm long; lemma to about half as long as the glumes, shining, awned, the awn surpassing the glumes.

Disturbed sites, often in moist areas; at widely scattered locations in Alaska and Yukon, where apparently introduced; widely distributed in temperate regions of the earth (*Alopecurus monspeliensis* L.).

PUCCINELLIA Parl.

Perennial tufted to stoloniferous or rhizomatous herbs; culms hollow; leaf sheaths open or partially closed; ligules membranous, glabrous or puberulent, leaf blades flat to folded or involute, lacking auricles; spikelets borne in more or less contracted to open panicles, 2–7-flowered, articulate above the glumes; glumes more or less keeled to rounded on the back, the first usually 1-nerved, the second commonly 3-nerved, awnless; palea shorter than, to equaling or longer than, the lemma; lemmas rounded on the back, 5-nerved, the nerves not converging apically, awnless; lodicules 2; stamens 3.

Fernald, M. L., and C. A. Weatherby. 1916. The genus *Puccinellia* in eastern North America. Rhodora 18:1–23.

Polypogon monspeliensis (L.) Desf. (× 0.3)

Swallen, J. R. 1944. The Alaskan species of *Puccinellia*. Jour. Wash. Acad. Sci. 34:16–23.

Puccinellia is a genus like *Poa*, whose species are difficult to distinguish. A part of the difficulty is based on the fact that some of the taxa are at least partly cleistogamous, and cleistogamous flowers tend to bear smaller anthers than do chasmogamous ones. Because of this, anther size, which has been utilized as a major diagnostic feature, probably does little more than separate cleistogamous from chasmogamous individuals. Vegetative reproduction and the presence of polyploid series within species tend to obscure further those characteristics utilized in segregation of entities. Also, because of the specialized habitat requirements for many species in the genus in saline or alkaline areas, fewer specimens are available for comparison of features. Because of these problems, the following key is tentative at best.

1a. Plants low, frequently with widely spreading stolons, seldom flowering; anthers mostly (1.5)1.7–2 mm long, mostly sterile; pistils vestigial or lacking; coastal and insular Alaska and northern Yukon. *P. phryganodes*

1b. Plants erect to decumbent, rarely or not at all with widely spreading stolons, usually fertile; anthers usually less than 1.5 mm long; pistils not or only rarely vestigial or lacking; distribution various. (2)

2a. Axis and branches of inflorescence smooth, lacking scabrous prickles, or sometimes with scattered prickles especially on the pedicels. (3)

2b. Axis and branches of inflorescence scabrous almost or quite throughout. (7)

3a. Lemma of second floret mostly 1.5–2 mm long; spikelets often purplish, the lemma tip yellow brown; plants mostly in the interior. *P. distans*

3b. Lemma of second floret mostly 2.5–4.3 mm long; spikelets various, but seldom bicolored; plants mostly of coastal regions. (4)

4a. Lemma thin, strongly nerved, that of the second floret mostly 2–3 mm long. *P. langeana*

4b. Lemma firm, the nerves obscure, that of the second floret commonly (2) 3–4.3 mm long. (5)

5a. Lemma of the second floret mostly (3)3.5–4.3 mm long; branches of the inflorescence ascending to spreading or reflexed. *P. andersonii*

5b. Lemma of the second floret mostly less than 3 mm long; branches of the inflorescence appressed or appressed-ascending or less commonly spreading. (6)

6a. Plants mostly 1–2(3) dm tall; panicle narrow, mostly 3–8(10) cm long, the branches appressed; lemmas entire-margined. *P. pumila*

6b. Plants mostly much more than 2 dm tall; panicle often open or over 10 cm long (or both), the branches appressed to spreading; lemmas often erose-ciliate. *P. nutkaensis*

7a. Lemma of second floret mostly 1.5–2 mm long; anthers 0.4–0.8 mm long; lower panicle branches usually reflexed. *P. distans*

7b. Lemma of second floret mostly more than 2 mm long; anthers often over 0.8 mm long (sometimes less); lower panicle branches spreading to ascending, seldom reflexed. (8)

8a. Lemmas 3–4.3 mm long; panicle branches bearing spikelets near the tips only; plants of coastal and insular regions. (9)

8b. Lemmas mostly less than 3 (to 3.4 in *P. arctica*) mm long; panicle branches with spikelets various; plants sometimes coastal, often in interior sites. (10)

9a. Panicle branches mostly stiffly ascending, the lowermost seldom less than 8 cm long. *P. lucida*

9b. Panicle branches mostly spreading or spreading-ascending, less commonly stiffly ascending, the lowermost seldom as much as 8 cm long. *P. andersonii*

10a. Lemma of second floret 1.8–2.3 mm long; plants densely tufted, the stems mostly stiffly erect; central eastern Alaska and southern Yukon. *P. nuttalliana*

10b. Lemma of second floret (2.1)2.4–3.2 mm long; plants densely to loosely tufted, often more or less decumbent-spreading; distribution various. (11)

11a. Lower portion of lemma usually distinctly hairy; plants of broad distribution in interior and northern Alaska and Yukon. *P. arctica*

11b. Lower portion of lemma glabrous or rarely with a few hairs; plants of coastal southern Alaska. *P. nutkaensis*

Puccinellia andersonii Swallen
Anderson Alkaligrass

Plants perennial, tufted and often more or less stoloniferous or subrhizomatous, mostly 1.5–5 dm tall; ligules mostly 1.7–3.2 mm long, glabrous or puberulent; leaf blades usually folded or involute, 0.5–2(3) mm broad; panicles 5–14 cm long, the branches ascending to spreading, glabrous or more commonly more or less scabrous; spikelets 5.3–10 mm long, (2)3–7-flowered, green or purplish, the pedicels with enlarged, tumid cells; glumes more or less erose to subentire, sometimes minutely ciliate, the first 1.6–2.2 mm long, the second 2.5–3 mm long; lemmas 3–4 mm long, acute to obtusish, sometimes erose-toothed, more or less hairy near the base; palea shorter than or subequal to the lemma; anthers 0.9–1.3 mm long.

Sea beaches, tidal flats, and salty marshes; in coastal and insular south-central and northwestern Alaska and to be sought elsewhere; endemic (*P. triflora* Swallen; *P. glabra* Swallen). Swallen treated the Alaskan species of *Puccinellia* (1944), and

named several critical new species as occurring within the Alaskan flora. *P. glabra* and *P. triflora* were maintained as distinct on the basis of such characteristics as divarication of branching of the panicles and number of florets per spikelet, both notoriously variable within this assemblage and within the genus as a whole. *P. andersonii* was supposed to differ in having lemmas not more than 3.5 mm long, whereas even in the type specimen some are as long as 4 mm. Furthermore, the maintenance of *P. andersonii* as distinct from *P. lucida* is difficult to justify. The scabrosity of the panicle branches is one of degree, varying from entirely smooth to scabrous throughout and apparently the only definitive feature is the length of the lower panicle branches.

Puccinellia arctica (Hook.) Fern. & Weath.
Arctic Alkaligrass

Plants perennial, tufted and seldom if ever stoloniferous or subrhizomatous, mostly 1–5.5(7.5) dm tall; ligules mostly 1–2.5 mm long, glabrous or puberulent; leaf blades folded, involute, or occasionally flat, 0.5–2.5 mm broad; panicles 5–15(21) cm long, the branches ascending to spreading, scabrous; spikelets 3.6–7 mm long, (2)3–6-flowered, green or purplish, the pedicels usually not with enlarged tumid cells; glumes usually minutely ciliate, the first 0.8–1.5(2.2) mm long, the second 1.5–2(3.3) mm long; lemmas 1.9–2.7(3.4) mm long, acute to obtusish, minutely ciliate and sometimes more or less toothed, hairy (sometimes densely so) towards the base; palea shorter than the lemma; anthers 0.5–1.1(1.5) mm long.

Stream banks, terraces, bars, spits, and roadsides; in most of continental and coastal western and northern Alaska and much of the Yukon; eastward to the Labrador (*P. borealis* Swallen). This entity belongs to an arctic complex of forms which have gone under several names. Among the names which might apply to the overall complex is *P. angustata* (R. Br.) Rand &

Redf. If that entity proves to be conspecific, as it might well be, then the proper name will be *P. angustata* since that name has priority.

Puccinellia distans (L.) Parl.

Plants perennial, tufted and seldom if ever stoloniferous or subrhizomatous, mostly 0.5–4(6.5) dm tall; ligules 0.6–1.5 mm long, usually more or less puberulent; leaf blades involute to folded, mostly less than 1.5 mm broad; panicles 5–19 cm long, the branches ultimately spreading or the lower ones strongly reflexed, scabrous; spikelets 3.2–5.7 mm long, 3–7-flowered, green to purplish, the pedicels not with enlarged tumid cells; glumes usually minutely ciliate, the first 0.7–1.4 mm long, the second 0.9–1.8 mm long; lemmas 1.7–2.3 mm long, obtusish, minutely ciliate and more or less toothed, hairy towards the base; palea about as long as, to slightly longer than, the lemma; anthers 0.4–0.8 mm long.

Shallow ponds, stream banks, sloughs, and roadsides; in most of continental Alaska and Yukon; widespread in North America; Eurasia (*P. hauptiana* [Krecz.] Kitagawa; *Atropis hauptiana* Krecz.; *P. interior* Sorens.; *P. vaginata* authors, not [?] [Lange] Fern. & Weath.). Our materials belong to a variable circumboreal complex which has been split previously into several critical entities. The lines of separation seem to be only arbitrary and without real morphological bases. Therefore, it seems that *P. distans* is represented in Alaska-Yukon at least by both indigenous and adventive Old World plants. The distinction is not clear.

Puccinellia langeana (Berlin) Sorens.
Arctic Alkaligrass

Plants perennial, tufted, seldom with a stoloniferous or subrhizomatous base, mostly 0.5–2.5 dm tall; ligules 0.5–2 mm long, glabrous; leaf blades involute or folded, mostly less than 1 mm broad; panicles 1.5–9 cm long, the branches appressed or appressed-ascending, smooth, not scabrous; spikelets 4–5.3 mm long, 3–5-flowered, green to purplish, the pedicels with enlarged tumid cells; glumes entire or more or less undulate, not ciliate, the first 0.9–1.4 mm long, the second 1.5–3 mm long; lemmas 2–3.1(3.4) mm long, obtuse to acutish, not or only sparsely and minutely ciliate, more or less hairy towards the base, strongly nerved; palea usually shorter than the lemma; anthers 0.3–0.7 mm long.

Beaches, rocky shores, spits, and lake shores; in coastal and insular western and northern Alaska and in the Aleutian Islands; eastward to Labrador and Greenland; circumpolar (*Glyceria langeana* Berlin; *G. paupercula* Holm in Fedde; *P. paupercula* [Holm] Fern. & Weath.; *P. alaskana* Scribn. & Merr.; *P. langeana* ssp. *alaskana* [Scribn. & Merr.] Sorens.; *P. langeana* ssp. *asiatica* Sorens.). Of the entities segregated within *P. langeana*, possibly the most striking is the *alaskana* phase. Its distinctive greenish spikelets whose parts average larger than those of the type phase are diagnostic and probably it should be retained at specific rank in a genus where characteristics tend to be tenuous at best.

Puccinellia lucida Fern. & Weath.

Plants perennial, tufted, sometimes with a stoloniferous or subrhizomatous base, mostly 4–7(9) dm tall; ligules 1.5–3 mm long, puberulent; leaf blades involute or folded or sometimes flat, mostly 1–4 mm broad; panicles 10–26 cm long, the branches appressed-ascending to spreading-ascending, scabrous; spikelets 7–13.5 mm long, 3–7 (10)-flowered, green to purplish, the pedicels lacking enlarged tumid cells; glumes minutely ciliolate, the first 1.3–2.4 mm long, the second 2–3.8 mm long; lemmas 3.0–4.9 mm long, obtuse to acutish, not or only sparingly ciliolate, more or less hairy towards the base; palea usually shorter than the lemma; anthers (0.8)1–1.5 mm long.

Beaches and tidal flats; in south-central and southeastern Alaska and reported from southern Yukon; southward to California and in Ontario and Quebec (*P. grandis*

Swallen). Our materials are not easily separable from those in eastern Canada.

Puccinellia nutkaensis (Presl) Fern. & Weath.

Pacific Alkaligrass

Plants perennial, tufted, often more or less stoloniferous or subrhizomatous at the base, mostly 2–7 dm tall; ligules 1–3.5 mm long, glabrous; leaf blades flat to folded or involute, 1–3 mm broad; panicles 4–19 cm long, the branches appressed-ascending to spreading, smooth or scabrous; spikelets 5–8.5 mm long, (2)4–6(7)-flowered, green or purplish, the pedicels usually with enlarged tumid cells; glumes minutely ciliate, the first 1.3–2.1 mm long, the second 2–3.2 mm long; lemmas 3–3.8(4.9) mm long, acutish, erose-ciliate to lobed, more or less hairy or sometimes glabrous at the base; palea subequal to or sometimes surpassing the lemma; anthers 0.7–1.5 mm long.

Sea beaches; in coastal and insular southern Alaska; southward to Washington (*Poa nutkaensis* Presl; *Puccinellia hulténii* Swallen; *P. kamtschatica* authors, not [?] Holmb.). This is the common sea beach alkaligrass of coastal regions. For the most part the inflorescence is narrow, with branches stiffly appressed-ascending, but occasional specimens, especially at or slightly below the level of high tide, bear panicles with spreading-ascending branches. These latter are the basis of P. *hulténii* and apparently represent only an ecological phase of *P. nutkaensis.*

Puccinellia nuttalliana (Schult.) A. S. Hitchc.

Plants perennial, densely tufted, rarely or not at all stoloniferous, mostly 1.5–6(10) dm tall; ligules 0.9–2(3) mm long, puberulent or glabrous; leaf blades involute, mostly less than 1.5 mm broad; panicles (3)7–20(30) cm long, the branches appressed-ascending to spreading, scabrous; spikelets 3.8–5.6 mm long, 3–7-flowered, green or purplish, the pedicels lacking en-

larged tumid cells, glumes sparingly and minutely ciliate, the first 1–1.5 mm long, the second 1.2–2 mm long; lemmas 1.7–2.3 mm long, acute to obtusish, usually erose and minutely ciliate, more or less hairy at the base; palea subequal to the lemma; anthers 0.7–1.2 mm long.

River flats, terraces, bars, roadsides, pond margins, lake shores, and sometimes on dryish slopes; southeastern continental Alaska and southern Yukon; eastward to Saskatchewan and southward to California, northern Mexico, Kansas, and Wisconsin (*Puccinellia agrostidea* Sorens.; *P. deschampsioides* Sorens.; *Poa airoides* Nutt.; *Puccinellia airoides* [Nutt.] Wats. & Coult.). *Puccinellia agrostidea* and *P. deschampsioides* do not differ markedly from *P. nuttalliana* in a strict sense. There seems to be a transition in anther size from that supposed to be diagnostic of *P. agrostidea* through that of both *P. nuttalliana* and *P. deschampsioides.* The degree of development of the keel and presence or absence of spines along the keel varies within the individual specimen and does not appear in any event to be definitive.

Puccinellia phryganodes (Trin.) Scribn. & Merr.

Creeping Alkaligrass

Plants perennial, tufted and mostly with well-developed elongate stolons, seldom flowering, mostly 0.5–1.5 dm tall; ligules 0.5–1 mm long, glabrous; leaf blades involute, usually less than 1.5 mm broad; panicles 1.5–3(5) cm long, the branches ascending to spreading, smooth; spikelets 5.7–8.4 mm long, 3–5-flowered, usually purplish, the pedicels with enlarged tumid cells; glumes entire, not ciliolate, the first 1.5–2.5 mm long, the second 2.5–4 mm long; lemmas 3.3–3.7 mm long, obtuse to acutish and often erose-toothed, not ciliolate, glabrous at the base; anthers 1.5–2.5 mm long, the second 2.5–4 mm long; lemmas 3.3–3.7 mm long, obtuse to acutish and often erose-toothed, not ciliolate, glabrous at the base; anthers 1.5–2 mm long.

Sea beaches and tidal flats; in insular and coastal western and northern Alaska and northern Yukon (?); eastward to Labrador; circumpolar (*Poa phryganodes* Trin.; *Puccinellia geniculata* authors, not [?] [Turcz.] Krecz.). This plant is ordinarily sterile, producing flowers only rarely and these seldom fertile.

Puccinellia pumila (Vasey) A. S. Hitchc.

Plants perennial, tufted, seldom or not at all stoloniferous, mostly 0.8–2 dm tall; ligules 0.8–2 mm long, glabrous; leaf blades involute, less than 1 mm broad; panicles mostly 3–8(10) cm long, the branches appressed or rarely spreading, smooth, not scabrous; spikelets 4.5–5.5 mm long, 3–6-flowered, green or purplish, the pedicels with enlarged tumid cells; glumes entire or more or less undulate, not ciliate, the first about 1.5 mm long, the second 1.5–3 mm long; lemmas 3–3.5 mm long, obtuse to acutish, not or only sparingly ciliate, strongly nerved; palea usually shorter than the lemma; anthers 0.7–1 mm long.

Sea beaches; in coastal south-central and southeastern Alaska; southward to California (*Glyceria pumila* Vasey). This might be altogether too near to *P. langeana* and probably would be best included within that species.

SCHIZACHNE Hack.

Perennial tufted herbs; culms hollow; leaf sheaths closed; ligules membranous; leaf blades flat to involute, lacking auricles; spikelets borne in open panicles, 3–6-flowered, articulate above the glumes; glumes unequal, the first 3-nerved, the second 5-nerved, awnless; palea shorter than the lemma; lemmas with 7–more convergent nerves, awned from a bifid apex; callus bearded; lodicules 2; stamens 3.

Schizachne purpurascens (Torr.) Swallen
False Melic

Plants tufted, mostly 6–10 dm tall; ligules mostly less than 1 mm long, glabrous; leaf blades flat or more or less involute, 2–5 mm broad, scabrous; panicles 8–14 cm long, the branches ascending to pendulous; spikelets 10–18 mm long (excluding awns), purplish; glumes acute, the first 4.5–6.5 mm long, the second 6–8.5 mm long; lemmas mostly 7.5–10 mm long, deeply bifid, awned, the curved awn to 10 mm long or more; callus strongly bearded; anthers 1.5–2 mm long.

Woods and slopes; in south-central Alaska and southern Yukon; eastward to eastern Canada and the northeastern states and southward to Mexico (*Trisetum purpurascens* Torr.). This is a most handsome grass.

SCOLOCHLOA Link. Nom. Cons.

Perennial rhizomatous herbs; culms hollow; leaf sheaths open; ligules membranous; leaf blades flat to more or less involute, lacking auricles; spikelets borne in open panicles, 3–4-flowered, articulate above the glumes; glumes unequal, the first 3-nerved, the second 5-nerved, awnless; palea about equaling the lemma; lemmas 7-nerved, the nerves convergent, unawned; callus strongly bearded; lodicules 2; stamens 3.

Scolochloa festucacea (Willd.) Link.
Sprangletop

Plants mostly 8–15 dm tall; ligules 2–6 mm long, lacerate; leaf blades flat to more or less involute, 4–10 mm broad; panicles 12–25 cm long, the branches ascending; spikelets compressed, mostly 7–10 mm long; glumes about 4–8 mm long; lemmas about 6 mm long; anthers 3–4 mm long.

Marshes, lake and stream margins; in Alaska known from central eastern portion along the Yukon River; from British Columbia to Manitoba and south to Oregon, Nebraska, and Iowa; Eurasia (*Arundo festucacea* Willd.).

SECALE L.

Annual tufted herbs; culms hollow; leaf sheaths open; ligules membranous; leaf blades flat, auriculate; spikelets borne sessile in terminal spikes, 1 per node, 2-flowered, borne flatwise on the rachis, this ultimately shattering; glumes subulate, 1-nerved; lemmas 5-nerved, curved, long-awned; lodicules 2; stamens 3.

Secale cereale L.
Rye

Plants 6–15 dm tall; ligules less than 1 mm long; leaf blades flat, 3–10 mm wide or more, scabrous; spikes 8–15 cm long, usually nodding; glumes shorter than the lemmas; lemmas curved, ciliate (strongly so along one side), awned, the awns mostly 1.5–7 cm long; anthers mostly 5–10 mm long.

Introduced in southeastern Alaska and to be expected elsewhere; widespread in temperate and subtropical regions of the earth. This is the cultivated rye of commerce.

SPHENOPHOLIS Scribn.

Perennial (or sometimes annual?) tufted herbs; culms hollow; leaf sheaths open; ligules membranous; leaf blades flat, lacking auricles; spikelets borne in slender panicles, usually 2-flowered, articulate below the glumes; glumes keeled, the first 1-nerved and narrow, the second plainly 3(5)-nerved and much broader, awnless; palea somewhat shorter than the lemma; lemmas obscurely 3-nerved, awnless; lodicules 2; stamens 3.

Sphenopholis intermedia (Rydb.) Rydb.
Wedgegrass

Plants tufted, mostly 2–8 dm tall; ligules 1–2.5 mm long, minutely ciliate and irregularly toothed, glabrous or puberulent; leaf blades flat, mostly 2–5 mm broad, scabrous; panicles mostly 7–15(20) cm long, slender, the branches erect or nearly so; spikelets 3–5 mm long; glumes unequal, the first 1.5–2.5 mm long, the second 2–2.5 mm long; lemmas mostly 2.5–3 mm long; anthers 0.5–0.7 mm long.

Meadows and pond and stream banks; in central Alaska (Manley Hot Springs) and northern British Columbia (Liard Hot Springs); eastward to Newfoundland and in nearly all of the contiguous states (*Eatonia intermedia* Rydb.; *Koeleria truncata* var. *major* Torr.; *Sphenopholis obtusata* var. *major* [Torr.] Erdman). This plant has not been collected in Alaska for many years and it might not now occur there (Tanana R., A. S. Hitchcock 4508, 28–29 July 1909, US).

STIPA L.

Perennial tufted herbs; culms hollow; leaf sheaths open; ligules membranous; leaf blades commonly involute, lacking auricles; spikelets borne in open to contracted panicles, 1-flowered, articulate above the glumes; glumes subequal, membranous, 3 (5)-nerved, acute to attenuate, awnless; lemma usually 5-nerved, narrow, firm or hardened, strongly convolute, awned, the prominent awn geniculate and twisted, the junction of body and awn evident; callus bearded; palea enclosed by the lemma and shorter than it; lodicules 2–3; stamens 3.

1a. Glumes 15–20 mm long or more; lemmas more than 8 mm long, with awns mostly over 10 cm long. S. *comata*
1b. Glumes less than 15 mm long; lemmas less than 8 mm long, with awns mostly less than 5 cm long. (2)

2a. Panicle open, with spreading branches, these bearing spikelets near the tip only; awns usually less than 25 mm long. S. *richardsonii*
2b. Panicle narrow, with erect or ascending branches, these usually bearing spikelets from near the base; awns commonly more than 25 mm long. S. *occidentalis*

Stipa comata Trin. & Rupr.
Needlegrass

Plants tufted, mostly 5–8(10) dm tall; ligules 1.5–5 mm long, puberulent; leaf blades involute, mostly 1–3 mm broad; panicles contracted, 7–20 cm long or more, the branches erect or nearly so, bearing spikelets from below the middle; glumes very long and slender, mostly 15–20 mm long or more; lemmas firm, the body 8–12 mm long, pubescent, the callus 3–4 mm long, the awn mostly 10–15 cm long, twisted and geniculate.

Roadsides; in the upper Yukon River Valley in southern Yukon, where apparently introduced; from British Columbia to Ontario and south to California, Arizona, Texas, and Indiana.

Stipa occidentalis Thurb. ex Wats.

Plants tufted, 4.5–10 dm tall; ligules 0.5–1 mm long, more or less puberulent; leaf blades involute, usually less than 1 mm broad; panicles slender, 8–21 cm long, the branches erect or nearly so, bearing spikelets from near the base; glumes mostly 9–11 mm long; lemmas firm, the body 5–8 mm long, pubescent, the callus 0.7–1.5 mm long, the awn mostly (20)25–35 mm long, twisted and geniculate.

Steep south-facing open slopes; in southern Yukon and adjacent northern British Columbia; widespread in western North America (*S. columbiana* Macoun; *S. minor* Vasey). Our materials are assignable to var. *minor* (Vasey) C. L. Hitchc.

Stipa richardsonii Link.

Plants tufted, 5–8(10) dm tall; ligules very short, mostly less than 0.5 mm long, glabrous or nearly so; leaf blades involute, usually less than 1 mm broad; panicles open, 7–15(20) cm long, the branches spreading, bearing spikelets near the tips only; glumes 7–10 mm long; lemmas firm, the body 5–6 mm long, pubescent, the callus 0.5–1 mm long, the awn 15–25 mm long, twisted and geniculate.

Steep south-facing slopes and bluffs; in southern Yukon; widespread in western North America.

TORREYOCHLOA Church

Perennial rhizomatous or stoloniferous herbs; culms hollow; leaf sheaths open; ligules membranous; auricles lacking; leaf blades flat; spikelets borne in rather loose panicles, articulate above the glumes, mostly 3–7-flowered; glumes more or less unequal, 1-nerved, shorter than the lowermost lemma; lemmas apparently 5-nerved or plainly 7-nerved, the nerves more or less parallel, not convergent at the apex, awnless; callus not hairy; lodicules 2; stamens commonly 3.

Torreyochloa pauciflora (Presl) Church
Weak Mannagrass

Plants rhizomatous, often rooting at the lower nodes, 2.5–13(14) dm tall; leaf sheaths smooth or more or less retrorsely scabrous, open; ligules 3–9 mm long; leaf blades flat, 3–15 mm broad, scabrous on one or both sides; panicle mostly 10–23 cm long, rather loose, the branches ascending to spreading or drooping; spikelets oblong to ovate in outline, compressed, 4.7–8 mm long, 3–7-flowered; first glume 0.8–1.3 mm long, the second 1.2–1.8 mm long; lemmas 5-nerved or plainly 7-nerved, the nerves often conspicuously scabrous.

Woods, thickets, marshes, and meadows; in coastal and insular south-central and southeastern Alaska and northern British Columbia; southward to California and New Mexico (*Glyceria pauciflora* Presl; *Panicularia pauciflora* [Presl] Kuntze; *Puccinellia pauciflora* [Presl] Munz). The assemblage of generic names within which this entity has been placed is an indication of the combination of characteristics which it possesses. The plant has numerous features of *Glyceria*, but has open sheaths and is placed with *Puccinellia* by some authors. *T. pauciflora* is distinguished from this latter genus by its usually much longer ligules, more prominent nerves of the lemma, and usually broader leaves.

TRISETUM Pers.

Perennial tufted herbs; culms hollow; leaf sheaths open; ligules membranous; leaf blades flat or more or less involute, lacking auricles; spikelets borne in spikelike to open panicles, usually 2–3-flowered, articulate above the glumes; glumes subequal, keeled, the first 1-nerved, the second 3-nerved, awnless; lemmas 5-nerved, keeled, bifid at the apex, awned, the awn inserted above the middle, geniculate; palea about as long as the lemma; callus hairy; lodicules 2; stamens 3.

1a. Panicles dense, spikelike, erect or nearly so; stems usually hairy below the panicle; our most common and most widespread *Trisetum*. *T. spicatum*

1b. Panicles more or less open to distinctly open, usually drooping or more or less nodding; stems glabrous below the panicle; plants of limited occurrence. (2)

2a. Panicles open, loose, drooping; ligules 1.5–3.7 mm long; plants of coastal south-central and southeastern Alaska. *T. cernuum*

2b. Panicles loose but not especially open, more or less nodding; ligules less than 1.5 mm long; plants of interior and coastal western Alaska. *T. sibiricum*

Trisetum cernuum Trin.
Nodding Oatgrass

Plants tufted, often decumbent and more or less stoloniferous at the base, mostly 3.5–10 dm tall; ligules 1.5–3.7 mm long; leaf blades flat, (3)4–8(10) mm broad, scabrous; panicles nodding, open, 10–25 cm long or more, the slender branches ascending to spreading, ultimately drooping, bearing spikelets towards the tip; spikelets 7–9 mm long (excluding awns), 2–3-flowered; glumes unequal, the first slender and 1.5–3.7 mm long, the second broader and more or less erose towards the tip, 2.7–4.8 mm long; lemmas 5.0–6.1 mm long; callus hairs about 1 mm long; awns mostly 7–12 mm long.

Woods, thickets, and marshes; in coastal south-central and southeastern Alaska from Prince William Sound eastward; southward to California, Idaho, and Montana (*Avena cernua* Kunth; *Avena nutkaensis* Presl; *T. nutkaense* [Presl] Scribn. & Merr. ex Davey).

Trisetum sibiricum Rupr.
Siberian Oatgrass

Plants tufted and apparently shortly rhizomatous, mostly 1.5–5(8) dm tall; ligules 1–1.5 mm long; leaf blades flat, 2–4 mm broad, scabrous; panicles more or less nodding, compact, finally somewhat open, the branches ascending to spreading, bearing spikelets from near the base; spikelets 5–9 mm long (excluding awns), (1)2–3-flowered; glumes unequal, the first slender and 2.5–3.6 mm long, the second broader and usually not erose near the tip, 4.5–5.5 mm long; lemmas 5–6 mm long; callus hairs 1–1.5 mm long; awns mostly 5–8 mm long.

Bluffs and marshes; in central eastern, northeastern, and west-central Alaska and western and northern Yukon; Asia (*T. sibiricum* var. *littorale* Rupr.; *T. sibiricum* ssp. *littorale* [Rupr.] Roshev.). Plants from Alaska and Yukon have been designated as variety or subspecies *littorale*, but the distinction is not clear since plants to 8 dm tall occur. Possibly the dwarf phases are only ecological races.

Trisetum spicatum (L.) Richter
Downy Oatgrass

Plants tufted, mostly 0.5–7 dm tall; ligules 0.8–1.5 mm long, ciliate; leaf blades flat to folded, 1–5 mm broad, scabrous; panicles usually erect, compact, spikelike, mostly 1–13 cm long, the branches very short, bearing spikelets from the base; spikelets 5–7 mm long, 2–3-flowered; glumes unequal, the first 3.8–5 mm long, the second 4.2–6 mm long, not erose near the tip; lemmas 4.2–5.7 mm long; callus hairs about 0.5 mm long; awns mostly 2–6 mm long.

Spits, tundra, heathlands, rock outcrops, bars, snow flushes, river banks, roadsides, landing strips, ridge tops, moraines, and open woods; in most of Alaska and Yukon; circumboreal; South America (*Aira spicata* L.; *Aira subspicata* L.; *Avena airoides* Koel.; *Avena mollis* Michx. not Salisb.; *T. molle* [Michx.] Kunth; *T. spicatum* var. *molle* [Michx.] Piper; *T. triflorum* Löve & Löve; *T. alaskanum* Nash; *T. spicatum*

Trisetum spicatum (L.) Richter (× 0.4).

var. *alaskanum* [Nash] Malte ex Louis-Marie; *T. spicatum* ssp. *alaskanum* [Nash] Hultén; *T. majus* Rydb.; *T. spicatum* ssp. *majus* [Rydb.] Hultén). This species has been segregated into a great many intricately critical subspecies (e.g. Hultén, E. 1959, Svensk. Bot. Tidskr. 53:203–28), of which four subspecies are reported from Alaska and Yukon. Mostly the diagnostic features are involved with variation in pubescence, relative length of spikelet parts, and differences in spike characteristics. None of these, taken either singly or in combination, appears to be definitive. Therefore it seems best to recognize our materials as portions of a polymorphic species.

TRITICUM L.

Annual tufted herbs; culms hollow; ligules membranous; leaf sheaths open; leaf blades flat, with well-developed auricles; spikelets borne in a terminal spike, flatwise to the rachis, 2–5-flowered, 1 per node, sessile, articulate above the glumes; glumes keeled, 3 (or more)-nerved, acute to awned; lemmas broad, keeled, several-nerved, these not convergent, acute to awned; lodicules 2; stamens 3.

Triticum aestivum L.
Wheat

Plants annual, mostly 8–15 dm tall; ligules to about 1 mm long; leaf blades flat, mostly 5–15 mm broad, smooth; spikes 5–12 cm long or more, awned or awnless; lemmas asymmetric, keeled, the keel smooth or scabrous.

Cultivated and escaping, not persisting, but constant reintroductions in feeds should allow the plant to continue as a waif in our flora; widespread in temperate regions of the earth.

IRIDACEAE
Iris Family

Perennial herbs from short rhizomes; leaves narrow, equitant; flowers regular or nearly so, perfect, subtended by spathaceous bracts; perianth of 6 segments in 2 series of 3 each; stamens 3, opposite the sepals; pistils 1; ovary inferior, 3-loculed; style 3-cleft; fruit a loculicidal capsule.

1a. Style branches large and petaloid; flowers large, usually over 5 cm wide; sepals and petals unlike. *Iris*

1b. Style branches not petaloid; flowers less than 2 cm broad; sepals and petals alike. *Sisyrinchium*

IRIS L.

Perennial herbs from short thick rhizomes; leaves sword shaped, large; flowers large and showy, in few-flowered terminal clusters; perianth bluish or purplish (rarely white), the sepals (falls) spreading or recurved, the petals (standards) erect or ascending, the tube prolonged beyond the ovary; styles 3, petaloid, arching over the 3 stamens; ovary pedicellate from within the involucre; capsule large, ellipsoid, many-seeded.

Iris setosa Pallas
Wild Iris, Flag

Plants densely tufted, (3)3.5–7 dm tall; basal leaves linear-lanceolate, 20–50 cm long, 0.5–1.5 cm wide; stem leaves 2–3, the stem usually once-branched; sepals bluish to purplish, 5–6 cm long, veined, the petals 2–4 cm long; style branches large, crested; capsules 3–4 cm long.

Floodplains, tidal flats, meadows, and lake shores; in much of Alaska south of the Brooks Range and in west-central Yukon; eastward to Newfoundland; Asia. Two poorly differentiated varieties are recognizable within our range.

1a. Bracts scarious; leaves less than 10 mm broad; plants of interior Alaska and Yukon (*I. setosa* ssp. *interior* [E. Anders.] Hultén). *I. setosa* var. *interior* E. Anders.

1b. Bracts slightly if at all scarious; leaves commonly 10–15 mm broad; plants of coastal and insular Alaska

Iris setosa Pallas (× 0.25).

(*I. arctica* Eastw.; *I. setosa* var. *platyrhyncha* Hultén). *I. setosa* var. *setosa*

SISYRINCHIUM L.

Grasslike, tufted, shortly rhizomatous perennials; flowers blue, borne in few-flowered umbels, subtended by paired, erect green bracts; perianth tube short or lacking; sepals and petals similar, spreading; style branches small, not covering the stamens; capsule globose, short-pedicellate.

Sisyrinchium angustifolium Mill.
Blue-eyed Grass

Plants 1–4 dm tall, the stems broadly to narrowly winged; leaves 4–25 cm long, commonly 1–5 mm broad; bracts greenish or purplish tinged; umbels 3–6-flowered; perianth blue to violet, 8–12(15) mm long; capsules subglobose to obovoid, 6–12 mm long.

Open grassy slopes, meadows, and tidal flats; in coastal and insular southern Alaska from the Aleutians eastward to the Panhandle and in southwestern Yukon; southward to Baja California and New Mexico and eastward across Canada (*S. littorale* Greene; *S. montanum* Greene).

JUNCACEAE
Rush Family

Annual or mostly perennial grasslike herbs; leaves alternate or basal, sheathing; flowers small, perfect, regular; sepals and petals each 3, similar, scalelike; stamens 3–6; pistils 1, ovary superior, 1–3-loculed; stigmas 3; fruit a capsule with 3–many seeds.

1a. Leaf sheaths open; capsule 1–3-loculed, with many seeds; plants glabrous. *Juncus*
1b. Leaf sheaths closed; capsule 1-loculed, with 3 seeds; plants glabrous or pubescent. *Luzula*

JUNCUS L.

Annual or perennial rhizomatous or caespitose herbs; stems round or flattened, solid or less commonly hollow; leaves 3-ranked, the sheaths open, the blades round or flattened, often septate within; flowers (1) few to many, borne singly or few to several in capitate clusters; perianth of six parts in two whorls; stamens 6 or 3 (rarely 1 or 2); pistils 1–3-loculed; seeds numerous.

Herman, F. J. 1964. The *Juncus mertensianus* complex in western North America. Leafl. West. Bot. 10:81–87.

1a. Plants annual; leaves less than 1 mm broad; rhizomes lacking. *J. bufonius*
1b. Plants perennial; leaves often more

Sisyrinchium angustifolium Mill. (× 0.4).

than 1 mm broad; rhizomes often present. (2)

2a. Lowermost bract of inflorescence terete and erect, seemingly representing a continuation of the stem axis, the inflorescence thus apparently lateral. (3)
2b. Lowermost bract of inflorescence flattened or variously channeled, not terete, erect to spreading, the inflorescence obviously terminal. (6)

3a. Lowermost bract 1–3 cm long; flowers 1–4 per stem. *J. drummondii*
3b. Lowermost bract usually over 5 cm

long; flowers (4)5–many per stem. (4)

4a. Perianth 4–6.5 mm long; stems 0.9–2.7 mm broad below inflorescence. *J. arcticus*

4b. Perianth 2–3.5(4) mm long; stems various, but if more than 0.9 mm broad then capsules sharply angled apically. (5)

5a. Involucral bract rarely half as long as the stem; stems 0.9–2 mm broad below the inflorescence; capsules sharply angled above. *J. effusus*

5b. Involucral bracts commonly more than half as long as the stem; stems 0.5–0.9 mm broad below the inflorescence. *J. filiformis*

6a. Plants scapose, the leaves all basal; heads solitary, 1–3(5)-flowered. (7)

6b. Plants with leafy stems, the leaves not all basal; heads more than one or with more than 5 flowers. (8)

7a. Involucral bract subequal to the head; capsule acute to obtuse apically; heads commonly 3-flowered. *J. triglumis*

7b. Involucral bract usually surpassing the head; capsule retuse apically; heads commonly 2-flowered. *J. biglumis*

8a. Leaf blades laterally flattened with one edge towards the stem, the leaves equitant; plants of coastal and insular southern Alaska. *J. ensifolius*

8b. Leaf blades terete or dorsiventrally compressed, the leaves not equitant; plants of various distribution. (9)

9a. Heads 4–more or the flowers loosely clustered; perianth 2.5–3.5 mm long (longer in some *J. supinformis*). (10)

9b. Heads 1–3; perianth 3.5–5 mm long or more (shorter in some *J. nodosus*). (14)

10a. Heads spherical; capsules attenuate apically. *J. nodosus*

10b. Heads hemispherical; capsules abruptly short beaked apically, not attenuate. (11)

11a. Flowers loosely arranged, not densely clustered in tight heads; perianth surpassing the capsule. *J. tenuis*

11b. Flowers densely aggregated into definite heads; perianth shorter than the capsule. (12)

12a. Stems often decumbent and rooting at the nodes, often with leaves proliferating from the heads; plants of coastal south-central and southeastern Alaska. *J. supiniformis*

12b. Stems erect or ascending, not proliferating leaves from the heads; plants with various distribution. (13)

13a. Outer perianth segments acute, the inner ones obtuse; plants of broad distribution. *J. alpinus*

13b. Outer perianth segments and the inner ones both acute; plants of southeastern Alaska. *J. articulatus*

14a. Capsules 5–8 mm long; seeds 2.5–3.5 mm long or more, appendaged at both ends; bracts, perianth, and capsules all brown. *J. castaneus*

14b. Capsules less than 5 mm long; seeds less than 2 mm long, not appendaged; bracts, perianth, and capsules various but sometimes all brown. (15)

15a. Heads commonly 2–3, spherical; capsules attenuate apically. *J. nodosus*

15b. Heads commonly solitary (rarely 2 or more), hemispherical; capsules abruptly beaked apically. (16)

16a. Heads 1–3(5)-flowered; leaves less than 1 mm broad. *J. stygius*

16b. Heads more than 5-flowered; leaves 1–3 mm broad (at least some). (17)

17a. Leaves flat, not septate; perianth segments 4–6 mm long. *J. falcatus*

17b. Leaves semiterete, septate; perianth segments 3–4 mm long. *J. mertensianus*

Juncus alpinus Vill.
Alpine Rush

Plants perennial from short rhizomes, the culms tufted, 1–4.5 dm tall, terete or nearly so; leaves 1–3 per stem, the blades terete; inflorescence terminal, much longer than the lower involucral bract, the branches ascending; heads commonly 5–many, 3–10-flowered, the flowers sessile or nearly so; perianth purplish brown to straw-colored, 2–2.5 mm long, the inner segments shorter and rounded, the outer acute; stamens 6, the anthers shorter than the filaments; capsule ovoid, subequal to the perianth; seeds minutely appendaged at each end.

Sandy banks and shores and moist sites near lakes, streams, and ponds, less commonly in shallow ponds; in most of Alaska and Yukon south of the 66th parallel, except for the Aleutians and the southern part of the Alexander Archipelago; eastward to the Atlantic and south to Idaho, Utah, Colorado, and Pennsylvania; Eurasia (*J. nodulosus* Wahl.; *J. alpinus* ssp. *nodulosus* [Wahl.] Hultén; *J. affinis* R. Br. ex Richards.; *J. richardsonianus* Schult.).

Juncus arcticus Willd.
Arctic Rush

Plants perennial from strong horizontal rhizomes, the culms arising along the rhizome in comblike fashion, 1–10 dm tall, terete or somewhat flattened; leaves all basal, commonly reduced to bladeless sheaths; inflorescence apparently lateral, the involucral bract erect, terete and sharply pointed, 2.5–20(25) cm long, the flowers congested or borne in a loose cluster; perianth 4–6.5 mm long, brownish to dark brown, the outer ones differing from the inner in size and often in texture; stamens 6, the anthers as long as the filaments or much longer; capsule more or less ovoid, about as long as the inner perianth segments; seeds minutely appendaged.

Tidal marshes, lake and pond shores, stream banks, and wet meadows; in most

Juncus arcticus Willd. (× 0.4).

of Alaska and Yukon; in most of temperate and arctic North America; circumboreal. Three more or less distinctive but apparently intergrading phases are recognizable among our materials. The following arbitrary key will allow for segregation of most of our materials.

1a. Plants coarse, the involucral bracts often more than 10 cm long; staminal filament about as long as the anther; plants of coastal and insular southern and southwestern Alaska, from the Aleutians eastward through the Panhandle (*J. balticus* Willd; *J. balticus* var. *littoralis* Engelm.; *J. arcticus* ssp. *littoralis* [Engelm.] Hultén; *J. balticus* var. *haenkei* [E. Meyer] Buch.; *J. balticus* ssp. *sitchensis* [Engelm.] Hultén; *J. arcticus* ssp. *sitchensis* Engelm.). *J. arcticus* var. *balticus* (Willd.) Trautv.

1b. Plants slender, the involucral bracts often less than 10 cm long; distribution various, but if coastal, then the bracts much less than 10 cm long or the staminal filament much shorter than the anther. (2)

2a. Inflorescence loosely flowered; staminal filament much shorter than the anther; plants of the southeastern third of Alaska and southern Yukon (*J. ater* Rydb.; *J. arcticus* ssp. *ater* [Rydb.] Hultén). *J. arcticus* var. *montanus* (Engelm.) Welsh comb. nov. (based on: *J. balticus* var. *montanus* Engelm. Trans. Acad. Sci. St. Louis 2:442. 1866).

2b. Inflorescence compactly flowered; staminal filament about as long as the anther; plants in most of continental and northwestern coastal Alaska and in the Yukon (*J. arcticus* ssp. *alaskanus* Hultén; *J. balticus* var. *alaskanus* [Hultén] Porsild). *J. arcticus* var. *alaskanus* (Hultén) Welsh comb. nov. (based on: *J. arcticus* ssp. *alaskanus* Hultén Lunds Univ. Arsskr. N. F. Avd. 2. 39:418. 1943).

Juncus articulatus L.

Plants perennial from short rhizomes, the culms tufted, 1–5 dm tall, terete or nearly so; leaves 1–3 per stem, the blades terete; inflorescence terminal, much longer than the lower involucral bract, the branches commonly spreading; heads 5–many, 6–12-flowered; perianth brown to straw-colored, 2.5–3 mm long, the segments all acute to acuminate; stamens 6, the anthers shorter than the filaments; capsule ovoid, subequal to or longer than the perianth; seeds minutely appendaged at each end.

Margins of ponds and lakes; in southeastern Alaska, where evidently rare; eastward to Newfoundland and south to California, Colorado, and New York; Eurasia.

Juncus biglumis L.
Two-flowered Rush

Plants perennial, tufted, from a short rhizome, the culms 0.2–3 dm tall, channeled along one side; leaves 1–4 per culm, all basal or nearly so, the blades terete; inflorescence terminal, commonly slightly surpassed by the lowermost bract, sessile; heads solitary, 2(3 or 4)-flowered; perianth purplish brown, 2.3–4.5 mm long, the segments acute to obtuse; stamens 6, the anthers shorter than the filaments; capsules obovoid, retuse apically, 3–5.5 mm long, much longer than the perianth; seeds minutely appendaged at each end.

Sphagnum mats, pond and lake margins, stream shores, and in shallow water; in most of Alaska and Yukon; east to the Atlantic and south in the mountains to British Columbia; circumpolar.

Juncus bufonius L.
Toad Rush

Plants annual, the stems solitary or tufted, from fibrous roots, the culms 2–3.5 dm tall, channeled, simple or more commonly branched; leaves 1–several per culm, basal and cauline, the blades channeled, narrowly linear-subulate; inflorescence terminal, commonly much branched; flowers solitary, subtended by scarious bracts; perianth greenish or scarious, 2.5–8.5 mm long, the segments lance-acuminate, with greenish midribs and broad hyaline margins; stamens usually 6, the anthers shorter than the filaments; capsules oblong-ellipsoid, obtuse apically, usually shorter than the perianth; seeds not appendaged.

Pond and stream margins, tidal flats, and wet places generally; in the southeastern two-thirds of Alaska and southern Yukon; widespread in North America; Eurasia and elsewhere (*J. ranarius* Perr. & Song.; *J. bufonius* var. *ranarius* [Perr. & Song.] Hayek).

Juncus castaneus J. E. Smith
Chestnut Rush

Plants perennial from elongate rhizomes, the culms solitary, 0.8–6.7 dm tall, terete or nearly so; leaves 3–5 per culm, cauline and basal, the blades channeled; inflorescence terminal, commonly surpassed by the lowermost bract; heads 1–3(5), (1)2–7-flowered; perianth purplish brown to brown, 2.9–4.8 mm long, the segments acute to acuminate; stamens 6, the anthers shorter than the filaments; capsules oblong-ellipsoid, 5–8 mm long, acute to obtuse and beaked apically, longer than the perianth; seeds appendaged at each end.

Moist meadows and bogs, in woods, heathlands, and tundra; in most of Alaska and Yukon; east to the Atlantic and south to New Mexico; circumpolar (*J. leucochlamys* Zinz. ex Krecz; *J. castaneus* ssp. *leucochlamys* [Zinz.] Hultén).

Juncus drummondii E. Meyer in Ledeb.
Drummond Rush

Plants perennial from short rhizomes, the culms tufted, 0.5–4.5 dm tall, terete or nearly so; leaves all basal, reduced to sheaths or the upper with bristlelike blades; inflorescence apparently lateral, commonly equaled or surpassed by the terete lowermost bract, the flowers solitary, usually 1–3 per inflorescence; perianth brownish, 4.8–6.5 mm long, the segments hyaline, acute to acuminate; stamens 6, the anthers sub-

equal to or longer than the filaments; capsules oblong-ellipsoid, 4–7 mm long, truncate to retuse apically, subequal to or longer than the perianth; seeds appendaged at each end.

Moist meadows, in open woods, heath, and alpine tundra; in south-central, southwestern, and southeastern Alaska and southern Yukon; south to California and New Mexico.

Juncus effusus L.
Bog Rush

Plants perennial, rhizomatous, the culms solitary or few, 4.5–5 dm tall, terete; leaves all basal, reduced to sheaths or the uppermost with bristlelike blades; inflorescence apparently lateral, compact to open, many-flowered, the lowermost bract 8–27 cm long; perianth greenish, 2–3.2 mm long, the segments lance-attenuate, with greenish or straw-colored broad midribs and narrow hyaline margins; stamens 3, the anthers subequal to the filaments; capsules ovoid to obovoid, 1.8–2.2 mm long, obtuse apically, usually shorter than the perianth; seeds not appendaged.

Marshy sites, pond margins, and muskegs; in coastal and insular southeastern Alaska; south to California and disjunctly eastward to the Atlantic.

Juncus ensifolius Wikstr.

Plants perennial, rhizomatous, the culms solitary or few, 1.5–6 dm tall, flattened; leaves cauline, ensiform (laterally flattened and equitant), the blades well developed; inflorescence terminal, the lowermost bract 0.5–10 cm long, shorter to longer than the inflorescence; heads (1)2–7 or the inflorescence compound and with numerous small heads, each few–many-flowered; perianth brown, (2)2.8–3.7 mm long, the segments lance-oblong to -attenuate, acute to attenuate; stamens 3, the anthers shorter than the filaments; capsules oblong, subequal to the perianth or shorter, obtuse apically; seeds not appendaged.

Marshy places, pond margins, and muskegs, in woods, heath, and alpine tundra; in coastal and insular southwestern, south-central, and southeastern Alaska; south to California and east to Utah.

Juncus falcatus E. Meyer

Plants perennial, rhizomatous, the culms solitary or few, 1–3 dm tall, terete or nearly so; leaves basal (few to several) and cauline (solitary), the blades all well developed; inflorescence terminal, the lowermost bract 0.5–5 cm long, flat; heads solitary or sometimes 2–3, 5–many-flowered; perianth brown, 3.5–6 mm long, the segments lanceolate, attenuate or acute apically; stamens 6, the anthers from shorter to longer than the filaments; capsules obovoid, 3.5–5 mm long, shorter than the perianth; seeds not appendaged.

Coastal marshes and tidal flats; in insular and coastal southwestern and southeastern Alaska; south to California; Asia (*J. falcatus* var. *alaskensis* Coville; *J. falcatus* ssp. *sitchensis* [Buch.] Hultén). Our material belongs to var. *sitchensis* Buch. in Engler.

Juncus filiformis L.

Plants perennial, rhizomatous, the culms more or less tufted, 0.7–4 dm tall, terete; leaves all basal, reduced to sheaths or the upper with bristlelike blades; inflorescence apparently lateral, the lowermost bract 3–20 cm long or more, often longer than the stem, terete; flowers few to several in a more or less open cluster; perianth greenish to brownish, (2.5)2.8–4.3 mm long, the segments lance-attenuate with broad hyaline margins; stamens 6, the anthers shorter than the filaments; capsules ovoid to obovoid, subequal to the perianth, obtuse apically; seeds not appendaged.

Wet sites, in meadows, heath, and woods; in much of southern Alaska south of the 66th parallel (except for the Aleutians) and southern Yukon; east to the Atlantic and south to Oregon, Utah, Wyoming, Michigan, and Pennsylvania; circumboreal.

Juncus mertensianus Bong.
Mertens Rush

Plants perennial, rhizomatous, the culms tufted, 0.5–3(4) dm tall, terete; leaves cauline (1–4) and basal (few to several), the blades well developed, half round; inflorescence terminal, the lowermost bract 0.8–3.5 cm long, shorter to longer than the head; heads solitary, the flowers several to many; perianth dark brown, 3–4.2 mm long, the segments lance-attenuate; stamens 6, the anthers shorter than the filaments; capsules obovoid, subequal to the perianth, obtuse apically; seeds apiculate at each end.

Bogs, muskegs, stream banks, and pond margins, in woods, heath, and alpine tundra; in interior south-central Alaska and southern Yukon and throughout coastal and insular southern Alaska; south to California and New Mexico.

Juncus nodosus L.

Plants perennial, rhizomatous, the culms arising singly, 1.5–5 dm tall, terete; leaves cauline (1–3) and basal (few), the blades well developed (half rounded); inflorescence terminal, usually exceeded by the lowermost bract; heads 3–many, many-flowered; perianth greenish or brownish, 3–4 mm long, the segments lance-subulate; stamens 6, the anthers subequal to or shorter than the filaments; capsule lance-ovoid, longer than the perianth; seeds minutely apiculate at each end.

Marshes; in interior central and coastal southeastern Alaska; eastward to the Atlantic and south to California, New Mexico, Nebraska, Ohio, and Pennsylvania.

Juncus stygius L.

Plants perennial, not rhizomatous, the culms tufted, 0.6–2(3) dm tall, filiform, terete; leaves 1–3 per stem, the blades filiform, terete; inflorescence terminal, commonly shorter than the lower involucral bract, unbranched; heads 1 or 2(4), 1–4-flowered; perianth pale with red midribs and hyaline margins, 3–4 mm long, the segments obtusish to acute; stamens 6, the anthers much shorter than the filaments; capsule ovoid, longer than the perianth; seeds appendaged at each end.

Bogs, mossy sites, and shallow pools; in south-central and southeastern Alaska; disjunctly eastward to Newfoundland; circumboreal (*J. stygius* ssp. *americanus* [Buch.] Hultén). Our material belongs to var. *americanus* Buch.

Juncus supiniformis Engelm.

Plants perennial, the stems generally tufted along slender rhizomes, 1–3 dm tall; leaves 2–4 per stem, the blades half round; inflorescence terminal, usually surpassing the lowermost involucral bract, commonly branched; heads 2–6, 3–several-flowered; perianth pale brown to dark brown, 3–4.5 mm long, the segments acute to acuminate; stamens 3 or 6, the anthers shorter than the filaments; capsule cylindrical, longer than the perianth; seeds shortly appendaged at each end.

Swamps, bogs, and ponds; in coastal and insular south-central and southeastern Alaska; south to California (*J. oreganus* Wats.).

Juncus tenuis Willd.
Slender Rush

Plants perennial, tufted, 1.5–5 dm tall, the culms terete or nearly so; leaves 1–3 per stem, basal or nearly so, the blades flattened; inflorescence terminal, commonly exceeded by the lowermost involucral bract, compact to open; heads 3–many, many-flowered; perianth greenish brown to tan, 4–5 mm long, the segments acuminate; stamens 6, the anthers shorter than the filaments; capsules ovoid, about equaling the perianth, usually retuse apically; seeds minutely appendaged at each end.

Wet soil along roads and in openings; in east-central and southeastern Alaska and southern Yukon; eastward to the Atlantic and south to Mexico; Eurasia, Australia, South America (*J. macer* S. F. Gray; *J. dudleyi* Wieg.).

Juncus triglumis L.

Plants perennial; culms densely tufted, 0.5–1(2) dm tall, terete or nearly so; leaves 1–3 per stem, basal or nearly so, the blades filiform, terete; inflorescence terminal, commonly about equaling the lower bract; heads solitary, 1–3(5)-flowered; perianth pale brown or whitish, 4–5 mm long, the segments obtuse; stamens 6, the anthers much shorter than the filaments; capsules cylindroid, subequal to or longer than the perianth, acute apically; seeds appendaged at each end.

Moist sites in arctic and alpine tundra and heathlands; in most of Alaska and Yukon; east to Newfoundland and south to Utah and Colorado; Greenland, Eurasia. Our materials are separable into two more or less distinctive varieties.

1a. Capsules subequal to the perianth; seeds mostly less than 1.8 mm long; plants of broad distribution (*J. triglumis* ssp. *albescens* [Lange] Hultén). *J. triglumis* var. *albescens* Lange

1b. Capsules longer than the perianth; seeds commonly more than 2 mm long; plants of west-central and northern Alaska. *J. triglumis* var. *triglumis*

LUZULA DC.

Perennial rhizomatous or caespitose herbs; stems terete or nearly so; leaves 3-ranked, the sheaths closed, the blades flat; flowers in spicate, umbellate, or paniculate clusters; perianth 6-parted, in 2 whorls; stamens 6; pistils 1-loculed, 3-seeded.

Hamet-Ahti, L. 1965. *Luzula piperi* (Cov.) M. E. Jones, an overlooked woodrush in western North America and eastern Asia. Aquilo, Ser. Botanica 3:11–21.

1a. Inflorescence with solitary (rarely 2–3) flowers terminating the ultimate branchlets; pedicels commonly more than 5 mm long (at least some). (2)

1b. Inflorescence with dense spicate or capitate clusters of flowers terminating the ultimate branchlets; pedicels commonly less than 1 mm long. (4)

2a. Inflorescence commonly simple; pedicels subumbellately arranged, arising from near a common point but of differing lengths. *L. rufescens*

2b. Inflorescence commonly few–many-branched; pedicels subumbellate or spicate to racemose. (3)

3a. Cauline leaves usually 2–3, rarely more than 3 mm broad; bracts and bracteoles of inflorescence long-ciliate. *L. wahlenbergii*

3b. Cauline leaves usually more than 3, commonly 6–10 mm broad or more; bracts and bracteoles merely erose or shredded, usually not long-ciliate. *L. parviflora*

4a. Lowermost bract of inflorescence well developed, leaflike, callous-thickened and rounded apically; bracts and bracteoles with few long-ciliate hairs or none; seeds carunculate. *L. campestris*

4b. Lowermost bract of inflorescence poorly developed, not leaflike or if so, then subulate apically; bracts and bracteoles commonly long-ciliate; seeds not carunculate. (5)

5a. Bracts longer than the flowers, silvery, conspicuous; inflorescence spicate, commonly nodding. *L. spicata*

5b. Bracts equaling or shorter than the flowers, conspicuous or inconspicuous; inflorescence of 1–several subcapitate clusters, erect or nodding. (6)

6a. Inflorescence of few to several subcapitate clusters, borne on elongate slender nodding or curved peduncles (at least some). (7)

6b. Inflorescence of 1–few sessile sub-capitate clusters, if branching, then the peduncles thick and erect or nearly so (rarely arching in some *L. confusa*). (8)

7a. Inflorescence of pale reddish brown to tan flower clusters, commonly with (1)2–3 main merely curved branches arising from brownish to straw-colored sheaths. *L. tundricola*

7b. Inflorescence of chestnut to purplish brown flower clusters, commonly with more than 3 main drooping branches, arising from chestnut to purplish brown sheaths. *L. arcuata*

8a. Inflorescence of 1 capitate flower cluster (rarely 2); basal leaf sheaths merely brownish or straw-colored; leaf blades flat, the leaves in a dense basal tuft. *L. nivalis*

8b. Inflorescence of 1–4 pedunculate flower clusters and usually 1 short pedunculate cluster or less commonly with only 1 capitate cluster; basal leaf sheaths purplish; leaf blades channeled at least apically, the leaves loosely tufted. *L. confusa*

Luzula arcuata (Wahl.) Wahl.
Alpine Woodrush

Plants tufted, shortly rhizomatous, 0.5–2.5 dm tall; leaf blades involute, channeled or flat, 1–4 mm broad, sparingly hairy or smooth; cauline leaves 1–3, the tips often brown or purplish; basal leaf sheaths brown to straw-colored or purplish; inflorescence of 3–10 or more capitate or spicate flower clusters borne on very slender drooping or arching branches; lowermost bract blade-less, sheathing, chestnut to purplish brown; bracts shorter than the flowers; perianth 1.8–3.5 mm long, brown, hyaline apically, attenuate; anthers 0.3–0.6 mm long, shorter than the filaments; capsules ovoid, from shorter to longer than the perianth; seeds about 1 mm long, brown, apiculate at each end.

Stream banks, ridge tops, and slopes, in tundra and heathlands; in most of Alaska and southern Yukon; southward to Washington; Eurasia. Our materials have been segregated into two varieties.

1a. Leaves involute or channeled, the sheaths often purplish at the base (*L. peringensis* Tolm.). *L. arcuata* var. *arcuata*

1b. Leaves flat, the sheaths commonly brown; our common phase (*L. unalaschkensis* [Buch.] Satake; *L. arcuata* ssp. *unalaschkensis* [Buch.] Hultén). *L. arcuata* var. *unalaschkensis* Buch.

Luzula campestris (L.) DC. ex DC. & Lam.

Plants tufted, 0.7–5.0 dm tall; leaf blades flat or channeled apically, 1.5–6.5 mm broad, long-hairy marginally; cauline leaves 2–4, the tips obtuse, callous-thickened and generally chestnut brown; basal leaf sheaths brownish to straw-colored; inflorescence of 1–5 densely congested capitate flower clusters, borne sessile or on stiffly erect or ascending branches; lowermost bract leaflike, the blade well developed, callous-tipped; bracts shorter than the flowers; perianth 2.2–4.7 mm long, chestnut brown to silvery white, hyaline apically, attenuate or acute; anthers 0.4–1 mm long or more, from shorter to longer than the filaments; capsules ovoid, shorter than or subequal to the perianth; seeds 1.3–1.7 mm long, brown, the cellular caruncle about half as long.

Wet to dry sites in tundra, heath, woods, and thickets; over much of Alaska and Yukon; east to the Atlantic and south to California; circumboreal. Four more or less completely intergrading varieties are recognizable within our materials. The following tentative key will allow for determination of most specimens.

1a. Perianth segments commonly 3–4.7 mm long; plants of coastal and insular southwestern, south-central, and southeastern sites. (2)

1b. Perianth segments commonly (1.6)

2.2–3 mm long; plants of coastal and interior sites. (3)

2a. Perianth greenish to straw-colored; plants of extreme southeastern Alaska (*L. comosa* E. Meyer; *L. campestris* var. *comosa* [E. Meyer] Fern. & Wieg; *L. campestris* ssp. *comosa* [E. Meyer] Hultén; *Juncus congestus* Thuill.). *L. campestris* var. *congesta* (Thuill.) E. Meyer

2b. Perianth chestnut to purplish brown (seldom otherwise); plants of coastal and insular Alaska, from the westernmost Aleutians eastward through southern and southeastern Alaska (*L. kobayasii* Satake; *L. multiflora* var. *kobayasii* [Satake] Sam.; *L. multiflora* ssp. *kobayasii* [Satake] Hultén). *L. campestris* var. *minor* (Satake) Welsh comb. nov. (based on *Luzula kobayasii* var. *minor* Satake Bot. Mag. 46:187. 1932).

3a. Inflorescence appearing pale due to prominent whitish bracts and bractlets, unbranched or with 1–2 elongate pedunculate clusters; plants rather broadly distributed in Alaska and Yukon (*L. frigida* Buch.; *L. multiflora* var. *frigida* [Buch.] Sam.). *L. campestris* var. *frigida* Buch.

3b. Inflorescence chestnut to purplish brown (sometimes pale), commonly with 2 or more long-pedunculate flower clusters; plants mostly of the southern half of Alaska and Yukon (*Juncus multiflorus* Ehrh.; *L. kjellmanniana* Miyabe & Kudo; *L. multiflora* var. *kjellmanniana* [Miyabe & Kudo] Sam.). *L. campestris* var. *multiflora* (Ehrh.) Celak

Luzula confusa Lindeb.
Northern Woodrush

Plants tufted, 0.6–3.7(4.2) dm tall; leaf blades channeled throughout or less commonly only apically, 1–3(7) mm broad, long-hairy marginally; cauline leaves (1) 2–3, the tips subulate, often purplish or brownish; basal leaf sheaths suffused with purple or purple brown; inflorescence of 1–4 densely congested capitate flower clusters, all subsessile, or commonly one cluster on a long suberect or ascending (rarely curved) peduncle; lowermost bract bladeless or with a slender aristate blade; bracts shorter than the flowers; perianth 1.7–2.5 mm long, the segments chestnut brown to brown, hyaline apically, acute to attenuate; anthers 0.5–0.7 mm long, about equaling the filaments; capsules ovoid, shorter than or subequal to the perianth; seeds 1–1.3 mm long, brown, minutely apiculate at each end.

Sandy and gravelly soils and humus, in arctic and alpine tundra and heathlands; in most of Alaska, except for coastal and insular southern regions and most of the Yukon; east to the Atlantic; circumboreal (*L. hyperborea* authors, not R. Br?.).

Luzula nivalis (Laest.) Beurl.
Snow Woodrush

Plants tufted, 0.4–2.0 dm tall; leaf blades flat throughout or more or less channeled apically, 1–4 mm broad, sparingly ciliate or quite smooth marginally; cauline leaves 1–2, the tips subulate, seldom purplish; basal leaf sheaths brown to straw-colored; inflorescence of one capitate subsessile erect flower cluster; lowermost bract bladeless or nearly so; bracts shorter than the flowers; perianth 1.5–2.4 mm long, chestnut brown, hyaline apically, acute; capsule ovoid, shorter to longer than the filaments; capsule ovoid, shorter than to surpassing the perianth; seeds 0.9–1.1 mm long, minutely apiculate at each end.

Meadows in peaty soil; in arctic and alpine tundra, north of the Brooks Range and in east-central Alaska and southern Yukon; circumpolar (*L. campestris* var. *nivalis* Laest.; *L. arctica* Blytt.). *L. nivalis* is a poorly understood entity in Alaska and Yukon. It shares features of both *L. confusa* and *L. tundricola*. Apparent transitional forms occur and these entities might best be considered as phases of *L. nivalis*.

Luzula parviflora (Ehrh.) Desv.
Small-flowered Woodrush

Plants tufted or the stems solitary or in small tufts from elongate rhizomes, 1.5–8.5 dm tall; leaf blades flat throughout or channeled apically, 8–11 mm broad, sparingly ciliate or quite glabrous marginally; cauline leaves 3–5, the tips subulate, often reddish or purplish; basal leaf sheaths brown or less commonly chestnut or purplish; inflorescence a congested to open nodding or spreading panicle, the flowers borne singly or in pairs on elongate pedicels; lowermost bract leaflike or the blade obsolete; bracts shorter than the flowers, erose and often somewhat ciliate; perianth 1.5–2.2 mm long, brownish, acute apically; anthers 0.4–0.6 mm long, commonly shorter than the filaments; capsule ovoid, as long as or longer than the perianth; seeds 1–1.4 mm long, brownish, apiculate at each end.

Lake shores, stream banks, bogs, thickets, meadows, and open woods, in tundra, heathlands, and forests; in most of Alaska and Yukon; eastward to the Atlantic and south to California, Arizona, New Mexico, Minnesota, and New York; circumboreal (*Juncus parviflorus* Ehrh.; *J. melanocarpus* Michx.; *L. parviflora* var. *melanocarpa* [Michx.] Buch.; *Juncoides piperi* Coville; *L. piperi* [Coville] M. E. Jones; *L. wahlenbergii* ssp. *piperi* [Coville] Hultén; *L. parviflora* ssp. *divaricata* sensu Hultén, not *L. divaricata* Wats.). *L. parviflora* is an extremely variable species, from which several infraspecific entities have been segregated. The phases apparently intergrade insensibly and hence segregation must be both arbitrary and artificial. Thus it seems best to recognize our materials as belonging to a single polymorphic entity.

Luzula rufescens Fisch. ex E. Meyer
Hairy Woodrush

Plants loosely tufted, 1.0–3.3 dm tall; leaf blades flat throughout, 1–3.5 mm broad, long-hairy marginally; cauline leaves 1–2, the obtuse tips callous-thickened and generally chestnut brown; basal leaf

Luzula parviflora (Ehrh.) Desv. (× 0.4).

sheaths brown; inflorescence an open sub-umbellate panicle with flowers solitary or paired on elongate pedicels; lowermost bract bladeless, bracts shorter than the flowers, subentire, not ciliate; perianth 1.5–3 mm long, the segments brownish with hyaline margins, acute to attenuate; anthers 0.3–0.7 mm long, shorter than the filaments;

capsule ovoid, longer than the perianth; seeds 2–2.5 mm long (including the caruncle), brown, the apical caruncle a third to two-thirds as long as the body.

Riverbanks, muskegs, hillsides, and marshes, in woods, thickets, and meadows, from low to high elevations; in most of interior Alaska and western Yukon; south to northwestern British Columbia; Asia.

Luzula spicata (L.) DC.
Spiked Woodrush

Plants tufted, 0.7–4.0 dm tall; leaf blades channeled, long-hairy marginally, 1.5–3 mm broad (folded); cauline leaves 1–3, the tips subulate, commonly purplish or chestnut; basal leaf sheaths brown to tan or chestnut, not purplish; inflorescence of 1 (2–3) congested capitate-spicate commonly nodding flower clusters, subsessile or when more than one, then one peduneculate; lowermost bract bladeless or with a slender subulate blade; bracts surpassing the flowers (at least some), the midrib often aristate-projecting, commonly long-ciliate; perianth 2.5–3 mm long, the segments brown to chestnut with hyaline margins, acuminate-aristate; anthers 0.3–0.6 mm long, shorter than the filaments or about equaling them; capsule ovoid, shorter to longer than the perianth; seeds 1.2–1.6 mm long, brown, slightly appendaged apically.

Alpine tundra, along streams and in meadows; in much of southern Alaska (except for western Aleutians and southern portion of Alexander Archipelago) and southern half of the Yukon; disjunctly eastward to the Atlantic and south to California, Utah, Colorado, and New York; Eurasia (*Juncus spicatus* L.). *L. spicata* apparently approaches some phase of *L. confusa*, rarely producing a stalked flower cluster similar to that entity. Otherwise the species is relatively uniform and distinctive.

Luzula tundricola Gorodk.
Tundra Woodrush

Plants tufted, (0.7)1.0–4.5 dm tall; leaf

blades flat except for the channeled apical portion, 1.5–4.5(5) mm broad, sparingly long-hairy or glabrous marginally; cauline leaves 1–5, the somewhat thickened tips obtuse to subulate, often chestnut or brown; basal leaf sheaths brown to straw-colored, not purplish; inflorescence of (1) 2–5 subcapitate or spicate flower clusters, all subsessile or commonly 1–3 clusters on very slender arching or nodding peduncles; lowermost bract bladeless; bracts shorter than the flowers; perianth 2–2.5 mm long, the segments commonly pale brown and hyaline, less commonly brown to chestnut, acute to attenuate; anthers 0.3–0.5 mm long, shorter than the filaments; capsules ovoid, commonly shorter than the perianth; seeds 0.9–1.1 mm long, brown, minutely apiculate at each end.

Alpine tundra, heathlands, and open woods, in dry to wet sites; in most of Alaska (except for south-central to southeastern portions) and in most of the Yukon; south to British Columbia and east to the Mackenzie; Asia (*L. arcuata* f. *latifolia* Kjellm. ex Nordenskj.; *L. nivalis* var. *latifolia* [Kjellm.] Sam.; *L. confusa* var. *latifolia* [Kjellm.] Buch.; *L. beeringiana* Gjaerevoll; *L. kamtschadalorum* authors, not [Sam.] Gorodk. [?]). The confusion with regard to affinities of *L. tundricola* is apparent in the list of synonyms. It approaches both *L. nivalis* and *L. confusa,* the latter the more closely. However, *L. tundricola* is rather easily distinguished by its flattened leaves (channeled at the apex), brownish leaf sheaths, and 2–4 slender arching peduncles and commonly solitary flower cluster.

Luzula wahlenbergii Rupr.
Wahlenberg Woodrush

Plants tufted and shortly rhizomatous, 10–40 cm long; leaf blades flat throughout or channeled apically, 1–3(6) mm broad, sparingly ciliate or quite glabrous; cauline leaves 2–3, the tips subulate, often purplish or brownish; basal leaf sheaths brown to chestnut or straw-colored; inflorescence an open, nodding or spreading panicle, the

flowers borne solitary or in 2s or 3s at the ends of filiform branches; lowermost bracts commonly bladeless; bracts shorter than the flowers, both shredded and long-ciliate; perianth 1.5–2.4 mm long, the segments brownish to chestnut with hyaline apex, acute; anthers 0.4–0.7 mm long, from shorter to slightly longer than the filaments; capsule ovoid, brown, commonly surpassing the perianth; seeds 1.2–1.6 mm long, brown, slightly appendaged apically.

Pond margins, wet slopes, stream banks, and moist meadows, in alpine and arctic tundra, heathlands, and open woods; in much of Alaska and southern Yukon (and to be expected elsewhere); eastward to Quebec and south to Washington; circumpolar (*L. wahlenbergii* ssp. *piperi* sensu Hultén, not *Juncoides piperi* Coville). This entity is easily mistaken for the much more common *L. parviflora.* The tendency to fewer, narrower leaves, smaller inflorescences, and more shredded and hairy bracts seems to allow adequate separation of the two entities.

JUNCAGINACEAE
Arrowgrass Family

Herbs of marshes or of maritime habitats, perennial, rhizomatous; leaves 2-ranked, basal or cauline, linear, sheathing at base; inflorescence a terminal spike or raceme; flowers perfect or imperfect, regular; perianth usually 6-segmented, in 2 series of 3; stamens commonly 6; pistils 1(3–6)-carpelled, or the carpels free except at the base; first capsular or follicular.

1a. Leaves arranged along the stem; carpels distinct or united only at the base. *Scheuchzeria*
1b. Leaves all basal, the inflorescence scapose; carpels united throughout their length. *Triglochin*

SCHEUCHZERIA L.

Plants perennial, the stems leafy; inflorescence a raceme; flowers few; perianth segments 6, in 2 similar series; stamens 6, the anthers linear, the filaments short; carpels usually 3, distinct or united only at the base; ovary usually 2-ovuled; fruit follicular.

Scheuchzeria palustris L.

Plants mostly subterranean (or submersed), the aerial stems commonly 1–2(4) dm tall, clothed at base with dead leaves; basal leaves 5–20 cm long or more, erect; cauline leaves reduced upwards, sheathing at the base, with an elongate ligule, the blade somewhat channeled; flowers 3–12, in the axils of well-developed bracts; pedicels 3–15 mm long or more; perianth greenish white, the segments to 3 mm long; follicles 5–8 mm long, divergent; seeds 4–5 mm long, dark brown.

Sphagnum bogs and lake margins; in south-central and southeastern Alaska and southern Yukon; east to Newfoundland and south to California, Idaho, Iowa, and New Jersey; Eurasia (S. *palustris* ssp. *americana* [Fern.] Hultén). Our materials belong to var. *americana* Fern.

TRIGLOCHIN L.

Plants perennial, the leaves all basal; inflorescence of terminal spicate racemes; flowers few to numerous; perianth segments 6, in 2 whorls of 3; stamens 6, the anthers oblong to elliptic, sessile or nearly so; carpels 3 or 6, these united, at length separating, sometimes only the inner 3 fertile and each 1-seeded; fruit capsular.

(*Note:* I consider the generic name to be feminine, not neuter).

Löve, A., and D. Löve. 1958. Biosystematics of *Triglochin maritimum* Agg. Nat. Can. 85:156–65.

1a. Fruit ovoid-oblong; fertile carpels 6, more or less cordate at the base. *T. maritima*

1b. Fruit linear-clavate; fertile carpels 3, subulate at the base. *T. palustris*

Scheuchzeria palustris L. (× 0.7).

Triglochin maritima L.
Maritime Arrowgrass

Plants 0.7–10.5 dm tall (rarely more) from a thick woody rhizome; leaves linear, 2–50 cm long, flattened or channeled, mostly 2–4 mm broad, obtuse apically, the sheath prominently hyaline margined; racemes several- to many-flowered, the flowers not subtended by bracts; pedicels 1–5 mm long; perianth segments greenish or yellowish, 1–2.2 mm long; fruit ovoid-oblong, mostly 4–6 mm long, deciduous, the axis terete.

Muskegs, saline meadows, tidal flats, and open woods; in much of Alaska (except for central, west-central, and insular southwestern portions) and in southern Yukon; widespread in North America; circumboreal.

Triglochin palustris L.
Marsh Arrowgrass

Plants 0.8–6.5 dm tall from short ascending rhizomes; leaves linear-filiform, 2–28 cm long, flattish, 0.3–2 mm broad, acutish apically, the sheath narrowly hyaline margined; racemes with few to many flowers, the flowers not subtended by bracts; pedicels 1–6 mm long; perianth segments yellowish or greenish, often suffused with purple, 1.5–2 mm long; fruit linear-clavate, commonly 8–10 mm long, the carpels separating at the base and remaining attached apically.

Fens, muskegs, pond margins, tidal flats, lake shores, and river banks; in most of Alaska south of the Brooks Range (except for the western Aleutians) and southern Yukon; eastward to the Atlantic and south to California, New Mexico, Iowa, and New York; circumboreal; Southern Hemisphere.

LEMNACEAE
Duckweed Family

Plants diminutive, aquatic, consisting of leafless flattened thalluslike stems (fronds), each bearing a single root from the lower surface; flowers from a sac- or flask-shaped

Triglochin palustris L. (× 0.4).

spathe, consisting of a single stamen (rarely 2) or a single pistil per spathe; fruit a 1–6-seeded utricle. The simplest and smallest of flowering plants, reproducing mostly by budding.

LEMNA L.

Plant bodies (fronds) consisting of flattened stems with a meristematic pouch on each side, from which arise vegetative and flower primordia; flowers monoecious, produced in a membranous spathe; staminate flowers usually 2 to each spathe, each flower consisting of a single stamen; fruit 1–3-seeded; not often found in flower.

1a. Fronds sessile or nearly so, commonly not remaining attached in long colonies, blades obovate to oblong. *L. minor*

1b. Fronds long-stalked, commonly remaining attached in long colonies, blades elliptic to lanceolate. *L. trisulca*

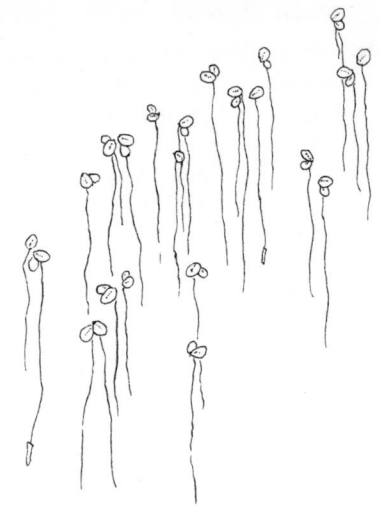

Lemna minor L.
Lesser Duckweed

Fronds floating on the surface, 2–4 mm wide, rounded to obovate-oblong, symmetrical, green or rarely reddish or purplish tinged, obscurely 3-nerved and often with a row of papillae on the midrib; fruit symmetrical, subturbinate; seed deeply and evenly 12–15-ribbed.

Stagnant ponds; in the southeastern quarter of Alaska (exclusive of the Panhandle) and southern Yukon; broadly distributed in North America; circumboreal.

Lemna trisulca L.
Ivy-leaved Duckweed

Fronds usually floating just below the surface, with several generations attached to each other, 6–10 mm long, 2–3 mm wide, elliptic to lanceolate, obscurely 3-nerved, denticulate at the apex; fruit asymmetrical; seeds ribbed.

Stagnant ponds; in the eastern half of continental Alaska (rarely westward) and most of the Yukon; widely distributed in North America; circumboreal.

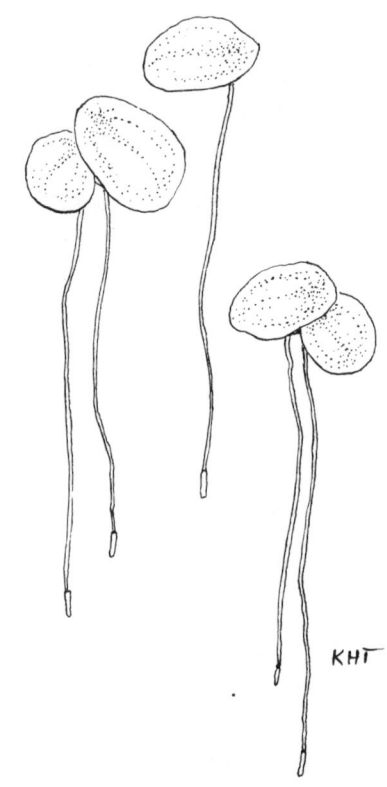

Lemna minor L. (× 1, below, ×10).

LILIACEAE
Lily Family

Perennial herbs from rhizomes, bulbs, or corms; stems erect; leaves basal or cauline, alternate or whorled; flowers regular or nearly so, usually showy, perfect (dioecious in *Asparagus*); perianth of 6 (4 in *Maianthemum*) segments in 2 series of 3 each, distinct or connate; stamens 6(4); pistils 1; ovary superior (2)3-loculed; styles 1 or 3; stigmas 2-3 or merely 2-3-lobed; fruit a capsule or a berry.

1a. Inflorescence umbellate; plants with the odor of onion. *Allium*

1b. Inflorescence racemose, paniculate, cymose, scapose, or the flowers axillary; plants not onion scented. (2)

2a. Stems much branched from thick tuberous roots; leaves scalelike, subtending filiform leaflike branchlets (phyllodia). *Asparagus*

2b. Stems simple or sparingly branched; underground parts various, usually not of thick tuberous roots; leaves not scalelike nor subtending phyllodia. (3)

3a. Plants (0.5)1-2.5 m tall; cauline leaves mostly 5-30 cm long, the bases sheathing; flowers in large panicles (these usually over 20 cm long) or rarely reduced to a spicate raceme. *Veratrum*

3b. Plants usually much less than 1 m tall; cauline leaves usually less than 15 cm long or absent, usually not sheathing. (4)

4a. Leaves all basal or essentially so; scapes with solitary white flowers (rarely 2-flowered), shorter than the leaves, or flowers several to many and blue to violet. (5)

4b. Leaves cauline, at least some (much smaller upward in *Lloydia, Tofieldia,* and *Zigadenus*); inflorescence various, but if a 1-flowered scape, the leaves linear. (6)

5a. Flowers solitary, white; leaves mostly 3-6 cm broad. *Clintonia*

5b. Flowers several to many, blue to violet; leaves commonly 0.5-1.5 cm broad; plants collected once along Haines Highway in southeastern Alaska where probably introduced. *Camassia quamash* (Pursh) Greene

6a. Flowers or flower clusters borne laterally in leaf axils. (7)

6b. Flowers or flower clusters borne in terminal inflorescences. (8)

7a. Perianth rotate, deeply wine-colored with greenish reflexed tips; plants mostly 2 dm tall or less. *Kruhsea*

7b. Perianth campanulate, yellowish white to pinkish; plants 3-10 dm tall. *Streptopus*

8a. Leaves mostly in 1-3 whorls, mostly above the middle of the stem; flowers dark purplish brown. *Fritillaria*

8b. Leaves alternate or basal, not whorled; flowers not dark purplish brown. (9)

9a. Perianth segments large, over 25 mm long; plants introduced. *Lilium*

9b. Perianth segments less than 15 mm long; plants indigenous. (10)

10a. Plants with leafy stems (basal leaves if present not conspicuously larger than the stem leaves). (11)

10b. Plants with stem leaves much reduced upwards, the basal leaves usually much larger. (12)

11a. Leaves cordate, petiolate; perianth segments 4. *Maianthemum*

11b. Leaves lanceolate to elliptic, subsessile or clasping; perianth segments 6. *Smilacina*

12a. Perianth segments 2.5-5 mm long; inflorescence a spicate raceme. *Tofieldia*

12b. Perianth segments mostly 6-12 mm long; inflorescence racemose or paniculate; plants bulbous; leaves not equitant. (13)

13a. Inflorescence 1–2-flowered; perianth white with purplish veins, the glands at the base not distinctive; styles 1. *Lloydia*

13b. Inflorescence several- to many-flowered; perianth cream to yellowish, the segments with broad greenish glands at the base; styles 3. *Zigadenus*

ALLIUM L.

Scapose bulbous plants with characteristic odor; leaves fleshy, hollow and quill-like or broad and flat; stems simple, erect; flowers in umbels, subtended by membranous bracts; petals and sepals free to partly united at the base, 1-nerved; stamens adnate to the base of the sepals and petals; styles filiform, usually deciduous; seeds black, 1–2 in each of the 3 locules; fruit a capsule.

1a. Leaves flat, 5–10 cm broad; flowers white; bulb scales persistent, reticulate; plants local in Aleutian Islands. *A. victorialis*

1b. Leaves fistulose, seldom over 2 cm broad (even when pressed); flowers pink or white; bulb scales membranous; plants widely distributed or cultivated. (2)

2a. Flowers pink; scapes not or only slightly inflated; plants indigenous, rarely cultivated. *A. schoenoprasum*

2b. Flowers white or pink; scapes inflated; plants cultivated. (3)

3a. Bulb well developed, conspicuously larger than the neck; flowers much shorter than the pedicels. *A. cepa*

3b. Bulb poorly developed, only slightly larger than the neck; flowers subequal to the pedicels. *A. fistulosum*

Allium cepa L.

Onion

Biennials (cultivated as annuals) with well-developed bulbs; leaves basal, hollow, glaucous; umbels borne on inflated scapes 3–10 dm tall; flowers pink to white, mostly 4–7 mm long, on pedicels mostly about 25 mm long; capsule about as long as the perianth.

Cultivated onion includes the var. *aggregatum* Don, which is the "multiplier onion," and var. *viviparum* Metz, which is the "top onion." Widely cultivated in Alaska and the Yukon.

Allium fistulosum L.

Welsh Onion

Tufted biennial or short-lived perennial from poorly developed bulbs; leaves basal, hollow, glaucous; umbels borne on inflated scapes 2.5–5 dm tall; flowers mostly about 10 mm long, subequal to the pedicels; capsules about as long as the perianth.

Cultivated plant for spring greens; in agricultural regions of Alaska.

Allium schoenoprasum L.

Chive

Tufted biennial or short-lived perennial from poorly developed clustered bulbs; leaves hollow, very narrow; umbels capitate on uninflated scapes 1.5–6(7.5) dm tall; pedicels much shorter than the flowers; flowers bright rose to purplish, 7–12 mm long; capsule about half as long as the perianth.

Stream banks, pond margins, lake shores, and meadows; in much of Alaska south of the Brooks Range (excluding the Aleutians) and Yukon south of the British Mountains; east to the Atlantic and south to Oregon, Wyoming, and New York; Eurasia (*A. sibiricum* L.). Plants of this species are indigenous to Alaska and the Yukon and have been designated as var. *sibiricum* (L.) Hartm. Cultivated chives belong to var. *schoenoprasum*.

Allium victorialis L.

Victory Onion

Biennials with the bulb scales persisting as a reticulate coat; leaves basal, the bases sheathing the lower part of the scape, mostly 5–10 cm broad and to 20 cm long; umbels capitate on uninflated scapes to 7.5 dm tall; flowers white, mostly 6–10 mm long.

Meadows on Attu Island; Eurasia. The Aleutian plant is known as ssp. *platyphyllum* Hultén.

ASPARAGUS L.

Erect perennial herbs from fascicled tuberous roots; leaves reduced to scales which subtend photosynthetic branchlets (phyllodia); flowers 1–4 in the axils of phyllodes; perianth with 6 yellowish green segments, about 6 mm long in staminate flowers, much shorter in pistillate ones; stamens 6; pistils 1; stigmas 3; fruit a berry.

Asparagus officinalis L.
Asparagus

Stems mostly 1.5–3 m tall; phyllodes filiform, to 20 mm long, 1 or more in the axil of each scale leaf; inflorescence of 2 (more rarely 1), 1-flowered racemes from the axils of phyllodia or branches; fruit a red few-seeded berry.

Cultivated food plant; escaped and persisting at Manley Hot Springs, Alaska.

CLINTONIA Raf.

Rhizomatous scapose perennials; leaves basal, broad, many-nerved; flowers solitary (rarely 2), borne on scapes; perianth of 6 similar distinct segments; stamens 6; pistils 1, 2–3-loculed; style slender; stigma obscurely 2–3-lobed; fruit a globose to ovoid berry.

Clintonia uniflora (Schult.) Kunth
Blue-bead

Plants mostly 1–2 dm tall; leaves 2–4, oblanceolate, more or less villous beneath, 10–20 cm long, 3–6 cm wide, acute at both ends; scape shorter than the leaves; flowers white, campanulate, the sepals and

petals about 2 cm long; berry about 1 cm long, 5–10-seeded.

Woods; in southeastern Alaska; southward to California and east to Montana (*Smilacina borealis* var. *uniflora* Schult.).

FRITILLARIA L.

Perennial bulbous plants with simple leafy stems; flowers campanulate, large, nodding; perianth segments 6, pistils 1, 3-loculed; styles 3, distinct almost to the base; fruit a loculicidal 3-valved capsule.

Fritillaria camtchatcensis (L.) Ker
Indian-rice, Black Lily

Plants 2–6 dm tall, from bulbs composed of several large fleshy scales subtended by numerous ricelike bulblets; leaves mostly in 2–3 whorls with a few scattered ones near the tip, lanceolate, blunt, 3–6(9) cm long, 0.7–3 cm wide; flowers 1–6, dark wine-colored, often almost black, tinged greenish yellow on the outside, 18–30 mm long; capsule obtusely angled, 2–3 cm long.

Clintonia uniflora (Schult.) Kunth (× 0.25).

Meadows, heathlands, tidal flats, and open woods; in or near coastal and insular southern Alaska (rarely some distance from coast in interior southern Alaska) and southern Yukon; Asia (*Lilium camtchatcense* L.)

KRUHSEA Reg.

Plant a low glabrous perennial from a rhizome; flowers solitary, borne laterally in leaf axils; perianth rotate, deeply wine-colored, 6-segmented, the greenish tips reflexed; stamens 6; pistils 1, the ovary 3-loculed; fruit a red globose berry.

Kruhsea streptopoides (Ledeb.) Kearney

Plants with simple stems 3–15(19) cm tall; leaves 4–8, sessile, ovate-lanceolate, acute, (20)25–50 mm long, 6–20 mm wide, somewhat clasping at the base; flowers borne singly in the axils of the upper 1–5 leaves, on recurved pedicels about 10 mm long; perianth segments 2–2.5 mm long; fruit about 4 mm in diameter.

Woods; in south-central to southeastern Alaska; south to Washington and Idaho; Asia (*Smilacina streptopoides* Ledeb.; *S. brevipes* Baker; *K. streptopoides* ssp. *brevipes* [Baker] Calder & Taylor; *K. streptopoides* var. *brevipes* [Baker] Fassett).

LILIUM L.

Perennial bulbous plants with erect leafy stems; bulbs scaly; flowers large, showy, arranged in terminal inflorescences; perianth with 6 distinct segments, each with a nectariferous groove or furrow at the base; stamens 6; pistils 1 with long style, the stigma 3-lobed; fruit a 3-loculed, loculicidal capsule.

1a. Flowers erect; perianth segments 2–4 cm broad. *L. maculatum*
1b. Flowers spreading or reflexed; perianth segments less than 2 cm broad. (2)

2a. Leaves single-nerved; style subequal

Kruhsea streptopoides (Ledeb.) Kearney (× 0.5).

to much longer than the ovary. *L. pumilum*

2b. Leaves several-nerved; style much shorter than the ovary. *L. callosum*

Lilium callosum Sieb. & Zucc.

Stems mostly 3–6 dm tall; leaves alternate, numerous, 2.5–10 cm long, 2–6(10) mm wide, several-nerved; flowers 2–several, nodding, bright orange red, sometimes obscurely dotted at the base; perianth segments sharply reflexed, 3–4 cm long, 6–10 mm wide; style shorter than the ovary.

Cultivated ornamental; in southern Alaska; introduced from the Old World.

Lilium maculatum Thunb.

Stems mostly 4.5–6 dm tall; leaves alternate, 5–10(14) cm long, (0.8)1–2.5 cm wide, several-nerved; flowers 1–several, erect, reddish orange, from scarcely to conspicuously spotted; perianth segments spreading, 7.5–10 cm long, 2–4 cm wide; style longer than the ovary.

Cultivated ornamental; in southern Alaska; introduced from the Old World.

Lilium pumilum DC.

Stems mostly 3–6 dm tall; leaves alternate, 5–10 cm long, 2–4 mm wide, 1-nerved; flowers few to several, nodding, bright red, rarely somewhat spotted; perianth segments reflexed, 2.5–3.5 cm long, 6–8 mm wide; style subequal to much longer than the ovary.

Cultivated ornamental; in southern Alaska; introduced from the Old World.

LLOYDIA Salisb.

Perennial bulbous herbs arising from rhizomes; stems erect, leafy, the leaves reduced upward; flowers 1–few in a terminal raceme; perianth segments 6, distinct, each with a transverse gland at the base; stamens 6; pistils 1, with a single style, the stigma 3-lobed; fruit a 3-loculed, loculicidal capsule.

Lilium pumilum DC. (× 0.4).

Lloydia serotina (L.) Wats.
Alp Lily

Plants erect, 5–15(20) cm tall; bulbs small, covered with a grayish fibrous coat; stems slender; basal leaves 2–8 cm long, mostly 1–2 mm wide; stem leaves alternate, reduced upwards, 1–4 cm long; flowers creamy white, 8–13 mm long, purple veined and tinged with rose on the back; capsule ovoid, about 8 mm long.

Alpine and arctic tundra and heathlands; in most of Alaska and the Yukon; south to Washington, Utah, and Colorado; Eurasia (*Bulbocodium serotinum* L.; *Anthericum serotinum* L).

MAIANTHEMUM Weber

Perennial rhizomatous plants with leafy stems; leaves usually 2–3, broad, many-nerved; flowers white, small, in a terminal spicate raceme; perianth segments 4, distinct, lacking glands at the base; stamens 4; pistils 1, the stigma 2-lobed or -cleft; fruit a 2-loculed berry with 1–2 seeds.

1a. Leaves distinctly ciliate marginally; stems and rachis of inflorescence hairy; plants of interior northern British Columbia. *M. canadense*
1b. Leaves merely scabrous marginally, not ciliate; stems and rachis of inflorescence glabrous; plants of coastal southern Alaska and coastal British Columbia. *M. dilatatum*

Maianthemum canadense Desf.

Stems 0.5–2(2.2) dm tall; leaves alternate, ovate to lanceolate, cordate basally, 3–10 cm long, 0.5–5 cm broad; racemes many-flowered, the pedicels 2–5 mm long, often fascicled; perianth segments, cream to white, 2–3 mm long, becoming reflexed; style stout; berry pale red, 3–4 mm in diameter.

Moist woods; at Liard Hot Springs, British Columbia; eastward to Newfoundland and south to South Dakota, Indiana, and New Jersey. Our material belongs to var. *interius* Fern.

Maianthemum dilatatum (Wood) Nels. & Macbr.
Deerberry

Stems (0.4)1 dm tall; leaves alternate, broadly cordate to sagittate, (2.0)5–15 cm

long, (1)3–10 cm wide or those of sterile stems up to 15 cm wide; racemes many-flowered, the pedicels 2–4 mm long, often fascicled; perianth segments cream to white, 2–3 mm long, becoming reflexed; style stout; berry spotted, becoming red on drying, globose, about 6 mm in diameter.

Woods, heathlands, and tundra; in coastal and insular southern Alaska from the Aleutians eastward through the Panhandle; south to California and Idaho (*Unifolium dilatatum* How.).

SMILACINA Desf.

Rhizomatous perennial herbs; stems erect or ascending, simple; leaves alternate, many-nerved, sessile or somewhat clasping; flowers in terminal racemes or panicles; perianth 6-segmented, white or greenish white, distinct or nearly so; stamens 6, ovary 3-loculed; style short; stigma (2–4)3-lobed; fruit a globose berry.

1a. Leaves 3 per stem; plants of northern British Columbia. *S. trifolia*
1b. Leaves commonly 5–9 or more; plants variously distributed. (2)

2a. Flowers in panicles; perianth segments about 2 mm long. *S. racemosa*
2b. Flowers in racemes; perianth segments 4–7 mm long. *S. stellata*

Smilacina racemosa (L.) Desf.
False Solomon-seal

Stems 3–10 dm tall from fleshy rhizomes; leaves 5–9 or more per stem, oblong-lanceolate, sessile or short-petioled, 7–20 cm long, pubescent below with short stiff hairs, the margins minutely ciliate; panicle densely many-flowered, 4–10 cm long; perianth segments about 2 mm long; fruit mottled, becoming red at maturity, 4–6 mm in diameter.

Woods; in extreme southeastern Alaska; eastward to the Atlantic and south to Arizona and Georgia (*Convallaria racemosa* L.).

Smilacina stellata (L.) Desf.

Stems 3–5 dm tall from a slender rhizome; leaves 5–9 or more per stem, sessile, pubescent beneath, lanceolate, 5–15 cm long, 2–4 cm wide; racemes 3–7 cm long, several-flowered; perianth segments 4–7 mm long; berries mottled, becoming red at maturity, 7–10 mm in diameter.

Tidal flats, open woods, and meadows; in south-central Alaska and southern Yukon; east to the Atlantic and south to California, Utah, Missouri, Indiana, and Virginia (*Convallaria stellata* L.).

Smilacina trifolia (L.) Desf.

Stems 1–4 dm tall from a rhizome; leaves sessile, 3(2–4) per stem, glabrous beneath, elliptic to oblong or lanceolate, 5–12 cm long, 1–4 cm broad; racemes 2.5–5 cm long, few- to several-flowered; perianth segments 3–4 mm long; berries dark red at maturity.

Moist woods; at Liard Hot Springs, British Columbia; eastward to Newfoundland and south to Michigan and New Jersey (*Convallaria trifolia* L.).

STREPTOPUS Michx.

Rhizomatous perennial herbs with leafy branching or simple stems; leaves alternate, thin, many-nerved, sessile or auriculate-clasping; flowers solitary or in pairs, borne laterally on the stem from the upper leaf axils; perianth campanulate, yellowish white, greenish, or pinkish, the tips at length spreading, 6-segmented; stamens 6; pistils 1, ovary 3-loculed; fruit a red berry.

1a. Leaves auriculate-clasping; flowers yellowish white (rarely pinkish); perianth segments recurved apically. *S. amplexifolius*
1b. Leaves merely sessile or only slightly clasping; flowers pinkish; perianth segments mostly straight. *S. roseus*

Streptopus amplexifolius (L.) DC.
Cucumber-root, Clasping Twisted-stalk

Plants with stems usually branched, 3–10 dm tall; leaves several, auriculate-clasping, ovate-lanceolate, 5–15 cm long, 2.5–5 cm wide; peduncles 2–5 cm long, 1–2-flowered; perianth 8–12 mm long, the segments recurved or spreading apically; stigma entire; fruit ovoid-ellipsoid, yellowish white to red, 10–18 mm long.

Woods and meadows; in much of Alaska south of the 65th parallel and in southern Yukon; disjunctly eastward to the Atlantic and south to California, Arizona, New Mexico, Minnesota, and North Carolina; Eurasia (*Uvularia amplexifolia* L.; *S. amplexifolius* var. *denticulatus* Fassett; *S. amplexifolius* var. *americanus* Schult.; *S. amplexifolius* ssp. *americanus* [Schult.] Löve & Löve; *S. amplexifolius* var. *papillatus* Ohwi; *S. amplexifolius* ssp. *papillatus* [Ohwi] Löve & Löve; *S. amplexifolius* var. *chalazatus* Fassett.). The fruit is edible and is used locally in jam and jelly.

Streptopus roseus Michx.

Plants with simple or less commonly branching stems 1–4 dm tall; leaves several, ovate to lance-ovate, (3.5)5–19 cm long, 0.8–3.5 cm wide, rounded to slightly clasping at the base; peduncles mostly 5–15(20) mm long, mostly 1-flowered; perianth 5–7 (8) mm long, the segments scarcely spreading apically; stigma 3-cleft; fruit globose, red, 7–9 mm in diameter.

Woods; in southeastern Alaska and northwestern British Columbia; south to Washington and in eastern North America (*S. curvipes* Vail; *S. roseus* ssp. *curvipes* [Vail] Hultén). Our plants belong to var. *curvipes* (Vail) Fassett.

TOFIELDIA Huds.

Perennial plants from short rhizomes; leaves slender, mostly basal, equitant or 2-ranked; flowers small, in terminal spikelike racemes, each subtended by a small involucre of 3 more or less united bractlets along the pedicel; perianth 6-segmented, persistent, the segments spreading, lacking basal glands, whitish, yellowish, or tinged with dark red; pistils 1, the styles 3, short and spreading; fruit a 3-loculed, septicidal capsule.

Hitchcock, C. L. 1944. The *Tofieldia glutinosa* complex of western North America. Amer. Midl. Nat. 31:487–98.

1a. Plants mostly over 15 cm tall; stems glandular-viscid below the inflorescence. *T. glutinosa*

1b. Plants mostly less than 15 cm tall; stems glabrous throughout. (2)

2a. Ovary and usually the petals tinged with red; stems usually with at least 1 enlarged leafy bract (frequently with 1–2 greatly reduced bracts in addition. *T. coccinea*

2b. Ovary and petals yellowish to green; stems usually with a single bract near the base. *T. pusilla*

Tofieldia coccinea Richards.
Northern Asphodel

Stems tufted, 0.4–1.0(1.2) dm tall; basal leaves densely tufted, equitant, (1)2–6 cm long, 0.2–0.4 cm wide, at least 1 stem leaf usually well developed; perianth mostly 1.5–3 mm long, whitish tinged with red; fruit 1.5–3 mm long, reddish tinged or dark red.

Tundra and heathland; in most of Alaska and western Yukon; east to Greenland; Asia.

Tofieldia glutinosa (Michx.) Pers.

Stems mostly 1.5–4.0 dm tall, viscid above with stalked glands; leaves 5–25 cm long, 0.2–0.6 mm wide, the stem with 2–4 leaves near the base; perianth mostly 3.2–5 mm long, yellowish green; fruit 4–7 mm long.

Muskegs, meadows, and open woods; in or near coastal southern Alaska from the Kenai Peninsula eastward through southeastern Alaska and southern Yukon; south to California and east to Saskatchewan (*Northecium glutinosum* Michx.; *T. occi-*

dentalis authors, not Wats.; *T. glutinosa* ssp. *brevistyla* Hitchc.).

Tofieldia pusilla (Michx.) Pers.
Scotch Asphodel, False Asphodel

Stems tufted, 0.5–2.0(2.5) dm tall; basal leaves densely tufted, equitant, (1)2–10 cm long, 0.2–0.5 cm wide, at least 1 poorly developed stem leaf usually present, this nearly basal; perianth mostly 1.5–3 mm long, yellowish white to greenish; fruit 1.5–3 mm long, green.

Arctic and alpine tundra and heathlands to open woods; in most of continental Alaska and most of the Yukon; eastward to Greenland; circumpolar (*Northecium pusillum* Michx.).

VERATRUM L.

Plants tall, stout, leafy, poisonous perennials from stout rhizomes; leaves broad, strongly veined and plaited, the bases sheathing; flowers greenish yellow to whitish, often bicolored, in large terminal panicles or rarely in spicate racemes; perianth segments 6, distinct, glandless or nearly so; stamens 6; pistils 1, the styles 3, persistent; fruit a 3-loculed septicidal capsule.

1a. Leaves glabrous beneath or pubescent only along the veins; inflorescence spicate or with lateral branches ascending. *V. album*

1b. Leaves pubescent over lower surface; inflorescence with drooping lateral branches. *V. eschscholtzii*

Veratrum album L.
European White Helebore

Stems mostly 0.5–1.5 m tall; leaves broadly ovate-lanceolate to elliptic, most 5–25 cm long and 3–18 cm wide, glabrous beneath or pubescent only along the veins; inflorescence a spicate raceme or a panicle with ascending to spreading branches; flowers whitish or yellowish green; perianth segments about 10 mm long; ovary subglabrous.

Heathlands and tundra; in the Seward Peninsula and western Aleutians; Eurasia (V. *oxysepalum* Turcz.; V. *album* ssp. *oxysepalum* [Turcz.] Hultén). Our material has been assigned to var. *oxysepalum* (Turcz.) Miyabe & Kudo.

Veratrum eschscholtzii Gray

Stems mostly 1–2.5 m tall; leaves broadly round-oval to ovate-lanceolate, narrower toward the inflorescence, glabrous above, pubescent, often densely so below, mostly 10–30 cm long, and 5–20 cm wide; inflo-

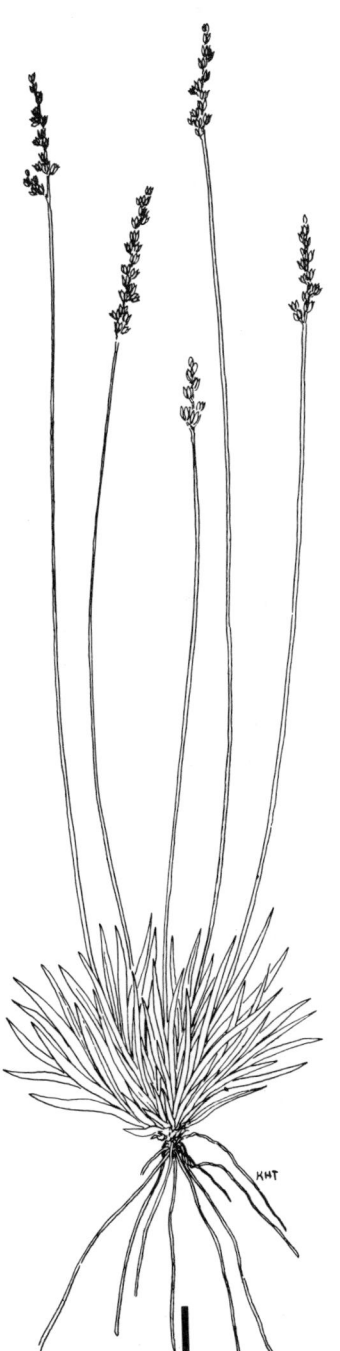

Tofieldia pusilla (Michx.) Pers. (× 0.9).

Veratrum eschscholtzii Gray (× 0.3).

rescence a large panicle in which at least the lower branches are pendulous and on which the pedicels turn up as the flowers open; flowers yellowish green; perianth segments 8–10(13) mm long; ovary often moderately pubescent.

Meadows, heathlands, and open woods; in southwestern, south-central and southeastern Alaska and southern Yukon; east to Mackenzie and Montana and south to California (*V. viride* ssp. *eschscholtzii* [Gray] Löve & Löve).

ZIGADENUS Michx.

Perennial bulbous poisonous plants; leaves linear, mainly basal, those on the stem reduced upwards; flowers in terminal racemes (rarely paniculate), perfect or polygamous; perianth withering and persistent, bearing 1–2 glands just above the narrowed base, 6-segmented; pistils 1; styles 3, distinct; fruit a 3-loculed septicidal capsule.

Zigadenus elegans Pursh
Elegant Death Camas

Plants (2)3–6(7) dm tall; basal leaves 8–30 cm long, 0.3–10(15) cm wide, slightly keeled; flowers greenish or yellowish white, borne in racemes or panicles; perianth segments 7–10 mm long, bearing a large obcordate gland at the base; capsule about 15 mm long.

Woods, heathlands, tundra, sandy slopes, dry hillsides, and rock outcrops; in most of Alaska (except for the southwestern portion) and most of the Yukon; eastward to Northwest Territory and the Great Lakes and south to Washington, Utah, and Colorado.

Zigadenus elegans Pursh (× 0.4).

ORCHIDACEAE
Orchid Family

Perennial herbs from corms, bulbs, rhizomes, or tuberous roots; leaves alternate, opposite, or basal, simple, entire, often sheathing; flowers perfect, irregular, solitary or in racemes or spikes; sepals 3, the lower sometimes united; petals 3 with one forming a prominent lip; stamens 1–2, adnate to the style and forming a column (gynandrium); pollen powdery or hanging together in waxy masses (pollinia), which are transferred to the stigmatic surface *en masse;* pistils 1; ovary inferior, mostly 3-loculed; fruit a many-seeded capsule.

1a. Leaves absent at flowering time; plants reddish or yellowish saprophytes with coralloid roots. *Corallorhiza*

1b. Leaves present at flowering time; plants green holophytes with roots various but not coralloid. (2)

2a. Fertile stamens 2; lip saccate, 1.2–3 cm long; leaves 2–several. *Cypripedium*

2b. Fertile stamens 1; lip usually neither saccate nor over 1.5 cm long (except in *Calypso*, which has a single leaf). (3)

3a. Lip of corolla (not of calyx) bearing a prominent spur at the base. (4)

3b. Lip of corolla lacking a spur at the base. (5)

4a. Flowers brightly colored with pink or purple; viscid disk of the pollinia enclosed in a pouchlike structure. *Orchis*

4b. Flowers greenish to white; viscid disk of pollinia free. *Habenaria*

5a. Leaves 1, folded lengthwise into prominently ribbed plaits; the lip saccate, 1.5–2.5 cm long. *Calypso*

5b. Leaves (1)2–several, variously folded or flat but not plaited; lip usually not saccate, less than 1.5 cm long. (6)

6a. Plants from small bulbous corms; pollinia smooth or waxy. *Malaxis*

6b. Plants from rhizomes or fascicled roots; pollinia granulose or powdery. (7)

7a. Leaves 2, opposite; flowers mostly less than 10 in a lax raceme. *Listera*

7b. Leaves several, alternate or basal; flowers mostly more than 10 in a compact or lax raceme or spike. (8)

8a. Roots from a creeping rhizome; basal leaves broadly elliptic to lanceolate or ovate, greenish above, pale beneath. *Goodyera*

8b. Roots from the base of the stem; basal leaves mostly narrowly lanceolate, not markedly different above and below. *Spiranthes*

CALYPSO Salisb.

Cormous herbs with fleshy roots or a coralloid rhizome (or both); leaves solitary, basal; flowers solitary, the scape with 1–2 sheathing scalelike leaves; sepals and lateral petals similar, spreading or ascending, pinkish, the lip large, saccate, spotted or lined brown purple, with two short spurs near the apex; column winged, petallike, the anther just below the summit; pollinia 2 in each sac.

Calypso bulbosa (L.) Oakes ex Thompson
Fairy Slipper, Calypso

Glabrous erect herbs 5–20(23) cm tall; leaf petiolate, the petiole 1–6 cm long, the blade cordate-ovate to elliptic, plicate, 2–6 cm long, 1.5–5.2 cm wide; flowers variegated with purple, pink, and yellow; sepals 12–23 mm long; petals 15–23 mm long; lip pendant, 15-25 mm long; capsule erect, ellipsoid, 2–3 cm long.

Mossy woods; rather rare except on some small islands, central and southern Alaska, and southern Yukon; east to Labrador and south to California, Arizona, Colorado, Minnesota and New York; Eurasia (*Cypripedium bulbosum* L.; *Cymbidium boreale* Sev.; *Calypso borealis* [Sev.] Salisb.; *C. bulbosa* f. *occidentalis* Holz.; *C. bulbosa* ssp. *occidentalis* [Holz.] Calder & Taylor; *C. occidentalis* [Holz.] Heller).

CORALLORHIZA (Hall.) Chat.

Brownish, purplish, or yellowish saprophytic plants from coralloid rhizomes; stems lacking foliage leaves, clothed with membranous sheaths; flowers in terminal racemes, yellowish, brownish, or purplish; sepals subequal, the lateral ones united with the foot of the column and often forming a short spur which is partly or wholly adnate to the top of the ovary; lip

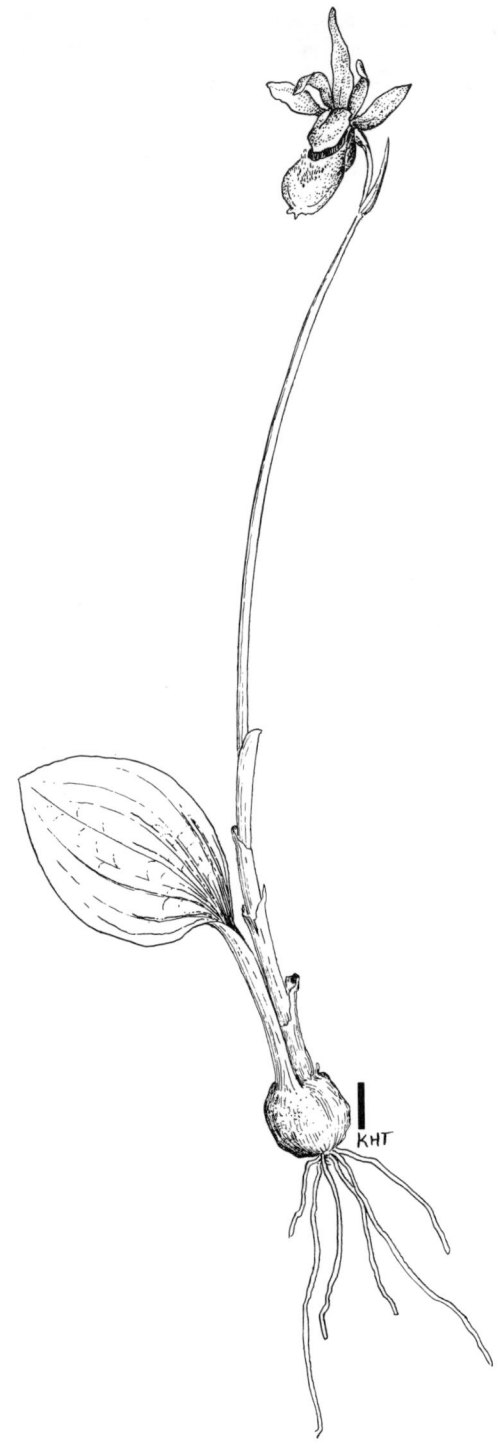

simple to 3-lobed; anther terminal, lidlike, with 4 waxy pollinia; capsule ellipsoid, pendant.

1a. Sepals 1-nerved; plants usually less than 2.5 dm tall; perianth 4.6–6.5 mm long. *C. trifida*
1b. Sepals 3-nerved; plants 1.5–5.5 dm tall; perianth 6–10 mm long. *C. mertensiana*

Corallorhiza mertensiana Bong.
Mertens' Coral-root

Erect glabrous leafless plants 1.5–5.5 dm tall; stems purple or brownish purple, with 2–3 sheathing bracts; flowers 1–25, purplish or greenish; sepals 3-nerved, 6–10.5 mm long; spur of sepal free from the ovary for about 1 mm; petals 3-nerved, 6–9 mm long; lip mottled reddish purple, 3-lobed, rounded to truncate or retuse at the apex; capsule 15–18 mm long.

Coniferous forest; in southeastern Alaska; south to California, Idaho, and Wyoming.

Corallorhiza trifida Chat.
Early Coral-root

Erect glabrous leafless plants 0.5–2.5(3.5) dm tall; stems pale yellow to deep yellow or greenish or brownish tinged, with 2–5 sheathing bracts, the bract apex often flaring and brownish; flowers 3–20, yellowish white to purplish, brownish, or greenish; sepals 1-nerved, 4.0–6.5 mm long; lip white, immaculate or maculate with purple, rounded to obtuse at the apex; capsule 8–12 mm long.

Woods and thickets; in much of southern Alaska and southern Yukon; eastward to Labrador and south to Oregon, Utah, Colorado, Indiana, and New Jersey; Eurasia.

CYPRIPEDIUM L.

Erect fibrous-rooted herbs from a rhizome; stems with 2 or more leaves; leaves plicate, prominently ribbed, the base sheathing; flowers showy, arranged in 1–several-flow-

Calypso bulbosa (L.) Oakes (× 0.6).

ered racemes; sepals spreading, free or more or less united; petals distinct, the lip sessile, inflated, saccate; column with a pair of fertile stamens; pollen granular; stigma somewhat 3-lobed; capsule ellipsoid.

1a. Leaves 2, subopposite to alternate; lip conspicuously purple blotched. *C. guttatum*
1b. Leaves 3 or more (or if only 2, then otherwise not as above), alternate; lip variously colored. (2)

2a. Sepals subequal to the lip; lip 1.2–2 cm long. *C. passerinum*
2b. Sepals 3–8 cm long, longer than or subequal to the lip; lip 2–3 cm long. (3)

3a. Lips yellow, purplish veined or spotted; plants of west-central and southern Yukon and adjacent British Columbia. *C. calceolus*
3b. Lips white, suffused or veined with purple; plants of coastal southeastern Alaska. *C. montanum*

Cypripedium calceolus L.
Lady's Slipper

Plants 1–3 dm tall or more from a slender rhizome; stems pubescent with short multicellular glandular hairs, the cross-walls yellowish, with (2)3–5 sheathing leaves; leaf blades elliptic to lanceolate, mostly 5–12 cm long, 3.5–6 cm broad; flowers solitary (rarely 2); sepals 3–5 cm long, (0.6)1–2 cm wide, greenish yellow to purplish or purplish veined, longer than or subequal to the lip; lip yellow, purplish veined, pouch- or slipper-shaped, mostly 2–3.5 cm long; capsule ellipsoid, glandular-hairy.

Open woods; in central western and southern Yukon and British Columbia; eastward to Newfoundland and south to Oregon, Arizona, New Mexico, Texas, Kansas, Louisiana, Alabama, and Georgia; Eurasia (*C. pubescens* Willd.; *C. parviflorum* Salisb.; *C. calceolus* var. *parviflorum* [Salisb.] Fern.; *C. calceolus* ssp. *parviflorum* [Salisb.] Hultén). Our materials belong to var. *pubescens* (Willd.) Correll.

Cypripedium guttatum Sw.
Spotted Lady's-slipper

Plants 1–3.5 dm tall from a slender rhizome; stems pubescent with multicellular nonglandular hairs, the cross-walls purplish with 2 subopposite to alternate leaves; leaf blades ovate to elliptic or lanceolate, 7–15

Cypripedium calceolus L. (× 0.4).

cm long, 3–6 cm wide; flowers solitary; sepals 1.3–2.8 cm long, shorter than or subequal to the lip; lip obovoid, 1.8–2.5(2.7) cm long, about 1–1.5 cm wide and deep white blotched with purple; capsule glandular-pubescent, strongly ribbed, reflexed.

Woods, heathlands, and meadows; in the Aleutians, Alaska Peninsula, and south-central to central and eastern Alaska, and southwestern to central Yukon; east to Mackenzie; Asia (*C. yatabeanum* Makino; *C. guttatum* ssp. *yatabeanum* [Makino] Hultén).

Cypripedium montanum Dougl. ex Lindl.
Mountain Lady's-slipper

Plants erect, 25–70 dm tall, from a stout rhizome; stems pubescent with multicellular hairs, the cross-walls yellowish, with 4–6 alternate leaves; leaf blades broadly ovate to elliptic-lanceolate, 5–16 cm long, 2.5–8 cm wide; flowers 1–3; sepals 3–6 cm long, brownish purple or dark green suffused with purple; lip globose, mostly white, tinged or veined with purple, 2–3 cm long, 1.5 cm wide and deep; capsule glandular-hairy, suberect, ellipsoid.

Open woods, often at high elevations; in southeastern Alaska and adjacent British Columbia; eastward to Alberta and Montana and south to California, Idaho, and Wyoming.

Cypripedium passerinum Richards.
Northern Lady's-slipper

Plants erect, 1.0–3.5 dm tall, from a slender rhizome; stems with 3–5 alternate leaves, pubescent with multicellular hairs, the cross-walls yellowish to colorless; leaf blades elliptic to ovate-lanceolate, 5–16 cm; lip obovoid, white to pink, often spotted with purple, 1–2 cm long, about 1 cm broad and deep; capsule erect to spreading obovoid, pubescent with multicellular hairs.

Woods and heathlands; from the Seward Peninsula eastward through central and southern Alaska and most of the Yukon; eastward to Quebec and southward to British Columbia, Alberta.

GOODYERA R. Br.

Scapose herbs from creeping rhizomes and fleshy fibrous roots; leaves basal or nearly so, ovate to lanceolate, dark green, often reticulate-veined or variegated with white; flowers white or cream-colored in 1-sided racemes on scapelike, bracted stems; lateral sepals distinct, the upper united with the lateral petals and forming a hood; lip deeply concave or saccate at the base, straight or recurved at the apex; anther with 2 pollinia; capsule erect, ovoid to ellipsoid.

1a. Perianth more than 5.5 mm long; leaves 4–10 cm long, 1.5–3.5 cm wide. *G. oblongifolia*
1b. Perianth less than 4 mm long; leaves 1–4.5 cm long, 0.6–2 cm wide. *G. repens*

Goodyera oblongifolia Raf.
Menzies' Rattlesnake Plantain

Erect scapose herb, the stem glandular-pubescent with multicellular hairs above, 1.0–4.5 dm tall; leaves usually oblong-elliptic, dark green or marked with white especially along the midvein, 4–11 cm long, 1.5–3.5 cm wide; flowers several to many, white, tinged with green; sepals 1-nerved, 6.5–10.5 mm long; lip 5–8 mm long, saccate; capsule about 1 cm long.

Woods; in southeastern Alaska; disjunctly eastward to Quebec and southward to California, Arizona, Minnesota, and New Hampshire.

Goodyera repens (L.) R. Br. ex Ait.
Lesser Rattlesnake Plantain

Erect scapose herb, the stems glandular-pubescent above with multicellular hairs, 1–4.5 dm tall; leaves ovate to oblong-elliptic, dark green with darker veins or the veins white margined, 1–4.5 cm long, 0.6–

2.0 cm wide; flowers several, white tinged with green; sepals 1-nerved, 3–3.5(4) mm long; lip 3–3.5 mm long, saccate; capsule about 6 mm long.

Woods; in central and south-central to eastern Alaska; eastward to Newfoundland and south to Arizona, New Mexico, South Dakota, Minnesota, Tennessee, and North Carolina; circumboreal. Two varieties are known from our region.

1a. Upper leaf surface green, the veins not conspicuously white margined, except the midvein sometimes white margined. *G. repens* var. *repens*

1b. Upper leaf surface conspicuously reticulate-veined; lateral veins white margined. *G. repens* var. *ophioides* Fern.

HABENARIA Willd.

Herbs of damp or wet sites from tuberous roots; stem leafy or merely bracted; leaves 1 or more, basal or cauline; flowers white or greenish, in spicate racemes; sepals free, the dorsal one erect or incurved and forming a hood; lip simple or 3-lobed, extended at the base into a spur; anther cells 2; pollen granular; capsules erect or spreading, cylindrical to ellipsoid.

1a. Lip unequally tridentate at the apex, the middle tooth small. *H. viridis*

1b. Lip entire at the apex. (2)

2a. Leaves 1–2 (rarely 3), basal or essentially so; stems without bracts or with a single one (rarely more) near the middle. (3)

2b. Leaves several, cauline or subbasal; stems leafy or with conspicuous bracts. (6)

3a. Perianth segments short, 1.5–2.5 mm long; spur bulbous. *H. chorisiana*

3b. Perianth segments mostly over 2.5 mm long; spur elongate. (4)

4a. Leaves 2 (rarely 3), large, orbicular to oval, reclining on the ground; spur over 14 mm long. *H. orbiculata*

4b. Leaves 1–2, small, obovate to linear-oblanceolate, erect or spreading; spur less than 12 mm long. (5)

5a. Spur subequal to the lip; lip linear to linear-lanceolate; plants widely distributed. *H. obtusata*

5b. Spur about twice as long as the lip; lip elliptic; plants of Aleutian Islands and coastal and insular south-central and southeastern Alaska. *H. behringiana*

6a. Leaves clustered at or near the base of the stem, usually withered at flowering; stem with numerous bracts; lip truncate at base or angled at each side; sepals 1-nerved. *H. unalascensis*

6b. Leaves cauline, persisting at flowering time; stem with foliaceous bracts; lip not truncate or angled at base; sepals 3-nerved. (7)

7a. Lip ovate-lanceolate, abruptly dilated at the base; flowers white, rarely greenish. *H. dilatata*

7b. Lip linear to broadly lanceolate, not dilated at the base; flowers usually greenish (sometimes purplish tinged). (8)

8a. Spur twice as long as the lip or longer. *H. behringiana*

8b. Spur subequal to the lip or shorter. (9)

9a. Spur bulbous to strongly saccate; lip linear to lanceolate; raceme usually laxly flowered, elongated. *H. saccata*

9b. Spur cylindrical, sometimes clavate; lip elliptical to lanceolate; raceme usually short and congested. *H. hyperborea*

Habenaria behringiana (Rydb.) Ames
Bering Bog-orchid

Plants erect, 0.7–1.8 dm tall; leaves solitary or rarely 2, arising near the middle of the stem, ovate-lanceolate to elliptic, 4–6 cm long, 1–2 cm wide, the single bract on the stem foliaceous, lanceolate; racemes few-flowered; flowers purplish; sepals 3-

nerved, 4–5 mm long; petals about 5 mm long; lip 5–8 mm long, elliptic to lanceolate, obtuse; spur slender, about 1 cm long.

Moist sites; in the western Aleutian Islands, endemic (*Platanthera tipuloides* var. *behringiana* [Rydb.] Hultén; *Limnorchis behringiana* Rydb.).

Habenaria chorisiana Cham.
Choris Bog-orchid

Plants erect, 0.7–2.5 dm tall; leaves subbasal, 2 or rarely only 1, ovate to lanceolate or elliptic, 4–6.5 cm long, 0.7–3.7 cm wide, the solitary stem bract (sometimes 2) lanceolate; racemes few- to several-flowered; flowers yellowish; sepals 1-nerved, 1.7–2.5 mm long; petals 3-nerved, 1.5–2 mm long; lip entire, about 1.5 mm long; spur bulbous, about 1 mm long.

Drier ridges in muskegs and in bogs; in coastal and insular southern Alaska from the Aleutians eastward to the Panhandle; south to British Columbia; Asia (*Platanthera chorisiana* [Cham.] Rchb.; *Limnorchis chorisiana* [Cham.] J. P. Anders.).

Habenaria dilatata (Pursh) Hook.
White Bog-orchid

Plants erect, 1.5–7(12.0) dm tall or taller; stems leafy; leaves few to several, linear to oblong or lanceolate, 10–30 cm long, 0.8–5.5 cm wide; racemes many-flowered; flowers white, yellowish, or greenish; sepals 3(4)-nerved, 3–9 mm long; petals 1–2-nerved, 4–8 mm long; lip strongly dilated at the base, 5–10 mm long; spur cylindrical, subequal to the lip.

Muskegs, woods, meadows, heathlands, rock outcrops, and bogs; in or near coastal and insular southern Alaska from the Aleutions eastward through the Panhandle and in southern Yukon; eastward to Newfoundland and south to California, Utah, Colorado, Minnesota, and New York (*Orchis dilatata* Pursh; *Platanthera dilatata* [Pursh] Lindl.; *P. dilatata* var. *angustifolia* Hook.; *P. dilatata* var. *chlorantha* Hultén; *H. dila-*

tata f. *chlorantha* [Hultén] B. Boi.; *H. dilatata* var. *albiflora* [Cham.] Correll). The varietal status of the white bog-orchid in Alaska needs clarification.

Habenaria hyperborea (L.) R. Br. ex Ait.
Northern Bog-orchid

Plants erect, 1.5–10 dm tall; stems leafy; leaves few to several, linear to oblong, lanceolate or elliptic, 5–30 cm long, 0.8–5 cm wide; racemes few- to many-flowered, lax to compact; flowers green or yellowish green; sepals 3-nerved, (2.5)3–9 mm long; petals 1–2-nerved, 3–7 mm long; lip linear to lanceolate, not conspicuously dilated at the base, (2)3–9(10) mm long; spur cylindrical, 3–8 mm long, usually shorter than the lip.

Stream beds, fens, lake margins, seeps, meadows, and open woods; in west-central to eastern and southern Alaska and southern Yukon; eastward to Newfoundland and south to California, Arizona, New Mexico, Nebraska, Minnesota, and New York; Greenland, Iceland. Two more or less distinctive phases of *H. hyperborea* occur in Alaska.

1a. Lateral sepals 4.5–6 mm long; lip 4.5–9 mm long; plants of coastal and insular southern Alaska from the Kenai Peninsula westward (*Orchis convallariifolia* Fisch.; *H. convallariifolia* (Fisch.) B. Boi.; *Platanthera convallariifolia* [Fisch.] Lindl.; *Limnorchis convallariifolia* [Fisch.] Rydb.; *P. convallariifolia* var. *dilatatoides* Hultén). *H. hyperborea* var. *viridiflora* (Cham.) Welsh comb. nov. (based on: *Habenaria borealis* var. *viridiflora* Cham. Linnaea 3:28, 1828).

1b. Lateral sepals 2.3–4 mm long; lip 2.5–4 mm long; plants of interior south-central to eastern and southeastern Alaska, less commonly in coastal southern region and in the Yukon (*Orchis hyperborea* L.; *Platanthera hyperborea* [L.] Lindl.; *Lim-*

norchis hyperborea [L.] Rydb.). *H. hyperborea* var. *hyperborea*

Habenaria obtusata (Banks) Richards.
Small Bog-orchid

Plants erect, 0.6–3.5 dm tall; stems naked or merely bracteate; leaves solitary or rarely 2, obovate to oblanceolate, 3–15 cm long, 0.8–4.5 cm wide; racemes few- to many-flowered; flowers greenish white; sepals 3-nerved, 3–7 mm long; petals 1-nerved, (3) 4–5.5 mm long; lip linear-lanceolate (5)6–10 mm long; spur slender, 3–8 mm long, subequal to the lip.

Woods, thickets, muskegs, and stream sides; in most of Alaska from the Brooks Range southward (except for the Aleutians) and most of the Yukon; eastward to Newfoundland and southward to Utah, Colorado, Minnesota, Illinois, and New York (*Orchis obtusata* Pursh; *Platanthera obtusata* [Pursh] Lindl.; *Lysiella obtusata* [Pursh] Britt. & Rydb.).

Habenaria orbiculata (Pursh) Torr.
Round-leaved Bog-orchid

Plants erect, 0.6–6 dm tall; leaves 2, basal, the scape with 1–several bracts, the blades orbicular, 7–25 cm long, 5–15 cm wide; racemes few- to several-flowered; flowers greenish white; sepals strongly nerved, 5–15 mm long; petals 5–12 mm long; lip linear-oblong, 10–20 mm long; spur cylindrical, 15–45 mm long, longer than the lip.

Woods; in extreme southeastern Alaska and northern British Columbia; disjunctly eastward to Newfoundland and south to Oregon, Idaho, Montana, and Georgia (*Orchis orbiculata* Pursh; *Platanthera orbiculata* [Pursh] Lindl.; *Lysias orbiculata* [Pursh] Rydb.).

Habenaria saccata Greene
Slender Bog-orchid

Plants erect, 1.5–10 dm tall or more; stems leafy; leaves several, narrowly lanceolate to elliptic, 4–15 cm long, 1–4 cm wide; racemes few- to many-flowered; flowers greenish; sepals 3-nerved, 3–6 mm long; petals 1–2-nerved, 3–5 mm long; lip linear to oblong, 4–8 mm long; spur cylindrical to bulbous, much shorter than or subequal to the lip.

Open woods, thickets, muskegs, and meadows; in or near coastal and insular southern Alaska from the easternmost Aleutians eastward through the Panhandle; southward to California and eastward to Arizona, New Mexico, Colorado, Wyoming, Montana, and Alberta (*Platanthera stricta* Lindl.; *Limnorchis stricta* [Lindl.] Rydb.; *P. saccata* [Greene] Hultén; *Limnorchis gracilis* [Lindl.] Rydb.; *Habenaria gracilis* [Lindl.] Wats.; *H. saccata* var. *gracilis* [Lindl.] B. Boi.).

Habenaria unalascensis (Spreng.) S. Wats.
Alaska Bog-orchid

Plants erect, scapose, 2.5–9.0 dm tall; stems tan to purplish brown, leafy at or near the base; leaves 2–4, oblanceolate to narrowly lanceolate, usually withered by flowering time, 7–15 cm long or more, 1–3(5) cm wide; racemes many-flowered; flowers white to yellowish green, often marked with purple; sepals 1-nerved, 2–4 mm long; petals 1-nerved, 2–4 mm long; lip ovate to lanceolate, 2.5–4.5 mm long; spur cylindrical, (1.5)3–5 mm long, about equal to the lip.

Muskegs, meadows, and woods; in coastal and insular southern Alaska from Unalaska eastward through the Panhandle; southward to California, Nevada, Utah, Colorado, South Dakota, and disjunctly to Quebec (*Spiranthes unalascensis* Spreng.; *Platanthera unalascensis* [Spreng.] Kurtz; *Piperia unalaschensis* [Spreng.] Rydb.).

Habenaria viridis (L.) R. Br.
Long-bracted Bog-orchid

Plants erect, 0.6–6.0 dm tall; stems leafy; leaves several, obovate to oblong or lanceolate, 4–15 cm long, 1–6.5 cm wide; racemes several- to many-flowered; flowers green to yellowish green; sepals 3–6 mm long; petals 3–5 mm long; lip 5–10 mm long, cuneate, 2–3-toothed apically, the central tooth often small, spur shorter than the lip, somewhat bulbous.

Tundra to heathlands, woods, and meadows; in western and south-central to coastal and insular southern Alaska and southern Yukon; eastward to Newfoundland and south to Utah, New Mexico, Nebraska, Iowa, North Carolina; Eurasia.

1a. Lowermost bracts 1–2 times longer than the flowers; leaves placed near stem base; plants of west-central Alaska (*Saltyrium viride* L.; *Peristylus islandicus* Lindl.; *Coeloglossum viride* var. *islandicum* [Lindl.] Schultz). *H. viridis* var. *viridis*

1b. Lowermost bract 2–several times longer than the flowers; leaves placed some distance above stem base; plants broadly distributed (*Orchis bracteata* Muhl.; *Coeloglossum viride* ssp. *bracteatum* [Muhl.] Hultén). *H. viridis* var. *bracteata* (Muhl.) Gray

LISTERA R. Br.

Slender inconspicuous woodland herbs; leaves 2, opposite, borne near the middle of the stem; flowers few to several in terminal racemes, greenish or purplish; sepals and petals distinct, similar and subequal; lip longer than the sepals and other petals; pollinia 2, powdery; capsule ovoid to obovoid.

1a. Lip narrow, deeply cleft into linear filiform lobes. *L. cordata*

1b. Lip oblong to narrowly cuneate or obovate, only slightly cleft or notched at the apex into oblong or rounded lobes. (2)

2a. Lip with auricles, about as broad basally as apically, lacking lateral teeth, the apex broadly notched; plants of broad distribution in eastern continental Alaska, rarely elsewhere. *L. borealis*

2b. Lip lacking auricles, broadest at the apex, with small teeth at the base, the apex retuse to entire, not broadly notched; plants coastal and insular. (3)

3a. Lip with a short slender claw, the lateral teeth very small; plants of the Aleutian Islands. *L. convallarioides*

3b. Lip sessile, the lateral teeth prominent; plants of southeastern Alaska. *L. caurina*

Listera borealis Morong
Northern Twayblade

Plants 0.6–2.5 dm tall; stems greenish to yellowish; leaves subopposite, elliptic to ovate, 1–6 cm long, 0.7–3 cm wide; racemes few-flowered; flowers green or yellowish green, the ovaries, pedicels, and rachis often pubescent with long white glandular hairs; sepals 1-nerved, 4–7 mm long; petals 1-nerved, 4–6 mm long; lip 8–13 mm long, 5–7 mm broad at the apex, broadly oblong, broadly notched at the widened apex, with two rounded lobes; capsule ovoid, to 5 mm long.

Muskegs, woods; in continental eastern to south-central (and disjunctly in west-central) Alaska and most of the Yukon; eastward to Labrador and southward to Utah and Colorado.

Listera caurina Piper
Western Twayblade

Plants 1.0–3.0(3.4) dm tall; stems slender; leaves 2, subopposite, suborbicular to ovate, (2)2.5–7 cm long, (1.4)1.8–4.5 cm broad; racemes few- to many-flowered; flowers greenish to yellowish, the ovaries, pedicels, and rachis often pubescent with multicellular hairs; sepals 1-nerved, 3–4 mm long; petals 1-nerved, 2.8–3.5 mm long; lip 4.5–6.5(8) mm long, 3.5 mm broad, cuneate, rounded or emarginate at the apex; capsule ovoid, to 7 mm long.

Woods; in southeastern Alaska; eastward to Newfoundland and south to California, Idaho, and Montana.

Listera convallarioides (Sw.) Nutt.
Broad-leaved Twayblade

Plants 1–3.5 dm tall; stems glandular-pubescent above; leaves 2, opposite or subopposite, broadly ovate to elliptic, 2–7 cm long, 1.5–6 cm broad; racemes many-flowered; flowers yellowish green; sepals 1-nerved, 4.5–6 mm long; petals 1-nerved, 4–5 mm long; lip 8–13 mm long, 5–7 mm wide near the apex, narrowly cuneate, with a shallow notch at the apex; capsule ovoid, to 4 mm long.

Moist sites along streams and lake shores; in the Aleutian Islands; disjunctly to British Columbia, east to Newfoundland and south to California, Arizona, Wyoming, Michigan, and New York (*Epipactis convallarioides* Sw.; *L. eschscholtziana* Cham.).

Listera cordata (L.) R. Br. ex Ait.
Heart-leaved Twayblade

Plants 0.7–2.5 dm tall; stems glabrous or slightly pubescent just above the leaves; leaves 2, opposite, ovate-cordate, 1–4(5.5) cm long, 1–4 cm wide; racemes few- to several-flowered; flowers yellowish green to purplish; sepals 2–3 mm long; petals 1.5–2.5 mm long; lip 3–6 mm long, linear-oblong, cleft deeply into 2 linear-lanceolate lobes; capsule ovoid, 4–6 mm long.

Woods and thickets, and in damp places along streams or in bogs; in the Aleutian Islands and in coastal southwestern, southern, and southeastern Alaska and southern Yukon; east to Newfoundland and south to California, Nevada, Utah, New Mexico, Michigan, and North Carolina; circumboreal.

MALAXIS Soland.

Slender inconspicuous herbs; leaves 1–5 on the lower part of the stem; flowers several to many in a raceme, whitish or greenish; sepals distinct or partially connate; lip usually uppermost in the flower, entire or variously lobed; pollinia 4, waxy; capsule ovoid to ellipsoid.

1a. Leaves 1 (rarely 2); lip auricled basally, the margin adjacent to the auricles verrucose-thickened. *M. monophyllos*
1b. Leaves 2–5; lip not auricled but often bearing lateral lobes near the base, not verrucose-thickened. *M. paludosa*

Malaxis monophyllos (L.) Sw.
White Adder's-tongue

Plants 1–2.5(3.3) dm tall, arising from a corm; leaves 1 or rarely 2 and subopposite, broadly ovate to elliptic or lanceolate, 1–10(15) cm long, (0.6)1–5 cm wide; racemes many-flowered; flowers small, pale greenish to yellowish green; sepals 1-nerved, 2–3 mm long; petals 1-nerved, 2–2.5 mm long; lip 3-lobed, 1.6–2.5(3.5) mm long, the central lobe tapering to a point; capsule ovoid, 3–6 mm long.

Gravelly beaches, open woods, meadows and tidal flats; in coastal and insular southern Alaska from the Aleutians eastward through the Panhandle; disjunctly from British Columbia and California eastward to northeastern North America; Eurasia (*Ophrys monophyllos* L.).

1a. Lip uppermost in the flower. *M. monophyllos* var. *monophyllos*
1b. Lip lowermost in the flower (*Microstylis brachypoda* Gray). *M. monophyllos* var. *brachypoda* (Gray) Morris & Ames

Malaxis paludosa (L.) Sw.
Bog Adder's-tongue

Plants 0.5–2.0 dm tall from a small corm; leaves 2–5 in a basal cluster, ovate to elliptic or obovate, 1–3 cm long, 0.2–1 cm wide; racemes several- to many-flowered; flowers yellowish green; sepals 2–2.5 mm long; petals 1–1.5 mm long; lip 1–2 mm long, uppermost in the flower, triangular; capsule ovoid, 2–4 mm long.

Muskegs and bogs; in south-central to southeastern Alaska; in disjunct localities eastward to the Great Lakes; Eurasia

(*Ophrys paludosa* L.; *Hammarbya palu-dosa* [L.] Kuntze).

ORCHIS L.

Scapose or leafy, mostly succulent herbs; leaves 1–few, basal or borne along the stem; flowers more or less showy, few to several in a raceme; sepals and petals distinct; lip simple or more or less lobed, produced at the base into a conspicuous spur; pollen cohering in waxy masses; capsule ellipsoid.

1a. Stem with 3–several leaves; floral bracts large, foliose. *O. aristata*
1b. Stem scapose, the single blade-bearing leaf basal; floral bracts slender, not foliose. *O. rotundifolia*

Orchis aristata Fisch. ex Lindl.
Fischer's Orchis

Plants 1–4 dm tall arising from fibrous and tuberous roots; leaves 3–more, oblanceolate to lanceolate, (4)5–15 cm long, (0.7) 1–5 cm wide; racemes few- to several-flowered; flowers magenta to violet purple or whitish; sepals 3–5-nerved, 9–13 mm long; petals 3-nerved, to 8 mm long; lip often purple-spotted, broadly ovate; spur 10–15 mm long; capsule to 12 mm long.

Meadows; in coastal and insular Alaska, from Prince William Sound westward through the Aleutians and on Nunivak Island (*Dactylorhiza aristata* [Fisch. ex Lindl.] Soo).

Orchis rotundifolia Banks ex Pursh
Round-leaved Orchis

Plants 1–3.5 dm tall, arising from slender rhizomes; leaves 1, nearly basal, orbicular to elliptic, 3–10 cm long, (1.2)2–7 cm wide; racemes 1–several-flowered; flowers pink to white; sepals 3–5-nerved, 6–10 mm long; petals 2–3-nerved, 5–6 mm long; lip 5–10 mm long, white, purple-spotted, 3-lobed; spur 3–7 mm long; capsule to 15 mm long.

Woods and muskegs; in central western to east-central and south-central Alaska and southern Yukon; eastward to Labrador and south to British Columbia, Montana, Wyoming, Minnesota, and New York; Greenland (*Amerorchis rotundifolia* [Banks] Hultén).

SPIRANTHES L. C. Richards.

Leafy herbs with tuberous-thickened or fleshy-fibrous roots; leaves alternate, reduced upwards, several; flowers in twisted spikes, white or cream-colored, small, spurless; sepals and petals distinct; lip concave, dilated at the reflexed apex; pollinia 2, powdery-granular; capsule ellipsoid to obovoid.

Spiranthes romanzoffiana Cham.
Hooded Ladies' Tresses

Plants erect, 0.8–3.5(5) dm tall; leaves several, mostly basal, linear to oblanceolate, 5–15(25) cm long, 0.3–1(1.5) cm wide; spike densely many-flowered, composed of 3 spiral ranks; flowers white or creamy white; sepals 7–10(14) mm long; petals 6–12 mm long; lip 5–11 mm long, 5 mm wide, constricted above the middle, recurved near the apex; capsule to 10 mm long.

Bogs, meadows, muskegs, and open woods; in much of Alaska south of the Brooks Range (except for coastal western regions) and in most of the Yukon south of the 65th parallel; eastward to Newfoundland and south to California, Utah, Colorado, and Pennsylvania.

POTAMOGETONACEAE
Pondweed Family

Perennial herbs of fresh water, rhizomatous; stems jointed, leafy, rooting at the nodes; leaves 2-ranked, alternate or opposite, sheathing at the base or sessile; flowers perfect or imperfect, in pedunculate axillary or terminal spikes; perianth of 4 distinct segments; stamens 4, attached to the perianth segments; pistils 4, superior; fruit an achene.

POTAMOGETON L.

Aquatic herbs; leaves alternate or the upper opposite, often dimorphic, the submersed thin and narrow, the floating broad and leathery; stipules prominent, sometimes sheathing; flowers sessile in pedunculate spikes, usually in whorls.

Fernald, M. L. 1932. The linear-leaved North American species of *Potamogeton*, Section Axillares. Mem. Gray Herb. 3: 1–183.

Ogden, E. C. 1943. The broad-leaved species of *Potamogeton* of North America north of Mexico. Rhodora 45:57–105, 119–63, 171–214.

1a. Submersed leaves with stipules adnate to the leaf base and forming a sheath around the stem, the blades arising from the summit of the sheath. (2)

1b. Submersed leaves with stipules free from the leaf base, the leaf blades or petioles attached directly to the node. (5)

2a. Blades of submersed leaves 3–8 mm broad, spinulose-serrulate apically, finely many-nerved, the leaves crowded on the stem. *P. robbinsii*

2b. Blades of submersed leaves 1–2 mm broad or less, entire, 1–3-nerved, the leaves not crowded on the stem. (3)

3a. Stipular sheaths of primary leaves more than twice as broad as the stem; spikes with 5–9 evenly spaced whorls. *P. vaginatus*

3b. Stipular sheaths of primary leaves commonly less than twice as broad as the stem; spikes with 2–5(6) whorls. (4)

4a. Sheathing stipules connate around the stem unless ruptured; fruit 1.8–2.5 mm long, beakless, the stigma sessile. *P. filiformis*

Spiranthes romanzoffiana Cham. (× 0.6).

4b. Sheathing stipules not connate around the stem; fruit 2.5–3.5 mm long, shortly beaked, the stigma on a very short style. *P. pectinatus*

5a. Leaves commonly more than 10 mm broad, sessile and rounded to clasping at the base, all submersed. (6)

5b. Leaves less than 10 mm broad or if broader then petiolate or not rounded to clasping at the base. (7)

6a. Leaves ovate to ovate-lanceolate, the largest ones less than 10 cm long; stipules soon shredded into white fibers 1–2 cm long; fruits 2.5–3.5 mm long. *P. perfoliatus*

6b. Leaves lance-oblong, the largest more than 10 cm long; stipules persistent, not shredded into fibers; achenes 4–5 mm long. *P. praelongus*

7a. Stems flattened, wing-margined, more than half as broad as the leaves; leaves all submersed, 2–5 mm broad, many-veined. *P. zosterifolius*

7b. Stems terete or only slightly flattened, not or only slightly wing-margined, variable in width; leaves all submersed or some floating. (8)

8a. Leaves more than 4 mm broad, at least some often dimorphic, the broader ones commonly floating. (9)

8b. Leaves less than 3.5 mm broad, linear, all submersed and of the same form. (12)

9a. Floating leaves with blades commonly 5 cm long or more, subcordate to rounded or obtuse basally with 19 or more nerves. *P. natans*

9b. Floating leaves with blades commonly less than 5 cm long, obtuse to cuneate basally (rounded in some *P. gramineus*, but then less than 19-nerved) or all leaves submersed and sessile. (10)

10a. Floating leaves, if any, not much different than the submersed ones; plants generally reddish tinged. *P. alpinus*

10b. Floating leaves much different from the submersed ones; plants generally green. (11)

11a. Submersed leaves linear, commonly 10–20 cm long, with a broad median strip of enlarged air chambers, the stripe 1 mm broad or more, the air chambers often more than 3 mm long. *P. epihydrus*

11b. Submersed leaves oblong-lanceolate, less than 10 cm long, lacking air chambers and hence lacking a median stripe. *P. gramineus*

12a. Submersed leaves, at least some, semi-terete, straight and stiffly spreading, without distinction of blade and petiole, soon deciduous. *P. natans*

12b. Submersed leaves all thin and flexible, with evident blades, more or less persistent. (13)

13a. Stipules conspicuous, commonly 10–20 mm long, strongly nerved, soon shredded into persistent white fibers; stems somewhat flattened. *P. friesii*

13b. Stipules conspicuous or inconspicuous, commonly 5–15 mm long, not strongly nerved or shredded or if shredded then lacking persistent white fibers; stems subterete or flattened. (14)

14a. Stipules connate around the stem; stems somewhat flattened and narrowly winged; nodes lacking paired globose yellowish to whitish glands situated on either side. *P. foliosus*

14b. Stipules with free margins; stems subterete or slightly flattened, not winged; nodes commonly with paired globose yellowish or whitish glands situated on either side. (15)

15a. Leaves with a narrow dark midvein and commonly 1 (or 2) pair of lateral veins; fruits 2–2.5 mm long. *P. berchtoldii*

15b. Leaves with a broad (to 0.3 mm) pale midvein and commonly 3–6 pairs of lateral veins; fruit (2.8)3–4 mm long. *P. subsibiricus*

Potamogeton alpinus Balbis
Northern Pondweed

Stems terete or nearly so, up to 10 dm long or more, commonly tinged reddish (as is the rest of the plant); submersed leaves (4.5)6–18(20) cm long, 5–12 mm broad, linear- to oblong-lanceolate, sessile, 7–9-nerved; floating leaves (usually lacking) transitional to the submersed ones, the blades 2.5–6 cm long, 0.5–2 cm broad, obovate to oblanceolate or elliptic, 7–15-nerved, tapering to the petioles; stipules membranous with freed edges, not adnate to the leaf or shredded; peduncles 3–12 (15) cm long, about as thick as the stem; spikes 1.5–3 mm long, with 5–9 whorls of flowers, crowded or the lower separated; achenes 3–4 mm long, obliquely obovoid, the dorsal keel prominent, the curved beak 0.4–0.6 mm long.

Lakes, ponds, and sluggish streams; in most of Alaska (except for the northern coastal plain and the western Aleutians) and most of the Yukon; east to the Atlantic and south to California, Colorado, Utah, and Pennsylvania; circumboreal (*P. tenuifolius* Raf.; *P. alpinus* ssp. *tenuifolius* [Raf.] Hultén). Our material belongs to var. *tenuifolius* (Raf.) Ogden.

Potamogeton berchtoldii Fieb.
Berchtold Pondweed

Stems subterete, up to 10 dm long, green; leaves all submersed, 2–4.5(6) cm long, 0.5–1.5(2) mm broad, sessile, usually 3-veined, the midvein broadened towards the base (due to lateral air chambers); stipules membranous with green edges, not adnate to the leaf, not shredded, soon deciduous, bordered at the base by glands; peduncles 0.8–3(5) cm long, very slender, about as thick as the stem; spikes 0.2–1 cm long, with 1–3 crowded or loose whorls of flowers; achenes 2–2.5 mm long, obliquely obovoid, the dorsal keel rounded, the oblique beak 0.4–0.6 mm long.

Ponds and lakes; in the southeastern three-quarters of Alaska and southern Yukon;

eastward to the Atlantic and southward to California, Nebraska, Indiana, and Delaware (*P. pusillus* authors, not L.).

Potamogeton epihydrus Raf.
Nuttall Pondweed

Stems somewhat flattened, up to 10 dm long or more, green; submersed leaves 5–20 cm long, (2)3–10 mm broad, sessile, 3–9-nerved, with a broad median band (commonly 1–1.5 mm broad) made up of air chambers to 3 mm long or more; floating leaves usually conspicuously different than the submersed, the blades 2.5–4.5(6) cm long, 8–20 mm broad, elliptic to oblong, with 15 or more nerves, tapering to flattened petioles usually shorter than the blade; stipules membranous, free from the leaf, not connate, not shredded; peduncles 2–8 cm long, about as thick as the stem; spikes mostly 1.5–4 cm long, densely flowered; achenes 3–4 mm long, obliquely obovoid, the dorsal keel sharp, the beak very short.

Ponds, lakes, and streams; in central southern and southeastern Alaska and in the westernmost Aleutian Islands; disjunctly eastward to the Atlantic and southward to California, Idaho, and Colorado in the west and to Iowa, Indiana, and Pennsylvania in the east (*P. nuttallii* Cham. & Schlecht.; *P. epihydrus* var. *nuttallii* [Cham. & Schlecht.] Fern.; *P. epihydrus* ssp. *nuttallii* [Cham. & Schlecht.] Calder & Taylor; *P. nuttallii* var. *ramosus* Peck; *P. epihydrus* var. *ramosus* [Peck] House).

Potamogeton filiformis Pers.
Filiform Pondweed

Stems subterete, up to 4–5 dm long, usually green or less commonly straw-colored; leaves all submersed, 5–12 cm long, 0.3–1 (1.5) mm broad, linear, sessile atop the connate-sheathing stipules, 1-nerved; stipules adnate to the leaf and connate-sheathing around the stem, the free upper portion ligulelike, not bordered at the base by glands; peduncles 3–15 cm long, very slender, about as thick as the stem; spikes 1–3 cm long, with 2–5 remote or crowded whorls of flowers; achenes 2–2.5 mm long, obovoid, the dorsal keel rounded, the stigma sessile, broad and low, not beaklike.

Lakes, ponds, streams, and tidal flats; in most of Alaska and southern Yukon; east to the Atlantic and south to California, Arizona, Colorado, Michigan, and Pennsylvania; circumboreal.

Potamogeton foliosus Raf.
Leafy Pondweed

Stems flattened and narrowly winged, up to 10 dm long, usually green; leaves all submersed, 3–10 cm long, commonly 1–2 mm broad, linear, sessile, 1–5-nerved; stipules membranous, free from the leaf, connate around the stem, not fibrous shredded, ultimately fragmented, lacking paired basal glands; peduncles 0.5–2 cm long, rather stout; spikes 0.3–0.5 cm long, with 2–3 crowded whorls of flowers; achenes 2–2.5 mm long, obliquely obovoid, the dorsal keel prominent and toothed, the beak 0.2–0.4 mm long.

Ponds and streams; in central to east-central and south-central Alaska; disjunctly eastward and southward over much of North America. This entity is apparently closely related to *P. berchtoldii*. More collections are necessary in order to evaluate the true range and infraspecific status of Alaskan materials.

Potamogeton friesii Rupr.
Fries Pondweed

Stems somewhat flattened, up to 10 dm long, green; leaves all submersed, 2–8 cm long, 1.5–3(3.5) mm broad, linear, sessile, 3–7 (commonly 5)-nerved; stipules white veined, conspicuous, free from the leaf, connate around the stem but soon ruptured, ultimately shredded into persistent white fibers, commonly with paired glands at the base; peduncles 1–5 cm long, slightly flattened and winged; spikes mostly 1–2 cm long, usually with 3–4 crowded to loose

whorls of flowers; achenes 2–2.5 mm long, obliquely obovoid, the dorsal keel rounded, the curved beak 0.4–0.6 mm long.

Ponds and lakes; in central and south-central Alaska and southern Yukon; in disjunct localities eastward to the Atlantic and south to Washington, North Dakota, Iowa, Indiana, and Virginia.

Potamogeton gramineus L.

Stems subterete, up to 10 dm long or more, commonly green; submersed leaves 2–10 cm long, 3–10 mm broad, narrowly lanceolate to oblong-lanceolate, sessile, 3–9-nerved, the midvein much less than 1 mm broad; floating leaves usually present, distinctly different from the submersed ones, the blades 1.5–6(7) cm long, 0.5–2.5 cm broad, elliptic to oblong or oblong-lanceolate, commonly 13–17-nerved, rounded to obtuse or acute basally, the petioles from shorter to longer than the blade; stipules brownish, thick, with membranous free margins, not adnate to the leaf, persistent; peduncles 2–11 cm long, club shaped, the upper end thicker than the stem; spikes 1–3.5 cm long, with 5–10 crowded whorls of flowers; achenes 2.5–2.9 mm long, obliquely obovate, the dorsal keel prominent, the slightly curved beak 0.2–0.4 mm long.

Ponds, lakes, and streams; in most of Alaska south of the Brooks Range and in southern Yukon; widely distributed in North America; circumboreal (*P. heterophyllus* authors, not Schreb.). Some phases of *P. gramineus* approach *P. natans* (see below), and poorly collected specimens lacking submersed leaves cannot always be easily assigned to one entity or the other. However, the smaller number of veins, fruit size, and beak length appear to be diagnostic.

Potamogeton natans L.
Floating Pondweed

Stems subterete, up to 10 dm long or more, commonly green or greenish; submersed

Potamogeton natans L. (× 0.25).

leaves 5–20 cm long or more, 1–2 mm broad, linear, not or only slightly differentiated into blade and petiole, soon deciduous; floating leaves 2.5–8.8 cm long, 0.8–4.5 cm broad, lance-elliptic to elliptic or ovate, commonly 19–25-nerved (or more), rounded to obtuse or subcordate basally, the petioles from shorter to much longer than the blades; stipules brownish or greenish, thick, with membranous free margins, not adnate to the leaf, persistent; peduncles 4–9 cm long, often broader than stem; spikes 1.5–4.5(5) cm long, with 10 or more crowded whorls of flowers; achenes 3–5 mm long, obliquely obovate, the dorsal keel rounded, the curved beak 0.5–0.8 mm long.

Lakes and ponds; in south-central and southeastern Alaska (and in southern Yukon?); eastward to the Atlantic and south to California, Arizona, New Mexico, Nebraska, Iowa, Indiana, and New Jersey; circumboreal.

Potamogeton pectinatus L.
Fennel-leaf Pondweed

Stems terete, up to 3–4 dm long, commonly greenish or straw-colored; leaves all submersed, 2.5–12 cm long, 0.2–1 mm broad, linear, sessile atop the sheathing stipules, 1(3)-nerved; stipules adnate to the leaf base, closely investing the stem but not connate, the upper free portion ligulelike, not bordered at the base by glands; peduncles 5–15 cm long, very slender, about as thick as the stem; spikes 1–5 cm long, with 3–5(6) unequally spaced clusters of flowers; achenes 2.5–3.5 mm long, obliquely obovoid, the dorsal keel rounded, the curved beak 0.3–0.5 mm long.

Ponds and lakes; from Kotzebue and Cape Thompson eastward along the Brooks Range and in the southeastern quarter of continental Alaska and southern Yukon; widely distributed in North America; Eurasia. This entity is easily mistaken for *P. filiformis* which it closely resembles.

Potamogeton perfoliatus L.
Clasping-leaf Pondweed

Stems terete, to 10 dm long or more, green to brownish; leaves all submersed, 2–7 cm long, 7–23 mm broad, ovate to lanceolate, sessile and cordate-clasping, 13–25-veined (or more); stipules white to brownish, prominently veined, soon shredded into white persistent fibers, not adnate to the leaf, not connate; peduncles 1–10 cm long or more, about as thick as the stem; spikes 1–4 cm long, with 6 or more closely crowded whorls of flowers; achenes 2.5–3.5 mm long, obliquely obovoid, the dorsal keel rounded, the almost straight beak 0.4–1 mm long.

Lakes, ponds, and streams; in most of Alaska south of the Brooks Range (except for the Panhandle, where it is to be sought) and most of the Yukon; widely distributed in North America; circumboreal (*P. perfoliatus* var. *richardsonii* Bennett; *P. richardsonii* [Bennett] Rydb.). Our material belongs to ssp. *richardsonii* (Bennett) Hultén.

Potamogeton praelongus Wulf.
White-stemmed Pondweed

Stems slightly flattened, to 20 dm long or more, greenish, reddish, or whitish; leaves all submersed, 10–15 cm long or more, 5–25 mm broad, oblong-lanceolate, sessile and somewhat clasping, with 3–7 main veins; stipules membranous, shortly adnate to the leaf base, not connate, not shredded; peduncles 7–20 cm long or more, often club shaped and broader than the stem; spikes (1)2–5 cm long, with 6 or more crowded whorls of flowers; achenes 4–5 mm long, obliquely obovoid, the dorsal keel prominent, the beak short.

Lakes; in disjunct sites from the Seward Peninsula and western Aleutians eastward to central, south-central, and northwestern Alaska and western Yukon; eastward to Newfoundland and south to Utah, Colorado, Nebraska, Indiana, and New York.

Potamogeton robbinsii Oakes
Robbins Pondweed

Stems terete, up to 10 dm long, greenish; leaves all submersed, 3–12 cm long, commonly 3–4 mm broad, linear-lanceolate, sessile atop the sheathing stipules, very finely many-nerved, 2-ranked and closely crowded on the stem; stipules adnate to the leaf base, the free portion strongly many-veined, ultimately shredded; peduncles 3–7 cm long, about as thick as the stems; spikes 0.7–2 cm long, with 2–4 compact to loose whorls of flowers; achenes 3.5–4 mm long, obovoid, the dorsal keel prominent, the beak conspicuous.

Lakes and ponds; in south-central Alaska (where evidently rare); disjunctly from British Columbia to Quebec and south to Idaho, Wyoming, Indiana, and Pennsylvania.

Potamogeton subsibiricus Hagstr.

Stems slightly flattened, up to 5 dm long,

green or greenish; leaves all submersed, 2–8 cm long, 1–1.5 mm broad, linear, sessile, with 9–17 nerves; stipules free from the leaf, not connate, whitish, membranous, not shredded, commonly with paired glands at the base; peduncles 2–5 cm long, about as thick as the stem; spikes 0.7–1 cm long, with 3–4 crowded whorls of flowers; achenes 2.9–4 mm long, obliquely obovoid, the dorsal keel rounded, the curved beak 0.3–0.5 mm long.

Lakes and ponds; from Kotzebue eastward through central and north-central Alaska to southern Yukon and the Mackenzie; east to Hudson Bay; Asia (*P. porsildiorum* Fern.).

Potamogeton vaginatus Turcz.
Sheathed Pondweed

Stems terete, up to 7 dm long or more, greenish or straw-colored; leaves all submersed, 2–15 cm long, 0.3–2 mm broad, linear to filiform, sessile atop the adnate sheathing stipules, 1–3-nerved; stipules adnate to the leaf base, loosely sheathing and 2–several times broader than the stem, not connate, often brownish, the upper free portion ligulelike, not bordered at the base by glands; peduncles 6–12 cm long, slender, about as thick as the stem; spikes 3–5 cm long with 5–9 evenly spaced clusters of flowers; achenes 2.8–3.3 mm long, obliquely obovoid, the dorsal keel obtuse, the stigma subsessile, the beak to 0.2 mm long.

Lakes, ponds, and streams; in most of Alaska and Yukon; east to Newfoundland and south to Oregon, Wyoming, Wisconsin, and New York; Eurasia (*P. interior* authors, not Rydb.?).

Potamogeton zosterifolius Schum.

Stems flattened and more or less winged, half as broad as the leaves or more, up to 8 dm long or more, greenish; leaves all submersed, 6–12 cm long or more, 2–4(5) mm broad, linear, sessile, with many veins; stipules not adnate to the leaf, not connate, whitish, firm, many-veined, eventually shredded, lacking paired glands at the base; peduncles 2–10 cm long, flattened, about as thick as the stem; spikes 1–2.5 cm long, with 3–5 crowded whorls of flowers; achenes 3.5–4.5 mm long, obliquely ob-ovoid, the dorsal keel obtuse to acute, more or less undulate-toothed, the slightly curved beak 0.4–0.7 mm long.

Lakes and ponds; in the southeastern third of Alaska (except for the Panhandle, where it should be expected) and to be sought in southern Yukon; eastward to Quebec and south to California, Idaho, Montana, Ne-braska, Illinois, and Virginia (*P. compressus* authors, not L.; *P. zosteriformis* Fern.). Our material belongs to ssp. *zosteriformis* (Fern.) Hultén.

RUPPIACEAE
Ditchgrass Family

Slender submersed branching aquatic pe-rennial herbs; leaves alternate or rarely op-posite, with sheathing adnate stipules; flow-ers perfect, enclosed in sheathing leaf bases, small, typically 2 per terminal spike; perianth lacking; stamens commonly 2, the anthers with a broad connective; pistils typically 4, the stigma broad and flat; fruit a drupelet.

RUPPIA L.

Stems filiform; leaves filiform; peduncles elongating and spirally coiled at maturity; flowers with 2 sessile anthers and 4 pistils sessile in flower and elevated on elongated stipes in fruit.

Fernald, M. L., and K. M. Wiegand. 1914. The genus *Ruppia* in eastern North America. Rhodora 16:119–27.

Ruppia maritima L.
Ditchgrass

Stems terete or nearly so, up to 8 dm long; leaves 2–20 cm long or more, linear to filiform, sessile atop the sheathing ad-nate stipules; stipules membranous, com-monly 2–several times broader than the stem, entirely adnate to the leaf base or the tips free for 1–2 mm, enclosing the flowers; flower spikes axillary, the pedun-cles elongating as the fruits develop and finally 3–20 cm long or more and straight or coiled; drupelets 1.5–3 mm long, obliquely ovoid.

Saline or brackish ponds and tidal flats; in coastal and insular southern Alaska; widely distributed in North America; Eur-asia, South America (*R. spiralis* Dumort; *R. occidentalis* Wats.). Proposed segregates have been based largely on the nature of the peduncle and on the shape, size, and beak characteristics of the fruit, but there does not appear to be any significance to these features and their recognition is pure-ly arbitrary.

SPARGANIACEAE
Burreed Family

Perennial aquatic herbs; leaves alternate,

Ruppia maritima L. (× 0.25).

sheathing at the base, erect or floating; plants monoecious; flowers imperfect, arranged in globose heads, the uppermost staminate, the lower ones pistillate; perianth of 3–6 linear-subulate scales; stamens usually 3–5; pistils 1, the ovary superior, 1(2)-loculed; styles 1(2), simple or forked; fruit a nutlet or achene.

SPARGANIUM L.

Aquatic plants; leaves linear with sheathing bases; flowers borne in pedunculate or sessile heads in the upper leaf axils or along the terminal portion of the stem.

Fernald, M. L. 1922. Notes on *Sparganium*. Rhodora 24:26–34.

Reveal, J. L. 1970. *Sparganium simplex* Huds., a superfluous name. Taxon 19: 796–97.

1a. Staminate heads solitary (rarely 2); leaves mostly 2–6(8) mm broad; anthers 0.3–0.7 mm long; achene beak 0.2–1.5 mm long. (2)

1b. Staminate heads 2 or more; leaves mostly 3–12 mm broad (see *S. angustifolium*); anthers often more than 0.8 mm long; achene beak more than 1.5 mm long. (3)

2a. Pistillate heads all axillary, the staminate head pedunculate; achenes tapering to a conical beak. *S. minimum*

2b. Pistillate heads, at least 1, borne above the leaf axils, the staminate head sessile; achenes rounded to obtuse and abruptly short beaked. *S. hyperboreum*

3a. Leaves 1.5–5(8) mm broad, the main veins on the lower surface 0.8 mm apart or less; mature pistillate heads 1–2 cm broad; achene beak (including the stigma about 1 mm long) about 2 mm long. *S. angustifolium*

3b. Leaves (5)6–12 mm broad, the main veins on the lower surface 0.8–2 mm apart; mature pistillate heads more than 2 cm broad; achene beak (in-

cluding the stigma about 1.5 mm long) more than 2 mm long. *S. emersum*

Sparganium angustifolium Michx.
Narrow-leaved Burreed

Stems 1.5–6 dm long or more, usually floating; leaves usually floating, 1.5–6(8) mm broad, the upper with dilated bases, the main veins mostly 0.2–0.8 mm apart; inflorescence with 2–4 pistillate heads and 2–5 staminate heads, the lower 1–2 pistillate heads 1–2 cm broad at maturity, peduncled and borne axillary or with some along the rachis; anthers about 1 mm long; stigmas about 1 mm long; achenes fusiform, the stipe 1–2 mm long, the body 2–3 mm long, the beak 1.5–2 mm long, including the style.

Ponds and sluggish streams; in much of Alaska south of the Brooks Range and southern Yukon; widely distributed in North America; Eurasia.

Sparganium emersum Rehmann
Emersed Burreed

Stems 1.5–6 dm tall, erect; leaves usually erect but sometimes floating, (5)6–12 mm broad, the upper with dilated scarious bases, the main veins mostly 0.8–2 mm apart; inflorescence with 3–5 pistillate heads and 2–8 staminate heads, the pistillate 2–3 cm broad at maturity, the lower ones peduncled and axillary or along the rachis; anthers 1–1.5 mm long; stigmas 1–2 mm long; achenes fusiform, the stipe 2–3.5 mm long, the body 3–5 mm long, the beak 3–4 mm long, including the stigma.

Ponds, lakes, and bogs; in west-central, south-central, central, and east-central Alaska and southern Yukon; east to Newfoundland and south to California, Colorado, and New England (*S. simplex* Huds.; *S. simplex* var. *multipedunculata* Morong; *S. multipedunculatum* [Morong] Rydb.). Our material belongs to var. *multipedunculata* [Morong] Reveal.

Sparganium emersum Rehmann (× 0.3).

Sparganium hyperboreum Laest.
Northern Burreed

Stems 1–2.5(3) dm tall, floating or erect; leaves erect or floating, 2–4(6) mm broad, the upper with dilated bases with more or less scarious margins, the main veins mostly 0.2–0.6(1) mm apart; inflorescence with 2–4 pistillate heads and 1(2) sessile staminate heads, the pistillate 0.8–1.5 cm broad at maturity, the lower ones peduncled and axillary or some along the rachis; anthers 0.3–0.7 mm long; stigmas 0.3–0.6 mm long; achenes fusiform, the stipe about 0.5 mm long, the body 4–5 mm long, rounded to obtuse apically, beakless or abruptly short beaked.

Ponds and lakes; in most of Alaska and northern and western Yukon; eastward to Newfoundland and Quebec; circumpolar (*S. williamsii* Rydb.).

Sparganium minimum Fries
Small Burreed

Stems 1.5–4.5 dm long or more, floating or emersed; leaves floating or erect, mostly 2–6 mm broad, the upper with dilated bases with narrow scarious margins, the main veins mostly 0.3–0.8 mm apart; inflorescence with 2–3 pistillate heads and 1(2) pedunculate staminate heads, the pistillate 0.8–1.2 cm broad at maturity, the lowermost ones peduncled or all sessile, axillary; anthers 0.3–0.6 mm long; stigmas 0.5–0.8 mm long; achenes fusiform, the stipe 0.5–0.8 mm long, the body 4–5 mm long, tapering apically to a beak about 1 mm long.

Ponds and lakes; in central western to eastern and south-central Alaska and western Yukon; eastward to Labrador and south to Utah, Montana, and Tennessee; Eurasia. It is evident that *S. minimum* is not common in Alaska or the Yukon. Since it can be confused with either *S. angustifolium* or *S. hyperboreum,* reports of *S. minimum* should be checked carefully.

TYPHACEAE
Cattail Family

Aquatic to semiaquatic monoecious perennial herbs with coarse rhizomes; leaves alternate, distichous, linear, sheathing at the base; flowers imperfect, borne in dense cylindrical spikes, the staminate above the pistillate, the pistillate pedicellate, interspersed with bracts and sterile flowers; perianth represented by bristles; staminate flowers with 2–5 stamens; pistils 1, the ovary long stipitate, 1-loculed, 1-carpelled, the stigmas 1; fruit a minute nutlet.

TYPHA L.

A single genus, with characteristics of the family.

Typha latifolia L.
Common Cattail

Plants 10–20 dm tall or more; leaves (6) 8–20 mm broad, the sheathing base with a scarious margin; pistillate spikes brown to blackish brown, mostly 20–30 mm thick at maturity, contiguous with the staminate spikes or nearly so; pollen grains in tetrads; fruit ellipsoidal, about 1 mm long.

Marshes and ponds; in central, south-central, and east-central Alaska and western Yukon; widely distributed in North America; Eurasia.

ZANNICHELLIACEAE
Horned Pondweed Family

Perennial submersed aquatic monoecious herbs with creeping rhizomes; stems slender, usually branched; leaves opposite or crowded at the nodes, with adnate, sheathing, usually ligulate stipules; flowers minute, imperfect, axillary, solitary or cymose; perianth of 3 scales or lacking; stamens 1; pistils 1–9, the ovary 1-loculed, 1-carpelled, the stigmas 1; fruit an achene.

ZANNICHELLIA L.

Stems slender; leaves opposite; flowers axillary; staminate flowers with 1 stamen, the filament slender; pistillate flowers sessile or short pedunculate subtended by a short hyaline bract, mostly with 3–5 pistils.

Zannichellia palustris L.
Horned Pondweed

Submersed herbs with slender stems up to 4 dm long or more; leaves opposite, 2–10 cm long, linear-filiform, 1-veined; staminate flowers with slender filaments; pistillate flowers sessile or short pedunculate; fruit short stipitate to subsessile, the body compressed, lunately curved, the keel often denticulate; styles less than half as long as the body.

Ponds and tidal flats, in brackish or fresh water; in coastal western and south-central

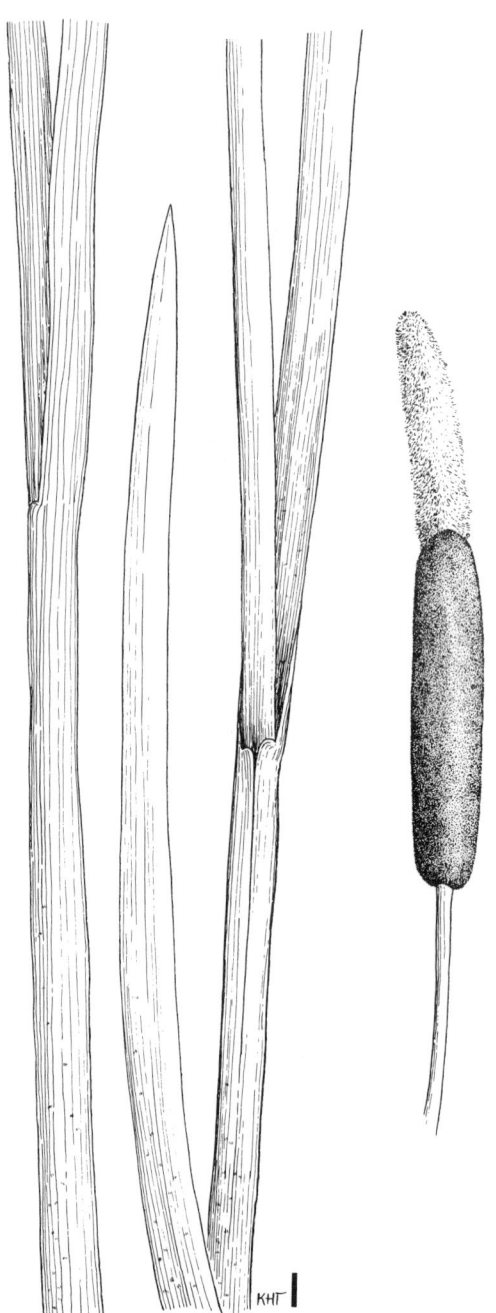

Typha latifolia L. (× 0.4).

Alaska (to be expected elsewhere); widely distributed in North America; Eurasia, Africa.

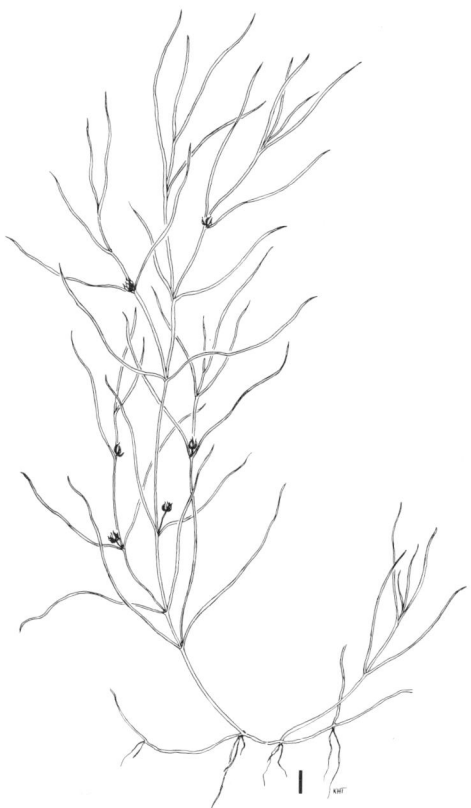

Zannichellia palustris L. (× 0.3).

ZOSTERACEAE
Eelgrass Family

Marine submersed or floating aquatic dioecious or monoecious perennials; leaves alternate, 2-ranked, linear and grasslike, sheathing at the base; flowers imperfect, arranged in 2 ranks on one side of a flattened spadix, this adnate to a thickened spathe terminated by a leaflike appendage; perianth lacking; staminate flower with a single stamen, enclosed by hyaline scales; pistils 1, the ovary 1-loculed, 2-carpelled, the stigmas 2; fruit an achene.

1a. Plants dioecious; spadix bordered by a conspicuous lobed margin. *Phyllospadix*
1b. Plants monoecious, with stamens and pistils on the same spadix; spadix bordered by a narrow entire margin. *Zostera*

PHYLLOSPADIX Hook.

Plants with a short thick rhizome and annual stems, dioecious; spadix pedunculate, the pistillate bordered by a series of lobes; anthers sessile; pistils sessile, the style short, the stigmas slender; fruit ovoid, beaked, with 2 acute basal outgrowths.

Phyllospadix scouleri Hook.
Surfgrass

Stems usually unbranched, 0.5–5 dm long, flattened; leaves linear, 2–4 mm broad, 30–150 cm long; pistillate plants with 1–2 spadices 2–5 cm long, these borne on short thick peduncles; spathe dilated and

Phyllospadix scouleri Hook. (× 0.3).

appendaged marginally, the apex leaflike; fruit 3–4 mm long, deeply lobed basally.

Intertidal and subtidal zones; in southeastern Alaska; southward to California.

ZOSTERA L.

Plants with elongate creeping rhizomes and annual stems, monoecious; spadices pedunculate, monoecious, with a narrow entiremargined spathe, lacking lobes; stamens and pistils arranged in 2 rows, with 1 stamen and 1 pistil at each level; anthers with elongate filaments; pistils sessile, the style short, the stigmas slender; fruit ellipsoid to cylindroid, rounded at the base.

Zostera marina L.
Eelgrass

Stems branching, up to 25 dm long, flattened; leaves linear, 2–10 mm broad or more, 20–100 cm long or more; spadix 3–8 cm long, enclosed by overlapping entire margins of the spathe, the spathe leaflike apically; fruit 3–4 mm long, ellipsoid to cylindroid, rounded at the base.

Subtidal or less commonly intertidal zones; in coastal and insular western, southwestern, south-central, and southeastern Alaska; south to California and along the Atlantic Coast; Eurasia (Z. *pacifica* Wats.; Z. *marina* var. *latifolia* Morong).

Glossary

A-, Ab-. Prefixes meaning without.

Abortion. Arrested development of an organ.

Abortive. Imperfectly developed.

Acaulescent. Stemless or apparently so.

Accrescent. Enlarging with age; especially said of sepals.

Achene. A small dry 1-seeded indehiscent fruit in which the ovary wall is free from the seed.

Acicular. Shaped like a needle.

Acrid. Sharp and irritating or biting to the taste.

Actinomorphic. With radial symmetry.

Acuminate. Tapering to the apex, the sides more or less concave.

Acute. Distinctly and sharply pointed, but not drawn out.

Ad-. Prefix meaning to or toward.

Adnate. With unlike parts congenitally grown together.

Adventitious. Developing in an unusual or abnormal position.

Adventive. Imperfectly naturalized; not native.

Affinity. The closeness of relationship between plants as shown by similarity of important organs.

Aggregate. Crowded into a dense cluster but not united.

Allopatric. Inhabiting distinct separate areas.

Alluvial. Pertaining to or composed of alluvium; soil, sand, gravel, or similar detrital material deposited by running water.

Alternate. An arrangement of leaves or other parts not opposite or whorled; placed singly at different heights on the axis or stem.

Ament. A catkin or spike of flowers usually bracteate, pendulous, and deciduous.

Amphibious. Capable of living on land or in water.

Amplexicaul. Clasping or embracing the stem, as a leaf.

Anastomosing. Netted; interveined; said of leaves marked by cross-veins forming a network; interlacing.

Androgynous. Hermaphroditic; having both male and female flowers in the same inflorescence, with the male above the female.

Aneuploid. An organism or cell having a chromosome number which is not an exact multiple of the haploid (n) or basic number.

Angiosperm. A plant with seeds enclosed in an ovary or pericarp.

Annual. Of one year's duration; completing its life cycle in one year.

Annulus. In ferns, the elastic organ which partially invests the sporangium and at maturity bursts it.

Anther. That portion of the stamen which bears the pollen.

Anthesis. The act of flowering.

Antrorse. Directed upward or forward.

Apetalous. Without petals or with a single perianth.

Apex. The tip, point, or angular summit of anything.

Aphyllopodic. Without leaves at the base.

Aphyllous. Without leaves.

Apical. Pertaining to the apex or tip.

Apices. Plural of apex.

Apiculate. Having a minute pointed tip.

655

Apomictic. The condition of apomixis.

Apomixis. Development of an individual plant from an egg without fertilization or fusion with pollen nucleus.

Appendage. An attached secondary part to a main structure.

Appressed. Lying flat against an organ.

Approximate. Drawn close together but not united.

Aquatic. Living in water.

Arachnoid. Cobwebby; composed of soft slender entangled hairs.

Arcuate. Moderately curved; bent like a bow.

Argenteus. Silvery.

Aristate. Awned; provided with a bristle at the end, rarely on the back edge.

Armed. Provided with any kind of strong and sharp defense, as of thorns, spines, prickles, barbs, etc.

Aromatic. Fragrant; spicy; pungent.

Articulate. Jointed; provided with nodes or joints or places where separation may naturally take place.

Ascend. To go or move upward; to rise.

Astringent. Binding; contracting; styptic.

Attenuate. Long-tapering; the sides straight.

Auricle. An ear-shaped lobe or appendage.

Auriculate. With earlike appendages.

Awl-shaped. Narrow and sharp-pointed; gradually tapering from base to a slender or stiff point.

Awn. A bristlelike appendage, especially on the glumes of grasses.

Axil. The upper angle formed between the axis and any organ that arises from it.

Axile. Belonging to the axis; said of placentae which are attached to the axis of an ovary.

Axillary. Situated in the axil.

Banner. The upper and usually largest petal in the papilionaceous corolla of a plant of the pea family; standard; vexillum.

Barbed. With rigid points or short bristles, usually reflexed like the barb of a fishhook.

Barbellate. Finely barbed.

Basal. At or pertaining to the base.

Basifixed. Attached or fixed by the base.

Beak. A long, prominent, and substantial projection; applied particularly to a prolongation of a fruit or carpel.

Berry. Any simple fruit having a pulpy or fleshy pericarp.

Bi-. A Latin prefix signifying two, twice, or doubly.

Bicolored. Two-colored.

Bidentate. Having two teeth.

Biennial. Of two seasons' duration from seed to maturity and death.

Bifid. Two-cleft or 2-lobed.

Bifurcate. Forked or 2-pronged.

Bilabiate. Two-lipped.

Bilobed. Two-lobed.

Binomial. The generic and specific name of an organism.

Biseriate. In 2 whorls or cycles, as perianth comprised of a calyx and a corolla.

Bladder. An inflated thin-walled structure.

Blade. The expanded part of a leaf or petal.

Bloom. The white, waxy, or pruinose covering on many fruits, leaves, and stems.

Body. Corpus; the main portion of a structure.

Boreal. Northern.

Bract. A modified leaf subtending a flower or belonging to an inflorescence.

Bracteate. With bracts.

Bracteole. A bractlet or small bract.

Bractlet. Bract borne on a secondary axis, as on the peduncle or even on a petiole.

Branch. A lateral division of the stem or axis of growth.

Branchlet. The ultimate divisions of a branch.

Bristle. A stiff hair.

Bud. An embryonic axis with its appendages.

Bulb. A subterranean leaf bud with fleshy scales or coats, like an onion.

Bulblet. A little bulb produced in the leaf axils, inflorescence, or other unusual places.

Bulbous. Having the character of a bulb.

Caducous. Falling off early or prematurely, as the sepals in some plants.

Caespitose. Growing in tufts.

Calcareous. Of or pertaining to calcium carbonate (limestone), as a calcareous soil.

Callous. Having the texture of a callus.

Callus. A hard prominence or protuberance, specifically the hardened base of a spikelet of grasses.

Calyces. Plural of calyx.

Calyx. The outermost circle of the floral envelopes.

Cambium. A layer, usually regarded as one cell thick, of persistent meristematic tissue, or a persistent meristematic layer which gives rise to secondary wood and secondary phloem.

Campanulate. Bell-shaped.

Canaliculate. Longitudinally channeled or grooved.

Canescent. Becoming hoary, usually with a gray pubescence.

Capillary. Hairlike; very slender.

Capitate. Headed; in heads; formed like a head; aggregated into a very dense or compact cluster.

Capsule. A dry dehiscent fruit made up of more than 1 carpel.

Carina. A keel; used either for the two combined lower petals of a papilionaceous flower or for a salient longitudinal projection on the center of the lower surface of an organ, as on the lemmas of many grasses.

Carinate. Keeled with 1 or more longitudinal ridges.

Carpel. A simple pistil; one unit of a compound pistil.

Carpellate. Possessing carpels.

Caryopsis. The grain or fruit of most grasses, with the seed coat grown fast to the pericarp.

Catkin. A scaly spike bearing apetalous, unisexual flowers; ament.

Caudex. The woody base of a perennial plant.

Caulescent. More or less stemmed or stem-bearing; having an evident stem above ground.

Cauline. Pertaining or belonging to the stem.

Chaff. Small membranous scales; degenerate bracts in many Compositae.

Chartaceous. Having the texture of writing paper.

Chlorophyll. The green coloring matter in the cells of autophytic plants.

Ciliate. Said of a margin fringed with hairs.

Ciliolate. Said of a margin fringed with small hairs.

Cinereous. Ash-colored; light gray.

Circinate. Coiled from the top downward; coiled into a ring, or partially so.

Circum. Near to; around.

Circumboreal. Around the northern hemisphere.

Circumpolar. Around the polar regions.

Circumscissile. Open or dehiscing along a horizontal line around the fruit or anther, the valve usually coming off like a lid.

Cladophyll. A branch assuming the form and function of a leaf.

Clambering. Vinelike, forming a mat or canopy over the undergrowth, often without the aid of tendrils or twining stems.

Clavate. Club-shaped; said of a long body thickened toward one end.

Claw. The long narrow petiolelike base of the petals or sepals in some flowers; the modified auricle of some grass leaves.

Cleft. Divided into lobes separated by narrow or acute sinuses which extend more than halfway to the midrib.

Cleistogamous. Having fertilization occur within the unopened flower.

Cleistogamy. The state of being cleistogamous.

Clone. The vegetatively produced progeny of a single individual.

Coetaneous. Of the same age or duration.

Comissural. Referring to or possessing a commissure.

Commissure. The place of joining or meeting, as the face by which one carpel joins another.

Compact. Close together.

Complete. A flower with sepals, petals, stamens, and pistils present.

Compressed. Flattened; especially flattened laterally.

Concave. Hollow, as the inside of a saucer.

Cone. The aggregation of scales on which seeds of spruce and other gymnosperms are borne; a strobilus.

Confluent. Blended into one; passing by degrees one into another.

Congested. Crowded.

Conic. Of, pertaining to, in the form of, or like a cone.

Connate. United congenitally or subsequently, as when like structures are joined.

Connective. That portion of a stamen that connects the two halves of an anther.

Connivent. Coming together or converging but not organically connected.

Conspectus. Survey; short general view.

Conspicuous. Attracting attention; striking.

Contiguous. Touching without fusion; used irrespective of whether the parts are like or unlike.

Contorted. Twisted or bent.

Convex. Having a more or less rounded surface.

Convolute. Said of floral envelopes in the bud in which one edge overlaps the next part, as sepal or petal or lobe, while the other edge or margin is overlapped by a preceding part; rolled up from the sides longitudinally.

Copiously. Plentifully; abundantly.

Coralloid. Having the appearance of coral.

Cordate. Heart-shaped; said of leaves having the petiole at the broader and notched end.

Coriaceous. Like leather.

Corm. A solid erect bulblike stem with scalelike leaves, usually subterranean.

Corolla. The inner floral envelope, composed of separate or connate petals.

Corona. Crown; coronet; any appendage or intrusion that stands between the corolla and stamens.

Corymb. Short and broad, more or less flattopped indeterminate flower cluster, the outer flowers opening first.

Corymbiform. Shaped like a corymb.

Corymbose. Arranged in corymbs.

Costa. A rib, as a midrib.

Cotyledon. Seed leaf; the primary leaf or leaves in the embryo.

Creeping. Running along the ground and rooting.

Crenate. Said of a margin with rounded or blunt teeth.

Crenulate. Finely crenate.

Crested. With elevated and irregular toothed ridge.

Crisped. Irregularly waved and twisted; kinky; curled.

Crown. Corona; the base of a tufted herbaceous perennial grass; the part of a stem at the surface of the ground.

Cruciform. Cross-shaped.

Cryptogam. Plants not bearing seeds, such as ferns and horsetails.

Culm. The jointed stem of grasses and sedges.

Cuneate. Wedge-shaped; triangular, with the narrow end at the point of attachment, as of leaves or petals.

Cusp. A pointed end; apex; beak.

Cuspidate. Tipped with a sharp rigid point.

Cuticle. A very thin waxy layer covering a plant.

Cyathia. Plural of cyathium.

Cyathium. The ultimate inflorescence in the genus *Euphorbia* consisting of unisexual flowers congested within a cup-shaped involucre.

Cylindrical. Elongated with a circular cross section.

Cyme. A broad, more or less flattopped determinate flower cluster, with central flowers blooming first.

Cymose. Bearing cymes or cymelike.

Cymule. A small cyme or a portion of one.

Deciduous. Not persistent; said of leaves falling in autumn or of floral parts falling after anthesis.

Decumbent. Reclining or lying on the ground, but with the end ascending.

Decurrent. Said of a leaf or leaf scar, part of which extends in a ridge down the twig below the point of insertion.

Decussate. In pairs, alternately crossing at right angles.

Deflexed. Bent or turned abruptly downward.

Dehiscent. That which dehisces or splits open, as the opening of anther or fruit along regular lines of suture.

Deliquescent. Dissolving or melting away; said of a stem which loses itself by repeated branching; opposed to excurrent.

Deltoid. Triangular, deltalike.

Dentate. Said of a margin with sharp teeth pointing outward.

Denticulate. Minutely or finely dentate.

Depauperate. Reduced or undeveloped; starved or stunted; said of small plants growing under favorable conditions.

Di-. Prefix in compounds meaning two or twice.

Diadelphous. Said of stamens formed in two groups through the union of their filaments.

Dichotomous. Branching by constantly forking in pairs.

Didymous. Found in pairs, as the fruits of Umbelliferae; divided into two lobes.

Didynamous. Said of four-stamened flowers with stamens in pairs, two long and two short.

Diffuse. Loosely branching or spreading; of open growth.

Digitate. Fingerlike; compound with the members arising from one point.

Diminutive. Below the average size; very small.

Dimorphic. Occurring in two forms.

Dioecious. Unisexual, the male and female elements in different plants.

Diploid. An organism or cell with two sets of chromosomes or two genomes.

Disarticulating. The parts separating at maturity.

Discoid. Having only disk flowers, as in some Compositae; or, in reference to stigma, disk-shaped.

Disjunctly. Said of noncontinuous distribution of plant taxa.

Disk. An enlargement or prolongation of the receptacle of a flower around the pistil, sometimes made up of coalesced nectaries or staminodia.

Dissected. Deeply divided or cut into many segments.

Dissimilar. Not similar; unlike.

Distal. The end opposite the point of attachment.

Distichous. Conspicuously two-ranked; in two rows.

Distinct. Separate; not united with parts in the same series.

Divaricate. Widely divergent.

Divided. Characterized by a lobing or segmentation which extends to the base.

Dorsal. Relating to the back or attached thereto; the surface turned away from the axis, which in a leaf is the lower surface.

Drupaceous. Resembling a drupe, possessing its character, or producing similar fruit.

Drupe. A fleshy one-seeded indehiscent fruit, with seed enclosed in a stony endocarp called a pit.

Drupelet. One drupe in a fruit made up of aggregate drupes, as in a raspberry.

E-, Ex-. In Latin-formed words usually denoting, as a prefix, that parts are missing.

Echinate. Armed with prickles.

Ecological. Pertaining to the relation of organisms to their environment.

Edaphic. Pertaining to or influenced by soil, rather than climatic conditions.

Eglandular. Without glands.

Ellipsoid. A solid body elliptic in outline.

Elliptic. A flat part of a body that is oval and narrows to rounded ends, widest at or about the middle.

Elongate. Stretched; lengthened.

Emarginate. With shallow notch at the apex.

Embryo. The rudimentary plant form in the seed.

Emergent. Protruding from the surface of water, said of some aquatic plants.

Endemic. Indigenous or native to.

Endocarp. The inner layer of a pericarp.

Endosperm. The albumen of a seed in angiosperms.

Ensiform. Sword-shaped, as in the leaf of the Iris.

Entire. Without toothing or division; with even margin.

Ephemeral. Persisting for one day only, as of some flowers.

Epi-. A Greek prefix signifying on or upon.

Epigynous. Borne on the ovary; said of floral parts in which the ovary is inferior.

Epithet. A single descriptive word or single descriptive phrase; in taxonomy, it is applied to the subdivisions of genera, to the second component of the name of species, and to the subdivisions of species.

Equitant. Folded over as if astride; used for conduplicate in which the leaves are folded together lengthwise in two ranks.

Erose. Irregularly toothed or eroded as though bitten or gnawed.

Estipulate. Without stipules.

Evanescent. Tending to evanesce; fleeting.

Evergreen. Remaining green during its dormant season; said of plants that are green throughout the year.

Ex-, see **E-.** Prefixes meaning without, not, lacking, from out.

Excurrent. Running through to the apex and beyond, as a mucro; a stem that remains central, the other parts being regularly disposed around it.

Exfoliating. Coming off in thin layers.

Exocarp. The outer layer of a pericarp.

Exserted. Sticking out; projecting beyond, as stamens from perianth; not included.

Falcate. Sickle- or scythe-shaped.

Farinose. Covered with a mealy, usually whitish substance.

Fascicle. A close cluster or bundle of flowers, leaves, stems, or roots.

Fastigiate. Parallel, clustered, and erect, as the branches of some trees.

Fen. Low land covered wholly or partly with water unless artificially drained.

Ferruginous. Rust-colored.

Fertile. Said of pollen-bearing stamens and seed-bearing fruits; capable of producing fruit.

Fetid. Having a disagreeable odor.

Few. Not many; of small number, usually two or three.

Fibrous. Having numerous woody fibers; having the appearance of fibers, as the roots of monocots.

-fid. Combining form denoting cleft; cut halfway to the middle.

Filament. The part of a stamen that supports the anther; threadlike structures.

Filiform. Threadlike; long and very slender.

Fimbriate. Fringed, the hairs longer or coarser as compared with ciliate.

Fistulose. Hollow cylindrical, as the leaf and stem of an onion.

Flabellate. Fan-shaped; dilated in a wedge shape, sometimes plaited.

Flange. A projecting edge or rim; edge flaring and conspicuous.

Fleshy. Succulent.

Flexuous. Bent alternately in different directions.

Floccose. Bearing tufts of woolly hairs.

Floret. The lemma and palea with included flower; also a small flower; one of a cluster.

Floriferous. Flower-bearing.

Flower. A modified stem concerned with the production of seeds in the angiosperms.

Foliaceous. Leaflike; said particularly of sepals, calyx lobes, and of bracts that resemble leaves in texture, size, or color.

Foliage. The leafy covering, especially of trees.

-foliate. A suffix meaning having leaves.

-foliolate. A suffix meaning having leaflets.

Foliose. Closely clothed with leaves; leafy; leaflike.

Follicle. A single-carpellate dry fruit dehiscing along one line or suture.

Follicular. Of or pertaining to a follicle.

Forage. Food for animals, especially domestic animals.

Forma. An infraspecific category (f.) ranked below a variety.

Free. Not joined to another organ.

Fruit. A mature ovary or ovaries with or without closely related parts.

Funnelform. With the tube gradually widening upward and passing insensibly into the limb.

Fusiform. Spindle-shaped; narrowed both ways from a swollen middle.

Galea. A hooked or helmet-shaped part of the perianth, usually the upper lip of an irregular corolla.

Gametophyte. The sexual stage in plants which bear sperm and egg; used particularly for the prothallus of ferns.

Gelatinous. Jellylike.

Geniculate. Abruptly bent so as to resemble the knee joint, as of awns and the lower nodes of some culms.

Gibbous. Swollen on one side as the glume; a pouchlike enlargement of the base of an organ, as of a calyx.

Glabrate. Nearly glabrous or becoming glabrous with maturity or age.

Glabrescent. About the same as glabrate.

Glabrous. Smooth, devoid of pubescence or hair in any form.

Gland. A secreting surface or structure; an appendage having the general appearance of such an organ.

Glandular. Having or bearing secreting organs or glands.

Glaucescent. Becoming sea green; somewhat glaucous.

Glaucous. Covered with a "bloom" or a whitish substance that rubs off, as of a plum or cabbage leaf.

Globose. Spherical; globular.

Globular. Spherical.

Glume. The chaffy two-ranked members of the inflorescence of grasses and similar plants; one of the two empty bracts at the base of a grass spikelet.

Glutinous. Covered with a sticky exudation.

Gynandrium. Adnate styles and stamens; a column.

Gynaecandrous. With staminate and pistillate flowers in the same spike, the pistillate at the apex.

Habit. The general appearance of a plant, whether erect, prostrate, climbing, etc.

Habitat. The kind of locality in which a plant grows.

Halophyte. A plant which grows in saline soil.

Haploid. An organism or cell having only one complete set (n) of chromosomes or one genome.

Hastate. Sagittate, with the basal lobes turned outward.

Head. A dense spherical or flattopped inflorescence of sessile flowers clustered on a common receptacle.

Heath. Any plant of the family Ericaceae, the heath family.

Heathlands. A tract of land with characteristic vegetation of low shrubs.

Hemi-. In Greek compounds, signifying half.

Hemispheric. Of, pertaining or belonging to, or like a hemisphere.

Herb. A plant naturally dying to the ground at the end of the growing season; without persistent stem above ground, and lacking definite woody, firm structure.

Herbaceous. Not woody; dying to the ground each year; said also of soft branches before they become woody.

Herbage. Vegetative parts of plant.

Hetero-. In Greek composition, signifying various or of more than one kind of form, as heterophyllous, with more than one kind of form of leaf.

Heterosporous. Producing two kinds of spores representative of two sexes, as in *Isoetes*, as well as seed plants.

Heterozygous. The state of an organism whose chromosomes do not carry identical members of any given pair of genes.

Hexa-. In Greek composition, signifying six.

Hilum. The scar or mark on a seed indicating the point of attachment.

Hip. The fruit of the rose; technically, a cynarrhodion.

Hirsute. With stiff or bristly hairs.

Hirsutulous. Slightly hirsute.

Hirtellous. Softly or minutely hirsute or hairy.

Hispid. Beset with rough hairs or bristles.

Hispidulous. Somewhat or minutely hispid.

Holophyte. A plant that manufactures its own food; a photosynthetic plant.

Holotype. The single specimen chosen as the basis for the original diagnosis or description of a taxon.

Homo-. In Greek compounds, signifying alike or very similar.

Homosporous. Producing spores of one kind only.

Horn. Stiff tapering appendage somewhat like the horn of a cow.

Humus. The earth, the soil, specifically that soil rich in decaying organic matter.

Hyaline. Thin and translucent or transparent.

Hybrid. A cross usually between two species of the same genus.

Hybridization. Interbreeding of species, races, varieties, etc. among plants.

Hydrophyte. A water plant, partially or wholly immersed.

Hygroscopic. Altering form or position through changes in humidity.

Hypanthium. A cup- or saucer-shaped enlargement or development of the receptacle which bear the calyx, corolla, and stamens at its apex.

Hypo-. In Greek composition, signifying below, under, beneath, lower.

Hypogaeous. Under the earth or soil.

Hypogynous. Borne on the receptacle or at the base of the ovary; said of the stamens or petals; not applicable to the ovary.

Imbricate. Partly overlapping like shingles on a roof, either vertically or laterally.

Immersed. Growing entirely under water.

Imperfect. Said of a flower with one of the sexes wanting.

In-. Prefix meaning in, within, among, at, into, onto, toward, during, on.

Incised. Cut sharply and irregularly, more or less deeply.

Included. Not protruding beyond the surrounding organ; not exserted.

Indehiscent. Not opening by valves or along regular lines.

Indigenous. Native to the country, not introduced.

Indumentum. A rather heavy hairy or pubescent covering.

Indurate. Hardened.

Indusium. The epidermal outgrowth covering the sori on ferns.

Inferior. Said of one organ when below another, as an inferior ovary with an adnate or superior calyx.

Inflorescence. Mode of flower bearing; technically less correct but much more common in the sense of a flower cluster.

Infra-. In combinations, signifying below.

Innovation. A basal offshoot from the main stem, shorter and less modified than a rhizome or stolon; in grasses an incomplete young shoot.

Insectivorous. Said of those plants which capture insects and presumably absorb nutriment from them.

Insipid. Without taste or savor; vapid.

Insular. Pertaining to an island.

Integument. The covering of an organ or body; the envelope of an ovule.

Inter-. In composition, signifying between, particularly between closely related parts or organs.

Intergrading. Merging gradually one with another through a continuous series of intermediate forms, kinds, or types.

Internodes. The part of the stem between two successive nodes.

Intra-. Prefix meaning within.

Introgressive. Referring to introgression or mutual exchange or unilateral flow of genes between closely related taxa, resulting in the formation of complex series of intermediate individuals.

Introrse. Turned or faced inward or toward the axis, as an anther facing toward the center of the flower.

Invest. To clothe, dress, or array.

Involucel. A secondary involucre; a small involucre about the parts of a cluster.

Involucrate. With an involucre.

Involucre. A cluster of bracts subtending a flower or inflorescence.

Involute. Rolled in from the edges, the upper surface within.

Irregular. Wanting in regularity of form; asymmetric, as a flower which cannot be halved in any plane, or one that is capable of bisection in one plane only; zygomorphic.

Isotype. A duplicate of the holotype specimen.

Keel. A central dorsal ridge; the united petals of a papilionaceous flower.

Labiate. Lipped; a member of the Labiatae.

Lacerate. Torn at the edge or irregularly cleft, as in some ligules.

Lacking. Deficient or wanting, something needed.

Laciniate. Cut into lobes separated by deep, narrow, irregular incisions.

Lanate. Clothed with woolly and interwoven hairs.

Lanceolate. Lance-shaped, rather narrow, tapering to both ends with the broadest part below the middle.

Lateral. On or at the side.

Latex. The milky juice of such plants as dandelion or lettuce.

Leaflet. A single division of a compound leaf.

Legume. The characteristic fruit of the Leguminosae family; usually a dehiscent fruit formed from one carpel with two lines of dehiscence; also used for any plant with this type of fruit.

Lemma. In grasses, the flowering glume; the lower of the two bracts immediately enclosing the flower.

Lenticel. A group of loose corky cells formed beneath the epidermis of woody plants, rupturing the epidermis and admitting gases to and from the inner tissues.

Lenticular. Shaped like a double convex lens.

Ligulate. With a ligule; strap-shaped or straplike.

Limb. The border or expanded part of a gamepetalous corolla, as distinct from the tube or throat; the lamina of a leaf or of a petal.

Linear. Long and narrow with margins parallel, or nearly so.

Littoral. Belonging to, or growing on the seashore.

Lip. Either the upper or lower division of a bilabiate or two-lipped corolla. Also the upper (but by twisting of the pedicel appearing to be the lower) petal in Orchidaceae.

Lobed. Bearing lobes; loosely used but technically cut in not over halfway to the base or midvein, the sinuses and apex of segments rounded.

Locule. Compartment or cell of an ovary or anther.

Loculicidal. With dehiscence on the back, between the partitions into the cavity.

Lodicule. A small scale outside the stamens in the flowers of grasses.

Loment. A flat legume which is constricted between the seeds, falling apart at the constrictions when mature into one-seeded joints.

Lousewort. Any of a genus of plants of the figwort family, formerly reputed to cause sheep feeding upon them to be subject to vermin, especially *Pedicularis*.

Lupine. Any member of the genus *Lupinus*.

Lyrate. Lyre-shaped, pinnatifid with the terminal lobe large and rounded, the lower lobes small.

Maculate. Blotched or spotted.

Malpighian. Pertaining to hairs with two prongs, attached at or near the center, like the head of a pick; dolabriform.

Many. Eleven or more; numerous.

Marcescent. Withering without falling off.

Margin. The edge of a leaf.

Maritime. Pertaining to the sea.

Mega-. Prefix, meaning large.

Megasporangium. The sporangium which produces the megaspores.

Megaspore. The more correct form of macrospore; the larger spore of heterosporous plants.

Megasporophyll. A sporophyll that bears megaspores, often produced in the axil of a bract; a carpel.

Membranous. Thin, more or less flexible, and translucent, like a thin membrane.

-merous. A suffix indicating division into parts; as flowers 5-merous, in which the parts of each kind or series are 5 or in 5s.

Mesic. Moderate, specifically as regards moisture requirements.

Mesocarp. The usually fleshy fruit layer between the endocarp and exocarp.

Mesophyte. A plant intermediate between hydrophytes and xerophytes; a plant of medium moisture requirement.

Micro-. In Greek compounds, signifying little, small.

Microsporangium. The receptacle in which the microspores develop.

Microspore. The smaller of the two kinds of spores in such plants as *Selaginella*.

Microsporophyll. A sporophyll bearing microsporangia; in angiosperms, the anther.

Midrib. The main rib or central vein of a leaf or leaflike structure.

Midvein. Same as midrib.

Monadelphous. Stamens united by their filaments into a tube or column.

Monocephalous. Bearing a single head or capitulum.

Monoecious. Having unisexual flowers with both sexes borne on the same plant.

Mottled. Marked with spots of different colors; dappled.

Mucilaginous. Slimy, composed of mucilage.

Mucronate. Furnished with a mucro (bristle-tipped).

Multi-. A Latin prefix meaning many.

Multiple fruit. One formed from several flowers crowded into a single unit on a common axis.

Muskeg. A bog characterized by an abundance of sphagnum moss and by tussocks.

Naked. Lacking some structure, appendage, or hairs which might ordinarily be expected to be present.

Naturalized. Having become thoroughly established in a region to which it is not indigenous.

Nectary. A nectar-secreting gland, often appearing as a protuberance, scale, or pit.

Nerve. In botany, a simple or unbranched vein or slender rib.

Neuter. Sexless, as a flower which has neither stamens nor pistils.

Nigrescent. Turning black.

Node. That point on a stem which normally bears a leaf or leaves.

Nodose. Knotty or knobby.

Nodulose. With little knobs or knots.

Nomenclature. The names of things in any science; in botany, restricted to the correct usage of scientific names in taxonomy.

Nudicaul. Having leafless stems.

Nut. An indehiscent 1-celled and 1-seeded hard and bony fruit, even if resulting from a compound ovary.

Nutlet. A small or diminutive nut; nucule.

Ob-. A Latin prefix signifying inversion.

Obconic. Conical, but attached at the narrower end.

Obcordate. Inversely heart-shaped, the notch being apical.

Oblanceolate. Inversely lanceolate, attached at the tapered end.

Oblique. Slanting, unequal sides.

Oblong. Longer than broad, with the margins nearly parallel.

Obovate. Reversed ovate, the distal end the broader.

Obovoid. A 3-dimensional figure of obovate outline.

Obsolete. Not evident or apparent; rudimentary; no longer used.

Obtuse. Blunt or rounded at the end.

Ochroleucous. Yellowish white, buff.

Ocrea. A legging-shaped or tubular structure formed by the union of two stipules, as in Polygonaceae.

Opposite. On both sides at the same level, as two leaves at the node and situated across the stem from each other.

Orbicular. Flat with a circular outline.

Ornamental. A plant cultivated essentially for decorative purposes.

Oval. Broadly elliptic with the width greater than half the length.

Ovary. That part of the pistil which contains the ovules.

Ovate. Shaped like a longitudinal section of a hen's egg, the broader end basal; applied to ovoid.

Ovoid. A solid that is oval (less correctly, ovate) in flat outline.

Ovule. That which becomes a seed after fertilization.

Oxytrope. Any member of the genus *Oxytropis*.

Palea. A chaffy scale or bract; the inner of the two bracts enclosing the grass flower.

Palmate. Lobed or divided or ribbed in a palmlike or handlike fashion; digitate.

Palustrine. Of or growing in marshes.

Panicle. A compound or branched raceme.

Paniculate. Having a panicle type of inflorescence.

Papilionaceous. Descriptive of the flower of many legumes having a standard or banner, wings, and keel; with a pealike flower; like a butterfly.

Papillate. Bearing minute pimplelike protuberances (papillae).

Papillose. Bearing papillae.

Pappus. The limb at the apex of the achene in Compositae, consisting of hairs, bristles, awns, or scales, presumably a modified calyx.

Parasite. An organism growing upon and obtaining nourishment from another; usually lacking chlorophyll in plants.

Parietal. Borne on or pertaining to the wall or inner surface of an ovary or fruit.

Parted. Divided by sinuses which extend nearly to the midrib.

Partition. A wall or dissepiment; a separated part or segment; the deepest division into which a leaf can be cut without becoming compound.

Paucity. Fewness; small number.

Pectinate. Pinnatifid with the segments narrow and arranged like the teeth of a comb; comblike.

Pedicellate. Borne on a pedicel.

Peduncle. A primary flower stalk supporting either a cluster or a solitary flower.

Pedunculate. Borne upon a peduncle.

Peltate. Shield-shaped, attached to the center or near the center, at least in some distance from the margin.

Pendulous. Drooping; hanging downward.

Penta-. Used as a prefix meaning *five*.

Pentagonal. Having five corners or angles.

Perennate. Lasting the whole year through;

self-renewing by lateral shoots from the base.

Perennial. A plant lasting three or more years.

Perfect. A flower with both functional stamens and pistils.

Perfoliate. Where the leaf has the stem apparently passing through it, or where opposite leaves are joined around the stem at their bases.

Perforate. Pierced with a hole or holes, or with pores.

Perianth. The floral envelope consisting of calyx and corolla however incomplete or modified.

Pericarp. The wall of the ripened ovary and therefore the wall of the fruit.

Perigynium. The more or less flask-shaped bract of *Carex* enclosing the achene.

Perigynous. Situated around but not attached to the ovary or its base directly; a flower with stamens and pistils on the calyx tube and the ovary superior.

Persist. To continue to exist; to recur constantly.

Persistent. Remaining attached after like parts ordinarily fall off.

Petal. One of the leafy expansions in the floral whorl called the corolla.

Petaloid. Like a petal or having a floral envelope resembling petals.

Petiolate. Having a petiole.

Petiole. The stalk to a leaf blade or to a compound leaf.

Petiolule. A small petiole; the petiole of a leaflet.

Phyllodes. Leaflike petiole having no blade.

Phyllopodic. With a leafy base.

Pilose. With soft hairs.

Pilosulous. Diminutive of pilose.

Pinna. The primary unit of a pinnately compound leaf.

Pinnate. Applied to a compound leaf when the leaflets are arranged along the rachis or stalk.

Pinnatifid. Pinnately lobed, cleft, or parted usually halfway into the midrib or more.

Pinnule. A secondary pinna or leaflet in a pinnately compound leaf.

Pistil. The female organ of a flower, consisting when complete of an ovary, style, and stigma.

Pistillate. Having pistals and no stamens; female.

Placenta. Any part of the interior of an ovary that bears ovules.

Plicate. Folded in plaits, usually lengthwise on the order of a folding fan.

Plumose. Hairs with side hairs along the main axis like the plume of a feather.

Pod. A dry dehiscent fruit.

Pollen. The microspores of seed plants, contained in the anther, which give rise to sperm nuclei.

Pollination. The transfer of pollen from the dehiscing anther to the receptive stigma.

Poly-. A Greek prefix meaning many, numerous.

Polychrome. Many-colored.

Polygamous. Bearing perfect and unisexual flowers on the same individual plant or on different individuals of the same species.

Polymorphic. With several or various forms, variable as to habit.

Polyploid. A plant with a chromosome complement of more than 2 sets of the haploid (n) number.

Pome. A fleshy indehiscent fruit with an inferior ovary and more than one locule, as an apple.

Population. The organisms, collectively, inhabiting an area or region.

Poricidal. Opening by pores.

Porrect. Directed outward and forward.

Precocious. Appearing or developing very early.

Prehensile. Clasping or grasping, as in tendrils.

Prickles. A small and weak spinelike body borne irregularly on the bark or epidermis.

Process. Any projecting appendage.

Procumbent. Lying or trailing on the ground, usually not rooting at the nodes.

Prostrate. Lying flat on the ground.

Proximal. Nearest the axis or the base.

Pruinose. With a waxy, powdery, usually whitish covering; glaucous.

Pseudo-. A prefix meaning false.

Pseudofasciculate. Apparently, but not actually joined; closely clustered.

Pseudoverticillate. Apparently, but not actually whorled.

Puberulent. With very short soft hairs; minutely pubescent.

Pubescent. Covered with short soft hairs or down; loosely, bearing hairs.

Pulvinate. Cushionlike, having short, very crowded stems.

Punctate. Dotted with depressions or with translucent internal glands or colored dots.

Pungent. Tipped with a sharp rigid point.

Pustular. Having slight blisterlike elevations.

Putative. Commonly thought or deemed; supposed; reputed.

Pyriform. Pear-shaped.

Pyxis. A capsule with circumcissile dehiscence, the top coming off as a lid.

Quadrangular. Four-angled.

Raceme. An inflorescence with pedicelled flowers borne along a more or less elongated axis with the younger flowers nearest the apex.

Racemose. Racemelike or bearing racemes.

Rachilla. A diminutive or secondary rachis or axis.

Rachis. The central elongated axis to an inflorescence or a compound leaf.

Radial. Arranged or having the parts arranged like rays; regular; actinomorphic.

Radiate. Spreading from or arranged around a common center; bearing rays.

Ray. Outer modified floret of some composites, with an extended or straplike flower axis on which some or all of the flower parts are borne.

Receptacle. That expanded portion of the axis which bears the floral organs; torus.

Recurved. Bent or curved downward or backward.

Reflexed. Abruptly curved or bent downward or backward.

Regular. A flower with all the members of each set alike in form, size, and color; radially symmetrical.

Reniform. Kidney-shaped; said of the form of some leaves.

Repent. Creeping, prostrate, and rooting at the nodes.

Replum. The septum of certain dry dehiscent fruits, persisting after the valves have fallen away.

Resin. An organic substance (hydrocarbon) insoluble in water, becoming hard when exposed to air.

Resupinate. Upside down or apparently so.

Reticulate. In the form of a network; leaf veins in a network.

Retrorse. Turned backward or downward.

Retuse. A rounded apex with a shallow notch.

Revolute. Rolled backward from each margin upon the lower side.

Rhizoid. In ferns, mosses, and liverworts, one of the rootlike filaments that attach the gametophyte to the substratum.

Rhizomatous. Having the characters of a rhizome or bearing rhizomes.

Rhizome. Any more or less elongated prostrate stem growing partly or completely beneath the surface of the ground, usually rooting at the nodes and becoming upcurved at apex.

Rhombic. Outline of an equilateral oblique-angled figure; 4-sided, like a diamond shape.

Rib. A primary vein, especially the central longitudinal or midrib.

Rosette. A cluster of spreading or radiating basal leaves.

Rostrate. With a beak, narrowed into a slender tip or point.

Rotate. Wheel-shaped, circular, and flat, applied to a gamopetalous corolla with a short tube.

Rotund. Rounded in outline, somewhat orbicular, but a little inclined toward the oblong.

Rudimentary. Imperfectly developed and nonfunctional.

Saccate. Bag-shaped; pouchy.

Sagittate. Shaped like an arrowhead with the basal lobes directed backward.

Saline. Of or pertaining to salt.

Salverform. A corolla with a long slender tube, abruptly flaring into a circular limb.

Samara. A dry indehiscent winged fruit.

Saprophyte. A plant deriving all of its nourishment from the bodies of decaying organisms.

Scaberulous. Slightly scabrous.

Scabrous. With short hairs; rough to the touch.

Scale. Any thin scarious body, usually a degenerative leaf, sometimes of epidermal origin.

Scape. Leafless peduncle arising from the ground; it may bear scales or bracts but not foliage leaves and may be 1–many-flowered.

Scapose. Bearing a scape or resembling one.

Scarious. Thin, dry, membranous and more or less translucent, not green.

Scar. A mark left on the stem by the separation of a leaf or on a seed by its detachment.

Schizocarp. A dry dehiscent fruit that splits into two halves, each half a mericarp, as in the Umbelliferae.

Scorpioid. Said of a coiled cluster in which the flowers are two-ranked and borne alternately at the right and left.

Scurfy. Covered with small branlike scales.

Secund. Said of parts or organs directed to one side only, usually by torsion.

Semi-. A Latin prefix meaning half.

Sepal. One of the parts of the outer whorl of the floral envelope or calyx, usually green in color.

Sepaloid. With the texture of or resembling a sepal.

Septate. Partitioned; divided by one or more partitions.

Septicidal. A capsule splitting down the septa and not through the locule.

Septum. Any kind of partition.

Sericeous. Silky, clothed with closely appressed, soft, straight pubescence.

Serotinus. Late coming, late to leaf, or flower, or to appear.

Serrate. With sharp teeth on the margin pointing forward.

Serrulate. Serrate with minute teeth.

Sessile. Without a stalk of any kind, as a leaf without a petiole.

Seta. A bristle or bristle-shaped body.

Setaceous. Bristlelike, with bristles.

Setose. Bristly, beset with bristles.

Several. Consisting of an indefinite number greater than two, but not very many.

Sheath. A tubular envelope, usually used for that part of the leaf of a sedge or grass that envelopes the stem.

Shrub. A woody perennial plant smaller than a tree and usually with several basal stems.

Silicle. The short fruit of certain Cruciferae usually less than 5 times longer than broad; silicule.

Silique. The fruit of Cruciferae usually more than 5 times longer than broad, with the two valves usually separating at maturity from the frame or replum.

Simple. Of one piece; not compound.

Sinuate. With a deep, wavy margin.

Sinus. The space or recess between two lobes or divisions of a leaf or other expanded organ.

Solitary. Borne singly or alone.

Sorus. A cluster of sporangia on a fern frond.

Spadix. A spike with a thick and fleshy axis, usually densely flowered with imperfect flowers.

Spathaceous. Spathelike.

Spathe. The bract or pair of bracts surrounding or subtending a flower cluster or spadix.

Spatulate. Broad and rounded at apex and tapering at base; flattened spoon-shaped.

Spicate. Like a spike, or disposed in a spike.

Spike. An inflorescence consisting of a central rachis bearing a number of sessile flowers.

Spine. A sharply pointed, rigid, deeply seated outgrowth from the stem, commonly a modified leaf axis.

Spinescent. Ending in a spine or sharp point; more or less spiny.

Spinose. Spinelike or with spines.

Spinulose. Minutely spiny; beset with small spines.

Spiny. Beset with spines.

Sporadic. Occurring here and there, without continuous range.

Sporangium. A structure within which spores are produced.

Spore. A cell which becomes free and capable of direct development into a new individual; the first cell of a gametophyte.

Sporophyll. A spore-bearing leaf, often highly modified.

Sporophyte. In ferns and seed plants, the foliaceous vegetative plant, as opposed to the gametophyte.

Spreading. Diverging nearly at right angles; nearly prostrate.

Spur. A tubular or saclike projection from a blossom, as of petal or sepal; it usually contains a nectar-secreting gland.

Squarrose. Spreading or recurved at the tip.

Stamen. The pollen-bearing organ of the flower.

Staminate. Having stamens and no pistil; male.

Staminode. A sterile stamen or any structure lacking an anther but corresponding to a stamen.

Stellate. Starlike or star-shaped with slender segments or hairs radiating out from a common center.

Sterile. Infertile and unproductive, as a flower without a pistil, a stamen without an anther or a leafy shoot without flowers.

Stigma. That part of the pistil that receives the pollen, usually at or near the apex of the pistil and mostly hairy, papillose or sticky.

Stipe. The stalklike support of a pistil; also the name for the petiole of a fern frond.

Stipitate. Provided with a stipe or with a slender stalklike base.

Stipulate. Provided with stipules.

Stipule. An appendage at the base of the petiole or leaf at each side of its insertion.

Stolon. A trailing shoot above ground rooting at the nodes; a runner.

Stoloniferous. With stolons or runners that take root.

Stomate. A small opening on the surface of a leaf through which gaseous exchange takes place.

Striate. With fine grooves, ridges or lines of color.

Strict. Very straight and upright.

Strigillose. Like strigose but hairs very short.

Strigose. With appressed, stiff, rather short hairs.

Strobilus. A structure characterized by imbricated bracts or scales, as a pine cone.

Stylar. Relating to the style.

Style. The usually stalklike part of a pistil connecting the ovary and stigma.

Stylopodium. A disklike enlargement at the base of the style as in Umbelliferae.

Sub-. A prefix meaning almost or below.

Subacute. Somewhat acute.

Subcapitate. Almost capitate.

Subcordate. Almost cordate.

Subcorymbose. Almost corymbose.

Subcylindrical. Almost cylindrical.

Subentire. Almost entire.

Subfoliaceous. Almost foliaceous.

Subrhizomatous. Almost rhizomatous.

Subscapose. Almost scapose.

Subshrub. A suffrutescent perennial (the stems basally woody), or a very low shrub often loosely treated as a perennial.

Subtend. To stand below and close to, as a bract below a flower or a leaf below a bud.

Subterranean. Below the surface of the earth, hypogaeous.

Subulate. Awl-shaped; narrowly triangular and tapering to a sharp point.

Succulent. Fleshy and full of juice.

Suffrutescent. Low-shrubby; applied to perennials, the lower part of the stems woody but the upper part herbaceous.

Suffuse. To overspread as with a fluid; tinge; tint.

Sulcate. Grooved or furrowed lengthwise.

Superior. Said of an ovary with floral parts inserted at its base.

Suture. A junction or seam of union; a line of opening or dehiscence.

Sympatric. Inhabiting one and the same area.

Sym-, syn-. Prefixes meaning together, with, joined.

Tailed. Said of anthers having caudal appendages.

Talus. A usually bouldery or gravelly slope.

Taproot. The primary descending root, forming a direct continuation from the radicle.

Tardily. Lately.

Taxa. Plural of taxon.

Taxon. A general term applied to any taxonomic element, population, or group.

Teeth. Plural of tooth.

Tendril. A rotating or twisting threadlike process or extension by which the plant grasps an object and clings to it for support.

Tenuous. Thin; slender.

Terete. Circular in transverse section.

Terminal. Proceeding from, or belonging to the end or apex.

Ternate. In threes.

Terrestrial. Growing in the soil in distinction from growing in water or other habitats.

Tetra-. In Greek composition, a prefix meaning four.

Tetrad. A group of four objects, as the four pollen grains formed from one pollen mother cell.

Tetradynamous. Having four long stamens and two short, as in Cruciferae.

Thorn. A sharp, usually aborted branch, simple or branched.

Throat. The opening or orifice into a gamepetalous corolla or perianth; the place where the limb joins the tube.

Tomentose. With tomentum; densely woolly or pubescent; with matted soft wool-like hairiness.

Tomentulose. Somewhat or delicately tomentose.

Tomentum. The covering of closely interwoven and tangled hairs in a tomentose surface.

Toothed. Dentate.

Trailing. Prostrate but not rooting.

Transverse. Across; at a right angle to the longitudinal axis.

Tree. A woody plant that produces one main trunk and a more or less distinct and elevated head.

Tri-. A prefix meaning 3 or 3 times.

Trichome. Any hairlike outgrowth of the epidermis, as a hair or bristle.

Tridentate. Three-toothed.

Trifoliolate. Having three leaflets.

Trifurcate. Having three forks or branches.

Triploid. An organism with three genomes or sets of chromosomes (3n).

Truncate. Ending abruptly, the base or apex nearly or quite straight across.

Trunk. The main axis of a tree below its branches.

Tuber. A short, thickened branch of a subterranean stem, beset with buds or "eyes."

Tubercle. A little tuber; a small rounded structure, often pimplelike.

Tuberculate. Furnished with knoblike excrescences or tubercles.

Tuberous. With tubers, tuberlike.

Tubular. Having the form of, or consisting of, a tube or tubes.

Tufted. Having a cluster of hairs or other slender outgrowths; stems in a very close cluster.

Turbinate. Top-shaped; inversely conical.

Turgid. Swollen or tightly drawn; said of a thin covering expanded by internal pressure.

Turion. A scaly, often thick and fleshy shoot produced from a bud on an underground rootstock.

Ubiquitous. Occurring everywhere.

Umbel. An inflorescence, more or less flattopped, in which all of the pedicels arise at the same point, like the ribs of an umbrella.

Unarmed. Destitute of prickles or other armature.

Umbellate. Umbelled; with umbels; pertaining to umbels.

Umbellet. A small or secondary umbel in a compound umbel.

Uncinate. Hooked near the apex or in the form of a hook.

Undulate. The margin gently wavy.

Uni-. A prefix meaning one.

Urn. The base of a pyxis.

Utricle. A small bladdery pericarp; any bladder-shaped appendage.

Valve. A separable part of a pod; the units or pieces into which a capsule splits or divides in dehiscing.

Valvate. Opening by valves, as in most dehiscent fruits and some anthers; parts of a flower bud that meet without overlapping.

Vascular. With vessels or ducts.

Vein. A strand of vascular tissue in a flat organ such as a leaf.

Velutinous. Velvety, due to a coating of fine soft hairs.

Venation. Veining; arrangement or disposition of veins.

Ventral. Belonging to the inner or axis side of an organ; the upper surface of a leaf.

Vernation. The disposition or arrangement of leaves in the bud.

Verrucose. Warty; covered with wartlike growths; tuberculate.

Versatile. Hung or attached near the middle and usually moving freely, as an anther attached crosswise on the apex of a filament and capable of turning.

Verticil. A whorl, or circular arrangement of similar parts round an axis.

Verticillate. Whorled, with two or more leaves at a node, cyclical.

Vestigial. Applied to the remnant or trace of an organ no longer fully developed, rudimentary.

Villosulous. Diminutive of villous.

Villous. With long, soft, somewhat wavy hairs.

Viscid. Glutinous, sticky or gummy to the touch.

Viviparous. Germinating or sprouting from seed or bud while attached to the parent plant.

Wart. A hard or firm excrescence.

Whorl. Three or more leaves or flowers at one node, in a circle.

Wing. Any membranous expansion attached to an organ; the lateral petal of a papilionaceous flower.

Woolly. Lanate, tomentose, clothed with long and tortuous or matted hairs.

Xero-. A combining form meaning dry.

Xerophyte. A plant which can subsist with a small amount of moisture, a desert plant.

Zygomorphic. Capable of division by only one plane of symmetry.

References

Anderson, J. P. 1916. Notes on the flora of Sitka, Alaska. Iowa Acad. Sci. 23: 427–82.

————. 1918. Plants of southeastern Alaska. Iowa Acad. Sci. 25:427–49.

————. 1919. Supplemental list of plants from southeastern Alaska. Iowa Acad. Sci. 26:328–31.

————. 1939. Plants used by the Eskimo in the northern Bering Sea and arctic regions of Alaska. Am. Jour. Bot. 26:714-16.

————. 1943a. Papers on the flora of Alaska–I. The genus *Cicuta*. Torreya 42:176–78.

————. 1943b. Two notable plant hybrids from Alaska. Iowa Acad. Sci. 50:155–57.

————. 1943-52. Flora of Alaska and adjacent parts of Canada. Iowa State Jour. Sci. 18:137–75, 381–445; 19:133–205; 20:213–57, 297–347; 21:363–423; 23:137–87; 24:219–71; 26:387–453.

————. 1947. Alaska and Yukon species of *Rubus* subgenus Cylactis Focke. Bull. Torrey Club 74:255–56.

————. 1959. Flora of Alaska and adjacent parts of Canada. Iowa State Univ. Press. 543 p.

Argus, G. W. 1969. New combinations in the *Salix* of Alaska and Yukon. Can. Jour. Bot. 47:795–801.

Bakewell, A. 1943. Botanical collections at the Wood Yukon Expeditions of 1939-41. Rhodora 45:305–16.

Beamish, K. I. 1955. Studies in the genus *Dodecatheon* of North America. Bull. Torrey Bot. Club 82:357–66.

Boivin, B. 1962. Études sur les *Oxytropis* DC. I — *Oxytropis deflexa* (Pallas) DC. Svensk. Bot. Tidskr. 56:496–500.

————. 1966a. Énumération des plantes du Canada. Nat. Can. 93:253–74.

————. 1966b. Énumération des plantes du Canada II—Lignidées. Nat. Can. 93:371–437.

————. 1966c. Énumération des plantes du Canada III—Herbidées, 1° partie; Digitatae: Dimerae, Liberae. Nat. Can. 93:583–646.

————. 1966d. Énumération des plantes du Canada IV—Herbidées, 2° partie; Connatae. Nat. Can. 93:989–1063.

————. 1967a. Énumération des plantes du Canada V—Monopsides (1ère partie). Nat. Can. 94:131–57.

————. 1967b. Énumération des plantes du Canada VI—Monopsides (2ème partie). Nat. Can. 94:471–528.

————. 1967c. Énumération des plantes du Canada VII—Resume statistique et régions adjacentes. Nat. Can. 94: 625–55.

————. 1967d. Études sur les Oxytropis DC.—II. Nat. Can. 94:73–78.

Bowden, W. M. 1959. Chromosome numbers and taxonomic notes on northern grasses. I. Triticeae. Can. Jour. Bot. 37:1143–51.

————. 1960a. Chromosome numbers and taxonomic notes on northern grasses. II. Tribe Festuceae. Can. Jour. Bot. 38:111–31.

————. 1960b. Chromosome numbers and taxonomic notes on northern grasses. III. Twenty-five genera. Can. Jour. Bot. 38:541–57.

————. 1961. Chromosome numbers

and taxonomic notes on northern grasses. IV. Tribe Festuceae: *Poa* and *Puccinellia*. Can. Jour. Bot. 39:123–37.

————. 1965. Cytotaxonomy of the species and interspecific hybrids of the genus *Agropyron* in Canada and neighboring areas. Can. Jour. Bot. 43:1441–48.

———— and W. J. Cody. 1961. Recognition of *Elymus sibiricus* L. from Alaska and the District of Mackenzie. Bull. Torrey Bot. Club 88:153–55.

Briggs, W. R. 1953. Some plants of Mount McKinley National Park, McGonegal Mountain area. Rhodora 55:245–52.

Calder, J. A. 1952. Notes on the genus *Carex* I: a new species of *Carex* from western Canada. Rhodora 54:246–50.

———— and R. L. Taylor. 1965. New taxa and nomenclatural changes with respect to the flora of the Queen Charlotte Islands, British Columbia. Can. Jour. Bot. 43:1387–99.

————. and R. L. Taylor. 1968. Flora of the Queen Charlotte Islands, British Columbia. I. Systematics of vascular plants. Can. Dep. Agr. Res. Mono. 4.

Cantlon, J. E., et al. 1959. A range extension in arctic Alaska for *Botrychium lunaria*. Amer. Fern Jour. 49:25–29.

Cody, Wm. J. 1960. Plants of the vicinity of Norman Wells, Mackenzie District, Northwest Territories. Can. Field-Nat. 74:71–100.

————. 1961. New plant records from the upper Mackenzie River Valley, Mackenzie District, Northwest Territories. Can. Field-Nat. 75:55–69.

————. 1963. Some rare plants from the Mackenzie Mountains, Mackenzie District, N. W. T. Can. Field-Nat. 77:226–28.

————. 1965. New plant records from northwestern Mackenzie District, N. W. T. Can. Field-Nat. 79:96–106.

————. 1971. A phytogeographic study of the floras of the continental Northwest Territories and Yukon. Nat. Can. 98:145–58.

———— and A. E. Porsild. 1968. Additions to the flora of the Northwest Territories, Canada. Can. Field-Nat. 82:263–75.

Cooper, W. S. 1924. The forests of Glacier Bay (Alaska), present, past, and yet unborn. Jour. Forest. 22(1):16–23.

————. 1930. The seed plants and ferns of the Glacier Bay National Monument, Alaska. Bull. Torrey Bot. Club 57:327–38.

————. 1939. Additions to the flora of Glacier Bay National Monument, Alaska, 1935-36. Bull. Torrey Bot. Club 66:453–56.

Coville, F. V. 1900. The tree willows of Alaska. Proc. Wash. Acad. Sci. 2:275–86.

————. 1901. The willows of Alaska. Proc. Wash. Acad. Sci. 3:297–362.

————. and F. Funston. 1895. Botany of Yakutat Bay, Alaska with a field report by F. Funston. Contr. U. S. Nat. Herb. 3:325–50.

Eastwood, A. 1947. A collection of plants from the Aleutian Islands. Leafl. West. Bot. 5:9–13.

————. 1957. A list of plants from Dall and Annette Islands, Alaska. Leafl. West. Bot. 7:102.

Gillett, G. W. 1960. A systematic treatment of the *Phacelia franklinii* group. Rhodora 62:205–22.

Gillett, J. M. 1968. The systematics of the Asiatic and American populations of *Fauria crista-galli* (Menyanthaceae). Can. Jour. Bot. 46:92–96.

Haglund, G. E. 1946. Contributions to the knowledge of the *Taraxacum* flora of Alaska and Yukon. Svensk. Bot. Tidskr. 40:325–61.

————. 1948. Further contributions to the knowledge of the *Taraxacum* flora of Alaska and Yukon. Svensk. Bot. Tidskr. 42:297–336.

————. 1949. Supplementary notes on the *Taraxacum* flora of Alaska and Yukon. Svensk. Bot. Tidskr. 43:107–16.

Harms, V. L. 1969. Range extensions for

some Alaskan aquatic plants. Can. Field-Nat. 83:253–56.

Heller, C. A. 1953. Wild edible and poisonous plants of Alaska. Univ. Alaska Ext. Bull. F-40. 87 p.

————. 1966. Wild flowers of Alaska. Portland, Oregon. 104 p.

Hitchcock, C. L., et al. 1955–69. Vascular plants of the Pacific Northwest. 5 vols. Univ. Washington Press.

Hodgson, H. J., and Wm. W. Mitchell. 1965. A new *Elymordeum* hybrid from Alaska. Can. Jour. Bot. 43:1355–58.

Hultén, E. 1936. New or notable species from Alaska. Contributions for the flora of Alaska I. Svensk. Bot. Tidskr. 30:515–28.

————. 1939. Two new species from Alaska. Contribution to the flora of Alaska II. Bot. Not. 1939:826–29.

————. 1940a. History of botanical exploration in Alaska and Yukon territories from the time of their discovery to 1940. Bot. Not. 1940:289–346.

————. 1940b. Two new species of *Salix* from Alaska. Contribution to the flora of Alaska III. Svensk. Bot. Tidskr. 34:374–76.

————. 1941–50. Flora of Alaska and Yukon, 1–10. Lunds Univ. Arsskr. N. F., Avd. 2, vol. 37–46. 1902 p.

————. 1946. New species of *Astragalus* and *Oxytropis* from Alaska and Yukon. Arkiv For Botanik 33:1–5.

————. 1957. De Amfiatlantiska Växterna. Svensk. Bot. Tidskr. 51:19–35.

————. 1958. The Amphi-Atlantic plants and their phytogeographical connections. Svensk. Vetenskapsakad. Handl. 7(71):1–340.

————. 1960. Flora of the Aleutian Islands. Ed. 2. Weinheim/Bergstr. 376 p.

————. 1962a. Flora and vegetation of Scammon Bay, Bering Sea coast, Alaska. Svensk. Bot. Tidskr. 56:36–54.

————. 1962b. The circumpolar plants I. Vascular cryptogams, conifers, monocotyledons. Svensk. Vetenskapsakad. Handl. 5:1–275.

————. 1964. The *Saxifraga flagellaris* complex. Svensk. Bot. Tidskr. 58:81–104.

————. 1966a. Contributions to the knowledge of flora and vegetation of the southwestern Alaska mainland. Svensk. Bot. Tidskr. 60:175–89.

————. 1966b. New species of *Arenaria* and *Draba* from Alaska and Yukon. Bot. Not. 119:313–16.

————. 1967. Comments on the flora of Alaska and Yukon. Svensk. Vetenskapsakad. Handl. Ser 2. 7(1):1–147.

————. 1968. Flora of Alaska and neighboring territories. Stanford Univ. Press. 1008 p.

————. 1973. Supplement to flora of Alaska and neighboring territories—a study of the flora of Alaska and the transberingian connection. Bot. Not. 126:459–512.

———— and H. St. John. 1931. The American species of *Lysichitum*. Svensk. Bot. Tidskr. 25:453–67.

Johnson, A. W., and L. A. Viereck. 1962. Some new records and range extensions of arctic plants from Alaska. Univ. Alaska Biol. Paper 6. 32 p.

————. et al. 1966. Vegetation and flora of the Cape Thompson-Ogotoruk Creek area, Alaska. pp. 277–354. *In:* Willomovsky, N. J., and J. N. Wolfe, eds. Environment of the Cape Thompson region, Alaska. U.S. Atomic Energy Comm. Div. Tech. Inform. 1250 p.

Johnson, P., and T. C. Vogel. 1966. Vegetation of the Yukon Flats region. U.S. Army Cold Reg. Res. Eng. Lab. Hanover, N. H., Res. Rep. 209. 53 p.

Jordal, L. H. 1951. Plants from the vicinity of Fairbanks, Alaska. Rhodora 53:156–59.

————. 1952. Some new entities in the flora of the Brooks Range region, Alaska. Rhodora 54:35–39.

Klebesadel, L. J., and Wm. W. Mitchell. 1964. An outbreak of horse poisoning from swamp horsetail (*Equisetum fluviatile* L.). Jour. Range Sci. 17:333–34.

Lawrence, Wm. E. 1947. Chromosome numbers in *Achillea* in relation to geographic distribution. Amer. Jour. Bot. 34:538–45.

Lepage, E. 1951. New and noteworthy plants in the flora of Alaska. Am. Midl. Nat. 46:574–79.

Löve, D., and N. J. Freedman. 1956. A plant collection from southwest Yukon. Bot. Not. 109:153–211.

Mitchell, Wm. W. 1965. Redefinition of *Bromus ciliatus* and *B. richardsonii* in Alaska. Brittonia 17:278–84.

———. 1967a. Taxonomic synopsis of *Bromus* section Bromopsis (Gramineae) in Alaska. Can. Jour. Bot. 45:1209-313.

———. 1967b. On the *Hordeum jubatum* — *H. brachyantherum* question. Madroño 19:108–10.

———. 1968a. Taxonomy, variation and chorology of three chromosome races of the *Calamagrostis canadensis* complex in Alaska. Madroño 19:235–46.

———. 1968b. Hybridization within the Triticeae of Alaska: a new × *Elyhordeum* and comments. Rhodora 70:467–73.

——— and H. J. Hodgson. 1965a. The status of hybridization between *Agropyron sericeum* and *Elymus sibiricus* in Alaska. Can. Jour. Bot. 43:855–59.

———. and H. J. Hodgson. 1965b. A new × *Agrohordeum* from Alaska. Bull. Torrey Bot. Club 92:403–7.

Mulligan, G. A. 1970. Cytotaxonomic studies of *Draba glabella* and its close allies in Canada and Alaska. Can. Jour. Bot. 48:1431–43.

———. 1971a. Cytotaxonomic studies of the closely allied *Draba cana, D. cinerea,* and *D. groenlandica* in Canada and Alaska. Can. Jour. Bot. 49:89–93.

———. 1971b. Cytotaxonomic studies in *Draba* species in Canada and Alaska: *D. ventosa, D. ruaxes* and *D. paysonii.* Can. Jour. Bot. 49:1435–60.

Murray, D. F. 1968. A plant collection from the Wrangell Mountains, Alaska. Arctic 21:106–9.

———. 1970. *Carex podocarpa* and its allies in North America. Can. Jour. Bot. 48:313–24.

———. 1971. Notes on the alpine flora of the St. Elias Mountains. Arctic 24:301–4.

Northstrom, T. E., and S. L. Welsh. 1970. Revision of the *Hedysarum boreale* complex. Great Basin Nat. 30:109–30.

Ohwi, J. 1965. Flora of Japan. Smithsonian Institution. 1067 p.

Packer, J. G. 1972. A taxonomic and phytographical review of some arctic and alpine *Senecio* species. Can. Jour. Bot. 50:507–18.

Pegau, R. E. 1972. Reflowering of some tundra plants in October near Nome Alaska. Can. Field-Nat. 86(2).

Polunin, N. 1940. Botany of the Canadian eastern arctic. Nat. Mus. Can. Bull. 92. 408 p.

———. 1959. Circumpolar arctic flora. Oxford Univ. Press. 514 p.

Porsild, A. E. 1937. Flora of the Northwest Territories. pp. 130–41. *In:* Canada's Western Northland.

———. 1938 Flora of Little Diomede Island in Bering Strait. Trans. Roy. Soc. Can. Ser. 3, Sect. 5. 32:21–38.

———. 1939. Contributions to the flora of Alaska. Rhodora 41:141–83, 199–254, 262–301.

———. 1943. Materials for a flora of the continental Northwest Territories of Canada. Sargentia 4:1–79.

———. 1944. Vascular plants collected in Kiska and Great Sitkin islands in the Aleutians by Lt. H. R. McCarthy and Capt. N. Kellas, August, September and October 1943. Can. Field-Nat. 58:130–31.

———. 1945. Alpine flora of the east slope of the Mackenzie Mountains, Northwest Territories. Nat. Mus. Can. Bull. 101. 35 p.

———. 1950. The genus *Antennaria*

in northwestern Canada. Can. Field-Nat. 64:1–25.

————. 1951a. Botany of southeastern Yukon adjacent to the Canol Road. Nat. Mus. Can. Bull. 121:1–400.

————. 1951b. Two new *Oxytropis* from arctic Alaska and Yukon. Can. Field-Nat. 64:1–25.

————. 1964a. Illustrated flora of the Canadian Arctic Archipelago. Nat. Mus. Can. Bull. 146. 218 p.

————. 1964b. *Potentilla stipularis* L. and *Draba sibirica* (Pall.) Thell. new to North America. Can. Field-Nat. 78: 92–96.

————. 1965. Some new or critical vascular plants of Alaska and Yukon. Can. Field-Nat. 79:79–90.

————. 1966. Contributions to the flora of southwestern Yukon Territory. Nat. Mus. Can. Bull. 216. 86 p.

————. 1972. The vascular flora of Limestone Hills, northern extension of the Ogilvie Mountains, Yukon Territory. Arctic 25:233–36.

———— and Wm. J. Cody. 1968. Checklist of the vascular plants of continental Northwest Territories, Canada. Plant Research Institute, Can. Dept. Agr. 102 p.

———— and H. A. Crum. 1961. The vascular flora of Liard Hot Springs, B.C., with notes on some Bryophytes. Nat. Mus. Can. Bull. 171:131–97.

Potter, L. 1962. Roadside flowers of Alaska. Thetford Center, Vermont. 590 p.

Raup, H. M. 1934. Phytogeographic studies in the Peace and upper Liard River regions, Canada. Contr. Arnold Arb. 6.

————. 1944. Expeditions to the Alaska Military Highway, 1943–44. Arnoldia 4:65–72.

————. 1947. The botany of southwestern Mackenzie. Sargentia 6:1–275.

————. 1959. The willows of boreal western America. Contr. Gray Herb. 185:1–95.

Reed, B. L. 1968. Geology of the Lake Peters area, northeastern Brooks Range, Alaska. U. S. Geol. Survey Bull. 1236. 132 p.

Rousi, A. 1965. Biosystematic studies of the species aggregate *Potentilla anserina* L. Ann. Botanici Fennici 2:47–112.

Scamman, E. 1940. A list of plants from interior Alaska. Rhodora 42:309–49.

Schaack, G. B. van. 1945. Flowers of Island X (Attu). Unpublished mss. 38 p.

Shacklette, H. S. 1966. Phytoecology of a greenstone habitat at Eagle, Alaska. U. S. Geol. Survey Bull. 1198–F. 36 p.

————. 1969. Vegetation of Amchitka Island, Aleutian Islands, Alaska. U. S. Geol. Survey Prof. Paper 648. 66 p.

Sharples, A. W. 1938. Alaska wild flowers. Stanford Univ. Press. 156 p.

Shetler, S. 1963. An annotated list of vascular plants from Cape Sabine, Alaska. Rhodora 65:208–24.

Sigafoos, R. S. 1958. Vegetation of northwestern North America, as an aid in interpretation of geological data. U. S. Geol. Survey Bull. 1061–E:165–85.

Spetzman, L. A. 1959. Vegetation of the arctic slope of Alaska. U.S. Geol. Survey. Prof. Paper 302–B:19–58.

Suda, Y., and G. W. Argus. 1969. Chromosome numbers of some North American arctic and boreal *Salix*. Can. Jour. Bot. 47:859–62.

Szczawinski, A. F. 1962. The heather family (Ericaceae) of British Columbia. British Columbia Provincial Mus. Handbook 19. 205 p.

Tatewaki, M. 1930–31. Notes on plants of the western Aleutian Islands collected in 1929, Parts 1–2. Trans. Sapporo Nat. Hist. Soc. 11:152–56; 12: 200–209.

Taylor, R. F. 1929. Pocket guide to Alaska trees. U. S. Dept. Agr. Misc. Publ. 55. 39 p.

Taylor, T. M. C. 1970. Pacific northwest ferns and their allies. Univ. Toronto Press. 247 p.

Thomas, J. H. 1957. The vascular flora of Middleton Island, Alaska. Contr. Dudley Herb. Stanford Univ. 5:39–56.

————. 1951a. A collection of plants from Point Lay, Alaska. Contr. Dudley Herb. Stanford Univ. 4:53–56.

————. 1951b. *Cochlearia officinalis arctica* in the vicinity of Point Barrow, Alaska. Rhodora 54:40–42.

Viereck, L. A., and J. M. Foote. 1970. The status of *Populus balsamifera* and *P. trichocarpa* in Alaska. Can. Field-Nat. 84:169–73.

———— and E. L. Little, Jr. 1972. Alaska trees and shrubs. U. S. Dept. Agr. Handbook 410. 265 p.

Welsh, S. L. 1968. Nomenclature changes in the Alaskan flora. Great Basin Nat. 28:147–56.

————. 1972. On the typification of *Oxytropis leucantha* (Pallas) Pers. Taxon 21:155–57.

———— and J. K. Rigby. 1971a. Botanical and physiographic reconnaissance of northern Yukon. Brigham Young Univ. Sci. Bull. Bio. Ser. 14(2):1–64.

———— and J. K. Rigby. 1971b. Botanical and physiographic reconnaissance of northern British Columbia. Brigham Young Univ. Sci. Bull. Bio. Ser. 14(4): 1–49.

Wiggins, I. L., and D. G. MacVicar. 1958. Notes on the plants in the vicinity of Chandler Lake, Alaska. Contr. Dudley Herb. Stanford Univ. 5:69–95.

———— and J. H. Thomas. 1962. A flora of the Alaskan arctic slope. Arct. Inst. N. Am. Spec. Publ. 4. 425 p.

Wight, W. F. 1908. A new larch from Alaska. Smithsonian Institution Misc. Collect. 50:174.

Williams, M. M. 1952. Alaska wild flower glimpses. Juneau: The Totem Press. 52 p.

Index

A R C T I C

Icy Cape

Cape Lisburne

Point
Hope

Noa

Noata

Kotzebue Sound

Kotz

Kotz

BERING STRAIT

Cape
Prince of Wales

Nome

St. Lawrence
Island

NORTON SOUND

Cape Romanzof

St. Matthew
Island

Bethel

Kusk

Nunivak
Island

Kuskokwim
Bay

Di

Cape Newenham

B

Pribilof
Islands

Attu
Island

B E R I N G

S
E
A

Agattu
Island

Fort Randall

Kiska
Island

Unimak Island

Amchitka
Island

Tanaga
Island

ALEUTIAN ISLANDS

Dutch Harbor

Atka
Island

Unalaska
Island

Adak
Island

Umnak

Umnak
Island

P